Foundations of
Cardiac Arrhythmias

FUNDAMENTAL AND CLINICAL CARDIOLOGY

Editor-in-Chief

Samuel Z. Goldhaber, M.D.

*Harvard Medical School
and Brigham and Women's Hospital
Boston, Massachusetts*

Associate Editor, Europe

Henri Bounameaux, M.D.

*University Hospital of Geneva
Geneva, Switzerland*

ADDITIONAL VOLUMES IN PREPARATION

Foundations of
Cardiac Arrhythmias

Basic Concepts and Clinical Approaches

edited by

Peter M. Spooner
National Institutes of Health
Bethesda, Maryland

Michael R. Rosen
Columbia University
New York, New York

MARCEL DEKKER, INC.　　　　　　　　NEW YORK • BASEL

ISBN: 0-8247-0266-2

This book is printed on acid-free paper.

Headquarters
Marcel Dekker, Inc.
270 Madison Avenue, New York, NY 10016
tel: 212-696-9000; fax: 212-685-4540

Eastern Hemisphere Distribution
Marcel Dekker AG
Hutgasse 4, Postfach 812, CH-4001 Basel, Switzerland
tel: 41-61-261-8482; fax: 41-61-261-8896

World Wide Web
http://www.dekker.com

The publisher offers discounts on this book when ordered in bulk quantities. For more information, write to Special Sales/Professional Marketing at the headquarters address above.

Current printing (last digit):
10 9 8 7 6 5 4 3 2 1

Series Introduction

Marcel Dekker, Inc., has focused on the development of various series of beautifully produced books in different branches of medicine. These series have facilitated the integration of rapidly advancing information for both the clinical specialist and the researcher.

My goal as editor-in-chief of the Fundamental and Clinical Cardiology series is to assemble the talents of world-renowned authorities to discuss virtually every area of cardiovascular medicine. In *Foundations of Cardiac Arrhythmias,* Peter M. Spooner and Michael R. Rosen have prepared a much-needed and timely book. Future contributions to this series will include books on molecular biology, interventional cardiology, and clinical management of coronary artery disease.

Samuel Z. Goldhaber

Foreword

Cardiac arrhythmias constitute a major clinical problem that presents itself in many ways, with the worst manifestation being sudden death. The development of the electrocardiogram by Einthoven nearly 100 years ago provided the first means for us to view their complexity and diversity. Yet, despite nearly a century of clinical observation of arrhythmias, it is only in recent years that we have witnessed significant progress in their treatment. Thus, it was interesting to note that one of the earliest discussions on the cardioversion of ectopic tachycardias, presented by Dr. Bernard Lown at the 1972 meeting of the Association of American Physicians, noted that "for the past 45 years, cardiac arrhythmia has been traditionally terminated by means of drugs, particularly quinidine." Although the utility of that approach remains problematic for most arrhythmias, much has changed and remarkable therapies have become available.

In his landmark book *Heart Disease*, published in 1931, Dr. Paul Dudley White devoted several pages to a consideration of cardiac arrhythmias. He then wrote:

> Notching or slurring of any of the components of the QRS complex may occur in slight degree normally or when there is considerable axis deviation in either direction, but it becomes marked only with ventricular premature beats, ventricular tachycardia, or intra-ventricular block.

Later in Dr. White's book, there are several chapters discussing abnormal cardiac rhythms due to unusual excitability and stimulation. Then a section follows on sudden death, which at the time was described as "an abrupt standstill of the heart, as yet unexplained."

That was 1931! As we now enter the next millennium, the scientific community brings with it an enormous trove of recent discoveries and much better understanding of the causes and clinical consequences of dysrhythmias and the enormous problem of sudden cardiac death.

This progress is the result of a convergence of several factors, foremost of which has been a considerable fundamental research effort coupled with a much better clinical understanding of the rhythm disorders. In fact, one could venture to say that this field today is a paradigm of what research should be: the merging of basic findings with their clinical applications. Biophysics, cell biology, cell signaling, and molecular biology have been powerful tools that are today increasingly supplemented by genetics. The genes determining signaling pathways, and the structural and functional identities of ion channels and gap junctions, are shedding new light on the pathogenesis of cardiac arrhythmias and, in the context of clinical findings, providing previously unanticipated directions toward prevention.

Is the problem solved? Of course not. But, as the editors of this volume point out in the first chapter, there is good indication that the course of research is well charted. However, they also underscore the need for continuing integration of basic and clinical research to ensure that outcomes are of direct benefit to the patient.

Students, veteran researchers, and clinicians could not dream of a more inspiring or knowledgeable cast of contributors to a major text. Indeed, the editors, Drs. Peter M. Spooner and Michael R. Rosen, have assembled a roster of authors whose expertise in their fields is recognized worldwide. Together, they present "the case" of cardiac arrhythmia in a unique and stimulating way, which will energize the field and, in the long run, benefit patients affected by these disorders.

Claude Lenfant, M.D.
Director, National Heart, Lung, and Blood Institute
National Institutes of Health
Bethesda, Maryland

Preface

Cardiac arrhythmias and the disciplines that facilitate their study encompass a vast area of scientific, medical, and technical development, impacting directly on one of the most frequent, severe, and costly causes of disability and mortality throughout the developed world. Our goal in designing this book was to provide students, teachers, and practicing professionals with a single source of information covering the fundamentals of progress in the arrhythmia sciences. The primary challenge to each contributor to the volume has been to facilitate the synthesis of basic and clinical research results into a solid foundation of fundamentals that promote understanding and can guide future progress. Hence, this is not just another introduction to the highly specialized world of the cardiac electrophysiologist. Rather, the volume is designed for the broad population of graduate scientists, engineers, molecular and cell physiologists, pharmacologists, and basic and clinical electrophysiologists and cardiologists concerned with understanding from the ground up how today's knowledge provides the clues for tomorrow's cures.

To ensure the translation of clinical and basic investigation and the integration of areas of conflicting opinion, we designated combinations of single or multiple authorships for individual chapters. Authors were selected based on their expertise, their unique contributions, and the often differing perspectives they have brought to their research. As editors we were often surprised and impressed with the innovative ways in which they met the challenge to work together. Many, if not all, of the resulting contributions reflect a real synthesis of ideas and observations that would have been impossible to achieve with standard approaches to authorship. It was also gratifying to see how frequently this different form of interaction altered the authors' own perspectives. We hope their efforts and insights, presented in this volume, are similarly helpful and contributory to our readers' goals.

In closing, we would like to acknowledge and express our sincere appreciation to those who contributed to the evolution of this book. Claude Lenfant, M.D., provided the vision that helped identify the need for such a work and we are much indebted for his continuing support. We also thank our publishers—first, the late Graham Garratt, who convinced us to undertake a project that might be unique and lasting, and subsequently Sandra Beberman, who so professionally guided us to the realization of a finished work. Our administrative assistants Michelle Cummings and Eileen Franey kept the project moving, and us on our toes. Suzanne Spooner and Tove Rosen were our strength through long weekends, late nights, and missed engagements. Their patience, encouragement, guidance, and advice played a major role in this effort that has greatly enriched both our personal and professional lives over the past three years. And finally, we thank the authors, who endured a profusion of editorial advice in preparing what we hope will be a continuing contribution to the science of arrhythmias

Peter M. Spooner
Michael R. Rosen

Contents

Contributors

Masood Akhtar, M.D.   Clinical Professor of Medicine, University of Wisconsin, and Director, Institute for Cardiac Rhythms, St. Luke's Medical Center, Milwaukee, Wisconsin

Christine M. Albert, M.D., M.P.H.   Assistant in Medicine, Cardiac Arrhythmia Service, Cardiovascular Division, Massachusetts General Hospital, Boston, Massachusetts

Charles Antzelevitch, Ph.D.   Executive Director, Masonic Medical Research Laboratory, Utica, New York

Gust H. Bardy, M.D.   Professor of Medicine, Division of Cardiology, Arrhythmia Services, Department of Medicine, University of Washington, Seattle, Washington

Luiz Belardinelli, M.D.   Executive Director, Department of Pharmacological Sciences, CV Therapeutics, Palo Alto, California

Charles I. Berul, M.D.   Associate in Cardiology and Assistant Professor of Pediatrics, Department of Cardiology, Children's Hospital and Harvard Medical School, Boston, Massachusetts

Penelope A. Boyden, Ph.D.   Professor of Pharmacology, College of Physicians and Surgeons, Columbia University, New York, New York

Arthur M. Brown, M.D., Ph.D.   Vice President, Research, and Professor, Rammelkamp Center for Education and Research, MetroHealth Campus, Case Western Reserve University, Cleveland, Ohio

Agustin Castellanos, M.D. Professor, Division of Cardiology, Department of Medicine, University of Miami School of Medicine, Miami, Florida

Leonard A. Cobb, M.D. Professor Emeritus, Department of Medicine, University of Washington, Seattle, Washington

Richard J. Cohen, M.D., Ph.D. Whitaker Professor of Biomedical Engineering, Harvard–MIT Division of Health Sciences and Technology, Massachusetts Institute of Technology, Cambridge, Massachusetts

Dario DiFrancesco, Ph.D. Professor, Department of General Physiology and Biochemistry, University of Milan, Milan, Italy

Stephen M. Dillon, Ph.D. Research Associate, University of Pennsylvania, and Presbyterian Medical Center, Philadelphia, Pennsylvania

Eugene Downar, M.D. Professor, Division of Cardiology, Department of Medicine, University of Toronto, Toronto, Ontario, Canada

Harry A. Fozzard, M.D. Otho S. A. Sprague Distinguished Service Professor of Medical Science, Emeritus, Departments of Neurobiology, Pharmacology, and Physiology, and Department of Medicine, The University of Chicago, Chicago, Illinois

Richard A. Gray, Ph.D. Assistant Professor, Department of Biomedical Engineering, University of Alabama at Birmingham, Birmingham, Alabama

H. Criss Hartzell, Ph.D. Professor, Department of Cell Biology, Emory University School of Medicine, Atlanta, Georgia

Heikki V. Huikuri, M.D. Professor of Medicine and Director of Academic Cardiology, Oulu University Hospital, Oulu, Finland

José Jalife, M.D. Professor and Chairman, Department of Pharmacology, SUNY Upstate Medical University, Syracuse, New York

Michiel J. Janse, M.D. Professor, Department of Cardiovascular Research, Academic Medical Center, Amsterdam, The Netherlands

Ronald W. Joyner, M.D., Ph.D. Professor, Department of Pediatrics, Children's Heart Center, Emory University School of Medicine, Atlanta, Georgia

Arnold M. Katz, M.D. Professor, Department of Medicine (Cardiology), University of Connecticut, Farmington, Connecticut

André G. Kléber, M.D. Professor, Department of Physiology, University of Bern, Bern, Switzerland

Ralph Lazzara, M.D. Director, Cardiac Arrhythmia Research Institute, and Professor of Medicine, University of Oklahoma Health Sciences Center, Oklahoma City, Oklahoma

Bruce B. Lerman, M.D.   Professor of Medicine and Chief, Division of Cardiology, Department of Medicine, New York Hospital–Cornell University Medical Center, New York, New York

Robert L. Lux, Ph.D.   Professor of Medicine, Cardiology Division, Department of Internal Medicine, University of Utah, Salt Lake City, Utah

Robert S. MacLeod, Ph.D.   Associate Professor of Bioengineering and the Nora Eccles Harrison Cardiovascular Research and Training Institute, University of Utah, Salt Lake City, Utah

Eduardo Marbán, M.D., Ph.D.   Michel Mirowski Professor of Cardiology; Professor of Medicine, Physiology, and Biomedical Engineering; and Director, Institute of Molecular Cardiobiology, The Johns Hopkins University, Baltimore, Maryland

Rahul Mehra, Ph.D.   Director, Atrial Fibrillation Research, Medtronic, Inc., Minneapolis, Minnesota

David M. Mirvis, M.D.   Professor, Department of Preventive Medicine, University of Tennessee, Memphis, Tennessee

Robert J. Myerburg, M.D.   Professor of Medicine and Physiology, and Director, Division of Cardiology, Department of Medicine, University of Miami School of Medicine, Miami, Florida

Denis Noble, Ph.D.   Professor, Department of Physiology, University of Oxford, Oxford, England

Robert Plonsey, Ph.D.   Department of Biomedical Engineering, Duke University, Durham, North Carolina

Amit Rakhit, M.D.   Clinical Fellow, Department of Cardiology, Children's Hospital and Harvard Medical School, Boston, Massachusetts

Richard B. Robinson, Ph.D.   Professor, Department of Pharmacology, Columbia University, New York, New York

Dan M. Roden, M.D.   Professor of Medicine and Pharmacology, and Director, Division of Clinical Pharmacology, Departments of Medicine and Pharmacology, Vanderbilt University School of Medicine, Nashville, Tennessee

Michael R. Rosen, M.D.   Gustavus Pfeiffer Professor of Pharmacology and Professor of Pediatrics, College of Physicians and Surgeons, Columbia University, New York, New York

Yoram Rudy   Director, Cardiac Bioelectricity Research and Training Center, Case Western Reserve University, Cleveland, Ohio

Jeremy N. Ruskin, M.D.   Director, Cardiac Arrhythmia Service, Cardiac Unit, Massachusetts General Hospital, Boston, Massachusetts

Jeffrey E. Saffitz, M.D., Ph.D.   Paul E. Lacy and Ellen Lacy Professor of Pathology, Department of Pathology, Washington University School of Medicine, St. Louis, Missouri

Michael Sanguinetti, Ph.D.   Professor, Division of Cardiology, Department of Internal Medicine, University of Utah, Salt Lake City, Utah

Benjamin J. Scherlag, Ph.D.   Professor of Medicine, University of Oklahoma Health Sciences Center, Oklahoma City, Oklahoma

Arnold Schwartz, Ph.D.   Director, Institute of Molecular Pharmacology and Biophysics, University of Cincinnati College of Medicine, Cincinnati, Ohio

Ketty Schwartz, Ph.D.   Director of Research, CNRS, INSERM U 523, Institut de Myologie, Paris, France

Madison S. Spach, M.D.   Professor of Pediatrics and Cell Biology, Department of Pediatrics, Duke University Medical Center, Durham, North Carolina

Peter M. Spooner, Ph.D.   Director, Arrhythmia, Ischemia, and Sudden Cardiac Death Research Program, Division of Heart and Vascular Diseases, National Heart, Lung, and Blood Institute, National Institutes of Health, Bethesda, Maryland

Harold C. Strauss, M.D.   Professor and Chairman, Department of Physiology and Biophysics, State University of New York, Buffalo, New York

Henk E. D. J. ter Keurs, M.D., Ph.D.   Professor of Medicine, Physiology and Biophysics, University of Calgary, Calgary, Alberta, Canada

Gordon F. Tomaselli, M.D.   Associate Professor, Department of Medicine, The Johns Hopkins University, Baltimore, Maryland

Jeffrey A. Towbin, M.D.   Professor, Cardiology Section, Department of Pediatrics, and Department of Molecular and Human Genetics, Baylor College of Medicine, Houston, Texas

David R. Van Wagoner, Ph.D.   Assistant Staff, Department of Cardiology, Cleveland Clinic Foundation, Cleveland, and Assistant Professor, College of Pharmacy, Ohio State University, Columbus, Ohio

Richard L. Verrier, Ph.D.   Associate Professor, Department of Medicine, Harvard Medical School, and Director, Institute for Prevention of Cardiovascular Disease, Beth Israel Deaconess Medical Center, Boston, Massachusetts

Marc A. Vos, Ph.D.   Associate Professor, Department of Cardiology, Academic Hospital Maastricht/Cardiovascular Research Institute Maastricht, Maastricht, The Netherlands

Albert L. Waldo, M.D.   Walter H. Pritchard Professor of Cardiology and Professor of

Medicine, Department of Medicine, Case Western Reserve University/University Hospitals of Cleveland, Cleveland, Ohio

Karl T. Weber, M.D.   Neuton Stern Professor of Medicine and Director, Division of Cardiovascular Diseases, University of Tennessee Health Science Center, Memphis, Tennessee

Mark J. Yeager, M.D., Ph.D.   Department of Cell Biology, The Scripps Research Institute, La Jolla, California

Douglas P. Zipes, M.D.   Distinguished Professor of Medicine, Pharmacology and Toxicology, and Director, Division of Cardiology and Krannert Institute of Cardiology, Department of Medicine, Indiana University School of Medicine, Indianapolis, Indiana

Adam Zivin, M.D.   Assistant Professor of Medicine, Division of Cardiology, Arrhythmia Services, Department of Medicine, University of Washington, Seattle, Washington

Foundations of
Cardiac Arrhythmias

1

Perspectives on Arrhythmogenesis, Antiarrhythmic Strategies, and Sudden Cardiac Death

PETER M. SPOONER

National Heart, Lung and Blood Institute, National Institutes of Health, Bethesda, Maryland

MICHAEL R. ROSEN

College of Physicians and Surgeons, Columbia University, New York, New York

I. INTRODUCTION

The chapters in this volume demonstrate the impressive progress that has been made in our understanding of cardiac arrhythmias, as well as the approaches used to study and control their occurrence. Given such advances, it may seem paradoxical that progress in translating this knowledge into improved medical therapies has been so slow and difficult. Much of this relates to the magnitude of our ignorance concerning those cardiac diseases which give rise to such arrhythmias. Moreover, from a scientific perspective, deciphering events in the transition from an organized, regular rhythm to the irregular, often chaotic states that characterize fibrillation and tachyarrhythmias has proved to be a subject as difficult and complex as any biomedical problem known today.

We have also only recently begun to appreciate the many interdependent concepts and approaches required to help translate knowledge of pathological cellular events into alterations in impulse initiation and conduction. For example, recent insights into the macroscopic, three-dimensional interactions implicated in the propagation of malignant ventricular tachycardias (Chap. 14) have only been possible because of the advent of methods for optically analyzing electrical excitation across the cardiac surface. The relevance of these observations, in terms of implications for new antiarrhythmic drug therapies (Chap. 26), could not have been appreciated without molecular and biochemical understanding of the regulation of cell-to-cell conduction (Chap. 8–10). Other important insights have emerged from widely diverse topics, including the discovery and identity of genes that determine the voltage-time course of the action potential in different cardiac tissues (Chap. 2, 3, and 6) and the role of specific mutations in those genes whose dysfunction results in inherited arrhythmias (Chap. 25). Population studies and trials are also beginning to provide insights into the relevance of high-frequency polymorphisms and rarer individual variations in both known and undiscovered molecules involved in cardiac control (Chap. 25). And epidemiological studies on sudden death populations (Chap. 20 and 29) have begun to suggest that we have yet to discover a significant proportion of the genetic determinants that govern whether and when some individuals are stricken with life-threatening arrhythmic events. Likewise, progress in implantation techniques (as well as new concepts in catheter design and electrical miniaturization)(Chaps. 21, 22 and 27), coupled with identification of better noninvasive markers of arrhythmia risk (Chap. 28) has been critical in recent successes with the use of pacemakers and defibrillators (Chap. 21, 22, and 29). Progress in this field can thus be appreciated to be extraordinarily dependent on frequently discordant advances at multiple technical and conceptual levels, and on the integration of results from different disciplines into a more powerful approach.

One of the messages evident in following the progression of ideas reviewed in this book is that, despite many successes, for the vast majority of life-threatening conditions, effective strategies for preventing disturbances of cardiac rhythm are largely unmet yet are among the most urgent public health needs today. Given the urgency of that situation and the wealth of new factual information emerging from multiple areas of research, we thought it would be useful to provide here a brief overview of the evolution of some of these more difficult issues. Our intent is to provide perspective on what, how, and why progress has occurred and how it is leading to new directions in the search for effective therapies.

II. THE PROBLEM OF SUDDEN CARDIAC DEATH

Prediction and prevention of sudden arrhythmic death in the most common forms of severe heart disease remain perhaps the primary scientific and medical challenge for investigators entering this field today. Given that cardiovascular disease (CVD) is generally accepted as the major cause of death in the developed world, the magnitude of the problem is enormous (1). It is also clear that this unseemly distinction is not likely to change without marked improvement in our ability to understand and prevent sudden death in populations in which it occurs most frequently. According to national statistics, including those developed by the American Heart Association (2) and others, between 40 to 50% of total U.S. mortality (more than 1,000,000 deaths per year) are due to CVD, about half involve arrhythmias, and more than 300,000 are directly attributable to ventricular tachycardia fibrillation occurring unexpectedly, usually within 1 to 2 h after onset of symptoms (3). Contributing to this total are more than 1,000,000 Americans who suffer a heart attack each year. Almost one-quarter of them do not survive and can be expected to die suddenly from fibrillation, asystole, and other causes on first occurrence. Coronary heart disease (CHD) is the most prevalent precursor to lethal arrhythmias and the largest factor precipitating CVD, heart attacks, and lethal arrhythmias in both male and female Americans. Approximately one-half of all patients who suffer a cardiac event as a result of CHD have no known risk factors and almost half have no prior history or symptoms of disease (4). Overall prevalence of CHD in the United States appears to be approximately 7,000,000 (2). Of those patients who die suddenly due to acute myocardial infarction, malignant tachy-arrhythmias, especially ventricular fibrillation, are most often the direct cause of death. One-year infarction recurrence rates are as high as 30% and almost one-third of initial survivors will succumb to premature arrhythmic death over the following years (5). Also, approximately 400,000 new cases of congestive heart failure are reported each year in the United States and, in 1996, approximately 1,000,000 Americans were diagnosed with NYHA Stage II–IV disease (2). Overall prevalence approaches 4,600,000 Americans and, within this population, the 5-year mortality rate is about 50%. Approximately 40 to 50% of heart failure deaths are believed to be due to fatal arrhythmias, and thus this population is the second highest contributor to total cardiovascular mortality (6). Altogether these arrhythmia-prone conditions—myocardial infarction, CHD, and heart failure—are the major cause of premature death in the United States, one that continues to surpass cancer. The toll, in terms of lives lost or seriously compromised, financial cost, utilization of medical resources, and family impact, is truly staggering.

Further, while worldwide incidences of ischemic and structural cardiac diseases are not known precisely, they undoubtedly represent an enormous multiple of U.S. numbers, especially in areas without resources for acute resuscitation and cardiac care where recent progress has made such a difference. Global prevalence is also increasing as risk factors for coronary heart disease and its "markers of excess" (smoking, poor diet, elevated cholesterol levels, increased body weight, physical inactivity, obesity, etc.), rise with economic improvements throughout developing nations. As these "hallmarks of progress" spread, they portend an even greater epidemic in cardiovascular morbidity, that is, unless there is rigorous progress in the application of therapies to control predisposing conditions like hypertension, diabetes, coronary atherosclerosis, plaque rupture and thrombosis, and development of new therapies to treat existing disease. Although recent surveys provide hope that rates of mortality from CHD (but not incidence of new disease) may in fact be leveling off in many countries (7), others argue that progress in acute medical care and secondary prevention (e.g., thrombolysis, angioplasty, bypass, and stenting in ischemia; use of renal agents, ACE inhibitors, and other drugs for structural disease) responsible for this decline in mortality merely delay, rather than reduce, long-term mortality. Nevertheless, it is encouraging to see a real decline in CHD mortality, even though progress in primary prevention in both developed and emerging nations is obviously lagging.

These worldwide events frame a major public health problem for the future. Unless there are significant advances in preventing ventricular fibrillation and malignant arrhythmias that are the most frequent causes of death in patients with common heart diseases—prior to their occurrence—little further progress in reducing this worldwide burden is likely to occur. Unfortunately, for most cardiac patients, arrhythmias are one of the most predominant life-threatening outcomes, as well as a consequence for which most forms of current therapy continue to be largely ineffective. A variety of considerations contribute to this dilemma. We are only beginning to understand arrhythmogenesis at the cellular level and have been only marginally successful in developing therapeutic strategies to modulate aberrant impulse initiation. Similarly, we have made little progress in controlling irregularities in impulse conduction other than by means of ablative or surgical excision.

We have, however, begun to understand that most life-threatening arrhythmias are the result of an extraordinarily complex dynamic, involving many different factors including temporal coincidence of at least three major elements: a compromised anatomical substrate; a primary triggering event, and self-perpetuating mechanisms for propagation (Fig. 1.1). With such an overlapping etiology, it has been particularly difficult to distinguish pat-

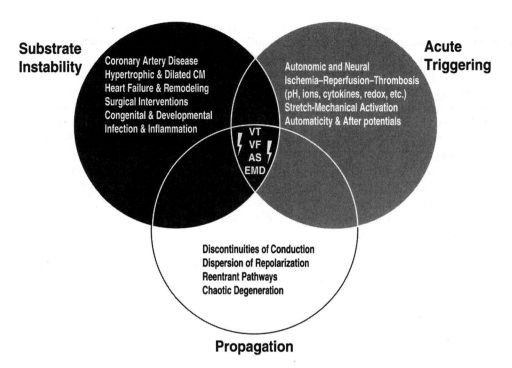

Substrate Instability

Coronary Artery Disease
Hypertrophic & Dilated CM
Heart Failure & Remodeling
Surgical Interventions
Congenital & Developmental
Infection & Inflammation

Acute Triggering

Autonomic and Neural
Ischemia–Reperfusion–Thrombosis
(pH, ions, cytokines, redox, etc.)
Stretch-Mechanical Activation
Automaticity & After potentials

VT
VF
AS
EMD

Discontinuities of Conduction
Dispersion of Repolarization
Reentrant Pathways
Chaotic Degeneration

Propagation

Figure 1.1   Interactions between arrhythmogenic factors affecting substrate instability (alterations in cardiac excitability), acute triggering (changes in transient conditions affecting either substrate stability or conduction), and propagation (alterations in wavefront direction and velocity). VT (ventricular tachycardia), VF (ventricular fibrillation), AS (asystole), and EMD (electromechanical coupling), resulting in a potentially lethal arrhythmia, require confluence of events in all three spheres. Individual pathological conditions (e.g., ischemia, heart failure) are multifactorial and may affect multiple elements to varying extents at different times throughout the course of different cardiac diseases. (Modified from Ref. 72.)

terns of cause from consequence and hence, identify optimal approaches. As Figure 1.1 suggests, the permutations and interactive nature of different elements of threat that combine to create a potentially lethal event can occur in an almost endless variety of ways in different patients at different times. Although Figure 1.1 lists some of these under just one major sphere of involvement, it is also apparent that most such conditions are themselves quite complex and can contribute to arrhythmias in many ways. In attempting to devise new methods of analysis and therapy, it is very important to keep in mind the dynamic nature of this myriad of interactions that occur in different diseases.

III. EARLY DEVELOPMENT OF ANTIARRHYTHMIC DRUGS

Another important factor contributing to the difficulties in treating arrhythmias is that progress in the development of drug therapies has been largely unsuccessful and good leads for safe, effective pharmacological strategies have been few and far between. Indeed, this very lack of progress appears to have created a climate in which commercial investments in the discovery of new antiarrhythmic compounds are likely to remain quite limited.

Given the importance of the problem, how is it that medical therapy has lagged so drastically where it is needed so desperately? It may help to review how we arrived at this position. Barely 50 years ago, only three drugs were considered effective antiarrhythmics in the United States: quinidine, procainamide, and digitalis. Digitalis, a derivative of the *foxglove* plant, had been used for more than 200 years based on its ability to improve the strength and regularity of the pulse and reduce edema in heart failure, or so-called "dropsical," patients (8). Years later it was recognized that dropsy was often accompanied by atrial fibrillation (9) and, in fact, digitalis is still given today for this combined clinical presentation. Quinine (10) was originally derived from the bark of the *Cinchona* tree and was used based on its ability to reduce fever and serve as an antimalarial. Early practitioners also noted that quinine preparations could sometimes normalize an irregular pulse. In 1853, quinidine, as opposed to quinine or other of the *Cinchona* alkaloids, was isolated by Pasteur and discovered to be effective in patients with an irregular pulse (11–13). Procaine was originally found to increase the electrical stimulation threshold of ventricular myocardium (14), and to have antiarrhythmic effects similar to quinidine. Because it was rapidly metabolized, longer lasting derivatives were sought and the result was the organic synthesis of procainamide, which also manifests less neural toxicity and remains in use today.

In the late 1950s and early 1960s, the number of available antiarrhythmics began to increase dramatically as tissue and cellular actions of these and related derivatives became known through the use of new electrophysiological techniques to study both animal and human tissue. The increase in available drugs, however, was not so much the result of innovative efforts to discover underlying pharmacological principles of antiarrhythmicity as it was an improved ability to recognize interesting and potentially useful properties of drugs under development for other purposes. Thus, lidocaine (15), originally developed as a local anesthetic, was tested for potential antiarrhythmic activity in the cardiac catheterization laboratory in 1950 (16), and subsequently further developed as an agent to treat ventricular arrhythmias. Its orally available derivatives, mexilitine and tocainide, were synthesized and tested in the mid-1980s. Flecainide and encainide were the result of attempts to increase the in vivo effectiveness of procainamide. Verapamil, shown to block L-type Ca-channel currents, was studied in the early 1960s as a coronary artery vasodilator to alleviate angina and as an antihypertensive (17). Its antiarrhythmic effects, for example on AV junctional arrhythmias, were noted during its early use. Phenytoin, which had been introduced as an antiepileptic in the late 1930s (18), was subsequently tested as an antiarrhythmic based on speculation that the mechanisms of myocardial infarction–induced arrhythmias might be similar to those for epileptic seizures (19).

Hence, early antiarrhythmic drug development frequently involved adaptation of medications in clinical use for other reasons, and new leads for effective compounds were most frequently obtained from careful clinical observations in the use of drugs for other diseases. Moreover, though highly inefficient, this empirical approach continues today. For example, azimilide, a new compound that blocks delayed rectifier K currents and is presently under investigation for several arrhythmias (20) was derived from research intended to improve antibacterial properties of nitrofurantoin. Even amiodarone, one of a few compounds that has shown promise for protecting against sudden death in subgroups of heart failure patients (21), was originally developed as an antianginal. Not surprisingly, the results of this approach, depending on which compound (even which stereoisomer) was tested in which disease setting, have led to mixed, and often confusing or contradictory data, a characteristic that continues to plague this work even today.

DRUG	Na Fast	Na Med	Na Slow	Ca	K	If	α	β	M₂	A1	Na-K ATPase	Left ventricular function	Sinus Rate	Extra-cardiac	PR	QRS	JT
Lidocaine	○											→	→	⊘			↓
Mexiletine	○											→	→	⊘			↓
Tocainide	○											→	→	●			↓
Moricizine	Ⓘ											↓	→	○		↑	
Procainamide		Ⓐ			⊘							↓	→	●	↑	↑	↑
Disopyramide		Ⓐ			⊘				○			↓	→	⊘	↑↓	↑	↑
Quinidine		Ⓐ			⊘		○		○			→	↑	⊘	↑↓	↑	↑
Propafenone		Ⓐ						⊘				↓	↓	○	↑	↑	
Aprindine		Ⓘ		○	○	○						→	→	⊘	↑	↑	→
Cibenzoline			Ⓐ	○	⊘				○			↓	→	○	↑	↑	→
Pirmenol			Ⓐ		⊘				○			↓	↑	○	↑	↑	↑→
Flecainide			Ⓐ		○							↓	→	○	↑	↑	
Pilsicainide			Ⓐ									↓→	→	○	↑	↑	
Encainide			Ⓐ									↓	→	○	↑	↑	
Bepridil	○			●	⊘							?	↓	○			↑
Verapamil	○			●						⊘		↓	↓	○	↑		
Diltiazem				⊘								↓	↓	○	↑		
Bretylium				●			◩	◩				→	↓	○			↑
Sotalol				●				●				↓	↓	○	↑		↑
Amiodarone	○			○	●		⊘	⊘				→	↓	●	↑		↑
Alinidine					⊘	●						?	↓	●			
Nadolol								●				↓	↓	○	↑		
Propranolol	○							●				↓	↓	○	↑		
Atropine									●			→	↑	⊘	↓		
Adenosine										□		?	↓	○	↑		
Digoxin										□	●	↑	↓	●	↑		↓

Relative potency of block: ○ Low   ⊘ Moderate   ● High
□ = Agonist   ◩ = Agonist/Antagonist
A = Activated state blocker
I = Inactivated state blocker

Figure 1.2 Important actions of drugs on membrane channels, receptors, and ion pumps and cardiac function in the heart. Most of these drugs have been marketed as antiarrhythmic agents, but some are not yet approved while others are no longer being used. For areas such as the clinical and electrocardiographic effects, the information available is so voluminous that the figure unavoidably includes some degree of subjectivity. Accordingly, the shading of the symbols and the direction of the arrows should not be taken as absolute. Moreover, the clinical information presented refers to the patient who does not have importantly compromised left ventricular function prior to drug administration. For the section on channels, receptors, and pumps, the actions of drugs on sodium (Na), calcium (Ca), potassium (I_k), and I_f channels are indicated. No attempt is made here to indicate effects on different channels within the Na, Ca, or K groups. Sodium channel blockade is subdivided into three groups of actions characterized by fast ($\tau < 300$ ms), medium ($\tau = 300$ to 1500 ms), and slow ($\tau > 1500$ ms) time constants for recovery from block. This parameter is a measure of use-dependence and predicts the likelihood that a drug will decrease conduction velocity of normal sodium-dependent tissues in the heart and perhaps the propensity of a drug for causing bundle branch block or proarrhythmia. The rate constant for onset of block might be even more clinically relevant. Blockade in the inactivated (I) or activated (A) state is indicated. Drug interaction with receptors [α, β, muscarinic subtype 2 (M₂)] and adenosine (A1) and drug effects on the sodium–potassium pump (Na/K ATPase) are indicated. Circles indicate blocking actions; unfilled squares indicate agonist ac-

IV. CONTEMPORARY STRATEGIES

As basic studies on arrhythmia mechanisms have advanced, we have come to know that many antiarrhythmic compounds (e.g., quinidine, procainamide, lidocaine, encainide, etc.), alter cardiac electrophysiology by changing fluxes through highly ion-specific membrane ion channels. As definition of these activities was clarified in the 1990s via innovative new methods to examine the molecular composition and biophysical properties of channels, a goal of antiarrhythmic development soon became that of trying to target the specific ion channels thought to carry various cardiac ion currents. The underlying strategy was that this might prevent development of reentrant conduction by selectively blocking, depolarizing (principally Na and Ca), or repolarizing (principally K) currents during the cardiac action potential cycle (22). Such pharmacological efforts have resulted over the past dozen years in literally hundreds of compounds being examined for potential antiarrhythmic activity (see Fig. 1.2). A much smaller number have obviously been tested in preclinical evaluations and fewer still have been examined for actual clinical utility. A major problem with this effort has been that, paradoxically, many of the agents developed were themselves proarrhythmic. Thus, while seemingly straightforward and specific at the ion channel and cellular levels, many of the new antiarrhythmic drugs (e.g., Fig. 1.2) were found to have much more complicated effects in vivo. The reasons were multiple and involved many different levels of effects, such as actions on channels in different tissues, effects of metabolites, and individual differences related to pharmacogenetic variations between patients. As a result, it is now widely acknowledged that it is all but impossible to predict biological responses to antiarrhythmics in intact individuals based simply on the projection of results, even the most optimistic ones, at only the cellular and subcellular levels (22,23).

Another dimension of this strategy arose when a number of drugs shown in Figure 1.2 were found not to be just ion channel blockers, but also to affect autonomic responses. The beginnings of this realization date to the mid-1950s when investigators pursuing state-of-the-art approaches to design better antihypertensives made important, but again, serendipitous observations. The goal was to synthesize α- and β-adrenergic receptor antagonists, and the successes here remain among the remarkable accomplishments in medicinal pharmacology. During clinical evaluation of several of these compounds it was found that patients treated with the beta-blocking drug propranolol appeared less likely to suffer acute arrhythmic death following myocardial infarction (24). Beta-antagonists were initially thought to exert this effect primarily by reducing heart rate, thus ameliorating destabilizing "triggers" (Fig. 1.1). Subsequent animal studies, including recent ones using cloned, heterologously expressed channel DNA, have since suggested that by altering activation pathways for Na, Ca, and K currents, beta-blockers (e.g., propranolol, nadolol, etc.; Fig. 1.2) also affect cardiac electrical stability (Fig. 1.1) as well as influencing extra-

---

tions; and filled/unfilled squares indicate combined agonist/antagonist actions. The intensity of the action is indicated by shading. The absence of a symbol indicates lack of effect. The use of a question mark (?) indicates uncertainty. The arrows in the clinical effect and electrocardiogram section indicate direction, no quantitative differentiation has been made between weak and strong effects. The effects listed for electrocardiogram, left ventricular function, sinus rate, and "extracardiac" are those that may be seen at therapeutic plasma levels. Deleterious effects that may appear with concentrations above the therapeutic range are not listed. (Modified with permission from Ref. 73 and 74.)

cardiac sites (e.g., neural pathways). Following these leads, the potential of autonomic-directed drugs to prevent lethal arrhythmias was explored and proven in the early 1980s with one of the most valuable tools in arrhythmia research—the large-scale, randomized, prospective clinical trial.

V. LARGE-SCALE TRIALS

The National Heart, Lung and Blood Institute's (NHLBI) *B*eta-Blocker *H*eart *A*ttack *T*rial (BHAT) was one of the first large-scale examples of this methodology. It tested the effects of propranolol in almost 4000 postinfarct patients (24) and, in the early 1980s, showed that beta-blockers have considerable value in the prevention of SCD. BHAT found that propranolol administered a short time following myocardial infarction resulted in an absolute reduction in mortality from ~10% (placebo) to ~6% (propranolol) over a 2-year follow-up period (Table 1.1). This was a relative reduction of ~35% and occurred with a coincident decrease in morbidity. In infarct patients with a history of congestive heart failure, SCD declined from ~10% (placebo) to 5% (propranolol), almost a 50% relative reduction. These landmark findings have since been repeatedly confirmed, making beta-blockade one of the most reliable, albeit underused (25), means of SCD prevention today.

Following this success, large-scale trials of similar design shortly became the "gold standard" for demonstrating antiarrhythmic efficacy in patients and the approach has remained of utmost value ever since. Detailed discussion of the design and approaches used in BHAT and similar studies, and their many important results over the intervening years, is well beyond the scope of this discussion. Nevertheless, the topic of antiarrhythmic trials is such a sufficiently rich and important subject that a separate book detailing the results of most of the important trials has been published recently (26). Because of their enormous impact on the field, and for the purpose of understanding the evolution of antiarrhythmic strategies, however, it is essential to review for just a moment some of the significant trials in our overview of strategies (Table 1.1).

One large-scale trial that has had enormous impact on treatment options began when the NHLBI initiated the *C*ardiac *A*rrhythmia *P*ilot *S*tudy (CAPS) to evaluate ion channel blockers that affect cardiac electrogenesis directly at the membrane level (27). By design, patient numbers were initially small, and a "surrogate" endpoint [each drug's ability to reduce electrophysiologically monitored premature ventricular complexes (PVCs)], was used to assess efficacy. After quite promising preliminary results, the full-scale *C*ardiac *A*rrhythmia *S*uppression *T*rial (CAST) was initiated to compare three Na-channel blockers (encainide, flecainide, and moricizine) with placebo in more than 2000 postinfarction patients (Table 1.1). Treatment was halted prematurely, however, when the Data and Safety Monitoring Board recognized in 1989 that drug-treated patients had higher total mortality than those taking a placebo. Despite clearly reducing PVCs, so-called "proarrhythmic" effects of each of these agents resulted in an actual increase in mortality (28,29). Unfortunately, the nature of this proarrhythmic action was never completely defined and remains a serious consideration even today (30). Nevertheless, a very important lesson was learned: "... suppression of PVCs and mildly symptomatic ventricular arrhythmias after infarction (supposedly valid surrogate endpoints), appears an unreliable strategy, one that could actually increase mortality" (30). Although there were significant caveats to this conclusion (e.g., appropriate primary and secondary endpoints, subgroup analysis, and patient selection), these results remain perplexing. Possible complications and explanations abound and are considered in detail in the volume on antiarrhythmic trials (26).

Table 1.1 Important Examples of Antiarrhythmia Sudden Cardiac Death Trials

Trial	Therapy tested	Results
A. Antiarrhythmic Drugs		
BHAT (1981)	Betaadrenergic antagonist (propranolol)	Reduction in post-MI mortality from 10% to 6% in 2 yrs.
CAST (1990)	Na-channel blockers (encainide, flecainide, moricizine)	Proarrhythmic increase in post-MI patients w/ PVCs from ~3% in 10–18 mon. Encainide RR[a] ~3.4. Flecainide ~ 4.4, Moricizine ~ 5.6.
SWORD (1996)	K-channel blocker (d-Sotalol)	Proarrhythmic mortality increase in post-MI patients from 2.7 to 4.6%.
DIAMOND (1998)	K-channel blocker (dofetilide)	No difference in all-cause mortality in low ejection fraction patients w/MI and heart failure
GESICA (1994)	Amiodarone	Reduction in mortality from 41 to 34% in heart failure patients.
EMIAT (1997)	Amiodarone	No difference in all-cause mortality: Reduction from 7% (placebo) to 4% (amiodarone) in arrhythmic mortality in patients w/ejection fraction < .40. Compliance < 60%.
B. Antiarrhythmic Devices		
MADIT (1996)	Drugs vs. ICD (Amio ± β-blocker)	ICDs prolong survival (>50% reduction in RR) Survival (0.84 vs. 0.70 at 2+ yrs) post-MI patients w/ejection fraction <.35 & nonsustained ventricular tachycardia
AVID (1998)	Drugs vs. ICD (Amio, rarely sotalol)	ICDs prolong survival (RR reduction = 30–40%) Survival (0.82 vs. 0.75 at 2+ yrs) Symptomatic SCD survivors, patients w/ejection fraction <.40 + sustained ventricular tachycardia
CABG-Patch (1998)	ICDs vs. no AA therapy	No survival benefit of ICDs in 3 yrs. in asymptomatic bypass recipients w/ejection fraction <.36 and abnormal SAECG

[a]RR, Relative Risk.

When CAST demonstrated that excess mortality was caused by drugs that block the fast inward current that generates the action potential upstroke, it raised the question of whether the fault lay with the drugs, the patients, or the therapeutic strategy. One issue was whether an alternative approach, of perhaps focusing on repolarizing K channels, rather than inward Na currents, might provide a better outcome. Many subsequent studies have since addressed this problem with drugs intended to inhibit the repolarizing K currents, I_{Kr} and I_{Ks}, with blockers of different and varying specificity. One recent example was the Survival With ORal D-Sotalol (SWORD) trial, which examined the ability of this agent to prevent SCD in high-risk patients with reduced ventricular function (31). Since the dextroisomer appeared more K-channel-specific, while the levo form also blocks β-receptors (Fig. 1.2), the trial was performed with d-sotalol. As in CAST, however, work was soon stopped prematurely when ongoing evaluation by the Safety Committee revealed significant proarrhythmia that compromised patient safety. At its conclusion, mortality rates were 4.6% in the treatment group and 2.7% in the placebo group, again an unacceptable result. More recently, some progress appears to have been made on the problem of proarrhythmia. Results with some of the most recent K-channel blockers have shown that it may be possible to develop I_{Kr} and I_{Ks} blockers that do not compromise patient safety to the extent of that associated with d-sotalol. For example, findings with dofetilide in the DIAMOND trial (32) and azimilide in ALIVE (20) indicate these agents may have favorable risk/benefit ratios. Unfortunately, at this stage beneficial effects on mortality have been difficult to demonstrate.

　　Another important approach illustrated in Table 1.1 has been a continuing series of trials focusing on another wide-spectrum drug—amiodarone—which has multiple effects on ion channels and adrenergic receptors, as well as other targets (Table 1.2). However, despite initially promising findings [e.g., in GESICA (21)], this drug's effectiveness in preventing SCD remains unproven. After more than a half-dozen trials [e.g., EMIAT (33); CAMIAT (34), etc.] and meta-analyses (35), it still seems unclear whether this drug may hold promise, and the only conclusion at this point is that, if there are beneficial effects, they must be small and demonstrable only in carefully selected study populations. The issue of carefully selecting drugs for specific subgroups of patients (e.g., based on ventricular performance) is becoming increasingly important and is treated in the volume on clinical trials (26).

VI.　MECHANISTIC APPROACHES TO THERAPY

What has been the outcome of these pharmacological explorations? Although it must be recognized that little overall progress in the prevention of arrhythmic death has been achieved, valuable lessons concerning appropriate strategies have indeed been learned. One of these is that attempts to prevent arrhythmogenesis with "pharmacologically unrefined" channel-blocking drugs may be *inherently* less efficacious than what can be achieved now with beta-blockers in high-risk populations (23). Whether better results can be obtained in more carefully selected populations remains an open question. Another issue is whether better targeting for particularly problematic or dysfunctional channels or currents, in only those tissues or cells responsible for initiating arrhythmias, can be more successful. As yet, good clues remain hard to find. The issue of whether there may be specific indicators to identify population subgroups that do respond to these types of drugs is also being pursued. In another exciting approach, several laboratories are working on the idea that instead of using drugs that simply block the channel pore, targeting of other sequence domains within the channel, particularly those that influence gating or inactivation, may be more effective (Chap. 26). Another potential tactic lies in designing cardiac-specific drugs that modulate regulatory interactions with accessory channel subunits, which in turn control channel activity and provide critical links with cell energy state and metabolism. Still another idea lies in exploiting information about differential intracardiac distributions of channels (36), to locally target disturbances in conducting tissues or the activation of electrogenic pathways normally silent in healthy cells, perhaps by precise pericardial or intramyocardial delivery. One approach now being tested with prototype drugs is to develop molecules that act, not as ion channel blockers, but as selective channel agonists or activators (37). Such agents could be especially useful in patients undergoing electrical and structural remodeling in chronic diseases where deficiencies in particular currents have been identified. One of the biggest challenges for the future will be to develop compounds that not only have specificity for channel proteins within the cardiac membrane, but also do not activate destabilizing mechanisms in this or other tissues to compromise patient safety. This is a particular concern because of the primary role of ion channels in the brain, vasculature, and skeletal muscle. Deleterious effects in these tissues and in heart may not be readily apparent in healthy individuals in whom such drugs are normally tested for safety, but could become manifest within the context of cells and tissues compromised with specific pathologies. Other problems such as choosing appropriate endpoints, surrogate markers, and complications such as reducing arrhythmic deaths but not all-cause mortality (33) appear to be issues that can be dealt with, especially as better noninvasive markers of arrhythmic risk are developed (Chap. 28). Unfortunately, the primary means to address them is likely to remain the large, full-scale clinical trial, an increasingly difficult, ethically challenging, and expensive proposition.

What are the events and principles that have led us to focus so hard on this difficult and frustrating approach to antiarrhythmic therapy? Most are based on the largely understated tactic that the most cardinal antiarrhythmic principle for therapy is to reduce aberrant propagation, thereby blocking reentrant conduction (Fig. 1.1). This reflects in large part successes achieved in the control of some arrhythmias by eliminating macroscopic conduction pathways with surgical or catheter ablation (Chap. 21 and 27). The concept goes back to fundamental work by a number of pioneering arrhythmologists in the early twentieth century. Notable among these was Mines (38), who not only reproduced the observation that a self-perpetuating circular excitation path could occur in cardiac tissue under experimental conditions (38), but accurately surmised the important implications of reentry in the pathology of common arrhythmias in humans (39).

Based on a largely accurate understanding of interrelationships between repolarization, refractoriness, and conduction velocity, achieved by Mines and other early workers (38–41), a useful series of postulates on reentrant conduction was developed, one that we continue to rely on today (Chap. 17). Critical among these is that termination of an arrhythmia, which propagates because of reentry, will occur on interruption of the anatomical pathway by which reexcitation occurs. Fifty years later, the potential of this idea was fully realized with the first demonstration that surgical ablation of accessory AV bypass tracts between atria and ventricles resolved arrhythmias associated with Wolff–Parkinson–White syndrome (42–44). This became the first demonstration of an interventional approach that has been highly successful and by which a number of atrial and ventricular arrhythmias can now be permanently terminated (45), albeit with some risk and considerable effort (Chap. 27).

In contrast to using macroscopic interventions like surgical or radiofrequency ablation to block propagation, channel-blocking strategies designed to prevent reentry have generally been aimed either at slowing conduction (e.g., with Na-channel blockers) or prolonging action potential duration (e.g., by inhibiting repolarizing K currents), with consequent and hopefully remedial effects on refractory periods. The underlying concept of preventing reentry by preventing reexcitation has been supported by theoretical, cellular, and intact animal electrophysiological studies, where the effects of modulating individual currents can be altered in desired or predictable ways. Clearly, the in vivo effects of most antiarrhythmic drugs on intact, diseased hearts have not had the same success as one would have hoped from a strictly channel-based interpretation of Mines' original principles.

VII. A NEW PARADIGM?

Insights into the complexities of ion channel function and the clinical expression of channel anomalies have emerged from recent clinical and molecular genetic studies on rare inherited arrhythmias associated with unexpected arrhythmic death. Moreover, the implications of this work go far beyond the relatively rare congenital arrhythmias and impact as well on multiple drug–channel interactions. Much of this began when pioneering work in the mid-1990s (46) revealed that defects in electrogenesis caused by mutations in genes coding for ion channel subunits underlie many hereditary forms of the long QT syndrome (LQTS) (47,48), some cases of idiopathic ventricular fibrillation (IVF) or Brugada's syndrome (49), and perhaps other congenital conditions (50) (Chap. 25). Figure 1.3 illustrates the range of mutations in channel subunits for the human cardiac Na protein (SCN5A), two human cardiac K channels (HERG and KvLQT1), and the K-channel accessory β-subunits, minK, and MiRP1, that have been discovered (51). Most of these have been identified as a result of linkage studies on patients with different forms of LQTS and Brugada's IVF and their diversity supports the concept that even subtle changes in channel function can contribute importantly to SCD. To design drugs that only alter specific functions of these proteins in the face of such sensitivity clearly remains a daunting task, although one that could markedly improve as our knowledge of channel structure does as well (52,53). The idea of attempting to rectify dysfunctional channels with directed therapies thus remains a valid strategy, but one that will require really innovative breakthroughs.

The possibility of using gene-directed drug treatment to prevent arrhythmias in such defined "single-current" congenital disorders as LQTS recently raised the very important question of whether this same paradigm might help provide direction for common arrhythmias in patients with acquired arrhythmias. Although functional changes in particular channels (e.g., a decrease in I_{to} in heart failure or I_{Kur} in atrial fibrillation) have frequently been associated with various disease conditions, there remain major difficulties in generalizing this model of arrhythmogenesis and its implications for therapy. First, even with the most virulent mutations, it appears alterations in almost any single-channel current may be insufficient to elicit a SCD event. There appears considerable redundancy in channels for both positive and negative charge and most patients with LQTS or Brugada mutations do not die at birth and only experience events when other factors interact in a particularly malignant constellation, resulting in an overlap of critical events as predicted in Figure 1.1 (5). Second, it appears we do not yet understand most aspects of what is called the "genotype–phenotype" problem (54). That is, what is the precise sequence of mechanisms that link clinical symptoms and arrhythmic events to particular changes in a channel protein at a particular time, in a particular interactive environment. We know, for example, that ventricular arrhythmias in LQT3 patients can arise because delayed Na entry via SCN5A channels prolongs the action potential (55), but how is it that an amino acid replacement in one channel domain, such as the fourth to last SCN5A transmembrane sequence, results in one clinical presentation, the LQTS syndrome, while a similar mutation several residues farther toward the carboxyl terminus of the same subunit, results in quite another (as in the Brugada syndrome) (see Fig. 1.3.)? Similarly, how is it that quite different mutations in the protein domain for a particular function (e.g., ion specificity) in one channel may result in a nearly identical clinical presentation as occurs in other situations as a result of alterations

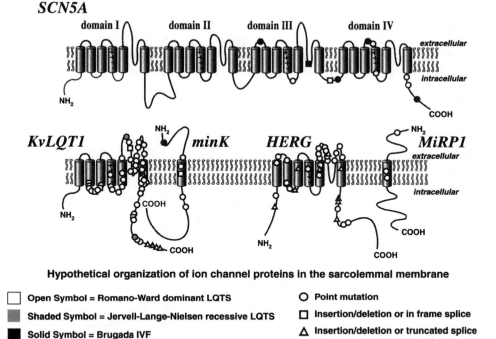

Figure 1.3 Hypothetical organization of major cardiac ion channel proteins in the cardiac membrane. The linear sequences shown here do not reflect the three-dimensional organization of functional channels. To form three-dimensional functional channels, the subunits shown here fold as a single unit (the sodium channel encoded by SCN5A) or as tetramers of the potassium channel subunits shown on the second line to form pore-containing complete structures. The pore is lined by the fifth and sixth transmembrane segments within major domains. The location of amino acid changes resulting from mutations known to exist as of 1999 are shown. Those in unfilled symbols are associated with the autosomal dominant (Romano-Ward) form of the long QT syndrome (LQTS); in gray, the autosomal recessive (Jervell-Lange-Nielsen) form of LQTS; and in black, Brugada syndrome (a congenital form of idiopathic ventricular fibrillation). Circles represent point mutations, squares are insertion/deletion events or splice alterations that leave the open reading frame intact, and triangles are insertion/deletion events or splicing errors that are predicted to result in proteins truncated at the locations shown. (Modified from Ref. 51.)

in very different sequence domains (e.g., voltage gating)? Even within families of patients with identical inherited LQT mutations, some individuals remain completely asymptomatic whereas others succumb at a young age (52). To attribute such differences to the diffuse concept of "incomplete penetrance" may simply be begging the question. Indeed, it is likely that other factors within the three types of elements necessary for arrhythmia development are involved (Fig. 1.1). Otherwise, factors such as age, sex, drugs, and environmental influences would not affect clinical presentation in ways that have now been amply documented (56). Independently inherited susceptibilities, mutations, or outcome-determining variations in other proteins thus seem likely to play a particularly important role, and it is worth noting that even minor variations in β-adrenergic receptor protein sequence have a major role in the survival of heart failure patients (57). Another example, also with clinical import, has been the finding that subtle, subclinical mutations in the accessory channel protein MiRP1 can influence susceptibility to environmental influences, such as antibiotics, and perhaps other non-QT-prolonging drugs and biologics (58).

There are also other indications that although the "channelopathy" paradigm reveals important clues regarding arrhythmia causes, there must be other influences, as yet undiscovered. Such evidence has come most recently from population and epidemiological studies. One case-controlled evaluation of the incidence of fatal arrhythmic events in first-degree relatives of SCD victims provided convincing evidence that previously recognized risk factors (i.e., blood pressure, lipid levels, diabetes, smoking, myocardial fibrosis, etc.) account for only somewhat more than half the familial risk of SCD in related individuals (59). Yet another study, in an even larger population, found that elevated relative inherited risks for primary SCD and fatal myocardial infarction appear not only to show a strong familial element, but are also, surprisingly, transmitted independently of each other (60). The eventual elucidation of the basis for these observations will be complex, but undoubtedly will help provide new insight into both acquired as well as inherited risks.

VIII. ELECTRICAL PREVENTION AND THERAPY

Meantime, the focus on antiarrhythmic drug development has been overshadowed by quite a different approach—actually a totally different type of technology. The implantable cardiac defibrillator (ICD), developed in the early 1980s, has become the mainstay strategy for SCD prevention. Its advantages have been demonstrated in direct comparison with pharmacological approaches in several, especially important, clinical trials (Table 1.1). The *M*ulticenter *A*utomatic *D*efibrillator *I*mplantation *T*rial (MADIT) (61) and the NHLBI's *A*ntiarrhythmics *V*ersus *I*mplantable *D*efibrillator (AVID) trial (62) demonstrated in the late 1990s a clear short-term survival advantage for ICDs in high-risk patients, as compared to therapy with amiodarone, sotalol, and other approved channel- or receptor-directed drugs. In AVID, 3-year survival rates of patients who had previous episodes of ventricular fibrillation or sustained ventricular tachycardia were approximately 75% with an ICD versus 64% with treatment with amiodarone—an absolute difference of 11% and a relative advantage of about 30%. Although overall costs for use of ICDs (vs. drugs) were higher (~ $25,000 additional per patient per year life saved), this increase in cost compared favorably with other forms of life-prolonging therapies (63).

Still, many problems remain and, although attractive and today essential, it also appears ICDs may not provide the optimal long-term answer. Devices, no matter how sophisticated, have finite failure and replacement rates, and at present most do not provide effective long-term protection. Thus, for example, although initial AVID results were encouraging, long-term mortality was less noticeably affected and about 40% of deaths in patients treated with an ICD were still arrhythmic. This may improve as such devices evolve technically. The problem is complicated, however, because we also do not yet know how to predict in which individuals or patient subgroups such devices are most likely to be effective. One trial (CABG-Patch), showed, for example, that for high-risk patients undergoing coronary artery bypass grafting, such devices are not effective at all in preventing SCD (64). Nevertheless, they are today's front-line prevention, and both the American Heart Association and the American College of Cardiology endorse ICD use for many patient categories (65). Their use and cost-effectiveness are therefore likely to increase over coming years until better pharmacological or ablative therapies emerge that do more than effect a postevent "rescue" of afflicted patients once an arrhythmia has occurred.

IX. DIRECTIONS FOR THE FUTURE

How do we move on from here and what are the research directions to improve SCD prevention? Progress is occurring on many fronts—toward better ways of identifying patient risk (Chap. 28); toward better electrical devices, detection algorithms, and improved indications for their use (Chap. 22); and toward better ideas for development of new pharmaceuticals (Chap. 26). Various combinations of different modes of therapy (e.g., drugs plus devices) are also beginning to be explored and initial results are promising (66). In addition, information from basic research is providing clues regarding new molecular and cellular mechanisms with as yet unknown potential. For example, basic studies suggest increases in intracellular Ca may be especially arrhythmogenic (Chap. 12). Thus, recent work on calmodulin kinase (CAM kinase) suggests that elevations in Ca occur in response to activation of the Na–Ca exchanger, and changes in this enzyme as might result in ischemia (68). The differential distribution of Na, Ca, K, and gap junction channel structural and regulatory subunits within different cardiac cell types and tissues, and their changes during disease or remodeling (Chap. 6 and 24), may also be especially important, but as yet we have little information as to their relevance in humans. Other possibilities include targeting different species of K, Ca, and Na channels, including cyclic nucleotide-gated channel proteins involved in pacemaker currents (Chap. 3 and 7). A quite different tack might involve targeting neural elements that play important roles in modulating triggering influences, like changes in autonomic tone (67). Other interesting clues are being discovered in studies on cardiac cell metabolism and on pathways of inter- and transcellular signaling. These may have significant implications for both triggering and propagation of arrhythmia. Interesting leads are also coming from work on inherited differences (mutations and polymorphic variations) in regulatory proteins that affect triggering or propagation (e.g., β_2-receptors) (56), or as yet only glimpsed possibilities. The latter would include changes in oxidative cell metabolites that influence channel regulatory subunits (69) or transcriptional factors that coordinately influence expression of genes for multiple, related channels or autonomic receptors (70).

As can be appreciated from this diversity of possibilities, the problem for future researchers will be to evaluate which of the hundreds of processes within the fundamental biology of the heart can best be exploited to reduce arrhythmic risk. To alleviate this risk, prevention must be the key goal for therapy. The challenge is to design approaches to intervene most appropriately in specific individuals or in a particular disease setting, prior to the onset of events. Based on recent progress, we anticipate that important new leads will come from an increasing focus on the role of genetic variation in inherited and acquired disease, an understanding of the importance of factors underlying new epidemiological observations, and on development of new molecular, biochemical, and biophysical indices of risk, such as receptor function or pharmacogenetic variations that determine drug sensitivity. Over the coming years our abilities here will clearly improve by orders of magnitude as a result of new knowledge obtained as the Human Genome Program progresses (71).

In closing, it is important to mention that future progress is also likely to require an increasing emphasis on integration of information from multiple types of research and development of a systems approach to the highly dynamic factors that underlie most arrhythmias (Fig. 1.4). The goal must remain development of individualized therapy for SCD prevention in patients with particular structural and functional deficits. Pharmacological or other approaches based on purely reductionist observations (i.e., laser narrow targeting of particular currents that reflect existing drug and disease classifications) will likely continue to prove inadequate. As Figure 1.4 emphasizes, progress will require new insights into factors regarding critical interactions among many underlying cellular, molecular, and metabolic elements, as well as a focus on their development within the whole organism. Understanding the dynamics and importance of the integrative aspects of arrhythmias will require approaches not readily predictable simply on the basis of the behavior of isolated components, as work on global propagation properties involved in reentry has already shown (Chap. 14). Integrative approaches represent some of this field's most exciting new directions and will be critical to future progress. We would encourage all those who share the goal of arrhythmia prevention to carefully consider potential future contributions of each of these major areas of promise—genetic variation, epidemiological clues, risk assessment and multilevel integration—in the context of the fundamental insights and knowledge that follow in the chapters of this book.

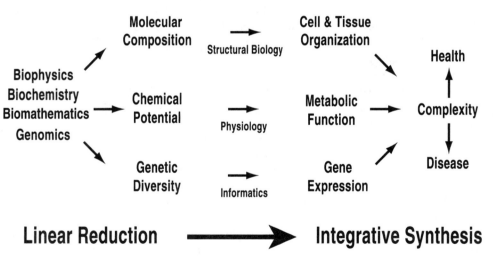

Figure 1.4 Pathways of reductionist and integrative approach to complex dynamic biological phenomena. This figure emphasizes multiple ways fundamental scientific disciplines (mathematics, physics, chemistry, genomics) can be directed at delineation of molecular, chemical, and genetic potential. In the context of various biological approaches (structural biology, physiology, and informatics), these approaches form our understanding of actual biological phenomena within the organism. Understanding the dynamic, synthetic interactions among those which contribute to the specific levels of complexity and the uniqueness of each individual is the goal of integrative approaches and represents the real key in understanding transitions between health and disease. Absent from the figure are interactions with various aspects of the environment and its impact on each unique biological function. With arrhythmias, interactions within various elements shown in Figure 1.1 are major determinants that alter how that complexity influences the origin and outcome of a potentially lethal event. (Based on discussions with Dr. David Robinson.)

REFERENCES

1. Myerburg RJ, Interian A, Jr, Mitrani RM, Kessler KM, Castellanos A. Frequency of sudden cardiac death and profiles of risk. Am J Cardiol 1997; 80:10F–19F.
2. American Heart Association. 1999 Heart and Stroke Statistical Update. Dallas, TX, 1999.
3. Zipes DP, Wellens HJJ. Sudden cardiac death. Circulation 1998; 98:2334–2351.
4. Hennekens CH. Increasing burden of cardiovascular disease: Current knowledge and future directions for research on risk factors. Circulation 1998; 97:1095–1102.
5. Baum RS, Alvarez H, Cobb LA. Survival after resuscitation from out-of-hospital ventricular fibrillation. Circulation 1974; 50:1231–1235.
6. Uretsky BF, Sheahan RG. Primary prevention of sudden cardiac death in heart failure: Will the future be shocking. J Am Coll Cardiol 1997; 30:1589–1597.
7. Rosamond WD, Chambless LE, Folsom AR, Cooper LS, Conwill DE, Clegg L, Wang C-H, Heiss G. Trends in the incidence of myocardial infarction and in mortality due to coronary heart disease, 1987 to 1994. N Engl J Med 1998; 339:861–867.
8. Withering W. An account of the foxglove, and some of its medical uses, etc. Birmingham: Robinson, 1785.
9. Senac JB de. Traité de la Structure du Coeur, de Son Action, et de Ses Maladies. Paris: Vincent, 1749.
10. Snell WE. Nathan Drake MD: A literary practitioner and his illness. Proc R Soc Med 1965; *58*:263.
11. Pasteur L. Note sur la quinidine. CR Acad Sci 1853; 36:26.
12. Pasteur L. Recherches sur les alcaloides des quiniquinas. CR Acad Sci 1853; 37:110.
13. Frey W. Weitere Enfahrungen mit Chinidin bei Absoluter Herzunregelmassigkeit. Wein Klin Wochenschr 1918; 55:849.
14. Mautz FR. The reduction of cardiac irritability by the epicardial and systemic administration of drugs as a protection in cardiac surgery. J Thorac Surg 1936; 5:612.
15. Lofgren N. Studies on Local Anesthetics. Xylocaine, A New Synthetic Drug. Stockholm: Ivar Haeggstroms, 1948.
16. Southworth JL, McKusick VA, Peirce EC, Rawson FL Jr. Ventricular fibrillation precipitated by cardiac catheterization. JAMA 1950; 143:717.
17. Haas H, Hartfelder G. Alpha-isoprophyl-α-[(N-methyl-N-hornoveratryl)-8-aminopropyl]-3,4-dimethoxyphenyl-acetonitril, eine Substanz mit Coronargefaberweiternden Eigenshaften. Arzneimittel Forsh 1962; 12:549.
18. Merritt HH, Putnam TJ. Sodium diphenylhydantoin in treatment of convulsive disorders. JAMA 1938; 111:1068.
19. Harris AS, Kokernot RH. Effects of diphenylhydantoin sodium (dilantin sodium) and phenobarbital sodium upon ectopic ventricular tachycardia in acute myocardial infarction. Am J Physiol 1950; 163:505.
20. Camm AJ, Karam R, Pratt CM. The azimilide post-infarct survival evaluation (ALIVE) trial. Am J Cardiol 1998; 81:35D–39D.
21. Doval HC, Nul DR, Grancelli HO, Perrone SV, Bortman GR, Curiel R, for the Grupo de Estudio de la Sobrevida en la Insuficiencia Cardiaca en Argentina (GESICA). Randomised trial of low-dose amiodarone in severe congestive heart failure. Lancet 1994; 344:493–498.
22. Task Force of the Working Group on Arrhythmias of the European Society of Cardiology. The Sicilian Gambit. Circulation 1991; 84:1831–1851.
23. Members of the Sicilian Gambit. The search for novel antiarrhythmic strategies. Eur Heart J 1998; 19:1178–1196.
24. Friedman LM, Byington RP, Capone RJ, Furberg CD, Goldstein S, Lichstein E, for the Beta-Blocker Heart Attack Trial Research Group. Effect of propranolol in patients with myocardial infarction and ventricular arrhythmia. J Am Coll Card 1986; 7:1–8.
25. Gottlieb SS, McCarter RJ, Vogel RA. Effect of beta-blockade on mortality among high-risk and low-risk patients after myocardial infarction. N Engl J Med 1998; 339:489–497.

26. Woosley RL, Singh SN, eds. Arrhythmia Treatment and Therapy: Evaluation of Clinical Trial Evidence. New York: Marcel Dekker, 2000.

27. Greene HL, Richardson DW, Hollstrom AP, McBride R, Capone RJ, Barker AH, Roden DM, Echt DS. Congestive heart failure after acute myocardial infarction in patients receiving antiarrhythmic agents for ventricular premature complexes (cardiac arrhythmia pilot study). Am J Cardiol 1989; 63:393–398.

28. Cardiac Arrhythmia Suppression Trial Investigators. Special Report. Effect of encainide and flecanide on mortality in a randomized trial of arrhythmia suppression after myocardial infarction. N Engl J Med 1989; 321:406–412.

29. Cardiac Arrhythmia Suppression Trial II Investigators. Effect of the antiarrhythmic agent moricizine on survival after myocardial infarction. N Engl J Med 1992; 327:227–233.

30. Friedman PL, Stevenson WG. Proarrhythmia. Am J Cardiol 1998; 82:50N–58N.

31. Waldo AL, Camm AJ, deRuyter H, Friedman PL, MacNeil DJ, Pauls JF, Pitt B, Pratt CM, Schwartz PJ, Veltri EP. Effect of d-Sotalol on mortality in patients with left ventricular dysfunction after recent and remote myocardial infarction, SWORD [Survival With *Oral* d-Sotalol]. Lancet 1996; 348:7–12.

32. Torp-Pederson C, Moller M, Block-Thomson PE, Kober L, Sandoe E, Egstrup K, Agner E, Carlsen J, Videbaek J, Marchant B, and Camm AJ. Dofetilide in patients with congestive heart failure and left ventricular dysfunction. N Engl J Med 1999; 341:857–865.

33. Julian DG, Camm AJ, Frangin G, Janse MJ, Munoz A, Schwartz PJ, Simon P. Randomized trial of effect of amiodarone on mortality in patients with left-ventricular dysfunction after recent myocardial infarction: European myocardial infarct amiodarone trial investigators: EMIAT. Lancet 1997; 349:667–674.

34. Cairns JA, Connolly SJ, Roberts R, and Gent M, for the CAMIAT Trial Investigators. Randomized trial of outcome after myocardial infarction in patients with frequent or repetitive ventricular premature depolarizations: CAMIAT. Lancet 1997; 349:675–682.

35. AMIODARONE Trials Meta-Analysis Investigators. Effect of prophylactic amiodarone on mortality after acute myocardial infarction and in congestive heart failure: Meta-analysis of individual data from 6500 patients in randomized trials. Lancet 1997; 350:1417–1424.

36. Dixon JE, Shi W, Wang H-S, MacDonald C, Yu H, Wymore R, Cohen IS, McKinnon D. Role of the Kv4.3 K$^+$ channel in ventricular muscle: A molecular correlate for the transient outward current. Circ Res 1996; 79:1–10.

37. Salata JJ, Jurkiewicz NK, Wang J, Evans BE, Orme HT, Sanguinetti MC. A novel benzodiazepine that activates cardiac slow delayed rectifier K currents. Mol Pharm 1998; 54:220–230.

38. Mines GR. On dynamic equilibrium in the heart. J Physiol (Lond) 1913; 46:350–383.

39. Mines GR. On circulatory excitation on heart muscles and their possible relation to tachyarrhythmia and fibrillation. Trans R Soc Can 1914; 4:43–53.

40. Garrey WE. The nature of fibrillatory contraction of the heart. Its relation to tissue mass and form. Am J Physiol 1914; 33:397–414.

41. Mayer AG. Rhythmical pulsation in scyphomedusae. Washington DC: Carnegie Institute Publication No. 47, 1906.

42. Durrer D, and Roos JP. Epicardial excitation of the ventricles in a patient with a Wolff-Parkinson-White syndrome (type B). Circulation 1967; 35:15–21.

43. Burchell HB, Frye RB, Anderson MW, McGoon DC. Atrioventricular and ventriculoatrial excitation in Wolff-Parkinson-White syndrome (type B). Circulation 1967; 36:663–672.

44. Cobb FR, Blumenschein SD, Sealy WC, Boineau JP, Wagner GS, Wallace AG. Successful surgical interruption of the bundle of Kent in a patient with Wolff-Parkinson-White syndrome. Circulation 1968; 38:1018–1029.

45. Morady F. Radio-frequency ablation as treatment for cardiac arrhythmias. N Engl J Med 1999; 340:534–544.

46. Keating MT. The long QT syndrome, a review of recent molecular genetic and physiologic discoveries. Medicine 1996; 75:1–5.

47. Priori SG, Bahranin J, Haver RNW, Haverkamp W, Jongsma HJ, Kleber AG, McKenna WJ, Roden DM, Rudy Y, Schwartz K, Schwartz PJ, Towbin JA, Wilde AM. Genetic and molecular basis of cardiac arrhythmias impact on clinical management, Parts I and II. Cirulation 1999; 99:518–528.

48. Priori SG, Bahranin J, Haver RNW, Haverkamp W, Jongsma HJ, Kleber AG, McKenna WJ, Roden DM, Rudy Y, Schwartz K, Schwartz PJ, Towbin JA, Wilde AM. Genetic and molecular basis of cardiac arrhythmias impact on clinical management, Part III. Circulation 1999; 99:674–681.

49. Chen Q, Kirsch GE, Zhang D, Brugada R, Brugada J, Brugada P, Potenza D, Moya A, Borggrefe M, Breithardt G, Ortiz-Lopez R, Wang Z, Antzelovitch L, O'Brien RE, Schulze-Bahr E, Keating MT, Towbin JA, Wang Q. Genetic basis and molecular mechanism for idiopathic ventricular fibrillation. Nature 1998; 392:293–296.

50. Schwartz PJ, Stramba-Badiale M, Segantini A, Austoni P, Bosi G, Giorgetti R, Grencini F, Marni ED, Perticone F, Rosti D, Salife D. Prolongation of the QT Interval and the Sudden Infant Death Syndrome. N Engl J Med 1998; 338:1709–1714.

51. Roden DM, Spooner PM. Inherited long QT syndromes: A paradigm for understanding arrhythmogenesis: proceedings of an NHLBI, NIH ORDR Workshop. J Cardiovasc Electrophys 2000; 10:1664–1683.

52. Doyle DA, Cabral M, Pfuetzner JH, Kuo RA, Gulbis JM, Cohen SL, Chait BT, MacKinnon R. The structure of the potassium channel: molecular basis of K conduction and specificity. Science 1998; 280:69–77.

53. Gulbis JM, Mann S, MacKinnon R. Structure of a voltage-dependent K channel beta subunit. Cell 1999; 97:943–952.

54. Priori SG, Nepolitano C, Schwartz PJ. Low penetrance in the long QT syndrome clinical impact. Circulation 1999; 99:529–533.

55. Bennet PB, Yazawa K, Makita N, George AL Jr. Molecular mechanism for an inherited cardiac arrhythmia. Nature 1995; 376:683–685.

56. Locati EH, Zareba W, Moss AJ, Schwartz PJ, Vincent GM, Lehmann MH, Towbin JA, Priori SG, Napolitano C, Robinson JL, Andrews M, Timothy H, Hall WJ. Age and sex-related differences in clinical manifestations in patients with congenital long-QT syndrome: Findings from the international LQTS registry. Circulation 1998; 97:2237–2244.

57. Ligget SB, Wagoner LE, Craft LL, Hornung RW, Holt BD, McIntosh TC, Walsh RA. The Ile B$_2$-adrenergic receptor polymorphism adversely affects the outcome of congestive heart failure. J Clin Invest 1998; 102:1534–1539.

58. Abbott GW, Sesti F, Splawski I, Buck ME, Lehmann MH, Timothy KW, Keating MT, Goldstein SAN. MiRP1 forms I$_{kr}$ potassium channels with HERG and is associated with cardiac arrhythmia. Cell 1999; 97:175–187.

59. Friedlander Y, Siscovick DS, Weinmann S, Austin MA, Psaty BM, Leuaitre RN, Arbogast P, Raghunathan TE, Cobb LA. Family history as a risk factor for primary cardiac arrest. Circulation 1998; 97:155–160.

60. Jouven X, Desnos M, Guerot C, Ducimetiere P. Predicting sudden death in the population: The Paris prospective study I. Circulation 1983; 99:1978.

61. Moss AJ, Hall WJ, Cannon DS, Daubert JD, Higgins SL, Klein H, Levine JH, Saksena S, Waldo AL, Wilber D, Brown MW, *Heo* M. Improved survival with an implanted defibrillator in patients with coronary disease at high risk for ventricular arrhythmia. N Engl J Med 1996; 335:1933–1940.

62. The Antiarrhythmics vs. Implantable Defibrillators (AVID) Investigators. A comparison of antiarrhythmic-drug therapy with implantable defibrillators in patients resuscitated from near-fatal ventricular arrhythmias. N Engl J Med 1997; 337:1576–1583.

63. Mushlin AI, Hall J, Zwanziger J, Gajary E, Andrews M, Marron R, Zou KH, Moss AT. The cost effectiveness of automatic implantable cardiac defibrillators: Results from MADIT. Circulation 1998; 97:2129–2135.

64. Bigger JT Jr, for the Coronary Artery Bypass Graft (CABG) PATCH Investigators. Prophylactic use of implanted cardiac defibrillators in patients at high risk for ventricular arrhythmias after coronary-artery bypass graft surgery. N Engl J Med 1997; 337:1569–1575.

65. Gregoratos G, Cheitlin MD, Conill A, Epstein AE, Fellows C, Ferguson TB, Freedman RA, Hlatky MA, Narcarolli GV, Saksena S, Schlant RC, Silka MJ. ACC/AHA guidelines for implantation of cardiac pacemakers and antiarrhythmia devices, executive summary. Circulation 1998; 97:1325–1335.

66. Pacifico A, Hohnloser SH, Williams JH, Tao B, Saksena S, Henry PD, Prystowsky EN. Prevention of implantable-defibrillator shocks by treatment with sotalol, d,1-sotalol implantable cardioverter-defibrillator study group. N Engl J Med 1999; 340:1855–1869.

67. Shi W, Wymore RS, Wang HS, Pan Z, Cohen IS, McKinnon D, Dixon JE. Identification of two nervous system-specific members of the erg potassium channel gene family. J Neurosci 1997; 17:9423–9429.

68. Wu Y, Roden DM, Anderson ME. Calmodulin kinase inhibition prevents development of the arrhythmogenic transient inward current. Circ Res 1999; 84:906–912.

69. Gulbis JM, Mann S, Mackinnon R. Structure of a voltage-dependent K channel beta subunit. Cell 1999; 97:943–952.

70. Zhou J, Morata M, Valverde P, Koren G. Upregulation of a Delayed Rectifying Outward Current in LQT Mouse Ventricular Myocytes. Circulation, in press.

71. Collins FS, Patrionas A, Jordon E, Chakravati A, Gesteland R, Walters L, et. al. New goals for the U.S. human genome project: 1998–2003. Science 1998; 282:682–689.

72. Zipes DP, Wellens HJJ. Sudden cardiac death. Circulation 1998; 98:2334–2351.

73. Kodama I, Ogawa S, Inoue H, Kasanuki H, Kato T, Mitamura H, Hiraoka M, Sugimoto T. Profiles of aprindine, cibenzoline, pilsicainide, and pirmenol in the framework of the Sicilian gambit. Jpn Circ J 1999; 63:1–12.

74. Schwartz PJ, Zaza A. The Sicilian gambit revisited, theory and practice. Eur Heart J 1992; 13:23F–29.

2

Cardiac Membrane and Action Potentials

DAN M. RODEN

Vanderbilt University School of Medicine, Nashville, Tennessee

I. INTRODUCTION

The cardiac action potential is the representation of the time dependence of voltage during the cardiac cycle. The shape and duration of this distinctive electrical signature is determined by the activity of individual ion currents flowing across the cell membrane during the action potential. Normal cardiac mechanical activity relies on the transmission of extracellular signals to the cell interior during the action potential to initiate contraction. Action potentials also provide excitatory stimuli for neighboring cells. Thus, the cardiac action potential underlies the normal spread of electrical activity and resultant contraction that characterize heart muscle. For scientists interested in the function of individual proteins (be they ion channels or others), the net effect of perturbed protein function is often conveniently evaluated at the level of the action potential. On the other hand, for the whole-heart physiologist, action potentials in single cells provide a convenient starting point from which the integrated activity of multiple cells can be evaluated. This central position of action potential physiology and pathophysiology in cardiac electrophysiology persists, and indeed has assumed new importance in a modern molecular context.

II. RECORDING CARDIAC ACTION POTENTIALS

The behavior of individual cardiac action potentials is sometimes inferred from recording the electrocradiogram from the body surface (for further details, see Chap. 13). Thus, prolongation of the QRS duration implies slowing of conduction within the ventricle. Similarly, prolongation of the QT interval implies prolongation of action potential duration, at least in some cells in the ventricle. While techniques are available to record action poten-

21

tials using extracellular techniques in situ, actual measurement of transmembrane potential is the gold standard. This chapter will describe the methods and their potential pitfalls in brief terms; the reader is referred to more specialized texts (1–3), including those accompanying contemporary electronic data acquisition hardware, for further theoretical and practical background.

A voltmeter is used to record the voltage (potential) across the membrane of a cell. Two inputs are required: an electrode to record voltage within the cell and an indifferent electrode, usually placed in the bath, as a reference. Glass microelectrodes are used to measure voltage changes across the cell membrane during an action potential. The electrodes are heated, drawn to a fine tip, and filled with a concentrated salt solution, most often 3 M KCl. The tip resistance is typically 10–20 MΩ. The glass microelectrode, mounted on the micromanipulator, is brought in apposition to a cardiac cell and the membrane is punctured by the electrode tip (Fig. 2.1). In most cells the membrane promptly seals around the microelectrode and the cell incurs no obvious damage.

The glass wall of the electrode provides an insulating medium, and thus the salt solution within the electrode acts like a conductor that is insulated except at the tip. The transmembrane potential measured by the amplifier may also include small components ("junction potentials") due to the interfaces among glass, KCl, and cytoplasm. Commercially available amplifiers and recording systems (usually digital) have appropriate frequency responses to faithfully record the rapid changes of voltage that occur during an action potential. The recording setup generally requires some form of shielding (e.g., a Faraday cage) to minimize ambient electrical noise, particularly that generated by nearby power lines and fluorescent lights.

III. VOLTAGE CLAMP

The recording of voltage differences across the cardiac cell membrane as a function of time (i.e., the action potential) is one rather simple application of the approach of using microelectrodes to gain insights into cardiac physiology. The action potential represents the

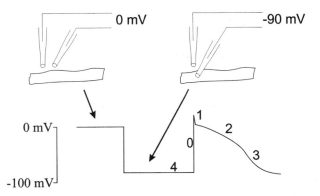

Figure 2.1 Recording a cardiac action potential. The amplifier records the voltage difference between the tip of the electrode and an extracellular reference electrode. With both electrodes in a tissue bath, the difference in potential is 0 mV. When one electrode punctures the cell, the resting potential (–90 mV in this case) is measured. The amplifier can then record action potentials occurring either spontaneously or after a stimulus delivered by the experimenter. The four different phases of the action potential are numbered according to common convention.

integrated effect of multiple distinct ion currents flowing across the membrane. As described in more detail in Chapter 3, the magnitude of a specific ionic current is determined by the number of ion channels underlying that current in the membrane, by the membrane potential, and by conductance, which is itself often a complex function of factors such as membrane voltage, time, or ligand binding. By holding voltage constant (voltage "clamping"), the time dependence of ionic current at a specific voltage can be directly observed (Fig. 2.2A). A variety of techniques, outlined below, can then be used to eliminate overlapping currents, allowing the investigator to analyze time and voltage dependence of individual ion currents to derive models of channel function; these have provided extraordinary insights into ion channel physiology and pathology.

The approach was originally developed by Hodgkin and Huxley (4) to measure sodium and potassium currents flowing across the membrane of the squid giant axon, and has since been adapted for use in many other preparations, including the heart. The fundamental principle is that current flowing across the cell membrane during an abrupt change in membrane potential consists of two components, the first due to charging the membrane capacitance and the second due to ion current flow. Thus, once the membrane capacitance has been charged, the magnitude of current injected by the voltage-clamp amplifier to maintain the membrane potential at the desired level is exactly equal to the membrane current (but of opposite sign).

The most commonly used contemporary application of the voltage-clamp experiment to cardiac tissue involves the patch-clamp technique (5). Here, a polished glass microelectrode is brought in close apposition to an isolated cell and a very high-resistance seal (Gigaohms) is allowed to form. Individual channels in the membrane patch isolated by

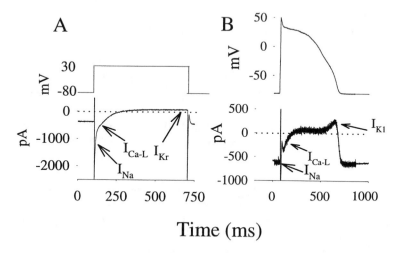

Time (ms)

Figure 2.2 Two types of voltage-clamp experiments in rabbit myocytes. (A) Current in response to a step voltage command (top) is shown. The major time-dependent currents recorded in this experiment are inward current through sodium and L-type calcium channels and outward I_{Kr}. (B) Currents recorded in response to a voltage-clamp command pulse that reproduces the action potential are shown. The major currents recorded here are I_{Na}, I_{Ca-L}, and I_{K1}, whose role in generating rapid terminal repolarization is well demonstrated here. As discussed in the text, both approaches have advantages and disadvantages. (Tracings courtesy of Mark E. Anderson and Yuejin Wu, Vanderbilt University.)

the high-resistance seal can then be voltage clamped. Alternatively, gentle pressure (e.g., suction) can be applied to rupture the patch, allowing the contents of the electrode to form a conductive pathway with the cell interior and thus the whole cell to be voltage clamped. With single-electrode voltage clamp applied to isolated patches or cells, the same electrode serves both to pass current and to monitor voltage.

Most voltage-clamp experiments use abrupt steps from one command potential to another. This approach allows the experimentalist to observe the time-dependent behavior of a specific ion current. However, voltage-clamp command pulses need not be square waves. The use of nonsquare-wave command pulses denies the experimentalist the ability of observing time-dependent behavior at a fixed voltage, but may offer other advantages. One example is the use of a slow ramp command pulse to assess multiple K^+ currents prior to and following exposure to a drug or other intervention (6). Another interesting approach is the use of a previously recorded action potential as the command pulse (7,8). This technique allows an estimate of the contribution of individual currents during each phase of the action potential, and inferences on the interrelationships among currents (Fig. 2.2B).

A. Experimental Problems in Voltage Clamp

One key requirement of the voltage-clamp technique is spatial control of membrane voltage. Current injection at a point in a membrane results in potential changes that decline exponentially with distance from that point, with the rate of decline determined by the length constant. The length constant of multicellular cardiac (or other) preparations depends on factors such as cell diameter and geometry, but may be ~100 μ, the length of a single cardiac myocyte (i.e., the change in voltage 100 μ from the point of injection will be ~37% of that seen at the point of injection). Thus, some mechanism must be incorporated into the method to allow spatial uniformity of voltage over the region of study. In the squid giant axon, which is a large biological preparation, it is possible to insert multiple wires axially into the cytoplasm to achieve this goal. One electrode measures voltage and one is used to inject current to maintain the voltage at the level desired by the investigator. For other large biological preparations, two glass microelectrodes are placed in close proximity to allow accurate and rapid control of the membrane voltage in the region between the two electrodes. To further minimize the problem of spatial control, experimentalists who attempt to voltage clamp multicellular cardiac tissues have used artificially shortened axial preparations (e.g., Purkinje fiber). Similar considerations apply to the use of two-electrode voltage clamp in *Xenopus* oocytes, which are approximately 1 mm in diameter. However, except under unusual conditions (e.g., attempts to record extremely large currents), spatial control is not generally a problem in voltage clamping single cardiac cells.

Unlike the situation with recording transmembrane voltages, the tip diameter of a patch electrode is typically large (~1 μ in diameter, tip resistances typically 1–3 MΩ). This allows current to pass rapidly and thereby to maintain the fidelity of the voltage clamp. As a consequence, the cell interior is dialyzed by the pipette electrode solution. This, in turn, requires that the investigator pay close attention to mimicking intracellular contents to the extent that is possible. In addition, shifts in voltage dependence of gating as a function of time after cell impalement and rundown of current is frequent and may be due to absence of key regulatory components in usual intracellular solutions. These can complicate assessment of the effects of drugs and of regulatory phenomena in the whole-cell voltage-clamp mode.

The lipid bilayer forming the cardiac cell membrane is the barrier that allows the interior and the exterior of the cell to maintain different compositions. Because of this lipid bilayer structure, the membrane acts as a capacitor, and current applied in response to a voltage-clamp command must first charge the membrane capacitance. This is a generally rapid event but can introduce errors into experiments in which currents with rapid activation kinetics (<1 ms) are studied. For example, if the desired voltage-clamp configuration is a perfect square wave, the time required to charge the membrane capacitance dictates that the actual membrane voltage achieved by current injection will not be perfectly "square," with the greatest error in the first few milliseconds after a desired step change in membrane voltage. This may introduce error in establishing the voltage dependence of activating current. Modern voltage-clamp amplifiers include circuitry that can partially compensate for the membrane capacitance (i.e., that can improve the response to a square-wave command).

Modern electronics accomplish voltage clamping by injecting current to maintain the command voltage (set by the experimentalist) equal to the voltage recorded at some site in the circuit assumed to be a near-equivalent to the actual membrane voltage. A common (usually small) source of error can arise as a result of high electrode resistance, which is in series with the cell membrane resistance across which the voltage is clamped. Because of this series resistance, attributable to the resistance of the electrode and of the cell membrane itself, the voltage is not actually clamped at the desired level. Error attributable to series resistance becomes a particular problem when large currents are recorded using a high-resistance electrode; in this situation, the experiment may need to be repeated using conditions to reduce the magnitude of the current (e.g., different voltages; different intra- and extracellular solutions) or larger electrodes.

B. Isolation of Currents in a Voltage-Clamp Experiment

An illustration of these difficulties and their potential solutions comes with a consideration of whole-cell voltage clamp of the cardiac sodium channel at 37°C and physiological extracellular and intracellular sodium concentrations. Under this condition, the current is too large to be clamped and its activation overlaps with the time course of charging of the membrane capacitance. Reduction in extracellular sodium is the most usual technique to reduce the amplitude of the sodium current to allow it to remain under control of the voltage clamp. To circumvent the problem of rapid activation, the current can be studied at reduced temperature, where its activation kinetics can be more readily separated from those of charging the membrane capacitance. Another approach is to use cell-attached patches, where the region of membrane whose capacitance must be charged to respond to the voltage-clamp command is very small (9).

To fully exploit the advantages of measuring current at a fixed voltage, it becomes desirable to develop methods to study specific currents in the absence of contamination by others. The approaches used to isolate membrane currents are illustrated by the techniques that can be used to eliminate the sodium current in a cardiac voltage-clamp experiment (e.g., in an experiment to study K^+ channels). First, sodium-channel-specific toxins, such as tetrodotoxin (TTX), can be used. Second, the composition of the extracellular solution can be altered to eliminate sodium (and other ions that might permeate through sodium channels) completely. Third, the preparation can be held (voltage clamped) at sufficiently depolarized voltages to inactivate the sodium channel. In practice, all three approaches are widely used to eliminate I_{Na}. Similar approaches (channel-specific toxins, altered intra- or extracellular solutions, inactivating prepulses) can be used to eliminate L-type or T-type calcium currents, and various components of outward K^+ current.

Another important consideration is the choice of tissue to be studied. Thus, an experimentalist interested in I_{Kr}, the rapid component of the delayed rectifier potassium current, might choose rabbit ventricular myocytes for study, since most of these cells lack the potentially contaminating slowly activating component I_{Ks}. In guinea pig, on the other hand, the two components (I_{Ks} and I_{Kr}) are both present (6). While these can be separated by component-specific blocking drugs or by differences in their voltage dependence of activation, it may be difficult to know at the beginning of an experiment how much of each component is present in an individual cell. Thus, if the question is one of drug block of a specific component, the guinea pig may be an unsuitable choice. If the question is one of interaction between components, or relative contributions of components to the action potential, then guinea pig might be an excellent choice, but rabbit would be unsuitable.

A logical extension of the approach in selecting specific cell types is to express channels in heterologous systems that ordinarily do not express any current whatsoever. The cloning of cDNAs encoding α-subunits for virtually all cardiac ion channels now allows the experimentalist to transfect mammalian cells or microinject *Xenopus* oocytes with appropriate DNA or RNA and to study the behavior of α-subunits in isolation and in physiological extracellular solutions without toxins. Transfections of mammalian cells can occasionally be too successful, in the sense that the currents are so large that the voltage cannot be adequately controlled; there are simple solutions for this problem, such as studying another cell line or transfecting with less DNA. Disadvantages to the use of heterologous expression systems are that intracellular signaling or other mechanisms modulating α-subunit function may well be absent in heterologous test systems, and action potentials cannot be recorded. In addition, some channel complexes are only assembled with coexpression of multiple ion channel or other proteins. While coexpression is possible in heterologous systems, this multimeric nature of ion channel complexes can be deceptive for scientists wishing to use such systems for cloning. For example, the recapitulation of I_{Ks} by expression of *minK* alone in *Xenopus* oocytes was initially a confusing finding because the protein does not resemble a typical voltage-gated ion channel α-subunit. This problem was resolved when it was found that the oocyte itself synthesizes an endogenous *XKvLQT1* α-subunit partner for exogenous *minK* and that expression of both proteins is required to recapitulate I_{Ks} in mammalian cells (10). Similarly, expression of both GIRK1 (also termed Kir 3.1) and GIRK4 (Kir 3.4) is required to recapitulate the acetylcholine-gated K^+ channel found in atrial cells. Expression of GIRK1 alone in *Xenopus* oocytes results in $I_{K\text{-}ACh}$ because the oocytes expresses an endogenous GIRK4 (11).

IV. THE CARDIAC MEMBRANE AT REST

When the transmembrane voltage of a ventricular cell at rest is measured, a value more negative than −80 mV is usually recorded. This negative voltage is thought to be maintained by negative charges on intracellular molecules that are so large that they cannot equilibrate into extracellular spaces. The extracellular sodium concentration is high, ~140 mM, while that within the cell is low, ~10 mM. Therefore, both a concentration gradient and an electrical gradient would be expected to drive sodium into cells at rest. This does not occur; the traditional way of describing the lack of sodium movement despite these

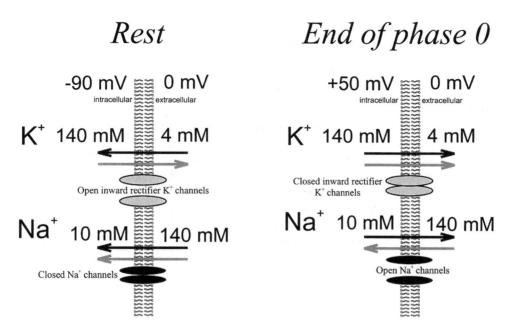

Rest   *End of phase 0*

Figure 2.3 Intracellular and extracellular sodium and potassium concentrations in a cardiac cell at rest (left) and at the end of phase 0 of the action potential (right). At rest, there is a concentration gradient (gray arrow) driving potassium out of cells and an electrical gradient (black arrow) driving it in. As described further in the text, there is an open permeation pathway (through inward rectifier channels) for potassium ions at rest to reach their electrochemical equilibrium potential as predicted by the Nernst equation. On the other hand, there is a large electrochemical gradient that would drive sodium into cells, but a permeation pathway is absent (sodium channels are closed). In the situation on the right, inward rectifier channels are closed, and sodium channels are open, and the equilibrium potential for sodium ions is approached.

gradients is that the resting membrane has very low permeability to sodium (Fig. 2.3). In more contemporary terms, this low permeability represents absence of any hydrophilic pathway (except for a small "leak") for sodium to travel along its electrochemical gradient to enter a cell. The membrane's low sodium permeability reflects the fact that those pathways, now recognized to be specific pore-forming, sodium-selective ion channels (the focus of the discussion in Chap. 3), are closed at rest.

Under physiological conditions, intracellular potassium concentration is high, ~140 mM, and the extracellular concentration is much lower, 3 to 5 mM. Under these conditions, the electrical gradient tends to maintain potassium within the cell, while the chemical gradient tends to drive it out. The potassium permeability of the resting cell membrane is high or, in more contemporary terms, the resting cell membrane includes a class of potassium channels (inward rectifiers such as I_{K1} and I_{K-ACh}) that are open and through which potassium can move according to its electrochemical gradient.

For any ion with electrochemical gradients such as these, there exists a membrane potential at which there will be no net ionic movement (i.e., net movement due to concentration and to electric gradients is zero). When the permeability of the membrane to one ionic species is much higher than it is to others, this equilibrium, or reversal, potential can be calculated using the Nernst equation:

$$E_{ion} = -\frac{RT}{F} \ln \frac{\text{intracellular ion concentration}}{\text{extracellular ion concentration}},$$

where E is the calculated reversal or equilibrium voltage, R is the universal gas constant, T is the absolute temperature, and F is Faraday's number. Simplifying and considering only potassium concentrations, the Nernst equation becomes $E_K = -61 \times \log ([K^+]_i/[K^+]_o)$, where $[K^+]_o$ is extracellular potassium concentration and $[K^+]_i$ is intracellular potassium concentration. For $[K^+]_i$ of 140 and $[K^+]_o$ of 4, an equilibrium voltage of -94 mV is calculated for potassium in the cell membrane at rest. This calculated equilibrium potential agrees very closely with the observed transmembrane voltage over a wide range of $[K^+]_o$ values (12). This finding indicates that the assumption that the membrane is permeable to a single species, potassium, at rest is viable. Importantly, it also indicates that resting membrane potential, which as described below is a crucial determinant of action potential configuration, is highly dependent on $[K^+]_o$.

V.   GENERAL DESCRIPTION OF THE CARDIAC ACTION POTENTIAL

The shape of an action potential recorded from a "generic" cell in the cardiac conducting system is shown in Figure 2.4(A), along with the ion currents that flow across the cell membrane to create this voltage signal. Variations in the relative magnitudes of the component ion currents then give rise to the region-specific action potential configurations described further below [Fig. 2.4(B)]. In most cells, an action potential is initiated when the membrane is depolarized, by charge flowing from an adjacent cell or by an investigator providing a depolarizing stimulus, to a threshold value. Subthreshold stimuli do not generate phase 0; they may depolarize the membrane, which then loses its charge and repolarizes.

A.   Phase 0

Under physiological conditions, excitation threshold in most cardiac cells is approximately -70 mV, corresponding to the threshold for activation of sodium channels (described further in Chap. 3). With suprathreshold stimuli, the membrane depolarizes sufficiently to open some sodium channels, allowing sodium ions to move along their electrochemical gradient. This inward movement of sodium rapidly results in further cellular depolarization, opening of additional sodium channels, and the characteristic abrupt depolarization of phase 0 of the action potential. Sodium channel flux during initiation of an action potential can be very large, exceeding 10^7 ions/s, and the equilibrium voltage for sodium ($\sim+70$ mV for $[Na^+]_i$ 10 mM and $[Na^+]_o$ 140 mM) may be approached at the end of phase 0 (Fig. 2.3). Measurement of the magnitude of the sodium current can be technically challenging (as described above), and the upstroke slope of phase 0 (dV/dt_{max} or V_{max}) is frequently used as an indirect measure of sodium current. V_{max} varies monotonically, although not entirely linearly, with I_{Na} (13). The total amplitude of the upstroke is another more indirect measure of the magnitude of the underlying sodium current. This description applies to tissues with sodium-channel-dependent action potentials (atrium, ventricle, infranodal conducting sys-

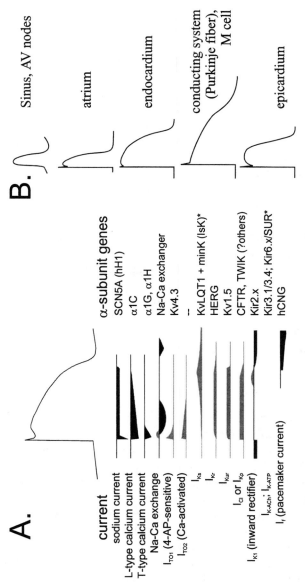

Figure 2.4 (A) The molecular genetic basis of the cardiac action potential. Shown here is an action potential with a configuration similar to one that might be recorded from a cell in the conducting system. The individual currents, and their time dependencies, that generate this time–voltage signature are shown below; the amplitudes are not to scale (the phase 0 sodium current, for example, is orders of magnitude larger than plateau currents). Inward currents are shown as solid areas and outward currents as gray ones. The genes whose expression generates the major structural (α) subunits for the channels that underlie these currents are indicated. For some of these, expression of ancillary (β) subunits, indicated by (*), is required to reproduce the major characteristics of the current in heterologous systems, and it is thus assumed that coexpression of these α- and β-subunits occurs in the heart. Function-modifying subunits have in fact been identified for most of the currents shown, but their roles in cardiac physiology remain uncertain. The varying action potential configurations in different regions of the heart reflect different expression or function of the individual components shown here. (B) Action potentials recorded from various regions of the heart. The different configurations reflect varying expression and/or function of the individual components shown in panel (A). Action potentials from the same region may vary considerably. In addition, there are striking changes with rate; for example, at slow rates, action potentials from the Purkinje system or M-cell layer prolong to a much greater extent than do those from other regions.

tem), where the magnitude of the phase 0 sodium current is the most important determinant of conduction velocity.

B.   Phases 1, 2, and 3

The opening of sodium channels and the accompanying abrupt membrane depolarization in turn produces both rapid sodium channel inactivation as well as opening of other ion channels, notably calcium channels and transient outward potassium channels. In some preparations, such as nerve, or arguably adult mouse or rat ventricle, this initial transient outward potassium current (I_{TO1} or more often termed I_{TO}) is sufficient to bring the membrane back to resting voltage within milliseconds (nerve) or tens of milliseconds (adult rat or mouse ventricle) (14). However, in most mammalian cardiac preparations, initial repolarization is then followed by the plateau of the action potential that is so characteristic of cardiac cells. During the plateau, which lasts several hundred milliseconds (depending on the specific cell type), multiple channels are open, allowing flow of inward (calcium, and under some circumstances, sodium and chloride, a repolarizing anionic current) and outward (potassium) currents along ion-specific electrochemical gradients. During phase 0, the membrane voltage may change by >120 mV in 1 ms, while membrane voltage may change by <20 mV over several hundred milliseconds during the plateau. Stated another way, both inward and outward currents are flowing during the plateau, but they are in a relative balance. Because the membrane resistance is very high, even a small perturbation in an individual current component during the plateau may lead to relatively large changes in membrane potential and hence in the duration of the action potential. This principle is particularly important in understanding diseases in which action potentials are prolonged and associated with arrhythmias, as discussed further in Chapters 16, 20, and 26.

The Nernst equation does not apply at the plateau, where the membrane is relatively permeant to multiple ion species. Here, the Goldman–Hodgkin–Katz equation may be used; one version is shown here, where P is permeability of a given ion:

$$E_{membrane} = 61 \times \ln \frac{(P_K \cdot [K]_o + P_{Na} \cdot [Na]_o)}{(P_K \cdot [K]_i + P_{Na} \cdot [Na]_i)}$$

Contemporary molecular and electrophysiological techniques have dissected multiple inward and outward currents that are activated by cellular depolarization and that flow during phase 2 and phase 3 in individual cardiac cells. The major inward currents are the L-type and T-type calcium currents (I_{Ca-L}, I_{Ca-T}), while the major outward currents are the ultrarapid, rapid, and slow-delayed rectifier components, I_{Kur}, I_{Kr}, and I_{Ks}, respectively. As a general principle, it is calcium-channel inactivation (which is slower than that for sodium channels) superimposed on time-dependent activation of delayed rectifier channels that eventually leads to a predominance of outward (repolarizing) currents over inward (depolarizing) current, and the initiation of repolarization. As repolarization proceeds, the membrane voltage becomes more negative and moves into a range in which outward current through inward rectifier channels (such as those responsible for I_{K1}) (Fig. 2.2B) develops, further accelerating normal repolarization. This regenerative process occurs over 100,000 times/day. Sodium overload is avoided by activation of sodium–potassium ATPase. This pump extrudes three sodium ions for each two potassium ions, and is therefore electrogenic, providing a small outward current. The role of this and other pumps and exchangers that act to maintain ionic homeostasis in the face of the large fluxes associated with the action potential are discussed further in Chapter 4.

One of the most abundant channel types in the heart is the ATP-inhibited potassium channel. With ATP depletion, these channels become active (15), hyperpolarizing the membrane to E_K (if it is depolarized) and shortening the action potential duration. One teleological view is that $I_{K\text{-ATP}}$ represents a defense against ischemia-related cellular depolarization and intracellular calcium overload. Since the channels are so abundant, even a very small open-channel probability may contribute repolarizing current, although whether this actually occurs under physiological conditions is controversial. Calcium-activated potassium currents have been described in multiple extracardiac tissues, but the extent to which these channels play a role in cardiac electrophysiology is controversial. A calcium-activated, rapidly inactivating chloride current (I_{TO2}) has been reported at depolarized potentials in rabbit ventricle; (16), intracellular calcium is also a well-recognized modulator of some of the currents mentioned above [e.g., $I_{Ca\text{-L}}$ (17), I_{Ks} (18)].

VI. ACTION POTENTIALS IN SPECIFIC REGIONS

The above description is most applicable to cells of the atrium, ventricle, and conducting system. One characteristic feature of cardiac electrophysiology is the striking variability of action potential duration and configurations among specific regions [Fig. 2.4 (B)], and indeed among cells within a specific region. The mechanisms determining this variability are an area of active investigation. Possibilities include not only variable expression of major structural pore-forming (α) subunits (and their function-modifying splice variants) (19–23), but also variable expression of ancillary (β) subunits as well as multiple posttranslational events, such as channel assembly and transport, possibly mediated by processes such as channel phosphorylation or expression of other regulatory proteins. These issues are discussed further in Chapters 5 and 6.

A. Sinus and AV Nodes

Action potentials in cells of the sinus and AV nodes differ considerably from the description provided above. Cells in the sinus node display a much decreased phase 0 upstroke slope and lack a robust sodium current. The other distinguishing feature of cells in the sinus node is pacemaker activity (i.e., depolarization during phase 4 that spontaneously brings the cell to threshold). Important components of the molecular basis for cardiac pacemaker activity have recently been defined and are discussed further in Chapters 3 and 7. Some cells in the AV junctional region display rapid phase 0 upstroke slopes, while others resemble those recorded in the sinoatrial node region. Phase 4 slope is greater in the sinoatrial node than in other tissues displaying pacemaker activity, and it is therefore this region that serves as the primary pacemaker for the heart. Phase 4 depolarization can also be observed in abnormal tissue (and pacemaker current occasionally recorded under unusual voltage-clamp conditions even in normal tissue) (24). Whether this represents activity of the same channels as underlie normal pacemaker activity in the sinus node is not clear.

B. Atrium

Action potentials in the atrium are shorter than those in the ventricle. This indicates that calcium currents are smaller and/or potassium currents are larger. Both L- and T-type calcium currents can be recorded in atrial cells, and all major time-dependent potassium currents (I_{TO}, I_{Kr}, I_{Ks}, and I_{Kur}) have been reported in human atrium. The acetylcholine-gated

K$^+$ channel is detected exclusively in atrium and may therefore contribute to shorter action potentials in atrium than in ventricle. Acetylcholine is well-known to provoke atrial fibrillation; one likely mechanism is activation of this channel with shortening (perhaps heterogeneously) of atrial action potentials, as discussed further below.

Considerable cell-to-cell heterogeneity in action potential duration configurations has been reported in human atrium, and variable levels of the transient outward current and/or delayed rectifier components have been seen. For example, in one study, a transient outward current was present in 29% of cells, delayed rectifier in 13%, and both components in 58% (25). Moreover, in dog atrium, action potentials and underlying currents were found to vary as a function of specific region within the atrium, suggesting a similar mechanism for the observed human data (26). This phenomenon appears to be one manifestation of the increasingly recognized phenomenon of variability among cells of a specific cardiac region in expression of ion channel genes, and thus in action potential duration and configuration. It is only now, with the development of molecular tools to examine cell-specific gene expression, that this is being defined and its mechanisms and consequences beginning to be explored. One study found that there was no potassium channel subtype that was uniformly expressed in all cells in different regions of the ferret heart (27). For some channel clones, such as Kv1.5, which is thought to underlie I_{Kur}, (28), mRNA expression was relatively uniform in atrium and ventricle. This contrasts with the situation in humans, where transcripts encoding Kv1.5 are much more abundant in atrium (29), and the corresponding current is observed in atrium but not ventricle. For other channel clones, there was considerably greater regional variation; *minK*, an ancillary subunit important for I_{Ks}, was found in 10 to 29% of working myocytes, but was detected in 34% of sinoatrial node cells. Another tool to study *minK* expression is genetically modified mice in which the *lacZ* reporter gene is under control of the *minK* promoter (30). In these animals, *lacZ* staining, indicating cells in which *minK* would ordinarily be expressed, was highly restricted, to the caudal atrial septum, the proximal conducting system, and the sinus node region. Whether such restriction also occurs in the human heart is unknown. Thus, the precedents from these electrophysiological and molecular studies strongly suggest that there is considerably more cell-to-cell heterogeneity in expression and function of cardiac ion channels than heretofore appreciated.

C. Ventricles and Conducting System

As in the atrium, the generic description of ventricular action potentials provided above require some modification as a function of region. In most large mammalian preparations, a prominent I_{TO} is recorded in epicardial cells but not in endocardial cells (31). In rat, I_{TO} is likely encoded by K4.2/4.3 heteromultimers, while in humans, Kv4.3 (or possibly Kv1.4 in endocardium) (32) is likely (33). Cells of the infranodal conducting system (common bundle of His, bundle branches, and beyond) display action potential characteristics that are somewhat different from those in epicardium and endocardium. Action potentials from this tissue are longer, and in particular prolong markedly with low [K$^+$]$_0$, slow rates, and QT-prolonging drugs (34), while when endocardial and epicardial cells are examined under these conditions, only modest action potential prolongation is observed (35). In ventricular muscle, inactivation of a large calcium current may be the primary driving force leading to initiation of terminal repolarization (phase 3), whereas the conducting system seems more sensitive to those maneuvers that affect K$^+$ currents (1). Although the mechanism for this behavior in the conducting system is uncertain, it does appear to be important in long QT-related arrhythmias, discussed further in Chapter 26. Another factor contributing to prolonged action potentials in the conducting system is inward sodium current flowing at the plateau. This can be inferred from the shortening of Purkinje action potentials seen with exposure to even low concentrations of the sodium-channel-specific toxin TTX (36). This inward current likely reflects the fact that curves describing the steady-state voltage dependence of I_{Na} activation and inactivation overlap slightly, defining a voltage "window" within which some sodium channels are in the conducting state. A similar calcium window current may contribute to marked action potential prolongation at very slow rates in this tissue (see also Chaps. 3, 12, and 16).

In many mammalian hearts, action potentials recorded from midmyocardium appear to have properties dissimilar to those of epicardium and endocardium, and quite similar to those in the conducting system. Specifically, cells in this "M-cell" region display phase 1 notches, as in epicardium. However, under conditions of repolarization stress (low $[K^+]_0$, slow rates, QT-prolonging drugs), M-cell action potentials also display marked action potential prolongation (31,35). Reduced I_{Ks} (37) (possibly as a result of greater expression of a dominant negative splice variant of *KvLQT1*, the major I_{Ks} α-subunit) (22,23) and/or a plateau component of I_{Na} (e.g., the "window" current described above) have both been implicated in action potential prolongation in the M-cell layer. Marked action potential prolongation in the M-cell layer may be another important component of arrhythmias related to action potential prolongation.

VII. THE CENTRAL ROLE OF THE ACTION POTENTIAL IN MODULATING CARDIAC ELECTROPHYSIOLOGY

The discussion above makes it clear that the action potential waveform represents the integrated activity of a multitude of individual channels and pumps, whose molecular identity is being elucidated. The way in which mutations or block of these individual components affect their function has provided fascinating biophysical and structural insights into the physiology and pathophysiology of individual molecular components. Nevertheless, it remains at the level of the action potential that integration of these changes occurs to affect overall cardiac function. This section will provide important examples of such integrated activity.

A. Action Potential Duration as a Primary Determinant of Cardiac Refractoriness

The phenomenon of refractoriness refers to the fact that when sufficiently premature stimuli are applied to cardiac preparations driven at a constant rate, no premature response is observed (i.e., the preparation is refractory) (Fig. 2.5, left panel). The term refractoriness is used by cardiac investigators in subtly different ways. For the clinical electrophysiologist or the physiologist studying the whole heart, tissue is refractory if a premature stimulus does not result in a propagated response in the whole heart. For the cellular physiologist, tissue is refractory if no premature action potential is observed; premature action potentials, on the other hand, may have sufficiently perturbed electrophysiological properties that they do not propagate. Compounding this semantic difficulty is the problem that very premature stimuli may produce abnormal responses whereas later (albeit still premature) stimuli may produce normal responses. The time window within which abnormal responses occur is often termed the relative refractory period. Responses to premature stimuli may also depend on the magnitude of the stimulus, so the terms absolute refractory period (during which no response is observed even at maximal stimulus strength) and effective re-

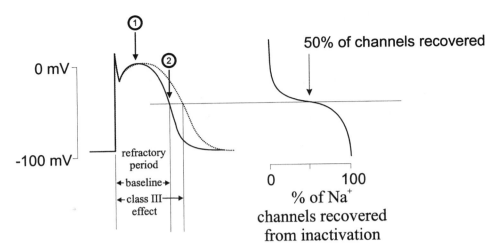

Figure 2.5 Refractoriness in a ventricular cell depends on recovery of sodium channels from inactivation. An action potential is shown on the left, and the voltage dependence of recovery of sodium channels from inactivation on the right. A premature stimulus delivered at the plateau (arrow 1) cannot generate an action potential because no sodium channels are available for reexcitation. As the action potential repolarizes (due to calcium channel inactivation and increasing outward current through multiple potassium channels such as I_{Ks}, I_{Kr}, I_{Kur}, and I_{K1}), sodium channels recover from inactivation. Once sufficient channels have recovered (50% in this example), a stimulus can generate an action potential, thereby defining the end of the refractory period (arrow 2). Note that potassium-channel-blocking drugs, which have no direct effect on sodium channel function, will prolong refractoriness by prolonging the action potential (dotted line).

fractory period (during which no response is observed at a fixed submaximal stimulus strength) are sometimes used in clinical electrophysiology.

When nerve or skeletal muscle is stimulated, very short action potentials result, and restimulation within tens of milliseconds then produces further action potentials. In cardiac muscle, on the other hand, restimulation within tens of milliseconds after phase 0 produces no effect. It is only when stimuli are applied during phase 3 of the action potential that a second action potential can be elicited (i.e., the preparation is no longer absolutely refractory). From a teleological point of view, this distinctively long refractory period in cardiac muscle prevents very rapid repetitive excitation.

The mechanism underlying this distinctively long refractory period of cardiac muscle is the interplay between the long plateau phase and the voltage dependence of sodium-channel inactivation (Fig. 2.5). At the depolarized potentials characteristic of the plateau, sodium channels are entirely inactivated, as discussed further in Chapter 3. Recovery of sodium channels from the inactivated to the closed (i.e., reexcitable) state is voltage dependent; the process also has a time dependence, but this is very short compared to the duration of an action potential. That is, as the action potential repolarizes, an increasingly large fraction of sodium channels becomes available for reexcitation, essentially as a function of the transmembrane voltage. A stimulus delivered when a sufficient proportion of sodium channels has become available will result in sodium channel reopening and an action potential. The duration and configuration of such a premature action potential will depend largely on the magnitude of the resultant depolarization due to sodium current as well as the extent to which calcium and potassium channels have become available (also a function of time and voltage). When the stimulus is applied prior to complete recovery of all sodium channels, the resulting magnitude of the depolarizing sodium current is less than that seen with a normal action potential upstroke. Whether such abnormal action potentials propagate into neighboring cells depends both on the magnitude of the sodium current as well as the geometry and electrical coupling characteristics of a particular preparation. As extrastimuli are applied at greater intervals following an initial phase 0, they are also, by the nature of the repolarization process, being applied to progressively more repolarized membranes (i.e., to membranes that include an increasingly large fraction of sodium channels available for excitation). Thus, very early extrastimuli produce no propagated responses, somewhat later extrastimuli may produce local but not propagating responses, still later extrastimuli may produce propagating responses with some delay, and extrastimuli applied after all sodium channels have recovered from inactivation (essentially equivalent to the end of the action potential) will result in normal propagation. Hence, although sodium current is not ordinarily a major component inward current during phases 2, 3, or 4 of the action potential, the changes in membrane potential occurring during this portion of the action potential exert direct and profound influences on the physiological behavior of sodium channels and thus on the electrophysiological properties of the whole heart.

This description has a number of important implications for the behavior of the heart in the presence of disease or drugs. First, interventions that prolong the action potential may have absolutely no direct effect on sodium channel function, but will nevertheless delay recovery from inactivation and hence prolong refractoriness. This has been termed the "class III" drug action. Second, drugs that interfere with sodium channel function may increase refractoriness by inducing a time-dependent component to I_{Na} recovery. That is, even when action potentials have been fully repolarized, sodium channels may be unavailable for reexcitation because they have not yet recovered from drug block. Drug unblocking during diastole then may become an important determinant of refractoriness. This is one manifestation of the general phenomenon of postrepolarization refractoriness, in which disease or drugs impose a significant time component on recovery from inactivation and hence prolong refractoriness. Third, this description does not apply to tissues, such as the AV node, in which sodium channels do not play as important a role in normal physiology. In the AV node, it is recovery of calcium channels following initial excitation that determines the ability of extrastimuli to produce propagated responses. However, calcium channels (unlike sodium channels) include not only voltage dependence, but also a prominent time-dependent component in recovery from the inactivated state. Thus, in AV nodal tissue, premature stimuli applied at the end of the action potential do not ordinarily result in normal responses because sufficient calcium channels have not yet recovered from inactivation. Fourth, when action potential prolongation occurs in a heterogeneous fashion, refractoriness may also become highly heterogeneous, setting the stage for reentrant excitation (see Chap. 17). Similarly, shortening of refractory periods (e.g., by stimulation of I_{K-ACh} or by loss of inward current) can lead to marked shortening of action potentials. Again, if this change occurs in only some regions, the conditions are ripe for reentrant excitation; this effect probably underlies atrial fibrillation occurring with vagal stimulation.

When an extrastimulus is applied to tissues with sodium-channel-dependent action potentials (atrium, ventricle, conducting system), the conduction velocity of the observed response is constant. On the other hand, when the same experiment is conducted in AV nodal tissue, the conduction velocity of the observed response is usually diminished with increasingly premature stimuli. This phenomenon is termed "decremental" conduction and is due to a combination of tissue geometry, electrical coupling properties, and I_{Ca-L} dependent propagation in tissues that depend on calcium channels for propagation. These issues, discussed further in Chapters 9, 11, and 17, are key to understanding mechanisms of reentrant arrhythmias that include the AV node as a portion of their circuits.

B.   Effects of Extracellular Potassium ([K$^+$]$_0$)

When [K$^+$]$_0$ is elevated, the membrane voltage depolarizes in accordance with the Nernst potential for potassium. This depolarization, in turn, results in inactivation of a proportion of sodium channels, resulting in decreased upstroke slope of phase 0 and slowed conduction. In addition, as the membrane voltage depolarizes, it is brought closer to the threshold for excitation, even as sodium channels are inactivated. This results in the apparent paradox that the stimulus strength required to produce an action potential may actually fall slightly as the membrane is modestly depolarized.

Simple electrochemical considerations would indicate that elevated [K$^+$]$_0$ should decrease outward current through potassium channels. However, for both I_{Kr} and I_{Kl}, elevated [K$^+$]$_0$ actually increases outward current through molecular mechanisms that are now being worked out and are discussed further in Chapter 3 (38,39). The result of this behavior is that, with elevated [K$^+$]$_0$, action potential duration shortens. On the other hand, low [K$^+$]$_0$ hyperpolarizes the resting voltage and produces marked action potential prolongation. The latter is particularly prominent at slow rates, presumably reflecting the fact that I_{Kr} plays a greater role in repolarization at slow rates and therefore its reduction by low [K$^+$]$_0$ has greater effect. These effects are thought to be important contributors to long-QT-related arrhythmias, which often arise at low [K$^+$]$_0$ and slow rates (Chap. 26).

Elevated [K$^+$]$_0$ is one component of the electrophysiological changes observed with ischemia. However, in this situation there are multiple other factors (e.g., intracellular ATP depletion, changes in pH, release of electrophysiologically active lipids from ischemic cells), each of which may alter the electrophysiological behavior of individual ion channels to contribute to the overall effects of ischemia (see also Chap. 20).

C.   Rate-Dependence of Action Potential Duration

It is now recognized that delayed rectifier current consists of at least two components, I_{Ks} and I_{Kr}. I_{Ks} results from coexpression of the *KvLQT1* gene with an ancillary subunit, *minK* (10,40), while I_{Kr} results from expression of HERG (41), possibly with other ancillary subunits (42,43). The amplitude of I_{Ks} is augmented by activation of intracellular signaling systems such as the β-adrenergic cascade, whereas I_{Kr} is not (44). In addition, in most tissues I_{Ks} activates and deactivates more slowly than I_{Kr}. Because I_{Ks} deactivation is incomplete when rate is rapid, residual I_{Ks} remains at the end of diastole, and subsequent action potentials are thereby shortened. At slow rates, on the other hand, I_{Ks} deactivation is more complete. As a consequence, I_{Kr} plays a relatively greater role in determining repolarization in slowly driven preparations. This framework forms the basis for the well-recognized phenomenon of "reverse use dependence," a term used to describe the fact that I_{Kr}-inhibiting drugs produce greater action potential prolongation at slow rates than at rapid ones (45).

The major effect of β-adrenergic stimulation on cardiac tissue is an increase in heart rate and an increase in an L-type calcium current (46). By the first principles outlined above, this increase in calcium current would ordinarily result in marked action potential prolongation, particularly at rapid rates seen with adrenergic stimulation. However, such an effect is not seen, presumably because β-adrenergic stimulation also results in an increase in the magnitude of outward repolarizing currents, particularly I_{Ks} and possibly a chloride current (47). Thus, a major role of I_{Ks} may be to defend against adrenergic-related increases in calcium current. Defects in I_{Ks} (as in the long-QT syndrome, discussed further in Chap. 26) are associated with arrhythmias that develop with adrenergic stimulation and may be inhibitable by beta-blockers.

D. Interdependence of Components of the Action Potential

The time and voltage dependencies of the ion channels that underlie the action potential impose a degree of interdependence on the components that has important consequences for arrhythmogenesis. For example, it is instructive to consider the potential consequences of inhibition of I_{TO} (Fig. 2.6). Under this condition, phase 1 would be abbreviated and phase 2 might therefore be initiated at a more depolarized membrane potential. This resetting of the start of phase 2 would, in turn, alter the magnitude of both inward current through L-type calcium channels as well as outward I_{Ks} and I_{Kr}. Depending on the extent of changes in these and other individual current components (and their baseline magnitudes, a function of particular cell type), action potentials could prolong or shorten. As a consequence, the time at which sufficient sodium channels had recovered from inactivation (thereby defining the end of the refractory period) would also be altered. This is a simple example of the way in which a change in a *single* component of the action potential might have multiple (often poorly predictable) consequences at the level of the individual action potential and in the whole heart. Another example arises from contemporary thinking about arrhythmia mechanisms in the long-QT syndrome (Chap. 26). Decreased I_{Ks} or

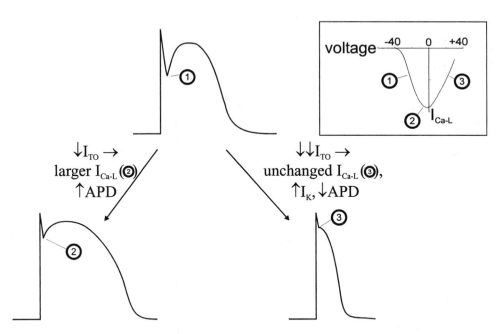

Figure 2.6   Interdependence of individual components of the action potential. Shown at the top is an action potential from a cell with a prominent I_{TO} (e.g., a ventricular epicardial cell). The possible effects of an I_{TO} blocker are shown at the bottom. In both cases, phase 1 ends at a more positive potential, the predicted effect of I_{TO} block. However, with a small reduction in I_{TO} (bottom left), the magnitude of the calcium current increases (position #2 on the current–voltage relation for I_{Ca-L} shown in the inset), prolonging the action potential duration. If the end of phase 1 occurs at even more positive potentials (bottom right), calcium current might in fact be smaller (position #3). In addition, potassium currents such as I_{Ks} would be larger. Thus, a shorter action potential might result.

I_{Kr} or increased plateau I_{Na} cause action potential prolongation in this disease. However, it now seems likely that a major consequence of this action potential prolongation is activation or reactivation of inward currents (through L-type calcium channels, T-type calcium channels, sodium–calcium exchange, or even the sodium channel) that actually initiates arrhythmias in this syndrome (48,49). Further, long-QT-related arrhythmias appear to be maintained by reentrant excitation due to much greater action potential prolongation in M cells than in epicardium or endocardium (50). This heterogeneity of repolarization produces heterogeneous prolongation of refractoriness setting up reentrant excitation.

These examples indicate that altering one ion current alters the time- and voltage-dependent behavior of other currents, in a complex and often cell-specific fashion. These principles extend beyond simple time and voltage dependence. For example, action potential prolongation may result in increased intracellular calcium. One consequence is increased contractile function (perhaps desirable), but it also may be that increased intracellular calcium promotes long-QT-related arrhythmias. There are many possible mechanisms for such an effect: examples include direct modulation of any of the currents involved in the abnormal repolarization process; calcium-mediated changes in intracellular signaling molecules (such as kinases or phosphatases) that then directly or indirectly modulate the function of repolarizing currents; altered cell–cell propagation by calcium-mediated modulation of gap junction function; or even conceivably by modulating the transcription of genes encoding ion channels or other proteins. Thus, the pleasingly simple action potential waveform actually reflects a complex interplay among its various molecular components. Unraveling those components may well identify not only new mechanisms for arrhythmias but also new drug targets for effective and safe antiarrhythmic therapy.

ACKNOWLEDGMENTS

This work was supported in part by grants from the United States Public Health Service (HL46681 and HL49989). Dr. Roden is the holder of the William Stokes Chair in Experimental Therapeutics, a gift from the Daiichi Corporation.

REFERENCES

1. Fozzard HA, Arnsdorf MF. Cardiac electrophysiology. In: Fozzard HA, Haber E, Jennings RB, Katz AM, Morgan HE, eds. The Heart and Cardiovascular System: Scientific Foundations, 2nd ed. New York: Raven Press, 1991:63–98.
2. ten Eick RE. Biophysical techniques for the study of cardiac tissue. In: Rosen MR, Janse MJ, Wit AL, eds. Cardiac Electrophysiology: A Textbook. Mt. Kisco: Futura, 1990:3–29.
3. Smith TG Jr, Lecar H, Redman S, Gage P. Voltage and Patch Clamping with Microelectrodes. Baltimore, MD: American Physiological Society, 1985.
4. Hodgkin AL, Huxley AF. A quantitative description of membrane current and its application to conduction and excitation in nerve. J Physiol (Lond) 1952;117:500–544.
5. Hamill OP, Marty A, Neher E, Sakmann S, Sigworth FJ. Improved patch clamp techniques for high-resolution current recording from cells and cell-free membrane patches. Pflügers Arch 1981;391:85–100.
6. Sanguinetti MC, Jurkiewicz NK. Two components of cardiac delayed rectifier K^+ current: Differential sensitivity to block by class III antiarrhythmic agents. J Gen Physiol 1990; 96:195–215.
7. Fischmeister R, DeFelice LJ, Ayer RKJ, Levi R, DeHaan RL. Channel currents during spontaneous action potentials in embryonic chick heart cells. The action potential patch clamp. Biophys J 1984;46:267–271.

8. Doerr T, Denger R, Doerr A, Trautwein W. Ionic currents contributing to the action potential in single ventricular myocytes of the guinea pig studied with action potential clamp. Pflugers Arch Eur J Physiol 1990;416:230–237.

9. Murray KT, Anno T, Bennett PB, Hondeghem LM. Voltage clamp of the cardiac sodium current at 37 degrees C in physiologic solutions. Biophys J 1990;57:607–613.

10. Sanguinetti MC, Curran ME, Zou A, Shen J, Spector PS, Atkinson DL, Keating MT. Coassembly of KvLQT1 and minK (IsK) proteins to form cardiac I_{Ks} potassium channel. Nature 1996;384:80–83.

11. Krapivinsky G, Gordon EA, Wickman K, Velimirovic B, Krapivinsky L, Clapham DE. The G-protein-gated atrial K$^+$ channel I_{KACh} is a heteromultimer of two inwardly rectifying K$^+$-channel proteins. Nature 1995;374:135–141.

12. Sheu SS, Korth M, Lathrop DA, Fozzard HA. Intra- and extracellular K+ and Na+ activities and resting membrane potential in sheep cardiac purkinje strands. Circ Res 1980;47:692–700.

13. Sheets MF, Hanck DA, Fozzard HA. Nonlinear relation between V_{max} and I_{Na} in canine cardiac Purkinje cells. Circ Res 1988;63:386–398.

14. Wang L, Duff HJ. Developmental changes in transient outward current in mouse ventricle. Circ Res 1997;81:120–127.

15. Deutsch N, Weiss JN. ATP-sensitive K$^+$ channel modification by metabolic inhibition in isolated guinea-pig ventricular myocytes. J Physiol Lond 1993;465:163–179.

16. Zygmunt AC, Gibbons WR. Calcium-activated chloride current in rabbit ventricular mycoytes. Circ Res 1991;68:424–437.

17. Kokubun S, Irisawa H. Effects of various intracellular Ca ion concentrations on the calcium current of guinea-pig single ventricular cells. Jpn J Physiol 1984;34:599–611.

18. Tohse N, Kameyama M, Irisawa H. Intracellular Ca^{2+} and protein kinase C modulate K$^+$ current in guinea pig heart cells. Am J Physiol 1987;253:H1321–H1324.

19. Kupershmidt S, Snyders DJ, Raes A, Roden DM. A K$^+$ channel splice variant common in human heart lacks a C-terminal domain required for expression of rapidly-activating delayed rectifier current. J Biol Chem 1998;273:27231–27235.

20. Lees-Miller JP, Kondo C, Wang L, Duff HJ. Electrophysiological characterization of an alternatively processed ERG K$^+$ channel in mouse and human hearts. Circ Res 1997;81:719–728.

21. London B, Trudeau MC, Newton KP, Beyer AK, Copeland NG, Gilbert DJ, Jenkins NA, Satler CA, Robertson GA. Two isoforms of the mouse Ether-a-go-go-Related Gene coassemble to form channels with properties similar to the rapidly activating component of the cardiac delayed rectifier K$^+$ current. Circ Res 1997;81:870–878.

22. Demolombe S, Bare I, Bliek J, Mohammad-Panah R, Loussouarn G, Mannens M, Wilde AA, Barhanin J, Charpentier F, Pereon Y, Escande D. The K$^+$ channel involved in the long QT syndrome 1 is a complex made of three different proteins. Circulation 1997;96:I-56 (abstr).

23. Demolombe S, Drouin E, Baro I, Charpentier F, Escande D, Pereon Y. Expression of KvLQT1 in human ventricular epicardial, midmyocardial, and endocardial layers. Circulation 1998;98:I-610 (abstr).

24. Yu H, Chang F, Cohen IS. Pacemaker current exists in ventricular myocytes. Circ Res 1993;72:232–236.

25. Wang Z, Fermini B, Nattel S. Delayed rectifier outward current and repolarization in human atrial myocytes. Circ Res 1993;73:276–285.

26. Feng J, Yue L, Wang Z, Nattel S. Ionic mechanisms of regional action potential heterogeneity in the canine right atrium. Circ Res 1998;83:541–551.

27. Brahmajothi MV, Morales MJ, Liu SG, Rasmusson RL, Campbell DL, Strauss HC. In situ hybridization reveals extensive diversity of K$^+$ channel mRNA in isolated ferret cardiac myocytes. Circ Res 1996;78:1083–1089.

28. Feng JL, Wible B, Li GR, Wang ZG, Nattel S. Antisense oligodeoxynucleotides directed against Kv1.5 mRNA specifically inhibit ultrarapid delayed rectifier K$^+$ current in cultured adult human atrial myocytes. Circ Res 1997;80:572–579.

29. Tamkun MM, Knoth KM, Walbridge JA, Kroemer H, Roden DM, Glover DM. Molecular cloning and characterization of two voltage-gated K$^+$ channel cDNAs from human ventricle. FASEB J 1991;5:331–337.

30. Kupershmidt S, Yang T, Anderson ME, Wessels A, Niswender KD, Magnuson MA, Roden DM. Replacement by homologous recombination of the minK gene with lacZ reveals restriction of minK expression to the mouse cardiac conduction system. Circ Res 1999; 84:146–152.

31. Antzelevitch C, Sicouri S, Lukas A, Nesterenko VV, Liu DW, Di Diego JM. Regional differences in the electrophysiology of ventricular cells: Physiological and clinical implications. In: Zipes DP, Jalife J, eds. Cardiac Electrophysiology: From Cell to Bedside, 2nd ed. Philadelphia: W.B. Saunders Co., 1995:228–245.

32. Näbauer M, Beuckelmann DJ, Überfuhr P, Steinbeck G. Regional differences in current density and rate-dependent properties of the transient outward current in subepicardial and subendocardial myocytes of human left ventricle. Circulation 1996;93:168–177.

33. Dixon JE, Shi WM, Wang HS, McDonald C, Yu H, Wymore RS, Cohen IS, McKinnon D. Role of the Kv4.3 K$^+$ channel in ventricular muscle: A molecular correlate for the transient outward current. Circ Res 1996;79:659–668.

34. Roden DM, Hoffman BF. Action potential prolongation and induction of abnormal automaticity by low quinidine concentrations in canine Purkinje fibers. Relationship to potassium and cycle length. Circ Res 1985;56:857–867.

35. Antzelevitch C, Sun ZQ, Zhang ZQ, Yan GX. Cellular and ionic mechanisms underlying erythromycin-induced long QT intervals and torsades de pointes. J Am Coll Cardiol 1996;28:1836–1848.

36. Coraboeuf E, Deroubaix E, Coulombe A. Effect of tetrodotoxin on action potentials of the conducting system in the dog heart. Am J Physiol 1979;236:HV561–HV561.

37. Liu DW, Antzelevitch C. Characteristics of the delayed rectifier current (I_{Kr} and I_{Ks}) in canine ventricular epicardial, midmyocardial, and endocardial myocytes: A weaker I_{Ks} contributes to the longer action potential of the M cell. Circ Res 1995;76:351–365.

38. Sanguinetti MC, Jurkiewicz NK. Role of external Ca^{2+} and K$^+$ in gating of cardiac delayed rectifier K$^+$ currents. Pflügers Arch 1992;420:180–186.

39. McAllister RE, Noble D. The time and voltage dependence of the slow outward current in cardiac Purkinje fibres. J Physiol 1966;186:632–662.

40. Barhanin J, Lesage F, Guillemare E, Fink M, Lazdunski M, Romey G. KvLQT1 and IsK (minK) proteins associate to form the I_{Ks} cardiac potassium current. Nature 1996;384:78–80.

41. Sanguinetti MC, Jiang C, Curran ME, Keating MT. A mechanistic link between an inherited and an acquired cardiac arrhythmia: HERG encodes the I_{Ks} potassium channel. Cell 1995;81:299–307.

42. Yang T, Kupershmidt S, Roden DM. Anti-minK antisense decreases the amplitude of the rapidly-activating cardiac delayed rectifier K$^+$ current. Circ Res 1995;77:1246–1253.

43. McDonald TV, Yu Z, Ming Z, Palma E, Meyers MB, Wang KW, Goldstein, SA, Fishman GI. A minK-HERG complex regulates the cardiac potassium current I(Kr). Nature 1997;388:289–292.

44. Sanguinetti MC, Jurkiewicz NK, Scott A, Siegl PKS. Isoproterenol antagonizes prolongation of refractory period by the class III antiarrhythmic agent E-4031 in guinea pig myocytes: Mechanism of action. Circ Res 1991;68:77–84.

45. Hondeghem LM, Snyders DJ. Class III Antiarrhythmic agents have a lot of potential, but a long way to go: Reduced effectiveness and dangers of reverse use-dependence. Circulation 1990;81:686–690.

46. Yue DT, Herzig S, Marban E. Beta-adrenergic stimulation of calcium channels occurs by potentiation of high-activity gating modes. Proc Natl Acad Sci 1990;87:753–757.

47. Harvey RD, Hume JR. Autonomic regulation of a chloride current in heart. Science 1989;244:983–985.

48. January CT, Moscucci A. Cellular mechanisms of early afterdepolarizations. Ann NY Acad Sci 1992;644:23–32.

49. Szabo B, Kovacs T, Lazzara R. Role of calcium loading in early afterdepolarizations generated by Cs^+ in canine and guinea pig Purkinje fibers. J Cardiovasc Electrophysiol 1995;6:796–812.
50. El-Sherif N, Chinushi M, Caref EB, Restivo M. Electrophysiological mechanism of the characteristic electrocardiographic morphology of torsade de pointes tachyarrhythmias in the long-QT syndrome: detailed analysis of ventricular tridimensional activation patterns. Circulation 1997;96:4392–4399.

3

Cardiac Ion Channels

HARRY A. FOZZARD

The University of Chicago, Chicago, Illinois

I. INTRODUCTION

Ion channels are scarce intrinsic membrane glycoproteins that form low-energy pathways for ions to move from one side of the lipid bilayer membrane to the other. These channels are characteristically selective for only one of the physiologically relevant ions, and they typically are gated open or shut by the membrane potential or by various ligands acting as extracellular or intracellular signaling agents. The cardiac action potential and its complex pattern of propagation results from orchestration of an array of these ion channels, in combination with the complex anatomy of the heart. In addition to contributing excitation and conduction, ion channels initiate and control cardiac contraction. These channels are the substrate for normal rhythm and for arrhythmias, and they are the targets or mediators of many important cardiac-active drugs.

Channel proteins are beautiful biophysical machines that have been characterized over many years by exquisitely sensitive electrical methods (1). Their modern study began with the classical work of Hodgkin and Huxley describing the ionic basis of the nerve action potential as time- and voltage-dependent opening and closing of Na and K channels. Ion channels are now the only complex molecules that can be studied by recording in real time the behavior of a single molecule. We can now combine this extensive biophysical information with the powerful tools of molecular biology to determine the primary structures of the channel proteins and the structural basis for their functions (2–4). These channels are cell constituents that have been highly conserved through evolution, descending from ancient bacterial ion channels. Given their central roles in cell function, it is not surprising that these channel proteins can function abnormally in disease in a wide variety of tissues. Some diseases result from genetic abnormalities in the channel DNA codes themselves, and others from abnormal expression or modulation of ion channels. Several of these genetic diseases are specific to the heart.

This chapter reviews our present understanding of channels as biophysical machines, including the key functions of permeation, selectivity, gating, and modulation. Rapidly emerging insights into channel structure are summarized, and, where possible, the relationships between structure and function are described. Both biophysical and molecular biological data can be combined to yield a picture of the three-dimensional structures of the proteins and their function as molecular machines. The many physiological roles of ion channels are discussed briefly here, and more extensively elsewhere in this book. Orchestration of ion channel currents into cardiac action potentials is found in Chapter 2. Modulation of cardiac channels in health and disease is then explored in Chapter 5. Chapter 6 considers what is known about the cell biology of cardiac ion channel proteins, including processing, assembly, and localization. This chapter focuses especially on the biophysical properties and structure–function of the magnificent protein machines themselves. With this information we may be able to understand the pathogenesis of several diseases, make more precise diagnoses, and design better therapeutic strategies.

II. GENERAL CHANNEL BIOPHYSICS AND STRUCTURE

A. Biophysical Functions of the Ion Channel Molecules

In order to perform their roles in electrogenesis, ion channel molecules have three essential functional properties: permeation, selectivity, and gating (Fig. 3.1). Permeation requires the provision of a low-energy pathway for hydrated ions across the low dielectric lipid bilayer, which is otherwise impermeable to ions. Selectivity allows the channels to discriminate between ions of similar size and charge, permitting only certain physiologically relevant ions to pass through their permeation paths. For example, the Na channel permits passage of Na, but not K or Ca. Gating is the process by which channels make transitions between closed and open conformations. Some are normally open and respond to stimuli by closing; others are closed and open on stimulation. Many of the ones that open on depolarization become inactivated after a time and must be polarized before they can open again (Fig. 3.2). They are typical of ion channels that are gated in response to changes in the membrane electric field (that is to say, they are voltage dependent). These channels

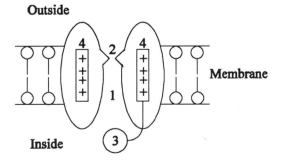

Figure 3.1 A biophysicist's concept of an ion channel. It is an intrinsic membrane protein, forming a low-energy path or pore for ions to cross the membrane lipid barrier (1). In the pore is a narrow region (2) called the selectivity filter. The channel has gates that open and shut (3), shown here as the inactivation ball. The gates are regulated by membrane potential or by ligand binding. In this model of a voltage-sensitive channel, a voltage sensor is shown as a charged structure within the protein (4).

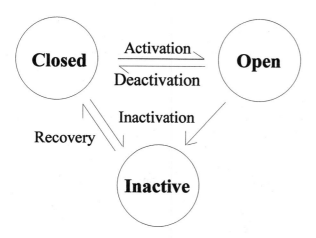

Figure 3.2 The three classes of ion channel gating states. When activated by either change in membrane voltage or by binding of a ligand, a channel may activate (open) and allow flow of ions through the pore down their electrochemical gradients. From the open state, the channel may deactivate to the closed state, where it is available to reopen. Often, channels "flicker" between closed and open states, leading to repeated openings and closings with the same stimulus. Alternatively, the open channel may inactivate. In that state it remains closed until it recovers, which requires reversal of the voltage change or dissociation of the ligand. The inactivated state may also be entered from the closed state. Biophysical evidence indicates that there are multiple states within each of these classes, so this diagram is oversimplified. Presumably, each state represents a specific conformation of the channel protein.

must have a sensor within their protein structure that monitors the electric field. Other channels are gated by intracellular ligands binding to specific sites on the channel.

The important primary functions of channels can be heavily modulated by extracellular and cytoplasmic processes. Some channels are gated by cytoplasmic second messengers such as G-proteins that are released by receptors for neurotransmitters or other extracellular signaling molecules. Others are regulated by such cytoplasmic messengers such as Ca and IP_3, or are modulated by phosphorylation through a protein kinase system. Many drugs and toxins act primarily on ion channels by blocking them or by altering their kinetics. The mechanisms of modulation by physiological second messengers or drugs/toxins are primarily changes in gating or permeation rates, with selectivity changes occurring rarely.

The critical physiological process that ion channels mediate is electrical current across the cell membrane. Net permeation of ions through channels constitutes membrane current. The ions are atoms or small molecules that are charged as a result of excess or deficiency of one or two electrons. Because they are small and highly charged, ions create a large electric field around themselves. In a physiological electrolyte solution the neutral, dipolar water molecules become oriented and tightly bound by the ion's electric field, creating an ion–water complex. It is this ion–water complex that must somehow pass through the ion channel. Physiologically important ions range in diameter from 2 to 3 Å, and the ion–water complex may be 6 to 8 Å in radius. Energy released during hydration can be 100 to 300 kcal/mol, so dehydration of ions for permeation is energetically highly unfavorable.

Biophysics is by definition quantitative and often the best way to describe biophysical processes is by mathematical notation. Net ion movement is driven by ion concentration and electric field gradients. They can be described by a variation of Ohm's law:

$$I_i = G_i (V_m - E_i),$$

where I_i is the membrane current of ion i; G_i is the membrane conductance for that ion; E_i is the electrochemical equilibrium for ion i; and V_m is the membrane potential. G_i is itself the product of γ, the single channel conductance; N, the number of channels; and p, the fraction of channels that are open at the time current is measured. The electrochemical equilibrium E_i is $RT/F \ln C_o/C_i$, where R is the gas constant; T is absolute temperature; F is the Faraday; C_o is the concentration of ion i in the outside solution; and C_i is its concentration in the inside solution.

Patch-clamp studies of single ion channels show that they have single-channel conductances in the range of fractions of a picoSiemen to several hundred picoSiemens. This means that with electrochemical gradients of around 100 mV, the net currents are in the range of picoamps (10^{-12} amps) (An amp is the current that flows through a resistance of 1 Ω because of a potential difference of 1 V. One amp is about 10 μM of charged particle per second.) For the voltage-gated Na channel that has an intermediate conductance, a picoamp of current represents the net movement of about 10 million Na ions per second. Because the pores of Na channels typically contain only one ion at a time, this places an upper limit of 100 ns on an individual ion's transit time through the pore, assuming that the channel is never empty and that there is only one-way traffic. For the higher conductance K channels, the limiting transit time is reduced to a maximum of less than 10 ns. Permeation in some channels approaches free diffusion rates, implying minimal interaction between the ion and the walls of the permeation path. In spite of this very short transit time, the channels can successfully discriminate between ions. For example, the K channel is about 1000 times more permeable to K than to Na. Membrane ion pumps also can have excellent selectivity. However, the high transport rates of channels (currents) are in striking contrast to membrane ion pumps. For example, the Na–K pump has a maximal rate of about 300 cycles per second, or 900 Na ions per second, instead of the millions per second of the Na channel. The much slower rate reflects the time required for biochemical transfer of energy from ATP to the protein and its consequent molecular movements required for ion transport. Some exchange transporters have much higher rates, and some ion channels, such as the pacemaker channel, transport at slow rates, leading to some overlap in rates between channels and transporters. The classical biophysical interpretation of the high permeation rates of ion channels is that very low-energy interactions between the ion and the channel are required. Either the ions pass through the channel without dehydration or the channel walls can substitute for the oxygens of water in an energetically equivalent fashion.

One simple selectivity mechanism is sieving. Sieving by size limitations could explain selectivity for small ions over large ones, but not the converse. Selectivity for a larger ion over a smaller one requires an intimate interaction between the pore and the ion. K channels can discriminate very well between the larger K ion (radius = 1.33 Å) and the smaller Na ion (radius = 0.95 Å). But the ions both have a single positive charge and differ in diameter by only about 0.4 Å. Such selectivity requires that the channel dehydrate the ion partially. The biophysical interpretation of this discrimination is that interaction of the ion with oxygen in its waters of hydration is replaced by interaction of the ion with equivalent oxygens of the amino acids lining the pore at some narrow region called the selectivity filter, so that the energy of the ion–oxygen interaction is minimally changed. Whatever the nature of the ion–channel interaction, it must not have a high affinity. Although its on-rate must be very high (up to the rate of free diffusion), its off-rate must also be very high in order to sustain the high net flux of ions. Another possible mechanism of selectivity is that some ions bind preferentially to the selectivity filter, blocking movement of any other ion through the pore. If that blocking ion can pass through occasionally and be displaced only by the same ion type, then permeation can occur, albeit at slow rates. This mechanism is thought to occur in Ca channels. For further discussion of these channel functional properties, see Ref. 1.

B. General Channel Protein Structure

Beginning in the early 1980s it became possible to purify and sequence parts of some channels, as the first step to cloning their coding DNA. In 1984, the first voltage-dependent channel was cloned—the Na channel (5), followed by the Ca channel (6) and many varieties of K channels during the next decade (7,8). In our discussion of the relation of structure to function we will focus on five functions: permeation, selectivity, gating, drug/toxin interaction, and modulation. The key features to be located in the structure are the voltage sensor, the pore lining and its selectivity filter, the gates, and the modulatory sites. Because ion channels are evolutionarily related, they often share structure–function features. This means that we can often draw on the entire body of information from all channels to understand each individual type.

Channel proteins are typically threaded through the membrane multiple times, so that part of the protein is intracellular, part extracellular, and part spans the membrane. The segments within the membrane bilayer are usually α-helical, which is a way for the amino acids to satisfy their need for hydrogen bonds in a lipid environment with few alternative hydrogen bond sources. In the principal subunits of voltage-gated channels, some transmembrane segments have a positively charged amino acid (usually arginine) at every third position in the helix, and this charged intramembrane element is the channel structure that senses the electric field in the membrane. The Na and Ca channels have 24 transmembrane segments, organized into four homologous, but not identical, domains of six transmembrane segments each, and with the N and C termini in the cytoplasm [Fig. 3.3(D)]. The domains are identified by Roman numbers I to IV, and the transmembrane segments in each domain are S1 to S6. The charged segment in each domain is S4. These four domains are closely packed in a circle, forming a pore at the center [Fig. 3.3(E)]. The smaller voltage-gated K-channel proteins have six transmembrane segments, and their membrane topology resembles one domain of the Na and Ca channels [Fig. 3.3(C)]. Four of them must assemble to form the pore. Various families of K-channel proteins can be identified by their amino acid homology and their ability to coassemble with homologous six-transmembrane-segment types. If all four channel proteins that form the channel are identical, then the structure is symmetrical. However, channels can be formed by different proteins within the same family, offering the opportunity for considerable diversity in their structure and function. The proteins forming some K-selective channels have only two transmembrane segments [Fig. 3.3(B)], similar to the six-transmembrane-segment type, but omitting the S1 to S4. Four such molecules must coassemble to form the pore.

The S5 to S6 extracellular segment in the six-transmembrane segment or domain folds back into the membrane partially. This segment, called the P-loop, contains the selectivity-determining sequences, and it forms the lining of the narrowest part of the pore. The P-loop selectivity regions are the most conserved parts of ion channels, and the

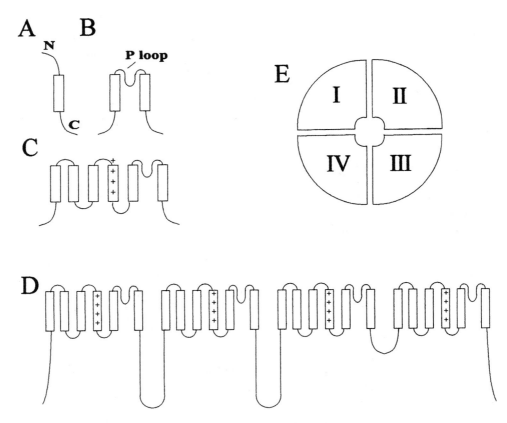

Figure 3.3   Illustrations of the topology of several membrane channel proteins. The barrel-like bars represent transmembrane segments, usually α-helices, which are threaded through the cell membrane. Above the bars is outside the cell membrane and below the bars is inside. (A) Single-transmembrane segment protein, with the N terminus outside (above the helix) and the C terminus on the inside (below the helix). This single-transmembrane segment protein is typical of several types of Na- and K-channel β-subunits. (B) A 2-transmembrane protein with topology typical of inward rectifier K channels. The N and C termini are intracellular and the extracellular loop folds partially back into the membrane to form part of the pore (P loop). (C) Six-transmembrane-segment protein with N and C termini intracellular, typical of voltage-gated K channels. The transmembrane segments are named S1 to S6. The P loop that forms part of the pore is the extracellular loop between S5 and S6. The S4 is a charged segment that forms the voltage sensor motif. (D) 24-Transmembrane-segment protein typical of the principal subunits of the Na and Ca channels. The six-transmembrane-segment units are internally homologous and are called domains (I–IV). Each domain resembles the voltage-gated K-channel motif, with a P loop between each S5 and S6. (E) Four K-channel subunits assemble around a central pore to form the channel, or one principal Na or Ca channel protein folds its four domains to form a pore.

K-channel P loop is nearly identical from bacteria to humans, an example of extraordinary evolutionary conservation.

The first K channels to be cloned were from a fruit fly lineage that manifested a characteristic behavioral mutation called "shaker." This type of channel protein has a long N-terminal segment in the cytoplasm containing a region of hydrophobic amino acids. This hydrophobic region interacts with itself to form a "ball" on the end of a "chain" of hydrophilic amino acids that connect it to the transmembrane part of the channel. The ball apparently functions as an inactivation gate by swinging into and blocking the inside mouth of the pore like a cork in a bottle (9). Other K channels without fast inactivation do not have this "ball-and-chain" structure. The Na channel has an analogous region on the intracellular linker between domains III and IV, forming that channel's intracellular inactivation gate (10). Other inactivation mechanisms for channels exist in the outer channel mouth, called C-type inactivation, and may represent a collapse of the pore itself. Activation gates have been difficult to locate in the channel structure. Recent clues point to the inner third of the transmembrane segment following the P loop (11,12).

Modulation sites abound on the intracellular regions of the channels, especially protein kinase A and protein kinase C phosphorylation sites and G-protein binding sites. Drug/toxin binding sites are found throughout these molecules, especially on the extracellular regions and in the pore itself. The outer surfaces of most ion channels are heavily glycosylated and negatively charged because of associated sialic acid residues. In general, no channels float freely in the lipid bilayer; rather, they are held in position by cytoplasmic attachments. Some channels, such as the Na channel, have intracellular structures that function as binding sites for ankyrin. These channels can be fixed in position in the membrane by the cytoskeleton. In noncardiac tissues, channels can be densely packed at specific locations in cells, such as Na and Ca channels at nerve synapses, or there can be polar distributions of ion channels in epithelial and endothelial cells that function to cause directional transcellular transport of ions. In the heart, the roles of these cytoskeletal attachments are not clear, but their disruption enzymatically can alter channel function. In addition, some of these attachments may keep certain enzymes located close to the channels, forming microdomains for efficient biochemical reactions involved in modulation.

III. CARDIAC POTASSIUM CHANNELS

A. Physiological Roles

All cells in the plant and animal kingdoms have K channels. Consequently, it is fair to conclude that K channels are crucial for basic cell function. In all animal cells these K channels are critical for the establishment and control of the resting membrane potential. In excitable muscle, nerve, and endocrine cells they have the special function of modulating excitability and contributing to repolarization of action potentials. Excitable and nonexcitable cells also use their resting potential to drive critical ion or ion-coupled transport systems. A particularly intriguing function of K channels is their ability to transduce metabolic information into electrical processes through their gating by adenine and guanine nucleotides such as ATP and ADP. For example, it is probably activation of this ATP-sensitive K channel that underlies the T-wave changes in cardiac ischemia. K channels are also primary or secondary targets for a large number of cardiac-active drugs, contributing to either the drug's therapeutic or its toxic actions.

The principal K channel that is open at the resting potential of heart muscle is the inward rectifier. This channel closes during depolarization to levels significantly more positive than the cell's resting potential [Fig. 3.4(A)]. If the extracellular K level is increased and the cell is depolarized, then the potential at which the channel begins to close changes, but closure is less complete. This behavior of closure upon depolarization is critical for heart cells, because it prevents excessive loss of K during the cardiac action potential plateau, while providing sufficient K conductance to stabilize the resting potential. The

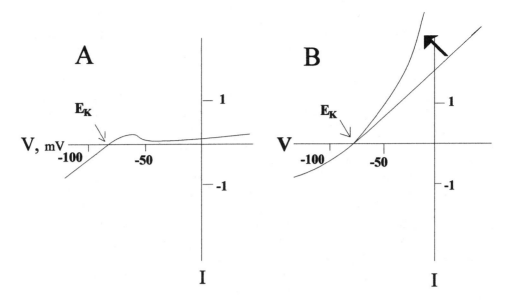

Figure 3.4 Current–voltage relationships of K channels. V is the membrane voltage coordinate and I is the ionic current coordinate. They cross at zero membrane potential and zero current. Displayed on the graphs are the I to V relationships of (A) inward rectifier K channels and (B) delayed rectifier K channels. The points where they cross the V axis represent the K electrochemical equilibrium potential (E_K), near the resting potentials of the cells. The slopes of the I to V relationships represent channel conductance. In (A), note that the conductance falls to a low value when the cell membrane is depolarized beyond its resting level. In (B), the more straight curve is the relationship shortly after depolarization; with time (arrow), the currents increase at each depolarized voltage.

channel reopens early during rapid repolarization, helping to terminate the plateau and assuring complete recovery of the resting potential. The mechanism for this unique gating behavior is not due to S4 voltage sensors, because these inward rectifier channels have only two transmembrane segments, not including S4. Rather, they are blocked during depolarization by small intracellular polyamines or by Mg (13). (See Chap. 2 for a discussion of the sequence of changes in ion channel gating that occur during the action potential.)

Certain inward rectifier channels have special roles to play in the heart. Acetylcholine-activated current in atrial and sinoatrial cells is the result of a G-protein second messenger. The vagal neurotransmitter acetylcholine (ACh) binds to the muscarinic receptor, releasing its associated heterotrimeric ($\alpha\beta\gamma$-subunits) GTP-binding protein (14). The $\beta\gamma$-component then binds to a special type of K channel that is normally closed, causing the channel to open. The resulting increase in K current hyperpolarizes atrial cells and shortens their action potentials, and it slows the SA node pacemaker by stabilizing the resting potential. Several other inward rectifier currents have been found in cardiac cells, including some activated by lipids.

At least three kinds of K channels are important during the action potential depolarization. First, most cardiac cells have a transient outward K channel, generating I_{to} that promotes initial repolarization after the Na current spike. The channel genes responsible for this current are thought to be *Kv1.4* (KCNA4) and *Kv4. 2/4.3* (KCND2/3). Second, there are two types of delayed rectifier channels, called fast (I_{Kf}) and slow (I_{Ks}) on the basis of their activation kinetics (Fig. 3.4B). The genes responsible for the delayed rectifier currents are HERG (for I_{Kf}) and *KvLQT1* (KCNQ1) (for I_{Ks}) in association with a β-subunit called *minK* (KCNE1) or another member of this β-subunit family. This family of β-subunits are transmembrane proteins with only one membrane segment [Fig. 3.2(A)]. These currents are the ones primarily responsible for repolarization of the action potential, and abnormalities of any of these currents significantly prolong the action potential.

Coupling of the electrical system to cardiac metabolism is accomplished by the K(ATP) channel. The channel is closed by normal levels of cytoplasmic ATP and ADP, and is opened by a rise in ATP/ADP ratio (15). The exact cytoplasmic levels of adenine nucleotides at which this channel activates have been difficult to resolve, because of the combined effects of ATP, ADP, pH, and possibly other factors. But density of this channel in ventricular muscle cells is so high and its single channel conductance is so large that opening of only a small percentage of available channels would shorten the action potential profoundly. It is likely that opening of this channel is the most important factor in action potential shortening seen in hypoxia/ischemia.

B. Structure and Its Correlation with Function

We have noted the two main types of K-channel structural patterns. The type with voltage-dependent gating has six transmembrane segments and the type without voltage dependence has two. The six transmembrane-segment type has a typical S4, found in all voltage-gated channels and a P loop between the fifth and sixth transmembrane segments, conferring K selectivity. They have 400 to 700 amino acids, depending on the lengths of their N- and C-terminal segments [Fig. 3.2(C)]. The two-transmembrane-segment types resemble their cousins, but with the first four transmembrane segments deleted, and they show inward rectification rather than voltage-dependent gating. The P loop in these channels is between their first and second transmembrane segments (the ones analogous to the S5 and S6 of the six-transmembrane-segment channels), and is highly homologous to the P loop in the six-transmembrane-segment types (Fig. 3.2B), establishing K selectivity for this type of channel.

1. The Voltage Sensor

The S4 voltage sensor has a characteristic motif, with positively charged amino acids—mostly arginines with a few lysines—at every third position. In a helical structure, which is likely for this segment, this means a barber-pole stripe of charge around the helix (2). Neutralization of the positive charges has dramatic effects on voltage sensing (16). The initial suggestion for how this charged structure could function as a membrane voltage sensor was that the positive charges were complemented by negative charges on adjacent helices in the protein. In response to a change in the electric field, the pairing of these salt bridges would shift and the S4 would twist and move out of the membrane, developing new pairing of charged residues. This mechanism poses energetic problems because the energy of a salt bridge is high (e.g., 3–10 kcal/mol) and the energy of the membrane electric field is relatively small. A typical sensitivity for channel conductance change as a function of membrane voltage in the range of gating is e-fold change for 4 to 5 mV change in field ($e \approx 2.72$). In addition, there are only a few negatively charged residues on the adjacent helices that are good candidates for this interaction. However, it has been shown that these S2 and S3 negative charges do interact in some way with the S4 charges, and they are at least necessary for proper folding of the protein during its synthesis (17). Direct study of critical residues on S4 have shown three important characteristics. First, neutralization of the charged residues in S4 have a profound effect on voltage sensing. Second, mutation of residues other than the charged ones are also important for voltage sensing. Presumably, these other residues, through their interactions with other parts of the protein, are important for the comformational changes of the S4 in response to voltage. Third, the charged residues become exposed on the surface of the protein, confirming that the S4 translocates toward the outside during sensing, or in some way shifts accessibility to ligands from the inside to the outside. How the sensor is coupled to the pore-associated gates is unclear at this point, but a plausible mechanism is by movement of the intracellular S4–S5 linker. Movement of these charges is also responsible for gating currents, which are capacity-like movements of the charged amino acids of the S4 segments that can be recorded electrically and are well correlated with channel gating.

2. Inactivation Gating

Most voltage-gated K channels both activate and inactivate upon depolarization, although sometimes the inactivation may be quite slow. More is known about the inactivation processes, so they will be discussed first. There are at least two types of inactivation in K channels, the rapid N type and the slow C type, and their molecular mechanisms are very different (9,18).

The Shaker family, represented in heart by Kv1.4 and Kv1.5 (KCNA4 and KCNA5 gene products) generate transient currents during depolarization, showing a rapid ball-and-chain-type inactivation process, as does the cardiac Kv4.2/3 (KCND2/3) type. However, biophysical studies of these K channels with voltage-clamp or single-channel analysis have shown no intrinsic voltage sensitivity of the inactivation process itself. It appears that all of the inactivation response that occurs upon depolarization is because it is coupled to the activation process. What this might mean in structural terms is that the conformation of the channel before activation is unfavorable for inactivation. But, during the voltage-dependent activation process, the inactivation system is either released or some protein conformation required for its function is achieved.

Early studies of the Shaker K channel revealed that removal of most of the N-terminal intracellular sequence abolished rapid inactivation, and this is the origin of the name for this type of inactivation (9). The N-terminal segment contains a large hydrophobic segment, connected to S1 by a hydrophilic tether. The hydrophobic segment was proposed to form a large "ball," and this has been confirmed by direct NMR studies of this peptide. The peptide sequence could be synthesized. and when it was injected it restored inactivation in a Shaker channel without the N terminus. The tether was also important, in that if it were shortened substantially the inactivation process was disabled. This rapid inactivation mechanism fits an original proposal derived from voltage-clamp studies of Na currents in squid axon that first led to the ball-and-chain model. When the inner mouth of the channel has a conformational change associated with activation, this ball can swing into the pore and occlude it (see Fig. 3.1). Recovery cannot occur until the ball detaches and the channel returns to its preactivation conformation.

Many voltage-gated K channels do not have this fast inactivation process, including some channels responsible for cardiac repolarizing, delayed rectifier currents. Mutation or removal of their N termini has no effect on inactivation. However, the Shaker channel without its N terminus still has a slow second inactivation process. Mutations in the C terminus affected this type of inactivation (18), resulting in its name of C type. Several clues have pointed to a conformational change in the external pore vestibule as the mechanism for slow C-type inactivation. First, mutation of residues just external to the conserved selectivity sequence alters slow inactivation (19). Second, binding of tetraethylammonium (TEA) to the external vestibule prevents or delays slow inactivation, as if it holds the channel open. It seems likely that this mechanism is important for the slow inactivation of the delayed rectifier KCNQ1 and *minK* channel that underlies I_K (slow) in the heart (20) and half of the identified channel defects of the long-QT syndrome. An important consequence of this mechanism is that drugs that interact with the outside of the channel could influence this inactivation process, with important effects on the action potential duration.

A particularly challenging gating problem is seen with HERG. This is the channel responsible for I_K (fast) in the heart, possibly in association with a *minK*-like β-subunit. HERG is the human version of the K channel responsible for the "ether-a-gogo" genetic phenotype in *Drosophila*. This channel shows rapid activation upon depolarization, but also very fast inactivation (21). Upon activation, most of the channels move rapidly to an inactivated state, so that the current on depolarization is not large (Fig. 3.5). However, upon repolarization the channels reenter the open state. Because deactivation from this open state is slow, there is a large "tail" current, which occurs upon repolarization after the initial activation–inactivation process. Consequently, this channel provides a large repolarizing current at the end of the cardiac action plateau, terminating the plateau in concert with I_K (slow) and the reactivation of the inward rectifier.

3. Activation

At the resting potential the delayed rectifier and transient outward types of K channels are mostly closed. Depolarization activates them by opening the pore in response to some movement of the S4 voltage sensor. The logical location for the activation gate is the inner mouth of the pore, where coupling can occur with fast inactivation. Mutation of residues in

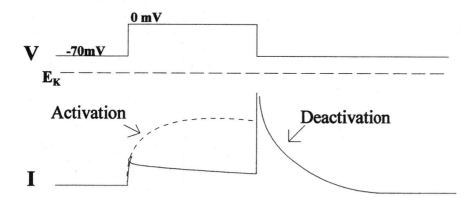

Figure 3.5 The response of the HERG channel to depolarization. If a step is made from −70 mV to 0 mV, the channel activates (the dashed line) and outward current increases. However, inactivation of the current is very rapid, so the current increases only a small amount. After repolarization to −70 mV, the channel recovers to the open state, with a large increase in outward K current. It then slowly deactivates to its original closed state at −70 mV. The channel does not contribute much current to the plateau, but as the action potential repolarizes, it provides a large current to terminate the plateau. This behavior is like an inward rectifier, but it results from an inactivation process faster than activation/deactivation.

the inner third of the S6 segments alters activation, pointing to this region as the gate (11). If the substitution is a Cys, this creates a covalent binding site for methansulfonates (MTS), which avidly develop a disulfide link to the substituted Cys. For these S6 Cys mutants, their reactivity requires opening of the channel, as if the residues are buried or protected by the activation gate, and not accessible to MTS until the voltage sensor has responded to depolarization and the activation gate is moved. Recent EPR studies show that residues in the inner third of the second transmembrane segment of a bacterial inward rectifier K channel (the segment that is analogous to the S6) move away from each other during activation (12). This does not prove that the inner thirds of the S6 helices are the gate because such movement could be secondary to a distant activation process, but it is convincing evidence of their involvement in the process.

4. Inward Rectification

Closure of the two-transmembrane-segment inward rectifier channels upon depolarization is not related to a voltage sensor, since these channels do not have S4 structures. As noted before, there appear to be two possibly related mechanisms for this process, block by Mg and block by polyamines (13). The heart channels affected by these mechanisms are the IK1 inward rectifier that is open at the resting potential, and, to a lesser degree, the muscarinic receptor and G-protein-activated K channel (GIRK) and the K(ATP) channel activated in hypoxia.

Millimolar concentrations of intracellular Mg are important for inward rectification. The channel has at least one divalent binding site. When this site is occupied by Mg, the channel is blocked or its single-channel conductance is altered. There are probably four subconductance states for this channel, and alteration of the equilibrium between them could change the current dramatically. Normally, Mg is prevented from binding by the electric field in the pore, but depolarization facilitates Mg block, and repolarization relieves it. The block can also be relieved by K entering the pore from the outside, and this relief may explain the shift in the voltage range of inward rectification in response to external K. This phenomenon has been proposed as a possible treatment for long-QT syndrome from genetic abnormalities of K channels. There is no evidence that this mechanism of K relief of Mg block exists for the delayed rectifier channels, so perhaps the increase in repolarization caused by increasing outside K may be by enhancing the inward rectifier contribution to terminating the plateau.

Even in the absence of intracellular Mg, the inward rectifier channels can close when in contact with the intracellular solution. The cytoplasm has several polyanion molecules of the spermine family with 1 to 4 positive charges. These small molecules can enter the channel upon depolarization and block the channel, and external K can also relieve their block. The relative importance of these two mechanisms is still being studied, but the involvement of these small molecules raises the possibility that cellular metabolism can influence these channels by alteration of the cytoplasmic concentrations of the spermine family. Intracellular levels of Mg are normally very stable, but this mechanism could be important in the arrhythmias of Mg deficiency or in the antiarrhythmic mechanism of Mg therapy.

5. The KcsA Crystal Structure

Although much can be inferred about rules for protein structure from x-ray analysis of soluble protein crystals, intrinsic membrane proteins have been more difficult to study. Only a few such proteins have been crystallized, such as bacterial rhodopsin, and the only structurally determined ion channels were those of bacterial toxins. Recently, Doyle and colleagues (22) published a landmark report of the crystal structure of the bacterial KcsA K channel to 3.2 Å resolution. This is a two-transmembrane segment K-selective channel with impressive amino acid sequence homology to eukaryotic K channels, so its structural properties are likely to be conserved in mammalian K channels. And because Na and Ca channels are very likely the result of K-channel gene duplications during evolution, these basic elements are likely to be found in those four-domain channels. This makes discussion of the KcsA structure particularly important.

The KcsA pore is formed by four of the second transmembrane helices (analogous to the S6) forming an inverted cone (Fig. 3.6). In the outer wide part of the cone the P loops are packed, each containing a helix-loop-strand conformation. The strand contains the triplet of amino acids Gly–Tyr–Gly (GYG) that is found in almost all K-selective channels (the exception being GFG, which is structurally almost identical). The strands with the signature sequence are densely packed to form a 14-Å-long narrow selectivity region. In contrast to the proposed structure of the Na-channel vestibule in which the side chains of the amino acids face the pore, the KcsA selectivity signature region is turned to form a pore lined by the backbone carbonyls of probably five successive residues, including the critical Gly–Tyr–Gly triplet. These sequences form a pore of fixed dimensions ~3 Å in diameter and containing probably two K ions. This structure is compatible with the Eisenman sequence selectivity theory (1) of a low-energy site for K, with amino acid carbonyls substituting for the K-ion waters of hydration. In order for the pore to be selective, it must have a fixed diameter. In the crystal structure, this signature strand region is locked rigidly into its fixed dimensions by the close packing of its side chains to the protein behind it—including especially the Tyr on the strand and the Trps of the N-end helices, so the intercar-

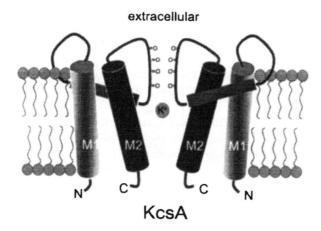

Figure 3.6 Schematic representation of the KcsA channel based on analysis of the x-ray study of the channel crystal. The view is from the side, showing only two of the four subunits. The N and C termini are inside the membrane. There are two transmembrane α-helices, M1 and M2. The pore is formed by tilted M2 helices, forming a "teepee" structure, with the M1 helices closer to the lipid bilayer. The outside P loop is composed of a helix-turn-strand structure. The selectivity filter is formed by the main chain carbonyl oxygens of the P-loop strand, with the side chains turned into the protein where they form stabilizing interactions with the P helix. (From Ref. 66; adapted from Ref. 22, with permission.)

bonyl oxygen spacing of the GYG sequence just fits the water–oxygen coordination of K (and Rb, which is also permeant), but not Na or Li. Beneath the narrow region is a large diameter "lake" large enough to include several free water molecules. The K ion could then enter the inner mouth of the pore in a hydrated form, until it reached the narrow signature region. The inner ends of the second (S6-like) helices form a narrow region at the inner pore mouth, and are likely to be involved in channel activation.

The motifs found in this ancient bacterial K channel with two transmembrane segments are likely to be inherited by mammalian K channels, with adaptations to different gating mechanisms. The theme of P loops and S6-equivalent lining of the pore will likely be found in Na and Ca channels as well, with the selectivity mechanism modified.

IV. THE CARDIAC SODIUM CHANNEL

A. Physiological Roles

Na channels underlie excitability and conduction of the action potential in mammalian atrial and ventricular muscle and in the His-Purkinje system. The channels are present in sinoatrial cells, but seem to play little or no physiological role in those tissues. Because the SA node normally operates at more depolarized potentials, the Na channels would be mostly inactivated. Na channels have not been found in the true AV nodal cells.

Typical atrial and ventricular cells have hundreds of these channels per μM of membrane, and His-Purkinje cells can have an even higher membrane density. Single-channel conductance is about 20 pS, depending on temperature and ion concentrations, and a typical single Na-channel current is 1 picoamp. Figure 3.2 illustrates the main classes of Na-channel gating states. The closed state of the channel is found at the resting potential (after recovery from any preceding activity). Depolarization by a pacemaker or a current flowing from an adjacent cell during action potential propagation causes some channels to open, letting some Na into the cell from the outside. This brings positive charge into the cell, further depolarizing it, and causing more channels to open. This process is called "regenerative" because once some channels open in response to depolarization, they admit positive charge that causes more depolarization, thereby promoting the opening of the remaining channels and producing maximal depolarization to near the Na electrochemical equilibrium potential. Normally, each Na channel has a probability of opening upon depolarization of only about 0.3 to 0.4, and then it opens only once or twice before inactivating. Recovery from the inactivated state requires repolarization to the resting potential. Recovery is usually quite fast at the normal resting potential, but is much slower at depolarized resting potentials such as those produced by high K_o or ischemia. An elegant mathematical model was developed by Hodgkin–Huxley to describe the squid nerve Na-channel gating, in which the rate constants for the switch from one state to the other (e.g., from the closed to the open) are described as voltage dependent. In this model, three processes called "m" must occur for the channel to activate (open), and one process called "h" must be reversed for the channel to inactivate (close).

The voltage dependency of activation and inactivation for the cardiac Na channels are usually more negative than Na channels found in the nervous system or in skeletal muscle (3). Threshold for activation is often as negative as –75 mV, and the midpoint of inactivation (voltage where 50% of the channels are inactivated) is typically –55 mV. The exact range is somewhat difficult to determine, because the experimental techniques used to study them often can influence the range, as if there is a cytoplasmic factor that regulates this range. There is usually a small crossover of the activation and inactivation curves, the size of which varies with conditions of measurement. This crossover produces a small, sustained inward "window" Na current over a small voltage range. This range for the window Na current includes the action potential plateau range, so it contributes a small amount of inward current to sustain the action potential plateau. Most of the voltage dependence of inactivation in Na channels is the result of coupling to the voltage dependence of activation, but inactivation in cardiac Na channels definitely has some voltage dependence independent of activation.

A host of drugs affect the cardiac Na channel, especially the local anesthetic drugs and certain marine toxins. A characteristic of cardiac Na channels is that they have an intermediate sensitivity to tetrodotoxin (TTX) and saxitoxin (STX), compared with channels in the nervous system where Na channels may be either very sensitive or completely resistant to these toxins.

B. Structure of the Channel

Na channels are composed of a large α-subunit and variably one or two small subunits (2). The α-subunit of about 2000 amino acids contains the permeability/selectivity path and the channel gates, although the gates can be modified by one or both subunits. Some 30 isoforms of the α-subunits from different tissues and from throughout the multicellular animal kingdom have been cloned so far. Those from different tissues in the same species are >90% identical, as are all the isoforms within mammalian species. A series of invertebrate isoforms from such species as squid and jellyfish have been cloned and identified as Na channels by homology (60–70% identity). They have been difficult to express in heterologous systems; consequently, we know much less about those isoforms.

Only one isoform of the Na channel α-subunit has been definitely found in adult atrial and ventricular cells (23,24). The sequence for a second type of molecule that is almost certainly a Na channel has been cloned from heart tissue, but seems to be associated with smooth muscle (25), rather than the cardiac myocytes themselves. Remember that about half of the cell mass in the whole heart is composed of cells other than myocytes. There are reports of some evidence of other Na-channel functional types, which have not been correlated with any cloned isoform of the channel.

The existence and roles of the β-subunits in cardiac cells has been difficult to resolve (26). The β2-subunit found in brain has never been identified in cardiac muscle. The β1-subunit is a small, single transmembrane segment protein (N-terminal outside). When expressed in *Xenopus* oocytes, the brain or skeletal muscle α-subunit isoforms produce currents with abnormally slow inactivation and recovery. In those isoforms, coexpression of the β1-subunit dramatically accelerates inactivation to approximately normal rates, alters recovery from inactivation, and shifts the voltage range of kinetics. It probably also alters stability of the α-subunit, because in heterologous systems it increases the membrane channel density. The identical β1-subunit sequence can be found in heart by Northern analysis or by PCR, but the message could be derived from noncardiac cells in the heart preparations. Fluorescent antibody studies have given conflicting results as to localization of β-subunits in cardiac myocytes. Coexpression of the messages for the cardiac α-subunit and the β1-subunit in *Xenopus* oocytes has some small effect on the channel kinetics, but does shift kinetics in the depolarizing direction and increases the density of current (27). One clue from studies on the development of skeletal muscle argues against a physiological role for the β1-subunit in heart. Embryonic skeletal muscle first expresses the cardiac muscle isoform, and with maturation it shifts to the adult skeletal isoform (28). Coincident with the expression of the adult skeletal isoform, these cells begin to express the β1-subunit. Definite conclusions about the physiological roles of the β1-subunit in heart, if any, must await further study. However, its depolarizing effect on the voltage range of kinetics (i.e., the voltage range of threshold) provokes the thought that it could be a cytoplasmic regulator of threshold for excitability.

C. The Permeation Path/Pore

Hydropathy analysis and location of antibody binding to specific epitopes has gradually led to a probably correct picture of the α-subunit's membrane topology. It has its N and C termini on the cytoplasmic side, and there are 24 transmembrane segments (Fig. 2D). These segments are composed of amino acids that have a strong tendency to form α-helices (3), but their secondary structures are not fully known. There is a pattern of internal homology within the transmembrane segments. It is possible to identify four domains of six transmembrane segments each. The domains are usually labeled Roman numerals I to IV. The transmembrane segments in each domain are labeled S1 to S6, and the S1 of each domain is homologous to the other S1, etc.

Most of the 2000 amino acid residues are cytoplasmic. The only long extracellular loop is between S5 and S6 of each domain. This segment contains a short hydrophobic region, which was first mistaken for a transmembrane segment, followed by a region rich in charged residues, mostly carboxyls. This charged region is extraordinarily well conserved in Na-channel molecules across the range from jellyfish to humans (29). This fact, combined with modeling based on protein structural rules, led to the proposal that this part of S5 to S6 loop was folded back into the membrane to form part of the permeation path and selectivity filter, and it is called the P loop.

The outer part of the permeation path, called the vestibule, is formed by a set of residues from each of the P loops (Fig. 3.7). The deepest point of this funnel-like vestibule is about one-third of the way through the membrane electric field. The funnel is formed of a stretch of about 8 to 10 amino acids with a sharp turn at its midpoint (30). The outer diameter of the funnel is about 12 Å and the depth is about 10 Å. The narrow inner mouth of the funnel is about 3×5 Å, and it is formed by one residue from each P loop—aspartate, glutamate, lysine, and alanine (DEKA, using the single letter code for the amino acids)

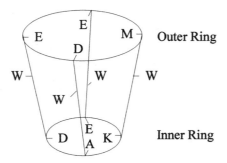

Figure 3.7 A model of the Na-channel outer vestibule, represented as a funnel. The inner ring, which is about 3×5 Å, is formed by four amino acid side chains—Asp (D), Glu (E), Lys (K), and Ala (A)—the DEKA selectivity filter motif. The outer ring, which is about 12 Å in diameter, is formed by four amino acid side chains—Glu (E), Glu (E), Met (M), and Asp (D). Each P strand has one Trp (W) that probably faces away from the pore and interacts with the protein interior to stabilize the vestibule structure.

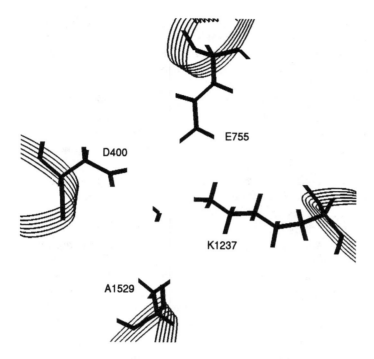

Figure 3.8   The Na-channel selectivity region. The P-loop backbones are shown as ribbons and the four DEKA amino acid side chains are shown as stick figures. The numbers of these amino acids are those of the skeletal muscle isoform. In the narrow part of the pore is a V-shaped water molecule.

(Fig. 3.8). The outer rim of the vestibule also has one residue from each P loop—two glutamates, a methionine, and an aspartate. In between is a tryptophan in each P loop that presumably stabilizes the structure. All of the charged residues in the vestibule are important for permeation.

The narrow DEKA ring (Fig. 3.8) is part of the selectivity filter, and it is the invariant structure in all isoforms of voltage-gated Na channels. The only exceptions are slight variations on the same theme: a DKEA sequence in jelly fish, an EDKA sequence in *C. elegans*, and an EEKE sequence in a rat brain isoform. The molecular mechanism of this selectivity filter remains elusive, but several suggestions can be made (31,32). The two or three negatively charged carboxyls can substitute for waters of hydration. The positively charged lysine is essential for selectivity of Na over K; even arginine will not maintain the selectivity. If lysine is neutralized, then the channel becomes permeable to Ca. Unraveling the complex physicochemical process that governs selectivity will be challenging.

The inner two-thirds of the pore are probably created by the four S6 helices. Part of the evidence for this is the role of S6 helices in the binding of local anesthetic drugs (see later). This role of S6 helices also is reasonable by analogy with the crystallized *KcsA* channel x-ray structure.

D. Gating

The cardiac Na channel has kinetic properties of activation and inactivation similar to nerve and skeletal muscle channels, and the gating currents appear so far to be indistinguishable from those in nerve. It has the typical S4 motif for voltage sensing, with minimal sequence differences. Much of the evidence for the roles of these S4 segments has been derived from K channels, but the structural similarity implies that the relations are the same. Domain I S4 of the Brain2 channel has been studied by successive neutralization of the positively charged residues (16). Their neutralization has a large effect on channel gating, with large shifts in the voltage range of gating directionally dependent on the location of the charge being neutralized. Domain IV S4 has been shown to be the primary voltage sensor for cardiac channel inactivation. Intramembrane residues on other helices are also likely to be important because of their proximity and interaction with the S4 segments. The present view is that S4 is in a large transmembrane pocket filled with water (33). There is a "gasket" near the middle of the pocket that avoids ion leakage through this pocket parallel to the pore. In response to a depolarization, the S4 segment moves or turns relative to the gasket, so that many of the charges are exposed to the outside.

The voltage range over which the S4 sensor moves is very sensitive to many residues in the S4 and neighboring structures. We do not yet understand which of these differences sets the physiological voltage range that is characteristic of each isoform. As we previously noted, the cardiac Na-channel kinetics are more negative than those of nerve (3), so that the channel is activated at more negative potentials and inactivates also at more negative potentials. This setting of the voltage range of kinetics is crucial for physiological function, because it, in association with the resting potential, determines the threshold for excitability. Shifts in either activation or inactivation could well be the pathological functional mechanism of causation of several lethal hereditary arrhythmias.

Cardiac Na channels display several types of inactivation. Fast inactivation can develop in milliseconds and recover equally quickly at many voltages. A second slow inactivation process can occur over seconds, with comparably slow recovery. These different states are important in state-dependent drug binding (see later), as well as in some diseases. Fast inactivation requires an intact cytoplasmic linker between domains III and IV, which is another of the highly conserved sequences in Na channels. One three-residue hydrophobic set of isoleucine–phenylalanine–methionine (IFM) is the most important, and deletion or replacement of these residues seriously compromises fast inactivation (10). It seems likely that IFM forms the core of a hydrophobic inactivation ball similar to that of Shaker-type K channels. Upon release by some movement of the domain IV S4, this cluster of amino acids can move into the inner mouth of the pore and block it. The blocking efficacy of the inactivation cluster is dependent on its "receptor," the site at the channel mouth to which it binds to close the permeation path. This receptor seems to include parts of the S6 helices (34) and the domain III S4 to S5 cytoplasmic loop (35). Perhaps the S6 part of the receptor is the molecular mechanism of coupling between activation and inactivation. The separate slow inactivation process does not involve the domains III to IV linker, and may be movement of residues in the outer vestibule, similar to the mechanism of C-type inactivation in K channels (36).

Several abnormalities of Na-channel function can result in disease by interfering with inactivation. About 10 to 15% of the patients with inherited long-QT syndrome and known channel defects have Na-channel mutations that result in incomplete inactivation, leading to a residual current that sets the stage for lethal arrhythmias.

E. Pharmacology/Toxicology

The cardiac Na channel is distinguished from other isoforms of the channel by its relative resistance to the marine toxins tetrodotoxin (TTX), saxitoxin (STX), and μ-conotoxin, and it can be blocked by clinically useful local anesthetic drugs, such as lidocaine. The mechanisms of these toxin and drug-binding interactions have been clarified considerably in the last few years, while simultaneously helping to understand the structure–function of the channel. We hope that understanding their mechanism will set the stage for design of drugs that have greater therapeutic efficacy.

TTX and STX have been used to locate the channel pore. These positively charged toxins bind with high affinity and perfect specificity to the outside mouth of the Na channel, probably to carboxyls at the Na site on the selectivity filter. Mutations in the P loops showed that all of the charged residues in the outer vestibule are important in toxin binding (37). As already noted, these mutations also altered single-channel conductance and, in some cases, selectivity. Based on the interactions, it has been possible to use the toxin structure to infer the structure of the channel mouth (30). The basis of the difference in cardiac sensitivity to TTX and STX is the absence of an aromatic residue in domain I just above the selectivity ring, which adds substantial energy to toxin binding to the nerve and skeletal muscle channels.

The local anesthetic class of drugs has a property called "use dependence." The effectiveness of these drugs in blocking Na currents is influenced by the frequency of action potentials. Clinically, this means that a drug concentration that has litle effect when the heart rate is 60 beats per minute is effective during a tachyarrhythmia with a heart rate of 150 to 200 beats per minute. Local anesthetic drugs bind to the inside of the channel pore with an affinity strongly affected by the channel's gating state (1). Lidocaine is a commonly used local anesthetic drug. It has a pK near the physiological level in extracellular solution, so it is partly neutral and partly positively charged. The neutral form can enter the cell (Fig. 3.9). The cytoplasmic drug then enters the inner mouth of the pore to a point just inside the selectivity filter. Several aromatic residues on the domains I and IV S6 helices influence local anesthetic block affinity (38), and their accessibility to the drug at different gating states may be the mechanism of use dependence. Specifically, block may be favored during depolarization and recovery from block after the end of the action potential is slowed, allowing accumulation of block with successive action potentials. The standard interpretation of this phenomenon is that drug binding stabilizes the inactivated state, but recently it has been shown convincingly that the critical IFM inactivation structure recovers at a normal rate in the drug-bound channel, even though the channel remains blocked by the drug (39). Different drugs bind with different affinity to various kinetic states, and their on- and off-rates appear to play an important role in their clinical usefulness. An example

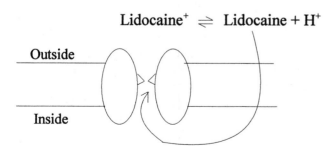

Figure 3.9 Lidocaine enters the cell in its neutral form. Lidocaine is partially charged at physiological pH. It reaches its binding site within the Na-channel pore from the inside, crossing the membrane mainly in its neutral form.

is the use of local anesthetic drugs with high affinity for the open state of the channel in the inherited long-QT syndrome variant that shows defects in Na-channel inactivation. As we understand better the molecular forces that control the blocking on- and off-rates of the family of local anesthetics, we should be able to design drugs that suit the clinical situation.

V.   THE CARDIAC CALCIUM CHANNELS

A.   Physiological Roles

Ca channels provide important currents in cardiac cells for excitation and conduction in the SA and AV nodes, for maintenance of the action potential plateau in atrial, ventricular, and His-Purkinje cells, and in activation and regulation of contraction (40). As the primary excitatory currents in the SA node, they also play a key role there in the SA node pacemaker process itself, and probably in pacemaker behavior of the His-Purkinje cells as well. They can sometimes maintain excitability in atrial and ventricular cells that have been depolarized by high extracellular K or ischemic injury, but the slow conduction of Ca-dependent action potentials is conducive to reentry arrhythmias. Recovery of Ca channels during long action potential plateaus can initiate depolarizations that probably underlie torsade de pointes ventricular tachycardia.

There are many types of Ca channels found in nerve, skeletal, heart, and smooth muscle and in some endocrine cells. The two types found in heart are L type and T type (41). Nerve cells have a preponderance of N and P/Q types. The nerve channels are essential for synaptic function, and their block by drugs acting on the cardiac channels would be highly undesirable. However, these types are not affected significantly by the L-type Ca-channel-active drugs (Ca-antagonist drugs) used for cardiac and vascular disease.

Ca channels can substitute for Na channels in providing the regenerative inward current for excitation and conduction of action potentials. SA and AV node cells have relatively few Na channels, and these are usually inactivated by the more positive resting potentials of the cells. Ca channels are activated by depolarization, similar to Na channels, but the activation process is slower. The intensity of peak Ca current in cardiac muscle cells is less than peak Na currents because the channel density in the surface membrane is much less and the single-channel conductance is also lower under physiological conditions. Consequently, conduction velocity of the I_{Ca}-derived action potentials is less in the SA and AV nodes.

In contrast to the case for Na current, the Ca current can also function as a potent second messenger. In heart, the principal target for Ca entering the cytoplasm is the ryanodine receptor channel. When Ca binds, this channel opens and releases Ca stored within the sarcoplasmic reticulum, which subsequently initiates interaction between actin and myosin to produce contraction. Under experimental conditions, the magnitude and duration of membrane I_{Ca} is an important determinant of the amount of Ca released from the sarcoplasmic reticulum and consequently the level of intracellular Ca achieved, thereby influencing contraction strength. In addition to interacting with the ryanodine receptor, Ca entering through channels can interact with a host of enzymes to influence metabolism. It can also interact with the Ca channel itself, as noted below.

The names of L type and T type explain the different kinetic behaviors of these channels (41). The dominant type in heart cells is the L type, so named because it is a *l*ong-lasting current. The smaller current is the T type, which is *t*ransient like the Na current. The L-type channel activates in the voltage range of −50 to −20 mV, and activation develops over several milliseconds, much slower than for Na channels. When intracellular Ca is buffered at its resting level, I_{Ca-L} decays under voltage clamp only very slowly. However, if intracellular Ca is allowed to rise to micromolar levels, as it typically does during Ca release from the sarcoplasmic reticulum, the L-type current decays substantially. This has been shown to result from binding of Ca to an intracellular site and consequent closure of the channel (42). When Ca in the cell is reduced by its uptake into the sarcoplasmic reticulum or is pumped out of the cell by the Na–Ca exchanger, then the channel is made reavailable for opening. It is not clear if the Ca-dependent reduction of current should be called inactivation, similar to that which develops in Na channels, or if it is an entirely different molecular event. Study of this process has been complicated by the difficulty in monitoring the level of Ca just beneath the membrane adjacent to the channel's Ca binding site. Ca diffuses slowly, so that Ca entering through the channel would achieve a local concentration near its own intracellular domains significantly higher than the general cytoplasmic level. Typically, Ca channels are located in membrane segments overlying and close to the ryanodine receptors, the sarcoplasmic release channel for Ca. Consequently, Ca released from the sarcoplasmic reticulum would also raise intracellular Ca near the channels to levels higher that the general cytoplasmic Ca, thereby producing closure of the Ca channels.

T-type Ca channels, in contrast, activate more quickly than L type and at voltages closer to the resting potential in the range of −70 to −60 mV, and they undergo a rapid inactivation that is voltage dependent (43), much like Na channels. Intracellular Ca has little effect on I_{Ca-T} inactivation. Both T-type and L-type channels can support excitation. Both also can activate contraction, but the principal source for excitation–contraction coupling is the L-type channel because of its greater density and more prolonged single-channel openings. However, the kinetic properties and the voltage range of activation for T-type channels make them more suitable for involvement in pacemaker activity.

The special kinetic properties of the L-type channel appear to underlie the phenomenon of early after-depolarization (EAD), which is thought to be the cellular basis of the ventricular tachycardia called torsade de pointes (44). EADs are reexcitations that spontaneously occur from prolonged action potential plateaus (Fig. 3.10). Most studies of EADs have been made in Purkinje cells, which typically have long action potentials, but the observations are probably applicable to atrial and ventricular cells as well. EADs originating at the end of the plateau usually start from potentials of −30 to −50 mV. This reexcitation can propagate and initiate a weak contraction. EADs are enhanced by BayK 8644, a L-type Ca-channel agonist, or adrenergic activity that enhances the Ca current, and are inhibited by low extracellular Ca and the dihydropyridine Ca-channel blockers. They occur over the voltage range where L-type channels can recover and be reactivated. The electrophysiological substrate for EADs is a plateau voltage range that permits recovery and reactivation of L-type channels, the "window" region. The plateau must be long enough for the cell's Ca transient to subside so that the channels can recover. Making the plateau higher or lower than this range, or shortening it, blocks EADs experimentally. Ca-channel blockers have not been particularly helpful clinically in the treatment of torsade de pointes, probably because of the dual drug effects of lower excitability, but faster recovery of the current during the plateau.

I_{Ca-T} is prominent in embryonic and neonatal atrial and ventricular cells. It diminishes in these cells with maturation, but the current remains prominent in adult His-Purkinje cells. Hypertrophy or various pathological injuries of the myocardium cause upregulation of the T-type current (45), making it a good candidate for a role in the arrhythmias associated with these abnormal processes.

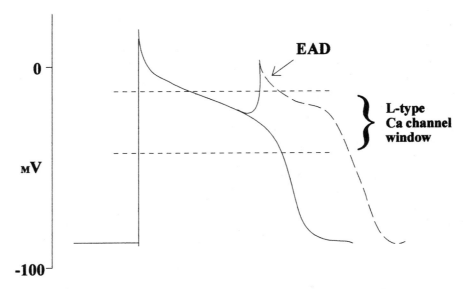

Figure 3.10   Early after-depolarization (EAD). The dashed lines indicate the approximate voltage range for Ca-channel window current, when the channels that were inactivated during the action potential may recover and be reactivated. Inactivation of the Ca current is primarily dependent on intracellular Ca, and its removal during relaxation allows the channels to recover enough for reactivation. The dashed second excitation is the EAD, which can occur if the plateau is long enough and in the correct voltage range for recovery and reactivation of the L-type Ca channel. The EAD can be prevented by changing the voltage range of the plateau or its duration.

B. Structure

The primary subunit of the L-type channel in the heart is called $\alpha 1C$, and the principal (and probably the only) subunit of the T-type channel has two or possibly three isoforms—$\alpha 1H$, $\alpha 1G$, and $\alpha 1I$. They are all similar in size and topology to each other and to the Na-channel α-subunit, with N and C termini on the intracellular side, 24 transmembrane segments organized into four homologous domains, and containing about 2000 amino acids (46). Between the S5 and S6 of each domain is the highly conserved P-loop region that contributes to the structure of the outer vestibule and its selectivity region. Each domain has the characteristic positively charged S4 transmembrane segment motif of voltage-dependent channels.

The L-type $\alpha 1C$ has four accessory subunits. The β-subunit binds to sites on the domains I–II linker and modifies kinetics (47). The $\alpha 2$-subunit is an extracellular protein that it is covalently linked to a small intramembrane δ-subunit. A γ-subunit is found in skeletal muscle, but not in heart. Although much remains to be learned about the roles of these subunits and their mechanisms of interaction with the $\alpha 1C$ principal subunit, they are clearly involved in modulating kinetics, in affecting expression, and perhaps targeting, and in mediating or regulating drug interactions and phosphorylation modulation.

The T-type isoforms were only recently cloned, and there is very little information about them at this time. $\alpha 1H$ and $\alpha 1G$ are both found in the heart, as well as the brain (48–50). The third type, $\alpha 1I$ has just been discovered and its role is uncertain. Both $\alpha 1H$ and $\alpha 1G$ generate characteristic transient Ca currents when expressed in heterologous cell systems. Although their kinetics are slightly different, they resemble native T-type currents closely. No evidence of accessory subunits has been found.

C. The Pore/Selectivity Region

The sequences of the P loops of all of the Ca-channel isoforms are nearly identical, and they are also highly homologous to the Na-channel P loops. Alignment of these P strands in a way similar to that of the Na channel predicts a ring of carboxyls at the P-loop turn point. Mutation of the Ca-channel carboxyl ring alters permeation, as expected if this ring is the critical selectivity region of the Ca channel (51,52). As already noted, if the DEKA selectivity ring of the Na channel is mutated to four carboxyls, the Na channels can conduct Ca, supporting the idea that the pore structures of these two channel types is similar.

Another property of the Ca-channel selectivity filter is if all divalent ions are removed from the outside solution, the channels can conduct Na very well. This Na current through the Ca channel can be blocked with only micromolar concentrations of Ca, demonstrating that the selectivity filter is a high-affinity Ca binding site (51). But if Ca blocks at micromolar concentrations, how can it ever conduct Ca efficiently? The best explanation at this point is that one Ca coordinates to all four carboxyls of the selectivity filter (53). When a second Ca approaches, it binds also and each Ca coordinates with two carboxyls. In this arrangement, the affinity is dramatically less, both because of fewer oxygen coordination sites and because of electrostatic repulsion between the two Ca ions. One or the other ion then moves away, one toward the inside and the other toward the outside, giving a 50% probability of Ca entering the cell with each encounter.

The T-type channel selectivity ring is structurally somewhat different (four aspartates instead of four glutamates), and, not surprisingly, its selectivity pattern and divalent block pattern are different. Comparison of the two Ca-channel types may help in resolving the mechanisms involved in selectivity and permeation in Ca channels.

D. Kinetics

Ca channels have the characteristic S4 motif of positively charged residues. These S4 segments are critical to activation of the L-type channel, but details are lacking. There is no apparent role for the cytoplasmic linker between domains III and IV in inactivation, in contrast to the role of this linker in inactivation of the Na channel. However, as already noted, Ca binds to the inside of the channel and leads to inactivation. Several sites in the C-terminal part of the channel protein have been identified as critical for this Ca-dependent inactivation process (54). The β-subunit binds to the cytoplasmic linker between domains I and II, and it appears to be essential for membrane expression and normal channel kinetics. The roles of the S4 segments in T-type channels have not yet been studied.

E. Modulation and Drug Interaction

Three types of "Ca-antagonist" drugs are used clinically, phenylalkylamines, benzothiazapines, and dihydropyridines. Their binding sites are separate but interactive, in that binding is not strictly competitive, but each drug alters the binding of the others (55). Reduction of current by these drugs is a pharmacological marker of L-type channels, including the cardiac and vascular smooth muscle isoform α1C, and also the skeletal muscle isoform α1S and the nerve/endocrine isoform α1D. These drugs do not affect the N and P/Q types, fortunately for the clinical usefulness of these drugs in cardiovascular disease. The drugs at higher levels do reduce current in T-type channels, but it is not likely that any of the therapeutic effects of the drugs are related to block of the T-type channel. Another drug, mebefridil, preferentially blocks the T-type channel, suggesting that that drug's efficacy in treating hypertension is because T-type channels are important for vascular smooth muscle.

The Ca-antagonist drugs all act to reduce probability of single L-type channel opening without change in single-channel conductance. Efficacy of drug-induced current reduction depends on the channel kinetic state and/or on voltage. In general, the drugs bind less well to the resting state, and are more effective with repeated activations, a use-dependent property like the action of local anesthetic drugs on Na channels. Partly because of the complexity of Ca-channel kinetics, the details of the state and voltage dependence of drug affinities are less clear than for local anesthetic drugs, although similar mechanisms are likely. These complex kinetic effects suggest that drug-induced reduction in current could be either a physical block of the pore or a disabling of the opening mechanism.

The sites of binding of these Ca-antagonist drugs on the α1-subunit are not fully resolved. It is clear that the midportion of S6 in domain IV is an important part of the binding site, analogous to the role of the domain IV S6 in local anesthetic drug binding in Na channels (56). In addition, the P-loop selectivity ring carboxyls also influence binding. There is also evidence of involvement of domain IIIS6, and perhaps other parts of domain III. Of particular interest is that certain dihydropyridines enhance channel opening, rather than inhibiting it. BayK 8644 is one example of such a drug. Its enantiomer reduces current, and the two drugs appear to bind to the same site on the channel. If that is true, then we must assume that it is not the physical presence of the drug at the binding site, but rather the way the drug affects the channel's function.

The cardiac L-type Ca channel was the first channel for which hormonal modulation was clearly demonstrated (57), establishing a paradigm for understanding action of membrane-directed hormones in the body. Binding of norepinephrine to the β-adrenergic receptor activates adenylyl cyclase via a G-protein. More cAMP further activates protein kinase A, which in turn phosphorylates some component of the channel. This can greatly increase I_{Ca-L}, and thereby increase conduction velocity in the AV node, spontaneous pacemaker activity of the SA node, and contraction in atrial and ventricular cells. There is no evidence for comparable effects of adrenergic modulation of the T-type channel. The site(s) of phosphorylation by the cAMP-activated protein kinase A have been difficult to identify in reconstituted systems, but they are likely to be on the C terminus of the α1-subunit. It is also possible that phosphorylation of other subunits also contributes to adrenergic effects.

Adrenergic modulation of cardiac Ca current follows the classical sequence of adrenergic action. Norepinephrine, released from sympathetic nerve endings in the heart, diffuses to the myocytes and binds to the seven-transmembrane-segment β-adrenergic receptors. These receptors have trimeric G-proteins bound to their intracellular surfaces. G-proteins are composed of α-, β-, and γ-subunits. The α-subunit is a GTPase, and the unstimulated receptor–G-protein complex retains the enzymatic product GDP, blocking further GTPase activity. When the receptor is occupied by norepinephrine (or any other of the receptor agonists), the G-protein α-subunit complex releases the GDP and binds a new GTP. In this form it dissociates from the receptor and from the $\beta\gamma$ complex. The α-subunit–GTP complex then binds to an adjacent membrane-bound adenylyl cyclase enzyme. Adenylyl cyclase converts ATP to cAMP, and the α-subunit–GTP complex stimulates it to increase the production of cAMP. The cAMP then diffuses to protein kinase A. When four cAMP are bound to the enzyme's regulatory subunit, the catalytic subunit is activated and phosphorylates its substrates. As described above, some component of the Ca channel,

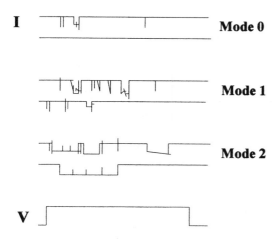

Figure 3.11   Kinetic modes of the L-type Ca channel. In response to a depolarization (bottom trace), the L-type channel opens. In the basal, unphosphorylated mode 0, the probability of the channel being open is very low. Moderate phosphorylation, typically by stimulation of the β-adrenergic receptors, increases the probability of being open after depolarization (mode 1). More extensive phosphorylation leads to mode 3, where prolonged channel openings occur, greatly increasing the current. Mode 0 is also favored by binding of the Ca-antagonist drugs, and mode 2 is produced by the agonist BayK 8644.

probably the α1-subunit, is phosphorylated, resulting in a change in its gating kinetics. This change is called a mode shift (58–60). The unstimulated mode O shows very low channel opening probability (Fig. 3.11). One phosphorylation shifts the channel to mode 1, with a substantial increase in opening probability and a significant increase in Ca current, with the attendant physiological effects of larger Ca current. A second phosphorylation shifts the channel to mode 2, in which the probability of channel closure is reduced. This leads to prolonged openings and a dramatic increase in Ca current. A set of phosphatase enzymes is present in the cell to dephosphorylate the channel complex and restore the unstimulated state. One of these is itself stimulated by cAMP. In summary, norepinephrine release initiates a cAMP-phosphorylation cascade that produces a transient increase in Ca current, affecting excitation, action potential duration, and contraction.

Study of Ca current modulation has been hampered by the experimental phenomenon of "run down," such that the whole cell current declines over several minutes. This process might be simply dephosphorylation of the channel, but direct proof of this has been elusive. If the membrane patch is removed from the cell so that easy access to the inside of the channel is obtained, then the channel activity disappears quickly. Some investigators have been able to recover normal activity with exposure of the inside of the membrane to the catalytic subunit of protein kinase A and ATP or treatment with blockers of phosphatases, but others have had difficulty confirming this. The relevance of rundown to in vivo behavior of the Ca channels is unclear, but it could play a role in reduced Ca current during ischemia.

Interestingly, the action of the dihydropyridines can be related to this modulatory mode switch. The Ca-antagonist drugs appear to shift the channel to mode 0, and agonists such as BayK 8644 appears to shift the channel to mode 2 (58). It is possible that the allosteric effects of phosphorylation are the reverse of those induced by drug binding, so they may bias the same molecular mechanism in opposite directions.

VI. PACEMAKER CHANNELS

The heart has the important property of generating its own electrical rhythm from specific pacemaker areas. Its pacemaker cells are characterized by a slow depolarization that begins immediately after repolarization of the prior action potential, rising slowly to threshold for the next action potential. Four variables influence the pacemaker rate (i.e., the interval between spontaneous action potentials): (1) duration of the preceding action potential; (2) potential achieved after repolarization; (3) threshold potential; and (4) rate of pacemaker depolarization. The latter three are pacemaker properties and are determined by the membrane currents underlying pacemaker behavior.

Normally, the primary pacemaker is the SA node, a discrete group of cells at the junction of the superior vena cava and the right atrium. Special atrial cells, such as those around the coronary sinus, also often manifest pacemaker behavior. Less often, AV nodal cells can show pacemaker behavior. The His-Purkinje cells also have latent pacemaker behavior, with much slower intrinsic rates. The excitatory currents of the SA-node and AV-node cells are Ca currents, with an unknown mix of T- and L-type currents. These Ca currents can be enhanced by sympathetic stimulation, which activates the adenylyl cyclase–cAMP–phosphorylation cascade. The result is a change in the threshold for the SA nodal action potential and acceleration of the pacemaker potential depolarization rate. In the His-Purkinje cells, the excitatory current is the Na current. Threshold for pacemaker-induced action potentials is the balance between the outward repolarizing K current near threshold and the barely activated Na-channel excitatory current. The primary SA-node pacemaker has less resting K current, partly because the inward rectifier channel (I_{K1}) is poorly expressed in those cells. This probably accounts for the reliability of the SA node pacemaker. The most dramatic effect of excitatory current changes affecting pacemaker rate is sympathetic–adrenergic effects on the open probability of the Ca channel in the SA node.

The pacemaker potential itself is a more-or-less steady and slow depolarization at potentials near the resting level—typically from –60 to –40 mV in SA-node cells and from –90 to –70 mV in His-Purkinje cells. In order for this depolarization to occur, there must be either a gradually increasing depolarizing current with a steady hyperpolarizing current or a gradually falling hyperpolarizing current with a steady depolarizing current, or both. It was early shown that this process is primarily time dependent after the preceding action potential repolarizes, because the process continues if the potential is held at its maximal level by voltage clamp during this interval. Release of the clamp does not start the pacemaker process, but rather the process jumps rapidly to the potential it would have had if the voltage had not been clamped.

A gradually increasing depolarizing current is generated by opening of a channel that is closed at depolarizing potentials, but opened at hyperpolarized potentials. This is the channel typically called I_f in heart, but called I_h in nerve. The channel is selective for monovalent cations, but distinguishes poorly between K and Na (61), and it has a reversal potential close to –20 mV. Because the channel opens at potentials close to the K equilibrium potential, the driving force on K is low and the contribution of outward K movement to the current is small. On the other hand, the potentials are distant from the Na equilibrium potential, so the predominant current through this channel is a depolarizing inward Na current. This channel is prominent in pacemaker cells such as the SA node, but it has also been reported in all cardiac cells when they are studied at sufficiently negative potentials (62). The single-channel conductance of the I_f/I_h channel is very small, and it has been difficult to study (61). One characteristic of value in studying its role in pacemakers is its sensitivity to block by Cd.

The role of the $I_f(I_h)$ channel in SA-node function has been controversial because experimentally the voltage range of its activation tends to be too negative for normal SA node pacemaker potentials. In addition, under some experimental conditions, exposure of the SA node to concentrations of Cd that are known to block this channel do not abolish SA-node pacemaker behavior. However, the voltage range of activation is shifted to more positive potentials during adrenergic stimulation. In contrast to its modulatory effects on Ca channels, cAMP binds to and modulates the channel directly, without phosphorylation (63). It is possible that the I_f current is normally not used for pacemaker function, but becomes important after sympathetic stimulation or when the membrane is hyperpolarized by vagal activity. Alternatively, it may be tonically modulated by cAMP in vivo and under those conditions it may function at typical pacemaker potentials of –60 to –40 mV. In any case, it must be an important player in the adrenergic acceleration of heart rate.

An alternative mechanism for SA-node pacemaker activity is that the preceding action potential activates a delayed rectifier K current, and that current declines slowly after repolarization. This declining K current, if coupled with a steady inward Na current (64), would result in depolarization to threshold. Evidence has been found for both currents. It must be noted that it is possible to make a pacemaker cell quiescent by elevating extracellular K. Purkinje cells are the most sensitive, and the SA node is the most resistant to elevated K. When K is subsequently lowered, the membrane begins progressively larger oscillations until it reaches threshold and begins regular pacemaking. Therefore, pacemaker behavior does not depend entirely on a preceding action potential. Given the crucial importance of a reliable cardiac pacemaker to survival, it is plausible that both the decaying K current mechanism and the $I_f(I_h)$ channel mechanism coexist, reinforcing each other.

Recently, a family of the I_f/I_h channels has been cloned. HAC1 is found in brain and heart, and HAC2 and HAC3 are found in brain only (65). Subsequently, several others of this family have been found. They are 6-transmembrane segment proteins with an S4 containing 10 positive charges organized differently from the typical S4 of Shaker-type K channels. The P loop contains the characteristic K-channel GYG motif, but with differences in adjacent residues, and it shows a P_{Na}/P_K ratio of about 0.3. The channel also has a cAMP binding site motif on the C terminus, and is activated by cAMP directly, without phosphorylation. Studies on these cloned HAC channels are just beginning. This should help resolve their roles in pacemaker tissues and perhaps facilitate the development of drugs that can modify pacemaker behavior.

VII. SUMMARY

Ion channels are intrinsic membrane proteins that regulate the movement of ions across the membrane down their electrochemical gradients. They generate membrane currents that underlie the cardiac resting and action potential. They are responsible not only for normal excitation and propagation, but also for cardiac arrhythmias. Ion channels can be characterized electrically in exquisite detail, and are remarkable biophysical machines. In the last decade many of the important ion channels have been cloned. Na and Ca channels consist of primary subunits of about 2000 amino acids, with four homologous domains of six transmembrane segments. K channels have two major types, a six transmembrane-segment type that resembles one domain of the Na- and Ca-channel proteins and a two-transmembrane type that resembles the last two segments of the larger type. These proteins can be expressed in heterologous systems, permitting study of the structural basis for their functions of permeation, selectivity, gating, modulation, and drug interaction. Genetic diseases of ion channels, and the underlying mechanisms of pathogenesis can often be understood on the basis of the known structure–function of the channels. This insight holds promise for rational drug design and targeted treatment for many arrhythmias.

REFERENCES

1. Hille B. Ionic Channels of Excitable Membranes, 2nd ed. Sunderland, MA: Sinauer, 1992.
2. Catterall WA. Cellular and molecular biology of voltage-gated sodium channels. Physiol Rev 1992; 72 (suppl): S15–S48.
3. Fozzard HA, Hanck DA. Structure and function of voltage-dependent sodium channels. Physiol Rev 1996; 76:887–926.
4. Marban E, Yamagishi T, Tomaselli GF. Structure and function of voltage-gated sodium channels. J Physiol 1998; 508:647–655.
5. Noda M, Shimizu S, Tanabe T, Takai T, Kayano T, Ikeda T, Takahashi H, Nakayama H, Kanaoka Y, Minamino N, Numa, S. Primary structure of electrophorus electricus sodium channel deduced from cDNA sequence. Nature 1984; 312:121–127.
6. Tanabe T, Takeshima H, Mikami H, Flockerzi V, Takahashi H, Kangawa K, Kojima M, Matsuo H, Hirose T, Numa S. Primary structure of the receptor for calcium channel blockers from skeletal muscle. Nature 1987; 328:313–318.
7. Papazian DM, Schwartz TL, Tempel BL, Jan YN, Jan, JY. Cloning of genomic and complementary DNA from Shaker, a putative potassium channel gene from *Drosophila*. Science 1987; 237:749–754.
8. Jan LY, Jan YN. Cloned potassium channels from eukaryotes and prokaryotes. Annu, Rev Neurosci 1997; 20:91–123.
9. Hoshi T, Zagotta WN, Aldrich RW. Biophysical and molecular mechanisms of Shaker potassium channel inactivation. Science 1990; 250:533–538.
10. West JW, Patton DE, Scheuer T, Wang Y, Goldin AL, Catterall WA. A cluster of hydrophobic amino acid residues required for fast Na-channel inactivation. Proc Natl Acad Sci 1992; 89:10910–10914.
11. Liu Y, Holmgren M, Jurman ME, Yellen G. Gated access to the pore of a voltage-dependent K channel. Neuron 1997; 19:175–184.
12. Perozo E, Cortes DM, Cuello LG. Three-dimensional architecture and gating mechanism of a K channel studied by EPR spectroscopy. Nature Struct Biol 1998; 5:459–469.
13. Nichols CG, Lopatin AN. Inward rectifier potassium channels. Annu Rev Physiol 1997; 59:171–191.
14. Wickham KD, Clapham DE. G-protein regulation of ion channels. Curr Opin Neurobiol 1995; 5:278–285.
15. Babenko AP, Aguilar-Bryan LA, Bryan J. A view of SUR/K 6.X K_{ATP} channels. Annu Rev Physiol 1998; 60:667–687.
16. Stuhmer W, Conti F, Suzuki H, Wang X, Noda M, Yahagi N, Kubo H, Numa S. Structural parts involved in activation and incativation of the sodium channel. Nature 1989; 339:597–603.
17. Seoh SA, Sigg D, Papazian DM, Bezanilla F. Voltage sensing residues in the S2 and S4 segments of the Shaker K channel. Neuron 1996; 16:1159–1168.
18. Choi KL, Aldrich RW, Yellen G. Tetraethylammonium blockade distinguishes two inactivation mechanisms in voltage-activated K channels. Proc Natl Acad Sci 1991; 88:5092–5096.
19. Liu Y, Jurman ME, Yellen G. Dynamic rearrangement of the outer mouth of a K channel during gating. Neuron 1996; 16:859–867.
20. Barhanin J, Lesage F, Guillemare E, Fink M, Romey G. KvLQT1 and IsK (minK) proteins associate to form the I_{ks} cardiac potassium current. Nature 1996; 384:78–80.

21. Smith PL, Baukrowitz T, Yellen G. The inward rectifier mechanism of the cardiac potassium current. Nature 1996; 379:833–836.

22. Doyle DA, Cabral JM, Pfuetzner RA, Kuo A, Gulbis JM, Cohen SL, Chait BT, MacKinnon R. The structure of the potassium channel: molecular basis of K conduction and selectivity. Science 1998; 280:69–76.

23. Rogart RB, Cribbs LL, Muglia LK, Kephart DD, Kaiser MW. Molecular cloning of a putative tetrodotoxin-resistant rat heart Na chanel isoform. Proc Natl Acad Sci 1989; 86:8170–8174.

24. Gellens ME, George AL, Chen L, Chahine M, Horn R, Barchi RL, Kallen RG. Primary structure and functional expression of the human cardiac tetrodotoxin-insensitive voltage-dependent sodium channel. Proc Natl Acad Sci 1992; 89:554–558.

25. George AL, Knittle TJ, Tamkun MM. Molecular cloning of an atypical voltage-gated sodium channel expressed in human heart and uterus. Proc Natl Acad Sci 1992; 89:4893–4897.

26. Goldin, AL. Accessory subunits and sodium channel inactivation. Curr Opin Neurobiol 1993; 3:272–277.

27. Makielski JC, Limberis JT, Chang SY, Fan Z, Kyle JW. Coexpression of $\beta 1$ with cardiac sodium channel α subunits in oocytes decreases lidocaine block. Molec Pharm 1996; 49:30–39.

28. Yang JS, Bennett PB, Makita N, George AL, Barchi RL. Expression of the sodium channel $\beta 1$ subunit in rat skeletal muscle is selectively associated with the tetrodotoxin-sensitive α subunit isoform. Neuron 1993; 11:915–922.

29. Goldin AL. Voltage-gated sodium channels. In: North R A, eds. Handbook of Receptors and Channels, Vol. II. Boca Raton, FL: CRC, 1994; 73–112.

30. Lipkind GM, Fozzard HA. A structural model of the tetrodotoxin and saxitoxin binding site of the Na channel. Biophys J 1994; 6:1–13.

31. Favre I, Moczudlowski E, Schild L. On the structural basis for ionic selectivity between Na, K, and Ca in the voltage-gated sodium channel. Biophys J 1996; 71:3110–3124.

32. Schlief T, Schonherr R, Imoto K, Heinemann SH. Pore properties of rat brain II sodium channels mutated in the selectivity filter domain. Eur Biophys J 1996; 25:75–91.

33. Yang N, George AL, Horn R. Molecular basis of charge movement in voltage-gated sodium channels. Neuron 1996; 16:113–122.

34. McPhee JC, Ragsdale DS, Scheuer T, Catterall WA. A critical role for transmembrane segment IVS6 of the sodium channel α subunit in fast inactivation. J Biol Chem 1995; 270:12025–12034.

35. Smith MR, Goldin AL. Interaction between the sodium channel inactivation linker and domain III S4–S5. Biophys J 1997; 73:1885–1895.

36. Todt H, Dudley SC, Kyle JW, French RJ, Fozzard HA. Ultra-slow inactivation in $\mu 1$ Na channels is produced by a structural rearrangement of the outer vestibule. Biophys J 1999; 76:1335–1345.

37. Terlau H, Heinemann SH, Stuhmer W, Pusch M, Conti F, Imoto K, Numa, S. Mapping the site of block by tetrodotoxin and saxitoxin of sodium channel-II. FEBS Lett 1991; 293:93–96.

38. Ragsdale DS, McPhee JC, Scheuer T, Catterall WA. Molecular determinants of state-dependent block of Na channels by local anesthetics. Science 1994; 265:1724–1728.

39. Vedantham V, Cannon SC. The position of the fast-inactivation gate during lidocaine block of voltage-gated Na channels. J Gen Physiol 1999; 113:7–16.

40. Pelzer D, Pelzer S, McDonald TF. Properties and regulation of calcium channels in muscle cells. Rev Physiol Biochem Pharmacol 1990; 114:107–207.

41. Bean BP. Two kinds of calcium channels in canine atrial cells. Differences in kinetics, selectivity, and pharmacology. J Gen Physiol 1985; 86:1–30.

42. Imredy JP, Yue DT. Mechanism of Ca-sensitive inactivation of L-type Ca channels. Neuron 1994; 12:1183–1194.

43. Hirano Y, Fozzard HA, January CT. Characteristics of L- and T-type Ca currents in canine cardiac Purkinje cells. Am J Physiol 1989; 256:H1478–H1492.

44. January CT, Moscucci A. Cellular mechanisms of early afterdepolarizations. Ann NY Acad Sci 1992; 644:23–32.

45. Nuss HB, Hauser SR. T-type Ca current is expressed in hypertrophied adult feline left ventricular myocytes. Circ Res 1993; 73:777–782.
46. Catterall WA. Molecular properties of sodium and calcium channels. J Bioenerg Biomembr 1996; 28:219–230.
47. Lacerda AE, Kim HS, Ruth P, Perez-Reyes E, Flockerzi V, Hofmann F, Birnbaumer L, Brown AM. Normalization of current kinetics by interaction between α1 and β subunits of the skeletal muscle dihydropyridine-sensitive Ca channel. Nature 1991; 352:527–530.
48. Cribbs LL, Lee J-H, Satin J, Zhang Y, Daud A, Barclay J, Williamson MP, Fox M, Rees M, Peres-Reyes E. Cloning and characterization of alpha1H from human heart, a member of the T-type calcium channel gene family. Circ Res 1998; 83:103–109.
49. Perez-Reyes E, Cribbs LL, Daud A, Lacerda AE, Rees M, Lee J-H. Molecular characterization of a neuronal low voltage-activated T-type calcium channel. Nature 1998; 391:896–900.
50. Lee J-H, Daud AN, Cribbs LL, Lacerda AE, Pereverzev A, Klockner U, Schneider T, Perez-Reyes E. Cloning and expression of a novel member of the low voltage-activated T-type calcium channel family. J Neurosci 1999; 19:1912–1921.
51. Elinor PT, Yang J, Sather WA, Zhang J-F, Tsien RW. Ca channel selectivity at a single locus for high-affinity Ca interactions. Neuron 1995; 15:1121–1132.
52. Yang J, Elinor PT, Sather WA, Zhang J-H, Tsien RW. Molecular determinants of Ca selectivity and ion permeation in L-type Ca channels. Nature 1993; 366:158–161.
53. Sather WA, Yang J, Tsien RW. Structural basis of ion channel permeation and selectivity. Curr Opin Neurobiol 1994; 4:313–323.
54. Zuhlke RD, Reuter H. Ca-sensitive inactivation of L-type Ca channels depends o multiple cytoplasmic amino acid sequences of the α1C subunit. Proc Natl Acad Sci 1998; 95:3287–3295.
55. Catterall WA, Streissnig J. Receptor sites for Ca channel antagonists. TIPS 1992; 13:256–262.
56. Hockerman GH, Johnson BD, Abbott MR, Scheuer T, Catterall WA. Molecular determinants of high affinity phenylalkylamine block of L-type calcium channels. J Biol Chem 1997; 272:18759–18765.
57. Reuter H. Calcium channel modulation by neurotransmitters, enzymes and drugs. Nature 1983; 301:569–574.
58. Hess P, Lansman JB, Tsien RW. Different modes of Ca channel gating behavior favored by dihydropyridine Ca agonists and antagonists. Nature 1984; 311:538–544.
59. Ochi R, Kawashima Y. Modulation of slow gating process of calcium channels by isoprenaline in guinea pig ventricular cells. J Physiol 1990; 424:187–204.
60. Ono K, Fozzard HA. Two phosphatase sites on the Ca channel affecting different kinetic functions. J Physiol 1993; 470:73–84.
61. DiFrancesco D. Pacemaker mechanisms in cardiac tissue. Annu Rev Physiol 1993; 55:455–472.
62. Yu H, Chang F, Cohen IS. Pacemaker current I_f in adult canine cardiac ventricular myocytes. J Physiol 1995; 485:469–483.
63. DiFrancesco D, Mangoni M. Modulation of single hyperpolarization-activated channels by cAMP in the rabbit sino-atrial node. J Physiol 1994; 474:473–482.
64. Hagiwara N, Irisawa H, Kasanuli H, Hosoda S. Background current in sino-atrial node cells of the rabbit heart. J Physiol 1992; 448:53–72.
65. Ludwig A, Zong X, Jeglitsch M, Hofmann F, Biel M. A family of hyperpolarization-activated mammalian cation channels. Nature 1998; 393:587–591.
66. Clapham DE. Unlocking family secrets: K channel transmembrane domains. Cell 1999; 97:574–550.

4

Cardiac Ion Pumps and Ion Exchangers

ARNOLD SCHWARTZ

University of Cincinnati College of Medicine, Cincinnati, Ohio

DENIS NOBLE

University of Oxford, Oxford, England

I. INTRODUCTION

Early in their evolution, cells must have conserved energy in several ways, but ATP is most certainly the primary molecule nature developed for supplying energy for the preservation and perpetuation of life. In protecting the early cell from the seas' high salt environment, a process of exclusion must have evolved in which ionic balance is maintained against a very high salt background. How might ATP accomplish this? First, this intriguing molecule is designed such that the terminal or γ-phosphate is linked to the vicinal or β-phosphate through an energy-rich bond. A simple splitting of the terminal P of ATP without intervening reactions would result in an energy liberation of about 10 Kcal/mol, causing a rise in body temperature. Indeed, this "simple splitting" model is utilized by nature in several ways, not the least of which is in the maintenance of a higher than ambient temperature in mammalian organisms. But the orderly process of the cellular production of ATP invested in mitochondria and in glycolysis is balanced with energy utilizing processes in a tight linkage to the machinery of the ionic cellular milieu. It is obvious that a "simple" ATP splitting (i.e., an "ATPase" reaction) would be valueless in maintaining osmotic balance of the cell. Accordingly, investigators initially hypothesized that a multistaged reaction sequence must be involved in the hydrolysis of ATP so that the released energy could provide the machinery for ionic balance. Indeed, it was postulated that ATPases in the cell membrane were intimately involved somehow in keeping internal Na low and K high and in transporting Na when this ion exits from the cell to the external environment and K enters from the exterior to the interior. The linkage between an ATPase activity and the movements of ions, or for that matter any energy-requiring process, has been a fascinating conundrum for over a hundred years (1).

What does this topic have to do with the heart and, in particular, cardiac arrhythmias? Clearly, the specifics of the heart beat, the maintenance of a rhythmic process, and an understanding of the complexities involved in events leading to rhythm disturbances are totally dependent on processes utilized to control a proper ionic milieu. Active ion transport and ion exchangers are the mechanisms nature has provided for these purposes. Nowhere is this more obvious than in the heart. Active ion transport not only maintains the ionic balance of the myocyte and promulgates impulse conduction, but in a remarkable series of reactions, links signal transduction to the control of cardiac contraction. The latter is accomplished in the heart through the multiple roles of Ca. By means of sophisticated instrumentation and Ca-sensitive dyes, it has been shown that intracellular Ca is exquisitely controlled in the cardiac cell. Its concentration varies during every beat from 10^{-8} M in diastole to 10^{-7} M in systole. This "concert" continues for years, uninterrupted, repeating every 600 ms in the human heart cell and even minor alterations result in cardiac diseases, including rhythm disturbances.

Linkage of this cycle with contraction as a result of the depolarization of the cell membrane is known as excitation–contraction coupling, with the "coupler" being Ca. Nature has used Ca to link almost all processes involving an excitatory event and many proteins, including those making up the sarcomere, have evolved specific Ca binding sites (e.g., troponin C) that, when occupied, cause conformational changes leading to specific, biological events. Other binding proteins such as calmodulin, the phosphatase calcineurin, etc., act as regulators of these processes as well. When Ca is removed, many Ca-dependent proteins inactivate, causing a termination and reversal of a particular biological event. It is obvious that the exquisite control of ionic Ca in cells is an essential element of the processes of life. Nature has invested an economical, though complex, major Ca regulatory mechanism in the Na- and K-dependent machinery of the heart cell. In addition to the membrane ATPases, it contains a well-orchestrated system of ion exchangers and intracellular regulators to maintain the coupling between excitation and contraction. This chapter deals with two of the major aspects of ionic control of Ca, Na, and K in the cardiac muscle cell, the Na/K ATPase system, and the Na/Ca exchange system; Figure 4.1 suggests in cartoon fashion how these processes may interdigitate to achieve this result.

A. The Membrane ATPase as a Driving Force for Monovalent Cations

In terms of evolution the ATP-driven pump system for Na and K homeostasis must have developed very early, since movements of these ions against a concentration and electrochemical gradient are absolute necessities for cellular stability, growth, differentiation, and excitability. It is well known that stimulation of a nerve produces an influx of Na concomitant with depolarization of the fiber and an increase in intracellular Na concentration (2). Many membrane depolarization processes are characterized by inward conduction of positive charges carried by Na through specific ion-gated channels. Recovery involves the outward movement of Na and this movement is regarded as an active transport process requiring energy because Na must move against a concentration and electrical gradient. Using giant axons from *Sepia officinalis* and form *Loligo forbesi*, (3), it has been reported that addition of cyanide, dinitrophenol, or azide reversibly inhibited the transport of Na out of the nerve and that these inhibitors acted in a concentration range that inhibited oxidative phosphorylation in mitochondria. Subsequent experiments proved that these inhibitors re-

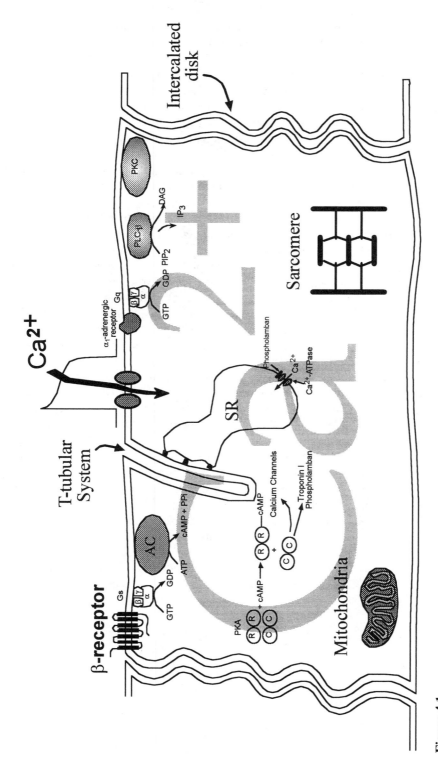

Figure 4.1

duced the content of high-energy phosphate compounds in the axoplasm. Others provided evidence showing that ATP must be involved. First, it was demonstrated that an ATP-splitting activity resided in the sheath of giant axons (4,5). It was further shown that both Ca- and Mg-dependent ATPases were present with the Ca portion in mitochondria and the Mg component in vesicles isolated from the nerve. There was evidence of a role for Na in "ATPase activity," showing that Mg-activated apyrase in brain homogenates responded to increasing concentrations of Na with an increase in activity (6).

A Danish scientist, J.C. Skou, concluded in the mid-1950s that there must be an energy-dependent process utilizing ATP in moving Na and K against their concentration gradients. Using homogenates of Crab nerve, the ATPase activity appeared due to activity in vesicles of plasma membrane. The activity required Mg in a 1:1 stoichiometry with ATP. Na stimulated ATPase activity and the addition of Na plus K produced a remarkable increase in activity. A detailed kinetic analysis provided data that were consistent with the intracellular concentration of Na and the extracellular concentration of K. Of considerable interest was the inhibition observed when Ca was added. The conclusion was that the "Crab nerve ATPase ... seems to fulfill a number of the conditions that must be imposed on an enzyme which is thought to be involved in the active extrusion of sodium ions from the nerve fiber" (7). Utilizing ingenious procedures for studying putative reaction sequences, including ATP–ADP and ATP–Pi exchange reactions, Skou reasoned further that the aforementioned ionic requirements the enzyme had to meet to be considered as the "active transporter" were in accord with all aspects that were known in the published literature. In one experiment, he added g-strophanthin (ouabain) to the enzyme and found that ouabain completely inhibited only the stimulated ATPase portion, in a concentration range the same as that shown for the red blood cell (8) and frog skin (9). How could the energy residing in ATP be used to move Na and K against their gradients? A multistaged complex reaction in which phosphorylation of the enzyme would result in a structural change, altering the affinity for intracellular Na, was postulated. When the enzyme is dephosphorylated, or perhaps when multiple conformers of phosphorylated intermediates are produced, the affinity for Na would be low and the affinity for K would be high. In this way, the ions could either sequentially or consecutively be transported in appropriate directions. This scheme for active ion transport first promulgated by Skou is shown in Figure 4.2.

In the scheme above, envision an enzyme (E) structurally placed in a cell membrane in such a way that one Mg ion would bind the terminal phosphate moiety of ATP to an inner membrane site in the presence of intracellular Na and a second Mg ion would bind to the adenine moiety of ATP when K is present. Ca would inhibit only at the second Mg site (competing with Mg). Ouabain might inhibit the enzyme by preventing dephosphorylation

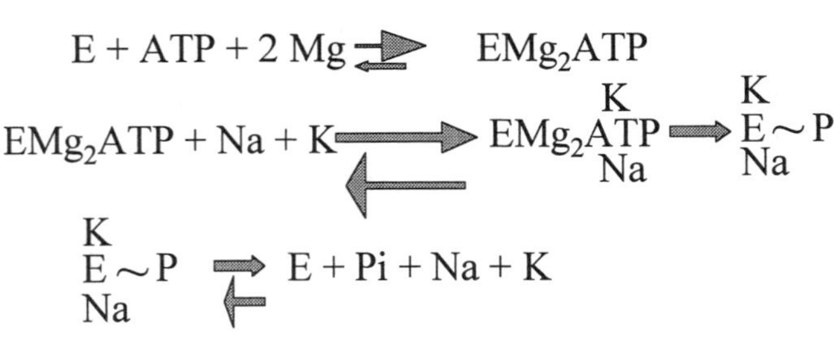

$$E + ATP + 2\,Mg \rightleftharpoons EMg_2ATP$$

$$EMg_2ATP + Na + K \rightleftharpoons EMg_2\underset{Na}{\overset{K}{ATP}} \Longrightarrow \underset{Na}{\overset{K}{E}} \sim P$$

$$\underset{Na}{\overset{K}{E}} \sim P \rightleftharpoons E + Pi + Na + K$$

Figure 4.2

triggered by K, or by interacting with some intermediate conformer of the phosphorylated enzyme somehow preventing the dephosphorylation process. Alternatively, ouabain might prevent a conversion of a particular phosphorylated intermediate state of the enzyme that may be insensitive to K to a state that is sensitive to K, perhaps accounting for the "competitive" interaction between cardiac glycosides and K. The enzyme with ATP and Mg attached would have a high affinity for Na and a low affinity for K, while some conformations of the phosphorylated intermediate state would have a high affinity for K and a low affinity for Na. The latter conformer could react with H_2O and undergo dephosphorylation, producing inorganic phosphate while Na and K were "transported" to the extracellular and intracellular compartments.

Thus the membrane ATPase studied in the test tube would require ATP and Mg in a stoichiometry of 2:1. Addition of Na and K simultaneously should produce a maximal increase in enzyme activity at a pH of about 7.0. The Na/K ATPase component should be completely inhibited by ouabain. An ATP–ADP exchange reaction should be inhibited by Na and not K, and might be insensitive to ouabain. Further, if the scheme is correct, the ATP–ADP exchange should be sensitive to ADP but not to P_i. An ATP–P_i exchange reaction would not require ADP, would require Na and K, and would be sensitive to ouabain. Theoretically, according to this scheme, and considering that all chemical reactions are reversible, it should be possible to "produce" a phosphorylated intermediate by adding inorganic phosphate to the membrane-bound enzyme in a test tube containing Mg. The reaction, in fact, was "reversed" by adding K and was stimulated by ouabain. The resultant E-P was indistinguishable from E-P formed in the forward direction (10).

Working with this scheme in late 1959, Skou showed it indeed fulfilled most of the requirements for an active monovalent cation transport enzyme, including the unusual reversibility of the third step. He also thought about the interesting effect of ouabain, recognizing as a physician that digitalis drugs were useful in stimulating the heart. Could it be that in some way this enzyme was involved? We shall see shortly how one of the present authors (AS) working with Skou began a study of this question that is still alive today. Meanwhile, despite an enormous amount of work on mechanisms through the present day we can add no significant new information that would modify the reaction above. Until we have more specific structural information (i.e., an accurate three-dimensional crystal analysis), significant refinement of mechanism will not be possible.

In thinking of the energetics of the process, it is important to emphasize that ATP is the essential primary driving force for almost all cellular energy-requiring reactions and it is estimated that a 70-kg adult at rest turns over around 50% of body weight in ATP per day; probably up to 800 kg (1 ton) of ATP per day when exercising (11).

B. Mechanisms of Digitalis Action

Cardiac glycosides (e.g., digitalis, ouabain) are alkaloids derived originally from the flowering plant, *Digitalis purpurea*. William Withering, an English physician, prepared infusions from its purple flowers and administered these to patients suffering from intense peripheral edema ("dropsy") and labored breathing. In 1785, Withering published in a beautiful monograph, several hundred cases describing the remarkable properties of this drug (12). He recognized that digitalis had a remarkable stimulatory action on the heart, yet reduced the pulse rate and produced copious amounts of fluid, which, upon excretion, in many cases totally relieved the symptoms and signs. For over 200 years, this drug has been the mainstay of the medical treatment of congestive heart failure and supraventricular arrhythmias. Digoxin, a purified derivative still partially synthesized from extracts of *Digitalis lanata*, is the most widely used of the cardiac glycosides. Ouabain is a water-soluble derivative that is most frequently used in experiments. Digitalis is still among the ten most prescribed drugs in the world, which is a "testimonial" to the fact that there is still no prevention or cure for heart failure and probably, at least for short-term therapy, digitalis used with caution is helpful. A recent long-term NHLBI clinical study revealed that while mortality is really not affected in a positive way, hospital stays were shortened and worsening heart failure, which is always the case with this disease, was actually reduced (13). The question then arises, could the beneficial effects of digitalis be due primarily to the inotropic action observed by Withering and others, or perhaps to some other effect? The aforementioned toxic action of digitalis is manifested as a series of effects that can best be attributed to an inhibition of Na/K ATPase (see above). When the serum level of the drug reaches a level above 2 to 10 nM, toxicity occurs, which is manifested by nausea, vomiting, peculiar eye color phenomena, rapid pulse, arrhythmias, and sudden death. It is difficult to reverse the lethal action of digitalis, although if caught soon enough, administration of a digoxin antibody whose affinity for the drug is extremely high can be beneficial.

To date, the best explanation for the toxic actions of digitalis is that ATPase becomes "overinhibited." But is the positive inotropic action (PIA) also due to an inhibition of the enzyme and, if so, is this action (viz., contractile force increase) related to any beneficial effect? In the past there has been much controversy on this point, but in a series of experiments beginning with a specific radioligand receptor binding of H^3-digoxin to a membrane-bound partially purified cardiac Na/K ATPase, and many subsequent studies, it is generally accepted that both the PIA and the toxic actions are due to a binding to and inhibition of Na/K ATPase (14–16). The interesting question currently posed is whether the beneficial actions, if any, are due to the inhibition of this enzyme. There are a number of intriguing experiments that describe a stimulatory action of digitalis in very low concentrations on Na/K ATPase (17), and in fact we suggested many years ago that stimulation of Na/K AtPase might be associated with a PIA. Note that the extremely low serum concentrations that are associated with beneficial therapeutic actions are probably too low to involve a significant inhibition of the enzyme. On the other hand, when a detailed analytical evaluation is carried out of the kinetics of digitalis binding to, and inhibition of, the cardiac Na/K ATPase, it appears that nanomolar concentrations of ouabain can in fact produce about a 20% inhibition of enzyme activity, without any visible stimulation (18). There is no doubt the drug binds specifically to Na/K ATPase and that the monovalent cation requirement for this binding and other characteristics provide compelling evidence that this enzyme is a receptor for the drug. Thus far, no other site has displayed receptor activity. On the other hand, what if the binding of digitalis and its action on the enzyme triggers a subsequent effect more closely associated with either inotropic or therapeutic effects? In point of fact, current opinion is that when digitalis is beneficial, it somehow may be associated with a neurohumoral action that leads to a diminution of heart rate, and perhaps an inhibition of negatively inotropic cytokines such as TNF-α and interleukins. It is suggested that an inotropic action is actually not desirable and may cause early demise.

It has been known for decades that low concentrations of cardiac glycosides inhibit the adrenergic system causing a decrease in norepinephrine (NE) and stimulation of cholinergic fibers. Packer was among the first to measure neurohumors in digitalis-treated patients and found elevated levels of serum catecholamines (19). In patients receiving digoxin, he and others found a diminution of NE. Packer postulated that any beneficial action of digitalis is due to an inhibition of "toxic" neurohumors known to cause deleterious effects on heart muscle. It is fascinating that digitalis is still causing much confusion, although continued research is uncovering other interesting actions of this very old drug. In a recent paper, it was reported that concentrations of ouabain below those that inhibit Na/K ATPase in most experiments, under certain conditions *stimulate* a Na channel by "slip-mode conductance" (20). The concentration–response curve was provocative in that the half-maximal stimulating action was in the nanomolar range, just what one would expect to find in the serum of digitalized patients. Although other interpretations are possible (21), it is possible that a heretofore unknown action of very low concentrations of cardiac glycosides may account for some beneficial action. On the other hand, as suggested above, any increase in contractility caused by the drug regardless of mechanism might be deleterious to an already compromised heart. Possible interactions between Na/K ATPase and the Na channel are an intriguing thought.

Why might one desire an increase in the force of contraction of a failing heart? Consider that heart failure is generally chronic, involves hypertrophy, and eventually becomes decompensated. Under such circumstances it would no longer be appropriate to use an inotropic intervention unless there is acute failure. Increasing heart contractility requires an increased energy expenditure, which is asking much of an already damaged heart. As Withering observed, patients who are acutely ill with major difficulties of breathing and massive edema are strikingly improved by digitalis treatment. But he never suggested that the drug be used for lifetime treatment. We know today that digitalis therapy provides a short-term benefit. In the long term, small doses seem to reduce the worsening of the disease and may modify quality of life to a small extent, but do not prolong life. The object of medical treatment is to improve the quality of life, reduce morbidity, extend life, and reduce hospital stay. Drugs such as ACE inhibitors and β-blockers do much to accomplish these desired effects. Digitalis is useful, but not for its action on contractility. In a provocative editorial, Stevenson pointed out, " . . . it has been evident for some time that inotropic therapy does not have a role in the treatment of most of the more than 4 million patients with heart failure"(22).

Taken together, our experience provides the inescapable conclusion that digitalis results in inotropic and toxic effects as well as short-term beneficial therapeutic effects. The only receptor for digitalis that has thus far been uncovered is Na/K ATPase. The actions of this intriguing drug can be summarized as follows:

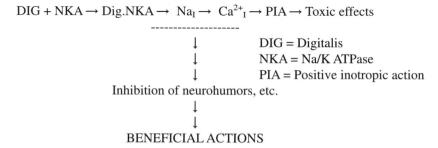

$$DIG + NKA \rightarrow Dig.NKA \rightarrow Na_I \rightarrow Ca^{2+}_I \rightarrow PIA \rightarrow Toxic\ effects$$

$\downarrow$ DIG = Digitalis

$\downarrow$ NKA = Na/K ATPase

$\downarrow$ PIA = Positive inotropic action

Inhibition of neurohumors, etc.

$\downarrow$

$\downarrow$

BENEFICIAL ACTIONS

This scheme suggests that digitalis binds to Na/K ATPase (NKA) and the complex so produced results in its slight inhibition leading to an increase in intracellular Na and through an action on the Na/Ca exchanger, an increase in intracellular Ca. Depending on the concentrations and compartmentation of the cations involved, any beneficial actions of the drug could be due to an inhibition of neurohumors such as norepinephrine, endothelin, NO, or cytokines, etc., that are destructive to heart muscle cells. Another possibility is that the digitalis.NKA complex in some unknown way directly triggers changes in neurohumors without alteration of Ca, something that is doubtful. The most logical mechanism for any therapeutically beneficial action is still via changes in Ca. Certainly one can imagine that inotropic and toxic actions would occur secondary to a higher intracellular Ca concentration produced by a greater inhibition of the enzyme.

II. MOLECULAR STUDIES

In order to understand the molecular function of pumps, it is necessary to deduce their structure. Since ion pumps consist of very large hydrophobic proteins, there was little progress in getting to the first step, the primary amino acid sequence, until 1985 when the Na/K ATPase from sheep kidney was cloned and the amino acid sequence of the catalytic subunit deduced from a complementary DNA (23). The 1016 amino acid protein was originally thought to have eight putative transmembrane segments and the apparent ouabain binding site was initially located at the extracellular junction of two transmembrane domains, linked to a phosphorylation site. More recent refinements suggest a ten-membrane spanning protein that has cation, ATP, and phosphorylation sites, and a "pocket" to which cardiac glycosides specifically bind (24,25). The catalytic subunit with a molecular mass of 113 kDa is thought to span the plasma membrane and in some way must by virtue of its tertiary conformation (which is at present unknown), bind and transport three Na ions outward and two K ions inward, at the expense of one ATP molecule. This "α" or main subunit is synthesized on the endoplasmic reticulum and transported to the plasma membrane and inserted with great precision. This remarkable feat may occur with the assistance of another closely attached and heavily glycosylated subunit, called the "β." This subunit may also be responsible for the maturation of the α-subunit on the ER and possibly for conferring stability to various conformations of the α-subunit thought to be vital to the functional activity of the enzyme. With an overall molecular mass of 55 kDa, the β-subunit seems to be required for cation transport and ATPase activity (26). Thus far at least three isoforms of the α-subunit, and three of the β-subunit, have been found. The α_1-isoform occurs in most tissues, the α_2 primarily in skeletal muscle, but also in the heart and brain, and the α_3 primarily in neural and cardiac tissues. The β_1 is expressed in practically all tissues while the β_2 is found mostly in neuronal tissue. A β_3 has been found in the membranes of oocytes of *Xenopus*, but not in other tissues. Stoichiometry of the two subunits in vivo is still unknown.

It is remarkable that nature has constructed a region of one protein subunit in such a way that a complex plant alkaloid, digitalis, binds with great specificity causing an inhibition of the enzyme. The simplest explanation put forth years ago is that plant chemicals interact with mammalian "receptors" because they mimic some endogenous principle that acts as a regulator. Further, receptors develop as a result of some endogenous ligand that has evolved into a specific agonist or antagonist. If these explanations are correct, it should be possible to isolate putative endogenous ligands. In the case of some plant alkaloids, such as the opiates, this postulate has proven to be correct. The digitalis story is, as yet, not this simple, for it has not yet been proven there is a highly specific endogenous ligand, despite hundreds of papers announcing a wide range of possibilities. Recently, evidence has been presented solidifying the presence of another component of the Na/K ATPase, dubbed the γ-subunit (27). This subunit may be an autoregulator of the enzyme, perhaps stabilizing a specific conformation of the protein. It is possible that digitalis might interact with the conformation and with the subunit.

Contemporary methods, such as site-directed mutagenesis, specific conformation, and site-detecting antibody epitope localizations have been employed to garner additional information concerning digitalis, cation binding, ATP, and phosphorylation sites. At present, it is thought that digitalis drugs bind tightly to the Na/K ATPase through an interaction with four of the ten membrane-spanning regions, the so-called H1–H2 and H4–H8 (25), and a variety of amino acids seem to be involved in ion binding (27). Purification and crystallization together with an accurate three-dimensional structure prediction will be necessary for more definitive structural identification.

A. Other Membrane ATPases

It is not surprising that the Na/K ATPase of the plasma membrane shares homology with other membrane ion transporters. As illustrated in Figure 1, the cardiac muscle cell contains a plasma membrane (PM) Ca ATPase and the SR contains a Ca ATPase different from the PM ATPase. The Golgi apparatus, the nuclear membrane, lysosomes, endosomes, secretory granules, storage granules, the inner mitochonrial membrane, chloroplasts, some bacteria, fungi, yeast, and a variety of plant membranes all contain membrane ATPases that direct the transport of K, Na, Ca, and H. At present, there are over 40 complete cDNA sequences, biochemical studies, gene isolation and characterization and chromosomal localization studies published on different ATPases (xx). Within these there appear to be two classes of proteins, the plasma membrane ATPases (PMCA), and organelle ATPases (SERCA). The former are generally stimulated by calmodulin and the latter are located at intracellular sites with specialized functions. These ion motive ATPases were grouped by Pedersen and Carafoli (11) into convenient categories designated P for covalent phosphorylation, V for vacuoles, such as found in yeast and the F type localized specifically in the mitochondria inner membrane and in bacteria and chloroplasts. The latter consist of a water-soluble F_1 moiety involved in catalytic activity, ATP synthesis or hydrolysis, and an F_0 component involved in proton transport. So, the F ATPases produce the ATP that is supplied to the various cellular systems that require and utilize ATP in a complex hydrolytic way (11).

B. Ion Exchangers and Their Role in Cardiac Arrhythmias

As we have seen, a key feature of the pump (Na/K ATPase) in cardiac cells is that it consumes energy in the form of ATP to create gradients of Na and K across the cell membrane. These gradients are in turn used as stores of energy to power other cell transport functions. Thus, the Na gradient is used by Na channels when they are opened to generate the inward flow of charge that underlies the rising phase of the action potential. Similarly, the K gradient is used when K channels open to cause repolarization. Such channels therefore function as simple gates that allow pump-created ion gradients to dissipate in a highly controlled way.

Ion exchangers use the ionic gradients established by the Na/K pump in a very different way: the movement of Na down its gradient is linked to the movement of other ions, often pumping them up their electrochemical gradients. Thus the Na–Ca exchanger uses the energy of the Na gradient to pump Ca up its gradient and in this mode it can extrude Ca from the cell. Similarly the Na–H exchanger can use the Na gradient to pump protons uphill.

Exchangers, therefore, do not simply form pores through which ions can move relatively freely. Instead, the proteins are configured in a way that ensures that specific ions must bind in a fixed stoichiometry before ion movement can occur. Thus, the Na–Ca exchanger requires that three Na ions must bind and be transported in one direction for each Ca bound and transported in the opposite direction. The Na–H exchanger binds one of each ion before moving them in opposite directions. The precise stoichiometry is important since it determines both the effectiveness of the ion pumping and whether the process is electrogenic. Thus, Na–Ca exchange is electrogenic and carries current in the direction of Na movement, but the Na–H exchange is electroneural. Later in this chapter we explain the significance of this electrogenicity for cardiac arrhythmias. Moreover, since three Na ions are used to transport each Ca ion, the Ca gradient that can be supported by a given Na gradient is much larger than it would be for lower stoichiometries.

C. Molecular Biology of the Na–Ca Exchanger

The cardiac Na–Ca exchanger is a member of a superfamily of exchangers that also includes exchangers that use the K ion gradient (e.g., in retinal cells). Here we will describe only the cardiac exchanger known as NCX1. Figure 4.3 shows the molecular topology of NCX1 as proposed by Philipson and Nicoll (28). They suggest there are nine transmembrane segments, with a very large intracellular loop between segments five and six. This loop contains 520 amino acids and represents more than half the size of the total protein. It also contains many of the regulatory sites, including a Ca binding site, a phosphorylation

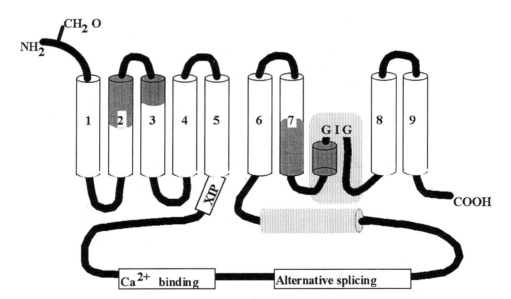

Figure 4.3 Proposed model for the Na–Ca exchanger containing nine transmembrane domains. The NH_2-terminus is glycosylated and extracellular. Earlier models predicted 11 transmembrane segments. Transmembrane segment six is likely to be a portion of the large intracellular loop (light-shaded cylinder). Hydrophobic segment nine does not span the membrane and is speculatively modeled to form a P-loop-like structure (light shading). The a-repeat regions in transmembrane segments two, three, and seven are shaded. Shown on the large intracellular loop are the endogenous XIP region, the binding site for regulatory Ca, and the region where extensive alternative splicing occurs. The large intracellular loop is not drawn to scale but encompasses almost 550 amino acids—more than half the length of the protein. (From Ref. 28.)

site, a Na-dependent inactivation site, which is also the site of action of an inhibitory peptide (XIP). The Ca regulatory site is distinct from the Ca transporting site.

D.   Stoichiometry of Na–Ca Exchanger

When the exchanger was first described in cardiac tissue (29), it was thought to be electroneutral, exchanging two Na ions for each Ca ion. Evidence that transport is in fact electrogenic has come from a variety of sources, both theoretical and experimental. The theoretical basis became clear when free intracellular levels of Na and Ca were first measured accurately in cardiac cells and were shown to be around 5 to 8 mM for Na and 100 nM for Ca. With this information it can be calculated that the energy gradient available to the exchanger would be insufficient to pump Ca out of the cell if the stoichiometry were 2:1. A minimum stoichiometry of 3:1 is required, which means that the exchanger must be electrogenic. Moreover, this stoichiometry also allows us to account for some of the slow inward ionic currents that might be attributable to Na–Ca exchange (see Ref. 30). The first direct experimental evidence for a stoichiometry of 3:1 came from the work of Reeves (30,31) in flux studies using reconstituted sarcolemmal vesicles, while the first clear evidence for the membrane current generated came from the work of Kimura et al. (32). Figure 4.4 shows current–voltage relations obtained for the exchanger at a variety of ion concentrations. Both the flux and ionic current studies show the exchanger is completely reversible. It can operate either in the forward mode, when it uses the Na gradient to ex-

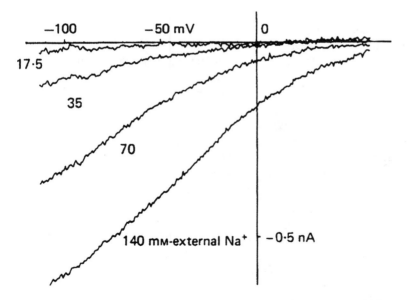

Figure 4.4   Current–voltage relations obtained from guinea pig ventricular cells in conditions where voltage-gated ion channels are blocked and most of the current is attributable to Na–Ca exchange. The current is inward at all voltages since the ion conditions are such that the exchanger works only in the forward mode (i.e., exchanging movement of Na inward for Ca moving outward). The current is a roughly exponential function of potential, with a tendency to become linear at strong negative potentials. These results show relations for various external Na concentrations between 140 and 17.5 mM. (From Ref. 32.)

trude Ca, or in the reverse mode, when it can carry Ca into the cell. The latter may even be sufficient to trigger contraction (33).

E. Kinetics of Na–Ca Exchanger

The kinetics of most ion channels have been determined using the membrane patch-clamp technique introduced by Neher and Sakmann. This approach is supremely effective since the currents carried by single channels are large enough to record their opening and closing as current jumps in the patch recording. This technique fails, however, in the case of ion exchangers since the charge carried by each turnover of the transporter is far too small. Full investigation of the kinetics of exchangers, therefore, required a new technique: the giant patch method introduced in 1989 by Hilgemann, in which the membrane area accessed by the patch electrode is large enough for exchanger currents to be detected (33). For the first time, this method enabled partial reactions of the Na–Ca exchanger to be recorded, and to determine which steps in the reaction sequence are voltage-dependent. Controlling the ion concentrations on both sides of the giant patch allows one to reduce certain steps in the reaction sequence to be either rate-limiting or absent. Recording the charge movements as the reaction steps up to the slowed or removed step then allows investigation of these 'partial' reactions. The results show a surprising degree of complexity in the reaction sequence, but the main conclusions are illustrated in Figure 4.5. Regarding the voltage-dependence of transport, the key conclusion is that Na translocation is electrogenic while Ca transport is electroneutral.

F. Contribution of Na–Ca Exchange to the Action Potential

Two important functional questions now arise. First, since the exchanger is electrogenic, which ion currents during normal electrical activity may be attributed to its activity? Second, quantitatively how much of the Ca extrusion during each beat may be attributed to the exchanger? As we shall see these questions are linked.

It is surprisingly easy to answer the first question. As shown in Figure 4.4, we know how much ion current is carried by the exchanger in various ionic conditions. These results can also be fitted by fairly simple equations for the transport process (see Ref. 35, 36). We also know the voltage time course during the action potential, the concentration of internal Na, and the time course of the intracellular Ca transient. This information is totally sufficient to calculate the activity of the exchanger and hence the ion current carried during the action potential. Figure 4.6 shows such a calculation in the case of the rabbit atrial action potential.

It can be seen there are three phases to exchanger activity. At rest, there is a very small inward current, indicating that the exchanger is contributing to Ca balance in the resting state by pumping Ca out to balance any inward Ca leak. At the beginning of the action potential, when the voltage becomes positive and intracellular free Ca is still low, the exchanger transiently reverses direction to operate in the outward mode. During this time it can contribute to the Ca entering to activate contraction (33). Finally, as intracellular Ca rises, the exchanger reverses direction again to operate strongly in the inward mode. During this phase the exchanger contributes the great majority of the inward current that maintains the late plateau phase of the atrial action potential.

There is considerable experimental evidence for this analysis. First, the late plateau in atrial action potentials (and in the case of the very similar ventricular action potentials in some species such as the rat) is abolished by substituting external Na by Li. The exchang-

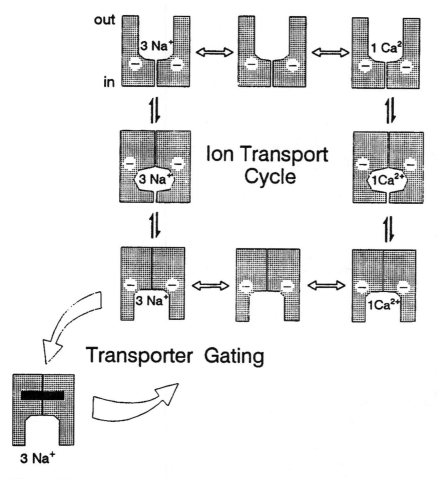

out

in

3 Na⁺

1 Ca²⁺

Ion Transport Cycle

3 Na⁺

1 Ca²⁺

3 Na⁺

1 Ca²⁺

Transporter Gating

3 Na⁺

Figure 4.5 Minimal ion transport cycle proposed for the Na–Ca exchanger. This scheme represents the cycle as consecutive, with the exchanger protein itself carrying two negative charges. In addition to the main transport cycle, the exchanger is also postulated to enter an inactive state whenever binding sites facing the cytoplasmic side are loaded with Na. (From Ref. 34.)

er cannot accept Li so this substitution abolishes the inward current. Second, buffering of intracellular Ca (e.g., by BAPTA), removes the signal that activates the inward mode of the exchanger and this also abolishes the late plateau (38).

How much flux does this current correspond to? Eisner (39) showed that if one integrates the ion current to obtain the total charge carried by the exchanger and compares this to the charge entering during each beat through Ca channels, the charge carried by the exchanger accounts for around 90% of the Ca extrusion necessary to balance Ca entry. The exchanger therefore functions as the main mechanism by which Ca balance is restored on a beat-to-beat basis. Hilgemann and Noble (36) came to a similar conclusion on the basis of comparing the model results with Ca fluxes calculated using extracellular Ca indicators.

This analysis is relatively easy in the case of action potentials showing a late low plateau since in such cases the Ca current and inward Na–Ca exchange hardly overlap. Once repolarization has occurred to around –40 mV (the level at the beginning of the late

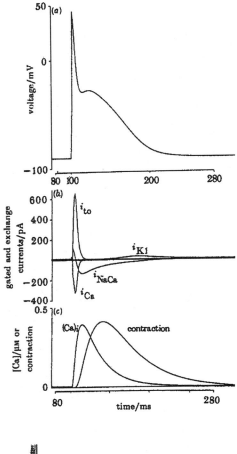

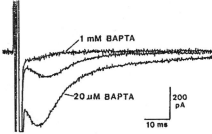

Figure 4.6 (Top) Time course of Na–Ca exchange computed in a model of the single atrial cell developed by Hilgemann and Noble (36). The exchanger first develops outward current during the initial depolarization. This phase corresponds to inward movement of Ca, and it may contribute to the Ca entry that triggers contraction. As intracellular Ca rises and the early stage of repolarization occurs, the current becomes inward, and so the exchanger starts to pump Ca out of the cell. This current is strong enough to maintain the late plateau of the action potential, as was originally shown by Mitchell et al. (37) in the case of rat ventricle. (Bottom) Experimental test of model predictions. These recordings show membrane current during voltage clamp to the level of the late plateau of the atrial action potential following a short, strong depolarization to trigger and then to deactivate the Na and Ca channel currents. The time course of the recorded inward current resembles that in the computer model shown above, and the current disappears when intracellular Ca is buffered by infusing BAPTA through the patch-clamp electrode. (From Ref. 38.)

plateau), the Ca channels rapidly close and nearly all the inward current during the remainder of the action potential is carried by Na–Ca exchange. However, this is not the situation with ventricular action potentials in species like the guinea pig, cat, dog, and human, where the plateau occurs at positive potentials. Here the flow of current through Ca channels greatly overlaps with that through the Na–Ca exchanger. Moreover, the high level of the plateau means that it takes longer for the combination of voltage and Ca changes to reach the point at which the initial outward phase of exchange activity switches to the inward mode. A simple dissection between phases of the action potential attributable to Ca current and to the exchanger is not then possible.

Can we nevertheless estimate the contribution of exchange current in this case? One way of achieving this is illustrated in Figure 4.7, which shows the results of an experiment in which a very rapid external solution switch (substituting Na ions with Li ions) at various stages during the action potential, results in action potential shortening. The experimental results compare well with those predicted by ventricular action potential models (40).

G. Role of Na–Ca Exchange in Arrhythmic Mechanisms

In 1976, Lederer and Tsien showed that in Purkinje fibers in which the Na pump had been blocked with cardiac glycosides, action potentials, or voltage-clamp steps are followed by an oscillatory inward current, which they called the transient inward current, I_{ti} (41). Previously, Ferrier and Moe (42) had shown the existence of transient depolarizations (now called delayed after-depolarizations) under similar conditions. If these depolarizations are sufficiently large, they can reach the threshold for initiation of further action potentials, so creating a focus of abnormal pacemaker activity and generating ectopic beats. One of the earliest theories for this process implicated the Na–Ca exchanger (43). This has turned out to be correct. The mechanism involved is now one of the best understood arrhythmic mechanisms and can be fully reconstructed using computer models.

The sequence of events is as follows. First, inhibition of the Na pump by cardiac glycosides, or by metabolic changes as in ischemia, for example, causes a rise in internal Na concentration, which may increase from its normal level of 5 to 8 mM to 15 to 20 mM. NMR studies of hearts in ischemic conditions have shown such an increase in the period of time leading up to the generation of ischemic arrhythmias (44).

This increase in internal Na reduces the energy gradient available to the Na–Ca exchanger, which therefore pumps less Ca out of the cell than enters during each beat. Ca therefore accumulates in the cell. Most of this excess Ca is sequestered by the sarcoplasmic reticulum (SR), which means that more Ca is available to be released during each contraction. This is one of the mechanisms of the well-known positive inotropic action of Na pump inhibition (see earlier section in this chapter).

Some of this excess Ca accumulates in the cytosol so that resting free intracellular Ca also rises, a condition known as Na–Ca overload. These two processes (i.e., accumulation of sequestered Ca and rise in free cytosol Ca) form a lethal combination. A sufficiently large rise of intracellular Ca triggers release of Ca from the SR. This is the process of Ca-induced Ca release that normally triggers SR release during each heart beat. But, in the abnormal conditions of Na–Ca overload, the process can occur spontaneously, instead of during the next wave of excitation. Moreover, since the SR is highly charged with Ca, the spontaneous release of Ca can be very large, and it can be repetitive, thereby generating an oscillation that may be continuous or damped depending on the degree of Na–Ca overload.

Each release of Ca from the SR during this process will activate the inward mode of

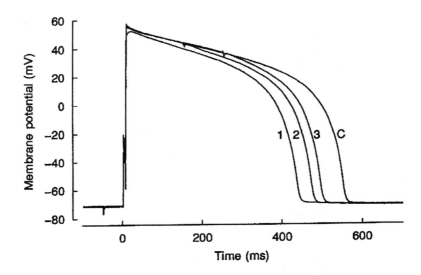

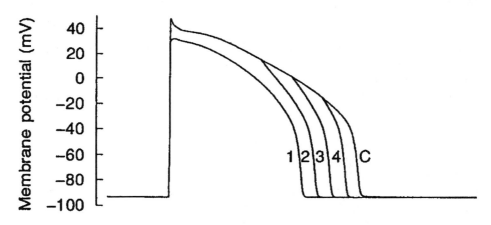

Figure 4.7 Experimental and theoretical evidence for activity of the Na–Ca exchanger during the high plateau of ventricular cells, as observed in many species, including guinea pig and humans. The top traces show experimental recordings made in isolated guinea pig ventricular cells during sudden external solution switches from 140 to 70 mM. The main change induced by this switch is a large reduction in the current carried by the Na–Ca exchanger (see also Fig. 4.4). The solution switch (which occurs within 50 ms) was applied at various times during the plateau. The earlier the switch, the greater the shortening of action potential duration. The lower traces show computations of the same effect using a model of the guinea pig ventricular action potential. (From Ref. 41.)

the Na–Ca exchange, so generating an inward depolarizing current. This is the mechanism of generation of I_{ti} (45). Figure 4.8 shows the reconstruction of this arrhythmic mechanism using the Hilgemann–Earm–Noble atrial cell model.

The conditions of Na–Ca overload were simulated by increasing internal Na to 15 mM, which is the level that would be reached after a 10-min period of inhibition of the Na/K pump. The first action potential is a triggered one using a stimulating current pulse

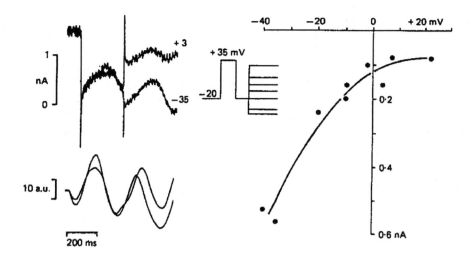

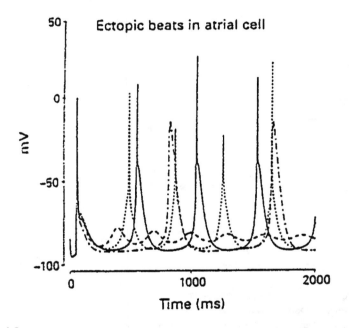

Figure 4.8 Ectopic beat mechanism attributable to intracellular Ca oscillations and consequent cyclic activation of the Na–Ca exchange. (Top) Identification of oscillatory current during Na–Ca overload in a guinea pig ventricular cell as originating from the Na–Ca exchanger. In this experiment, Ca oscillations occurred following repolarization from +35 mV to –20 mV. Top record shows membrane current; bottom record shows cell contraction. During the inward current oscillation, the membrane was suddenly clamped to various potentials in order to derive the current–voltage relation of the mechanism generating the current. This is plotted on the right. It clearly resembles that for the Na–Ca exchanger (compare with Fig. 4.4) (45). (Bottom) Computed ectopic activity in model of single atrial cell following Na overload by raising intracellular Na from 8 mM to 16.34 mM. The oscillatory current then generated is sufficient to initiate action potentials. The precise frequency of the oscillations depends on the binding constant for Ca activation of Ca release from the sarcoplasmic reticulum. (From Ref. 46, where the further details of the computations are given.)

mimicking the normal spread of excitation. The subsequent action potentials are sponta-neous ones generated as each oscillation of intracellular Ca activates inward Na–Ca ex-change current to depolarize the membrane to threshold.

Would these depolarizations be sufficiently strong to trigger conducted waves of ex-citation in the mass of cardiac tissue? This is an important question. Ectopic beats can only trigger life-threatening reentrant arrhythmias if they propagate. A purely local depolariza-tion restricted to a small region of damaged or ischemic tissue would not be dangerous.

This question has been answered using supercomputers to represent massive net-works of cardiac tissue in which small regions of Na–Ca overload were inserted (46). Fig-ure 4.9 shows the results of these computations.

The results are expressed in terms of the minimal size of regions that must be in a Na

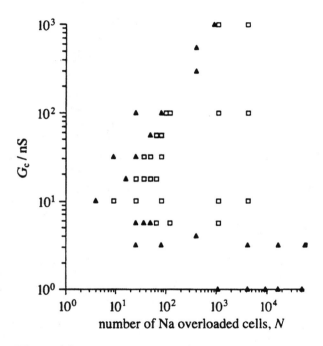

Figure 4.9 Incorporation of arrhythmic mechanism illustrated in Figure 4.8 into large networks of cardiac cells. These computations were done in networks of up to one million cells interconnected with resistances between near neighbors representing nexus junctions. Na–Ca overload was then in-corporated into regions of various sizes to determine the minimum size required for the Ca oscilla-tion to generate sufficient current carried by then Na–Ca exchange to trigger a fully propagated ec-topic beat. The results were obtained at various levels of nexal conductance from 10^3 nS (representing normal tissue) to 10 nS and below (representing the extensive cell decoupling that oc-curs in ischemic tissue). Filled symbols show combinations that failed to initiate propagated conduc-tion of the ectopic beat. Open symbols represent combinations of parameters that allowed full con-duction to occur so that the overloaded region invaded the normal regions. The results show that surprisingly small regions of Na–Ca overload are sufficient to ensure propagation. At normal levels of intercellular connection, for example, 1000 cells are sufficient (representing around 1 mm^2 of damaged tissue). At "ischemic" levels of connectivity, even smaller numbers are sufficient. The Na–Ca exchange current triggered by Ca oscillations is therefore a robust arrhythmic mechanism. (From Ref. 46.)

overload state in order to trigger propagated ectopic beats. Clearly, if this region is too small, the Na–Ca exchanger will not be able to generate sufficient current to excite adjacent normal regions. The ectopic beat is then aborted and does not propagate. The minimal size required to ensure propagation of ectopic beats depends on a number of factors, including the density of intercellular connections. But, in all cases, a surprisingly small region of overloaded cells is sufficient to ensure propagation. At normal levels of intercellular connectivity around 1000 cells are sufficient. When intercellular connectivity is reduced, as happens in ischemic tissue, even small areas are sufficient. These computations have been repeated in three-dimensional reconstructions and they show that regions as small as 1 to 3 mm^3 can initiate ectopic beats. Therefore, this mechanism of arrhythmia is a very powerful one.

H.   Role of Na–Ca Exchange in the Arrhythmias of Ischemia

Na overload frequently occurs in ischemic tissue and the question therefore naturally arises whether Na–Ca exchange may play a role in the initiation of fatal arrhythmias in this disease state. Ch'en et al. (47) have recently used modeling of cardiac metabolism to attempt to unravel this sequence of events. Figure 4.10(A) shows the major events computed during a 10-min period of total ischemia. As observed experimentally, the level of ATP is hardly changed for several minutes, while breakdown of creatine phosphate occurs. During this period, the contraction of the modeled tissue falls, internal Na rises, and at around 4 min a period of arrhythmia occurs as each stimulus evokes multiple beats. The mechanism has been illustrated previously: intracellular Ca oscillations generate cyclic activation of inward Na–Ca exchange current. Figure 4.10(B) shows that this model also succeeds in reproducing arrhythmias that can occur during reperfusion. To mimic reperfusion, the level of ATP was restored to its original level of 7 mM and maintained there. Intracellular Na recovers as the Na pump is reactivated. At the same time the SR Ca pump is strongly activated by the high level of cytosol-free Ca to produce a supercharged Ca concentration within the SR. The results are that when the level of Na enters the range where intracellular Ca oscillations occur, they are even stronger than during the period of ischemia itself. This provides an explanation for the well-known strength of reperfusion arrhythmias.

I.   Role of Na–Ca Exchange in Acid-Induced Arrhythmia

The transport processes that control acidity in the heart are now well understood. Figure 4.11 illustrates the main features. There are two acid extruding transporters: the Na–H exchanger and the $Na–HCO_3$ cotransporter. Acid loading is mediated by $Cl-HCO_3$ exchange and by a putative Cl–OH exchanger (48). This transport system has also been successfully modeled and the main results are shown in Figure 4.12. In each case, the result of a sudden acid load is computed. Reducing pH_i from 7.1 to 6.9 induces a modest rise in intracellular Na as the Na-dependent transporters use the energy of the Na gradient to extrude protons or bring HCO_3 ions into the cell. There is also a biphasic effect on contraction as protons inhibit contraction and then the rise in Na produces its well-known positive inotropic effect as the rise of Na works through the Na–Ca exchanger to produce a rise in Ca. This biphasic effect is also observed experimentally. A fall in pH_i to 6.7 also induces a biphasic change in contraction, but the rise in intracellular Na is now great enough to induce Ca-overload arrhythmias. These are even stronger when pH_i is reduced to 6.5, which is within the range of severe ischemia.

This mechanism of arrhythmogenesis is the basis of a cardioprotective drug, HOE6,

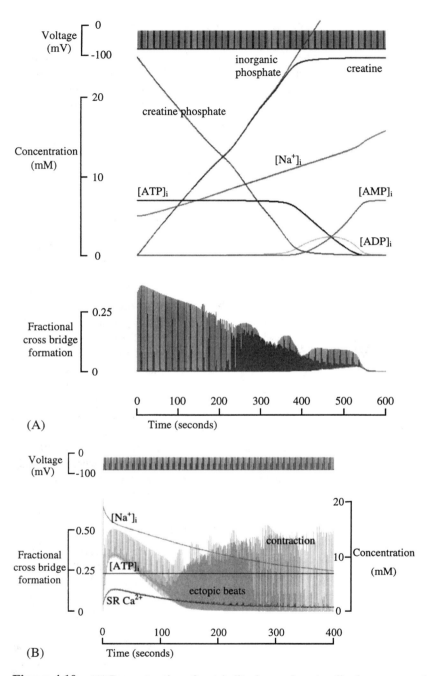

Figure 4.10 (A) Reconstruction of metabolite, ion, and contractile changes occurring during the first 5 min of total ischemia. The top trace in each panel shows stimulation via a voltage-clamp pulse at 0.1 Hz, while the middle traces represent various metabolites and ion concentrations. The bottom trace is cell contraction, calculated as fraction of cross-bridge formation. The computation reproduces all the major changes observed experimentally, including a period where ectopic beat arrhythmias are triggered, represented here by multiple contractile responses to each stimulus. These computed arrhythmias are attributable to the Ca oscillation/Na–Ca exchange mechanism discussed in

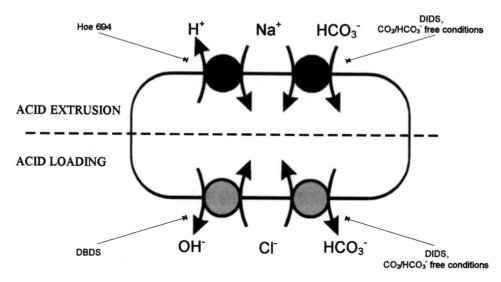

Figure 4.11 Transporters involved in controlling pH in cardiac cells, together with their inhibitors. The Na–H exchange and the Na–HCO₃ cotransporter are involved in acid extrusion, using the Na gradient to move protons or bicarbonate ions across the cell membrane. Na–H exchange is blocked by HOE 694, while DIDS can be used to block Na–HCO₃. Two transporters are involved in acid loading: Cl-HCO₃ and a putative Cl–OH exchanger. These are the transporters modeled in the results shown in Figure 4.12.

694, a Na–H exchange inhibitor. Figure 4.12(D) shows that the model succeeds in reproducing this cardioprotective effect. When Na–H exchange is inhibited, the rise in intracellular Na is more gradual and less severe, since only the Na–HCO₃ cotransporter is now functional. With a less severe rise in internal Na, Na–Ca-overload arrhythmias would be expected to be less severe or even suppressed.

This cardioprotective action of Na–H exchange inhibitors is currently one of the more promising lines of drug intervention in ischemia (49). The remarkable ability to accurately model such effects illustrates the need and power of quantitative, integrative studies in situations where many different protein transporters, enzymes, metabolites, etc., are involved in the function of intact cardiac cells. Such methods not only serve to unravel underlying mechanisms of arrhythmias, but they can, as illustrated here, also be used to suggest and test strategies for arrhythmia prevention.

---

this chapter. (B) Continuation of the computation with immediate restoration of metabolite levels, which are then held constant (corresponding to reperfusion). Note that the arrhythmic episode following reperfusion is even more severe than that during ischemia. The mechanism of this process is that, following restoration of ATP levels, and during the period of continued Ca overload, the sarcoplasmic reticulum becomes supercharged with Ca so that Ca oscillations, when they occur, are even more severe. (From Ref. 47.)

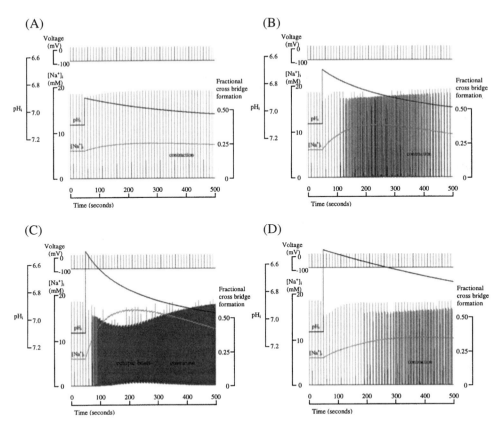

Figure 4.12 Computations of the changes in contraction and internal Na following acid loading of a model cardiac cell. The top trace in each panel shows stimulation via a voltage-clamp pulse at 0.1 Hz, while the middle trace represents internal pH. The acid load increases successively from the resting pH of 7.1 to (A) 6.9, (B) 6.7, and (C) 6.5. The trace that starts to rise on acid loading is internal Na. The greater the acid load, the greater the rise in internal Na. The bottom trace represents contraction, represented as cross-bridge formation. Note the negative inotropic effect seen immediately following the change in pH. This effect is seen experimentally and is the well-known acid inhibition of Ca binding to troponin. Subsequently, contraction increases as internal Na rises (this is exactly the same mechanism as occurs during Na rise following Na pump inhibition, as discussed earlier in this chapter). At a certain level of internal Na, Ca oscillations occur which generate multiple contractions in (B) and (C). The effect of blocking Na–H exchange (mimicking the effect of HOE694) is shown in (D). This greatly slows the recovery of pH, and the rise of internal Na, thus reducing the occurrence of Ca oscillation arrhythmias. This is thought to be the mechanism of the cardioprotective action of HOE694. (From Ref. 47.)

REFERENCES

1. Boyer PD. Ann Rev Biochem 1997; 66: 717–749.
2. Keynes RD. J Physiol (Lond) 1951; 114:119.
3. Hodgkin AL, Keynes RD. J Physiol (Lond) 1955; 128: 28.
4. Libet B. Fed Proc 1948; 7: 72 (abstr).
5. Abood LG, Gerard RW. J Cell Comp Physiol 1954; 43: 379.

6. Utter MF. J Biol Chem 1950; 185: 499.
7. Skou JC. Biochim Biophys Acta 1957; 23: 394–401.
8. Schatzmann HJ. Helv Physiol Acta 1953; 11: 346.
9. Koefoed-Johnson V. Acta Physiol Scand 1957; 42(suppl) 87.
10. Lindenmeyer GE, Laughter AH, Schwartz A. Arch. Biochem Biphys 1968; 127: 187–192.
11. Pedersen PE, Carafoli E. TIBS 1987; 12: 146–189.
12. Withering W. An Account of the Foxglove and Some of Its Medical Uses. Birmingham, AL: The Classics of Medicine Library, Division of Gryphon Editions, Ltd., 1979.
13. The Digitalis Investigation Group. N Engl J Med 1997; 336: 525–533.
14. Schwartz A, Lindenmeyer GE, Allen JC. Pharmacol Rev 1975; 27: 3–134.
15. Schwartz A. Circ Res 1976; 39: 2–7.
16. Van Winkle WB, Schwartz A. Annu Rev Physiol 1976; 38: 247–272.
17. Noble D. Cardiovasc Res 1980; 9: 495–514.
18. Michael L, Pitts BJR, Schwartz A. Science 1978; 200: 1287–1289.
19. Packer M. N Engl J Med 1993; 329: 201–202.
20. Santana LF, Gomez AM, Lederer WJ. Science 1998; 279: 1027–1033.
21. Nuss HB, Marban E. Sodium channels do not conduct Ca in response to phosphorylation: Direct evidence against "slip-mode conductance." Science, in press, 1999.
22. Stevenson LW. N Engl J Med 1998; 339: 1848–1850.
23. Shull GE, Schwartz A, Lingrel JB. Nature 1985; 316: 691–695.
24. Lingrel JB, Kuntzweiler T. J Biol Chem 1994; 269: 19659–19662.
25. Croyle ML, Woo AL, Lingrel JB. Eur J Biochem 1997; 248: 488–495.
26. Therien AG, Karlish SJD, Blostein R. J Biol Chem 1999; 274: 12252–12256.
27. Lingrel JB, Arguello JM, Van Huysse J, Kuntzweiler TA. Ann NY Acad Sci 1997; 834: 194–206.
28. Phillipson KD, Nicolls DA. Ann Rev Physiol (in press).
29. Reuter H, Seitlz N. J Physiol 1968; 195: 451–470.
30. Noble D. J Physiol 1984; 353: 1–50.
31. Reeves JP, Philipson KD. In: Sodium-Calcium Exchange. OUP 1989: 27–53.
32. Kimura J, Miyamae S, Noma A. J Physiol 1987; 384: 199–222.
33. Hobai IA, Levi A. Cardiovasc Res 1999 (in press).
34. Hilgemann D. Pflugers Archiv (Eur J Physiol) 1989; 415: 247–249.
35. DiFrancesco D, Noble D. Phil Trans R Soc B 1985; 307: 353–398.
36. Hilgemann D, Noble D. Proc R Soc B 1987; 230: 163–205.
37. Mitchell MR, Powell T, Terrar DA, Twist VW. J Physiol 1987; 391: 545–600.
38. Earm YE, Ho WK, So IS. Proc R Soc B 1999; 240: 61–81.
39. Negretti N, Varro A, Eisner DA. J Physiol 1995; 486: 581–591.
40. LeGuennecx, Noble D. J Physiol 1994; 478: 493–504.
41. Lederer W, Tsien RW. J Physiol 19xx; 262: 73–100.
42. Ferrier GR, Moe G. Circ Res 1973; 33: 508–515.
43. Tsien RW, Kass RS, Weingart R. J Exp Biol 1979; 81: 205–215.
44. Clarke K, et al. Mag Res Med 1991; 18: 15–27.
45. Fedida D, Noble D, Rankin AC, Spindler AJ. J Physiol 1987; 392: 523–542.
46. Winslow R, Kimball A, Varghese A, Noble D, Adlakha C, Hoythya A. Proc R Soc B 1993; 254: 55–61.
47. Ch'en F F-T, Vaughan-Jones RD, Clarke K, Noble D. Prog Biophys Mol Biol 1998; 69: 515–538.
48. Leem CH, Vaughan-Jones RD. J Physiol 1998; 509: 487–496.
49. Scholz W, Albus U. Cardiovasc Res 1995; 29: 184–188.

5

Regulation of Cardiac Ion Channels and Transporters

H. CRISS HARTZELL

Emory University School of Medicine, Atlanta, Georgia

ARNOLD M. KATZ

University of Connecticut, Farmington, Connecticut

I. INTRODUCTION

According to Claude Bernard, the "primary condition for freedom and independence of existence" is the ability to maintain a constant "internal environment" (milieu intérieur) within our bodies (1). This constancy, called "homeostasis" by Walter Cannon (2), is made possible by an array of regulatory responses that meet the many challenges encountered during our active, often hazardous, lives. Homeostatic mechanisms, while providing adaptive short-term responses to physiological stresses, can become maladaptive in pathological conditions. For example, deleterious effects of these mechanisms become clinically important in heart failure, where prolonged and inappropriate utilization of the hemodynamic defense reaction that protects the body from a fall in cardiac output contributes to disability and death, often by generating lethal arrhythmias. This dichotomy, between short-term benefit and long-term harm, characterizes many of the systems that regulate cardiac ion channels and transporters.

This chapter focuses on the participation of cardiac ion channels, ion exchangers, and ion pumps in the hemodynamic defense reaction, an integrated neurohumoral response that is called upon, for example, by underfilling of the systemic arterial system or a fall in blood pressure. This response, which causes salt and water retention by the kidney and constriction of vascular smooth muscle, has important effects on the heart (Table 5.1). In short-term emergencies, such as occur during exercise or following hemorrhage, the hemodynamic defense reaction helps the body to maintain blood pressure and tissue perfusion. However, when employed for long periods, the same response becomes maladaptive and can exacerbate disability and worsen prognosis.

II. THE HEMODYNAMIC DEFENSE REACTION

The body's response to a fall in blood pressure is detailed in a brilliant essay by Peter Harris (3), who contrasts the neurohumoral mechanisms that are activated by exercise and hemorrhage with those evoked when arterial filling is reduced by a damaged cardiac pump (i.e., in the patient with chronic heart failure). Although all are examples of the same coordinated hemodynamic defense reaction, their long-term consequences are very different (Table 5.2). During exercise, the challenge is to provide the large increase in cardiac output required to perfuse the active skeletal muscles. In this setting, where muscle vasodilatation markedly lowers peripheral resistance, the hemodynamic defense reaction overcomes the tendency for blood pressure to fall, largely by constricting arterioles supplying tissues other than the exercising muscles, and stimulating the heart. Cardiac output is increased by tachycardia, which is brought about by activation of pacemaker currents in the sinus node, and by an inotropic response of working myocardial cells that is due in part to increased opening of plasma membrane calcium channels (Table 5.1). Similar mechanisms operate in the patient who has sustained a severe hemorrhage, where blood loss reduces cardiac output by decreasing venous return. In this setting, the challenge is to maintain systemic blood pressure at levels sufficiently high to perfuse the brain and heart. Once again,

Table 5.1 The Hemodynamic Defense Reaction: A Neurohumoral Response to Decreased Arterial Filling

Mechanism	Short-term, adaptive		Long-term, maladaptive
Vasoconstriction	↑Afterload Maintain blood pressure Maintain cardiac output		↓Cardiac output ↑Cardiac energy demand Cardiac necrosis
Increased cardiac Adrenergic drive	↑Contractility ↑Relaxation ↑Heart Rate	Maintain Cardiac Output	↑Cytosolic calcium ↑Cardiac energy demand Cardiac necrosis Arrhythmias, sudden death
Salt and water retention	↑Preload Maintain cardiac output		Edema, anasarca Pulmonary congestion

Table 5.2 Three Conditions That Evoke the Hemodynamic Defense Reaction

Condition	Duration	Challenge	Response
Exercise	Min/h	Increase perfusion to exercising muscles	Selective vasoconstriction Cardiac stimulation
Hemorrhage	h	Loss of vascular volume	Vasoconstriction Cardiac stimulation Salt and water retention
Heart Failure	Lifetime (Progressive)	Damage to heart Impaired pumping	Vasoconstriction Cardiac stimulation Salt and water retention

this challenge is met by peripheral vasoconstriction and cardiac stimulation, along with salt and water retention by the kidney (Table 5.1). Neither of these challenges lasts very long. Intense exercise generally ends after several minutes, or at most a few hours, while recovery is either well underway soon after a severe hemorrhage or the patient dies.

Similar responses characterize the hemodynamic defense reaction in patients with chronic heart failure, a clinical syndrome caused by a variety of diseases that damage the cardiac pump and so reduce the heart's ability to maintain a normal cardiac output. Although the duration of the neurohumoral response in heart failure is much longer than that seen during exercise and following hemorrhage, all of these hemodynamic challenges activate the same signal transduction cascades, and so evoke similar responses by cardiac ion channels, ion exchangers, and ion pumps. Among the actions of the mediators of the hemodynamic defense reaction are powerful effects on voltage-gated ion channels that control the cardiac action potential. In heart failure, the resulting abnormalities of impulse generation and propagation are of considerable clinical significance as approximately 40% of these patients die suddenly and unexpectedly, most likely of a lethal arrhythmia. Arrhythmias also play an important role in morbidity and mortality following acute myocardial infarction, where abnormal impulse generation and conduction are exacerbated by the neurohumoral response to the fall in blood pressure commonly seen in this condition.

III. MEDIATORS OF THE HEMODYNAMIC DEFENSE REACTION

The three major components of the hemodynamic defense reaction listed in Table 5.1—vasoconstriction, cardiac stimulation, and fluid retention—are mediated by signaling cascades that are activated, and in some cases inhibited, by an array of extracellular "messengers." The signaling molecules that act upon the heart include peptides such as angiotensin II, which is generated both systemically and locally, atrial natriuretic peptide (ANP), which is released by the atrial myocardium, and endothelin, which is produced by vascular endothelial cells (Table 5.3). Probably the most important mediators of cardiac stimulation are the β-adrenergic agonists epinephrine and norepinephrine. The norepinephrine that is released locally by sympathetic nerve endings in the heart has especially potent effects on cardiac ion channels (discussed later in this chapter).

Additional classes of signaling molecules, including prostaglandins and nitric oxide,

Table 5.3 Some Mediators of the Hemodynamic Defense Reaction

Mediator	Effect on the Heart
Peptide and Protein	
Angiotensin II	+
Endothelin	+
Catecholamine	
Norepinephrine	
α_1-receptors	+
α_2-receptors (central)	–
β-receptors	++

+ weak stimulatory effect, ++ strong stimulatory effect, – inhibitory effect.

participate in the hemodynamic defense reaction, but rather little is known of their effects on cardiac ion channels (4). Cytokines and peptide growth factors, which generally activate tyrosine kinases, have important effects on cardiac function through their effects on myocyte growth; however, short-term effects of these peptides on cardiac ion channels in heart have not been clearly demonstrated. Because channels in other systems are modulated by tyrosine kinase phosphorylation, it seems likely that these mechanisms also exist in heart.

A. Peptide Mediators

1. Angiotensin

Angiotensin II, which has important vasoconstrictor actions, also modifies the performance of the heart. This peptide, which is formed by enzymatic cleavage of an inactive precursor called angiotensinogen, can be synthesized systemically by enzymes that circulate in the blood, or formed locally by tissue enzymes. The circulating and tissue systems appear to serve different functions: the former is probably most important in regulating vasomotor tone, while the tissue system exerts local trophic effects that modulate gene expression. The effects of angiotensin II are mediated by at least two receptor subtypes, designated AT_1 and AT_2, which not only differ qualitatively, but can oppose one another (5,6). The AT_1 receptors, which predominate in the adult heart, mediate the vasoconstrictor actions of angiotensin II (6) and increase cardiac contractility (6–10). The inotropic effects of angiotensin II, which are due in part to increased calcium currents across the plasma membrane, are relatively minor.

2. Endothelin

In the late 1980s, a potent vasoconstrictor was isolated from endothelial cells and, because of its source, was named endothelin (ET) (11). In addition to its prominent vasoconstrictor effects, binding of this peptide to ET-A receptors increases myocardial contractility (12,13). As in the case of the heart's response to angiotensin II, the inotropic effects of endothelin are weak.

B. Catecholamines

By far the most important mediator of the cardiovascular response in the hemodynamic defense reaction is norepinephrine that is released by sympathetic stimulation. The receptors that mediate the cellular actions of norepinephrine are divided into two classes: α-adrenergic receptors, which are responsible for the vasoconstrictor effects of sympathetic stimulation, and β-adrenergic receptors, which cause both vasodilation and cardiac stimulation (14). Both the α- and β-adrenergic receptors include several subclasses (Table 5.4), a diversity that allows norepinephrine to initiate a variety of cardiovascular responses. Binding of this catecholamine to peripheral α_1-adrenergic receptors on vascular smooth muscle evokes a powerful vasoconstrictor response. Binding of this catecholamine to preganglionic α_2-adrenergic receptors in the central nervous system inhibits sympathetic outflow and so causes vasodilation and attenuates the cardiac actions of norepinephrine; this indirect counter-regulatory effect is not considered further in this chapter.

The powerful cardiac effects of norepinephrine are primarily mediated by β-adrenergic receptors that increase contractility (inotropy), accelerate relaxation (lusitropy), and increase heart rate (chronotropy). As discussed below, many of these effects are initiated

Table 5.4   Major Receptor Subtypes That Mediate Actions of Norepinephrine

α_1-Adrenergic receptors
   Smooth muscle contraction—vasoconstriction
   Sodium retention by the kidneys
   Positive (α1A) or negative (α1B) chronotropy in heart
α_2-Adrenergic receptors
   Central inhibition of sympathetic activity
   Vasodilation
   Cardiac inhibition
β_1-Adrenergic receptors
   Major effect: smooth muscle relaxation—vasodilatation
   Minor effect: Cardiac stimulation—positive inotropy, lusitropy, and chronotropy
β_2-Adrenergic receptors
   Cardiac stimulation—positive inotropy, lusitropy, and chronotropy

when norepinephrine activates signal transduction systems that modify a variety of cardiac ion channels, ion pumps, and ion exchangers.

1. α_1-Adrenergic Receptors

In addition to their prominent role in mediating the vasoconstrictor response to sympathetic stimulation, α_1-receptor activation has a positive inotropic effect on the heart (15,16); this cardiac response, however, is generally minor, and in some species absent. In humans, although α_1-adrenergic receptor activation increases contractility, this inotropic response is weak and, in the human ventricle where the density of α_1-adrenergic receptors is low, of little clinical significance (17). As discussed below, these effects are mediated by a family of GTP binding proteins called G-proteins, which activate phospholipase-coupled intracellular signal transduction systems.

2. β-Adrenergic Receptors

As is the case for most of the receptors that mediate the responses to extracellular messengers, there are several β-adrenergic receptor subtypes. Two of them, the β_1- and β_2-receptors, play an important role in cardiovascular regulation. The relative amount of the two subtypes differs depending on cardiac tissue, species, pathophysiological state, and age or developmental stage. The β_2-receptors normally predominate in vascular smooth muscle, where they mediate a relaxing effect of norepinephrine that is generally overwhelmed by the constrictor effects of α_1-adrenergic receptor activation. In mammalian ventricular muscle, the most important cardiac subtype is the β_1-receptor, although β_2-receptors are also found. Both receptor subtypes exert their effects by G-protein-coupled signal transduction (see below). Binding of norepinephrine to the β_1- and β_2-receptors has been reported to yield quite different responses. The effects of β_1-receptor stimulation increase inotropy and lusitropy, and are mediated mainly by protein kinase–activated phosphorylations that occur in the cytosol, whereas β_2-receptor-mediated phosphorylation may be localized to the plasma membrane and the vicinity of calcium channels (18–21). Similar differences in the effects of these β-adrenergic receptor subtypes have been found in the human heart, although these appear to be less marked than in the rat (22).

IV. SIGNAL TRANSDUCTION PATHWAYS THAT MEDIATE THE HEMODYNAMIC DEFENSE REACTION

The signals that are mediated by the extracellular messengers discussed above begin when these molecules bind to specific receptors in the plasma membrane. These receptor proteins allow the cell to respond to the arrival of a regulatory molecule at the cell surface by initiating an intracellular signal that alters cardiac function. For example, the sympathetic neurotransmitters (norepinephrine released from sympathetic nerves and epinephrine released from the adrenal gland) bind to β_1- and β_2-adrenergic receptors to increase heart rate and myocardial excitability and contractility, and to α_1-receptors that cause vasoconstriction. Acetylcholine (Ach), the major parasympathetic neurotransmitter, has opposing effects that slow the heart, decrease myocardial excitability and contractility, and cause peripheral vasodilation.

The increase in contractile force initiated by activation of β-adrenergic receptors is brought about largely by stimulation of adenylyl cyclase, an enzyme that synthesizes the intracellular second messenger cyclic AMP (cAMP). cAMP, by an allosteric effect, then activates cAMP-dependent protein kinase, an enzyme that phosphorylates a variety of effector systems involved in regulating cell excitability or contractility. These effector systems include a variety of sarcolemmal ion channels, regulatory proteins in the sarcoplasmic reticulum that control cellular Ca fluxes, as well as proteins of the contractile apparatus (23,24). The increase in contractile force initiated by α_1-adrenergic receptors is brought about by a cAMP-independent mechanism that is not yet fully understood. Part of the mechanism may involve activation of a phospholipase C-catalyzed signaling pathway (see below). In addition, α_1-receptor activation decreases K currents and thus prolongs the Ca influx by signal transduction systems that are not well understood (25).

Acetylcholine, the parasympathetic neurotransmitter, decreases contractile force and frequency by binding to muscarinic receptors whose actions generally oppose those of β-adrenergic receptor-stimulated cAMP production. The muscarinic receptors, which are coupled to different G-proteins than the β-receptors, also activate cAMP-independent, membrane-delimited pathways that regulate ion channel function.

We have learned a tremendous amount in the last decade about the mechanisms by which the transmitter/hormone receptors are coupled to their effector systems. This signal transduction process usually involves the G-proteins, which bind and hydrolyze guanine nucleotides (26,27). The ability of the G-proteins to hydrolyze GTP, and exchange GDP for GTP, regulates the activity of these coupling proteins. Figure 5.1 provides an overview of G-protein-mediated signal transduction in the heart: ion channels can be directly modulated by G-proteins, by second messengers generated by G-protein-coupled enzymes, and by G-protein-initiated phosphorylation cascades.

V. G-PROTEIN-COUPLED SIGNAL TRANSDUCTION

A. The G-Protein Cycle

The G-proteins that couple receptors to enzymes such as adenylyl cyclase and phospholipase C, or to ion channels are heterotrimeric membrane proteins located on the cytoplasmic side of the plasma membrane. They consist of an α-subunit, which binds guanine nucleotides, and two smaller β- and γ-subunits. Figure 5.2 depicts the activation cycle of the heterotrimeric G-proteins. In the resting or inactive state, GDP is tightly bound to the α-subunit. Hormone binding to the receptor induces a conformational change in the receptor

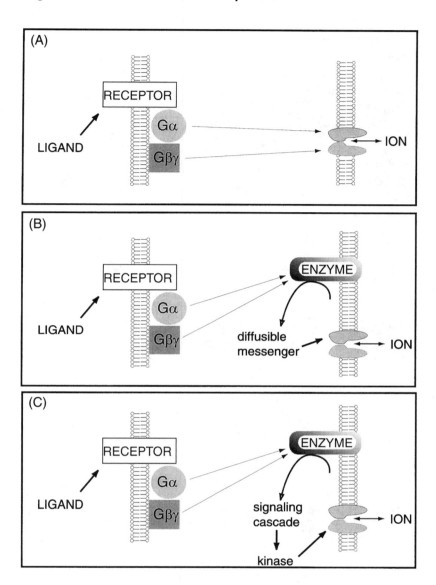

Figure 5.1   General mechanisms of regulation of ion channels in heart by G-protein-coupled pathways. (A) Ion channels can be regulated directly by binding of α- or βγ-subunits of G-proteins. (B) Both α- and βγ-subunits of G-proteins can activate enzymes, such as adenylyl cyclase and phospholipase C, which produce diffusible messengers that directly regulate the channels. (C) The second messengers produced in (B) can stimulate kinases that regulate the channel via phosphorylation.

that stimulates its interaction with the G-protein. When the receptor binds to a G-protein, the affinity of the α-subunit for GDP decreases significantly and the GDP dissociates. Because the ratio of cytosolic GTP to GDP is high, the empty site becomes quickly occupied by GTP. Once GTP binds, the α-subunit changes conformation to a state having low affinity for both the receptor and the βγ-subunits (which remain tightly bound to each other and function as a dimer) (28). Both the dissociated α-subunit and the βγ-dimer then regulate

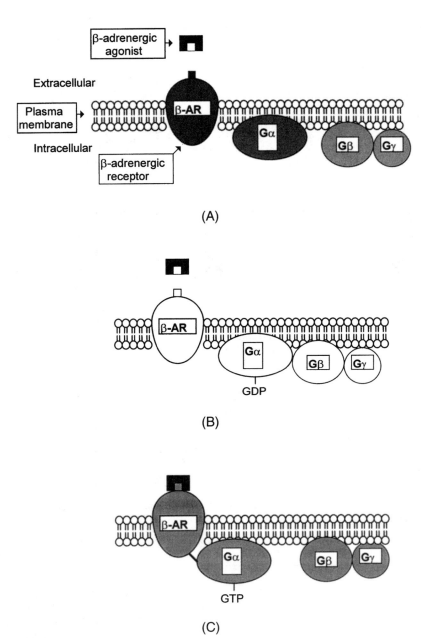

Figure 5.2 G-Protein-coupled signal transduction. (A) Four proteins mediate signal transduction by β-adrenergic agonists: the β-adrenergic receptor (β-AR) and the three components of the G-proteins (Gα, Gβ, and Gγ). (B) In the inactive system, when the receptor is not bound to the agonist, Gα is bound to GDP and to the Gβ–Gγ complex. In this state, the receptor and all of the G-protein subunits are inactive (unshaded). (C) Binding of the agonist to the receptor activates the latter, which is able to bind to Gα; this leads to replacement of bound GDP with GTP and dissociation of the Gβ–Gγ complex from the Gα-subunit. As a result, all of these components become activated (shaded).

COLOR PLATES

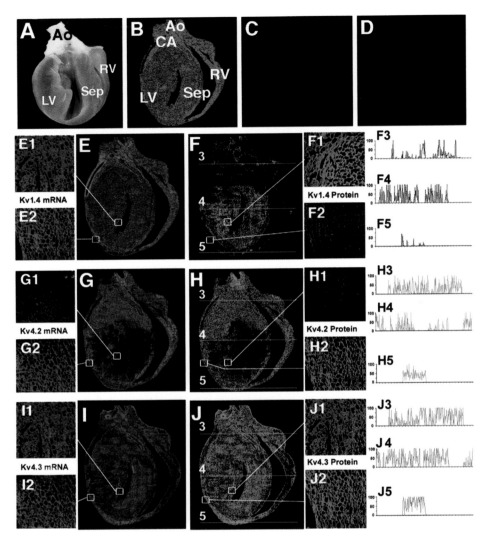

Figure 6.3 Comparison of Kv1.4, Kv4.2, and Kv4.3 α-subunit mRNA transcript and protein distribution in ferret ventricular sagittal sections. Relative mRNA levels were detected by fluorescence in situ hybridization (FISH) and are indicated by the red fluorescence. Relative protein levels were detected by indirect immunofluorescence (IF) and are indicated by the green fluorescence. Results shown were obtained from adjacent sagittal sections. Panel (A) provides a view of a sagittal cut through the ventricle and great vessels (AO=aorta; CA=coronary artery; RV=right ventricle; LV=left ventricle; Sep=septum). Panels (B) and (C) are sections that have been probed with antisense and sense probes to TnIc and serve as positive and negative controls to this transcript. Panel (D) represents a negative control signal to Kv4.2/Kv4.3 using a secondary antibody (see Ref. 149) in the absence of primary antibodies. FISH results are shown in Panels (E), (G), and (I) and IF results are shown in Panels (F), (H), and (J). The smaller numbered panels correspond to enlarged sections obtained from selected regions indicated by the white boxes. Note the general correspondence between the Kv4.2 and Kv4.3 mRNA and protein levels. On the other hand, Kv1.4 mRNA and protein levels do not correlate. Relative fluorescence intensity profiles of Kv1.4, Kv4.2, and Kv4.3 antibodies are measured in transverse sections obtained from the indicated levels in the basal, midventricular, and apical regions of the ventricle. The marked variation in relative intensity profiles measured from the different transverse sections indicates that protein levels vary between different small domains within adjacent regions as well as varying between different regions of the heart (regional localization). (Reproduced by permission from Ref. 149.)

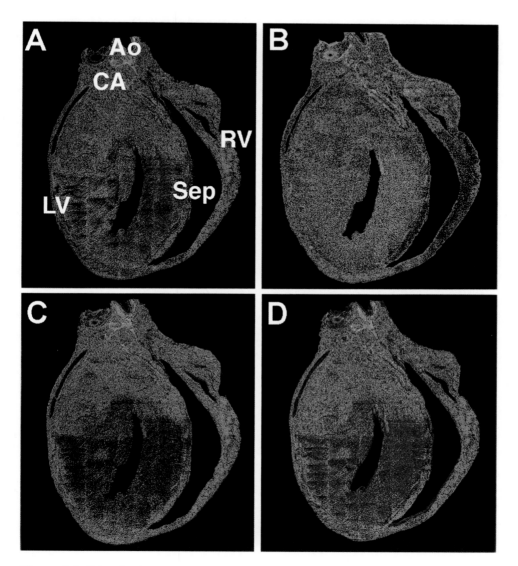

Figure 6.4 Colocalization and distribution of Kv1.4, Kv4.2, Kv4.3 α subunits in ferret ventricular sagittal sections determined from direct IF. Panel A shows Kv1.4 (red) and Kv4.2 (green) colocalized (yellow to orange) mainly in the apical LV endocardial region and septum. Panel B shows Kv1.4 (red) and Kv4.3 (green) also colocalized (yellow to orange) mainly in the apical LV epicardial region and septum. Panel C shows Kv4.2 (green) and Kv4.3 (red) also colocalized (yellow to red) in the apex. Panel D shows overall distribution patterns of Kv1.4 (blue), Kv4.2 (green), and Kv4.3 (red) α subunits. Colocalization of Kv4.2 and Kv4.3 is indicated by yellow to orange and Kv1.4 and Kv4.3, purple to pink. (Reproduced by permission from Ref 149.)

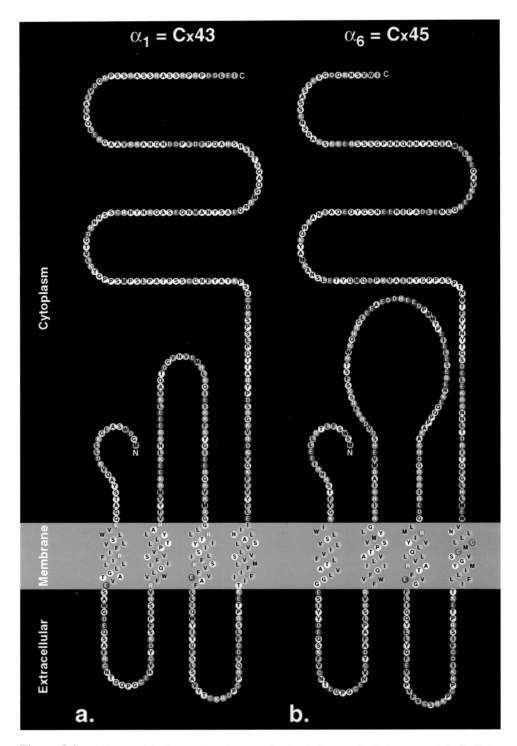

Figure 8.1 Folding models for gap junction proteins in the heart: Cx43 (α_1 connexin), Cx45 (α_6 connexin), Cx40 (α_5 connexin), and Cx37 (α_4 connexin). The amino acid sequences were deduced from cDNA analysis (16,29,63,134), and the residues are coded as follows: hydrophobic in yellow, acidic in red, basic in blue, and cysteine in green. Hydropathy analysis predicts that the polypeptide spans the lipid bilayer of the plasma membrane four times. Among the connexins there is

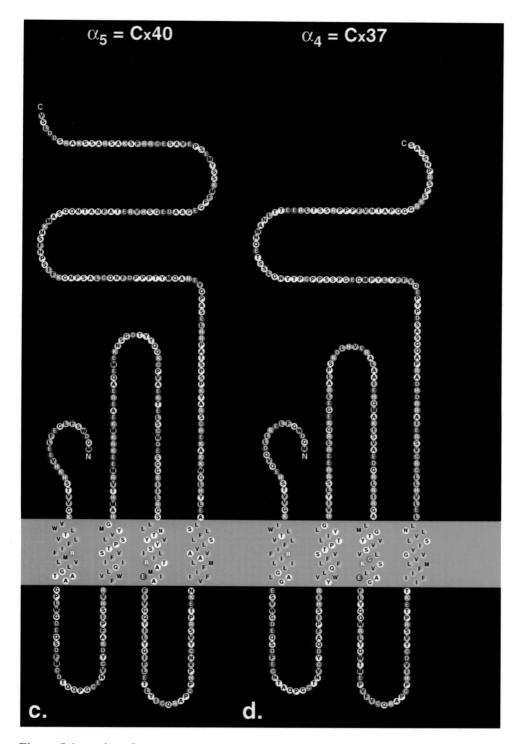

Figure 8.1 continued conservation in the amino acid sequence of the four hydrophobic domains, refererred to as M1, M2, M3 and M4. Note the three cysteine residues (shown in green) located in each of the extracellular loops (designated E1 and E2) that are also conserved among the connexins. The cytoplasmic loop between M2 and M3 and the carboxy-tail are the most divergent in size and sequence.

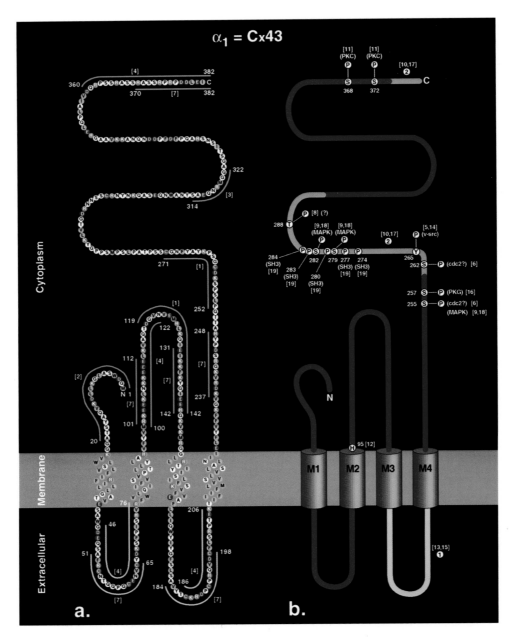

Figure 8.2 (a) Folding model for Cx43 (α_1 connexin). The amino acid sequence of Cx43 was deduced from cDNA analysis (29), and the residues are coded as in Figure 8.1. Of all cardiac connexins, the topology of Cx43 has been most thoroughly studied. The predicted locations of the extracellular and cytoplasmic regions have been confirmed with site-directed antibodies (blue and yellow bars indicate cytoplasmic and extracellular epitopes, respectively). (b) Locations of selected functionally important residues and domains as determined by mutagenesis and chimera studies. Indicated sites of covalent modification are based on consensus sequences, modification of synthetic peptides in vitro, and mutagenesis studies. The significance of the functional domains (circled numbers) and specifically mutated residues are as follows: Domain 1 (yellow) is the predominant determinant for specificity of heterotypic interactions between Cx43 (α_1), Cx50 (α_8), and Cx46 (α_3). This result is based on studies of chimeras of Cx50 (α_8) and Cx46 (α_3). Domains 2 (blue) impart high

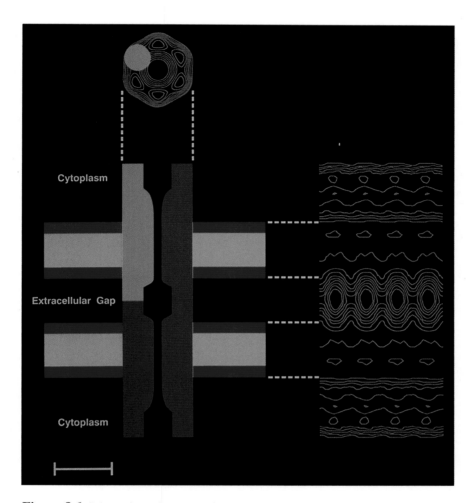

Figure 8.6 Schematic scale model of the cardiac gap junction membrane channel based on the planar density map (shown above the model) and the edge-on density map perpendicular to the membrane plane (shown to the right of the model) (scale bar = 50Å). The connexon has a diameter of ~65Å and is formed by a ring of 6 Cx43 subunits. The Cx43 subunit (solid) has a diameter of ~25Å and is highly asymmetric with an axial ration of 4:1 to 5:1. The hexameric connexons of adjacent cells are in register to define the central channel. The schematic model is in the same orientation as the edge-on density (right) derived by image analysis of the micrograph in Figure 8.5. The relationship between the 50Å thick lipid bilayers and the edge-view is indicated by dashed lines. The model shows the location of the 10Å thick lipid headgroup region and the 30Å hydrophobic membrane interior containing the lipid alkyl chains. The detailed shape and transmembrane structure of the channel are depicted here in a stylized fashion. (Modified from Ref. 5 and reproduced with permission of Current Biology Ltd.)

Figure 8.2 continued

sensitivity to pH gating of Cx43. Gating by pH is also influenced by the charge on His[95]. The SH3 domains of v-Src bind to proline-rich motifs and a phosphorylated tyrosine at position 265 of Cx43 (α_1). The numbered references are as follows: 1 (7), 2 (8), 3 (135), 4 (9), 5 (69), 6 (136), 7 (10), 8 (68), 9 (70), 10 (84), 11 (71), 12 (85), 13 (48), 14 (72), 15 (50), 16 (73), 17 (86), 18 (74), 19 (75). (Modified from Ref. 5 and reproduced with permission of Current Biology Ltd.)

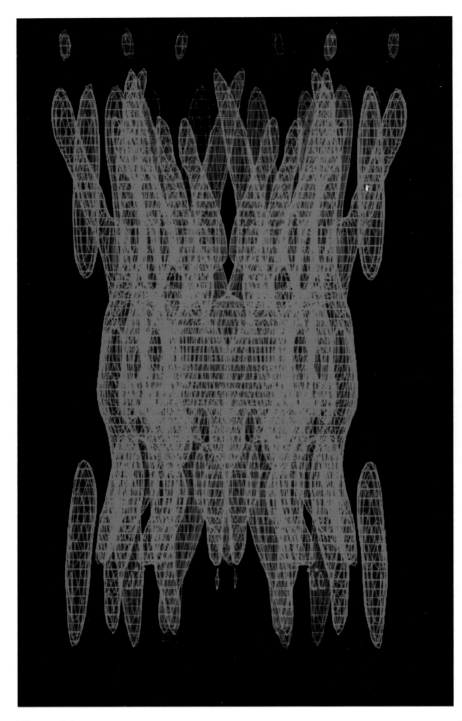

Figure 8.8 Side view of the three-dimensional density map of a recombinant cardiac gap junction channel (14). The dodecameric channel formed by a C-terminal truncation mutant of Cx43 has a length of ~150Å. The outer boundary of the map shows that the diameter of the channel is ~70Å within the two membrane regions and then narrows in the extracellular gap to a diameter of ~50Å. The rodlike features in the map correspond to transmembrane α-helices.

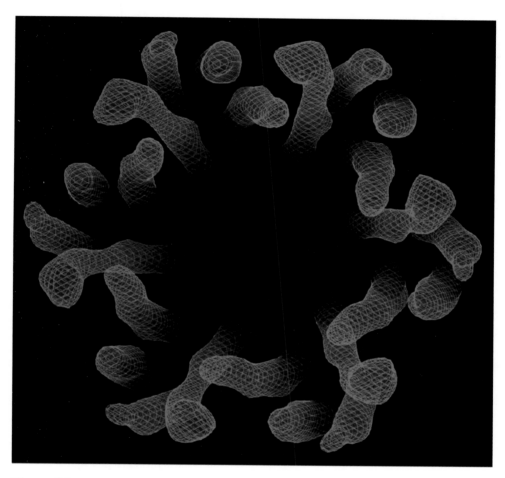

Figure 8.9 A top view looking toward the extracellular gap of the three-dimensional map of a recombinant cardiac gap junction channel. For clarity, only the cytoplasmic and most of the membrane-spanning regions of one connexon are shown. The rod-shaped features are resolved well and reveal the packing of twenty-four transmembrane α-helices within each connexon.

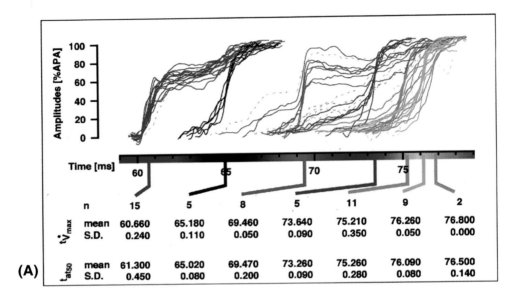

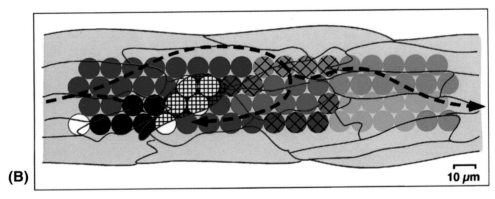

Figure 10.5 Characteristics of microscopic impulse propagation during ultraslow conduction induced by gap junctional uncoupling. (A) Plots of action potential upstrokes recorded simultaneously during activation of the strand. Upstrokes that occur at nearly the same time are color coded and the respective activation times determined from either the point in time of occurrence of maximum dV/dt or occurrence of 50% depolarization (t_{at50}) are shown below the graph. (B) A local activation map. The recording sites producing the upstrokes shown in (A) are color coded accordingly and superimposed on a schematic drawing of the preparation showing the cell borders. At the site of the white disks, no changes in transmembrane voltage were recorded (electrically uncoupled cell with hatched outline). The cross-hatched disks indicate the locations of signals for which it was not possible to assign unambiguous activation times because of notched upstrokes (dashed signals in part A). Arrows indicate qualitatively the direction of activation that advanced in a stepwise fashion along the preparation (overall conduction velocity ≈ 1.1 cm/s). (From Ref. 19.)

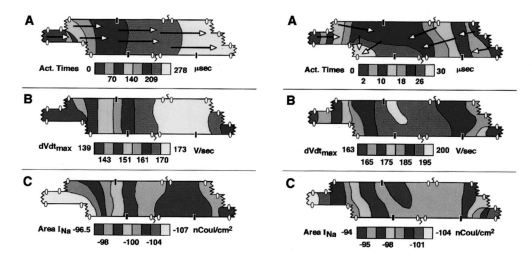

Figure 10.6 Left panels show intracellular two-dimensional spatial relations of excitation spread, maximum dV/dt, and time integral of the Na current (area I_{Na}) calculated for a single model myocyte during longitudinal propagation. Right panels show the same relationships for the same model cell during transverse propagation. (From Ref. 29.)

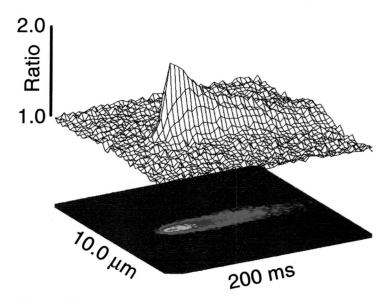

Figure 12.4 Properties of an average Ca^{2+} spark of a ventricular trabecula (n=79). Three-dimensional contour plot indicates the average amplitude of a spark (see text for more detail). (Reproduced with permission from Ref. 153.)

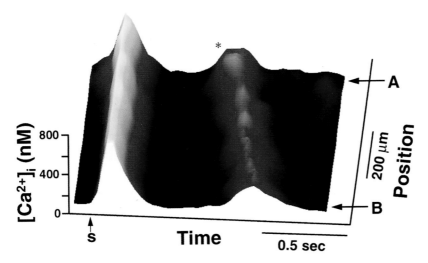

Figure 12.7 Spatial and temporal changes in Ca_1 in normal RV trabecula during the last electrically stimulated twitch (S) and a triggered propagated contraction (TPC). Note that after the end of the clearly uniform stimulated Ca_1 transient, a smaller Ca_1 transient propagates from site A in the muscle to site B in the muscle. The Ca^{2+} waves appear to travel at a constant velocity along this trabecula. (Reproduced with permission from Am J Physiol 1998; 274: H266-H276.)

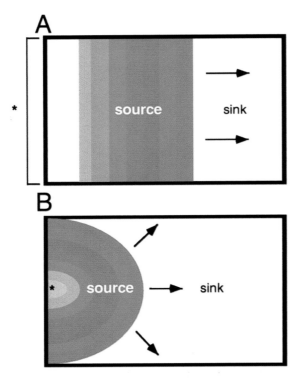

Figure 14.16 Role of wavefront curvature in establishing sink:source relationships. (A) Hypothetical example of a plane wave initiated by stimulation (asterisk) of the entire left border of the tissue. The wave propagates from left to right. An appropriate balance between amount of excited tissue at the wavefront (source) and amount of tissue to be excited by that wavefront (sink) allows for rapid and uniform propagation. (B) In the case of a curved wave initiated by a point source, sink $\geq$ source. Therefore, propagation is slower than for the plane wave case.

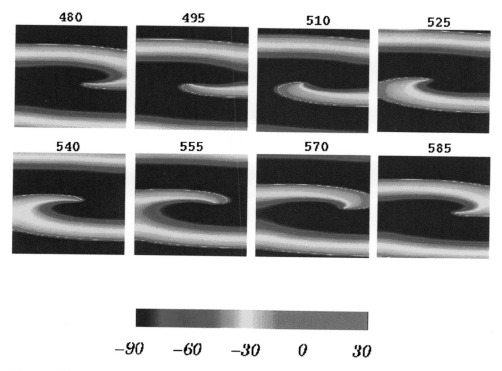

Figure 14.20 Spiral wave activity in a 2 x 2-cm anisotropic cardiac sheet (modified LRd model) with a refractory period of 147 ms. Ratio of horizontal to vertical velocity during plane waves is 4:1. Numbers on top of each panel indicate time (ms) after spiral wave initiation by cross-field stimulation protocol. The color bar indicates membrane potential distribution in millivolts. In this continuous ion model, membrane channels are distributed uniformly at the surface of the sheet. The model incorporates seven ion currents. The conductance of each current as well as steady-state and time constants for each gating variable are given in Ref. 1. The systems of equations was solved using a finite-element method and a semi-implicit integration scheme as described in detail in Ref. 48. The sheet consisted of 3600 cubic elements with 16 nodes per element, which implies an internode spacing of 111 mm. (Reprinted from Ref. 48, with permission.)

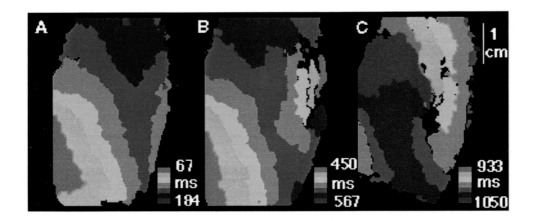

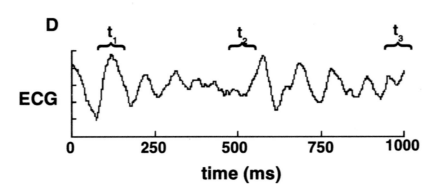

Figure 14.26 Video-imaging experiment showing spiral wave activity on the epicardial surface of the isolated rabbit heart. Isochrone maps from the surface of the free wall of the left ventricle at three different time intervals. White denotes earliest activation times and dark gray denotes late activation. (A) time t_1, a "V-shaped" collision pattern is evident on the surface. (B) An altered collision pattern. (C) A spiral wave pattern with a rotation period of 117 ms. (D) Simultaneous horizontal ECG. Notice the similarity to the simulation results shown in Figure 14.14. (Modified from Ref. 84, with permission.)

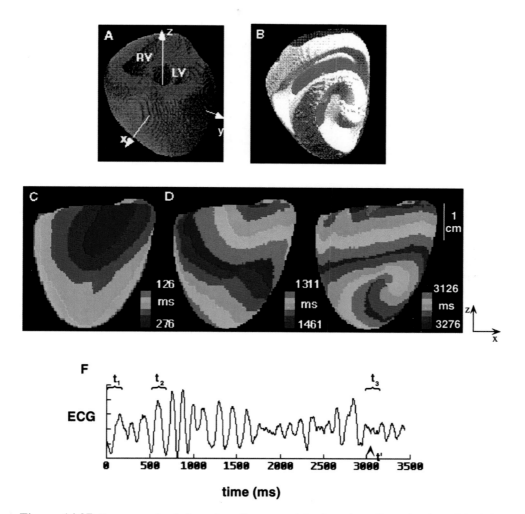

Figure 14.27 Computer simulation of scroll wave activity in a three-dimensional model of the heart. (A) Realistic geometry of right (RV) and left (LV) ventricles. (B) Three-dimensional isosurface of activity (variable analogous to normalized membrane potential) at time t_3 reveals a (scroll wave) shown in white (the red heart was made transparent to visualize the three-dimensional structure). (C) to (E) Isochrone maps from the right ventricular epicardial surface at three different time intervals. White denotes earliest activation times and dark gray denotes late activation. (C) At time interval t_1 a V-shaped collision pattern is evident on the surface. (D) At time interval t_2 the scroll wave filament has moved to the lateral surface of the left ventricle and the waves emanating from the scroll wave filament move upward in a "V" pattern and also propagate from the bottom. (E) At time interval t_3 a spiral wave pattern with a (clockwise) rotation period of 125 ms is manifest on the RV surface. (F) A simultaneous horizontal lead (x) ECG displaying irregular period and morphology. (Modified from Ref. 84, with permission.)

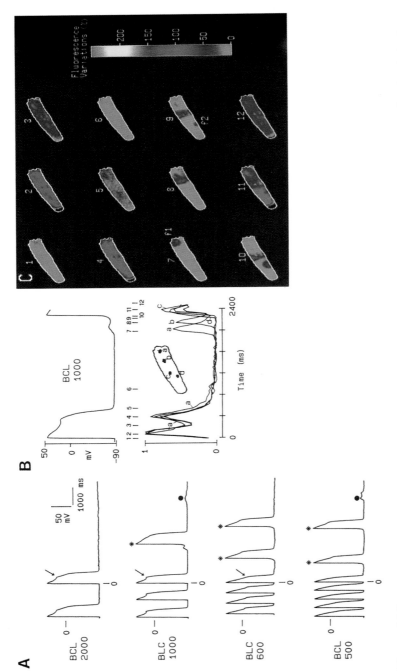

Figure 16.9 Tracings showing simultaneous presence of early after-depolarizations (EADs) (arrows) and delayed after-depolarizations (DADs) (black circles) coupled to the same beat. (A) Decrease in basic cycle length [(BCL) ms] is accompanied by shortening of the coupling interval (CI) and an increased likelihood of triggered beats (asterisks); the EAD CI did not significantly change but its activation voltage decreased and it disappeared at a BCL of 500 ms. (B) Electrical signal and normalized fluorescence signal from four different cell sites of the last driven action potential at a BCL of 1000 ms. Sites are evident in the cell profile. (C) Twelve images in pseudocolor, taken as indicated by the vertical bars over the fluorescence signal. CA²⁺ transient was synchronous during both the evoked action potential (image 2) and the EAD (image 4), but not during the DAD. A localized Ca²⁺ release (f1) originated in region a (image 7) and the DAD developed; when the wave moved away from the end of the cell and propagated toward region b (image 8), the DAD decreased. While the wave reached region c, a second focus (f2) started in region d (image 9), reached threshold, and triggered an action potential (image 11). (From Ref. 31.)

the activities of target proteins, in other words, two intracellular messengers are generated. The cycle is completed when the α-subunit hydrolyzes the GTP to GDP and the α- and βγ-subunits reassociate. Typically, the average lifetime of GTP on a Gα-protein is several seconds. X-ray crystal structures of G-proteins (29–31) have revealed that five noncontiguous peptide stretches form the GTP pocket, probably located on the cytoplasmic face of the Gα-protein; these regions of primary structure are highly conserved among the Gα-subunits.

B.   G-Protein Diversity in Heart

Heterotrimeric G-proteins are grouped into subtypes, originally based on the function of the α-subunits. For example, the α-subunit of the G_s family stimulates adenylyl cyclase and the α-subunit of the G_i family inhibits adenylyl cyclase. Toxin sensitivity provides another criterion for G-protein classification. ADP-ribosylation by pertussis or cholera toxins affects G-protein activity: $Gα_s$ is constitutively activated by cholera toxin whereas the interaction of the heterotrimeric G_i- and G_o-proteins with their upstream receptors are blocked by pertussis toxin ADP-ribosylation of the α-subunits.

Molecular cloning has demonstrated at least 17 α-subunit genes, 4 β-genes, and 7 γ-genes in mammals (27). Although heterotrimeric G-proteins were historically named by their α-subunits, the combination of an α-subunit with different βγ-subunits can result in G-protein complexes with different receptor and effector specificities. The α-subunits can be classified into families based on sequence homologies, which fortunately correspond generally to their function. α-Subunit diversity also comes about as the result of alternative RNA splicing. This is probably best studied in the case of $Gα_s$. Four isoforms of $Gα_s$ are encoded by a single gene and are generated by alternative splicing at exons 3 and 4. On SDS-polyacrylamide gel electrophoresis, these proteins migrate as two major bands with molecular weights of ~45 kD and ~52 kD. In adult ventricle, both forms are present, but the 45-kD form predominates. The functional significance of the different isoforms is unknown, but the 52-kD form seems to exchange GTP for GDP more rapidly than the 45-kD form. Heart also expresses $Gα_{i2}$ and $Gα_{i3}$ ($Gα_{i2}$ is the predominant form in ventricle). There is little, if any, $Gα_{i1}$ present. $Gα_o$ is present in the atrium (at a concentration about half of that of $Gα_i$), but there are only low levels of $Gα_o$ in ventricle. Members of the G_q class of α-subunits, $Gα_q$, $Gα_{11}$, and $Gα_{15}$, are also expressed in heart.

C.   Covalent Modifications of Heterotrimeric G-Proteins

The heterotrimeric G-proteins are tightly associated with the plasma membrane largely by virtue of covalent modifications of the proteins (32–34). The γ-subunits of G-proteins are modified by attachment of a 20-carbon isoprenoid (geranylgeranyl) and by carboxyl methylation on the C-terminal cysteine (following proteolytic removal of three C-terminal amino acids). It is known that γ-subunit prenylation plays an important role in membrane association because preventing prenylation, either by site-directed mutagenesis of the cysteine residue or by blocking the synthesis of isoprenoids, reduces their association with the membrane. Furthermore, activated α-subunits, dissociated from βγ, remain associated with the membrane. As with the γ-subunits, α-subunit hydrophobicity is increased by covalent modification. The N-terminal glycine of several Gα-subunit subtypes ($Gα_o$, $Gα_i$, and $Gα_z$) is myristoylated, which increases the affinity of the α-subunit for the βγ-dimer. Myristoylation may also facilitate association of the α-subunit with the membrane, independent of its association with βγ, and is required for activity of the α-subunit. In addition, many Gα-subunits can be palmitoylated by a thioester linkage to a cysteine near their amino terminus. Unlike myristoylation, which is cotranslational and irreversible, palmitoylation can occur post-translationally.

D. Enzymatic Cascades

Receptor activation of G-proteins ultimately regulates cardiac excitability by altering sarcolemmal ion channel function. G-protein-mediated regulation of ion channels is either direct, through interaction of a G-protein subunit with the ion channel, or indirect, via the enzymatic production, liberation, or destruction of second messengers, such as cAMP, IP_3, or Ca (Fig. 1). The enzymatic pathways that regulate the levels of these second messengers include adenylyl cyclase, phospholipases C-β, A_2, and D, and phosphodiesterase. The indirect pathways can involve multiple steps that often include phosphorylation.

1. Adenylyl Cyclase

Adenylyl cyclase (AC) catalyzes the formation of cAMP from ATP (Fig. 5.3). cAMP is an allosteric activator of cAMP-dependent protein kinase (PKA). Phosphorylation of proteins (enzymes and ion channels) by PKA modulates their activity. There are at least nine mammalian isoforms of AC (35,36). The cloned mammalian adenylyl cyclases are integral membrane proteins with 12 predicted transmembrane domains, in two groups of six. There are two large cytoplasmic domains, one between the two sets of transmembrane domains and the other at the cytoplasmic carboxy terminus. The predominant isoforms that are expressed in heart are types V and VI. Interestingly, these two isoforms, unlike others that have been studied, are inhibited by micromolar concentrations of Ca. Although all of the isoforms of adenylyl cyclase are stimulated by $Gα_s$ and are activated in vitro with the stable GTP analogue GTPγS, G-protein βγ-subunits regulate only types I (inhibition), II, and

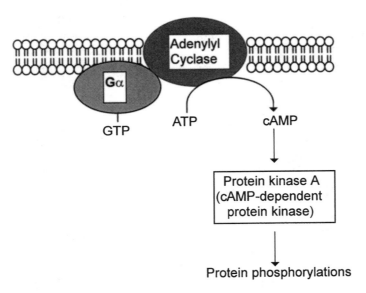

Figure 5.3   Activated $Gα_S$ exerts many of its effects by activating the enzyme adenylyl cyclase, which forms cyclic AMP (cAMP) from ATP. This nucleotide activates cyclic-AMP-dependent protein kinases that phosphorylate a variety of proteins that lie within the cytosol.

IV (stimulation). Although type I adenylyl cyclase is inhibited by G_i via direct binding of the $\beta\gamma$-subunits, other adenylyl cyclase isozymes are inhibited by binding the α-subunits of G_i. The modulatory effects of G-protein subunits on adenylyl cyclase may depend not only on the isozyme type, but also on the activator ($G\alpha_s$, Ca, etc.).

2. Phospholipases

Phospholipases in heart, which are also G-protein regulated (37) (see Fig. 5.4) cleave glycerophospholipids to yield a variety of different signaling molecules. The phospholipase C (PLC) family of enzymes most commonly cleaves phosphatidylinositol(5,6)bis-phosphate (PIP_2) to produce diacylglycerol (DAG), which remains associated with the plasma membrane and inositol 1,4,5-trisphosphate (IP_3), which is released into the cytosol (Fig. 5.5). DAG in conjunction with phosphatidylserine and Ca^{2+} is involved in activating protein kinase C and targeting it to the membrane, whereas IP_3 binds to and activates the IP_3 receptor, which is a ligand-gated calcium channel in the membrane of the endoplasmic reticulum. Opening the IP_3-receptor intracellular calcium channels releases calcium into the cytosol from a pool sequestered in an intracellular membrane system called the sarco(endo)plasmic reticulum. The activities of numerous cytosolic proteins are affected by changes in cytosolic calcium. The PLC-IP_3-Ca release is relatively slow, and so is unlikely to participate in beat-to-beat regulation, but it may play a role in regulating diastolic compliance.

Seven isozymes of mammalian PLC have been identified: three are regulated by G-proteins and others are regulated by tyrosine phosphorylation and/or Ca. In heart, the PLC isoforms that predominate are PLC-β1 and PLC-β2. These isoforms are stimulated by the $\beta\gamma$-subunits of a G_i protein (38). This stimulation is blocked by pertussis toxin.

The physiological role of PLC in heart is not entirely clear. In brain and smooth muscle, IP_3 plays a central role in calcium signaling, but in heart the major pathway for calcium release in cardiac myocytes involves the ryanodine receptor, which is gated by Ca, not IP_3. However, IP_3-induced calcium release may play a role in cardiac function. α_1-Adrenergic stimulation triggers IP_3-induced calcium accumulation in rat ventricular muscle. This intracellular messenger, along with Ca, is probably important in regulating cell growth and the hypertrophic response to overload.

Vagal and muscarinic stimulation usually slow the heart and decrease excitability and contractility. These effects are mediated in part by inhibition of adenylyl cyclase, and in part by effects on K channels. There is considerable evidence that Ach also causes positive chronotropic and inotropic effects that may be mediated by PLC (39). The stimulatory effects are particularly apparent in cardiac tissues from animals treated with pertussis toxin, but are also present in untreated tissues, particularly in embryonic tissue, or when high concentrations of muscarinic agonists are used. Although the mechanism is not understood, the positive effects of Ach correlate with phospholipase activation. It has been suggested that the stimulatory effects of Ach and vagal stimulation provide a mechanism for autoregulation of the inhibitory effects of the vagus nerve. Phospholipase A_2 (PLA_2) releases the fatty acid from the second position on the glycerol backbone. This fatty acid is usually arachidonic acid, the substrate for the synthesis of many bioactive compounds, including the steroid hormones, prostaglandins and leukotrienes. α-Adrenergic-receptor stimulation activates PLA_2 via a G-protein-dependent pathway. G-protein-mediated release of arachidonic acid has been proposed to play a role in the activation of cardiac inwardly rectifying K channels. Furthermore, arachidonic acid has potent effects on cardiac Ca currents. Arachidonic acid has been shown to be released from heart during ischemia.

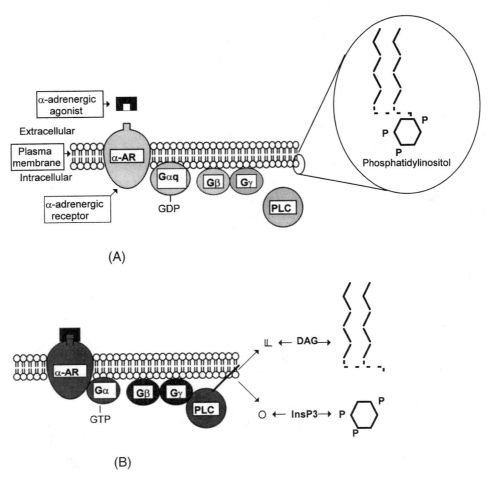

(A)

(B)

Figure 5.4 Phospholipase-coupled signal transduction. (Top) Five proteins can mediate signal transduction by such extracellular messengers as angiotensin II and α-adrenergic agonists: these are the receptor, depicted here as an α-adrenergic receptor (α-AR), Gα, the Gβγ complex, and the lipolytic enzyme phospholipase C (PLC). (A) In the inactive system, when the receptor is not bound to the agonist, Gα is bound to GDP and all components of the G-protein as well as PLC are inactive (lightly shaded). (B) Binding of the agonist to the receptor activates the G-protein (usually G_q) by accelerating the replacement of bound GDP with GTP on Gα. This results in dissociation of the Gα and Gβγ complex. The latter activates PLC. All of the activated signaling molecules are shaded. Activated PLC hydrolyzes phosphatidylinositol, a membrane phospholipid, which yields diacylgylcerol (DAG) and inositol triphosphate ($InsP_3$). Each of the latter represents an intracellular messenger, as depicted in Figure 5.5.

VI. ION CHANNELS CONTROLLING THE HEART BEAT

The membrane potential of a cell is determined by the sum of all currents flowing into and out of the cell. By convention, inward current is considered to depolarize and outward current to hyperpolarize the cell membrane. During each action potential, ion channels cycle between different states: depolarization converts closed voltage-gated Na^+ and Ca^{2+} channels to their open state ("activation"). The opening of these channels is quickly followed

Inositol Trisphosphate   Diacylglycerol

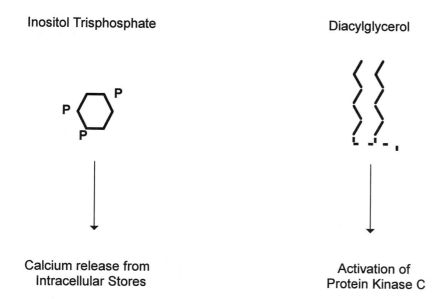

Calcium release from   Activation of
Intracellular Stores   Protein Kinase C

Figure 5.5 Inositol triphosphate and diacylglycerol, the two products released by phospholipase C cleavage of membrane inositol phospholipids, have different effects. The former binds to receptors that open calcium channels on intracellular membranes and so increases cytosolic calcium. The latter activates members of a family of lipid-dependent protein kinases called protein kinase C, which phosphorylate substrates that participate in signal transduction.

by the transition to a nonconducting, "inactivated" state. The rate of recovery from this inactivation is one of the determinants of the duration of the refractory period of the action potential. In this state, the channels cannot be reopened until they have cycled back to the closed state, which normally occurs during diastole. Repolarization occurs as a result of the activation of outward currents carried by the outward movement of K^+ and inward movement of Cl^- that bring the membrane back toward the resting potential. These inward and outward currents alternate in time because the outward currents are activated slowly in response to the depolarization generated by the inward currents and because the inward currents "inactivate" or turn off a short time after they are activated (see Chaps. 2 and 3).

Different regions of the heart are specialized for different functions and express different complements of ion channels. Heart rate is determined by the primary pacemaker cells of the sinoatrial (SA) node. Several ion channels of the SA node are specialized for generating spontaneous pacemaker depolarizations. This spontaneous activity is then transmitted by a wave of depolarization that spreads first over the atria, and then is transmitted to the ventricles though the AV node, His bundle, bundle branches, and the His-Purkinje system. Arrival of this wave of at the ventricular myocardium initiates the forceful contraction that propels blood through the vasculature. The ventricular myocardium has a complement of ion channels that function to control the concentration of intracellular Ca^{2+}, which regulates contraction and transmembrane fluxes of Na, K, and Cl that determine excitability and impulse propagation.

At least 12 different ion channels interact to generate action potentials in different regions of the heart (see Chap. 6). The functioning of these channels is highly interdependent because changes in membrane potential caused by the opening of one channel will affect the opening, inactivation, and reactivation of other channels, and because ions that enter via one channel can bind to and alter the function of other ion channels. The ion channels responsible for the SA node pacemaker and for controlling ventricular contraction are described in more detail below and elsewhere in this text.

A. Ionic Mechanisms of Pacemaking in the Sinoatrial Node

Spontaneous, rhythmic depolarization of the sinoatrial node, which is the primary pacemaker region of the heart, is the result of a complex interplay between several ion currents, including T- and L-type Ca^{2+} currents ($I_{Ca.T}$ and $I_{Ca.L}$), the background sodium current (I_b), the delayed rectifier (I_K), the hyperpolarization-activated inward current (I_f), and the Na^+–Ca^{2+} exchange current (I_{NaCa}) (40–42) (see Chap. 7).

B. Currents in Ventricular Myocardium

The depolarization that is generated in the SA node is transmitted to the atrial and ventricular muscle. The wave of electrical depolarization that is conducted to myocardium then activates inward Na^+ and Ca^{2+} currents responsible for the upstroke and plateau, respectively, of the ventricular action potential. This is followed by activation of outward currents that are responsible for repolarization (43–45).

The fast Na^+ current, I_{Na}, is activated at membrane potentials more positive than –70 mV. The depolarization produced by current through Na^+ channels, which is responsible for the upstroke of the action potential, then opens voltage-gated L-type Ca^{2+} channels. The inward current that flows though the latter is directly responsible for the plateau of the action potential and for initiating contraction. A fraction of the Ca^{2+} that enters adult cardiac myocytes via these channels activates the contractile apparatus; however, most of this activator Ca^{2+} stimulates Ca^{2+} release from the sarcoplasmic reticulum by a process called calcium-induced calcium release (CICR). During depolarization, an additional Ca^{2+} influx may occur via the Na^+–Ca^{2+} exchanger.

Repolarization of the ventricular action potential is then initiated by several temporally overlapping processes. Both the Na^+ and Ca^{2+} channels inactivate. As these inward currents decay, outward currents serve to bring the membrane potential back to diastolic potential. Phase 1 of the repolarization in some species is initiated by a transient outward current (I_{to}) that appears to represent an important determinant of action potential duration. This is followed in phase 2 by activation of the delayed rectifier current, I_K. In the final stages of repolarization, current flowing through background inward rectifier channels (I_{K1}) contribute to outward current as the inward rectification declines.

VII. CHANNEL REGULATION

Virtually all of these cardiac ion channels are regulated by one or more of the signal transduction pathways as described in Table 5.5 (46,47). Obviously, the precise functional consequences depend upon the neural pathways that are activated, the transmitters, receptors, and signal transduction pathways involved, and the ion channels affected in different regions of the heart. For example, activation of neural pathways that are targeted to the sinoatrial node might primarily have effects on heart rate, whereas secretion of epinephrine by the adrenal glands might have effects more on force of ventricular contraction.

Neurotransmitters and hormones affect the function of ion channels by remarkably diverse mechanisms. Several channels are regulated by phosphorylations that are cat-

Table 5.5   Examples of Ion Channels Regulated by β-Adrenergic Receptors in Heart

Ion channel	Signal	Effect
Fast Na channel	cAMP-dependent phosphorylation	Shift in voltage dependence
		Change in ion selectivity
L-type Ca channel	cAMP-dependent phosphorylation	Increase in channel availability
		Increase in P_o
Delayed rectifier K	cAMP-dependent phosphorylation	Increase in maximal current
Hyperpolarization activated (pacemaker)	Direct effect of cAMP, also cAMP-dependent phosphorylation	Shift in activation curve to more positive potentials
Chloride current (CFTR)	Direct effect of cAMP	cAMP-gated channel

alyzed by a number of protein kinases, including a cAMP-dependent protein kinase that is activated by β-adrenergic stimulation. Stimulation of the channels by cAMP-dependent protein kinase can be mediated by different mechanisms including, for example, increases in the probability that the channel will be opened in response to a voltage step (as occurs with L-type Ca channels), shifts in the voltage dependence of activation (as occurs with the pacemaker current I_f), or possibly even changes in ion selectivity (as has been reported to occur with Na channels). Channels can also be regulated directly by diffusible second messengers. For example, the pacemaker current I_f is regulated directly by cAMP, several K, Cl, and Ca channels are Ca regulated, and an inwardly rectifying K channel is regulated by ATP. Several channels, including the muscarinic inwardly rectifying K channel and the ATP-sensitive K channel, are regulated by G-proteins. Examples of these diverse regulatory mechanisms are provided below.

A.   Regulation of I_{Ca} by cAMP-Dependent Phosphorylation

Many ion channels in heart are regulated by G-protein-coupled second messenger cascades, the best known being the voltage-gated, L-type Ca channel (48–50). Activation of β-adrenergic receptors stimulates cAMP synthesis by activation of adenylyl cyclase via G_s, activation of cAMP-dependent protein kinase (PKA), and phosphorylation of a protein that is commonly thought to be a subunit of the L-type calcium channel. Phosphorylation of the channel increases integrated Ca flux through the channel as a result of complex changes in channel gating. The largest fraction of the increase in I_{Ca} produced by β-agonists is clearly due to cAMP-dependent phosphorylation. The phosphorylation hypothesis has been verified by a large number of laboratories. It is well known that β-agonists stimulate adenylyl cyclase and increase cAMP levels in cardiac cells. The effect of β-agonists on I_{Ca} is mimicked by cAMP or cAMP analogs, by phosphodiesterase inhibitors that inhibit cAMP breakdown, and by intracellular cAMP-dependent protein kinase. AMP-PNP, a competitive inhibitor of phosphorylation, reduces β-adrenergic stimulation of I_{Ca}. Also, ATPγS, a substrate used by protein kinases to thiophosphorylate proteins "irreversibly," potentiates the stimulation and slows the washout of the effect of β-agonists. Further evidence supporting the cAMP hypothesis is that inhibition of phosphatases potentiates the effects of β-agonist stimulation. Interestingly, the regulation of L-type Ca channels by cAMP-dependent protein kinase appears to require a macromolecular signaling complex composed of the Ca channel, the kinase, and a scaffolding protein (51,52). The scaffolding protein is a member of the AKAP (A-kinase associated protein) family of proteins. These proteins have binding sites for different signaling enzymes including kinases and phosphatases and serve to bring these signaling molecules together into a signaling complex.

Single-channel conductance of L-type Ca channels is not altered by β-adrenergic agonists, which suggests that there is no change in the channel pore. Channel gating, however, is affected in a complex manner (49). β-Agonists have been shown to increase mean time a channel remains in the open state once opened and also to increase the probability that the channel will open in response to a depolarization. Several models have been proposed to account for the effects of phosphorylation, but the model we prefer is one proposed by Tsien and coworkers (53), who suggested that the Ca channel has multiple phosphorylation sites and that one site regulates the availability of channels: that is, this site must be phosphorylated in order for the channel to be capable of opening in response to voltage. In addition, there are other phosphorylation sites that control fast gating.

B. Muscarinic Ach Effects on I_{Ca}

Parasympathetic nerves release acetylcholine (Ach), which activates muscarinic receptors to antagonize the effects of β-adrenergic stimulation. The predominant muscarinic receptor coupled to adenylyl cyclase inhibition in heart is the M_2-subtype, although other subtypes may be coupled to the phosphatidyl inositol pathway. Activation of M_2-muscarinic receptors by Ach activates an inhibitory GTP-binding protein, $G_{\alpha i}$, that in turn inhibits adenylate cyclase. As adenylyl cyclase activity is inhibited, intracellular cAMP concentration drops and Ca channels are dephosphorylated, presumably by protein phosphatases 1 and 2A. Because Ach has no effect on basal I_{Ca} and only decreases I_{Ca} stimulated by previous activation of AC, it is generally concluded that Ach acts on I_{Ca} solely by adenylyl cyclase inhibition.

C. Regulation of Na Channels by cAMP-Dependent Phosphorylation

Cardiac Na channels are also regulated by cAMP-dependent phosphorylation, but the effects on Na-channel activity differ markedly from those described above for the Ca channel. Under physiological conditions, the effect of cAMP-dependent phosphorylation on Na current is voltage dependent such that the current can either be stimulated or reduced, depending on the specific voltage-clamp protocol used in experimental studies (54). However, the observation that the rate of rise of the action potential is reduced in the presence of cAMP suggests that physiologically the Na current is reduced by cAMP-dependent phosphorylation. Furthermore, it has recently been reported that the ionic selectivity of cardiac Na channels changes in response to cAMP-dependent phosphorylation (55). The channels become much more permeant to Ca than in the absence of β-agonists: the P_{Ca}/P_{Na} ratio changes from 0 to ~1.3. Ca, which enters via the Na channel and can stimulate additional Ca release from the sarcoplasmic reticulum by CICR. This suggests that, unlike Ca channels, the Na-channel pore can be modified by phosphorylation. The idea that the selectivity of a channel can be modulated by something as simple as phosphorylation is a heretical idea and these data remain an area of controversy in the field.

D. Regulation of the Pacemaker Current I_f Directly by cAMP

β-Adrenergic stimulation increases the frequency of cardiac contraction. This is due partly to a shift in the activation curve for I_f to more positive potentials, which allows the depolarizing effect of this inward current to begin earlier as the SA node repolarizes from the previous action potential. Pacemaker depolarization is also accelerated by increased inward Ca and outward K currents; the former speeds depolarization while the latter shortens action potential duration (42). The regulation of I_f has been somewhat confusing, because in Purkinje cells the shift in the activation curve appears to be mediated by cAMP-dependent phosphorylation, whereas in the sinoatrial node it appears to be mediated by the direct action of cAMP on the channel (41,56). This channel has recently been cloned and has been found to have ~28% amino acid identity to cyclic nucleotide-gated channels and exhibits a cyclic nucleotide binding domain (57). Although heterologously expressed cloned channels are regulated directly by cAMP, different clones of the I_f channel turn out to have different sensitivities to cAMP. It may be that the Purkinje fiber channel is a low-affinity form that is regulated by phosphorylation, whereas the sinoatrial node channel is a high affinity cAMP-regulated channel.

E. Regulation of the ATP-Sensitive K Channel

Although this channel was first discovered in heart tissue, the precise function of the ATP-sensitive K channel in heart remains unknown. The channel may play an important role in shortening the action potential in an energy-starved heart, as occurs in myocardial ischemia. This channel is normally closed, but opens when the intracellular ATP/ADP ratio falls (58,59). The opening of this channel during ischemia, by shortening action potential duration and so reducing the entry of activator Ca, could reduce contractility and Ca overload in energetically compromised cells. It has been shown that activation of K_{ATP} channels is likely to play a role in ischemic preconditioning, a mechanism whereby prior brief ischemic episodes protect against a subsequent period of severe ischemia. However, opening of these channels could also promote serious arrhythmias.

ATP-sensitive K channels are regulated in an extremely complex way. They are inhibited by ATP and stimulated by ADP on the cytosolic side of the channel. In addition, channel activity may be dependent on phosphorylation or membrane phospholipids such as phosphatidylinositol 4,5-bisphosphate, and is regulated directly by G-protein α-subunits. Decreases in intracellular pH also stimulate channel activity. The ATP-sensitive K channel has recently been cloned and is a heteromultimer composed of an inwardly rectifying K channel, $K_{ir}6.2$, and the sulfonylurea receptor (SUR). Studies on heterologously expressed and mutated ATP-sensitive K channels are now providing insights into the regions involved in regulation of this channel.

F. Direct Regulation of Ion Channels via G-Proteins

1.   G-Protein Regulation of $I_{K.ACh}$

ACh not only affects ion channels indirectly, by interfering with cAMP production, but it also affects function via a membrane-delimited pathway. In supraventricular tissue, ACh released from the vagus nerve causes an increase in potassium conductance of an inwardly rectifying channel that leads to hyperpolarization. The clearest demonstration that diffusible second messengers are not involved in activation of this current is that Ach can activate the channels in excised patches of membrane. Several lines of evidence show that channel activation involves a G-protein (60). First, activation of the K^+ channel by ACh requires intracellular GTP and is blocked by pertussis toxin. Second, ACh irreversibly activates $I_{K.ACh}$ when the cell is perfused internally with stable, nonhydrolyzable analogs of GTP. Finally, although Ach-activated K^+ channels disappear quickly in inside-out excised patches deprived of GTP, the activity could be restored (when Ach is present at the extracellular surface) by addition of GTP to the cytoplasmic membrane face. The effect of GTP is blocked by pertussis toxin. Activation of the muscarinic K channel clearly involves the $\beta\gamma$-subunit of a G-protein because both native and recombinant $\beta\gamma$-subunits activate these channels in excised patches and $\beta\gamma$-subunits bind to the GIRK4 subunit of the inwardly rectifying channel (61,62).

2. Direct G-Protein Regulation of L-Type Ca Channels

It has been suggested that a small fraction of the stimulation of I_{Ca} produced by β-agonists is due to direct effects of $G\alpha_S$. Direct effects of $G\alpha_S$ on skeletal muscle and cardiac calcium channels have been demonstrated in cell-free systems: G-proteins have effects on survival of Ca channels in excised patches and gating of channels incorporated into lipid bilayers. Furthermore, biochemical evidence suggests that G-protein α-subunits may bind to L-type Ca channels. Although it was reported by one lab that the response to β-agonists was biphasic, these data have been seriously questioned (63). On balance, there seems to be no compelling evidence to support a direct $G\alpha_s$ effect on cardiac L-type Ca channels at the present time.

VIII. REGULATION OF TRANSPORTERS

Compared to the amount of information available regarding channel regulation, our knowledge of transporter regulation is minimal. Plasma membrane Na–Ca exchange is a major mechanism for controlling cytosolic Ca in cardiac cells, but it is not clear whether this transporter is regulated by neurohormones. However, its activity is stimulated by cytosolic Ca and ATP, and inactivated by Na (64). Thus, one would expect that its exchange activity would be altered by different pathological and physiological states The exchanger is also stimulated by phosphatidylinositol 4,5 bisphosphate (65) and α-adrenergic-receptor stimulation (66). The Na–K pump is also regulated by β-adrenergic stimulation by cAMP-dependent phosphorylation (67), but a full discussion of this enzyme is beyond the scope of this discussion. Na–H exchange is also an important transporter in the cardiovascular system, but its regulation by autonomic mechanisms is not thoroughly understood.

IX. REGULATION OF CHANNELS BY NOVEL OR POORLY UNDERSTOOD MECHANISMS

Although we have gained a good understanding of many specific features of channel regulation, our global understanding of these processes remains limited. We know that several ion channels are affected by the free radical gas nitric oxide, and by such fatty acids as arachidonic acid, phospholipids, and leukotrienes, but we remain reasonably ignorant of how these signals function physiologically or pathologically, and how their effects on channels are brought about. Growth factors and cytokines can also influence ion channels, but it remains unclear whether most of their effects are mediated directly through transcriptional or translational regulation of channel genes or their regulators or whether the signaling pathways stimulated by these factors might also have shorter term actions directly on ion channel proteins. These are important questions as there is increasing evidence that cytokines are activated in pathological states including the failing heart (68). With regard to mechanisms of regulation that involve changes in protein synthesis, most ion channels are heteromultimeric proteins. It is a common observation that assembly of pore-forming subunits with different subtypes of accessory subunits have important effects on channel biophysical properties (such as activation or inactivation kinetics) and channel regulation (69). Thus, channel properties can be modulated by mix-and-match assembly of different subunits. Finally, transcripts of ion channels are frequently alternatively spliced. Alternative splicing of channels can have dramatic effects on channel function, as has been demonstrated for effects of flip and flop variants of the inotropic glutamate receptors on channel gating and pharmacology (70). Whether these mechanisms are involved in regulation of cardiac ion channels remains to be determined.

X. CONCLUSIONS

The mechanisms reviewed in this chapter provide only an introduction to the complex, fascinating, and rapidly expanding field of cardiac signal transduction, which is mediated by pathways that "notify" individual cardiac myocytes and the heart as a whole of changes in the external world. The heart, as described in this chapter, responds by interpreting these signals as a challenge to modify its function. The most important of the physiological stimuli discussed above are those mediated by sympathetic activation, for example, during exercise or emotional stress. In the sympathetic response, the requirement that the heart be made aware of the body's need for a large increase in cardiac output is met by signals that augment pump function and increase heart rate. When these physiological mechanisms are called upon repeatedly, as occurs during training, they are accompanied by a growth response that produces a form of physiological hypertrophy. Unfortunately, if the stress is sustained, a different form of growth occurs; this is the pathological hypertrophy that plays an important role in the pathophysiology of heart failure (71).

In signaling a short-term increase in cardiac output during exercise, many signals work together. One simple feature of this response is an increase in end-diastolic volume, brought about when vasodilatation in the exercising muscles increases flow from the arteries into the veins, and so the venous return to the heart. The resulting increase in preload, by stretching the cardiac fibers, increases their ability to do work through the operation of Starling's Law of the Heart. The heart's response to exercise also involves several of the mechanisms discussed in detail in the preceding pages. Most important are increases in heart rate, contractility, and relaxation, all of which are initiated when norepinephrine, released from sympathetic nerve endings, binds to β-adrenergic receptors on the myocyte plasma membrane. This is usually the first in a multistep cascade of signaling, in which each step is controlled by a different messenger or chemical reaction.

One way to view these signal transduction systems is as a series of ponds, linked by waterfalls, along a swiftly flowing river. In each step, an incoming signal (water in this bucolic analogy) fills the downstream pond, causing an overflow that transmits a signal to the next pond in the cascade. Unlike the river, where a single substance (water) flows from one pond to another, in biological signaling, different substances are sent from one "pond" to the next.

One might reasonably ask why so many different steps are needed to link the arrival of the extracellular messenger (norepinephrine) at the cell surface to responses that increase heart rate and myocardial contractility. The answer probably lies in the ability of these multistep cascades to enhance the control of cell signaling. The many steps that link sympathetic stimulation to changes in ion channel function, which at first glance might seem almost perversely complex, actually provide the cell with many opportunities. Were signaling to proceed by a simple linear cascade, where one step couples to a single downstream reaction, the response would be stereotyped, limited in scope, and essentially "all or none." The fact that biological signaling pathways generally branch, loop both forward and backward, and interact with other pathways provides rich opportunities for control;

they allow a response to be graded ("fine tuned"), to be amplified, and to use feedback loops in preventing the response from going out of control ("runaway signaling"). The "intra-pathway" and "extra-pathway" cross-talk of these signaling mechanisms thus allows responses to be graded, both to match the needs of the circulation and to remain within the limitations of the cell. These pathways also provide for amplifications that allow a response to develop rapidly. Understanding these control mechanisms, however, presents a daunting challenge; it is our hope that this chapter will help the reader meet this challenge.

REFERENCES

1. Bernard C. Leçons sur les phénomènes de la vie communs aux animaux et aux vegetaux. JF Fulton, tr. Paris: Ballièe, 1878.
2. Cannon WB. The Wisdom of the Body. New York: WW Norton, 1932.
3. Harris P. Evolution and the cardiac patient. Cardiovasc Res 1983; 17:313–319, 373–378, 437–445.
4. Han X, Kobzik L, Severson D, Shimoni Y. Characteristics of nitric oxide-mediated cholinergic modulation of calcium current in rabbit sino-atrial node. J Physiol 1998; 509:741–754.
5. Mukoyama M, Hosada K, Suga S, Saito Y, Ogawa Y, Shirakami G, Jougasaki M, Obata K, Yasue H, Kambayashi Y, Inouye K, Imura H. Brain natriuretic peptide (BNP) as a novel cardiac hormone in humans—evidence for a exquisite dual natriuretic peptide system, atrial natriuretic peptide and brain natriuretic peptide. J Clin Invest 1991; 87: 1402–1412.
6. Timmermans PBMWM, Wong PC, Chiu AT, Herblin WF, Benfield P, Carini DJ, Lee RJ, Wexler RR, Saye JAM, Smith RD. Angiotensin II receptors and angiotensin II receptor antagonists. Pharmacol Rev 1993; 45:205–251.
7. Koch-Weser J. Myocardial actions of angiotensin. Circ Res 1964; 14:337–344.
8. Freer RS, Pappano A, Peach MJ, Bing KT, McLean MJ, Vogel SM, Sperelakis N. Mechanism for the positive inotropic effect of angiotensin II on isolated cardiac muscle. Circ Res 1976; 39:178–183.
9. Kass RS, Blair ML. Effect of angiotensin II on membrane current in cardiac Purkinje fibers. J Mol Cell Cardiol 1981; 13:797–809.
10. Ikenouchi H, Barry WH, Bridge JHB, Weinberg EO, Apstein CS, Lorell BH. Effects of angiotensin II on intracellular Ca^{2+} and pH in isolated beating rat hearts loaded with the indicator indo-1. J Physiol (Lond) 1994; 480:203–215.
11. Yanagisawa M, Kurihara H, Kimura S, Tomobe Y, Kobayashi M, Mitsui Y, Goto K, Masaki T, Yazaki Y. A novel potent vasoconstrictor peptide produced by vascular endothelial cells. Nature 1988; 331:411–415.
12. Suzuki T, Hoshi H, Mitsui Y. Endothelin stimulates hypertrophy and contractility of neonatal rat cardiac myocytes in a serum-free medium. FEBS Lett 1989; 268:149–151.
13. Ishikawa T, Yanagisawa M, Kurihara H, Goto K, Masaki T. Positive inotropic effects of novel potent vasoconstrictor peptide endothelin on guinea pig atria. Am J Physiol 1988; 255:H970–973.
14. Ahlquist RP. A study of the adrenotropic receptors. Am J Physiol 1948; 153:586–600.
15. Skomedal T, Aass H, Osnes J, Fjeld NB, Klingen G, Langslet A, Semb G. Demonstration of an alpha-adrenoreceptor-mediated inotropic effect of norepinephrine in human atria. J Pharmacol Exp Ther 1985; 233:441–446.
16. Endoh M, Blinks JR. Actions of sympathomimetic amines on Ca^{2+} transients and contractions of rabbit myocardium: reciprocal changes in myofibrillar responsiveness to Ca^{2+} mediated through α- and β-adrenoreceptors. Circ Res 1988; 62:247–265.
17. Lee HR. α_1-Adrenergic receptors and heart failure. In: Gwathmey JK, Briggs GM, Allen PD. Heart Failure. Basic Science and Clinical Aspects. New York: Marcel Dekker, 1993: 211–234.
18. Skeberdis VA, Jurevicius J, Fischmeister R. Beta-2 adrenergic activation of L-type Ca^{++} current in cardiac myocytes. J Pharmacol Exp Ther 1997; 283:452–461.

19. Xiao RP, Lakatta EG. Beta-1 adrenoreceptor stimulation and beta-2 adrenoreceptor stimulation differ in their effects on contraction, cytosolic Ca^{2+}, and Ca^{2+} current in single rat ventricular cells. Circ Res 1993; 73:286–300.

20. Xiao RP, Hohl C, Altschuld R, Jones L, Livingston B, Ziman B, Tantini B, Lakatta EG. Beta-2 adrenoreceptor-stimulated increase in cAMP in rat heart cells is not coupled to changes in Ca^{2+} dynamics, contractility, or phospholamban phosphorylation. J Biol Chem 1994; 269:19151–19156.

21. Xiao RP, Ji X, Lakatta EG. Functional coupling of the beta-2 adrenoreceptor to a pertussis toxin-sensitive G protein in cardiac myocytes. Mol Pharmacol 1995; 47:322–329.

22. Bristow MR, Hershberger RE, Port JD, Minobe W, Rasmussen R. β_1 and β_2 Adrenergic receptor mediated adenylate cyclase stimulation in non-failing and failing human ventricular myocardium. Mol Pharmacol 1989; 35:295–303.

23. Hartzell HC. Regulation of cardiac ion channels by catecholamines, acetylcholine and second messenger systems. Prog Biophys Mol Biol 1988; 52:165–247.

24. Katz AM. Cyclic AMP effects on the myocardium: A man who blows hot and cold with one breath. J Am Coll Cardiol 1983; 2:143–149.

25. Fedida D, Braun AP, Giles WR. α_1-Adrenoceptors in myocardium: functional aspects and transmembrane signaling mechanisms. Physiol Rev 1993; 73:469–487.

26. Gilman AG. Guanine nucleotide-binding regulatory proteins and transmemebrane signaling. In: Fidia Research Foundation Neuroscience Award Lectures. New York: Raven Press, 1998.

27. Quarmby LM, Hartzell HC. Molecular biology of G proteins and their role in cardiac excitability. In: Zipes DP, Jalife J, eds. Cardiac Electrophysiology: From Cell to Bedside. Philadelphia: Saunders, 1994.

28. Lee E, Taussig R, Gilman AG. The G226A mutant of $Gs\alpha$ highlights the requirement for dissociation of G protein subunits. J Biol Chem 1992; 267:1212–1218.

29. Hamm HE. The many faces of G protein signaling. J Biol Chem 1998; 273:672.

30. Hamm HE, Gilchrist A. Heterotrimeric G proteins. Curr Opin Cell Biol 1996; 8:196.

31. Lambright DG, Sondek J, Bohm A, Skiba NP, Hamm HE, Sigler PB. The 2.0 Angstrom crystal structure of a heterotrimeric G protein. Nature 1996; 379:319.

32. Higgins JB, Casey PJ. The role of prenylation in G-protein assembly and function. Cell Signaling 1996; 8:433–437.

33. Morales J, Fishburn CS, Wilson PT, Bourne HR. Plasma membrane localization of Galpha(z) requires two signals. Mol Biol Cell 1998; 9:14.

34. Resh MD. Regulation of cellular signaling by fatty acid acylation and prenylation of signal transduction proteins. Cell Signaling 1996; 8:412.

35. Iyengar R. Molecular and functional diversity of mammalian Gs-stimulated adenylyl cyclases. FASEB J 1993; 7:768–775.

36. Krupinski J, Lehman TC, Frankenfield CD, Zwaagstra JC, Watson PA. Molecular diversity in the adenylylcyclase family. J Biol Chem 1992; 267:24858–24862.

37. Morris AJ, Scarlatta S. Regulation of effectors by G-protein α and $\beta\gamma$ subunits. Recent insights from studies of the phospholipase C-β isozymes. Biochem Pharmacol 1997; 54:429–435.

38. Schnabel P, Gas H, Nohr T, Camps M, Bohm M. Identification and characterization of G protein-regulated phospholipase C in human myocardium. J Mol Cell Cardiol 1996; 28:2426–2435.

39. Pappano AJ. Vagal stimulation of the heartbeat: Muscarinic receptor hypothesis. J Cardiovasc Electrophysiol 1991; 2:262–273.

40. Campbell D, Rasmusson R, Strauss H. Ionic current mechanisms generating vertebrate primary cardiac pacemaker activity at the single cell level:an integrative view. Annu Rev Physiol 1992; 54:279–302.

41. DiFrancesco D. Pacemaker mechanisms in cardiac tissue. Annu Rev Physiol 1993; 55:455–472.

42. Irisawa H, Brown HF, Giles W. Cardiac pacemaking in the sinoatrial node. Physiol Rev 1993; 73:197–227.

43. Carmeliet E. K$^+$ channels and control of ventricular repolarization in the heart. Fundam Clin Pharmacol 1993; 7:19–28.

44. Katz AM. Cardiac ion channels. N Engl J Med 1993; 328:1244–1251.

45. Noble D. The surprising heart: a review of recent progress in cardiac electrophysiology. J Physiol (Lond) 1984; 353:1–50.

46. Hartzell HC. Regulation of cardiac ion channels by catecholamines, acetylcholine and second messenger systems. Prog Biophys Mol Biol 1988; 52:165–247.

47. McDonald TF, Pelzer S, Trautwein W, Pelzer DJ. Regulation and modulation of calcium channels in cardiac, skeletal, and smooth muscle cells. Physiol Rev 1994; 74:365–507.

48. Hartzell HC. Filling the gaps in Ca channel regulation. Biophys J 1993; 65:1358–1359.

49. Hartzell HC, Duchatelle-Gourdon I. Structure and neural modulation of cardiac channels. J Cardiovasc Electrophysiol 1992; 3:567–578.

50. Trautwein W, Hescheler J. Regulation of cardiac L-type calcium current by phosphorylation and G proteins. Annu Rev Physiol 1990; 52:257–274.

51. Gao T, Yatani A, Dell'Acqua ML, Sako H, Green SA, Dascal N, Scott JD, Hosey MM. cAMP-dependent regulation of cardiac L-type Ca^{2+} channels requires membrane targeting of PKA and phosphorylation of channel subunits. Neuron 1997; 19:196.

52. Gray PC, Johnson BD, Westenbroek RE, Hays LG, Yates IIJ, Scheuer T, Catterall WA, Murphy BJ. Primary structure and function of an A kinase anchoring protein associated with calcium channels. Neuron 1998; 20:1026.

53. Tsien RW, Bean BP, Hess P, Lansman JB, Nilius B, Nowycky MC. Mechanisms of calcium channel modulation by β-adrenergic agents and dihydropyridine calcium agonists. J Mol Cell Cardiol 1986; 18:691–710.

54. Arita M, Muramatsu H, Ono K, Kiyosue T. β-adrenergic regulation of the cardiac Na$^+$ channel. In: Morad M, Ebashi S, Trautwein W, Kurachi Y, eds. Molecular Physiology and Pharmacology of Cardiac Ion Channels and Transporters. Boston: Kluwer, 1996: 53–62.

55. Santana LF, Gomez AM, Lederer EJ. Ca^{2+} flux through promiscuous cardiac Na$^+$ channels: slip-mode conductance. Science 1998; 279:1027–1033.

56. DiFrancesco D, Tortora P. Direct activation of cardiac pacemaker channels by intracellular cyclic AMP. Nature 1991; 351:145–147.

57. Clapham DE. Not so funny anymore: pacing channels are cloned. Neuron 1998; 21:5–7.

58. Hiraoka M, Furukawa T. Functional modulation of cardiac ATP-sensitive K$^+$ channels. News Physiol Sci 1998; 13:131–137.

59. Terzic A, Jahangir A, Kurachi Y. Cardiac ATP-sensitive K$^+$ channels: regulation by intracellular nucletides and K$^+$ channel-opening drugs. Am J Physiol 1995; 269: C525–C545.

60. Kurachi Y. G protein regulation of cardiac muscarinic potassium channel. Am J Physiol 1995; 269:C821–C830.

61. Krapivinsky G, Krapivinsky L, Wickman K, Clapham DE. Gβγ binds directly to the G protein-gated K$^+$ channel, I$_{KACh}$. J Biol Chem 1995; 270:29059–29062.

62. Wickman KD, Iniguez-Lluhi JA, Davenport PA, Taussig R, Krapivinsky GB, Linder ME, Gilman AG, Clapham DE. Recombinant G-protein beta gamma-subunits activate the muscarinic-gated atrial potassium channel. Nature 1994; 368:255–257.

63. Hartzell HC, Fischmeister R. Direct regulation of cardiac Ca^{2+} channels by G proteins: neither proven nor necessary? TIPS 1992; 13:380–385.

64. Philipson KD, Nicoll DA. Sodium-calcium exchange. Curr Opin Cell Biol 1992; 4:678–683.

65. Hilgemann DW, Ball R. Regulation of cardiac Na$^+$-Ca^{2+} exchange and K$_{ATP}$ potassium channels by PIP$_2$. Science 1996; 273:956–959.

66. Stengl M, Mubagwa K, Carmeliet E, Flameng W. Phenylephrine-induced stimulation of Na$^+$/Ca^{2+} exchange in rat ventricular myocytes. Cardiovasc Res 1998; 38:703–710.

67. Gao J, Cohen IS, Mathias RT, Baldo GJ. The inhibitory effect of beta-stimulation on the Na/K pump current in guinea pig ventricular myocytes is mediated by a cAMP-dependent pathway. Pflugers Arch 1998; 435:479–484.

68. Kapadia S, Oral H, Lee J, Nakano M, Taffet GE, Mann DL. Hemodynamic regulation of tumor necrosis factor-β gene and protein expression in adult feline myocardium Circ Res 1997; 81:187–195.

69. Isom LL, De Jongh KS, Catterall WA. Auxiliary subunits of voltage-gated ion channels. Neuron 1994; 12:1183–1194.

70. Partin KM, Fleck MW, Mayer ML. AMPA receptor flip/flop mutants affecting deactivation, desensitization, and modulation by cyclothiazide, aniracetam, and thiocyanate. J Neurosci 1998; 16:6634–6647.

71. Katz AM. Heart Failure: Pathophysiology, Molecular Biology, Clinical Management. Philadelphia: Lippincott/Williams and Wilkins, 2000.

6

Molecular Diversity and Distribution of Cardiac Ion Channels

HAROLD C. STRAUSS

State University of New York, Buffalo, New York

ARTHUR M. BROWN

Case Western Reserve University, Cleveland, Ohio

I. INTRODUCTION

The ordered depolarization and repolarization that occurs with each cardiac cycle represents a coordination of pacemaker activity, activation, and repolarization throughout the heart. For decades these physiological variables have been studied in vivo and in vitro using a variety of increasingly sophisticated techniques. The more recent introduction of voltage-clamp, cellular and molecular biological techniques into the cardiac electrophysiology field has enabled us to establish the ionic, molecular, and cellular basis of many electrophysiological properties of the heart. Such studies reveal a highly integrated system that includes the well-known specialized regions of the heart, and the "working" atrial and ventricular myocardium. While the absence of specialization within the ventricular myocardium implied uniformity of function, recent studies have demonstrated a remarkable heterogeneity of function throughout both the atrial and ventricular myocardium. This heterogeneity is very important as it likely contributes to the pathophysiology of reentrant arrhythmias. This chapter reviews the cellular and molecular basis of the diversity of electrophysiological function within the heart.

II. TRANSMEMBRANE ACTION POTENTIALS AND ION CURRENTS

Transmembrane action potential recordings have demonstrated differences in action potential waveforms from the SA node, crista terminalis, Bachmann's bundle, pectinate muscle and atrial roof, AV node, His bundle, Purkinje fibers, and subendocardial, mid, and subepicardial layers of the ventricle (1–5). Differences in rate of rise and amplitude of phase 0, and shape, time course and voltage of phases 1–4 of the action potential have been noted in recordings from these different regions. These differences in action potential waveform indicate that the molecular basis of electrogenesis cannot be uniform throughout the heart and have been detailed using voltage-clamp and molecular biology techniques. Voltage-clamp techniques were used to isolate, identify, and characterize the properties of the many inward and outward currents in cardiac myocytes responsible for the different phases of the action potential (Fig. 6.1) (6–42). More recently, these techniques have been used to examine the ionic basis of the differences in action potential morphology throughout the heart by studying myocytes isolated from different anatomical regions of the heart. We will begin with a consideration of the ionic basis of the "generic" action potential. Although differences in the individual ionic currents between animal species are well recognized, this chapter will not review this material in any detail, but instead will focus on the generic properties of the currents.

A. Depolarizing Currents

In the atrial and ventricular myocardium and His-Purkinje system the fast Na^+ current (I_{Na}) is primarily responsible for phase 0 of the action potential (6–8). The high threshold, dihydropyridine-sensitive L-type Ca^{2+} current ($I_{Ca,L}$) makes a nominal contribution to phase 0 in these cells. However, because of its slower inactivation, $I_{Ca,L}$ makes a significant contri-

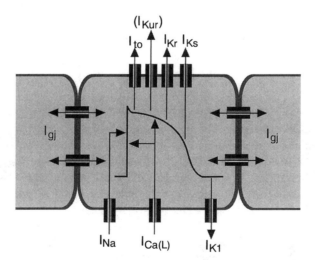

Figure 6.1 A transmembrane action potential is superimposed on a simplified scheme of a myocardial cell. The cell is connected to two neighboring cells via gap junction channels (I_{gj}). Contribution of the different inward (I_{Na}, $I_{Ca,L}$) and outward currents (I_{to}. I_{Kur}, I_{Kr}, I_{Ks}, I_{K1}) to the transmembrane action potential is approximated by the arrows. The time course of the different currents is discussed in further detail in the text. I_{Kur} is enclosed in parenthesis because the current has only been detected in atrial cells. (Reproduced by permission from Ref. 5a.) I_{Na} = Na^+ current; $I_{Ca,L}$ = L-type Ca^{2+} current; I_{to} = transient outward K^+ current; I_{Kur} = ultrarapid component of the delayed rectifier K^+ current; I_{Kr} = rapid component of the delayed rectifier K^+; I_{Ks} = slow component of the delayed rectifier K^+ current; I_{K1} = inward rectifier K^+ current.

bution to phases 1 and 2 of the action potential (9–16). It is also an important contributor to phase 0 in SA and AV nodal cells and is essential for conduction in the SA and AV nodes (14). Under basal conditions, it makes a modest contribution to phase 4 in SA nodal cells (32,39,40). $I_{Ca,L}$ is essential for excitation–contraction coupling because it is the major pathway for Ca^{2+} entry, which in turn produces Ca^{2+}-induced Ca^{2+} release from the SR. The low threshold, T-type Ca^{2+} current ($I_{Ca,T}$), makes a significant contribution to the early part of repolarization in the atrium and to phases 4 and 0 in the SA node (31,32,40,42).

B. Repolarizing Currents

In the atrial and ventricular myocardium, a minimum of six distinct K^+ currents have been identified in cardiac myocytes (Fig. 6.1). The calcium-independent transient outward K^+ current (I_{to1}) is a prime contributor to phase 1 and the early part of phase 2 (18–24). An additional calcium-dependent transient outward K^+ current (I_{to2}) was identified in some studies (18–24). The ultrarapid component of the delayed rectifier K^+ current (I_{Kur}), which is only detectable in atrial myocytes, begins in phase 1 and contributes to the remainder of the action potential plateau (25,26). The rapid and slow components of the delayed rectifier K^+ current (I_{Kr}, I_{Ks}) make an increasing contribution throughout phase 2 and produce the early part of phase 3 (27–30,33,34). The inward rectifier K^+ current (I_{K1}) produces the terminal part of repolarization and is a major determinant of the resting potential in phase 4 (30). In addition, two related inward rectifier K^+ currents are the ligand-gated ATP-sensitive K^+ current ($I_{K,ATP}$) and the muscarinic-gated K^+ current ($I_{K,ACh}$) (35–38). $I_{K,ATP}$ is inhibited by physiological levels of ATP within the cell and requires significant ATP depletion to activate the channel. $I_{K,ACh}$ is activated by acetylcholine (ACh) binding to M2 receptors in the SA and AV nodes and atrium (38). Another current that is cation, but not K^+ selective, is the hyperpolarization-activated inward cation current (I_h or I_f), observed in all cardiac pacemaker cells (39–43). I_f is recognized to be an important contributor to the pacemaker potential in SA nodal cells. The corresponding gene is a member of the cyclic nucleotide-gated family of ion channels and is related to the voltage-gated potassium channel family (43). Other cationic and anionic currents flowing through ion channels, exchangers, and transporters also contribute to the cardiac action potential but will not be considered further (42,44).

III. ION CHANNELS AND THEIR GENES

The elucidation of the molecular basis of these various currents began with the cloning of members of the evolutionarily conserved, voltage-dependent Na^+-, Ca^{2+}-, and K^+-channel superfamily of genes (42,44). The putative structure of each functioning Na^+-, Ca^{2+}-, and K^+-channel unit is a symmetrical arrangement of four repeats (Na^+, Ca^{2+} channels) or subunits (K^+ channels), each consisting of six transmembrane-spanning segments. In the case of the Na^+ and Ca^{2+} channels, the structure is monomeric; the pore-forming α-subunit consists of four repeats of the six-transmembrane-segment region. In contrast, the putative structure of each K^+ channel is tetrameric, consisting of four separate pore-forming voltage-sensing α-subunits that associate to form a channel. As for each repeat of Na^+ and Ca^{2+} channels, each K^+-channel α-subunit contains six putative transmembrane-spanning segments (see Chap. 3). It has been suggested that Na^+ and Ca^{2+} channels arose by gene duplication from a simpler K^+-channel subunit (42).

Each of the repeats in the Na^+- and Ca^{2+}-channel α-subunits and each K^+-channel α-subunit monomer contains domains that subserve essential channel functions (42,44). A membrane-spanning segment enriched with positive charges, S4, serves as the putative voltage sensor and transduces changes in transmembrane potential into gating changes (45–49). A segment linking membrane-spanning segments S5 and S6 forms the outer part of the ion conduction pathway or pore (so-called P region), while the C-terminus part of S6 contributes to the inner pore region (50–52).

The P region is highly conserved in K^+ channels including a stripped-down pH-dependent K^+ channel from the bacterium *Streptomyces lividans*, KcsA. This channel has been crystallized and the structure of the pore has been solved at 3.2 Å resolution (52). The structure provides insight into the essential properties of K^+ conduction and the large amount of structure–function data accumulated in recent years. The model shows that the channel is a tetramer, with the selectivity filter near the outer mouth of the pore. The charge on potassium ions is stabilized by the water molecules in the hydrophilic cavity located internal to the filter and by the dipole moment of short α-helices that point to the cavity. At its cytoplasmic end, the pore may be occluded by the C-terminal ends of the second of two transmembrane-spanning α-helices which may form the activation gate. In voltage-sensitive K^+ (Kv) channels the activation gate would correspond to the C-terminal ends of S6.

Inactivation in K^+ channels results from the occlusion of the inner mouth by the N-terminal region (ball) or from a constriction of the pore. For some K^+ channels, the positively charged stretch of amino acids in the N terminus may produce the fast inactivation referred to as N-type inactivation. Residues in the C-terminal region of the S5–S6 linker, and the N-terminal end of the pore and S6 regions may be involved in a generally slower type of inactivation resulting from pore constriction and is referred to as C-type (53–56) (see Chap. 3). N- or C-type inactivation may be the only mechanism of inactivation in Kv channels or they both may occur in the same channel. For Na^+ channels, a conserved region in the linker between repeats 3 and 4 acts as an inactivation hinge and pore lid, and other parts of the channel (S6 and S4–S5) also contribute to inactivation (57–62). For Ca^{2+} channels, the domain corresponding to inactivation has not yet been assigned, although a putative Ca^{2+}-binding motif called an EF-hand and a nearby consensus calmodulin-binding isoleucine-glutamine (IQ) motif in the carboxy terminus of the $α_{1C}$ channel subunit have been implicated in Ca^{2+}-dependent inactivation (63,64). The structural basis of inactivation in the Ca^{2+} channel has not been determined. In addition, ancillary membrane-bound or cytoplasmic subunits have been identified for many of the known ion channels (65–67). Each of these subunits associates with varying degrees of specificity to α-subunits and appears to be important in modulating the gating properties and expression of the hetero-oligomeric channel (*vida infra*).

A. Na Channels

The eel electroplax voltage-gated Na^+ channel was the first ion channel to be cloned (68). Subsequently, three mammalian orthologs were cloned from rat brain and at the present time 11 mammalian voltage-dependent Na^+-channel genes have been cloned (60). The genes encode tissue-specific isoforms that are expressed mainly in brain, peripheral nerve, heart, and skeletal muscle (60,69,70). The human heart Na^+-channel gene is *SCN5A* and is expressed in denervated skeletal muscle as well (71) (Tables 6.1 and 6.2). Na^+-channel cDNA expressed in non-native cells (heterologously expressed) generates an inward current that has many of the features of the native Na^+ current. While the human heart Na^+-channel cDNA expresses currents that are similar to the brain and skeletal muscle Na^+ channels, it is about 1000-fold less sensitive to tetrodotoxin (60,71,72). This difference is due to the presence of a *cys* in place of a *tyr* or *phe* in the P region of the first repeat of the

Table 6.1 Nomenclature of Ion Channels and Corresponding Genes

cDNA	Gene
Kv1	KCNA
Kv2	KCNB
Kv3	KCNC
Kv4	KCND
HERG	KCNH
KvLQT	KCNQ
Hyperpolarization-Activated Cyclic-Nucleotide Gated (HACNG)	HCN

Table 6.2 Correspondence of Cardiac Ionic Currents with Channel Subunits and Genes

Current	Channel subunit	Gene
I_{Na}	α-subunit	SCN5A
	β-subunit	SCN5B
$I_{Ca,L}$	α1-subunit	CACNA1C
	α2/δ-subunit	CACNA2
	β2-subunit	CACNB2
$I_{Ca,T}$	α1H (Ca$_v$T.2) α-subunit	CACNA1H
I_{to-epi}	Kv4.2 α-subunit	KCND2
	Kv4.3 α-subunit	KCND3
$I_{to-endo}$	Kv1.4 α-subunit	KCNA4
	Kvβ1.2 β-subunit	KCNAB1
I_{Kr}	HERG1 α-subunit	KCNH2
	HERG1B α-subunit	KCNH2
	MiRP1	KCNE2
I_{Ks}	KvLQT1 α-subunit	KCNQ1
	minK	KCNE1
I_{Kur}	Kv1.5 α-subunit	KCNA5
I_f	HACNG1 α-subunit	HCN1
	HACNG2 α-subunit	HCN2
	HACNG4 α-subunit	HCN4

heart Na^+ channel. Two ancillary subunits, β_1 and β_2, have been cloned (60). Coexpression of the rat brain β_1-subunit with the brain α-subunit in *Xenopus* oocytes substantially modifies both the magnitude and gating properties of the current expressed. Because coexpression of the β_1-subunit with the cardiac Na^+-channel isoform appeared to have nominal effects on the expressed current, the role of the cardiac β_1 in heart has been less clear. However, recent data have suggested that the ancillary β_1-subunit does associate with the cardiac α-subunit, as coexpression of both α- and β-subunits is necessary to detect phenotypic changes resulting from a mutation in *SCN5A* gene that was detected in an extended family with hereditary long-QT syndrome (73).

B. Ca Channels

The L-type dihydropyridine-sensitive cardiac Ca^{2+} channel is a hetero-oligomer of four distinct subunits, probably in a 1:1:1:1 stoichiometry: α_1, which is the functional channel and is encoded by *CACNA1C* gene; α_2-δ which is encoded by another gene, *CACNA2*, and is proteolytically processed into α_2 and δ, which are covalently linked, and the β_2-subunit encoded by *CACNB2* gene (Tables 6.1 and 6.2). The α_2-subunit is extracellular and anchored to the membrane by δ (74–76); α_2 is heavily glycosylated and the glycosylation enhances expression of the α_1-subunit and modifies gating (75). The β_2-subunit has even greater effects on expression and gating (76). The T-type channel is probably encoded by *CACNA1H* gene which is expressed primarily in heart, although *CACNA1G*, which is expressed primarily in brain, is also expressed in heart (77–79).

C. K-Channel α-Subunits

The largest representation of K^+-channel α-subunits in the heart is from mammalian orthologs of the *Drosophila* family of K^+ channels. These channels were cloned using cDNA probes from select regions of *Drosophila* K^+ channels and screening of cDNA libraries prepared from heart tissue of different animal species and humans (80–82). The numerous Kv channel genes indicate that there are many more relevant genes than identified K^+ currents, and as a result the linkage between the genes and K^+ currents described above has not been fully established. Four related, but distinct, subfamilies have been cloned: Shaker (Kv1, KCNA1), Shab (Kv2, KCNA2), Shaw (Kv3, KCNA3), and Shal (Kv4, KCNA4) (83) (Tables 6.1 and 6.2). Transcripts from Kv1–Kv4 subfamilies are represented in the murine, rat, and ferret heart and many are also present in human heart as well (84–87). Recently, a related set of Kvs5–9 have been cloned from mammals (88–90). These subunits are electrically silent when expressed alone, but Kvs 5,6,8, and 9 may coassemble with other Kv channels and alter their function and expression (87–90). Only Kvs 5 and 6 are represented in heart (87). Linkage studies of patients with familial long-QT-interval syndrome have recently called attention to two additional members of the family of voltage-dependent K^+ channels, KvLQT1 and the more distantly related *HERG* (*h*uman *e*ther-à-go-go *r*elated *g*ene) (91–98). K^+-channel monomers within each subfamily assemble as a homo- or heteromultimer to form a functional channel. For example, in the Kv1 subfamily, Kv1.4 can coassemble with Kv1.1, Kv1.2, and Kv1.5 to form heteromultimers (99,100).

The phylogenetic relationship between representative members of the K^+ channel family is indicated in the dendrogram shown in Figure 6.2. Kv subfamilies 1 through 6 plus 9 are more closely related to each other than to either KvLQT1 or *HERG*. The surprisingly high degree of sequence homology between Kv2.1 and Kv5.1 may explain why Kv5.1 preferentially inhibits the expression of Kv2.1 channels.

One elusive gene product that has been recently cloned is the channel related to the cyclic-nucleotide-gated channel class found in photoreceptors and olfactory cells, termed the *H*yperpolarization-*A*ctivated, *C*yclic *N*ucleotide-*G*ated Channel (HA-CNG1), which only bears 28% identity to the CNG family of channels found in photoreceptors and olfactory cells. It has a putative topology suggesting that each α-subunit consists of six transmembrane-spanning segments. Five separate HA-CNG cDNA clones have been found in the mouse heart; however, only two have been adequately characterized and appear to be related to I_h or I_f (43,101–103).

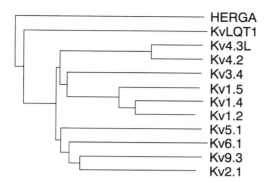

Figure 6.2 A simplified dendrogram of voltage-dependent K⁺ channel genes (six-transmem-brane-spanning segment) described in heart. Phylogenetic tree of cardiac voltage-gated K channels. The protein sequences for each of the above human channels were aligned using the Clustal algorithm (Ref. 100a) as implemented in multiple sequence alignment (InforMax). The phylogenetic tree was then generated using the Neighbor Joining method (Ref. 100b) also using multiple sequence alignment. Note that the members of the Kv1 family are more closely related to each other than to other family members. Kv subfamilies 1 through 6 plus 9 are more closely related to each other than to either KvLQT1 or *HERG*.

In contrast, the Kir K⁺-channel family members are not truly voltage dependent, are distantly related to the voltage-dependent K⁺ channels, and each α-subunit consists of only two transmembrane segments separated by a linker that contains a P region that is homologous to K⁺ channels (43,44,104) (see Chap. 3). Members of this family form the inwardly rectifying K⁺ currents found in heart. Recent data suggest that: (1) Kir2.x family members (IRK) encode I_{K1}; (2) $I_{K,ATP}$ is a multimeric complex of inwardly rectifying K⁺ channel subunits (Kir6.2) and the sulfonylurea receptor (SUR1); and (3) $I_{K,ACh}$ is a hetero-multimer of two inwardly rectifying K⁺-channel subunits, GIRK1 (Kir3.1) + GIRK4 (CIR or Kir3.4) (43,44,105–111). Recently a new family of K⁺ channels originally discovered in *Caennorhabdytis elegans* has been shown to have mammalian orthologs. The prototype is the TWIK, *T*andem of P-domains in a *W*eak *I*nwardly rectifying *K*⁺ channel (112–114). Structurally these channels resemble a tandem of two Kirs but there is little homology at the amino acid level. These channels and some isoforms have been identified in mammalian heart (108) and they may help set the resting potential.

D. K⁺-Channel Ancillary Subunits

Cytoplasmic β-subunits coassemble with α-subunits of the Kv1 subfamily and at least four different β-subunit genes have been isolated. In mammals, there are the three alternatively spliced forms of Kvβ1, Kvβ1.1, Kvβ1.2, Kvβ1.3, as well as Kvβ2, Kvβ3, and Kvβ4 (65–67,116–118). The C-terminal regions of the four mammalian Kvβ proteins are highly conserved and differences arise from divergent N termini (67). Interaction with Kv1 α-subunits appears to occur between the highly conserved NAB domain in the N terminus of the α-subunit (which is also the site for α_1/α_1 coassembly) and the C terminus of the β-subunit (67). The N termini of Kvβ1 subunits produce inactivation by blocking the inner mouth of the pore region of the Kvα-subunits, and can substantially modify both activation and inactivation properties (66,122–124). The C termini produce significant increases in α-subunit expression through a chaperone-like effect, but unlike the chaperones they remain strongly bound to the α-subunits (125). In addition to the complexity of heteromultimeric assembly of α-subunits, data suggest that for Kvβ2 the major α–β complexes in brain appear to have a stoichiometry of $\alpha_4\beta_4$ (126). This raises the possibility that β-subunits also form heteromultimers (119–121) and the formation of α–α, α–β, and β–β heteromultimers, where the different α's and β's may be greater than one (122), may contribute to the functional diversity of K$^+$-channel function in vivo. Furthermore, cotranslational assembly of the α/β-hetero-oligomer is unknown and a least two possibilities exist: cotranslation of each α with each β followed by tetramerization or separate multimerization of αs and βs followed by assembly of the hetero-oligomer.

Kvβ2 subunits have been crystallized recently and the structure resolved at 2.8 Å (126). The resultant structure is tetrameric and each subunit is an oxidoreductase containing NADP$^+$ in its active site. This result confirms an earlier inference that noted the strong homolgy in sequence between Kvβs and oxidoreductases (127). The results point to a possible connection between the redox state of the cell and K$^+$-channel activity.

An unrelated membrane subunit previously called minK or IsK appears to be a member of a larger family, referred to as MinK-related peptides (*MiRP*) (128). MinK is a small membrane-bound peptide that was originally believed to independently form a channel but is now believed to function as an ancillary subunit that associates with KvLQT1 to produce I_{Ks}. Recently another member of this family, *MiRP1*, was identified and shown to coassemble with HERG and form a channel whose phenotype closely resembles I_{Kr}. Mutations in the *MiRP1* gene may be associated with inherited and acquired cardiac arrhythmias (128).

IV.   REGIONAL DIFFERENCES IN ACTION POTENTIALS AND IONIC CURRENTS

Cardiac transmembrane action potential recordings show substantial differences in different regions of the heart. For example, the notch of the action potential and the underlying I_{to} is particularly illustrative of this point (Fig. 6.1). The "notch" or spike-and-dome morphology in the transmembrane action potentials in the epicardium and endocardium of the left ventricle has been analyzed extensively (4,5,129,130). Differential responses to changes in rate between these two regions, and changes during restitution and to compounds such as 4-aminopyridine, led investigators to propose that different regions of the heart have distinct cellular electrophysiological properties. The parallels between the behavior of the spike-and-dome morphology of the action potential and the transient outward current also led investigators to propose that the transient outward current was more prominent in the epicardium than in the endocardium. In addition, a subpopulation of cells (M cells) has been identified in the deep subepicardial layers of the canine ventricle. These M cells display longer action potential duration (phase 2), a steeper rate dependence of action potential duration, and a pharmacological responsiveness that differs from adjacent epicardial and endocardial layers (4,5,129,130). More recently, it has been shown that specific ion currents are also not uniformly expressed throughout the heart. In addition, nonuniformity of sarcolemmal ion channel localization within individual myocytes has been observed, presumably reflecting differences in cell sorting of the different ion channel proteins.

A. Na⁺ Channels

Nonuniformity of expression of different individual channels occurs between different regions of the heart and within individual myocytes. For Na^+ channels, relatively little information in this area is available. The absence of an effect of TTX on action potentials recorded from cells in the center of the sinus node, the site of the primary pacemaker, suggests that I_{Na} density is markedly reduced in the primary pacemaker region of the SA node (131). Nonuniformity of Na^+ channel distribution also occurs within individual myocytes, as immunofluorescence data indicate that they are preferentially localized at the intercalated disk instead of being uniformly distributed along the sarcolemma (132). The functional significance of this distribution pattern has been discussed by Spach (133). During excitation spread, the greatest load inside each cell occurs in membrane patches next to the intercalated disks at the ends of cells. Hence, the increased Na^+ channel expression and, as a result, Na^+ current density near the gap junction serves to increase the safety factor for propagation.

B. Ca²⁺ Channels

L- and T-type Ca^{2+} channels are distributed very differently in heart. L-type channels, which are intimately involved in E–C coupling, are expressed in the ventricles, atria, and nodal tissues (134). They provide for propagation of the cardiac impulse through the AV node and depolarizing current in SA nodal cells. On the other hand, T-type Ca^{2+} channels, which are minimally involved in E–C coupling, have a more restricted distribution mainly in SA nodal, atrial, and Purkinje cells (135–138). Ca^{2+} entry via T-type Ca^{2+} channels appears to be a less effective signal to the SR than Ca^{2+} entry via the L-type Ca^{2+} channel (138). On the other hand, T-type Ca^{2+} channels contribute to pacemaker activity in SA nodal cells as the activation threshold for these channels falls within a voltage range that overlaps with the pacemaker potential in SA nodal cells (135,136).

C. Kv Channels

I_{to1} has been extensively studied in myocytes isolated from the nonspecialized atrial and ventricular myocardium of many different species including humans (139–149). While I_{to1} is present in nearly all working myocyte types, the current density and phenotype does not appear to be uniform. The density of I_{to1} is greater in epicardial than endocardial cells from the left ventricle of patients, although the range in the epicardial/endocardial ratios is wide, ranging between 1.6 and 4.0 in different studies (143,145,146,148). In part, such differences may reflect the heterogeneous distribution of K^+ channels discussed below. Such differences in current density are not confined to the ventricle, but also have been reported between different cell types of the right atrium and, in fact, I_{to1} may not be present in all atrial myocytes (147,150). The phenotype of and response to the *Heteropoda* toxin 2 for I_{to1} differs in cells obtained from the epicardial and endocardial regions (145,149,151). Specifically, differences in inactivation kinetics and, more importantly, large differences in recovery from inactivation have been reported, with recovery kinetics in I_{to1} in endocardial cells being roughly two orders of magnitude slower than in epicardial cells (~50 vs. ~3000 ms at 22°C). These differences in properties of the currents strongly suggest that different combinations of genes encode for I_{to} in the epicardial and endocardial cells.

Differences in the current density of other K^+ currents have been noted between different regions of the heart. While I_{Kr} has been described in SA and AV nodal cells, and atrial and ventricular myocytes, the current is not universally found in myocytes from these regions. Variation in the magnitude of the I_{Kr} tail currents has been reported in ventricular myocytes isolated from the human heart (34,150–153). Similarly, the density of I_{Ks} has been reported to be lower in the midmyocardial region than in the endocardial and epicardial regions (130). In addition, the density of I_{Ks} is lower in atrial than in ventricular myocytes and in some species, such as cat, there is little or no I_{Ks} as compared to I_{Kr} (34).

Another difference in K$^+$ current expression that is of potential therapeutic importance has been reported for I_{Kur}. I_{Kur} is detected in atrial myocytes but not in human ventricular myocytes (146,154). A selective blocker or modulator of this channel might therefore be very useful for the treatment of atrial arrhythmias. Finally, the current density of I_{K1}, which helps to stabilize the resting potential, is widely recognized to be higher in ventricular than in atrial myocytes and lowest in SA nodal myocytes (155,156). The absence of I_{K1} in SA nodal cells allows other relatively small inward currents to contribute to diastolic depolarization.

In sum, there is an abundance of phenotypical data to indicate that there are marked differences in genes encoding for the different K$^+$ currents between different regions of the heart. Arrhythmias occurring in patients with ischemic heart disease can originate in or result from diseased tissue in well-circumscribed regions of the heart and therefore are amenable to therapy that can be delivered to localized regions, such as radiofrequency ablation or surgery. Should the distribution of critical K$^+$ channels fall within these circumscribed regions, selective pharmacological targeting could offer improved therapeutic outcomes with a lower risk of proarrhythmic events than current therapy with nonselective antiarrhythmic agents.

V. REGIONAL DIFFERENCES IN ION CHANNEL (Na, Ca, AND Kv) GENE EXPRESSION

Dixon and McKinnon used the RNase protection assay to examine the expression of fifteen different K$^+$ channel transcripts in rat atrial and ventricular muscle (84). Amongst this group, mRNA levels of Kv1.2, Kv1.4, Kv1.5, Kv2.1, and Kv4.2 were found to be relatively high. The abundance of the remaining transcripts ranged between low to undetectable. In addition, these investigators established that Kv4.2 mRNA was more abundant in the epicardial than in the endocardial region of the left ventricle (157).

One limitation of this approach is that RNA isolated from whole hearts and measured by RNase protection results in analysis of RNA from all cells present in the sample analyzed, including endothelial, vascular, and nerve cells. Fluorescent in situ hybridization (FISH) performed on specific isolated myocyte types determines the fraction of myocytes expressing the particular transcript in question, and addresses issues related to heterogeneity of transcript expression in the heart (85). Shortcomings of this method include the uncertain relationship between the number of dispersed cells and the number of cells in the intact heart, the degradation of mRNA during cell isolation, and the presence of single-stranded DNA.

In one study of the ferret heart, the most widely distributed K$^+$ transcripts in atrial and ventricular myocytes were found to be Kv1.5 and Kv1.4 (85). Other Kv transcripts were also expressed in the atria and ventricles. As anticipated, Kv1.1, Kv1.3, and Kv1.6 were rare in myocytes isolated from the atria and ventricles. On the other hand, mRNA coding for Kv2 and Kv3 channels was expressed in myocytes from all regions of the heart.

Kv2.1 transcript was two- to threefold more abundant than Kv2.2, and Kv3.4 transcript expression was threefold more common in ventricular than atrial myocytes. All ventricular myocytes expressed some member of the Kir 2,3, or 6 family. As expected, Kir was nearly absent from the SA node.

Analysis of mRNA transcript distribution both in samples of tissue and isolated cardiac myocytes is also limited by the potential lack of correlation between mRNA and protein expression (i.e., while the lack of message reliably predicts lack of protein expression, the presence of message need not necessarily signify functional protein expression). To address these concerns, investigators have begun to use both the fluorescent in situ hybridization and immunolocalization techniques to identify where K$^+$ channel transcripts are expressed in heart, whether differences in voltage-gated K$^+$ channel expression exists between major anatomical regions of the heart, and to determine the extent of uniformity of gene expression between myocardial cells within the major anatomical regions of the heart. Such studies have been performed on *ERG*, Kv1.4, Kv4.2, and Kv4.3 (149,158) and are discussed further in the following sections (Figs. 6.3 and 6.4).

VI. RELATIONSHIP BETWEEN GENE PRODUCTS AND CARDIAC CATION CURRENTS

Assignment of electrical currents to specific channel cDNAs requires demonstration of similarities in biophysical properties between the native and the heterologously expressed current. It also requires that the native current be modulated by an intervention that is specific for the proposed gene product, such as function-altering, isoform-specific, antibodies or toxins, sequential coimmunoprecipitation antisense cRNA, dominant-negative subunits, transgenic constructs, or gene knockouts (154). In addition, both the native current and the gene product must be expressed within the cell type of the area of the heart studied. In the case of Na$^+$ and Ca^{2+} channels the assignment may be more straightforward given the limited number of subunits involved. However, for K$^+$ channels, the problem is much more complex because the number of candidate genes is large, their expression in different regions of the heart is variable, and they may assemble different $\alpha_4\beta_4$ combinations. The typical electrophysiological/biophysical analysis of a K$^+$ channel homotetramer is generally incomplete, since native K$^+$ channels are likely formed from heteromeric coassembly of α-subunits within subfamilies, and coassembly of different α- and β-subunits. As a result, knowledge of the distribution patterns of different gene products in in heart has the potential to shorten the list of candidate genes that are responsible for the regional differences in ionic currents.

VII. PHYSIOLOGY, MOLECULAR BIOLOGY, AND MUTATIONS OF ION CURRENTS

A. I_{Na}

Despite the difference in TTX sensitivity between the brain and cardiac Na channels, their gating and permeation properties and pharmacological sensitivity appear to be very similar (60). Although two Na$^+$ channel genes, *SCN5A* and *SCN6A*, have been identified in heart, only *SCN5A* has been demonstrated to be functional. The major difference between *SCN5A* and the skeletal muscle *SCN4A* or neuronal *SCNs1-3A* is TTX sensitivity as described previously. The third most common form of hereditary long-QT syndrome (LQTS), LQT3, has been linked to *SCN5A* (159) (Table 6.2). The disease is inherited as an

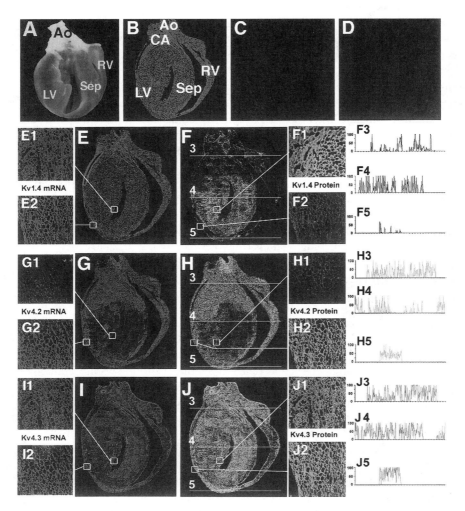

Figure 6.3 Comparison of Kv1.4, Kv4.2, and Kv4.3 α-subunit mRNA transcript and protein distribution in ferret ventricular sagittal sections. Relative mRNA levels were detected by fluorescence in situ hybridization (FISH) and are indicated by the red fluorescence. Relative protein levels were detected by indirect immunofluorescence (IF) and are indicated by the green fluorescence. Results shown were obtained from adjacent sagittal sections. Panel (A) provides a view of a sagittal cut through the ventricle and great vessels (AO = aorta; CA = coronary artery; RV = right ventricle; LV = left ventricle; Sep = septum). Panels (B) and (C) are sections that has been probed with antisense and sense probes to TnIc and serve as positive and negative controls to this transcript. Panel (D) represents a negative control signal to Kv4.2/Kv4.3 using a secondary antibody (see Ref. 149) in the absence of primary antibodies. FISH results are shown in Panels (E), (G), and (I) and IF results are shown in Panels (F), (H), and (J). The smaller numbered panels correspond to enlarged sections obtained taken from selected regions indicated by the white boxes. Note the general correspondence between the Kv4.2 and Kv4.3 mRNA and protein levels. On the other hand, Kv1.4 mRNA and protein levels do not correlate. Relative fluorescence intensity profiles of Kv1.4, Kv4.2, and Kv4.3 antibodies are measured in transverse sections obtained from the indicated levels in the basal, midventricular, and apical regions of the ventricle. The marked variation in relative intensity profiles measured from the different transverse sections indicates that protein levels vary between different small domains within adjacent regions as well as varying between different regions of the heart (regional localization). (Reproduced by permission from Ref. 149.) (See also color plate.)

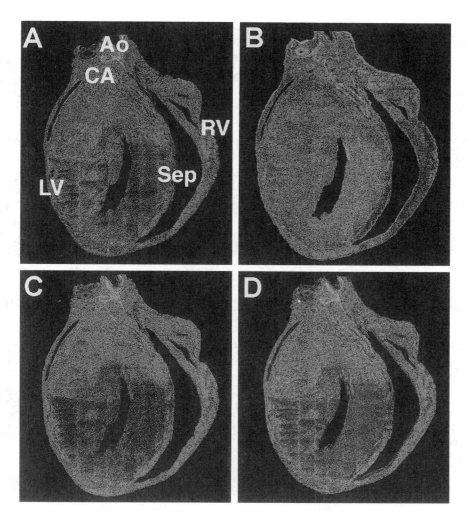

Figure 6.4 Colocalization and distribution of Kv1.4, Kv4.2, Kv4.3 α subunits in ferret ventricu-lar sagittal sections determined from direct IF. Panel A shows Kv1.4 (red) and Kv4.2 (green) colo-calized (yellow to orange) mainly in the apical LV endocardial region and septum. Panel B shows Kv1.4 (red) and Kv4.3 (green) also colocalized (yellow to orange) mainly in the apical LV epicardial region and septum. Panel C shows Kv4.2 (green) and Kv4.3 (red) also colocalized (yellow to red) in the apex. Panel D shows overall distribution patterns of Kv1.4 (blue), v4.2 (green) and Kv4.3 (red) α subunits. Colocalization of Kv4.2 and Kv4.3 is indicated by yellow to orange and Kv1.4 and Kv4.3, purple to pink. (Reproduced by permission from Ref. 149.) (See also color plate.)

autosomal dominant trait and all of the mutations that have been tested functionally have to date shown gain of function, which explains the dominant transmission (160). Channel inactivation is the property that is impaired by the mutation and as a result an increase in late current (gain of function) is the usual result. One consequence of the functional stud-ies has been the demonstration that the increased steady-state current is more sensitive to Na^+ channel blockers, such as mexiletine, than is the peak current (161). Mexiletine was selected because of its preferential binding to the prolonged inactivated state of the Na^+

channel, which makes this drug particularly effective in blocking the increased steady-state current in the mutant channel. This result has been translated into the therapy of patients with LQT3 (162), where the inward current that produces the long-QT syndrome can be blocked without any detectable effect on the transient inward current required for propagation of the impulse (see Chap. 9).

Interestingly, several recently described mutations have not produced phenotypes that can readily explain the long-QT interval and the propensity toward ventricular arrhythmias. One such mutation in the α-subunit (D1790G) has not produced any detectable cellular phenotype. Coexpression of the mutated α-subunit with the human β1-subunit is required to show an abnormal phenotype (a hyperpolarizing shift in the inactivation curve), but there was no evidence for any gain of function (i.e., an enhanced window current) (73). A similar situation applies to missense mutations that link *SCN5A* to familial idiopathic ventricular fibrillation (163). The cellular phenotype expressed by the mutations is an increased recovery from inactivation, which by itself offers no ready explanation for the increased susceptibility to ventricular arrhythmias.

B. I_{Ca}

Channelopathies involving Ca^{2+} channels thus far have not been described for *CACNA1C*, *CACNA2*, or *CACNB2*, the subunits that compose the cardiac L-type-Ca^{2+} channel. The absence of such data implies that the mutations that have occurred either have failed to cause a gain or loss of function or had lethal effects.

C. I_{to1}

Differences in peak current density and electrophysiological/biophysical properties of I_{to1}, the rapidly activating and inactivating calcium-independent transient outward K^+ current, in the subendocardial and subepicardial regions of the ventricle indicate that there are at two least different I_{to1} phenotypes (143,145,146,149). The most marked difference in gating kinetics has been in the kinetics of recovery from inactivation, where cells from the subepicardial region display a rapid recovery from inactivation, while in cells from the subendocardial region, recovery is generally much slower. In addition, the differential action of *Heteropoda* toxin 2 on I_{to1} in endocardial and epicardial cells (149) further supports the presence of at least two distinct proteins that form I_{to1} in the heart. The very slow recovery kinetics of I_{to1} in the endocardial cells measured at 21 to 22°C implies that this current makes a minimal contribution to repolarization at normal heart rates. However, when similar measurements were performed at 35 to 37°C, values for the time constant of recovery from inactivation ranged between 490 and 840 ms, implying that this current system could make a meaningful contribution to repolarization in endocardial cells in vivo (146,148).

Homotetramers of Kvα 1.4, 4.2, and 4.3 produce rapidly activating and inactivating currents and are realistic candidate genes for I_{to1} (24,142,149,164). The gating kinetics of Kv4.2/4.3 and the pharmacological properties (4-AP, HpTx2) of Kv4.2 previously discussed suggest that these genes produce native I_{to1} in epicardial cells (149,164,165). The distribution pattern of Kv1.4 protein in the ventricle in conjunction with its slow recovery kinetics and insensitivity to HpTx2 makes it likely that this gene is responsible for the native I_{to1} in subendocardial myocytes of the left ventricle of both the human and ferret heart (24,100,143,145,149).

Kv4.2 transcript and protein have been reported in mouse, rat, ferret, and dog, but not the human heart (157). Kv4.3 transcript and protein has been reported in rat, ferret, dog, and human heart (166). Among the members of the Kv4 subclass, the Kv4.2 transcript is expressed in the greatest percent of ferret cardiac myocytes (149). While Kv4.2 transcript is more abundant in the epicardium than in the endocardium of the rat and canine left ventricle, this has not been the case for Kv4.3 (157). As a result, the absence of a difference in distribution of Kv4.3 protein between the epicardial and endocardial regions of the heart is inconsistent with the reported differences in I_{to} density seen between these two regions of the human left ventricle (144,145). The regional heterogeneity of Kv4.2/4.3 in the ferret ventricle may shed some light on this topic. In sagittal sections of the ferret heart, Kv4.2 protein was localized primarily to the RV free wall, epicardial layers of the LV in the apex and base, but was markedly reduced or absent in the apical LV endocardium. On the other hand, Kv4.3 protein was more uniformly distributed throughout the LV apical, mid, and basal regions, but was also concentrated in the epicardial layers of the LV and virtually absent in the RV (149). These data call attention to the need for accurately identifying the sites from which samples are taken for biochemical, in situ hybridization and immunofluorescence studies and provide guidance for the evaluation of potential changes in ion channel expression in the human heart in different disease states.

One further interesting facet is the marked discrepancy between the distribution of the Kv1.4 transcript and protein in the heart. Kv1.4 transcript is abundant in atrial and ventricular myocytes; however, Kv1.4 protein is reported as absent in rat ventricular myocytes (167,168) or expressed at a low level in the endocardium and midmyocardium (free wall and septum) of the ferret ventricle (149). Whether the inhibition of translation can be removed to increase Kv1.4 protein expression in the ventricle and be used therapeutically in patients with long-QT syndrome is unkown.

In attempting to explain the marked disparity between the distribution of Kv1.4 transcript and protein, the observations of Shi et al. (125) may be relevant. These investigators noted that Kvβ2 associates with the Kv1.2α-subunit early in channel biosynthesis, and that Kvβ2 exerts chaperonelike effects on associated Kv1.2α-subunits, by mediating the biosynthetic maturation and surface expression of this channel in COS cells. As Kvβ1 and Kvβ2 have been demonstrated to associate with Kv1.4 as well as Kv1.2 subunits, it is plausible that coexpression of one of these ancillary β-subunits may be necessary for expression of Kv1.4 protein in cardiac myocytes. Channelopathies involving I_{to} K$^+$ channels thus far have not been described for Kv1.4, Kv4.3, and Kvβ1.1-3 genes that may form the human cardiac I_{to} K$^+$ channel.

D. I_{Kr}

The selective blockade of I_{Kr} by E-4031 and dofetilide allowed investigators to separate two components of the delayed rectifier current in cardiac muscle (27,28,169,170). The more rapidly activating component, I_{Kr}, initially believed to have relatively rapid activation kinetics, has a rapid inactivation process that complicated the analysis of activation. Voltage-clamp protocols subsequently demonstrated that I_{Kr} activation was slow relative to other delayed rectifier K$^+$ channels (33,170). In addition, the more rapid inactivation process, which accounts for the rectification properties of I_{Kr}, was shown to recover quickly and to have its own intrinsic voltage sensitivity (31,171). The rapid inactivation process competes with the activation process at depolarized potentials. As a result, there is very little current during the action potential plateau; however, the rapid recovery from inactivation enables the current to increase markedly during phase 3 of the action potential.

Homomultimers of heterologously expressed *ERG* produce currents whose gating kinetics closely resemble those of native I_{Kr} (172). Slower deactivation kinetics and different pharmacological properties, such as differing dofetilide and E-4031 sensitivities between ERG and the native I_{Kr}, were thought to reflect differences between the particular gene product being tested and that which exists in the heart, or the absence of the appropriate ancillary subunit (172,173). In fact, a gene encoding a MinK-related peptide 1 (MiRP1) that assembles with *HERG* has been cloned (128). The mixed complex formed from *HERG* and MiRP1 more closely resembles the native current in terms of its electrophysiological and pharmacological properties. Although three N-terminal isoforms of the mouse *ERG1* clone have recently been identified (174,175) only two have been detected in humans; they are *HERG1* and *HERG1b* (176). *HERG1b* is an alternative splice variant with a markedly shorter N terminus and is expressed in heart. The human (H) and mouse (M) homologs of ERG have very similar electrophysiological properties. *MERG1b* deactivates rapidly and the heteromultimer formed by coassembly of *MERG1a* and *MERG1b* has deactivation kinetics that are more rapid than those of *MERG1a* or *HERG1*. An additional C-terminal splice variant of *HERG1* (*HERG USO*) has been reported, which when coexpressed with HERG1 shows different electrophysiological properties than the *HERG* homotetramer (177). Thus, if multiple *HERG* gene products are expressed robustly in the ventricle, then the differences in activation and inactivation gating properties between the different splice variants of *HERG* predict that the electrophysiological properties of I_{Kr} might not be uniform in heart cells. It is of interest that such differences in gating properties of I_{Kr} have not been described.

Another factor complicating the evaluation of the role of *ERG* in heart is that ERG mRNA is expressed in about half of the myocytes of the ferret heart, whereas KvLQT1 mRNA is somewhat more common (85). ERG protein was heterogeneous in its distribution and most abundant in the outer cell layers of the ventricle, particularly in the apical and midmyocardial sections. Interestingly, the transmural distribution of ERG protein is not uniform in the left ventricular free wall. Differences in ERG protein expression also were seen in the ferret right atrium, where ERG protein was most abundant in the medial right atrium, especially in the trabeculae of the atrial appendage and in the SA nodal cells. Its detection in SA nodal cells may help explain the bradycardias seen in some patients with familial long-QT interval syndrome (158).

Mutations in *ERG* have been linked to the second most common form of hereditary LQTS in patients, LQT2 (Table 6.2). Sufficient numbers of patients have been analyzed to show that the mutations in its gene, *KCNH1*, are scattered through the open reading frame (92–94). Expression studies have shown that most of the mutations markedly reduce expression of the channel and exert a dominant negative effect on wild-type channels, thereby explaining the dominant mode of transmission (93). The dominant negative effect is explained by the putative tetrameric structure of the channel with one or two deficient and dysfunctional subunits being sufficient to produce "poisoning" and reduction in expression of transcript and protein. These mutations in *HERG1* were sufficient to prolong the QT interval in these patients, indicating that the expression of this protein is sufficiently broad to affect repolarization throughout the ventricle. Recently, several novel mutations were identified that, when coexpressed with wild-type channels, shifted the gating properties of the channel indicating that the abnormal phenotype resulted from altered protein function, not the level of expression (178).

Another series of mutations in the amino-terminal region accelerated deactivation (179). Recent analysis of the crystal structure of the NH terminus demonstrated that it possesses a Per-Arnt-Sim (PAS) domain. Mutations in the NH terminus were proposed to disrupt the PAS domain and interfere with its interaction with S4-S5 linker (180), indicating that this tertiary structure was an important determinant of the rate of deactivation.

HERG is also a very frequent target in acquired LQTS. Nonselective class I and III antiarrhythmic drugs, nonsedating antihistamines, antigastrointestinal medications, second-generation antipsychotic drugs, and macrolide and conazole antibiotics all block the channel at concentrations that correlate with drug levels that cause toxicity in patients. It has been speculated that patients with other mutations of *HERG* or other repolarizing K$^+$ currents, which ordinarily do not cause a lengthening of repolarization, may nonetheless show an increased sensitivity to K$^+$ channel blockade and an increased risk of cardiotoxic drug effects. At this time no such linkage has been demonstrated with this gene; however, a linkage has been suggested with *KCNE1* (181).

E. I_{Ks}

The other component of the delayed rectifier is the E-4031 and dofetilide-insensitive component, which activates and deactivates more slowly than I_{Kr}, shows minimal inward rectification, and is inhibited by azimilide (28,29,182). I_{Ks} is an important component of the delayed rectifier K$^+$ current recorded at positive potentials in the ventricle, and, as a result, makes an increasingly important contribution to the plateau phase of repolarization with either an increase in frequency or plateau duration (29,34). Heterologous expression of KvLQT1 alone produces a current that activates far more quickly than I_{Ks} (94,96,182,183). However, coexpression with the ancillary subunit minK produces currents with gating kinetics and a pharmacological profile that are very similar to the endogenous current (29,94,96,182). MinK may also increase the expression of KvLQT1, but recent experiments show that the hetero-oligomeric channel expresses significantly larger single-channel currents with a higher opening probability (182). In addition, minK eliminates or greatly slows inactivation of KvLQT1 (183). A splice variant of KvLQT1 having an N-terminal truncation has been identified (KvLQT2) (184,185). The truncated cDNA isoform reduces I_{Ks} expression when it is coinjected with the full-length cDNA isoform into *Xenopus* oocytes (184). The question of whether such a mechanism is operative in the intact heart has not yet been resolved, since it has not been determined whether both KvLQT1 and KvLQT2 are expressed in the same myocyte.

Regional heterogeneity of KvLQT1 transcript expression is also likely as the transcript is expressed in just over half of ferret cardiac myocytes (85). Surprisingly, minK mRNA is only expressed in 10 to 18% of ferret ventricular myocytes and 21 to 29% of ferret atrial myocytes. MinK associates with both ERG and KvLQT1 and increases the magnitude of the current over that seen with expression of the α-subunit alone. The relatively sparse distribution of minK mRNA may account for some of the variability in I_{Ks} levels between the different regions of the ventricular wall that have been reported by Antzelevitch and his colleagues.

Mutations in *KCNQ1* cause the LQT1 variant of familial LQTS (94,95,97) (Table 6.2). This is the most common form of the disease and, as is the case for LQT2, the reported mutations are scattered throughout the open reading frame (Table 6.3). Also, dominant negative suppression of wild-type KvLQT1 has been demonstrated. Mutations in *KCNE1* cause one of the rare forms of hereditary LQTS, LQT5. As for LQTS, 1 and 2 dominant negative suppression has been demonstrated. Biallelic mutations involving both *KCNQ1* and *KCNE1* cause a recessive form of hereditary LQTS that is associated with deafness. The latter is due to loss of function in the secretory epithelium in the inner ear (182).

Table 6.3   Summary of Mutations in Familial Long-QT Syndrome

Summary

Locus	LQT1	LQT2	LQT3	LQT5	All
Gene	KvLQT1	HERG	SCN5A	minK	genes
	KCNQ1	KCNH2	SCN5A	KCNE1	
Location	11p15.5	7q35–36	3p21	21q22	total
Total mutations found	78(49)	81(69)	13(8)	3(3)	175(129)

Distribution by position

Position					
Extracellular	0	6	1	1	8
Transmembrane segment	32	14	4	0	50
Pore	9	12	0	N/A	21
Intracellular	37	49	8	2	96
Total	78	81	13	3	175

Numbers in parentheses denote the number of mutations that are distinct.
Keating and colleagues have studied more than 400 unrelated individuals with arrhythmia susceptibility. They screened these individuals for mutations in KvLQT1 (*KCNQ1*), minK (*KCNE1*), HERG (*KCNH2*) and SCN5A genes. The number of mutations identified with each type of familial long-QT syndrome does not correlate with the prevalence of the disease, as the most prevalent form of the disease is LQT1. Mutations in specific genes have now been identified by Keating et al. in 175 individuals. Mutations in these genes were not detected in 222 individuals. No distinct distribution pattern has emerged; mutations are scattered throughout the protein-encoding exons of the gene. Almost all of the K$^+$ channel mutations that have been studied decrease current.
Source: Ref. 187.

F. I_{Kur}

A rapidly activating, sustained outward current occurs during depolarization in atrial, but not ventricular, myocytes. I_{Kur} is distinguished by virtue of its relatively high sensitivity to 4-aminopyridine (K_D 5.8–49 µM), but lack of sensitivity to other K$^+$ channel blockers (TEA, dendrodotoxin, Ba^{2+}) (26). Its rapid activation and minimal inactivation indicates that this current makes an important contribution to repolarization in atrial myocytes. The heterologously expressed Kv1.5 clone produces currents that have gating and pharmacological properties that closely resemble I_{Kur} (e.g., Kv1.5 current is the most sensitive to 4-AP among the Kv channels with a K_D for 4-AP comparable to that reported for native I_{Kur}).

The distribution of Kv1.5 protein was studied by Mays et al. (154), who used antibodies directed against two distinct channel epitopes. They determined that while the Kv1.5 protein was present in human atrial myocytes, it was also present in ventricular myocytes. Additional data confirming the linkage between the Kv1.5 protein and I_{Kur} was provided by Feng et al. (186), who demonstrated that I_{Kur} was selectively decreased in atrial myocytes by antisense oligonucleotides. On the other hand, the presence of the Kv1.5 protein in ventricular myocytes is at variance with the observation that the native I_{Kur} is only found in atrial myocytes. The molecular basis of this difference in I_{Kur} expression has not been described. Nonetheless, the differential expression of an important K$^+$ current between the atria and ventricles is a very significant observation with important therapeutic potential. It suggests that antiarrhythmic drugs could be designed to have effects against atrial arrhythmias, without any effect on ventricular repolarization, thereby reducing risk to patients.

G. I_f

I_f is found in many cell types of the heart; however, its properties are not uniform throughout the heart. In all cardiac pacemaker cells, I_f slowly activates on hyperpolarization and more rapidly deactivates on depolarization. Present data suggest that it is a mixed cation current (Na^+ and K^+) with a reversal potential of -30 to -15 mV. I_f is controlled by numerous signal transduction pathways, such as the β-adrenergic receptor signaling pathway involving G-proteins and cyclic AMP, which, when activated, increase I_f. The expressed HA-CNG1 protein and I_f have many similar properties (101–103). Both the clone and native current are approximately four times more permeant to K^+ than Na^+ and are not permeable to divalent cations and anions. Both expressed clones are activated by hyperpolarization, and their conductance is increased by intracellular cyclic AMP and cyclic GMP. Similar effects of pharmacological agents on I_f and the clones have been observed, including the blocking action of millimolar extracellular Cs^+. The activation of this channel by hyperpolarization is similar to that seen with the voltage-gated, inwardly rectifying plant K^+ channel, KAT1, and is surprising given the presence of a positively charged S4 region in this channel. Unfortunately, there are no immunolocalization data for HA-CNG1 proteins in the heart.

VII. SUMMARY

The elucidation of the molecular basis of the different ionic currents in the heart is progressing rapidly and has significantly advanced our understanding of the electrophysiological and biophysical data. It has become apparent that an additional variable in the heart that needs to be addressed is the nonuniform distribution of channel proteins and their modulators. The heterogeneity of channel protein is greater than was anticipated many years ago, and the distribution patterns as well as the underlying mechanisms are just beginning to be addressed. The immunolocalization data from the ferret heart indicate that this variability in channel expression appears to be a combination of variability among cells as well as regional sublocalization.

The distinct distribution on patterns of the channel proteins that have been studied, Kv1.4, Kv4.x, and *HERG*, call attention to many new and important questions. Do other ion channels and associated subunits also display distinct pattens of distribution? Are distinct patterns of ion channel distribution present in the human heart? What are the mechanisms of these distinct patterns of distribution? How do these patterns of expression contribute to cardiac electrophysiological properties and pathophysiology of cardiac arrhythmias? Since reentrant arrhythmias are frequently maintained or originate in circumscribed regions of the heart, can the distinct distribution patterns of ion channels be exploited to selectively deliver therapy to these localized regions? Determination of the patterns of localization of K^+ channel proteins within the human heart will have important ramifications with regard to our understanding of the molecular basis of normal and abnormal function, as well as for antiarrhythmic therapy.

ACKNOWLEDGMENT

We would like to thank Dr. Michael J. Morales for preparing the dendrogram shown in Figure 6.2.

REFERENCES

1. Hoffman BF, Cranefield PF. Electrophysiology of the Heart. New York: McGraw-Hill, 1960:42–210.
2. Paes de Carvalho A, de Mello WC, Hoffman BF. Electrophysiological evidence for specialized fiber types in rabbit atrium. Am J Physiol 1959; 196:483–488.
3. Wagner ML, Lazarra R, Weiss RM, Hoffman BF. Specialized conducting fibers in the interatrial band. Circ Res 1966; 18:502–518.
4. Antzelevitch C, Sicouri S, Lukas A, Krishnan SC, Di Diego JM, Gintant GA, Liu D-W. Heterogeneity within the ventricular wall. Electrophysiology and pharmacology of epicardial, endocardial and M cells. Circ Res 1991; 69:1427–1449.
5. Antzelevitch C, Sicouri S, Lukas A, Nesterenko VV, Liu D-W, Di Diego JM. Regional differences in the electrophysiology of ventricular cells. Physiological and clinical implications. In: Zipes DP, Jalife J, eds. Cardiac Electrophysiology. From Cell to Bedside. Philadelphia: WB Saunders, 1995:228–245.
5a. Jongsma HJ. Sudden cardiac death: a matter of faulty ion channels? Curr Biol 1998; 8:R568–R571.
6. Brown AM, Lee KS, Powell T. Sodium current in single rat heart muscle cells. J Physiol (Lond) 1981; 318:479–500.
7. Makielski JC, Sheets MF, Hanck DA, January CT, Fozzard HA. Sodium current in voltage clamped internally perfused canine cardiac Purkinje cells. Biophys J 1987; 52:1–11.
8. Mitsuiye T, Noma A. Exponential activation of the cardiac Na^+ current in single guineapig ventricular cells. J Physiol (Lond) 1992; 453:261–277.
9. Reuter H. The dependence of the slow inward current in Purkinje fibers on the extracellular calcium-concentration. J Physiol (Lond) 1967; 192:479–492.
10. Rose WC, Balke CW, Wier WG, Marban E. Macroscopic and unitary properties of physiological ion flux through L-type Ca channels in guinea-pig heart cells. J Physiol (Lond) 1992; 456:267–284.
11. Campbell DL, Giles WR, Shibata EF. Ion transfer characteristics of the calcium current in bull-frog atrial myocytes. J Physiol (Lond) 1988; 403:239–266.
12. Campbell DL, Giles WR, Hume JR, Noble D, Shibata EF. Reversal potential of the calcium current in bull-frog atrial myocytes. J Physiol (Lond) 1988; 403:267–289.
13. Campbell DL, Giles WR, Robinson K, Shibata EF. Inactivation of calcium current in bull-frog atrial myocytes. J Physiol (Lond) 1988; 403:287–315.
14. Campbell DL, Rasmusson RL, Strauss HC. Ionic current mechanisms generating vertebrate primary cardiac pacemaker activity at the single cell level: an integrative view. Annu Rev Physiol 1992; 54:279–302.
15. Campbell DL, Giles WR, Robinson K, Shibata EF. Studies of the sodium-calcium exchanger in bull-frog atrial myocytes. J Physiol (Lond) 1988; 403:317–340.
16. Campbell DL, Strauss HC. Regulation of Ca^{2+} currents in the heart. In: Means AR, ed. Calcium regulation of cellular function. (Advances in Second Messenger and Phosphorylation Research 30). New York: Raven Press, 1995:25–88.
17. Schulze D, Kofuji P, Hadley R, Kirby MS, Kieval RS, Doering A, Niggli E, Lederer WJ. Sodium/calcium exchanger in heart muscle: molecular biology, cellular function, and its special role in excitation-contraction coupling. Cardiovasc Res 1993; 27:1726–1734.
18. Josephson IR, Sanchez-Chapula J, Brown AM. Early outward current in rat ventricular cells. Circ Res 1984; 54:157–162.
19. Escande D, Coulombe A, Faivre JF et al. Two types of transient outward currents in adult human atrial cells. Am J Physiol 1987; 252:H142–H148.
20. EF Shibata, Drury T, Refsum H, Aldrete V, Giles WR. Contributions of a transient outward current to repolarization in human atrium. Am J Physiol 1988; 257:H1773–H1781.
21. Tseng GN, Hoffman BF. Two components of transient outward current in canine ventricular myocytes. Circ Res 1989; 64:633–647.

22. Hiraoka M, Kawano S. Calcium-sensitive and insensitive transient outward current in rabbit ventricular myocytes. J Physiol (Lond) 1989; 410:187–212.

23. Campbell DL, Rasmusson RL, Qu Y, Strauss HC. The calcium-independent transient outward potassium current in isolated ferret right ventricular myocytes. I. Basic characterization and kinetic analysis. J Gen Physiol 1993; 101:571–601.

24. Campbell DL, Rasmusson RL, Comer MB, Strauss HC. The cardiac calcium-independent transient outward potassium current: kinetics, molecular properties, and role in ventricular repolarization. In: Zipes DP, Jalife J, eds. Cardiac Electrophysiology. From Cell to Bedside. Philadelphia: WB Saunders, 1995:83–96.

25. Fedida D, Wible B, Wang Z et al. Identity of a novel delayed rectifier current from human heart with a cloned K channel current. Circ Res 1993; 73:210–216.

26. Wang Z, Fermini B, Nattel S. Sustained depolarization-induced outward current in human atrial myocytes. Evidence for a novel delayed rectifier K^+ current similar to Kv1.5 cloned channel currents. Circ Res 1993; 73:1061–1076.

27. MC Sanguinetti, Jurkiewicz NK. Two components of cardiac delayed rectifier K^+ current: differential sensitivity to block by Class III antiarrhythmic agents. J Gen Physiol 1990; 96:195–215.

28. JR Balser, Bennett PB, Roden DM. Time-dependent outward current in guina pig ventricular myocytes. Gating kinetics of the delayed rectifier. J Gen Physiol 1990; 96:835–863.

29. Kass RS. Delayed potassium channels in the heart. Cellular, molecular, and regulatory properties. In: Zipes DP, Jalife J, eds. Cardiac Electrophysiology. From Cell to Bedside. Philadelphia: WB Saunders, 1995:74–82.

30. Muraki K, Imaizumi Y, Watanabe M, Habuchi Y, Giles WR. Delayed rectifier K^+ current in rabbit atrial myocytes. Am J Physiol 1995; 269:H524–532.

31. Bean BP. Two kinds of calcium channels in canine atrial cells. J Gen Physiol 1985; 86:1–30.

32. Hagiwara N, Irisawa H, Kameyama M. Contributions of the two types of calcium currents to the pacemaker potentials of rabbit sino-atrial node cells. J Physiol (Lond) 1988; 395:233–253.

33. Liu S, Rasmusson RL, Campbell DL, Wang S, Strauss HC. Activation and inactivation kinetics of an E-4031-sensitive current from single ferret atrial myocytes. Biophys J 1996; 70:2704–2715.

34. Gintant GA. Two components of delayed rectifier current in canine atrium and ventricle. Does I_{Ks} play a role in reverse rate dependence of Class III agents? Circ Res 1996; 78:26–37.

35. Noma A. ATP-regulated K^+ channels in cardiac muscle. Nature 1983; 305:147–148.

36. Lederer WJ, Nichols CG. Regulation and function of adenosine triphosphate-sensitive potassium channels in the cardiovascular system. In: Zipes DP, Jalife J, eds. Cardiac Electrophysiology. From Cell to Bedside. Philadelphia: WB Saunders, 1995:103–115.

37. Breitweiser GE, Szabo G. Mechanism of muscarinic receptor-induced K^+ channel activation as revealed by hydrolysis-resistant GTP analogues. J Gen Physiol 1988; 91:469–493.

38. Yamada M, Inanobe A, Kurachi Y. G protein regulation of potassium ion channels. Phamacol Rev 1998; 50:723–757.

39. DiFrancesco D. Pacemaker mechanisms in cardiac tissue. Annu Rev Physiol 1993; 55:455–72.

40. DiFrancesco D, Mangoni M, Maccaferri G. The pacemaker current in cardiac cells. In: Zipes DP, Jalife J, eds. Cardiac Electrophysiology. From Cell to Bedside. Philadelphia: WB Saunders, 1995:96–103.

41. DiFrancesco D. Dual allosteric modulation of pacemaker (f) channels by cAMP and voltage in rabbit SA node. J Physiol (Lond) 1999; 515:367–376.

42. Hille B. Ionic channels of excitable membranes, 2nd ed. Sunderland: Sinauer, 1992.

43. Clapham DE. Not so funny anymore: pacing channels are cloned. Neuron 1998; 21:5–7.

44. Aidley DJ, Stanfield PR. Ion channels. Molecules in action. Cambridge: Cambridge University Press, 1996.

45. Stuhmer W, Conti F, Suzuki H, Wang X, Noda M, Yahagi N, Kubo H, Numa S. Structural parts involved in activation of the sodium channel. Nature 1989; 339:597–603.

46. Papazian DM, Timpe LC, Jan YN, Jan LY. Alteration of voltage-dependence of Shaker potassium channel by mutations in the S4 sequence. Nature 1991; 349:305–310.

47. Aggarwal SK, MacKinnon R. Contribution of the S4 segment to gating charge in the Shaker K$^+$ channel. Neuron 1996; 16:1169–1177.

48. Larsson HP, Baker OS, Dhillon DS, Isacoff EY. Transmembrane movement of the Shaker K$^+$ channel S4. Neuron 1996; 16:387–397.

49. Starace DM, Stefani E, Bezanilla F. Voltage-dependent proton transport by the voltage sensor of the Shaker K$^+$ channel. Neuron 1997; 19:1319–27.

50. Yellen G, Jurman ME, Abramson T, MacKinnon R. Mutations affecting internal TEA blockade identify the probable pore-forming region of a K$^+$ channel. Science 1991; 251:939–942.

51. Hartmann HA, Kirsch GE, Drewe JA, Taglialatela M, Joho RH, Brown AM. Exchange of conduction pathways between two related K$^+$ channels. Science 1991; 251:942–944.

52. Doyle DA, Cabral JM, Pfuetzner RA, Kuo A, Gulbis JM, Cohen SL, Chait BT, MacKinnon R. The structure of the potassium channel: molecular basis of K$^+$ conduction and selectivity. Science 1998; 280:69–77.

53. Hoshi T, Zagotta WN, Aldrich RW. Two types of inactivation in Shaker K$^+$ channels: effects of alterations in the carboxy-terminal region. Neuron 1991; 7:547–56.

54. Zagotta WN, Hoshi T, Aldrich RW. Restoration of inactivation in mutants of Shaker potassium channels by a peptide derived from ShB. Science 1990; 250:568–571.

55. Liu Y, Jurman ME, Yellen G. Dynamic rearrangement of the outer mouth of a K$^+$ channel during gating. Neuron 1996; 16:859–867.

56. Rasmusson RL, Morales MJ, Wang S, Liu S, Campbell DL, Brahmajothi MV, Strauss HC. Inactivation of voltage-gated cardiac K$^+$ channels. Circ Res 1998; 20:739–50.

57. Hartmann HA, Tiedeman AA, Chen SF, Brown AM, Kirsch GE. Effects of III–IV linker mutations on human heart Na$^+$ channel inactivation gating. Circ Res 1994; 75:114–122.

58. Patton DE, West JW, Catterall WA, Goldin AL. Amino acid residues required for fast Na$^+$-channel inactivation: charge neutralizations and deletions in the II–IV linker. Proc Natl Acad Sci USA 1992; 89:10905–10909.

59. Makitka, XX, Bennett PB, George AL Jr. Multiple domains contribute to the distinct inactivation properties of human heart and skeletal muscle Na$^+$ channels. Circ Res 1996; 78:244–252.

60. Roden DM, George AL Jr. Structure and function of cardiac sodium and potassium channels. Am J Physiol 1997; 273 (Heart Circ Physiol 42): H511–H525.

61. McPhee JC, Ragsdale DS, Scheuer T, Catterall WA. A critical role for transmembrane segment IVS6 of the sodium channel alpha subunit in fast inactivation. J Biol Chem 1995; 270:12025–12034.

62. Tang L, Kallen RG, Horn R. Role of an S4–S5 linker in sodium channel inactivation probed by mutagenesis and a peptide blocker. J Gen Physiol 1996; 108:89–104.

63. Peterson BZ, DeMaria CD, Yue DT. Calmodulin is the Ca^{2+} sensor for Ca^{2+}-dependent inactivation of L-type calcium channels. Neuron 1999; 22:549–558.

64. Zühlke RD, Pitt GS, Deisseroth K, Tsien RW, Reuter H. Calmodulin supports both inactivation and facilitation of L-type calcium channels. Nature 1999; 399:159–161.

65. Rettig J, Heinemann SH, Wunder F, Lorra C, Parcej DN, Dolly JO, Pongs O. Inactivaiton properties of voltage-gated K$^+$ channels altered by presence of beta-subunit. Nature 1994; 369:289–94.

66. Morales MJ, Castellino RC, Crews AL, Rasmusson RL, Strauss HC. A novel beta subunit increases rate of inactivation of specific voltage-gated potassium channel alpha subunits. J Biol Chem 1995; 270:6272–6277.

67. Xu J, Li M. Auxilliary subunits of Shaker-type potassium channels. Trends Cardiovasc Med 1998; 8:229–234.

68. M Noda, Shimizu S, Tanabe T, Takai T, Kayano T, Ikeda T, Takahashi H, Nakayama H, Kanaoka Y, Minamino N, Kangawa K, Matsuo H, Raftery MA, Hirose T, Inayama S,

Hayashida H, Miyata T, Numa S. Primary structure of Electrophorus electricus sodium channel deduced from cDNA sequence. Nature 1984; 312:121–127.

69. Smith MR, Smith RD, Plummer NW, Meisler MH, Goldin AL. Functional analysis of the mouse Scn8a sodium channel. J Neurosci 1998; 18:6093–6102.

70. Plummer NW, Galt J, Jones JM, Burgess DL, Sprunger LK, Kohrman DC, Meisler MH. Exon organization, coding sequence, physical mapping, and polymophic intragenic markers for the human neuronal sodium channel gene SCN8A. Genomics 1998; 54:287–296.

71. George AL Jr, Varkony TA, Drabkin HA, Han J, Knops JF, Finley WH, Brown GB, Ward DC, Haas M. Assignment of the human heart tetrodotoxin-resistant voltage-gated Na⁺ channel alpha-subunit gene (SCN5A) to band 3p21. Cytogenet Cell Genet 1995; 68:67–70.

72. Satin J, Kyle JW, Chen M, Bell P, Cribbs LL, Fozzard HA, Rogart RB. A mutant of TTX-resistant cardiac sodium channels with TTX-sensitive properties. Science 1992; 256:1202–1205.

73. An RH, Wang XL, Kerem B, Benhorin J, Medina A, Goldmit M, Kass RS. Novel LQT-3 mutation affects Na⁺ channel activity through interactions between alpha- and betal-subunits. Circ Res 1998; 83:141–6.

74. Catterall WA, Seager MJ, Takahashi M. Molecular properties of dihydropyridine-sensitive calcium channels in skeletal muscle. J Biol Chem 19xx; 263:3533–3538.

75. Hullin R, Biel M, Flockerzi V, Hofmann F. Tissue-specific expression of calcium channels. Trends Cardiovasc Med 1993; 3:48–53.

76. Felix R, Gurnett CA, DeWaard M, Campbell KP. Dissection of functional domains of the voltage-dependent Ca²⁺ channel alpha2delta subunit. J Neurosci 1997; 17:6884–6891.

77. Perez-Reyes E, Schneider T. Molecular biology of calcium channels. Kidney Int 1995; 48:1111–1124.

78. Cribbs LL, Lee JH, Yang J, Satin J, Zhang Y, Daud A, Barclay J, Williamson MP, Fox M, Rees M, Perez-Reyes E. Cloning and characterization of alpha1H from human heart, a member of the T-type Ca²⁺ channel gene family. Circ Res 1998; 13:103–109.

79. Perez-Reyes E, Cribbs LL, Daud A, Lacerda AE, Barclay J, Williamson MP, Fox M, Rees M, Lee JH. Molecular characterization of a neuronal low-voltage-activated T-type calcium channel. Nature 1998; 391:896–900.

80. Roberds SL, Tamkun MM. Cloning and tissue-specific expression of five voltage-gated potassium channel cDNAs expressed in rat heart. Proc Natl Acad Sci USA 1991; 88:1798–1802.

81. Tamkun MM, Knoth KM, Walbridge JA, Kroemer H, Roden DM, Glover DM. Molecular cloning and characterization of two voltage-gated K⁺ channel cDNAs from human ventricle. FASEB J 1991; 5:331–337.

82. Takumi T, Ohkubo H, Nakanishi S. Cloning of a membrane protein that induces a slow voltage-gated potassium current. Science 1988; 242:1042–1045.

83. Covarrubias M, Wei A, Salkoff L. Shaker, Shal, Shab and Shaw express independent K⁺ current systems. Neuron 1991; 7:763–773.

84. JE Dixon, McKinnon D. Quantitative analysis of potassium channel mRNA expression in atrial and ventricular muscle of rats. Circ Res 1994; 75:252–260.

85. MV Brahmajothi, Morales MJ, Liu S, Rasmusson RL, Campbell DL, Strauss HC. In situ hybridization reveals extensive diversity of K⁺ channel mRNA in isolated ferret cardiac myocytes. Circ Res 1996; 78:1083–1089.

86. Han W, Yue L, Nattel S. Unusual properties of transient outward current in canine cardiac Purkinje cells point to a unique molecular basis. Circulation 1998; 98: (suppl I) I–611.

87. Kramer JW, Post MA, Brown AM, Kirsch GE. Modulation of potassium channel gating by co-expression of Kv2,1 with regulatory Kv5.1 or Kv6.1 alpha-subunits. Am J Physiol 1998; C:1501–1510.

88. Hugnot JP, Salinas M, Lesage F, Guillemare E, Weille J de, Heurteaux C, Mattei MG, Lazdunski M. Kv8.1, a new neuronal potassium channel subunit with specific inhibitory properties towards Shab and Shaw channels. EMBO J 1996; 15:3322–3331.

89. Salinas M, Weille J de, Guillemare E, Lazdunski M, Hugnot JP. Modes of regulation of shab K⁺ channel activity by the Kv8.1 subunit. J Biol Chem 1997; 272:8774–8780.

90. Salinas M, Duprat F, Heurteaux C, Hugnot JP, Lazdunski M. New modulatory alpha subunits for mammalian Shab K⁺ channels. J Biol Chem 1997; 272:24371–24379.

91. Warmke JW, Ganetzky B. A family of potassium channel genes related to eag in Drosophila and mammals. Proc Natl Acad Sci USA 1991; 91:1560–1562.

92. Sanguinetti MC, Jiang C, Curran ME, Keating MT. A mechanistic link between an inherited and an acquired cardiac arrhythmia: HERG encodes the IKr potassium channel. Cell 1995; 81:299–307.

93. Sanguinetti MC, Curran ME, Spector PS, Keating MT. Spectrum of HERG K⁺-channel dysfunction in an inherited cardiac arrhythmia. Proc Natl Acad Sci (USA) 1996; 93:2208–2212.

94. Sanguinetti MC, Curran ME, Zou A, Shen J, Spetor PS, Atkinson DL, Keating MT. Coassembly of K(V)LQT1 and minK (IsK) proteins to form cardiac I(Ks) potassium channel. Nature 1996; 384:80–83.

95. Splawski I, Tristani-Firouzi M, Lehmann MH, Sanguinetti MC, Keating MT. Mutations in the hminK gene cause long QT syndrome and suppress I$_{Ks}$ function. Nature Genetics 1997; 17:338–340.

96. Barhanin J, Lesage F, Guillemare E, Fink M, Lazdunski M, Romey G. K(V)LQT1 and IsK (minK) proteins associate to form the I$_{Ks}$ cardiac potassium current. Nature 1996; 384:78–80.

97. Chouabe C, Neyroud N, Guicheney P, Lazdunski M, Romey G, Barhanin J. Properties of KvLQT1 K⁺ channel mutations in Romano-Ward and Jervell and Lange-Nielsen inherited cardiac arrhythmias. EMBO 1997; 16:5472–5479.

98. Romey G, Attali B, Chouabe C, Abitbol I, Guillemare E, Rarhanin J, Lazdunski M. Molecular mechanism and functional significance of the minK control of the KvLQT1 channel activity. J Biol Chem 1997; 272:16713–16716.

99. M Sheng, Liao YJ, Jan YN, Jan LY. Presynaptic A-current based on heteromultimeric K⁺ channels detected in vivo. Nature 1993; 365:72–75.

100. Po S, Roberds S, Snyders DJ, Tamkun MM, Bennett PB. Heteromultimeric assembly of human potassium channels. Circ Res 1993; 72:1326–1336.

100a. Thompson JD, Higgins DG, Gibson TJ. CLUSTAL W: improving the sensitivity of progressive multiple sequence alignment through sequence weighting, position-specific gap penalties and weight matrix choice. Nucleic Acids Res 1994; 22:4673–4680.

100b. Studier JA, Keppler KJ. A note on the neighbor-joining algorithm of Saitou and Nei. Mol Biol Evol 1988; 5:729–731.

101. Santoro B, Liu DT, Yao H, Bartsch D, Kandel ER, Siegelbaum SA, Tibbs GR. Identification of a gene encoding a hyperpolarization-activated pacemaker channel of brain. Cell 1998; 29:717–729.

102. Gauss R, Seifert R, Kaupp UB. Molecular identification of a hyperpolarization-activated channel in sea urchin sperm. Nature 1998; 393:583–587.

103. Ludwig, A, Zong X, Jeglitsch M, Hofmann F, Biel M. A family of hyperpolarization-activated mammalian cation channels. Nature 1998; 393:587–591.

104. Kubo Y, Baldwin TJ, Jan YN, Jan LY. Primary structure and functional expression of a mouse inward rectifier potassium channel. Nature 1993; 362:127–133.

105. N Inagaki, Gonoi T, Clement JP, Namba N, Inazawa J, Gonzalez G, Aguilar-Bryan L, Seino S, Bryan J. Reconstitution of I$_{KATP}$: an inward rectifier subunit plus the sulfonyl receptor. Science 1995; 270:1166–1170.

106. N Inagaki, Gonoi T, Clement JP, Wang CZ, Aguilar-Bryan L, Bryan J, Seino S. A family of sulfonylurea receptors determines the pharmacological properties of ATP-sensitive K⁺ channels. Neuron 1996; 16:1011–1017.

107. Babenko AP, Gonzalez G, Aguilar-Bryan L, Bryan J. Reconstituted human cardiac KATP channels: functional identity with the native channels from the sarcolemma of human ventricular cells. Circ Res 1998; 83:1132–43.

108. Kubo Y, Reuveny E, Slesinger PA, Jan YN, Jan LY. Primary structure and functional expression of a rat G-protein-coupled muscarinic potassium channel. Nature 1993; 364:802–806.

109. Krapivinsky G, Gordon EA, Wickman K, Velimirovic B, Krapivinsky L, Clapham DE. The G-protein-gated atrial K$^+$ channel I$_{KACh}$ is a heteromultimer of two inwardly rectifying K$^+$-channel proteins. Nature 1995; 374:135–141.

110. Corey S, Krapivinsky G, Krapivinsky L, Clapham DE. Number and stoichiometry of subunits in the native atrial G-protein-gated K$^+$ channel, I$_{KACh}$, J Biol Chem 1998; 273:5271–5278.

111. Corey S, Clapham DE. Identification of native atrial G-protein-regulated inwardly rectifying K$^+$ (GIRK4) channel homomultimers. J Biol Chem 1998; 273:27499–27504.

112. Lesage F, Guillemare E, Fink M, Duprat F, Lazdunski M, Romey G, Barhanin J. TWIK-1, a ubiquitous human weakly inward rectifying K$^+$ channel with a novel structure. EMBO J 1996; 15:1004–1011.

113. Lesage F, Lauritzen I, Duprat F, Reyes R, Fink M, Heurteaux C, Lazdunski M. The structure, function and distribution of the mouse TWIK-1 K$^+$ channel. FEBS Lett 1997; 402:28–32.

114. Chavez RA, Gray AT, Zhao BB, Kindler CH, Mazurek MJ, Mehta Y, Forsayeth JR, Yost CS. TWIK-2, a new weak inward rectifying member of the tandem pore domain potassium channel family. J Biol Chem 1999; 274:7887–7892.

115. Kim D, A Fujita, Y Horio, Y Kurachi. Cloning and functional expression of a novel cardiac two-pore background K$^+$ channel (cTBAK-1). Circ Res 1998; 82:513–518.

116. Scott VE, Rettig J, Parcej DN, Keen JN, Findlay JB, Pongs O, Dolly JO. Primary structure of a beta subunit of alpha-dendrodotoxin-sensitive K$^+$ channels from bovine brain. Proc Natl Acad Sci 1994; 91:1637–1641.

117. Majumder K, Biasi M De, Wang Z, Wible BA. Molecular cloning and functional expression of a novel potassium channel beta-subunit from human atrium. FEBS Lett 1995; 361:13–16.

118. Leicher T, Bahring R, Isbrandt D, Pongs O. Coexpression of the KCNA3B gene product with Kv1.5 leads to a novel A-type potassium channel. J Biol Chem 1998; 273:35095–35101.

119. Accili EA, Kiehn J, Yang Q, Wang Z, Brown AM, Wible BA. Separable Kvβ subunit domains alter expression and gating of potassium channels. J Biol Chem 1997; 272:25824–25831.

120. Xu J, Li M. Kvβ2 inhibits the Kvβ1-mediated inactivation of K$^+$ channels in transfected mammalian cells. J Biol Chem 1997; 272:11728–11735.

121. Accili EA, Kiehn J, Wible BA, Brown AM. Interactions among inactivating and noninactivating Kvβ subunits, and Kvα1.2, produce potassium currents with intermediate inactivation. J Biol Chem 1997; 272:28232–28236.

122. Shamotienko OG, Parcej DN, Dolly JO. Subunit combinations defined for K$^+$ channel Kv1 subtypes in synaptic membranes from bovine brain. Biochemistry 1997; 36:8195–8201.

123. Accili EA, Kuryshev YA, Wible BA, Brown AM. Separable effects of human Kvβ1.2 N- and C-termini on inactivation and expression of human Kv1.4. J Physiol (Lond) 1998; 512:326–336.

124. Morales MJ, Wee JO, Wang S, Strauss HC, Rasmusson RL. The N-terminal domain of a K$^+$ channel β subunit increases the rate of C-type inactivation from the cytoplasmic side of the channel. Proc Natl Acad Sci USA 1996; 93:15119–15123.

125. Shi G, Nakahira K, Hammond S, Rhodes KJ, Schechter LE, Trimmer JS. β subunits promote K$^+$ channel surface expression through effects early in biosynthesis. Neuron 1996; 16:843–852.

126. Gulbis JM, Mann S, MacKinnon R. Structure of a voltage-dependent K$^+$ channel beta subunit. Cell 1999; 97:943–952.

127. McCormack T, McCormack K. Shaker K$^+$ channel beta subunits belong to an NAD(P)H-dependent oxidoreductase superfamily. Cell 1994; 79:1133–135.

128. Abbott GW, Sesti F, Splawski I, Buck ME, Lehmann MH, Timothy KW, Keating MT, Goldstein SA. MiRP1 forms I$_{Kr}$ potassium channels with HERG and is associated with cardiac arrhythmia. Cell 1999; 97:175–187.

129. Sicouri S, Antzelevitch C. A subpopulation of cells with unique electrophysiological properties in the deep subepicardium of the canine ventricle. The M cell. Circ Res 1991; 68:1729–1741.

130. Liu D-W, Antzelevitch C. Characteristics of the delayed rectifier current (I_{Kr} and I_{Ks}) in canine ventricular epicardial, midmyocardial, and endocardial myocytes. A weaker I_{Ks} contributes to the longer action potential of the M cell. Circ Res 1995; 76:351–365.

131. Kodama I, Nikmaram MR, Boyett MR, Suzuki R, Honjo H, Owen JM. Regional differences in the role of the Ca^{2+} and Na^+ currents in pacemaker activity in the sinoatrial node. Am J Physiol 1997; 272:H2793–806.

132. Cohen SA. Immunocytochemical localization of rH1 sodium channel in adult rat heart atria and ventricle: presence in terminal intercalated disks. Circulation 1996; 94:3083–3086.

133. Spach MS. Discontinuous cardiac conduction: Its origin in cellular connectivity with long-term adaptive changes that cause arrhythmias. In: Spooner PM, Joyner RW, Jaliife J, eds. Discontinuous Conduction in the Heart. Armonk, NY: Futura Publishing, 1997:5–51.

134. Dolphin AC. L-type calcium channel modulation. Adv Second Messenger Phosphoprotein Res 1999; 33:153–77.

135. Doerr T, Denger R, Trautwein W. Calcium currents in single SA nodal cells of the rabbit heart studied with action potential clamp. Pflugers Arch 1989; 413:599–603.

136. Y Hirano, HA Fozzard, CT January. Characteristics of L- and T-type Ca^{2+} currents in canine cardiac Purkinje cells. Am J Physiol 1989; 256:H478–H492.

137. Vassort G, Alvarez J. Cardiac T-type calcium current: Pharmacology and roles in cardiac tissues. J Cardiovasc Electrophysiol 1994; 5:376–393.

138. Zhou Z, January CT. Both T- and L-type Ca^{2+} channels can contribute to excitation-contraction coupling in cardiac Purkinje cells. Biophys J 1998; 74:1830–1839.

139. Fedida D, Giles WR. Regional variations in action potentials and transient outward current in myocytes isolated from rabbit left ventricle. J Physiol (Lond) 1991; 442:191–209.

140. Furukawa T, Myerburg RJ, Furukawa N, Bassett AL, Kimura S. Differences in transient outward currents of feline endocardial and epicardial myocytes. Circ Res 1990; 67:1287–1291.

141. Benitah JP, Gomez AM, Bailly P, DA Ponte J-P, Berson G, Delgado C, Lorente P. Heterogeneity of the early outward current in ventricular cells isolated from normal and hypertrophied rat hearts. J Physiol (Lond) 1993; 469:111–138.

142. Wettwer E, Amos G, Gath J, Zerkowski H-R, Reidmeister J-C, Ravens U. Transient outward current in human and rat ventricular myocytes. Cardiovasc Res 1993; 27:1662–1669.

143. Wettwer E, Amos GJ, Posival H, Ravens U. Transient outward current in human ventricular myocytes of subepicardial and subendocardial origin. Circ Res 1994; 75:473–482.

144. Näbauer M, Beuckelmann DJ, Erdmann E. Characteristics of transient outward current in human ventricular myocytes from patients with terminal heart failure. Circ Res 1993; 73:386–394.

145. Näbauer M, Beuckelmann DJ, Überfuhr P, Steinbeck G. Regional differences in current density and rate-dependent properties of the transient outward current in subepicardial and subendocardial myocytes of human left ventricle. Circulation 1996; 93:168–177.

146. Amos GJ, Wettwer E, Metzger F, Li Q, Himmel HM, Ravens U. Differences between outward currents of human atrial and subepicardial ventricular myocytes. J Physiol (Lond) 1996; 491:31–50.

147. Yamashita T, Nakajima T, Hazama H, Hamada E, Murakawa Y, Sawada H, Omata M. Regional differences in transient outward current density and inhomogeneities of repolarization in rabbit right atrium. Circulation 1995; 92:3061–3069.

148. Li G-R, Feng J, Yue L, Carrier M. Transmural heterogeneity of action potentials and I_{to1} in myocytes isolated from the human right ventricle. Am J Physiol 1998; 275:H369–H377.

149. Brahmajothi MV, Campbell DL, Rasmusson RL, Morales MJ, Trimmer JS, Nerbonne JM, Strauss HC. Distinct transient outward potassium current (I_{to}) phenotypes and distribution of

fast-inactivating potassium channel alpha subunits in ferret left ventricular myocytes. J Gen Physiol 1999; 113:581–600.

150. Wang Z, Fermini B, Nattel S. Delayed rectifier outward current and repolarization in human atrial myocytes. Circ Res 1993; 73:276–285.

151. Sanguinetti MC, Johnson JH, Hammerland LG, Kelbaugh PR, Volkmann RA, Saccomano NA, Mueller AL. Heteropodatoxins: Peptides isolated from spider venom that block Kv4.2 potassium channels. Mol Pharmacol 1997; 51:491–498.

152. Veldkamp MW, van Ginneken ACG, Opthof T, Bouman LN. Delayed rectifier channels in human ventricular myocytes. Circulation 1995; 92:3497–3504.

153. Konarzewska H, Peeters GA, Sanguinetti MC. Repolarizing K^+ currents in nonfailing human hearts. Similarities between right septal subendocardial and left subepicardial ventricular myocytes. Circulation 1995; 92:1179–1187.

154. Mays DJ, Foose JM, Philipson LH, Tamkun MM. Localization of the Kv1.5 K^+ channel protein in explanted cardiac tissue. J Clin Invest 1995; 96:282–292.

155. Shibata EF, Giles WR. Ionic currents that generate the spontaneous diastolic depolarization in individual cardiac pacemaker cells. Proc Natl Acad Sci USA 1985; 82:7796–7800.

156. Giles WR, Imaizumi Y. Comparison of potassium currents in rabbit atrial and ventricular cells. J Physiol (Lond) 1988; 405:123–145.

157. Dixon JE, Shi WM, Wang HS, McDonald C, Yu H, Wymore RS, Cohen IS, McKinnon D. Role of the Kv4.3 K^+ channel in ventricular muscle: a molecular correlate for the transient outward current. Circ Res 1996; 79:659–668.

158. Brahmajothi MV, Morales MJ, Reimer KA, Strauss HC. Regional localization of ERG, the channel protein responsible for the rapid component of the delayed rectifier, K^+ current in the ferret heart. Circ Res 1997; 81:128–135.

159. Wang Q, Shen J, Splawski I, Atkinson D, Li Z, Robinson JL, Moss AJ, Towbin JA, Keating MT. SCN5A mutations associated with an inherited cardiac arrhythmia, long QT syndrome. Cell 1995; 80:805–811.

160. Wang DW, Yazawa K, George AL Jr, Bennett PB. Characterization of human cardiac Na^+ channel mutations in the congenital long QT syndrome. Proc Natl Acad Sci USA 1996; 93:13200–13205.

161. Priori SG, Napolitano C, Cantu F, Brown AM, Schwartz PJ. Differential response to Na^+ channel blockade, beta-adrenergic stimulation, and rapid pacing in a cellular model mimicking the SCN5A and HERG defects present in the long-QT syndrome. Circ Res 1996; 78:1009–1015.

162. Schwartz PJ, Priori SG, Locati EH, Napolitano C, Cantu F, Towbin JA, Keating MT, Hammoude H, Brown AM, Chen LS. Differential response to Na^+ channel blockade, beta-adrenergic stimulation, and rapid pacing in a cellular model mimicking the SCN5A and HERG defects present in the long-QT syndrome. Circ Res 1996; 78:1009–15.

163. Chen Q, Kirsch GE, Zhang D, Brugada R, Brugada J, Brugada P, Potenza D, Moya A, Borggrefe M, Breithardt G, Ortiz-Lopez R, Wang Z, Antzelevitch C, O'Brien RE, Schulze-Bahr E, Keating MT, Towbin JA, Wang Q. Genetic basis and molecular mechanism for idiopathic ventricular fibrillation. Nature 1998; 392:293–296.

164. Tseng GN, Jiang M, Yao JA. Reverse use dependence of Kv4.2 blockade by 4-aminopyridine. J Pharmacol Exp Ther 1996; 279:865–76.

165. Campbell DL, Qu Y, Rasmusson RL, Strauss HC. The calcium-independent transient outward potassium current in isolated ferret right ventricular myocytes. II. Closed state reverse use-dependent block by 4-aminopyridine. J Gen Physiol 1993; 101:603–26.

166. Takimoto K, Li D, Hershman KM, Li P, Jackson EK, Levitan ES. Decreased expression of Kv4.2 and novel Kv4.3 K^+ channel subunit mRNAs in ventricles of renovascular hypertensive rats. Circ Res 1997; 81:533–539.

167. Barry DM, Nerbonne JM. Myocardial potassium channels: electrophysiological and molecular diversity. Ann Rev Physiol 1996; 58:363–394.

168. Barry DM, Trimmer JS, Merlie JP, Nerbonne JM. Differential expression of voltage-gated K^+ channel subunits in adult rat heart. Relation to functional K^+ channels? Circ Res 1995; 77:361–9.

169. Verheijck EE, van Ginneken AC, Bourier J, Bouman LN. Effects of delayed rectifier current blockade by E-4031 on impulse generation in single sinoatrial nodal myocytes of the rabbit. Circ Res 1995; 76:607–15.

170. Shibasaki T. Conductance and kinetics of delayed rectifier potassium channels in nodal cells of the rabbit heart. J Physiol (Lond) 1987; 387:227–250.

171. Johnson JP Jr, Mullins FW, Bennett PB. Human ether-à-go-go-related gene K+ channel gating probed with extracellular Ca^{2+}: Evidence for two distinct voltage sensors. J Gen Physiol 1999; 113:565–580.

172. Wang S, Liu S, MJ Morales, HC Strauss, RL Rasmusson. A quantitative analysis of the activation and inactivation linetics of HERG expressed in Xenopus oocytes. J Physiol (Lond) 1997; 502:45–60.

173. Spector PS, Curran ME, Keating MT, Sanguinetti MC. Class III antiarrhythmic drugs block HERG, a human cardiac delayed rectifier K^+ channel. Open-channel block by methanesulfonanilides. Circ Res 1996; 78:499–503.

174. Lees-Miller JP, Kondo C, Wang L, Duff HJ. Electrophysiological characterization of an alternatively processed ERG K^+ channel in mouse and human hearts. Circ Res 1997; 81:719–726.

175. London B, Trudeau MC, Newton KP, Beyer AK, Copeland NC, Gilbert DJ, Jenkins NA, Satler CA, Robertson GA. Two isoforms of the mouse ether-a-go-go-related gene coassemble to form channels with properties similar to the rapidly activating component of the cardiac delayed rectifier K^+ channel. Circ Res 1997; 81:870–878.

176. Splawski I, Shen J, Timothy KW, Vincent GM, Lehmann MH, Keating MT. Genomic structure of three long QT syndrome genes: KVLQT1, ERG, and KCNE1. Genomics 1996; 51:86–97.

177. Kupershmidt S, Snyders DJ, Raes A, Roden DM. A K^+ channel splice variant common in human heart lacks a C-terminal domain required for expression of rapidly activating delayed rectifier current. J Biol Chem 1998; 272:27231–27235.

178. Nakajima T, Furukawa T, Tanaka T, Katayama Y, Nagai R, Nakamura Y, Hiraoka M. Novel mechanism of HERG current suppression in LQT2: shift in voltage dependence of HERG inactivation. Circ Res 1998; 83:415–22.

179. Morais JH, Lee A, Cohen SL, Chait BT, Li M, MacKinnon R. Crystal structure and functional analysis of the HERG potassium channel N terminus: a eukaryotic PAS domain. Cell 1998; 95:649–655.

180. Chen J, Zou A, Splawski I, Keating MT, Sanguinetti MC. Long QT syndrome-associated mutations in the per-arnt-Sim (PAS) domain of HERG potassium channels accelerate channel deactivation. J Biol Chem 1999; 274:10113–10118.

181. Wei J, Yang IC-H, Tapper AR, Murray KT, Viswanathan P, Rudy Y, Bennett PB, Norris K, Balser J, Roden DM, George AL Jr. KCNE1 polymorphism confers risk of drug-induced long QT syndrome by altering kinetic properties of IKs potassium channels (abstr). Circulation 2000; 102 (in press).

182. Barhanin J, Attali B, Lazdunski M. IKS, a slow, and intriguing cardiac K^+ channel and its associated long QT diseases. Trends Cardiovasc Med 1998; 8:207–214.

183. Tristani-Firouzi M, Sanguinetti MC. Voltage-dependent inactivation of the human K^+ channel KvLQT1 is eliminated by association with minimal K^+ channel (minK) subunits. J Physiol (Lond) 1998; 510:37–45.

184. Jiang M, Tseng-Crank J, Tseng G-N. Suppression of slow delayed rectifier current by a truncated isoform of KvLQT1 cloned from normal human heart. J Biol Chem 1997; 272:24109–24112.

185. Nakamura M, Watanabe H, Kubo Y, Yokoyama M, Matsumoto T, Sasai H, Nishi Y. KQT2, a

new putative potassium channel family produced by alternative splicing. Isolation, genomic structure, and alternative splicing of the putative potassium channels. Receptors Channels 1998; 5:255–71.

186. Feng J, Wible B, Li GR, Wang Z, Nattel S. Antisense oligodeoxynucleotides directed against Kv1.5 mRNA specifically inhibit ultrarapid delayed rectifier K$^+$ current in cultured adult human atrial myocytes. Circ Res 1997; 80:572–579.

187. Splawski I, Shen J, Timothy KW, Lehmann MH, Priori S, Robinson JL, Moss AJ, Schwartz PJ, Towbin JA, Vincent GM, Keating MT. Spectrum of mutations in long QT syndrome genes KvLQT1, HERG, SCN5A, KCNE1, and KCNE2. Circulation 2000; 102 (in press).

7

Sinoatrial Node and Impulse Initiation

RICHARD B. ROBINSON

Columbia University, New York, New York

DARIO DIFRANCESCO

University of Milan, Milan, Italy

I. INTRODUCTION

Pacemaking originates in a specialized region of the heart, the sinoatrial node (SAN). In the human heart, the SAN is located in the terminal groove lateral to the superior vena-caval–atrial junction (1). In contrast, in the rabbit heart the SAN is located superficially to the crista terminalis of the right atrium, in the area surrounding the superior vena cava orifice. Because the rabbit SAN forms a thin superficial sheet that is relatively easy to isolate from surrounding atrial tissue (2), it is commonly employed in electrophysiological studies and much of our present information on its function was obtained in studies on the rabbit. A comparison of the two is illustrated in Figure 7.1.

The SAN action potential characteristically differs from that of working myocardium in two ways. First, the SAN action potential has a maximal upstroke velocity (i.e., rate of maximal depolarization) that is quite slow compared to those of ventricular myocytes (2–4 V/s in the central node versus ~200 V/s in ventricular myocytes) with depolarizing current being largely carried by Ca^{2+}. Second, the cells of the SAN have the unique ability to generate spontaneous action potentials at a constant rate. These are characterized by a slow diastolic (or "pacemaker") depolarization phase, which brings the membrane potential from the most hyperpolarized level reached after an action potential [maximum diastolic potential (MDP)] up to the threshold level required to initiate a new action potential. These differences are illustrated in Figure 7.2.

The characteristic transmembrane action potentials of the SAN reflect the fact that the ion currents underlying them differ from those in the ventricle. These differences relate to both the population of currents present in the SAN versus those in the ventricle and the

151

functional characteristics of the individual ion currents (i.e., magnitude, voltage, and time dependence). Further, the electrical characteristics of the SAN region are heterogeneous, with the action potential exhibiting a different configuration in central versus peripheral regions of the node. This chapter will first focus on the characteristics of the central SAN region. We will consider the individual ionic currents present in this area and their responsiveness to drugs and other agents that interact with autonomic receptors. Differences in cells from the peripheral region of the SAN, and the physiological implications of those differences, will then be addressed.

II. PACEMAKER ACTIVITY IN THE SA NODE

The rhythmic, spontaneous impulse initiation of SAN pacemaker cells results from a diastolic depolarization that is initiated immediately after repolarization of the preceding action potential [Fig. 7.2(A)]. This "slow diastolic" or "pacemaker" depolarization is typical of "automatic" cells and essential to their function. Since any depolarizing process requires a net inward current to be generated, the presence of slow diastolic depolarization implies an inward current being activated just prior to termination of the preceding action potential.

A great deal of research has gone into identifying the inward currents responsible for generation of spontaneous activity and other currents activated during the diastolic depolarization. The information gathered to date allows a detailed description of the mechanistic determinants of pacemaker activity, although the relative contributions of various com-

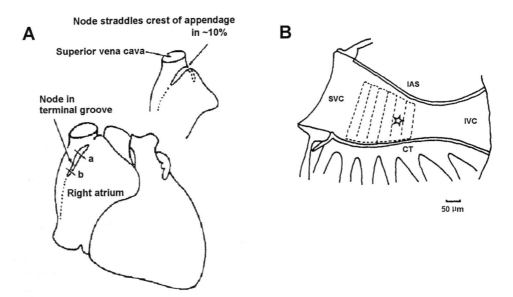

Figure 7.1 Anatomy of the SA node. (A) Location of the sinus node in the human heart. In most hearts the SAN is located in the terminal groove lateral to the superior cavoatrial junction, but in 10% of hearts it is a horseshoe-shaped structure straddling the node. Lines labeled (a) and (b) mark the body and tail of the node, respectively. (Reprinted from Ref 1.) (B) Illustration of the location of the SAN in a rabbit heart. The region generally dissected for SAN myocyte isolation is shown by the series of dashed lines, with the star marking the central nodal area. SVC: superior vena cava; IVC: inferior vena cava; IAS: interatrial septum; CT: crista terminalis. (Reprinted from Ref. 2.)

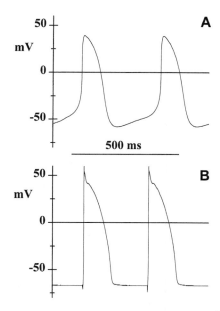

Figure 7.2 (A) Spontaneous action potentials recorded from a SAN myocyte of adult rabbit heart. Temperature was 34°C. (B) Action potential recorded from a rabbit ventricular myocyte, paced at a rate comparable to the spontaneous rate of the SAN myocyte. Temperature was 35°C. (Ventricular record provided courtesy of Mr. Thai Pham.)

ponents is still being debated (3–6). Currents characteristically expressed in pacing myocytes are schematically illustrated in Figure 7.3. These include:

1. A hyperpolarization-activated inward current, termed "pacemaker" (I_f) current, whose properties determine initiation and modulate rate control of the slow (diastolic) depolarization phase subsequent to action potential repolarization.
2. A delayed K^+ current that is responsible for repolarization and whose decay contributes to diastolic depolarization; it possibly results from the contribution of two components, one sensitive to the drug E-4031, such as the rapid (I_{Kr}) component in the ventricle, and one insensitive to E-4031, such as the slow ventricular (I_{Ks}) component (9).
3. Two Ca^{2+} currents (an L type and a T type), both contributing to the rapid depolarization phase of the action potential and to the last fraction of the slow depolarization phase.
4. Other time-independent and pump- or exchange-related components, including Na/Ca exchange current (see below and Ref. 7).

Examples of the major time-dependent currents recorded from a rabbit SAN myocyte are shown in Figure 7.4. When the membrane potential is changed from –35 mV to –45 or –55 mV, as it would be during repolarization after an action potential, a slowly increasing inward current, I_f, is observed [Fig. 7.4(A)]. When the membrane potential is then depolarized to –5 mV, as it would be during initiation of an action potential, a large and transient inward current, I_{Ca}, is observed [Fig. 7.4(A); displayed on expanded timebase in Fig. 7.4(B)], followed by a sustained outward current, I_K [Fig. 7.4(A)].

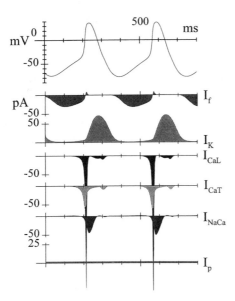

Figure 7.3 Computer simulation of ion currents active in a spontaneously active SAN myocyte. Top trace: action potential; bottom traces: time course of ion, exchange, and pump currents contributing to action potential generation, as indicated. The computation was performed by using the OXSOFT Heart computer program (version 4.2) as derived from the original SAN simulation of DiFrancesco and Noble (7). The parameters used were as in Ref. 8, with the following modifications: half activation of I_f: –64 mV; maximal delayed K^+ current: 300 pA; maximal pump current: 15 pA. The background Na^+ conductance was set to zero.

The paragraphs below summarize some of the mechanisms governing the slow diastolic depolarization responsible for impulse initiation in the SAN, and the basic properties of the major ion currents that play a role in this process.

III. SLOW DEPOLARIZATION: ION CURRENTS INVOLVED IN PACEMAKING

A. I_f Current

The I_f current of the SAN is a nonspecific cation current that is carried by Na^+ and K^+. The mixed cation permeability generates a current reversal potential near –20 mV, such that in physiological settings the current is inward and slowly activates upon hyperpolarization in a voltage range comprising that of the diastolic depolarization (–60 to –40 mV) (Fig. 7.3, second panel from top). Since the initial description in 1979 (10), its properties have been established with some detail at the multicellular (11,12), single-cell (2,13–16), single-channel (17,18), and most recently, molecular levels (19–22). I_f has features well suited for generating a depolarizing process in response to hyperpolarization entering the range of I_f activation, and, as described below, contributes essentially to the modulatory action of neurotransmitters on heart rate (4).

The study of I_f and its contribution to pacemaking is complicated by the fact that only a small amount of current is required to drive the diastolic depolarization. Also, I_f shows decay with time under some experimental conditions [referred to as "run-

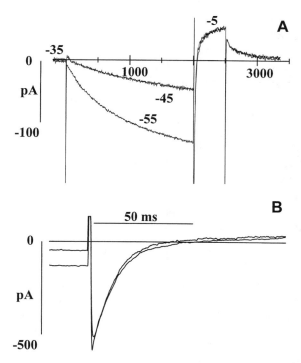

Figure 7.4 Time-dependent currents recorded in a rabbit SAN cell. (A) Hyperpolarizing steps from the holding potential of –35 mV elicit activation of the inward I_f current (2-s steps to –45 and –55 mV as indicated). Depolarization to –5 mV elicits onset of the "slow" inward current, consisting mostly of the L-type Ca^{2+} current, and of the "delayed" K^+ current, originally termed I_K (composed mostly of the rapid component I_{Kr}). Upon return to –35 mV, deactivation of the delayed K^+ current results in a decaying outward tail. Temperature 34°C. (B) Expanded time base of portion of traces at –5 mV showing activation and inactivation of the Ca^{2+} current.

down" (14)]. Nonetheless, the following is clear: in some studies of multicellular preparations or single cells, I_f activates on hyperpolarization from holding potentials of about –35 to –40 mV (10,11,16,23). These voltages are well within the diastolic range for SAN. Still other studies of similar preparations have shown activation occurring only on hyperpolarization from the –50 to –65 mV range (2,12), suggesting a much more peripheral role for this pacemaker current. In addition to the aforementioned run-down, a possible cause for this variability among studies is the control of I_f activation by intracellular cAMP (24); in settings where this second messenger is maintained at a physiological concentration range, I_f is activated in the more positive voltage range.

B. The Delayed K^+ Current

The delayed rectifier is the primary repolarizing current in the SAN, since there is little inward rectifying current (25,26), a fact that contributes to the high-input resistance of these cells and the requirement for such a small net inward current to generate the depolarization. It activates with a slow time course following action potential initiation and deactivates during repolarization (Fig. 7.3, third panel from top). Several recent studies have stressed the contribution of I_K and, in particular, of the rapid I_{Kr} component (defined as E-4031-sensitive current) to diastolic depolarization in the rabbit (27–29). At the same time, an E-4031-insensitive component has been noted, particularly with respect to contributions in the plateau range of voltages and repolarization (27). Further, potential species specificity is suggested by the observation that in the guinea pig sinus node the delayed rectifier appears to be predominantly the slower I_{Ks} (30). Differences in relative contribution of I_{Kr} and I_{Ks} would be expected to impact on both the contribution of the delayed recitfier to basal automaticity (since the two components exhibit distinct deactivation kinetics) and modulation by autonomic agonists (see below).

The time course of I_K deactivation following repolarization would result in a decreasing outward current. For this to account for the required net inward current during diastole, it must be combined with an inward current. The interplay between I_f activation and I_K decay during diastolic depolarization is such as to generate a nearly constant net inward current, which is the required condition for a depolarization of constant rate (7).

C. Background Inward Current

Some investigators have suggested that a background current, rather than I_f, might supply the required inward current in conjunction with a decaying I_K. A background current is by definition a time-independent current that changes in magnitude instantaneously with a change in voltage and retains that new value as long as the voltage remains constant. As such, it is difficult to distinguish from artifactual leakage current between the recording electrode and the cell. On the basis of indirect evidence extrapolated from data in other tissues and for consistency with previous models (7,31,32), a background component, inward in the pacemaker range and Na^+-dependent, has often been assumed and included in the reconstruction of nodal electrical activity. A Na^+-dependent current is observed in ion-substitution experiments when channel blockers are used to abolish all time-dependent components (33). However, the properties of this current contrast with evidence that no net inward current is recorded down to about −65 mV in beating cells (15). To date, no single-channel measurements have confirmed the existence of a "background" component with properties compatible with whole-cell data, leaving the nature of the net background current in SAN uncertain (34). The computation in Figure 7.3 was performed assuming no background inward current.

D. Ca^{2+} Currents

Early multicellular studies indicated that a Ca^{2+}-dependent current was elicited upon depolarization beyond a threshold of about −50 mV (35). It was originally proposed that this current was responsible for pacemaker generation (36,37) but this was not supported by current-clamp experiments, which indicated that Ca^{2+} entry can contribute to only the last fraction of pacemaker depolarization (38). Later, two Ca^{2+} inward components (L type and T type) were reported (39). The larger, long-lasting (L) component ($I_{Ca,L}$) activates near −30 mV, while the transient (T) component ($I_{Ca,T}$) activates at about −50 mV. Both Ca^{2+} currents are activated on depolarization with fast activation and, for the T component, inactivation kinetics, and their properties are appropriate for generation of the rapid action potential depolarization and upstroke (Fig. 7.3, fourth and fifth panels from top).

The T component of calcium flux has also been proposed to contribute to diastolic depolarization, on the basis of evidence that Ni^{2+}, a blocker of $I_{Ca,T}$, slows the diastolic depolarization rate (39). Currents activated on depolarization, such as the Ca^{2+} currents, can only contribute to diastolic depolarization if there is a sustained current in the diastolic potential range. Such a sustained current can result from either incomplete inactivation or a "window" component, defined as a voltage range where the activation and inactivation curves overlap (in other words, a voltage range where a small percentage of channels are open but not inactivated at steady state). While activation and inactivation curves for the L component overlap over a relatively wide range at depolarized voltages, there is disagreement as to whether overlap occurs for the T component (39,40). Further, evidence that a Ca^{2+}-dependent "background" component is only observed positive to –45 mV (15) argues against a substantial T-type window component contributing to diastolic depolarization in SAN cells.

Another potential inward current has been suggested by Guo et al. They have reported a "sustained inward current" in the diastolic range of voltages of both rabbit and guinea pig SAN myocytes, which may add to the inward components contributing to phase 4 depolarization (41,42). This current has peculiar characteristics in that its pharmacological and modulatory properties are indistinguishable from those of $I_{Ca,L}$, and yet it appears to be carried by Na^+. Since Na^+ permeation through L-type Ca^{2+} channels is known to occur only when external Ca^{2+} is lowered to micromolar concentrations (43), these data do not seem compatible with the idea that the "sustained inward current" corresponds to a special mode of permeation through L-type Ca^{2+} channels. Until single-channel data are collected, however, this remains an intriguing possibility.

E. Na/Ca Exchanger and Intracellular Calcium

The Na/Ca exchanger normally functions to extrude Ca from the cell in exchange for the entry of Na. Since one divalent Ca^{2+} ion is exchanged for every three monovalent Na^+ ions, the result is a net inward current in response to the Ca influx and SR Ca release occurring during an action potential (Fig. 7.3, sixth panel from top). The concept that this current, or other $[Ca]_i$-sensitive mechanisms, might contribute to SAN automaticity has been the focus of several recent studies. In cat myocytes, Lipsius and colleagues concluded that such a current contributed to automaticity in latent atrial pacemakers, but was less of a factor in primary SAN myocytes (44,45). Other laboratories have reported negative chronotropic effects of ryanodine in rabbit SAN myocytes (46,47). Since ryanodine depletes SR Ca stores, it would be expected to reduce Na/Ca exchange current, but both these studies identified additional effects of ryanodine on other currents (e.g., $I_{Ca,L}$) that could explain the negative chronotropic effect. In addition, an action potential clamp study has provided evidence of Ca-dependent modulation of current(s) active during diastolic depolarization (48). Thus, while the extent to which Na/Ca exchange current contributes to diastolic depolarization in primary SAN myocytes is uncertain, it appears likely that some Ca-dependent processes are present during this period.

IV. AUTONOMIC REGULATION OF ADULT SAN AUTOMATICITY

The SAN spontaneously generates action potentials at a constant rate, but also responds to the changing needs of the organism by increasing and decreasing heart rate in response to extrinsic signals. The most common signals are provided by the autonomic neurotransmitters acetylcholine (ACh) and norepinephrine. While the number of currents potentially contributing to diastolic depolarization and automaticity is large, the subset of currents that also are potential targets for autonomic agonists is more limited.

A. $I_{K,ACh}$

The action of vagal stimulation on heart rate was shown in early studies to be associated with membrane hyperpolarization and an increased K^+ permeability (49–51). A K^+ current activated by ACh ($I_{K,ACh}$) has been described in both amphibian and mammalian hearts, and its properties have been characterized in detail in intact cardiac tissue (52,53), single-cell (54,55), and single-channel (56–58) studies, as well as molecular reconstitution experiments (59). Activation of the current tends to accelerate repolarization and hyperpolarize membrane potential. Moreover, by generating a net outward current, it would tend to decrease diastolic depolarization and slow heart rate.

However, activation of $I_{K,ACh}$ is not the only mechanism, and is most likely not even the primary mechanism, by which vagal stimulation slows heart rate. In SAN cells, ACh has an additional strong inhibitory action on the current I_f, which occurs at 20-fold lower concentrations than those activating $I_{K,ACh}$ (60) (see Ref. 61 for an opposing view). Measurement of the dose–response relationships for I_f inhibition and $I_{K,ACh}$ activation indicates that in the concentration range up to 0.01–0.03 µM ACh acts selectively on I_f. Since at these concentrations ACh is indeed effective in slowing spontaneous rate, without inducing the membrane hyperpolarization characteristic of increasing $I_{K,ACh}$ [as is also true with moderate vagal stimulation (62)], I_f inhibition appears to play a key role in vagal control of rhythm.

B. Ca^{2+} Currents

The Ca^{2+} current also is inhibited by ACh in various cardiac preparations (63). However, the Ca^{2+} current of intact SAN is "remarkably insensitive" to ACh (64). Thus, in experiments where I_f and $I_{Ca,L}$ were recorded simultaneously, ACh inhibition of $I_{Ca,L}$ occurred at concentrations higher than those required for I_f inhibition and activation of $I_{K,ACh}$, (half-inhibition of $I_{Ca,L}$ estimated at ~2900 µM) and the maximal inhibition that could be experimentally achieved was only some 31% (65). Figure 7.5 [panels (B) and (D)] illustrates the differential sensitivity of I_f and $I_{Ca,L}$ to inhibition by ACh. During exposure to higher ACh concentrations or intense vagal stimulation, $I_{Ca,L}$ would indeed be expected to be inhibited, and $I_{K,ACh}$ would be activated. Under these conditions, the heart is at risk from sinus node conduction block or impulse failure. Interestingly, one situation where this could conceivably occur is early in development when the parasympathetic nerves have reached the heart, cholinesterase levels may be low (66), and sympathetic innervation is immature. Such a setting would optimize cholinergic inhibition and minimize sympathetic activation of currents. Here, the heart may be protected by a tetrodotoxin-sensitive Na current present only in the young SAN (67). The $I_{K,ACh}$-dependent hyperpolarization would relieve the partial inactivation of the Na current and thereby increase the safety factor for automatic impulse initiation by SAN cells.

The adrenergic agonists norepinephrine and epinephrine act at both α- and β-adrenergic receptors. However, unlike the situation in Purkinje fibers and ventricle, where α-adrenergic activation affects electrophysiological activity (68), there appears to be no α-adrenergic effect on the SAN (69). In contrast, β-adrenergic stimulation has several actions on ion currents participating in the generation of electrical activity of SAN cells. Stimulation of the Ca^{2+} current has been reported in multicellular preparations (10,70) and in single myocytes, where only the L component appears to be modulated (39). As in ventricular myocytes (71,72), activation of adenylyl cyclase and cAMP production are involved in this action. Epinephrine increases the fully activated Ca^{2+} current of SAN without appar-

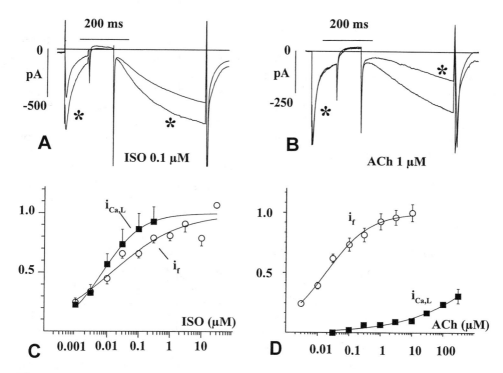

Figure 7.5 Action of isoproterenol and acetylcholine on Ca^{2+} current and I_f in rabbit SAN myocytes. (A) Traces recorded during depolarizing steps to 0 mV from a holding potential of –40 mV to elicit activation of the Ca^{2+} current, followed by hyperpolarizations to –100 mV to activate I_f, before and during superfusion with isoproterenol 0.1 µM. (B) A similar protocol before and during superfusion with acetylcholine 1 µM. The Ca^{2+} current is the 5-µM nifedipine-sensitive current. Isoproterenol enhances both the Ca^{2+} current and I_f, whereas ACh only inhibits I_f at this concentration. (C) Dose–response relationship of current increase by isoproterenol (ISO). On vertical axis, 1.0 signifies maximum increase. (D) Dose–response relationship of current decrease by acetylcholine (ACh). On vertical axis, 1.0 signifies maximum decrease. (Modified from Ref 65.)

ently affecting its kinetics (70), in agreement with evidence in other cardiac preparations (73). Also in agreement with results in different cardiac cells (74,75), single-channel studies in nodal myocytes have shown that epinephrine increases the probability of L-channel opening without modification of its conductance (39). As the T component is not modulated by β-adrenergic stimulation, and the L component is activated at voltages more positive than –30 mV (39), neither current is likely to play a major role in sympathetic rhythm modulation in pacemaker tissue. However, some effect of modification of Ca^{2+} currents on spontaneous rate is likely, given the finding that Ca^{2+} current activation occurs during the last phase of pacemaker depolarization (32,65).

C. Delayed Rectifier Current

Catecholamines have also been reported to activate the I_K current in multicellular SAN preparations, although single-cell studies are still lacking (10,70). Dependence upon β-adrenergic stimulation has been characterized for delayed K^+ currents in other cardiac cell types, and appears to involve protein kinase A–dependent phosphorylation (76–78). Interestingly, we have recently demonstrated that β-adrenergic stimulation of I_K in ventricular epicardial myocytes is age dependent, with the current being unresponsive in the newborn but not the adult (79). This raises the possibility that there could be a developmental element to the modulation of I_K in the SAN by adrenergic agonists. In addition, the observation that the guinea pig SAN delayed rectifier is largely I_{Ks} rather than I_{Kr} (30) raises questions of species differences in autonomic modulation of this current, since I_{Ks} is more responsive to catecholamines than I_{Kr} (80). In the human SAN, the relative contribution of the two components of I_K, and thus the extent to which this current is modulated by sympathetic innervation, is unknown and work is needed here to evaluate the relative importance of this current.

D. I_f Current

Modulation of the hyperpolarization-activated current by epinephrine and the relevance of this mechanism to the sympathetic control of rhythm have been known since the current was first studied in the SAN (10). In the SAN (14) and Purkinje fibers (81), I_f activation by β-adrenergic stimulation occurs via a positive shift of the voltage dependence of the current activation curve and kinetics. This causes acceleration of spontaneous activity by increasing the slope of diastolic depolarization, which shortens the time required to reach threshold for action potential generation. The dose–response relation for the action of isoproterenol on I_f and $I_{Ca,L}$ are in fact similar [Fig. 7.5 (A) and (C)], but the contribution of the latter seems to be significant only during the terminal phase of diastolic depolarization (65).

Particularly interesting is the mechanism by which autonomic agonists modulate I_f. Although cAMP dependent, I_f is not phosphorylation dependent in rabbit SAN; rather, there is direct nucleotide interaction with the channel (24). This action contrasts with that of cAMP on other cardiac and noncardiac currents, such as the L component of the Ca^{2+} current, and can be demonstrated in cell-free, inside-out macro patches by direct superfusion of cAMP on the intracellular side of the channel (24). Voltage-ramp measurements indicate that cAMP shifts the current activation curve maximally by about 14 mV, without modifying the fully activated I-V relation. Further, single-channel data in inside-out patches show that cAMP acts by increasing the probability of channel opening, but does not affect the single-channel conductance (82). These results agree with evidence from whole-cell experiments where the intracellular cAMP concentration is changed by autonomic receptor stimulation or other maneuvers modifying cAMP synthesis and/or hydrolysis (14) and provide a molecular basis for the I_f-mediated chronotropic action of neurotransmitters.

In summary, the voltage dependence of I_f is such that the current will be active during diastolic depolarization and autonomic agonists act to shift that voltage dependence in an excitatory (norepinephrine) and inhibitory (ACh) direction. I_f responds to ACh at concentrations lower than any other SAN current, and responds to norepinephrine at concentrations at least as low as other currents. Thus, modulation of I_f by neurotransmitters is a major contributor to the neural regulation of heart rate.

V. MOLECULAR BIOLOGICAL IDENTIFICATION OF ION CHANNELS

The use of molecular biological techniques has allowed the sequencing and cloning of several cardiac ion channels (for review, see Chap. 3). This approach has strongly accelerated the understanding of the basic molecular mechanisms underlying properties such as permeability and kinetics, as well as channel modulation by second messengers and/or G-proteins. The properties of native ion channels of tissues or isolated cells can therefore be compared with those of cloned channels. Interestingly, a strict comparison of the properties of native channels and their clones has proven to be essential for optimal clone identification. An unexpected finding has been that ion channels in the heart, as well as in other tissues, are often heteromultimers, composed of several different subunits whose assembly is a necessary condition for correct function. This applies, for example, to the ACh-activated K^+ channel (59) or the rapid (I_{Kr}) (83) and slow (I_{Ks}) (84,85) components of the voltage-dependent K^+ current. It is possible that the requirement of assembly of several different subunits for correct channel function applies to most channel types.

A. Potassium Currents

The ACh-activated K^+ channel is composed of two distinct inwardly rectifying subunits, products of a GIRK1 gene and a CIR gene (59). Coassembly of the two subunits greatly augments the current expressed in *Xenopus* oocytes and is required to reproduce specific properties of the native channel, such as a correct dependence of channel activity upon agonist concentration and coupling with G-proteins (59). Particularly during high parasympathetic activity, the contribution of this channel to the diastolic potential can be significant (60).

Several cardiac channels involved in pacemaker activity have been cloned, including cardiac depolarization-activated K^+ channel clones corresponding to delayed K^+ currents. Coassembly of the product of the KvLQT1 clone and the protein minK (IsK) forms the I_{Ks} K^+ channel (84,85). Although KvLQT1 alone forms a functional channel, the kinetic and biophysical features of the native current are best reproduced when the two clones (KvLQT1 and minK) are coexpressed. Interest in the delayed K^+ current clones has greatly increased with the finding that they are linked to different forms of the congenital long-QT syndrome (86).

The major subunit of the "rapid" K^+ current I_{Kr} (87) is encoded by another gene, referred to as ERG. Some of the main properties of I_{Kr} channels, such as rectification with a negative slope conductance, dependence on extracellular K^+, block by lanthanum ions and rapid current inactivation, are reproduced by heterologously expressed ERG clones (87,88). However, ERG current kinetics are slower than I_{Kr} kinetics, and expressed clones are not blocked by E-4031 and other I_{Kr}-blocking drugs, indicating that other subunits may combine to form the channel (87). Indeed, I_{Kr} appears to be regulated by heteromultimers composed of ERG subunits and the protein minK (83) or related proteins like MiRP1 (MinK-related protein 1). A preliminary study has found that coexpression of ERG and MiRP1 restores E-4031 sensitivity and speeds kinetics, resulting in a current more like the native I_{Kr} (89).

In situ hybridization data indicate that both ERG and KvLQT1 transcripts are found in SA nodal tissue of the ferret (90,91). Regional distribution of subsidiary subunits such as minK and MiRP1 within the SAN and periphery remains unresolved. Further, species variability remains an issue with respect to the delayed rectifier. Kinetic data and studies on sensitivity to the specific I_{Kr} blocker E-4031 indicate that a major component of the delayed K^+ current of rabbit nodal tissue is I_{Kr} (27,29); in contrast, in the guinea pig, little E-4031-sensitive current is recorded (30). Since only I_{Ks} appears to be regulated by phosphorylation, a different balance in the expression of the I_{Ks} and I_{Kr}-type components in different species may underlie differences in the mode of regulation of these channels and of the duration of the action potential. As stated earlier, the relative expression of the two components of delayed rectifier in human SAN is unknown.

B. Ca²⁺ Currents

The subunit composition of L-type Ca^{2+} channels is thought to comprise a pore-forming α_1 subunit (α_{1C} in cardiac tissue), an extracellular α_2 subunit linked by disulfide bridges to a transmembrane δ-subunit, to form $\alpha_2\delta$-compound, and an intracellular β-subunit (92,93). The major determinant of the L-type Ca^{2+} channel phenotype is the α_1-subunit. Coexpression of the α_1- and α_2-subunits confers to expressed channels the major kinetic and pharmacological properties of native cardiac L-type Ca^{2+} channels (94). A recent, important result has been the cloning of cDNA coding for several new α_1-subunits whose heterologous expression yields a current with properties similar to those of native T channels (95–98). Although these subunits are expressed mostly in the brain, several (α_{1G}, α_{1H}) are also present in the heart (95,96). As more is learned about the biophysical and pharmacological characteristics of each of these subunits, it will be interesting to see the level of expression of each in SAN tissue. Understanding the extent of expression could provide a direct way to test the hypothesis that T channels contribute substantially to pacemaker activity (39).

C. Pacemaker Current

Four different pacemaker channel isoforms, termed HCN (from Hyperpolarization- and Cyclic-Nucleotide-gated), have been identified in mammals. These isoforms are differentially expressed in brain, cardiac, and other tissues (20–22) HCN channels belong to a superfamily of channels that includes cyclic-nucleotide-gated (CNG) and voltage-dependent K^+ channels (99). The HCN2 isoform appears to be preferentially expressed in the heart (20).

Expressed HCN channels have properties similar to the native I_f current. It may be interesting to note that the HCN2 (cardiac) isoform appears to have a greater cAMP dependence than the HCN1 (neuronal) isoform (20,22); this agrees with experimental evidence indicating that in rabbit SAN myocytes the maximal cAMP-induced shift of the I_f activation curve is over 14 mV (82), whereas smaller shifts are reported in neurons ((100) e.g., 4–6 mV). It will be interesting to see if other differences between hyperpolarization-activated currents in different preparations, such as slow versus fast kinetics (the latter observed in some types of neurons; e.g., Ref. 101), are also accounted for by differences in the properties of isoforms.

VI. SAN HETEROGENEITY

The action potential characteristics and corresponding ionic currents discussed above are primarily characteristic of the central region of the SAN, which is the usual site of impulse initiation. In fact, the SAN is both morphologically and electrophysiologically heterogeneous, and this heterogeneity has implications for both normal activity and the response to autonomic agonists and pharmacological agents.

Central SAN cells exhibit a characteristic action potential, with a slower rate of rise and lower amplitude than those of the periphery. Figure 7.6 compares the action potential shape of cells from four different regions in the node, progressing from most central [region 1, Fig. 7.6(A)] to peripheral [region 3, Fig. 7.6(C)] and to atrial [region 4, Fig. 7.6(D)]. It is clear that there is a transition in action potential characteristics from center to periphery, but that all nodal regions—unlike the atrial tissue—share the characteristic of a slow diastolic potential (102). One might assume that the rate of diastolic depolarization would be greatest in cells from the center, where the heartbeat originates. In fact this is not the case. When small pieces of tissue were isolated from different regions of the node, the more peripheral samples were found to beat at a higher intrinsic rate (103). This suggests that electrotonic interactions with surrounding atrial tissue may suppress the greater automaticity of the peripheral nodal region, allowing the center to drive the heart rate.

The fact that small pieces of tissue (in which electrotonic interactions are minimal) from the center and periphery have different inherent rates indicates that the underlying ion currents differ regionally. This has not been directly confirmed by isolating myocytes from different regions and comparing ion currents. However, differences were found between large and small cells, with larger cells having the higher intrinsic rate (104). This is interesting, since cells in the center of the node are smaller than those of the periphery (102). The difference in intrinsic rate between isolated large and small SAN myocytes indicates that there must be differences in the underlying ion currents. In fact, larger (suggesting more peripheral) cells had a higher density of the pacemaker current I_f, and a greater density of rapid Na^+ current I_{Na} (104). Consonant with this observation, studies in intact tissue have suggested that automaticity in the peripheral region of the node is more Na^+-dependent while that in the center is more Ca^{2+}-dependent (105). That the central region of the node serves as the dominant pacemaker probably reflects the influence of electrotonic interactions of peripheral nodal cells with adjacent and more polarized atrial myocytes, thereby suppressing the tendency to phase 4 depolarization of the peripheral SAN cells.

The heterogeneity of ion currents throughout the node and the influence of electrotonus with atrial tissue have implications for the characteristics of impulse initiation during neural stimulation, exposure to pharmacological interventions, and disease. Any intervention that selectively enhances and/or inhibits currents that are dominant in one region, or alters cell-to-cell communication, has the potential to shift the site of origin of the impulse within the node (for review, see Ref. 106). Possible regional heterogeneity in gap junction isoform distribution (107), the extent of sympathetic and parasympathetic innervation, and myocyte neurotransmitter receptor density add further levels of complexity.

Evidence exists that spontaneous changes in the site of impulse initiation and/or exit from the node are common in the human (108) as well as in animal models (109). Further, when the node is surgically removed from the canine heart, latent pacemakers in the atrium (rather than in the AV node) take over to drive the heart (110). These latent atrial pacemakers have their own unique mix of ion current characteristics, including a small and slowly activating I_f (111), a T-type Ca current that contributes to late diastolic depolarization (112), and a significant Na/Ca exchange current (44).

In summary, action potential parameters, cycle length, and the underlying ion currents vary with cell size in the sinus node. Since anatomical studies indicate that peripheral cells are larger than central cells, the assumption is that the intrinsic ion currents of the larger cells contribute to the distinct electrophysiological character of the peripheral SAN. However, studies have not yet been done to specifically isolate cells from distinct regions of the node for comparison with the ion currents.

VII. CONCLUSION

Although the issue of the relative importance of various ion currents to the generation of pacemaker depolarization in the sinoatrial node is still debated, their electrophysiological

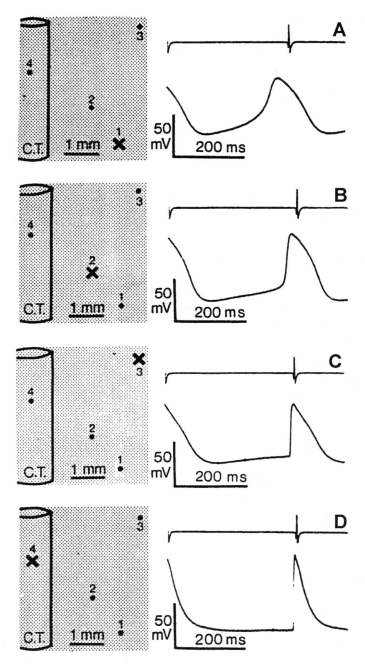

Figure 7.6 Regional differences in action potential configuration within the sinoatrial node. The left portion of each panel is a schematized drawing of the sinus node region adjacent to the crista terminalis (CT) with the recording site indicated (X) in each case (see Fig. 7.1 for anatomical reference). The right portion of each panel depicts representative action potentials from the central node (panel A, region 1) to progressively more peripheral regions (panels B and C, regions 2 and 3, respectively), and from atrial tissue in the crista terminalis (panel D, region 4). Note the progressive decrement in slope of phase 4 diastolic depolarization from regions 1 to 3, and the absence of any diastolic depolarization in the atrial action potential (region 4). (Modified from Ref. 102.)

properties have been described in some detail, and a quickly growing set of complementary data on their molecular properties is presently being gathered. This is likely to provide a more complete, molecular-based understanding of the different phenomena underlying pacemaker generation and its control by neurotransmitters in physiological and pathological settings.

As highlighted here, these will also include regional heterogeneity in ion channel function, heterogeneity in cell coupling, and in the distribution of autonomic receptors and cell-signaling cascades, as well as the pattern of innervation. In thinking of pathological situations, we must also not forget the potential contribution of pacemakers outside the node. All these components have the potential to affect either intrinsic rate or modulation of that rate by extrinsic regulators: The result is a complex, dynamic, and robust impulse initiation process.

REFERENCES

1. Anderson RH, Ho SY. The architecture of the sinus node, the atrioventricular conduction axis, and the internodal atrial myocardium. J Cardiovasc Electrophysiol 1998; 9: 1233–1248.
2. Denyer JC, Brown HF. Rabbit sino-atrial node cells: isolation and electrophysiological properties. J Physiol (Lond) 1990; 428: 405–424.
3. Irisawa H, Brown HF, Giles W. Cardiac pacemaking in the sinoatrial node. Physiol Rev 1993; 73: 197–227.
4. DiFrancesco D. Pacemaker mechanisms in cardiac tissue. Ann Rev Physiol 1993; 55: 455–472.
5. DiFrancesco D. Cardiovascular controversies: The pacemaker current, i_f, plays an important role in regulating SA node pacemaker activity. Cardiovasc Res 1995; 30: 307–308.
6. Vassalle M. Cardiovascular controversies: The pacemaker current, i_f, does not play an important role in regulating SA node pacemaker activity. Cardiovasc Res 1995; 30: 309–310.
7. DiFrancesco D, Noble D. A model of cardiac electrical activity incorporating ionic pumps and concentration changes. Phil Trans R Soc Lond B 1985; 307: 353–398.
8. DiFrancesco D. The hyperpolarization-activated (I_f) current: autonomic regulation and the control of pacing. In: Morad M, Ebashi S, Trautwein W, Kurachi Y, eds. Molecular Physiology and Pharmacology of Cardiac Ion Channels and Transporters. Dordrecht: Kluwer Academic Publishers, 1996:31–37.
9. Sanguinetti MC, Jurkiewicz NK. Delayed rectifier outward K^+ current is composed of two currents in guinea pig atrial cells. Am J Physiol 1991; 260: H393–H399.
10. Brown HF, DiFrancesco D, Noble SJ. How does adrenaline accelerate the heart? Nature 1979; 280: 235–236.
11. Brown HF, DiFrancesco D. Voltage-clamp investigations of membrane currents underlying pacemaker activity in rabbit sino-atrial node. J Physiol (Lond) 1998; 308: 331–351.
12. Yamagihara K, Irisawa H. Inward current activated during hyperpolarization in the rabbit sinoatrial node cell. Pflugers Arch 1980; 385: 11–19.
13. Nakayama T, Kurachi Y, Noma A, Irisawa H. Action potential and membrane currents of single pacemaker cells of the rabbit heart. Pflugers Arch 1984; 402: 248–257.
14. DiFrancesco D, Ferroni A, Mazzanti M, Tromba C. Properties of the hyperpolarizing-activated current (i_f) in cells isolated from the rabbit sino-atrial node. J Physiol (Lond) 1986; 377: 61–88.
15. DiFrancesco D. The contribution of the 'pacemaker' current (i_f) to generation of spontaneous activity in rabbit sino-atrial node myocytes. J Physiol (Lond) 1991; 434: 23–40.
16. van Ginneken ACG, Giles W. Voltage clamp measurements of the hyperpolarization-activated inward current i_f in single cells from rabbit sino-atrial node. J Physiol (Lond) 1991; 434: 57–83.

17. DiFrancesco D. Characterization of single pacemaker channels in cardiac sino-atrial node cells. Nature 1986; 324: 470–473.
18. DiFrancesco D, Tromba C. Channel activity related to pacemaking. In: Piper HM, Isenberg G, eds. Isolated Adult Cariomyocytes, Volume II: Electrophysiology and Contractile Function. Boca Raton: CRC Press, 1989:97–115.
19. Santoro B, Grant SGN, Bartsch D, Kandel ER. Interactive cloning with the SH3 domain of N-src identifies a new brain specific ion channel protein, with homology to Eag and cyclic nuceotide-gated channels. Proc Natl Acad Sci USA 1997; 94: 14815–14820.
20. Santoro B, Liu DT, Yao H, Bartsch D, Kandel ER, Siegelbaum SA, Tibbs GR. Identification of a gene encoding a hyperpolarization-activated pacemaker channel of brain. Cell 1998; 93: 1–20.
21. Gauss R, Seifert R, Kaupp UB. Molecular identification of a hyperpolarization-activated channel in sea urchin sperm. Nature 1998; 393: 583–587.
22. Ludwig A, Zong X, Jeglitsch M, Hofmann F, Biel M. A family of hyperpolarization-activated mammalian cation channels. Nature 1998; 393: 587–591.
23. DiFrancesco D. The hyperpolarization-activated current, i_f, and cardiac pacemaking. In: Rosen MR, Janse MJ, Wit AL, eds. Cardiac Electrophysiology: A Textbook. New York: Futura, 1990:117–132.
24. DiFrancesco D, Tortora P. Direct activation of cardiac pacemaker channels by intracellular cyclic AMP. Nature 1991; 351: 145–147.
25. Giles WR, van Ginneken ACG, Shibata EF. Ionic currents underlying cardiac pacemaker activity: a summary of voltage-clamp data from single cells. In: Nathan RD, ed. Cardiac Muscle: The Regulation of Excitation and Contraction. New York: Academic, 1986:1–27.
26. Irisawa H. Membrane currents in cardiac pacemaker tissue. Experientia 1987; 43: 1131–1240.
27. Ito H, Ono K. A rapidly activating delayed rectifier K^+ channel in rabbit sinoatrial node cells. Am J Physiol 1995; 269: H443–H452.
28. Ono K, Ito H. Role of rapidly activating delayed rectifier K^+ current in sinoatrial node pacemaker activity. Am J Physiol 1995; 269: H453–H462.
29. Verheijck EE, van Ginneken ACG, Bourier J, Bouman LN. Effects of delayed rectifier current blockade by E-4031 on impulse generation in single sinoatrial nodal myocytes of the rabbit. Circ Res 1995; 76: 607–615.
30. Anumonwo JM, Freeman LC, Kwok WM, Kass RS. Delayed rectification in single cells isolated from guinea pig sinoatrial node. Am J Physiol 1992; 262: H921–H925.
31. Noble D, Noble SJ. A model of sino-atrial node electrical activity based on a modification of the DiFrancesco-Noble (1984) equations. Proc R Soc Lond B 1984; 222: 295–304.
32. Brown HF, Kimura J, Noble D, Noble SJ, Taupignon A. The slow inward current, i_{si}, in the rabbit sino-atrial node investigated by voltage clamp and computer simulation. Proc R Soc Lond B 1984; 222: 305–328.
33. Hagiwara N, Irisawa H, Kasanuki H, Hosoda S. Background current in sinoatrial node cells of the rabbit heart. J Physiol (Lond) 1992; 448: 53–72.
34. Denyer JC, Brown HF. Pacemaking in rabbit isolated sino-atrial node cells during Cs block of the hyperpolarization-activated current i_f. J Physiol (Lond) 1990; 429: 401–409.
35. Irisawa H, Yamagihara K. The slow inward current of the rabbit sino-atrial nodal cell. In: Zipes DP, Bailey JC, Elharrar V, eds. The Slow Inward Current and Cardiac Arrhythmias. The Hague: Nijhoff, 1980:265–284.
36. Yamagihara K, Irisawa H. Potassium current during the pacemaker depolarization in rabbit sinoatrial node cell. Pflugers Arch 1980; 388: 255–260.
37. Yamagihara K, Noma A, Irisawa H. Reconstruction of sinoatrial node pacemaker potential based on voltage clamp experiments. Jpn J Physiol 1980; 30:841–857.
38. Brown HF, Kimura J, Noble D, Noble SJ, Taupignon A. The ionic currents underlying pacemaker activity in rabbit sino- atrial node: experimental results and computer simulations. Proc R Soc Lond B 1984; 222: 329–347.

39. Hagiwara N, Irisawa H, Kameyama M. Contribution of two types of calcium currents to the pacemaker potentials of rabbit sino-atrial node cells. J Physiol (Lond) 1988; 395: 233–253.

40. Fermini B, Nathan RD. Removal of sialic acid alters both T- and L-type calcium currents in cardiac myocytes. Am J Physiol 1991; 260: H735–H743.

41. Guo J, Ono K, Noma A. A sustained inward current activated at the diastolic potential range in rabbit sino-atrial node cells. J Physiol (Lond) 1995; 483: 1–13.

42. Guo J, Mitsuiye T, Noma A. The sustained inward current in sino-atrial node cells of guinea-pig heart. Pflugers Arch 1997; 433: 390–396.

43. Hess P, Lansman JB, Tsien RW. Calcium channel selectivity for divalent and monovalent cations. Voltage and concentration dependence of single channel current in ventricular heart cell. J Gen Physiol 1986; 88: 293–319.

44. Zhou Z, Lipsius SL. Na^+-Ca^{2+} exchange current in latent pacemaker cells isolated from cat right atrium. J Physiol (Lond) 1993; 466: 263–285.

45. Bassani RA, Bassani JW, Lipsius SL, Bers DM. Diastolic SR Ca efflux in atrial pacemaker cells and Ca-overloaded myocytes. Am J Physiol 1997; 273: H886–H892.

46. Satoh H. Electrophysiological actions of ryanodine on single rabbit sinoatrial nodal cells. Gen Pharmacol 1997; 28: 31–38.

47. Li J, Qu J, Nathan RD. Ionic basis of ryanodine's negative chronotropic effect on pacemaker cells isolated from the sinoatrial node. Am J Physiol 1997; 273: H2481–H2489.

48. Zaza A, Micheletti M, Brioschi A, Rocchetti M. Ionic currents during sustained pacemaker activity in rabbit sino-atrial myocytes. J Physiol (Lond) 1997; 505 (Pt 3): 677–688.

49. Hoffman BF, Suckling EE. Cardiac cellular potentials: effect of vagal stimulation and acetylcholine. Am J Physiol 1953; 173: 312–320.

50. del Castillo J, Katz B. Production of membrane potential changes in the frog's heart by inhibitory nerve impulses. Nature 1955; 175: 1035.

51. Hutter O, Trautwein W. Vagal and sympathetic effects on the pacemaker fibres in the sinus venosus of the heart. J Gen Physiol 1956; 39: 715–733.

52. Giles W, Noble SJ. Changes in membrane currents in bullfrog atrium produced by acetylcholine. J Physiol (Lond) 1976; 261: 103–123.

53. Ten Eick R, Nawrath H, McDonald TF, Trautwein W. On the mechanism of the negative inotropic effect of acetylcholine. Pflugers Arch 1976; 36: 207–213.

54. Pfaffinger PJ, Martin JM, Hunter DD, Nathanson NM, Hille B. GTP-binding proteins couple cardic muscarinic receptors to a K channel. Nature 1985; 317: 536–538.

55. Breitwieser GE, Szabo G. Uncoupling of cardiac muscarinic and beta-adrenergic receptors from ion channels by a guanine nucleotide analogue. Nature 1985; 317: 538–540.

56. Sakmann B, Noma A, Trautwein W. Acetylcholine activation of single muscarinic K channels in isolated pacemaker cells of the mammalian heart. Nature 1983; 303: 250–253.

57. Soejima M, Noma A. Mode of regulation of the ACh-sensitive K-channel by the muscarinic receptor in rabbit atrial cells. Pflugers Arch 1984; 400: 424–431.

58. Kurachi Y, Nakajima T, Sugimoto T. Acetylcholine activation of K channels in cell-free membrane of atrial cells. Am J Physiol 1986; 251: H681–H684.

59. Krapivinsky G, Gordon EA, Wickman K, Velimirovic B, Krapivinsky L, Clapham DE. The G-protein-gated atrial K^+ channel $I_{K,ACh}$ is a heteromultimer of two inwardly rectifying K^+-channel proteins. Nature 1995; 374: 135–141.

60. DiFrancesco D, Ducouret P, Robinson RB. Muscarinic modulation of cardiac rate at low acetylcholine concentrations. Science 1989; 243: 669–671.

61. Boyett MR, Kodama I, Honjo H, Arai A, Suzuki R, Toyama J. Ionic basis of the chronotropic effect of acetylcholine on the rabbit sinoatrial node. Cardiovasc Res 1995; 29: 867–878.

62. Hirst GDS, Bramich NJ, Edwards FR, Klemm M. Transmission at autonomic neuroeffector junctions. Trends Neurosci 1992; 15: 40–46.

63. Hartzell HC. Regulation of cardiac ion channels by catecholamines, acetylcholine and second messenger systems. Prog Biophys Mol Biol 1988; 52: 165–247.

64. Noma A, Trautwein W. Relaxation of the ACh-induced potassium current in the rabbit sinoatrial node cell. Pflugers Arch 1978; 377: 193–200.
65. Zaza A, Robinson RB, DiFrancesco D. Basal responses of the L-type Ca^{2+} and hyperpolarization-activated currents to autonomic agonists in the rabbit sino-atrial node. J Physiol (Lond) 1996; 491: 347–355.
66. Danilo P, Jr, Binah O, Hordof AJ. Chronotropic and dromotropic actions of acetylcholine on the developing fetal heart. Dev Pharmacol Ther 1993; 20: 231–238.
67. Baruscotti M, DiFrancesco D, Robinson RB. A TTX-sensitive inward sodium current contributes to spontaneous activity in newborn rabbit sino-atrial node cells. J Physiol (Lond) 1996; 492: 21–30.
68. Rosen MR, Robinson RB, Cohen IS, Steinberg SF, Bilezikian JP. Alpha-adrenergic modulation of cardiac rhythm in the developing heart. In: Sperelakis N, ed. Physiology and Pathophysiology of the Heart. Hingham: Kluwer Academic Publishers, 1995:457–465.
69. Hewett KW, Rosen MR. Developmental changes in the rabbit sinus node action potential and its response to adrenergic agonists. J Pharmacol Exp Ther 1985; 235: 308–312.
70. Noma A, Kotake H, Irisawa H. Slow inward current and its role mediating the chronotropic effect of epinephrine in the rabbit sinoatrial node. Pflugers Arch 1980; 388: 1–9.
71. Osterrieder W, Brum G, Hescheler J, Trautwein W, Hofmann F, Flockerzi V. Injection of subunits of cyclic AMP-dependent protein kinase into cardiac myocytes modulates Ca current. Nature 1982; 298: 576–578.
72. Kameyama M, Hescheler J, Hofmann F, Trautwein W. Modulation of Ca current during the phosphorylation cycle in the guinea-pig heart. Pflugers Arch 1986; 407: 123–128.
73. Reuter H, Scholz H. The regulation of Ca conductance of cardiac muscle by adrenaline. J Physiol (Lond) 1977; 264: 49–62.
74. Reuter H. Calcium channel modulation by neurotransmitters, enzymes and drugs. Nature 1983; 301: 569–574.
75. Cachelin AB, De Peyer JE, Kokubun S, Reuter H. Calcium channel modulation by 8-bromocyclic AMP in cultured heart cells. Nature 1983; 304: 462–464.
76. Brown HF, Noble SJ. Effects of adrenaline on membrane currents underlying pacemaker activity in frog atrial muscle. J Physiol (Lond) 1974; 238: 51P–53P.
77. Giles W, Nakajima T, Ono K, Shibata EF. Modulation of the delayed rectifier K^+ current by isoprenaline in bull-frog atrial myocytes. J Physiol (Lond) 1989; 415: 233–249.
78. Yazawa K, Kameyama M. Mechanism of receptor-mediated modulation of the delayed outward potassium current in guinea-pig ventricular myocytes. J Physiol (Lond) 1990; 421: 135–150.
79. Charpentier F, Liu O-Y, Rosen MR, Robinson RB. Age-related differences in β-adrenergic regulation of repolarization in canine epicardial myocytes. Am J Physiol 1996; 271: H1174–H1181.
80. Sanguinetti MC, Jurkiewicz NK, Scott A, Siegl PK. Isoproterenol antagonizes prolongation of refractory period by the class III antiarrhythmic agent E-4031 in guinea pig myocytes. Mechanism of action. Circ Res 1991; 68: 77–84.
81. Tsien RW. Effects of epinephrine on the pacemaker potassium current of cardiac Purkinje fibers. J Gen Physiol 1974; 64: 293–319.
82. DiFrancesco D, Mangoni M. Modulation of single hyperpolarization-activated channels (i_f) by cAMP in the rabbit sino-atrial node. J Physiol (Lond) 1994; 474: 473–482.
83. McDonald TV, Yu Z, Ming Z, Palma E, Meyers MB, Wang K-W, Goldstein SAN, Fishman GI. A minK-HERG complex regulates the cardiac potassium current I_{kr}. Nature 1997; 388: 289–292.
84. Sanguinetti MC, Curran ME, Zou A, Schen J, Spector PS, Atkinson DL, Keating MT. Coassembly of K_v LQT1 and minK (IsK) proteins to form cardiac I_{ks} potassium channel. Nature 1996; 384: 80–83.
85. Barhanin J, Lesage F, Guillemare E, Fink M, Lazdunski M, Romey G. K_v LQT1 and IsK (minK) proteins associate to form the I_{ks} cardiac potassium current. Nature 1996; 384: 78–80.

86. Rosen M.R. Long QT syndrome patients with gene mutations. Circulation 1995; 92: 3373–3375.

87. Sanguinetti MC, Jiang C, Curran ME, Keating MT. A mechanistic link between an inherited and an acquired cardiac arrhythmia: HERG encodes the I_{kr} potassium channel. Cell 1995; 81: 299–307.

88. Trudeau MC, Warmke JW, Ganetzky B, Robertson GA. HERG, a human inward rectifier in the voltage-gated potassium channel family. Science 1995; 269: 92–95.

89. Abbott GW, Sesti F, Splawski I, Buck ME, Lehmann MH, Timothy KW, Keating MT, Goldstein SAN. MiRP1 forms IKr potassium channels with HERG and is associated with cardiac arrhythmia. Cell 1999; 97:175–187.

90. Brahmajothi MV, Morales MU, Rasmusson RL, Campbell DL, Strauss HC. Heterogeneicity in K^+ channel transcript expression detected in isolated feret cardiac myocytes. Pacing Clin Electrophysiol 1997; 20: 388–396.

91. Brahmajothi MV, Morales MU, Reimer KA, Strauss HC. Regional localization of ERG, the channel protein responsible for the rapid component of the delayed rectifier K^+ current in the ferret heart. Circ Res 1997; 81: 128–135.

92. Randall AD. The molecular basis of voltage-gated Ca^{2+} channel diversity: is it time for T? J Membrane Biol 1998; 161: 207–213.

93. Dolphin A.C. Mechanisms of modulation of voltage-dependent calcium channels by G-proteins. J Physiol (Lond) 1998; 506: 3–11.

94. Mikami A, Imoto K, Tanabe T, Niidome T, Mori Y, Takeshima H, Narumiya S, Numa S. Primary structure and functional expression of the cardiac dihydropyridine-sensitive calcium channel. Nature 1989; 340: 230–233.

95. Perez-Reyes E, Cribbs LL, Daud A, Lacerda AE, Barclay J, Williamson MP, Fox M, Rees M, Lee JH. Molecular characterization of a neuronal low-voltage activated T-type calcium channel. Nature 1998; 391: 894–900.

96. Cribbs LL, Lee JH, Yang J, Satin J, Zhang Y, Daud A, Barclay J, Williamson MP, Fox M, Rees M, and Perez-Reyes E. Cloning and characterization of alpha lH from human heart, a member of the T-type Ca^{2+} channel gene family. Circ Res 1998; 83: 103–109.

97. Lee JH, Daud AN, Cribbs LL, Lacerda AE, Pereverzev A, Schneider T, Perez-Reyes E. Cloning and expression of a novel member of the low voltage-activated T-type calcium channel family. J Neurosci 1999; 19: 1912–1921.

98. Talley EM, Cribbs LL, Lee JH, Daud A, Perez-Reyes E, Bayliss DA. Differential distribution of three members of a gene family encoding low voltage-activated (T-type) calcium channels. J Neurosci 1999; 19: 1895–1911.

99. Clapham DE. Not so funny anymore: pacing channels are cloned. Neuron 1998; 21: 5–7.

100. McCormick DA, Pape HC. Noradrenergic and serotoninergic modulation of a hyperpolarization-activated cation current in thalamic relay neurones. J Physiol (Lond) 1990; 431: 319–342.

101. Maccaferri G, Mangoni M, Lazzari A, DiFrancesco D. Properties of the hyperpolarization-activated current in rat hippocampal CA1 pyramidal cells. J Neurophysiol 1993; 69: 2129–2136.

102. Masson-Pevet M, Bleeker WK, Mackaay AJC, Bouman LN, Houtkooper JM. Sinus node and atrium cells from the rabbit heart: A quantitative electron microscopic description after electrophysiological localization. J Mol Cell Cardiol 1979; 11: 555–568.

103. Opthof T, van Ginneken ACG, Bouman LN, Jongsma HJ. The intrinsic cycle length in small pieces isolated from the rabbit sinoatrial node. J Mol Cell Cardiol 1987; 19: 923–934.

104. Honjo H, Boyett MR, Kodama I, Toyama J. Correlation between electrical activity and the size of rabbit sinoatrial node cells. J Physiol (Lond) 1996; 496: 795–808.

105. Kodama I, Nikmaram MR, Boyett MR, Suzuki R, Honjo H, Owen JM. Regional differences in the role of the Ca^{2+} and Na^+ currents in pacemaker activity in the sinoatrial node. Am J Physiol 1997; 272: H2793–H2806.

106. Schuessller RB, Boineu JP, Bromberg BI. Origin of the sinus impulse. J Cardiovasc Electrophysiol 1996; 7: 263–274.

107. Davis LM, Kanter HL, Beyer EC, Saffitz JE. Distinct gap junction protein phenotypes in cardiac tissues with disparate conduction properties. J Am Coll Cardiol 1994; 24: 1124–1132.

108. Brody DA, Woolsey MD, Arzbaecher RC. Application of computer techniques to the detection and analysis of spontaneous P-wave variations. Circulation 1967; 36: 359–371.

109. Boineau JP, Schuessler RB, Mooney CR, Wylds AC, Miller CB, Hudson RD, Borremans JM, Brockus CW. Multicentric origin of the atrial depolarization wave: the pacemaker complex. Relation to dynamics of atrial conduction, P-wave changes and heart rate control. Circulation 1978; 58: 1036–1048.

110. Randall WC, Talano J, Kaye MP, Euler D, Jones S, Brynjolfsson G. Cardiac pacemakers in absence of the SA node: responses to exercise and autonomic blockade. Am J Physiol 1978; 234: H465–H470.

111. Zhou Z, Lipsius SL. Properties of the pacemaker current (I_f) in latent pacemaker cells isolated from cat right atrium. J Physiol (Lond) 1992; 453: 503–523.

112. Zhou Z, Lipsius SL. T-type calcium current in latent pacemaker cells isolated from cat right atrium. J Mol Cell Cardiol 1994; 26:1211–1219.

8

Intracardiac Cell Communication and Gap Junctions

JEFFREY E. SAFFITZ

Washington University School of Medicine, St. Louis, Missouri

MARK J. YEAGER

The Scripps Research Institute, La Jolla, California

I. INTRODUCTION

Impulse propagation in an electrically excitable tissue requires a means of conducting action potentials throughout a multicellular array. In the nervous system, this is accomplished by release of chemical neurotransmitters at synapses. In skeletal muscle, each individual multinucleated myotube is electrically activated by its own motor neuron and excitation spreads within the confines of a continuous sarcolemma. In contrast, cardiac and smooth muscle are composed of individual cells each invested with an insulating lipid bilayer that would effectively prevent intercellular current flux were there not specialized cell–cell junctions to serve this purpose. In these tissues, ions flow from one cell to another via gap junctions, specialized regions of the plasmalemma containing transmembrane protein assemblies that adjoin in the extracellular space to create aqueous pores that directly link the cytoplasmic compartments of neighboring cells. A gap junction consists of an array of from tens to thousands of closely packed channels that permit intercellular passage of ions and small molecules up to ~1 kDa in molecular weight. Gap junctions are ubiquitous throughout the animal kingdom. They facilitate intercellular exchange of molecular information in virtually all multicellular tissues. They form during the earliest stages of embryonic development and are thought to play important roles in the spread of morphogens, signaling molecules and ions. Intercellular spread of ions via gap junctions occurs in both nonelectrically excitable tissues (e.g., spread of Ca waves in a layer of epithelial cells) and electrically excitable tissues (e.g., smooth muscle and myocardium) that propagate action potentials.

171

The cloning and sequencing of genes encoding gap junction channel proteins have spawned great advances in knowledge of the molecular structure, distribution, and potential functional specializations of gap junction channels in the heart. This chapter summarizes current understanding of the structure and function of gap junctions in the mammalian heart. Potential roles of gap junctions in cardiac development and the electrophysiological consequences of expression of specific gap junction channel proteins by functionally distinct cardiac tissues in the normal heart are considered. Mechanisms responsible for alterations in the expression, distribution, and function of gap junction channel proteins and the potential role of such alterations in cardiac remodeling, ventricular dysfunction, and arrhythmogenesis are also reviewed.

II. GAP JUNCTION CHANNELS ARE COMPOSED OF CONNEXINS

Gap junction channels are composed of members of a multigene family of proteins referred to as connexins. More than a dozen unique connexins have been cloned (reviewed in Refs. 1–3). These proteins are identified using the abbreviation Cx for connexin followed by the molecular weight of the specific protein (1). Based on phylogenetic considerations, the connexins have been divided into two types, designated as α and β (3,4). Hydropathy analysis predicts that connexins have four hydrophobic domains (designated M1, M2, M3, and M4) and two extracellular loops (designated E1 and E2). The N terminus, M2–M3 loop and C-terminal tail are all predicted to be localized on the cytoplasmic side of the plasmalemma. The α-type connexins have a larger M2–M3 cytoplasmic loop and a larger C-terminal cytoplasmic tail than the β-type connexins. Channel structure has been most extensively characterized for Cx43 (α1), Cx32 (β1), and Cx26 (β2) (reviewed in Ref. 5 and updated in Ref. 6). The topology of connexins expressed in the cardiovascular system, shown in Figures 8.1 and 8.2, conforms to the general topology of the α-type connexins.

Cx43 is the best characterized cardiac connexin. Its topology has been determined using protease cleavage and peptide antibody labeling directed to selected sites in the protein sequence (Fig. 8.2) (7–10). Primary antibodies directed against synthetic peptides can be labeled with gold-conjugated secondary antibodies and visualized by electron microscopy, thereby localizing specific regions of the connexin molecule. For example, gap junctions containing Cx43 become heavily decorated on their cytoplasmic surfaces by peptide antibodies directed to cytoplasmic domains (Fig. 8.3). With this approach, it has been demonstrated that the N-terminal and C-terminal domains and the M2–M3 loop are all located on the cytoplasmic side of the membrane. Based on their degree of sequence homology with Cx43, it is presumed that Cx40, Cx45, and Cx37 have a similar membrane homology (Fig. 8.1).

III. GAP JUNCTION PLAQUES ARE COLLECTIONS OF MANY CHANNELS

Gap junction channels tend to form aggregates in the plasma membrane, referred to as plaques, which contain from tens to thousands of channels. When cells are subjected to treatments that break the plasmalemmal into small fragments, regions of the membrane containing gap junction plaques have a greater density than other regions and can be isolated by centrifugation on a sucrose density gradient. When the isolated plaques are treated with low concentrations of detergent, membrane lipids are extracted and the channels become packed even tighter in the membrane and can form ordered two-dimensional crystals (Fig. 8.4). Occasionally, edge-on views of gap junctions may be viewed by electron

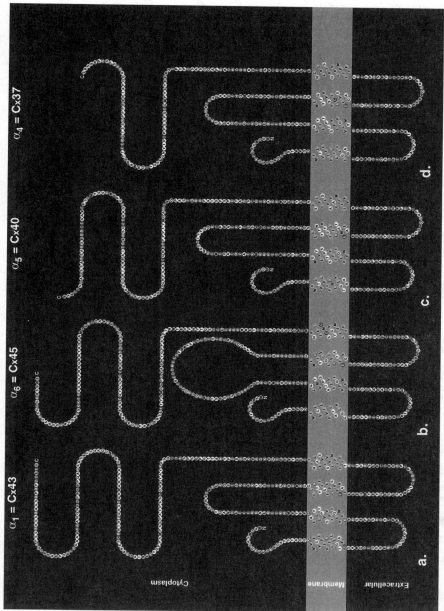

Figure 8.1 Folding models for gap junction proteins in the heart: Cx43 (α_1 connexin), Cx45 (α_6 connexin), Cx40 (α_5 connexin) and Cx37 (α_4 connexin). The amino acid sequences were deduced from cDNA analysis (16,29,63,134), and the residues are coded as follows: hydrophobic in yellow, acidic in red, basic in blue, and cysteine in green. Hydropathy analysis predicts that the polypeptide spans the lipid bilayer of the plasma membrane four times. Among the connexins there is conservation in the amino acid sequence of the four hydrophobic domains, referred to as M1, M2, M3 and M4. Note the three cysteine residues located in each of the extracellular loops (designated E1 and E2) that are also conserved among the connexins. The cytoplasmic loop between M2 and M3 and the carboxy-tail are the most divergent in size and sequence. (See also color plate.)

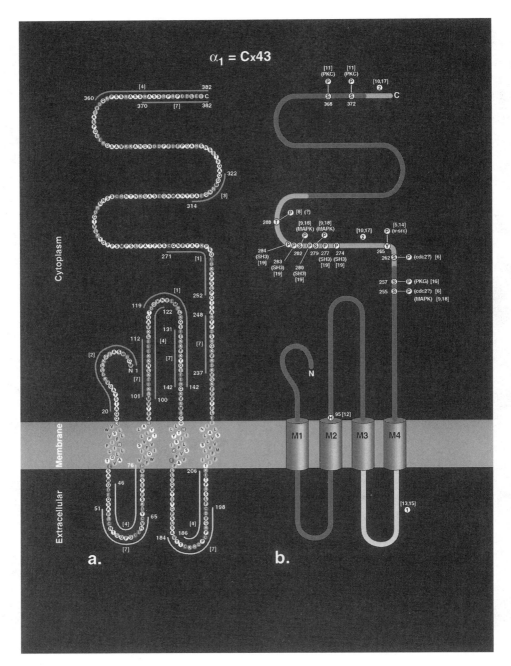

Figure 8.2   (a) Folding model for Cx43 (α_1 connexin). The amino acid sequence of Cx43 was deduced from cDNA analysis (29), and the residues are coded as in Figure 8.1. Of all cardiac connexins, the topology of Cx43 has been most thoroughly studied. The predicted locations of the extracellular and cytoplasmic regions have been confirmed with site-directed antibodies (bars indicate cytoplasmic and extracellular epitopes). (b) Locations of selected functionally important residues and domains as determined by mutagenesis and chimera studies. Indicated sites of covalent modification are based on consensus sequences, modification of synthetic peptides in vitro, and mutagenesis studies. The significance of the functional domains (circled numbers) and specifically mutated residues are as follows: Domain 1 is the predominant determinant for specificity of heterotypic inter-

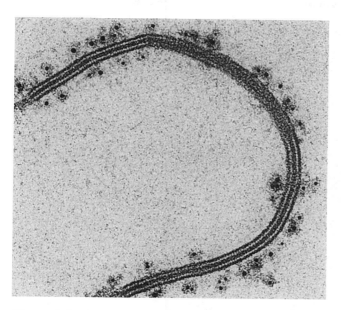

Figure 8.3 Convention transmission electron microscopy of ultrathin sections has served as a keystone for defining the "morphological signature" of gap junctions. The characteristic appearance is septalaminar, with stain exclusion in the hydrophobic domains of the lipid bilayer and stain accumulation in the extracellular gap and on the cytoplasmic faces. Cardiac gap junctions are ~250Å thick. Immunogold labeling by site-specific peptide antibodies directed against a cytoplasmic peptide in Cx43 supports the folding model shown in Figure 8.2 (5).

microscopy when the plaques are folded 90° with respect to the surface of the supporting grid. Figure 8.5 shows an orthogonal edge view of a gap junction that has been "negatively" stained with a heavy metal stain (in this case, uranyl acetate) and visualized by electron microscopy. The electron-dense stain has been excluded by the two lipid bilayer regions, but has accumulated in the extracellular vestibules of the channels where it is seen as repeating densities.

IV. CONNEXONS ARE FORMED BY A HEXAMERIC CLUSTER OF SUBUNITS

An individual intercellular channel is created by stable, noncovalent interactions of two hemichannels referred to as connexons, each located in the plasma membranes of adjacent cells. Computer analysis and image enhancement can be used to average all of the individual channel images in the well-ordered two-dimensional crystalline domains illustrated in Figure 8.4 to derive a map of a connexon that shows greater detail. The top view of the im-

---

actions between Cx43 (α_1), Cx50 (α_8) and Cx46 (α_3). This result is based on studies of chimeras of Cx50 (α_8) and Cx46 (α_3). Domains 2 impart high sensitivity to pH gating of Cx43. Gating by pH is also influenced by the charge on His[95]. The SH3 domains of v-Src bind to proline-rich motifs and a phosphorylated tyrosine at position 265 of Cx43 (α_1). The numbered references are as follows: 1 (7), 2 (8), 3 (135), 4 (9), 5 (69), 6 (136), 7 (10), 8 (68), 9 (70), 10 (84), 11 (71), 12 (85), 13 (48), 14 (72), 15 (50), 16 (73), 17 (86), 18 (74), 19 (75). (Modified from Ref. 5 and reproduced with permission of Current Biology Ltd.) (See also color plate.)

Figure 8.4   Transmission electron micrograph of a cardiac gap junction membrane plaque formed by hundreds of donut-like channels. This biochemically isolated membrane sheet has been treated with detergents to concentrate the channels into crystalline domains, the largest of which is ~50 μm^2. Such crystalline patches are amenable to analysis in order to derive an averaged density map (Figure 8.6, top). Scale bar = 1000Å. (Reproduced from Ref. 137 with permission of Acta Crystallographica.)

age-enhanced channel (top of Fig. 8.6) shows that each connexon is formed by six connexin subunits. The schematic model in Figure 8.6 is an interpretation of the edge-on map, which was derived by image analysis of the micrograph shown in Figure 8.5. Each connexon spans the lipid bilayer. The end-to-end interaction between two hexameric connexons occurs in the extracellular "gap" between the membranes of adjacent cells, thus giving rise to the term "gap junction." The thickness of cardiac gap junctions is ~250 Å, the lipid bilayers have a thickness of ~50 Å, and the extracellular gap is ~40 Å thick. The connexons appear to extend ~20 Å into the extracellular space and ~55 Å into the cytoplasmic space.

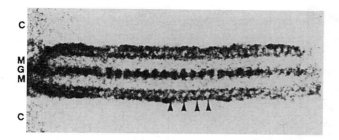

Figure 8.5 Orthogonal edge view of a cardiac gap junction membrane two-dimensional crystal negatively stained with uranyl acetate. The cytoplasmic, lipid membrane, and extracellular gap regions are indicated by the letters C, M and G. The repeating densities identify stain that has accumulated in the extracellular vestibule. The arrows indicate repeating stain excluding areas in the cytoplasmic membrane face which appear to suggest ordered packing of the protein in this region. The total thickness is ~250Å.

V. TRANSMEMBRANE α-HELICES CAN BE VISUALIZED BY ELECTRON CRYOCRYSTALLOGRAPHY

The electron micrographs shown in Figures 8.3 to 8.5 are of specimens that have been dried and stained with heavy metal to enhance electron microscopic contrast. Image analysis of such micrographs is limited to about 15 Å resolution. Biological specimens can be quick-frozen so that the aqueous phase does not crystallize but instead becomes vitrified like glass. Native biological structure is preserved in this frozen-hydrated state because the samples have not been dried or stained. This approach has been applied to analysis of plaques isolated from cells that express only a single recombinant connexin (11). Figure 8.7 shows images obtained from analysis of such a plaque containing pure Cx43. A projection density map derived from negatively stained two-dimensional crystals of recombinant Cx43 [Fig. 8.7(c)] closely resembles the maps of native cardiac gap junction membranes (10,12). Diffraction patterns of frozen-hydrated two-dimensional crystals [Fig. 8.7(b)] extend to much higher resolution than negatively stained crystals [Fig. 8.7(a)]. A map at 17 Å resolution [Fig. 8.7(d)] derived from analysis of frozen-hydrated two-dimensional crystals shows substantially more detail than the map of negatively stained and dried crystals [Figure 8.7(c)]. In particular, at 17 Å radius, the channel is lined by circular densities with the characteristic appearance of transmembrane α-helices that are oriented roughly perpendicular to the membrane plane (13). A similar appearance for densities at 33 Å radius suggests the presence of α-helices at the interface with the membrane lipids. The two rings of α-helices are separated by a continuous band of density at a radius of 25 Å, which may arise from the superposition of projections of additional transmembrane α-helices and polypeptide density arising from the extracellular and intracellular loops within each connexin subunit.

VI. THE DODECAMERIC CHANNEL IS FORMED BY 48 TRANSMEMBRANE α-HELICES

To further explore the transmembrane architecture of gap junction channels, three-dimensional maps with resolution cutoffs of 7.5 Å in the membrane plane and 21 Å perpendicular to the membrane have been derived from frozen-hydrated, tilted two-dimensional crys-

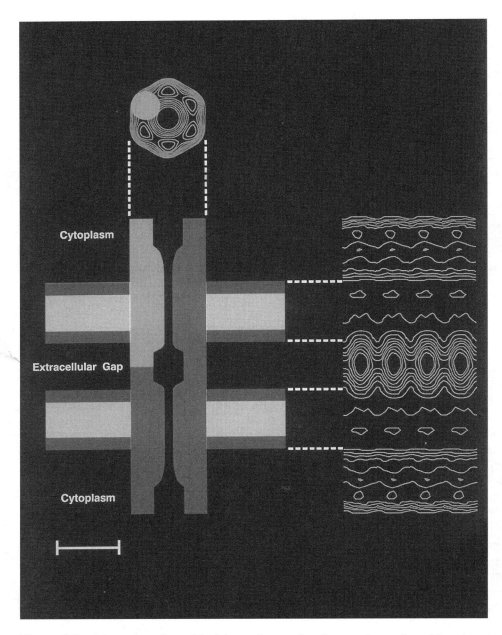

Figure 8.6 Schematic scale model of the cardiac gap junction membrane channel based on the planar density map (shown above the model) and the edge-on density map perpendicular to the membrane plane (shown to the right of the model) (scale bar = 50Å). The connexon has a diameter of ~65Å and is formed by a ring of 6 Cx43 subunits. The Cx43 subunit (solid) has a diameter of ~25Å and is highly asymmetric with an axial ratio of 4:1 to 5:1. The hexameric connexons of adjacent cells are in register to define the central channel. The schematic model is in the same orientation as the edge-on density (right) derived by image analysis of the micrograph in Figure 8.5. The relationship between the 50Å thick lipid bilayers and the edge-view is indicated by dashed lines. The model shows the location of the 10Å thick lipid headgroup region and the 30Å hydrophobic membrane interior containing the lipid alkyl chains. The detailed shape and transmembrane structure of the channel are depicted here in a stylized fashion. (Modified from Ref. 5 and reproduced with permission of Current Biology Ltd.) (See also color plate.)

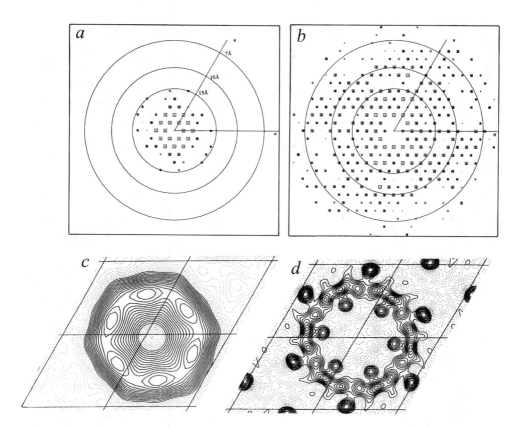

Figure 8.7 Computed diffraction patterns (a and b) and projection density maps (c and d) for gap junctions containing a recombinant form of rat heart Cx43 in which the majority of the C-terminal domain has been deleted (truncation at Lys 263) (12). (a) Negatively stained crystals display diffraction to ~15Å resolution. (c) The projection density map shows the hexameric connexon, similar to maps at comparable resolution from rat heart gap junctions (Figure 8.6, top; (10, 137). (b) When unstained crystals are examined using electron cryomicroscopy, the diffraction patterns extend to 7Å resolution. (d) The projection density map at 7Å resolution displays three major features: a ring of circular densities centered at radius 17Å interpreted as α-helices that line the channel, a ring of densities centered at 33Å radius interpreted as α-helices that are exposed to the lipid and a continuous band of density at 25Å radius separating the two groups of helices. The hexagonal lattice has parameters a = b = 79Å and γ = 120Å. The spacing between grid bars is 40Å. (Modified from Ref. 5 and reproduced by permission of Current Biology Ltd.)

tals of channels formed by mutant recombinant Cx43 that lacks the ~13-kDa carboxy-tail (see Fig. 8.2) (14). A side view of the map (Fig. 8.8) shows that the mutant channel has a thickness of ~150 Å compared with ~250 Å for gap junction channels formed by the full-length Cx43 protein. The reduced length is consistent with the loss of the ~13-kDa C terminus. The outer diameter of the mutant channel within the membrane region is ~70 Å and the diameter narrows to ~55 Å within the extracellular gap. A top view of the map (Fig. 8.9) shows that there are 24 rod-shaped densities per connexon. Because each connexon is formed by six subunits, the map confirms that each connexin has four transmembrane α-helices as originally predicted by hydropathy analysis. Because two connexons dock end

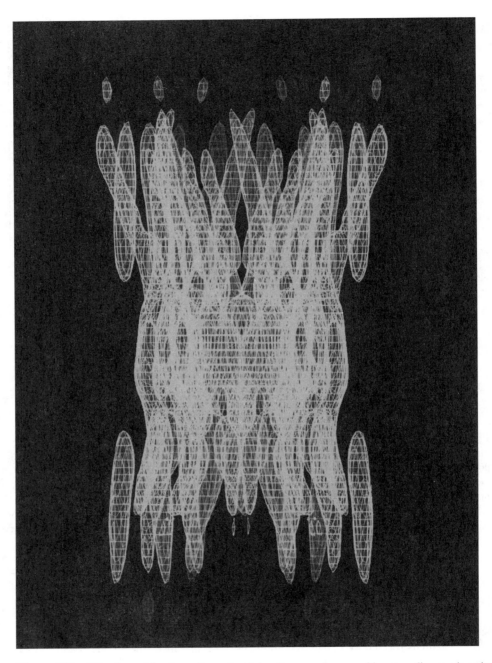

Figure 8.8 Side view of the three-dimensional density map of a recombinant cardiac gap junction channel (14). The dodecameric channel formed by a C-terminal truncation mutant of Cx43 has a length of ~150Å. The outer boundary of the map shows that the diameter of the channel is ~70Å within the two membrane regions and then narrows in the extracellular gap to a diameter of ~50Å. The rodlike features in the map correspond to transmembrane α-helices. (See also color plate.)

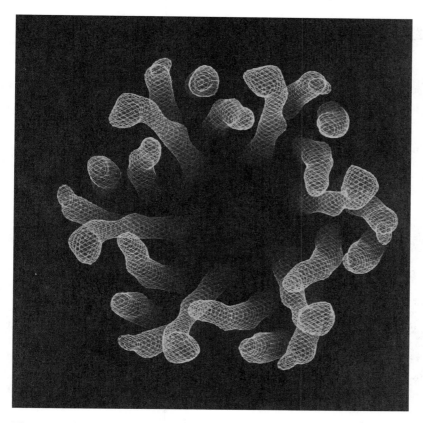

Figure 8.9   A top view looking toward the extracellular gap of the three-dimensional map of a re-combinant cardiac gap junction channel. For clarity, only the cytoplasmic and most of the mem-brane-spanning regions of one connexon are shown. The rod-shaped features are resolved well and reveal the packing of twenty-four transmembrane α-helices within each connexon. (See also color plate.)

to end to form the intercellular channel, the dodecamer contains 48 transmembrane α-he-lices. The close packing of the α-helices confers considerable structural stability.

VII.   MULTIPLE CONNEXINS ARE EXPRESSED IN THE HEART IN TISSUE-SPECIFIC PATTERNS

Messenger RNAs encoding Cx37, Cx40, Cx43, Cx45, Cx46, and Cx50 have been detected in homogenates of mammalian heart muscle (15–20). Not all of these proteins are neces-sarily expressed by cardiac myocytes. For example, expression of Cx37 in the heart is con-fined to the endothelium of the coronary vasculature (18). Cx50 was originally localized to the region of the atrioventricular valves in the early postnatal rat heart and may be ex-pressed by an interstitial cell (19). Other proteins such as Cx43 and Cx40 are expressed by both cardiac myocytes and vessel wall cells (endothelium and/or smooth muscle). It has been unequivocally established that mammalian cardiac myocytes express Cx43, Cx45, and Cx40, but different tissues of the heart express different patterns of these connexins.

Table 8.1 Immunolocalization of Connexins in the Heart

	Cx43	Cx40	Cx45	Cx37
Atrium				
Myocardium	+	+	+	−
Endothelium	−	+	−	+
Endocardium	−	+	?	+
Ventricle				
Myocardium	+	−	+	−
Endothelium	−	+	−	+
Endocardium	−	+	+	+
Conducting system				
SA node	+?	+?	+	
AV node	+?	?	?	
His/Purkinje system	+	+	+	

+ = present; − = absent; ?= conflicting results and/or species differences; SA = sinoatrial; AV = atrioventricular.

Cx46, a protein expressed abundantly in the mammalian lens, has been detected by Northern blot analysis in homogenates of adult rat heart (20). Recent evidence suggests that Cx46 may be expressed in certain specialized myocytes of the sinus node (21). Table 8.1 summarizes current knowledge about connexin expression patterns in the heart based on immunohistochemical studies.

The major cardiac gap junction protein Cx43 is expressed in atrial and ventricular myocytes and in the distal His-Purkinje system (22–24). Cx43 has been identified in the rabbit and canine sinus nodes (25,26), but not in bovine sinus or atrioventricular nodes (27,28). Originally cloned and sequenced from rat heart by Beyer et al. (29), Cx43 has been characterized in several mammalian species (including human) as well as in birds and amphibians. Its sequence is highly conserved in mammals (>97% amino acid identity). Although it is the most abundant cardiac connexin, Cx43 is expressed in numerous tissues including vascular endothelium, vascular and uterine smooth muscle, lens epithelium, fibroblasts, astrocytes, pancreatic islet cells, ovarian granulosa cells, and macrophages (2,30).

The mammalian protein Cx40 was originally identified as a homologue of avian Cx42, which was first cloned from a chick embryo cDNA library (15). Cx40 has been cloned from rat, mouse, and human sources and exhibits somewhat greater sequence variability than the other connexins. Expression of Cx40 in the heart is more restricted than Cx43 (23,24,31–34). Cx40 is expressed by atrial myocytes and by the His-Purkinje fibers of the atrioventricular conduction system but not by adult ventricular myocytes. Expression of Cx40 by myocytes of the sinus and atrioventricular nodes is less well characterized, but Cx40 appears to be present in at least some areas within the canine sinus node (26). It is also expressed in endothelial cells of the coronary vasculature (31). Expression of Cx40 may be regulated developmentally. In mice, for example, Cx40 appears to be expressed in the embryonic left ventricle but then undergoes down-regulation during the early postnatal period (35).

Cx45 was also originally cloned and sequenced from a chick embryo cDNA library (15) and, thereafter, from a human source by genomic cloning (36). Cx45 appears to be expressed in relatively low levels in atrial myoctes and the atrioventricular conduction system of mouse and rat hearts (37). Recent evidence suggests that it is widely expressed during cardiac morphogenesis in the mouse heart but then becomes down-regulated (38). Whether it is expressed to an appreciable extent in adult mammalian ventricular myocytes is not clear at the present time (37). In general, connexin expression patterns in the human heart are similar to those observed in mouse, rat, and canine hearts.

Despite uncertainties about the exact patterns of connexin expression in different tissues of the heart and possible species differences in these patterns, there is no doubt that individual cardiac myocytes express multiple proteins. For example, individual atrial myocytes and specialized myocytes in the His-Purkinje system express three gap junction channel proteins: Cx43, Cx40, and Cx45. Multilabel confocal microscopy and immuno-electron microscopy have shown that multiple connexins colocalize within gap junctions (17). There is no evidence to suggest that different connexins are segregated into different gap junctions or within subregions of individual junctions. Little is known, however, about the distribution of multiple connexins within junctions in terms of the molecular composition of individual channels. This is currently a subject of intense research interest.

VIII. CARDIAC CONNEXINS FULFILL DIFFERENT TISSUE COUPLING FUNCTIONS

Expression of multiple connexins by different cardiac tissues raises major questions about the specific functional roles of individual connexins in the heart and the potential biological advantages provided by expression of multiple gap junction channel proteins. Insights into these questions have been revealed recently in studies of mouse models in which connexin expression has been manipulated by gene targeting strategies. In 1995, Reaume et al. (39) reported that targeted deletion of the gene encoding Cx43 in mice (Cx43 –/– mice) results in neonatal death due to a complex malformation of the conotruncal (right ventricular outflow tract) region of the heart. Interestingly, similar malformations have been observed in transgenic mice in which cardiac neural crest cells were selectively targeted either to overexpress Cx43 or express a dominant negative Cx43 construct (40,41). These results suggest that disturbances in intercellular communication resulting in either gain or loss of function in critical tissues during development can lead to similar types of cardiac malformations.

The fact that Cx43 –/– mice develop to term and have beating hearts in utero suggests that other connexins can provide adequate intercellular coupling to support development. Because Cx43 –/– mice die as neonates, it has been difficult to analyze whole heart electrophysiology in these animals. However, it appears that mice which are heterozygous for a Cx43 null allele (Cx43 +/– mice) have a subtle phenotype related to diminished expression of Cx43. Although Cx43 +/– mice appear indistinguishable from their wild-type littermates, the velocity of paced beats measured on the epicardial ventricular surface in isolated, perfused hearts is slow in Cx43 +/– compared with wild-type mice (42,43). In similar preparations, no difference in atrial conduction velocity was observed even though both ventricular and atrial tissues normally express abundant Cx43 and its expression in Cx43 +/– mice is diminished by ~50% in both tissues (43). These results suggest that Cx43 is a major coupling protein in ventricular but not atrial muscle. Recently, two groups have independently created mouse models with targeted deletion of the Cx40 gene (44,45). Phenotypes in Cx40 –/– mice include sinoatrial, intra-atrial, and atrioventricular conduction disturbances and increased vulnerability to induction of atrial arrhythmias (44–46). These findings suggest that Cx40 plays an important role as a coupling protein in tissues in which it is expressed. Nevertheless, there may be considerable redundancy in the specificity of functions fulfilled by different connexins. For example, no overt abnormalities in cardiac structure or function have been observed in mice in which the liver gap junction protein, Cx32, was substituted for one or both Cx43 alleles (47). With anticipated development and analysis of additional mouse models, the specific functional roles of individual connexins in normal cardiac physiology and arrhythmogenesis will be defined in detail.

IX. MULTIPLE CONNEXINS MAY FORM HYBRID CHANNELS

Because individual differentiated cells generally express more than a single species of connexin, the possibility exists for the formation of individual channels composed of more than one isoform. The theoretical combinations of connexins in hybrid channels and a currently used system of nomenclature are illustrated in Figure 8.10. An individual connexon is referred to as homomeric if it consists of subunits of only a single type of connexin. Alternatively, a connexon composed of two or more connexin isoforms is referred to as a heteromeric connexon. The two connexons that form a single, complete channel may each have an identical connexin composition. In this case, the resultant channel is referred to as a homotypic channel. If the connexons have disparate connexin compositions, then the resultant channel is described as a heterotypic channel.

The fact that individual cells typically express two or more connexins means that the potential number of unique combinations of heteromeric connexons and heterotypic channels is truly vast. As described in the next section, homomeric, homotypic channels (i.e., channels, composed entirely of a single connexin species) exhibit distinct biophysical properties. Therefore, formation of various types of hybrid channels could greatly expand the potential for functional specialization in gap junction channels. At the present time, little is known about the natural occurrence or biological significance of hybrid channels in the heart or other differentiated tissues. Much is known, however, about the potential to form channels composed of multiple connexins, based largely on transfection studies in *Xenopus* oocytes. The eggs of the African clawed toad *Xenopus laevis* are approximately 1 mm in diameter and can be easily injected with nucleic acid sequences encoding specific connexins. Such transfected cells produce connexin proteins in abundance and, if pairs of cells are juxtaposed physically, form gap junction channels whose biophysical proper-

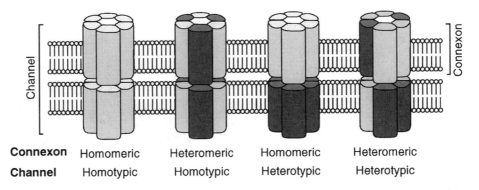

| Connexon | Homomeric | Heteromeric | Homomeric | Heteromeric |
| Channel | Homotypic | Homotypic | Heterotypic | Heterotypic |

Figure 8.10 Nomenclature for connexons and gap junction channels formed by 1 or more connexins. (Modified from Ref. 138.)

ties can be quantified in detail. Using this strategy, it has been shown that some, but not all, homomeric connexons can form heterotypic channels. For example, White et al. (48) have shown that Cx43, Cx46, and Cx50, the multiple connexins expressed in the vertebrate lens, form hybrid channels selectively. Homomeric, homotypic channels were always detected when *Xenopus* oocytes transfected with the same connexins were brought together. However, not all combinations of homomeric, heterotypic channels were observed in oocytes transfected with different connexins. Connexons composed of Cx50 in one oocyte did not form heterotypic channels with an oocyte expressing Cx43. However, functional Cx46/Cx43 and Cx46/Cx50 homomeric, heterotypic channels did occur and their biophysical properties were distinct from those of the corresponding homotypic channels. The second extracellular domain (E2) was specifically implicated as a determinant of formation of stable homomeric, heterotypic channels (48).

Analysis of hybrid channels in *Xenopus* oocytes or other cell types transfected with connexins that are naturally expressed in the heart has revealed that homomeric Cx40 connexons cannot form functional heterotypic channels with homomeric Cx43 connexons (49–51). However, Cx40 and Cx43 may form heteromeric connexons (52). Recent evidence suggests that Cx43 and Cx45 can form both heteromeric connexons and homomeric, heterotypic channels (53,54). These findings have potential implications for intercellular coupling in regions of the heart such as sinus node–atrial myocyte junctions or Purkinje fiber–ventricular myocyte junctions in which pairs of coupled cells may express different connexin phenotypes. However, the functional significance, if any, of hybrid channels in normal cardiac physiology or pathophysiology is unknown at the present time.

X. CHANNELS COMPOSED OF DIFFERENT CONNEXINS ARE FUNCTIONALLY DISTINCT

Rigorous electrophysiological characterization of the biophysical properties of gap junction channels is challenging. Many of the powerful methods used to measure single-channel events in membrane patches in isolated cells must be modified for the study of gap junction channels. For example, except under special circumstances, functional gap junction channels can only be studied in cell pairs. Furthermore, whereas other types of ion channels may be dispersed in the membrane such that only a few channels are included in the portion of the membrane covered by a patch pipette, cells coupled at gap junctions are typically connected by dozens, if not hundreds or thousands, of channels. Analysis of single-channel events, therefore, requires the use of uncoupling agents to create conditions in which only a few channels remain open and can be studied in relative isolation. Because differentiated cells typically express multiple connexins that colocalize in individual gap junctions, it has been difficult to know the specific connexin composition of single channels being analyzed. Naturally occurring gap junction channels in differentiated mammalian cells generally exhibit a wide range of single-channel properties that may reflect both the diversity of hybrid channels and potential post-translational modifications of channel proteins (e.g., phorphorylation) that can affect the biophysical properties of the channels. Because of these complexities, much of what is known about the functional properties of single gap junction channels has come from studies of homomeric, homotypic channels produced by transfection of "communication-deficient" cells such as *Xenopus* oocytes or cultured cell lines that have little or no endogenous connexin expression. The results of this approach have demonstrated that channels formed by different connexins

Table 8.2 Electrophysiological Properties of Gap Junction Channels Formed from Different Cardiovascular Connexins

Connexin	Main state conductance (pS)[a]	Voltage sensitivity[b]	Ionic selectivity[c] (anion/cation)
Cx43	90–115	+	0.13
Cx40	150–160	++	0.29
Cx45	~25	+++	0.20
Cx37	300–350	+++	0.43

[a]Conductance values vary with buffer and salt concentrations. All connexins form channels with different substates.
[b]Voltage sensitivity refers to the relationship between transjunctional voltage potential and conductance.
[c]All connexins have been shown to allow intercellular diffusion of dye molecules and are, therefore, relatively nonselective.
Source: Data from Refs. 55–57, 59–61, 62, 63, and 64.

exhibit distinct biophysical properties, including unitary conductances, pH dependence, voltage-dependence, and selective permeability to ions and small molecules such as fluorescent dyes.

Table 8.2 summarizes selected electrophysiological properties of channels formed by cardiac connexins. Main conductance states measured in Cx43 channels range from 90 to 115 pS (55–58). Cx43 channel conductance is regulated by phosphorylation of the Cx43 protein (55,58). Cx43 channels are less sensitive to changes in transjunctional voltage than channels composed of Cx40, Cx45, or Cx37 (56,57). Both mammalian Cx40 and its avian homolog, Cx42, form channels with larger conductances (150 to 160 pS) than Cx43 channels (57,59–61). In contrast, Cx45 channels from both avian and mammalian sources exhibit a much lower main state conductance of ~25 pS and are highly sensitive to transjunctional voltage (57,62). Cx37 channels have the greatest main state conductance (300 to 350 pS) of any of the cardiovascular connexins and are also highly sensitive to transmembrane voltage (63,64).

Homomeric, homotypic channels formed by different connexins also differ in other functional properties that are presumably determined by divergent sequences in their C-terminal intracellular domains. For example, the degree of cytoplasmic acidification required to achieve channel closure and the sensitivity to lipophilic uncoupling agents differs among channels formed by different connexins (57). Some homomeric, homotypic channels exhibit differential permeabilities to ions and molecules of different sizes and charges. For example, Cx43 channels are permeable to both ions and fluorescent dyes such as Lucifer yellow and 6-carboxy-fluorescein. In contrast, Cx45 channels allow intercellular flow of ions but do not pass larger dye molecules (57). Clearly, the original concept that gap junctional channels are nonselective pores that indiscriminantly permit passage of ions and small molecules has undergone significant revision. Although the occurrence and biological significance of hybrid channels in naturally occurring tissues has not been elucidated, the prevailing sentiment among gap junction biologists is that hybrid channels almost certainly exist in nature and account, at least in part, for the wide range of single-channel properties observed in cell pairs that express multiple connexins. Presumably, multiple connexins have evolved to enhance fine control of intercellular communication and perhaps to provide a mechanism for maintaining both preferential communication pathways and communication boundaries within complex multicellular structures.

XI. INTERCELLULAR COMMUNICATION AT GAP JUNCTIONS IS REGULATED BY COMPLEX MECHANISMS

The molecular determinants that confer distinct biophysical properties on connexin channels are incompletely understood, but major advances have been achieved. Phosphorylation of both Cx43 and Cx45 affects channel function and possibly channel assembly in gap junctions (55,58,65–75). Increasing evidence suggests a direct relationship between phosphorylation of Cx43 and unitary conductance state, but the relationship between the extent and sites of Cx43 phosphorylation and channel function is complex. Alterations in Cx43 phosphorylation affect both macroscopic junctional currents and single-channel behavior in neonatal rat ventricular myocytes and other cells (55,58). Cx43 mutants with progressive deletions of the carboxy terminus form channels with different unitary conductances (76) in keeping with the hypothesis that this portion of the protein is a principal determinant of unitary conductance. However, deletion of all putative phosphorylation sites from Cx43 does not prevent formation of functional channels (77). Musil and Goodenough (67) observed differential solubility of phosphorylated Cx43 isoforms in selected detergents and showed that the abundance of different phosphorylated forms of Cx43 correlated with the assembly of gap junctional plaques. The distribution and extent of Cx45 phosphorylation may be less complex than for Cx43, but less is known about the functional implications of different phosphorylated isoforms of Cx45. Cx45 phosphorylation apparently affects protein stability. A recent study has shown that mutation of specific serine residues in Cx45 has a significant impact on its turnover kinetics (78). Several protein kinases, including protein kinase C, cGMP-dependent protein kinease, mitogen-activated protein kinase, and v-Src have been shown to phosphorylate Cx43 in vitro or in transfected cells engineered to overexpress a specific kinase (58,66,68–75). However, little is known about the endogenous kinases and phosphatases responsible for maintaining high basal levels of phosphorylated connexins or enzymes mediating changes in connexin phosphorylation under pathophysiological conditions.

Changes in the intracellular milieu play an important role in regulating intercellular coupling under both physiological and pathophysiological conditions. Channel conductance properties change in response to exogenous cyclic nucleotides and other signaling molecules. The specific signal transduction pathways responsible for these changes have not been elucidated in detail. Probably, many of these effects are ultimately mediated by changes in phosphorylation. Gap junction channels close in response to increasing levels of intracellular Ca^{2+} and H^+ which act synergistically to promote uncoupling (79,80). Intracellular concentrations of these ions required to fully uncouple cells are probably achieved only under conditions of severe injury. Gap junction channel function is also modulated by fatty acids and lipid metabolites of ischemia that may contribute to uncoupling during early stages of myocardial infarction (81–83).

Biochemical and molecular mechanisms of pH gating of Cx43 channels have been elucidated in mutagenesis experiments (84–87). For example, replacement of histidine 95 in the cytoplasmic loop domain of Cx43 with hydrophobic residues such as methionine or phenylalanine or with the charged basic residue arginine or the noncharged residue glutamine results in nonfunctional channels. Replacement of histidine 95 with the acidic residue aspartate or the polar uncharged residue tyrosine yields functional channels but with diminished susceptibility to uncoupling in response to acidification. Transfection studies involving expression of a 17-amino acid peptide from the carboxy terminus of Cx43 suggest that pH gating is mediated by a particle–receptor interaction in which the carboxy terminus acts as a gating particle that can bind noncovalently to a receptor domain, perhaps in the intracellular loop, and, thereby, lead to channel closure in response to acidification (87).

XII. CONNEXINS TURN OVER RAPIDLY

Although gap junctions may be relatively stable structures in terms of their number, size, and distribution in normal differentiated tissues, increasing evidence suggests that the flux of connexins into and out of gap junctions is highly dynamic. Laird et al. (65) originally reported that Cx43 turns over rapidly ($t_{1/2}$ ~2 h) in cultured neonatal rat ventricular myocytes. This observation has been confirmed in neonatal rat ventricular myocytes (88) and in other cells (67) and extended to include Cx45, which also turns over rapidly in cultured neonatal rat ventricular myocytes ($t_{1/2}$ ~3 h) (88). Although these high turnover rates could be a special feature of cultured cells, it appears that connexins also turn over rapidly in adult tissues. For example, the liver gap junction proteins Cx32 and Cx26 have half-lives of 2 to 3 h in both cultured hepatocytes and liver tissue in vivo (89). Cx43 has a half-life of 1.3 h in isolated, perfused adult rat hearts analyzed by metabolic labeling and pulse-chase experiments (90). These observations suggest that trafficking of connexins into and out of gap junctions occurs at an astonishing rate. Such rapid turnover of protein in junctions may provide a mechanism by which cells adjust their levels of intercellular coupling. Interventions that affect connexin turnover could facilitate remodeling of conduction pathways and have significant consequences regarding the relative and total amounts of individual connexin proteins present in myocytes.

The determinants of connexin turnover are not fully characterized. Studies of connexin degradation in cultured cells, including cardiac myocytes, have revealed that Cx43 is degraded by both endosomal and proteasomal proteolysis pathways (91–93). Both pathways also appear to be responsible for Cx43 degradation in the adult rat heart (90). It has been reported that inhibition of endosomal proteolysis in the rat heart leads to the accumulation of phosphorylated Cx43, whereas inhibition of proteasomal degradation causes nonphosphorylated Cx43 to accumulate (90). These observations are consistent with previous studies suggesting that changes in connexin phosphorylation patterns play an important role in targeting proteins for degradation.

XIII. GAP JUNCTIONS ARE CONTAINED WITHIN THE INTERCALATED DISK

There is an intimate spatial relationship between sites of intercellular electrical coupling at gap junctions and regions within the intercalated disk responsible for mechanically coupling cardiac myocytes (94–96). Gap junctions are typically located at or near the ends of working atrial or ventricular myocytes where the largest intercalated disks connect neighboring cells. When ventricular myocardium is viewed by transmission electron microscopy, two different groups of gap junctions can be identified based on their locations within the intercalated disk. A population of small gap junctions is located within segments of the intercalated disk composed mainly of fascia adherens junctions (open arrow in Fig. 8.11). Many of the remaining junctions are located between offset segments of the adhe-

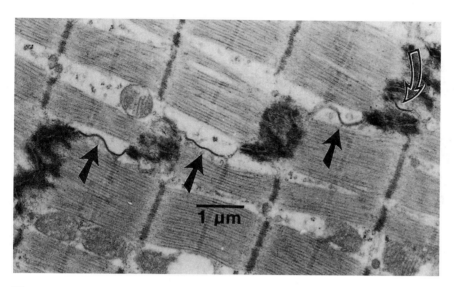

Figure 8.11   A transmission electron micrograph of an intercalated disk connecting two canine ventricular myocytes. The tissue has been sectioned in a plane parallel to the longitudinal axis of the myocytes. The disk contains multiple adhesive components, seen as electron dense structures capping the ends of rows of sarcomeres and offset by units of sarcomere length. Gap junctions (solid arrows) are located between the adhesive junctions and are oriented in a plane parallel to the long axis of the cells. A smaller gap junction (open arrow) is located entirely within an adhesive component. (From Ref. 97.)

sive components (solid arrows in Fig. 8.11). In the longitudinal plane of section, these gap junction profiles are oriented parallel to the long cell axis and juxtaposed perpendicularly to adhesive components of the intercalated disk. As seen in Figure 8.11, these gap junctions have profile lengths that occur in multiples of sarcomere length (inter Z-band length) because each individual adhesive segment typically caps the end of 1 to 3 sarcomeres and, thus, are themselves offset by multiples of sarcomere length.

The majority of gap junction profiles seen in ultrathin sections of normal canine ventricular myocardium cut in longitudinal or transverse planes of section are small, discoid structures with diameters <1.0 μm (97). Another population of junctions, located between offset adhesive regions of the intercalated disk, have longer profile lengths when viewed in the transverse plane versus the longitudinal plane. Although they are not numerous, these long gap junctions probably play an important role in intercellular current transfer based on their size and location within intercalated disks. For example, the longest 10% of gap junction profiles measured in cross-sections of canine ventricular myocardium constitute nearly 40% of aggregate gap junction length (97). The largest of these junctions in ventricular myocytes may have surface areas in excess of 8 μm². Reconstructions of gap junction distributions at the ends of ventricular myocytes based on transmission electron microscopy of ultrathin tissue sections and scanning electron microscopy of isolated ventricular myocytes led to a model first proposed in 1989 (Fig. 8.12) (94). In this model, adhesive elements of the intercalated disk are shown as fingerlike projections that interdigitate with neighboring cells to form fascia adherens and other types of mechanical junctions. Small gap junctions are located within these adhesive segments. Larger junctions are

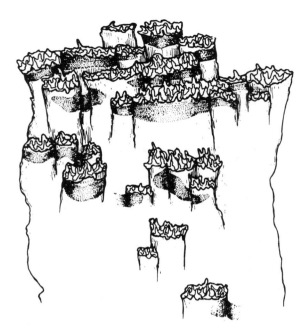

Figure 8.12 Diagram showing the relationship between large gap junctions (shown as stippled patches on the cell surface membrane) and adhesive components (shown as fingerlike projections) of the intercalated disk at the end of a typical ventricular myocyte. The largest gap junctions have a rectangular or ribbonlike shape. They are located in membrane regions between offset shelves of adhesive elements. The long axis of the large gap junctions is oriented perpendicular to the long axis of the myocyte. (From Ref. 94.)

shown as rectangular stippled patches on the cell surface membrane between offset adhesive segments of the intercalated disk. The largest and longest of these gap junctions have a ribbonlike shape with their long axes oriented perpendicular to the long axis of the cell. This model has been confirmed in confocal immunofluorescence images of ventricular muscle stained with anti-gap junction protein antibodies (98,99). When the ends of individual immunostained myocytes are viewed in cross sections, large ribbonlike gap junctions can be seen forming a well-defined ring of junctions outlining the periphery of the cell. The interior aspects of the cell terminus contain numerous, mainly punctate, spots of specific immunofluorescent signal that presumably represent small discoid junctions located deep within the adhesive component of the intercalated disk.

The spatial distribution of gap junctions within large terminal intercalated disks appears to be well suited to the functional requirements of the heart. The close proximity to the nonjunctional sarcolemma (i.e., sarcolemmal membrane adjacent to the extracellular space) of a ring of large gap junctions encircling the terminal intercalated disk probably ensures that sufficient current density is delivered to a localized region of the sarcolemma of an adjacent cell for safe conduction. One consequence of this arrangement is that the dense packing of protein in the lipid bilayer makes these large gap junctions stiff and susceptible to fragmentation in response to shear stress. Shear forces could be significant in large junctions situated between contracting ventricular myocytes. However, packaging these junctions between points of mechanical stabilization at adhesive junctions of adjacent cells, and their configuration in a ribbonlike shape with the long axis of the ribbon oriented perpendicular to the long axis of the cell, likely minimizes shear stress during the cardiac cycle.

XIV. THE SPATIAL DISTRIBUTION OF GAP JUNCTIONS VARIES IN DIFFERENT CARDIAC TISSUES

Although they are concentrated at the ends of myocytes where the largest intercalated disks are located, gap junctions occur at various points along the surface of a typical ventricular myocyte (Fig. 8.13). This distribution of junctions along the cell length provides opportunities for myocytes to be electrically coupled in complex three-dimensional patterns in which individual cells are connected to many neighbors in varying degrees of end-to-end and side-to-side apposition. This pattern is evident in thin slices of ventricular tissues (Fig. 8.14).

Because intercellular current transfer is possible only at points where two cells share gap junctions, it follows that the number and spatial distribution of gap junctions are important determinants of the pattern of current spread in myocardium. Reconstructions of the three-dimensional distribution of intercellular connections at gap junctions have revealed that functionally distinct cardiac tissues are coupled in different patterns. In general, these patterns appear to be consistent with the degree of anisotropy and the velocity of conduction in each tissue.

For example, both the velocity of conduction and the extent to which conduction velocity varies in longitudinal and transverse directions (anisotropy) are moderate in ventricular myocardium. Typical longitudinal conduction velocities in the canine left ventricle are 0.6 to 0.7 m/s and transverse velocities are approximately 0.2 m/s, yielding an anisotropy ratio of roughly 3:1 (100,101). Reconstructions of intercellular connections in the canine left ventricle have revealed that individual ventricular myocytes are coupled to an average of 11.3 neighbors (Fig. 8.15) (34). Approximately half of the attached neighbors are connected to a typical ventricular myocyte in a purely or predominantly side-to-side orientation (referred to as type I and type II connections, respectively, as illustrated in Fig. 8.15)

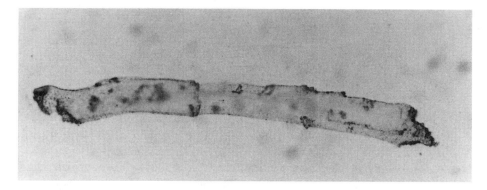

Figure 8.13 An isolated canine left ventricular myocyte stained with an anti-gap junction channel protein antibody to delineate the distribution of gap junctions on the cell surface. The largest gap junctions are concentrated at the ends of the cell but numerous smaller junctions arise at various points along the length of the cell. (From Ref. 139.)

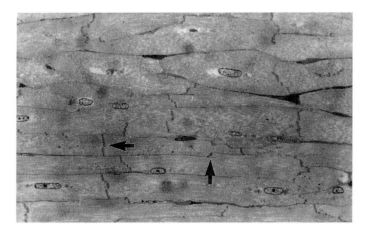

Figure 8.14 A 1-μm-thick section of canine left ventricle showing the complex packing of myocytes in ventricular tissue. Individual cells are linked by intercalated disks and gap junctions that connect cells in end-to-end (horizontal arrows) and side-to-side (vertical arrows) orientations. (From Ref. 97.)

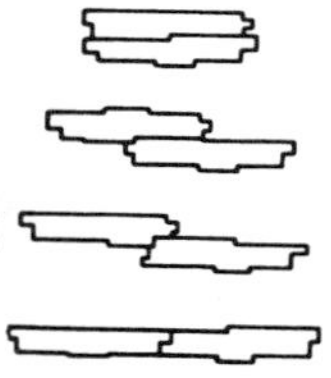

		Left Ventricle	Crista Terminalis	Sinoatrial Node
I		$3.3 \pm 1.4\ (29)^{*}$	$0.8 \pm 0.6\ (12)$	$0.7 \pm 0.5\ (14)$
II		$2.0 \pm 0.7\ (18)$	$0.7 \pm 0.6\ (11)$	$1.4 \pm 1.0\ (30)$
III		$2.1 \pm 0.9\ (19)$	$1.1 \pm 0.7\ (17)$	$2.1 \pm 0.9\ (44)$
IV		$3.9 \pm 1.1\ (34)^{*}$	$3.8 \pm 0.7\ (60)^{*}$	$0.6 \pm 0.7\ (12)$
		11.3 ± 2.2 cells connected to each myocyte	6.4 ± 1.7 cells connected to each myocyte	4.8 ± 0.7 cells connected to each myocyte

Figure 8.15 Diagram of the number and spatial orientation of cellular connections in canine left ventricle, crista terminalis, and sinus node. The numbers in parentheses indicate the percentages of each type of interconnection. Data are expressed as mean ± standard deviation, and asterisks indicate significant differences ($p < 0.01$) between numbers of connections of each type in the ventricle or crista terminalis compared with the sinus node. (From Ref. 34.)

and the remaining half are connected in a purely or mainly end-to-end fashion (type III and type IV connections). The extensive degree of overlap in the packing of ventricular myocytes is evident in a comparison of actual and "effective" length-to-width ratios. The actual single-cell length-to-width ratio, measured in isolated canine ventricular myocytes, is approximately 6:1, but the "effective" length-to-width ratio of ventricular myocytes in tissue, defined as the ratio of the number of cells crossed by equidistant test lines oriented parallel versus perpendicular to the long cell axis, is only 3.4:1, reflecting the significant degree of overlap of cells in tissue. This "blueprint" of intercellular connections is consistent with the moderate degree of anisotropy typical of ventricular conduction. Activation wavefronts moving through a sheet of ventricular myocardium have numerous opportunities to propagate across intercellular junctions in both longitudinal and transverse directions. Because of the elongated shape of the cells, however, wavefronts traveling in the transverse direction must cross more intercellular junctions per unit distance traveled and, thus, would encounter greater resistance and propagate more slowly than wavefronts traveling an equal distance in the longitudinal direction. Anisotropy of conduction velocities in the ventricle, therefore, appears to be determined as much by the shape of the myocytes as by an anisotropic distribution of intercellular junctions.

In contrast to ventricular muscle, conduction in the crista terminalis, a discrete bundle of atrial myocardium that conducts impulses from the sinus node to the AV junction, is rapid (~1 m/s) and highly anisotropic. The ratio of longitudinal to transverse conduction velocity in the crista may be as high as 10:1 compared with only 3:1 in the ventricle (102). Whereas ventricular myocytes are connected on average to 11.3 neighbors, individual myocytes of the crista terminalis are connected to an average of only 6.4 neighbors (Fig. 8.15). However, the great majority of interconnections in the crista terminalis occur between cells oriented in end-to-end apposition. Nearly 80% of the interconnections in the crista terminalis occur between myocytes juxtaposed in a purely (type IV junctions) or mainly (type III junctions) end-to-end orientation (Fig. 8.15).

Like ventricular myocytes, cells of the crista terminalis also have an elongated shape. Myocyte length-to-width ratios are equivalent in the crista and the ventricle, but myocytes of the crista terminalis are more smoothly cylindrical and have relatively few gap junctions along their lengths (34). The functional consequences of the pattern of intercellular connections in the crista seem clear. Transverse propagation in the crista terminalis would certainly be impeded by the relative paucity of connections between cells in side-to-side apposition and the resultant increase in resistance to current propagation in the transverse direction. Highly anisotropic and rapid longitudinal conduction velocity may also be facilitated by more limited electrotonic spread (dissipation) of current in the transverse direction in the crista. Rapid conduction in the crista could also be related to the presence of high-conductance Cx40 channels in atrial, but not ventricular, myocardium.

Reconstructions of the extent and spatial orientation of intercellular connections have revealed yet another entirely different "blueprint" in the sinus node. A typical canine sinus node myocyte is connected to an average of only 4.8 neighbors (Fig. 8.15) (103). Gap junctions in the sinus node occur almost exclusively at small, simple intercalated disks located on cytoplasmic projections arising at various points along the node myocytes. This arrangement of small disks and junctions on cytoplasmic projections provides opportunities for complex packing patterns in which most node myocytes are interconnected in varying degrees of both side-to-side and end-to-end orientation. Nearly 75% of interconnections between sinus node cells are of mixed side-to-side and end-to-end orientation (types II and III), whereas this packing orientation occurs in only 37% and 40%, respectively, of ventricular and crista terminalis connections (Fig. 8.15).

Ultrastructural measurements have revealed that virtually all sinus node junctions are small (mean profile length of 0.17 μm) and fall within a narrow size range. In fact, the aggregate gap junction profile length per 100 μm of intercalated disk length is 12.7 and 3.2 times greater in ventricular and crista terminalis myocytes, respectively, than in sinus node myocytes, and the total amount of gap junction profile length calculated as a function of myocyte area is nearly 27 times greater in the ventricle and 5 times greater in the crista terminalis than in the sinus node (103). Not only are sinus node myocytes connected by intercalated disks to many fewer neighbors than ventricular and crista myocytes, the intercellular connections in the central portion of the sinus node contain relatively few and exceptionally small gap junctions. These structural features may contribute to the slow, nonuniform conduction typical of the sinus node region.

XV. GAP JUNCTION DISTRIBUTIONS CHANGE DURING CARDIAC DEVELOPMENT

The distribution of gap junctions in fetal and neonatal ventricular myocytes is strikingly different than that observed in myocytes of the adult ventricle (19,22,104). Neonatal ventricular myocytes are interconnected by numerous tiny, punctate gap junctions and fascia adherens junctions distributed uniformly over the cell surface (104). With postnatal maturation, there is progressive redistribution of both gap junctions and fascia adherens components of the intercalated disk toward the adult pattern in which large junctions at or near the ends of the cells are associated with large intercalated disks. In the human heart, this transition is not completed until approximately 6 years of age (104). A similar transition in the distribution of ventricular myocyte gap junction occurs in the rat heart (19,22).

The electrophysiological significance of different gap junction distributions at different developmental stages is unclear. Changes in gap junction distribution may help account for the observation that the interval required to depolarize the ventricles increases by only ~20% during a period of rapid growth in which heart weight increases by 16-fold (105). Redistribution of gap junctions may underlie age-related increases in conduction velocity in Purkinje fibers (106) and enhanced anisotropy in atrial muscle (107). It has also been suggested that the neonatal pattern may play a role in both hyperplastic and hypertrophic cardiac growth and help the heart maintain its overall geometry and contractile function during an interval of rapid enlargement (104).

XVI. CELL-TO-CELL COMMUNICATION IS ALTERED IN CARDIAC DISEASE

Alterations in gap junctions probably contribute importantly to the pathogenesis of heart disease but much remains to be learned about disease mechanisms. Mutations in Cx43 at putative phosphorylation sites have been reported, although not confirmed, in children with cardiac malformations characterized by disturbances in the normal left–right asymmetry of the heart (108). Mutations in other connexins associated with specific disease states have also been delineated recently. Mutations in Cx32, a protein expressed in the liver, but also in Schwann cells responsible for myelinating peripheral nerves, have been implicated in the pathogenesis of X-linked Charcot-Marie-Tooth disease (109), a form of hereditary neuropathy with demyelination. Mutations in Cx26, another protein expressed abundantly in the liver but also in other sites such as the inner ear, have been identified in kindreds affected by nonsyndromic forms of hereditary deafness (110). And mutations in the lens protein Cx50 have been implicated in the development of cataracts (111).

Although there is probably not a direct genetic basis underlying altered intercellular coupling in most forms of heart disease, both acute and chronic myocardial responses to injury are associated with pathophysiologically relevant alterations in cell-to-cell communication. Changes in the amount of connexin expression and the distribution of gap junctions occur in several forms of heart disease. The functional implications of these changes have not been elucidated in detail but it is likely that they affect anisotropy of conduction and potentially alter the spatial and temporal coordination of electrical activation in the heart. Altered coupling may also have important implications for cardiac metabolism and contractile function.

XVII. CONNEXIN EXPRESSION IS UP-REGULATED DURING THE ACUTE HYPERTROPHIC RESPONSE

Compensatory hypertrophic growth of cardiac myocytes in response to a moderate increase in load is characterized by increased protein content, changes in cell structure, and increased cardiac performance. Fundamental features of the hypertrophic response include increased synthesis of contractile proteins, assembly of new sarcomeres, and improved contractile function. Studies in vitro and in vivo suggest that compensatory hypertrophic growth may also be associated with increased connexin levels, increased numbers of gap junctions, and enhanced intercellular coupling. For example, long-term (24 h) exposure of neonatal rat ventricular myocyte cultures to cAMP increases the tissue content of Cx43 by approximately twofold and increases the number of gap junctions interconnecting the cells (112). These changes are associated with a significant increase in conduction velocity without apparent changes in active membrane properties (112). Cultured neonatal rat ventricular myocytes exposed for 24 h to angiotensin II also exhibit a twofold increase in Cx43 content and an increase in the number of gap junctions (113).

Remodeling of conduction pathways during early, compensatory responses to increased load in vivo may also be an active process involving enhanced connexin expression and rearrangements of gap junction distributions. For example, connexin expression is enhanced during early stages of hypertrophy induced by renovascular hypertension in guinea pigs (114). Bastide et al. (31) have also shown that expression of Cx40 is enhanced in Purkinje fibers of the rat when hypertrophy is induced by hypertension.

XVIII. CONNEXIN EXPRESSION IS DOWN-REGULATED IN CHRONIC HEART DISEASE

The hypertrophic response is a dynamic continuum in which progressive changes in gene expression and the structure of cells and extracellular matrix occur during the transition from a phase of compensated structural and functional adaptation to an increasingly maladaptive state culminating in heart failure. Ventricular conduction delay, often reflected as prolongation of the QRS interval in the surface electrocardiogram, is a general feature of chronic left ventricular hypertrophy in humans. Conduction velocity first increases in hypertrophied ventricles but then decreases as hypertrophy becomes more severe (115–117). Decrements in conduction velocity may be related to increases in extracellular resistance caused by interstitial fibrosis (118–120) and increases in intercellular resistance due to decreased connexin expression (97,99,118,120,121).

Recently, it has been demonstrated immunohistochemically that Cx43 expression is reduced in segments of "hibernating" myocardium in patients with chronic ischemic heart disease (122). Interestingly, the reduced gap junction protein expression appears to be attributable mainly to a loss of the larger gap junctions normally seen at the major intercalated disks at the ends of cells (122). These results and others suggest that reduced gap junction channel protein levels occur as a general rule in chronic myocardial disease states, including healed myocardial infarction (97,99,118,120,121), chronic hibernation (122), end-stage disease in chronic aortic stenosis (99), and even with aging (107,119). Changes in gap junction distribution in chronic forms of heart disease are only one component of a more generalized, stereotypical response of cardiac myocytes to chronic injury (123–127). This response is characterized most conspicuously by partial or nearly complete loss of sarcomeres, accumulation of glycogen, and disorganization and loss of sarcoplasmic reticulum (123–127), in addition to a reduction in gap junction protein expression and loss of large gap junctions (97,99,122). These stereotypical morphological changes, referred to as myocytolysis, are universally observed, at least to some extent, in dysfunctional wall segments in patients with chronic ischemic heart disease (123) and diverse forms of cardiomyopathy in patients with heart failure (123–126). A similar picture has been reported in experimental models of heart failure or chronic arrhythmias such as chronic, pacing-induced atrial fibrillation (127). Probably the most frequent setting in which myocytolysis occurs is in regions of viable but structurally altered myocytes bordering healed myocardial infarcts which, as described below, are sites of altered gap junction distribution and diminished coupling that likely contribute to reentrant arrhythmogenesis.

XIX.   CHANGES IN GAP JUNCTION DISTRIBUTION CREATE ARRHYTHMIA SUBSTRATES

The best-studied disease setting in which alterations in gap junction distribution have been closely linked to reentrant arrhythmias is myocardial infarct healing. During the inflammatory and reparative phases of infarct healing, viable myocytes at the edges of the infarct scar develop complex structural alterations involving both cardiac myocytes and the extracellular matrix. A common pattern of structural alteration in peri-infarct tissue is accumulation of interstitial bundles of collagen oriented parallel to the long axis of groups of cardiac myocytes (97,128–131). This "substrate" has been observed in regions identified by activation mapping to be sites of slow conduction, conduction block, and complex fractionated electrograms (128–132). Ultrastructural measurements in a healed canine left ventricular infarct model have shown that epicardial border zone myocytes in bundles separated by interspersed collagen are connected by smaller gap junctions than normal myocytes (97). Myocytes in the epicardial border zone lack the subset of very large gap junctions seen in normal myocytes. Furthermore, the number of cells connected to a single canine ventricular myocyte is reduced by nearly half in epicardial border zone regions but the loss of intercellular connections does not occur in a spatially uniform distribution (97). In fact, the mean number of border zone myocytes connected to one another in side-to-side configuration is reduced by 75%, whereas connections between epicardial border zone cells in end-to-end orientation are reduced by only 22% (97). The predicted pathophysiological consequences of these structural alterations are consistent with observations made in both experimental animals and human arrhythmia mapping studies. Longitudinal propagation through remodeled regions would be expected to remain relatively rapid because end-to-end connections are preserved. Ventricular tachycardia is typically induced and maintained when wavefronts activate these critical regions in a direction transverse to the long myocyte axis (128–131). Macroscopic propagation in the transverse direction is likely to be greatly impaired because side-to-side connections are selectively disrupted and wavefronts are forced to zig–zag through the tissue (131) until they reenter postrefractory tissue and initiate the next beat of the tachycardia. The complex pathways followed by such wavefronts probably account for the slow, heterogeneous conduction properties and the presence of fractionated electrograms and late ("diastolic") potentials in border zone regions (120,128–131). Remodeling of gap junctions in peri-infarct myocytes and loss of side-to-side connections have been observed in canine infarcts only 4 days after infarction (120) before significant fibrosis has occurred, and also in healed canine (97) and human (131) infarcts.

One intriguing aspect of the process of gap junction remodeling after myocardial infarction is the apparent recapitulation of a "fetal" arrangement of junctions by viable myocytes at the edges of the acute infarct. Adult cardiac myocytes bordering regions of acute infarction undergo dramatic rearrangement of their gap junctions in which multiple, small junctions become uniformly dispersed over the cell surface in a pattern strikingly similar to that observed in early development (120). Thus, it appears that in response to near-lethal injury at the immediate edges of ischemic infarcts, myocytes recapitulate the fetal pattern of gap junctions, perhaps in a manner analogous to the reexpression of fetal isoforms of some contractile proteins and enzymes during induction of the hypertrophic response. The responsible mechanisms have not been defined nor have the functional consequences of this striking pattern change been elucidated directly. However, strong circumstantial evidence links this pattern to regions of abnormal conduction critical to the pathogenesis of reentrant ventricular tachycardia (120,128–131). It is likely that dynamic changes in both connexin synthesis and degradation play a role in this process.

The potential role of gap junction remodeling in the pathogenesis of other types of arrhythmias such as atrial fibrillation is unknown. No consistent patterns of remodeling of intercellular junctions or change in connexin expression have been identified in patients with chronic atrial fibrillation. A recent report suggests, however, that the distribution of Cx40, but not Cx43, is altered in the atria of goats with chronic atrial fibrillation (133). This observation raises intriguing questions about potential mechanisms regulating differential expression of multiple connexins in heart disease.

REFERENCES

1. Beyer EC, Paul DL, Goodenough DA. Connexin family of gap junction proteins. J Membr Biol 1990; 116:187–194.
2. Willecke K, Hennemann H, Dahl E, Jungbluth S, Heynkes R. The diversity of connexin genes encoding gap junction proteins. Eur J Cell Biol 1991; 56:1–7.
3. Kumar NM, Gilula NB. Molecular biology and genetics of gap junction channels. Semin Cell Biol 1992; 3:3–16.
4. Gimlich RL, Kumar NM, Gilula NB. Differential regulation of the levels of three gap junction mRNAs in Xenopus embryos. J Cell Biol 1990; 110:597–605.
5. Yeager M, Nicholson BJ. Structure of gap junction intercellular channels. Curr Opin Struct Biol 1996; 6:183–192.
6. Yeager M. Structure of cardiac gap junction intercellular channels. J Struct Biol 1998; 121:231–245.
7. Beyer EC, Kistler J, Paul DL, Goodenough DA. Antisera directed against connexin43 peptides react with a 43-kD protein localized to gap junctions in myocardium and other tissues. J Cell Biol 1989; 108:595–605.
8. Yancey SB, John SA, Lal R, Austin BJ, Revel J-P. The 43-kD polypeptide of heart gap junctions: immunolocalization, topology, and functional domains. J Cell Biol 1989; 108:2241–2254.
9. Laird DW, Revel J-P. Biochemical and immunochemical analysis of the arrangement of connexin43 in rat heart gap junction membranes. J Cell Science 1990; 97:109–117.
10. Yeager M, Gilula NB. Membrane topology and quaternary structure of cardiac gap junction ion channels. J Mol Biol 1992; 223:929–948.

11. Kumar NM, Friend DS, Gilula NB. Synthesis and assembly of human α1 gap junctions in BHK cells by DNA transfection with the human α1 cDNA. J Cell Sci 1995; 108:3725–3734.

12. Unger VM, Kumar NM, Gilula NB, Yeager M. Projection structure of gap junction membrane channel at 7Å resolution. Nature Struct Biol 1997; 4:39–43.

13. Unwin PN, Henderson R. Molecular structure determination by electron microscopy of unstained crystalline specimens. J Mol Biol 1975; 94:425–440.

14. Unger VM, Kumar NM, Gilula NB, Yeager M. Three-dimensional structure of a recombinant gap junction membrane channel. Science 1999; 283:1176–1180.

15. Beyer EC. Molecular cloning and developmental expression of two chick embryo gap junction proteins. J Biol Chem 1990; 265:14439–14443.

16. Kanter HL, Saffitz JE, Beyer EC. Cardiac myocytes express multiple gap junction proteins. Circ Res 1992; 70:438–444.

17. Kanter HL, Laing JG, Beyer EC, Green KG, Saffitz JE. Multiple connexins colocalize in canine ventricular myocyte gap junctions. Circ Res 1993; 73:344–350.

18. Reed KE, Westphale EM, Larson DM, Wang HZ, Veenstra RD, Beyer EC. Molecular cloning and functional expression of human connexin37, an endothelial cell gap junction protein. J Clin Invest 1993; 91:997–1004.

19. Gourdie RG, Green CR, Severs NJ, Thompson RP. Immunolabeling patterns of gap junction connexins in the developing and mature rat heart. Anat Embryol 1992; 185:363–378.

20. Paul DL, Ebihara L, Takemoto LJ, Swenson KI, Goodenough DA. Connexin 46, a novel lens gap junction protein, induces voltage-gated currents in nonjunctional plasma membrane of Xenopus oocytes. J Cell Biol 1991; 115:1077–1089.

21. Verheule S. Distribution and physiology of mammalian cardiac gap junctions. Doctoral Thesis, University of Utrecht, 1999.

22. van Kempen MJ, Fromaget C, Gross D, Moorman AF, Lamers WH. Spatial distribution of connexin43, the major cardiac gap junction protein, in the developing and adult rat heart. Circ Res 1991; 68:1638–1651.

23. Kanter HL, Laing JG, Beau SL, Beyer EC, Saffitz JE. Distinct patterns of connexin expression in canine Purkinje fibers and ventricular muscle. Circ Res 1993; 72:1124–1131.

24. Davis LM, Kanter HL, Beyer EC, Saffitz JE. Distinct gap junction protein phenotypes in cardiac tissues with disparate conduction properties. J Am Coll Cardiol 1994; 24:1124–1132.

25. Anumonwo JMB, Wang H-Z, Trabka-Janik E, Dunham B, Veenstra RD, Delmar M, Jalife J. Gap junctional channels in adult mammalian sinus nodal cells: Immunolocalization and electrophysiology. Circ Res 1992; 71:229–239.

26. Kwong KF, Schuessler RB, Green KG, Boineau JP, Saffitz JE. Differential expression of gap junction proteins in the canine sinus node. Circ Res 1998; 82:604–612.

27. Oosthoek PW, Viragh S, Mayen AEM, van Kempen MJA, Lamers WH, Moorman AFM. Immunohistochemical delineation of the conduction system. I. The sinoatrial node. Circ Res 1993; 73:473–481.

28. Oosthoek PW, Viragh S, Lamers WH, Moorman AFM. Immunohistochemical delineation of the conduction system. II. The atrioventricular node and the Purkinje fibers. Circ Res 1993; 73:482–491.

29. Beyer EC, Paul DL, Goodenough DA. Connexin43: a protein from rat heart homologous to a gap junction protein from liver. J Cell Biol 1987; 105:2621–2629.

30. Goodenough DA, Goliger JA, Paul DL. Connexins, connexons, and intercellular communication. Annu Rev Biochem 1996; 65:475–502.

31. Bastide B, Neyses L, Ganten D, Paul M, Willecke F, Traub O. Gap junction protein connexin40 is preferentially expressed in vascular endothelium and conductive bundles of rat myocardium and is increased under hypertensive conditions. Circ Res 1993; 73:1138–1149.

32. Gros D, Jarry-Guichard T, Ten Velde I, de Maziere A, van Kempen JA, Davoust J, Briand JP, Moorman AFM, Jongsma HJ. Restricted distribution of connexin40, a gap junctional protein, in mammalian heart. Circ Res 1994; 74:839–851.

33. Gourdie RG, Severs NJ, Green CR, Rothery S, Germroth P, Thompson RP. The spatial distribution and relative abundance of gap-junctional connexin40 and connexin43 correlate to functional properties of components of the cardiac atrioventricular conduction system. J Cell Sci 1993; 105:985–991.

34. Saffitz JE, Kanter HL, Green KG, Tolley TK, Beyer EC. Tissue-specific determinants of anisotropic conduction velocity in canine atrial and ventricular myocardium. Circ Res 1994; 74:1065–1070.

35. Delorme B, Dahl E, Jarry-Guichard T, Briand J-P, Willecke K, Gros D, Théveniau-Ruissy M. Expression pattern of connexin gene products at the early developmental stages of the mouse cardiovascular system. Circ Res 1997; 81:423–437.

36. Kanter HL, Saffitz JE, Beyer EC. Molecular cloning of two human cardiac gap junction proteins, connexin40 and connexin45. J Mol Cell Cardiol 1994; 26:861–868.

37. Coppen SR, Dupont E, Rothery S, Severs NJ. Connexin45 expression is preferentially associated with the ventricular conduction system in mouse and rat heart. Circ Res 1998; 82:232–243.

38. Alcoléa S, Théveniau-Ruissy M, Jarry-Guichard T, Marics I, Tzouanacou E, Chauvin J-P, Briand J-P, Moorman AFM, Lamers WH, Gros DB. Downregulation of connexin 45 gene products during mouse heart development. Circ Res 1999; 84:1365–1379.

39. Reaume AG, de Sousa PA, Kulkarni S, Langille BL, Zhu D, Davies TC, Jeneja SC, Kidder GM, Rossant J. Cardiac malformation in neonatal mice lacking connexin43. Science 1995; 267:1831–1834.

40. Ewart JL, Cohen MF, Meyer RA, Huang GY, Wessels A, Gourdie RG, Chin AJ, Park SMJ, Lazatin BO, Villabon S, Lo CW. Heart and neural tube defects in transgenic mice overexpressing the Cx43 gap junction gene. Development 1997; 124:1281–1292.

41. Huang GY, Wessels A, Smith BR, Linask KK, Ewart JL, Lo CW. Alteration in connexin 43 gap junction gene dosage impairs conotruncal heart development. Develop Biol 1998; 198:32–44.

42. Guerrero PA, Schuessler RB, Davis LM, Beyer EC, Johnson CM, Yamada KA, Saffitz JE. Slow ventricular conduction in mice heterozygous for a Cx43 null mutation. J Clin Invest 1997; 99:1991–1998.

43. Thomas SA, Schuessler RB, Berul CI, Beardslee MA, Beyer EC, Mendelsohn ME, Saffitz JE. Disparate effects of deficient expression of connexin43 on atrial and ventricular conduction: Evidence for chamber-specific molecular determinants of conduction. Circulation 1998; 97:686–691.

44. Kirchhoff S, Nelles E, Hagendorff A, Krüger O, Traub O, Willecke K. Reduced cardiac conduction velocity and predisposition to arrhythmias in connexin 40-deficient mice. Curr Biol 1998; 8:299–302.

45. Simon AM, Goodenough DA, Paul DL. Mice lacking connexin40 have cardiac conduction abnormalities characteristic of atrioventricular block and bundle branch block. Curr Biol 1998; 8:295–298.

46. Hagendorff A, Schumacher B, Kirchoff S, Lüderitz B, Willecke K. Conduction disturbances and increased atrial vulnerability in connexin 40-deficient mice analyzed by transesophageal stimulation. Circulation 1999; 99:1508–1515.

47. Schumacher B, Hagendorff A, Plum A, Willecke K, Jung W, Lüderitz B. Electrophysiological effects of connexin43-replacement by connexin32 in transgeneous mice. PACE 1999; 22:858 (abstr).

48. White TW, Bruzzone R, Wolfram S, Paul DZ, Goodenough DA. Selective interactions among the multiple connexin proteins expressed in the vertebrate lens: The second extracellular domain is a determinant of compatibility between connexins. J Cell Biol 1994; 125:879–892.

49. Bruzzone R, Haefliger JA, Gimlich RL, Paul DL. Connexin 40, a component of gap junctions in vascular endothelium, is restricted in its ability to interact with other connexins. Mol Biol Cell 1993; 4:7–20.

50. White TW, Paul DL, Goodenough DA, Bruzzone R. Functional analysis of selective interactions among rodent connexins. Mol Biol Cell 1995; 6:459–470.

51. Elfgang C, Eckert R, Lichtenberg-Frate H, Butterweck A, Traub O, Klein RA, Hulser D, Willecke K. Specific permeability and selective formation of gap junction channels in connexin-transfected HeLa cells. J Cell Biol 1995; 129:805–817.

52. He DS, Burt JM. Function of gap junction channels formed in cells co-expressing connexin40 and 43. In: Werner R, ed. Gap Junctions. Amsterdam: IOS Press, 1998:40–44.

53. Moreno AP, Fishman GI, Beyer EC, Spray DC. Voltage dependent gating and single channel analysis of heterotypic channels formed by Cx45 and Cx43. Prog Cell Res 1995; 4:405–408.

54. Koval M, Geist ST, Westphale EM, Kemendy EM, Civitelli R, Beyer EC, Steinberg TH. Transfected connexin 45 alters gap junction permeability in cells expressing endogenous connexin 43. J Cell Biol 1944; 130:987–995.

55. Moreno AP, Sáez JC, Fishman GI, Spray DC. Human connexin43 gap junction channels—Regulation of unitary conductances by phosphorylation. Circ Res 1994; 74:1050–1057.

56. Brink PR, Ramanan SV, Christ GJ. Human connexin43 gap junction channel gating: evidence for mode shifts and/or heterogeneity. Am J Physiol 1996; 271:C321–C331.

57. Veenstra RD. Size and selectivity of gap junction channels formed from different connexins. J Bioenerg Biomembr 1996; 28:317–337.

58. Moreno AP, Fishman GI, Spray DC. Phosphorylation shifts unitary conductance and modifies voltage dependent kinetics of human connexin43 gap junction channels. Biophys J 1992; 62:51–53.

59. Beblo DA, Wang H-Z, Beyer EC, Westphale EM, Veenstra RD. Unique conductance, gating, and selective permeability properties of gap junction channels formed by connexin 40. Circ Res 1995; 77:813–822.

60. Bukauskas FF, Elfgang C, Willecke K, Weingart R. Biophysical properties of gap junction channels formed by mouse connexin40 in induced pairs of transfected human HeLa cells. Biophys J 1995; 68:2289–2298.

61. Beblo DA, Veenstra RD. Monovalent cation permeation through the connexin40 gap junction channel. Cs, Rb, K, Na, Li, TEA, TNA, TBA and effects of anions Br, Cl, F, acetate, aspartate, glutamate, and NO_{t3}. J Gen Physiol 1997; 104:509–522.

62. Veenstra RD, Wang H-Z, Beyer EC, Brink PR. Selective dye and ionic permeability of gap junction channels formed by connexin45. Circ Res 1994; 75:483–490.

63. Reed KE, Westphale EM, Larson DM, Wang H-Z, Veenstra RD, Beyer EC. Molecular cloning and functional expression of human connexin37, an endothelial cell gap junction protein. J Clin Invest 1993; 91:997–1004.

64. Veenstra RD, Wang H-Z, Beyer EC, Ramanan SV, Brink PR. Connexin37 forms high conductance gap junction channels with subconductance state activity and selective dye and ionic permeabilities. Biophys J 1994; 66:1915–1928.

65. Laird DW, Puranam KL, Revel JP. Turnover and phosphorylation dynamics of connexin43 gap junction protein in cultured cardiac myocytes. Biochem J 1993; 273:67–72.

66. Lau AP, Hatch Pigott V, Crow DS. Evidence that heart connexin43 is a phosphoprotein. J Mol Cell Cardiol 1991; 23:659–663.

67. Musil LS, Goodenough DA. Biochemical analysis of connexin43 intracellular transport, phosphorylation, and assembly into gap junctional plaques. J Cell Biol 1991; 115:1357–1374.

68. Goldberg GS, Lau AP. Dynamics of connexin43 phosphorylation in pp60v-src-transformed cells. Biochem J 1993; 295:735–742.

69. Swenson KI, Piwnica-Worms H, McNamee H, Paul DL. Tyrosine phosphorylation of the gap junction protein connexin43 is required for the pp60v-src-induced inhibition of communication. Cell Regul 1990; 1:989–1002.

70. Kanemitsu MY, Lau AF. Epidermal growth factor stimulates the disruption of gap junctional communication and connexin43 phosphorylation independent of 12-0-tetradecanoylphorbol

13-acetate-sensitive protein kinase C: the possible involvement of mitogen-activated protein kinase. Mol Biol Cell 1993; 4:837–848.

71. Sáez JC, Nairn AC, Czernik AJ, Spray DC, Hertzberg EL. Rat connexin43: regulation by phosphorylation in heart. In: Hall JE, Zampighi GA, Davis RM, eds. Progress in Cell Research, Vol 3: Gap Junctions. Amsterdam: Elsevier. 275–282.

72. Loo LWM, Berestecky JM, Kanemitsu MY, Lau AF. pp60src-mediated phosphorylation of connexin43, a gap junction protein. J Biol Chem 1995; 270:12751–12761.

73. Kwak BR, Sáez JC, Wilders R, Chanson M, Fishman GI, Hertzberg EL, Spray DC, Jongsma HJ. cGMP-dependent phosphorylation of connexin43: Influence on gap junction channel conductance and kinetics. Pflügers Arch 1995; 430:770–778.

74. Warn-Cramer BJ, Lampe PD, Kurata WE, Kanemitsu MY, Loo LWM, Eckhart W, Lau AF. Characterization of the mitogen activated protein kinase phosphorylation sites on the connexin43 gap junction protein. J Biol Chem 1996; 271:3779–3786.

75. Kanemitsu MY, Loo LWM, Simon S, Lau AF, Eckhart W. Tyrosine phosphorylation of connexin43 by v-Src is mediated by SH2 and SH3 domain interactions. J Biol Chem 1997; 272:22824–22831.

76. Fishman GI, Moreno AP, Spray DC, Leinwand LA. Functional analysis of human cardiac gap junction channel mutants. Proc Natl Acad Sci USA 1991; 88:3525–3529.

77. Lash JA, Critser BS, Pressler ML. Cloning of gap junctional protein from vascular smooth muscle and expression in two-cell mouse embryos. J Biol Chem 1990; 265:13113–13117.

78. Hertlein B, Butterweck A, Haubrich S, Willecke K, Traub O. Phosphorylated carboxyl terminal serine residues stabilize the mouse gap junction protein connexin45 against degradation. J Membrane Biol 1998; 162:247–257.

79. Burt JM. Block of intercellular communication: interaction of intracellular H^+ and Ca^{2+}. Am J Physiol 1987; 352:C607–C612.

80. White RL, Doeller JE, Verselis VK, Wittenberg BA. Gap junctional conductance between pairs of ventricular myocytes is modulated synergistically by H^+ and Ca^{++}. J Gen Physiol 1990; 95:1061–1075.

81. Burt JM, Massey KD, Minnich BN. Uncoupling of cardiac cells by fatty acids: Structure-activity relationship. Am J Physiol 1991; 260:C439–C448.

82. Hirschi KK, Minnich BN, Moore LK, Burt JM. Oleic acid differentially affects gap-junction mediated communication in heart and vascular smooth muscle cells. Am J Physiol 1993; 265:C1517–C1526.

83. Wu J, McHowat J, Saffitz JE, Yamada KA, Corr PB. Inhibition of gap junctional conductance by long-chain acylcarnitines and their preferential accumulation in junctional sarcolemma during hypoxia. Circ Res 1993; 72:879–89.

84. Liu SG, Taffet S, Stoner L, Delmar M, Vallano ML, Jalife J. A structural basis for the unequal sensitivity of the major cardiac and liver gap junctions to intracellular acidification—the carboxyl tail length. Biophys J 1993; 64:1422–1433.

85. Ek JF, Delmar M, Perzova R, Taffet SM. Role of histidine95 on pH gating of the cardiac gap junction protein connexin43. Circ Res 1994; 74:1058–1064.

86. Morley GE, Taffet SM, Delmar M. Intramolecular interactions mediate pH regulation of connexin43 channels. Biophys J 1996; 70:1294–1302.

87. Calero G, Kanemitsu M, Taffet SM, Lair AF, Delmar M. A 17mer peptide interferes with acidification-induced uncoupling of connexin43. Circ Res 1998; 82:929–935.

88. Darrow BJ, Laing JG, Lampe PD, Saffitz JE, Beyer EC. Expression of multiple connexins in cultured neonatal rat ventricular myocytes. Circ Res 1995; 76:381–387.

89. Traub O, Look J, Dermietzel R, Brummer F, Hulser D, Willecke K. Comparative characterization of the 21-kD and 26-kD gap junction proteins in murine liver and cultured hepatocytes. J Cell Biol 1989; 108:1039–1051.

90. Beardslee MA, Laing JG, Beyer EC, Saffitz JE. Rapid turnover of connexin43 in the adult rat heart. Circ Res 1998; 83:629–635.

91. Laing JG, Beyer EC. The gap junction protein connexin43 is degraded via the ubiquitin pro-teasome pathway. J Biol Chem 1995; 270:26399–26403.

92. Laing JG, Tadros P, Westphale EM, Beyer EC. Degradation of connexin43 gap junctions in-volves both the proteasome and the lysosome. Exp Cell Res 1997; 236:483–492.

93. Laing JG, Tadros PN, Green K, Saffitz JE, Beyer EC. Proteolysis of connexin43-containing gap junctions in normal and heat-stressed cardiac myocytes. Cardiovasc Res 1998; 38:711–718.

94. Hoyt RH, Cohen ML, Saffitz JE. Distribution and three-dimensional structure of intercellular junctions in canine myocardium. Circ Res 1989; 64:565–574.

95. Severs NJ. The cardiac gap junction and intercalated disc. Int J Cardiol 1990; 26:137–173.

96. Page E. Cardiac gap junctions: In: Fozzard HA, Haber E, Jenings RB, Katz AM, Morgan HE, ed. The Heart and Cardiovascular System. New York: Raven Press, 1992:1003–1047.

97. Luke RA, Saffitz JE. Remodeling of ventricular conduction pathways in healed canine infarct border zones. J Clin Invest 1991; 87:1594–1602.

98. Gourdie RG, Green CR, Severs NJ. Gap junction distribution in adult mammalian myocardi-um revealed by an antipeptide antibody and laser scanning confocal microscopy. J Cell Sci 1991; 99:41–55.

99. Peters NS, Green CR, Poole-Wilson PA, Severs NJ. Reduced content of connexin43 gap junc-tions in ventricular myocardium from hypertrophied and ischemic human hearts. Circulation 1993; 88:864–875.

100. Draper MH, Mya-Tu M. A comparison of the conduction velocity in cardiac tissue of various mammals. Q J Exp Physiol 1959; 44:91–109.

101. Kadish AH, Shinnar M, Moore EN, Levine JH, Balke CW, Spear JF. Interaction of fiber ori-entation and direction of impulse propagation with anatomic barriers in anisotropic canine myocardium. Circulation 1988; 78:1478–1494.

102. Spach MS, Miller WT, Geselowitz DB, Barr RC, Kootsey JM, Johnson EA. The discon-tinuous nature of propagation in normal canine cardiac muscle. Evidence for recurrent dis-continuities of intracellular resistance that affect the membrane currents. Circ Res 1981; 48:39–54.

103. Saffitz JE, Green KG, Schuessler RB. Structural determinants of slow conduction in the ca-nine sinus node. J Cardiovasc Electrophysiol 1997; 8:738–744.

104. Peters NS, Severs NJ, Rothery SM, Lincoln C, Yacoub MH, Green CR. Spatiotemporal rela-tion between gap junctions and fascia adherens junctions during postnatal development of hu-man ventricular myocardium. Circulation 1994; 90:713–725.

105. Zak R. Development and proliferative capacity of cardiac muscle cells. Circ Res 1974; 34(suppl II):II17–II26.

106. Rosen MR, Legato MJ, Weiss RM. Developmental changes in impulse conduction in the ca-nine heart. Am J Physiol 1981; 240:H546–H554.

107. Dolber PC, Spach MS. Structure of canine Bachmann's bundle related to propagation of exci-tation. Am J Physiol 1989; 257:H1446–H1457.

108. Britz-Cunningham SH, Shah MM, Zuppan CW, Fletcher WH. Mutations of the connexin43 gap junction gene in patients with heart malformations and defects of laterality. N Engl J Med 1995; 332:1323–1329.

109. Bergoffen J, Scherer SS, Wang S, Scott MO, Bone LJ, Paul DL, Chen K, Lensch MW, Chance PF, Fischbeck KH. Connexin mutations in X-linked Charcot-Marie-Tooth disease. Science 1993; 262:2039–2042.

110. Kelsell DP, Dunlop J, Stevens HP, Lench NJ, Liang JN, Parry G, Mueller RF, Leigh IM. Con-nexin26 mutations in hereditary non-syndromic sensorineural deafness. Nature 1997; 387:80–83.

111. Shiels A, Mackay D, Ionides A, Berry V, Moore A, Bhattacharya S. A mis-sense mutation in the human connexin50 gene (GJA8) underlies autosomal dominant "Zonular Pulverulent" cataract, on chromosome 1q. Am J Hum Genet 1998; 62:526–532.

112. Darrow BJ, Fast VG, Kléber AG, Beyer EC, Saffitz JE. Functional and structural assessment of intercellular communication: increased conduction velocity and enhanced connexin expression in dibutyryl cAMP-treated cultured cardiac myocytes. Circ Res 1996; 79:174–183.

113. Dodge SM, Beardslee MA, Darrow BJ, Green KG, Beyer EC, Saffitz JE. Effects of angiotensin II on expression of the gap junction channel protein connexin43 in neonatal rat ventricular myocytes. J Am Coll Cardiol 1998; 32:800–807.

114. Peters NS, del Monte F, MacLeod KT, Green CR, Poole-Wilson PA, Severs NJ. Increased cardiac myocyte gap-junctional membrane early in renovascular hypertension. J Am Coll Cardiol 1993; 21:59A (abstr).

115. Winterton SJ, Turner MA, O'Gorman DJ, Flores NA, Sheridan DJ. Hypertrophy causes delayed conduction in human and guinea pig myocardium: accentuation during ischaemic perfusion. Cardiovasc Res 1994; 23:47–54.

116. Cooklin M, Wallis WRJ, Sheridan DJ, Fry CH. Changes in cell-to-cell electrical coupling associated with left ventricular hypertrophy. Circ Res 1997; 80:765–771.

117. McIntyre H, Fry CH. Abnormal action potential conduction in isolated human hypertrophied left ventricular myocardium. J Cardiovasc Electrophysiol 1997; 8:887–894.

118. Peters NS. New insights into myocardial arrhythmogenesis: distribution of gap junctional coupling in normal, ischaemic and hypertrophied human hearts. Clin Sci 1996; 90:447–452.

119. Spach MS, Dolber PC. Relating extracellular potentials and their derivatives to anisotropic propagation at a microscopic level in human cardiac muscle. Circ Res 1986; 58:356–371.

120. Peters NS, Coromilas J, Severs NJ, Wit AL. Disturbed connexin43 gap junction distribution correlates with the location of reentrant circuits in the epicardial border zone of healing canine infarcts that cause ventricular tachycardia. Circulation 1997; 95:988–996.

121. Smith JH, Green CR, Peters NS, Rothery S, Severs NJ. Altered patterns of gap junction distribution in ischemic heart disease: an immunohistochemical study of human myocardium using laser scanning confocal microscopy. Am J Pathol 1991; 139:801–821.

122. Kaprielian RR, Gunning M, Dupont E, Sheppard MN, Rothery SM, Underwood R, Pennell DJ, Fox K, Pepper J, Poole-Wilson PA, Severs NJ. Downregulation of immunodetectable connexin43 and decreased gap junction size in the pathogenesis of chronic hibernation in the human left ventricle. Circulation 1998; 97:651–660.

123. Maes A, Flameng W, Nuyts J, Borgers M, Shivalkar B, Ausma J, Bormans G, Schiepers C, De Roo M, Mortelmans L. Histological alterations in chronically hypoperfused myocardium. Correlation with PET findings. Circulation 1994; 90:735–745.

124. Ausma J, Schaart G, Thoné F, Shivalkar B, Flameng W, Depré C, Vanoverschelde J-L, Ramaekers F, Borgers M. Chronic ischemic viable myocardium in man: Aspects of dedifferentiation. Cardiovasc Pathol 1995; 4:29–37.

125. Borgers M, Thoné F, Wouters L, Ausma J, Shivalkar B, Flameng W. Structural correlates of regional myocardial dysfunction in patients with critical coronary artery stenosis: chronic hibernation? Cardiovasc Pathol 1993; 2:237–245.

126. Schaper J, Froede R, Hein ST, Buck A, Hashizume H, Speiser B, Friedl A, Bleese N. Impairment of the myocardial ultrastructure and changes of the cytoskeleton in dilated cardiomyopathy. Circulation 1991; 83:504–514.

127. Ausma J, Wijffels M, Thoné F, Wouters L, Allessie M, Borgers M. Structural changes of atrial myocardium due to sustained atrial fibrillation in the goat. Circulation 1997; 96:3157–3163.

128. Gardner PI, Ursell PC, Fenoglio Jr JJ, Wit AL. Electrophysiologic and anatomic basis for fractionated electrograms recorded from healed myocardial infarcts. Circulation 1985; 72:596–611.

129. Ursell PC, Gardner PI, Albala A, Fenoglio JJ, Wit AL. Structural and electrophysiological changes in the epicardial border zone of canine myocardial infarcts during infarct healing. Circ Res 1985; 56:436–451.

130. Dillon SM, Allessie MA, Ursell PC, Wit AL. Influences of anisotropic tissue structure and

reentrant circuits in the epicardial border zone of subacute canine infarcts. Circ Res 1988; 63:182–206.

131. DeBakker MJT, van Capelle FJL, Janse MJ, Tasseron S, Vermeulen JT, de Jonge N, Lahpor JR. Slow conduction in the infarcted human heart. "Zigzag" course of activation. Circulation 1993; 88:915–926.

132. Janse MJ, Wit AL. Electrophysiological mechanisms of ventricular arrhythmias resulting from myocardial ischemia and infarction. Physiol Rev 1989; 69:1049–1169.

133. van der Velden HM, van Kempen MJ, Wijffels MC, van Zijverden M, Groenewegen WA, Allessie MA, Jongsma HJ. Altered pattern of connexin40 distribution in persistent atrial fibrillation in the goat. J Cardiovasc Electrophysiol 1998; 9:596–607.

134. Kanter HL, Saffitz JE, Beyer EC. Molecular cloning of two human cardiac gap junction proteins, connexin40 and connexin45. J Mol Cell Cardiol 1994; 26:861–868.

9

Impulse Conduction
Continuous and Discontinuous

CHARLES ANTZELEVITCH

Masonic Medical Research Laboratory, Utica, New York

MADISON S. SPACH

Duke University Medical Center, Durham, North Carolina

I. INTRODUCTION

Every beat of the heart is initiated and regulated by electrical impulses produced by the flow of ions across the membrane of individual cardiac cells. This process, when repeated in each adjoining cell of the heart, causes an orderly spread of electrical activity (conduction) and a synchronous contraction of the myocardium. Like other excitable tissues, cardiac cells are electrically connected through low-resistance pathways known as gap junctions. These pathways facilitate the spread of the electrical impulse from one cell to the next, ensuring efficient activation and pumping action.

Under normal conditions, impulse conduction appears continuous, although at a microscopic level discontinuities can be discerned as the impulse courses through each succeeding gap junction. In disease states, these microscopic discontinuities are greatly amplified, leading to delays in impulse conduction and disorganization of the orderly activation of the heart; thus abnormal rhythms arise that can render the heart inefficient or totally ineffective as a pump.

In this chapter, we will examine the basis for continuous and discontinuous impulse conduction in various tissues of the heart and explore the role of discontinuous conduction in the development of cardiac arrhythmias.

II. DETERMINANTS OF NORMAL AND DISCONTINUOUS CONDUCTION

Impulse conduction in cardiac tissues involves the transmission of local circuit current through gap junctions that electrically couple the cells that make up the cardiac syncytium. Local circuit current or electrotonic current flows whenever a voltage difference occurs between two sites within the syncytium. This current is governed by Ohm's law such that its intensity is directly proportional to the voltage gradient between the two sites and inversely proportional to the resistance to the flow of local circuit current through the intracellular (gap junctional and myoplasmic resistance) and extracellular space.

A. Electrotonus and Local Circuit Current

Simply defined, electrotonus is a voltage change attributable to the flow of current through the resistance and capacitance of a structure. In excitable cells, it is the potential change produced by current acting to discharge the capacity of the membrane. A true electrotonic potential is a response whose characteristics are dictated by the passive cable properties of the preparation and one in which active generator properties of the membrane play no role. The mathematical derivations describing the cable properties that delineate electrotonic behavior in excitable tissues are similar to those formulated in the mid-19th century by Lord Kelvin to describe the decrement in the signal carried by the trans-Atlantic telegraph cable. Modifications of these "cable" equations have been applied to both nerve and cardiac tissues (for reviews, see Refs. 1–3).

The basis for electrical interaction between two excitable cardiac cells is illustrated in Figure 9.1. The equivalent circuit of each cell is represented by a membrane resistance (R_m) and capacitance (C_m) in parallel. R_m depends on how effectively the protein ion channels embedded in the lipid bilayer conduct ions such as sodium, potassium, calcium, and chloride. C_m is a property of the lipid bilayer whose hydrophobic center excludes ions and thereby serves as a dielectric across which capacitative charge can develop. The two cells are connected by a low-resistance pathway denoted as R_i, representing the resistance of the gap junctions that join the intracellular space of the two cells and, to a lesser extent, the resistance of the myoplasm of each cell.

Excitation of cell A causes an increase in the conductance of the membrane to sodium (gNa). Sodium ions flow down their electrochemical gradient and displace the negative charges on the inside of the membrane. In cardiac cells, as with many other excitable cells, the increase in gNa is regenerative such that the initial influx of Na^+ causes a depolarization that in turn leads to a further increase in gNa. As cell A depolarizes due to the flow of ions across the membrane (ion current, I_i), a voltage difference develops between it and cell B. This potential difference causes current to flow across the gap junction between the two cells. This current may be carried by any ion species capable of traversing the gap junctions and is believed to be in large part carried by K^+ ions (the predominant cation inside the cell). The magnitude of the current is directly proportional to the voltage difference between cells A and B and inversely proportional to the resistance of the gap junction and myoplasm (according to Ohm's law, $I = V/R$). The current acts to displace negative charges from the inside of the membrane of cell B, which in turn causes positive ions on the outside of the membrane to flow toward cell A. Although no ions move across the membrane of cell B, the cell is depolarized by what is termed a capacitive current (I_c). If the depolarization is sufficient to bring the membrane to its threshold potential, sodium will enter through the sodium channels and an action potential will be generated in cell B.

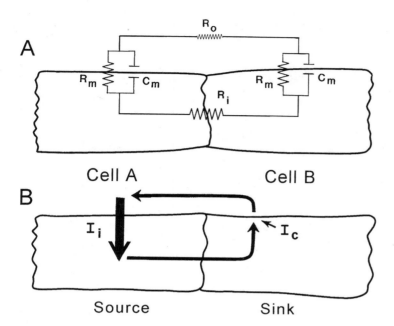

Figure 9.1   Basis for electrotonic communication between excitable cells. (A) Equivalent electrical circuit of two adjoining cells. R_m = membrane resistance; C_m = membrane capacitance; R_o = extracellular resistance; R_i = intracellular resistance to the flow of local circuit current. (B) Components
of local circuit current. I_i = ionic current; I_c = capacitive current; total membrane current, $I_m = I_i + I_c$.
(Modified from Ref. 40, with permission.)

The circular current flow described comprises the local circuit current or electrotonic current; it serves as the basis for impulse conduction and all cell-to-cell electrical interactions
[Fig. 9.1(B)].

Total membrane current is therefore the sum of the ion (I_i) and capacitive (I_c)
currents:

$$I_m = I_i + I_c. \tag{1}$$

Where I_i is dependent on the gating and maximum conductance of ion channels within the
membrane (see Chaps. 2 and 3) and I_c is dependent on membrane capacity and the rate of
change of membrane voltage:

$$I_c = C_m \cdot \frac{dV_m}{dt}. \tag{2}$$

B.   Sink and Source Factors

It is generally accepted that the determinants of conduction in heart can be divided into
source and sink factors where the source is represented by cells already activated or an extrinsic stimulus and the sink is composed of tissue awaiting activation (3). For conduction
to succeed, the intensity of the local circuit current provided by the source must exceed the
threshold current (I_{th}) requirements of the sink. The point in the cardiac cycle at which the
source current exceeds I_{th} marks the end of the refractory period of the system [Fig.
9.2(A)]. The electrical coupling between cardiac cells is normally such that the cell-to-cell

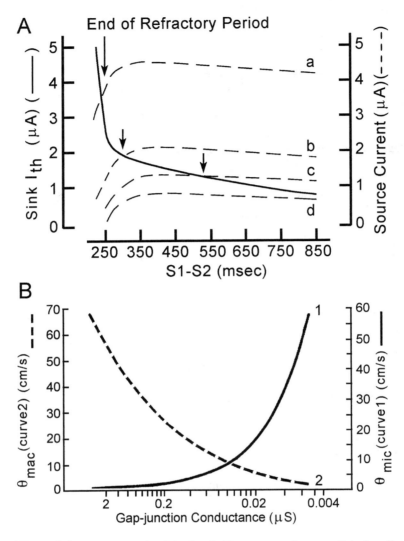

Figure 9.2 (A) Schematic of the threshold current requirements (I_{th}) of cardiac tissue awaiting activation (sink) and the availability of current provided by tissue already activated (source current) at different phases of the cardiac cycle (S_1–S_2 interval) and different levels of electrotonic interaction between source and sink. (Based on data obtained from canine Purkinje fibers.) Conduction is possible only when the source current provided to the sink exceeds the I_{th} of the sink. The crossover of the two relationships marks the end of the refractory period of the sink (vertical arrows). With progressive electrical uncoupling of source and sink, the intensity of source current available to activate the sink is diminished resulting in the development of conduction delays, postrepolarization refractoriness, and impulse conduction failure. Complete conduction block occurs when the intensity of source current reaching the sink is insufficient to meet the threshold current requirements of the sink. (B) Effects of variations in gap junctional conductance on microscopic and macroscopic velocities. Macroscopic average velocity over many cells (dashed line) decreases whereas microscopic intracellular velocity (solid line) increases with decreasing gap junction conductance. (Modified from Refs. 40 and 41, with permission.)

delay is minimal and conduction velocities as high as 2 to 4 m/s are achieved. Figure 9.2 illustrates the effects of altering electrical coupling. When the coupling between source and sink is normal, the source current available to meet the threshold requirement of the sink is relatively high (curve a) and the refractory period (indicated by the vertical arrows) is approximately 250 ms, coinciding with the terminal repolarization phase of the action potential (canine Purkinje fiber). When the availability of source current is diminished due to poor electrical coupling or other factors, the refractory period can extend well beyond full repolarization of the action potential, a phenomenon known as postrepolarization refractoriness (trace c) (4–9). A further decrease in source current leads to conduction block or failure of the cell to excite its neighbors (trace d). When the availability of source current is diminished, apparent conduction velocity can decrease to levels as low as 0.003 m/s [Fig 9.2(B)]. In summary, under normal conditions the source current greatly exceeds I_{th}. As the source/sink ratio approaches 1, major delays can be expected, leading to discontinuities in conduction. When the ratio drops below 1, conduction block occurs.

A pathophysiological reduction in the source/sink ratio can occur as a result of:

1. An increase in R_i, caused by either a decrease in gap junctional conductance or the imposition of real or functional inexcitable zone.
2. A reduction in the amplitude of inward currents and action potential amplitude at the source.
3. A reduction of the excitability of the sink (increased I_{th}).

An increase in R_i, that is, an increase in the resistance to the flow of local circuit current between cells or between two active (excitable) regions of tissue, can greatly alter conduction characteristics. An increase in effective resistance to the flow of electrotonic current can result from (1) a decrease in gap junctional conductance or (2) the imposition of an inexcitable zone at a time when gap junctional resistance is normal or even reduced.

When a decrease in gap junctional conductance is homogeneous among all cells of the tissue, conduction is slowed but remains macroscopically continuous. Conduction velocity (θ) under these conditions is directly proportional to the length constant (λ) and inversely proportional to the membrane time constant (τ_m, defined below):

$$\theta \propto \lambda/\tau_m$$

When conduction becomes discontinuous due to imposition of an inexcitable zone (illustrated by the schematic in Fig. 9.3), an action potential elicited by stimulation of the proximal (P) side of the preparation conducts normally up to the border of the inexcitable zone. Although active propagation of the impulse stops at this point, the local circuit current generated by the proximal segment continues to flow axially, encountering the resistance of successive gap junctions as it traverses the inexcitable zone. Transmembrane recordings from the first few inexcitable cells show responses not very different from the action potential recorded in the neighboring excitable cells. This occurs despite the fact that no ions may be moving across the membrane of the inexcitable cells. The responses recorded in the inexcitable zone result from the discharge of the membrane capacity of cells in that region by local circuit current flowing through the tissue; they are the electrotonic images of activity generated in the proximal segment. The resistive–capacitive properties of the tissue lead to a progressive slowing of the rate of rise of upstroke voltage (dV/dt) and an exponential decline in the amplitude of the membrane potential recorded along the length of the inexcitable cable, according to the following equation

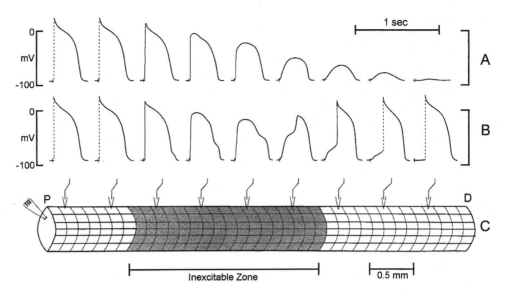

Figure 9.3 Discontinuous conduction (B) and conduction block (A) in a Purkinje strand with a central inexcitable zone (C). The schematic illustration is based on transmembrane recordings obtained from canine Purkinje fiber–sucrose gap preparations. An action potential elicited by stimulation of the proximal (*P*) side of the preparation conducts normally up to the border of the inexcitable zone. Active propagation of the impulse stops at this point, but local circuit current generated by the proximal segment continues to flow through the preparation encountering a cumulative resistance (successive gap junctions). Transmembrane recordings from the first few inexcitable cells show a response not very different from the action potentials recorded in the neighboring excitable cells, in spite of the fact that no ions may be moving across the membrane of these cells. The responses recorded in the inexcitable region are the electrotonic images of activity generated in the proximal excitable segment. The resistive–capacitive properties of the tissue lead to an exponential decline in the amplitude of the transmembrane potential recorded along the length of the inexcitable segment and to a slowing of the rate of change of voltage as a function of time. If, as in panel (B), the electrotonic current is sufficient to bring the distal excitable tissue to its threshold potential, an action potential is generated after a step delay imposed by the slow discharge of the capacity of the distal (*D*) membrane by the electrotonic current (foot-potential). Active conduction of the impulse therefore stops at the proximal border of the inexcitable zone and resumes at the distal border after a step delay that may range from a few to tens or hundreds of milliseconds. (Modified from Ref. 40, with permission.)

$$V_x = V_o \, e^{-x/\lambda}. \tag{3}$$

The distance over which the amplitude of an electrotonic potential decays to 1/e (36.8%) of its peak value (amplitude at the beginning of inexcitable cable) is defined as the length constant (λ) and can be described mathematically as proportional to the $\sqrt{R_m/R_i}$, or equal to the square root of the ratio of transmembrane resistance (r_m) to the sum of the inside (r_i) and outside (r_o) axial resistances:

$$\lambda = \sqrt{r_m/r_i + r_o}, \tag{4}$$

where *r* values are normalized for unit length (1 cm).

In cardiac tissues, λ ranges between 0.4 and 2.0 mm. The time taken for the electrotonic potential to reach $1-1/e$ (63.2%) of its steady-state value defines another characteristic cable constant known as the membrane time constant (τ_m). Mathematically, τ_m is described as the product of R_m and C_m:

$$\tau_m = R_m * C_m. \tag{5}$$

If the electrotonic current encounters excitable tissue before dissipating totally, it may bring the distal excitable tissue to its threshold potential [Fig. 9.3(B)]. The result is a step delay in conduction imposed by the slow discharge of the capacity of the distal (D) tissue by the electrotonic current emerging from the inexcitable gap, manifested as a footpotential preceding the action potential upstroke. Active conduction therefore stops at the proximal border of the inexcitable zone and resumes at the distal border after a step delay that may be on the order of tens or hundreds of milliseconds. Conduction under these conditions is described as discontinuous or "saltatory."

A reduction in the amplitude of net inward current and action potential amplitude at the source has the same effect on conduction as an increase in axial resistance. Both reduce the availability of source current needed to meet the I_{th} of the sink.

Finally, a reduction in the excitability of the sink current (increased I_{th}), secondary to a decrease in the availability of net inward, can reduce the source/sink ratio and lead to slowing of conduction and development of postrepolarization refractoriness. Referring back to Figure 9.2(A), we can appreciate the fact that an upward shift of the excitability curve (I_{th} vs. $S1-S2$) has a similar effect on refractoriness as reduced availability of source current.

C. Modulation of Excitability of the Sink and Electrotonic Inhibition

The excitability of the sink can be influenced by a wide variety of factors ranging from ion channel modulators to discrete subthreshold events. Our discussion in this section will be limited to the influence of subthreshold depolarizations on excitability and how this can affect conduction in the heart. Figure 9.4 shows the inhibitory effects of subthreshold depolarizations on the excitability of canine Purkinje fibers. The degree of inhibitory influence of a subthreshold depolarization is shown to be a sensitive function of both the amplitude of the conditioning event and its temporal relationship to the test pulse (10). These observations suggest that the electrotonic potential of a blocked impulse can influence the excitability of the sink and thus the conduction of the next impulse. An example of this phenomenon, termed "electrotonic inhibition," is illustrated in Figure 9.5. The two transmembrane recordings were obtained from the proximal and distal segments of a canine Purkinje fiber–sucrose gap preparation. Propagation of responses elicited by stimulation of the proximal segment (P) during basic drive (S_1) at a basic cycle length (BCL) of 1500 ms was relatively prompt (P–D interval: 90 ms). An interpolated stimulus delivered to the proximal segment (S_2) was used to elicit proximal responses at progressively longer S_1-S_2 intervals, once after every fifteenth basic cycle. The first beat in each panel represents the last of a train of 15 beats. In panel A, an interpolated response introduced at an S_1-S_2 interval of 600 ms fails to propagate to the distal site but produces an electrotonically mediated subthreshold depolarization and a conspicuous delay in the conduction of the next basic impulse (S'_1). In panel B, when the interpolated response is introduced 100 ms later, it is again blocked, but the resulting subthreshold depolarization in the distal segment now causes transmission failure of the next basic beat. Further increase of the S_1-S_2 interval (panel C) is attended by augmentation of the first subthreshold event and diminution of the second (basic response) until at an S_1-S_2 interval of 1100 ms the interpolated beat suc-

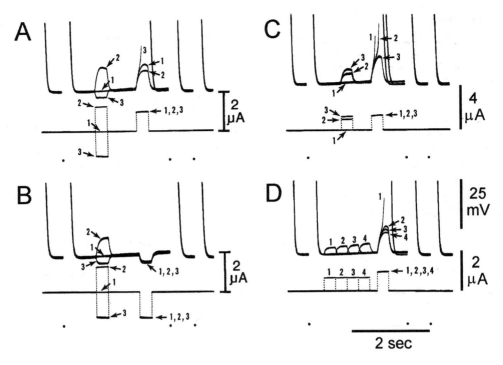

Figure 9.4 Effects of depolarizing and hyperpolarizing conditioning pulses on subsequent responses to current pulses of either polarity. Each panel is a composite of three or more sweeps of the oscilloscope recorded during pauses of 2600 ms interposed between trains of 10 beats each (BCL = 600 ms). Transmembrane activity (top trace) was recorded from the distal segment of a canine Purkinje fiber–sucrose gap preparation at high gain (only bottom third of action potential is shown). The proximal segment of the preparation was inactivated (depolarized with 20 mM KCl) and constant current pulses were applied across the gap. The bottom trace is the current monitor. Responses obtained with each successive sweep are numerically labeled. (A) The membrane response to a depolarizing test pulse was diminished by a depolarizing conditioning pulse but enhanced but a hyperpolarizing conditioning pulse. (B) The electrotonic response to a hyperpolarizing test pulse was little affected by a conditioning pulse of either polarity. (C) Inhibition of the response to a depolarizing test pulse is greater when the amplitude of the conditioning pulse is larger. (D) Inhibition is accentuated when the conditioning response is temporally closer to the test pulse. (Modified from Ref. 10, with permission of the American Physiological Society.)

cessfully conducts across the inexcitable gap, thus defining the end of the effective refractory period (ERP) of the system.

Conduction across poorly coupled regions of myocardium is a sensitive function of rate (see Refs. 11 and 12). The frequency-dependent characteristics of impulse conduction under these conditions are due in part to the rate dependence of the intensity of local circuit current provided by the source as well as rate-dependent changes in the excitability of the sink (9). Because subthreshold responses can modulate the excitability of the sink, the overall rate dependence of conduction is determined not only by impulses successfully conducted, but also by the electrotonic manifestation of impulses blocked at the proximal border of a poorly coupled zone. The subthreshold depolarization of the blocked beat can delay conduction of the subsequent beat or cause it to block. When the subthreshold depo-

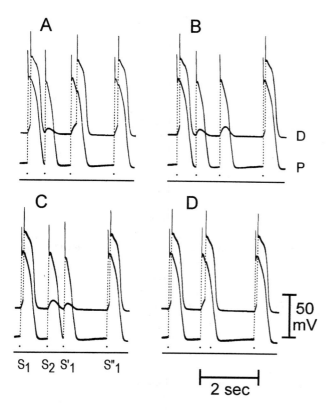

Figure 9.5 Electrotonic inhibition. The two upper traces are transmembrane recordings from the proximal (*P*) and distal (*D*) segments of a canine Purkinje fiber–sucrose gap preparation. Bottom trace is the stimulus marker. Basic stimuli (S_1) were applied to proximal segment at cycle length of 1500 ms and interpolated extrastimuli (S_2) were introduced at progressively longer S_1–S_2 intervals, one after every fifteenth basic beat. The first beat in each panel is the basic beat which, under steady-state conditions, propagates across the inexcitable segment (sucrose gap) with a delay of 90 ms. At an S_1–S_2 interval of 600 ms (panel A), the interpolated response (P_2, second response), although blocked, caused a delay in the conduction of the next basic beat (P'_1) through its electrotonic manifestation at the distal site. When the interpolated beat was introduced later in diastole, its electrotonic image at the distal site increased in amplitude and caused conduction failure of the next basic beat (panel B: S_1–S_2 = 700; panel C: S_1–S_2 = 950 ms). At an S_1–S_2 of 1100 ms (panel D), the premature beat succeeded in conducting across the inexcitable segment, thus defining the end of the effective refractory period of the system. (From Ref. 10, with permission.)

larizations occur close enough in time, the inhibitory influence of each event on the next may cause complete conduction block in a preparation that, at slower stimulation rates, displays almost normal conduction, as in the example illustrated in Figure 9.6. This manifestation of electrotonic inhibition has also been demonstrated in atrial tissues (13,14) and appears to be a characteristic of all excitable cardiac cells.

Thus, electrotonic inhibition can lead to complete suppression of conduction across anisotropic or poorly coupled regions of myocardium. This phenomenon has also aided our understanding of the characteristics of impulse propagation in tissues such as the AV node, where conduction is thought to be discontinuous in some regions. Here, electrotonic

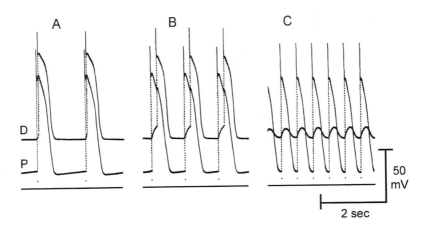

Figure 9.6 Tachycardia-induced complete conduction block as a consequence of the cumulative effects of electrotonic inhibition (Wedensky inhibition). The two transmembrane recordings were obtained from the proximal (*P*) and distal (*D*) segments of a canine Purkinje fiber–sucrose gap preparation. Bottom trace is the stimulus marker. (A) BCL = 1500 ms, 1:1 transmission with little delay. (B) BCL = 1050 ms, 1:1 transmission with long conduction delays. (C) BCL = 500 ms, complete block. (From Ref. 10, with permission of the American Physiological Society.)

inhibition has been suggested to be the mechanism underlying concealed conduction. Lewis and Master (15) were the first to show that impulses blocked in the AV node have an effect on the conduction of subsequent beats, providing evidence for a phenomenon later termed "concealed conduction" by Langendorf (16). Concealed conduction has since been invoked to explain a number of phenomena observed both clinically and experimentally (10,13,14).

D. The Effect of Electrical Load

The issue of electrical load is an important one. If cell A in Figure 9.1 were connected to a dozen cells instead of one, it would "experience" the electrical load of the neighboring cells. The local circuit current generated by cell A would be diluted among 12 cells. As a consequence, far less current would be available to meet the current threshold requirement of each cell and conduction of the impulse from cell A to the others (cells B_{1-12}) would fail. If we consider cells B_{1-12} as the sink, we can conclude that in order to effect conduction the source current must be augmented. This can be achieved by increasing the membrane currents generated by cell A or more practically by having more than one cell serve as the source. In fact, this is how impulses normally propagate in the heart, as groups of cells alternately serve as sources and sinks. When a mismatch in load occurs, discontinuities in conduction are to be expected. As will be discussed later in this chapter, a mismatch in load can occur as a result of an abrupt change in geometry of the tissue under physiological as well as pathophysiological conditions.

E. Discontinuous Conduction and Reentry

Discontinuous conduction recorded from a fiber whose central segment is rendered inexcitable is illustrated in Figure 9.7. The two traces are transmembrane recordings from the

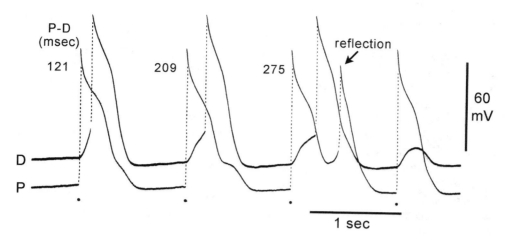

Figure 9.7 Delayed transmission and reflection across an inexcitable gap created by superfusion of the central segment of a Purkinje fiber with an "ion-free" isotonic sucrose solution. The two traces were recorded from proximal (*P*) and distal (*D*) active segments. *P–D* conduction time (indicated in ms in the upper portion of the figure) increased progressively with a 4:3 Wenckebach periodicity. The third stimulated proximal response was followed by a reflection. (From Ref. 12, with permission.)

proximal (*P*) and distal (*D*) segments of an isolated canine Purkinje fiber preparation in which a narrow inexcitable zone was created by superfusion of the central fiber segment with an "ion-free" isotonic sucrose solution. Stimulation of the proximal segment elicits an action potential that propagates normally up to the proximal border of the sucrose gap. Although active propagation across the gap is not possible because of the ion-depleted extracellular milieu, local circuit current continues to flow through the intercellular low-resistance pathways (Ag/AgCl electrodes placed in the two outer compartments provide an extracellular shunt pathway for the return of the local circuit current). The electronic current, much diminished upon emerging from the gap, slowly discharges the capacity of the distal (*D*) tissue. The resulting depolarizations manifest as either subthreshold responses (last distal response) or as footpotentials that bring the distal excitable element to its threshold potential. Active impulse propagation therefore stops and then resumes after a delay that may be as long as several hundred milliseconds. When anterograde (*P → D*) transmission time is long enough to permit recovery of refractoriness at the proximal site, electrotonically mediated transmission of the impulse in the retrograde direction reexcites the proximal segment, thus generating a closely coupled extrasystole by a mechanism known as reflection or reflected reentry (7,8,11,13,14,17,18).

Electrical uncoupling between segments of cardiac tissue giving rise to discontinuities of impulse conduction can occur under a variety of conditions. Any agent or agency capable of suppressing the active generator properties of cardiac tissues can diminish excitability to the point of rendering a localized region functionally inexcitable and thus creating a discontinuity in the active propagation of the advancing wavefront. Exam-ples include an ion-free (sucrose gap), ischemic, or high K^+ environment, (7,8,10,13,14,17, 19–21) as well as electrical blocking (depolarizing) current (22), localized pressure (10,23), and localized cooling (23). Inhibition of the inward currents using sodium and/or calcium blockers can also create discontinuities of active impulse conduction when applied to localized segments.

The resistive barriers created secondary to the development of inexcitable regions are fundamentally similar to those encountered in nonuniform anisotropy where apparently very slow conduction occurs in the direction perpendicular to the long axis of the muscle fibers, especially when bundles of muscle are separated by connective tissue strands (23–28). The important role of anisotropy will be discussed later in the chapter.

Conduction delays on the order of tens or hundreds of milliseconds can occur when electrotonic communication between source and sink is weak. With gradual uncoupling, conduction characteristics generally become progressively more sensitive to changes in the active and passive membrane properties of both the source and sink (9). At low levels of electrotonic communication (see Fig. 9.2, curves b and c), slight changes in the intensity of the source current, as reflected by the action potential amplitude, duration, or maximum rate of rise of the action potential upstroke, $\dot{V}_{max}$ (9,29), as well as small changes in the threshold current requirement of the sink, can produce remarkable changes in conduction characteristics, leading to the development of long conduction delays, postrepolarization refractoriness, and block (9,10).

Nonuniform recovery of refractoriness and geometric factors also play an important role in determining impulse conduction characteristics. Disparity in the recovery of neighboring tissues from refractoriness permits unidirectional block and may contribute to the development of functionally inexcitable or refractory zones as well as lines of block in response to premature beats (30). Disparity of local refractoriness can also contribute to a major slowing of impulse propagation and thus to reentry, as illustrated in Figure 9.8. The three transmembrane traces shown were recorded along the length of a thin canine Purkinje fiber preparation bathed in Tyrode's solution containing 20 mM K^+ and 0.3 μM epinephrine. Epinephrine was added to increase the inward calcium current, the primary charge carrier responsible for the upstroke of the action potential in the setting of K-induced depolarization. The resting potentials were: –53 to –55 mV. The action potential duration (APD_{90}) increased from a value of 155 ms at cell 1 to 190 ms at cell 3. With stimulation applied near cell 1 at a basic cycle length of 1000 ms (first and last responses), propagation is relatively prompt and unencumbered. A premature impulse elicited with an extrastimulus applied at an S_1–S_2 interval of 180 ms propagates relatively promptly to cell 2 but is blocked between sites 2 and 3 where action potential duration is longest [Fig. 9.8(A)]. With an S_1–S_2 of 190 ms, conduction between sites 2 and 3 succeeds but with a step delay of 120 ms, long enough to permit reflection of the impulse and the generation of an extrasystole [Fig. 9.8(B)]. With an S_1–S_2 of 210 ms, conduction between cells 2 and 3 also occurred with a step delay, but one too short to permit reflected reentry [Fig. 9.8(C)]. The discontinuity in conduction occurred at the site where refractoriness would be expected to be longest. The results suggest that local differences in refractoriness can contribute to the development of block and prominent step delays in the conduction of impulses at "functionally" inexcitable zones, thus setting the stage for reflection (and circus movement reentry).

The influence of geometry on impulse conduction characteristics is also well appreciated. Regions at which the cross-sectional area of interconnected cells increases abruptly are known to be potential sites for the development of unidirectional block or delayed conduction. Slowing or block of conduction generally occurs when the impulse propagates in the direction of increasing diameter, because the local circuit current provided by the advancing wavefront is insufficient or barely sufficient to charge the capacity of the larger

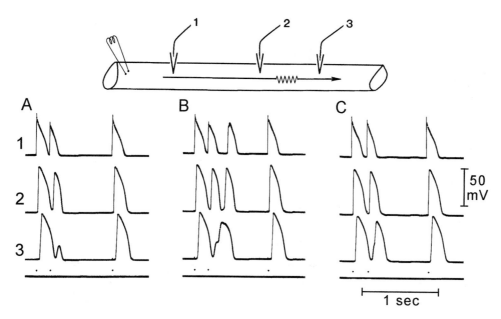

Figure 9.8 Disparity of local refractoriness contributes to conduction delay and reflection in a canine Purkinje fiber bathed in 20 mM K$^+$ and 0.3 μM epinephrine. The diagram at the top shows the arrangement of stimulating and recording electrodes. In the lower panels, the upper three traces are transmembrane recordings from sites 1–3 and the lower trace is the stimulus marker. (A) Application of an extrastimulus at an S_1–S_2 = 180 ms elicits a premature impulse that propagates to site 2, but is blocked between sites 2 and 3. (B) An extrastimulus applied at an S_1–S_2 interval of 190 ms conducts successfully between sites 2 3 with a step delay of 120 ms, which is long enough to permit reflection of the impulse. (C) An extrastimulus applied at a longer S_1–S_2 interval (210 ms) conducts between sites 2 and 3 with a delay too short to permit reflection. (From Ref. 18, with permission.)

volume of tissue ahead and thus bring the larger mass to its threshold potential. The Purkinje–muscle junction is one example of a site at which unidirectional block and conduction delays are commonly observed. The preexcitation (Wolff–Parkinson–White) syndrome is another example, where a thin bundle of tissue (Kent bundle) inserts into a large ventricular mass, creating very significant load mismatch.

Figure 9.9 illustrates an example in which the geometry of the preparation contributes to delayed conduction, block, and reentry of premature beats in a normal Purkinje fiber preparation. The preparation, shown schematically at the top of the figure, was obtained from the right ventricle of a canine heart. With stimulation at a BCL of 1000 ms, conduction of the basic beat (first beat in each panel) was unimpeded and the action potential durations at sites 1, 2, and 3 were 300, 330, and 340 ms, respectively (panel A). A premature beat elicited with an extrastimulus applied at an S_1–S_2 interval of 230 ms conducted to site 1 but was blocked near the branch point before reaching sites 2 and 3 (panel B). When the S_1–S_2 interval was increased by 15 ms (245 ms), the premature wavefront was once again blocked at site 2 but succeeded in activating site 3 after a delay of 125 ms (panel C). With a slight (5 ms) abbreviation of the S_1–S_2 interval to 240 ms, the premature wavefront blocked before reaching site 2 and the activation of site 3 was further delayed (145 ms). The delay was now long enough to permit reentrant excitation of the proximal

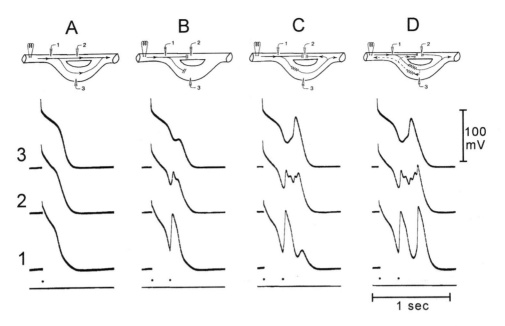

Figure 9.9 Contribution of geometry to delayed conduction, block, and reentry of premature beats in an isolated canine Purkinje preparation. The schematic drawing of the preparation at the top of each panel shows the location of stimulating and recording electrodes and the proposed pathways of impulse propagation. In the lower part of each panel, the top three traces are transmembrane recordings from sites 1–3 and the bottom trace is the stimulus marker. (A) Basic beats at a BCL of 1000 ms. (B) Application of an extrastimulus at an S_1–S_2 = 230 ms elicits a premature response that conducts to site 1 but is blocked near the branch point before reaching sites 2 and 3. (C) A slight increase in the S_1–S_2 interval to 245 ms elicits a premature beat that once again blocks at site 2 but succeeds in activating site 3 after a 125-ms delay. (D) A premature beat elicited at an S_1–S_2 = 240 ms also blocks before reaching site 2, but activates site 3 after a longer delay (145 ms). The delay is sufficient to permit reentrant excitation of the proximal site (site 1). (From Ref. 18, with permission.)

site. The data suggest that retrograde conduction of the impulse (dotted line) may have occurred along the same path as anterograde conduction thus giving rise to a reflected reentry. An alternative interpretation is that the impulse traveled around the preparation in a counterclockwise direction returning through site 2, thus causing reentry through a circus movement mechanism. Regardless of the mechanism involved, it appears clear that the geometry of the preparation (branching leading to a near doubling of the total cross-sectional area) as well as the gradient of action potential duration contributed to the unidirectional block and the major *step* delay in conduction of the impulse, both of which were required for reentry to occur.

Nonuniform recovery of refractoriness giving rise to discontinuous conduction is also encountered in ventricular myocardium, as in the examples illustrated in Figures 9.10 and 9.11. Figure 9.10 shows traces recorded from an arterially perfused left ventricular wedge model of the LQT3 form of the long-QT syndrome. The sea anemone toxin ATX-II, through its actions to augment late sodium current, dramatically increases transmural dispersion of repolarization by producing a much greater prolongation of the action potential of the midmyocardial M cell than in epicardial or endocardial cells. Premature stimulation

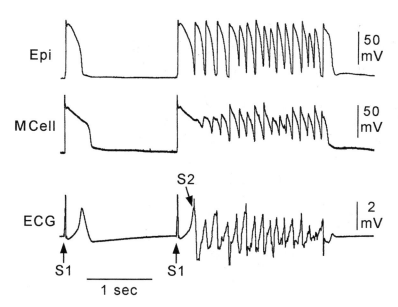

Figure 9.10 Polymorphic ventricular tachycardia displaying features of torsade de pointes (TdP) in an arterially perfused canine left ventricular wedge model of the LQT3 form of the long-QT syndrome. Each trace shows action potentials simultaneously recorded from M and epicardial (Epi) cells together with a transmural ECG. The preparation was paced from the endocardial surface at a BCL of 2000 ms (*S*1). ATX-II produced very significant transmural dispersion of repolarization (first grouping). A single extrastimulus (*S*2) applied to the epicardial surface at an *S*1–*S*2 interval of 320 ms activates the M region after a long delay and initiates TdP (second grouping). (Modified from Ref. 32, with permission.)

applied to the epicardium, the site of briefest refractoriness, elicits a response that propagates into the M region with a long step delay, allowing for the establishment of a presumably reentrant polymorphic ventricular tachycardia exhibiting the characteristics of torsade de pointes (31–37).

Accentuation of transmural electrical heterogeneity can also occur as a result of a disproportionate abbreviation of the epicardial action potential as demonstrated in a model of the Brugada syndrome, illustrated in Figure 9.11 (32,38,39).

III. DISCONTINUOUS PROPAGATION AT A MICROSCOPIC LEVEL: RECURRENT DISCONTINUITIES OF RESISTANCE PRODUCED BY THE CELL-TO-CELL CONNECTIONS

At a microscopic size scale, the spread of excitation in cardiac muscle occurs by a process that is discontinuous in nature (advancing in steps with abrupt changes) due to recurrent discontinuities of intracellular resistance (r_i) that affect the membrane currents (24). Anatomically, the recurrent r_i discontinuities are produced primarily by the gap junctions and the discrete boundaries of cardiomyocytes. Therefore, in this section we will emphasize the distribution of the gap junctions in cardiac bundles composed of parallel fibers. A detailed account of the origin of ideas about discontinuous conduction in cardiac muscle can be found in a recent book (42). We will concentrate on resistive discontinuities that are

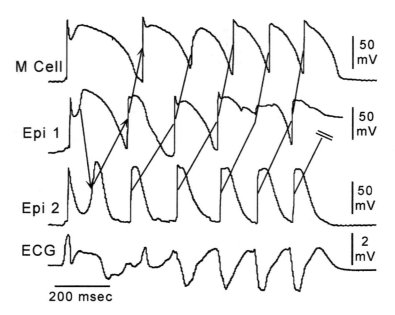

M Cell

Epi 1

Epi 2

ECG

50 mV

50 mV

50 mV

2 mV

200 msec

Figure 9.11   Pressure-induced phase 2 reentry and VT in an arterially perfused right ventricular wedge model of the Brugada syndrome. Shown are transmembrane action potentials simultaneously recorded from two epicardial (Epi 1 and Epi 2) and one M region (M) sites, together with a transmural ECG. Local application of pressure near Epi 2 results in loss of the action potential dome at that site but not at Epi 1 or M sites. The dome at Epi 1 then reexcites Epi 2, giving rise to a phase 2 reentrant extrasystole that propagates into adjoining cells with a very long delay, sufficient to trigger a short run of ventricular tachycardia. (Modified from Ref. 32, with permission.)

located in the intracellular circuit (R_i of Fig. 9.1), in part because only minimal information is available about resistive discontinuities in the restricted extracellular space in between cells beneath the surface of cardiac bundles (43,44) [i.e., interstitial space (R_o of Fig. 9.1)]. The interstitial resistance varies inversely with the volume of interstitial space. Consequently, discontinuities of interstitial resistance occur at sites where there is an abrupt change in the volume of interstitial space, such as at sites where there is an abrupt transition from a smaller to a larger area of the interstitium in which there has been deposition of collagen (43,44). We mention extracellular interstitial discontinuities here because the concept of discontinuous conduction was introduced based on the r_a hypothesis (24), which involves the following: the axial resistance (r_a) depends on the intracellular and extracellular resistances, the size and shape of the cells, the relative volumes of intra- and extracellular space (cell packing), and the resistance and distribution of the cell-to-cell connections. These structural factors produce discontinuities of axial resistance at microscopic and macroscopic levels and, in turn, the resistive discontinuities create nonuniformities of electrical loading that have an important influence on the propagation of excitation in normal and abnormal cardiac muscle (45).

In this section, the experimental examples are based on data from human and canine atrial and ventricular muscle (working myocardium) in which there is general similarity in both species of the geometry of the cells and the cellular distribution of the gap junctions. Abnormal conduction that is proarrhythmic is presented by showing examples from human atrial muscle bundles in which aging has altered the microstructural properties of the atrium (e.g., the incidence of atrial fibrillation increases considerably with aging).

Before proceeding to microstructural discontinuities, however, we present the following five principles as a basis for linking the normal properties of discontinuous conduction to the manner by which these properties change and thereby produce abnormal anatomical substrates that are important in the genesis of arrhythmias:

1. The properties of discontinuous propagation are different along different axes (i.e., anisotropy) in cardiac bundles, which normally contain parallel fibers. The anisotropic behavior occurs as a result of the geometry of elongated cells and the distribution of their gap junctions.

2. Discontinuities of propagation are more prominent during transverse propagation across cardiac fibers than during longitudinal propagation along the fibers.

3. Resistive discontinuities produce anisotropic differences in cellular loading that result in directional differences in the maximum rate of rise of the action potential ($\dot{V}_{max}$). That is, in the presence of constant sarcolemmal membrane ion–channel properties, sites at which there is an increase in electrical load produce a decrease the value of $\dot{V}_{max}$ and a lower safety factor of conduction. Conversely, when the electrical load is decreased, there is an increase in $\dot{V}_{max}$ and an increase in the safety factor of conduction (45).

4. Remodeling of cardiac microstructure with aging and structural heart disease causes discontinuous propagation to become proarrhythmic due to anisotropic changes in conduction velocity and in the safety factor of conduction.

5. An extracellular potential waveform with multiple deflections (fractionation) provides a way to detect anisotropic structural discontinuities that are proarrhythmic.

A. The Dual Nature of Normal Cardiac Conduction—Continuous at a Macroscopic Size Scale and Discontinuous at the Microscopic Level

Atrial and ventricular bundles from normal children and young adults have an abundance of side-to-side and end-to-end electrical connections between cells. Such cardiac bundles have "uniform" anisotropic properties, meaning that the cells are well coupled electrically in all directions. Figure 9.12(A) shows an isochrone map that demonstrates the spread of excitation following the onset of excitation at a point in normal mature ventricular muscle with uniform anisotropic properties. The spread of excitation occurs at a size scale of several millimeters (macroscopic level), with characteristic directional differences in the propagation velocity in relation to the orientation of the fibers. Fast propagation (widely spaced isochrones) occurs along the longitudinal axis of the fibers (LP) and slower propagation (closely spaced isochrones) occurs in the transverse direction across the fibers (TP). The longitudinal velocity (0.51 m/s) and the transverse velocity (0.17 m/s) differ by a factor of 3. An LP/TP velocity ratio of 1.5 to 5 is typical of uniform anisotropic cardiac bundles. The directional differences in propagation velocity are due to the fact that cardiac bundles have a lower axial resistance along the (fast) long axis of the fibers and a higher axial resistance in the (slow) transverse direction.

At the macroscopic size scale, the smooth-appearing spread of excitation depicted by the isochrones in Figure 9.12(A) is similar to that expected in a continuous medium

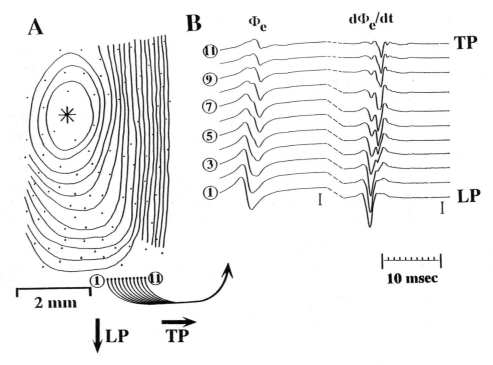

Figure 9.12 Discontinuous anisotropic conduction in uniform anisotropic myocardium. (A) Spread of excitation at a macroscopic level in normal mature ventricular muscle. The isochrone map of excitation spread is shown for a point stimulus (*) in a typical canine left ventricular epicardial preparation. Isochrones, which represent the excitation wave, were drawn at 1-ms intervals based on the excitation times measured at the locations of the scattered dots. The row of small solid circles at the bottom indicates the location of a microarray containing 11 extracellular electrodes. (B) Original extracellular waveforms (Φ_e) and the first derivatives of the extracellular waveforms ($d\Phi_e/dt$) recorded with the microarray. Although the original waveforms appear rather smooth in contour (as would occur in an continuous medium), their first derivatives show multiple notches. The notches in the $d\Phi_e/dt$ waveforms signify underlying discontinuities in propagation at a microscopic level. Note that the notches in the first derivative waveforms increase in going from LP to TP, which signifies that TP is more discontinuous than LP. Vertical calibration lines: $\Phi_e = 5$ mV; $d\Phi_e/dt$ 5 = V/s. (From Ref. 49, with permission.)

with different resistances along different axes. In a syncytium with anisotropic properties, the directional differences in propagation velocity can be accounted for by the following equation for the conduction velocity in a continuous cable:

$$\theta \propto k(1/\sqrt{r_a}), \tag{7}$$

where θ is the effective conduction velocity, k represents a constant cell size with the same excitatory membrane properties (including the membrane capacitance) at all locations, and r_a is the axial resistance in the direction of propagation (45–47). As we shall see, the theoretical and practical implications of Eq. (7) are considerable. First, Eq. (7) indicates that the action potential can be made to propagate as rapidly or as slowly as is desired by choosing the appropriate axial resistance—the higher the value of the axial resistance, the lower the conduction velocity, and vice versa. Second, if conduction is continuous in nature, the time course of the transmembrane action potential (V_m) should not change when there are differences in conduction velocity due to differences in axial resistance along different axes [k in Eq. (7)].

The original unipolar extracellular waveforms (Φ_e) of Figure 9.12(B) demonstrate that the amplitudes of the extracellular waveforms have the same directional dependence as occurs with the propagation velocity. (For purposes of clarity, it should be noted here that "unipolar" extracellular waveforms are recorded as the difference in potential between two sites located far apart. The recording electrode is located near the origin of the excitatory currents and the indifferent electrode is located so far away that the excitatory currents have no detectable effect.) A large-amplitude, smooth positive–negative deflection occurs along the longitudinal axis of fast conduction [Fig. 9.12(B), position 1, longitudinal propagation (LP), and a smaller deflection occurs with slower transverse propagation (TP) [Fig. 9.12(B), position 11]. Equation (7) has been used to provide a quantitative basis for the low amplitude of the extracellular waveforms along the slow transverse (high-resistance) axis of the fibers and for the large amplitude of the extracellular waveforms in the fast longitudinal (low-resistance) direction of the fibers (48). Thus, Figure 9.12 illustrates that by inspecting the amplitude of extracellular waveforms, one can obtain an index of the relative conduction velocities at different sites i.e., large Φ_e waveforms occur with fast conduction and small Φ_e waveforms occur with slow conduction.

The preceding demonstrates some of the powerful applications of continuous medium theory (1,45–47) in accounting for directional differences in conduction velocity and in the extracellular waveforms (48) in uniform anisotropic cardiac bundles. In a continuous anisotropic medium, the first derivatives of the original Φ_e waveforms also should produce curves that are smooth in contour if the underlying propagation process is continuous. However, as illustrated in the associated first derivative extracellular ($d\Phi_e/dt$) waveforms of Figure 9.12(B), in going from longitudinal to transverse propagation irregularities (notches) appear in the derivative waveforms (49). Notches in the $d\Phi_e/dt$ waveforms should not occur in a continuous medium, and they signify discontinuities of propagation at a smaller size scale. Specifically, even though the original extracellular waveforms appear smooth in contour, their derivatives make visible the irregularities (changes in slope) that result from the underlying propagation events being asynchronous or discontinuous. Consequently, the first derivative waveforms of Figure 9.12(B) illustrate two major features of discontinuous propagation in normal cardiac bundles:

1. The discontinuities of propagation become more prominent as the direction of propagation shifts progressively from the fast longitudinal axis of the fibers to occur along the transverse axis across the fibers.
2. The spread of currents generated by the excitation of multiple groups of cells in the region of a recording electrode affects the extracellular potential waveform at a single site (50). In a syncytium that behaves like a continuous medium, the cells are well coupled electrically in all directions. Due to the high degree of electrical coupling, the sequential excitation of small groups of cells occurs as a smooth, synchronized process in the path of propagating depolarization. In turn, the synchronized process produces smooth Φ_e and $d\Phi_e/dt$ extracellular waveforms. On the other hand, notches in the derivative waveforms provide a way to detect small changes in the slope of the extracellular waveforms that reflect underlying discontinuities (delays) in the propagation process. The small local de-

lays result in the asynchronous firing of small groups of cells. In order for different groups of cells to fire asynchronously, they must be separated from one another by an elevated resistance ("loose" coupling) which, in this case, is related to the anisotropic distribution of the gap junctions. Since the discontinuities are maximal during TP and minimal during LP, it follows that the cells are more loosely coupled in a side-by-side manner and they are more tightly coupled in an end-to-end manner (45).

B. Directional Differences in $\dot{V}_{max}$ Due to Microscopic Discontinuities

Figure 9.13 illustrates the effects of discontinuous anisotropic propagation on the rate of rise of the intracellular action potential in epicardial ventricular muscle of an adult dog. In practice, the transmembrane potential V_m is obtained by measuring the intracellular potential (Φ_i) and the extracellular potential (Φ_e) just outside at the membrane surface ($V_m = \Phi_i - \Phi_e$). Because the extracellular potentials were so small in this case, we use V_m and Φ_i interchangeably. The four action potential upstrokes (Φ_i) and their first derivatives ($d\Phi_i/dt$) were recorded while the microelectrode tip remained at the same impalement site. The direction of conduction was altered to occur in each direction along the longitudinal axis of the fibers (LP1, LP2) and in each direction along the transverse axis of the fibers (TP1, TP2). Note that the maximum rate of rise of the action potential has a greater value during slow conduction along the transverse axis of the fibers and a smaller value during fast conduction along the longitudinal axis of the fibers, a characteristic of adult working myocardium of mammals (24,49). The decrease in $\dot{V}_{max}$ with increasing velocity is the opposite of what is expected from changes in membrane properties in a continuous medium (51) e.g., increases in the activation of Na^+ channels increase both $\dot{V}_{max}$ and the velocity (24).

At this point, one can ask whether the recognition that the microstructure is anisotropic helps one to understand why the average value of $\dot{V}_{max}$ is greater during slow transverse than during fast longitudinal propagation. A quantitative explanation requires a computer model that contains an electrical representation (circuit) of the associated network of myocytes and their gap junctions. Accordingly, the major features of the cellular networks of normal adult canine ventricular and atrial muscle (which are similar to those of the human) are illustrated in the drawing of Figure 9.14(A). Note that the gap junctions are located primarily at the ends of the myocytes with only a few gap junctions along the lateral borders of each cell. Consequently, the normal adult distribution of the gap junctions provides a high degree of electrical coupling (low resistance) between cells in an end-to-end manner, but there is loose coupling (elevated resistance) between cells in a side-to-side direction. This anatomical arrangement of the gap junctions is consistent, then, with our interpretation of the derivatives of the extracellular potential waveforms of Figure 9.12(B)—smooth derivatives occur during LP while notches appear in the derivatives during TP due to the asynchronous firing of cells separated by a high resistance.

For the interested reader, Ref. 52 provides a quantitative analysis using the 2-D cellular model represented in Figure 9.14(A), the results of which account for the characteristic anisotropic variations in $\dot{V}_{max}$ shown in Figure 9.13. Based on those quantitative results, we present the following qualitative explanation as to why the distribution of the gap junctions of adult working myocardium produces $\dot{V}_{max}$ results that change in the right direction: a specified set of membrane properties will produce the largest $\dot{V}_{max}$ when the membrane is in the form of an isopotential patch (no extra load). When the membrane is distributed in the form of a continuous cable, each patch of membrane as it is depolarizing has to supply

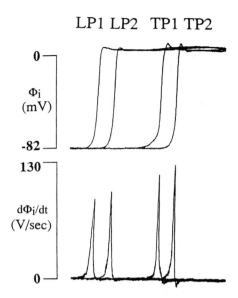

Figure 9.13   Upstrokes of the intracellular action potential Φ_i (top) and the first derivatives of Φ_i (bottom). LP1 and LP2 indicate conduction along the longitudinal axis of the fibers in each direction, and TP1 and TP2 indicate conduction along the transverse axis across fibers in each direction. The in vitro measurements were performed in the epicardium of normal adult canine left ventricular muscle. (From Ref. 49, with permission.)

charging current for the next piece of membrane downstream (an extra load), thus slowing its own depolarization. It is reasonable, then, that $\dot{V}_{max}$ might be larger for transverse propagation than for longitudinal propagation. In the transverse direction, the membrane is divided up into relatively isolated patches by the high resistance of the gap junctions compared to the cytoplasm, whereas in the longitudinal direction, the membrane distribution more closely approximates a continuous cable. During transverse propagation, then, activity would seem to halt momentarily, or hesitate, before moving on to the next unit of cells along the propagation path. Thus, the action potential would be expected to be closer to a membrane action potential (patch) during transverse propagation (24,52).

C. Normal Developmental Changes: Effects of Remodeling the Distribution of Gap Junctions from Birth to Maturity

Normal developmental changes are pertinent to arrhythmias because pathological reiteration of some of the developmental patterns of the gap junctions (53–55) occurs when heart disease and aging alter normal mature atrial and ventricular muscle. Therefore, before proceeding to changes of discontinuous propagation that occur in abnormal anatomical substrates, which are proarrhythmic, it is necessary to consider (1) the geometry of cardiac myocytes and the cellular distribution of gap junctions at birth; and (2) the manner in which cell size and the distribution of the gap junctions change to arrive at the normal coupling pattern of the mature heart [Fig. 9.14(A)].

The drawing of Figure 9.14(B) illustrates the size of the cells and the cellular distribution of the gap junctions in the canine left ventricle at birth. The details fit those of other neonatal mammalian hearts, including the human heart (54,55). At birth, the gap junc-

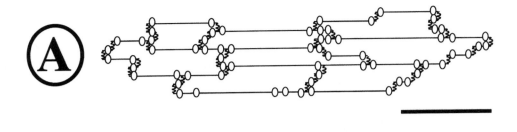

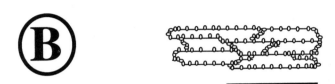

Figure 9.14   Cellular networks representing normal adult and neonatal ventricular muscle. (A) Normal mature ventricular muscle, in which the average value of TP $\dot{V}_{max}$ is greater than that of LP $\dot{V}_{max}$. Gap junctions are located at the ends of the elongated adult cells at the intercalated disks; the transverse irregular lines represent plicate gap junctions within the adhesive component (fascia adherens) of the intercalated disk, and the ovals represent interplicate gap junctions juxtaposed to the fascia adherens (56,57). (B) Normal neonatal ventricular muscle, in which the average values of LP $\dot{V}_{max}$ and TP $\dot{V}_{max}$ are not significantly different. Punctate gap junctions (small ovals) are distributed along the entire surface of the small elongated myocytes. The outlines of the myocytes represent reconstructions from measurements from photomicrographs of disaggregated myocytes from adult (52) and neonatal dog hearts. The distribution of the gap junctions in (A) represents that described by Hoyt et al. (56) and Gourdie et al. (57). The neonatal distribution of gap junctions was reconstructed from photomicrographs of neonatal canine ventricular muscle sections stained immunohistochemically for connexin 43, the major gap junction protein in ventricular muscle. Bars = 50 μm.

tions are distributed along the entire perimeter of small cells. With normal developmental hypertrophy, however, the gap junctions shift from along the lateral borders of cardiac myocytes and become localized at the intercalated disks at the ends of the much larger adult cells (56–58). Consequently, the loss of side-to-side connections between cells produces increasing electrical isolation of the cells from their lateral neighbors, while simultaneously the increased number of end-to-end connections produces increased coupling between cells juxtaposed end to end. Interestingly, the LP/TP velocity ratio remains similar in adult and neonatal canine ventricular muscle, in the range of 2.7 to 3.5 (44), although the LP and TP velocities are lower in the neonate than in its adult counterpart. Also, in the neonate, the value of $\dot{V}_{max}$ changes at different sites along any given axis of conduction, as well as at the same site when the direction of conduction is altered (44), just as occurs in the adult canine ventricle.

Adult and neonatal ventricular muscle differ, however, in their anisotropic patterns of the mean value of $\dot{V}_{max}$. In contrast to the average value of TP, $\dot{V}_{max}$ being greater than LP $\dot{V}_{max}$, as occurs in adult ventricular muscle (Fig. 9.13), there is no significant difference in the mean values of $\dot{V}_{max}$ during LP and TP in neonatal canine ventricular muscle (44). Of additional interest is that the diffuse cellular distribution of the gap junctions illustrated in the cells of Figure 9.14(B) is similar to the pattern that occurs in the small neonatal rat heart cells in synthetic monolayers studied by Fast et al. (59). Electrophysiologically, these synthetic monolayers are similar to neonatal canine ventricular muscle in that there is no significant difference in the mean values of $\dot{V}_{max}$ during LP and TP.

The dissimilar LP–TP $\dot{V}_{max}$ relationships in adult versus neonatal muscle indicate that the differences in the microstructure at the two different ages produce different anisotropic loading effects. That is, normal developmental hypertrophy produces a relative decrease in electrical load ($\uparrow\dot{V}_{max}$) on an increasing area of membrane within each cell during transverse propagation compared to longitudinal propagation. Data based on computer models (such as those represented in Fig. 9.14) are needed for a quantitative analysis of the mechanisms involved when there are changes in both cell size and in the distribution of the gap junctions. However, the following description shows that normal developmental remodeling of the microstructure can produce results that change in the right direction: as the cells enlarge secondary to developmental hypertrophy, the gap junctions along the borders of the cells migrate to the ends of the cells at the intercalated disks in association with an increase in the area of membrane within each cell. These combined structural changes, which result in increasing electrical isolation of enlarging cells from their lateral neighbors, simultaneously produce two electrophysiological changes during TP: (1) progressively larger units of membrane produce an increase in total current within each cell as the cell is excited within a very short time interval [21 μs (52)]; and (2) the lateral delays between cells become progressively longer as a reflection of the loss of side-to-side connections from birth to maturity. Both of these electrical changes result in a progressive increase in the "boundary" effect (60) of the lateral borders of each cell, which increases $\dot{V}_{max}$. These boundary effects of the lateral cell borders on the time course of depolarization of the membrane during TP are similar to those of a membrane (isolated patch) action potential (52) or a collision of excitation waves (60), both of which increase $\dot{V}_{max}$ considerably.

On the other hand, in the longitudinal direction, the increases in cell size are associated with an increase in gap junction plaques at the ends of the cells (increased coupling). Due to the maintenance of tight electrical coupling between the ends of the myocytes from birth to maturity, the membrane along the longitudinal axis of the fibers better approximates the form of a continuous cable, with much smaller delays in impulse transfer from one cell to the next. Thus, at all stages of development, as each patch of membrane depolarizes during longitudinal propagation, it has to supply current to downstream membrane within its own cell and to downstream cells (extra load), thus slowing depolarization of its own action potential.

D. Aging of the Atrium and Healing After Myocardial Infarction Set the Stage for Reentrant Arrhythmias

Two of the major clinical categories of reentrant arrhythmias involve the following: (1) an increased incidence of ventricular reentrant tachycardia following the healing of myocardial infarcts (61); and (2) an escalating incidence of atrial fibrillation with aging (e.g., people over 75 years of age exhibit a prevalence of atrial fibrillation greater than 10%) (62). An intriguing feature of atrial fibrillation in the aging population is that this arrhythmia occurs in some elderly people who do not have any other sign of heart disease (63).

The following question therefore arises: what do healing of ventricular infarction and aging of the atrium have in common? Even though these two reentrant arrhythmias affect different cardiac chambers, they share the same adaptive microstructural response— remodeling of the distribution of the gap junctions occurs in both conditions to produce the general arrangement represented in the drawing of Figure 9.15. First, there is reappearance of gap junctions along the lateral borders of most of the cells (64–66) [i.e., a change in the distribution of the gap junctions to an arrangement similar to that of the neonate as shown in Fig. 9.14(B)]. Second, associated with the enhanced number of lateral connections between most of the myocytes, the cells become divided into small groups due to the development of elongated areas of expanded interstitial space. Widening of the interstitium is due to the accumulation of interstitial fluid (61,66) or to the depostion of collagen (50,67), which forms septa around small groups of cells (microfibrosis). These enlarged interstitial spaces represent locations at which there cannot be direct contact between the laterally juxtaposed cells. Therefore, in the absence of gap junctions connecting cells side to side at these locations, the electrophysiological significance of the collagenous septa is that they mark areas where there is no direct cell-to-cell current between laterally adjacent cells, which results in nonuniform anisotropic properties (26,68).

E. Proarrhythmic Effects of Microfibrosis Due to Very Slow Transverse Conduction

To illustrate the role of discontinuous conduction in the creation of reentrant circuits, we will show the functional effects of the microstructural changes to nonuniform anisotropy in atrial bundles (Fig. 9.15). As noted previously, these properties develop with aging in human atrial bundles. As to experimental models, we know of no animal models that mimic the extensive development of collagenous septa that occurs with aging in human atrial bundles. However, less extensive collagenous septa that create nonuniform anisotropic conduction occur in prominent atrial bundles of adult dogs (e.g., Bachmann's interatrial bundle, the crista terminalis, and the limbus of the fossa ovalis) (68). Figure 9.16 shows the events associated with the conduction of normal action potentials in a nonuniform anisotropic atrial bundle (crista terminalis) of an adult dog (69). The events shown also occur in atrial pectinate bundles from humans over 60 years of age. The isochrone activation sequence of Figure 9.16(A) shows that LP is considerably faster than TP, as expected. However, in contrast to the activation sequence of the uniform anisotropic bundle illustrated in Figure 9.12(A), the close spacing of the isochrones (1 ms apart) in the transverse direction in Figure 9.16(A) indicates that very slow conduction (0.1 m/s) occurs across the fibers. In atrial bundles with nonuniform anisotropic properties from human subjects between 50 and 70 years of age, the average effective TP velocity is 0.08 m/s (67,69). The term "effective velocity" is used to indicate that the path of excitation spread at a microscopic level is complex with a "zigzag" course in the small groups of cells that are separated (uncoupled) laterally over variable distances. These events are reflected in the complex multiphasic ("fractionated") extracellular potential waveforms shown in Figure 9.16(B) (50,67). Thus, during transverse propagation the small deflections of the "fractionated" extracellular waveforms indicate the presence of insulated electrical boundaries (no lateral cytoplasmic continuity) for variable lengths along the small groups of cells. Bundles with fractionated electrograms of nonuniform anisotropy have LP/TP velocity ratios in the range of 9 to 15, the high ratios being accounted for primarily by the very slow transverse effective conduction velocities.

The complexity of transverse excitation spread at a microscopic level is averaged in the isochrone representation of transverse propagation at a macroscopic level in Figure 9.16(A). However, in nonuniform anisropic tissues fast longitudinal propagation continues as an overall smooth process with smooth extracellular potential waveforms [Fig. 9.16(B)].

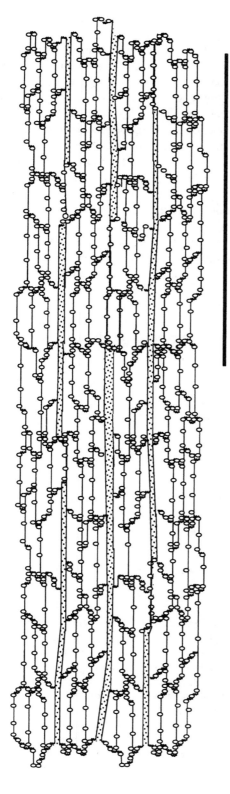

Figure 9.15 Representation of abnormal cardiac microstructure with nonuniform anisotropic properties. The general arrangement of the gap junctions is that which occurs in the atrium during aging (50,67) and in the ventricle following healing of ventricular infarction (64–66). A common feature is the reappearance of punctate gap junctions (ovals) along the lateral borders of the cells. This feature is accompanied by intermittent areas of expanded interstitial space (stippled areas) due to connective tissue or edema. The cellular network is based on histological results of Ref. 67 and immunohistological studies of atrial bundles from patients 60 to 75 years of age (unpublished). Bar = 0.5 mm.

The smooth course of longitudinal conduction is due to the maintenance of the continuity of cell-to-cell connections along the long axis of the fibers (Fig. 9.15). In older human atrial bundles, therefore, the dominant effect of remodeling the microstructure is the development of elongated areas of connective tissue in the interstitium (microfibrosis). Whereas the reappearance of the neonatal distribution of gap junctions along the lateral cell borders should enhance lateral conduction, very slow conduction in the transverse direction occurs due to the lateral separation of groups of cells by connective tissue (Fig. 9.15).

The importance of slow conduction is that low conduction velocities are necessary to produce reentry. To establish a reentrant circuit, conduction block must occur in one pathway (one-way block) while conduction is maintained in a second pathway. In the pathway in which propagation continues, the excitatory impulse must move slowly enough around a barrier, which usually is formed by totally refractory tissue, to permit time for recovery of excitability to occur in the area of initial block. The impulse that continues to propagate around the barrier eventually "reenters" the area in which initial one-way block occurred. At this point, if the membrane of that area has sufficiently recovered its excitability (reactivation of the fast Na^+ current), propagation will continue around the circuit and a persistent reentrant circuit (arrhythmia) will be established. If the impulse propagating around the barrier moves at a high velocity, however, the impulse will reenter the area of initial one-way block before sufficient time has elapsed for repolarization to reactivate the sodium current. In this case, conduction fails and no persistent reentrant circuit will be established. Thus, a low conduction velocity is proarrhythmic because it enhances the establish-

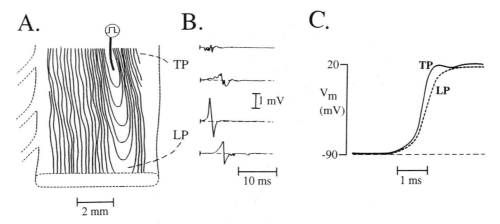

Figure 9.16 Events associated with nonuniform anisotropic conduction of normal action potentials. The data were obtained in a nonuniform anisotropic atrial bundle (crista terminalis) from an adult dog. The events shown are similar to those that occur with normal action potentials in atrial pectinate bundles from human subjects greater than 60 years of age. (A) Isochrone map of the spread of excitation from a point source of stimulation. (B) Characteristic extracellular waveforms associated with nonuniform anisotropic properties [i.e., multiphasic or fractionated waveforms occur during very slow transverse propagation (TP) and larger biphasic smooth waveforms occur during fast longitudinal propagation (LP)]. (C) Isochrone map (A) and the extracellular waveforms (B) were produced by the propagation of fast upstroke action potentials that arose from resting potentials of –90 mV. Note that the rate of depolarization during very slow transverse propagation (TP) is more rapid than the upstroke during fast longitudinal propagation (LP). (Modified from Ref. 69, with permission.)

ment of reentrant circuits, as well as provides for the maintenance of a reentrant circuit once it is established. The effect of microfibrosis on maintaining a reentrant circuit are especially important with regard to the high incidence of reentrant arrhythmias (e.g., atrial fibrillation) with aging of the atria. Here, microfibrosis that occurs with aging provides the major known mechanism (anisotropic resistive discontinuities) for creating the very low effective velocities necessary for a reentrant circuit to be small enough to occur within some regions of the atria.

Prior to the realization that recurrent resistive discontinuities produce discontinuous conduction in cardiac muscle (24), the low conduction velocities (< 0.1 m/s) necessary for reentry were considered to be caused by slow-response action potentials (70). The upstrokes of slow-response action potentials are produced by slow inward membrane currents, which occur primarily via the Ca^{2+} channels (71–73). Therefore, the question arises as to whether the very low effective conduction velocities during TP in nonuniform anisotropic tissue are caused by (1) an elevation of the effective axial resistance (r_a) along the transverse axis of the fibers, which can be accounted for by Eq. (7) or (2) the slow inward current. Figure 9.16(C) shows that both fast longitudinal and very slow transverse conduction occur with fast upstroke action potentials generated from resting (take-off) potentials of –90 mV. At resting potentials in the range of –80 and –90 mV, the fast Na^+ current is fully activated (72–74). However, as the take-off potential shifts from –80 mV to –60 mV (which occurs with premature beats), there is progressive inactivation of the fast Na^+ current, which decreases the excitability of the membrane and reduces the propagation velocity in each direction from its normal value. At more depolarized take-off potential between –60 and –50 mV, the sarcolemmal membrane is totally inexcitable. However, with resting potentials in the range of –50 to –30 mV, the slow inward current can be activated to produce slow upstroke action potentials (70) without turn-on of the fast sodium current. These slow upstroke action potentials occur in the AV node (75) and in very depressed tissues (76). To date, however, the very slow transverse conduction observed in nonuniform anisotropic tissues of aging atrial bundles (50,67) and in healing infarcts (61) has been associated with fast upstroke action potentials generated by the fully activated or partially inactivated fast sodium current (i.e., normal action potentials or premature impulses initiated at take-off potentials more negative than –65 mV) (61,67).

We wish to point out, however, that prominent delays of conduction across sites of elevated gap junction resistance create important interactions between the Ca^{2+} and Na^+ currents in fast upstroke action potentials that have normal resting potentials. When there are lengthy delays (> 0.5 ms) in transfer of the upstroke across such sites, these currents have a combined role. First, the fast sodium current generates the upstroke. As the upstroke approaches its peak value, there is turn-on of the prolonged L-type Ca^{2+} current (77), which maintains the transmembrane potential near its peak value after the Na^+ current is inactivated. Maintaining a maximum depolarized value is important to ensure that a prominent potential gradient occurs across the area of delay. That is, the prolonged potential gradient across the site of delay results in sufficient current flow during the delay to depolarize the capacitance of the membrane on the other side and activate its sodium current (78).

F. Reentry Effects of Microfibrosis Due to Unidirectional Longitudinal Block of Premature Impulses

Most reentrant arrhythmias are initiated by a premature beat that occurs during the rapid repolarization phase of the preceding action potential. This event results in the take-off potential of the premature action potential having a value less negative than -80 mV. In turn, as the take-off potential of earlier premature impulses occurs at progressively less negative values in the range of -80 to -65 mV, inactivation of the fast Na^+ current increases until propagation ceases at take-off potentials of approximately -65 mV. Because the upstrokes of early premature beats propagate with partial inactivation of the fast Na^+ current, there is a reduction in the excitability of the membrane which, in turn, lowers $\dot{V}_{max}$ and decreases the safety factor of conduction. The safety factor is a dimensionless parameter that indicates the margin of safety with which the action potential propagates relative to the minimum requirements for sustained propagation (26). The precise relationships that exist between the safety factor, $\dot{V}_{max}$, and variations in the availability of the Na^+ and Ca^{2+} depolarization ion currents have not been established for the different loading conditions that exist during LP and TP in anisotropic cardiac bundles. However, the complex interrelationships of these variables involve one overall principle—the lower the value of $\dot{V}_{max}$, the lower the safety factor of conduction.

In cardiac bundles with nonuniform anisotropic properties, the average value of $\dot{V}_{max}$ is greater during slow TP than fast LP [Fig. 9.16(C)], a relationship that also occurs in normal mature uniform anisotropic bundles (Fig. 9.13). Thus, the lower average value of $\dot{V}_{max}$ during LP is consistent with a larger current load on the membrane and a lower safety factor for conduction during fast LP than during slow TP (24). In such bundles, propagation should fail first in the longitudinal direction of lower $\dot{V}_{max}$ (greater load) as the depolarization current is reduced by early premature beats. However, the occurrence of unidirectional longitudinal conduction block of premature impulses is rare in tightly coupled uniform anisotropic bundles. That is, as the premature stimulus interval is progressively shortened, propagation ceases in all directions simultaneously. On the other hand, as illustrated in Figure 9.17, in older human atrial bundles with nonuniform anisotropic properties (microfibrosis), early premature impulses routinely produce unidirectional longitudinal decremental conduction to block, while simultaneously very slow but stable transverse propagation continues with effective conduction velocities as low as 0.04 m/s (67).

Thus, the resistive discontinuities associated with microfibrosis in aging atrial bundles produce the two requirements needed to initiate a reentrant circuit—unidirectional block and very slow conduction. Indeed, Figure 9.18 illustrates that in an atrial bundle with microfibrosis from a 64-year-old patient, this combination resulted in reentry within a small area. With an early premature beat, decremental conduction to block occurred during LP, while very slow discontinuous transverse conduction persisted to initiate reentry within the single atrial bundle (67).

G.   Significance of Discontinuous Propagation

At this point, the major significance of discontinuous conduction is that it is providing new insights to previously unrecognized arrhythmia mechanisms produced by microscopic discontinuities. For example, it is now apparent that resistive discontinuities play a major role in producing the most commonly encountered arrhythmias, such as atrial fibrillation with aging and reentrant arrhythmias after healing of ventricular infarction. The dual behavior of discontinuous conduction also provides a way to begin to resolve a poorly understood major clinical problem—premature beats and Na^+ channel blocking drugs that alter conduction in either an antiarrhythmic or a proarrhythmic manner (79). That is, when the depolarizing Na^+ current is partially inactivated by premature beats or drugs, the microstruc-

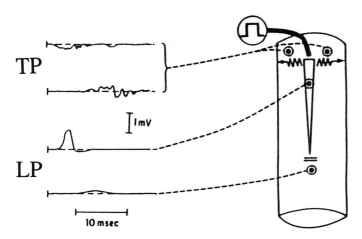

Figure 9.17   Representative propagation response to premature action potentials in older human atrial bundles with microfibrosis (nonuniform anisotropic properties). In the longitudinal direction (LP), the extracellular waveforms indicate decremental conduction to block; this is represented in the drawing of the bundle by an open triangle. In the transverse direction, the irregular (fractionated) extracellular waveforms with multiple deflections indicate that stable, but very slow, transverse propagation (TP) continued with markedly discontinuous conduction. The excitation sequence during transverse propagation was so complex that it was not possible to construct an isochrone map of TP. Therefore, in the drawing of the bundle, the "sawtooth" curves denote the irregular course of excitation spread. The atrial bundle is from a 64-year-old patient. (Modified from Ref. 67, with permission.)

ture has either a protective or a proarrhythmic effect. Which of these electrophysiological responses occurs is dependent upon whether the discontinuities of the microstructure prevent or produce the two events required to initiate a reentrant circuit—very slow conduction and unidirectional block (67).

1. Protective Mechanism of Discontinuous Propagation with Normal Microstructure

Although a premature beat is the usual triggering event for reentrant arrhythmias in hearts with a pathological anatomical substrate, premature beats occur quite commonly in normal hearts but do not lead to arrhythmias. Thus, a long-standing question has been: what is the normal protective structural mechanism that provides immunity against reentrant arrhythmias when premature beats occur? Most reentrant arrhythmias can be viewed as a loss of the natural structural immunity to reentry. Therefore, it has become important to identify normal microstructural mechanisms as a corollary of ionic current mechanisms that provide protection against reentrant arrhythmias. Accordingly, the following two electrical effects of normal discontinuous conduction play a major role in providing immunity to reentrant arrhythmias (52,67).

1.   The irregular arrangement of the cells and the gap junctions of normal cardiac muscle [Fig. 9.14(A)] result in microscopic propagation being stochastic in nature as well as discontinuous (52). That is, instead of being orderly, stable, and uniform as it appears at the larger macroscopic size [Fig. 9.12(A)], normal propagation at a microscopic level is seething with change and disorder in the sense

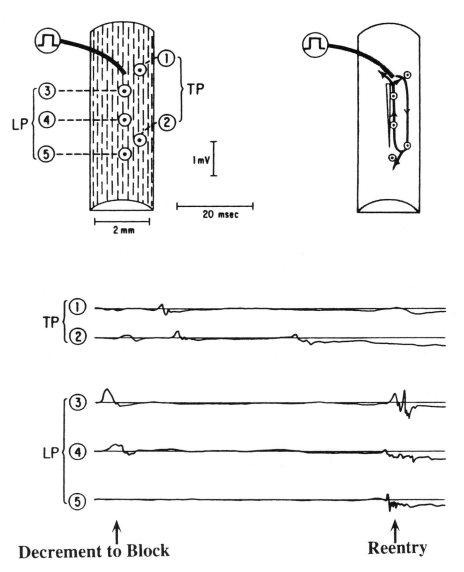

Figure 9.18 Reentrant effects of the development of microfibrosis (nonuniform anisotropic properties) with aging in human atrial bundles. The upper left drawing shows the locations at which each extracellular waveform was recorded. The upper right drawing shows the perimeter of reentrant circuit indicated by solid lines with arrows; the open elongated triangle denotes the initial decremental conduction to block in the longitudinal direction. The preparation is from a 64-year-old patient. (Modified from Ref. 67, with permission.)

of continuously varying excitatory events (e.g., variations in $\dot{V}_{max}$ as illustrated in Fig. 9.13). Functionally, the irregular conduction events at a microscopic level provide a considerable protective effect against arrhythmias by reestablishing the general trend of wavefront movement after small variations in excitation events occur (52). This feature maintains the direction of wavefront movement which, in turn, keeps the effective conduction velocity well above that of the very slow conduction needed to initiate and maintain reentrant circuits within the size of the structures involved.

2. In normally coupled tissues with uniform anisotropic properties, the upstroke of an action potential propagating in one direction has a considerable effect on depolarization moving in another direction in nearby cells (presumably by means of electrotonic interactions). Thus, when progressively earlier premature beats produce failure of propagation in any direction from the site of origin, failure occurs simultaneously in all directions (67). Since simultaneous propagation failure in all directions prevents the occurrence of unidirectional block, a reentrant circuit cannot be established following premature beats.

2. Implications of Microfibrosis with Proarrhythmic Discontinuous Conduction

One can view reentrant arrhythmias that occur with structural heart disease in the following perspective: there is loss of the mechanisms provided by normal cardiac microstructure that prevent initial unidirectional block and very slow conduction of premature impulses. As we have illustrated, this loss occurs commonly with the development of microfibrosis (80) during aging in atrial bundles (50) and following healing of infarcted tissue (61). An atrial bundle from a 62-year-old patient is shown in Figure 9.19 to illustrate the morphology of the microfibrosis associated with the proarrhythmic events of very slow transverse propagation, initial one-way block, and reentry within a small area (Figs. 9.16–9.18). The key feature is the loss of side-to-side electrical connections between small groups of cells, as marked by the areas of interstitium in which there are prominent collagenous septa.

Identification of the areas of very slow conduction with the structural mechanisms necessary for reentry will become increasingly important in the future. A practical question arises, therefore, as to how these proarrhythmic areas can be recognized. Here, detailed analysis of extracellular waveforms with multiple deflections (fractionation) provides a way to identify and assess areas of anisotropic structural discontinuities that are proarrhythmic. The fractionated waveforms illustrated in Figures 9.16–9.18 are associated with collagenous septa (microfibrosis) as shown in Figure 9.19. An important feature is that this structural morphology is reflected in the extracellular waveforms, and the spatial frequency of the collagenous septa can be calculated during transverse conduction (81). Because the effective conduction velocity θ equals dx/dt, we also know that dx equals $\theta\, dt$. Thus, the lateral distances separating the discontinuities (collagenous septa) can be obtained from (1) the effective transverse conduction velocity and (2) the time differences between each consecutive small deflection in the fractionated waveforms. For the low effective conduction velocity θ of the TP extracellular waveforms of Figure 9.16(B), for example, the calculation indicates that the distances separating the collagenous septa vary between 40 to 70 μm, which is consistent with the morphology of the bundle in Figure 9.19. We use this example to emphasize the need for future research on extracellular waveforms as a practical way to link clinical studies in patients with basic experimental and computer modeling studies.

In the pathological states associated with most reentrant arrhythmias, there are associated changes from normal in the distribution of the gap junctions (64–66). Consequently, molecular and genetic interventions that alter gap junction proteins, collagen, and other microstructures provide a new preventive or therapeutic approach for these arrhythmias (82). At present, however, advances at the molecular level have exceeded our experimental ability to analyze the electrophysiological implications of the data. The difficulty is that

Figure 9.19 Collagenous septa in an atrial bundle with nonuniform anisotropic electrical properties. The collagenous septa (gray to black) in this longitudinal section are thick and long and thus isolate adjacent cells and groups of cells (white). The preparation is from a 62-year-old patient. Bar = 200 μm. Note that the lateral distances between collagenous septa are approximately 30 to 70 μm. (From Ref. 67, with permission.)

of evaluating interactions between the different components of cardiac microstructure to account for phenomena observed at two quite different size scales (i.e., anisotropic variations at a microscopic level and the associated effective velocities at the considerably larger macroscopic level). Consequently, mathematical models with an electrical representation of the associated microstructure are becoming increasingly necessary (52). For both clinical and basic scientists, this area provides an important challenge to link experimental and computer modeling studies about the dual effects of discontinuous conduction to the prevention and treatment of reentrant arrhythmias.

ACKNOWLEDGMENTS

Supported by grants HL47678 (CA) and HL 50537 (MS) from the National Institutes of Health, American Heart Association–New York State Affiliate, the North Carolina Supercomputing Center, and the Masons of New York State and Florida.

REFERENCES

1. Jack JJB, Noble D, Tsien RW. Electrical Current Flow in Excitable Cells. Oxford: Oxford Press, 1975.
2. Hodgkin AL. Conduction of the Nervous Impulse. Liverpool: Liverpool University, 1964.
3. Fozzard HA. Conduction of the action potential. In: Berne RM, Sperelakis N, Geiger SR, eds. Handbook of Physiology. Sect.2. The cardiovascular system. Bethesda, MD: American Physiological Society, 1979:335–356.
4. Cranefield PF, Hoffman BF. Conduction of the cardiac impulse. II. Summation and inhibition. Circ Res 1971; 28: 220–233.
5. Cranefield PF, Klein HO, Hoffman BF. Conduction of the cardiac impulse. I. Delay, block and one-way block in depressed Purkinje fibers. Circ Res 1971; 28: 199–219.
6. Cranefield PF, Wit AL, Hoffman BF. Conduction of the cardiac impulse III. J Gen Physiol 1972; 59: 227–246.
7. Antzelevitch C, Jalife J, Moe GK. Characteristics of reflection as a mechanism of reentrant arrhythmias and its relationship to parasystole. Circulation 1980; 61: 182–191.
8. Antzelevitch C, Moe GK. Electrotonically-mediated delayed conduction and reentry in relation to "slow responses" in mammalian ventricular conducting tissue. Circ Res 1981; 49: 1129–1139.
9. Davidenko JM, Antzelevitch C. Electrophysiological mechanisms underlying rate-dependent changes of refractoriness in normal and segmentally depressed canine Purkinje fibers. The characteristics of post-repolarization refractoriness. Circ Res 1986; 58: 257–268.
10. Antzelevitch C, Moe GK. Electrotonic inhibition and summation of impulse conduction in mammalian Purkinje fibers. Am J Physiol 1983; 245: H42–H53.
11. Antzelevitch C, Lukas A. Reflection and reentry in isolated ventricular tissue. In: Dangman KH, Miura DS, eds. Basic and Clinical Electrophysiology of the Heart. New York: 1990.
12. Antzelevitch C. Clinical applications of new concepts of parasystole, reflection, and tachycardia. Cardiol Clin 1983; 1: 39–50.
13. Lukas A, Antzelevitch C. Reflected reentry, delayed conduction, and electrotonic inhibition in segmentally depressed atrial tissues. Can J Physiol Pharmacol 1989; 67: 757–764.
14. Antzelevitch C, Lukas A, Litovsky SH. Reflection as a subclass of reentrant cardiac atrial arrhythmias. In: Attuel P, Coumel P, Janse MJ, eds. The Atrium in Health and Disease. Mt. Kisco, NY: Futura Publishing Co., 1989:43–64.
15. Lewis T, Master AM. Observations upon conduction in the mammalian heart. AV conduction. Heart 1925; 12: 209–269.

16. Langendorf R. Concealed A-V conduction: the effect of blocked impulses on the formation and conduction of succeeding impulses. Am Heart J 1948; 35: 542–552.

17. Rozanski GJ, Jalife J, Moe GK. Reflected reentry in nonhomogeneous vnetricular muscle as a mechanism of cardiac arrhythmias. Circulation 1984; 69: 163–173.

18. Antzelevitch C. Electrotonic Modulation of Conduction and Automaticity. In: Janse HJ, Meijler FL, Van Der Tweel LH, eds. Proceedings of the Royal Academy of Arts and Sciences of the Netherlands. Amsterdam: 1990.

19. Antzelevitch C, Bernstein MJ, Feldman HN, Moe GK. Parasystole, reentry, and tachycardia: A canine preparation of cardiac arrhythmias occurring across inexcitable segments of tissue. Circulation 1983; 68: 1101–1115.

20. Janse M, Van Capelle FJL. Electrotonic interactions across an inexcitable region as a cause of ectopic activity in acute regional myocardial ischemia. A study in intact porcine and canine hearts and computer models. Circ Res 1982; 50: 527–537.

21. Jalife J, Moe GK. Excitation, conduction, and reflection of impulses in isolated bovine and canine cardiac Purkinje fibers. Circ Res 1981; 49: 233–247.

22. Ferrier GR, Rosenthal JE. Automaticity and entrance block induced by focal depolarization of mammalian ventricular tissues. Circ Res 1980; 47: 238–248.

23. Downar E, Waxman MB. Depressed conduction and unidirectional block in Purkinje fibers. In: Wellens HJ, Lie KI, Janse MJ, eds. The Conduction System of the Heart. Philadelphia: Lea and Febiger, 1976:393–409.

24. Spach MS, Miller WT, Geselowitz DB, Barr RC, Kootsey JM, Johnson EA. The discontinuous nature of propagation in normal canine cardiac muscle. Evidence for recurrent discontinuities of intracellular resistance that affect the membrane currents. Circ Res 1981; 48: 39–54.

25. Spach MS, Kootsey JM, Sloan JD. Active modulation of electrical coupling between cardiac cells of the dog. Circ Res 1982; 51: 347–362.

26. Spach MS, Josephson ME. Initiating reentry: the role of nonuniform anisotropy in small circuits. J Cardiovasc Electrophysiol 1995; 5: 182–209.

27. Hofer E, Urban G, Spach MS, Schafferhofer I, Mohr G, Platzer D. Measuring activation patterns of the heart at a microscopic size scale with thin-film sensors. Am J Physiol 1994; 266: H2136–H2145.

28. Spach MS. Anisotropic structural complexities in the genesis of reentrant arrhythmias. Circulation 1991; 84: 1447–1450.

29. De Bakker JMT, Van Capelle FJL, Janse MJ, et al. Reentry as a cause of ventricular tachycardia in patients with chronic ischemic disease: Electrophysiologic and anatomic correlation. Circulation 1988; 77: 589–606.

30. Gilmour RF, Salata JJ, Zipes DP. Rate-related suppression and facilitation of conduction in isolated canine cardiac Purkinje fibers. Circ Res 1985; 57: 35–45.

31. El-Sherif N, Chinushi M, Caref EB, Restivo M. Electrophysiological mechanism of the characteristic electrocardiographic morphology of torsade de pointes tachyarrhythmias in the long-QT syndrome. Detailed analysis of ventricular tridimensional activation patterns. Circulation 1997; 96: 4392–4399.

32. Antzelevitch C, Yan GX, Shimizu W, Burashnikov A. Electrical heterogeneity, the ECG, and cardiac arrhythmias. In: Zipes DP, Jalife J, eds. Cardiac Electrophysiology: From Cell to Bedside. Philadelphia: W.B. Saunders Co., 1998:1–34.

33. Shimizu W, Antzelevitch C. Cellular basis for the electrocardiographic features of the LQT1 form of the long QT syndrome: Effects of -adrenergic agonists, antagonists and sodium channel blockers on transmural dispersion of repolarization and torsade de pointes. Circulation 1998; 98:2314–2322.

34. Yan GX, Antzelevitch C. Cellular basis for the normal T wave and the electrocardiographic manifestations of the long QT syndrome. Circulation 1998; 98:1928–1936.

35. Yan GX, Shimizu W, Antzelevitch C. The characteristics and distribution of M cells in arterially-perfused canine left ventricular wedge preparations. Circulation 1998; 98:1921–1927.

36. Shimizu W, Antzelevitch C. Sodium channel block with mexiletine is effective in reducing dispersion of repolarization and preventing torsade de pointes in LQT2 and LQT3 models of the long-QT syndrome. Circulation 1997; 96: 2038–2047.

37. Sicouri S, Antzelevitch D, Heilmann C, Antzelevitch C. Effects of sodium channel block with mexiletine to reverse action potential prolongation in in vitro models of the long QT syndrome. J Cardiovasc Electrophysiol 1997; 8: 1280–1290.

38. Antzelevitch C. The Brugada syndrome. J Cardiovasc Electrophysiol 1998; 9: 513–516.

39. Yan GX, Antzelevitch C. Cellular basis for the Brugada Syndrome. Circulation 1999; 100:1660–1666.

40. Antzelevitch C. Electrotonus and reflection. In: Rosen MR, Janse MJ, Wit AL, eds. Cardiac Electrophysiology: A Textbook. Mount Kisco, NY: Futura Publishing Company, Inc., 1990:491–516.

41. Shaw R, Rudy Y. Gap junctions and the spread of electrical excitation. In: De Mello WC, Janse MJ, eds. Heart Cell Communication in Health and Disease. Norwell, MA: Kluwer Academic Publishers, 1998:125–147.

42. Spooner PM, Joyner RW, Jalife J, eds. Discontinuous Conduction in the Heart. Armonk, NY: Futura Publishing Co., 1997.

43. Spach MS, Heidlage JF, Dolber PC. The dual nature of anisotropic discontinuous conduction in the heart. In: Zipes DP, Jalife J, eds. Cardiac Electrophysiology: From Cell to Bedside. Philadelphia: W.B. Saunders Co., 1999:213–222.

44. Spach MS, Heidlage JF, Dolber PC, Barr RC. Extracellular discontinuities in cardiac muscle: evidence for capillary effects on the action potential foot. Circ Res 1998; 83: 1144–1164.

45. Spach MS. Discontinuous cardiac conduction: its origin in cellular connectivity with long-term adaptive changes that cause arrhythmias. In: Spooner P, Joyner RW, Jalife J, eds. Discontinuous Conduction in the Heart. Armonk, NY: Futura Publishing Co. Inc, 1997:5–51.

46. Hodgkin AL. A note on conduction velocity. J Physiol (Lond) 1954; 125:221–224.

47. Clerc L. Directional differences of impulse spread in trabecular muscle from mammalian heart. J Physiol (Lond) 1976; 255:335–346.

48. Spach MS, Miller WT, Miller-Jones E, Warren R, Barr RC. Extracellular potentials related to intracellular action potentials during impulse conduction in anisotropic canine cardiac muscle. Circ Res 1979; 45:188–204.

49. Spach MS, Heidlage JF, Darken ER, Hofer E, Raines KH, Starmer CF. Cellular $\dot{V}_{max}$ reflects both membrane properties and the load presented by adjoining cells. Am J Physiol 1992; 263: H1855–H1863.

50. Spach MS, Dolber PC. Relating extracellular potentials and their derivatives to anisotropic propagation at a microscopic level in human cardiac muscle: evidence for electrical uncoupling of side-to-side fiber connections with increasing age. Circ Res 1986; 58:356–371.

51. Tasaki I, Hagiwara S. Capacity of muscle fiber membrane. Am J Physiol 1957; 188:423–429.

52. Spach MS, Heidlage JF. The stochastic nature of cardiac propagation at a microscopic level. An electrical description of myocardial architecture and its application to conduction. Circ Res 1995; 76:366–380.

53. Gourdie RG, Litchenberg WH, Eisenberg LM. Gap junctions and heart development. In: De Mello WC, Janse MJ, eds. Heart Cell Communication in Health and Disease. Boston: Kluwer Academic Publishers, 1998:19–43.

54. Gourdie RG, Green CR, Severs NJ, Thompson RP. Immunolabeling patterns of gap junction connexins in the developing and mature rat heart. Anat Embryol 1992; 185:363–378.

55. Peters NS, Severs NJ, Rothery SM, Lincoln C, Yacoub MH, Green CR. Spatiotemporal relationship between gap junctions and fascia adherens junctions during postnatal development of human ventricular myocardium. Circulation 1994; 90:713–725.

56. Hoyt RH, Cohen ML, Saffitz JE. Distribution and three-dimensional structure of intercellular junctions in canine myocardium. Circ Res 1989; 64:563–574.

57. Gourdie RG, Green CR, Severs NJ. Gap junction distribution in adult mammalian myocardium

revealed by anti-peptide antibody and laser scanning confocal microscopy. J Cell Sci 1991; 99:41–55.

58. Dolber PC, Beyer EC, Junker JL, Spach MS. Distribution of gap junctions in dog and rat ventricle studied with a double-label technique. J Mol Cell Cardiol 1992; 24:1443–1457.

59. Fast VG, Darrow BJ, Saffitz JE, Kléber AG. Anisotropic activation spread in heart cell monolayers assessed by high-resolution optical mapping. Role of tissue discontinuities. Circ Res 1996; 79:115–127.

60. Spach MS, Kootsey JM. Relating the sodium current and conductance to the shape of transmembrane and extracellular potentials by simulation: effects of propagation boundaries. IEEE Trans Biomed Eng 1985; 32:743–755.

61. Wit AL. A model of arrhythmias that may necessitate a new approach to antiarrhythmic drug development. In: Rosen MR, Palti Y, eds. Lethal Arrhythmias Resulting from Myocardial Ischemia and Infarction. Boston: Kluwer Academic Publishers, 1988:199–214.

62. Lake FR, Cullen KJ, de Klerk NH, McCall MG, Rosman DL. Atrial fibrillation and mortality in an elderly population. Aust N Z J Med 1989; 19:321–326.

63. Werkö L. Atrial Fibrillation. In: Olsson SB, Allessie MA, Campbell RWF, eds. Atrial Fibrillation: Mechanisms and Therapeutic Strategies. Armonk, NY: Futura Publishing Co., Inc., 1994:1–13.

64. Luke RA, Saffitz JE. Remodeling of ventricular conduction pathways in healed canine infarct border zones. J Clin Invest 1991; 87:1594–1602.

65. Smith JH, Green CR, Peters NP, Rothery S, Severs NJ. Altered patterns of gap junction distribution in ischemic heart disease. An immunohistochemical study of human myocardium using laser scanning confocal microscopy. Am J Pathol 1991; 139:801–821.

66. Peters NS, Coromilas J, Severs NJ, Wit AL. Disturbed connexin 43 gap junction distribution correlates with the location of reentrant circuits in epicardial border zone of healing canine infarcts that cause ventricular tachycardia. Circ Res 1997; 95:988–996.

67. Spach MS, Dolber PC, Heidlage JF. Influence of the passive anisotropic properties on directional differences in propagation following modification of the sodium conductance in human atrial muscle. A model of reentry based on anisotropic discontinuous propagation. Circ Res 1988; 62:811–832.

68. Spach MS, Miller III WT, Dolber PC, Kootsey JM, Sommer JR, Mosher CE, Jr. The functional role of structural complexities in the propagation of depolarization in the atrium of the dog. Cardiac conduction disturbances due to discontinuities of effective axial resistivity. Circ Res 1982; 50:175–191.

69. Spach MS. The stochastic nature of cardiac propagation due to the discrete cellular structure of the myocardium. Int J Bifurcation Chaos 1996; 6:1637–1656.

70. Cranefield PF. The Conduction of the Cardiac Impulse. The Slow Response and Cardiac Arrhythmias. Mt. Kisco, NY: Futura Publishing Co., 1975:304.

71. Beeler GW, Reuter H. Reconstruction of the action potential of ventricular fibres. J Physiol (Lond) 1977; 268:177–210.

72. Luo C-H, Rudy Y. A model of the ventricular cardiac action potential. Depolarization, repolarization, and their interaction. Circ Res 1991; 68:1501–1526.

73. Luo C-H, Rudy Y. A dynamic model of the cardiac ventricular action potential. I. Simulations of ionic currents and concentration changes. Circ Res 1994; 74:1071–1096.

74. Ebihara L, Johnson EA. Fast sodium current in cardiac muscle: a quantitative description. Biophys J 1980; 32:779–790.

75. Paes de Carvalho A, Hoffman BF, de Paula Carvalho M. Two components of the cardiac action potential. I. Voltage-time course and the effect of acetycholine on atrial and nodal cells in the rabbit. J Gen Physiol 1969; 54:607–635.

76. Cranefield PF, Klein HO, Hoffman BF. Conduction of the cardiac impulse. I. Delay, block, and one-way block in depressed Purkinje fibers. Circ Res 1971; 28:199–219.

77. Shaw RM, Rudy Y. Ionic mechanisms of propagation in cardiac tissue. Roles of the sodium and

L-type calcium current during reduced excitability and decreased gap junction coupling. Circ Res 1997; 81:727–741.

78. Sugiura H, Joyner RW. Action potential conduction between guinea pig ventricular cells can be mediated by calcium current. Am J Physiol 1992; 262:H1591–H1602.
79. Cardiac Arrhythmia Suppression Trial (CAST) Investigators. Preliminary report: Effect of encainide and flecainide on mortality in a randomized trial of arrhythmia suppression after myocardial infarction. N Engl J Med 1989; 321:406–412.
80. Weber KT, Brilla CG, Janicki JS. Myocardial fibrosis: functional significance and regulatory factors. Carrdiovasc Res 1993; 27:341–348.
81. Spach MS, Dolber PC. The relation between discontinuous propagation in anisotropic cardiac muscle and the "vulnerable period" of reentry. In: Zipes DP, Jalife J, eds. Cardiac Electrophysiology and Arrhythmias. Orlando, FL: Grune & Stratton, Inc., 1985:241–252.
82. Spach MS, CF Starmer. Altering the topology of gap junctions: a major therapeutic target for atrial fibrillation. Cardiovasc Res 1995; 30:336–344.

10

Networks
Fundamental Properties and Models

RONALD W. JOYNER

Emory University School of Medicine, Atlanta, Georgia

ANDRÉ G. KLÉBER

University of Bern, Bern, Switzerland

I. INTRODUCTION

Recent advances in experimental methods have permitted theoretical and experimental studies of impulse propagation in diverse systems, ranging from cell pairs to complex networks of cells. Of particular importance to the subject matter of this chapter is what we have learned regarding spatial inhomogeneity in the membrane properties of individual cells as well as in electrical coupling among cells. We shall discuss simple cell pairs, strands of cells, and then more complex structures, showing how the general features of the electrical interactions are preserved as the structure becomes more complex.

II. ACTION POTENTIAL PROPAGATION ALONG LINEAR CELLULAR STRUCTURES

A. Continuous Propagation

Early models of impulse propagation were derived from studies carried out in nonmyelinated giant axons of sepia (1) and subsequently taken as a paradigm for cardiac propagation. The continuous propagation model represents cardiac tissue as a syncytial structure in which the cytoplasmic and gap junctional resistors are lumped into a single so-called internal resistor, r_i, and the cell membranes are represented by a cylindrical surface circumscribing the internal resistive continuum (2,3). If the radius of such a cylinder is assumed to remain constant along the axis, there is continuity in time and space. Consequently, the membrane potential distribution along the x axis at a given time t corresponds to the action potential recorded as a function of t, if t is substituted by $t = s/\Theta$, where Θ corresponds to the conduction velocity. This model has provided reasonable predictions of the effects of metabolic changes (ischemia, hypoxia) and antiarrhythmic drugs on conduction (4–6). However, for many situations, especially in the case of propagation across tissue discontinuities (trabeculated structures, conduction across the atrioventricular junction or midmural ventricular layers) and in electrically uncoupled tissue, the applicability of the continuous propagation model is limited. A major feature of the continuous propagation model is the so-called square root relationship between conduction velocity and internal resistance r_i, on one hand, and $(\mathrm{d}V/\mathrm{d}t)_{max}$ of the action potential on the other (3). These independent relationships also predict that the resistive properties of the tissue (represented by r_i) affect propagation independently of the depolarizing ion currents (represented by $(\mathrm{d}V/\mathrm{d}t)_{max}$). Yet interdependence and interaction between resistive network properties and ion channel function is a major feature of discontinuous conduction, thereby illustrating the limited applicability of the continuous propagation model.

B. Discontinuous Propagation

Discontinuous propagation models incorporate the passive and active properties of individual cells interacting with one another rather than accepting the uniform syncytial behavior that characterizes the continuous propagation model. As such, discontinuous propagation models tend to reflect the anatomical and physiological properties of the heart more accurately than continuous propagation models. The simplest case of discontinuous propagation is that which occurs between two coupled excitable cells, in which case the coupling resistance, R_c (Ohms) or its reciprocal, the coupling conductance, G_c (Siemens) represents the aggregate pathway for current flow from one cell to the other through gap junctions, as illustrated in Figure 1(A). With this representation, the effective size of either cell can be altered experimentally by scaling the coupling currents. Data obtained using this "coupling clamp" circuit (7) to produce a known value of resistive coupling between two isolated cells (8) have shown progressive delays in action potential propagation as the coupling resistance was increased, with maximal values of coupling resistance of about 150 MΩ for two coupled guinea pig ventricular cells but successful conduction for values of coupling resistance up to 2500 MΩ for propagation from a sinoatrial node cell to an atrial cell (9). These differences can be explained by the much greater input resistance of the atrial cells. Figure 10.1(B) and 10.1(C) shows an example of action potential conduction between two ventricular cells at a coupling resistance of 143 MΩ, with the action potentials of the two cells displayed in (B) and the time delay for propagation as a function of coupling resistance displayed in (C). As the coupling resistance is varied from 50 to 200 MΩ, the conduction delay varies from 2 ms to 25 ms (8). However, the ability of two cells to propagate action potentials can be affected in complex ways by the coupling conductance. Methods have also been developed to replace one of the real cells with a real-time solution of a theoretical model of a particular cell type (10). Figure 10.2 shows results obtained by coupling a theoretical model cell with intrinsic automaticity (11) to a real rabbit ventricular cell (12). For the normal size of the sinoatrial node cell model, increasing cell coupling converts the spontaneous pacing of the model cell to quiescence because of the impedance mismatch of the low-input resistance ventricular cell. For increased sizes of the model cell (size 5 or greater, which is equivalent to specifying a group of five or more cells closely coupled and working as a functional unit) increases in cell coupling show a pro-

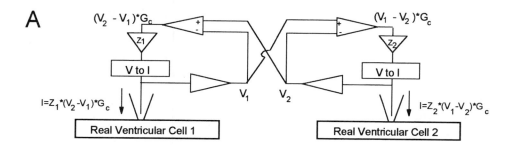

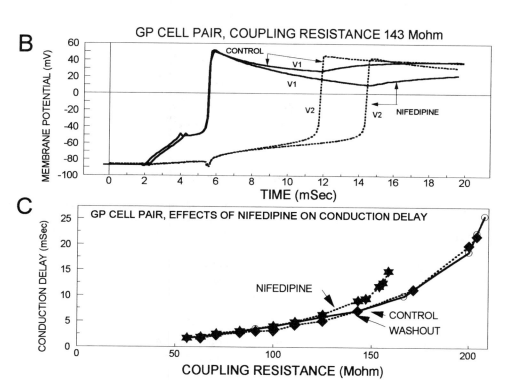

Figure 10.1 (A) Schematic diagram of how the current flow between two cardiac cells can be represented by an equivalent electrical circuit. (B) Results obtained with simultaneous recordings from two isolated guinea pig ventricular cells coupled by a "coupling clamp circuit" with a resistance of 143 MΩ. Cell 1 (V1, solid lines) receives a repetitive stimulus and then propagates an action potential to cell 2 (V2, dotted lines). Results are shown in the "control" solution and in a nifedipine solution to partially inhibit L-type calcium currents. (C) Summary of the dependence of conduction delay on coupling resistance in the normal solution, the nifedipine solution, and after washout of the nifedipine solution. (From Ref. 8.)

gression from pacing of the model cell but not driving the ventricular cell, successful pacing of the model cell and driving of the ventricular cell, and then failure of pacing of the model cell. Thus, even at the level of a pair of cells of different membrane properties, the anatomical properties that organize cardiac cells into functional groups can be seen to play a critical role in enabling propagation.

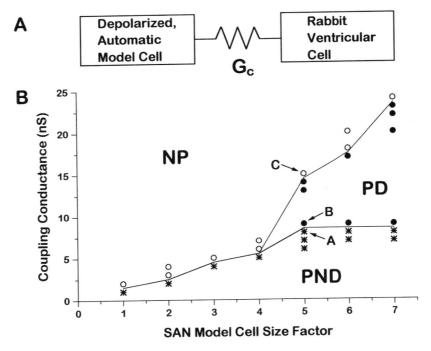

Figure 10.2 Summary of results obtained by coupling a real-time simulation of a sinoatrial node cell to a real isolated rabbit ventricular cell. The nodal cell model was used at a variable size, representing a well-coupled aggregate of 1,2, . . . 7 nodal cells. For each set of values of size and coupling conductance, the results were of three possible outcomes: pacing of the nodal cell without driving of the ventricular cell ("pace but not drive" (PND)—asterisks); failure of the nodal cell to pace ("not pace" (NP)—open circles); or successful pacing of the nodal cell with also successful driving of the ventricular cell ("pace and drive" (PD)—filled circles). (From Ref. 12.)

Theoretical simulations of discontinuous propagation in a strand consisting of model cardiac cells separated by a gap junctional conductance, g_c, have been extensively studied (13). The main differences from the continuous propagation model is that electrical current in the intracellular compartment is flowing through two electrical conductances in series—the cytoplasmic conductance, g_{cyto}, and the junctional conductance, g_c (14). If the parameters for the active and passive electrical properties are within the range of normal, the conduction time across the cell is approximately the same as the conduction time across the gap junctions at the cell end. Moreover, comparison of velocities from the continuous model with the discontinuous model yields similar values, indicating that this distinction is not relevant for normal macroscopic propagation in linear tissue (linear strands or propagation with parallel isochrone lines or planes in 2-D and 3-D tissue, respectively).

These simulated results were confirmed in cultured strands of well-coupled neonatal rat myocytes, as shown on Figure 10.3 (15). In this technique, special coverslips are prepared with the glass surface partially coated in a particular geometrical pattern with a substance that inhibits the attachment and growth of dissociated cells. When these coverslips are then incubated with a suspension of cells, the cells attach and grow in the desired spatial pattern (16). In such strands, average propagation time along a 30-μm distance within the cytoplasm was 38 μs, and average propagation time across an end-to-end cell connec-

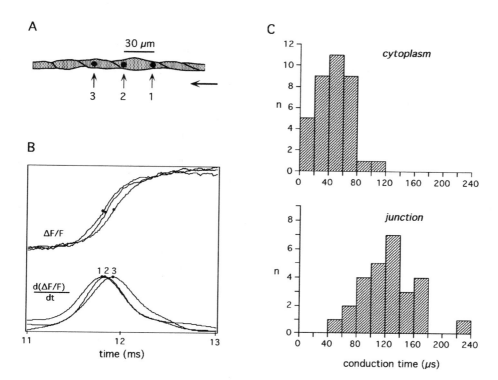

Figure 10.3 Impulse propagation in one-dimensional cell chains. (A) Diagram of a portion of a cell chain. (B) Potential-dependent fluorescence change (ΔF/F) recorded by the diodes in (A) and their time derivatives. (C) Histograms of conduction times for all experiments. Average cytoplasmic conduction time (38 ± 25 µs, $n = 37$) was markedly shorter then the average junctional conduction time (118 ± 40 µs, $n = 27$). The mean difference, attributed to the conduction delay induced by the gap junctions, amounted to 80 µs, which is 51% of the overall conduction time. (From Ref. 15.)

tion was 118 µs, indicating a delay at the gap junctions of 80 µs, or of about 50% of overall conduction time. Although this delay may seem to exert a highly significant effect on the subcellular pattern of propagation, it is confined to single cell chains. The situation is more complex in 2-D tissue, as shown in Figure 10.4. Mapping longitudinal propagation in a cell strand (consisting of approximately six cells in parallel) lengthened cytoplasmic conduction time and shortened the conduction delay across the cell border if compared to the results obtained in single cell chains. This resulted in a reduction of the junctional delay from 50% to 20%. Thus lateral cell apposition makes propagation more smooth. This effect is explained by flow of electrotonic current through lateral gap junctions, which partially cancels the role of the end-to-end junctions in delaying conduction (15).

While in modeling studies there is no major difference between continuous and discontinuous propagation in tissue with normal electrical parameters and network properties, a marked difference is unmasked with progressing cell-to-cell uncoupling, a process that is likely to be relevant for arrhythmogenesis in acute ischemia (acute uncoupling) and in advanced heart failure (where there is remodeling of gap junctions) (17,18). In simulated linear strands of model cells (18), effects on conduction velocity of increasing cell-to-cell uncoupling have been compared to effects of decreasing depolarizing ion currents. Although decreasing the inward current was able to modestly lower velocity, cell-to-cell uncoupling

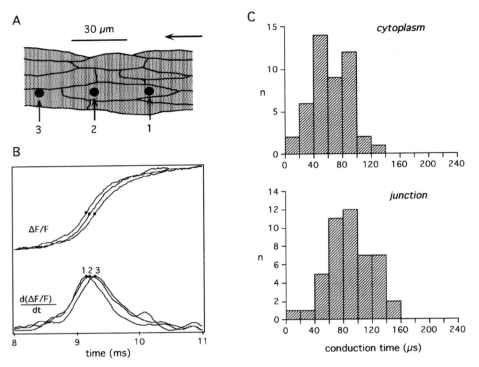

Figure 10.4 Impulse propagation in two-dimensional cell chains. (A) Diagram of a wide cell chain (4–6 cells in parallel). (B) Potential-dependent fluorescence change (ΔF/F) recorded by the diodes in (A) and their time derivatives. (C) Histograms of conduction times for all experiments. Average cytoplasmic conduction time (57 ± 26 μs, $n = 46$) was shorter then the average junctional conduction time (89 ± 39 μs, $n = 48$). The gap junctional delay amounted to 32 μs, which is 22% of the overall conduction time. (From Ref. 15.)

in the presence of normal excitability produced very slow conduction velocities on the order of a few centimeters per second. For very slow conduction to occur, cell-to-cell uncoupling needs to be extreme (i.e., g_c has to decrease over a hundredfold or more). Moderate changes in coupling resistance or changes in gap junction expression by 2- to 10-fold, for example, induce only small-to-moderate changes in conduction velocity. Recent experimental work, carried out with high-resolution optical mapping in patterned myocyte cultures, has confirmed these very low velocities and revealed some additional features of very slow conduction in uncoupled tissue, as illustrated in Figure 5 (19). Advanced cell-to-cell uncoupling in small patterned strands was associated with a marked inhomogeneity in conduction. This inhomogeneity was due to the fact that, in contrast to the computer simulations, gap junctions are nonuniformly distributed and occur at distinct sites in the cell's perimeter and at changing densities. A similar decrease of gap junctional conductance at the molecular level may therefore lead to total uncoupling between certain cells and maintained coupling between other cells. As a consequence, during advanced cell-to-cell uncoupling, cell strands were excited in "unit elements" that consisted of one or more cells. The delays between these cells increased markedly, from a few tenths of milliseconds to two or more milliseconds. Occasionally, cells were found to be inexcitable during propagation and completely sealed off by total uncoupling. These cells develop full action po-

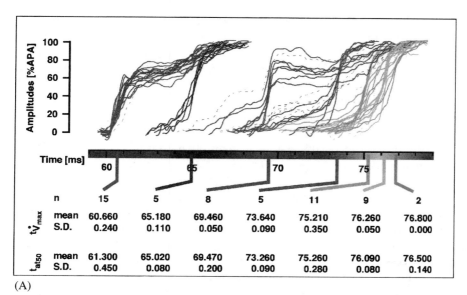

(A)

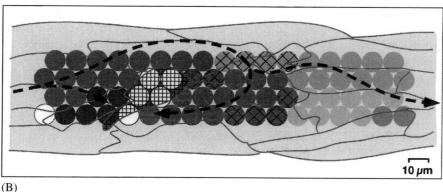

(B)

Figure 10.5 Characteristics of microscopic impulse propagation during ultraslow conduction induced by gap junctional uncoupling. (A) Plots of action potential upstrokes recorded simultaneously during activation of the strand. Upstrokes that occur at nearly the same time are color coded and the respective activation times determined from either the point in time of occurrence of maximum dV/dt or occurrence of 50% depolarization (t_{at50}) are shown below the graph. (B) Local activation map. The recording sites producing the upstrokes shown in (A) are color coded accordingly and superimposed on a schematic drawing of the preparation showing the cell borders. At the site of the white disks, no changes in transmembrane voltage were recorded (electrically uncoupled cell with hatched outline). The cross-hatched disks indicate the locations of signals for which it was not possible to assign unambiguous activation times because of notched upstrokes (dashed signals in part A). Arrows indicate qualitatively the direction of activation that advanced in a stepwise fashion along the preparation (overall conduction velocity ≈ 1.1 cm/s). (From Ref. 19.) (See also color plate.)

tentials when excited with field stimulation, which indicates that possible depression of excitability by the uncoupling agent played a very minor role. The propagation pattern of Figure 10.5 shows highly irregular activation with partially retrograde activation of single cells or cellular clusters due to the fact that single cells block the impulse. Estimation of the excitation wavelength (λ) in such cases suggests that such microscopic retrograde pathways are unlikely to cause reexcitation. At a velocity of 1 cm/s and an estimated refractory period of 100 ms, λ amounts to 1 mm. This is orders of magnitude longer than the pathway observed in Figure 10.5. Thus, despite the fact that propagation velocity can decrease to as much as 1 to 3 cm/s, the smallest possible reentry pathways will amount to 1 to 3 mm (which is equivalent to a circumference of a circle with a diameter of 320–960 μm) and are expected to involve a relatively large number of cells. This result again illustrates the importance of the anatomical features of the cardiac syncytium and the coordinated activity of groups of heart cells.

One of the most distinct and important features of uncoupled networks is the interaction between network properties and ion currents. The large conduction delays that are present between the successively depolarized cell groups shown in Figure 10.5 indicate that the upstrokes of a given cell group occur at a time when the previous group is already depolarized to its early plateau. The ion depolarizing current responsible for the generation of local current circuits in propagation during this time is $I_{Ca,L}$. Success of propagation in the state of advanced cell-to-cell uncoupling therefore becomes $I_{Ca,L}$-dependent in a way equivalent to success of action potential transfer between two cells, as shown by the effects of nifedipine in Figure 10.1(C) and (D) to increase the conduction delay between two coupled cells and to decrease the maximal allowable delay for conduction (8); similar effects have been observed in multicellular arrays of cells (19). This phenomenon may be of considerable significance in understanding the way in which changes in autonomic tone, which is partly expressed as a modulation of $I_{Ca,L}$, or the use of drugs that directly alter $I_{Ca,L}$ (e.g., calcium channel blockers) or modulate autonomic tone (e.g., beta-blockers) may alter the propagation of action potentials specifically at regions where action potential conduction is discontinuous.

III. PROPAGATION IN TWO-DIMENSIONAL AND THREE-DIMENSIONAL NETWORKS

A. General Considerations

Multidimensional cardiac tissue differs in conduction behavior from linear models in several respects, which are all related to specific structural features and which will be discussed in detail as follows.

1.   The unit element of cardiac function, the myocyte, is a longitudinal element of varying length and diameter changing with postnatal development and species. As an example, average adult canine myocytes are 122 μm in length and 23 μm in width (20,21). This longitudinal shape or structural anisotropy necessarily bears a functional counterpart. The cytoplasmic distance in the transverse direction is reduced and therefore the overall resistance of the intracellular space is higher in transverse versus longitudinal direction. In addition to cell shape, anisotropy is highly influenced by the pattern of gap junction distribution, which also varies with age, tissue type, and pathology. As a consequence of cell shape and gap junction distribution, the propagation velocity (V_T) in the transverse direction is reduced compared to the longitudinal velocity (V_L). Values for the so-called anisotropic velocity ratio (V_L/V_T) vary with the type of tissue from approximately 2.7 in the ventricle to as much as 12 in the atrial crista terminalis (5,22–24).

2.   Cardiac tissue is arranged in trabecula, strands, bundles, and sheets that are in turn separated by connective tissue strands (25,26). Transferring this situation into a mod-

el of excitable elements connected by resistive elements would mean that the pattern of connectivity would be determined by large zones of absent cell-to-cell contact, according to the shape and extension of the connective matrix, in addition to the connectivity determined by the cellular network. Normally, connective tissue structures are aligned with myocytes. Therefore, their effect to act as resistive obstacles for a propagating wave is much more relevant to transverse than to longitudinal propagation. When such barriers occur, they alter the normal uniform anisotropy of the tissue into a spatially inhomogeneous pattern that has been termed "discontinuous anisotropy" (27).

3. The biophysical rules determining the relationship between propagation velocity, shape of the action potential upstroke, and activation of ion currents are very similar at any scale (i.e., similar phenomena can be observed to occur at the level of cell pairs, strands of cells, two-dimensional syncytia of cells, and spatially inhomogeneous networks that include macroscopic connective tissue obstacles). Moreover, changes in ion channel function and/or cell-to-cell coupling can affect the role of connective tissue discontinuities in conduction and vice versa. The effects of myocardial ischemia or hypertropy in altering the spatial pattern of cell-to-cell coupling thus may be seen as a "remodeling" of the anatomical substrate, which may then serve to facilitate the occurrence of arrhythmias.

B. Cellular Networks

The electrical behavior of 2-D networks has been studied in computer models and in cell cultures. Simulation studies (28,29) have revealed interesting features of direction-dependent conduction, which became apparent when the resolution was increased to reveal the role of the cellular architecture in propagation. Based on earlier histological analyses (21), one study (29) simulated propagation in a corresponding cellular network in which each cell consisted of up to 36 excitable elements. Figure 10.6 shows the simulated changes in the activation times, maximal upstroke velocity $(dV/dt)_{max}$, and the time integral of I_{Na} (area I_{Na}) during longitudinal (left panels) and transverse (right panels) propagation with an anisotropic velocity ratio of 3.1. During longitudinal propagation, as a consistent rule, area I_{Na} (which represents not only the amount of charge entering the cell, but also the quantity of sodium ions that must then be pumped back out of the cell by use of cellular energy stores) was larger at the beginning of a cell and got activated to a lesser degree when the wavefront approached the cell end. Inversely, $(dV/dt)_{max}$ was lower at the cell pole that was activated first and higher at the cell pole that was activated last. Local conduction velocity was always lowest at the pole that was activated first and fastest toward the cell end. In a given cell, the upstream and downstream impedance of the wavefront depended on the shapes, sizes, and connectivity of the cells surrounding this location. Since the distribution of these structural parameters depends on the pattern of connectivity, the exact pattern of the changes in $(dV/dt)_{max}$, area I_{Na}, and activation varied in each simulated cell and with wavefront direction. During conduction in the transverse direction $(dV/dt)_{max}$, area I_{Na} and activation depended on the arrangement of the small transverse gap junctions on opposite cell borders. The smallest values for $(dV/dt)_{max}$ and the highest for area I_{Na} were found immediately beyond a gap junction. Inversely, the largest values for $(dV/dt)_{max}$ and the smallest for area I_{Na} were found at sites remote from insertions of gap junctions, where microcollisions of waves occurred. As discussed below, these changes at the subcellular level follow the same rules that characterize discontinuous conduction at the more macroscopic level. One important point from these studies is that the geometrical arrangement of cells and cell junctions in a network of cells, combined with the direction of propagation, can

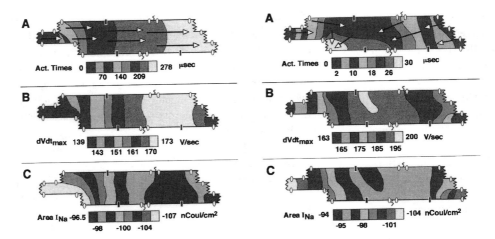

Figure 10.6   Left panels show intracellular two-dimensional spatial relations of excitation spread, maximum dV/dt, and time integral of the Na current (area I_{Na}) calculated for a single model myocyte during longitudinal propagation. Right panels show the same relationships for the same model cell during transverse propagation. (From Ref. 29.) (See also color plate.)

significantly alter the action potential waveform and the ion accumulation for sodium ions in a similar way to that discussed above for alterations in the entry of calcium ions.

It has not been possible as yet to verify the subcellular characteristics of propagation experimentally in all the details. However, some of the typical simulated phenomena of Figure 10.6 were experimentally shown in patterned anisotropic cell cultures assessed with high-resolution optical mapping of membrane potential. If such cultures are grown on a collagen matrix with a specific geometrical pattern, they form an anisotropic network with fusiform cells and a gap junction distribution pattern corresponding to the neonatal type (i.e., with a regular spacing of equally sized gap junctions around the cell perimeter). The anisotropic velocity ratio of such cultures (1.9–2.1) is in the lower range of values observed in ventricles in vivo (30). Two examples of experimentally observed conduction discontinuities at a cellular level are described in Figure 10.7 (31). In part (A), the central myocyte expressed only one detectable gap junction at the left cell pole. As a consequence, conduction was blocked at the upper cell margin and propagated around this myocyte. The myocyte itself was excited across the remaining gap junction at the left cell pole. In part (B), a small intercellular cleft separated a cell at the longitudinal perimeter from its neighbors along a distance of approximately 50 μm. During transverse conduction, this produced excitation at the left and right cell poles only, with collision of wavefronts in the middle of the cell. As a consequence, (dV/dt)$_{max}$ of the action potential upstrokes was low at the cell poles and high at the collision sites.

Although both computer simulations and experimental results suggest that the cardiac cellular structure significantly affects the action potential shape, transmembrane flow of ion current, and cellular activation patterns, several caveats remain.

1.   Recent experimental studies have suggested that some of the ion channels are preferentially introduced into the surface membranes closely adjacent to gap junctions. Remodeling of gap junctions is also associated with remodeling of the pattern of membrane channel expression. A preferential expression of I_{Na} channels at gap junctions would be expected to affect impulse propagation at the

A

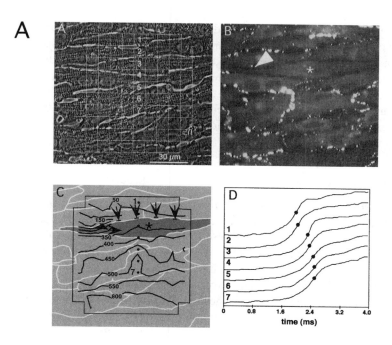

B

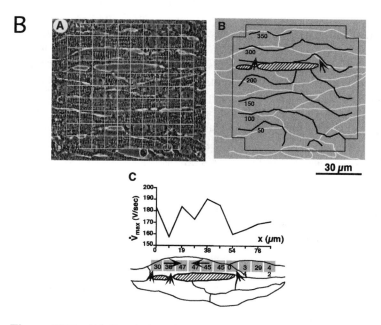

Figure 10.7 (A) Correlation of activation spread with gap junction distribution. Panels (A) and (B) are phase contrast and fluorescence images (immunolabeling of gap junctional protein Cx43), respectively. Asterisk depicts a central myocyte with almost no Cx43 expression. Panels (C) and (D) show isochronal activation map and selected optical recordings. (B) Anisotropic activation in a region with a small intercellular cleft. Panels A, B, and C show an image of the cell culture and the photodiode array, an isochronal map during transverse activation, and distribution of maximum dV/dt, respectively. (From Ref. 31.)

cellular level (32). It might be anticipated that such a preferential location would partially cancel the effect of the discontinuous distribution of gap junctions.

2. The change in the shape of the action potential upstroke with changing direction of impulse spread appears not to be a consequence of the cellular anisotropy and discontinuity alone. Thus, it was shown that the initial portion of the action potential can be affected by both the presence of a superficial fluid layer (volume conductor) (33) and electrical interactions between myocardial cells and the capillary vascular cells (34).

C.   Propagation in Networks Containing Resistive Obstacles

Macroscopic cardiac structure is highly discontinuous at the level of the atria, the atrioventricular junction, and the ventricles. Structural complexities affecting propagation and extracellular waveforms have been previously described in the atria and in the Purkinje fibers (24,35–37) and at the Purkinje-ventricular muscle junction (38). More recent work on the morphology of the ventricles suggests that similar discontinuities may play a role in midmural ventricular layers that are highly anisotropic and discontinuous with small fiber bundles connecting layers of excitable tissue otherwise separated by a collagen matrix (26). Several theoretical and experimental studies have shown that the biophysical processes determining propagation are common to all these macroscopic discontinuities. Discontinuities that have been investigated include: (1) a linear region of reduced coupling that can produce a line of block around which the action potential can propagate (39–41); (2) a connective tissue sheet or linear connective tissue structure interrupted by a narrow region of excitable tissue ('isthmus" or "gate') (42,43); and (3) an abrupt tissue expansion consisting of a cell strand emerging to a large tissue area (44–46). Variations of this latter geometry and their functional consequences have been described in experimental and theoretical studies (45,47). The common major biophysical behavior of all these macroscopic obstacles in a network of excitable cells is due to the fact that the excitatory current in the head of the wavefront is dispersed by the change in tissue geometry (so-called "current-to-load mismatch"). As a consequence of this dispersion, the density of excitable current flowing into the nonexcited elements ahead of a wavefront may decrease to an extent such that propagation is critically slowed and eventually becomes blocked.

Figure 10.8 shows the results of a simulation study assessing conduction propagating through an isthmus (45). Just prior to the isthmus, local conduction velocity is increased. This increase is an effect of collision of the lateral wavefronts with the obstacle (see below). The faster action potential upstrokes at the collision sites are fed back electrotonically to the gate and transiently accelerate conduction. Beyond the gate, conduction is locally slowed down to a minimum value that is located at some distance from the gate. This minimum is due to the curvature of the wave at this site, which creates current-to-load mismatch. Conduction slowing beyond the gate is dependent on gate width. Very similar results have been obtained in simulations of abrupt tissue expansions in which retrograde propagation was always successful and showed acceleration at the transition from the large area into the strand, while propagation was blocked in the anterograde direction due to the very high degree of impedance to load mismatch. Occurrence of block was critically dependent on the geometry at the transition site. If the transition was abrupt, the critical width amounted to approximately 200 μm in 2-D tissue and to 540 μm in 3-D tissue. With a small so-called "taper" at the transition (which is always present to some degree in cell cultures), the critical width decreased to 30 μm in 2-D tissue. This value corresponds

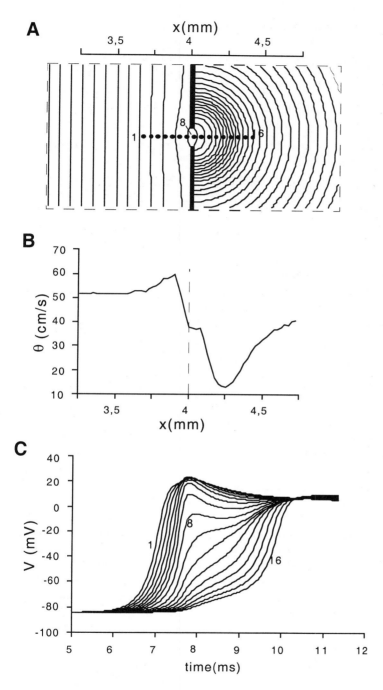

A

B

C

Figure 10.8 Simulated impulse conduction across an isthmus. Isthmus length = 50 μm, isthmus width = 150 μm. (A) Isochrone map of activation, isochrone intervals of 0.15 ms. The distance, x, is measured from the left end of the strand. (B) Conduction velocity (θ) profile across the transition along a row of 16 recorded points. (C) Action potentials (V) at the selected points depicted in (A). (From Ref. 45.) Simulations done with a ventricular cell model with $gNa_{max} = 23$ mS/cm^2, $R_x = R_y = 450$ Ω· cm. (From Ref. 59.)

closely to the values observed in experiments and indicated a close fit of the simulated data with the experiments.

Propagation around pivoting points (e.g., in anisotropic reentry) follows rules very similar to conduction across geometrical expansion and gates. The wavefront turning around the pivoting point assumes a spiral form and the formation of a convex curvature is associated with conduction slowing. This slowing has several important consequences related to the initiation or behavior of arrhythmias (41).

1. Theoretically, the curvature of the turning wave in the immediate vicinity of the pivoting point cannot decrease beyond a critical value. Therefore, the tissue immediately adjacent to the pivoting point remains inexcitable. This produces a so-called wave break or a phase singularity (i.e., a single location where refractory tissue, fully activated tissue, and resting tissue merge).

2. Slowing of propagation at the pivoting point is expected to change the head-to-tail interactions (the interactions between the leading edge of excitation and the regions that are in a state of partial recovery from inactivation) in reentrant circuits and to create or widen an excitable gap.

3. If a critical relation between excitation wavelength and conduction velocity around the pivoting point is met (relatively low state of excitability and/or low degree of cell-to-cell coupling), the turning waves may detach from the pivoting point and create spiral waves (so-called "wave shedding")(43). In other situations, pivoting points may act as anchors of moving spirals and stabilize reentry (48). As mentioned above, an essential feature of discontinuous conduction is the observation that cellular electrical properties and network properties are mutually interactive. The flow of ion current and the degree of cell-to-cell coupling affect the role of tissue discontinuities in producing conduction slowing and block and in initiating or anchoring spiral waves. Conversely, discontinuities in tissue architecture affect the ion currents participating in maintaining propagation.

D. Effect of I_{Na} Inhibition on Macroscopic Discontinuities

Figure 10.9(A) and (B) depict the results from simulated propagation across abrupt tissue expansions where the maximal sodium conductance, g_{Na}, is lowered either in the expanded region (A) or in the narrow region (B) to simulate partial inhibition of Na current (45). Reducing g_{Na} increases the width at which conduction block across the abrupt expansion takes place (i.e., it enhances the effect of a discontinuity to slow or block conduction). This effect is likely to explain the experimental observation that conduction block at an isthmus or gate develops with increasing stimulation or excitation frequency (42). It predicts that the probability of occurrence of unidirectional conduction block at tissue discontinuities will increase in pathophysiological settings associated with a high rate (e.g., an established tachycardia) or with inhibition of I_{Na} (acute ischemia) and underlines the role of tissue discontinuities in contributing to spatial and temporal instabilities of propagation.

E. Effects of Cell-to-Cell Uncoupling on Macroscopic Discontinuities

Both simulations and experimental studies have shown that increasing gap junctional conductance in discontinuous tissue may exert complex effects on propagation. Thus, in tissue with discontinuous architecture, propagation may be blocked unidirectionally in the states of high and of very low cell-to-cell coupling while conduction may be bidirectional in an

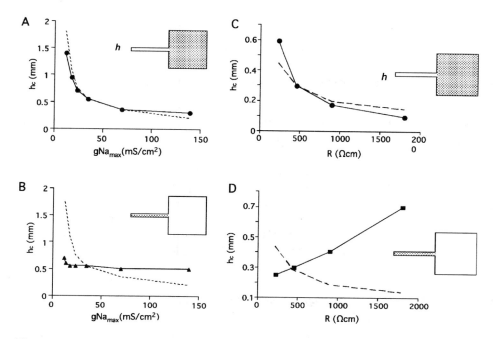

Figure 10.9 (A) and (B) Dependence of critical strand width, h_c, on spatially nonuniform changes in maximum sodium conductance, gNa_{max}. Dashed curves correspond to the control conduction when excitability was changed uniformly throughout the model while solid curves and symbols correspond to spatially inhomogeneous changes. Shaded areas in the inset indicate the parts of the model where gNa_{max} was altered: (A) in the large area or (B) in the strand. (C) and (D) Dependence of critical strand width, h_c, on spatially nonuniform changes in intracellular resistivity. Dashed curves correspond to the control when resistivity was changed uniformly throughout the model, while solid curves and symbols correspond to spatially inhomogeneous changes. Shaded areas in the inset indicate the parts of the model where resistivity was altered: (C) in the large area or (D) in the strand. (From Ref. 45.)

intermediate state of cell-to-cell coupling. The explanation is given by computer simulations of propagation across abrupt tissue expansions (46), illustrated in Figure 10.9(C) and (D), where the gap junctional resistors in longitudinal and transverse direction were changed separately. Increasing cell-to-cell coupling in the small strand slows conduction and favored formation of unidirectional block at the insertion to the large area, while increasing resistance in the large tissue area assured propagation across the same expansion (49). This latter effect was exclusively due to the increase of resistance in the vertical direction. (i.e., the resistance that opposes lateral dispersion of local current and therefore reduces current-to-load mismatch) (49,50). If one defines the "safety factor" for propagation as the ratio of the electrical charge produced by the excitation process of a given cell or region to the charge required for activation of the same cell or region (18), then the finding that partial cell-to-cell coupling can *increase* the safety factor of macroscopic propagation across tissue discontinuities (e.g., in fibrosis) complicates the interpretation of results related to gap junctional remodeling as an arrhythmogenic factor. Cell-to-cell uncoupling, on one hand, decreases conduction velocity and the wavelength of excitation and, therefore, can favor formation of reentrant circuits. On the other hand, cell-to-cell uncoupling may render conduction safer and more stable by decreasing the probability of occurrence of unidirectional blocks. Which of these effects will predominate in a particular geometric setting is not yet understood.

F. Effects of Macroscopic Discontinuities on Ion Currents

The interaction of macroscopic resistive discontinuities with ion currents is due to the fact that resistive obstacles may either accelerate or delay local activation. In the first case (e.g., in wavefront collision), flow of electrotonic current into the membrane may be so fast that it can interfere with the activation of I_{Na}. In the second case, current-to-load mismatch may delay activation between closely adjacent tissue to a point where flow of I_{Na} is no longer sufficient to excite the cells downstream. In this case, propagation depends on flow of $I_{Ca,L}$ and becomes sensitive to inhibitors or enhancers of $I_{Ca,L}$. Thus, unidirectional block at abrupt tissue expansions could be produced by application of nifedipine and reversed by Bay-K, an enhancer of $I_{Ca,L}$ (51), an effect that had also been observed for conduction between two coupled ventricular myocytes (52) and emphasizes the potential importance of drugs affecting $I_{Ca,L}$ on macroscopic propagation spread.

IV. RULES GOVERNING PROPAGATION IN NETWORKS ARE SCALE INDEPENDENT

The theoretical and experimental findings of work discussed above underline the multiple complexities present at the cellular and the more macroscopic levels. They illustrate the fact that the effects of changing structure (i.e., electrical network properties) cannot be separated from the effects of changing cellular electrical properties (e.g., ion channel function). In addition, the phenomena described in the previous sections suggest close similarities among events occurring at a cellular scale and events occurring at a more macroscopic scale. This suggests that the biophysical rules governing these processes are common to all these phenomena and, therefore, independent of scale. Interestingly, such rules have been established by theoretical work many years before the experimental data in real tissue could be obtained (13,53–57). Figure 10.10(A) shows a simple discontinuous excitable structure consisting of a linear chain of excitable elements. A set of N elements is connected by a resistor of relatively constant low value, r_i, and all the sets are interconnected by a resistor of a relatively high value R_i' (13). For a given total number of excitable elements, changing R_i, leaving r_i constant, and adjusting N modifies the degree of discontinuity without changing the overall resistance R_i (serial resistance of all r_i and R_i' resistors). Importantly, this arrangement of elements, which corresponds to a linear, resistive model, can easily be extended to a "geometrical" 2-D (or 3-D) model, as illustrated on Figure 10.10(B). In principle, all the structures discussed in the first parts of this chapter can be superimposed on the schemes depicted in Figure 10.10. This similarity helps to focus the discussion of underlying biophysical rules to three processes.

1.   Events occurring at the transition of a low- to a high-resistance region, which corresponds to sites where local current converges in the geometrical models. Terms such as "collision" and "partial collision" have been used to describe comparable processes. Examples of such partial collisions are apparent at the cellular level during longitudinal propagation (event occurring at the distal cell pole) (Fig. 10.6), at the site of local block (Fig. 10.7), and during collision of a wave with the resistive obstacles at either side of an isthmus or gate (Fig. 10.8). In all three examples, a local increase in propagation velocity is observed

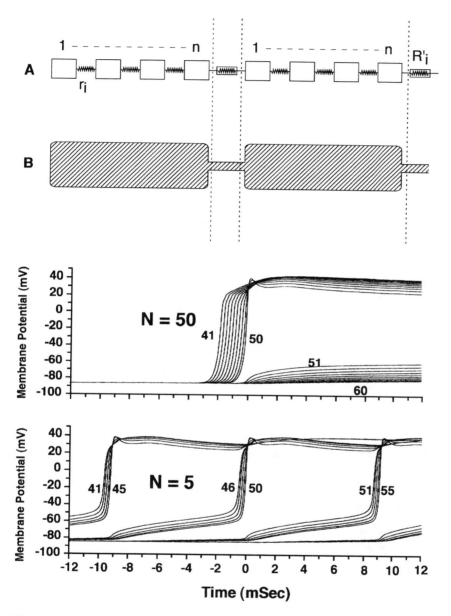

Figure 10.10 Diagram of a linear strand in which the coupling resistance between adjacent cells is spatially inhomogeneous and (B) diagram showing how this strand can be conceptually extended to represent propagation in a multidimensional structure (see text). The lower panels show simulations of a linear strand of 100 cardiac cells in which the coupling resistance between adjacent cells is spatially inhomogeneous, as diagrammed in Figure 10(A) and (B). In the upper graph (bottom panel) the strand is assumed to have a coupling resistance of 1 MΩ (1000 nS) except at the junction to the right of every 50th cell ($n = 50$), where the coupling resistance is increased to 20 MΩ (50 nS). Action propagation fails at this resistive barrier. In the lower graph (bottom panel), additional barriers have been inserted, such that now there is an increase in coupling resistance to 20 MΩ at every 5th ($n = 5$) cell and propagation now succeeds in a discontinuous fashion throughout the strand.

2.   Events occurring at the transition of a high- to a low-resistance region, or in geometrical terms, events occurring at a site of current-to-load mismatch with dispersion of local current. The effect of dispersion of local current on the shape of the action potential and ion current corresponds to the effect shown during impulse conduction through the gate (Fig. 10.8) (45). In essence, the changes in action potential shape, flow of depolarizing ion current, and in local activation are opposite to the changes during collision. As a consequence of dispersion, conduction velocity increases locally, the upstroke velocity of the action potential decreases and may get biphasic, and the amount of I_{Na} flow increases locally. The changes in action potential shape are due to the fact that more local current than in steady state is drained into the dispersing sink downstream, and to the fact that the slow inflow of electrotonic current allows I_{Na} at that site enough time to fully activate. Examples of current dispersion can be observed in Figure 10.9 at the macroscopic level and in both Figures 10.2 and 10.6 at the cellular level

3.   Interactions between subsequent discontinuities. Theoretical studies (13) showed that discontinuities not only produce local conversion or dispersion sites for local current, but that there is interaction between successive resistive obstacles. An example of this effect is shown in the lower panels of Figure 10.10, which shows a simulation of a linear strand of 100 elements, with each element represented by a model ventricular cell (58) of length 100 μm. Each cell is electrically coupled to its neighbors by a resistance of one MΩ(1000 nS), except at specified locations where the cell-to-cell resistance is raised to 20 MΩ (50 nS). If we use a spacing between these specific sites of relative uncoupling of 50 cells ($N = 50$, 5-mm spacing), then propagation fails at the first barrier. If we then introduce more uncoupling into the strand, with placement of the high resistance between every fifth cell ($N = 5$, 0.5-mm spacing), then conduction is discontinuous but successful. Thus, the spatial distribution of the uncoupling has a large effect on the success of propagation, allowing successful propagation to be produced by adding additional uncoupling. This interaction is essential to the behavior of discontinuous systems. What is the consequence of this complex behavior of a discontinuous structure? Again, it is discernible at both the cellular as well as the macroscopic level in the results shown in the previous section. At the cellular level, its counterpart relates to the observation that increasing the degree of discontinuity by decreasing electrical cell-to-cell coupling leads initially to an increase in the safety factor for conduction until block occurs at high levels of uncoupling (18). At the macroscopic level, this behavior is reflected by the observation that propagation across a macroscopic discontinuity is maintained at an intermediate degree of cell-to-cell uncoupling, but is blocked at both a normal stage of coupling and at a very high degree of uncoupling (49)

V.   SUMMARY

The above considerations about excitable networks with a discontinuous structure, such as heart muscle, show that the basic rules governing propagation are relatively simple and applicable to the observed phenomena at any structural level. The complexities are introduced by the fact that the effects at all these levels are interactive, and with respect to their arrhythmogenic potential, partially opposite. Especially in states where fibrosis increases and/or cell-to-cell uncoupling occurs and/or membrane ion current flow decreases, these interactions can produce propagation phenomena that are difficult to predict. If one adds the complex interactions of different membrane ion channels, which determine the shape of the action potential, its duration, and its refractoriness, the predictability of such system behavior becomes even more difficult. Sophisticated computer modeling, taking into account the experimental and theoretical findings at all levels, will be helpful in the future to predict and explain the electrical behavior of such networks. However, the actual testing of hypotheses must be done in experimental models, such as directly coupled isolated cells, patterned tissue cultures, and optical or electrical recordings from intact tissue with histological determinations of tissue anatomy and cell-to-cell couplings in order to fully understand these complex interactions.

ACKNOWLEDGMENTS

This work was supported by the Swiss National Science Foundation, the Swiss Heart Foundation, and the National Heart, Lung and Blood Institute.

REFERENCES

1. Hodgkin AL. A note on conduction velocity. J Physiol 1954; 125:221–224.
2. Jack JJB, Noble D, Tsien RW. Electric Current Flow in Excitable Cells. Oxford: Clarendon Press, 1975.
3. Walton MK, Fozzard H. The conducted action potential: Models and comparison to experiments. Biophys J 1983; 44:9–26.
4. Buchanan JW, Saito T, Gettes LS. The effects of antiarrhythmic drugs, stimulation frequency, and potassium-induced resting, membrane potential changes on conduction velocity and dV/dt-max in guinea pig myocardium. Circ Res 1985; 56:696–703.
5. Kleber AG, Janse MJ, Wilms-Schopmann FJG, Wilde AAM, Coronel R. Changes in conduction velocity during acute ischemia in ventricular myocardium of the isolated porcine heart. Circulation. 1986; 73:189–198.
6. Riegger CB, Alperovich G, Kleber AG. Effect of oxygen withdrawal on active and passive electrical properties of arterially perfused rabbit ventricular muscle. Circ Res 1989; 64:532–541.
7. Joyner RW, Sugiura H, Tan RC. Unidirectional block between isolated rabbit ventricular cells coupled by a variable resistance. Biophys J 1991; 60:1038–1045.
8. Sugiura H, Joyner RW. Action potential conduction between guinea pig ventricular cells can be modulated by calcium current. Am J Physiol 1992; 263:H1591–604.
9. Joyner RW, Kumar R, Golod DA, Wilders R, Jongsma HJ, Verheijck EE, Bouman LN, Goolsby WN, Van Ginneken ACG. Electrical interactions between a rabbit atrial cell and a nodal cell model. Am J Physiol 1998; 274:H2152–H2162.
10. Wilders R, Kumar R, Joyner RW, Jongsma HJ, Verheijck EE, Golod D, van Ginneken AC, Goolsby WN. Action potential conduction between a ventricular cell model and an isolated ventricular cell. Biophys J 1996; 70:281–295.
11. Wilders R, Jongsma HJ, van Ginneken AC. Pacemaker activity of the rabbit sinoatrial node. A comparison of mathematical models. Biophys J 1991; 60:1202–1216.
12. Wagner MB, Golod D, Wilders R, Verheijck EE, Joyner RW, Kumar R, Jongsma HJ, van Ginneken AC, Goolsby WN. Modulation of propagation from an ectopic focus by electrical load and by extracellular potassium. Am J Physiol 1997; 272:H1759–69.
13. Joyner RW. Effects of the discrete pattern of electrical coupling in propagation through an electrical syncytium. Circ Res 1982; 50:192–200.
14. Rudy Y, Quan WL. A model study of the effects of the discrete cellular structure on electrical propagation in cardiac tissue. Circ Res 1987; 61:815–823.
15. Fast VG, Kleber AG. Microscopic conduction in cultured strands of neonatal rat heart cells measured with voltage-sensitive dyes. Circ Res 1993; 73:914–925.
16. Rohr S, Scholly DM, Kleber AG. Patterned growth of neonatal rat heart cells in culture. Morphological and electrophysiological characterization. Circ Res 1991; 68:114–130.

17. Kleber AG, Riegger CB, Janse MJ. Electrical uncoupling and increase of extracellular resistance after induction of ischemia in isolated, arterially perfused rabbit papillary muscle. Circ Res 1987; 61:271–279.

18. Shaw RM, Rudy Y. Ionic mechanisms of propagation in cardiac tissue. Roles of the sodium and L-type calcium currents during reduced excitability and decreased gap junction coupling. Circ Res 1997; 81:727–741.

19. Rohr S, Kucera JP, Kleber AG. Slow conduction in cardiac tissue, I: effects of a reduction of excitability versus a reduction of electrical coupling on microconduction. Circ Res 1998; 83:781–794.

20. Luke RA, Beyer EC, Hoyt RH, Saffitz JE. Quantitative analysis of intercellular connections by immunohistochemistry of the cardiac gap junction protein connexin 43. Circ Res 1989; 65:1450–1457.

21. Hoyt RH, Cohen ML, Saffitz JE. Distribution and three-dimensional structure of intercellular junctions in canine myocardium. Circ Res 1989; 64:563–574.

22. Clerc L. Directional differences of impulse spread in trabecular muscle from mammalian heart. J Physiol 1976; 255:335–346.

23. Goodman D, Steen AB, Dam RT. Endocardial and epicardial activation pathways of the canine right atrium. Am J Physiol 1971; 220:1–11.

24. Spach MS, Miller WT, Geselewitz DB, Barr RC, Kootsey JM, Johnson EA. The discontinuous nature of propagation in normal canine cardiac muscle. Circ Res 1981; 48:39–54.

25. Sommer JR, Scherer B. Geometry of cell and bundle appositions in cardiac muscle: light microscopy. Am J Physiol 1985; 248:H792–H803.

26. LeGrice IJ, Smaill BH, Chai LZ, Edgar SG, Gavin JB, Hunter PJ. Laminar structure of the heart: ventricular myocyte arrangement and connective tissue architecture in the dog. Am J Physiol 1995; 269:H571–H582.

27. Spach MS, Josephson ME. Initiating reentry: the role of nonuniform anisotropy in small circuits. J Cardiovasc Electrophysiol 1994; 5:182–209.

28. Leon LJ, Roberge FA, Vinet A. Simulation of two-dimensional anisotropic cardiac reentry: effects of the wavelength on the reentry characteristics. Ann Biomed Eng 1994; 22:592–609.

29. Spach MS, Heidlage JF. The stochastic nature of cardiac propagation at a microscopic level. Electrical description of myocardial architecture and its application to conduction. Circ Res 1995; 76:366–380.

30. Fast VG, Kleber AG. Anisotropic conduction in monolayers of neonatal rat heart cells cultured on collagen substrate. Circ Res 1994; 75:591–595.

31. Fast VG, Darrow BJ, Saffitz JE, Kleber AG. Anisotropic activation spread in heart cell monolayers assessed by high-resolution optical mapping. Role of tissue discontinuities. Circ Res 1996; 79:115–127.

32. Cohen SA. Immunohistochemistry of rat cardiac sodium channels. Circulation 1991; 84:83(abstr).

33. Suenson M. Interaction between ventricular cells during the early part of excitation in the ferret heart. Acta Physiol Scand 1985; 125:81–90.

34. Spach MS, Heidlage JF, Dolber PC, Barr RC. Extracellular discontinuities in cardiac muscle: Evidence for capillary effects on the action potential foot. Circ Res 1998; 83:1144–1164.

35. Spach MS, Miler WT, Dolber PC, Kootsey JM, Sommer JR, Mosher CE. The functional role of structural complexities in the propagation of depolarization in the atrium of the dog. Cardiac Conduction disturbances due to discontinuities of effective axial resistivity. Circ Res 1982; 50:175–191.

36. Spach MS, Lieberman M, Scott JG, Barr RC, Johnson EA, Kootsey JM. Excitation sequences of the atrial septum and the AV node in isolated hearts of the dog and rabbit. Circ Res 1971; 29:156–172.

37. Spach MS, Barr RC, Serwer GS, Johnson EA, Kootsey JM. Collision of excitation waves in the dog purkinje system: Extracellular identification. Circ Res 1971; 29:499–511.

38. Veenstra RD, Joyner RW, Rawling DA. Purkinje and ventricular activation sequences of canine papillary muscle. Effects of quinidine and calcium on the Purkinje-ventricular conduction delay. Circ Res 1984; 54:500–515.

39. Schalij MJ, Lammers WJ, Rensma PL, Allessie MA. Anisotropic conduction and reentry in perfused epicardium of rabbit left ventricle. Am J Physiol 1992; 263:H1466–H1478.

40. Dillon SM, Allessie MA, Ursell PC, Wit AL. Influences of anisotropic tissue structure on reentrant circuits in the epicardial border zone of subacute canine infarcts. Circ Res 1988; 63:182–206.

41. Fast VG, Kleber AG. Role of wavefront curvature in propagation of cardiac impulse. Cardiovasc Res 1997; 33:258–271.

42. Cabo C, Pertsov AM, Baxter WT, Davidenko JM, Gray RA, Jalife J. Wave-front curvature as a cause of slow conduction and block in isolated cardiac muscle. Circ Res 1994; 75:1014–1028.

43. Cabo C, Pertsov AM, Davidenko JM, Baxter WT, Gray RA, Jalife J. Vortex shedding as a precursor of turbulent electrical activity in cardiac muscle. Biophys J 1996; 70:1105–1111.

44. Rohr S, Salzberg BM. Characterization of impulse propagation at the microscopic level across geometrically defined expansions of excitable tissue: multiple site optical recording of transmembrane voltage (MSORTV) in patterned growth heart cell cultures. J Gen Physiol 1994; 104:287–309.

45. Fast VG, Kleber AG: Block of impulse propagation at an abrupt tissue expansion: evaluation of the critical strand diameter in 2- and 3-dimensional computer models. Cardiovasc Res 1995; 30:449–459.

46. Fast VG, Kleber AG. Cardiac tissue geometry as a determinant of unidirectional conduction block: assessment of microscopic excitation spread by optical mapping in patterned cell cultures and in a computer model. Cardiovasc Res 1995; 29:697–707.

47. Rohr S, Salzberg BM. Multiple site optical recording of transmembrane voltage (MSORTV) in patterned growth heart cell cultures: assessing electrical behavior, with microsecond resolution, on a cellular and subcellular scale. Biophys J 1994; 67: 1301–1315.

48. Gray RA, Jalife J, Panfilov A, Baxter WT, Cabo C, Davidenko JM, Pertsov AM. Nonstationary vortexlike reentrant activity as a mechanism of polymorphic ventricular tachycardia in the isolated rabbit heart. Circulation 1995; 91: 2454–2469.

49. Rohr S, Kucera JP, Fast VG, Kleber AG. Paradoxical improvement of impulse conduction in cardiac tissue by partial cellular uncoupling. Science 1997; 275:841–844.

50. Joyner RW, Ramza BM, Tan RC, Matsuda J, Do TT. Effects of tissue geometry on initiation of a cardiac action potential. Am J Physiol 1989; 256:H391–403.

51. Rohr S, Kucera JP. Involvement of the calcium inward current in cardiac impulse propagation: induction of unidirectional conduction block by nifedipine and reversal by Bay K 8644. Biophys J 1997; 72:754–766.

52. Joyner RW, Kumar R, Wilders R, Jongsma HJ, Verheijck EE, Golod DA, van Ginneken AC, Wagner MB, Goolsby WN. Modulating L-type calcium current affects discontinuous cardiac action potential conduction. Biophys J 1996; 71:237–245.

53. Joyner RW, Ramon F, Morre JW. Simulation of action potential propagation in an inhomogeneous sheet of coupled excitable cells. Circ Res 1975; 36:654–661.

54. Joyner RW, Picone J, Rawling D, Veenstra R. Propagation through electrically coupled cells: Effects of regional changes in membrane properties. Circ Res 1983; 53:526–534.

55. Joyner RW, Veenstra R, Rawling D, Chorro A. Propagation through electrically coupled cells: Effects of a resistive barrier. Biophys J 1984; 45:1017–1025.

56. Spach MS, Heidlage JF, Darken ER, Hofer E, Raines KH, Starmer CF. Cellular V_{max} reflects both membrane properties and the load presented by adjoining cells. Am J Physiol 1992; 263:Pt 2):H1855–63.

57. Spach MS, Kootsey JM. Relating the sodium current and conductance to the shape of transmembrane and extracellular potentials by simulation: effects of propagation boundaries. IEEE Trans Biomed Eng 1985; 32:743–755.

58. Zeng J, Laurita KR, Rosenbaum DS, Rudy Y. Two components of the delayed rectifier K+ current in ventricular myocytes of the guinea pig type. Theoretical formulation and their role in repolarization. Circ Res 1995; 77:140–152.
59. Luo CH, Rudy Y. A dynamic model of the cardiac ventricular action potential. I. Simulations of ionic currents and concentration changes. Circ Res 1994; 74:1071–1096.

11

Atrioventricular Conduction

RALPH LAZZARA and BENJAMIN J. SCHERLAG

University of Oklahoma Health Sciences Center, Oklahoma City, Oklahoma

LUIZ BELARDINELLI

CV Therapeutics, Palo Alto, California

I. ATRIOVENTRICULAR CONDUCTION

In this chapter we trace the evolution of previously held concepts of AV nodal function and introduce some of the more recent findings from experimental and clinical studies. The remarkable success of clinical electrophysiologists in curing arrhythmias involving the atrioventricular (AV) node and perinodal tissues (1–5) has provided the impetus for new basic research regarding the function of the various tissues and cell types comprising the AV junction and their relation to atrioventricular conduction. Studies suggest the bases for new concepts of normal AV conduction and may aid in understanding the mechanisms for AV junctional arrhythmias, particularly AV nodal reentrant tachycardia (AVNRT). Evidence has been taken from both in vivo and in vitro investigations, and a substantial portion deals with myocytes isolated from AV junctional tissues, particularly the various ion currents that mediate their special functions.

A. The AV Node

The anatomy of the AV node was initially described by Tawara in 1905 (6), and confirmed by Keith and Flack in 1907 (7), as a half-oval structure consisting of interwoven fibers located at the apical end of the triangle of Koch. The functions ascribed to the slowing effect of the cardiac impulse through the AV node have been: (1) allowing the slower mechanical transmission of blood from atria to ventricles to precede the relatively rapid excitation of the ventricles via the His-Purkinje system, thereby providing optimal efficiency for ventricular ejection; (2) a filtering function by which inordinately rapid ventricular responses can be prevented during atrial tachyarrhythmias, particularly atrial fibrillation.

The electrophysiological mechanism for this function to slow conduction began to be addressed with the advent of single-cell microelectrode recordings from the compact AV node and perinodal zones starting in the late 1950s and early 1960s (8,9). Alanis and his associates (10) and Paes de Carvalho and Almeida (11) showed that there are three types of cells found in the AV node and perinodal zones. These cells, called AN, N, and NH cells, were categorized mainly based on their electrophysiological characteristics rather than their specific anatomic location. Electrophysiologically, these cell types showed distinct differences from atrial cells on one side and His bundle cells on the other. AN, N, and NH cells respond to increasing rate or premature stimulation with progressive conduction slowing either due to loss of source current or increasing resistance between cells. It is possible that both these factors are responsible for this effect, which is called "decremental conduction" (12). The N region was shown to be the locale in which the greatest slowing of conduction occurred and most reports indicated that these N cells comprise a large part of the compact AV node (Fig. 11.1). The AN cells that were originally considered as part of the node have now been shown to arise from the transitional tissues of the slow and fast pathway or posterior or distal AV nodal inputs, and the N region and the NH cells were in-

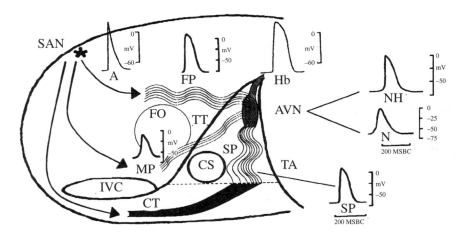

Figure 11.1   A schematic representation of the right atrium and A-V junction as seen in a right anterior oblique view. The atrial inputs from the sino-atrial node (SAN) arrive at the compact A-V node via the transitional tissues (wavy, parallel lines), which include the fast pathway (FP), the mid pathway (MP), and the slow pathway (SP). Anatomically, these transitional cell groups have been described as the superficial, deep transitional cells (28,29) and the posterior nodal extension (30). Note that the AN cells composing these transitional tissues all show some plateau (phase 2), yet each has distinctively different resting membrane potentials (side scale) and differ from the higher resting potentials atrial (A) and His bundle (Hb) cells. The cells of the atrioventricular node (AVN) consist mainly of N cells with low resting potentials along with the NH cells that are intermediate, in terms of resting potential, with relation to the adjacent Hb. The compact AVN and Hb lie at the apex of Koch's triangle, which consists of the angle made by the tendon of Todaro (TT) and the tricuspid annulus (TA). The base of the triangle (dashed line) runs from the coronary sinus (CS) ostium to the TA and demarcates the juncture of the crista terminalis (CT) and the SP as it passes the opening of the inferior vena cava (IVC).

terposed between the N and His bundle cells (Fig. 1). Thus, all these cell types—AN, N, and NH—represent the substrate for the normal slowing of conduction characteristic of responses such as: delay of AV conduction leading to first-degree heart block (i.e., P-R interval greater than 200 ms); or with further increases in heart rate, Wenckebach-type second-degree AV block (i.e., progressive increase in P-R or A-H interval leading to dropped ventricular beats) and, finally, 2:1 AV block observed with even further increasing heart rate. It should be mentioned that atrial or ventricular pacing at increasing rates most commonly invokes these forms of heart block both for AV (antegrade) as well as VA (retrograde) conduction. On the other hand, increasing the heart rate through the action of the sympathetic nervous system (e.g., during exercise) usually results in shortened AV conduction time even at very rapid rates due to the effect of adrenergic neurotransmitters that enhance AV conduction.

The property of decremental conduction can also explain, in part, another feature sometimes seen in the AV node, particularly in response to atrial premature beats or, more commonly, in response to closely coupled ventricular beats that conduct into the AV node. As a result of a high degree of nonuniform, decremental conduction among closely adjacent cells, local block and dissociation of conduction can occur. This may result in reentry manifesting as nonstimulated reciprocal activation of the atrium, usually referred to as echo or reciprocal beats (13).

B. Dual AV Nodal Physiology

Mendez and Moe (14) suggested that, under the influence of closely coupled premature atrial stimuli, the AV node dissociates into two dichotomous pathways, alpha and beta (Fig. 11.2). Furthermore, one pathway (in this case, beta) has a longer refractory period; therefore, a premature atrial beat (A$_2$) would block in this pathway while the dissociated impulse conducting slowly through the other pathway (in this case, alpha) could arrive at the distal end of the beta pathway when it had recovered. Reentry and retrograde conduction thus propagates back to the atria as an echo beat (E). It seems natural to assume that a perpetuation of this type of reentrant circuit studied initially in rabbit hearts could be responsible for the sustained form of reentry (viz., AVNRT observed clinically). However, utilizing the same rabbit preparation other investigators found that sustained AVNRT involves the AV node only as part of the reentrant circuit, rather than wholly containing the circuit. Watanabe and Dreifus (15) and Mazgalev et al. (16), when mapping with microelectrodes during sustained reentry, were able to determine that the circuit extended posteriorly between the AV node and the coronary sinus ostium. In relation to these findings, Janse et al. (17) found that early or closely coupled premature beats extended the area of reduced action potentials (i.e., decremental conduction, outside the compact AV node and even beyond the anterior border of the tendon of Todaro) (Fig. 11.3). Since AN cells, but not atrial cells, show decremental conduction, particularly in response to premature stimulation, AN cells must be located beyond the borders of the compact AV node (10). These data suggest that there might be at least two different locations in the AV junction at which reentry could occur: (1) within the compact AV node giving rise to single or, less frequently, multiple echo beats; and (2) a posterior reentrant circuit composed of AN or transitional cells and utilizing only a part of the AV node as a turn-around site. Such a circuit could give rise to single or sustained reentry; the sustained manifesting as AVNRT.

The data from isolated tissue studies suggested that evidence for two pathways might be exploited clinically. These would be expected to manifest themselves as discon-

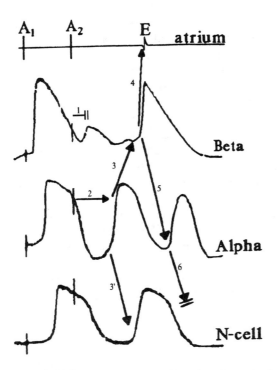

Figure 11.2 Reentry recorded by microelectrodes in the rabbit A-V node induced by an atrial premature depolarization (A_2) following the regular driven stimulus (A_1, top trace). The premature stimulus, A_2, blocks at the beta cell (#1, —⫲) but conducts slowly to the Alpha cell (#2), which then produces propagation (#3) to activate the beta cell. The last, having had time to recover can now be activated and conducts (#4) as a reentrant or echo beat (E) in the atrium. At the same time activation continues downstream (#5) reactivating the alpha cell but arriving at the N cell region too early leading to block (#6). (From Ref. 14.)

tinuous curves describing atrio-Hisian conduction time as a function coupling intervals of atrial premature depolarizations. In fact, utilizing His bundle recordings, clinical investigators determined that a series of premature beats with decreasing coupling intervals could induce dissociated curves describing AV nodal function in patients with AVNRT (Fig. 11.4) (18). As the coupling of the atrial premature beats decrease (shortening of A_1–A_2 interval) the AV nodal conduction time (determined from the A–H interval) gradually prolong (filled circles, fast pathway conduction). However, over a critical range of cycle lengths, a short decrease in A_1–A_2 (e.g., 10 ms) led to a sudden jump of A_1–H_2 of ≥50 ms. This new curve (open circles, slow pathway) continued to show greater AV nodal delay until AVNRT ensued. These findings were interpreted as the counterparts of the behavior of the Mendez, Moe α, and β pathways described above (i.e., block in the pathway with a longer refractory period and slow conduction in the other pathway, allowing recovery and reentry and leading to echo beats).

Taken together with other evidence obtained in clinical studies, the widely accepted hypothesis became that the reentrant circuit responsible for AVNRT was for the most part contained within the AV node. A key study was performed by Sung et al. in 1981 (19). They found that patients who manifested dual conduction curves during ventricular pacing

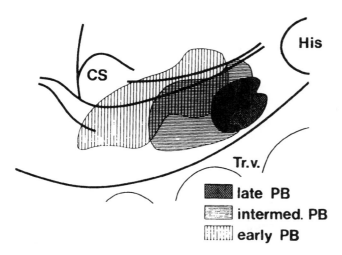

Figure 11.3 The use of premature atrial stimulation to map areas of reduced action potential amplitude indicative of AN, N, or NH cell characteristics. Late coupled premature beats (late PB) show block in the N and NH regions of the compact A-V node. Premature beats with intermediate (intermed PB) coupling show areas with reduced action potential amplitude extending mostly posteriorly, whereas microelectrode mapping of early coupled beats (early PB) lead to extension of reduced action potential amplitude in cells (most likely AN) beyond the tendon to Todaro (double line from CS to His) in the mid and anteroseptal region. (From Ref. 17.)

showed earliest atrial activation near the apex of Koch's triangle when conducting over the fast pathway. However, during conduction over the slow pathway, earliest atrial activation occurred at or near the coronary sinus ostium. Thus the atrial entrance/exit for the fast AV nodal pathway appeared to be near the site at which the His bundle potential was recorded, whereas the atrial entrance/exit for the slow AV nodal pathway was located at the base of Koch's triangle between the coronary sinus ostium and the tricuspid annulus.

A strategy to treat patients with AVNRT then became a matter of producing just enough AV nodal damage near the AV node so as to effectively perturb the slow and/or fast pathways, but not enough to induce a degree of heart block of clinical consequence. It was the cardiac surgeons who first succeeded in damaging the perinodal zone to achieve termination and cure of AVNRT without significant impairment of AV nodal conduction (20,21). This procedure was superseded when clinical electrophysiologists achieved the same result using radiofrequency current applied to the perinodal zone through strategically placed electrode catheters. Radiofrequency catheter ablation was aimed at AV nodal "modification" based on the assumption that the critical portion of the reentrant circuit was within the AV node proper. Thus ablative lesions were directed at the area close to the AV node, that is, near the fast pathway entrance/exit, close to the apex of Koch's triangle (22). Although the resulting cures of AVNRT were gratifying, the incidence of complete heart block, though small in number, was cause for concern.

The cure of AVNRT by radiofrequency ablation was also puzzling when clinical electrophysiologists discovered that lesions placed at the base of Koch's triangle between the coronary sinus ostium and the tricuspid annulus were equally effective in ablating AVNRT (1,2). The possibility that this success might be due to AV node "modification" was dismissed because the distance between the lesions and the AV node itself was con-

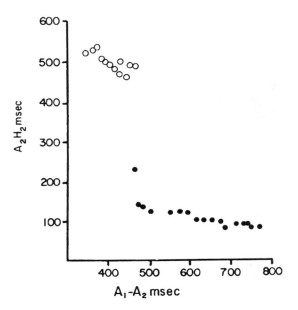

Figure 11.4 A-V nodal function curve in a patient with dual A-V nodal pathways determined by introduction of premature atrial depolarizations (A_2) with progressive shortening of the coupling interval to the basic drive (A_1, abscissa). As coupling interval (A_1-A_2) decreases, atrioventricular conduction increases, measured as the A_2H_2 interval (ordinate). In this case, at an A_1-A_2 of 450 ms, the A_1-H_2 conduction time jumps from 220 to 440 ms. The discontinuity represents refractoriness in the fast pathway (filled circles) leading to conduction over the slow pathway (open circles). See text for further discussion. (From Ref. 18.)

siderable. As a result, the incidence of complete AV block was less than 1% when this "slow pathway" ablation was performed.

C. Recent Studies on the Histological and Functional Anatomy of the AV Junction

In addition to the salutary effects that ablation of the "fast or slow pathways" provided for patients with AVNRT, there were, in addition, electrophysiological features that could be used for basic investigations of AV junctional physiology. Specifically, fast pathway ablation induced prolongation of the A–H interval at a given atrial rate, whereas the cycle length or pacing rate at which second-degree Wenckebach-type block occurred (typified by progressive prolongation of the P–R interval until a P wave failed to propagate to the ventricle) was not significantly affected. In contrast, ablation of the slow pathway in the posterior-inferior region of Koch's triangle induced no change in A–H interval, but Wenckebach cycle length was significantly prolonged. These specific effects implicate damage to inputs to the AV node since damage to the compact AV node itself affected *both* A–H prolongation and varying degrees of Wenckebach, 2:1, and complete AV block (23).

To further elaborate upon these specific responses to fast and slow pathway ablation, several basic studies were reported in which similar radiofrequency current ablations in the normal dog heart were induced at the same anatomical sites for fast and slow pathway ablations as in patients. It was found that such lesions consistently reproduced the same distinctive and separate electrophysiological alterations of A–H conduction and Wenckebach cycle length as described above (24–27). These experimental studies were carried further so that combined fast and slow pathway ablations were also performed. Surprisingly, the incidence of complete heart block was rare. Histological examinations showed that the focal lesions did not damage the compact AV node or His bundle. These studies cast further doubts on the concept that modification of the compact AV node was involved in fast and slow pathway interruption leading to termination of AVNRT. In addition, the possibility of more than two AV nodal inputs also challenged the dual AV nodal input hypothesis (17).

D. Transitional Cell Inputs to the Compact AV Node

What appeared most puzzling to electrophysiologists and anatomists was that even though catheter ablation of either the fast or slow pathway was consistently successful in curing AVNRT "the substrate(s) involved remains largely unknown" (28). Attention began to turn toward the transitional cells (Fig. 11.5) originally described as fibers that are narrower than atrial muscle cells and are arranged in a parallel direction connecting the atrium in the region of the coronary sinus ostium to the head of the compact AV node (6). These observations were extended to include anterior–superior transitional cells running in a craniocaudal direction in animal hearts and posterior–anterior in the horizontally positioned human heart. Anatomical studies had previously demonstrated transitional cells designated as deep and superficial atrial to compact AV nodal connections (29,30).

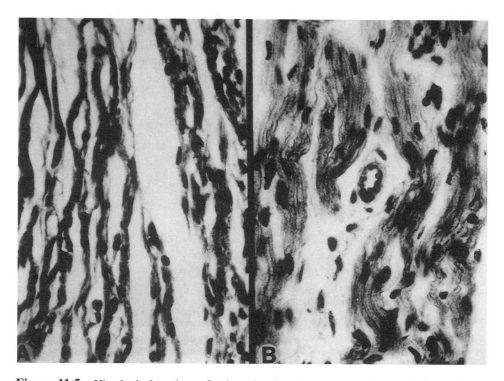

Figure 11.5 Histological sections of atrium showing the arrangement size and shape of transitional cells compared to regular atrial myocardium. (A) Transitional cells show characteristic long strands of spindle shaped, smaller diameter cells. (B) Atrial myocardial cells which are shorter and wider than transitional cells. For functional differences see text. (From Ref. 29.)

Recent studies in humans and rabbit have provided evidence that the posterior nodal extension constitutes the slow pathway in both species (31,32). Additional evidence was provided by studies in the dog heart showing parallel fibers consisting of transitional cells in the posterior–inferior AV junction which, when the slow pathway was ablated, showed direct evidence of damage in histological studies. These lesions consistently induced significant prolongation of the cycle length at which Wenckebach type 2° block occurred without affecting the A–H interval measured at a constant pacing rate (24), which is entirely consonant with the clinical criteria for slow pathway ablation (4,5). Other studies (24,27) in the normal dog heart provided clear evidence that discrete ablation of the fast pathway as well could be achieved adjacent to the apex of Koch's triangle, the site described for ablation of the fast pathway in clinical reports. In addition to meeting all the clinical criteria for fast pathway ablation, histological studies showed no damage to the AV node or His bundle but damage to transitional cells along the anterior limbus of the fossa ovalis connecting the atrium to the compact AV node (27).

If indeed the fast and slow pathways can be located and identified as the transitional cells in the anterior–superior and posterior–inferior AV junction, respectively, and if these exist in normal hearts, what mechanisms are necessary for the occurrence of AVNRT using transitional tissues? Some insight may be obtained from the work cited above (15,16). In these experimental studies, longitudinal dissociation of the posterior–inferior perinodal zone of the AV junction provided the substrate for reentrant tachycardias in some rabbit hearts. More recently, it was reported that in the rabbit hearts studied with microelectrode mapping techniques, approximately 25% showed inducible AVNRT localized in the slow pathway (i.e., posterior–inferior transitional cell AV nodal input) (33). In addition to localizing the dissociated slow pathways in transitional cells situated along the tendon of Todaro or adjacent to the tricuspid annulus, these workers demonstrated the presence of a "band of well-polarized but poorly excitable cells" that separated the two functionally disparate transitional cell groups (solid line, Fig. 11.6). In effect there was a line of block extending from the area of the coronary sinus ostium to, and in some cases through, the compact AV node to the upper portions of the His bundle. In Figure 6, a schematic representation of this form of longitudinal dissociation is depicted as a circuit utilizing the transitional cells of the slow pathway, SP_1 and SP_2, rotating around a line of block. The pivot points engage the compact AV node anteriorly and the region of the coronary sinus ostium posteriorly. The existence of this peripheral circuit does not preclude reentry occurring within the compact AV node, utilizing dissociated α and β pathways (Fig. 6 insert, lower right). Such a circuit wholly contained within the compact AV node would explain echo beats elicited by premature atrial or ventricular stimulation. Recent experimental studies in the rabbit heart, in which connections between the compact node and its transitional cell inputs were disconnected, still showed inducible echo beats in response to premature His stimulation (34). Another confirmatory study utilized microelectrode mapping and clearly implicated the N, NH, and Hb regions as the substrate for echo beats induced by premature His bundle stimulation (35). The possibility of other reentry circuits could be hypothesized based on functional longitudinal dissociation of the fast pathway, FP_1, and FP_2 (Fig. 6). Indeed, these circuits have been demonstrated in the rabbit heart (35) using microelectrode mapping and in a clinical study of 11 patients with an unusual form of AVNRT. In the latter report, the reentry circuit appeared to be located at superior rather than inferior right atrium. However, in these latter cases, the involvement of the AV node was open to question (36).

In summary, new clinical and experimental studies have provided evidence that mul-

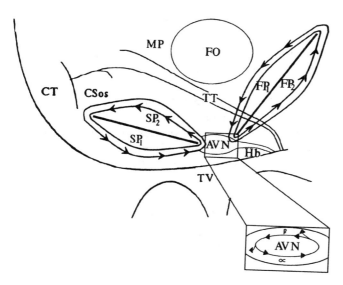

Figure 11.6 Schematic representation of the multiple reentrant circuits that could account for A-V node reentry leading to echo beats or AVNRT. The circuits for echo beats are confined mainly to the compact A-V node utilizing the dissociated α and β pathway described by Mendez and Moe (13). This is depicted in the insert (lower right). On the other hand, the common reentrant circuit responsible for AVNRT in the rabbit, and possibly in humans, requires dissociation of the posterior-inferior or slow pathway (SP₁, SP₂), which circulates around a line of block (solid line) stretching between turnaround sites in the A-V node (AVN) and the region near the coronary sinus ostium (CSos). Other possible reentrant circuits may occur in the dissociated fast pathway (FP₁, FP₂) just anterior to the fossa ovalis (FO). See text for further discussion. Other abbreviations: CT = crista terminalis; MP = mid pathway transitional cell input to the AVN; TT = tendon of Todaro; TV = tricuspid valve; Hb = His bundle.

tiple reentrant circuits exist in the AV junction. Echo beats appear to utilize dissociated pathways within the compact AV node, whereas AVNRT in all its various forms utilizes dissociated pathways within transitional tissues and the compact AV node as critical elements for sustained reentry.

II. CHARACTERISTICS OF ISOLATED SINGLE AV NODAL CELLS: MORPHOLOGY, MEMBRANE POTENTIAL, AND CURRENTS

In this section, the term AV nodal cell(s) refers to single myocytes isolated by enzymatic dissociation from the AV node region of rabbit hearts that have electrophysiological properties similar to those found in cells from the compact node (N zone) of rabbit multicellular AV nodal preparations. There is no reliable method to isolate and to select specific single myocytes from the subregions of the AV node, namely, AN, N, and NH zones.

Early attempts to isolate single AV nodal myocytes yielded rounded cells when they were exposed to physiological salt solutions containing normal calcium (37–41). These "AV nodal" cells stored in Ca-free solution had an average length and width of 93 ± 23 μm and 14.6 μm, but when superfused with normal Tyrode solution, changed from rod to round shape or oblate spheroid (41). The maximum diastolic potential, action potential amplitude, and maximum rate of rise of the upstroke were -65.1 ± 9.1 mV, 100.9 ± 9.3 mV, and 7.9 ± 4.0 V/S, respectively (39). Thus, these early studies yielded AV nodal cells that assumed an abnormal shape when exposed to Ca and had resting and action potential amplitudes that were greater than those recorded from intact AV node preparations of rabbit hearts (Fig. 2) (42).

In recent years, a number of laboratories have succeeded in isolating from the rabbit heart Ca-tolerant single AV nodal cells that maintain their in vivo rod or ellipsoid shape (43–48). In the presence of normal Ca, some AV nodal myocytes are quiescent, whereas others beat spontaneously, but they all respond in a predictable manner to M_2 muscarinic, A_1 adenosine, and β_1-adrenergic receptor agonists (43–45,49,50). Electrophysiological and morphological characteristics of rabbit isolated quiescent single atrial, AV nodal, and ventricular myocytes are summarized in Table 11.1. The action potential configuration of isolated single AV nodal myocytes (Fig. 11.7) is similar to those of slow action potentials recorded from the compact node (N zone) of multicellular AV nodal preparations (51,52). The resting membrane potential (-40 to -60 mV), the maximum rate of rise (<15 V/s), and

Table 11.1   Properties of Dispersed Rabbit Cardiomyocytes

	Atrial	AV nodal	Ventricular
Resting potential (mV)	-75 ± 2	-58 ± 2 (23) [39]	-82 ± 3 [70]
AP Amplitude (mV)	110 ± 3 [70]	85 ± 1 (23) [39]	125 ± 4 [70]
V_{max} (*V/s*)	>60	<15 [39,91]	>100
Cell input resistance (MΘ)	41 ± 18.2 (7) [44]	776 ± 283 (13) [44]	<40 [91]
Inward current$_{max}$ (nA)	>10	≤6 [91]	>10
Inward rectifier K^+ current l_{K1}	Yes [44]	No [44]	Yes [70]
ACh- and Ado-activated K^+ current $l_{KACh,Ado}$	Yes	Yes	No (or small)
Pacemaker current (l_f)	No	Yes	No
Response to high [K]$_o$ (*e.g.*, 8 mM)	Depolarization [55]	No change [55]	Depolarization [55]
Pharmacology (ACh,Ado)	Hyperpolarization Shortening APD	Hyperpolarization Shortening APD ↓ AP amplitude ↓ Rate of rise of AP	No effect
Morphology *Relaxed*	Rod-shaped Elongated Clear striation	Ellipsoid-shaped with smooth surface Less striated	Rectangular-shaped Clear striation
Contracted	Medium size, ball-shaped with rough surface	Small size, ball-shaped with shining smooth surface	Large size, ball-shaped with rough surface
Dimensions (μm)	170 ± 42 (21)(L) [44] 17 ± 1.8(21)(W) [44]	113 ± 9 (21)(L) [44] 8.8 ± 1.4(21)(W) [44]	137.3 ± 1.9(90)(L) [68] 29.2 ± 0.7(90)(W) [68]

Values are mean ± SEM or range (minimum of maximum) reprinted from Ref. 45 unless otherwise indicated. Number in parentheses represent the number of AV nodal cells studied. Numbers in brackets refer to references. Atrial and ventricular myocyte data represent experiments on four to seven myocytes from at least three different preparations. Abbreviations: AP = action potential; V_{max} = maximum velocity of rate of rise; ACh = acetylcholine; Ado = adenosine; APD = AP duration; and ↓ = decrease.

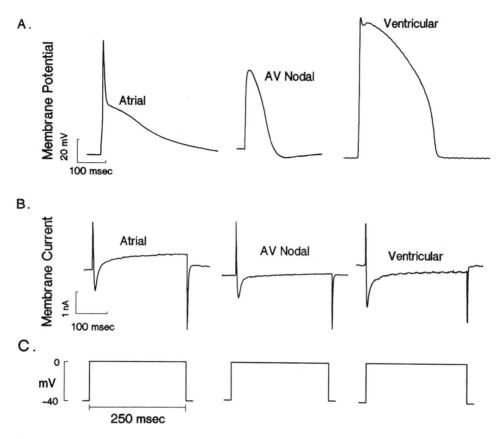

Figure 11.7 (A) Action potentials and (B) L-type Ca inward currents recorded from an atrial, AV nodal, and ventricular myocyte isolated from the same rabbit heart. Membrane potentials and currents were recorded using the whole-cell voltage clamp technique. (C) Voltage clamp protocol.

the amplitude (65 to 90 mV) of the action potential of the isolated single AV nodal cells are within the range of values reported for N cells of intact AV node preparations of rabbit hearts (52). Further, their membrane capacitance and input resistance are approximately 40 pF and 700 to 1000 MΩ, respectively (43,44). The high input resistance of the AV nodal cells has important physiological implications in that small changes in membrane currents (and conductance) cause significant changes in membrane potential and thereby affect cell excitability.

In beating rabbit AV nodal cells, spontaneous action potentials occur at a rate of 188 ± 47 min^{-1}, with maximum diastolic potentials of –58 ± 10 mV (44), and a maximum rate of rise of the action potential of 7.4 ± 0.9 V·s^{-1} (43). The action potential amplitude and overshoot are 83 ± 15 mV and 25 ± 9 mV, respectively (44). These values are similar to those summarized in Table 11.1 for quiescent AV nodal cells and for nonquiescent cells reported by other investigators (43,46). More importantly, the action potential characteristics of spontaneously beating isolated single AV nodal cells are also similar to those reported for the intact AV node preparations, including their beating rate (51,53,54).

Variability in action potential shape among enzymatically dispersed single cells from

AV node preparations has been reported (43,44,46). Some cells have action potentials with smooth overshoots, whereas others from the same preparation have "notched" overshoots. This and other differences in action potential configuration may reflect the different origin (i.e., area within the AV node region) of the myocytes. Preparations of isolated single cells from the AV node are inevitably mixed with at least atrial, AN, N, and NH cells. Identification of normal isolated single AV nodal cells requires rigorous electrophysiological and morphological criteria, which are not always possible. Several criteria, morphological and electrophysiological, to distinguish isolated single AV nodal from atrial and ventricular myocytes have been proposed (43–46). Some of the distinguishing properties and differences among these cells isolated from rabbit hearts are summarized in Table 11.1. More recently, Martynyuk et al. (55) showed that isolated rabbit single AV nodal cells, in contrast to atrial and ventricular myocytes, do not depolarize when exposed to elevated (8 mM) extracellular κ ($[K]_o$). This lack of response of isolated single AV nodal cells to changes in $[K]_o$ is consistent with the well-known insensitivity of the AV node to higher $[K]_o$ (56). The determination of the response of the cells to high $[K]_o$ may prove to be a simple and reliable test for distinguishing normal AV nodal cells from other types of myocytes present in the preparation.

A. Membrane Currents

A number of ion currents that are activated during the action potential have been identified in single cells obtained from the rabbit AV node. The characteristics of the inward and outward currents of rabbit isolated single AV nodal cells are each described below. The most salient differences between the membrane currents of AV nodal, atrial, and ventricular myocytes are also discussed.

1. Inward Sodium Current (I_{Na})

At a normal resting or maximum diastolic potential of –40 to –60 mV, I_{Na} channels are expected to be inactivated. Inactivation of I_{Na} channels at these low membrane potentials is a possible explanation, other than the absence of Na channels, for the lack of I_{Na} in isolated single AV nodal cells (46). Evidence that Na channels are in fact present derives from reports in which depolarizing steps elicit I_{Na} in some AV nodal myocytes when the holding potential is set at –90 mV (44,47). Assuming that these are "true" AV nodal cells (and neither transitional nor atrial cells), this observation is reminiscent of I_{Na} of sinoatrial nodal cells (57) and suggests that Na channels are present in AV nodal cells. A recent study of the distribution of Na-channel-specific protein expression in the AV node using a Na-channel-specific antibody disputes this conclusion (58). Immunofluorescent labeling of Na channels is low or absent in the compact node, whereas transitional and lower nodal regions display high levels of immunofluorescence that are comparable to that of atrial and ventricular myocytes (58). This finding indicates that Na channels are absent (or in small number) in the cells of N zone, but are present in AN and NH zones of the AV node. Regardless, because of the low resting and/or diastolic membrane potential, it is likely that I_{Na} does not play a significant functional role in the total depolarizing inward current of AV nodal myocytes, and hence the action potential upstroke and impulse propagation in the compact node (N zone) are not dependent on I_{Na}.

2. Inward Ca Current

Given the lack of significant I_{Na} contribution to the upstroke of the action potential, inward Ca current generates the upstroke of the action potential and determines the conduction of the impulse through the AV node. Like atrial and ventricular myocytes, AV nodal cells have two inward Ca current components: $I_{Ca,T}$ and $I_{Ca,L}$, both of which are activated at approximately –60 mV and –40 mV, respectively, and their maximal amplitude is reached at approximately –30 and +10 mV, respectively (59). The $I_{Ca,T}$ has fast activation and inactivation kinetics, which are strongly voltage dependent (57,60). Reactivation of $I_{Ca,T}$ is much slower than that of $I_{Ca,L}$ (59,60).

3. T-type Calcium Current ($I_{Ca,T}$)

To our knowledge, the study of Liu et al. (61) is the only published report on $I_{Ca,T}$ of isolated single AV nodal cells. The $I_{Ca,T}$ of isolated single AV nodal cells was activated at –50 mV, reached a maximum amplitude between –35 to ~30 mV, and contributed about one-third to one-half of the total calcium current (61). In AV nodal cells, neither the density of $I_{Ca,T}$ nor its kinetics (e.g., steady-state activation and inactivation curves) have been completely determined. Nevertheless, using a conditioning pulse voltage clamp protocol, Liu et al. (61) found that the time course of reactivation of $I_{Ca,T}$ was monoexponential with a time constant of 405 ms. These authors showed that a time-dependent partial inactivation of $I_{Ca,T}$ plays a major role in the electrotonic inhibition of excitability of AV nodal cells, and proposed that this is the underlying ion basis of concealed AV nodal conduction (61).

Future experiments are needed to define the role of $I_{Ca,T}$ in the upstroke phase of the action potential of AV nodal cells, and in the action(s) of neurotransmitters and pharmacological agents in AV nodal conduction. The activation threshold for $I_{Ca,T}$ is approximately 20 mV more negative than $I_{Ca,L}$, but is within the voltage range of the diastolic potential of AV nodal cells. The likelihood of activation of $I_{Ca,T}$ in these cells should be greater in vivo because the AV node is electrically coupled to atrial and ventricular cells. Coupling of AV nodal myocytes to atrial cells causes hyperpolarization of the nodal myocytes (62), and should decrease the steady-state inactivation of $I_{Ca,T}$. Finally, hyperpolarizing agents (e.g., acetylcholine, adenosine) may increase the relative contribution of $I_{Ca,T}$ to the total Ca current of AV nodal cells.

4. L-Type Ca Current ($I_{Ca,L}$)

The action potentials of cells of the intact AV node, as well as AV nodal conduction, are depressed and/or abolished by inorganic (e.g., manganese) and organic (e.g., verapamil) Ca channel antagonists, but not by tetrodotoxin (TTX) that blocks fast Na channels (63,64). This has led to the proposal that the inward Ca current (I_{Ca}), and not I_{Na}, is responsible for the upstroke phase of the action potential of AV nodal cells (65). The I_{Ca} differs from the I_{Na} in that it is activated at more positive membrane potential (–40 mV), has slower activation and inactivation kinetics, and is blocked by Ca antagonist agents such as verapamil, diltiazem, and nifedipine (59). First recordings of the I_{Ca} of AV node cells were made in multicellular preparations (53,66,67) or in isolated, single, rounded AV nodal cells (39). Notwithstanding the fact that the experimental conditions for recording ion currents in multicellular preparations are far from perfect, the characteristics of slow I_{Ca} recorded in multicellular AV node preparations are to a great extent similar to the $I_{Ca,L}$ of rabbit isolated single rod- and spindle-shaped AV nodal myocytes (45,49,68). $I_{Ca,L}$ activates at –30 mV, reaches a maximal amplitude at +10 mV, and has a "bell-shaped" current-voltage relationship (45,49,68). The current density of $I_{Ca,L}$ in isolated single AV nodal myocytes has been reported to be 9.3 pA/pF (68), a value somewhat smaller than that reported for atrial or ventricular myocytes (59). Representative examples of $I_{Ca,L}$ from rabbit isolated single atrial, AV nodal, and ventricular myocytes are shown in Figure 7. As is the case for other cardiac cells, inactivation of $I_{Ca,L}$ of AV nodal cells is voltage dependent and biexponential

(68). The time course of recovery of $I_{Ca,L}$ from inactivation is also biexponential (with time constants of 194.7 and 907.4 ms) (68). The steady-state activation and inactivation curves for $I_{Ca,L}$ show half-maximal activation at –3.6 mV and half-maximal inactivation at –25.8 mV (68). The ion (Ca and Na) selectivity of $I_{Ca,L}$ of isolated single AV nodal myocytes has yet to be determined. In rabbit multicellular AV node preparations, Akiyama and Fozzard (52) estimated that the relative permeability to Ca and Na (P_{Ca}/P_{Na}) of AV nodal cell membrane at a time of maximal overshoot is ~70, and this P_{Ca}/P_{Na} ratio is due to the selectivity of Ca channels rather than to a mixture of I_{Na} and I_{Ca}. Although the "slow" Ca channels are more permeable to Ca than Na ions, Na ions account for a significant part of the slow inward current (52). Thus, it is not surprising that lowering extracellular Na depresses intranodal conduction (69), even though TTX has little or no effect on the upstroke of the action potential of AV nodal cells (63). These properties of $I_{Ca,L}$ are largely responsible for the slow upstroke velocity of the action potentials and excitability of AV nodal cells. This, in turn, may contribute to the postrepolarization refractoriness of AV nodal cells, slow conduction, and rate-dependent AV node conduction disturbances.

5. The Inward Rectifier Current (I_{K1})

The κ current I_{K1} is responsible for the resting membrane potential and possibly for the late repolarization phase of the action potential of atrial and ventricular myocytes (70), but not of AV nodal cells (44,55,71). That is, in contrast to atrial and ventricular, but similar to SA nodal myocytes, in AV nodal cells the inward rectifier current I_{K1} is negligible or absent (43,44,55). The representative current–voltage relationships from a rabbit single AV nodal and atrial myocytes depicted in Figure 8 illustrate that I_{K1} is small or absent in AV nodal cells. Additional evidence that I_{K1} is negligible in AV nodal but present in ventricular myocytes is derived from recordings of single I_{K1} channel unitary current showing that the maximum number of open I_{K1} channels is on average 2.3 per patch in ventricular myocytes, but only 0.03 to 0.06 in AV nodal cells (71). This lack or low density of I_{K1} is responsible for the low resting conductance and high input membrane resistance of AV nodal cells.

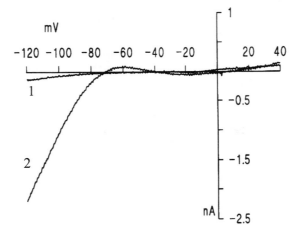

Figure 11.8 Current–voltage relationship obtained from a rabbit single AV nodal (1) and atrial (2) myocyte in response to a voltage ramp (~24 mV/s from a holding potential of –45 mV). Note that this inward rectifier K current (I_{K1}) is negligible in the AV nodal cell. (Modified with permission from Ref. 44.)

6. Delayed Rectifier K+ Current (I_K)

The delayed rectifier κ current (I_K) plays an important role in the repolarization phase of the action potential (72). This current may also contribute to the diastolic depolarization of SA nodal cells, and hence pacemaker activity (73). In atrial and ventricular myocytes, I_K is composed of at least two components—a rapid activating component referred to as I_{Kr} and a slow activating component referred to as I_{Ks} (72,74). According to Nakayama et al. (39), in rabbit isolated single AV nodal myocytes (rounded cells), I_K is activated at voltage potentials positive to −50 mV, saturates at +20 mV, and the amplitude at a holding potential of −40 mV is 61.3 ± 21.3 pA. In rod- and spindle-shaped AV nodal myocytes, Howarth et al. (75) found that I_K can be activated at membrane potentials more positive than −30 mV with maximal activation occurring at membrane potentials between +10 to +30 mV, and the density of 0.88 ± 0.08 pA/pF was similar to that of ventricular myocytes (0.76 to 0.07 pA/pF). In AV nodal myocytes, I_K is rapidly activated and sensitive to the I_{Kr}-blocking drug E-4031, indicating that this current is primarily composed of the rapid activating component I_{Kr} (75). The I_{Ks} component of I_K appears not to be present in AV nodal cells (75). The time course of deactivation of I_K is significantly faster in AV nodal (480 ms) than in ventricular (230 ms) myocytes (75). The role of this outward current system in the electrophysiological properties of single AV nodal cells (e.g., rate-dependent activation failure), and hence AV nodal conduction and pacemaker activity is far from understood.

7. Transient Outward Current (I_{to})

The transient outward current, I_{to}, activates rapidly upon depolarization and inactivates with slower time course (76). This current system consists of several distinct types of ion channels (e.g., Cl and K). I_{to} is mostly responsible for the early phase of repolarization and contributes to the plateau height and total duration of the action potential (76,77). Nakayama and Irisawa (38) observed a transient outward current in quiescent but not spontaneously beating multicellular AV nodal cell preparations. These authors proposed that the activation of I_{to} may have a role in keeping AV nodal cells quiescent—that is, as latent pacemakers. To our knowledge, the only published study of I_{to} in isolated single AV nodal cells is that of Munk et al. (46). These authors found that I_{to} is present in 93% of rod- and 42% of ovoid-shaped AV nodal cells and had properties similar to those described for I_{to} recorded in rounded AV nodal cells (38). The I_{to} density of rod-shaped AV nodal cells is 5.1 ± 1.9 pA/pF, a value that is considerably smaller than the 23 ± 4.2 pA/pF of atrial myocytes (46). The contribution of I_{to} to the configuration of the AV nodal action potential is doubtful (46).

8. The Hyperpolarization-Activated Current (I_f)

The hyperpolarization-activated current (I_f) contributes to the diastolic depolarization of SA nodal myocytes (78). This current is insensitive to Ba, sensitive to Cs, and blocked by zatebradine in a use-dependent manner (43–45). The presence of this current in AV nodal myocytes and their functional role is not as clear as it is in SA nodal myocytes. Kokubun et al. (53) reported that I_f is present in only one-fifth of spontaneously beating multicellular AV nodal preparations. In rounded isolated single AV nodal cells, I_f was detected in four of the seven cells studied (39). In spontaneously beating single AV nodal cells, Hancox et al. (43) observed I_f in only 10 to 20% of the cells, and the amplitude of the current was only significant at membrane potentials more negative than the pacemaker potential. On the other hand, in another study using beating AV nodal myocytes, I_f was observed in 90% of the cells (47). Similarly, Munk et al. (46) found that in 95% of the ovoid AV nodal myocytes I_f could be activated at membrane potentials of –60 and –90mV, whereas in rod-shaped AV nodal myo-cytes I_f could be activated in 88% of the cells at potentials more negative than –100 mV. The discrepancies in the percentage of AV nodal cells expressing I_f may in part be due to the type (transitional vs. compact node) and the condition of the cells from which recordings are made. Regardless as to whether I_f is present in AV nodal cells, its contribution to the pace-maker potential and/or excitability of these cells remains to be determined.

9. Acetylcholine- and Adenosine-Activated K Current ($I_{KACh,Ado}$)

$I_{KACh,Ado}$ flows through a distinct population of K channels coupled to muscarinic (M_2) and adenosine (A_1) receptors by a pertussis toxin-sensitive G-protein (81). This G protein-gated K (K_G) channel, a member of the family of inward-rectifier K channels, has been cloned and found to be composed of two subunits referred to as GIRK1 and GIRK4 (82–84). In the heart the GIRK1 and GIRK4 proteins are more abundantly expressed in the atria than in the ventricles (82,83). In atrial myocytes, ACh- and Ado-regulated K_G chan-nels have a unitary conductance of 45 ± 4 pS and a mean open time of 1.4 ms (81). The biophysical properties of this K channel in atrial and SA nodal myocytes, and its modula-tion by various factors and receptor agonists, have been widely reported (81,85–87). In isolated single AV nodal cells there are only a few studies on $I_{KACh,Ado}$ (44,45,55,88). The presence of ACh-activated single K channels and whole-cell I_{KAch} in rabbit isolated single AV nodal cells (nodal "cardioballs") were first reported by Sakman et al. (88). The unitary conductance of these ACh-activated K channels was 35 ± 3 pS and their mean open time was 1.8 ms (88). Of interest, spontaneous openings (i.e., in the absence of ACh in the recording pipette) of K channels in membrane patches of AV nodal cells were often ob-served; these "resting" K channel currents have gating and conductance properties similar to those activated by ACh (88). In quiescent and nonquiescent rod- and ellipsoid-shaped single AV nodal cells, both ACh (43) and adenosine (44,45,55) cause hyperpolarization of membrane potential, decrease action potential amplitude and duration, and slow the beat-ing rate. Adenosine activates an outward current that is inwardly rectifying and has rever-sal potentials of –105 ± 2 mV and –63 ± 4 mV at $[K]_o$ of 2.7 and 12.7 mM, respectively (44). These reversal potentials are similar to the calculated equilibrium potentials for potassium (assuming $[K]_I = 141$ mM), suggesting that this outward current is carried by K ions (44). This outward current, which is blocked by Ba (55), is equivalent to the I_{KAdo} of atrial (85,89) and SA nodal myocytes (86,87). Activation of $I_{KAch,Ado}$ by adenosine and pre-sumably ACh in AV nodal cells is in great part responsible for the changes in the resting and action potential (44,45,55). In addition, this current is likely to indirectly contribute to the effects of adenosine to decrease excitability and increase the refractory period of AV nodal cells (45). Recently, Martynyuk et al. (55) showed that activation of I_{KAdo} in AV nodal cells is markedly potentiated by elevated $[K]_o$. This potentiation of I_{KAdo} by $[K]_o$ ap-pears to be responsible for the enhanced depressant effect of adenosine on AV nodal con-duction in the setting of hyperkalemia (55). In summary, $I_{KAch,Ado}$ activation plays a major role in the control and modulation of AV nodal function by vagal tone and the nucleoside adenosine.

B. Mechanisms of Rate-Dependent Activation Failure

The mechanisms underlying rate-dependent AV nodal conduction delay and block are still not fully understood. Studies in rabbit isolated single AV nodal cells have provided new in-sights into the genesis of this phenomenon (90,91). Rate-dependent activation failure of isolated single AV nodal cells has been proposed to play a mechanistic role in conduction

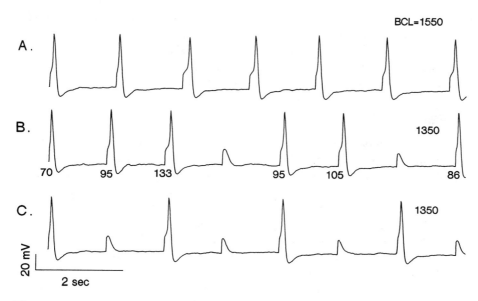

Figure 11.9 Rate-dependent activation failure (Wenckebach-like periodicity) of response of a rabbit isolated single AV nodal cell. All records are from same cell stimulated at two basic cycle lengths (BCL) of 1550 and 1350 ms. (A) Each stimulus is followed by an action potential. (B) Responses (i.e., action potentials) are periodically absent. Time from beginning of stimulus to upstroke of the action potential (activation delay, denoted by numbers at bottom of each action potential) increases progressively before failure (no action potential following a stimulus). (C) Record obtained a few minutes after (B) showing a stable S:R ratio of 2:1. Calibration bars apply to all panels.

delay and rate-dependent properties of the AV node (91). Accordingly, the Wenckebach-like periodicity of action potentials of isolated single AV nodal cells (Fig. 11.9) provides evidence that the rate dependence of impulse conduction through the AV node is due at least in part to intrinsic properties of individual AV nodal cells. Postrepolarization refractoriness, a phenomenon whereby the refractory period outlasts the repolarization phase of the action potential, is due to the slow recovery of the excitability of single AV nodal cells. The membrane current(s) responsible for this phenomenon in AV nodal cells remains to be determined. Hoshino et al. (91) suggested that at least I_K, I_{K1}, and I_{Ca} are involved in the postrepolarization refractoriness of AV nodal cells. The slow deactivation of I_K was shown to be the cause of the delayed recovery of the excitability of guinea pig ventricular myocytes, and in turn of the postrepolarization refractoriness and Wenckebach periodicity observed in these cells (90). This observation, and an analytical model of rate-dependent processes of AV nodal cells developed by these authors (91), have led them to propose that the slow time course of deactivation of I_K is in part responsible for the diastolic changes of AV nodal cell excitability. In keeping with this interpretation are the findings showing that the time constant of deactivation of I_K in AV nodal cells is such that this current could contribute to the changes in cell excitability during diastole (40,75). A role for I_{K1} in the phenomenon of postrepolarization refractoriness and excitability of AV nodal cells is unlikely because this current is negligible or absent in isolated single AV nodal cells (see above). On the other hand, $I_{Ca,L}$ is a prime candidate to mediate this phenomenon because this current is in great part responsible for the maximal upstroke velocity of the action potential of AV nodal cells, which is a major determinant of cell excitability and impulse conduction.

The involvement of $I_{Ca,T}$ and other ion current systems present in AV nodal cells (e.g., I_{to}, I_f) in the rate-dependent activation failure remains to be determined. Regardless, the time- and rate-dependent changes in excitability of single AV nodal cells is likely to be a major cause of second-degree AV block; however, additional studies will be necessary to elucidate the ion mechanisms responsible for this phenomenon.

III. SUMMARY AND CONCLUSIONS

Over most of this century, many studies of the AV node have described phenomenologically its morphological and electrophysiological features with a goal of clarifying its functions as a critical element in atrioventricular conduction to slow the transmission of the impulse from atria to ventricles and to serve as a high-rate filter to protect the ventricles. In recent years, there has been a revived interest in the AV node, in part because of the development of the clinical procedure of radiofrequency catheter ablation to eliminate AV node reentrant tachycardia that has stimulated studies of the atrial inputs to the AV node. The nature of the multiple inputs and their properties are receiving renewed interest. In addition, the properties of the AV node are receiving clarification by studies in isolated AV nodal cells of various channels and currents including gap junctions. Notable findings include the apparent absence of Na channels and the K current I_{K1} from cells from the compact node. Clarification of electrophysiological properties at this more molecular level is in its infancy but already an understanding of certain aspects of integrated AV nodal functions has been achieved, such as the diminished excitatory current (absent Na channel) and the reduced resting potentials (absent I_{K1}) resulting in slow conduction. Accelerating progress is to be anticipated.

REFERENCES

1. Jackman WM, Beckman KJ, McCelland JH, Wang X, Friday KJ, Roman CA, Moulton KP, Twidale N, Hazlitt HA, Prior MI, Oren J, Overholt ED, Lazzara R. Treatment of supraventricular tachycardia due to atrioventricular reentry by radiofrequency catheter ablation of slow pathway conduction. N Engl J Med 1992; 327:313–318.
2. Haissaguerre M, Gaita F, Fischer B, Commenges D, Montserrat P, d'Ivernois C, Lemetayer P, Warin J. Elimination of atrioventricular nodal tachycardia using discrete slow potentials to guide application of radiofrequency energy. Circulation 1992; 85:2162–2175.
3. Jazayeri MR, Hempe SL, Sra JS, Dhala AA, Blanck Z, Deshpande SS, Avitall B, Drum DP, Gilbert CJ, Akhtar M. Selective transcatheter ablation of the fast and slow pathways using radiofrequency energy in patients with atrioventricular nodal reentrant tachycardia. Circulation 1992; 85:1318–1328.
4. Jazayeri MR, Akhtar M. Electrophysiological behavior of atrioventricular node after selective fast or slow pathway ablation in patients with atrioventricular nodal reentrant tachycardia. PACE 1993; 16(II):623–628.
5. Kay GN, Epstein AE, Dailey SM, Plumb VJ. Role of radiofrequency ablation in the management of supraventricular arrhythmias: Experience in 760 consecutive patients. J Cardiovasc Electrophysiol 1993; 4:371–389.
6. Tawara S. Die topographic und histologie der bruckenfasern: Ein beitrag zur lehre von der bedeutung der Purkinjeschen faden, "Das Reizleitungssystem des Saugetierherzens." Zentralbl F Physiol 1906; 3:70–76.
7. Keith A, Flack M. The form and nature of the muscular connections between the primary division of the vertebrate heart. J Anat Physiol 1907; 41:172–189.
8. Hoffman BF, Paes de Carvalho A, DeMello WC. Transmembrane potentials of single fibers of the atrio-ventricular node. Nature 1958; 181:66.

9. Matsuda K, Hoshi T, Kameyama S. Action potentials of the A-V node. Tohoku J Exp Med 1958; 68:8.

10. Alanis J, Lopez E, Mandoki JJ, Pilar G. Propagation of impulses through the atrioventricular node. Am J Physiol 1959; 197(6):1171–1174.

11. Paes de Carvalho A, Almeida DF. Spread of activity through the A-V node. Circ Res 1960; 8:801–809.

12. Hoffman BF, Cranefield PF. The atrioventricular node. In: Electrophysiology of the Heart. New York: McGraw-Hill, 1960: 132–174.

13. Moe GK, Preston JB, Burlington H. Physiologic evidence for a dual A-V transmission system. Circ Res 1956; 4:357–375.

14. Mendez C, Moe GK. Demonstration of a dual A-V conduction system in the isolated rabbit heart. Circ Res 1966; 19:378–393.

15. Watanabe Y, Dreifus LS. Inhomogenous conduction in the A-V node: A model for reentry. Am Heart J 1965; 70:505–514.

16. Mazgalev T, Dreifus LS, Bianchi J, Michelson EL. The mechanism of AV junctional reentry: Role of the atrionodal junction. Anat Rec 1988; 201:179–188.

17. Janse MJ, Van Capelle FJL, Anderson RH, Touboul P, Billette J. Electrophysiology and structure of the atrioventricular node of the isolated rabbit heart. In: Wellens HJJ, Lie KS, Janse MJ, eds. The Conduction System of the Heart. Leiden: HE Stenfert Kroese, 1976:296–315.

18. Denes P, Wu D, Dhingra RC, Chuquima R, Rosen KM. Demonstration of dual A-V nodal pathways in patients with paroxysmal supraventricular tachycardia. Circulation 1973; 48:549–555.

19. Sung R, Waxman H, Saksena S, Juma Z. Sequence of retrograde atrial activation in patients with dual atrioventricular nodal pathways. Circulation 1981; 64:1059–1067.

20. Ross DL, Johnson DC, Denniss AT, Cooper MJ, Richards DA, Uther JB. Curative surgery for atrioventricular junctional ("AV Nodal") reentrant tachycardia. J Am Coll Cardiol 1985; 6:1383–1392.

21 Cox JL, Holman WL, Cain ME. Cryosurgical treatment of atrioventricular node reentrant tachycardia. Circulation 1987; 76:1329–1336.

22. Huang SKS, Bharati S, Graham AR, Gorman G, Lev M. Chronic incomplete atrioventricular block induced by radiofrequency catheter ablation. Circulation 1989; 80:951–961.

23. Scherlag BJ, Antz M, Otomo K, Tondo C, Patterson E, Lazzara R, Jackman WM. Atrial inputs as determinants of atrioventricular nodal conduction: re-evaluation and new concepts. Cardiologia 1995; 40:753–761.

24. Antz M, Scherlag BJ, Patterson E, Otomo K, Tondo C, Pitha J, Gonzalez M, Jackman WM, Lazzara R. Electrophysiology of the right anterior approach to the atrioventricular node: Studies in vivo and in the isolated perfused dog heart. J Cardiovasc Electrophysiol 1997; 8:47–61.

25. Hirao K, Scherlag BJ, Poty H, Otomo K, Tondo C, Antz M, Patterson E, Jackman WM, Lazzara R. Electrophysiology of the atrio-AV nodal inputs and exits in the normal dog heart: Radiofrequency ablation using an epicardial approach. J Cardiovasc Electrophysiol 1997; 8:904–915.

26. Tondo C, Scherlag BJ, Otomo K, Antz M, Patterson E, Arruda M, Jackman WM, Lazzara R. Critical atrial site for ablation of pacing-induced atrial fibrillation in the normal dog heart. J Cardiovasc Electrophysiol 1997; 8:1255–1265.

27. Antz M, Scherlag BJ, Otomo K, Pitha J, Tondo C, Patterson E, Jackman WM, Lazzara R. Evidence for multiple atrio-AV nodal inputs in the normal dog heart J. Cardiovasc Electrophysiol 1998; 9:395–408.

28. Becker A. Atrioventricular nodal anatomy revisited. Learning Center Highlights. Am Coll Cardiol 1994; 19:6.

29. Becker AE, Anderson RH. Morphology of the human atrioventricular junction area. In: Wellens HJJ, Lie KI, Janse MJ, eds. The Conduction System of the Heart. Leiden: HE Stenfert Kroese, 1976: 263.

30. Anderson RH, Becker AE, Brechenmacher C. The human atrioventricular junctional area. A morphological study of the AV node and bundle. Eur J Cardiol 1975; 3:11–25.
31. Inoue S, Becker AE. Posterior extensions of the human compact atrioventricular node. Circulation 1998; 97:188–193.
32. Medkour D, Becker AE, Khalife K, Billette J. Anatomic and functional characteristics of a slow posterior AV nodal pathway: Role in dual-pathway physiology and reentry. Circulation 1998; 98:164–174.
33. Patterson E, Scherlag BJ. Longitudinal dissociation within the posterior AV nodal input of the rabbit: A substrate for AV nodal reentry. Circulation 1999; 99:143–155.
34. Loh P, de Bakker JMT, Hocini M, Thibault B, Janse MJ. High resolution mapping and dissection of the triangle of Koch in canine hearts: Evidence for subatrial reentry during ventricular echoes. PACE 1997; 20:1080.
35. Patterson E, Scherlag BJ. Multiple reentrant pathways incorporating the compact AV node in superfused rabbit AV preparations. PACE 1998; 21:919.
36. Iesaka Y, Takahashi A, Goya M, Soijima Y, Okamoto Y, Fujiwara H, Aonuma K, Nogame A, Hiroe M, Marumo F, Hiraoka M. Adenosine-sensitive atrial reentrant tachycardia originating from the atrioventricular nodal transitional area. J Cardiovasc Electrophysiol 1997; 8:854–864.
37. Kakei M, Noma A. Adenosine-5-triphosphate-sensitive single potassium channel in the atrioventricular node cell of the rabbit heart. J Physiol 1984; 352:265–284.
38. Nakayama T, Irisawa H. Transient outward current carried by potassium and sodium in quiescent atrioventricular node cells of rabbits. Circ Res 1985; 57:65–73.
39. Nakayama T, Kurachi Y, Noma A, Irisawa H. Action potential and membrane currents of single pacemaker cells of the rabbit heart. Pflugers Arch 1984; 402:248–257.
40. Shibasaki T. Conductance and kinetics of delayed rectifier potassium channels in nodal cells of the rabbit heart. J Physiol 1987; 387:227–250.
41. Tanaguchi J, Kokubun S, Noma A, Irisawa H. Spontaneously active cells isolated from the sino-atrial and atrio-ventricular nodes of the rabbit heart. Jpn J Physiol 1981; 31:547–558.
42. Anderson RH, Janse MJ, van Capelle FJL, Billette J, Becker AE, Durrer D. A combined morphological and electrophysiological study of the atrioventricular node of the rabbit heart. Circ Res 1974; 35:909–922.
43. Hancox JC, Levi AJ, Lee CO, Heap P. A method for isolating rabbit atrioventricular node myocytes which retain normal morphology and function. Am J Physiol 1993; 265:H755–H766.
44. Martynyuk AE, Kane KA, Cobbe SM, Rankin AC. Adenosine increases potassium conductance in isolated rabbit atrioventricular nodal myocytes. Cardiovasc Res 1995; 30:668–675.
45. Wang D, Shryock JC, Belardinelli L. Cellular basis for the negative dromotropic effect of adenosine on rabbit single atrioventricular nodal cells. Circ Res 1996; 78:697–706.
46. Munk AA, Adjemian RA, Zhao J, Ogbaghebriel A, Shrier A. Electrophysiological properties of morphologically distinct cells isolated from the rabbit atrioventricular node. J Physiol 1996; 493.3:801–818.
47. Habuchi Y, Nishio M, Tanaka H, Yamamoto T, Lu L-L, Yoshimura M. Am J Physiol 1996; 271:H2274–H2282.
48. Han X, Kobzik L, Balligand J-L, Kelly RA, Smith TW. Nitric oxide synthase (NOS3)-mediated cholinergic modulation of Ca^{2+} current in adult rabbit atrioventricular nodal cells. Circ Res 1996; 78:998–1008.
49. Martynyuk AE, Kane KA, Cobbe SM, Rankin AC. Nitric oxide mediates the anti-adrenergic effect of adenosine on calcium current in isolated rabbit atrioventricular nodal cells. Pflugers Arch -Eur J Physiol 1996; 431:452–457.
50. Martynyuk AE, Kane KA, Cobbe SM, Rankin AC. Role of nitric oxide, cyclic GMP and superoxide in inhibition by adenosine of calcium current in rabbit atrioventricular nodal cells. Cardiovasc Res 1997; 34:360–367.
51. Hoffman BF, Paes de Carvalho A, de Mello WC, Cranefield PF. Electrical activity of single fibers of the atrioventricular node. Circ Res 1959; 7:11–18.

52. Akiyama T, Fozzard HA. Ca and Na selectivity of the active membrane of rabbit AV nodal cells. Am J Physiol 1979; 236:C1-C8.

53. Kokubun S, Nishimura M, Noma A, Irisawa H. Membrane currents in the rabbit atrioventricular node cell. Pflugers Arch 1982; 393:15–22.

54. West TC, Toda N. Response of the A-V node of the rabbit to stimulation of intracardiac cholinergic nerves. Circ Res 1967; 20:18–31.

55. Martynyuk AE, Morey TE, Belardinelli L, Dennis DM. Hyperkalemia enhances the effect of adenosine on $I_{K,Ado}$ in rabbit isolated AV nodal myocytes and on AV nodal conduction in guinea pig isolated heart. Circulation 1999; 99:312–318.

56. Watanabe Y, Dreifus LS. Interactions of lanatoside C and potassium on atrioventricular conduction in rabbits. Circ Res 1970; 27:931–940.

57. Denyer JC, Brown HF. Rabbit sino-atrial node cells: Isolation and electrophysiological properties. J Physiol 1990; 428:405–424.

58. Petrecca K, Amellal F, Laird DW, Cohen SA, Shrier A. Sodium channel distribution within the rabbit atrioventricular node as analysed by confocal microscopy. J Physiol 1997; 501;2:263–274.

59. McDonald TF, Pelzer S, Trautwein W, Pelzer DJ. Regulation and modulation of calcium channels in cardiac, skeletal, and smooth muscle cells. Physiol Rev 1994; 2:365–445.

60. Vassort G, Alvarez J. Cardiac T-type calcium current: Pharmacology and roles in cardiac tissues. J Cardiovasc Electrophysiol 1994; 5:376–393.

61. Liu Y, Zeng W, Delmar M, Jalife J. Ionic mechanisms of electrotonic inhibition and concealed conduction in rabbit atrioventricular nodal myocytes. Circulation 1993; 88:1634–1646.

62. Spitzer KW, Sato N, Tanaka H, Firek L, Zaniboni M, Giles WR. Electrotonic modulation of electrical activity in rabbit atrioventricular node myocytes. Am J Physiol 1997; 273:H767–H776.

63. Zipes DP, Mendez C. Action of manganese ions and tetrodotoxin on atrioventricular nodal transmembrane potentials in isolated rabbit hearts. Circ Res 1973; 32:447–454.

64. Wit AL, Cranefield PF. Verapamil inhibition of the slow response. A mechanism for its effectiveness against reentrant A-V nodal tachycardia. Circulation 1974; 50:3–146.

65. Meijler FL, Janse MJ. Morphology and electrophysiology of the mammalian atrioventricular node. Physiol Rev 1988; 68:608–647.

66. Noma A, Irisawa H, Kokubun S, Kotake H, Nishimura M, Watanabe Y. Slow current systems in the A-V node of the rabbit heart. Nature 1980; 285:228–229.

67. Ho WK, Kim WG, Earm YE. Membrane currents in the atrioventricular node of the rabbit. Seoul J Med 1986; 27:211–231.

68. Hancox JC, Levi AJ. L-type calcium current in rod- and spindle-shaped myocytes isolated from rabbit atrioventricular node. Am J Physiol 1994; 267:H1670–H1680.

69. Watanabe Y. Effects of calcium and sodium concentrations on atrioventricular conduction: Experimental study in rabbit hearts with clinical implications on heart block and slow calcium channel blocking agent usage. Am Heart J 1981; 102:883–891.

70. Giles WR, Imaizumi Y. Comparison of potassium currents in rabbit atrial and ventricular cells. J Physiol 1988; 405:123–145.

71. Noma A, Nakayama T, Kurachi Y, Irisawa H. Resting K conductances in pacemaker and non-pacemaker heart cells of the rabbit. Jpn J Physiol 1984; 34:245–254.

72. Sanguinetti MC, Keating MT. Role of delayed rectifier potassium channels in cardiac repolarization and arrhythmias. News Physiol Sci 1997; 12:152–157.

73. Noma A. Ionic mechanisms of the cardiac pacemaker potential. Jpn Heart J 1996; 37:673–682.

74. Sanguinetti MC, Jurkiewicz NK. Two components of cardiac delayed rectifier K+ current: differential sensitivity to block by class III antiarrhythmic agents. J Gen Physiol 1990; 96:195–215.

75. Howarth FC, Levi AJ, Hancox JC. Characteristics of the delayed rectifier K current compared in myocytes isolated from the atrioventricular node and ventricle of the rabbit heart. Pflugers Arch Eur J Physiol 1996; 431:713–722.

76. Kass RS, Freeman LC. Potassium channels in the heart: Cellular, molecular, and clinical implications. Trends Cardiovasc Med 1993; 3:149–159.

77. Hiraoka M, Sawanobori T, Kawano S, Hirano Y, Furukawa T. Functions of cardiac ion channels under normal and pathological conditions. Jpn Heart J 1996; 37:693–707.

78. DiFrancesco D. The onset and autonomic regulation of cardiac pacemaker activity: relevance of the f current. Cardiovasc Res 1995; 29:449–456.

79. DiFrancesco D. Some properties of the UL-FS 49 block of the hyperpolarization activated current (I_f) in sino-atrial node myocytes. Pflugers Arch 1994; 427:64–70.

80. Goethals M, Raes A, van Bogaert P-P. Use-dependent block of the pacemaker current I_f in rabbit sinoatrial node cells by zatebradine (UL-FS 49): On the mode of action of sinus node inhibitors. Circulation 1993; 88:2389–2401.

81. Kurachi Y, Nakajima T, Sugimoto T. On the mechanism of activation of muscarinic K+ channels by adenosine in isolated atrial cells: involvement of GTP-binding proteins. Pflugers Arch 1986; 407:264–274.

82. Kubo Y, Reuveny E, Slesinger PA, Yan YN, Yan LY. Primary structure and functional expression of a rat G-protein-coupled muscarinic potassium channel. Nature 1993; 364:802–806.

83. Dascal N, Schreibmayer W, Lim NF, Wang W, Chavkin C, DiMagno L, Labanca C, Kieffer BL, Gaveriaux-Ruff C, Trollinger D, Lester HA and Davidson N. Atrial G protein-activated K+ channel: expression cloning and molecular properties. Proc Natl Acad Sci USA 1993; 90:10235–10239.

84. Krapivinsky G, Gordon EA, Wickman K, Velimirovic B, Krapivinsky L, Clapham DE. The G-protein-gated atrial K+-channel I_{KAch} is a heteromultimer of two inwardly rectifying K+-channel proteins. Nature 1995; 374:135–141.

85. Belardinelli L, Isenberg G. Actions of adenosine and isoproterenol on isolated mammalian ventricular myocytes. Circ Res 1983; 53:287–297.

86. Belardinelli L, Giles WR, West A. Ionic mechanisms of adenosine actions in pacemaker cells from rabbit heart. J Physiol (Lond) 1988; 405:615–633.

87. Zaza A, Rocchetti M, DiFrancesco D. Modulation of the hyperpolarization-activated current (I(f)) by adenosine in rabbit sinoatrial myocytes. Circulation 1997; 94:734–741.

88. Sakamann B, Noma A and Trautwein W. Acetylcholine activation of single muscarinic K+ channels in isolated pacemaker cells of the mammalian heart. Nature 1983; 303:250–253.

89. Visentin S, Wu S-N, Belardinelli L. Adenosine-induced changes in atrial action potential: contribution of Ca and K currents. Am J Physiol 1990; 258:H1070–H1078.

90. Delmar M, Glass L, Michaels DC, Jalife J. Ionic basis and analytical solution of the Wenckebach phenomenon in guinea pig ventricular myocytes. Circ Res 1989; 65:775–788.

91. Hoshino K, Anumonwo J, Delmar M, Jalife J. Wenckebach periodicity in single atrioventricular nodal cells from the rabbit heart. Circulation 1990; 82:2201–2216.

12

Ca²⁺ and Arrhythmias

HENK E. D. J. TER KEURS

University of Calgary, Calgary, Alberta, Canada

PENELOPE A. BOYDEN

Columbia University, New York, New York

I. INTRODUCTION

This chapter discusses the role of cellular Ca^{2+} in cardiac arrhythmias and its interactions with other components of the working cardiac myocyte. We describe what is known about the mechanisms involved and the exquisitely complex movement of this ion that someday may permit better therapies to treat or prevent arrhythmias.

II. OVERVIEW OF Ca²⁺ TRANSPORT IN THE CARDIAC CELL

Structural aspects of the myocyte are illustrated in Figure 12.1, a longitudinal electron microscopic section through a trabecula from the right ventricle of a rat heart. The cell border is delineated by a glycoprotein layer overlying the sarcolemma, which invaginates into the cell near the Z lines of the myofibrils. These transverse tubuli (T-tubuli) are rich in dihydropyridine-sensitive Ca^{2+} channels (DHPR) and Na/Ca exchange proteins. The T-tubuli contact a longitudinal network of tubules—the sarcoplasmic reticulum (SR)—which is a prominent Ca^{2+} storing organelle. Terminal cisternae of the SR abutting the T-tubuli contain Ca^{2+} channels with a high affinity for ryanodine receptors (RyR) that are involved in Ca^{2+} release from the SR. The RyR form ultrastructurally recognizable junctional "foot proteins" in close proximity to the DHPR. The longitudinal SR envelops the myofibrils and is densely covered by Ca^{2+} pump molecules, which drive Ca^{2+} into the SR where it is buffered by binding to calsequestrin. The rate of activity of the SR-Ca^{2+} pump depends on the free cytosolic $[Ca^{2+}]_i$. The Ca^{2+} sensitivity of the pump is controlled by the degree of phosphorylation of the regulatory protein, phospholamban, in the SR membrane. The con-

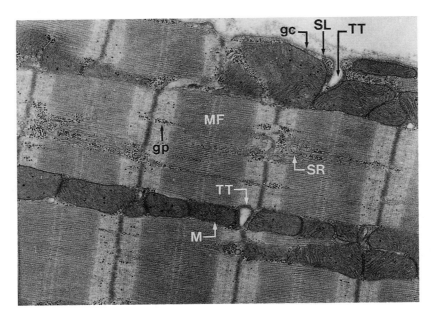

Figure 12.1 Electron micrograph of cardiac muscle illustrating structural components of excitation contraction coupling. SL = sarcolemmal membrane; TT = transverse tubuli; gc = glycocalyx; SR = sarcoplasmic reticulum; MF = myofibrils; gp = glycogen particles; and M = mitochondria. The diameter of the myocytes permits rapid electrical conduction of action potentials along the cell. Action potentials are carried into the core of the cells along the transverse tubuli. Each sarcomere in every myofibril is accompanied by closely adjacent sarcoplasmic reticulum. Thus, Ca^{2+} ions need to diffuse over <1 μm in order to reach the most centrally located contractile proteins after release from the terminal cisternae, permitting rapid synchronous activation of the sarcomeres. Close proximity of the mitochondria allows rapid ATP supply as well as participation of the mitochondria in Ca^{2+} homeostasis. (Reproduced from Can J Cardiol 1993; 9: 1F–11F, with permission.)

tractile proteins arranged in sarcomeres in the myofibrils occupy almost 60% of the intracellular space. Mitochondria adjacent to the sarcomeres occupy much of the remainder of the cell.

The ultrastructure of Purkinje and atrial cells differs from ventricular myocytes, in that they have no T-tubuli and the SR is in two forms, subsarcolemmal junctional SR and corbular SR located in the core of the cell. The diameter of a Purkinje cell is 30 to 40 μm, that of a normal ventricular myocyte is 20 to 25 μm, and that of an atrial cell is 15 μm.

A model [see Fig. 12.2(A)] of excitation contraction coupling (ECC) has been developed to help explain Ca^{2+} movements in the cardiac cell (10,126). During the action potential, Ca^{2+} ions enter the cell through the DHPR or L-type Ca^{2+} channels. Most of this Ca^{2+} is immediately bound to the Ca^{2+} pump in the SR. Ca^{2+} ions entering via DHPR channels in the T-tubuli also trigger the release of Ca^{2+} from the RyR in the terminal cisternae. In the rat (12), the ratio of RYR to L-type Ca^{2+} channels receptors is 7:1, while the ratio of superficial L-type Ca^{2+} channels versus T-tubular channels is 1:3 (150). It is thought one T-tubular L-type channel faces ~10 RYRs. Ca^{2+}-induced Ca^{2+} release (CICR) is proportional to the Ca^{2+} content of the SR and dictates the force of cardiac contraction by activating contraction of the sarcomere. This contraction is short-lived due to the rapid elimina-

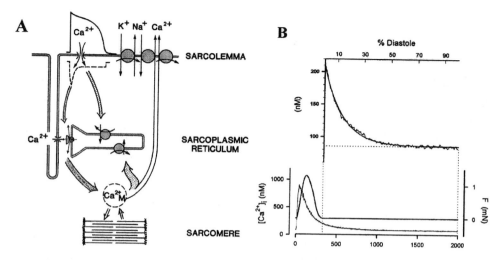

Figure 12.2 (A) Diagram illustrating components of normal excitation contraction coupling in ventricular myocardium. Sarcolemmal transport proteins (e.g., ion channels and/or pumps) mediate Ca^{2+} influx or Ca^{2+} efflux. Shown are the L-type Ca^{2+} channels indicated by pore openings below the plateau phase of the action potential (upper left), the Na/K Pump, and the Na/Ca transport proteins of sarcolemma. Various SR membrane proteins are Ca^{2+} sensitive and provide Ca^{2+}-induced Ca^{2+} release (ryanodine receptor, RYR) or Ca^{2+} reuptake into SR (Ca^{2+} ATPase pump). Ca^{2+} released from SR binds to proteins of the sarcomere (Ca^{2+}M) and contraction is initiated. (B) Lower tracings are actual force (thick black line) and Ca_i transient (thin black line) recordings of the electrically stimulated trabecula. Upper tracing illustrates the slow change in Ca_i occurring during the diastolic period (between vertical dotted lines). (Panel A modified from Can J Cardiol 1993; 9:1F–11F and Panel B reproduced from J Physiol 1997; 502: 661–677, with permission.)

tion of Ca^{2+} ions from the cytosol. About two-thirds of the Ca^{2+} is resequestered by the SR; the remainder leaves the cell mostly via the low-affinity, high-capacity Na/Ca exchanger, while a low-capacity, high-affinity Ca^{2+} pump in the cell membrane lowers the cytosolic Ca^{2+} level further during the diastolic interval (24) [Fig. 12.2(B)]. In the steady state the sum of the Ca^{2+} effluxes through the membrane balances influx during the action potential. It follows, then, that the Ca^{2+} content of the SR depends on the heart rate and on the duration of the action potential. Furthermore, a fraction of the Ca^{2+} involved in activation of the heartbeat recirculates into the SR and becomes available for activation of the next beat. Thus, force of a subsequent heartbeat will depend on the force of the previous one. In addition, it takes time for the Ca^{2+} release process to recover completely from the last release, such that sequestered Ca^{2+} can again be released from the SR. Therefore, the force of the heartbeat will also depend on heart rate, duration of diastole, and duration of the action potential (155).

It is possible to load the SR excessively with Ca^{2+} ions and this may occur following damage of cardiac cells (108) or following exposure to interventions that increase intracellular Ca^{2+} levels (digitalis, high external Ca^{2+}, high stimulus rate, etc.). SR-Ca^{2+} overload is defined as the condition in which the SR releases Ca^{2+} spontaneously. Spontaneous, uncoordinated Ca^{2+} release between heartbeats can be observed as spontaneous contractions of small groups of sarcomeres in cells of the myocardium, and gives rise to fluctuations of the light-scattering properties of the muscle (79). Spontaneous Ca^{2+} release increases the diastolic force generated by the contractile filaments and, in so doing, reduces Ca^{2+} release during the next heart beat. Spontaneous release of Ca^{2+} ions is likely to lead to cell depolarization as a result of activation of Ca^{2+}-dependent channels or by electrogenic Na/Ca exchange.

Because of differences in the structure of the Purkinje cell, coupling of excitation of the cell membrane with Ca^{2+} release in the core of these large cells differs substantially from that in myocytes. The action potential in these cells precedes rapid Ca^{2+} entry into the subsarcolemmal space. The latter induces Ca^{2+} release by the corbular SR, which then propagates into the core of the cell (13).

III. MOLECULAR BUILDING BLOCKS OF EXCITATION CONTRACTION COUPLING

The prominent L-type (L for long-lasting I_{CaL}) and lesser T-type (T for transient I_{CaT}) Ca^{2+} currents were initially described in neuronal tissues but subsequently recognized to exist in the various tissues of the heart. Multiple cardiac Ca^{2+} channels were first described in canine atrial cells (8). Subsequently, I_{CaL} and I_{CaT} have been recorded in tissues of most species. However, within the same species it appears that the density of I_{CaL} and I_{CaT} currents varies depending on the location of the myocyte within the heart. High densities of both the L- and the T-type channels have been described in rabbit SA node cells (55) and latent atrial pacemaker cells (160). To date, there have been no conclusive data showing T-type Ca^{2+} currents in AV node cells but studies of myocytes from canine ventricles revealed a large peak T/L current density ratio in Purkinje cells from both free-running fiber bundles and the subendocardium of the left ventricle, while myocytes dispersed from mid and epicardial layers have a smaller T/L current ratio.

Cardiac L- and T-type Ca^{2+} channels are different in several biophysical properties: (1) their voltage range of activation; T-channel activation occurs at more negative voltages than the L channel (e.g., in 5-mM $[Ca^{2+}]_o$ the threshold for activation is -50 and -30mV for T and L, respectively) (142); (2) their voltage range of inactivation; in 5-mM $[Ca]_o$ the T channel is inactivated by depolarizations positive to -70 mV while the L channel remains fully available for activation at potentials positive to -40 mV; (3) mechanism of inactivation; the T channel inactivates solely by membrane depolarization; with L channels, both membrane depolarization and Ca^{2+} ions participate in the inactivation process. Due to its relatively low density and rapid inactivation, the relative total amount of Ca^{2+} influx via I_{CaT} in the working heart is thus probably small compared to that of I_{CaL}. Voltage-dependent inactivation of the L-type Ca^{2+} current even occurs with channel proteins incorporated into lipid bilayers (122). In fact, inactivation of L-type Ca^{2+} current can occur at voltage steps where activation is absent.

The molecular structure of the L-type Ca^{2+} channel structure is thought to consist of a combination of an α_1C-subunit with β_2-, a α_2/δ- and γ-subunits. The γ-subunit is exclusively expressed in skeletal muscle. The α_1C-subunit alone is sufficient to express L-type channel activity; however, when coexpressed with the β-subunit, peak currents increase fourfold. This may occur by an accelerated opening of the pore and reduction in the rate of channel closure (81,115). Furthermore, the β_2-subunit causes a shift in the activation curve, a slowing of activation, and an enhancement of the inactivation process. Phosphorylation of the β_2-subunit may be involved in the increased I_{CaL} seen with β-adrenergic stimulation (54). Coexpression of the α_2/δ-with the α_1C-subunit also results in an increase in current, suggesting a role for the α_2/δ-subunit in the formation of functional L-type Ca²⁺ channels.

The molecular basis of neuronal and cardiac T-type Ca²⁺ channels have also been defined recently (116). In both cases, the α-subunits (α_1H, α_1G) have high sequence identity with the α_1C-subunit (116). Importantly, in terms of possible targets for pharmacological modulation, intracellular loop motifs involved in β-subunit binding (84,119) or Ca²⁺ binding (35) of the L-type α_1C protein are missing in both the α_1G and α_1H proteins.

The cardiac Na/Ca exchange protein transports Ca²⁺ ions across the sarcolemma in exchange for Na⁺ ions and is important in maintaining Ca²⁺ homeostasis in the myocyte. Normally, Na/Ca exchange works in the so-called "forward mode" (i.e., extruding Ca²⁺ ions in exchange for extracelluar Na⁺ ions). Reverse-mode operation of the Na/Ca exchanger can provide additional Ca²⁺ influx into the cell. In the forward mode, Ca²⁺ ions are being transported out against their electrochemical gradient; therefore, energy is needed. It is generally accepted that it is the Na⁺ ion distribution that provides this energy. Stoichiometric determinations have shown that three Na⁺ ions are transported for one Ca²⁺ ion and thus the exchanger is electrogenic. Under normal conditions, the reversal potential of the Na/Ca exchanger has been shown to be about –30 mV (78). Accordingly, negative to this potential, Na⁺ is moving in and Ca²⁺ out and inward current is generated; positive to this potential the Na/Ca exchanger works in reverse mode and outward current is generated (70). The Na/Ca exchanger protein itself has been shown to have distinct regions that are involved in the Na/Ca translocation process, while other regions are involved in regulation by Na⁺ and Ca²⁺ ions (99). Ca²⁺-dependent regulation of exchange activity is via a high-affinity binding site that is distinct from Ca²⁺ transport site (97). Binding of the regulator Ca²⁺ decreases Na⁺-dependent inactivation of the exchanger (59).

Stretch-activated Ca²⁺ channels have been described in both atrial and ventricular cells of several species, but not as yet in humans (65). The atrial channel is permeable to monovalent cations and Ca²⁺ ions (77) and thus can provide a source of Ca²⁺ influx. In single cells and isolated tissues, stretch has been observed to lead to a gradual increase in [Ca]$_i$ (48), as well as increases in the inositol phosphates IP₃ and IP₄, both of which may modulate [Ca]$_i$ levels (34). In adult guinea pig cells, large stretch-induced [Ca]$_i$ changes are blocked by streptomycin, a blocker of mechanosensitive transduction currents in hair cells, but not by ryanodine or the toxin tetradotoxin. Interestingly, streptomycin also blocks stretch-induced atrial tachyarrhythmias in the isolated heart (109), presumably by inhibiting a mechanosensitive cation channel in atrial myocytes (48).

IV. THE SR Ca²⁺ ATPase PUMP

Two Ca²⁺ ions are usually transported by the cardiac SR Ca²⁺ pump for each ATP molecule consumed (135), though other stoichiometries have been reported. ATP binds to high-affinity binding sites on the cytoplasmic side of the pump, and the terminal phosphate of ATP is then transferred to aspartate-351 on the pump protein while the bound Ca²⁺ ions become "occluded." ATP hydrolysis of the protein alters the structure such that Ca²⁺ cannot return to the cytoplasmic side and its phosphorylation reduces the Ca²⁺ affinity such that Ca²⁺ can then be released into the lumen of the SR. The cardiac Ca²⁺-pump protein appears similar to that from slow-twitch muscle (15,16) and has 10 membrane-spanning regions. Most of the protein is on the cytoplasmic side of the SR membrane, including the β-strand, phosphorylation sites, nucleotide binding sites, stalk domains, and a hinge region. The crucial high-affinity Ca²⁺ binding sites were initially thought to reside in the highly anionic stalk region; however, recent data suggest that these are within transmembrane regions M4–M6 and M8 (28). It is likely that in the membrane these transmembrane spans may be arranged in a cylinder to form an ion channel through the SR bilayer (93).

The rate of the cardiac SR Ca^{2+}-pump is regulated by phosphorylation of the protein phospholamban (31). In the dephosphorylated state, phospholamban interacts with the SR Ca^{2+} pump near the phosphorylation site, inhibiting pump activity. Phosphorylation of phospholamban removes the inhibitory effect and increases Ca^{2+} pumping rate. Phospholamban is a homopentamer; each monomic subunit has 52 amino acids within one hydrophobic and one hydrophilic domain. A structural model has been proposed showing the pentamer forming a hydrophilic pore through the SR membrane, with phosphorylation sites on the cytoplasmic surface (128). Phospholamban is phosphorylated by cAMP-dependent protein kinases (31). Studies on intact perfused hearts show that β-adrenergic stimulation increases protein kinase A phosphorylation, reduces the K_m for Ca^{2+}, and thus accelerates relaxation of the muscle. Ca^{2+} calmodulin-dependent protein kinases and protein kinase C (68) can also phosphorylate phospholamban (148). This phosphorylation increases the maximal transport rate of the Ca^{2+} pump. The cardiac Ca^{2+} pump has two ATP binding sites, a high-affinity site ($K_d \sim 1$ μM), which is the substrate site, and a second, lower affinity site ($K_d \sim 200$ μM) that serves a regulatory role (39). The major substrate for the Ca^{2+} pump is probably Mg-ATP, but other nucleotides can also be used.

V.   THE RYANODINE-SENSITIVE SR-Ca²⁺ RELEASE CHANNELS

There are two kinds of Ca^{2+} release channels in the SR membrane, a Ca^{2+}-activated channel and an inositol triphosphate (IP_3R)–activated channel. It is thought that the major mechanism regulating Ca^{2+} release in cardiac cells is Ca^{2+}-induced Ca^{2+} release (CICR). CICR requires that Ca^{2+} ions entering through L-type Ca^{2+} channels bind to the SR-Ca^{2+} release channel and cause opening of a high-conductance channel, allowing rapid Ca^{2+} efflux from the SR. Studies of the SR-Ca^{2+} release channel have been greatly accelerated by the recognition that ryanodine, a plant alkaloid, is a selective and specific binding ligand for this channel. The ryanodine receptor functionally constitutes the Ca^{2+} release channel of the SR, and structurally forms the foot linking the T-tubules to the SR (Fig. 12.1). The recognition of selective ryanodine binding has allowed purification of several isoforms (RyR1, RyR2, RyR3) from skeletal and cardiac muscle (66,82). Most of what is known about the RyR comes from electrophysiological experiments on the channels after they have been incorporated into lipid bilayers (83). The RyR2 opening probability increases up to nM concentrations of $[Ca^{2+}]_i$ and then decreases at higher (>μM) $[Ca^{2+}]_i$. In addition, increase of SR luminal $[Ca^{2+}]$ causes a marked increase in the open probability of the Ca^{2+} release channel (164). Recently, it has been shown that (26a) substitution of alanine for glutamine in the putative transmembrane sequence of RyR3 reduced Ca^{2+} sensitivity of the channel 10,000 fold. Location of the Ca^{2+} sensor in the transmembrane domain of RYR is attractive because it might confer the observed Ca^{2+} sensitivity to the channel both at the cytoplasmic and the luminal side of the SR membrane. The channel has a high Ca^{2+} conductance, but can also conduct other divalent cations, such as Ba^{2+} and Mg^{2+} (10), as well as monovalent ions in the absence of Ca^{2+} (130). Compared to the sarcolemmal Ca^{2+} channel, the SR-Ca^{2+} release channel has lower selectivity for Ca^{2+} and tenfold higher conductance (10). The ability of Ca^{2+} ions to cause release depends on the $[Ca^{2+}]_i$, the rate of rise of $[Ca^{2+}]_i$ (45), as well as the presence of nucleotides and Mg^{2+}. RyR channels close rapidly as a result of deactivation (56) or a decrease in Ca^{2+} influx.

The RyR is a homotetramer with a monomer molecular weight of ~320 to 450 kDa

(66,67,82) and its three-dimensional architecture resembles the junctional "feet" observed in the muscle (146). Several studies have elucidated the sites for modulation of CICR by various agents (e.g., ATP, Mg^{2+}, caffeine, calmodulin Ca^{2+}, ATP, ryanodine, temperature, voltage, ruthenium red, pH) on the RyR protein and on its associated "chaperone" proteins (17). Smaller modulatory proteins that have been found to copurify with the RyR are triadin, junction (53), and FKBP12. The latter is required for normal function of RyR2 and plays a key role in coupled gating between neighboring RyR2 channels (95). An immunosuppressant agent, FK06, binds to the FKBP12 protein, presumably inhibiting its modulation of RyR1, thereby increasing Ca_i transients by increasing the rate of release from the SR (101). Exposure to the immunosuppressant cyclosporin A increases spontaneous Ca^{2+} release by the SR in intact cardiac muscle, suggesting an increase of the opening probability of the RyR channel (6). Modulation of RyR-mediated Ca^{2+} release by calcineurin may be an important pathway for control of the Ca^{2+} release process (139). Ryanodine shifts the cardiac SR-Ca^{2+} release channel to a stable subconducting state where it no longer responds to Ca^{2+}, ATP, or Mg^{2+}, and high concentrations of ryanodine (>100 μM) appear to lock the channel in a closed state (102).

VI. COUNTERCURRENTS

The presence of large Ca^{2+} fluxes through the SR membrane requires the existence of other channels that allow large countercurrents to protect against electrical instability of SR membrane. A large-conductance (150 pS) K^+ channel exists in SR membrane and provides counter ion transport for Ca^{2+} release (1). In addition, a large-conductance (120 pS) Cl^- channel exists in SR membrane that can also be permeable to Ca^{2+} (133). Interestingly, this Cl^- channel's activity is altered with phosphorylation and some have suggested that phospholamban modulates its conductance (36).

VII. SPONTANEOUS Ca²⁺ RELEASE

Spontaneous SR-Ca^{2+} release was first observed by Fabiato (46) in mechanically skinned fibers. He found spontaneous oscillatory contractions were initiated by loading the SR using low concentrations of Ca^{2+} insufficient alone to induce Ca^{2+} release. The observation that skinned myocyte fragments started to contract in an oscillatory fashion led to the concept that a heavily Ca^{2+}-loaded (or Ca^{2+}-overloaded) SR emits a spontaneous Ca^{2+} release. The importance of this phenomenon is that spontaneous contractions, caused by cytosolic $[Ca^{2+}]_i$ oscillations (151), are accompanied by spontaneous oscillations in current and membrane potential in both single myocytes as well as multicellular cardiac preparations (74,79). Agents that reduce Ca^{2+} load of the SR (e.g., ryanodine, caffeine, EGTA) abolish spontaneous $[Ca^{2+}]_i$ oscillations as well as the oscillatory potentials, current, and contractions (134). Therefore, it is thought that spontaneous $[Ca^{2+}]_i$ oscillations are not secondary to transmembrane potential changes, but given the particular initiating conditions, may cause depolarizations and even automatic activity (22).

VIII. IP₃-DEPENDENT Ca²⁺ RELEASE

The IP₃ receptor (IP₃R) has been identified by immunohistolocalization techniques in myocytes and, while density appears less than that of RyR2, it appears significantly higher in Purkinje cells (51). Most studies show it located to a region of the intercalated disk (47). Three isoforms have been identified, with IP₃R2 occurring in working cardiac muscle

(117) and IP$_3$R2 in the Purkinje fiber system (51). The role of IP$_3$-induced Ca^{2+} release in cardiac contraction coupling is unknown, but IP$_3$R2 from ferret ventricle, when incorporated into planar bilayer, is Ca^{2+}-selective, IP$_3$-activated, blocked by heparin, and not altered by ryanodine (117). Interestingly, in skinned cardiac fibers IP$_3$ can induce tension oscillations and enhance submaximal, caffeine-induced CICR (162). IP$_3$ does not increase the Ca^{2+} sensitivity of the Ca^{2+} release channel, but still enhances Ca^{2+} oscillations from SR of skinned fibers (162). Presumably this is because luminal Ca^{2+} ions can bind to cytosolic IP$_3$R sites and modulate function. Ca^{2+} waves, which occur in cardiac cells, depend on the regenerative production of a diffusible molecule that triggers Ca^{2+} release from adjacent SR stores (see below). Cytosolic Ca^{2+} may be one such ion, but IP$_3$ could also serve as a propagating signal in cardiac cells, as has been suggested for some nonexcitable cells (20).

IX. MITOCHONDRIAL Ca^{2+} TRANSPORT

Ca^{2+} enters mitochondria down a large electrochemical gradient (~180 mV) set up by proton extrusion linked to electron transport down the respiratory chain. Its transport is blocked competitively by physiological concentrations of Mg^{2+} and by ruthenium red and lanthanides (10). Ca^{2+} extrusion occurs mainly via Na/Ca and Na/H exchangers and is thus dependent on Na$^+$ concentrations (52). Ca^{2+} uptake by the mitochondria is too slow to affect muscle relaxation significantly (11), but may have an important role in the regulation of [Ca^{2+}]$_i$ during mechanical restitution of force in cardiac muscle. Recent data suggest mitochondrial Ca^{2+} uptake is apparent only after a progressive Ca^{2+} load (cytosolic threshold ~30 to 500 nM) and is sensitive to the mitochondrial Ca^{2+} uniport blocker, Ru360 (161). Mitochondria can also accumulate a large amount of Ca^{2+} under pathological conditions (e.g., acute ischemia) (24). Also, when Ca^{2+} overload occurs, mitochondria will temporarily compensate by taking up large amounts of Ca^{2+}, which may help prevent cell damage. However, Ca^{2+} accumulation by mitochondria can diminish ATP production and may eventually compromise the mitochondria.

X. SARCOLEMMAL Ca^{2+} BINDING

Interactions between Ca^{2+} ions and the sarcolemma are pivotal in the cells' feedback mechanisms and one additional process that affects this is determined by the cell's buffering systems. One of these is binding to phospholipids, mostly the phosphatidylserines and phosphatidylinositols of the cell membrane. The density of phosphatidylserine and phosphatidylinositol (118) permits binding of ~1.2 mmol/L cellular space (149). The K$_d$ of Ca^{2+} binding (~10 nM) allows these phospholipids to act as a dynamic buffer during the contractile cycle. Hence feedback of subsarcolemmal [Ca^{2+}] on protein function in the sarcolemma depends critically on this buffer system. Given the low K$_d$ of this buffer, it would be expected that in Ca^{2+} overloaded cells the buffer may saturate and cease to buffer [Ca^{2+}] variations near the sarcolemma.

XI. THE CARDIAC CYCLE: CYTOSOLIC Ca^{2+} TRANSIENTS AND FORCE DEVELOPMENT

Figure 12.2(B) shows force development and the estimated cytosolic [Ca^{2+}] as a function of time in ventricular muscle. These results are representative of contractions at long and short end-systolic sarcomere lengths (SL) (i.e., at the extremes of the function curve of cardiac muscle). The figure shows that the time course of the transient increase in Ca^{2+} is independent of muscle length. Full activation of the contractile system requires saturation of all Ca^{2+} sites on troponin C (Tn-C), with simultaneous binding of additional Ca^{2+} to calmodulin (154). Hence, even activation of the muscle at only 25% of its maximum, such as in this figure, is accompanied by significant Ca^{2+} turnover. Changes in the kinetics of the Ca^{2+} transient with stretch are consistent with the hypothesis that the force–length relationship is determined principally by the length-dependent sensitivity of the contractile system, which resides in the relation between Ca^{2+} affinity of Tn-C and stretch (58,76). This also implies that in the stretched myocardium more Ca^{2+} is bound (62). It is understandable why the peak amplitude of the $[Ca^{2+}]_i$ transient of the stretched muscle is identical to that of the short muscle, if one assumes that the larger amount of Ca^{2+} released by the SR is balanced by the larger amount of Ca^{2+} bound to the contractile filaments at longer SL. The molecular mechanism underlying length dependence of Ca^{2+} binding to Tn-C remains unknown, but one hypothesis is that force exerted on the actin filament deforms the Tn-C molecule, thus retarding the dissociation of Ca^{2+} from Tn-C. This effect is thought to be length-dependent since the number of myosin crossbridges that can attach to actin increases with SL. Thus, the mechanical load on a sarcomere may influence the dissociation of Ca^{2+} from Tn-C, and it has been shown that rapid removal of an external load causes a robust additional $[Ca^{2+}]_i$ transient (64). This phenomenon may become important when the EC coupling properties of the myocardium are nonuniform, as in disease. Thus, the relaxation phase of the $[Ca^{2+}]_i$ transient depends both on the rate of Ca^{2+} binding and dissociation with Tn-C, and on the rate of Ca^{2+} binding to the sarcolemmal Na/Ca exchanger and the Ca^{2+}-pump of the SR, as well as on the rate of removal of Ca^{2+} ions by these transporters.

XII. Ca²⁺ SPARKS IN NORMAL CARDIAC MUSCLE

Spontaneous release of Ca^{2+} from the SR in single ventricular cells can be demonstrated with fluorescence techniques as spatially discrete, local increases in $[Ca^{2+}]_i$, termed Ca^{2+} sparks (26). Ca^{2+} sparks are also triggered during voltage-clamp pulses (91) and during action potentials where they have been termed evoked sparks or local $[Ca^{2+}]$ transients. Evoked Ca^{2+} sparks are probably triggered by Ca^{2+} entering via single L-type Ca^{2+} channels (21,92). An important consideration is that Ca^{2+} sparks may also trigger each other to produce Ca^{2+} waves, which propagate through the cell (25). Whatever their fundamental nature, Ca^{2+} sparks evoked by L-type Ca^{2+} currents are believed to summate, spatially and temporally, constituting the electrically evoked whole-cell $[Ca^{2+}]_i$ transient that couples excitation to contraction (152). Similar to the situation with Ca^{2+} sparks, Ca^{2+} waves have been recorded only in single isolated cells, although waves of sarcomere shortening, limited to single cells, have been reported in multicellular preparations. In addition, rapidly propagating Ca^{2+} waves accompanied by propagating contractions, have been recorded in trabeculae with focal damage.

Confocal images of microscopically quiescent trabeculae are illustrated in Figure 12.3. Ca^{2+} sparks are readily visible in the full-frame image as spatially localized bright regions and in the line-scan images as localized transient changes in fluorescence in microscopically quiescent muscle (153). Scan images of Ca^{2+} waves are apparent as regions of elevated $[Ca^{2+}]_i$ that move at constant velocity. Ca^{2+} sparks are common with ~10% of these being generated from repeatedly firing single sites. The average spacing between Ca^{2+} spark sites is about 2 μm, roughly intervals of one sarcomere length. Ca^{2+} sparks larger than 2 μm occur at ~10% of the frequency of single sparks [Fig. 12.3(B)]. Ca^{2+} sparks

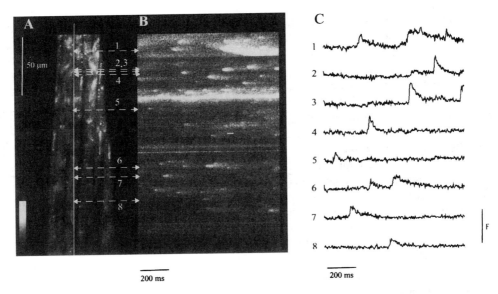

Figure 12.3 Ca^{2+} sparks during prolonged quiescence in a normal rat trabecula. (A) Full frame confocal microscopic image of trabecula loaded with fluorescent Ca^{2+} indicator Fura 2. Level of fluorescence intensity indicates elevations of Ca_i. The solid white vertically oriented line through the image indicates the single line that was scanned 512 times at a rate of 500 Hz to produce the line scan image. (B) Line scan image shows the fluorescence along the scan line of (A) as a function of time. Intensity calibration is the same as in (A). In the line scan image, an area of persistent high brightness occurs in the nuclear region, small sparks of transient increases of brightness are recognizable elsewhere in the muscle (~25 are evident in this scan). In one area, a high level of brightness occurs that moves both in time (along x axis) and in space (~10 μm along y axis). (C) Line plots (fluorescence as a function of time) at eight different sites in the muscle. The numbers of (A) and (B) correspond to the numbers of the line plots in (C). Line plots are obtained by averaging the fluorescence from 5 pixels (1.36 μm) at the place indicated by tips of arrows in (B). Note the variation in the amplitude of the local Ca_i transients (sparks) at the different sites. In line plot 1, the Ca^{2+} wave of (B) appears as a local Ca_i transient of prolonged duration. (Reproduced with permission from Ref. 153.)

appear similar in time course and spatial spread in unstimulated muscle, compared to single isolated cardiac cells. Figure 12.4 shows an averaging of single Ca^{2+} sparks. Triggering of some of these by a local rise of $[Ca^{2+}]_i$, [Fig. 12.3(C)] has also been described in isolated cells. The peak amplitude occurs at ~200 nmol/L, below the level at which crossbridges are activated in intact trabeculae. Also, Ca^{2+} sparks are usually spatially restricted, suggesting that the Ca^{2+} concentration in the myofilament space during and after the peak of the Ca^{2+} spark must have been substantially lower than 170 nM. This makes it even more unlikely that crossbridges are activated by individual sparks.

XIII. MICROSCOPIC Ca^{2+} WAVES IN NORMAL MUSCLE

Slowly traveling Ca^{2+} waves occur rather rarely in trabeculae and appear comparable to those in single cells (137). They occur several seconds following a twitch with an average frequency of ~2.5-Hz per cell, and an extent ≤~4 sarcomere lengths. More often than not, these waves propagate in only one direction. Figure 12.5 shows two waves that start at the

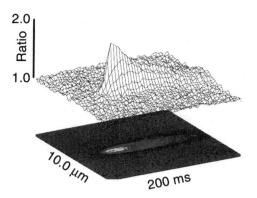

Figure 12.4 Properties of an average Ca^{2+} spark of a ventricular trabecula ($n = 79$). Three-dimensional contour plot indicates the average amplitude of a spark (see text for more detail). (Reproduced with permission from Ref. 153.) (See also color plate.)

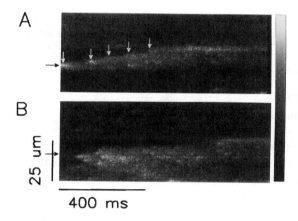

Figure 12.5 Enlarged confocal image depicting the characteristics of line scans during propagation of one Ca^{2+} wave (A) and during initiation and propagation of another (B) in normal muscle. In (A), the Ca^{2+} wave has an asymmetric appearance, as if it encountered a border or failed to propagate in one direction. In (B), the wave begins as a "V," indicating equal propagation in both directions; however, this wave stops propagating. The black arrows in both (A) and (B) mark the same position in the two scans, indicating that the two waves started at the same place. The white arrows indicate the position of sparks at the leading edge of the wave in (A). (Reproduced with permission from Ref. 153.)

same site in a cell; one is propagated in only one direction (panel A), whereas the other propagates in both directions (panel B). This suggests that if these waves start at a gap junction, their propagation into one or both cells connected to the gap junction would be dictated by chance. It is striking that Ca^{2+} waves show Ca^{2+} sparks on their leading edge, with a distance along the edge occurring at about 1 sarcomere length. This suggests that Ca^{2+} sparks may be present at the leading edge of Ca^{2+} waves (25). One group of investigators suggested that in single myocytes, Ca^{2+} sparks may provide the regenerative mechanism for a Ca^{2+} propagation wave from one terminal cisterna to another (25). In this case, the trigger for Ca^{2+} spark generation during a propagated wave would consist of Ca^{2+} arriving from an adjacent Ca^{2+} spark site (25). If Ca^{2+} release from the site is proportional to the rate of rise and the absolute $[Ca^{2+}]_i$ reached, one would anticipate that waves would propagate at a constant velocity, since the same process would repeat itself at each following site.

XIV. $[Ca^{2+}]_i$ DEPENDENT REGULATION OF ION TRANSPORT

The transient elevation of $[Ca^{2+}]_i$ following the start of an action potential can affect ion channel function and alter the time course of action potential voltage changes during each cardiac cycle. This could occur by Ca^{2+} directly binding to specific channel proteins or by influencing the activity of other proteins, which could then modulate ion channel behavior. At high $[Ca^{2+}]_i$, inactivation of the L-type Ca^{2+} current may be due to a Ca^{2+}-induced reduction of the open probability of the channel (38) or result from direct effects on Ca^{2+} binding sites on the α-subunit of the channel. Ca_i-dependent inactivation of the L-type Ca^{2+} current has been recognized for many years (86); however, the mechanism of how influx and subsequent release of Ca^{2+}, but not other ions, hastens Ca^{2+} current decay is unknown. One hypothesis is that Ca^{2+} ions bind to the α-subunit of the L-type Ca^{2+} channel protein. A consensus Ca^{2+} binding motif (EF hand) is thought to be located near the inner mouth of the proposed channel and is required for current-induced inactivation of channel activity (110), although this idea has recently been challenged (159) in favor of another site (131,159). Other recent data have suggested that there may be a mediating role for calmodulin in Ca^{2+}-dependent inactivation of the L-type channel. In fact, a calmodulin binding motif has been identified on the $\alpha_1 C$-subunit (163).

Once the L-type Ca^{2+} channel is activated and inactivated, it follows a predictable time course as it recovers from inactivation, repriming for the next stimulus. This recovery process is voltage and Ca_i-dependent. Voltage-dependent recovery has a reasonably fast phase and slow or very slow phases (14). Importantly, recovery from Ca_i-induced inactivation may occur at positive plateau potentials and depends on both SR and Na/Ca exchanger function (129). Presumably it is this recovery from Ca_i-dependent inactivation that allows the L-type Ca^{2+} channels to reopen, allowing Ca^{2+} influx during early after-depolarizations (see below).

Current generated by the Na/Ca exchanger protein ($I_{Na/Ca}$) also depends on $[Ca^{2+}]_i$, because of the contribution of $[Ca^{2+}]_i$ to the diffusion gradient for Ca^{2+} ions. In normal myocytes, the time course of decline of inward $I_{Na/Ca}$ repolarization is related to the time course of the spatially averaged $[Ca^{2+}]_i$ transient (40). In myocytes from diseased hearts that have abnormal Ca^{2+} cycling, $I_{Na/Ca}$ could contribute substantially to both altered outward and inward currents. Therefore, in myocytes from hypertrophied or failing hearts, where the relaxation phase of Ca_i transients may be slowed, more slowly decaying inward Na/Ca exchanger currents would occur during diastole. Under conditions where Ca^{2+} channel function is decreased [e.g., following coronary artery occlusion (2)] and the Na/Ca exchange activity is increased, the large Ca^{2+} influx seen upon depolarization could be carried by the Na/Ca exchanger (90). Outward currents generated by Na/Ca exchanger could be both sustained and oscillatory during a maintained depolarization. Thus, in myocytes from diseased hearts $I_{Na/Ca}$ exchange could contribute significantly to the total time of repolarization, particularly when $[Na^+]_i$ is increased.

Ion currents mediating repolarization can also be modified by changes in $[Ca^{2+}]_i$. The open probability of the cardiac delayed rectifier channel, I_K, is increased with an increase in $[Ca]_i$ (125,138), producing enhanced outward currents with increasing $[Ca^{2+}]_i$.

Elevation of [Ca]$_i$ above 10 nM also enhances I_{Ks} without an effect on current voltage relationship (147). Noise analysis has shown that [Ca]$_i$ increases the probability of opening of I_{Ks} channels (138). What remains unclear is whether this K⁺ channel modulation is due to Ca²⁺ binding to a channel protein or activation of a Ca²⁺-dependent signaling molecule (111). β-Adrenergic stimulation of the slow component of human I_K (hI_{Ks}) is mediated by an increase in [Ca²⁺]$_i$ (144). Direct Ca²⁺ injection into oocytes expressing a mouse I_{Ks} clone (63) also enhances current amplitude. Ca²⁺ influx may also contribute to a "dynamic" rectification of the inwardly rectifying K⁺ current, I_{K1}, since both the probability of opening of the channel in subconductance states and rectification of I_{K1} appear to be [Ca²⁺]$_i$-dependent (100). Interestingly, cytochalasin, but not colchicine, removes this Ca²⁺-dependent effect, suggesting a role for cytoskeletal actin filaments in rectification of this channel.

Transient outward currents reflect the sum of a K⁺ current through a voltage-dependent, [Ca²⁺]$_i$-independent channel (I_{to1}), and one through a [Ca²⁺]$_i$-dependent Cl⁻ conducting channel (I_{to2}). In normal canine and feline myocytes, the amplitude of I_{to2} is small relative to the voltage-dependent, 4-aminopyridine-sensitive I_{to1} (143). This [Ca²⁺]$_i$-dependent Cl⁻ channel is important in normal cardiac repolarization and could be involved in arrhythmogenesis (61), yet little is known about its physiology. Recently, a low-conductance, Ca²⁺-activated Cl⁻ channel was described in canine myocytes (29). This current had relatively low Ca²⁺ sensitivity but can carry significant current (29,75) and be activated upon depolarization after I_{CaL}-induced Ca²⁺ release from the SR, as well as following caffeine-induced Ca²⁺ release. The presence of a [Ca²⁺]$_i$-dependent Cl⁻ current in normal human cells remains controversial (43,87), but it appears that human atrial cells can express a [Ca²⁺]$_i$-dependent nonspecific cation channel (49). Two apparently different Ca²⁺-dependent cation channels have also been identified in adult and neonatal ventricular cells. Ca²⁺ is needed for activation of both channels and each appears equally permeable to Na⁺, K⁺, Li⁺, and Cs⁺ (30,41,80,60).

XV. ARRHYTHMOGENESIS: Ca²⁺ AND AUTOMATICITY

Automatic (nondriven) electrical activity occurs in various regions of the normal heart. The term "normal automaticity" usually refers to this type of activity in sinoatrial (SA) nodal cells, latent atrial pacemaker cells, and Purkinje fibers. Recent data combining voltage clamp with Ca²⁺ ion imaging techniques suggests Ca²⁺ ions have a significant role in modulating the slope of phase 4 depolarization and thus automatic firing rates of these cell types. In SA nodal cells, ryanodine, which reduces conductance of the Ca²⁺ release channel, slows the final phase of depolarization and thus pacemaker activity of cat, guinea pig, and rabbit SA nodal cells (88). Whether ryanodine reduces the contribution of Ca$_i$-dependent Na/Ca exchange currents, or Ca$_i$-dependent T-type Ca²⁺ currents to phase 4 depolarization in SA node cells is not known. Nevertheless, some evidence suggests that Ca$_i$, at least in part, modulates SA node activity. In cells from the pacemaker region of cat atria, an important late diastolic component of nondriven rhythmic activity depends on the release of Ca²⁺ from the SR (123). A slow SR leak of Ca²⁺ during diastole provides persistent Ca²⁺ extrusion via the Na/Ca exchanger, which in turn generates inward current and atrial cell depolarization. In normal cat atrial and ventricular myocytes, the rate of Ca²⁺ leak from the SR is very low and thus no diastolic depolarization occurs (7).

Using normal Purkinje fibers several laboratories (19,120) have shown that the polarized individual canine Purkinje cell lacks normal automaticity in the absence, as well as in the presence, of catecholamines (120). This is unlike adult sinus node cells where the individual cell shows normal automaticity and, under voltage-clamp conditions, I_f is prominent. Moreover, the minimal element needed for automaticity is an aggregation of Purkinje cells suggesting that factors traveling between myocytes in aggregates (e.g., Ca^{2+}) may play a role here as well. Diseased Purkinje as well as diseased human atrial fibers that become chronically depolarized show nondriven electrical activity that does not depend on an initiating beat. In one study, rates of firing of abnormal foci were strongly modulated by agents that affect SR function (42), but there have been no studies to show concomitant changes in $[Ca^{2+}]_i$ linked to abnormal automatic activity.

XVI. Ca^{2+} AND TRIGGERED ACTIVITY

While the role of propagating Ca^{2+} waves in delayed after-depolarizations (DADs) is reasonably well accepted (see below), some EADs in ventricular myocytes appear not to be due to regional increases in $[Ca^{2+}]_i$, or propagating Ca^{2+} waves. Rather, during EADs, fluorescence transients show synchronous changes throughout the myocyte (104). This supports the idea that a change in membrane potential primarily causes increases in $[Ca^{2+}]_i$ during an EAD. In some models, EADs have been shown to depend on a Ca^{2+} "window current" (69) while in others changes in Ca^{2+} loading and Na/Ca exchange current (145) and/or CAM-kinase activity (156) may be involved. Although it is known that elevated $[Ca^{2+}]_i$ can inactivate L-type Ca^{2+} currents, its predominant effect is to further enhance Ca^{2+} currents through activation of a Ca^{2+}/calmodulin-dependent CaM kinase, and recent data suggest that CaM kinase inhibitors can suppress clofilium-induced EADs (156).

Triggered arrhythmias induced by DADs can lead to premature firing and early generation of a subsequent impulse. DADs arise from a spontaneous increase in $[Ca^{2+}]_i$, leading to a transient inward current, and activation of contractile filaments (74). A small $[Ca^{2+}]_i$ transient, assumed to be due to spontaneous Ca^{2+} release from the SR, leads to a transient inward current. Hence, a sufficiently large Ca^{2+} load on the SR would create an unstable state where spontaneous Ca^{2+} release could become so large that the resulting transient inward current would depolarize the cells sufficiently to trigger a new action potential, which could then propagate as a triggered arrhythmia.

Spontaneous Ca^{2+} release from the SR has been well documented in both isolated dispersed cells and in cardiac trabeculae using confocal microscopy. Figure 12.6 shows a regional Ca^{2+} wave after an action potential–induced synchronous Ca^{2+} transient in a myocyte accompanied by an after-contraction and a DAD. In this case, the spontaneous Ca^{2+} wave emerged at the center of the myocyte and spread in both directions (105). Typically, the interval between the last stimulation and the onset of the first Ca^{2+} wave shortens and the probability of multiple foci of Ca^{2+} waves increases when the stimulus frequency or $[Ca^{2+}]_o$ is increased (23). These observations are consistent with the concept that an increase in $[Ca^{2+}]_i$ causes a transient net inward current and results in a DAD. Any of the $[Ca^{2+}]_i$-dependent currents described above could be involved in the generation of net inward current.

Ca^{2+} waves usually start at one end of a myocyte, where one might envisage gap junctions, and spread at a constant velocity in all directions (137), although they may also propagate in spirals around intracellular organelles (89). Their amplitude and width are usually fairly constant during propagation and the velocity is typically about 100 μm/s in quiescent cells (137). In isolated myocytes, they occur randomly with a frequency that varies from <0.1 to ~5 Hz, although remarkably stable intervals between spontaneous Ca^{2+} waves can be observed. Frequency increases with increased SR Ca^{2+} loading, as does the

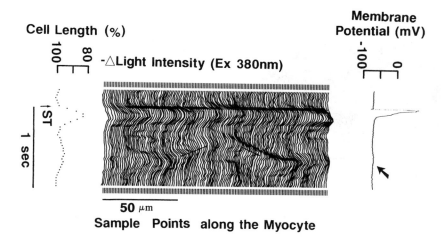

Figure 12.6 Spatial changes in fluorescence signals during a delayed after-depolarization in a single guinea pig myocyte superfused with potassium-free Tyrodes solution. Upper panel shows transmembrane potential recording; middle panel shows spatiotemporal changes in Fura 2 signals; and bottom panel shows cell length changes. Note that the action potential shown in the upper panel is followed by a delayed after-depolarization (at arrow). Note in the middle panel that the synchronous fluorescent transients elicited by the action potential are followed by focal transients that emerge spontaneously at the center of the myocyte and then propagate in opposite directions. These propagating patterns of fluorescent transients are Ca²⁺ waves. In the bottom panel, it is seen that the spontaneous Ca²⁺ waves are accompanied by a small contraction. ST = electrical stimulation. (Reproduced with permission from Ref. 105.)

number of initiating foci (23). When Ca²⁺ waves start from two or more foci within a myocyte, the waves appear to collide and $[Ca^{2+}]_i$ declines without evidence of further propagation, demonstrating refractoriness of the propagation mechanism (9). Thus Ca²⁺ waves are the consequence of a process with a "refractory period." If an action potential is elicited during propagation of a Ca²⁺ wave, the amplitude of the Ca²⁺ transient and the accompanying twitch are reduced. This decrease is more pronounced if the interval with the preceding Ca²⁺ wave transient is short (106), suggesting indirectly that the spontaneous transient and twitch generation share similar mechanisms of Ca²⁺ cycling.

When cardiac muscle is damaged locally, such as by microelectrode impalement, Ca²⁺ waves start near the damaged region and propagate in a coordinated fashion into adjacent tissue (32). Several observations suggest that after-contractions in multicellular preparations occur as the combined result of mechanical effects and elevated cellular Ca²⁺ levels owing to the regional damage. This may give rise to premature beats, as well as triggered arrhythmias. One unique aspect of these after-contractions is that they appear to be initiated by stretch and release of the damaged region during the regular twitch and can propagate into neighboring myocardium: hence, the term triggered propagated contractions (TPCs). Damage-induced TPCs may, therefore, serve as a mechanism that couples regional damage to initiation of premature beats and arrhythmias in the adjacent myocardium. Figure 12.7 shows a characteristic TPC in a rat cardiac trabecula. Displacement of the TPC occurs at a velocity of propagation (V_{prop}) that varies from 0.1 to 15 mm/s (108) and is correlated tightly with the amplitude of the twitch preceding the TPC, suggesting that the Ca²⁺ load of the SR dictates V_{prop}. In contrast, sarcomere stretch, which increases

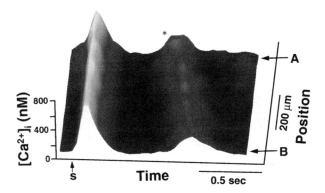

Figure 12.7   Spatial and temporal changes in Ca$_i$ in normal RV trabecula during the last electrically stimulated twitch (S) and a triggered propagated contraction (TPC). Note that after the end of the clearly uniform stimulated Ca$_i$ transient, a smaller Ca$_i$ transient propagates from site A in the muscle to site B in the muscle. The Ca^{2+} waves appear to travel at a constant velocity along this trabecula. (Reproduced with permission from Am J Physiol 1998; 274: H266–H276.) (See color plate.)

twitch force for any level of loading of the SR, does not increase V_{prop} (33). Studies of the effects of interventions, such as varied [Ca^{2+}]$_o$, or Ca^{2+} channel agonists and antagonists, also support the idea that the Ca^{2+} load is the main determinant of V_{prop}. Interventions that cause a leak of Ca^{2+} (caffeine and ryanodine) increase V_{prop}, suggesting that V_{prop} also depends on the diastolic cytosolic Ca^{2+} level (103). Finally, the rate of initiation of TPCs is tightly correlated with V_{prop} when the Ca^{2+} load of the SR is modulated, suggesting that triggering and propagation processes share closely related mechanisms.

XVII.   MECHANISMS UNDERLYING PROPAGATED Ca^{2+} WAVES

Initiation: In mechanically skinned cells in which the SR is intact, excessive SR Ca^{2+} loading causes spontaneous Ca^{2+} release (44). The mechanisms whereby this increases opening of the SR-Ca^{2+} channel are uncertain, but could involve altered channel sensitivity to luminal [Ca^{2+}]. Localization of the Ca^{2+} sensor in the transmembrane domain of the channel protein would make it suitable as a sensor to either altered luminal or cytosolic [Ca^{2+}]. Intact cells with a high SR-Ca^{2+} load show similar behavior. Hence, the oscillatory character of a triggered arrhythmia in myocardium with a high cellular Ca^{2+} load may be due to further increases in Ca^{2+} entry during the arrhythmia, causing even more Ca^{2+} loading, etc. Consequently, as soon as the release process has recovered, the overloaded SR again releases another fraction of Ca^{2+}. The requirement that the Ca^{2+} release mechanism must recover first would explain the delay between after-contractions and after-depolarizations due to the preceding beat.

As noted above, TPCs arise in damaged regions of cardiac muscle as a result of increased Ca^{2+} entry, which in turn induces Ca^{2+} overload and asynchronous activity (108). Another consequence of injury is that Ca^{2+} can diffuse via gap junctions into intact adjacent cells. Entry of Ca^{2+} by this type of mechanism near an ischemic area of myocardium can continue, as long as these junctions remain open. Spontaneous activity in the damaged zone is thus usually random and accompanying Ca^{2+} transients are small and do not propagate through the muscle, but, they can cause Ca^{2+} overload and spontaneous activity in bordering cells. Spontaneous SR-Ca^{2+} release and subsequent contractions can then increase resting tension and hence twitch force (79). Thus, twitch force of the damaged cells and cells of the neighboring border zone is less than that of the central region of the trabeculae. So, during an electrically evoked twitch, contraction of the central region of the trabeculae stretches the damaged region. During the rapid relaxation phase of the twitch, this stretched, damaged region then shortens suddenly and the stretch or quick release of damaged ends of trabeculae can then trigger TPCs (Fig. 12.8). TPCs always start shortly after rapid shortening of damaged areas, suggesting that it is actually the shortening during relaxation that initiates a TPC. The observation (4,64) that rapid shortening of a contracting muscle releases Ca^{2+} from myofilaments thus provides a candidate mechanism for initiation of TPCs. Ca^{2+} ions that dissociate from the contractile filaments due to the quick release of the damaged areas during relaxation could initiate a TPC if Ca^{2+}-induced Ca^{2+} release has recovered sufficiently to allow amplification of the initial Ca^{2+} transient in the damaged region.

Propagation: The observation that Ca^{2+} waves travel at a constant velocity and with constant amplitude through an isolated myocyte or multicellular preparation is an important clue about the mechanism of propagation of these waves. Diffusion of Ca^{2+} alone would clearly be too slow by at least two to three orders of magnitude and would be accompanied by a decline of the observed wave amplitude. Propagation of electrical activity is much faster (1 m/s for the action potential in ventricular myocardium) and thus electrotonic conduction is too fast (~0.1 m/s) to be compatible with the observed values of V_{prop} in trabeculae. One mechanism that has been proposed is diffusion of Ca^{2+} due to local increases in [Ca^{2+}]$_i$ and subsequent CICR from adjacent SR, similar to the waves propagated by Ca^{2+}-sparks (Fig. 12.8). The transition from nonpropagating sparks to propagating sparks and a Ca^{2+} wave may be related to an increase in Ca^{2+} sensitivity of the SR-Ca^{2+} release channel as a consequence of greater SR-Ca^{2+} loading (26). Propagation with a constant velocity is thought to be consistent with CICR propagating by Ca^{2+} diffusion to adjacent sarcomeres or cells (3). Such a model is supported by work on saponin-skinned muscle fibers, which also exhibit propagating local contractions after rapid local exposure of the fiber to a Ca^{2+}-containing solution, suggesting that an intact SR, but not the cell membrane, is essential (32).

The characteristics of Ca^{2+} waves and TPCs in trabeculae are quite similar. Neither spontaneous activity in single myocytes nor TPCs in trabeculae require an intact sarcolemma and both are abolished by agents that interfere with SR-Ca^{2+} loading or release. On the other hand, at first glance, there is a striking difference in propagation velocity. The velocity of Ca^{2+} waves in unstimulated cells is about ten times lower than V_{prop}. However, TPCs are generated in cardiac muscle preparations at short intervals after the twitch such that their properties are affected by residual binding of Ca^{2+} to intracellular ligands. This is in contrast to the situation in myocytes where the moment of appearance of a Ca^{2+} wave following the twitch is both random and usually later. Hence, elimination of Ca^{2+} from ligands during late diastole, after the Ca^{2+} extrusion processes have done their work, should reduce V_{prop} (108).

Backx et al. investigated which parameters of Ca^{2+} diffusion and CICR are required for the high V_{prop} in muscles by modeling the behavior of a myofibril and the SR during a sudden focal Ca^{2+} release (3). From this work we learned that Ca^{2+} transients propagate through the cytosol at a rate modified by binding to troponin, calmodulin and Ca^{2+} sequestration by the SR, as well as by the rate of Ca^{2+} release from adjacent release sites of the SR. This combination of changes might be expected to result from loading of cardiac cells

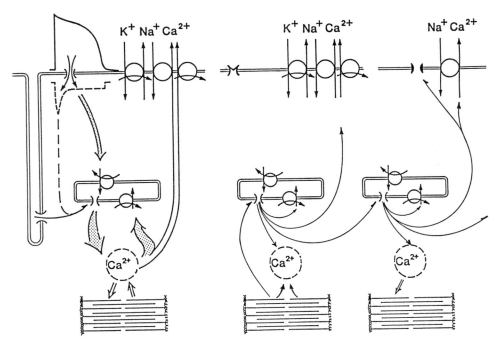

Figure 12.8 Diagram of the excitation–contraction coupling system in the cardiac cell, and as its role during TPCs. The left panel shows events during the twitch. During the action potential, a large transient Ca^{2+} influx enters the cells followed by a maintained component of the slow inward current (dashed line). Ca^{2+} entry does not lead directly to force development as the Ca^{2+} that enters is rapidly bound to binding sites on the SR. The rapid influx of Ca^{2+} via the T tubuli is thought to induce release of Ca^{2+} from a release compartment in the SR, by triggering opening of Ca^{2+} channels in the terminal cisternae, thus activating the contractile filaments to contract. Rapid relaxation follows because the cytosolic Ca^{2+} is sequestered rapidly in an uptake compartment of the SR and partly extruded through the cell membrane by the Na^+/Ca^{2+} exchanger and by the low-capacity, high-affinity Ca^{2+} pump. This process loads the SR. It is important to note that the process of Na^+/Ca^{2+} exchange is electrogenic so that Ca^{2+} extrusion through the exchanger leads to a depolarizing current. The middle panel shows events near a damaged region during triggering of TPC. Rapid shortening occurs during relaxation of the twitch, following stretch by the normal and therefore stronger myocardium, during contraction. This rapid release of the sarcomeres leads to dissociation of Ca^{2+} from the contractile filaments during the relaxation phase. The SR is enough recovered to respond to the increase in $[Ca^{2+}]_i$ by Ca^{2+}-induced Ca^{2+} release, leading to an after-contraction. The resultant elevation of $[Ca^{2+}]_i$ also causes diffusion of Ca^{2+} to adjacent sarcomeres. The right panel shows that arrival of diffusing Ca^{2+} after release in the damaged region leads to Ca^{2+}-induced Ca^{2+} release by the SR in adjacent sarcomeres. Ca^{2+} diffuses again to the next sarcomere, while causing a local contraction, as well as a delayed after-depolarization (DAD) due to electrogenic Na^+/Ca^{2+} exchange and activation of Ca^{2+}-sensitive, nonselective channels in the sarcolemma. Diffusion of Ca^{2+} along its gradient maintains the propagation of the TPC. (Reproduced with permission from Cardiovasc Res 1999; 40: 444–455.)

with Ca^{2+} during repetitive stimulation, as well as exposure to high $[Ca^{2+}]_o$ or Ca^{2+} agonists (103,108). Although these observations provide a reasonable framework to explain propagated Ca^{2+} waves, this is still only a working model and many questions remain unanswered. For example, the mechanism of propagation of a Ca^{2+} wave from one cardiac cell to another has received little attention. It has been reported that there is apparently continuous propagation of a Ca^{2+} wave from one cell to another with no delay or change of velocity at cell–cell junctions (136). On the other hand, it has been noted that in myocytes without Ca^{2+} overload, a local increase in $[Ca^{2+}]_i$ does not propagate (112), and that a Ca^{2+} wave induced by local application of caffeine decreases in both amplitude and velocity as it propagates along the cell (141).

TPCs are accompanied by a depolarization similar to a DAD and the duration of depolarization correlates exactly with the time during which the TPCs travel through the trabeculae. The amplitude of the after-depolarizations also correlates exactly with the amplitude of the TPCs (32). This correspondence suggests that the depolarizations are elicited by a Ca^{2+}-dependent current of the same duration as the $[Ca^{2+}]_i$ transient wave (74). In the small trabeculae used for TPC studies, this depolarization can be recorded over a distance of a few millimeters without much decrement due to electrotonic conduction. Local heating of trabeculae causes a TPC to stop at the site of heating, while the concomitant depolarization is still measurable at a distance of about a millimeter distal to the heating site again as a result of electrotonic conduction of the DAD (32). This indicates that the depolarization is not the source of the TPC, but rather is induced by the TPC. Further, a TPC accompanied by a DAD can become sufficiently large to elicit an action potential and associated muscle twitch, as shown in Figure 12.9. An action potential triggered by an initial TPC may add so much Ca^{2+} to the cell that a triggered arrhythmia starts. Triggered arrhythmias indeed occur in a damaged muscle when the Ca^{2+} load of the SR is large at room temperature, usually during the first hour after damage has occurred. In such a case, the full-blown arrhythmia is usually preceded by the repeated occurrence of single premature beats. At 37°C, the time span over which these damage-related events occur in human trabecula is much shorter and the TPCs, which cause the premature beats, disappear in 10 min or less. Under these conditions, it is likely their occurrence is limited by rapid closure of gap junctions as a result of persistently elevated Ca^{2+} levels in damaged cells. In addition, the pH in these cells may be low due to the enormous metabolic load resulting from the intense load on ion transport, and the lowered pH may promote gap junction closure. Gap junctional conductance is regulated acutely by changes in pH, Ca^{2+}, cAMP, and cGMP levels (158). Thus, Ca^{2+} ions can be both flowing through gap junctions as well as altering their conductance. In the first case, increases in Ca^{2+} ions released upon RyR activation can travel as a wave (see Fig. 12.7) across a cell and propagate to adjoining cells via gap junctions. How the wave crosses the gap junction is unknown, but extracellular segments of Cx43 protein components of the gap junction appear critical for wave propagation (140).

XVIII. Ca²⁺ AND REENTRY

Two important determinants of reentrant excitation are continued aberrant impulse propagation and unidirectional block. Both theoretical and experimental models suggest that under some circumstances the L-type Ca^{2+} current and $[Ca^{2+}]_i$ can affect cardiac impulse propagation. Several studies using patterned cultures of neonatal myocytes and adult cardiac cell pairs have emphasized that the L-type Ca^{2+} current of the myocyte in a region of a current-to-load mismatch can become essential for continued impulse propagation. Re-

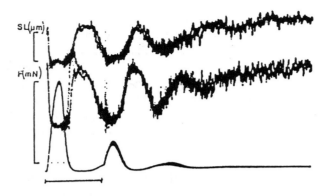

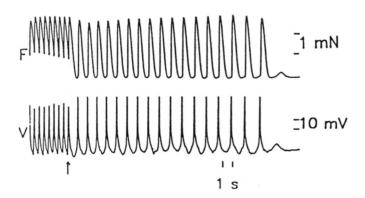

Figure 12.9 Spontaneous contractions cause development of twitches and arrhythmias. (A) An example of the development of a spontaneous twitch that is triggered by a propagating after-contraction. Note the acute increase of the rate of rise of force during the development of the first after-contraction (arrow), and also the similarity of the time course of the subsequent twitch and that of the electrically elicited twitch. (Reproduced with permission from Ref. 108.) (B) Force (F) and membrane potential (V) recordings during a train of conditioning stimuli (ending at the arrow) and a subsequent triggered arrhythmia. Note the initial slow upstroke in both force and membrane potential of triggered twitches, suggestive of an underlying TPC and DAD. The triggered arrhythmia terminated spontaneously with an increase in the interval between triggered beats, followed by a TPC and DAD. (Reproduced with permission from Daniels M.C.G. Mechanism of Triggered Arrhythmias in Damaged Myocardium. Utrecht, The Netherlands: The University of Utrecht, Ph.D dissertation 1991.)

versal of a block in propagation can be accomplished using Bay K8644, an agonist of the L-type Ca^{2+} current (72). The critical relationship of this finding to discontinuities of conduction in arrhythmogenic substrates, such as occurs in acute ischemia or healing sections of previously infarcted myocardium, however, is not clear. Thus, mechanisms of "pseudoblock" during reentrant arrhythmias in the healing, infarcted heart may depend not only on the redistribution of gap junctions, but also on the loss of functional L-type Ca^{2+} channels and its subsequent impact on intracellular Ca^{2+} cycling. In electrophysiological studies to map reentrant excitation in ventricles of a previously infarcted heart, an L-type Ca^{2+} current agonist is antiarrhythmic (18), presumably because its increases depressed L-type Ca^{2+} currents and $[Ca^{2+}]_i$ in border zone myocytes that contribute to initiation of reentrant arrhythmias.

XIX. Ca²⁺ AND ELECTROMECHANICAL ALTERNANS

T-wave alternans, as well as QT alternans, are body surface eletrocardiographic descriptors characterized by alternating, beat-to-beat changes in T-wave morphology or in QT interval. Both measures appear to be predictors of arrhythmic events and have been linked to spatial and temporal heterogeneity of repolarization of action potentials in the heart, and frequently, to spatial and temporal changes in intracellular Ca^{2+} (121). Mechanical alternans is accompanied by Ca_i transient alternans and both are sensitive to agents that affect SR function, such as ryanodine (124). Mechanical alternans can be explained on the basis of APD alternans in some models of ECC, such as that in Figure 12.2, which predicts two types of alternans. The first is alternans in which APD and force increase and decrease together (i.e., in-phase alternans) and the second, in which changes in APD occur out of phase with force. As discussed above, it is believed that the amount of Ca^{2+} released during any cardiac beat depends on the Ca^{2+} influx during the preceding action potential. This is true for myocytes with normal SR function; in cells with diminished SR Ca^{2+} storage ability, APD determines concomitant Ca^{2+} release directly. Acute changes in $[Ca^{2+}]_i$ affect membrane currents, and thus provide a feedback mechanism that controls APD of the concurrent beat. At high heart rates, where alternans is usually found, relaxation may be incomplete. As a result of the elevation in $[Ca^{2+}]_i$, Ca^{2+} binding to ligands in the cytoplasm, including the SR-Ca^{2+} pump, would be occupied and the effect of variation of Ca^{2+} influx would be more pronounced in terms of force development and feedback to sarcolemmal channels.

In a recent study discordant electrical alternans, described as the situation when membrane repolarization alternates with an opposite phase, between groups of neighboring cells (114). This type of alternans was directly linked to the formation of unidirectional lines of block and reentrant ventricular fibrillation. Whether mechanical alternans accompanies or causes these arrhythmogenic beats remains unknown at this time. Although multiple explanations for alternans can be envisaged based on models such as that in Figure 12.2, a more precise explanation awaits further study.

XX. Ca²⁺ AND EXCITATION–TRANSCRIPTION COUPLING

Recent studies suggest that transcription of different ion channel proteins is greatly affected by disease. For instance, experimental myocardial hypertrophy, and insults that prolong APD, result in a substantial increase in Kv1.4 K^+ channel mRNA levels in cultured rat myocytes. This gene regulation is reversed on regression of hypertrophy (96). There is no evidence as to whether this enhanced transcription results in an enhancement of functional channel proteins in the hypertrophied cell, but Tseng et. al. (71) studied the effects of myocardial infarction on K^+ channel expression using RNase protection assays in the dog. mRNA levels for several K^+ currents (KvLQT1, IsK, and ERG) were all reduced in epicardial tissues within 2 days after occlusion. By day 5, KvLQT1 transcripts were recovered, but IsK and ERG levels remained reduced. These findings are consistent with a reduction of the K^+ currents I_{Ks} and I_{Kr} in border zone myocytes. In the rat, there is a significant reduction in Kv4.2 channel protein levels in noninfarcted, regionally hypertrophied tissues with no changes in Kv2.1 or Kv1.5 levels 3 days after occlusion (157). After 3 to 4 weeks, mRNAs of Kv1.4, Kv2.1 and Kv4.2 all appeared significantly decreased, with no changes in either Kv1.2 or Kv1.5 levels (50). Quantitative analysis of mRNA levels in normal and failing human ventricles, including some following myocardial infarction, indicate that steady-state mRNAs for Kv4.3 and HERG are decreased, while no mRNA changes were noted in Kv1.2, Kv1.4, 1.5, 2.1, or I_{K1} (73).

Changes in intracellular Ca^{2+} and hormonally induced changes in cAMP levels have both been linked to altered transcription of mRNAs encoding different ion channel proteins (127). In some studies, increases in intracellular Ca^{2+} produced by pacing caused a fall in mRNA, as well as the density of rat skeletal muscle Na^+ channel proteins or decreased density of the fast cardiac Na^+ current (27). On the other hand, forskolin-induced changes in cAMP increased Na^+ channel mRNA. Additionally, cAMP presumably through a protein kinase system decreases the rate of transcription of the K^+ channel gene Kv1.5 (107).

Catecholamines appear to regulate the level of L-type Ca^{2+} channel expression in cardiac cells. In particular, α-adrenergic agonists decrease, while β-agonists increase, L-type Ca^{2+} channel mRNA and function (94). In the latter studies, the change in the level of mRNA was correlated with a change in functional channel density. More direct contact of sympathetic neurons with cardiocytes also increases Ca^{2+} channel expression (113). The dependence of these changes on intracellular Ca^{2+} changes is not yet known, but we do know Ca^{2+} is an intracellular messenger that can quickly affect many cell surface proteins.

The recognition that Ca^{2+} might also regulate long-term expression of channels is an important step in unraveling the role of this key signaling molecule yet determining how Ca^{2+} could trigger a stimulus-specific nuclear response is still an enigma. In cardiac cells, Ca^{2+} changes in the cytoplasm can occur within milliseconds. Moreover, Ca^{2+} can be extruded from the cell within hundreds of milliseconds. So what is the signal? Recent evidence suggests that it is not only the amplitude of the Ca^{2+} response but also the interval between $[Ca^{2+}]_i$ transients that determine the overall impact on the nucleus and thus the nuclear stimulus-specific response. For instance, studies have shown that the permeability of the nuclear pore is controlled by $[Ca^{2+}]_i$ and second messengers, for example, IP_3 (132). It appears that depletion of cisternal Ca^{2+} stores in the nuclear membrane triggers a conformational change in the nuclear pore complex (85). In neurons, after Ca^{2+} signals are transmitted to the nucleus, they activate transcription of various factors. Two principal Ca^{2+}-dependent responsive DNA regulatory elements—CRE (cAMP response element) and SRE (serum response element)—have been identified (5). Further work has shown that signals activated by cytosolic Ca^{2+} target SRE, while nuclear Ca^{2+} signals control gene expression through CRE (57). In neurons, increased nuclear Ca^{2+} concentration controls the activities of the transcriptional activator CREB and the transcriptional coactivator CBP and this process is mediated by CaM kinase activation (98). Calmodulin-dependent protein kinase II can apparently decode the frequency of Ca^{2+} transients into distinct levels of kinase activity (37). Equally as important, the stability of a phosphorylated nuclear CREB is dynamically regulated by nuclear phosphatases that may be under the control of the Ca^{2+} binding protein calcineurin. Thus, in cardiac cells, as in neurons, there may be at least two pathways in the cell's response to Ca^{2+} transients—an early and a late response. Which signaling pathway becomes activated seems to be dictated by the duration and magnitude and the interval of the $[Ca^{2+}]_i$ increase.

XXI. SUMMARY

Ca^{2+} ions play a key role in almost all functional aspects of cardiac cells and are intimately involved in the development and propagation of arrhythmias. The nature of the many mechanisms by which Ca^2 exerts its multitude of effects is by no means fully understood, nor is their role in arrhythmogenesis. The structure of cardiac cells enables rapid electrical conduction, as well as rapid activation of the contractile system, even though diffusion of Ca^{2+} is slow. Nature has, therefore, provided amplification stations between the sarcolemma and the myofibrils so that the delivery of Ca^{2+} can be accelerated. Ca^{2+} sensitivity of many proteins in the cardiac cell is sufficiently high that activation of these proteins occurs at $[Ca^{2+}]_i$ levels only slightly above those of diastolic level. Furthermore, $[Ca^{2+}]_i$ also profoundly affects electrical processes at the surface membrane. It is therefore plausible that instability of Ca^{2+} transport systems contribute as well to the mechanisms that lead to arrhythmias. Ca^{2+} also plays an important role in the long-term life of the cardiac cell by affecting both the composition and cellular distribution of proteins that dictate the cell's phenotype. Turnover of these proteins is so fast that it is likely that the very factors that determine the initiation of an arrhythmia may themselves change the cardiac cell phenotype and thus alter the cell's future response to the same factors. Solving the nature of these intricate and dynamic interactions promises to be an important area of research essential to understanding of the nature of arrhythmias.

ACKNOWLEDGMENTS

Supported by grants HL-58860 from the National Heart, Lung and Blood Institute, National Institutes of Health, Bethesda, Maryland, MRC and Alberta Heritage Foundation for Medical Research, and a NATO Collaborative Research Grant.

REFERENCES

1. Abramcheck CW, Best PM. Physiological role and selectivity of the in situ potassium channel of the sarcoplasmic reticulum in skinned frog skeletal muscle fibers. J Gen Physiol 1989; 93: 1–21.
2. Aggarwal R, Boyden PA. Diminished calcium and barium currents in myocytes surviving in the epicardial border zone of the 5 day infarcted canine heart. Circ Res 1995; 77: 1180–1191.
3. Backx PH, de Tombe PP, van Deen JHK, Mulder BJM, ter Keurs HEDJ. A model of propagating calcium-induced calcium release mediated by calcium diffusion. J Gen Physiol 1989; 93: 963–977.
4. Backx PH, Gao WD, Azan-Backx MD, Marban E. The relationship between contractile force and intracellular $[Ca^{2+}]$ in intact rat cardiac trabeculae. J Gen Physiol 1995; 105: 1–19.
5. Bading H, Ginty DD, Greenberg ME. Regulation of gene expression in hippocampal neurons by distinct calcium signaling pathways. Science 1993; 260: 181–186.
6. Banijamali HS, ter Keurs MHC, Paul LC, Ter Keurs HEDJ. Excitation contraction coupling in rat heart: influence of cyclosporin A. Cardiovasc Res 1993; 27: 1845–1854.
7. Bassani RA, Bassani JWM, Lipsius SL, Bers DM. Diastolic SR Ca efflux in atrial pacemaker cells and Ca-overloaded myocytes. Am J Physiol 1997; 273: H886–H892.
8. Bean BP. Two kinds of calcium channels in canine atrial cells: Differences in kinetics, selectivity, and pharmacology. J Gen Physiol 1985; 86: 1–30.

9. Berlin JR, Cannell MB, Lederer WJ. Cellular origins of the transient inward current in cardiac myocytes. Role of fluctuations and waves of elevated intracellular calcium. Cir Res 1989; 65: 115–126.

10. Bers D. Sarcoplasmic reticulum calcium uptake, content and release. In: Bers D, ed. *Excitation-Contraction Coupling and Cardiac Contractile Force*. Boston, London: Kluwer Academic, 1991:93–116.

11. Bers DM, Bridge JHB. Relaxation of rabbit ventricular muscle by Na-Ca exchange and sarcoplasmic Ca-pump: Ryanodine and voltage sensitivity. Circ Res 1989; 65: 334–342.

12. Bers DM, Stiffel VM. Ratio of ryanodine to dihydropyridine receptors in cardiac and skeletal muscle and implications for EC coupling. Am J Physiol 1993; 264: C1587–C1593.

13. Boyden PA, Pu J, Pinto JMB, ter Keurs HEDJ. Ca^{2+} transients and Ca^{2+} waves in Purkinje cells. Circ Res 2000; 86:448–455.

14. Boyett MR, Honjo H, Harrison SM, Zang WJ, Kirby MS. Ultraslow voltage dependent inactivation of the calcium current in guinea pig and ferret ventricular myocytes. Pflugers Arch 1994; 428: 39–50.

15. Brandl CJ, deLeon S, Martin DR, MacLennan DH. Adult forms of the Ca^{2+} ATPase of sarcoplasmic reticulum. J Biol Chem 1987; 262: 3768–3774.

16. Brandl CJ, Green NM, Korczak B, MacLennan DH. Two Ca^{2+} ATPase genes: Homologies and mechanistic implications of deduced amino acid sequences. (Univ Calgary, Thesis) 1986; 44: 597–607.

17. Burne HR, Sanders DA, McCormick F. The GTPase superfamily: conserved structure and molecular mechanism. Nature 1991; 349: 117–127.

18. Cabo C, Schmitt H, Coromilas J, Ciaccio EJ, Wit AL. A new mechanism of antiarrhythmic drug action: Increasing inward L type calcium current can stop reentrant ventricular tachycardia in the infarcted canine heart. Circulation 2000 (in press).

19. Callewaert G, Carmeliet EE, Vereecke J. Single cardiac Purkinje cells: General electrophysiology and voltage-clamp analysis of the pace-maker current. J Physiol 1984; 349: 643–661.

20. Camacho P, Lechleiter JD. Calreticulin inhibits repetitive intracellular Ca^{2+} waves. Cell 1995; 82: 765–771.

21. Cannell MB, Cheng H, Lederer WJ. The control of calcium release in heart muscle. Science 1995; 268: 1045–1049.

22. Capogrossi MC, Houser SR, Bahinski A, Lakatta EG. Synchronous occurrence of spontaneous localized calcium release from the sarcoplasmic reticulum generates action potentials in rat cardiac ventricular myocytes at normal resting membrane potential. Circ Res 1987; 61: 489–503.

23. Capogrossi MC, Lakatta EG. Frequency modulation and synchronization of spontaneous oscillations in cardiac cells. Am J Physiol 1985; 248: H412–H418.

24. Carafoli E. Intracellular calcium homeostasis. Ann Rev Biochem 1987; 56: 395–433.

25. Cheng H, Lederer MR, Lederer WJ, Cannell MB. Calcium sparks and [Ca++]i waves in cardiac myocytes. Am J Physiol 1996; 270: C148–C159.

26. Cheng H, Lederer WJ, Cannell MB. Calcium sparks: elementary events underlying excitation-contraction coupling in heart muscle. Science 1993; 262: 740–744.

26a. Chen SR, Ebisawa K, Li X, Zhang L. Molecular identification of the ryanodine receptor Ca^{2+} sensor. J Biol Chem 1998; 273:14675–14678.

27. Chiamvimonvat N, Kargacin ME, Clark RB, Duff HJ. Effects of intracellular calcium on sodium current density in cultured neonatal rat cardiac myocytes. J Physiol 1995; 483: 307–318.

28. Clarke DM, Loo TW, Inesi G, MacLennan DH. Location of high affinity Ca^{2+}-binding sites within the predicted transmembrane domain of the sarcoplasmic reticulum Ca^{2+}-ATPase of sarcoplasmic reticulum. Nature 1989; 339: 476–478.

29. Collier ML, Levesque PC, Kenyon JL, Hume JR. Unitary Cl-Channels activated by cytoplasmic Ca^{2+} in canine ventricular myocytes. Circ Res 1996; 78: 936–944.

30. Colquhoun D, Neher E, Reuter H, Stevens C. Inward current channels activated by intracellular Ca in cultured cardiac cells. Nature 1981; 294: 752–754.

31. Colyer J. Translation of Ser 16 and Thr 17 phosphorylation of phospholamban into Ca^{2+} pump stimulation. Biochem J 1996; 15: 201–207.

32. Daniels MCG, Fedida D, Lamont C, ter Keurs HEDJ. Role of the sarcolemma in triggered propagated contractions in rat cardiac trabeculae. Circ Res 1991; 68: 1408–1421.

33. Daniels MCG, ter Keurs HEDJ. Spontaneous contractions in rat cardiac trabeculae: Trigger mechanism and propagation velocity. J Gen Physiol 1990; 95: 1123–1137.

34. Dassouli A, Sulpice JC, Roux S, Crozatier B. Stretch induced inositol triphosphate and tetrakisphosphate production in rat cardiocytes. JMCC 1993; 25: 973–982.

35. de Leon M, Wang Y, Jones L, Perez-Reyes E, Wei X, Soong TW, Snutch TP, Yue DT. Essential Ca^{2+}-binding motif for Ca^{2+}-sensitive inactivation of L-type Ca^{2+} channels. Science 1995; 270: 1502–1506.

36. Decrouy A, Juteau M, Rousseau E. Examination of the role of phosphorylation and phospholamban in the regulation of the cardiac sarcoplasmic reticulum Cl-channel. J Membr Biol 1995; 146: 315–326.

37. DeKonicnck P, Schulman H. Sensitivity of CaM kinase II to the frequency of Ca^{2+} oscillations. Science 1998; 279: 227–230.

38. de Leon M, Wang Y, Jones L, Perez-Reyes E, Wei X, Soong TW, Snutch TP, Yue DT. Essential Ca^{2+} binding Motif for Ca^{2+} sensitive inactivation of L type Ca^{2+} channels. Science 1995; 270: 1502–1506.

39. de Meis L, Vianna AL. Energy interconversion by the Ca-dependent ATPase of the sarcoplasmic reticulum. Ann Rev Biochem 1979; 48: 275.

40. Egan TM, Noble D, Noble SJ, et al. Sodium-calcium exchange during the action potential in guinea pig ventricular cells. J Physiol 1989; 411: 639–661.

41. Ehara T, Noma A, Ono K. Calcium-activated non-selective cation channel in ventricular cells isolated from adult guinea-pig hearts. J Physiol 1988; 403: 117–133.

42. Escande D, Coraboeuf E, Planche C. Abnormal pacemaking is modulated by sarcoplasmic reticulum in partially depolarized myocardium from dilated right atria in humans. J Mol Cell Cardiol 1987; 19: 231–241.

43. Escande D, Coulombe A, Faivre JF, Deroubaix E, Coraboeuf E. Two types of transient outward currents in adult human atrial cells. Am J Physiol 1987; 252: H142–H148.

44. Fabiato A. Spontaneous versus triggered contractions of calcium tolerant cardiac cells from the adult rat ventricle. Basic Res Cardiol 1985a; 80(suppl 2): 83–88.

45. Fabiato A. Time and calcium dependence of activation and inactivation of calcium induced release of calcium from the sarcoplasmic reticulum of a skinned cardiac Purkinje cell. J Gen Physiol 1985b; 85: 247–290.

46. Fabiato A, Fabiato F. Calcium-induced release of calcium from the sarcoplasmic reticulum of skinned cells from adult human, dog, cat, rabbit, rat, and frog hearts and from fetal and newborn rat ventricles. Annl NY Acad Sci 1978; 307: 491–522.

47. Fitzgerald M, Anderson KE, Woodcock EA. Inositol-1,4,5 triphosphate IP3 and IP3 receptor concentrations in heart tissues. Clin Exp Pharmacol Physiol 1998; 21: 257–260.

48. Gannier F, White E, Garnier D, LeGuennec J-Y. A possible mechanism for large stretch induced increase in Ca$_i$ in isolated guinea pig ventricular myocytes. Cardiovasc Res 1996; 32: 158–167.

49. Gannier F, White E, Lacampagne A, Garnier D, Le Guennec JY. Streptomycin reverses a large stretch induced increase in intracellular calcium in isolated guinea pig ventricular myocytes. Cardiovascular Res 1994; 28: 1193–1198.

50. Gidh-Jain M, Huang B, Jain P, El-Sherif N. Differential expression of voltage gated K channel genes in left ventricular remodeled myocardium after experimental myocardial infarction. Circ Res 1996; 79: 669–675.

51. Gorza L, Schiaffino S, Volpe P. Inositol 1,4,5-triphosphate receptor in heart: evidence for

its concentration in Purkinje myocytes of the conduction system. J Cell Biol 1993; 121: 345–352.

52. Gunter TE, Pfeiffer HJ. Mechanisms by which mitochondria transport calcium. Am J Physiol 1990; 258: C755–C786.

53. Guo W, Jorgensen AO, Jones LR, Campbell KP. Biochemical characterization and molecular cloning of cardiac triadin. J Biol Chem 1996; 271: 458–465.

54. Haase H, Karczewski P, Beckert R, Krause EG. Phosphorylation of the L type calcium channel beta subunit is involved in beta adrenergic signal transduction in canine myocardium. FEBS Lett 1993; 335: 217–222.

55. Hagiwara N, Irisawa H, Kameyama M. Contribution of two types of calcium currents to the pacemaker potentials of rabbit sino-atrial node cells. J Physiol 1988; 395: 233–253.

56. Han S, Schiefer A, Isenberg G. Ca^{2+} load of guinea pig ventricular myocytes determines efficacy of brief Ca^{2+} currents as trigger for Ca^{2+} release. J Physiol 1994; 480: 411–421.

57. Hardingham GE, Chawla S, Johnson CM, Bading H. Distinct functions of nuclear and cytoplasmic calcium in the control of gene expression. Nature 1997; 385: 260–265.

58. Hibberd MG, Jewell BR. Calcium and length dependent force production in rat ventricular muscle. J Physiol 1982; 329: 527–540.

59. Hilgemann DW, Matuoska S, Nagel GA, Collins A. Steady state and dynamic properties of cardiac NaCa exchange. Sodium dependent inactivation. J Gen Physiol 1992; 100: 905–932.

60. Hill JAJ, Coronado R, Strauss HC. Reconstitution and characterization of a calcium-activated channel from the heart. Circ Res 1988; 62: 411–415.

61. Hiraoka M, Kawano S. Calcium-sensitive and insensitive transient outward current in rabbit ventricular myocytes. J Physiol 1989; 410: 187–212.

62. Hofmann PA, Fuchs F. Bound calcium and force development in skinned cardiac muscle bundles: effect of sarcomere length. J Mol Cell Cardiol 1988; 20: 667–677.

63. Honore E, Attali B, Romey G, Heurteaux C, Ricard P, Lesage F, Lazdunski M, Barhanin J. Cloning, expression pharmacology and regulation of a delayed rectifier K channel in mouse heart. EMBO J 1991; 10: 2805–2811.

64. Housmans PR, Lee NKM, Blinks JR. Active shortening retards the decline of the intracellular calcium transient in mammalian heart muscle. Science 1983; 221: 159–161.

65. Hu H, Sachs F. Stretch-activated ion channels in the heart. J Mol Cell Cardiol 1997; 29: 1511–1523.

66. Inui M, Saito A, Fleischer S. Isolation of the ryanodine receptor from cardiac sarcoplasmic reticulum and identity with the feet structures. J Biol Chem 1987a; 262: 15637–15642.

67. Inui M, Saito A, Fleischer S. Purification of the ryanodine receptor and identity with feet structures of junctional terminal cisternae of sarcoplasmic reticulum from fast sketetal muscle. J Biol Chem 1987b; 262: 1740–1747.

68. Iwasa Y, Hosey MM. Phosphorylation of cardiac sarcolemma proteins by the calcium-activated phospholipid-dependent protein kinase. J Biol Chem 1984; 259: 534–540.

69. January C, Riddle JM. Early afterdepolarizations: Mechanisms of induction and block. A role for the L type Ca current. Circ Res 1989; 64: 977–990.

70. Janvier NC, Boyett MR. The role of Na Ca exchange current in the cardiac action potential. Cardiovasc Res 1996; 32: 69–84.

71. Jiang M, Yao J-A, Wymore RS, Boyden PA, Tseng G-N. Suppressed transcription and function of delayed rectifier K channels in post myocardial infarction canine ventricle. Circulation 1998; 98: 1–818.

72. Joyner RW, Kumar R, Wilders R, Jongsma HJ, Verheijek EE, Golod DA, van Ginneken ACG, Wagner MB, Goolsby WN. Modulating L type calcium current affects discontinuous cardiac action potential conduction. Biophys J 1996; 71: 237–245.

73. Kaab SH, Duc J, Ashen D, Nabauer M, Beuckelmann DJ, Dixon J, McKinnon D, Tomaselli GF. Quantitative analysis of K+ channel mRNA expression in normal and failing human ventricle reveals the molecular identity of Ito. Circulation 1996; 94: 1–592.

74. Kass RS, Tsien RW, Weingart R. Ionic basis of transient inward current induced by strophanthidin in cardiac Purkinje fibers. J Physiol 1978; 281: 209–226.

75. Kawano S, Hirayama Y, Hiraoka M. Activation mechanism of Ca²⁺ sensitive transient outward current in rabbit ventricular myocytes. J Physiol 1995; 486: 593–604.

76. Kentish JC, Ter Keurs HEDJ, Ricciardi L, Bucx JJ, Noble MIM. Comparison between the sarcomere length force relationships of intact and skinned trabeculae from rat right ventricle. Circ Res 1986; 58: 755–768.

77. Kim D. A mechanosensitive K⁺ channel in heart cells. Activation by arachidonic acid. J Gen Physiol 1992; 100: 1021–1040.

78. Kimura J, Miyamae S, Noma A. Identification of sodium-calcium exchange current in single ventricular cells of guinea pig. J Physiol 1987; 384: 199–222.

79. Kort AA, Lakatta EG. Calcium-dependent mechanical oscillations occur spontaneously in unstimulated mammalian cardiac tissues. Circ Res 1984; 54: 396–404.

80. Koster OF, Szigeti GP, Beuckelmann DJ. Characterization of a Cai dependent current in human atrial and ventricular cardiomyocytes in the absence of Na+ and K+. Cardiovasc Res 1999; 41: 175–187.

81. Lacerda AE, Kim HS, Ruth P, Perez Reyes E, Lockerzi F, Hofmann F, Birnbaumer L, Brown AM. Normalization of current kinetics by interaction between alpha 1 and beta subunits of the skeletal muscle dihydropyridine sensitive Ca²⁺ channel. Nature 1991; 352: 527–530.

82. Lai FA, Anderson K, Rousseau E, Liu Q, Meissner G. Evidence for a Ca²⁺ channel within the ryanodine receptor complex from cardiac sarcoplasmic reticulum. Biochem Biophys Res Commun 1988; 151: 441–449.

83. Lai FA, Erickson HP, Rousseau E, Liu Q, Meissner G. Purification and reconstitution of the calcium release channel from skeletal muscle. Nature 1988; 331: 315–319.

84. Lambert RC, Maulet Y, Mouton J, Beattie R, Volsen S, De Waard M, Feltz A. T-type Ca²⁺ current properties are not modified by Ca²⁺ channel β subunit depletion in nodosus ganglion neurons. J Neurosci 1997; 17: 6621–6628.

85. Lee AM, Dunn RC, Clapham DES-BL. Calcium regulation of nuclear pore permeability. Cell Calcium 1998; 23: 91–101.

86. Lee KS, Marbán E, Tsien RW. Inactivation of calcium channels in mammalian heart cells: joint dependence on membrane potential and intracellular calcium. J Physiol 1985; 364: 395–411.

87. Li GR, Feng J, Wang Z, Fermini B, Nattel S. Comparative mechanisms of 4 aminopyridine resistant Ito in human and rabbit atrial myocytes. Am J Physiol 1995; 269: H463–H472.

88. Li J, Qu J, Nathan RD. Ionic basis of ryanodine's negative chronotropic effect on pacemaker cells isolated from the sinoatrial node. Am J Physiol 1997; 273: H2481–H2489.

89. Lipp P, Niggli E. Microscopic spiral waves reveal positive feedback in subcellular calcium signaling. Biophys J 1993; 65: 2272–2276.

90. Litwin SE, Bridge JHB. Enhanced NaCa exchange in the infarcted heart. Implications for excitation contraction coupling. Circ Res 1997; 81: 1083–1093.

91. Lopez-Lopez JR, Shacklock PS, Balke CW, Wier WG. Local stochastic release of Ca²⁺ in voltage clamped rat heart cells: visualization with confocal microscopy. J Physiol 1994; 480: 21–29.

92. Lopez-Lopez JR, Shacklock PS, Balke CW, Wier WG. Local Ca²⁺ transients triggered by single L type Ca²⁺ channel currents in cardiac cells. Science 1995; 268: 1042–1045.

93. MacLennan DH, Brandl CJ, Korczak B, Green NM. Calcium ATPase: Contribution of molecular genetics to our understanding of structure and function. In: Hille B, Frambrough DM, eds. Proteins of Excitable Membranes. New York: John Wiley & Sons, Inc., 1987:287–300.

94. Maki T, Gruver EJ, Toupin D, Marks AR, Davidoff A, Marsh JD. Catecholamines regulate calcium channel expression in rat cardiomyocytes: distinct alpha and beta adrenergic regulatory mechanisms. Circulation 1993; 88: 1276.

95. Marx SO, Ondrias K, Marks AR. Coupled gating between individual skeletal muscle calcium release channels. Science 1998; 281: 818–821.

96. Matsubara H, Suzuki J, Murasawa S, Inada M. Kv1.4 mRNA regulation in cultured rat heart myocytes and differential expression of Kv1.4 and Kv1.5 genes in myocardial development and hypertrophy. Circulation 1993; 88: 186.

97. Matsuoka S, Nicoll DA, Hryshko LV, Levitsky DO, Weiss JN, Philipson KD. Regulation of cardiac NaCa exchanger by Ca^{2+}. Mutational analysis of the Ca^{2+} binding domain. J Gen Physiol 1995; 105: 403–420.

98. Matthews RP, Guthrie CR, Wailes LM, Zhao X, Means AR, McKnight GS. Calcium/calmodulin dependent protein kinase types II and IV differentiately regulate CREB dependent gene expression. Mol Cell Biol 1994; 16: 6107–6116.

99. Matuoska S, Nicoll DA, He Z, Philipson KD. Regulation of cardiac Na/Ca exchanger by endogenous XIP region. J Gen Physiol 1997; 109: 273–286.

100. Mazzanti M, Assandri R, Ferroni A, DiFrancesco D. Cytoskeletal control of rectification and expression of four substates in cardiac inward rectifier K channels. FASEB J 1996; 10: 357–361.

101. McCall E, Li L, Satoh H, Shannon TR, Blatter LA, Bers DM. Effects of FK-506 on contraction and Ca^{2+} transients in rat cardiac myocytes. Circ Res 1996; 79: 1110–1121.

102. Meissner G. Ryanodine activation and inhibition of the Ca release channel of sarcoplasmic reticulum. J Biol Chem 1986; 261: 6300–6306.

103. Miura M, Boyden PA, ter Keurs HEDJ. Ca^{2+} waves during triggered propagated contractions in intact trabeculae: determinants of the velocity of propagation. Circ Res 1999; 84: 1459–1468.

104. Miura M, Ishide N, Numaguchi H, Takishima T. Diversity of early afterdepolarizations in guinea pig myocytes; Spatial characteristics of intracellular Ca^{2+} concentration. Heart Vessels 1995; 10: 266–274.

105. Miura M, Ishide N, Oda H, Sakurai M, Shinozaki T, Takishima T. Spatial features of calcium transients during early and delayed afterdepolarizations. Am J Physiol 1993; 265: H439–H444.

106. Miura M, Ishide N, Sakurai M, Shinozaki T, Takishima T. Interactions between calcium waves and action potential-induced calcium transients in guinea pig myocytes. Heart Vessels 1994; 9: 79–86.

107. Mori Y, Matsubara H, Folco E, Siegel A, Koren G. The transcription of a mammalian voltage-gated potassium channel is regulated by cAMP in a cell-specific manner. J Biol Chem 1993; 268: 26482–26493.

108. Mulder BJM, de Tombe PP, ter Keurs HEDJ. Spontaneous and propagated contractions in rat cardiac trabeculae. J Gen Physiol 1989; 93: 943–961.

109. Nazir S, Dick DJ, Lab MJ. Mechanoelectric feedback and arrhythmia in the atrium of the isolate, Langendorff-perfused guinea pig hearts and its modulation by streptomycin. J Physiol 1995; 483: 24–25.

110. Neely A, Olcese R, Wei X, Birnbaumer L, Stefani E. Calcium-dependent inactivation of a cloned cardiac calcium channel alpha (1) subunit (alpha 1c) expressed in Xenopus oocytes. Biophys J 1994; 66: 1895–1903.

111. Nitta J, Furukawa T, Marumo F, Sawanobori T, Hiraoka M. Subcellular mechanism for calcium-dependent enhancement of delayed rectifier potassium current in isolated membrane patches of guinea pig ventricular myocytes. Circ Res 1994; 74: 96–104.

112. O'Neill SC, Mill JG, Eisner DA. Local activation of contraction in isolated rat ventricular myocytes. Am J Physiol 1990; 258: C1165–C1168.

113. Ogawa S, Barnett JV, Sen L, Galper JB, Smith TW, Marsh JD. Direct contact between sympathetic neurons and rat cardiac myocytes in vitro increases expression of functional calcium channels. J Clin Invest 1992; 89: 1085–1093.

114. Pastore JM, Girouard SD, Laurita KR, Akar FG, Rosenbaum DS. Mechanism linking T wave alternans to the genesis of cardiac fibrillation. Circulation 1999; 99: 1385–1394.

115. Perez Garcia MT, Kamp TJ, Marbán E. Functional properties of cardiac L type calcium channels transiently expressed in HEK293 cells. Role of alpha 1 and beta subunit. J Gen Physiol 1995; 105: 289–305.

116. Perez Reyes E, Cribbs LL, Daud A, Lacerda AE, Barclay J, Williamson MP, Fox M, Rees M, Lee JH. Molecular characterization of a neuronal low voltage activated T type calcium channel. Nature 1998; 391: 896–900.

117. Perez PJ, Ramos-Franco J, Fill M, Mignery GA. Identification and functional reconstitution of the type2 inositol 1,4,5-trisphosphate receptor from ventricular cardiac myocytes. J Biol Chem 1997; 272: 23961–23969.

118. Post JA, Langer GA, Op den KAmp, JAF, Verkleij AJ. Phospholipid asymmetry in cardiac sarcolemma. Analysis of intact cells and "gas dissected" membranes. Biochim Biophys Acta 1988; 943: 255–266.

119. Pragnell M, De Waard M, Mori Y, Tanabe T, Snutch TP, Campbell KP. Calcium channel beta-subunit binds to a conserved motif in the I–II cytoplasmic linker of the alpha 1—subunit. Nature 1994; 368: 67–70.

120. Robinson RB, Boyden PA, Hoffman BF, Hewett KW. The electrical restitution process in dispersed canine cardiac Purkinje and ventricular cells. Am J Physiol 1987; 253: H1018–H1025.

121. Rosenbaum DS, Jackson LE, Smith JM, Garna H, Ruskin JN, Cohen RJ. Electrical alternans and vulnerability to ventricular arrhythmias. Engl J Med 1994; 330: 235–241.

122. Rosenberg RL, Hess P, Reeves JP, Smilowitz H, Tsien RW. Calcium channels in planar lipid bilayers: insights into mechanism of ion permeation and gating. Science 1986; 231: 1564–1566.

123. Rubenstein DS, Lipsius SL. Mechanisms of automaticity in subsidiary pacemakers from cat right atrium. Circ Res 1989; 64: 648–657.

124. Rubenstein DS, Lipsius SL. Premature beats elicit a phase reversal of mechanoelectrical alternans in cat ventricular myocytes: A possible mechanism for reentrant arrhythmias. Circulation 1995; 91: 201–214.

125. Scamps F, Carmeliet EE. Delayed K current and external K in single Purkinje cells. Am J Physiol 1989; 257: C1086–C1092.

126. Schouten VJA, Deen JK, deTombe PP, Verveen AA. Force interval relationship in heart muscle of mammals. A calcium compartment model. Biophys J 1987; 51: 13–26.

127. Sherman SJ, Chrivia J, Catterall WA. Cyclic adenosine 3′5′-monophosphate and cytosolic calcium exert opposing effects on biosynthesis of tetrodotoxin sensitive sodium channels in rat muscle cells. J Neurosci 1985; 5: 1570–1576.

128. Simmerman HKB, Collins JA, Theiber JL, Weger AD, Jones LR. Sequence analysis of phospholamban. Identification of phosphorylation sites and two major structural domains. J Biol Chem 1986; 261: 13333–13341.

129. Sipido KR, Callewaert G, Porciatti F, Vereecke J, Carmeliet E. Ca$_i$ dependent membrane currents in guinea pig cells in the absence of Na/Ca exchange. Pflugers Arch 1995; 430: 871–878.

130. Smith JS, Imagawa T, Fill M, Ma J, Campbell DL, Coronado R. Purified ryanodine receptor from rabbit skeletal muscle is the calcium-release channel of sarcoplasmic reticulum. J Gen Physiol 1988; 92: 1–26.

131. Soldatov NM, Oz M, O'Brien KA, Abernathy DR, Morad M. Molecular determinants of L type Ca^{2+} channel inactivation. J Biol Chem 1998; 273: 957–963.

132. Stehno-Bittel L, Perez-Terzic C, Clapham D. Diffusion across the nuclear enevlope inhibited by depletion of the nuclear calcium store. Science 1995; 270: 1835–1838.

133. Sukhareva M, Morrisette J, Coronado R. Mechanism of chloride dependent release of Ca^{2+} in the sarcoplasmic reticulum of rabbit skeletal muscle. Biophys J 1994; 76: 75–765.

134. Sutko JL, Kenyon JL. Ryanodine modification of cardiac muscle responses to potassium-free solutions. J Gen Physiol 1983; 82: 385–404.

135. Tada M, Yamada M, Kadoma M, Inui M, Ohmori F. Calcium transport by cardiac sarcoplasmic reticulum and phosphorylation of phospholamban. Mol Cell Biochem 1982; 46: 74–95.

136. Takamatsu T, Minamikawa T, Kawachi H, Fujita S. Imaging of calcium wave propagation in

guinea plg ventricular cell pairs by confocal laser scanning microscopy. Cell Struct Funct 1991; 16:341–346.

137. Takamatsu T, Wier WG. Calcium waves in mammalian heart: quantification of origin magnitude waveform and velocity. FASEB J 1990; 4: 1519–1525.

138. Tohse N, Kameyama M, Irisawa H. Intracellular Ca and protein kinase C modulate K current in guinea pig heart cells. Am J Physiol 1987; 253: H1321–H1324.

139. Tong Q, Mumby MC, Hilgemann DW. Type 2 phosphatidylinositol 4' kinase regulates cardiac Na/Ca exchange via memberane anchored cyclic AMP kinase and calcineurin. *Biophys J* 1999; 76: A151.

140. Toyofukyu T, Yabuki M, Otsu K, Kuzuya T, Hori M, Tada M. Intercellular calcium signaling via gap junction in connexin43 transfected cells. J Biol Chem 1998; 273: 1519–1528.

141. Trafford AW, Lipp P, O'Neill SC, Niggli E, Eisner DA. Propagating calcium waves initiated by local caffeine application in rat ventricular myocytes. J Physiol 1995; 489: 319–326.

142. Tseng G-N, Boyden PA. Multiple types of Ca currents in single canine Purkinje myocytes. Circ Res 1989; 65: 1735–1750.

143. Tseng G-N, Hoffman BF. Two components of transient outward current in canine ventricular myocytes. Circ Res 1989; 64: 633–647.

144. Tseng G-N, Yao J-A, Tseng-Crank J. Modulation of K channels by coexpressed human alpha 1c-adrenoceptor in Xenopus oocytes. Am J Physiol 1997; 272: H1275–H1286.

145. Volders PG, Kulcsar A, Vos MA, Sipido KR, Wellens HJ, Lazzara R, Szabo B. Similarities between early and delayed afterdepolarizations induced by isoproterenol in canine ventricular myocytes. Cardiovasc Res 1997; 34: 348–359.

146. Wagenknecht T, Grassucci R, Frank JS, Saito A, Inui M, Fleischer S. Three-dimensional architecuture of the calcium channel/foot structure of sarcoplasmic reticulum. Nature 1989; 338: 167–170.

147. Walsh KB, Arena JP, Kwok W-M, Freeman L, Kass RS. Delayed-rectifier potassium channel activity in isolated membrane patches of guinea pig ventricular myocytes. Am J Physiol 1991; 260: H1390–H1393.

148. Wegener AD, Simmerman HKB, Lindemann JP, Jones LR. Phospholamban phosphorylation in intact ventricles. Phosphorylation of serine 16 and threonine 17 in response to adrenergic stimulation. J Biol Chem 1989; 264: 11468–11474.

149. Wendt-Gallitelli MF, Isenberg G. Total and free myoplasmic calcium during a contraction cycle: X ray microanalysis in guinea pig ventricular myocytes. J Physiol 1991; 435: 349–372.

150. Wibo M, Bravo G, Godfraind T. Postnatal maturation of excitation contraction coupling in rat ventyricle in relation to the subcellular localization and surface density of 1,4 dihydropyridine and ryanodine receptors. Circ Res 1991; 68: 662–673.

151. Wier WG, Cannell MB, Berlin JR, Marbán E, Lederer WJ. Cellular and subcellular heterogeneity of intracellular calcium concentration in single heart cells revealed by fura-2. Science 1987; 235: 325–328.

152. Wier WG, Egan TM, Lopez-Lopez JR, Balke CW. Local control of excitation contraction coupling in rat heart cells. J Physiol 1994; 474: 463–471.

153. Wier WG, ter Keurs HEDJ, Marbán E, Gao WD, Balke CW. Ca²⁺ 'sparks' and waves in intact ventricular muscle resolved by confocal imaging. Circ Res 1997; 81: 462–469.

154. Wier WG, Yue DT. Intracellular calcium transients underlying the short term force interval relationship in ferret ventricular myocardium. J Physiol 1986; 376: 507–530.

155. Wohlfart B. Relationships between peak force, action potential duration and stimulus interval in rabbit myocardium. Acta Physiol Scand 1979; 106: 395–409.

156. Wu Y, Roden DM, Anderson ME. CaM kinase inhibtion prevents development of the arrhythmogenic transient inward current. Circ Res 1999; 84:906–912.

157. Yao J-A, Jiang M, Fan J-S, Zhou Y-Y, Tseng G-N. Heterogeneous changes in K⁺ currents in rat ventricles 3 days after myocardial infarction. Cardiovasc Res 1999; 44:132–145.

158. Yeager M. Structure of Cardiac Gap junction membrane channels. In: Spooner PM, Joyner RW, Jalife J, eds. Discontinuous Conduction in the Heart. Armonk: Futura, 1997:161–184.

159. Zhou J, Olcese R, Qin N, Noceti F, Birnbaumer L, Stefani E. Feedback inhibition of Ca²⁺ channels by Ca²⁺ depends on a short sequenec of the C treminus that does not include the Ca²⁺ binding function of a motif with similarity to Ca²⁺ binding domains. PNAS 1997; 94: 2301–2305.

160. Zhou Z, Lipsius SL. T-type calcium currents in latent pacemaker cells isolated from cat right atrium. J Mol Cell Cardiol 1994; 26: 1211–1219.

161. Zhou Z, Matlib MA, Bers DM. Cytosolic and mitochondrial Ca²⁺ signals in patch clamped mammalian ventricular myocytes. J Physiol 1998; 507: 379–403.

162. Zhu Y, Nosek TM. Inositol trisphosphate enhances Ca²⁺ oscillations but not Ca²⁺-induced Ca²⁺ release from cardiac sarcoplasmic reticulum. Pflugers Arch 1991; 418: 1–6.

163. Zuhlke RD, Pitt GS, Deisseroth K, Tsien RW, Reuter H. Calmodulin supports both inactivation and facilitation of L type calcium channels. Nature 1999; 399: 159–162.

164. Sitsapesan R, Williams AJ. Regulation of current flow through ryanodine receptors by luminal Ca²⁺. J Membr Biol 1997; 159:179–185.

13

The Electrocardiogram and Electrophysiological Analysis

DAVID M. MIRVIS

University of Tennessee, Memphis, Tennessee

ROBERT PLONSEY

Duke University, Durham, North Carolina

I. OVERVIEW

Previous chapters of this text have rigorously presented the anatomical and physiological features of the *cardiac electrical generator*, that is, the characteristics of ion fluxes across cell membranes. In this chapter, we will explore how the electrical forces produced by this biological generator result in signals that can be registered on the electrocardiogram. The steps or processes that lead to an electrocardiographic recording are depicted in Figure 13.1.

First, an extracellular cardiac electrical field is generated by ion fluxes across cell membranes and between adjacent cells. These ion currents are synchronized by the cardiac activation sequence, as determined by patterns of cell-to-cell conduction and the properties of specialized cardiac conduction tissues. The net result is a cardiac electrical field in and around the heart that varies with time during the cardiac cycle. The properties of this field at each instant reflect, in a very complex manner, the state of activation or recovery of segments of the heart.

This electrical field must pass through numerous other structures before reaching the body surface. These tissues, including the heart muscle itself as well as the pericardium, lungs, blood, skeletal muscle, subcutaneous tissues and skin, form the *volume* conductor in which the biological generator sits and through which the cardiac field must pass. The components of the volume conductor differ from each other in their passive electrical

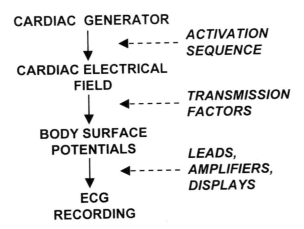

Figure 13.1 Schematic representation of the factors resulting in the recording of the electrocardiogram. The major path leading to the electrocardiogram is denoted by solid lines, while factors influencing or perturbing this path are shown with dotted lines. (Modified from Mirvis DM. Body Surface Electrocardiographic Mapping. Boston: Kluwer Academic Publishers, 1988.)

properties, such as in resistance. These differences perturb the cardiac electrical field as it passes through them. Thus, the cardiac potentials recorded on the body surface reflect both the active effects of the cardiac generator and the passive effects of the volume conductor contents known as *transmission factors*.

The cardiac field, once manifest on the skin, must be detected and processed to yield an electrographic recording. This stage involves the use of *electrodes* configured to produce *leads*. The outputs of these leads are amplified, filtered, and displayed by a variety of electronic devices. Each of these components has characteristics that also alter the signal in either desired or undesired ways. Because each of these steps has a significant impact on the potentials that are finally recorded, a clear understanding of each step is vital to appreciate the values and limitations of the final product—the *electrocardiogram*.

II. GENERATION OF THE CARDIAC ELECTRICAL FIELD DURING ACTIVATION

A. Definition of Current Direction

Transmembrane ion currents and their associated action potentials described in prior chapters are ultimately responsible for the extracellular potential field that is recorded as an electrocardiogram. Current may be analyzed as if carried by positively or negatively charged ions. Through a purely arbitrary choice, electrophysiological currents are considered as the movement of positive charge. A positive current moving in one direction is equivalent to a negative current of equal strength moving in the opposite direction. For example, both an *inward* flow of positive current and an *outward* flow of negative charge across a membrane constitute an *inward* current. In this chapter, we will consider currents to be carried by positive ions, which is consistent with conventions.

B. Generation of Extracellular Potential Fields

The conceptual sequence of events converting a transmembrane action potential into the cardiac electrical field is depicted in Figure 13.2 (1). Panel (A) shows a cylindrical, linear cardiac fiber that is 20 mm in length. The difference between intracellular and extracellular voltages (the *transmembrane potential* or V_m) is recorded at a site near the middle of the fiber (point X_0) by a voltmeter. Panel (B) depicts a 10-ms recording of V_m from the electrodes shown in panel (A) during which the cell is stimulated by an impulse applied to the left margin of the fiber (point $X = 0$). The fiber generates a classic action potential, with the transmembrane potential reversing from negative to positive over time. Time point t_0 marks the midpoint of phase 0 of the action potential, corresponding to the activation time of site X_0.

Panel (C) plots the intracellular potential as a function of distance along the fiber length as the impulse propagates from left to right along the fiber shown in panel (A). Sites to the left of point X_0 have already been activated, whereas those to the right of that point remain in a resting state. Hence, intracellular potentials to the right of X_0 are negative and those to the left are positive. Near site X_0, the site currently undergoing activation, the potentials reverse polarity over a short distance.

Panel (D) displays the direction and magnitude of transmembrane currents along the fiber. As expected from prior discussions of transmembrane currents, current flow is inwardly directed in fiber regions that are undergoing activation (i.e., to the left of point X_0) and outwardly directed in neighboring zones still at rest (i.e., to the right of X_0).

The outward current flow is a *current source* and the inward current flow is a *current sink*. As a good approximation, we can localize these currents to the sites of maximal current flow, as depicted in panel (E). In the fiber shown in the figure, the source and sink are separated by distance d, usually 1.0 mm or less. As activation proceeds along the fiber, extracellular current flows from source to sink as the source–sink pair moves to the right at the speed of propagation for the particular type of fiber.

C. Generation of a Current Dipole

Two point sources of equal strength but of opposite polarity located very near to each other, such as the current source and the current sink depicted in panel (E), may be represented as a *current dipole*. Such a dipole is fully characterized by three parameters—strength or *dipole moment*, location, and orientation. In this case, the location of the dipole is the activation site (point X_0) and its orientation is in the direction of activation (i.e., from left to right along the fiber). Dipole moment equals the product of the magnitude of the charge and the separation of the point sources, and is proportional to the rate of change of intracellular potential with respect to distance along the fiber (i.e., action potential shape).

A current dipole produces a characteristic potential field. As current flows from source to sink, the potential field that is generated is perpendicular to the direction of current flow (Fig. 13.3). Positive potentials are projected ahead of and negative potentials are projected behind the moving dipole. The potential recorded at any site in this field distant from the dipole is proportional to the dipole moment, inversely proportional to the square of the distance from the dipole to the recording site, and proportional to the cosine of the angle between the axis of the dipole and a line drawn from the dipole to the recording site (Fig. 13.4).

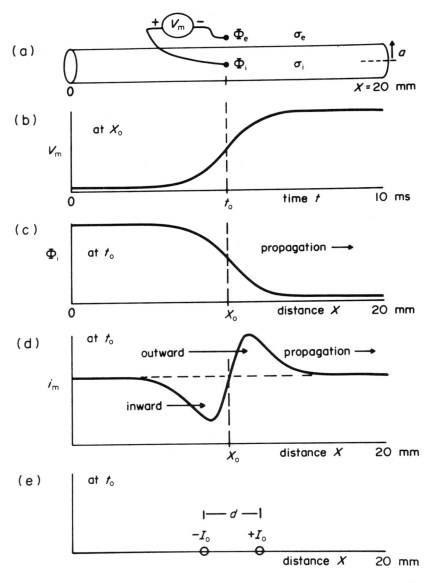

Figure 13.2 Example of potentials and currents generated by the activation of a single cardiac fiber. (a) Intracellular (Φ_i) and extracellular (Φ_e) potentials are recorded with a voltmeter (V_m) from a fiber 20 mm in length with a radius a and intracellular and extracellular conductivities σ_i and σ_e, respectively. The fiber is stimulated at site $X = 0$ and propagation proceeds from left to right. The recording site is X_o. (b) The upstroke (phase 0) of the transmembrane action potential recorded at site X_o. Time t_o represents the midpoint of the upstroke. (c) Plot of intracellular potential (Φ_i) at t_o along the length of the fiber depicted in (a). Positive potentials are recorded from activated tissue to the left of site X_o and negative ones are registered from not yet excited areas to the right of site X_o. (d) Membrane current (i_m) flows along the length of the fiber at time t_o. The outward current is the depolarizing current that propagates ahead of the activation site X_o, while an inward one is the activated sodium current that flows behind the site X_o. (e) Representation of the sites peak of inward and outward current flow in (d) as two point sources, a sink ($-I_o$) and a source ($+I_o$), separated by distance d. (Reproduced with permission from Ref. 1, with permission.)

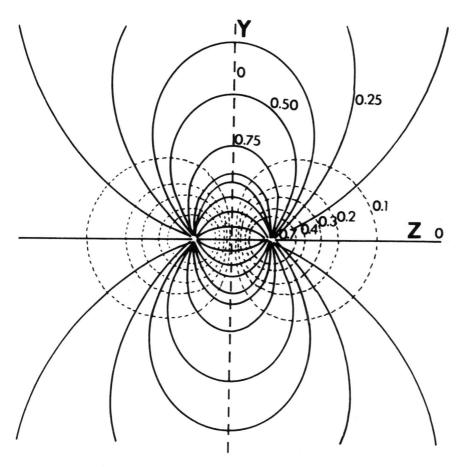

Figure 13.3 The potential field generated by a dipole oriented along the Z axis. Solid lines represent current flows and dotted lines represent potential lines. (From Plonsey R. The biophysical basis for electrocardiography. In: Liebman J, Plonsey R, Gillette PC, eds. Pediatric Electrocardiography. Baltimore: Williams and Wilkins, 1982.)

D.   Activation Wavefronts

This example from one cardiac fiber may be generalized to the more realistic case in which multiple adjacent fibers are activated in synchrony (i.e., an *activation wavefront* exists) (Fig. 13.5). Activation of each fiber creates a current dipole oriented in the direction of activation (A). In a sheet of fibers undergoing activation, each fiber generates a dipole oriented in the direction of activation (B).

The net effect of all the dipoles in this wavefront or layer is the vector sum of the effects of all the simultaneously active dipoles. The effects of all dipoles in the sheet of muscle may then be represented as a single dipole called an *equivalent dipole* (C). This single dipole is located in the geometric center of the activating sheet of myocardium, is oriented perpendicular to the plane of the wavefront or layer in the direction of activation, and has a moment that is proportional to the area of the activating zone. This model of an activation wavefront containing multiple dipoles is often called a *double layer*, referring to the fi-

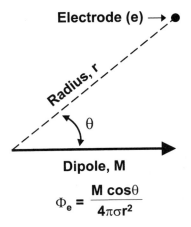

$$\Phi_e = \frac{M \cos\theta}{4\pi\sigma r^2}$$

Figure 13.4   The formula for computing the electrical potential (ϕ_e) recorded by an electrode e, in terms of the distance from the dipole to the electrode (radius, r), the angle between the radius to the electrode and the dipole (θ), dipole moment (M), and extracellular conductivity (σ_e). (Modified from Mirvis DM. Body Surface Electrocardiographic Mapping. Boston: Kluwer Academic Publishers, 1983.)

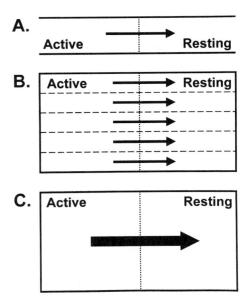

Figure 13.5   The relationship between the equivalent dipole of a single fiber and that of an activation front. Each fiber in a myocardial sheet generates a dipole (B) similar to that of an individual fiber (A). All of these dipoles may be lumped into a single equivalent dipole approximating the overall activation wavefront (C).

nite thickness of the layer of the myocardium undergoing activation. The single resultant dipole is a good approximation when the extent of the double layer is less than the distance to the measurement site (on the torso).

Thus, an activation front propagating through the heart may be represented by a single dipole that projects positive potentials ahead of it and negative potentials behind it. If more than one area of the heart is undergoing activation at the same instant, as commonly occurs during ventricular activation (2), each can be represented by a separate dipole. The net potential field is then determined by the vector sum of all the equivalent dipoles. This relation between the direction of the activation wavefront and the cardiac electrical field is critical to understanding the genesis of the electrocardiogram.

E. The Solid Angle Theorem

One important and commonly used method of estimating the potentials expected at some point removed from an activation front is an application of the *solid angle theorem*. A *solid angle* is a geometric measure of the size of a region when viewed from a distant electrode, and is the area on the surface of a sphere of unit radius cut by lines drawn from a recording electrode to all points around the boundary of the region of interest. This region may be a wavefront, an area of ischemia, or any other region on or in the heart. This relationship is depicted in Figure 13.6.

The solid angle theorem states that the potential recorded by a remote electrode (Φ) is defined by the relationship

$$\Phi = (\Omega/4\pi)(V_{m2} - V_{m1})K, \tag{1}$$

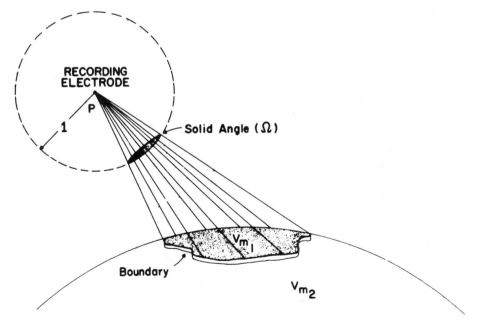

Figure 13.6 Schematic representation of the solid angle theorem for estimating the potential generated by an activation front and recorded at electrode site P in terms of the solid angle (Ω) and the potentials on either side of the active layer, V_{m1} and V_{m2}. (Modified from Ref. 3, with permission.)

where Ω is the solid angle, $V_{m2} - V_{m1}$ is the potential difference across the boundary under study, and K is a constant reflecting differences in intracellular conductivity.

This construct indicates that the recorded potential is composed of the product of two factors (3). First, the solid angle reflects spatial parameters, including the size of the boundary of the region under study and the distance from the electrode to that boundary. The potential will rise as the boundary size increases and as the distance to the electrode shrinks. Note that the solid angle depends solely on the size of the boundary of the activating zone, not on its geometry nor on the shape of the contained myocardium. This condition is based upon the usual expectation that the double layer is uniform.

A second set of parameters includes nonspatial factors. These include the potential difference across the surface and the intracellular and extracellular conductivities. Examples of nonspatial effects include factors, such as ischemia, that change transmembrane action potential shapes or that alter conductivity. Thus, the potential generated by an activation front is determined by both spatial and nonspatial parameters.

This theorem does have significant conceptual limitations. It assumes, for example, that all sites within the boundary have identical electrophysiological properties and that the recording electrode is located within an infinite and unbounded volume conductor. Neither is true for real recordings. However, to the extent that this approach is valid, it may be more accurate than replacing a double layer with a single, resultant dipole.

III. THE CARDIAC ELECTRICAL FIELD DURING VENTRICULAR RECOVERY

The cardiac electrical field during recovery differs in fundamental ways from that described for activation. These include differences in the orientation of the potential gradients and current flows, as well as in the manner in which recovery spreads through cardiac tissue.

A. Recovery Potentials and Current Flows

Intercellular potential differences and, hence, the direction of longitudinal current flows during recovery are (in the idealized case of equal action potential durations along the fiber) the opposite of those described for activation. As a cell undergoes recovery, its intracellular potential becomes progressively more negative. Hence, for two adjacent cells, the intracellular potential of the cell whose recovery has progressed further will be more negative than an adjacent, less recovered cell. Intracellular (positive) currents will then flow from the less to the more recovered cell, and extracellular currents will flow in the opposite direction [Fig. 13.7 (B)]. An equivalent dipole can then be constructed for recovery, just as for activation. Its orientation will point from less to more recovered cells [Fig. 13.7(A)].

B. Differences Between Activation and Recovery

Key differences can be described between activation and recovery forces in the single fiber where propagation is uniform. First, the recovery dipole is oriented away from the activation front (i.e., in the direction opposite that of the activation dipole). Whereas during activation intracellular currents flow from activated to resting cells (i.e., in the direction of propagation), during recovery intracellular currents flow from more completely recovered to less completely recovered cells (i.e., the direction opposite to that of propagation), as illustrated in Figure 13.7(A).

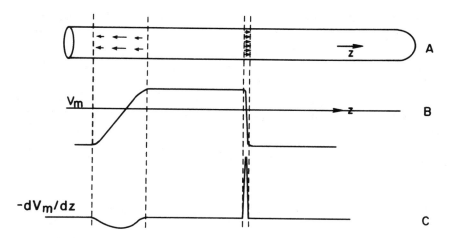

Figure 13.7 Schematic representation of the differences between activation and recovery forces in a linear fiber. (A) A linear fiber in which activation proceeds from left to right. The region undergoing activation (short arrows representing dipoles) and recovery (longer arrows) are marked. Recovery occurs over a larger region but dipole strengths (corresponding to dipole density) are lower than during activation. (B) Transmembrane potential (V_m) along the fiber, indicating regions of activation and recovery. (C) Rate of change of transmembrane potential with respect to distance along the fiber. Rates of change are lower during recovery than during activation corresponding to the difference in dipole density. (From Plonsey R. The biophysical basis for electrocardiography. In: Liebman J, Plonsey R, Gillette PC, eds. Pediatric Electrocardiography. Baltimore: Williams and Wilkins, 1982.)

The recovery dipole density will also differ from that of the activation dipole density. As described previously, the strength of the activation dipole density is proportional to the negative of the rate of change of transmembrane potential with respect to distance along the fiber. Rates of change in potential during the recovery phases of the action potential (phases 1 through 3) are considerably slower than during activation (phase 0), so that recovery dipole density will be less than that during activation. However, the *total* strength or moment of the recovery dipole will be the same as that of the total activation dipole since these quantities depend upon the total charge on the membrane (i.e., from resting potential to peak potential and from peak potential to resting potential).

A third difference between activation and recovery is the rate of movement of the activation and the recovery dipoles. Activation, because it is rapid (as fast a 1 ms in duration), occurs over only a small distance along the fiber. Recovery, in contrast, lasts 100 ms or longer. If the propagation and recovery velocity in the fiber were 50 cm/s, activation would occur in a 0.5-mm segment, but recovery would occur in a 5-cm section if duration of recovery were 100 ms [Fig. 13.7(C)]. Potential mapping studies have shown, for example, that whereas only discrete portions of the heart are undergoing activation at any one instant, the whole heart undergoes recovery simultaneously (4).

For a uniform fiber, these differences result in a characteristic electrocardiographic difference between activation and recovery patterns [Fig. 13.7(C)]. All other factors being equal (an assumption that is often not true, as described below), repolarization forces will (1) be of opposite polarity to activation forces (because the dipole is oriented in the opposite direction); (2) have a lower amplitude than activation forces (because the dipole moment density is lower); and (3) have a longer duration (because of the longer duration of recovery forces). Thus, electrocardiographic patterns generated by stimulation of a linear fiber with uniform recovery properties during recovery (i.e., the ST-T wave) may be expected to be of opposite polarity, lower amplitude, and longer duration than those due to activation (the QRS complex) [Fig. 13.7(C)].

That the T wave is, in reality, of the same polarity as the QRS complex is due to transmural differences in action potential duration (5). Refractory periods, reflecting action potential durations, progressively decrease from the endocardium to the epicardium (Fig. 13.8). Because the difference in refractory period durations exceeds transmural conduction time, epicardial cells that are activated last will complete recovery ahead of endocardial cells that are activated first. Hence, the intracellular currents flowing from less to more recovered cells will flow from the endocardium toward the epicardium, and result in an equivalent dipole directed toward the epicardium (i.e., in the same direction as the equivalent dipole generated by transmural activation). Thus, during recovery, the activation sequence interacts with the variation in cellular properties to determine the extracellular potentials so that QRS and ST-T waves have the same polarity.

IV. THE ROLE OF TRANSMISSION FACTORS

The activation fronts described in the previous section lie within a very complex three-dimensional physical environment—the *volume conductor*—through which the cardiac electrical field spreads. The structures that are present within the volume conductor modify the cardiac electrical field in significant ways so that the potentials on the body surface are quite different from what they would be in a homogeneous medium. Furthermore, as a consequence of the decreases in potential amplitudes with increasing distances from the source, surface potentials have an amplitude of only 1% of the amplitude of transmembrane potentials. As a result, there is also smoothing of the details of the electrical field

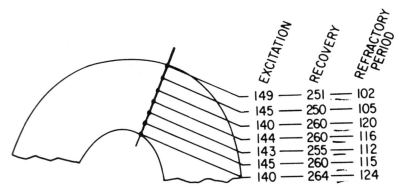

Figure 13.8   Schematic representation of transmural activation times, recovery times marking the end of the local action potential, and refractory periods reflecting the duration of the local action potential. Time measurements are in milliseconds from an arbitrary reference time. Activation (excitation) times progressively increase but refractory periods decrease from endocardium to epicardium. Because the transmural difference in refractory period duration (22 ms) is greater than the transmural difference in activation times (9 ms), epicardial cells complete recovery before endocardial cells. (From Ref. 5, with permission.)

with distance so that surface potentials have only a general spatial relation to the underlying cardiac events.

A. Biophysical versus Biological Effects on Electrical Fields.

The modifying effect of these physical structures is a biophysical one, dependent upon the physical properties of the structures and the related laws of physics. This is in contrast to the biological cardiac generators, whose output is dependent upon cellular structure and physiological and biochemical processes. Thus, electrocardiographic potentials on the body surface are dependent upon both biological and biophysical properties (6). These biophysical effects may be called *transmission factors* to emphasize their effects on the transmission of the cardiac electrical field throughout the body.

B. Types of Transmission Factors

The transmission factors that are present in the body may be grouped into four broad categories—*cellular factors*, *tissue or cardiac factors*, *organism or extracardiac factors*, and *purely physical factors* (7).

1. Cellular Factors

Cellular factors determine the intensity of current fluxes that result from local transmembrane potential gradients and include intracellular and extracellular resistances and the concentration of relevant ions, especially the sodium cation. For example, smaller cardiac cell sizes, higher intracellular resistances, and lower ion concentrations reduce the magnitude of current flow resulting from a given transmembrane potential gradient. A lower current flow corresponds to a lower equivalent dipole moment and, hence, to reduced extracellular potentials.

2. Tissue Factors

Tissue factors affect the relation of one cardiac cell to another. These properties are related to the *cable properties* of a fiber that describe the contribution of the passive conducting medium on the shape of the action potential and its conduction velocity.

Two major tissue factors are (1) cardiac *anisotropy* (i.e., the property of cardiac tissue that results in greater current flow and more rapid propagation along the fiber direction than transversely); and (2) the presence of connective tissue between cardiac fibers that disrupts effective electrical coupling between adjacent fibers so that transverse conduction occurs through irregular and circuitous paths around and within activation fronts. These factors have been discussed in detail in Chapter 9.

The major effect of tissue factors is to alter the current paths and, hence, potential fields. This leads to changes in the amplitude and configuration of recorded electrograms and electrocardiograms. Recording electrodes oriented along the long axis of a cardiac fiber will register larger potentials than will electrodes oriented perpendicular to the long axis. Waveforms recorded from fibers with little or no intervening connective tissue will be narrow in width and smooth in contour, whereas those recorded from tissues with abnormal fibrosis will be prolonged and heavily fractionated (8).

3. Organism Factors

Organism factors include all the tissues and structures that lie between the activation region and the body surface. These include the ventricular walls, the intracardiac and intrathoracic blood volumes, the pericardium, the lungs, and the skeletal muscle and subcutaneous fat layers of the thorax. These tissues alter the cardiac field because of the differences in the electrical resistivity of adjacent tissues (i.e., the presence of *electrical inhomogeneities* within the torso). For example, intracardiac blood has a much lower resistivity (162 Ω·cm) than do the lungs (2150 Ω·cm) or heart muscle (563 Ω·cm). When the cardiac field crosses the boundary between two tissues with differing resistivity, the field is altered.

The significance of these organism or extracardiac factors has been quantified by simulation techniques. In one simulation model known as the *eccentric spheres model*, the torso is represented by a series of eccentric spheres (Fig. 13.9) (9). Each sphere represents one tissue and, in the mathematical simulation, is assigned the biophysical properties of that tissue. By varying the properties of individual layers, the effects of various physiological or pathological conditions on the cardiac electrical field can be quantified.

One organism factor that is commonly studied is the effect of the mass and conductivity of the blood found within the cardiac chambers on the electrocardiogram (10). The conductivity of blood is changed by anemia, for example, in which nonconductive cellular elements are replaced with more conductive plasma, leading to an increase in overall conductivity. The size of the intracavitary blood mass is increased as ventricular chamber size increases with ventricular dilitation.

According to the *Brody effect* (10), the high conductivity of the intracavitary blood mass increases the potentials generated by dipoles oriented radial to the boundary of the mass of blood and the myocardium, but decreases the potentials generated by dipoles oriented tangential to that boundary. Because activation of the ventricular walls proceeds in a transmural direction that is radial to the interface between intracavitary space and the myocardium, the major effect of the intracavitary blood is to increase the magnitude of potentials sensed by body surface electrodes. Factors that increase the conductivity or size of the blood mass (e.g., with anemia and ventricular dilitation) would then be expected to increase the magnitude of the potentials recorded. This effect has been documented in simulation studies using the eccentric spheres model (Fig. 13.10).

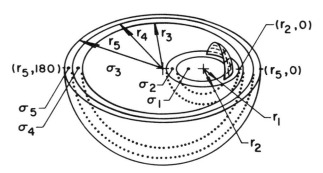

Figure 13.9   The eccentric spheres model used to simulate the effect of organism or extracardiac transmission factors. The sphere representing the heart is located eccentrically within a larger sphere representing the torso. The active myocardial area is marked by + and – signs. Regions with conductivities σ_1 through σ_5 represent intracavitary blood, heart muscle, lungs, skeletal muscle, and fat, respectively. Radii r_1 and r_2 represent distances from center of the heart to the endocardium and epicardium, respectively, and radii r_3, r_4, and r_5 are the distances from the center of the torso to parts of the chest wall. (From Ref. 9, with permission.)

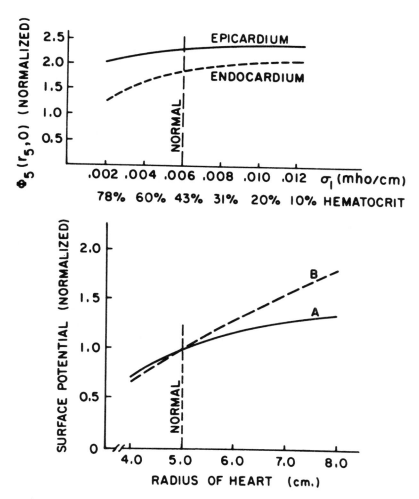

Figure 13.10 Effect of changes in the content (changes in hematocrit, top) and volume (ventricular dilatation, bottom) of intracavitary blood on body surface potentials in the eccentric spheres model. Decreasing hematocrit (top) increased surface potentials generated by either an endocardial or an epicardial source. Increasing the volume of intracavitary blood (bottom) by increasing the radius of the heart, either with maintaining the size of the activation front constant (line A) or with increasing the size of the front in proportion to the increase in radius (line B), likewise increased surface potentials. (From Rudy Y. The effects of the thoracic volume conductor (inhomogeneities) on the electrocardiogram. In: Liebman J, Plonsey R, and Rudy Y, eds. Pediatric and Fundamental Electrocardiography. Boston: Martiinus Nijhoff, 1987.)

As described in an earlier section, these changes in potential occur without any change in the characteristics of the biological cardiac generator, and reflect biophysical rather than physiological effects. Any change in the physiological properties of the heart by, for example, the pathological cause of the ventricular enlargement, would produce additional changes in the electrocardiographic recordings. Thus, biophysical and pathophysiological events interact to produce the cardiac field.

4. Physical Factors

Other transmission factors reflect fundamental laws of physics. First, changes in the distance between the heart and the recording electrode reduce potential magnitude in accord with the *inverse square law* (i.e., amplitude decreases in proportion to the square of the distance). A related factor is the effect of *eccentricity*. The heart is located eccentrically within the chest, lying closer to the anterior than to the posterior torso. Similarly, the right ventricle and the anteroseptal left ventricle are located closer to the anterior chest than are other parts of the left ventricle and the atria. Therefore, electrocardiographic voltages will be higher anteriorly than posteriorly, and voltages projected from the anterior left ventricle will be greater on the body's surface than those generated by the posterior ventricular regions.

Another physical factor is *cancellation*. More than one region of the ventricle undergoes activation simultaneously during much of the cardiac cycle (2,4). As described above, each activation wavefront can be represented by a dipole reflecting the location, orientation, and size of the wavefront. The net effect is the vector sum of all of these dipoles. Because some of these dipoles may be oriented opposite to others, the effects of one wavefront may be cancelled in part or in whole by that of another. For example, potentials generated by a wavefront moving to the left will be partially or wholly canceled by potentials generated by a second wavefront moving to the right. Similarly, forces generated by two wavefronts located near each other and with similar spatial orientations may meld to give the appearance of a single wavefront (11). In humans, over two-thirds of all cardiac electrical activity during ventricular activation is canceled (12).

A final effect is due to the presence of a finite, physical boundary surrounding the volume conductor—the skin. The potentials recorded on the surface of a bounded volume conductor are from two to three times larger than those recorded the same distance from the cardiac source but in an infinitely large, unbounded volume conductor.

V. RECORDING ELECTRODES AND LEADS

Potentials generated by the cardiac electrical generator and modified by transmission factors are sensed by electrodes placed on the torso, on the heart, or within the myocardium that are configured to form various types of electrocardiographic leads. Electrodes have physical properties that affect their ability to faithfully sense the potentials, and the various lead systems are designed to be sensitive to different aspects of the cardiac field. These will be described in this section.

A. Bipolar and Unipolar Lead Configurations

Electrocardiographic leads may be subdivided into two general types—*bipolar leads* and *unipolar leads*. A bipolar lead consists of two electrodes placed at two different sites. The leads sense the difference in potential between these two sites. The actual potential at either electrode is not known; only the difference between them is recorded. One electrode is designated as the positive input; the potential in the other, or negative electrode, is subtracted from the potential in the positive electrode to yield the bipolar potential.

Unipolar leads, in contrast, measure the absolute electrical potential at one site. To do so requires a *reference site* (i.e., a site at which the potential either is or is deemed to be zero). The reference site may be a location far away from the active electrode within an experimental preparation or, as in clinical electrocardiography, a specially designed electrode configuration as described below. The unipolar recording is then the potential sensed by a single electrode at one site—the *recording* or *active* or *exploring electrode*—in relation to the designated zero or reference potential.

These two lead types have important differences when applied to electrocardiography and electrophysiology. A bipolar lead composed of two closely spaced electrodes will be very sensitive to electrophysiological events occurring very near the electrode pair, but insensitive to those occurring in more remote portions of the heart (13). The more remote the event, the more it will affect the potentials at both electrode sites equally, so that the difference in potential between the two electrodes that is registered by the bipolar lead will approach zero. In the case of activation fronts, a close bipolar lead will detect a wavefront only when it is passing between or very near the two electrodes.

The output of the bipolar lead resembles a sharp spike as the wavefront passes the electrode pair (Fig. 13.11). The timing of the peak of the bipolar spike marks the timing of activation of the adjacent myocardium. Factors affecting the morphology of the bipolar spike include the distance between the poles (the closer the electrodes are to each other, the more insensitive the lead will be to remote events), the orientation of the axis between the poles in relation to the direction of the wavefront (the more nearly perpendicular are the two axes, the greater will be the amplitude), the distance from the poles to the wavefront (the closer the wavefront, the greater the amplitude), and fiber orientation (with taller spikes along the direction of myocardial fibers) (14).

In addition, the morphology of the spike can provide information about the properties of the underlying myocardium (15). For example, myocardial ischemia leads to low-amplitude potentials. Chronic scarring results in fractionated, multiphasic, and prolonged spikes reflecting the circuitous path of the activation front through scarred tissue.

A unipolar lead, in contrast, is very sensitive to remote events as well as those occurring in the immediate vicinity of the single exploring electrode. Indeed, the shape and

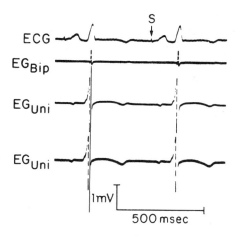

Figure 13.11   Body surface electrocardiogram (ECG) during paced rhythm (s = stimulus artifact), two unipolar recordings from adjacent electrodes on a plunge electrode (EG_{Uni1} and EG_{Uni2}), and the bipolar recording constructed from the two unipolar recordings (EG_{Bip}). Unipolar recordings demonstrate complex multiphasic waves while the bipolar recording is characterized by a narrow spike. (From Kupersmith J. Electrophysiologic mapping during open heart surgery. Prog Cardiovasc Dis 1976; 19:167, with permission.)

amplitude of unipolar recordings may be primarily determined by remote events. The result is a broad multiphasic complex.

Timing of local events can be determined, however, from unipolar recordings by the timing of the sharp, negative deflection or the *intrinsic deflection*. If one considers an activation wave or double-layer propagating, say, from left to right under an epicardial unipolar electrode, one can apply the solid angle theorem to evaluate the expected signal. As the wave approaches the electrode, the solid angle it subtends increases (with the actual magnitude dependent upon the size of the wavefront). As the wave reaches the electrode, the solid angle is maximum, being equal to $+\pi$ [see Eq. (1)]. As it continues to the right, past the electrode, the solid angle changes abruptly to $-\pi$ and then decreases in absolute magnitude as the distance from the electrode to the wave increases. Consequently, there is a rapid change in signal polarity at the electrode, corresponding to the intrinsic deflection. The timing of the change in polarity marks the timing of the arrival of the activation front with little ambiguity, and the actual slope of the change on potential is dependent upon wave thickness, its speed of propagation, and, possibly, electrode size if it is large.

This concept is illustrated in Figure 13.12. A unipolar lead records potentials on the epicardium as activation proceeds from the endocardium to the epicardium. As the wavefront enters the subjacent ventricular wall (18 ms) and moves across the ventricular wall, epicardial potentials rise. Once the wavefront reaches the epicardium (42 ms), that is, the recording site, the unipolar lead records an abrupt fall in potential. This deflection has been called the *intrinsicoid deflection*, having a significance analogous to that of the intrinsic deflection described above.

B. Direct Cardiac Recording Systems

Electrophysiological recordings may be made from electrodes placed directly on the epicardium, from electrodes implanted within the myocardium, and from electrodes located within the cardiac chambers. Epicardial electrodes of various sizes and shapes may be unipolar or bipolar, depending on the specific need. Single or single pairs of electrodes may be placed on a hand-held probe or mounted on intravascular catheters, 60 or more electrodes may be enmeshed in a "sock" electrode placed around the entire heart, or multiple electrodes may be mounted on devices passed transvascularly into the cardiac chambers. For a comprehensive recording of epicardial activity, a minimum of 64 electrodes is needed, although as many as 1124 have been used (13). With newer optical techniques, the number of sites examined may reach 16,000.

Plunge electrodes, with numerous electrodes located on the shaft of a needle that is plunged through the ventricular wall (Fig. 13.13), are commonly used to record intramural extracellular potentials. Typically, 10 to 30 electrodes are placed on a single needle. A full, three-dimensional intramural map of cardiac activity requires a minimum of 128 such electrodes, as depicted in Fig. 13.14.

Recording from two adjacent electrodes on the plunge electrode results in a bipolar lead, whereas potentials from each electrode (with reference to a remote electrode) will yield a unipolar recording (Fig. 13.11). Timing of activation of tissue near each electrode or electrode pair can be determined from either configuration, as detailed above. Unipolar configurations permit recording of potential amplitudes, as well as timing, so that intramural and epicardial potential distributions can be determined as well as the timing of excitation. The relative advantages of each will be considered later in this chapter.

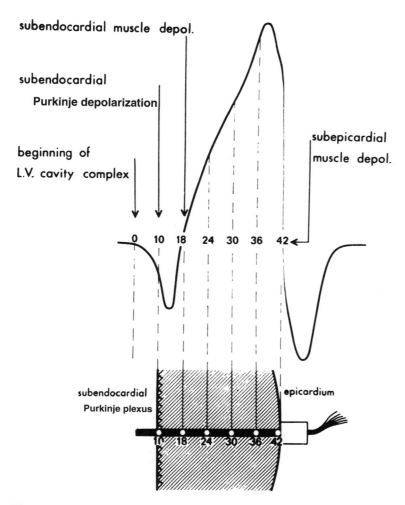

Figure 13.12   Unipolar recording from an epicardial electrode (top) of a transmural plunge electrode (bottom) during activation of the left ventricular free wall. Events labeled on the waveform are timed relative to the onset of endocardial ventricular activation. Activation of the muscle directly subjacent to the epicardial electrode is associated with the timing of the most rapid fall in recorded voltage. (Reprinted from Durrer D. Electrical aspects of human cardiac activity: a clinical-physiological approach to excitation and stimulation. Cardiovasc Res 1968; 2:1, with permission from Elsevier Science.)

C.   Body Surface Electrocardiographic Lead Systems

The standard 12-lead electrocardiogram is derived from three types of leads—three *bipolar limb leads* (leads I, II, and III), *six precordial unipolar leads* (leads V_1 through V_6), and three modified unipolar limb leads (*the augmented limb leads*, aVR, aVL, and aVF). The locations of the electrodes and the definition of the positive and negative inputs for each lead are listed in Table 13.1.

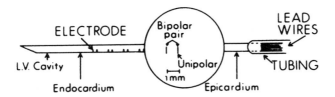

Figure 13.13 Schematic representation of a plunge electrode for recording intramural potential. Electrodes are in pairs, separated by 1.0 mm. A bipolar recording may be made by registering the potential difference between members of a pair or a unipolar recording may be registered by comparing the potential at one electrode to a remote reference electrode. (From Claydon FJ, Pilkington TC, Ideker RE. Classification of heart tissue from bipolar and unipolar potentials. IEEE Trans Biomed Eng 32:513, 1985, with permission.)

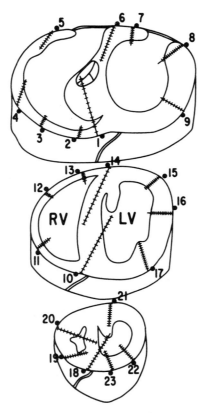

Figure 13.14 Locations of 23 intramural plunge electrodes for recording detailed activation and recovery pattern. Each plunge electrode contains 15 electrodes placed 1–2 mm apart. (From Ref. 4, with permission.)

Table 13.1 Location of Electrodes and Lead Configurations for the Standard
12-Lead Electrocardiogram

Lead type, lead	Positive input	Negative input
Unipolar limb leads		
Lead I	Left arm	Right arm
Lead II	Left leg	Right arm
Lead III	Left leg	Left arm
Augmented unipolar limb leads		
aVR	Right arm	Left arm + left leg
aVL	Left arm	Right arm + left leg
aVF	Left leg	Left arm + left arm
Precordial leads		
V_1	Right sternal margin, 4th intercostal space	Wilson central terminal
V_2	Left sternal margin, 4th intercostal space	Wilson central terminal
V_3	Midway between V_2 and V_4	Wilson central terminal
V_4	Left midclavicular line, 5th intercostal space	Wilson central terminal
V_5	Left anterior axillary line, same level as V_4	Wilson central terminal
V_6	Left midaxillary line, same level as V_4	Wilson central terminal

1. Bipolar Limb Leads

The bipolar limb leads record the potential differences between two limbs, as detailed in
Table 13.1. As bipolar leads, the output measures the potential difference between the positive and negative inputs and limbs. The electrical connections for these leads are such that
the potential in lead II equals the sum of potentials sensed in lead I and lead III. This relation is known as *Einthoven's law*.

2. Unipolar Precordial Leads and the Wilson Central Terminal

The unipolar precordial leads register the potential at each of the six designated torso sites
in relation to a theoretical zero reference potential. To do so, the exploring electrode for
recording the precordial leads is connected to the positive input of the recording system.

The negative or reference input is composed of a *compound electrode* (i.e., a configuration of more than one electrode connected electrically) known as the *Wilson central terminal* (16). This is formed by combining the outputs of the left arm, right arm, and left leg
electrodes through 5000-Ω resistances. The result is that each precordial lead registers the
potential at a precordial site with reference to the average potential on the three limbs. The
potential recorded by the Wilson central terminal remains relatively constant during
the cardiac cycle, so that it functions as reference potential. The output of a precordial lead
is then determined predominantly by changes in the potential at the precordial site.

3. Augmented Unipolar Limb Leads

The three augmented limb leads, aVR, aVL, and aVF, are modified unipolar leads. The exploring electrode is on the right arm (aVR), the left arm (aVL), or the left foot (aVF). It is the reference electrode that is modified. Instead of consisting of a full Wilson central terminal composed of outputs from three limb electrodes, the reference potential is the mean potential of the two limb electrodes not including the electrode used for the exploring electrode (17). Thus, for lead aVL, the exploring electrode is on the left arm and the reference electrode is the mean output of the electrode on the right arm and the left foot. Similarly, for lead aVF, the reference potential is the mean of the output of the two arm electrodes.

This modified reference system was designed to increase the amplitude of the output. Outputs tended to be small, in part because the same electrode potential was included in both the exploring and reference potential inputs. Eliminating this duplication results in a theoretical increase in amplitude of 50%.

4. Other Lead Systems

Other lead systems may be used for specific purposes. For example, additional right precordial unipolar leads may be used to assess right ventricular lesions and a vertical parasternal bipolar pair may facilitate detection of P waves for diagnosing arrhythmias. Precordial and anterior–posterior thoracic electrode arrays of up to 150 or more electrodes may be used to display the spatial distribution of body surface potentials as *body surface isopotential maps* (18), as described below. Isopotential maps permit analysis of abnormalities not seen in temporal patterns of the standard electrocardiogram or that may be manifest only from areas of the chest not recorded from by the six standard precordial leads (19).

D. Lead Theory: Lead Vectors and Lead Fields

A lead can be represented as a vector known as the *lead vector*. For a simple bipolar lead, the lead vector is directed from the negative electrode toward the positive one. For a unipolar lead, the vector is directed from the reference electrode site to the exploring electrode location. If the reference electrode for the unipolar lead is a compound electrode such as the Wilson central terminal, the origin of the lead vector lies on the axis connecting the electrodes that make up the compound electrode. That is, for lead aVL, the vector points from the midpoint of the axis connecting the right arm and left leg electrodes toward the left arm. For the precordial leads, the lead vector points from the center of the torso to the precordial electrode site. Lead vectors for the three limb leads, the three augmented unipolar limb leads, and the unipolar precordial leads are shown in Figure 13.15.

Instantaneous cardiac activity may also be approximated as a single dipole representing the vector sum of the various active wavefronts. This vector is known as the *heart vector*. This vector sum is only an approximation that distorts the information contained in the double-layer distribution, but is useful in providing a first approximation to the cardiac source.

The potential sensed in a lead is the dot or scalar product of the heart vector and the lead vector, or

$$V_L = \mathbf{L} \cdot \mathbf{H}, \tag{2}$$

where $\mathbf{L}$ is the lead vector, $\mathbf{H}$ is the heart vector, and V_L is the potential recorded in lead L.

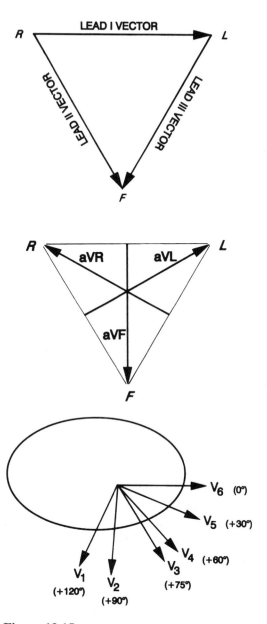

Figure 13.15 Lead vectors for the three bipolar limb leads (top), augmented unipolar limb leads aVR, aVL, and aVF (middle), and for the precordial unipolar leads (bottom). Vectors for bipolar limb leads point from the negative to the positive electrode of each pair. For augmented unipolar limb leads, the vector points from a locus midway between the two electrodes making up the reference lead to the active extremity electrode. For precordial leads, the vector is directed from the middle of the torso to the precordial electrode sites.

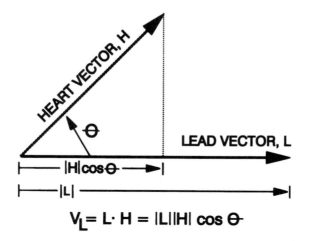

$$V_L = L \cdot H = |L||H| \cos \Theta$$

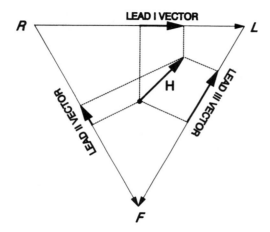

Figure 13.16 Geometric representation of the lead vector concept. (Top) Depiction of the equation for computing the potential in a lead as the dot product of the lead vector and the heart vector. (Bottom) The projection of a heart vector (**H**) on the lead vectors of the three limb leads.

The dot product equals the length of the projection of the heart vector on the lead vector multiplied by the length of the lead vector, or

$$V_L = |L||H|(\cos \theta), \tag{3}$$

where L and H equal the length of the lead and heart vectors, respectively, and θ is the angle between the two vectors. This relation is depicted in Figure 13.16 for the three limb leads.

An additional complexity occurs because the lead vector varies with the location of the heart vector. The result is a *lead vector field*, rather than a single lead vector for each electrocardiographic lead. The lead vector field (Fig. 13.17) depicts the lead vector for a particular lead for all locations of the heart vector. The potential recorded by an electrode then equals the dot product of the heart vector and the lead vector field at the location of the heart vector.

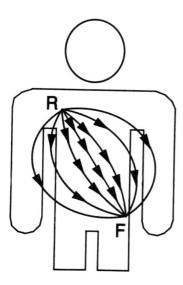

Figure 13.17 The lead vector field for lead II. Lines connect sites with equal lead vector strengths, with the number of arrowheads indicating the relative strength of each line. R and F represent the right arm and left foot electrode sites, respectively. (Modified from Mirvis DM. Body Surface Electrocardiographic Mapping. Boston: Kluwer Academic Publishers, 1988.)

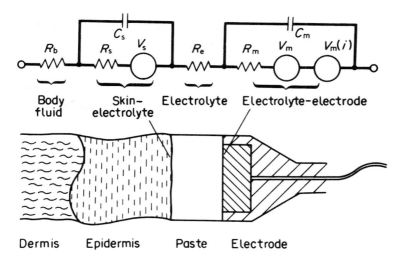

Figure 13.18 Physical model of the skin-electrode interface (bottom) and the equivalent electronic circuit (top). The circuit includes resistances (R_b, R_s, R_e, and R_m), capacitances (C_s and C_m), and voltages (V_s, V_m, and $V_{m(i)}$). (From Ref. 20, with permission.)

Because this field varies in strength for different locations of the heart vector, a lead will register different potentials for a heart vector (i.e., cardiac activity, of constant strength) that is located in different sites within the heart. This means that leads have regional sensitivity (i.e., they are more sensitive to cardiac activity in certain areas of the heart than in others).

A final complexity is the common occurrence of electrical activity in more than one cardiac region. Because the different sites have different lead vector field strengths and orientations, the recorded potential is not the simple sum of the potentials generated by each. Rather, the result is the sum of the dot products of the heart vectors and the lead vector field for each.

VI. ELECTROPHYSIOLOGICAL AND ELECTROCARDIOGRAPHIC DISPLAY SYSTEMS

A. Electrode Characteristics

Electrodes placed on or within various tissues and structures to sense the cardiac electrical field are not passive devices that merely detect the field. Rather, they are complex systems that alter the potentials that they sense (20). The complexities of skin electrodes are shown in Figure 13.18, and include the properties of the dermal and epidermal layers of the skin, the electrolytic paste applied to the skin, and the electrode itself. The net effect is a complex electrical circuit produced by these different components and the interfaces between them that includes resistances, capacitances, and voltages. Each of these modify the cardiac field potentials that reach the electrodes themselves before they are displayed as an electrocardiogram.

B. Amplifier Characteristics

Another group of factors that modifies the cardiac field includes the characteristics of the electronic systems used to amplify, filter, and digitize the recorded signals. The major features include *gain, frequency response, coupling, and digitizing rate.*

1. Gain

Electrocardiographic amplifiers are *differential amplifiers* (i.e., they amplify the difference between two inputs). For bipolar leads, the differential output is the difference between the two active leads; for unipolar leads the difference is between the exploring or active lead and the reference lead. The major advantage of this differential configuration is the significant reduction of electrical noise that is sensed by both inputs and hence canceled. The standard amplifier gain for routine electrocardiography is 1000, but may vary from 500 (*half-standard*) to 2000 (*double standard*). For direct cardiac recordings, gain settings are not standardized but reflect the specific requirements of the study at hand and the electronic devices being used.

2. Frequency Response

Amplifiers respond differently to the range of signal frequencies included in an electrophysiological signal. The *bandwidth* of an amplifier defines the frequency range over which the amplifier accurately amplifies the input signals. Waveform components with frequencies above or below the bandwidth may be artifactually reduced or increased in amplitude, or they may be delayed in transmission in relation to other frequencies.

In addition, recording devices include high- and low-pass filters that intentionally reduce the amplitude of specific frequency ranges of the signal. For example, this may be done to reduce the effect of body motion or line-voltage frequencies (i.e., 60-Hz interference). For routine electrocardiography, the standards of the American Heart Association require a bandwidth of 0.05 to 100 Hz (21).

Filters are also incorporated into digital recording systems to avoid *aliasing*. If frequencies greater than one-half the digital sampling rate (see below) are permitted to be recorded, the higher frequencies are folded back onto the lower frequency components, producing distortions. Thus, signals are filtered at a frequency less than one-half the sampling rate before digitization.

3. Coupling

Amplifiers for routine electrocardiography include a capacitor stage between the input and the output pole (i.e., they are *capacitor coupled*). The major impact of this configuration is the blocking of DC voltages while permitting flow of AC signals. Because the electrocardiographic waveform may be analyzed as an AC signal (that accounts for the waveform shape) that is superimposed upon a DC baseline (that determines the actual voltage levels of the recording), this coupling has significant effects on the recording process. First, unwanted DC potentials, such as those produced by the electrode interfaces, are blocked from affecting the electrocardiogram. Second, and more importantly, the elimination of the DC potential level from the final product means that the electrocardiographic potentials are not calibrated against an external reference level. Rather, electrocardiographic potentials are measured in relation to an internal standard. Thus, amplitudes of electrocardiographic waves are measured in millivolts relative to another portion of the waveform, for example, the TP segment (occurring between the end of the T wave and the onset of the P wave), rather than in relation to ground potentials. Finally, capacitive coupling decreases the possibility that dangerous ground loops will include current flow to the patient, especially in the case of component failure in the power supply or when the patient is connected to multiple devices.

4. Digitizing Rate

An additional property is the *digitizing* or *sampling rate* for computerized systems. The rate must be high enough to adequately sample brief signals such as notches in QRS complexes or brief bipolar spikes, but not so fast as to introduce artifacts, include high-frequency noise, or require excessive digital storage capacities (22,23). The lower the sampling rate, the lower will be the precision and the accuracy of estimating local activation times and the less reliable will be the reproduction of waveform morphologies. In general, the sampling rate should be at least twice the frequency of highest frequencies of interest in the signal being recorded.

Standard electrocardiography is most commonly performed with a sampling rate of 500 Hz, with each sample representing a 2-ms period. Suggested sampling rates for direct cardiac recordings vary from as low as 250 Hz for activation time mapping, to as high as 2000 Hz for epicardial recordings and 15,000 Hz for recording from the specialized conduction system if waveform morphology is to be accurately registered (14,22).

C. Display Formats

Cardiac potentials may be processed for display in numerous formats. The most common of these formats includes the *classical scalar electrocardiogram, isopotential maps*, and *isochrone maps*.

1. Scalar Recordings

Scalar recordings, such as the standard electrocardiographic display, depict the potential recorded from one lead, either unipolar or bipolar, as a function of time (e.g., as in the upper panel of Fig. 13.11). For standard electrocardiography, amplitude is displayed on a

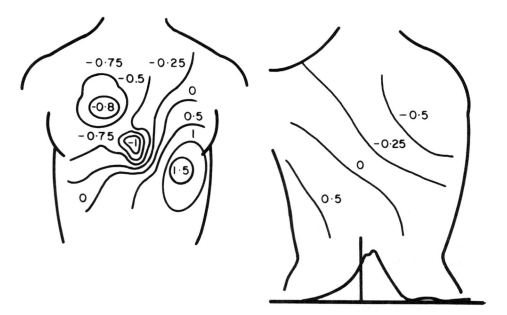

Figure 13.19 Body surface potential maps depicting the distribution of potentials at one instant, marked on the scalar waveform, on the surface of the anterior (left) and posterior (right) torso. Lines connect sites on the torso with equal potential relative to the Wilson central terminal. (From Liebman J, Plonsey R, Gillette PC. Pediatric Electrocardiography. Baltimore: Williams and Wilkins, 1982, with permission.)

scale of 1-mV to 10-mm vertical displacement and time as 200 ms/cm on the horizontal scale.

2. Isopotential Maps

Isopotential maps (Fig. 13.19) are constructed from potentials recorded from multiple unipolar leads and displayed as a series of *isopotential lines*. These lines connect sites at equal potential relative to the reference potential at one instant during cardiac activity. A different isopotential map can thus be constructed for each instant in time during cardiac activation and recovery. Body surface maps have two major advantages over scalar recordings (19). First, they permit analysis of the spatial distribution of potentials on the body's surface rather than only temporal information. Cardiac abnormalities may often produce spatial changes in the cardiac field that are not detectable from analysis of only temporal patterns. Second, because body surface mapping uses electrodes placed on right and posterior regions of the torso not sampled in routine electrocardiography, it may detect abnormalities projected only to these torso sites.

3. Isochrone Maps

Isochrone maps depict the sequence of activation of the heart as determined from multiple unipolar or bipolar leads on the epicardial surface of the heart or within intramural tissues (Fig. 13.20) An isochrone map outlines the locations of recording sites in relation to anatomic landmarks on the heart, and presents local activation at each site as a number.

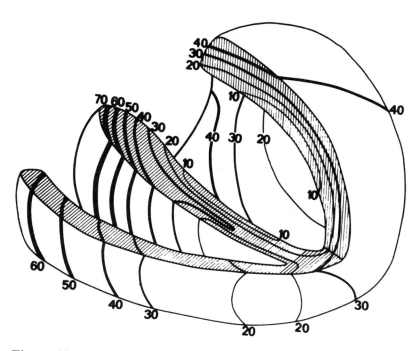

Figure 13.20 An isochrone map showing the sequence of activation of the human heart. Isochrone lines connect sites undergoing activation at the same instant, as noted in milliseconds after the earliest detectable evidence of ventricular activation. Portions of the free walls of the ventricles and of the septum have been removed. (From Durrer D. Electrical aspects of human cardiac activity: a clinical-physiologic approach to excitation and stimulation. Cardiovasc Res 1968; 2:1, with permission.

Lines connect sites activated at the same point in time after the earliest detected electrical activity (time 0).

Potential and isochrone maps differ in several fundamental features (4). Isochrone maps provide information only about activation times; no information is presented about the strength of a wavefront, only about its position at a given instant. Potential maps, in contrast, provide intensity information as potential amplitudes. Isochrone maps can only be applied to activation forces since, as described above, repolarization is a nonpropagating event. Potential maps, in contrast, can be applied to study both activation and recovery.

VII. THE ELECTROCARDIOGRAM

The waveforms and intervals that make up the standard electrocardiogram are displayed in the diagram in Figure 13.21 The P wave is generated by activation of the atria, the PR interval represents the duration of atrioventricular conduction, the QRS complex is produced by activation of both ventricles, and the ST-T wave reflects ventricular recovery. Although a detailed description of the waveforms of the electrocardiogram is beyond the scope of this chapter, two features associated with cardiac arrhythmias will be presented.

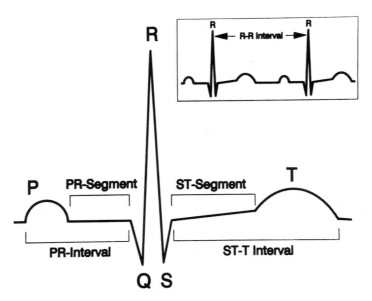

Figure 13.21   The waves and intervals of the normal electrocardiogram. (Modified from Mirvis DM, ed. Body Surface Electrocardiographic Mapping. Boston: Kluwer Academic Publishers, 1988.)

A.   QT Interval

The QT *interval* extends from the onset of the QRS complex to the end of the T wave and includes the full duration of ventricular activation and recovery. The timing of the onset of ventricular activation and the end of ventricular recovery can be assessed by the *activation recovery interval*, measured from the instant of the maximum negative rate of change of voltage (i.e., –dV/dt) during the QRS complex to the corresponding instant during the T wave (24).

Two important features of the QT interval are its rate sensitivity and its variation in different electrocardiographic leads. The normal QT interval, like the duration of the action potentials it reflects, changes in direct relation to cycle length (i.e., the QT interval shortens as heart rate rises and lengthens as heart rate falls). Various formulas have been reported to correct for this rate sensitivity to yield a *corrected QT interval*. The most commonly used formula is the *Bazett formula*, defined as

$$QT_c = k \sqrt{RR}, \tag{4}$$

where QT_c is the corrected QT interval, k is a constant equal to 0.397 for men and 0.415 for women, and RR is the RR interval.

The second characteristic of the QT interval is that the measurement differs in different leads, leading to a measure of *QT interval dispersion* (25,26). The QT interval has been shown to vary significantly from lead to lead (25). Because the QT interval is a measure of action potential duration, the difference in QT interval durations among the ECG leads may then quantify the inhomogeneity of recovery times throughout the heart (26). Such inhomogeneity, in turn, reflect a tendency to ventricular arrhythmias in conditions such as congenital long-QT-interval syndrome, acute myocardial infarction, and congestive heart failure. The significance of this phenomenon in addition to, or in association with, prolongation of the QT interval will be considered in a subsequent chapter.

B. Ventricular Gradient

The concept of the *ventricular gradient* was developed to emphasize recovery properties (27). In the case of simple linear cardiac fiber with uniform action potential shapes, recovery forces would be the simple opposite of activation forces, so that the sum of potentials generated by activation (the QRS complex) and those generated by recovery (the ST-T wave) would be zero. According to the concept of the ventricular gradient, differences in recovery properties in regions of the heart cause the sum of QRS and ST-T potentials to differ from zero, so that the size of the QRST area is a reflection of the extent of regional differences in recovery properties. Since the QRST area is dependent essentially on recovery properties alone and is not affected by activation sequence, the ventricular gradient should be independent of the ventricular activation sequence. Since the QRST area is determined primarily by dispersion of recovery, it may reflect vulnerability to arrhythmias (28), as will be discussed in a later chapter.

REFERENCES

1. Barr RC. Genesis of the electrocardiogram. In: Macfarlane PW, Lawrie TDV, eds. Comprehensive Electrocardiography. New York: Pergamon Press, 1989.
2. Durrer D, van Dam RTh, Freud GF, Janse MJ, Meijler FL, Arzbaecher RC. Total excitation of the human heart. Circulation 1970; 41:899–912.
3. Holland RP, Arnsdorf MF. Solid angle theory and the electrocardiogram: physiologic and quantitative interpretations. Prog Cardiovasc Dis 1977; 19:431–457.
4. Spach MS, Barr RC. Ventricular intramural and epicardial potential distributions during ventricular activation and repolarization in the intact dog. Circ Res 1975; 37:243–257.
5. Abildskov JA. The sequence of normal recovery of excitability in the dog heart. Circulation 1975; 52:442–446.
6. Mirvis DM. Physiology and biophysics in electrocardiography. J Electrocardiol 1996; 29:175–177.
7. Lepeschkin E. Physiological influences on transfer factors between heart currents and body surface potentials. In: Nelson CV, Geselowitz DB, eds. The Theoretical Basis of Electrocardiography. Oxford: Clarendon Press, 1976.
8. Spach MS, Dolber PC. Relating extracellular potentials and their derivatives to anisotropic propagation at a microscopic level in human cardiac muscle. Circ Res 1986; 58:356–371.
9. Rudy Y, Plonsey R. A comparison of volume conductor and source geometry effects on body surface and epicardial potentials. Circ Res 1980; 46:283–291.
10. Brody DA. A theoretical analysis of intracavitary blood mass influence on the electrocardiogram. Circ Res 1956; 4:731–738.
11. Abildskov JA, Burgess MJ, Lux RL, Wyatt RF. Experimental evidence for regional cardiac influence in body surface isopotential maps of dogs. Circ Res 1976; 38:386–391.
12. Abildskov JA, Klein RM. Cancellation of electrocardiographic effects during ventricular excitation. Circ Res 1962; 11:247–251.
13. Ideker RE, Smith WM, Blanchard SN, Reiser SL, Simpson EV, Wulf PD, Danieley ND. The assumptions of isochronal cardiac mapping. Pacing Clin Electrophysiol 1989; 12:456–478.
14. Breithardt G, Shenasa M, Bermann M, Borggrefe M, Haverkamp W, Hindricks G, Vogt B, Reinhardt L. Precision and reproducibility of isochronal electrical cardiac mapping. In: Shenasa M, Borgreffe M, Breithardt G, eds. Cardiac Mapping. Mt Kisco: Futura, 1993.
15. Biermann M, Shenasa M, Borggrefe M, Hindricks G, Haverkamp W, Breithardt G. The interpretation of cardiac electrograms. In: Shenasa M, Borgreffe M, Breithardt G, eds. Cardiac Mapping. Mt Kisco: Futura, 1993.

16. Wilson FN, MacLeod AG, Barker PS. Electrocardiographic leads which record potential variations of the heart at a single point. Proc Soc Exp Biol Med 1932; 29:1011–1012.
17. Goldberger E. A simple indifferent, electrocardiographic electrode of zero potential and a technique of obtaining augmented, unipolar extremity leads. Am Heart J 1942; 23:483–492.
18. Taccardi B. Distribution of heart potentials on the thoracic surface of normal human subjects. Circ Res 1963; 12:341–352.
19. Mirvis DM. Current status of body surface electrocardiographic mapping. Circulation 1987; 75:684–688.
20. Zywietz C. Technical aspects of electrocardiogram recording. In: Macfarlane PW, Lawrie TDV, eds. Comprehensive Electrocardiography. New York: Pergamon Press, 1989.
21. Pipberger HV, Arzbaecher RC, Berson AS, Briller SA, Brody DA, Flowers NC, Geselowitz DB, Lepeschkin E, Oliver GC, Schmitt OH, Spach MS. Recommendations for standardization of leads and of specifications for instruments in electrocardiography and vectorcardiography. Circulation 1975; 52:11–31.
22. Barr RC, Spach MS. Sampling rates required for digital recording of intracellular and extracellular potentials. Circulation 1977; 55:40–48.
23. Bailey JJ, Berson AS, Garson A, Horan LG, Macfarlane PW, Mortara DW, Zywietz C. Recommendations and specifications in automated electrocardiography: bandwidth and digital signal processing. Circulation 1990; 81:730–739.
24. Millar CK, Kralios FA, Lux RL. Correlation between refractory periods and activation-recovery intervals from electrograms: effects of rate and adrenergic interventions. Circulation 1985; 72:1372–1379.
25. Mirvis DM. Spatial variation in QT intervals in normal subjects and in patients with acute myocardial infarction. J Am Coll Cardiol 1985; 5:625–631.
26. Day CP, McComb JM, Campbell RWF. QT dispersion: an indicator of arrhythmic risk in patients with long QT intervals. Br Heart J 1990; 63:342–344.
27. Wilson FW, Macleod AG, Barker PS, Johnston FR. The determination and the significance of the areas of the ventricular deflections of the electrocardiogram. Am Heart J 1934; 10:46–61.
28. Urie PM, Burgess, MJ, Lux RL, Wyatt RF, Abildskov JA. The electrocardiographic recognition of cardiac states at high risk of ventricular arrhythmias. An experimental study in dogs. Circ Res 1978; 42:350–358.

14

Global Behaviors of Cardiac Activation

YORAM RUDY

Case Western Reserve University, Cleveland, Ohio

JOSÉ JALIFE

SUNY Upstate Medical University, Syracuse, New York

I. INTRODUCTION

Propagation of excitation in cardiac tissue is determined by the relationship between the availability of depolarizing charge (source) and the amount of charge required for successful propagation (sink). This relationship reflects a complex interaction between membrane ion currents that generate the action potential and the structural properties of the tissue that determine the electric load on a depolarizing cell. Both membrane factors (e.g., excitability, defined in terms of sodium current availability) and structural factors (e.g., conductance and distribution of gap junctions) can be affected by pathology. In this chapter, we describe processes and principles that govern cardiac excitation and arrhythmia. We do so at progressively increasing levels of integration and complexity, starting with the single cardiac cell. Extensive use is made throughout the chapter of mathematical models and computer simulations. Theoretical model simulations are designed to be both reductionist and integrative. In the reductionist mode, modeling is used to determine the role of component processes in observed integrated behavior (e.g., the role of individual ion channels in generating the whole cell action potential; the kinetics of specific ion channels and the role of gap junctions during slow conduction in the multicellular tissue). In the integrative mode, observed component behavior (e.g., modification of channel function by drug or mutation) is introduced into an integrated model (the whole cell, the multicellular tissue) to evaluate and predict its global electrophysiological consequences. This application is of particular importance since many components are studied experimentally in isolation, away from their physiological environment (e.g., cloned ion channels in expression systems). Similar to simplified experimental preparations such as membrane patches, isolated cells, and isolated tissue, theoretical models of single cells, one-dimensional strands, and two- and three-dimensional tissue are extremely helpful in providing mechanistic insights and quantitative characterization of fundamental phenomena that determine excitation of the whole heart. These principles can then be applied to the analysis of global cardiac excitation. The models in this chapter are introduced at progressively increasing levels of complexity from the single cell to the whole heart, providing description of phenomena with a step-by-step increase in the level of integration.

II. CARDIAC EXCITATION BEGINS IN THE SINGLE CELL

The process of cardiac excitation occurs at several levels of integration and of increasing complexity. Cardiac muscle is constructed of many individual cells (myocytes) that are interconnected by tubular protein structures (gap junctions) that allow cells to communicate electrically (see Chap. 8). The process of excitation involves generation of an electrical impulse, the action potential, by individual cells and its conduction from cell to cell through the gap junctions. The resulting propagating wave of electrical activation triggers contraction of the heart and synchronizes its blood-pumping action. At the cellular level, the electrical action potential triggers mechanical contraction by inducing a transient increase of the intracellular calcium concentration which, in turn, carries the contraction message to the contractile elements of the cell. The process of excitation–contraction coupling is discussed in Chapter 12.

A. Action Potentials Are Generated by a Dynamic Balance of Membrane Currents

The single myocyte is the building block of cardiac tissue. Generation of the action potential by a single myocyte involves complex interactions between membrane ion channels and the cellular environment. The biophysical basis of action potential generation is described in Chapters 2 to 5. Here we use a mathematical model of the ventricular myocyte [the Luo–Rudy (LRd) model] (1–4) to demonstrate that this process involves a dynamic interplay between different ion channels, pumps, and exchangers. A schematic diagram of the LRd cell model is shown in Figure 14.1, together with an action potential and the accompanying calcium transient generated by the model. Experimentally measured action potential and calcium transient (5) are shown for comparison.

In Figure 14.2 we describe the role played by various ion currents in generating the action potential and in determining its duration and morphology. The action potential and calcium transient are shown, together with the time course of selected ion currents during the action potential. Once the cell is excited to threshold, the fast inward sodium current, I_{Na}, activates and depolarizes the membrane at a very fast rate (maximum rate of 382 V/s), generating the fast rising phase of the action potential. Subsequently, the inward L-type calcium current, $I_{Ca(L)}$, is activated to depolarize the membrane, supporting the plateau of the action potential against the outward repolarizing currents I_{Kr} and I_{Ks}. $I_{Ca(L)}$ displays a "spike and dome" morphology during the action potential. It reaches its early peak value of -4.92 μA/μF in 2.74 ms, whereas I_{Na} reaches its peak value of -381 μA/μF (two orders of magnitude larger) in 1 ms, a time when $I_{Ca(L)}$ is still very small (only -0.84 μA/μF). Therefore, the early spike of $I_{Ca(L)}$ contributes very little to the rising phase of the action potential; it plays an important role in triggering calcium release from the sarcoplasmic reticulum to generate the calcium transient. It is the dome portion of $I_{Ca(L)}$ that acts to maintain the action potential plateau. Both I_{Kr} and I_{Ks} are outward currents that act to repolarize the

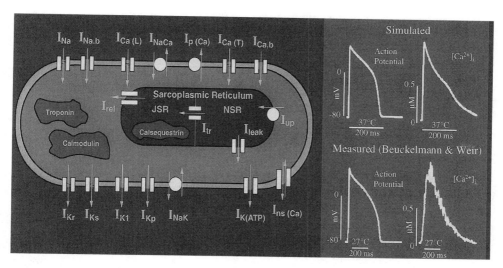

Figure 14.1 Schematic diagram of the Luo–Rudy (LRd) ventricular cell model (left). Action potential and calcium transient, $[Ca^{2+}]_i$, simulated by the model are shown on the right in comparison with experimental data measured by Beuckelmann and Wier. (Adapted from Ref. 5, with permission.) In the LRd model, the time and voltage dependence of the major ion channels are formulated with Hodgkin–Huxley-type formalism with additional $[K^+]_0$-dependent conductances of I_{Kl}, I_{Kr}, and $I_{K(ATP)}$, $[Ca^{2+}]_i$-dependent conductance of I_{Ks}, and $[Ca^{2+}]_i$-dependent inactivation of $I_{Ca(L)}$. I_{Na}, fast sodium current; $I_{Ca(L)}$, calcium current through L-type calcium channels; $I_{Ca(T)}$, calcium current through T-type calcium channels; I_{Kr}, fast component of the delayed rectifier potassium current; I_{Ks}, slow component of the delayed rectifier potassium current; I_{Kl}, inward rectifier potassium current; I_{Kp}, plateau potassium current; $I_{K(ATP)}$, ATP-sensitive potassium current; I_{NaK}, sodium–potassium pump current; I_{NaCa}, sodium-calcium exchange current; $I_{P(Ca)}$, calcium pump in the sarcolemma; $I_{Na,b}$, sodium background current; $I_{Ca,b}$, calcium background current; $I_{ns(Ca)}$, nonspecific calcium-activated current; I_{up}, calcium uptake from the myoplasm to network sarcoplasmic reticulum (NSR); I_{rel}, calcium release from junctional sarcoplasmic reticulum (JSR); I_{leak}, calcium leakage from NSR to myoplasm; I_{tr}, calcium translocation from NSR to JSR. Calmodulin, Troponin and Calsequestrin are considered here as calcium buffers. Details of the LRd model are provided in Refs. 1–4.

membrane and bring it back to its rest potential. At the very beginning of the plateau, I_{Kr} increases faster than I_{Ks} due to its faster kinetic properties. However, I_{Ks} is larger in magnitude during most of the plateau, reflecting the greater density of I_{Ks} channels in the membrane and the inward rectification property of I_{Kr} that limits its magnitude. I_{NaCa} is initially a relatively small outward current (it operates in the "reverse mode" to extrude Na^+ and bring in Ca^{2+} with a stoichiometry of 3:1). It then reverses direction and operates in the "direct mode" to extrude Ca^{2+}, becoming a significant inward current that acts to slow the rate of repolarization during the late phase of the action potential and to prolong its duration. Finally, there is a large increase of I_{Kl}, which together with I_{Kr} and I_{Ks}, repolarizes the membrane back to its rest potential.

B. Single Cells: Source of Arrhythmogenic Activity

Figure 14.2 and the related discussion demonstrate how the precise time course and dynamic balance of inward and outward currents determine the morphology of the action po-

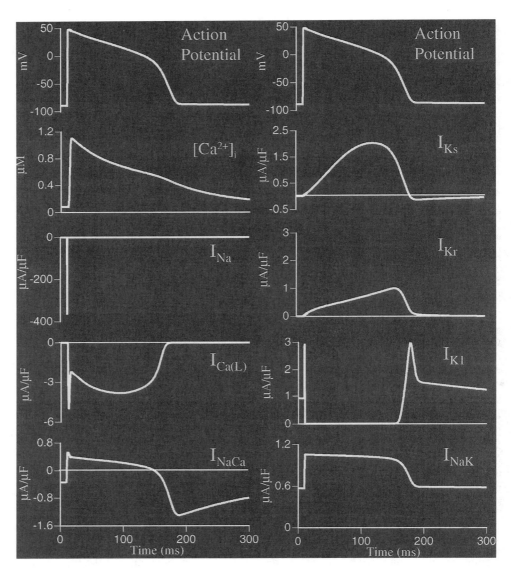

Figure 14.2 Major ion currents during the action potential. Shown are the action potential (repeated at top of both columns for reference), the calcium transient (free calcium in the myoplasm during the action potential, $[Ca^{2+}]_i$), and selected ion currents that determine the action potential morphology (current symbols are defined in Fig. 14.1). All quantities are simulated using the LRd model. (Compiled from Refs. 2 and 3, with permission.)

tential. Any shift of this balance can lead to an abnormal action potential morphology with possible arrhythmogenic consequences. Using the LRd model, we have shown that reactivation of $I_{Ca(L)}$ provides the mechanistic basis of arrhythmogenic early after-depolarizations (EADs) that occur during the plateau phase of a prolonged action potential (plateau EADs) (6–8). In contrast, EADs that occur during the late repolarization phase of the action potential result from activation of I_{NaCa} secondary to spontaneous calcium release from the sarcoplasmic reticulum (9). This I_{NaCa}-based mechanism also underlies the formation of delayed after-depolarizations (DADs) which occur following an action potential (9,10). EADs, DADs, and their role in cardiac arrhythmias are considered in Chapter 16.

In the rhythmically beating heart, cardiac cells are subject to periodic stimulation from their neighbors. The rate of periodic stimulation is modulated by neural inputs and other influences. During abnormally fast rhythms (tachycardia, flutter, fibrillation), a cell might not be able to respond to every stimulus with an action potential, thereby failing to maintain a 1:1 ratio between the number of stimuli and the number of successful responses. The inability to follow in a 1:1 fashion could result in local excitation failures that decrease the spatiotemporal organization of the cardiac activation process, exerting an arrhythmogenic influence. A periodic, rate-dependent activation failure in cardiac tissue is termed "Wenckebach periodicity" and was originally characterized clinically as a propagation-related phenomenon in the AV node (11). In this clinical setting, there is progressive prolongation of the PR interval in a series of beats (usually 3 to 8) with the final sinus beat in the sequence failing to propagate to the ventricle. More recently, it has been demonstrated that Wenckebach periodicity is an intrinsic property of single cardiac cells (1,12). Figure 14.3 shows the response of a cell to periodic stimulation at a basic cycle length (BCL) ranging from 160 to 2000 ms. Simulated behavior using the Luo–Rudy cell model (1) is shown in the left panels (A,B); corresponding experiments (12) are shown in the right panels (C,D). Figures 14.3(A) and (C) show activation patterns for different BCLs. The behavior is summarized in a staircase plot of response-to-stimulus (R:S) ratios versus BCL [14.3(B) (D)]. As the BCL of periodic stimulation decreases from 2000 ms (pacing rate increases), a monotonic decrease of the R:S ratio is observed. At fast pacing rates, the cell cannot maintain a 1:1 stimulus-to-response (S:R) ratio. Below a certain BCL, only every other stimulus elicits a successful response and an action potential, establishing a 2:1 pattern of stimulus to response. With further decrease of BCL (increase of pacing rate), only every third stimulus is able to elicit an excitatory response and a 3:1 pattern is established. Thus, the staircase pattern demonstrates phase locking at increasing integer S:R ratios (1:1, 2:1, 3:1, 4:1, etc.), with sharp transitions between these values through many noninteger S:R ratios (e.g., 8:7, 7:6, 6:5). The theoretical simulations (1) show that the excitation failure results from a progressive beat-to-beat decrease of membrane excitability due to residual activation (incomplete deactivation) of the delayed rectifier potassium current, I_K (the sum of I_{Kr} and I_{Ks}).

Figure 14.3 shows the response of a cardiac cell to periodic stimulation at extracellular potassium concentration $[K]_o = 5.4$ mM. The response pattern is even more complex at lower $[K]_o$. In Figure 14.4, the R:S staircase plot is shown for $[K]_o = 4$ mM. It displays a less regular behavior that deviates from the monotonically decreasing pattern of the staircase in Figure 14.3. A stable 1:1 pattern is followed by a region of nonmonotonic behavior as BCL is decreased below 458 ms (right arrow). This nonmonotonic region covers a range of BCLs from 458 to 415 ms. Below 415 ms, another regular region of a 2:1 pattern is obtained, followed by another region of nonmonotonic behavior (BCLs from 205 to 190 ms, left arrow). Note that the same patterns repeat for different regions of BCL. For example, a 1:1 pattern is observed for a BCL of 500, 438, and 428 ms. These regions are separated by regions of 2:1 patterns. A similar nonmonotonic behavior has been observed experimentally (12).

As the BCL is reduced below 205 ms, irregular activation patterns are observed. The response patterns are unstable and a regular periodicity is not established for more than 250 consecutive stimuli in the simulated behavior (similar aperiodic activity has been observed in experiments at BCL = 200 ms for recording periods that encompassed 100 or

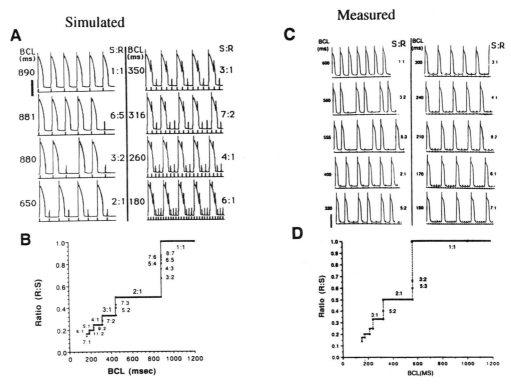

Figure 14.3 Periodic rate-dependent patterns of cellular responses at $[K^+]_0 = 5.4$ mM. (A) and (B) show simulated behavior using the Luo–Rudy model (1). (C) and (D) show corresponding measured behavior recorded by Chialvo et al. (12). (A) and (C) provide examples of activation patterns. Numbers on the left indicate basic cycle length of periodic stimulation (BCL); ratios on the right are the resulting stimulus-to-response (S:R) patterns. The train of stimuli is indicated below each pattern. (B) and (D) show the staircase plot of response-to-stimulus (R:S) ratios as a function of BCL. The numbers next to the staircase indicate the S:R ratios. (From Refs. 1 and 12, with permission.)

more stimuli) (12). The simulation studies reveal that the unstable patterns and chaotic responses of the single cell result from a complex interplay between the kinetic behavior of the repolarizing currents, I_K and I_{K1}. The interested reader is referred to Reference 1 for a detailed study of the ionic mechanism that underlies this behavior. The important message is that during fast rhythms (tachycardia), cardiac cells, due to their intrinsic properties, might develop complex rhythm patterns that decrease the spatiotemporal organization of the excitation process, thereby aggravating the arrhythmia.

III. PROPAGATION OF EXCITATION

In Section II, we demonstrated that the excitatory activity of the single cardiac cell is determined by a complex interplay between many interacting processes. These processes include the flow of different ions through membrane ion channels, pumps, and exchangers that are subject to regulatory processes and that are modulated through interactions with the membrane potential and with the dynamically changing ion environment of the cell

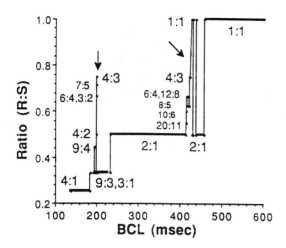

Figure 14.4 Simulation of cellular responses to periodic stimulation at low $[K^+]_0 = 4$ mM. Note the nonmonotonic behavior of the R:S staircase (arrows). Format is the same as in Figure 14.3(B). (From Ref. 1, with permission.)

(e.g., the calcium transient). The integrated response of the cell is the outcome of these highly interactive processes. Under abnormal conditions, we have demonstrated that cells can generate arrhythmogenic EADs and DADs (6–10). As shown in Section II, the cellular responses during an arrhythmia can be very complex and contribute to the arrhythmogenic behavior of the heart (see Fig. 14.4). In this section we increase the level of integration by allowing the cell to interact with other cells in a multicellular fiber.

A. Action Potential Conduction and Electrical Communication Between Cells

While substantial integration of component processes occurs at the level of the cell, a much higher level of integration occurs in the multicellular cardiac tissue and in the intact heart, where cells interact through gap junctions to generate a propagating wave of excitation. In this section, we use theoretical simulations of action potential conduction in a multicellular tissue model (13–15) (Fig. 14.5) to explain the principles of impulse propagation in the heart. Related experimental studies are described in Chapter 10.

In the multicellular tissue, an excited cell is electrically loaded by unexcited neighboring cells. The higher membrane potential of the excited, depolarized cell generates a spatial potential gradient that, following Ohm's law, results in a flow of current to the neighboring cells through intercellular gap junctions. Thus, the charge generated by the membrane ion currents in the excited cell is divided between discharging local membrane capacitance and further depolarizing that cell, and depolarizing neighboring cells toward their excitation threshold. In this sense, the excited cell serves as a source of electric charge, while the loading, nonexcited cells constitute an electric sink. In this context of source–sink relationship, we define an important index of propagation success or failure termed the safety factor (SF) for conduction (15). SF is defined as the ratio of charge generated for the multicellular tissue by a cell's excitation to the minimal amount of charge required to cause the excitation. SF > 1 indicates that more charge was produced by the cell than required to cause excitation. The fraction of SF > 1 indicates the margin of safety.

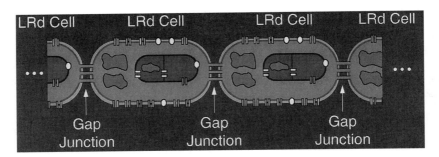

Figure 14.5 Theoretical multicellular cardiac fiber composed of a series LRd cells interconnected by resistive gap junctions. Fiber is 7 mm (70 cells) long. Conduction is initiated by an external stimulus applied to the proximal end of the fiber on the left (cell 1). Cellular parameters are studied for cells in the middle of the fiber, and conduction velocity is computed between cells 20 and 50.

When SF falls below 1, the charge requirements are not met and conduction fails. This definition of SF, computed at a particular cell, takes into consideration charge lost and gained with respect to the entire multicellular tissue. Simulations (presented in Figs. 14.8 and 14.9) use the safety factor concept to characterize action potential propagation under a variety of conditions.

Excitation of the multicellular fiber by a propagating action potential results in both membrane and intracellular responses (e.g., the transmembrane action potential and the intracellular calcium transient). Figure 14.6(A) shows the spatial profile of a computed action potential (AP) that propagates down the fiber under normal conditions of membrane excitability and gap junctional coupling. The normal conduction velocity is 54 cm/s; the AP profiles are shown at time t and $t + 3$ ms. The distance covered by the propagating AP in 3 ms is 1.6 mm (16 cells). Ahead of the AP upstroke there is a slow electrotonic depolarizing "foot" that precedes the local fast depolarization phase. The sharp upstroke to the left (upstream) of the foot is due to fast depolarization of local membrane by I_{Na}. Further upstream—to the left of the sharp upstroke, late upstroke and early plateau—depolarization is maintained by the L-type calcium current, $I_{Ca(L)}$.

The propagating action potential of Figure 14.6(A) triggers a chain of events that result in calcium release from the sarcoplasmic reticulum. Figure 14.6(B) shows the spatial calcium transient that accompanies the action potential of Figure 14.6(A). At each time instant (t and $t + 3$ ms) the action potential upstroke (wavefront) is about 25 cells ahead of the calcium transient. This spatial lag reflects the period of calcium accumulation that is required to trigger calcium release via the calcium-induced–calcium-release process (16) and the dynamics of calcium release by the sarcoplasmic reticulum (see Chap. 12). An important property of the excitatory process is the mutual interaction between the action potential and the calcium transient during propagation. Calcium entry through $I_{Ca(L)}$ during the late AP upstroke triggers the calcium transient. The resulting elevated $[Ca^{2+}]_i$ acts to reduce $I_{Ca(L)}$ via calcium-dependent inactivation (2,17) in a negative feedback fashion. It also serves to activate I_{NaCa} in its direct mode and to increase I_{Ks} conductance (3,18). Thus, the action potential triggers the calcium transient which, in turn, modulates ion processes that determine the AP. As will be shown below, $I_{Ca(L)}$ becomes an important current to sustain slow conduction caused by reduced gap junction coupling. Under such conditions, the calcium transient through its calcium-dependent inactivation of $I_{Ca(L)}$ may play an important role in action potential propagation.

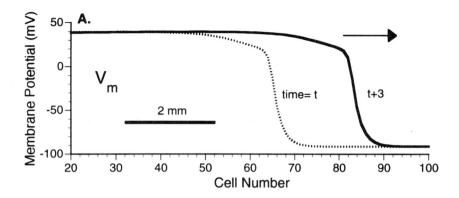

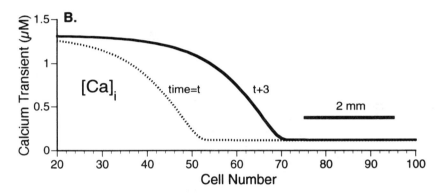

Figure 14.6 Action potential propagating down the multicellular fiber of Figure 14.5. (A) Head of action potential traveling from left to right at time = t ms (dashed line) and 3 ms later ($t + 3$, solid line). (B) Corresponding spatial profile of the intracellular calcium transient at the two time instants. The calcium transient lags behind the action potential upstroke. the spatial spread of calcium is not as sharp as the action potential upstroke due to calmodulin and troponin buffering of free myoplasmic calcium. (From Ref. 15, with permission.)

In addition to membrane factors and the intracellular environment, it is intuitively clear that gap junctions play an important role in propagation of the action potential since they control the flow of ions between cells. Figure 14.7 shows action potential profiles in two neighboring cells under conditions of normal gap junctional coupling (A) and a marginal (tenfold) decrease in coupling (B). For normal intercellular coupling, the gap junctional conductance between cells ($g_j = 2.5\ \mu S$) is roughly equal to the myoplasmic conductance of the entire cell. The result is a similar conduction time of 0.09 ms between cells as in crossing the entire cell length. Note, however, that the gap junctional dimension is only 80 Å, whereas the cell length is 100 μm. On a macroscopic scale of many cells, the result is an apparent uniform spread of excitation [Figure 14.7(A)]. With only tenfold reduction in gap junctional coupling [Figure 14.7(B)], a large (0.5 ms) conduction delay is introduced at each intercellular junction, while the entire cell depolarizes almost simultaneously as one unit. The almost simultaneous depolarization of the cell is due to increased confinement of depolarizing charge to the cell when intercellular coupling is reduced, decreasing loading effects and charge leakage out of the cell and maximizing charge avail-

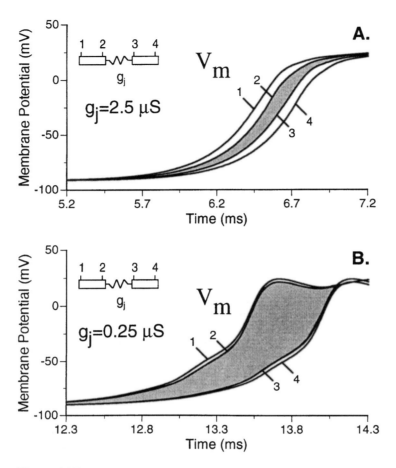

Figure 14.7 Transmission of action potential through intercellular gap junction. Action potential upstrokes from the edge elements (see inset) of two adjoining cells for intercellular conductance of 2.5 μS (A) and intercellular conductance of 0.25 μS (B). For control coupling (A), intercellular conduction delay (shaded) is approximately equal to intracellular (myoplasmic) conduction time. A tenfold decrease in intercellular conductance (B) increases intercellular conduction time and decreases intracellular conduction time dramatically, resulting in gap junction dominance of overall conduction velocity. (From Ref. 15, with permission.)

ability for local depolarization. Note [Fig. 14.7(B)] that propagation under these conditions is nonuniform (discontinuous), with long delays at gap junctions. In fact, the macroscopic conduction velocity over many cells is determined by the gap junctional delays rather than by the (negligible) time spent in traveling across individual cells. Of course, membrane processes are still essential for generation of the action potential, but its velocity of propagation is dominated by the gap junctions. This observation identifies the gap junctions as a potentially important target for antiarrhythmic intervention.

B. Reduced Membrane Excitability and the Occurrence of Conduction Block

Reduced membrane excitability is present during conditions of acute ischemia (19–22), tachycardia, and treatment with antiarrhythmic agents that block I_{Na}(23). Figure 14.8(A) shows conduction velocity and safety factor as a function of membrane excitability. Because excitability is determined by availability of sodium channels, membrane excitability is reduced by lowering g_{Na}, the maximum conductance of I_{Na}. Both velocity and SF decrease monotonically with decreasing excitability, indicating that as conduction slows it also becomes less safe. However, SF decreases slowly in the range of moderate excitability reductions. For extreme reduction of I_{Na} availability (<11.25%), the depolarizing charge needed to reach threshold increases dramatically and SF drops precipitously toward 1. For SF < 1, minimal depolarizing charge requirements for sustaining conduction are not met and propagation fails. The rapid decrease of SF when excitability is highly depressed reflects the nonlinear, "all-or-none" behavior of the cell membrane. Note that the minimum conduction velocity before failure is 17 cm/s, about one-third the control velocity (54 cm/s) with a fully excitable membrane.

Figures 14.8(B) and (C) explore the role of I_{Na} and $I_{Ca(L)}$ in action potential (AP) propagation under conditions of depressed membrane excitability. In Figure 14.8(B), SF is shown with (solid line) and without (dotted line) the contribution of $I_{Ca(L)}$. For most values of membrane excitability, the presence of $I_{Ca(L)}$ does not influence SF. Only at extreme reduction of excitability (<30%), $I_{Ca(L)}$ augments SF slightly, delaying somewhat the occurrence of conduction failure. Figure 14.8(C) shows the membrane potential, V_m, during the AP rising phase with (solid line) and without (dotted line) $I_{Ca(L)}$, for a severely depressed membrane (20% I_{Na} availability). The slightly higher amplitude of the AP with $I_{Ca(L)}$ acts to increase the electrotonic driving force (voltage gradient) and the depolarizing current to the downstream fiber that is still undergoing depolarization. The resulting increased supply of depolarizing charge slightly augments the SF of the extremely depressed membrane. Quantitatively, however, the effect is very small. The bar graph inset of Figure 14.8(C) shows relative charge generated by I_{Na} and $I_{Ca(L)}$ to depolarize the downstream fiber under these conditions. The computed charge ratio $Q_{Na}:Q_{Ca}$ is 75:1. Thus it can be concluded that conduction is predominantly maintained by I_{Na}, even when its availability is greatly reduced under conditions of extremely depressed membrane excitability.

C. Reduced Intercellular Coupling and Very Slow Conduction

It was shown in Figure 14.7 that, under conditions of decreased intercellular coupling, propagation of the action potential is a discontinuous process with long delays at the gap junctions. Uncoupling of cells due to reduced gap junctional conductance has been demonstrated to occur in mammalian heart tissue during chronic ischemia and infarction (24). In the normal myocardium, intercellular coupling is smaller in the directions transverse to the fibers, rather than along the fibers, resulting in anistropic conduction that is slower in the transverse direction (25) (Chaps. 8–10). Figure 14.9(A) shows conduction velocity and SF as a function of gap junctional coupling as simulated by the model. As gap junctional conductance (g_j) decreases, conduction velocity decreases monotonically. This behavior is more dramatic but qualitatively similar to that caused by reduced membrane excitability [Fig. 14.8 (A)]. However, in contrast to the reduced excitability case, SF does not decrease with the decrease in coupling and in conduction velocity. In fact, SF increases and attains a maximum value (close to 3) at $g_j = 0.023$ μS, a conductance that is about 100 times smaller than the normal value of 2.5 μS. At this level of reduced coupling, conduction is very slow (about 1/15 of normal velocity) but, paradoxically, very safe. Due to the high SF, extremely slow conduction velocities can be sustained by a fiber with greatly reduced intercellular coupling. The minimum velocity before block is 0.26 cm/s, a 200-fold decrease from the normal value for a well-coupled fiber. In comparison, the minimum velocity be-

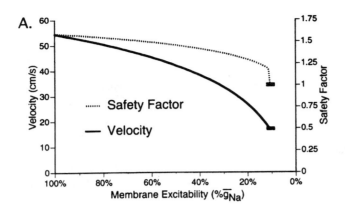

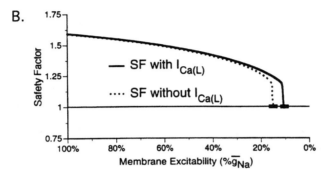

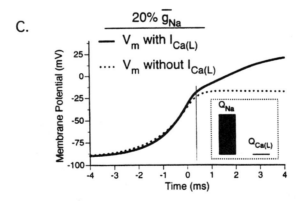

Figure 14.8 Conduction during decreased membrane excitability. (A) Conduction velocity (solid line) and safety factor (SF) of conduction, (dotted line) versus sodium channel availability (% g_{Na}). (B) SF computed over a range of membrane excitability with (solid line) and without (dotted line) $I_{Ca(L)}$. SF without $I_{Ca(L)}$ begins to diverge from SF with $I_{Ca(L)}$ at 30% g_{Na}. (C) High temporal resolution action potential upstrokes from a cell in the middle of the fiber of Figure 14.5 at greatly reduced excitability of only 20% g_{Na}, with (solid line) and without (dotted line) $I_{Ca(L)}$. Inset bar graph shows relative charge, Q (current integrated over time) generated by I_{Na} and $I_{Ca(L)}$ in the period from the middle cell's excitation to excitation of the adjoining cell (marked by thin vertical line at time = 0.37 ms). This charge computation reflects ion source charge during conduction. Even at severely reduced g_{Na}, charge contributed by I_{Na} is much larger than charge contributed by $I_{Ca(L)}$. Horizontal bars in (A) and (B) indicate conduction failure (note that conduction cannot be sustained when SF < 1, i.e., when the minimal charge requirements are not met). (From Ref. 15, with permission.)

fore block due to reduced membrane excitability is a relatively high 17 cm/s, one-third of control. Therefore, one can view reduced excitability as a prescription for conduction failure and reduced coupling as a prescription for very slow, stable propagation. The increase in SF at reduced coupling is due to greater confinement of depolarizing current to the depolarizing cell, with less shunting of current and loss of charge to the downstream fiber (reduced load). The fact that reduced coupling can support extremely slow velocities (<1 cm/s) suggests that this phenomenon is implicated in microreentrant arrhythmias where sustenance of reentry requires very slow conduction (24) (Chap. 17).

Figure 14.9 (B) through (D) explore the relative importance of I_{Na} and $I_{Ca(L)}$ in supporting slow conduction due to decreased gap junctional coupling. Figure 14.9(B) shows SF with (solid line) and without (dotted line) the contribution of the $I_{Ca(L)}$, for the entire range of intercellular coupling. In contrast to the case of reduced excitability [Figure 14.8(B)], $I_{Ca(L)}$ plays an important role in the setting of reduced coupling. Without $I_{Ca(L)}$, propagation fails at $g_j = 0.0197$ μS. With $I_{Ca(L)}$, slower propagation is sustained at much lower levels of coupling, and block occurs only when g_j is reduced to 0.0056 μS (more than threefold decrease relative to the value without $I_{Ca(L)}$). The important contribution of $I_{Ca(L)}$ during very slow, discontinuous conduction is further illustrated in (C) and (D). For $g_j = 0.02$ μS (C) the delay between local depolarization (fast rising phase at 0 ms) and fast depolarization of the downstream neighboring cell (indicated by a thin vertical line in the figure) is 3.6 ms. This long delay extends into the plateau of the source cell, during which $I_{Ca(L)}$ is the major inward membrane current that generates depolarizing charge (I_{Na} is already inactivated). Consequently, charge contribution from $I_{Ca(L)}$ becomes significant; in fact, for $g_j = 0.02$ μS Q_{Na}: $Q_{Ca} = 1.47{:}1$, indicating almost equal depolarizing charge contribution from I_{Na} and from $I_{Ca(L)}$ and an equal importance of $I_{Ca(L)}$ in supporting conduction. With further reduction of gap junction coupling [(D), $g_j = 0.006$ μS], role reversal occurs and $I_{Ca(L)}$ becomes the major source of depolarizing charge (Q_{Na};$Q_{Ca} = 0.26$). Therefore, unlike the I_{Na} dominance during reduced excitability, $I_{Ca(L)}$ becomes the major source of depolarizing charge and the most significant current to sustain conduction in the situation of highly reduced coupling [I_{Na} is still needed, but only to depolarize the membrane into the activation range of $I_{Ca(L)}$]. It follows that under such circumstances (e.g., in the setting of a chronic infarct, where uncoupling is present), $I_{Ca(L)}$ constitutes a potential target for antiarrhythmic intervention.

IV. ASYMMETRY OF ELECTRICAL PROPERTIES: REENTRANT EXCITATION

The previous section dealt with membrane changes (reduced excitability) and tissue structural changes (reduced gap junctional coupling) that can result in slow conduction and conduction failure. Slow conduction and asymmetrical conduction failure (unidirectional block) (26) are conditions that lead to the development of reentry, a major cause of clinical arrhythmias (Chap. 17). In its simplest form, reentry involves propagation of the action potential in a closed fixed pathway around an inexcitable anatomical obstacle. This form of rotating excitation is known as "circus-movement" reentry. Compared to the propagation in a linear array of cells that was examined in Section III, reentrant propagation in a closed pathway involves interaction between the head and tail of the action potential that adds complexity to the situation. Faithful to this chapter's objective to characterize cardiac excitation at progressively increasing levels of complexity and integration, we proceed to

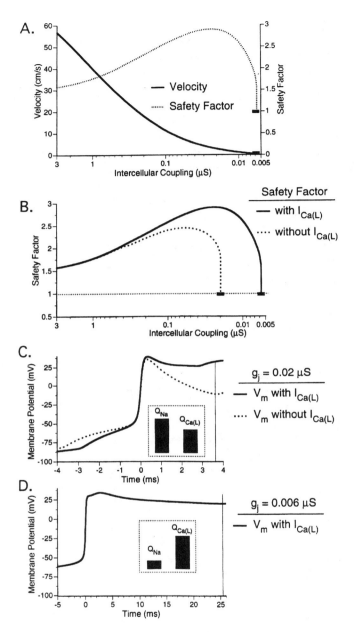

Figure 14.9   Conduction during decreased intercellular coupling. (A) Conduction velocity (solid line) and safety factor (SF) for conduction (dotted line) versus gap junction conductance, g_j. Although velocity decreases in a monotonic fashion with decreased coupling, the behavior of SF is biphasic (an increase followed by a decrease). Conduction velocity decreases two-hundred-fold to 0.26 cm/s before block occurs. SF increases initially, reflecting greater confinement of current for depolarizing local membrane (reduced loss to downstream fiber due to reduced g_j). (B) SF computed over a range of intercellular coupling with (solid line) and without (dotted line) $I_{Ca(L)}$. SF computed without $I_{Ca(L)}$ begins to diverge from SF with $I_{Ca(L)}$ at a fivefold decrease in intercellular coupling ($g_j = 0.5$ μS). Conduction fails without $I_{Ca(L)}$ at three times the conductance for which block occurs when $I_{Ca(L)}$ is present. (C) High temporal resolution action potential upstrokes from the middle cell of the fiber at $g_j =$

characterize important aspects of reentry in a fixed pathway (27,28). More complex forms of reentry where the restrictions of a fixed pathway and an inexcitable obstacle are removed (e.g., spiral waves) will be described in later sections.

A. Parameters Used to Describe Circus-Movement Reentry

Circus-movement reentry is the best understood type of rotating wave in the heart. It may be described as an electrical impulse propagating over a unidimensional circuitous pathway (i.e., a ring) (Fig. 14.10) and returning to its site of origin to reactivate that site. As originally postulated by George Mines (29), the complex anatomy of the heart, together with the heterogeneous distribution of tissue types and hence electrophysiological properties (i.e., excitability, wave speed, and refractory period), provide an excellent substrate for the establishment of circus-movement reentry. Because of such complexities, a prematurely initiated propagating wave may encounter refractory tissue and block in a particular direction, a phenomenon known as unidirectional conduction block. As illustrated in Figure 14.10, if unidirectional conduction block occurs, the wave may propagate indefinitely in the opposite direction around the ring. Under these conditions, the rotation period (T) is determined by the conduction speed (C) and the perimeter of the circuit ($P = 2\pi R$). If the ring is completely homogeneous, and C is constant, then

$$T = P/C. \tag{1}$$

Since the wavefront is always followed by a finite tail of refractoriness, it follows that the perimeter of the ring in Figure 14.10 must be larger than the spatial extent of the excitation wave. It was postulated many years ago by Wiener and Rosenblueth (30) that, during reentry around a ring of homogeneous excitable tissue, the spatial extent of the excited region, also known as the wavelength (WL), is the product of the conduction speed (cm/s) and the refractory period (RP) (s). If the tissue is normally polarized and excitable, then the action potential duration (APD) can be used as a measure of the refractory period and wavelength is given by

$$WL = C \times APD. \tag{2}$$

In this ring model of reentry (Fig. 14.10), the tissue in the circuit that is not occupied by the propagating wave can be excited and is called the spatial excitable gap (EG); its length is easily obtained by

$$EG = P - WL. \tag{3}$$

This type of reentry has been shown to explain the mechanism underlying supraventricular tachycardias associated with preexcitation, or Wolff–Parkinson–White syndrome, in which an anomalous bundle of muscle (i.e., an accessory pathway) connecting atria and

---

0.020 μS with (solid line) and without (dotted line) $I_{Ca(L)}$. Inset bar graph shows relative charge, Q (current integrated over time) generated by I_{Na} and $I_{Ca(L)}$ in the period from the middle cell's excitation to excitation of the adjoining cell (marked by a thin vertical line at time = 3.6 ms). The long intercellular delay extends into the plateau of the source cell, during which $I_{Ca(L)}$ is the major inward membrane current. At the reduced coupling of (C), charge contributed by $I_{Ca(L)}$ approaches that contributed by I_{Na}. At extreme levels of uncoupling, illustrated in (D), charge contribution from $I_{Ca(L)}$ exceeds that from I_{Na} by up to an order of magnitude. (From Ref. 15, with permission.)

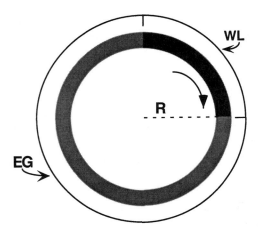

Figure 14.10 Circus-movement reentry around a ring of excitable tissue. In this scheme, a reentrant wave is seen propagating clockwise. The spatial extent of the excited region, shown in black, is called the wavelength (WL). The perimeter of the ring is given by $P = 2\pi R$. Since $P>$WL, there is an excitable gap (EG) and the wave propagates indefinitely around the ring.

ventricles bypasses the atrioventricular node and allows for rapid clockwise or counterclockwise rotating activity between atria and ventricles (20).

B. Asymmetry in Membrane Excitability

To simulate reentry in a closed pathway, the multicellular one-dimensional fiber of Figure 14.5 was formed into a ring that included up to 1500 cells [Fig. 14.11(A)]. A necessary condition for the initiation of reentry is unidirectional block. Unidirectional block and reentry were initiated in the model by a combination of a primary stimulus and a premature stimulus [Fig. 14.11(A)]. The action potential elicited by the primary stimulus at the top cell (cell 1) propagated down both branches of the ring and, due to symmetry, collided at the bottom cell. By applying a properly timed premature stimulus at a point in the left branch during the repolarization phase of this propagating action potential, a premature action potential was elicited. It propagated a short distance, decrementally, in the anterograde direction (same direction as the primary action potential) before block occurred. In the retrograde direction, the premature action potential propagated around the ring fiber and returned to its point of initiation. Thus, unidirectional block was induced and reentry established. The unidirectional block results from asymmetry in the state of membrane excitability due to the passage of the primary action potential.

To further characterize the asymmetry of membrane excitability and its underlying ionic mechanism, we define a vulnerable window (TW) as the time interval during the repolarization phase of a propagating action potential during which unidirectional block and reentry can be induced [Fig. 14.11(B)]. The vulnerable window can also be represented as a distance in the space domain (SW) or as a range of membrane potentials in the voltage domain (VW). Outside this window, it is impossible to induce unidirectional block and reentry; an action potential elicited by a premature stimulus either propagates or blocks in both directions. When a premature stimulus is applied inside the window, the membrane generates a critical sodium current, giving rise to an action potential that propagates incre-

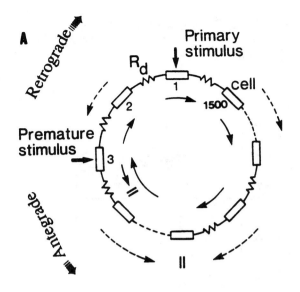

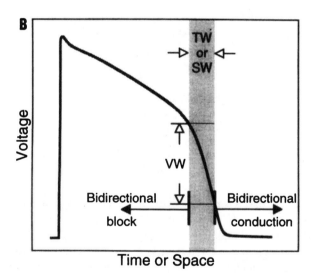

Figure 14.11 (A) Ring-shaped cable model. The ring model consists of up to 1500 cells, each 100 μm in length and 16 μm in diameter. A primary stimulus was applied at the top of the ring, while a premature stimulus was applied at a site in the left branch during the relative refractory period of the propagating action potential (during TW) (B). Dashed arrows outside the ring represent propagation of the primary action potential. Due to symmetry, it blocks at the bottom (middle) cell of the ring. Solid arrows inside the ring represent propagation of the premature (reentrant) action potential. It blocks in the antegrade direction (= symbol) and reenters in the retrograde direction. R_d = gap junction resistance. (From Ref. 27, with permission.) (B) Schematic representation of the vulnerable window during the refractory period of a propagating action potential. TW, SW, and VW represent the vulnerable window in the time domain, space domain, and voltage domain, respectively. (From Ref. 28, with permission.)

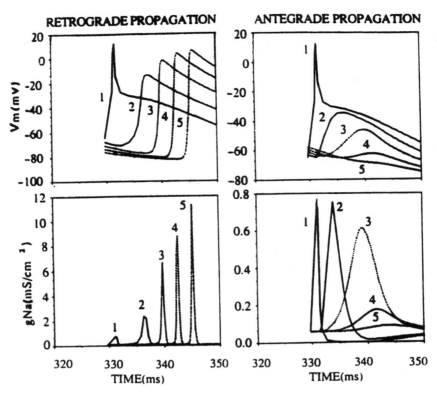

Figure 14.12 Antegrade decremental propagation (right) and retrograde incremental propagation (left) induced by a premature stimulus in the vulnerable window. V_m = membrane potential (mV); g_{Na} = sodium channel conductance (mS/cm^2). The parameters are displayed at increasing distances (curves 1 to 5) from the site of premature stimulation in both directions. Note different g_{Na} scales in left and right panels. (From Ref. 27, with permission.)

mentally in the retrograde direction (where tissue is progressively more recovered from activation by the propagating action potential) and decrementally in the anterograde direction (where tissue is progressively less excitable due to activation by the propagating action potential). This is because in the retrograde direction the tissue is progressively more recovered as the distance from the window increases in this direction, while in the anterograde direction the membrane is progressively less excitable as the distance from the window increases.

Figure 14.12 shows an example of propagation induced by a premature stimulus in the vulnerable window. The primary stimulus was applied to cell 1 at $t = 0$ ms. The premature stimulus was applied in the wake of a propagating action potential to cell 250 at time = 331 ms. In the retrograde direction, the action potential propagated a distance of 40 cells before reaching the region of fully excitable membrane. The conduction velocity was only 19.5 cm/s from cell 250 to cell 240. As the distance increased, the conduction velocity gradually increased, reaching 38.2 cm/s from cell 220 to cell 210. In the anterograde direction, propagation gradually diminished as a result of decreased membrane excitability in this direction. The action potential gradually decreased in amplitude and in conduction velocity (17.5 cm/s from cell 250 to cell 255 and 9.7 cm/s from cell 255 to cell 260). Note the graded nature of electrical excitation induced in the vulnerable window. In the retrograde direction, g_{Na} (sodium channel conductance which determines membrane excitability) recovers slowly from curve 1 (0.78 mS/cm^2) to curve 2 (2.4 mS/cm^2), reflecting the time course of sodium channel recovery from inactivation. In the anterograde direction, there is a sharp decrease in g_{Na} from curve 3 (0.62 mS/cm^2) to curve 4 (0.18 mS/cm^2), reflecting a sharp decrease of sodium channel activation. This asymmetry of sodium channel availability and membrane excitability results in the development of unidirectional conduction block.

Figure 14.13 is an example of sustained reentry in the ring model of Figure 14.11. A primary stimulus is applied at cell 1 (star symbol) and a premature stimulus at cell 40 (plus sign). It propagates decrementally in the anterograde direction for a distance of only five cells before it blocks. In the retrograde direction, the premature action potential propagates around the ring fiber, returns to cell 40, reenters, and continues to propagate around the ring in a sustained fashion. The duration of the reentrant action potential is initially very short and then increases during a transient period until stable propagation with a duration of 120 ms (less than half the 270-ms duration of the fully developed action potential) is obtained. In other simulations, under conditions of greater head–tail interaction of the reentrant action potential, stable beat-to-beat alternans (long-short pattern) of action potential duration is observed (27,28).

C. Asymmetry of Myocardial Structure

The simulations above focus on asymmetry of membrane excitability and its role in induction of reentry. In the section on action potential propagation we demonstrated that, in addition to membrane factors, tissue structural factors (e.g., cell-to-cell communication through gap junctions) play an important role in excitation and conduction. In the following simulations, we expand this concept to reentrant excitation. Two structural asymmetries are examined in the context of reentry: nonuniformities of fiber cross-section and nonuniformities in the degree of cellular coupling.

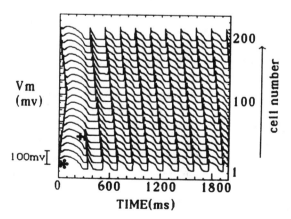

Figure 14.13 Sustained reentry induced in the retrograde direction in the ring model of Figure 14.11. Location (cell number) and time of application of primary stimulus are marked by asterisk; location and time of application of premature stimulus is marked by plus sign. (From Ref. 27, with permission.)

1.   Fiber Cross-Section

The complex architecture of cardiac tissue involves geometrical nonuniformities. An important example is a change in the size (expansion, narrowing) of the excitable tissue that supports conduction. Expansions occur at Purkinje–muscle junctions, at the insertion site of an accessory pathway in the WPW syndrome, and in chronic infarcts where narrow strands connect islands of surviving tissue, to mention a few examples. Sites at which the cross-section of excitable tissue suddenly increases may be sites for unidirectional block. An action potential propagating in a strand of small cross-section might not supply sufficient current (charge) for induction of propagation in the fiber with large cross-section. In other words, there is a mismatch between a small source and a large sink (load) that may cause the safety factory for conduction to decrease below 1, leading to conduction failure. We simulated this situation by introducing nonuniformities in the fiber radius, as shown in Figure 14.14. There were three fibers with radii of 4, 8, and 16 µm. Unidirectional block could occur at the junction of two fibers with different diameters when propagation traveled from a small-diameter fiber to a large-diameter fiber. For creation of unidirectional block, the ratio of diameters of connected fibers must be sufficiently high. Conduction was not blocked at either the junction between fiber I and fiber II or the junction between fiber II and fiber III, because the ratio of diameters (1:2) of the connected fibers was not high enough. Unidirectional block occurred at the junction between fiber I and fiber III when the impulse propagated from fiber I to fiber III. Note that the ratio of diameters of fibers I and III is 1:4. The stimulus was applied at cell 40 as indicated by St in Figure 14.14(A). The action potential propagated in both the counterclockwise and clockwise directions. In the clockwise direction, the action potential propagated decrementally (about 20 cells) and completely blocked before it reached the junction between fibers I and III. In the counterclockwise direction, the action potential propagated through all the junctions, returned to the point of initiation, and kept propagating. Thus, sustained reentry was obtained. For this type of nonuniform ring, there is a favorable direction of circus movement (counterclockwise).

We noticed that there is a spatial gap on the ring at which a single stimulus could induce reentry. This gap is located close to the unidirectional block site on the small fiber [shown in Fig. 14.14(A) by a large arrow]. The existence of a finite gap results from the requirement that the tissue be excitable at the time of arrival of the (counterclockwise) reentrant action potential at the I–III junction. Figure 14.14(B) shows the propagating action potentials along the ring fiber of Figure 14.14(A). Note the decremental propagation (D) in the clockwise direction from cell 10 (stimulus site) to cell 1 where block occurs. It is important to emphasize that in the heart, geometrical asymmetries during propagation can be structural or functional. An increase in cross-section (load) experienced by the propagating wavefront can be anatomical in nature, due to branching of fibers, Purkinje–muscle junctions, or a small surviving strand merging into a larger island of surviving tissue in an infarct. It can also be functional in nature; for example, high-curvature convex segments of the wavefront encounter an increased cross-section of unexcited tissue (load) due to the fanning-out effect associated with the curvature. This situation arises in the pivot points of a reentry pathway or at the core of a spiral wave. The concept of curvature in the context of spiral wave reentry is developed further below.

2.   Cellular Coupling

In the heart, nonuniformities of gap junctional coupling can occur in a variety of situations. It is a property of normal myocardium, where coupling is greater along fibers than trans-

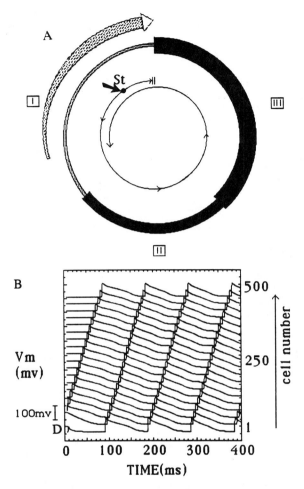

Figure 14.14 (A) Ring fiber with nonuniform distribution of fiber cross-sections. Unidirectional block could occur at junction of two fibers with different diameters when propagation is from small-diameter fiber to large-diameter fiber. Sustained circus movement could be induced by a single stimulus (St) applied within gap (large arrow) without the need for a premature stimulus. (B) Propagating action potentials along the ring fiber of (A). Note decremental propagation in the clockwise direction from cell 10 to cell 1 (marked by D next to left axis). (From Ref. 27, with permission.)

verse to the fibers (anisotropy). Consequently, a wavefront experiences a change in load upon changing its direction of propagation relative to the fiber axis. Nonuniformities in coupling are also likely to be present in an infarct or during chronic ischemia due to the spatial nonuniformity of the injury. The simulations predict that nonuniform distribution of gap junctional resistance can provide conditions for induction of unidirectional block and reentry by a single stimulus without the need for premature stimulation. In Figure 14.15, there are three segments (identified as I, II, and III) with gap junctional conductances of 0.1, 1.0, and 0.3 µS, respectively. Unidirectional block may occur at the junction of two fibers with different gap junctional conductances when propagation is from the fiber with lower gap junctional conductance to the fiber with higher gap junctional conductance. The

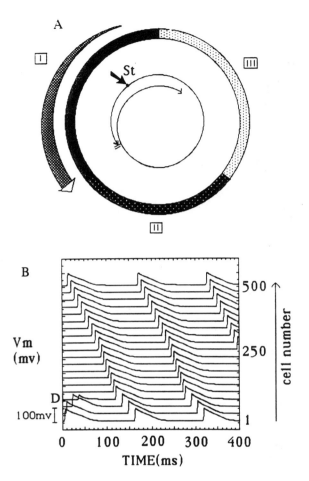

Figure 14.15 (A) Ring fiber with nonuniform distribution of gap junctional conductances. Unidirectional block could occur at junction of two fibers with different gap junction conductances when propagation is from fiber with lower gap junction conductance to fiber with higher gap junction conductance. Sustained circus movement could be induced by a single stimulus (St) applied within gap (large arrow) without the need for a premature stimulus. (B) Propagating action potentials along the ring fiber of (A). Note decremental propagation in the counterclockwise direction from cell 55 to cell 75 (marked by D). (From Ref. 27, with permission.)

high gap junctional conductance causes leakage of current and loss of charge to the downstream fiber, which constitutes a large electrical load. The increase in load lowers the safety factor at the transition from the less coupled to the better coupled fiber and could lead to conduction failure.

As in the case of the ring with fibers of nonuniform radii (Fig. 14.14), the ratio of gap junctional conductance in adjacent segments is critical for the induction of unidirectional block. Propagation does not block at junctions I–III and III–II because this ratio (1:3 and 3:10, respectively) is not small enough. However, block occurs at junction I–II, at which the ratio of gap junctional conductances is 1:10. The stimulus was applied at cell 10 [St in Fig. 14.15(A)]. The favorable direction of propagation is clockwise. The gap in which reentry could be induced by a single stimulus is located close to the site of unidirectional block [large arrow in Fig. 14.15(A)]. Figure 14.15(B) shows the propagating action potentials along the ring fiber of Figure 14.15(A). Note the decremental propagation (D) in the counterclockwise direction from cell 55 to cell 75, where block occurs.

V. TWO-DIMENSIONAL WAVE PROPAGATION

In previous sections, the discussion centered on the characteristics of propagation in unidimensional cables of cardiac cells coupled electrically through gap junctions. We now focus our attention on propagation in two dimensions.

A. Plane Waves Versus Curved Waves

When propagation of the impulse occurs in what is, effectively, two-dimensional cardiac muscle, as in the case of thin preparations shaved from the subepicardial muscle of sheep hearts (31,32), or that of the rabbit ventricles treated by endocardially applied cryoablation (33), additional factors come into play. Most importantly, the shape (i.e., curvature) of the wavefront is thought to be a major determinant of the success or failure of propagation (34–36). The more convexly curved a wavefront is, the lower its velocity of propagation. Beyond a certain critical curvature, propagation of the wavefront cannot proceed. This concept of critical curvature is very much related to the existence in two-dimensional cardiac muscle of a *liminal area* for propagation, equivalent to the liminal length in a one-dimensional cable. This is easily appreciated in the schematic diagram of Figure 14.16(B), which illustrates the fact that the relative area of tissue to be excited (the sink) ahead of a convexly curved wavefront (the source) is larger than the area forming the sink in front of an equivalent plane wave (A).

B. The Concept of Critical Curvature

Based on the theory of wave propagation in excitable media, investigators (37,38) predicted that there should be a relationship between the curvature of the wavefront and its propagation speed. This has been confirmed experimentally in cardiac muscle (35,36,39). Indeed, recently published video imaging studies in isolated cardiac tissue, in combination with computer simulations using ionic model representations of cardiac excitation (35) have demonstrated that conduction velocity in two-dimensional cardiac muscle is steeply dependent on the curvature of the wavefront, and that there is a critical wavefront curvature beyond which propagation stops (see below). This relationship may be observed readily when following the propagation of the cardiac excitation wavefront across an artificially produced isthmus in a thin piece of atrial or ventricular tissue. In 1971, De la Fuente et al. (40) studied propagation through a narrow isthmus in canine atrial tissue. Although their study did not address the issue of wavefront curvature, they demonstrated that there was a range of isthmus widths of 0.5 to 1 mm at which conduction block occurred. Cabo et al. (35) investigated the characteristics of propagation through an isthmus, as well as the influence of anisotropy in both computer simulations and experiments in thin slices of isolated ventricular tissue. In the diagram of Figure 17(A), we reproduce Cabo's numerical experiment in which an isthmus was created in a two-dimensional array of excitable cardiac cells by placing an impermeable screen with a hole at the center of the array. Propagation was initiated by planar stimulation in the proximal side (left) with the resulting wavefront being parallel to the screen. The size of the isthmus was reduced gradually (not

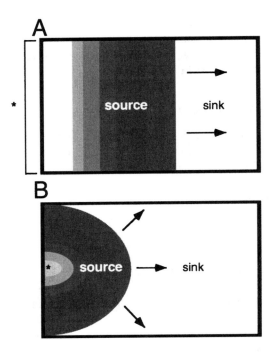

Figure 14.16   Role of wavefront curvature in establishing sink:source relationships. (A) Hypothetical example of a plane wave initiated by stimulation (asterisk) of the entire left border of the tissue. The wave propagates from left to right. An appropriate balance between amount of excited tissue at the wavefront (source) and amount of tissue to be excited by that wavefront (sink) allows for rapid and uniform propagation. (B) In the case of a curved wave initiated by a point source, sink ≥ source. Therefore, propagation is slower than for the plane wave case. (See also color plate.)

shown) until the wave failed to propagate from the proximal to the distal side. The critical width of the isthmus (i.e., the width at which propagation failed), was shown to be a function of the direction of propagation (longitudinal, 0.2 mm; transverse, 0.6 mm). Isthmus widths larger than critical resulted in wave diffraction, with the result that wavefronts propagating distally changed their velocity depending on their curvature.

Figure 14.17 (A1) shows simulation results in the form of an activation map for longitudinal propagation, with isochrone lines representing the position of the wavefront every 5 ms as it moved from left to right. The velocity of propagation of the planar wave was 41 cm/s. Distal to the isthmus, diffraction resulted in a steeply curved wavefront. The changes in curvature and conduction velocity across the relatively narrow (0.25 mm) isthmus are shown in panel A1. The local conduction velocity along a line in the direction of propagation through the center of the isthmus is presented graphically in (A2). Proximal to the isthmus, the velocity of the wavefront increased as it approached the isthmus. The velocity decreased to a minimum (25 cm/s) after the isthmus and then progressively increases again toward the value of the velocity of a planar wave. Panel (A3) shows the changes in wavefront curvature as a function of space. Clearly, the changes in the curvature correlated very well with changes in the conduction velocity (35): because of the high resistivity at the barrier, the increase in the conduction velocity just before the isthmus corresponded to that of a wavefront of negative curvature (concave). On the other hand, the

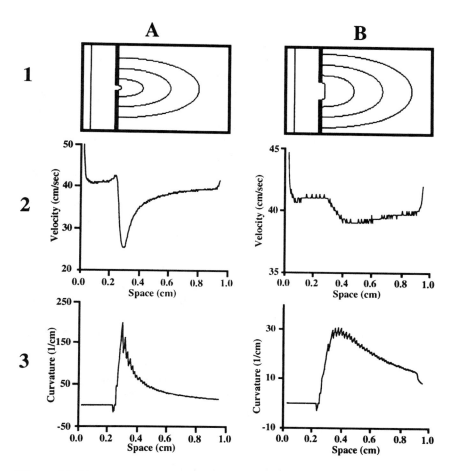

Figure 14.17 Computer simulation (LRd model) showing longitudinal propagation through narrow (width, 250 μm; A1 through A3;) and wide (width, 950 μm; B1 through B3) isthmuses. The isthmus was located 0.25 cm from the left border (A1 and B1); electrical activity was initiated by planar stimulation of the left border of the matrix. A1 and B1 show isochronal maps for each condition; the spacing between isochrones was 5 ms. A2 and B2 are graphs showing the changes in the velocity of propagation along a line in the direction of propagation (left to right) through the middle of the isthmus. A3 and B3 are graphs showing the changes in the curvature of the wavefront along the same line. The simulations correspond to a case with normal excitability. (Modified from Ref. 35, with permission.)

decrease in the conduction velocity beyond the isthmus (right) corresponded to wavefronts of positive curvature (convex). The minimum velocity occurred at a distance of 0.5 mm distal to the isthmus and coincided with the site of maximum curvature.

Propagation through a wider isthmus (950 μm) is presented in Figure 14.17(B1–3). Qualitatively, the results are similar to those shown in Figure 14.17(A1–3): velocity and curvature were again related to each other. However, there were some quantitative differences: minimum velocity and maximum curvature occurred further away from the isthmus and over a wider range of distances.

The critical size of the isthmus for propagation was strongly dependent on the excitability of the medium. In fact, if the maximum value of the sodium conductance of all cells in the array was decreased, there was a decrease in the velocity of a planar wave (longitudinal conduction velocity = 31 cm/s) and an increase in the size of the critical isthmus.

It is important to stress that, in this model, conduction block occurs at an appreciable distance away from the isthmus. In fact, close scrutiny of Figure 14.17 shows that, right at the isthmus, propagation velocity was normal because the wavefront was flat; the minimum velocity of propagation occurred at some distance beyond the isthmus where the convex curvature of the wavefront was maximal. At a further distance away from the isthmus, the velocity was still slower than that of a planar wave because the wavefront was curved. Thus, according to Cabo's results, the most important effect of the isthmus was to diffract the planar wavefront into an elliptic curvature (35).

Experimental results (35) confirmed the validity of the simulations; at nominal excitability, the critical isthmus for longitudinal propagation in sheep ventricular muscle was estimated to be somewhat larger (i.e., between 0.5 and 1 mm) than in the simulations (0.2 mm). In this regard, Fast and Klèber (41) studied propagation through narrowing groups of cells in patterned rat myocyte cultures. They found that an action potential emerging from a strand five to eight cells wide always propagated to a large growth area, albeit at a reduced velocity. The latter would suggest that, in this monolayer cell culture model, the critical isthmus was less than 65 to 130 μm. As discussed by Cabo et al. (35) a number of factors may contribute to such differences, including species, tissue excitability, anisotropic ratio, and length of the tissue occupied by the isthmus. Overall, however, the studies provide strong support to the contention that, in cardiac muscle, wavefront curvature may be a cause of slow conduction and block. As confirmed by Cabo's experimental results, the local changes in curvature themselves were responsible for the corresponding changes in velocity. It follows from these results that, since a large curvature of the wavefront causes a reduction in velocity, propagation may be possible only for wavefronts whose curvature is less than critical.

C. Wavefront–Wavetail Interactions and Wavebreaks

The concept of wavebreak (38,42,43) is a useful tool in the understanding of the mechanisms of initiation and maintenance of reentrant arrhythmias. As previously shown in computer simulations (43,44) and in excitable media (45), under appropriate conditions of excitability, the interaction of a wavefront with an obstacle can lead to fragmentation of that wavefront and vortex formation. This is illustrated diagrammatically in Figure 14.18, which shows three snapshots in the hypothetical time evolution of a wavebreak created by an anatomical obstacle (e.g., a scar) in a two-dimensional sheet of cardiac muscle. When the excitability of the sheet is high (not shown), upon circumnavigating the obstacle, the two broken ends of the wavefront would come together and fuse to recover the previous shape of the wavefront. By contrast, as shown in Figure 14.18, when the excitability is somewhat lower, the free ends of the wavebreak do not converge. They move in opposite directions, thus causing fragmentation of the wavefront, with formation of two curled wavelets (vortices) that begin to rotate in opposite directions. The end result is functional figure-of-eight reentry (46). This phenomenon is know as vortex shedding and has been demonstrated to occur in cardiac muscle (47).

Another consequence of the formation of a wavebreak is an appreciable distortion of the spatial (wavelength) and temporal (action potential duration) extents of the excited state. As illustrated by the diagrams presented in Figure 14.19, during the propagation of a wave initiated by a linear source (i.e., a plane wave) (A), or by a point source (i.e., a circular

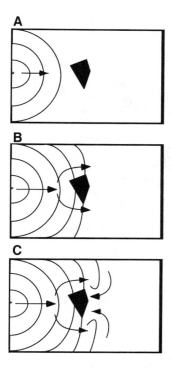

Figure 14.18 Schematic representation of vortex shedding as a result of wavebreak formation by the interaction of a wavefront with an anatomical obstacle. (A) Wave of excitation is initiated by point stimulation (asterisk) on left border of a two-dimensional piece of cardiac muscle. (B) Upon reaching the obstacle, the wavefront breaks into two parts that circumnavigate the obstacle. (C) Under appropriate conditions of low excitability, the two wavefronts detach from the obstacle. The two wavebreaks formed curl and begin to rotate forming two counterrotating vortices.

wave) (B), the wavefront is always followed by a recovery band of finite dimensions. The distance between the front and the tail corresponds to the wavelength (WL) of excitation. In other words, the edge of the wavefront and the edge of the wavetail never meet each other. In contrast, as shown in Figure 14.19(C), broken waves have a "dangling" end, the wavebreak or phase singularity (black dot), at which the wavefront fails to activate the tissue ahead. Instead, it rotates around a small region (i.e., the core). Thus, the wavebreak forms a phase singularity that acts effectively as the pivoting point (the rotor) and forces the wave to become a vortex (i.e., acquire a spiral shape) as it rotates around the core. Moreover, under these conditions, the spatial extent of the excited region [stippled region in Fig. 19(C)], as well as the action potential duration, decreases gradually toward the phase singularity. As such, the wavelength is no longer constant but depends on distance from the core. Computer simulations using ionic formulations of cardiac excitation have provided testable explanations to such phenomena (48). Briefly, the idea is that, upon the initiation of reentry, an exceedingly large wavefront curvature near the core results in an area of functional block inside a small rim of tissue surrounding the core. Any section of the wavefront entering the rim will undergo decremental propagation and die out before ever reaching the very center of the core. Consequently, the tissue in the core, although excitable, is never excited. Under these conditions, voltage gradients are established between the core and its surroundings,

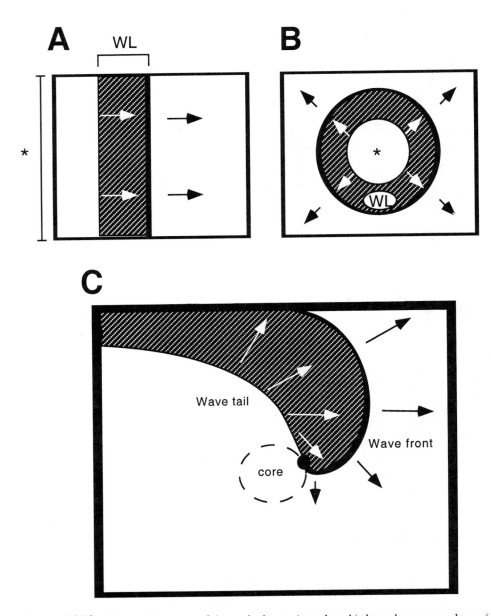

Figure 14.19   The spatial extent of the excited state (wavelength) depends on wave shape. (A) Planar wave initiated by linear stimulation (asterisk) of the entire left border of a piece of homogeneous excitable sheet of cardiac muscle. (B) Circular wave initiated by point stimulation of the center of the sheet. Note that wavelength (WL) is constant for both (A) and (B). (C) Spiral wave shape produced by pivoting of wavebreak (i.e., phase singularity; black dot) around the core. Note that, in this case, WL is not constant but increases from core to periphery.

whereby the relatively large current sink effect exerted by the core acts to abbreviate APD and thus shorten wavelength in the tissue that surrounds it (48). The latter effect would explain why traditional concepts, which define wavelength as the product of refractory period times conduction velocity, and apply very well to anatomical reentry (see Fig. 14.10), cannot be used to predict the properties or behavior of vortexlike reentry.

VI. ROTATING WAVES IN TWO DIMENSIONS

A. Two-Dimensional Vortexlike Reentry: Rotors and Spirals

Recent advances in the understanding of rotating waves in two-dimensional excitable media in general, and of cardiac arrhythmias in particular, have come from studies that apparently have nothing to do with the heart. Thus, it has recently become clear that vortexlike reentry, defined as a rapidly spiraling electrical wavefront, is not unique to abnormal cardiac rhythms. Phenomena that have very much in common with vortexlike reentry have been observed in many biological systems, including the brain and retina (49), the social amoeba *dyctiostelium* (50) and calcium waves in *Xenopus laevis* oocytes (51), single ventricular myocytes (52), and cardiac cell monolayers (53), as well as in autocatalytic chemical reactions (54). The common feature of all these systems is "excitability" since, just as in myocardial tissue, in any of these systems, a local excitation above a certain threshold results in the initiation of a propagating wave akin to the cardiac action potential. These systems are called excitable media (55) and are governed by a mathematical description that is very similar to the well-known models of cardiac propagation (1,2). We should point out, however, that there are enormous differences in time and space scales in the var-

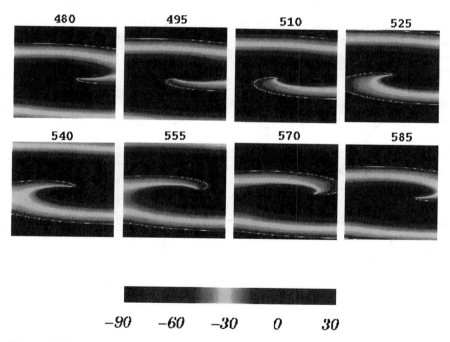

Figure 14.20 Spiral wave activity in a 2 × 2-cm anisotropic cardiac sheet (modified LRd model) with a refractory period of 147 ms. Ratio of horizontal to vertical velocity during plane waves is 4:1. Numbers on top of each panel indicate time (ms) after spiral wave initiation by cross-field stimulation protocol. The color bar indicates membrane potential distribution in millivolts. In this continuous ionic model, membrane channels are distributed uniformly at the surface of the sheet. The model incorporates seven ionic currents. The conductance of each current as well as steady-state and time constants for each gating variable are given in Reference 1. The system of equations was solved using a finite-element method and a semi-implicit integration scheme as described in detail in Reference 48. The sheet consisted of 3600 cubic elements with 16 nodes per element, which implies an internode spacing of 111 μm. (Reprinted from Ref. 48, with permission.) (See also color plate.)

ious excitable media; for example, conduction velocity of calcium waves in the oocyte is ~30 μm/s, whereas in heart muscle the normal velocity of action potential wave propagation is ~0.3 to 0.5 m/s. Yet, in all types of excitable media that have been studied thus far, it has been demonstrated that a particular perturbation of the excitation wave may result in vortexlike activity (Fig. 14.20). During such activity in two-dimensional media, the excitation wave acquires the shape of a spiral (3–6) and is called a spiral wave or vortex, organized by its rotor pivoting around the core. As in the case of functional reentry, spiral waves do not require an anatomical obstacle. In the example of Figure 14.20, taken from a computer simulation in an anisotropic cardiac sheet (48), the spiral is elongated, roughly orientating horizontally along the fiber direction.

The hypothesis that functional reentry is the result of a rotor giving rise to vortices of electrical waves was formulated many years ago (30). Subsequently, in 1973, Allessie (56) demonstrated the validity of the hypothesis in isolated cardiac muscle and, more recently, a newly developed high-resolution video-imaging technique has enabled direct visualization of the excitation vortices and the quantitative study of their dynamics (31,32).

B. Drifting Spirals Give Rise to Complex Excitation Patterns

Traditional electrophysiological concepts see functional reentry as a strictly stationary phenomenon: an "arc of block" in the center of a reentrant loop acts as a functional obstacle around which the reentrant excitations revolve (46,57). Such concepts do not provide any mechanism for reentry drift. In contrast, the concept of rotors applies not only to stationary reentry but to drifting reentry as well. In fact, similar to other vortices, like hurri-

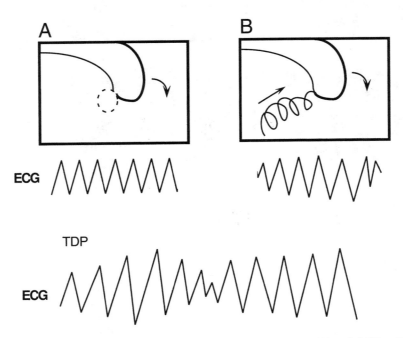

Figure 14.21   Diagrammatic representation of stationary (A) and drifting (B) spiral waves. In (A), the spiral wave rotates around a stationary core (top) and results in a monomorphic ECG pattern (middle). In (B), the spiral has drifted from the bottom left corner of the sheet (top). Consequently, the ECG pattern (middle) is polymorphic. The bottom trace depicts a hypothetical pattern of torsade de pointes (TdP), which would result from a relatively long episode of spiral drift (B).

canes and tornados, rotating waves in the heart would tend to drift [Fig. 14.21(B)] rather than stay in the same place [Fig. 14.21(A)]. It has been shown in media other than the heart that drifting may be either intrinsic (58) or appear as a result of heterogeneity of the medium (59–61), external electrical field (62), or boundary effects (61). Often, drifting rotors hit the border of the medium and disappear (63). Sometimes they become attached to heterogeneities (obstacles) giving rise to stable rhythms (64) whose periodicity depends in part on the size of the obstacle around which they rotate.

The concept of drifting rotors provides new insight into the mechanisms of variability and termination of reentry (31,32,65). Drifting of a reentrant circuit consistent with the theory of wave propagation in excitable media has been demonstrated in isolated pieces of thin (~500-μm thick) ventricular epicardial muscle (31,32). However, in the latter case, the observation of drifting was not the result of drug effects or any other external influence but was caused by intrinsic heterogeneities in the epicardial muscle itself. Optical mapping experiments (31,32) have demonstrated that drifting spirals can yield beat-to-beat changes in ventricular complexes resulting in a polymorphic pattern that mimicks that of so-called torsade de pointes (TdP) (see Fig. 14.21) (66). Moreover, as illustrated in Figure 14.22(B), occasionally it has been observed that drifting spirals become attached (anchored) to small obstacles in their path (e.g., an artery or a scar), in which case the ECG pattern [Fig. 14.22(C)] changes from polymorphic, when the spiral is drifting [Fig. 14.22(A)], to monomorphic, when the spiral becomes anchored [Fig. 14.22(B)] and begins to rotate around the obstacle. Finally, sustained rotating activity has been amply demonstrated in the atrium, and the possibility that the dynamics of drifting and anchoring rotors may be involved in the conversion of atrial flutter to atrial fibrillation and vice-versa has been considered by some authors (67,68).

C. Are Wavebreaks Important in the Mechanisms of Fibrillation?

Over the last several years, evidence has accumulated in favor of the idea that cardiac fibrillation may find its explanation in the behavior of generic models of excitable media (69). The basic postulates are that: (1) the seemingly irregular electrical activity that is responsi-

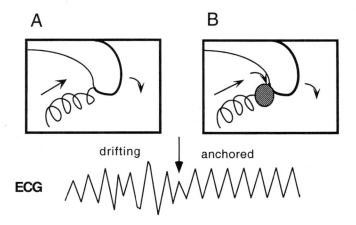

Figure 14.22   Drifting and anchoring of spiral waves. (A) Spiral was initiated at bottom left corner. It drifted toward the center. (B) Drifting spiral finds a rounded obstacle in the center of sheet and begins to rotate around it. (C) In the ECG, drifting and anchoring manifest as conversion from polymorphic to monomorphic tachycardia.

ble for fibrillation may be explained on the basis of self-organized rotors giving rise to vortices of electrical waves; and (2) the formation of rotating waves is the result of wavebreaks of the propagating wavefronts (see Fig. 14.18). The dynamics of such broken wavefronts are dependent on the conditions of excitability and refractoriness of the tissue, which will establish the final outcome after a wavebreak has formed. Exploration of the significance of wavebreaks in the context of propagation of electrical waves in cardiac muscle has just recently begun. From studies in other excitable media (44,45), we hypothesize that under normal conditions of excitability and cycle length, collision of an excitation wave with an obstacle in the cardiac muscle should not cause a wavebreak. It is only when the excitability of the tissue is relatively low that the broken wavefront may become laterally unstable (i.e., its broken ends do not progress laterally) after its interaction with an appropriate anatomical or functional obstacle. But what constitutes an appropriate obstacle? The complicated anatomical structure of the normal atria and ventricles is probably an important substrate for the establishment of reentrant arrhythmias. For example, the natural orifices of the caval and pulmonary veins, the atrioventricular rings, and the highly heterogeneous tissue-type distributions and geometrical arrangements of cardiac cells in the atrium have all been implicated in the mechanisms of arrhythmias. However, sustained arrhythmias do not occur under normal circumstances and can only be induced by high-frequency or premature stimulation and facilitated by the presence of hormones or neurotransmitters. On the other hand, there has been ample speculation about the histopathological substrate for reentrant arrhythmias in cardiac tissues of patients with congestive heart failure or myocardial ischemia and/or infarction (70). Sclerotic patches, diffuse fibrotic displacement of the cardiac muscle, or both, are commonly found, particularly in elderly patients who are most vulnerable to these arrhythmias. In addition, in some of these patients, the onset of arrhythmias has been ascribed to increased dispersion of refractoriness (i.e., temporal differences in the duration of refractory periods between neighboring cells or tissues), secondary to uneven chamber enlargement (55). Such increased dispersion would set the stage for functionally determined obstacles that would interfere with the propagation of wavefronts. Wavebreaks and lateral instabilities may result from such interference and may be one of the initiating causes of reentrant activation and fibrillation. In fact, it is well known that, in heart muscle and other excitable media, wavebreaks leading to rotating wave initiation may be a consequence of the interaction of a wavefront with a wavetail (32,69). During stable vortexlike reentry, the broken wavefront rotates around the functional core without short-cutting the circuit. However, broken ends of the wavefront do not always lead to reentrant activity, and it is possible that the unstable and fragmented activation fronts that are frequently observed during the apparently turbulent activity that characterizes atrial or ventricular fibrillation are, in fact, waves that have been broken by interaction with other wavefronts or with obstacles. As a result of their lateral instability, some such waves may shrink and undergo decremental conduction; other broken waves may propagate unchanged until annihilated by collision with another wave or boundary; and still other waves may undergo curling at their broken end to create new rotors and spiral waves. The final result would be the fragmentation of the wavefronts into multiple independent daughter wavebreaks giving rise to new wavebreaks, which sometimes may lead to new rotors and new wavebreaks, and so on, in a self-perpetuating motion that characterizes fibrillatory activity.

VII. PROPAGATION IN THREE-DIMENSIONAL CARDIAC MUSCLE

Cardiac muscle is spatially three-dimensional. Thus the activity observed on the surface of the ventricles can only be used as a rough approximation to the real-life situation. An appropriate three-dimensional representation of the activity is needed for a meaningful study of the mechanisms of tachycardia and fibrillation in the whole heart. Frazier et al. (71) carried out a detailed study of transmural propagation in the right ventricular outflow tract of the dog. These investigators convincingly demonstrated that, in the presence of the deeper layers of myocardium, epicardial surface propagation of an impulse initiated by point stimulation of that surface was different from that which would be manifest if the tissue were two-dimensional. While, as expected from anisotropic cardiac muscle, the activation fronts near the epicardial site of stimulation were elliptic, most were asymmetrical and had folds and undulations. In addition, the ellipses in each plane rotated clockwise toward the endocardium. The rotation was less than expected from the transmural rotation of the fibers (72). More recently, Taccardi et al. (73) confirmed that complex anatomical factors significantly affected three-dimensional propagation in the ventricles. Stimulation of the epicardium of the left ventricle of the dog resulted in ellipsoidal wavefronts propagating transmurally toward the endocardium and rotating more and more in the different planes. However, upon reaching the endocardium, the impulse propagated rapidly across that layer and then moved from endocardium to epicardium in such a way that activation of some regions distant from the stimulus site occurred as epicardial breakthroughs emerging from the deeper layers of myocardium. Such an extremely complex pattern of three-dimensional propagation is explained in part by the intricate anatomical structure of the ventricles. In addition to the elongated shape of, and asymmetrical connections among, myocardial fibers, which results in anisotropy of propagation, the muscle fiber axis rotates transmurally as much as 120 degrees in some areas (72,74,75). Moreover, excitation patterns are also affected by the Purkinje fiber network in the subendocardium and by macroscopic discontinuities and connective tissue septa separating muscle bundles.

A. Plane Wave Propagation

Unfortunately, experimental tools available today do not permit a thorough quantitative understanding of the multiple factors involved in the propagation of the cardiac impulse across the three-dimensional myocardium. Therefore, investigators have developed simplified mathematical models in which to study three-dimensional cardiac impulse propagation (76–79). One such model is based on so-called eikonal curvature equations that track the location of the action potential front, while ignoring the fine ionic mechanisms involved in propagation. This front tracking method is derived from the original work of Christian Huygens who, in the 17th century, suggested that light travels along rays that can be tracked perpendicularly to the wavefront surface. The method can be used when the propagated signal has an identifiable front. As applied to the heart, the eikonal curvature equation views the action potential upstroke location as a curve or surface in the myocardial wall, and seeks to determine its location in time, while ignoring the quantitative details of the upstroke or the rest of the action potential. We illustrate some of their results in Figures 23 and 24; the interested reader should consult the original papers for the mathematical formulations. In Figure 14.23, we have redrawn results of Keener and Panfilov (77), who carried out simulations of propagation of a plane wave in a 5×5-cm slab of ventricular muscle, with 1-cm thickness and 120-deg rotational anisotropy, from endocardium to epicardium. A plane wave was defined as any wave whose intersection with any constant layer of the myocardial wall between endocardium and epicardium was a straight line. Plane waves were initiated by stimulation of the leftmost endocardial border of the slab. Wavefronts propagated first transmurally toward the epicardium and then toward the right border. They moved as two-dimensional surfaces in three-dimensional

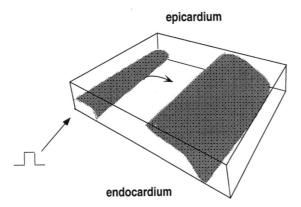

Figure 14.23 Schematic representation of computer simulation data (77) showing the propagation of two plane waves in a three-dimensional slab of anisotropic cardiac tissue in which fiber orientation undergoes a gradual rotation of 120 deg from endocardium (bottom) to epicardium (top). The slab's dimensions are $5 \times 5 \times 1$ cm; the speed of propagation along fibers is 0.5 cm/s; the ratio of plane wave velocities in the longitudinal and transverse directions is 3:1. Plane waves were initiated by stimulation of the left endocardial border. Note that velocity is faster in the epicardium than in the endocardium as a result of the fact that in the epicardium the wavefront moves along fibers, whereas propagation in the endocardium occurs across fibers. In the transmural direction waves are not planar because of the rotation of fiber orientation.

space; in the transmural direction, these waves were not planar because of the tissue anisotropy.

B.   Point Stimulation and Concentric Wave Propagation

The data in Figure 14.24 were also taken from the article by Keener and Panfilov (77). As shown by the cartoon in the inset, for these simulations they used the same slab of muscle to simulate three-dimensional propagation of a wavefront emanating from a point stimulus applied to the epicardium (asterisk) in the center of the slab. Isochronal maps (10-ms intervals) are shown from five constant levels in the z axis, from epicardium ($z = 0$) through endocardium ($z = 1.0$). Keener and Panfilov (77) have pointed to some interesting features in this figure. First of all, it is clear from the isochrone maps that the expanding oblong wavefronts do not follow the direction of fiber orientation (dotted lines). For example, on the epicardium ($z = 0$) the long axis of the wavefront is about 5 deg from the fiber direction and on the endocardial surface ($z = 1.0$) it is rotated about 30 deg from the fiber direction. From epicardium to endocardium there is approximately a 60-deg rotation of the oblong isochrone maps with the 120 deg of fiber rotation (with respect to $z = 0$), which corresponds well to previous experimental estimates in the dog heart (71).

C.   Scroll Wave Propagation

In all types of excitable media that have been studied thus far, it has been demonstrated that a particular perturbation of the excitation wave may result in vortexlike activity (80). Recently, the application of nonlinear dynamics theory to the study of wave propagation in the heart (69,81), together with high-resolution mapping techniques (31,32), has enabled investigators to demonstrate spiral wave activity on the surface of ventricular muscle. This

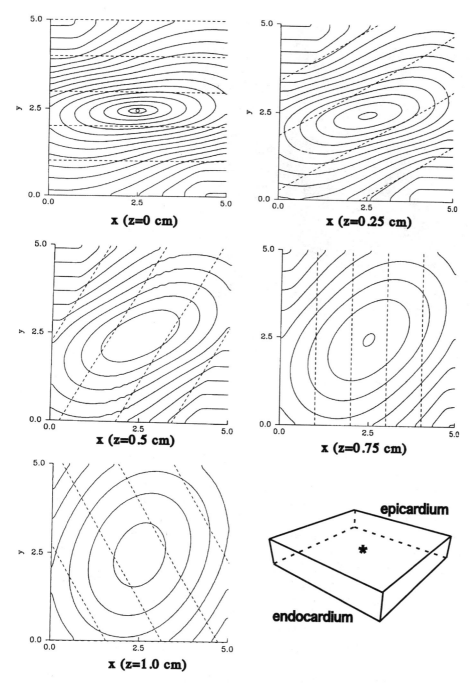

Figure 14.24 Computer simulation of propagation of wavefronts in a three-dimensional tissue slab of $5 \times 5 \times 5$ cm, after point stimulation at the epicardium (top). Rotational anisotropy from epicardium to endocardium is $2\pi/3$. Wavefronts were initiated at 10-ms intervals. Maps are from cross-sectional views of the wavefront locations at different depths. Broken lines indicate fiber direction at each plane. (Modified from Ref. 77, with permission.)

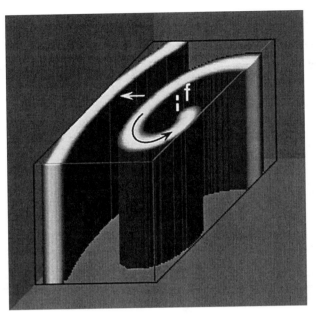

Figure 14.25   A snapshot of a three-dimensional scroll wave in a computer simulation (86) using the FitzHugh–Nagumo equations. The model consisted of an array of $48 \times 48 \times 48$ elements connected to each other by a diffusion term. The cuboidal array was anisotropic in that element size in the x axis was twice as large as those in the y and z axes. Nonexcited areas are transparent. The rotation axis f (filament) is shown by the broken line. White arrow indicates direction of propagation. (Modified from Ref. 86, with permission.)

has led to the application of new experimental and numerical approaches to the study of the two- and three-dimensional spatiotemporal patterns of excitation that result in cardiac rhythm disturbances (31,32,82,83). However, since the myocardium is spatially three-dimensional, the activity observed on the surface of the ventricles can only be used as a rough approximation to the real-life situation. Thus an appropriate three-dimensional representation of the activity is needed for a meaningful study of the mechanisms of arrhythmias in the whole heart. The theory of wave propagation in excitable media provides some additional tools which allow such a representation (84). If, in a hypothetical experiment, one were to stack many thin sheets of cardiac muscle one on top of the other, it might be possible to observe the three-dimensional representation of the spiral wave activity occurring synchronously in all the stacked sheets. The three-dimensional electrical object formed in such an experiment would be a scroll wave. Connecting the cores of all stacked spirals to one another results in a filament (f) around which the scroll wave rotates (Fig. 14.25).

There is some information in the electrophysiological literature about whether the myocardium is able to sustain three-dimensional scroll waves. For example, the experiments of Frazier et al. (85) obtained transmural recordings of the activation patterns of the right ventricular outflow tract of the in situ canine heart during and immediately after the application of cross field stimulation (32), through long orthogonally located electrodes sutured to the epicardial surface over the recording electrodes. The isochrone maps obtained from epicardium, midmyocardium, and endocardium demonstrated vortexlike ac-

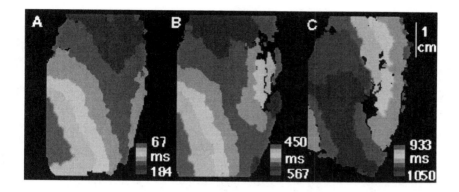

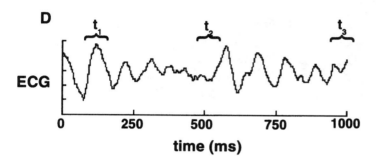

Figure 14.26 Video-imaging experiment showing spiral wave activity on the epicardial surface of the isolated rabbit heart. Isochrone maps from the surface of the free wall of the left ventricle at three different time intervals. White denotes earliest activation times and dark gray denotes late activation. (A) time t_1, a "V-shaped" collision pattern is evident on the surface. (B) An altered collision pattern. (C) A spiral wave pattern with a rotation period of 117 ms. (D) Simultaneous horizontal ECG. Notice the similarity to the simulation results shown in Figure 14. (Modified from Reference 84, with permission.) (See also color plate.)

tivity (period 110 ms) throughout the thickness of the outflow tract wall (85), suggesting that the stimulation protocol resulted in the formation of a scroll wave whose filament was nearly perpendicular to the surface (86).

Recently, video-imaging technology that permits recording from two orders of magnitude more sites than traditional methods, has been used to study spiral waves on the surface of the whole heart (83,84). Similar to Chen et al. (87) and Frazier et al. (85), stimulation protocols derived from spiral wave theory were used to initiate scroll waves whose filaments remained perpendicular to the heart surface. Spiral waves were observed on the heart surface of the rabbit heart and these spiral waves tended to drift (Fig. 14.26). The degree of movement of the spiral waves related to the irregularity of the electrocardiograms (ECGs) with the episodes displaying the largest movement resulting in the largest changes in the ECG (82–84). These results confirmed the predictions of Winfree (69), showing that a spiral wave moving over a large portion of the heart gave rise to undulating ECGs characteristic of torsade de pointes, as well as ventricular fibrillation.

To determine whether the spiral waves observed on the epicardial surface of the ventricles corresponded in fact to three-dimensional scroll waves spanning the entire ven-

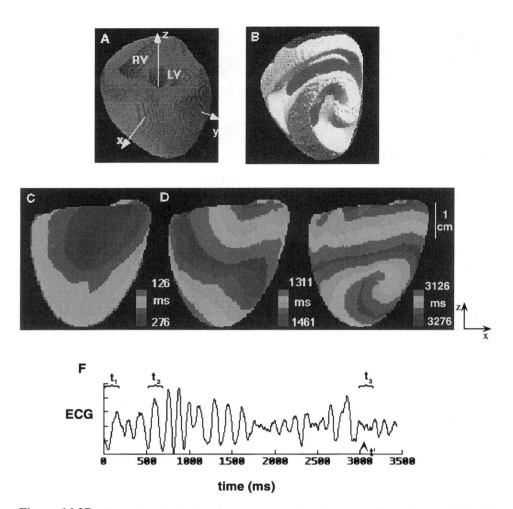

Figure 14.27 Computer simulation of scroll wave activity in a three-dimensional model of the heart. (A) Realistic geometry of right (RV) and left (LV) ventricles. (B) Three-dimensional isosurface of activity (variable analogous to normalized membrane potential) at time t_3 reveals a (scroll wave) shown in white (the heart was made transparent to visualize the three-dimensional structure). (C) to (E) Isochrone maps from the right ventricular epicardial surface at three different time intervals. White denotes earliest activation times and dark gray denotes late activation. (C) At time interval t_1 a V-shaped collision pattern is evident on the surface. (D) at time interval t_2 the scroll wave filament has moved to the lateral surface of the left ventricle and the waves emanating from the scroll wave filament move upward in a "V" pattern and also propagate from the bottom. (E) at time interval t_3 a spiral wave pattern with a (clockwise) rotation period of 125 ms is manifest on the RV surface. (F) A simultaneous horizontal lead (x) ECG displaying irregular period and morphology. (Modified from Ref. 84, with permission.) (See also color plate.)

tricular wall, computer simulations were carried out based on an electrophysiological computer model of the right and left ventricles of the canine heart, which was developed using anatomical geometrical data (88). The dynamics of the excitable tissue were represented using the FitzHugh–Nagumo equations (89), and the results obtained were scaled such that the average period of activation of an individual site during fibrillation matched the rotation period (133 ms) of spirals recorded from the surface of the rabbit heart (82–84).

In Figure 14.27, we have reproduced modeling predictions of vortexlike activity in the whole heart (84). (A) shows a snapshot of the ventricular model with realistic three-dimensional morphology and spatial coordinates. In (B), the simulated heart was made transparent to present an image of a scroll wave spanning the thickness of the ventricular wall at time t_3 (see below). In (C), at time t_1, a three-dimensional scroll wave spanning the left ventricular wall was hidden from view. It was manifest on the right ventricular epicardial surface as a collision of two upwardly moving waves in a "V" pattern. This pattern resulted from the waves emanating from the reentrant source on the left ventricle wrapping around the heart. In (D), at time t_2, the scroll wave had moved to the posterior surface and the pattern on the right ventricular surface was altered with a central breakthrough band (white) giving rise to V-shaped wavefronts moving upward and to the right, and S-shaped wavefronts moving downward and to the left. Finally, in (E), at time t_3, the three-dimensional scroll moved to the wall of the right ventricle and was manifest on the external ventricular surface as a spiral wave. (F) illustrates an ECG in the x direction showing complex patterns of activation that cannot be distinguished from ventricular fibrillation.

Although the spatial patterns on the epicardial surface and the myocardial wall exhibited organized scroll wave activity, the signal at individual sites was irregular, as the scroll wave moved throughout the heart. The simulated ECG (82–84) obtained during this numerical experiment displayed patterns characteristic of fibrillation. Overall, the experimental and numerical results, illustrated respectively in Figures 14.26 and 14.27, provide strong support to the idea that complex patterns of excitation during polymorphic ventricular tachycardia and ventricular fibrillation are the result of three-dimensional scroll wave dynamics.

VIII. DRIFTING SPIRAL WAVES APPLICABLE TO VF IN THE HUMAN HEART

Although it is generally believed that fibrillation in the human heart is the result of large numbers of wandering wavelets, the detailed mechanisms of fibrillation remain unclear. As discussed above, recent advances in optical mapping technology have made it possible to demonstrate that, in small mammalian hearts such as those of the mouse and the rabbit, even a single scroll wave drifting throughout the ventricles can give rise to electrocardiographic patterns that are indistinguishable from VF. However, it is unlikely that given the larger mass of the human heart, only one drifting spiral can prevail. It seems more feasible that self-organization in the case of the larger human heart occurs in the form of a relatively small number of drifting scroll waves that interact with each other giving rise to complex spatiotemporal patterns. Therefore, video imaging of Langendorff-perfused hearts of larger mammals such as sheep, or even humans, should provide insight into the number of spirals and their movement during fibrillation. Recent studies in the sheep heart (90) suggest that within a given VF episode of 5 s, the number of rotating waves varies between 2 and 5. The area of the core around which these rotors reverberate is approximately 0.135 cm^2 and the rotation period is ~114 ms. In addition, estimates indicate that the rotor density on the ventricular surface is approximately one per 3.8 cm^2 (area of recording, 13.4 cm^2). Why is this important? Perhaps the most immediate practical application involves incorporating spiral wave theory to reduce the energy of the field required to terminate fibrillation. In 1990, Krinsky et al. (91) studied how point stimuli, as well as uniform and nonuniform field application, affect the dynamics of spiral waves in both two- and three-dimensional mathematical models. Krinsky et al. (91) suggest that the timing of the stimuli and the position of the spiral waves play a crucial role in defibrillation. Further studies using numerical methods have shown that spirals drift in the direction of an applied electric field (92). It has also been shown in mathematical models that the application of a periodic uniform electric field at a rate similar to the spiral wave rotation period results in a resonant drift of the spiral (93). In fact, a low-energy defibrillator has been suggested utilizing a feedback mechanism and the effect of resonant drift (93). Although it is not clear if resonant drift will occur in heart tissue, it has been shown to exist in chemical media (94).

ACKNOWLEDGMENTS

This work was supported by grants R01-HL49054 and R37-HL33343 (Y.R.) and P01-H139707 (J.J.) from the National Heart, Lung and Blood Institute, National Institutes of Health, Bethesda, Maryland.

This chapter summarizes findings described in previously published articles from our laboratories (see reference list). Our thanks to Gregory Faber and Prakash Viswanathan for their help in preparing figures and to Kathryn Shields and Brenda Hudson for their help in preparing the manuscript.

REFERENCES

1. Luo C, Rudy Y. A model of the ventricular cardiac action potential: depolarization, repolarization and their interaction. Circ Res 1991; 68: 1501–1526.
2. Luo C, Rudy Y. A dynamic model of the cardiac ventricular action potential: I. simulations of ionic currents and concentration changes. Circ Res 1994; 74: 1071–1096.
3. Zeng J, Laurita KR, Rosenbaum DS, Rudy Y. Two components of the delayed rectifier K^+ current in ventricular myocytes of the guinea pig type: theoretical formulation and their role in repolarization. Circ Res 1995; 77: 1–13.
4. Shaw RM, Rudy Y. Electrophysiologic effects of acute myocardial ischemia: a theoretical study of altered cell excitability and action potential duration. Cardiovasc Res 1997; 35: 256–272.
5. Beuckelmann DJ, Wier WG. Mechanism of release of calcium from sarcoplasmic reticulum of guinea-pig cardiac cells. J Physiol (Lond) 1988; 405: 233–255.
6. Zeng J, Rudy Y. Early afterdepolarizations in cardiac myocytes: mechanism and rate dependence. Biophys J 1995; 68: 949–964.
7. Cranefield PF, Aronson RS. Cardiac Arrhythmias: The Role of Triggered Activity and Other Mechanisms. New York: Futura Publishing Co, Inc, 1988.
8. January CT, Riddle JM. Early afterpolarizations: mechanism of induction and block, a role for L-type Ca^{2+} current. Circ Res 1989; 64: 977–990.
9. Luo C, Rudy Y. A dynamic model of the cardiac ventricular action potential: II. afterdepolarizations, triggered activity and potentiation. Circ Res 1994; 74: 1097–1113.
10. Rosen MR. The concept of afterdepolarization. In: Rosen MR, Janse MJ, Wit AL, eds. Cardiac Electrophysiology: A Textbook. New York: Futura Publishing Co, Inc; 1990:267–282.
11. Wenckebach KF. Zur Analyse des Unregelmassingen Pulses: II. Uber Den Regelmassig Intermittirenden Puls. Z F Klin Med 1899; 37: 475–488.
12. Chialvo DR, Michaels DC, Jalife J. Supernormal excitability as a mechanism of chaotic dynamics of activation in cardiac purkinje fibers. Circ Res 1990; 66: 525–545.
13. Rudy Y, Quan W. A model study of the effects of the discrete cellular structure on electrical propagation in cardiac tissue. Circ Res 1987; 61: 815–823.
14. Rudy Y, Quan W. Propagation delays across cardiac gap junctions and their reflection in extracellular potentials: a simulation study. J Cardiovasc Electrophysiol 1991; 2: 299–315.

15. Shaw RM, Rudy Y. Ionic mechanisms of propagation in cardiac tissue: roles of sodium and L-type calcium currents during reduced excitability and decreased gap-junction coupling. Circ Res 1997; 81: 727–741.

16. Fabiato A. Simulated calcium current can both cause calcium loading in and trigger calcium release from the sarcoplasmic reticulum of a skinned canine cardiac purkinje cell. J Gen Physiol 1985; 85: 291–320.

17. Lee KS, Marban E, Tsien RW. Inactivation of calcium channels in mammalian heart cells: joint dependence on membrane potential and intracellular calcium. J Physiol (Lond) 1985; 364: 395–411.

18. Nitta J, Furukawa T, Marumo F, Sawanobori T, Hiraoka M. Subcellular mechanism for Ca^{2+}-dependent enhancement of delayed rectifier K^+ current in isolated membrane patches of guinea pig ventricular myocytes. Circ Res 1994; 74: 96–104.

19. Shaw RM, Rudy Y. Electrophysiologic effects of acute myocardial ischemia: a mechanistic investigation of action potential conduction and conduction failure. Circ Res 1997; 80: 124–138.

20. Gettes LS, Reuter H. Slow recovery from inactivation of inward currents in mammalian myocardial fibres. J Physiol (Lond) 1974; 240: 703–724.

21. Kodama I, Wilde A, Janse MJ, Durrer D, Yamada K. Combined effects of hypoxia, hyperkalemia and acidosis on membrane action potential and excitability of guinea-pig ventricular muscle. J Mol Cell Cardiol 1984; 16: 247–259.

22. Veenstra RD, Joyner RW, Wiedmann RT, Young ML, Tan RC. Effects of hypoxia, hyperkalemia, and metabolic acidosis on canine subendocardial action potential conduction. Circ Res 1987; 60: 93–101.

23. Bigger JT Jr, Hoffman BF. Antiarrhythmic drugs. In: Gilmam AG, Rall TW, Nies AS, Taylor P, eds. The Pharmacological Basis of Therapeutics. New York: Pergamon Press, 1990:840–873.

24. Wit AL, Janse MJ. The Ventricular Arrhythmias of Ischemia and Infarction: Electrophysiological Mechanisms. New York: Futura Publishing Co, 1992.

25. Spach MS, Dolber PC, Heidlage JF, Kootsey JM, Johnson EA. Propagating depolarization in anisotropic human and canine cardiac muscle: apparent directional differences in membrane capacitance: a simplified model for selective directional effects of modifying the sodium conductance on V_{max}, tau foot, and the propagation safety factor. Circ Res 1987; 60: 206–219.

26. Shaw RM, Rudy Y. The vulnerable window for unidirectional block in cardiac tissue: characterization and dependence on membrane excitability and cellular coupling. J Cardiovasc Electrophysiol 1995; 6: 115–131.

27. Quan W, Rudy Y. Unidirectional block and reentry of cardiac excitation—a model study. Circ Res 1990; 66: 367–382.

28. Rudy Y. Reentry: insights from theoretical simulations in a fixed pathway. J Cardiovasc Electrophysiol 1995; 6: 294–312.

29. Mines GR. On circulating excitation on heart muscles and their possible relation to tachycardia and fibrillation. Trans R Soc Can 1914; 4: 43–53.

30. Wiener N, Rosenblueth A. The mathematical formulation of the problem of conduction of impulses in a network of connected excitable elements, specifically in cardiac muscle. Arch Ins Cardiol Mex 19xx; 16: 205–265.

31. Davidenko JM, Pertsov AM, Salomonsz R, Baxter W, Jalife J. Stationary and drifting spiral waves of excitation in isolated cardiac muscle. Nature 1991; 355: 349–351.

32. Pertsov AM, Davidenko JM, Salomonsz R, Baxter WT, Jalife J. Spiral waves of excitation underlie reentrant activity in isolated cardiac muscle. Circ Res 1993; 72: 631–650.

33. Schalij MJ. Anisotropy and ventricular tachycardia. Doctoral Thesis, University of Limburg, Maastricht, The Netherlands, 1988.

34. Zykov VS. Analytical evaluation of the dependence of the speed of an excitation wave in two-dimensional excitable medium on the curvature of its front. Biophizica 1980; 25: 888–892.

35. Cabo C, Pertsov AM, Baxter WT, Davidenko J M, Gray RA, Jalife J. Wave-front curvature as a cause of slow conduction and block in isolated cardiac muscle. Circ Res 1994; 75: 1014–1028.

36. Fast VG, Klèber AG. Role of wavefront curvature in propagation of cardiac impulse. Cardiovasc Res 1997; 33: 258–271.

37. Winfree AT. When Time Breaks Down. Princeton, NJ: Princeton University Press, 1987.

38. Zykov VS. Simulation of Wave Processes in Excitable Media. New York: University Press, 1987.

39. Knisley S, Hill B. The importance of wavefront curvature for longitudinal propagation and anisotropy in myocardium. J Am Coll Cardiol 1992; 19: 90A.

40. de la Fuente D, Sasyniuk B, Moe GK. Conductance through a narrow isthmus in isolated canine atrial tissue: a model of the W-P-W syndrome. Circulation 1971; 44: 803–809.

41. Fast VG, Klèber AG. Cardiac tissue geometry as a determinant of unidirectional conduction block: assessment of microscopic excitation spread by optical mapping in patterned cell cultures and in a computer model. Cardiovasc Res 1995; 29: 697–707.

42. Krinsky VI. Mathematical models of cardiac arrhythmias (spiral waves). Pharm Ther B 1978; 3: 539–555.

43. Krinsky VI. Self-Organization: Autowaves and Structures Far from Equilibrium. Berlin: Springer-Verlag, 1984.

44. Starobin JM, Zilbeter YI, Rusnak EM, Starmer CF. Wavelet formation in excitable cardiac tissue: the role of wavefront-obstacle interactions in initiating high-frequency fibrillatory-like arrhythmias. Biophys J 1996; 70: 581–594.

45. Agladze K, Keener JP, Müller SC, Panfilov A. Rotating spiral waves created by geometry. Science 1994; 264: 1746–1748.

46. El-Sherif N. The figure-8 model of reentrant excitation in the canine post infarction hearts. In: Zipes DP, Jalife J, eds. Cardiac Electrophysiology and Arrhythmias. New York: Grune and Stratton, Inc., 1985:363–372.

47. Cabo C, Pertsov AM, Davidenko JM, Baxter WT, Gray RA, Jalife J. Vortex shedding as a precursor of turbulent cardiac electrical activity in cardiac muscle. Biophys J 1996; 70: 1105–1111.

48. Beaumont J, Davidenko N, Davidenko J, Jalife J. Spiral waves in two-dimensional models of ventricular muscle: formation of a stationary core. Biophys J 1998; 75: 1–14.

49. Gorelova NA, Bures J. Spiral waves of spreading depression in the isolated chicken retina. Neurobiology 1983; 14: 353–363.

50. Gerisch G. Standienpezifische Aggregationsmuster bei Distyostelium Discoideum. Wilhelm Roux Archiv Entwick Org 1965; 156: 127–144.

51. Lechleiter J, Girard S, Peralta E, Clapham D. Spiral calcium wave propagation and annihilation in Xenopus laevis oocytes. Science 1991; 252: 123–126.

52. Lipp P, Niggli E. Microscopic spiral waves reveal positive feedback in subcellular calcium signaling. Biophys J 1993; 65: 2272–2276.

53. Bub G, Glass L, Publicover NG, Shrier A. Rotors in cultured cardiac myocyte monolayers. Proc Natl Acad Sci *(USA)* 1998; 95: 10283–10287.

54. Winfree AT. Spiral waves of chemical activity. Science 1972; 175: 634–636.

55. Winfree AT. Varieties of spiral wave behavior: An experimentalistís approach to the theory of excitable media. Chaos 1991; 1: 303–334.

56. Allessie MA, Bonke FIM, Schopman FJC. Circus movement in rabbit atrial muscle as a mechanism of tachycardia. Circ Res 1973; 33: 54–62.

57. Dillon SM, Allessie MA, Ursell PC, Wit AL. Influence of anisotropic tissue structure on reentrant circuit in the epicardial border zone of subacute canine infarcts. Circ Res 1988; 63: 182–206.

58. Skinner GS, Swinney HL. Periodic to quasiperiodic transition of chemical spiral rotation. Physica D 1991; 48: 1–16.

59. Ermakova EA, Pertsov AM, Shnol EE. On the interaction of vortices in two-dimensional active media. Physica D 1989; 40: 185–195.

60. Rudenko AN, Panfilov AV. Drift and interaction of vortices in two-dimensional heterogeneous active medium. Stud Biophys 1983; 98: 183–188.

61. Pertsov AM, Ermakova EA. Mechanism of the drift of spiral wave in an inhomogeneous medium. Biophysics 1988; 33: 338–341.
62. Steinbock O, Schütze J, Müller SC. Electric-field-induced drift and deformation of spiral waves in an excitable medium. Phys Rev Lett 1992; 68: 248–251.
63. Yermakova YA, Pertsov AM. Interaction of rotating spiral waves with a boundary. Biophysics 1986; 31: 932–940.
64. Zou X, Levine H, Kessler DA. Interaction between a drifting spiral and defects. Phys Rev 1993; E47: R800–R803.
65. Fast VG, Pertsov AM. Shift and termination of functional reentry in isolated ventricular preparations with quinine-induced inhomogeneity in refractory period. J Cardiovasc Electrophysiol 1992; 3: 255–265.
66. Davidenko JM. Spiral wave activity a common mechanism for polymorphic and monomorphic ventricular tachycardias. J Cardiovasc Electrophysiol 1993; 4: 730–746.
67. Ortiz J, Niwano S, Abe H, Rudy Y, Johnson NJ, Waldo AL. Mapping the conversion of atrial flutter to atrial fibrillation and atrial fibrillation to atrial flutter: insights into mechanisms. Circ Res 1994; 74: 882–894.
68. Ikeda T, Wu FJ, Uchida T, Hough D, Fishbein MC, Mandel WJ, Chen PS, Karagueuzian HS. Meandering and unstable reentrant wave fronts induced by acetylcholine in isolated canine right atrium. Am J Physiol 1997; 273: H356–370.
69. Winfree AT. Electrical turbulence in three-dimensional heart muscle. Science 1994; 266: 1003–1006.
70. Callans DJ, Josephson ME. Ventricular tachycardias in the setting of coronary artery disease. In: Zipes DP, Jalife J, eds. Cardiac Electrophysiology: From Cell to Bedside. Philadelphia: W.B. Saunders Co., 1995:732–743.
71. Frazier DW, Krassowska W, Chen PS, Wolf PD, Danieley ND, Smith WM, Ideker RE. Transmural activations and stimulus potentials in three-dimensional anisotropic canine myocardium. Circ Res 1988; 63: 135–146.
72. Streeter D. Gross morphology and fiber geometry of the heart. Handbook of Physiology Vol. 1: The Heart, Section 2: The Cardiovascular System. Baltimore, MD: Williams & Wilkins, 1979:61–112.
73. Taccardi B, Macchi E, Lux RL, Ershler PR, Spaggiari S, Baruffi S, Vyhmeister Y. Effect of myocardial fiber direction on epicardial potentials. Circulation 1994; 90: 3076–3090.
74. Greenbaum RA, Ho SY, Gibson DG, Becker AE, Anderson RH. Left ventricular fiber architecture in man. Br Heart J 1981; 45: 248–263.
75. Hunter PJ, Smail BH. The analysis of cardiac function: a continuum approach. Prog Biophys Mol Biol 1988; 52: 101–164.
76. Keener JP. An eikonal-curvature equation for action potential propagation in myocardium. J Math Biol 1991; 29: 629–651.
77. Keener JP, Panfilov AV. Three-dimensional propagation in the heart: The effects of geometry and fiber orientation on propagation in myocardium. In: Zipes DP, Jalife J, eds. Cardiac Electrophysiology: From Cell to Bedside. Philadelphia: W. B. Saunders Co., 1995:335–347.
78. Colli Franzone P, Guerri L, Taccardi B. Spread of excitation in a myocardial volume: Simulation studies in a model of anisotropic ventricular muscle activated by point stimulation. J Cardiovasc Electrophysiol 1993; 4: 144–160.
79. Colli Franzone P, Guerri L, Taccardi B. Potential distributions generated by point stimulation in a myocardial volume. Simulation studies in a model of anisotropic ventricular muscle. J Cardiovasc Electrophysiol 1993; 4: 438–458.
80. Winfree AT. Evolving perspectives during 12 years of electrical turbulence. Chaos (Focus Issue: Fibrillation in Normal Ventricular Myocardium) 1998; 8: 1–19.
81. Winfree AT. Theory of spirals. In Zipes DP, Jalife J, eds. Cardiac Electrophysiology: From Cell to Bedside. Philadelphia: W. B. Saunders Co., 1995:379–389.
82. Gray RA, Jalife J, Panfilov AV, Baxter WT, Cabo C, Davidenko JM, Pertsov AM. Nonstation-

ary vortexlike reentrant activity as a mechanism of polymorphic ventricular tachycardia in the isolated rabbit heart. Circulation 1995; 91: 2454–2469.

83. Gray RA, Jalife J, Panfilov AV, Baxter WT, Cabo C, Davidenko J, Pertsov AM. Mechanisms of cardiac fibrillation. Science 1995; 270: 1222–1225.

84. Jalife J, Gray R. Drifting vortices of electrical waves underlie ventricular fibrillation in the rabbit heart. Acta Physiol Scand 1996; 157: 123–131.

85. Frazier DW, Wolf PD, Wharton JM, Tang ASL, Smith WM, Ideker RE. Stimulus-induced critical point: Mechanism for the electrical initiation of reentry in normal canine myocardium. J Clin Invest 1989; 83: 1039–1052.

86. Pertsoy AM, Jalife J. Three-dimensional vortex-like reentry. In: Zipes DP, Jalife J, eds. Cardiac Electrophysiology: From Cell to Bedside. Philadelphia: W. B. Saunders Co., 1995:403–410.

87. Chen PS, Wolf PD, Dixon EG, Danieley ND, Frazier DW, Smith WM, Ideker RE. Mechanism of ventricular vulnerability to single premature stimuli in open chest dogs. Circ Res 1988; 62: 1191–1209.

88. Nielsen PMF, Le Grice IJ, Smaill BH, Hunter PJ. Mathematical model of geometry and fibrous structure of the heart. Am J Physiol 1991; 260: H1365–H1378.

89. Panfilov AV, Keener JP. Reentry in an anatomical model of the heart. Chaos, 1995; 5: 681–689.

90. Gray RA, Pertsov AM, Jalife J. Spatial and temporal organization during cardiac fibrillation. Nature 1998; 392: 75–78.

91. Krinsky VI, Biktashev VN, Pertsov AM. Autowave approaches to cessation of reentrant arrhythmias. Ann NY Acad Sci 1990; 591: 232–246.

92. Pumir A, Plaza F, Krinsky VI. Effect of an externally applied electric field on excitation propagation in the cardiac muscle. Chaos 1994; 4: 1–9.

93. Biktashev VN, Holden AV. Resonant drift of an autowave in a bounded medium. Phys Lett 1993; 181: 216–224.

94. Perez-Muñizuri V, Aliev R, Vasiev B, Krinsky VI. Electric current control of spiral wave dynamics. Physica D 1992; 56: 229–234.

15

Mapping and Invasive Analysis

ROBERT L. LUX and ROBERT S. MACLEOD

University of Utah, Salt Lake City, Utah

MASOOD AKHTAR

University of Wisconsin and St. Luke's Medical Center, Milwaukee, Wisconsin

I. INTRODUCTION

The field of electrocardiographic mapping evolved out of visual observations of the motion of the heart during normal and abnormal rhythms and the development of the technology to record electric potentials from multiple sites on the heart and body surface. The list of nineteenth- and early twentieth-century physicians and scientists who contributed to the study of arrhythmias, particularly the conditions that cause them, includes MacWilliam, Mines, Garrey, Lewis, Wiggers, and many others (1–5). To a large extent, their ability to understand mechanisms depended on their ability to link mechanical to electrical phenomena despite very limited technical capabilities and sparse knowledge of electrophysiology at the cellular level. The advent of the electronic and then the computer age removed many of the technical barriers and the field of electrocardiographic mapping emerged in the early 1960s, largely through the work of Taccardi (6). For the first time, many body-surface ECGs could be recorded and the potential distribution observed as a function of time. Computers then made it possible to perform the necessary signal processing, numerical calculations, and visualization tasks required to analyze cardiac electrical activity. The use of mapping techniques simplified the interpretation of electrocardiographic phenomena by providing a more complete picture of electrical activity of the heart than could be inferred from a few, sparsely sampled ECGs. In effect, the technique permitted electrocardiography to move beyond the study of empirical links between a few voltage-time signals and disease states to the characterization of the spatiotemporal behavior of bioelectric sources.

This same technology was soon applied to the heart in experimental studies, which then allowed direct measurement of cardiac electrograms (EGs) on the epicardial surface or within the wall of the muscle by means of transmural needle electrodes. The early work of many pioneers in this area (7–16) provided new insights and impetus for cardiac mapping to develop to its current status as a widely used experimental and clinical technique.

The field of cardiac mapping can be subdivided into two categories. The first is potential mapping, in which the main objectives are to analyze electric potential in order to understand or infer the nature and time course of cardiac sources in response to interventions and disease. The second category is activation (or recovery) time mapping, which has the more focused goal of observing and tracking the sequences of depolarization associated with the spread of excitation and repolarization in the heart. For the diagnosis and analysis of arrhythmias, activation mapping is more useful as it provides the information required to follow the course of activation in time and in space. In this chapter, we present a brief review of the electrical and electrophysiological basis for cardiac mapping, the fundamental technical and technological aspects that make it possible, and finally, a view toward the future of this technique.

II. PRINCIPLES OF ARRHYTHMIA MAPPING

A. Electrocardiographic Potentials and Activation Times

The biophysical basis for all arrhythmia mapping is the dynamic distribution of electrocardiographic potentials generated by the propagation of action potentials throughout the excitable myocardium. Cellular depolarization and its propagation from cell to cell are complex phenomena covered in other chapters of this book. The essential result for cardiac mapping is the close juxtaposition of regions of the heart that are depolarized and neighboring regions still at rest. The resulting potential differences create currents that flow throughout all electrically conductive tissues in the body and produce time- and space-varying voltage differences that are measurable from the heart as well as from the body surface (17–20).

Figure 15.1 shows a sequence of instantaneous potential distributions recorded from the lateral wall of the left ventricle of a normal canine heart using an array of 1200 electrodes embedded in a nylon sock that fits over the ventricles. Also included in the figure are two unipolar epicardial EGs recorded from the indicated locations. The potential distributions shown are from time instants early, mid, and late during depolarization (QRS complex of the ECG waveform) of the ventricles following initiation of activation by electrical stimulation at a point near the center of the ellipsoidal pattern visible in the left panel. Of note is the progression of the sharp boundary between positive (leading) and negative (trailing) regions away from the site of stimulation. Another feature of note is that the QRS deflections in EGs in the vicinity of the stimulus site are mostly negative, while those farther from the stimulus site show a biphasic morphology with an early positive followed by a later negative wave. At locations even further from the stimulus site, the QRS waves of the EGs become almost entirely positive.

The transition between the positive and negative waves in EGs, called the *intrinsic deflection* of the QRS, defines the time at which the wavefront associated with propagating activation moves past the recording site. By determining the time of this EG transition, referred to as the *activation time*, for enough sites it is possible to describe the sequence of spread of activation as a contour map. In this map, lines connecting regions with identical activation times are called isochrones of activation time and the collection of isochrones over a single heartbeat is called an activation isochrone or activation sequence map. Figure 15.2 shows the activation map associated with the isopotential maps presented in Figure

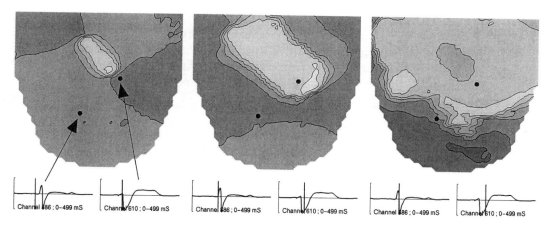

Figure 15.1   Potential distributions from the anterior epicardium of a dog recorded with a 1200-electrode sock. Each panel contains the potential distributions at a different time instant, indicated by the vertical lines in the electrograms below each panel. Lighter shades indicate more negative potentials while darker shades mark regions of more positive potentials. Dark circles mark the locations of the electrograms shown below each map. (B. Taccardi, unpublished data.)

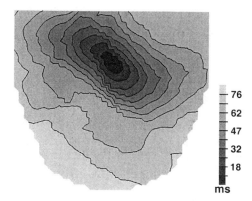

Figure 15.2   Activation isochrone maps from the beat shown in the previous figure. The scale bar beside the map indicates the range of activation times and their corresponding gray shades. (B. Taccardi, unpublished results.)

15.1. This example also shows the link between activation time and spatial distribution of potential. The isochrone line for each value of time—originally derived from the voltage signals—also marks the location of the steepest spatial gradient in the isopotential map from that same instant in time. This provides a second definition of the activation wavefront as the location of steepest gradient of the potential for each instant in time.

   Activation isochrone lines, however, actually reflect the intersection of a three-dimensional surface of propagating depolarization with the physical surface at which the measurements are made. It is the determination of this three-dimensional activation wavefront that is the fundamental objective of arrhythmia mapping. Two-dimensional measurements from, for example, the epicardial and endocardial surfaces or catheters in the heart chambers and vessels, provide projections of activation wavefronts. To measure three-dimensional wavefronts directly requires arrays of multipole plunge electrodes that are inserted into the heart. The goal of many research projects in cardiac mapping is to establish the links between fundamental characteristics of propagation and their reflection in measurable signals.

B. Activation Sequence

Electrical activation of the heart normally proceeds from an initial depolarization in the sinoatrial (SA) node located in the right atrium near the superior vena cava, from which it spreads throughout both atria. To reach the ventricles, activation then passes through a sequence of specialized conduction tissues—the atrioventricular (AV) node, bundle of His, and then the left and right bundle branches—and then spreads very rapidly through the network of Purkinje fibers that lines the subendocardium to finally reach the working myocardium. For a normally initiated and conducted beat, the ventricles are activated in a consistent pattern determined by the locations of the terminal ends of the Purkinje fibers. There are some beat-to-beat variations in the details of this sequence caused by changes in mechanical load, heart rate, or other influences of the autonomic nervous system, but the more striking feature is the consistency of the pattern of activation over time.

In arrhythmias, this normal sequence is disrupted. For example, premature beats that are initiated within the ventricles propagate from a point source through the working myocardium, eventually reaching one or more terminal points of the Purkinje fibers, and then spreading rapidly through the remainder of the conduction system. The resulting activation surface initially has the contour of a single ellipsoid but then breaks up into multiple surfaces, each of which is initiated at the terminal ends of the conduction system. Another example is monomorphic reentrant ventricular tachycardias (VTs), which arise from stable loops of propagating excitation. In this case, the moving activation surface is thought to remain relatively coherent, but may follow a complex path that includes regions of slowed conduction. Such VTs are stable and exhibit an activation sequence that often rotates around anatomical obstacles or regions of functional block (as we will demonstrate in Fig 15.7). Polymorphic VT, on the other hand, does not have a stable pattern and the activation surface wanders through the ventricles, changing from beat to beat (Fig. 15.15 demonstrates examples of both). The technical objectives of mapping during arrhythmias are to measure and track this abnormal activation surface, to identify regions of normal and abnormal conduction, and to develop models that explain the nature of the arrhythmias and suggest appropriate therapy. There are many challenges to this task that derive from the complexity and transient nature of propagation, as well as the appearance of regions of complete loss of excitability that often accompany ischemia, infarction, or heart failure.

An electrophysiological feature unique to the myocardium is the long refractory period that follows each depolarization. While nerve cells remain refractory for just a few milliseconds, refractoriness in cardiac cells lasts several hundred milliseconds, corresponding to the plateau and repolarization phases of the cardiac action potential. Refractoriness plays a protective role in cardiac propagation because it ensures that once activated, tissue cannot be restimulated for a period of time long enough to ensure that activation has passed entirely out of the region. By this means, each activation (contraction) is followed by a period of quiescence (relaxation) that permits proper recovery of the cells and adequate filling of the chambers of the heart with blood from the venous circulation. The sequence of repolarization is governed by the sequence of activation and the intrinsic action potential duration of cells. In the context of arrhythmia mapping, changes in repolarization, especially heterogeneous distributions of refractory times, are thought to play a key role in the initiation and maintenance of reentrant arrhythmias, as is discussed in Chapter 17.

C. Conduction and Conduction Block

A critical component of cardiac arrhythmia mapping is the correct identification and characterization of regions of the heart that exhibit either partial or complete block of conduction. The reasons for such conduction blocks include regional heart disease and infarction, acutely compromised perfusion of the tissue, or abnormal response of refractoriness to variations in cycle length. The mechanisms of block range from completely inactive cells, to cells with reduced action potential amplitude and duration, and to cells that have a prolonged refractory period and are thus inexcitable following a short preceding cycle length. Cells may also remain individually active, but become decoupled from their neighbors in response to changes in extracellular or intracellular environment, such as occurs with reduced pH or elevated Ca^{2+} concentration. Regardless of the cause, conduction block in a region containing otherwise normally active cells during one beat will leave the region *more* excitable than its neighbors on the subsequent beat. The resulting disparity of repolarization then creates an environment conducive to reentrant arrhythmias.

The detection of conduction block from cardiac maps, usually indicated as the absence of well-defined activation times within a region, is crucial in understanding the mechanisms responsible for specific arrhythmias. Importantly, regions of abnormal conduction usually coincide with regions critical to the initiation and maintenance of the arrhythmias. However, due to the complexity of disease, there is often more than one such region, and suppression of abnormal electrophysiological influences in one may unmask others, so that the nature of the arrhythmia will change.

D. Circuits and Pathways

We have described reentry as a mechanism of cardiac arrhythmias, but not yet discussed the interplay among activation, heterogeneous repolarization, and conduction block that combine to form the pathways of continuous reactivation that lead to arrhythmias. Reentrant pathways often originate as a consequence of conduction block, especially those that allow activation to pass in one direction but not the other (i.e., unidirectional block). Regions that contain a heterogeneous mix of viable and damaged tissue may be a substrate for unidirectional block, sometimes in the form of narrow strands of viable cells embedded in otherwise necrotic tissue. These same conditions may also result in propagation delays that are severe enough that the activation emerging from the damaged zone finds tissue at least partially recovered from a previous activation. If the stimulation from the delayed activation is strong enough and the recovered tissue large enough, a subsequent ectopic activation will result. Under suitable conditions of cardiac geometry, conduction velocity, and rates of recovery, this single ectopic event can initiate a closed loop of propagation through the ventricles that can lead to a sustained reentrant tachycardia (see Fig. 15.7). In most clinical cases, the location of the critical exit sites from the heterogeneous regions is in the subendocardium, a result of the fact that this region is the most likely to be affected by myocardial infarction. A significant minority of arrhythmia cases, however, appears to originate in the subepicardial region (21). The goals of arrhythmia mapping then become the localization of the pathways of reentry and especially the critical sites of origin of arrhythmias, using methods that examine both the endocardial and epicardial regions of the heart.

E. Measurement Spaces

As we have stated, activation sequences in the heart, whether during normal or abnormal propagation, can be considered to consist of one or more three-dimensional surfaces expanding through the myocardium. Since the objective of arrhythmia mapping is to characterize these surfaces and to determine the critical pathways and circuits that allow a specific arrhythmia to occur, it is important to make measurements that permit adequate assessment of the surface. For some arrhythmias that are well localized in the subendocardial region, it may be adequate to use relatively simple curvilinear arrays of electrodes mounted in a catheter. Other cases that require larger coverage or more spatial detail in sampling may demand more extensive arrays of electrodes mounted on epicardial socks or ventricular balloons. Even more detailed mapping of the heart volume might require the use of multielectrode needles plunged into the ventricular walls.

In experimental mapping of normal or abnormal propagation, the use of dense (1- to 2-mm spacing) arrays of electrodes mounted on nylon meshes permits detailed analysis of waveform, potential distribution, and activation pathway information. Similarly, large numbers of multielectrode plunge electrodes can be used to define the surfaces of propagation within the myocardium itself. Of course, the more sites studied, the greater the cost in terms of acquisition hardware and time, as well as the computational capabilities required to display and interpret the data. As a result of the costs and the invasive nature of the measurements, such methods are not of particular value in clinical settings.

For the clinical assessment of arrhythmias, invasive techniques similar to those used in animal experiments were employed during the 1980s and early 1990s when open-chest surgery and direct cardiac surface mapping provided the only means for detecting and treating arrhythmias. However, recent advances in catheter-based techniques now provide a means both to assess activation sequence and to interrupt reentry by radiofrequency ablation without surgery and have reduced the mortality and morbidity associated with arrhythmias (22). The newest approaches use multiple electrodes mounted on catheter "baskets" that expand in the ventricular or atrial cavities and directly measure the endocardial surface potentials and activation times (23,24). A further method of intracavitary mapping uses noncontact multielectrode probes that measure potentials from within the ventricular cavities. Estimating the endocardial potentials from the noncontact probes requires additional signal-processing steps that include the numerical solution of an equation for the electric field in the blood volume (25–27).

III. TECHNOLOGICAL METHODS OF ARRHYTHMIA MAPPING

While the technical capabilities of cardiac mapping systems have historically placed significant limits on such parameters as the maximum number of recording channels, sample frequency, and the duration of continuous recordings, contemporary computer-based acquisition hardware minimizes these obstacles. The management, basic signal processing, and visualization of mapping data have also become much more straightforward than in the past. The technical challenges of today have shifted much more into the areas of automated analysis, feature extraction, and quantitative characterization of mapping data. Here we provide a brief overview of some of the essential aspects of acquiring, analyzing, interpreting, and displaying arrhythmia mapping data for both experimental and clinical applications. Further sources of technical information on mapping include a number of excellent comprehensive reviews (28–31).

A. Lead Systems and Electrode Arrays

There are two basic approaches to measuring cardiac activity, known as unipolar and bipolar leads. Unipolar leads represent the potential difference between one site on or near the heart and a reference, or *indifferent*, electrode typically located remotely from the heart. The actual electrodes consist of a small metal pellet, disk, or uninsulated length of wire, placed in direct contact with myocardium. The electrode is connected via a wire to the noninverting input of an electronic differential amplifier while the inverting (negative) inputs of all the amplifiers are connected to the indifferent electrode. Thus, the potential measured by each amplifier represents the difference of voltage between an individual electrode recording site and the reference site. The main advantages of unipolar leads is that they sense potential relative to a common electrode and thus provide the necessary information for potential mapping. Their major disadvantage is that a unipolar signal contains both local information and also attenuated fields from remote events and measurement noise. The signals in Figure 15.1 are unipolar EGs.

Bipolar leads were devised to increase the sensitivity to local activity while reducing distant effects. They consist of two, closely spaced electrodes, with one connected to the inverting and the other to the noninverting input of an amplifier. Thus the signal measured represents the voltage difference between two neighboring recording sites—the algebraic difference between the unipolar signals that would have been recorded from the separate poles using an arbitrary, indifferent reference. The important feature of recordings from closely spaced bipolar electrode pairs is that similarities in the unipolar signals are removed via electronic subtraction and leave only those aspects of the unipolar signals that are different. Thus bipolar leads are most sensitive to local electrocardiographic information and are insensitive to distant activity. As a further consequence, bipolar leads emphasize the passage of the activation wavefront and may improve detection of activation. In essence, bipolar electrodes function as spatial differentiators of electrocardiographic potential. One weakness of bipolar leads is that they do not share a common reference and thus cannot be compared to other leads in an isopotential map. Figure 15.3 shows two pairs of unipolar EGs recorded from closely spaced electrodes as well as their differences, which are the bipolar signals that would have been recorded from them.

A fundamental question in any mapping application is the number of electrodes required together with their locations. The answer is based on a combination of features such as the desired biomedical goal, the accessibility of the heart to measurements, the types of electrodes available, and the capabilities of the acquisition system. The question of the minimum required spatial resolution has not been resolved completely but current consensus indicates a spacing of ≈ 2.5 mm is required to capture the details of cardiac activation (32). A selection of typical electrode configurations of either unipolar or bipolar leads includes the following:

Epicardial arrays sewn into nylon socks that cover some or all of the ventricles
Rigid plaque arrays with regular electrode spacing that cover 1- to 10-cm^2 areas of the ventricles or atria
Sets of transmural plunge needles with 3 to 12 electrodes in each needle
Catheters with 1 to 16 electrodes that are placed in the ventricles, atria, or coronary veins

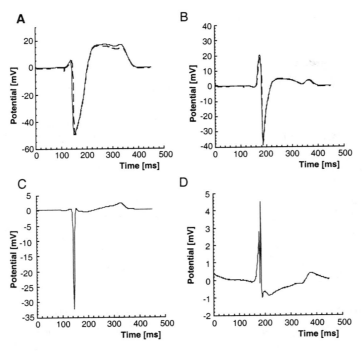

Figure 15.3   A sample of unipolar electrograms and the bipolar leads formed from their difference. (A) and (B) each contain two very similar unipolar electrograms from neighboring sites in a 1200-lead epicardial array. (C) and (D) contain the bipolar leads formed by subtracting the two unipolar signals in (A) and (B), respectively. Note the different vertical scales in each plot. (B. Taccardi, unpublished data.)

Basket catheters consisting of 4 to 6 strands, each containing 4 to 8 electrodes for insertion into the right or left atrium

An inflatable, multielectrode balloon or large bore catheter with 10 to 80 electrodes that can be introduced into the ventricles through an incision.

Figure 15.4 contains diagrams of a variety of electrode configurations for cardiac mapping including a 10-pole plunge needle, a multielectrode catheter, a basket catheter, and an inflatable endocardial electrode array.

Because mapping depends explicitly on spatial relationships among signals at different sites, it is necessary to determine the locations of all the electrodes in an array. When the electrodes are regularly spaced as in, for example, a rigid epicardial plaque array or a plunge needle, only a small number of locations needs to be determined in order to reconstruct the complete geometry. For irregularly spaced arrays or a collection of individual catheters, it is necessary to specify the location of each electrode individually. In clinical applications, electrode locations are typically specified relative to anatomical landmarks by means of fluoroscopy or by intracardiac echocardiography (33). A recent innovation is a system that uses external magnetic field emitters and miniature magnetic field sensors mounted on a catheter to measure both electrode location and individualized endocardial geometry (34). Similar approaches may be available for experimental use but, typically,

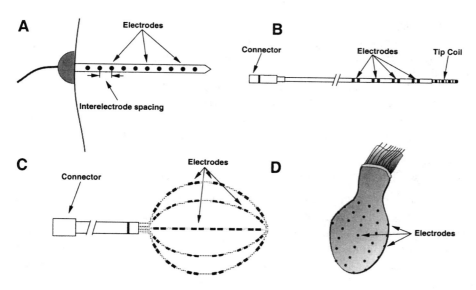

Figure 15.4 A sample of electrode arrays common used in cardiac mapping. (A) Plunge needle with 10 individual electrodes; (B) multielectrode catheter arranged in four bipolar pairs; (C) basket catheter with six wands, each containing six bipolar pairs; and (D) inflatable sock array for surgical insertion into the left ventricle.

access to the heart is sufficient for direct measurements of electrode locations using either mechanical or electromagnetic digitization devices.

B. Recording Systems

Typical recording systems for acquisition and analysis in arrhythmia mapping include high impedance amplifiers, sample-and-hold and multiplexing circuits, analog-to-digital converters, computer interfaces, and software to store the streams of digitized values. The amplifiers translate the millivolt level signals acquired by the electrodes to signals in the range of 1 to 10 V. Sample-and-hold circuits are used to synchronously capture the voltages from all amplifier outputs to ensure that the sample times of all signals from the electrodes are identical. The sampling rate of a mapping system establishes its time resolution and must be high enough to capture the most rapid fluctuations (highest frequencies) of interest in the signals and yet as low as possible to minimize the overall required acquisition capacity—the product of sampling rate and the number of individual recording channels determines both the necessary throughput and storage demands of an acquisition system. Typical sampling rates for cardiac mapping range from 1000 to 5000 samples/s (equivalent to 1–0.2-ms between samples) with some experimental systems supporting up to 50,000 samples/s. Multiplexing circuits are used to merge the signals from a set of input channels into a data stream that can then pass through a single analog-to-digital (A/D) converter. A/D converters then translate the continuous spectrum of analog voltages into a discrete, digital representation, typically using 12 or 16 bits to span the full range of voltages. The result is a voltage amplitude coded into one of 4096 (2^{12}) or 65,536 (2^{16}) discrete values. Thus each EG becomes a numerical sequence of equally spaced time samples, with each sample consisting of a discrete number representing the voltage at a single electrode. The digital outputs from all the A/D converters are typically multiplexed once again so that all the time samples from all channels merge into a single stream that can then be passed to a computer. Computer interfaces and appropriate "real-time" software capture the sequences of sampled, multiplexed, and digitized EGs in computer memory or disk for immediate display and quality control as well as off-line analysis, interpretation, and review. Early systems, whether for experimental research or clinical application, were bulky, costly, and limited in the number of signals (channels) they could record as well as the duration over which they could accumulate data. Present systems are compact, relatively inexpensive, and run on small workstations, desktop, or laptop computers with sampling capacities as high as 2000 channels (35).

C.   Activation Time Measurements

As described above, times of activation at each recording site are the primary measurements needed to characterize and interpret the dynamic changes of electrocardiographic signals recorded during arrhythmias. The algorithms used depend on factors such as the type of signals recorded, the amplitude of each EG, the level of electrical noise contaminating each signal, and the nature of the activation (*e. g.*, normal, ectopic, or reentrant). For unipolar EGs, the time of minimum dV/dt (time of the most rapid downstroke of the intrinsic deflection of the QRS) is known to correspond to the time of most rapid upstroke of the action potentials in myocardial cells near the sampling electrode (11). For unipolar electrodes having a small contact area (corresponding to tens of cells), numerical algorithms to estimate signal derivatives yield well-defined, sharp spikes that define activation times unambiguously to within one or two sample intervals. For bipolar EGs, signal morphology varies depending on the orientation of the electrodes relative to the activation wavefront, forming either triphasic or biphasic QRS deflections. As a result, the algorithm for determining activation time must also vary with signal morphology (36). For triphasic EGs, the time of the central peak yields the best estimate of activation time, while for biphasic signals, the zero crossing between the positive and negative components is most accurate (36).

When recording signals from myocardium that is either ischemic or necrotic (scarred from prior infarction or other cardiac disease), there are complications that introduce ambiguities in the measurement of activation times. Unipolar EGs recorded from ischemic or necrotic tissue show slower intrinsic deflections and/or intrinsic deflections having a jagged character, resulting in multiple peaks in the estimated derivative (dV/dt) (37). EGs recorded from bipolar leads in similar tissue may exhibit similar ambiguities with wide spikes or multiple, "fractionated" waves and no obvious single time of activation. Both abnormal EG morphologies are associated with disrupted conduction in the region of the recording site that results from nonuniform propagation of depolarization in the region, secondary to consequences of infarction or disease. One approach to extracting activation times in cases of ambiguity in the temporal derivative, at least for regular electrode grids, is to estimate the *spatial* gradient of the potential field for each instant in time and use its maximum as a marker of the path of activation (38). Another approach under similar conditions is to match the recorded cardiac maps to a theoretical model of propagation and extract activation time as a parameter using a fitting procedure (39). This is one example of a more general strategy suggested by Ideker *et al.* of using measured data to estimate parameters of the underlying activation by solving an appropriately formulated inverse problem (29). There always remain some instances in which activation cannot be adequately detected, or may not even exist in the region sensed by the electrode, such as in the case of conduction blocks. EGs, especially those from unipolar leads, may contain inflections of reduced amplitude from neighboring regions and adequate detection of these may be critical to interpreting the arrhythmia and understanding its mechanisms.

D. Visualization Analysis and Interpretation

In Figures 15.1 to 15.3 we have already illustrated the basic means for visualizing electro-cardiographic mapping, namely, electrograms and their morphology, isopotential maps, and isochrone or activation sequence maps. It is clear that adequate spatial sampling of the electrocardiographic signals and appropriate display of the information provide the means to understand and interpret the sequence of depolarization, whether normal or abnormal. During complex arrhythmias, when electrogram features become ambiguous as a consequence of regional block, slow conduction, fractionation arising from colliding wave-fronts, or reentry itself, sampling density, robust analysis tools, and flexible visualization methods become even more critical for interpreting the activation sequences in the measured maps.

One common visualization technique in clinical electrophysiology is to present all the individual EG signals in a column ordered either by physical proximity or the sequence of activation, an example of which is shown in Figure 15.5. By comparing activation times and then correlating with the underlying electrode/heart geometry, it is possible to determine the location of the ectopic focus or identify some part of the reentrant pathway.

A more directly spatial method of viewing activation is to project sequences of isopotential maps or activation maps on a two- or three-dimensional rendering of the actual measurement geometry. Modern graphics workstations now make it possible to view long sequences of maps or a sliding window of activation time maps in real-time animations, controlled by the viewer (40,41). Such facilities permit a thorough examination of both the temporal and spatial aspects of the spread of activation and can be crucial in re-constructing complex reentrant loops.

Interpretation of reentry requires that individual components of the wavefronts be identified and differentiated from other components that may travel close by, but later in

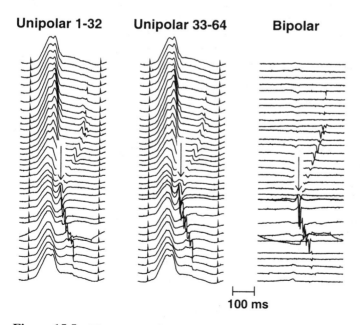

Figure 15.5 Electrograms from two parallel rings of electrodes on the atrial base of the heart of a WPW patient. (Unpublished data from University of Utah clinical study.)

time and/or in different directions. A useful visual aid is to identify those areas or lines that separate noncontiguous wavefronts as these often serve as anchors around which reentrant waves will circulate. In Figure 15.7, the thick lines indicate such regions of block. Locating block requires comparison of activation times from neighboring measurement sites; if the difference exceeds a threshold that is based on electrode spacing and expected propagation velocities, then there is likely to be block between the two sites.

IV. EXPERIMENTAL APPLICATIONS OF ARRHYTHMIA MAPPING

In order to illustrate mapping techniques, we present several examples from experimental studies. Figure 15.1 contains activation information from unipolar EGs, recorded from a flexible nylon sock containing 1200 electrodes at about 2-mm spacing that covers a portion of the left ventricle. An obvious feature of the map is the elliptically shaped region of negative potentials (lighter gray shades) that expands with time from the stimulus site. The orientation of the ellipse is a function of local fiber direction and the "bulging" of the ellipse with time reflects electrotonic interactions with activation wavefronts in the deeper fibers (19). Another important feature of this map that would be undetectable without high-resolution cardiac mapping are the sites of "breakthrough," the small islands of activation visible in the right panel of the figure that emerge from under the epicardium and precede the coordinated propagation over the epicardial surface that follows within a few milliseconds. Breakthroughs are thought to originate from depolarization of the terminal ends of the Purkinje fibers in the subendocardium, which then propagate outwardly toward the epicardium, emerge there, and then join to form ever more complex wavefronts of activation.

The high-resolution potential distributions in Figure 15.6 show data from a 21×25 flexible plaque array of unipolar electrodes (spaced on 2-mm centers). The isopotential lines show the spread of activation at three instants following stimulation from a site within the array. The left and middle panels illustrate typical patterns of early ectopic activation in which an elliptically shaped central activation region (lighter gray shades in the figure) has a region of positive potential (darker shades) at either end of the ellipse—this figure contains only one of these positive regions. The rightmost panel of this figure also shows the appearance of a breakthrough as an isolated island of negative potential within an otherwise unexcited region.

An example of a classic form of ventricular tachycardia from a reentrant loop around a region of block that has been studied extensively is shown in Figure 15.7 (42). Thin lines indicate the isochrones of activation and reveal the pathway of the wavefront as it moves around the upper line of block indicated by the heavy lines. Reentry occurs when the wavefront reaches the right edge of the map (190 ms) and then reactivates tissue that has recovered sufficiently from the previous beat (10 ms).

V. CLINICAL APPLICATION OF INTRACAVITARY MAPPING

Since its introduction in the late 1960s, cardiac mapping has played a rapidly growing role in the diagnosis and treatment of clinical arrhythmias (43,44). Mapping of both the epicardium and the endocardium is often combined with programmed electrical stimulation, a technique that uses direct stimulation of the heart to elicit electrical abnormalities (45–52). In the remainder of this chapter, we illustrate the use of electrophysiological mapping as a means of detecting and differentiating arrhythmias in patients.

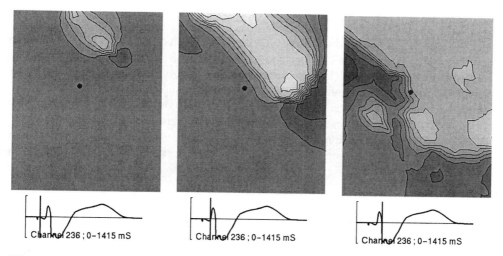

Figure 15.6 Potential distributions from a grid of epicardial electrodes. Recording electrodes were in a 21 × 25, regularly spaced (2 mm) mesh placed on the right ventricle. The location of the electrogram at the bottom of each panel is indicated by the black circle in the maps and the vertical line marks the time instant of each map. (B. Taccardi, unpublished data.)

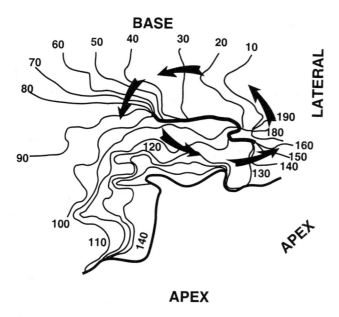

Figure 15.7 Activation isochrone maps from an epicardial array showing reentry. Thin lines are activation isochrones labeled with the time of each while the thick lines mark lines of conduction block. (Adapted from Ref. 42 with permission.)

A. Routine Electrophysiological (EP) Evaluation

The first recording of a His bundle electrogram using an electrode catheter was published in 1969 (43). An electrode catheter was placed across the atrioventricular junction to record from the His bundle (Fig. 15.8). This method was quickly accepted as a gold standard to separate the PR segment into two components of supra-His and infra-His conduction. The atria to His interval (A–H) correlates with AV nodal conduction, while the time from the onset of His deflection to ventricular activation represents the overall His–Purkinje system (HPS) conduction, frequently labeled as the HV interval (Fig. 15.9). In addition to the His bundle electrogram, other recordings are frequently obtained from the right bundle and right and left atria (the latter via a catheter in the coronary sinus), and the right

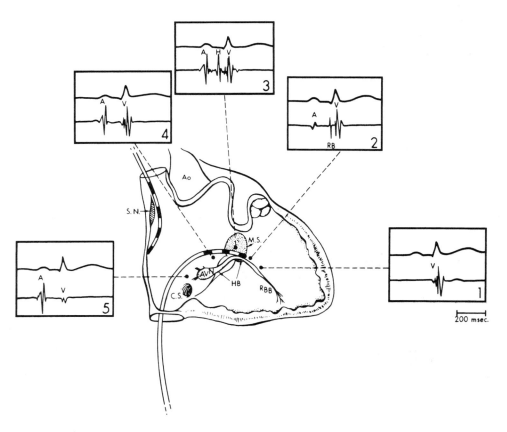

Figure 15.8   Intracardiac recordings from the region of AV junction. An electrode catheter introduced percutaneously and placed across the tricuspid valve. Bipolar intracavitary recordings from five sites (1–5) are depicted. A surface ECG is also shown. Note that when the catheter is in the right ventricle, only a ventricular electrogram (corresponding to the QRS) is recorded (position 1). As the catheter is withdrawn, a small atrial (A) and a right bundle (RB) deflection are noted. At position 3 is the typical His (H) potential recording that is sandwiched between the A and V. Withdrawal to location 4 indicates disappearance of His deflection and A and V are seen when the catheter is in the right atrium (position 5). RBB = right bundle branch; AO = aorta; MS = membranous septum; CS = coronary sinus; AVN = atrioventricular node; SN = sinus atrial node; HB = His bundle; RBB = right bundle branch.

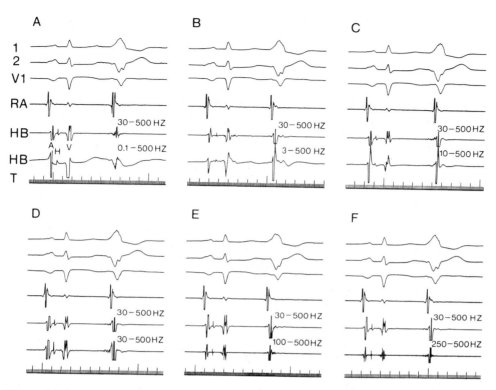

Figure 15.9 Effect of filtering frequency on HB electrograms. A sinus and premature ventricular complex is shown in all six panels. While the usual filtering frequency of 30 to 500 Hz is kept constant on HRA (right atrium) and top HB (His bundle) tracings, the high-pass filter is changed from 0.1 (unfiltered) to 250 Hz in the lower HB trace (A–F). The A, V, and H can be recognized in all panels, loss of low-frequency signals and, consequently, progressively sharper deflections are noted (A–F). Time lines at 10,100, and 1000 ms are at the bottom. (Reproduced with permission from Akhtar M. Techniques of electrophysiologic testing. In: Schlant R, Alexander W, Fuster V, eds. Hurst's The Heart: Arteries and Veins, 9th ed. City: Publisher, 1998.)

and left ventricles. Routine electrophysiological study typically utilizes just three catheters [i. e., RA (right atrium), HB (His bundle), and RV (right ventricle)]. In the presence of an accessory pathway, a coronary sinus catheter is also important to examine left atrial and/or left ventricular activation.

The main purpose of intracardiac recordings is to determine the timing of various electrical events, rather than the characteristics of any particular signal. For this reason, the intracardiac signals are filtered such that most of the low-frequency components are eliminated for clearer separation of A, H, and V electrograms. The usual filtering frequency is 30 to 40 Hz for the high-pass and 400–500 Hz for the low-pass filter (Fig. 15.2), which is vastly different from the surface ECG waveform, which is essentially unfiltered.

In addition to providing intracardiac electrocardiographic recordings, multipolar electrode catheters are also used for local electrical stimulation of tissue. Stimulation can be performed via any device that is capable of delivering rectangular electrical impulses of adjustable voltage and duration. The typical impulse is less than 2 ms in duration and roughly twice the diastolic threshold, (i. e., typically less than 1 and 2 mA for the ventricles and atria, respectively).

Conduction across the atria, AV nodes, and the HPS in the anterograde and retrograde directions can be studied with constant cycle length pacing of the atria or ventricles, respectively, usually from sites in the right atrium and right ventricle. During constant cycle length pacing the refractory periods of atrioventricular and ventricular atrial conduction systems can be studied with the use of extrastimuli (Fig. 15.10). Intravenous drugs are commonly used in conjunction with EP studies, either to learn about the effect of a drug on cardiac conduction and/or refractory period or to facilitate or suppress induction of arrhythmia. The most common agent used to facilitate the induction of tachycardia is intravenous isoproterenol (53). Intravenous procainamide is frequently used to unmask AV conduction abnormalities in the HPS and also to check for suppression of inducibility of ventricular tachycardia (54).

Electrophysiological evaluation is carried out in a variety of symptomatic tachycardias. We present a sample here under the headings of supraventricular tachycardia, ventricular tachycardia, and a more generic category, wide QRS tachycardia.

B. Supraventricular Tachycardias

1. Preexcitation Syndromes

The most common variety of ventricular preexcitation is the so-called Wolff–Parkinson–White (WPW) syndrome, which is typified by a short PR interval, initial slurring of the body surface QRS (delta wave) (Fig. 15.11), and a history of tachycardia (45,46). The accessory pathway in WPW is a muscle-to-muscle bridge directly connecting the atria with the ventricles and bypassing the normal AV conduction system (i. e., the AV node and the HPS). Since conduction in these accessory pathways is faster than that in the AV node, the impulse reaches the ventricle ahead of the expected arrival through the normal pathway, hence the shorter PR interval and the term preexcitation (Fig. 15.11). The onset of the resulting R wave is premature, but its shape is broadened because conduction within the accessory pathway and the ventricular muscle at the insertion point is slower than that through the HPS. The resulting slurred onset of the QRS is referred to as the delta wave.

Because individuals with WPW have both a normal and an accessory pathway they are prone to tachycardia along a circuit consisting of impulse traveling from atria to ventricle via one and returning to the atria through the other pathway. This is the best-documented example of a reentrant phenomenon in the human heart. When the impulse reaches the ventricles through the normal pathway and returns via the accessory pathway (AP), it is called orthodromic AV reentry; the reverse is known as antidromic reentry (Fig. 15.12).

Supraventricular Tachycardias (Reentrant)

In the presence of an AP:

> AP as essential part of circuit (orthodromic and antidromic AV reentry, atriofascicular antidromic tachycardia)
> AP as passive bystander (atrial tachycardia, atrial flutter, atrial fibrillation, AV nodal reentry)

In the absence of an AP:

> AV nodal reentry common, uncommon, and intermediate varieties
> Atrial tachycardias, atrial flutter, atrial fibrillation

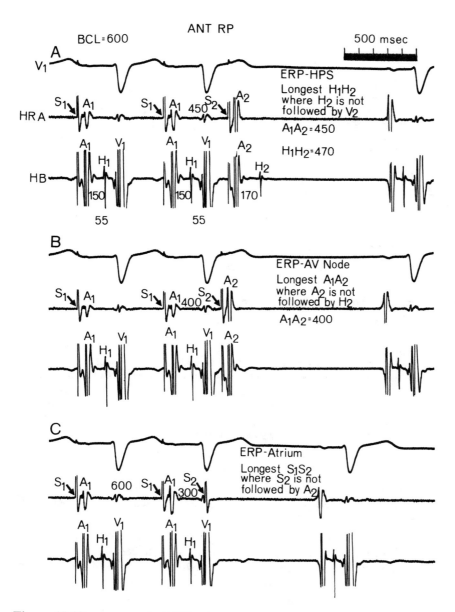

Figure 15.10 Anterograde (ANT) refractory period determination in the clinical settings. During basic or constant cycle length (BCL, S1 S1, or A1, A1) of six to eight cycles (only the last one is shown) a premature atrial stimulus (S2) is delivered at progressively shorter coupling (S1, S2, A1, A2) (A–C). The last complex in each panel is sinus. (A) The premature impulse fails to reach the ventricles (no QRS) and the site of block is in the His Purkinje system (i.e., H2 but no V2). The definition of effective refractory period is outlined in all panels. ERP of the AV node (i.e., the longest A1, A2 where the A2 does not reach the His bundle) is illustrated in (B) and that of the atrium in (C). S = stimulus artifact; S1, A1, H1, V1 = designations for the basic drives; S2, A2, H2, V2 = the premature impulse; ERP = effective refractory period; VI = surface ECG; HRA = high right atrial electrogram; HB = His bundle. Similar electrogram designations are used throughout the text. (Reproduced with permission from Akhtar M. Techniques of electrophysiologic testing. In: Schlant R, Alexander W, Fuster V, eds. Hurst's The Heart: Arteries and Veins, 9th ed. New York: Health Professions Division/McGraw Hill, 1998.)

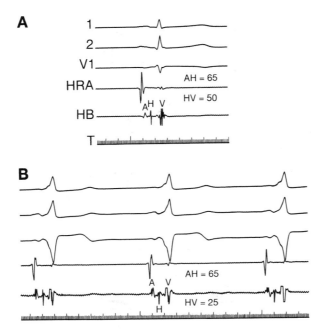

Figure 15.11 Ventricular preexcitation. Electrogram tracings from top to bottom in each panel are surface ECG leads 1, 2, and V1, HRA, and HB. (A) a normal complex without any ventricular preexcitation where the AH (AV nodal conduction time) is 65 ms and the HV (His Purkinje system conduction time) is 50 ms. The HV, measured from onset of the HB deflection to the earliest ventricular activation in either the surface or intracardiac tracing, is much shorter in (B). This shortens the PR interval (ventricular preexcitation). The initial part of the QRS is slurred due to muscle-to-muscle spread of activation front over the accessory pathway (the delta wave).

Mapping the AV ring during surgery for the purpose of ablating the accessory pathway(s) provides the necessary temporal information to locate the earliest site of ventricular activation following atrial stimulation. Figure 15.5 contains an example of electrograms recorded from an array of two parallel rows of 32 electrodes mounted on a band of nylon mesh that covered the atria above the AV ring of a patient undergoing treatment for WPW. Each column in the figure contains all the electrograms from one of the rings, ordered as they were on the sock, a form of display known as a "ladder diagram." The EGs in the left two columns are unipolar leads while those in the rightmost column are bipolar EGs obtained by taking the differences of the two paired unipolar signals. The top and bottom of the display correspond to anterior regions of the AV ring and the middle (vertically) corresponds to the posterior aspect of the AV ring. Ventricular pacing spikes are clearly visible near the beginning of each unipolar electrogram, followed by the large ventricular QRS complexes and then the retrogradely conducted P waves. The special value of the ladder diagram is that it reveals clearly the close temporal proximity of the QRS complexes, but also the disparity in the timing of the P-wave spikes. The diagram clearly shows that the earliest atrial response (marked by the arrows) occurred on the lower third of the display, corresponding to a right-sided pathway (that was subsequently successfully ablated). The EGs in the right column illustrate the utility of bipolar leads to suppress remote elec-

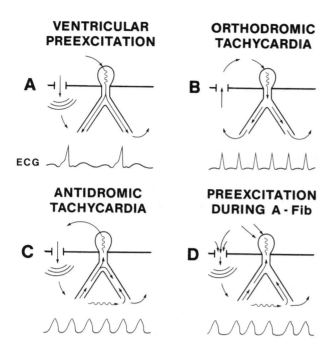

VENTRICULAR PREEXCITATION

ORTHODROMIC TACHYCARDIA

ANTIDROMIC TACHYCARDIA

PREEXCITATION DURING A-Fib

ECG

Figure 15.12 WPW syndrome. (A) Ventricular preexcitation (Panel A) and various arrhythmias noted in WPW are schematically shown. Note that during orthodromic reentry (B) there is no evidence of ventricular preexcitation (since the AP is only used in the retrograde direction), whereas the QRS complexes are preexcited during antidromic AV reentry (C) and atrial fibrillation (D). As expected, the R-R intervals are irregular during atrial fibrillation. (Reproduced with permission from Akhtar M, The Examination of the Heart: Part 5, The Electrocardiogram. Dallas: American Heart Association, 1993.)

trical activity (stimulus and "distant" QRS) and enhance the features arising locally in the atrial tissue above the A-V ring.

2. Atrioventricular Nodal Reentry

In the absence of ventricular preexcitation, the most common type of supraventricular tachycardia is atrioventricular nodal (AVN) reentry (55). Here the circuit of reentry is localized above the His bundle in the general region of the AV node. The basis of this tachycardia is dual AV nodal physiology such that the anteriorly located fibers conduct faster and have a longer refractory period, while the reverse is true of posteriorly located fibers. These are often referred to as fast and slow pathways, respectively (Fig. 15.13). Several types of AVN reentry can be recognized. The most common form is associated with slow pathways conducting anterogradely while the fast pathway is used retrogradely (toward the atria) (55). When an impulse reaches the His bundle, its travel times anterogradely toward the ventricle and retrogradely to the low septal atrium take about the same time, such that the QRS and P wave occur simultaneously (short HA). On the surface ECG, therefore, the P wave is obscured. However, on the intracardiac electrogram, the atrial activation of the His bundle occurs first and may precede the local V electrogram [Fig. 15.14 (A)]. If

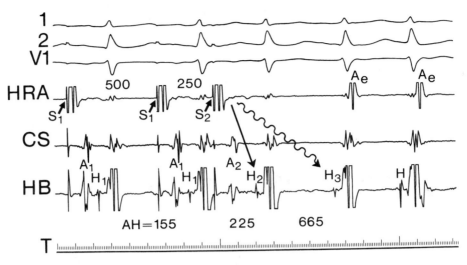

Figure 15.13   Dual AV nodal physiology and AVN reentry. The occurrence of two narrow QRS complexes in response to a single premature atrial complex (A2) and the onset of common AVN reentry is illustrated. A2 reaches the His over the fast and slow pathway, respectively, and then returns to the atria through the fast pathway.

there is any doubt about the timing of atrial activation, recordings from other positions within the right atrium, away from the AV junction, can clearly show the atrial activation uncontaminated by ventricular depolarization.

In a small percentage (approximately 10%), the sequence of activation in the two pathways is reversed [Fig. 15.14 (B)] (55). In this case the AH is shorter than the H–A interval (long HA and RP intervals) and the retrograde P wave clearly can be seen outside the QRS and often following the T wave. Even more infrequent is AV nodal reentry in which the AH and HA intervals have similar durations. Long-RP-interval tachycardia is an ECG pattern sometimes noted with other mechanisms. Among them are atrial tachycardia from the low interatrial septum and a slowly conducting retrograde AP forming the return route to the atrium rather than the slow AVN conducting pathway. The latter can be distinguished from AVN reentry of the uncommon type by timed ventricular premature stimuli to preexcite the atria when the HB is refractory, as described above.

3.   Atrial Tachyarrhythmias

All forms of AP and AVN reentry are lumped together under the category of AV junctional tachycardias. A common clinical problem is the separation of AV junctional tachycardias from regular atrial tachycardia. The vast majority of AV junctional tachycardias are reentrant in nature while arrhythmias originating in the atria can be reentrant or originate from other mechanisms such as abnormal automaticity and/or triggered activity. Relatively few EP maneuvers are available to discern reentrant from nonreentrant atrial tachycardia in the clinical setting. Tachycardias that can be readily initiated and terminated with premature atrial complexes are generally classified as reentrant. Tachycardias due to automaticity cannot be induced with atrial extrastimuli, but they may respond to the phenomenon of overdrive suppression. Prognosis using intravenous isoproterenol is also inconclusive be-

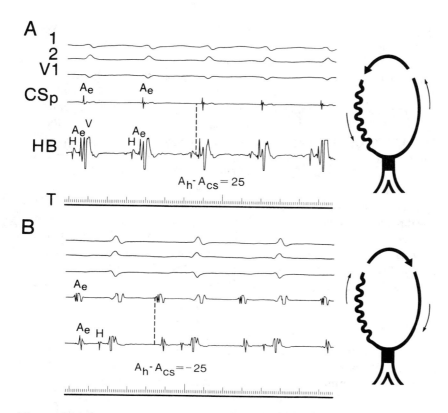

Figure 15.14 Common and uncommon AVN reentry. (A) Common and (B) dis-uncommon AVN reentry—along with schematic representation of anterograde and retrograde pathways. The earliest activation of retrograde A (Ae) is noted on the HB (i.e., anterior) during common AVN reentry (A) while the proximal CS (OS of CS) is the earliest point of activation during uncommon AVN reentry (i.e., posterior location of atrial breakthrough via the slow AV nodal pathway) (B).

cause it may trigger spontaneous episodes of automatic tachycardias, but can also facilitate electrical induction of reentry (56) or triggered activity.

Atrial (rather than AV junctional) origin of tachycardias is suggested by:

1. Atrial activation sequence during the tachycardia that is different from that noted during VA conduction induced by ventricular pacing.
2. Absence of retrograde (VA) conduction with ventricular pacing when the tachycardia can be electrically induced with atrial pacing.
3. Lack of effect of AV nodal block induced with intravenous adenosine (6 or 12 mg), suggesting that atrial tachycardia is independent of the AV node (57).
4. Abolition of the tachycardia by focal atrial tissue ablation away from the AV junction.

Other examples of atrial tachyarrhythmias are atrial flutter and atrial fibrillation (AF). The latter is the most common symptomatic clinical arrhythmia, afflicting more than two million Americans, with more than 160,000 new cases every year. The mechanism of AF is controversial, but it appears that both focal firing and multiple reentrant circuits in both atria may be involved in different patients. EP evaluation for atrial fibrillation is appropriate under two conditions. The first occurs when regular supraventricular tachycardia or atrial flutter is thought to be the trigger for AF, in which case elimination of these could prevent atrial fibrillation. The second reason for EP study of AF arises when atrial tissue ablation is contemplated. Although at the present time the scope of atrial ablation is limited unless AF is of focal origin, which is difficult to determine clinically, it is anticipated that with improving mapping technology AF will become a curable arrhythmia at least in some cases (58,59). Atrial flutter, on the other hand, can be cured now and is therefore frequently evaluated in the EP laboratory. The common forms of atrial flutter localize to the right atria with a circuit that can be mapped (60,61). The route of impulse propagation is through the isthmus (*i. e.*, an area of slow conduction between the entry of inferior vena cava into the right atrium and the tricuspid annulus). From there, the impulse travels up the interatrial septum to the anterolateral right atrium to enter the isthmus. Using several atrial recordings, the direction of reentry can be fairly well mapped out with the atrial activation sequence being ostium of the coronary sinus (CS) → anterior septum (HBE) → right atria (HRA) → anterolateral right atria → the region of the isthmus. Common atrial flutter results in a negative flutter wave in an inferior lead (the so-called sawtooth appearance). A reversal in the direction of the excitation produces counterclockwise or uncommon flutter with posterior or upright flutter waves in leads II, III, and AVF.

C. Ventricular Tachycardia

At least two forms of ventricular tachycardia (VT) must be separated in clinical practice (62). Monomorphic VT refers to identical QRS morphologies for each complex [Fig. 15.15(A)] whereas polymorphic VT implies a constantly changing morphology [Fig. 15.15(B)]. The bulk of the effort regarding EP evaluation of VT has been centered around the monomorphic variety and most of these tachycardias are reentrant in nature. The substrate of reentry is usually myocardial scarring, due to prior myocardial infarction or fibrosis from any etiology. Another mechanism is reentry within a diseased HPS, such that the reentry circuit incorporates the bundle branch, the His bundle, and the intervening myocardial septum (63,64). VT may also occur in an otherwise normal heart in either the right or left ventricle. Here we deal primarily with analysis and mapping of myocardial VT (fibrosis) and bundle branch reentry (reentry within the HPS).

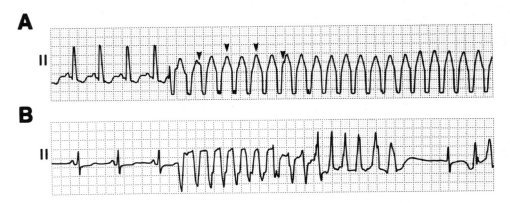

Figure 15.15 Monomorphic and polymorphic ventricular tachycardia. Note a uniform morphology of QRS with each complex in (A), whereas a constantly changing configuration is seen in (B). A long QT interval can be appreciated following the pause (QT of the complex after termination of VT). This combination is referred to as torsade de pointes.

Ventricular Tachycardias
Monomorphic VT:

> Myocardial reentry (fibrosis)
> Bundle branch reentry
> Miscellaneous (RV outflow, idiopathic LV–VT)

Polymorphic VT:

> Prolonged QT interval (congenital and acquired forms—torsade de pointes)
> Normal QT interval (myocardial hypertrophy and/or diffuse fibrosis)
> Normal QT interval (ischemia)

1. Myocardial VT

The best model for this type of reentry is seen in myocardial infarcts, particularly those that have healed. Surviving myocardial fibers connecting the healthy myocardium through scarred tissue can provide an ideal milieu for this arrhythmia. It is well known that the speed of electrical propagation is much faster along the longitudinal axis of muscle fibers compared to the transverse. When surviving myocardial cells creating these bridges through scarred myocardium have transverse orientation, this results in an area of slow conduction. Emergence of the impulse from this area of slow conduction initiates a new QRS complex as the impulse reenters the other end of the slow conduction zone to perpetuate the process (65,66). This is obviously an oversimplification since extensive scarring can create a fairly complex substrate with multiple actual and potential circuits. EP findings and commonly performed maneuvers to locate the critical area of slow conduction and also to apply curative ablative therapy include:

> Fractionated electrograms recorded from the area of scarring indicate the presence of abnormal conduction (67).
> Locating the earliest electrical activation preceding the surface QRS complex helps indicates the point of exit from the zone of slow conduction.
> Recording potentials from the area of slow conduction that are sandwiched between the QRS complexes, the so-called mid-diastolic potential (Fig. 15.16). When the mid-diastolic potential to the onset of QRS interval is equal to the stimulus to QRS interval from the same site, it suggests that the catheter tip is in the line of impulse propagation through the critical zone of slow conduction. A high degree of success for ablation at this site is expected. A longer stimulus to QRS interval suggests that the catheter tip is within a blind loop connected to the critical area of slow conduction but somewhat outside the main circuit.
> Pacing along the line of slow conduction to produce QRS morphology identical to that of the spontaneous VT [Fig. 15.17 (A) and (B)]. If pacing is initiated during VT, entrainment may occur in which the rate is similar to the overdrive pacing rate, yet the morphology of the paced QRS is identical to that during VT. During this type of entrainment, the interval between the stimulus and the QRS complex provides some clue regarding the proximity of pacing sites relative to the site of exit [Fig. 15.17 (B)].

2. Bundle Branch Reentry

Bundle branch reentry (BBR) accounts for up to 40% of inducible monomorphic VTs in patients with idiopathic dilated cardiomyopathy and patients with valvular disease (63,68). Because the incidence of ischemic cardiomyopathy is significantly greater than other

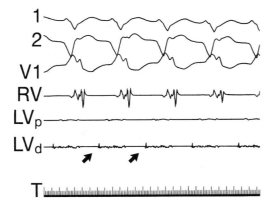

Figure 15.16 Fractionated electrograms during myocardial VT. During myocardial VT in a patient with prior myocardial infarction, electrograms are obtained from the area of slow conduction. A fractionated electrogram is preceded by a sharp diastolic deflection from the zone of slow conduction within the VT circuit. Ablation at the site of diastolic potential is often successful in control of VT.

forms of myopathy, BBR is not uncommon in this setting. However, myocardial VT is far more frequent than BBR in the presence of prior myocardial infarct, particularly in the absence of ventricular dilatation.

The main prerequisite for BBR to occur is a conduction abnormality within the HPS, so that the baseline HV is always prolonged (normal HV ranges from 35–55 ms) (63,68,69). Several other features of BBR need examination before a definitive diagnosis can be made. BBR (rather than myocardial VT) is suggested by the following observations:

1. Since the impulse can only reach the ventricles through the HPS in BBR, the His bundle activation always precedes the QRS with an HV interval equal to or longer than the HV interval during sinus or other supraventricular impulse (Fig. 15.18).
2. The sequence of His to bundle branch activation preceding the QRS is appropriate for the direction of impulse propagation. For example, when the QRS morphology has a left bundle branch block pattern, the His bundle activation precedes anterograde right bundle depolarization (Fig. 15.18). The reverse is the case if the VT has a right bundle branch block configuration, in which case the left bundle branch potential will follow the His deflection.
3. A spontaneous or induced block into the HPS in the anterograde or retrograde direction always terminates the VT (Fig. 15.18).
4. Whenever there is a change in the cycle length of the VT, the H–H (or RB–RB/LB–LB) cycle length change precedes the subsequent V–V cycle length change. In myocardial VT, in contrast, a V–V cycle length change will occur before the H–H variation.
5. Ventricular pacing at the rate of the VT does not produce a visible His bundle deflection, indicating that the stable His potential sandwiched between the VT QRS complexes is not due to bystander HPS delay secondary to a myocardial VT, but is indeed pertinent to subsequent ventricular activation.

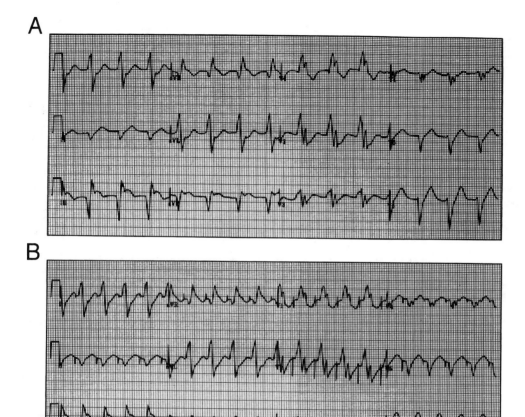

Figure 15.17 Pacing within the area of VT circuit. (A) Spontaneous VT on a 12-lead ECG. Identical QRS morphology is reproduced by pacing in the left ventricle with a stimulus to QRS interval of 80 ms suggesting pacing closer to the exit point of impulse from the zone of slow conduction.

6. Ablation of the right and/or left bundle eliminates the VT. It is important to point out that at times, interfascicular reentry between the two fascicles of the left bundle may exist in these patients with extensive HPS disease. Ablation of one of the fascicles will be necessary to eliminate this type of tachycardia.

3. Right Ventricular Dysplasia

Myocardial VT also occurs in the absence of obvious left ventricular pathology, an example of which is right ventricular (RV) dysplasia (70), in which the right ventricular myocardium is gradually replaced by fatty deposition. VT due to myocardial reentry, not too dissimilar to other forms of fibrosis, is the main clinical manifestation. Due to thinning of the right ventricular myocardium, ablative therapy is not the initial choice. Class III antiarrhythmic drugs are tried first with implantable cardiac defibrillator (ICD) as the next choice both for overdrive VT termination and prevention of arrhythmic sudden death if the drug therapy is not effective or not tolerated.

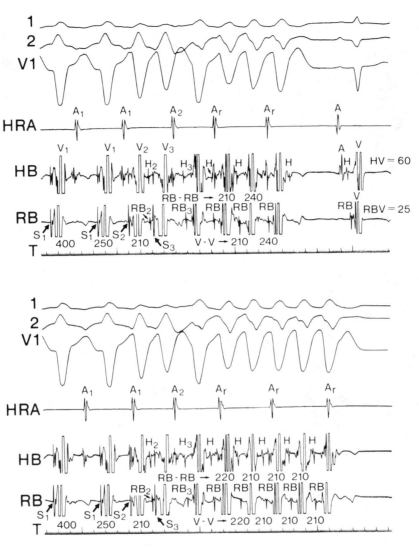

Figure 15.18   Bundle branch reentrant (BBR) VT. Onset of BBR VT is shown in the two panels with a left bundle branch block configuration. Many of the features of BBR are seen: (1) typical LBBB morphology on surface ECG suggesting ventricular activation via the right bundle; (2) retrograde 2:1 VA block suggesting VT; (3) sequence of HB–RB preceding the QRS same as in sinus (last complex is panel A); (4) HB–V, RB–V longer than sinus complex; (5) anterograde block in the HPS (His but no RB in panel A) and retrograde HPS block [i.e., no H after the QRS in (B) terminate the VT in both cases. Ar = retrograde atrial activation.]

4.   Other Forms of Myocardial VT

Two other forms of myocardial VTs deserve attention (62,71). The RV outflow is the site of exercise-related VT that may arise from reentrant or automatic mechanisms. When the VT is inducible with programmed electrical stimulation and after isoproterenol infusion, activation mapping is an effective method to determine the origin. Reproduction of the VT configuration on the 12-lead ECG with pacing (pace-mapping) is helpful when the VT cannot be reproduced or the induction is inconsistent. Beta-blockers, antiarrhythmic drugs, and radiofrequency ablation are the therapeutic choices. Due to the infrequent occurrence of arrhythmic death in this population, ICD therapy is seldom recommended.

The left ventricular inferoapical region is the origin of so-called idiopathic VT, probably due to reentry utilizing the peripheral Purkinje system (71). A sharp Purkinje potential is often recorded and precedes the ventricular activation. The exact mechanism (*i. e.*, reentry versus triggered activity) is not always certain; perhaps both forms exist. Many patients respond to intravenous and oral verapamil and catheter ablation in close proximity to the earliest Purkinje potential has a high success rate in controlling the VT.

5. Polymorphic VT

Polymorphic VT can occur with or without associated long QT interval. In the setting of QT interval prolongation, a VT termed *torsade de pointes* can be seen (72) [Fig. 15.15 (B)], of which there are both congenital and acquired varieties. It appears that their onset may be due to early after-depolarization while continuation may be reentry related. Traditional EP evaluation is seldom helpful in these syndromes and therapeutic approaches include beta-blockers for the congenital form and withdrawal of offending agents for the acquired form. The incidence of sudden arrhythmic death can be reduced with the use of ICDs in this population.

Polymorphic VT in the absence of QT prolongation is frequently due to acute or active myocardial ischemia and therefore anti-ischemic therapy is the treatment of choice (62,73). If complete revascularization cannot be accomplished, ICD therapy and beta-blockers are often added. Where ischemia is not present and the QT is normal, a polymorphic VT is likely to be due to reentrant mechanisms from extensive scarring, in which case it can be treated as monomorphic VT in the same setting. However, a diagnosis of polymorphic VT due to fibrosis should only be arrived at after QT prolongation and acute ischemia have been excluded. Polymorphic VT does not provide an ideal ablation substrate and ICD therapy should be the first choice, particularly in the setting of extensive scarring and the associated poor LV function.

VI. FUTURE OF ARRHYTHMIA MAPPING

As in the past, the future of invasive arrhythmia therapies will be closely tied to technological advances in electrodes, acquisition systems, and computer-based analysis. One area of particular promise is in the integration of electrical measurements with fast, quantitative anatomical imaging. Systems now under development will permit the acquisition of cavitary potentials, anatomical reconstruction of the heart and its chambers, and geometric registration of the recording electrode arrays, thus permitting rapid and accurate interpretation of arrhythmic events. Ultrasound and electromagnetic sensing provide the means to develop three-dimensional imaging techniques that can locate electrode geometries within mapping arrays relative to recognizable features of cardiac anatomy. Small, inexpensive workstations provide the necessary computing power to integrate electrical and anatomical imaging information, as well as the means of visualizing arrhythmia characteristics that aid in interpretation.

One example of a mapping approach that has emerged because of advanced technology is pace mapping. The basis of this technique is that body surface potential distributions that arise from pacing specific cardiac regions have characteristics that are consistent from patient to patient, yet unique to the site of pacing and may thus be cataloged in an atlas (74–76). Thus, when a monomorphic VT is observed on the body surface, its characteristic signature may be reproduced by pacing from regions near the exit site of the reentrant arrhythmia. This permits both a faster localization of the critical sites of the arrhythmia, as well as a means to confirm the location of the ablation catheter.

An example of the impact of new electrode design on arrhythmia mapping is the growing use of multielectrode venous catheters that facilitate detailed, percutaneous electrical measurements of the epicardium near the coronary vessels (77–80). Estimation and interpolation strategies based on catheter mapping may provide means to interpret and understand arrhythmic events from a limited number of recording sites accessible through routine catheterization (81–83).

Clearly the future of arrhythmia mapping is directed at improving the speed, accuracy, and reliability with which electrical and anatomical images may be acquired, integrated, and interpreted in the setting of arrhythmic events. The objective of these newer strategies is to achieve improvements in diagnosis and therapy while further reducing the time, cost, patient discomfort, and risk associated with the procedures.

ACKNOWLEDGMENTS

We gratefully acknowledge the contribution of Dr. Bruno Taccardi for providing data for several of the figures and his reading and valuable comments on the manuscript. Experimental data used in this study were acquired as part of research supported by the National Institutes of Health under grant HL43276 (Dr. Taccardi). Other support for this work comes from the Nora Eccles Treadwell Foundation and the Richard A. and Nora Eccles Harrison Fund for Cardiovascular Research.

REFERENCES

1. Mines GR. On dynamic equillibrium in the heart. J Physiol 1913; 4:349–383.
2. Garrey WE. The nature of fibrillary contraction of the heart—its relation to tissue mass and form. Am J Physiol 1914; 3:397–414.
3. Mines GR. On circulating excitations in the heart muscles and their possibilitie relation to tachycardia and fibrillation. Trans R Soc Can 1914; 14:43–52.
4. Lewis T, Feil HS, Stroud WD. Observations upon flutter and fibrillation. Part II. the nature of auricular flutter. Heart 1920; 7:191–245.
5. Wiggers CJ, Wégria R. Ventricular fibrillation due to single, localized induction and condenser shocks applied during the vulnerable phase of ventricula systole. Am J Physiol 1940; 128:500–505.
6. Taccardi B. Distribution of heart potentials on the thoracic surface of normal human subjects. Circ Res 1963; 1:341–351.
7. Durrer D, van Dam RT, Freud GE, Janse MJ, Meijler FL, Arzbaecher RC. Total excitation of the isolated human heart. Circulation 1970; 41:899–912.
8. Kléber AG, Janse MJ, van Capelle FJL, Durrer D. Mechanism and time course of ST- and TQ-segment changes during acute regional myocardial ischemia in the pig heart determined by extracellular and intracellular recordings. Circ Res 1978; 42:603–613.
9. Spach MS, Barr RC, Lanning CF, Tucek PC. Origin of body surface QRS and T-wave potentials from epicardial potential distributions in the intact chimpanzee. Circulation 1977; 55:268–278.
10. Spach MS, Barr RC. Origin of epicardial ST-T wave potentials in the intact dog. Adv Cardiol 1978; 21:15–18.
11. Spach MS, Miller WT, Miller-Jones E, Warren RB, Barr RC. Extracellular potentials related to intracellular action potentials during impulse conduction in anisotropic canine cardiac muscle. Circ Res 1979; 4:188–204.

12. Taccardi B, G Marchetti. Distribution of heart potentials on the body surface and in artificial conducting media. In: Taccardi B, Marchetti G, eds. International Symposium on the Electrophysiology of the Heart. New York: Pergamon Press, 1965:257–280.

13. Taccardi B. Changes in cardiac electrogenesis following coronary occlusion. In: Coronary Circulation and Energetics of the Myocardium. Basel: S. Karger, 1966:259–267.

14. Wit AL, Bigger JT. Possible electrophysiological mechanisms for lethal arrythmias accompanying myocardial ischemia and infarction. Circulation 1975; 51,52(suppl II):III-96–III-115.

15. Ideker RE, Bandura JP, Larsen RA, Cox JW, Keller FW, Brody DA. Localization of heart vectors produced by epicardial burns and ectopic stimuli. Circ Res 1975; 36:105–112.

16. Laxer C, Ideker RE, Pilkington TC. The use of unipolar epicardial QRS potentials to estimate myocardial infarction. IEEE Trans Biomed Eng 1985; BME-32:64–67.

17. Arisi G, Macchi E, Baruffi S, Spaggiari S, Taccardi B. Potential fields on the ventricaular surface of the exposed dog heart during normal excitation. Circ Res 1983; 5:706–715.

18. Arisi G, Macchi E, Corradi C, Lux RL, Taccardi B. Epicardial excitation during ventricular pacing: Relative independence of breakthrough sites from excitation sequence in canine right ventricle. Circ Res 1992; 71(4):840–849.

19. Taccardi B, Macchi E, Lux RL, Ershler PR, Spaggiari S, Baruffi S, Vyhmeister Y. Effect of myocardial fiber direction on epicardial potentials. Circulation 1994; 90:3076–3090.

20. Lux RL, Ershler PR, Taccardi B. Measuring spatial waves of repolarization in canine ventricles using high resolution epicardial mapping. J Electrocardiol 1996; 29(suppl):130–134.

21. Kaltenbrunner W, Cardinal R, Dubuc M, Shenasa M, Nadeau R, Tremblay G, Vermeulen M, Savard P, Page PL. Epicardial and endocardial mapping of ventricular tachyarrhythmia in patients with myocardial infarction. Is the origin of the tachycardia always subendocardially localized? Circulation 1991; 84:1058–1071.

22. Jackman WM, Xang X, Friday KJ, Roman CA, Moulton KP, Beckman KJ, McClelland JH, Twidale N, Hazlitt HA, Prior MI. Catheter ablation of accessory atrioventricular pathways (Wolff-Parkinson-White syndrome) by radiofrequency current. N Engl J Med 1991; 324(23): 1605–1611.

23. Fitzpatrick A, Chin M, Stillson C, Lesh M. Successful percutaneous deployment, pacing and recording from a 64-polar, multi-strut "basket" catheter in the swine left ventricle. PACE 1994; 17:482.

24. Eldar M, Ohad DG, Goldberger JJ, Rotstein Z, Hsu S, Swanson DK, Greenspon AJ. Transcutaneous multielectrode basket catheter for endocardial mapping and ablation of ventricular tachycardia in the pig. Circulation 1997; 96:2340–2437.

25. Khoury DS, Taccardi B, Lux RL, Ershler PR, Rudy Y. Reconstruction of endocardial potentials and activation sequences from intracavity probe measurements. Circulation 1995; 91: 845–863.

26. Lui ZW, Ershler PR, Taccardi B, Lux RL, Khoury DS, Rudy Y. Noncontact endocardial mapping: Reconstruction of electrocardiograms and isochrones from intracavitary probe potentials. J Cardiovasc Electrophysiol 1997; 8:415–431.

27. Schilling RJ, Peters NS, Davies DW. Simultaneous endocardial mapping in the human left ventricle using a noncontact catheter: comparison of contact and reconstructed electrograms during sinus rhythm. Circulation 1998; 9:887–898.

28. Ideker RE, Smith WM, Wolf P, Danieley ND, Bartram FR. Simultaneous multichannel cardiac mapping systems. *PACE* 1987; 10:281–291.

29. Ideker RE, Smith WM, Blanchard SM, Reiser SL, Simpson EV, Wolf PD, Danielei ND. The assumptions of isochronal cardiac mapping. PACE 1989; 12:456–478.

30. Berbari EJ, Lander P, Geselowitz DB, Scherlag BJ, Lazzara R. The methodology of cardiac mapping. In: Shenasa M, Borggrefe M, Breithardt G, ed. Cardiac Mapping. Futura Publishing Co., Inc., 1993:63–77.

31. Downar E, Parson ID, Mickleborough LL, Cameron DA, Yao LC, Waxman MB. On-line epicardial mapping of intraoperative ventricular arrhythmias: Initial clinical experience. J Am Coll Cardiol 1984; 4:703–714.

32. Pieper CF, Pacifico A. The epicardial field potential in dog: Implications for recording site density during epicardial mapping. PACE 1993; 16:1263–1274.

33. Chu E, Kalman JM, Kwasman MA, Jue JCY, Fitzgerald PJ, Epstein LM, Schiller NB, Yock PG, Lesh MD. Intracardiac echocardiography during radiofrequency catheter ablation of cardiac arrhythmias in humans. J Am Coll Cardiol 1994; 24(5):1351–1357.

34. Gepstein L, Hayam G, Ben-Haim SA. A novel method for nonfluoroscopic catheter-based electroanatomical mapping of the heart: In vitro and in vivo accuracy results. Circulation 1997; 95: 1611–1622.

35. Martel S, Lafontaine S, Bullivant D, Hunter IW, Hunter PJ. A hardware object-oriented cardiac mapping system. In: Proceedings IEEE Eng Med Biol Soc 17th Ann Int Conf 1995;1647.

36. Ndrepepa G, Caref EB, Yin H, El-Sherif N, Restivo M. Activation time determination by high-resolution unipolar and bipolar extracellular electrograms. J Cardiovasc Electrophysiol 1995; 6(3):174–188.

37. Bern M, Eppstein D. Mesh generation and optimal triangulation. In: Hwang FK, Du DZ, ed. Computing in Euclidean Geometry. New York: World Scientific, 1992.

38. Lux RL, Green LS, MacLeod RS, Taccardi B. Assessment of spatial and temporal characteristics of ventricular repolarization. J Electrocardiol 1994; 27(suppl):100–104.

39. Ellis WS, Eisenberg SJ, Auslander DM, Dae MW, Zakhor A, Lesh MD. Deconvolution: A novel signal processing approach for determining activation times from fractionated electrograms and detecting infarcted tissue. Circulation 1996; 94:2633–2640.

40. MacLeod RS, Johnson CR, Matheson MA. Visualization of cardiac bioelectricity—a case study. In: Proc IEEE Visualization 92. IEEE CS Press, 1992:411–418.

41. MacLeod RS, Johnson CR, Matheson MA. Visualizing bioelectric fields. IEEE Comp. Graph. Appl. 1993; 13(4):10–12.

42. Wit AL, Janse MJ. The Ventricular Arrhythmias of Ischemia and Infarction: Electrophysiological Mechanisms. Mount Kisco, NY: Futura Publishing, 1993.

43. Goldreyer BN, Bigger JT, Jr. Spontaneous and induced reentrant tachycardia. Ann Intern Med 1969; 70(1):87–98.

44. Scherlag BJ, Lau SH, Helfant RH, Berkowitz WD, Stein E, Damato AN. Catheter technique for recording His bundle activity in man. Circulation 1969; 39(1):13–8.

45. Durrer D, Schoo L, Schuilenburg RM, Wellens HJ. The role of premature beats in the initiation and the termination of supraventricular tachycardia in the Wolff-Parkinson-White syndrome. Circulation 1967; 36(5):644–62.

46. Gallagher JJ, Gilbert M, Svenson RH, Sealy WC, Kasell J, Wallace AG. Wolff-Parkinson-White syndrome: the problem, evaluation, and surgical correction. Circulation 1975; 51(5): 767–85.

47. Akhtar M, Damato AN, Ruskin JN, Batsford WP, Reddy CP, Ticzon AR, Dhatt MS, Gomes JA, Calon AH. Antegrade and retrograde conduction characteristics in three patterns of paroxysmal atrioventricular junctional reentrant tachycardia. Am Heart J 1978; 95(1):22–42.

48. Wellens HJ, Duren DR, Lie KI. Observations on mechanisms of ventricular tachycardia in man. Circulation 1976; 54(2):237–44.

49. Wilber DJ, Garan H, Finkelstein D, Kelly E, Newell J, McGovern B, Ruskin JN. Out-of-hospital cardiac arrest: use of electrophysiologic testing in the prediction of long-term outcome. N Engl J Med 1988; 318(1):19–24.

50. Josephson ME, Horowitz LN, Spielman SR, Waxman HL, Greenspan AM. Role of catheter mapping in the preoperative evaluation of ventricular tachycardia. Am J Cardiol 1982; 49(1): 207–20.

51. Mirowski M. The automatic implantable cardioverter-defibrillator: an overview. J Am Coll Cardiol 1985; 6(2):461–6.

52. Jackman WM, Wang XZ, Friday KJ, Roman CA, Moulton KP, Beckman KJ, McClelland JH, Twidale N, Hazlitt HA, Prior MI. Catheter ablation of accessory atrioventricular pathways (Wolff-Parkinson-White syndrome) by radiofrequency current. N Engl J Med 1991; 324(23): 1605–11.

53. Jazayeri MR, Van Wyhe G, Avitall B, McKinnie J, Tchou P, Akhtar M. Isoproterenol reversal of antiarrhythmic effects in patients with inducible sustained ventricular tachyarrhythmias. J Am Coll Cardiol 1989; 14(3):705–714.

54. Waxman HL, Buxton AE, Sadowski LM, Josephson ME. The response to procainamide during electrophysiologic study for sustained ventricular tachyarrhythmias predicts the response to other medications. Circulation, 1983; 67(1):30–7, 1983.

55. Akhtar M, Jazayeri MR, Sra J, Blanck Z, Deshpande S, Dhala A. Atrioventricular nodal reentry. clinical, electrophysiological, and therapeutic considerations. Circulation 1993; 88(1): 282–95.

56. Niazi I, Naccarelli G, Dougherty A, Rinkenberger R, Tchou P, Akhtar M. Treatment of atrioventricular node reentrant tachycardia with encainide: reversal of drug effect with isoproterenol. J Am Coll Cardiol 1989; 13(4):904–10.

57. DiMarco JP, Sellers TD, Berne RM, West GA, Belardinelli L. Adenosine: electrophysiologic effects and therapeutic use for terminating paroxysmal supraventricular tachycardia. Circulation 1983; 68(6):1254–63.

58. Lesh MD, Kalman JM, Olgin JE, Ellis WS. The role of atrial anatomy in clinical atrial arrhythmias. J Electrocardiol 1996; 29(suppl):101–113.

59. Olgin J, Kalman J, Maguire M, Chin M, Stillson C, Lesh MD. Electrophysiologic effects of long linear atrial lesions placed under intracardiac echo guidance. PACE 1996; 19(4, Part II):581.

60. Cosio FG, Lopez-Gil M, Goicolea A, Arribas F, Barroso JL. Radiofrequency ablation of the inferior vena cava-tricuspid valve isthmus in common atrial flutter. Am J Cardiol 1993; 71(8): 705–9.

61. Schumacher B, Lewalter T, Wolpert C, Jung W, Luderitz B. Radiofrequency ablation of atrial flutter. J Cardiovasc Electrophysiol 1998; 9(8 (suppl)):S139–45.

62. Akhtar M. Clinical spectrum of ventricular tachycardia. Circulation. 1990; 82(5):1561–73.

63. Caceres J, Jazayeri M, McKinnie J, Avitall B, Denker ST, Tchou P, Akhtar M. Sustained bundle branch reentry as a mechanism of clinical tachycardia. Circulation 1989; 79(2):256–70.

64. Blanck Z, Dhala A, Deshpande S, Sra J, Jazayeri M, Akhtar M. Bundle branch reentrant ventricular tachycardia: cumulative experience in 48 patients. J Cardiovasc Electrophysiol 1993; 4(3):253–62.

65. Josephson ME, Harken AH, Horowitz LN. Endocardial excision: a new surgical technique for the treatment of recurrent ventricular tachycardia. Circulation 1979; 60(7):1430–9.

66. Miller JM, Harken AH, Hargrove WC, Josephson ME. Pattern of endocardial activation during sustained ventricular tachycardia. J Am Coll Cardiol 1985; 6(6):1280–7.

67. Josephson ME, Horowitz LN, Farshidi A. Continuous local electrical activity. A mechanism of recurrent ventricular tachycardia. Circulation 1978; 57(4):659–65.

68. Narasimhan C, Jazayeri MR, Sra J, Dhala A, Deshpande S, Biehl M, Akhtar M, Blanck Z. Ventricular tachycardia in valvular heart disease: facilitation of sustained bundle-branch reentry by valve surgery. Circulation 1997; 96(12):4307–13.

69. Blanck Z, Jazayeri M, Dhala A, Deshpande S, Sra J, Akhtar M, Bundle branch reentry: a mechanism of ventricular tachycardia in the absence of myocardial or valvular dysfunction. J Am Coll Cardiol 1993; 22(6):1718–22.

70. Fontaine G, Frank R, Rougier I, Tonet J, Gallais Y, Farenq G, Lascault G, Lilamand M, Fontaliran F, Chomette G. Electrode catheter ablation of resistant ventricular tachycardia in arrhythmogenic right ventricular dysplasia: experience of 15 patients with a mean follow-up of 45 months. Heart Vessels 1990; 5(3):172–87.

71. Nakagawa H, Beckman K, McClelland J, Wang X, Lazzara R, Arruda M, Santoro I, Hazlitt HA, Abdalla I, Singh A, Gossinger H, Sweidan R, Hirao K, Widman L, Jackman W. Radiofrequency ablation of idiopathic left ventricular tachycardia guided by a purkinje potential. Pace 1993; 16:161.

72. Jackman WM, Clark M, Friday KJ, Aliot EM, Anderson J, Lazzara R. Ventricular tachyarrhythmias in the long QT syndromes. Med Clin North Am 1984; 68(5):1079–109.

73. Tchou P, Atassi K, Jazayeri M, McKinnie J, Avitall B, Akhtar M. Etiology of polymorphic ventricular tachycardia in the absence of prolonged QT. J Am Coll Cardiol 1989; 13:21A.
74. SippensGroenewegen A, Spekhorst H, van Hemel NM, Kingma JH, Hauer RNW, Janse MJ, Dunning AJ. Body surface mapping of ectopic left and right ventricular activation: QRS spectrum in patients without structural heart disease. Circulation 1990; 82:879–896.
75. SippensGroenewegen A, Spekhorst H, van Hemel NM, Kingma JH, Hauer RN, Janse MJ, Dunning AJ. Body surface mapping of ectopic left ventricular activation. QRS spectrum in patients with prior myocardial infarction. Circ Res 1992; 71(6):1361–78.
76. SippensGroenewegen A, Spekhorst H, van Hemel NM, Kingma JH, Hauer RN, Bakkerde CA, Grimbergen CA, Janse MJ, Dunning AJ. Localization of the site of origin of postinfarction ventricular tachycardia by endocardial pace mapping. body surface mapping compared with the 12-lead electrocardiogram. Circulation 1993; 88(5 Pt 1):2290–306.
77. D'Avila A, Brugada P. Letter to the editor. *PACE* 1994; 17:1832–1833.
78. Stellbrink C, Diem B, Schauterte P, Ziegert K, Hanrath P. Transcoronary venous radiofrequency catheter ablation of ventricular tachycardia. J Cardiovasc Electrophysiol 1997; 8:916–921.
79. de Paola AA, Melo WD, Tavora MZ, Martinez EE. Angiographic and electrophysiological substrates for ventricular tachycardia mapping through the coronary veins. Heart (CEN) 1998; 79(1):59–63.
80. Zickler P. Cardiac mapping. Biomed Instrum Technol (BTI) 1997; 31(2):173–5.
81. Macleod RS, Kuenzler RO, Taccardi B, Lux RL. Estimation of epicardial activation maps from multielectrode venous catheter measurements. *PACE* 1998; 21(4):595.
82. MacLeod RS, Ni Q, Kuenzler RO, Taccardi B, Lux RL, Brooks DH. Spatiotemporal analysis of cardiac electrical activity. In: Matthews MB, ed. The Thirty-Second Asilomar Conference on Signals, Systems, and Computers. Piscataway, NJ: IEEE Press, 1998:309–313.
83. Kuenzler RO, MacLeod RS, Taccardi B, Ni Q, Lux RL. Estimation of epicardial activation maps from intravascular recordings. J Electrocardiol 1999; 32:77–92.

16

Automaticity and Triggered Activity

MARC A. VOS

Academic Hospital Maastricht/Cardiovascular Research Institute Maastricht, Maastricht, The Netherlands

BRUCE B. LERMAN

New York Hospital–Cornell University Medical Center, New York, New York

I. INTRODUCTION

Arrhythmias can occur due to abnormalities in impulse initiation, impulse conduction, or a combination of the two. Abnormal impulse initiation is further subclassified as due to abnormal automaticity or triggered activity. The focus of this chapter is to review the cellular mechanisms and clinical manifestations of abnormal impulse initiation.

II. AUTOMATICITY

Rhythmic (pacemaker) activity is an inherent property of different cell types [Fig. 16.1 (A)]. Under physiological conditions, the ability to generate spontaneous phase 4 depolarization (normal automaticity) is restricted to cells within the sinoatrial (SA) node, atrioventricular (AV) node, and/or His–Purkinje system. However, the pacemaker current I_f has been identified throughout the heart, even in atrial and ventricular myocytes, where it activates at extremely negative membrane potentials (see Chap. 6). Whether pacemaker activity in tissues such as these is expressed clinically has not yet been documented. There is a normal hierarchy in the frequency of the initiated action potentials, with the sinus node being the dominant pacemaker. Automaticity in the distal conduction system (or working myocardium) may compete with that in the sinus node on the basis of enhanced normal automaticity or abnormal automaticity.

III. ABNORMAL AUTOMATICITY

Under pathological conditions, a decrease in the resting membrane potential (RMP) may occur, which can lead to spontaneous phase 4 depolarization in all cell types, including the working myocardium (1). Abnormal automaticity is defined as spontaneous impulse initiation in cells that are not fully polarized (Fig. 16.1). Therefore, it is a definition based on the level of RMP, which in pathological settings in extranodal tissues can range from –80 to as low as –30 mV. The disturbances in the normal ion balance leading to abnormal automaticity may result from perturbations in various currents (e.g., reduction in the inwardly rectifying potassium current I_{k1}). Like normal automaticity, abnormal automaticity is enhanced by catecholamine stimulation.

Three types of abnormal automaticity have been identified: those which occur at high, intermediate, and low membrane potentials (1). In Table 16.1, the specific characteristics of the three forms have been depicted showing that the rate response to overdrive pacing and the efficacy of drugs vary considerably among the subtypes. Moreover, at the low membrane potential, it may transiently accelerate after overdrive pacing. As shown in Figure 16.2, suppression of automatic ventricular tachycardia (VT) with pacing indicates

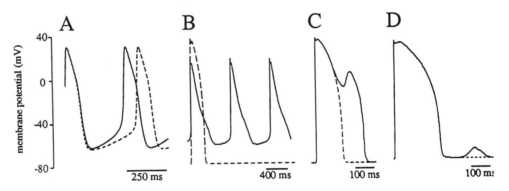

Figure 16.1 Examples of (A) enhanced automaticity; (B) abnormal automaticity; (C) early after-depolarization; and (D) delayed after-depolarization. Dashed lines represent action potentials under control conditions. (Reproduced with permission from the thesis of A. Verkerk, Ph.D. "Cellular Mechanisms of Arrhythmias," University of Amsterdam, The Netherlands, 1998.)

Table 16.1 Characteristics of Subtypes of Abnormal Automaticity

			Overdrive	Drugs	
	RMP	Rate	suppression	Lidocaine	Verapamil
High	>–70 mV	Slow	++	++	–
Intermediate	–61 to –70 mV	Intermediate	–/+	–/+	–/+
Low	<–60 mV	Fast	–	–	++

(–), No response; (–/+), variable response; (++), effective.

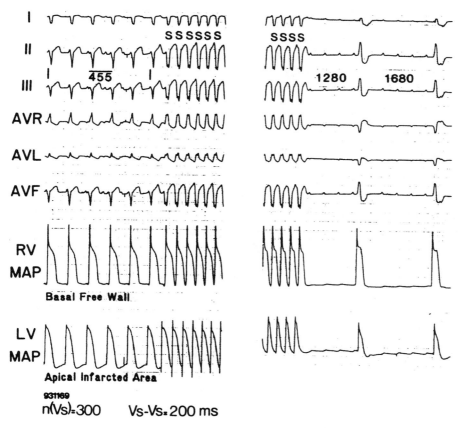

I

II   S S S S S S   S S S S

III   455   1280   1680

AVR

AVL

AVF

RV MAP

Basal Free Wall

LV MAP

Apical Infarcted Area

931169

n(Vs)=300   Vs-Vs=200 ms

Figure 16.2 Abnormal automaticity in a dog with AV block. Six surface ECG leads and two endocardial monophasic action potential (MAP) recordings are shown. Ventricular tachycardia (VT) is present (left panel), which originates around the border of an infarction (24 h after a left anterior artery occlusion) in the left ventricle (LV). On the LV MAP catheter, a clear diastolic phase 4 upslope is observed during VT, suggesting that the VT is due to abnormal automaticity. In contrast, the normal right ventricular (RV) MAP shows a flat baseline. (Right panel) VT is temporarily suppressed by prolonged overdrive pacing (300 stimuli (S) at an interstimulus interval of 200 ms), which causes the diastolic line to become flat. (Illustration prepared in collaboration with Dr. S.H.M. de Groot and Mrs. H.D.M. Leunissen.)

that the arrhythmia is related to either the high or the intermediate RMP forms of abnormal automaticity (1,2).

One well-defined experimental example of abnormal automaticity is VT that occurs in subendocardial Purkinje fibers after ischemia/infarction (3). During the subacute phase of repolarization (24–72 h following coronary occlusion), automatic arrhythmias arise from the borders of the infarction. In the Harris dog model, the characteristics of abnormal automaticity have been confirmed both in vivo and in vitro (2). More recently, the monophasic action potential (MAP) catheter demonstrated abnormal automaticity in this canine model (e.g., Fig. 16.2). Unfortunately, the MAP technique does not allow quantification of the RMP.

A.   Clinical Manifestations

A representative clinical example of an automatic rhythm is VT that is precipitated by exercise in patients without structural heart disease. The tachycardia may arise from either ventricle and can present as monomorphic or polymorphic VT.

   This form of VT is thought to represent adrenergically mediated automaticity because programmed stimulation cannot initiate or terminate the arrhythmia, whereas the tachycardia is induced with catecholamine stimulation and is sensitive to β-blockade (4). The tachycardia is not likely due to abnormal automaticity dependent on a low resting membrane potential (~ –60 mV) because calcium channel blockers are ineffective (4) and the rhythm does not demonstrate acceleration during overdrive pacing. The tachycardia is also transiently suppressed (up to 20 s) but not terminated by adenosine (4,5) (Fig. 16.3).

   The cellular mechanism governing this arrhythmia and the anatomical substrate is poorly delineated. However, since adenosine's effects on adrenergically mediated automaticity arising from Purkinje fibers depend on the level of the resting membrane potential, with greater effect observed at more negative potentials (~ –80 to –90 mV) and a diminished or absence of effect at more positive potentials, it is likely that these automatic rhythms arise from cells that are nearly fully polarized (6). This is also consistent with the clinical observation that automatic VT can be transiently suppressed with overdrive pacing. The membrane current hypothesized to mediate this arrhythmia in Purkinjes fibers and ventricles is the pacemaker current I_f. This net inward cation current is activated by hyperpolarization (see Chap. 7). Catecholamines modulate the rate in automatic cells by increasing cAMP synthesis and shifting the I_f activation curve such that I_f is activated at less negative membrane potentials (7). Adenosine appears to attenuate I_f through an inhibition of cAMP synthesis, an antiadrenergic mechanism (8) having much in common with that of vagal stimulation.

IV.   AFTER-DEPOLARIZATIONS

In cardiac cells, oscillations of membrane potential that occur during or after the action potential and depend for their initiation on the preceding action potentials are referred to as after-depolarizations (9). They are generally divided into two subclasses: early and delayed after-depolarizations (EADs and DADs, respectively), [Fig. 16.1 (C) and (D)]. When an after-depolarization achieves a sufficient amplitude and threshold potential is reached, a triggered response is evoked. Under appropriate circumstances, this process may become iterative, resulting in a sustained triggered rhythm.

A.   Early After-Depolarizations and Arrhythmogenesis

EADs can create transmural or interventricular inhomogeneities of repolarization (10) that can lead to marked changes in the ECG, as evidenced by the occurrence of different T-wave morphologies or TU-waves (Fig. 16.4). There is growing evidence that EADs and EAD-induced triggered activity play a significant role in the initiation of certain clinical arrhythmias (11). Examples include polymorphic ventricular tachycardia or torsade de pointes arrhythmias (TdP) that occur on the basis of a genetic defect (congenital) or due to drugs that prolong the duration of repolarization (acquired form). An example of TdP as a consequence of the congenital long-QT syndrome is shown in Figure 16.4.

   An EAD can appear during the plateau (phase 2) and/or repolarization (phase 3) phase of the action potential. A variety of conditions have been associated with the occur-

A. Isoproterenol 1 µg/min (60 sec)

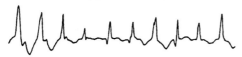

B. Isoproterenol 1 µg/min (95 sec)

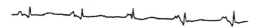

C. Adenosine 9 mg

D. Adenosine + 20 sec

E. Propranolol 2.5 mg

F. Verapamil 4 mg

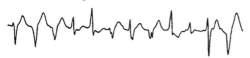

⊢——⊣ = 400 msec

Figure 16.3   Initiation of idiopathic automatic ventricular tachycardia (VT) in a patient during isoproterenol infusion. (A) Sinus rhythm recording 60 s after beginning isoproterenol infusion. (B) Sustained polymorphic VT (identical to the patient's clinical arrhythmia developed 95 s after initiation of isoproterenol infusion (1 µg/min)). (C) Adenosine transiently suppressed automatic VT (for approximately 5 s). This was followed by the emergence of multiform ventricular extrasystoles until sustained VT resumed. (D) Resumption of automatic VT 20 s after initial effects of adenosine were demonstrated. (E) Termination of automatic VT with propranolol. (F) Lack of effect of verapamil on automatic VT. (From Ref. 4.)

rence of EADs in single myocytes, particularly in midmyocardial cells (M cells) or Purkinje fibers (Fig. 16.5). Distinction between phase 2 and phase 3 EADs is often based on the takeoff potential of the EAD, being less negative than −35 mV for phase 2 and ≥ −35 mV for phase 3 or late EADs. Sometimes both forms of EADs appear simultaneously in the same cell.

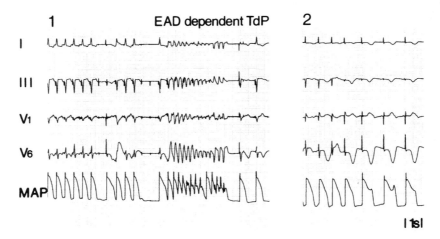

Figure 16.4 Relationship between early after-depolarizations (EADs) and torsade de pointes (TdP). In this illustration, four surface ECG leads and a MAP recording are shown in a patient with congenital long-QT syndrome. In panel 1, faulty pacemaker function creates sinus pauses of variable duration that induce TdP. It can also be seen that immediately following the pause the T wave changes dramatically (particularly in lead V6), which is accompanied by an EAD inscription on the MAP recording. This is also illustrated in panel 2. During slowing of the ventricular rhythm, an EAD develops in beats 5 to 7, accompanied by associated T-wave changes. (Illustration prepared in collaboration with Drs. A.P.M. Gorgels and K. den Dulk.)

Critically prolonging repolarization, either by reducing outward currents, increasing inward currents, or a combination of the two is normally required for the manifestation of EAD-induced ectopic activity. During the plateau phase of an EAD, the delicate equilibrium between inward and outward currents is interrupted by a depolarizing current (11). Propagation of an EAD or EAD-induced triggered activity from a focus to the surrounding cells or regions of the heart will generate an ectopic beat (Fig. 16.5).

The mechanism of propagation of phase 2 EADs has not been entirely delineated and may include true propagation or reflection of the electrotonic difference (prolonged repolarization-dependent reexcitation) (12,13). Alternatively, prolongation of action potential duration by an EAD can create spatial and temporal inhomogeneities in repolarization and a substrate that favors reentrant arrhythmias.

Evidence derived from voltage-clamp experiments shows that $Ca_{(L)}$ window current can be responsible for generation of EADs, especially during phase 2 of the action potential (14). More recent data suggest the possible involvement of Na/Ca exchange, especially under circumstances of uncontrolled release of calcium from the sarcoplasmic reticulum. Fluorescence microscopy studies in which Ca transients and APDs are measured simultaneously in myocytes (15) indicate that a second calcium transient precedes the generation of an EAD (Fig. 16.6). Since experimental evidence suggests that EADs occur by a mechanism independent of calcium-induced calcium release through the $Ca_{(L)}$ channel (16), we and others have attributed the phenomenon of acceleration-induced EADs to an increase in free intracellular calcium (17–19).

A wide variety of drugs can produce EADs, EAD-related ectopic activity, or even TdP (20). What all these agents have in common is that they excessively prolong repolarization via an increase in inward (Na^+ or Ca^{2+}) currents and/or a decrease in outward (K^+)

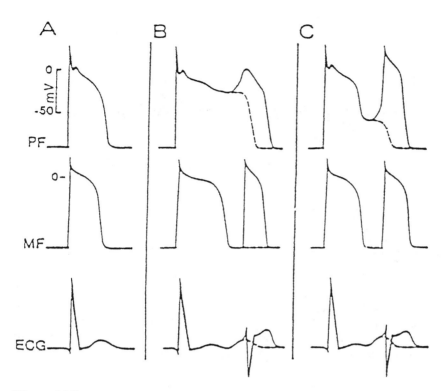

Figure 16.5 Diagramatic representation of transmembrane action potentials from a Purkinje fiber (PF) and an adjacent muscle fiber (MF) and the surface ECG to illustrate the changes associated with the development of prolonged repolarization, EADs, and an EAD-triggered action potential. (A) Control; (B) phase 2 EAD. First, the action potential duration (APD) prolongs and the plateau flattens in the PF (dotted line). Prolongation of the APD in the MF is much less. The EAD may or may not succeed in conducting to the adjacent MF. The ECG shows the appearance of a U wave (dotted line) that corresponds in time with the terminal part of the prolonged Purkinje fiber APD and/or with the EAD. Successful conduction of the EAD will generate a ventricular premature complex arising from the U wave. (C) Phase 3 EAD. There is an interruption and retardation of phase 3 (dotted line) before a depolarization triggers the new action potential that may be conducted to generate a premature ventricular beat. (From El-Sherif, Arch Mal Coeur 1991; 84:227–234.)

currents. These drugs include so-called class Ia (quinidine and procainimide) and class III (sotalol, ibutilide) antiarrhythmics (Fig. 16.7), as well as a variety of noncardiac drugs, including macrolide antiobiotics (e.g., erythromycin) and certain nonsedating antihistamines (e.g., terfenadine and astemizole). More recently, it has been shown that drugs or toxins that delay sodium current inactivation, such as anthopleurin A (21), or increase calcium current directly (BAYK 8644) or indirectly (catecholamines), also give rise to EADs.

1. Clinical Manifestations of TdP

TdP is a form of polymorphic VT that oscillates around the ECG isoelectric line in the presence of a prolonged QT interval. It is often preceded by a specific sequence of beats (i.e., short–long–short). However, TdP can also occur without preceding ectopic beats or even when the heart rate suddenly accelerates. The ectopic beats usually arise within the

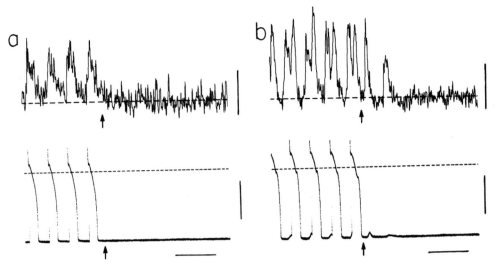

Figure 16.6 The relationship between calcium transients (upper panels) and EADs and DADs recorded with microelectrodes in myocytes (lower panels). (a) Under baseline conditions, there is no electrical or mechanical activity postpacing (arrow) (b) After administration of isoprenaline (right panel), double-peak Ca transients can be seen within the pacing train responsible for the EADs, while two Ca transients can be seen postpacing which are accompanied by DADs. (From Ref. 15.)

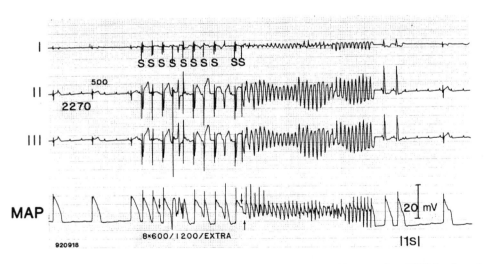

Figure 16.7 Acceleration-induced early after-depolarizations. Three surface ECG leads and a MAP recording are shown in a dog with chronic complete AV block after d-sotalol (2 mg/kg). Pacing is performed (S) using a basic train of eight beats with an interstimulus interval of 600 ms, followed by a delay of 1200 ms and a final stimulated beat (extra). During the pacing train, it can clearly been seen that not all ventricular beats are stimulated beats. The 2nd, 4th, 5th, and 7th are not properly conducted. In some cases, ectopic beats arise interfering with stimulation (e.g., between the 4th and the 5th). Pacing-induced deflections (EADs in the 3rd) that generate new action potentials (see the ectopic beat in the pause of 1200 ms) eventually lead to the TdP. These EADs are referred to as acceleration-induced EADs. (From Ref. 17.)

QT(U) interval. TdP often terminates spontaneously although electrical cardioversion is sometimes necessary. A variety of clinical associations have been described: congenital long-QT syndrome, hypertrophy, heart failure, hypokalemia, bradycardia, intracellular calcium overload, drugs that delay repolarization, catecholamines, and ischemia/reperfusion.

In recent years, different genes have been identified encoding specific cardiac ion channels (11) [e.g., HERG for I_{kr}, minK and KVLQT1 for I_{ks} (22), and SCN5A for I_{Na}]. Whether congenital anomalies of the ion channels described also have relevance for acquired TdP is not yet clear (23).

Magnesium is often used to treat TdP (24). More specific therapies based on the specific etiology include potassium channel openers, such as pinacidil, or drugs that block the sodium channel (lidocaine) or calcium current (verapamil). Agents that interact with intracellular calcium handling have also been shown to suppress EADs and EAD-related TdP such as ryanodine and flunarizine (18,19).

2. EAD Recordings in the Intact Heart

Although there are a number of techniques to measure repolarization in the intact human heart, only the MAP recording technique can provide visualization of EADs. These "humps" in the MAP were first recorded by Gravilescu et al. and later by Bonatti et al. (25,26). Since then, a number of reports have provided evidence that EADs and EAD-related ectopic activity are present in patients with documented TdP.

So-called pseudo EADs may occur due to electrotonic interactions between neighboring cells or regions with disparate repolarization phases that give rise to deflections simulating EADs. To optimize the registration of EADs in MAP recordings, we use the following criteria: (1) the MAP catheter should be deployed so as to ensure contact with the endocardium with appropriately applied pressure; (2) in the control state, the MAP contour should be smooth, with a minimum amplitude of 15 mV, and should possess a stable baseline (DC coupling of the amplifier); (3) consecutive beats should have an identical MAP morphology and duration; and (4) EADs recorded by the MAP catheter (e.g., after "class III" agents) should be abolished by interventions that cause EAD suppression in vitro (e.g., continuous rapid pacing).

B. Delayed After-Depolarizations

DADs refer to oscillations in membrane potential that occur after repolarization and during phase 4 of the action potential. In contrast to automatic rhythms that originate de novo during spontaneous diastolic depolarization, DADs are induced by and dependent on the preceding action potential.

DAD-induced triggered action potentials have been studied in isolated preparations under a variety of conditions: drugs that increase intracellular Ca^{2+} (e.g., cardiac glycosides), catecholamines, elevated $[Ca^{2+}]_0$, low $[K^+]_0$, and rapid pacing (27). Typically, during the plateau phase of the action potential, Ca^{2+} enters the cell. The increase in $[Ca^{2+}]_i$ triggers release of Ca^{2+} from the sarcoplasmic reticulum (SR) (calcium-induced calcium release), further elevating $[Ca^{2+}]i$ and initiating contraction. Relaxation occurs through sequestration of Ca^{2+} by the SR. Spontaneous oscillatory release of Ca^{2+} from Ca^{2+} overloaded SR activates a transient inward current, I_{Ti}, which gives rise to a DAD. I_{Ti} is generated by the Na^+-Ca^{2+} exchanger (I_{NaCa}) and/or a nonspecific Ca^{2+}-activated current ($I_{ns(Ca)}$) (28,29) (Fig. 16.8). At plateau potentials $I_{ns(Ca)}$ and I_{NaCa} are typically outward currents.

EADs generated during the plateau phase of the action potential (phase 2) differ in

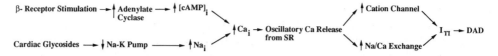

Figure 16.8   Schema for delayed after-depolarizations (DADs). (Modified from Ref. 38.)

mechanism from DADs since ryanodine, an inhibitor of SR Ca^{2+} release, and intracellular Ca^{2+} chelators abolish DADs but have no effect on EADs (30). It appears that recovery from inactivation and reactivation of L-type Ca^{2+} channels is necessary for generation of EADs rather than Ca^{2+} release from the SR since calcium agonists potentiate EADs and are suppressed by Ca^{2+} channel blockers (13). $[Ca^{2+}]i$ transients studied using fluorescence microscopy show that EADs elicited with isoproterenol are associated with a secondary increase in intracellular Ca^{2+}, and occur synchronously throughout the cell. In contrast, Ca^{2+} transients that give rise to DADs result from focal spontaneous release of Ca^{2+} from the SR, which propagates as a Ca^{2+} wave throughout the cell (31) (Fig. 16.9). Faster pacing rates result in multifocal Ca^{2+} transients. Of note, late phase 3 EADs may be similar in mechanism to DADs since the spatiotemporal characteristics of spontaneous Ca^{2+} release associated with these after-depolarizations resemble those of DADs (14,31) (Fig. 16.6).

DADs can originate from Purkinje fibers, sinus node, myocardial, mitral valve, and coronary sinus tissues. In ventricular myocytes, DADs appear to originate preferentially from M cells (10,32). These cells are localized to the sub- and midmyocardium, occupy approximately 30% of the left ventricle, and have characteristics of both Purkinje and myocardial cells. Similar to the Purkinje fiber, the M-cell action potential exhibits rate dependence (i.e., the action potential disproportionately prolongs with slowing of the pacing rate). However, like myocardial cells, M cells demonstrate an absence of spontaneous phase 4 automaticity. DADs have been shown to be preferentially elicited in M cells with digitalis, $[Ca^{2+}]_0$, or the Ca^{2+} agonist Bay K 8644 (Fig. 16.10). DADs often show the following characteristic responses to pacing techniques: the amplitude and coupling interval of DADs are dependent on the duration and rate of the preceding pacing stimuli, and this relationship is usually linear. In general, at a fixed pacing rate, the coupling interval of the triggered beat to the preceding beat is constant; as the drive rate increases, the coupling interval decreases. Once successive triggered beats are elicited and tachycardia develops, the coupling interval between the triggered beats can increase, decrease, or remain the same, depending on the rate of tachycardia. In digitalis-induced triggered activity, the rate usually stabilizes within the first 20 beats (27).

In the case of catecholamine-induced DADs, subthreshold DADs occur singly. The decrease in drive cycle length potentiates DADs because more Na^+ (and Ca^{2+}) enters the cell during rapid depolarization, further contributing to Ca^{2+} loading of the cell. In contrast, digitalis-induced DADs may occur as two or more subthreshold oscillations. As the pacing cycle length decreases, the amplitude of the first DAD may increase and reach threshold at a critical length; however, with further decreases in pacing cycle length the amplitude of the first DAD may diminish and no longer achieve threshold, whereas a second elicited DAD may be present and reach threshold. This can produce a wide range of coupling intervals from the last paced beat to the first triggered beat (33) (Fig. 16.11).

Overdrive pacing of tachycardias due to DADs may be associated with abrupt termination or termination preceded sequentially by acceleration and deceleration (34). Reen-

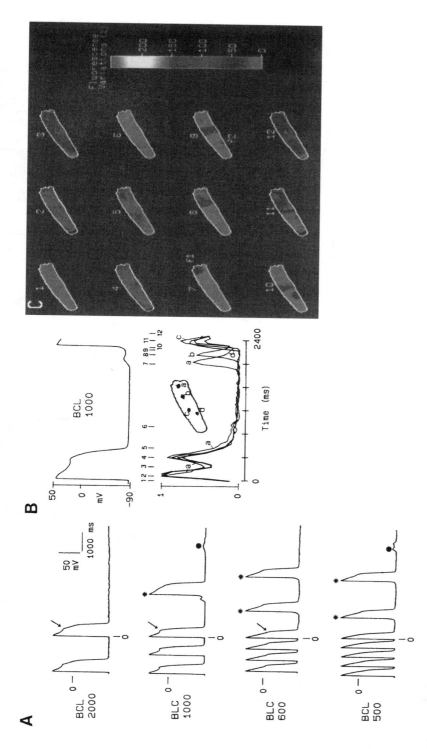

Figure 16.9 Tracings showing simultaneous presence of early after-depolarizations (EADs) (arrows) and delayed after-depolarizations (DADs) (black circles) coupled to the same beat. (A) Decrease in basic cycle length [(BCL) ms] is accompanied by shortening of the coupling interval (CI) and an increased likelihood of triggered beats (asterisks); the EAD CI did not significantly change but its activation voltage decreased and it disappeared at a BCL of 500 ms. (B) Electrical signal and normalized fluorescence signal from four different cell sites of the last driven action potential at a BCL of 1000 ms. Sites are evident in the cell profile. (C) Twelve images in pseudocolor, taken as indicated by the vertical bars over the fluorescence signal. Ca^{2+} transient was synchronous during both the evoked action potential (image 2) and the EAD (image 4), but not during the DAD. A localized Ca^{2+} release (f1) originated in region a (image 7) and the DAD developed; when the wave moved away from the end of the cell and propagated toward region b (image 8), the DAD decreased. While the wave reached region c, a second focus (f2) started in region d (image 9), reached threshold, and triggered an action potential (image 11). (From Ref. 31.) (See also color plate.)

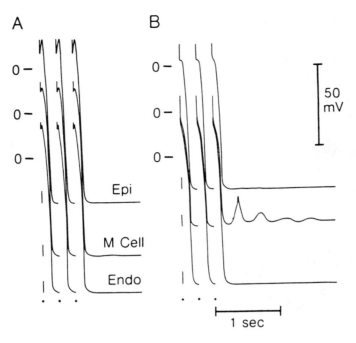

Figure 16.10   Relationship between digitalis-induced delayed after-depolarizations and myocardial cell type. Each panel shows transmembrane activity recorded from separate epicardial (Epi), endocardial (Endo), and M-cell preparations. Shown are the last several beats of a train of 10 basic responses elicited at a basic cycle length of 250 ms. Each train was followed by a 3-s pause. (A) Control; (B) acetylstrophanthidin (AcS) (10ms7 g/mL). Acetylstrophanthidin induced prominent delayed after-depolarizations in the M cell but not in epicardium or endocardium. (From Ref. 32.)

trant arrhythmias may show a similar response, with the exception that slowing is not typically observed before termination.

Although DAD-mediated triggered activity has been well characterized in isolated preparations, a major challenge has been to identify clinical analogues of these arrhythmias. The significance of DADs as a mechanism of arrhythmogenesis will ultimately depend on the definitive demonstration that this arrhythmia is responsible for some forms of clinical tachycardia. Most in vivo efforts have focused on studying intact dog models exposed to digitalis intoxication and have relied primarily on the response to programmed stimulation and extracellular recordings of DADs to confirm the diagnosis (35,36). The inherent limitation with these approaches is that the induction and responses of the arrhythmia to programmed stimulation in the clinical laboratory are often sufficiently irreproducible to preclude meaningful conclusions. The search for other means to identify triggered clinical arrhythmias, in particular those due to catecholamine cAMP stimulation, has led to the development of an electropharmacological matrix (see below) that, while not definitively conclusive, provides evidence that cAMP-mediated triggered activity may be responsible for some clinical arrhythmias (37).

The prototypical clinical arrhythmia that is thought to be due to cAMP-mediated triggered activity (and is DAD-dependent) is idiopathic right ventricular outflow tract tachycardia (RVOT) (38), which segregates into two phenotypes: paroxysmal stress-in-

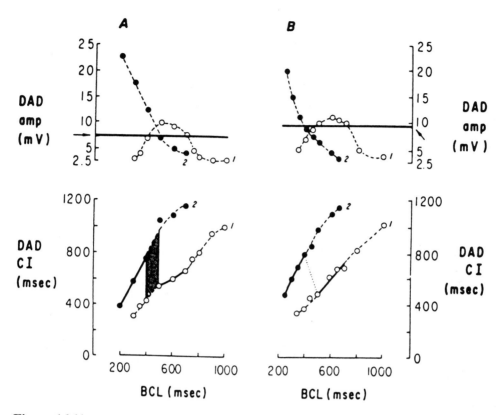

Figure 16.11 The relationship of delayed after-depolarization amplitude (DAD amp) and coupling interval (DAD CI) to basic cycle length (BCL) in two digitalis-toxic Purkinje fibers (A and B), each having two delayed after-depolarizations in sequence. The arrows and horizontal lines identify the threshold potential. (A) As drive cycle length decreases the first after-depolarization that occurs (DAD1) attains threshold at BCL = 700 ms and would initiate one or a series of triggered action potentials. As BCL decreases, so would the coupling intervals of impulses initiated by DAD1. As the BCL decreases further and DAD1 begins to decrease in its amplitude, the second delayed after-depolarization (DAD2) attains threshold and initiates an action potential. As a result, over a wide range of BCL either the first or the second delayed after-depolarization reaches threshold and fires. Considering the coupling intervals for the beats initiated by these delayed after-depolarizations, DAD1 will fire at basic cycle lengths of 700 to 400 ms, and the resultant action potentials will have coupling intervals of about 650 to 500 ms. As BCL becomes sufficiently short such that DAD2 attains threshold and fires (basic cycle length = 400 to 200 ms), the coupling interval for the resultant beat will be about 800 to 400 ms (about twice the drive cycle length). Panel B presents another possibility (a different Purkinje fiber). Here, as BCL decreases to about 500 ms, the amplitude of the DAD1 becomes subthreshold. DAD2 does not attain threshold until the basic cycle length is less than 400 ms. Within this range of basic cycle lengths (i.e., 400 to 500 ms), driving the preparation would result in no arrhythmia. Hence, a "gap" of 100 ms would be seen during which no arrhythmia occurs (broken line). At longer BCL (700 to 500 ms) an arrhythmia having a coupling interval of 700 to 475 ms would be seen. At shorter BCL (about 400 to 225 ms), the coupling interval for the arrhythmia, now resulting from the second delayed after-depolarization, would be about 800 to 450 ms. (From Ref. 33.)

duced VT and repetitive monomorphic VT (RMVT). These forms of arrhythmia represent polar ends of the spectrum of clinical VT due to cAMP-mediated triggered activity (39). RMVT occurs during rest and is characterized by frequent ventricular extrasystoles, ventricular couplets, and salvos of nonsustained VT with intervening sinus rhythm. In contrast, paroxysmal stress-induced VT usually occurs during exercise or emotional stress and is a sustained arrhythmia. Common to both groups is the absence of structural heart disease, similar tachycardia morphology (LBBB, inferior axis), and similar site of origin (RVOT), although the tachycardia can occasionally originate from the left ventricle as well (40). There can be considerable overlap between these two subtypes of VT.

1. Electropharmacological Matrix

The electrophysiological rationale for the matrix is outlined in Figure 16.12 and the responses are summarized in Table 16.2. The following discussion of the clinical application of this matrix is provided to demonstrate how information derived from studies of basic mechanisms can be used to help understand the arrhythmias occurring in patients. Since activation of adenylyl cyclase and Ca entry via $I_{Ca(L)}$ is critical to the development of cAMP-mediated triggered activity, the triggered arrhythmia would be expected to be sensitive to a constellation of electropharmacological perturbations, including beta-blockade, calcium channel blockade (verapamil), vagal maneuvers, and adenosine.

Beta-blockade is often effective as adjunctive, but not primary, therapy of reentrant VT since it rarely prevents induction or termination of tachycardia. In contrast, both automatic VT and VT due to cAMP-mediated triggered activity terminate or slow in response

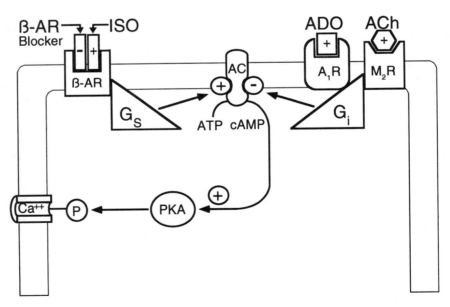

Figure 16.12 Schematic representation of the cellular model for pharmacological probes used to identify cAMP-mediated tachycardia resulting from intracellular calcium overload. AC = adenylyl cyclase; ACh = acetylcholine; ADO = adenosine; A_1R = adenosine A_1 receptor; β-AR = β-adrenergic receptor; ISO = isoproterenol; G_s = stimulatory G protein; G_i = inhibitory G protein; M_2R = muscarinic cholinergic receptor; PKA = protein kinase A. (From Ref. 39.)

Table 16.2   Electropharmacological Matrix

	Reentry	Automaticity	cAMP triggered activity
Catecholamine stimulation	Facilitates/no effect	Facilitates	Facilitates
Induction with rapid pacing	Facilitates/no effect	No effect	Facilitates
Overdrive pacing	Terminates/accelerates	Transiently suppresses	Terminates/accelerates
Beta-blockade	No effect/rarely terminates	Terminates	Terminates
Vagal maneuvers	No effect[a]	Transiently suppresses	Terminates
Calcium channel blockade	No effect	No effect	Terminates
Adenosine	No effect	Transiently suppresses	Terminates

Source: Ref. 37.
Automaticity refers to arrhythmias that arise from spontaneous phase 4 depolarization from nearly fully repolarized cells. Abnormal automaticity (which arises from cells with resting membrane potentials ≤–70 mV) is not included in this table because it has not conclusively been shown to be a cause of clinical arrhythmias.
[a]An exception is intrafascicular reentry, which is sensitive to verapamil.

to beta-blockers (4,39). Vagal maneuvers transiently suppress (but do not terminate) automatic VT, have no effect on reentrant VT (4,38), but terminate cAMP-mediated triggered activity (4,38,39). Since the direct release of ACh by carotid sinus pressure or potentiation of endogenous ACh with the cholinesterase inhibitor edrophonium can also terminate this form of VT, it is thought that Valsalva's effects are mediated by activation of the muscarinic receptor (M_2) (4) and its coupling to the inhibitory guanine nucleotide binding protein (G_i), which decreases stimulated levels of cAMP (41).

Verapamil terminates VT due to triggered activity, but is also effective in experimental ischemic-related reentry, intrafascicular reentry, and abnormal automaticity (39,42,43). Adenosine's effects on VT, with rare exceptions, are mechanism-specific, terminating that which is due to cAMP-mediated triggered activity.

Adenosine's electrophysiological effects are activated through its binding to the adenosine A_1 cell surface receptor and are mediated by a pertussis toxin–sensitive G-protein (G_i). G_i couples the A_1 receptor with its intracellular effectors on the inner surface of the cell membrane [i.e., an ion channel that activates the inward rectifying current I_{KAdo} (confined to supraventricular tissue) or adenylyl cyclase], which decreases stimulated levels of intracellular cAMP throughout the heart, thus producing an antiadrenergic effect (44). The G-protein is a heterotrimer, composed of three subunits—α, β, and γ. In the inactive state, guanosine diphosphate (GDP) is bound to the α-subunit. When adenosine binds to the A_1 receptor, GDP dissociates from the α-subunit and is replaced by guanosine triphosphate (GTP), which then activates the G-protein (41).

In the absence of β-adrenergic stimulation, adenosine has no effect on ventricular resting membrane potential, action potential amplitude and duration, or levels of intracellular cAMP. However, c-AMP-mediated changes in the action potential are attenuated or completely reversed by adenosine (Fig. 16.13). Voltage-clamp studies in ventricular myocytes demonstrate that adenosine abolishes cAMP-stimulated increases in $I_{Ca(L)}$ (46,47). This inhibition is dependent on adenosine attenuating/abolishing catecholamine-induced prolongation of the available state of the calcium channel (48). I_{Ti} is also reduced in response to adenosine, secondary to adenosine's effects on $I_{Ca(L)}$ (47). The antiadrenergic ef-

A.

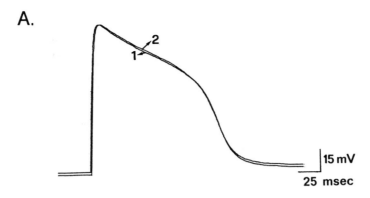

B.

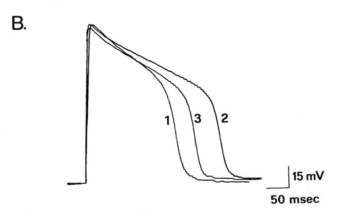

Figure 16.13 Demonstration of the direct and indirect antiadrenergic effects of adenosine on isolated bovine ventricular myocytes. (A) The action potential designated 1 is the control. Action potential 2 shows the absence of a direct effect of adenosine (50 μM). (B) Action potential 1 is the control action potential in the same preparation as shown in (A). Action potential 2 shows an increase in action potential duration and elevation of the plateau in response to 1 nM of isoproterenol. Action potential 3 shows the antiadrenergic effects of adenosine (50 μM). (From Ref. 45.)

fects of adenosine have not only been demonstrated in guinea pig, rabbit, and bovine ventricular myocytes but also in the human His–Purkinje system and ventricular myocardium (5,49).

The specificity of adenosine's effects on DADs, I_{Ti}, and triggered activity has been extensively studied in cellular preparations (Fig. 16.14). Adenosine abolishes DADs and I_{Ti} induced by isoproterenol and forskolin (an adenylate cyclase activator), but not those induced by dibutyryl cAMP. This suggests that the antiadrenergic effects of adenosine are mediated at a level proximal to cAMP (i.e., at the level of adenylyl cyclase) (47). Since the effects of adenosine on DADs are abolished by specific A_1-receptor antagonists and by pertussis toxin, which inhibits G_i through ADP ribosylation (47) (Fig. 16.15), the effects of adenosine on DADs and triggered activity are likely mediated through the A_1 receptor and the signal transduction protein, G_i. The clinical effects of adenosine (and verapamil) in a patient with VT presumed to be due to cAMP-mediated triggered activity are shown in Figure 16.16.

A. CONTROL

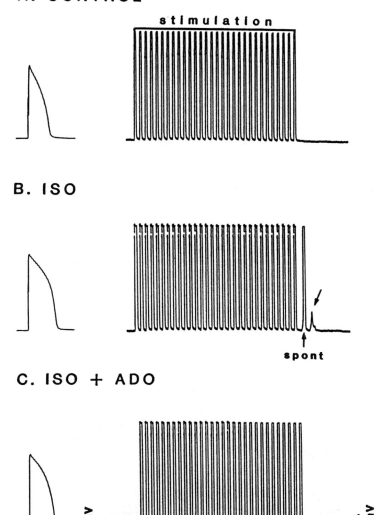

B. ISO

C. ISO + ADO

Figure 16.14 Abolition of delayed after-depolarizations (DAD) with adenosine. (A) Transmembrane potential from a single guinea pig ventricular myocyte during a drive train (stimulation). (B) In the presence of isoproterenol (ISO), the drive train triggers a DAD that reaches threshold, followed by a subthreshold DAD. (C) Abolition of the DAD with adenosine. (Modified from Ref. 47.)

Adenosine's effects are specific with regard to the etiology of the DAD since it is ineffective in attenuating DADs and I_{Ti} induced by cAMP-independent mechanisms (i.e., digitalis-related inhibition of Na^+, K^+, ATPase) (47,50). It is also unlikely that adenosine has any effect on DADs and triggered activity mediated through stimulation of the $alpha_1$-adrenergic–inositol–trisphosphate signaling cascade (51). Furthermore, adenosine has no effect on phase 2 EADs induced by quinidine or the calcium channel activator Bay K8644

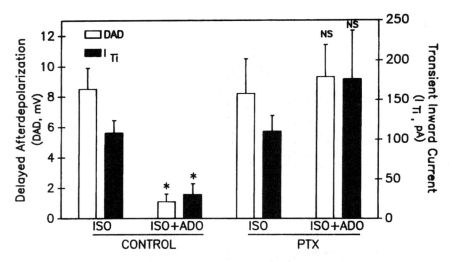

Figure 16.15   Bar graph showing lack of effect of adenosine (ADO) on isoproterenol (ISO)-induced delayed after-depolarizations (DADs) and transient inward current (I_{Ti}) in ventricular myocytes from pertussis toxin (PTX)-treated guinea pigs. Each bar represents the mean ±SEM. *Values significantly different ($p < 0.05$) from ISO alone. NS, differences between ISO + ADO vs. ISO alone were not significant. (From Ref. 47.)

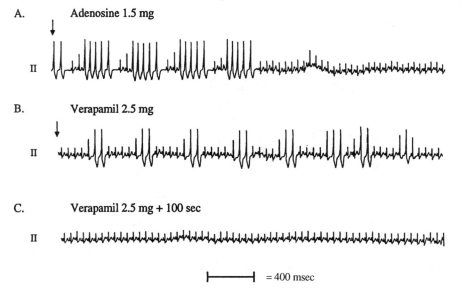

Figure 16.16   (A) ECG recording showing termination of incessant repetitive monomorphic ventricular tachycardia (VT) by adenosine. The vertical arrow indicates the completion of adenosine administration and a saline flush. (B) Administration of verapamil during incessant repetitive monomorphic VT. Vertical arrow indicates completion of verapamil infusion. (C) Termination of VT 100 s after verapamil administration. Surface lead II is shown. (From Ref. 39.)

(47). Finally, of particular interest, is that adenosine has no effect on catecholamine-facilitated reentry due to structural heart disease (52). Consistent with this finding, adenosine has only a minimal antiadrenergic effect on depressed action potentials in infarcted Purkinje fibers or in partially depolarized ventricular or Purkinje cells (6,46). The mechanism for adenosine's desensitization in these tissues is unknown, but possibilities include down-regulation of the A_1 adenosine receptor or uncoupling of the receptor from its G protein. Finally, some data suggest that adenosine's antiadrenergic effects are mediated by a cAMP-independent process involving A_1-receptor coupling to protein-specific phosphatases (53). This observation is contradicted, however, by the fact that the electrophysiological effects of adenosine on I_{Ca} and I_{Ti} in cAMP-dependent triggered activity are related to increased cAMP levels (47) and that adenosine does not abolish triggered activity induced by dibutyryl cAMP. The latter should occur if adenosine's antiarrhythmic effects are due to dephosphorylation of proteins associated with ion channel gating rather than to inhibition of adenylate cyclase (47).

Several compounds besides adenosine have been proposed to have specificity for identifying triggered activity. Doxorubicin, an anthracycline antibiotic, suppresses VT due to digitalis-induced triggered activity but has no effect on infarct-related or ouabain-mediated automatic arrhythmias (54). The effects of doxorubicin on reentrant arrhythmias or cAMP-dependent triggered arrhythmias have not been studied. Furthermore, because of organ toxicity, doxorubicin is not an appropriate antiarrhythmic drug for clinical use.

Flunarizine, a compound that decreases intracellular Ca^{2+} through an unspecified mechanism has also been proposed to have mechanism-specific antiarrhythmic effects. For instance, flunarizine has no effect on reentrant or automatic rhythms but terminates DAD-dependent triggered activity (55,56). However, unlike adenosine, flunarizine is also effective in terminating ouabain-mediated triggered activity and abolishing EADs (18).

An important unresolved issue regarding VT due to cAMP-mediated triggered activity centers on the biochemical/molecular substrate for this arrhythmia. Nearly all clinical forms of VT are due to an anatomical substrate. However, this arrhythmia occurs in the absence of anatomical abnormalities and originates from a discrete focus (<1 cm diameter) in the RVOT. It has been hypothesized that a mutation in the various signal transduction pathways involving cAMP may provide the electrophysiological substrate. To this end, a somatic cell mutation was recently identified in the inhibitory G protein $G_{\alpha i2}$ in a patient with adrenergically mediated idiopathic RVOT tachycardia (which was adenosine-insensitive) (57). A point mutation (F200L) in the GTP binding domain of $G_{\alpha i2}$ was detected in a biopsy sample from the arrhythmogenic focus. This mutation increased intracellular cAMP and inhibited suppression of cAMP by adenosine (Fig. 16.17). This mutation may potentiate VT by eliminating the antiarrhythmic effects of endogenous adenosine, leading to an increase in intracellular cAMP. As shown in Figure 17, it may also increase the concentration of intracellular cAMP in response to β-adrenergic stimulation. No mutation was identified in $G_{\alpha i2}$ in myocardial tissue obtained from regions remote from the site of tachycardia origin or from peripheral lymphocytes.

V. CONCLUDING REMARKS

Although much is known about the cellular mechanisms of abnormal impulse formation, the emerging elaboration of the molecular mechanisms responsible for this category of arrhythmogenesis has provided a level of understanding and integration not previously appreciated. It is anticipated that the continued molecular characterization of EAD- and

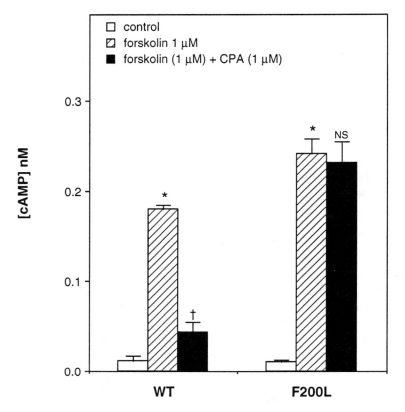

Figure 16.17 Effects of forskolin and cyclopentyladenosine (CPA) on cAMP levels in CHO cells stably transfected with human adenosine A_1 (hA_1R) receptor and either native (WT) or mutant (F200L) $G_{\alpha i2}$. Differences in cAMP levels between control and forskolin were significant at $p < 0.0001$ for both WT- and F200L-transfected cells (*). Differences in cAMP levels between forskolin and forskolin + CPA were significant at $p < 0.0001$ (†) for WT-transfected cells, but not significant (NS) for F200L-transfected cells. (From Ref. 57.)

DAD-mediated arrhythmias over the next decade will provide not only improved methods for preclinical diagnosis but also the basis for the development of novel antiarrhythmic therapies.

ACKNOWLEDGMENTS

This work was supported in part by grants from the National Institute of Health (RO1 HL56139) and the Netherlands Organization for Scientific Research (NWO #950.10.647).

REFERENCES

1. Dangman KH, Hoffman BF. Studies on overdrive stimulation of canine cardiac Purkinje fibers: maximal diastolic potential as a determinant of the response. J Am Coll Cardiol 1983; 2:1183–1190.
2. Vos MA, Gorgels APM, Leunissen HDM, Havenith MG, Kriek E, Smeets JLRM, Wellens HJJ. Programmed electrical stimulation and drugs identify two subgroups of ventricular tachycar-

dias occurring 16–24 hours after occlusion of the left anterior descending artery. Circulation 1992; 85:747–755.

3. Janse MJ, Wit AL. Electrophysiologic mechanisms of ventricular arrhythmias resulting from myocardial ischemia and infarction. Physiol Rev 1989; 69:1049–1168.

4. Lerman BB. Response of nonreentrant catecholamine-mediated ventricular tachycardia to endogenous adenosine and acetylcholine: evidence for myocardial receptor-mediated effects. Circulation 1993; 87:382–390.

5. Lerman BB, Wesley RC, DiMarco JP, Haines DE, Belardinelli L. Antiadrenergic effects of adenosine on His-Purkinje automaticity: Evidence for accentuated antagonism. J Clin Invest 1988; 82:2127–2135.

6. Rosen MR, Danilo P, Weiss RM. Actions of adenosine on normal and abnormal impulse initiation in canine ventricle. Am J Physiol 1983; 83;H715–H721.

7. DiFrancesco D, Angoni M, Maccaferri G. The pacemaker current in cardiac cells. In Zipes DP, Jalife J, eds. Cardiac Electrophysiology: From Cell to Beside. Philadelphia: WB Saunders Co., 1995: 96–103.

8. Belardinelli L, Shryock JC, Song Y. Ionic basis of the electrophysiological actions of adenosine on cardiomyocytes. FASEB J 1995; 9:359–365.

9. Cranefield PF, Aronson RS. Cardiac arrhythmias: The role of triggered activity and other mechanisms. Mount Kisco, NY: Futura, 1988.

10. Antzelevitch C, Sicouri S. Clinical relevance of cardiac arrhythmias generated by afterdepolarizations. Role of M cells in the generation of U-waves, triggered activity and torsade de pointes. J Am Coll Cardiol 1994; 23: 259–277.

11. Roden DM, Lazzara R, Rosen M, Schwartz PJ, Towbin J, Vincent GM. Multiple mechanisms in the long QT syndrome: current knowledge, gaps, and future directions. Circulation 1996; 94:1996–2012.

12. Brugada P, Wellens HJJ. Early afterdepolarizations: role in conduction block, "prolonged repolarization dependent reexcitation," and tachyarrhythmias in the human heart. PACE 1985; 8: 889–896.

13. Surawicz B. Electrophysiologic substrate of torsade de pointes: dispersion of repolarization or early afterdepolarizations? J Am Coll Cardiol 1989; 14:172–184.

14. January CT, Riddle JM. Early afterdepolarizations: mechanism of induction and block: a role for the L-type window current. Circ Res 1989; 64:977–990.

15. Yamada KA, Corr PB. Effects of β-adrenergic receptor activation on intracellular calcium and membrane potential in adult cardiomyocytes. J Cardiovasc Electrophysiol 1992; 3: 209–224.

16. Volders PG, Kulcsar A, Vos MA, Sipido KR, Wellens HJ, Lazzara R, Szabo B. Similarities between early and delayed afterdepolarizations induced by isoproterenol in canine ventricular myocytes. Cardiovasc Res 1997; 34:348–359.

17. Vos MA, Verduyn SC, Gorgels AP, Lipcsei GC, Wellens HJJ. Reproducible induction of early afterdepolarizations and torsade de pointes arrhythmias by d-sotalol and pacing in dogs with chronic atrioventricular block. Circulation 1995; 91:864–872.

18. Burashnikov A, Antzelevitch C. Acceleration induced action potential prolongation and early afterdepolarizations. J Cardiovasc Electrophysiol 1998; 9:934–948.

19. Verduyn SC, Vos MA, Gorgels AP, Zande van der J, Leunissen JD, Wellens HJ. The effect of flunarizine and ryanodine on acquired torsades de pointes arrhythmias in the intact canine heart. J Cardiovasc Electrophysiol 1995; 6:189–200.

20. Hohnloser SH, Singh BN. Proarrhythmia with class III antiarrhythmic drugs: definition, electrophysiologic mechanisms, incidence, predisposing factors, and clinical implications. J Cardiovasc Electrophysiol 1995; 6:920–936.

21. El-Sherif N, Caref EB, Yin H, Restivo M. The electrophysiological mechanism of ventricular arrhythmias in the long QT syndrome. Tridimensional mapping of activation and recovery patterns. Circ Res 1996; 79:474–492.

22. Duggal P, Vesely MR, Wattanasirichaigoon D, Villafane J, Kaushik V, Beggs AH. Mutation of

the gene for IsK associated with both Jervell and Lange Nielsen and Romano-Ward forms of long-QT syndrome. Circulation 1998; 97:142–146.

23.   Donger C, Denjoy I, Berthet M, Neyroud N, Cruaud C, Bennaceur M, Chivoret G, Schwartz K, Coumel P, Guicheney P. KVLQT1 C-terminal missense mutation causes a form fruste long-QT syndrome. Circulation 1997; 96:2778–2781.

24.   Banai S, Tzivoni D. Drug therapy for torsade de pointes. J Cardiovasc Electrophysiol 1993; 4:206–210.

25.   Gavrilescu S, Luca C. Right ventricular monophasic action potentials in patients with long QT syndrome. Br Heart J 1978; 40:1014–1018.

26.   Bonatti V, Rolli A, Botti G. Recording of monophasic action potentials of the right ventricle in long QT syndromes complicated by severe ventricular arrhythmias. Eur Heart J 1983; 4:168–179.

27.   Wit AL, Rosen MR. Afterdepolarizations and triggered activity. In: Fozzard HA, Haber E, Jennings RB, Katz AM, Morgan HE, eds. The Heart and Cardiovascular System. New York: Raven Press, 1992: 2113–2140.

28.   Han Z, Ferrier GR. Contribution of $Na^+ - Ca^{2+}$ exchange to stimulation of transient inward current by isoproterenol in rabbit cardiac pukinje fibers. Circ Res 1995; 76:664–674.

29.   Luo C-H, Rudy Y. A dynamic model of the cardiac ventricular action potential. II. Afterdepolarizations, triggered activity, and potentiation. Circ Res 1994; 74:1097–1113.

30.   Marban E, Robinson SW, Wier WG. Mechanisms of arrhythmogenic delayed and early afterdepolarizations in ferret ventricular muscle. J Clin Invest 1986; 78:1185–1192.

31.   DeFerrari GM, Viola MC, D'Amato E, Antolini R, Forti S. Distinct patterns of calcium transients during early and delayed afterdepolarizations induced by isoproterenol in ventricular myocytes. Circulation 1995; 91:2510–2515.

32.   Sicouri S, Antzelevitch C. Afterdepolarizations and triggered activity develop in a select population of cells (M cells) in canine ventricular myocardium: the effects of acetylstrophantidin and Bay K 8644. PACE 1991; 14:1714–1720.

33.   Rosen MR, Reder RF. Does triggered activity have a role in the genesis of cardiac arrhythmias? Ann Intern Med 1991; 94:794–801.

34.   Moak JP, Rosen MR. Induction and termination of triggered activity by pacing in isolated canine Purkinje fibers. Circulation 1984; 69:149–162.

35.   Vos MA, Gorgel A, Leunissen J, van Deursen R, Wellens H. Significance of the number of stimuli to initiate ouabain-induced arrhythmias in the intact heart. Circ Res 1991; 68:38–44.

36.   SH de Groot S, Vos MA, Gorgels A, Leunissen J, van der Steld BJ, Wellens H. Combining monophasic action potential recordings with pacing to demonstrate delayed afterdepolarizations and triggered arrhythmias in the intact heart. Value of diastolic slope. Circulation 1995; 92:2697–2704.

37.   Lerman BB, Stein KM, Markowitz SM. Adenosine-sensitive ventricular tachycardia: a conceptual approach. J Cardiovasc Electrophysiol 1996; 7:559–569.

38.   Lerman BB, Belardinelli L, West GA, Berne RM, DiMarco JP. Adenosine-sensitive ventricular tachycardia: evidence suggesting cyclic AMP-mediated triggered activity. Circulation 1986; 74:270–280.

39.   Lerman BB, Stein K, Engelstein ED, Battleman DS, Lippman N, Bei D, Catanzaro D. Mechanism of repetitive monomorphic ventricular tachycardia. Circulation 1995; 92:421–429.

40.   Lerman BB, Stein KM, Markowitz SM. Mechanisms of idiopathic left ventricular tachycardia. J Cardiovasc Electrophysiol 1997; 8:571–583.

41.   Gilman AG. G proteins: transducers of receptor-generated signals. Annu Rev Biochem 1987; 56:615–649.

42.   El-Sherif N, Lazzara R. Reentrant ventricular arrhythmias in late myocardial infarction period. 7. Effect of verapamil and D-600 and the role of the "slow channel." Circulation 1979; 60:605–615.

43.   Belhassen B, Rotmensch HH, Laniado S. Response of recurrent sustained ventricular tachycardia to verapamil. Br Heart J 1981; 46:679–682.

44. Lerman BB, Belardinelli L. Cardiac electrophysiology of adenosine: basic and clinical concepts. Circulation 1991; 83:1499–1509.

45. Lerman BB, Belardinelli. Effects of adenosine on ventricular tachycardia. In: Pellig A, Michelson EL, Dreifus LS, eds. Cardiac Electrophysiology and Pharmacology of Adenosine and ATP: Basic and Clinical Aspects. New York: Alan R. Liss, 1987: 301–314.

46. Isenberg G, Belardinelli L. Ionic basis for the antagonism between adenosine and isoproterenol on isolated mammalian ventricular myocytes. Circ Res 1984; X:309–325.

47. Song Y, Thedford S, Lerman BB, Belardinelli L. Adenosine-sensitive afterdepolarizations and triggered activity in guinea pig ventricular myocytes. Circ Res 1992; 70:743–753.

48. Kato M, Yomaguchi H, Ochi R. Mechanism of adenosine-induced inhibition of calcium current in guinea pig ventricular cells. Circ Res 1990; 67:1134–1141.

49. Nunain SO, Garratt C, Paul V, Debbas N, Ward DE, Camm AJ: Effect of intravenous adenosine on human atrial and ventricular repolarization. Cardiovasc Res 1992; 26:939–943.

50. Belardinelli L, Song Y. Adenosine and ATP regulated ion currents in cardiomyocytes. In: Godfraind T, Mancia O, Abbracchio MP, Aguilar-Bryan L, Govonl S, eds. Pharmacological Control of Calcium and Potassium Homeostasis: Biological, Therapeutical and Clinical Aspects. Dordrecht, The Netherland Kluwer Academic Publishers, 1995; 65–72.

51. Endoh M, Yamashita S. Adenosine antagonizes the positive inotropic action mediated via β-, but not □-adrenoceptors in the rabbit papillary muscle. Eur J Pharmacol 1980; 65:445–448.

52. Lerman BB, Stein KM, Markowitz SM, Mittal S, Slotwiner D. Catecholamine facilitated reentrant ventricular tachycardia: uncoupling of adenosine's antiadrenergic effects. J Cardiovasc Electrophysiol 1999; 10:17–26.

53. Gupta RC, Neumann J, Durant P, Watanabe AM. A_1-adenosine receptor-mediated inhibition of isoproterenol-stimulated protein phosphorylation in ventricular myocytes. Evidence against a cAMP-dependent effect. Circ Res 1992; 72:65–74.

54. Le Marec H, Spinelli W, Rosen MR. The effects of doxorubicin on ventricular tachycardia. Circulation 1986; 74:881–889.

55. Park JK, Danilo P, Rosen MR. Effects of flunarizine on impulse formation in canine Purkinje fibers. J Cardiovasc Electrophysiol 1992; 3:306–314.

56. Vos MA, Gorgels APM, Leunissen HDM, vander Nagel T, Halbertsma FJJ, Wellens HJJ. Further observations to confirm the arrhythmia mechanism specific effects of flunarizine. J Cardiovasc Pharmacol 1992; 19:682–690.

57. Lerman BB, Dong B, Stein KM, Markowitz SM, Linden J, Catanzaro D. Right ventricular outflow tract tachycardia due to a somatic cell mutation in G protein subunit αi2. J Clin Invest 1998; 101:2862–2868.

17

Reentry

MICHIEL J. JANSE

Academic Medical Center, Amsterdam, The Netherlands

EUGENE DOWNAR

University of Toronto, Toronto, Ontario, Canada

I. REENTRY

During normal sinus rhythm, each electrical wave becomes extinct after sequential activation of atria and ventricles, and a new impulse must arise in the sinus node to permit subsequent activation. Under special conditions, an excitation wave can be blocked in circumscribed areas, rotate around these zones, and reenter the site of original excitation in repetitive cycles. This is called reentrant excitation, or circus movement. According to the nature of the area of block around which the reentrant wave propagates, a distinction has been made between anatomical and functional reentry.

II. ANATOMICAL REENTRY

Studies early in the twentieth century clearly defined the mechanisms by which an electrical impulse can continue to propagate and reexcite tissue it has previously excited (1–3). The simplest model of reentry in cardiac tissue was described by Mines in 1913 (1) and is illustrated in Figure 17.1. Mines defined one of the basic requirements for the initiation of reentry—the establishment of unidirectional block. He also realized that the initiation and maintenance of reentry were dependent on both conduction velocity and the duration of the refractory period. In Figure 17.1 (A) and (B), excitation propagates in one direction only. In (A), "if the rate of propagation is rapid as compared with the duration of the wave, the whole circuit will be in the excited state at the same time, and the excitation wave will die out." However, as in (B), when "the wave is slower and shorter . . . the excited state will have passed off at the region where the excitation started before the wave of excitation reaches this point. . . . Under these circumstances, the wave of excitation may spread a second time over the same tract of tissue" (1). In this example, the revolution time exceeds the absolute and relative refractory periods, and therefore the reentrant circuit contains a fully excitable gap. This implies that impulses originating outside the reentrant circuit can penetrate into the circuit and influence the reentrant rhythm. Because the normal pacemakers of the heart are likely to be overdrive suppressed during a reentrant tachycardia (see Chap. 16), such impulses are usually initiated artificially, either mechanically (chest thump) or by electrical stimulation. The effect of such impulses on a reentrant circuit with an excitable gap is illustrated in Figure 17.2 (4). In (A), the impulse enters the reentrant circuit at the end of the relative refractory period and propagates in both antegrade and retrograde directions. In the retrograde direction, the wave collides with the circulating wavefront and both waves annihilate, but in the antegrade direction the wave continues to propagate (B). As a result, the arrhythmia is reset (i.e., propagation continues to circulate with its former frequency but with a phase shift determined by the penetrating impulse). As shown in (C), when the impulse enters the circuit early in the relative refractory period, it fails to propagate in the antegrade direction because it meets progressively less excitable tissue. In the retrograde direction, it meets increasingly recovered tissue and is able to propagate until it meets the circulating wave and terminates the arrhythmia (D). When the heart is paced at a regular rate faster than the rate of the wave rotation, the situation depicted in (A) and (B) may be perpetuated: every paced impulse blocks the circulating wavefront, but also enters the antegrade pathway and maintains the circulating wave. The heart, therefore, follows the pacing rate, but when pacing is stopped, the original rhythm resumes. This phenomenon is called "transient entrainment" (5) and is an important diagnostic tool in determining whether a tachycardia is in fact caused by reentry. Of course, overdrive pacing may also terminate a reentrant tachycardia, according to the principle illustrated in (C) and (D). This is an important therapeutic tool used in "antitachycardia pacing" (see Chap. 22).

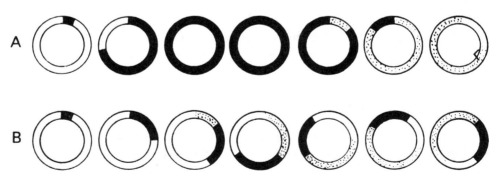

Figure 17.1   Mines' diagram to explain that reentry will occur if conduction is slowed and the refractory period duration is decreased. A stimulated impulse leaves in its wake absolutely refractory tissue (black area) and relatively refractory tissue (stippled area). (A) and (B) The impulse conducts in one direction only due to unidirectional block of the region to the left of the stimulation site. (A) Because of fast conduction and a long refractory period, the tissue is still absolutely refractory when the impulse has returned to its site of origin. (B) Because of slow conduction and a short refractory period, the tissue has recovered excitability by the time the impulse has reached the site of origin and the impulse continues to circulate. (Reproduced with permission from Ref. 1.)

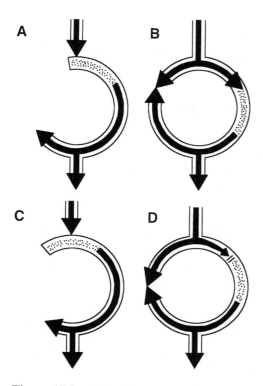

Figure 17.2 Effect of a premature impulse entering a reentrant circuit. The black and dotted areas show the absolute and relative refractory periods, respectively (A) and (B) Premature impulse enters the circuit at the end of relative refractory period and spreads in two directions. In the retrograde direction, the premature impulse collides with and annihilates the circulating wave; in the anterograde direction, the premature impulse advances resetting the tachycardia. (C) and (D) Premature impulse reaches the circuit closer to the state of absolute refractoriness. The impulse again collides with and annihilates the retrograde wave while failing to propagate in the anterograde direction, thereby terminating the tachycardia. (Reproduced with permission from Ref. 4.)

III. TACHYCARDIAS IN THE WOLFF–PARKINSON–WHITE SYNDROME

Mines was the first to consider reentrant excitation as a cause for arrhythmias in humans. In 1913 he wrote: "I venture to suggest that a circulating mechanism of this type may be responsible for some cases of paroxysmal tachycardia as observed clinically" (1). One year later, after having read an article by Kent (6) in which a human heart was described with an extra connection between atria and ventricles on the right side of the heart, Mines repeated his view:

> I now repeat this suggestion in the light of the new histological demonstration of Stanley Kent that the muscular connection between auricles and ventricles is multiple. Suppose that for some reason an impulse from the auricle reached the main A–V bundle but failed to reach this "right lateral" connection. It is possible then that the ventricle would excite the ventricular end of this lateral connection, not finding it refractory as normally it would be at such a time. The wave spreading then to the auricle might be expected to circulate around the path indicated (2).

This was written 16 years before Wolff, Parkinson, and White described the clinical syndrome that now bears their name (7), 18 years before Holzmann and Scherf ascribed the abnormal electrocardiogram in these patients to preexcitation of the ventricles during sinus rhythm via an accessory atrioventricular bundle (8), 19 years before Wolferth and Wood published the diagram shown in Figure 17.3 (9), and 53 years before the first studies in patients with the Wolff–Parkinson–White (WPW) syndrome employing intraoperative mapping and programmed stimulation via intracardiac catheters proved Mines' predictions to be correct (10–12). None of these papers quoted Mines.

A clinical example of preexcitation and the initiation of circus-movement tachycardia in a patient with a left lateral accessory connection is shown in Figure 17.4. The first three beats are the last of a train of atrial paced beats at a cycle length of 500 ms. These three beats are all preexcited as shown by a positive delta wave in V1 and the extremely short AV interval on the coronary sinus electrograms particularly on CS 7,8. The vertical line and the short arrows indicate the response to a premature atrial stimulus introduced at a coupling interval of 270 ms. At this degree of prematurity, the accessory tract is still refractory and the short arrows show pure atrial electrograms on the CS recordings. The atrial impulse conducts through the normal AV node and His bundle after a physiological delay and produces a normally activated (nonpreexcited) ventricular complex as seen in the surface ECG signals (beat A). This normally excited ventricular beat is able to activate the ventricular insertion of the accessory tract and conduct back retrogradely to the left atrium with so little delay that, again, ventricular and atrial electrograms (indicated by the long arrows) appear as one complex in the coronary sinus. The AV node by now has recovered and allows excitation to pass antegradely into the ventricles, producing the second beat (beat B) of a sustained orthodromic tachycardia at a cycle length of 300 ms. The patient was permanently cured by catheter ablation with RF energy applied to the mitral annulus at the site of the accessory tract (in proximity to CS 5,6).

Less commonly, circus-movement tachycardia proceeds down the accessory tract producing global preexcitation and back to the atria passing in a reverse manner through the His bundle and AV node. This antidromic tachycardia manifests with a wide complex anomalous QRS due to global preexcitation and can be mistaken for ventricular tachycardia. In approximately 10% of WPW patients there are two accessory AV nodal connections

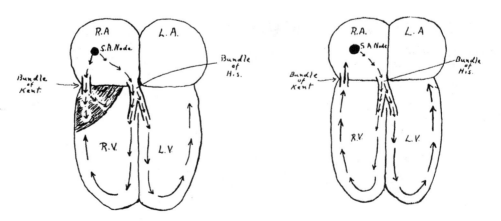

Figure 17.3   Mechanism of preexcitation and of circus-movement tachycardia as originally predicted by Mines and published by Wolferth and Wood in 1933. (Reproduced with permission from Ref. 9.)

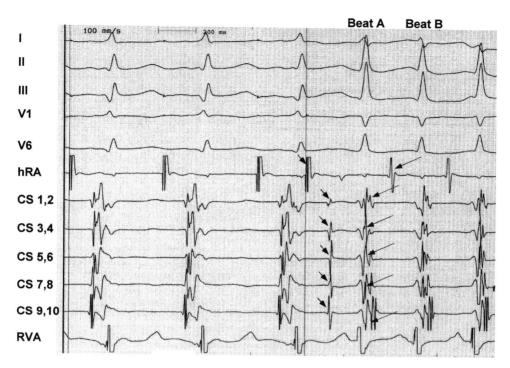

Figure 17.4 Preexcitation and initiation of a circus-movement tachycardia (orthodromic) in a patient with a left lateral accessory connection. Top five tracings are surface ECG leads 1, 2, 3, V1, V6. Lower tracings are all intracardiac bipolar electrograms from high right atrium (HRA) distal to proximal coronary sinus (CS 1,2 through CS 9,10) and right ventricular apex (RVA). See text for discussion.

(13). In such instances, circus-movement tachycardia can follow a multitude of circuits depending on which of the three (two anomalous and one normal AV connection) pathways are activated and which direction is taken. Figure 17.5 depicts six different circuits. The complexity seen in the examples in Figure 17.5 provides unique insights into the range of reentrant phenomena in a clinical setting. In practice, it is not uncommon for there to be a single accessory tract capable of conducting only in a reverse direction (from ventricle to atrium). Such cases never show preexcitation on the ECG and are therefore called concealed tracts in contradistinction to tracts that conduct anomalously into the ventricles producing manifest preexcitation. Manifest preexcitation occurs with an incidence of one to three per thousand persons (14). Approximately one-third of all accessory tracts are concealed.

That the bundle branches constitute a portion of the reentrant circuit introduces yet another variable that can affect circus-movement tachycardia. The abrupt initiation of any supraventricular tachycardia may catch either the right or left bundle in a relatively refractory state. This may not only give the tachycardia a right or left bundle branch block morphology but in the case of an accessory tract it may slow the cycle length of the tachycardia by forcing an enlargement of the reentrant circuit (see Fig. 17.6). As first noted by Coumel and Attuel (15), the coexistence of an accessory connection and a bundle branch block in the same ventricle causes excitation to proceed down the contralateral bundle branch, then across the intraventricular septum into the ipsilateral ventricle. Bundle branch

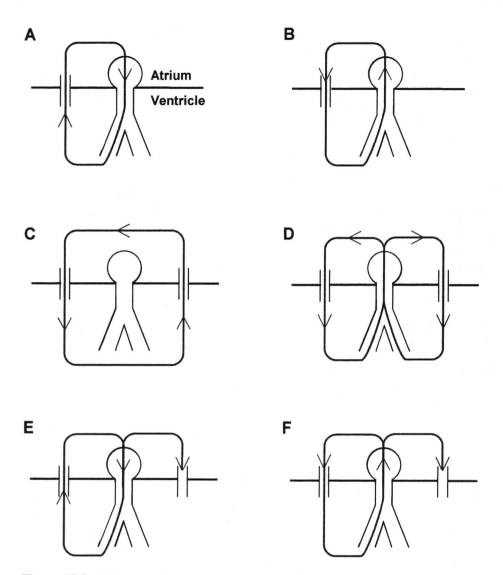

Figure 17.5 Schematic of a selection of six different circus-movement tachycardias that can be seen in preexcitation. (A) Orthodromic tachycardia such as the example in Figure 17.4. Activation proceeds normally over the AV node and His bundle into the ventricles and returns by retrograde conduction through the accessory tract back to the atria. (B) Antidromic tachycardia with ventricular activation being exclusively via the accessory tract (causing global preexcitation mimicking ventricular tachycardia) and returning to the atria by retrograde conduction through the normal AV connection. (C)–(F) Effect of two accessory tracts. (C) Global preexcitation via one tract with retrograde conduction to the atria via the second tract. The normal AV connection is excluded from active participation in the circuit. (D) Simultaneous preexcitation by two accessory tracts with a retrograde limb of this figure-8 tachycardia being formed by the normal AV node. (E) Orthodromic tachycardia with passive preexcitation via the second accessory tract. (F) Antidromic tachycardia with passive additional preexcitation from the second accessory tract.

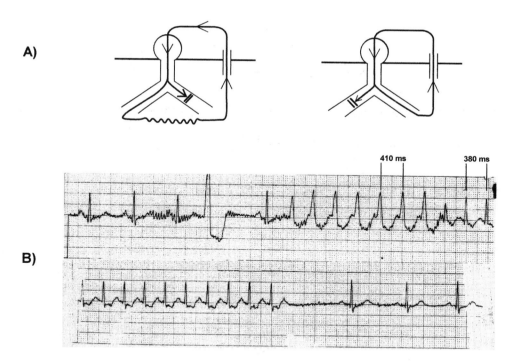

Figure 17.6 Effect of functional bundle branch block on an orthodromic tachycardia. (A, right) Schematic of an orthodromic tachycardia in which the circuit utilizes a concealed left accessory tract. The right bundle does not form part of the circuit so that appearance of a functional right bundle branch block would not affect the tachycardia. By contrast, development of a functional left bundle branch block would force the circuit to bypass the block by conducting down the right bundle and back to the left ventricle through the left-sided bypass tract. (A, left) Situation is shown schematically. (B) Clinical example. The upper ECG rhythm strip shows the onset of an orthodromic tachycardia in a patient with a concealed left lateral accessory connection. The lower strip shows termination of this tachycardia. Note that the onset of the tachycardia is marked by a functional bundle branch block that disappears after seven beats. For the reasons shown in the upper panels, the tachycardia cycle length is slower by 30 ms (410 vs. 380 ms) during the left bundle branch block.

block in the contralateral ventricle will of course not affect the reentrant circuit and will not cause a change in cycle length of the tachycardia.

IV. BUNDLE BRANCH REENTRY

The long reentrant circuit composed of the bundle branches and distal ventricular myocardium is another example of an anatomical circuit causing ordered reentry in which there is a single, rotating wavefront (16–18). Figure 17.7 shows an example of "concealed" bundle branch reentry in a normal canine heart. When a premature stimulus was applied to the right side of the interventricular septum, the stimulated impulse was blocked in the right bundle branch, which had a longer refractory period than the ventricular muscle. It still conducted through the septal myocardium toward the left bundle branch, through which the impulse traveled retrogradely toward the His bundle and subsequently anterogradely through the right bundle, activating the bundle at the recording site 97 ms after the stimu-

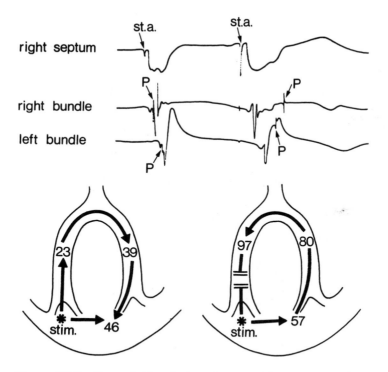

Figure 17.7   Concealed bundle branch reentry. Upper traces are bipolar electrograms from the right side of the interventricular septum and the right and the left bundle in an isolated, Langendorff-perfused canine heart. Stimuli were applied to the right septal surface (St.a. = stimulus artifact). The last of a series of basal stimuli (first beat shown) is followed by a premature stimulus 210 ms later. In the recordings of the bundle branches, the Purkinje spike (P) precedes the myocardial complex during the basic beat, and activity in the right bundle precedes activity in the left bundle. During the premature beat, the Purkinje spikes occur later than the myocardial complexes, and the activity in the left bundle precedes activity in the right bundle. In the lower panels, diagrams depict the activation sequence of the basic and the premature impulse. Numbers are activation times in milliseconds (time zero is the stimulus artifact). Double bars indicate conduction block. (Reproduced with permission from Ref. 16.)

lus artifact. In this case, reexcitation of the septal myocardium did not occur because the ventricular refractory period had not yet ended. However, Lyons and Burgess demonstrated overt bundle branch reentry in the normal canine heart (18), and this can be elicited as well by programmed stimulation in the normal human heart. Usually, only one or two successive reentrant responses caused by bundle branch reentry follow the stimulated premature impulse. However, sustained ventricular tachycardia involving bundle branch reentry can occur when cardiac disease slows intraventricular conduction (19,20). This type of tachycardia can be prevented by surgical or electrical ablation of the right bundle branch (21).

Commonly, pathological processes will block retrograde conduction via the right bundle branch. This will not be manifest on the surface ECG. However, antegrade conduction through the zone of retrograde block may still be possible and so the scene is set for a bundle branch reentrant ventricular tachycardia [Fig. 17.8 (A)]. A spontaneous ventricular

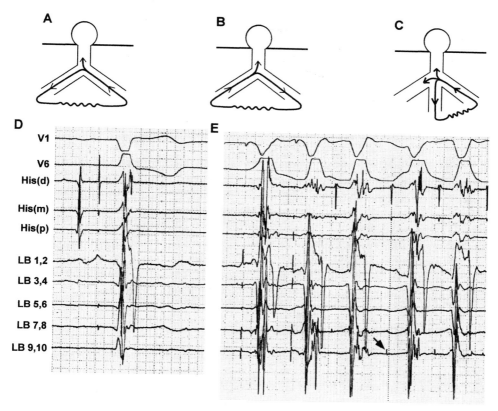

Figure 17.8 Ventricular tachycardia due to bundle branch reentry. (A)–(C) Three common circuits. (A) Most common variant with excitation proceeding antegradely down the right bundle across the septum and then retrogradely up the left bundle and back to the right bundle. The surface ECG would show a left bundle branch block morphology. (B) Activation proceeds down the left bundle across the septum and retrogradely along the right bundle to reactivate the left bundle. The ECG would show a right bundle branch block morphology. (C) Reentry circuits confined to the left ventricle. Activation proceeds down one fascicle of the left bundle through ventricular muscle then retrogradely along another fascicle to complete the circuit. The right bundle plays no significant role. (D)–(E) Clinical examples of the schematic in (C) with left bundle branch reentry. (D) Recorded in sinus rhythm. (E) Initiation of a left bundle branch tachycardia. The top two tracings are of surface leads V1 and V6. Three bipolar electrograms show distal to proximal His electrograms [His (d), His (m), and His (p)] and five electrograms are shown recorded from the region of the left bundle branch (LB 1,2 distal through LB 9,10 proximal). Note with the onset of the tachycardia there are both His and left bundle potentials before each ventricular complex. Note also that the left bundle branch potential (indicated by the arrow) precedes the His potential in keeping with retrograde activation of the His bundle. Ablation of the right bundle branch had no effect on the tachycardia while ablation of the left bundle resulted in cure of the tachycardia.

ectopic beat arising in the right ventricle will enter the distal portion of the right bundle only to falter at the site of block, leaving the proximal portion in an excitable state. Excitation will continue to spread by ventricular conduction through the septum into the left ventricle. There the left bundle branch system will be invaded retrogradely, activating the His bundle and thence down the right bundle through the zone of block only to start the whole cycle again as the second beat of a sustained ventricular tachycardia. Such a tachycardia manifests a left bundle branch block morphology on the surface ECG.

Less commonly, the circuit may operate in the reverse direction due to retrograde block in the left bundle branch [Fig. 17.8 (B)]. The surface ECG in this case reveals a right bundle branch block morphology. Still less frequently, bundle branch tachycardia can be confined to the fascicles of the left bundle branch system. The circuit proceeds retrogradely through a region of block in one fascicle to the common left bundle then antegradely down the remainder of the left bundle into the ventricular muscle to reenter to the original fascicle retrogradely again [Fig. 17.8(C)].

The bundle branch tachycardias are of more than passing interest since they are usually malignant presenting clinically as syncope or sudden death in more than 75% of instances (22). The underlying pathological process is coronary artery disease or dilated cardiomyopathy in 90% of cases. It has been reported that as much as one-third of all ventricular tachycardias in dilated cardiomyopathies are due to bundle branch reentry (19). Clinical recognition of the mechanisms of these ventricular tachycardias is extremely important because they are eminently curable with catheter ablation of the appropriate bundle branch. In 90% of cases, this means producing a complete block within the right bundle branch—a simple target to identify and to ablate with radiofrequency energy.

V. VENTRICULAR TACHYCARDIA

In the above examples the bundles of the specialized conducting system formed a relatively simple discrete reentrant pathway that could be traversed in either direction, resulting in two different ventricular tachycardias. Detailed mapping studies (23–25) have shown that in healed infarcts the arrhythmogenic substrate that forms the reentrant pathway can be extremely heterogeneous. The substrate ranges from small single tracts to an extensive subendocardial sheetlike complex, manifestations of which may be seen over more than 20% of the entire left ventricular endocardial surface.

In Figure 17.9, the endocardial excitation of the left ventricle is depicted in an isolated, blood-perfused human heart. The heart was from a patient with a myocardial infarction undergoing cardiac transplantation. The endocardial surface of the left ventricle is depicted as though a cut were made along the posterior descending coronary artery, from base to apex, and the walls then folded outward. The reentrant pathway followed a counterclockwise course and involved tissue overlying the infarct. Electrical activity could be recorded throughout the complete cycle of the tachycardia, which was 264 ms. By making histological sections of the anterolateral wall and part of the adjacent interventricular septum, it could be demonstrated that there was continuity of myocardial bundles from one side of the infarct to the other (Fig. 17.10).

An example of a sheetlike return complex is seen in Figure 17.11, which shows left ventricular activation maps in a polar projection. Panels (A) and (B) depict in 12-ms isochrones the excitation sequence of two consecutive beats of a ventricular tachycardia. Each beat started with an elliptical band of excitation high on the left ventricular septum and ended at an inferoapical region 228 ms later. The next beat of the tachycardia also started on the septum at 408 ms. There appeared to be a large spatial (7 cms) and a large temporal (180 ms) separation between the end of one beat and the beginning of the next. Close scrutiny of the local electrogram revealed diastolic potentials on virtually a full quadrant of the total endocardium linking the terminal activation of one beat and the beginning of the next.

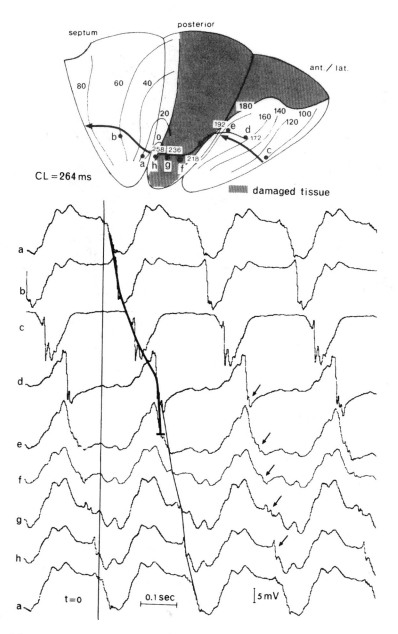

Figure 17.9 (Top) Endocardial activation pattern of one cycle of a sustained ventricular tachycardia induced in a Langendorff-perfused human heart with extensive inferoposterior infarction. Isochrones, constructed from endocardial electrograms recorded with a balloon electrode, are in milliseconds with respect to the time reference ($t = 0$). Heavy arrows indicate the main spread of subendocardial activation. Figures next to the black dots indicate activation times. (Bottom) Endocardial electrograms recorded at sites indicated in the top panel. The heavy printed line connects times of activation of subendocardial tissue at the sites a to e. At site d, the main deflection is followed by a second response of small amplitude (arrow). At sites e to h, signals mainly reflect remote activity, but in all signals small responses are present (arrows). The course of these small responses is indicated by the light printed line. (Reproduced with permission from Ref. 23.)

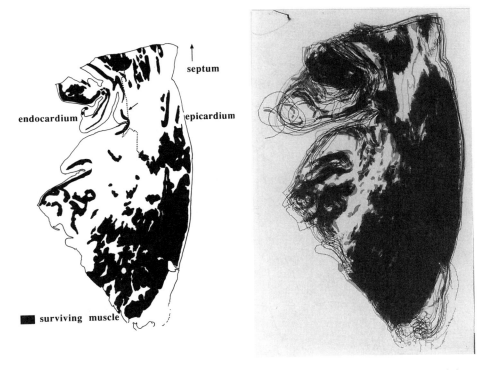

Figure 17.10 (Left) Drawing of one section of the left ventricular wall in which surviving myocardial tissue (black) formed part of the reentrant circuit. The scar tissue is indicated by white. (Right) Superimposed drawings of a number of sections showing continuity of viable myocardium from one side of the infarct to the other. (De Bakker JMT, Tasseron S, unpublished results.)

There may be multiple connections between the surviving sheetlike complex and the surrounding healthy myocardium. These connections may offer alternative routes of entry into and exit out of the return sheet complex. Figure 17.12 is from a patient with two paths of entry into a common return complex. It shows the effect of intermittent spontaneous block in one of these entrance paths. The top panel shows ECG lead a VL of a ventricular tachycardia characterized by unusual intermittent pauses in which the tachycardia cycle lengths increased spontaneously by about 100 ms for one beat only without a change in configuration. The systolic activation pattern (not shown) was consistent with a figure-8 pattern with early systolic activation appearing on electrode row 12. This pattern remained unchanged for beats of long as well as short cycle lengths. The middle two panels show diastolic activation during short and long cycle length beats. There were two inputs into a common return path across the apex, one from row 1 and one from row 3. The more common shorter cycle lengths were sustained by activation from row 1 reaching electrode 13.8 at 198 ms (relative to an arbitrary reference on electrode 5 of row 1). When this input failed, electrode 13.8 was activated 65 ms later at 255 ms from row 3. As a consequence, earliest systolic activation of the ensuing beats started at 290 ms instead of 193 ms. The lower panel shows bipolar electrograms from electrode row 1 during a transition from short (412 ms) to long (508 ms) cycles. Note the diastolic potentials on electrodes 1.6 and 1.7. In the short cycle, the potential on 1.6 precedes that on 1.7, but with the long cycle the

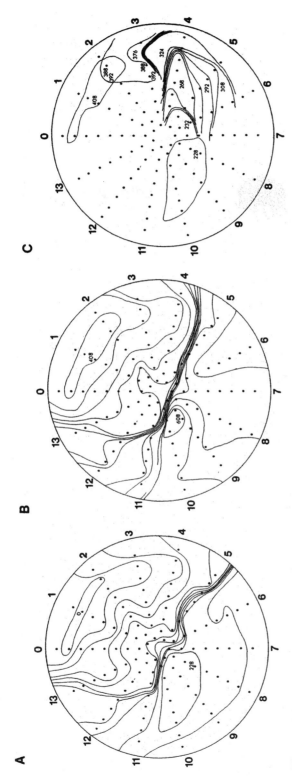

Figure 17.11 Left ventricular endocardial activation maps shown in a polar projection. Fourteen electrode rows (0–13) extend from the base (periphery) to the apex (center). Row 0 is parallel to the course of the left anterior descending coronary artery. (A) and (B) 12-ms isochrones of two consecutive beats of a ventricular tachycardia. The first beat begins with an elliptical band of activation (0 ms) across electrode rows 0, 1, and 2 and ends at 288 ms in an inferoapical region. (B) The next beat starts at 408 ms and follows a similar spread of activation. (C) Activation sequence of diastolic potentials linking the end of the first beat and the start of the second. Note how diastolic activation extends over a full quadrant of the total endocardium (Reproduced with permission from Ref. 24.)

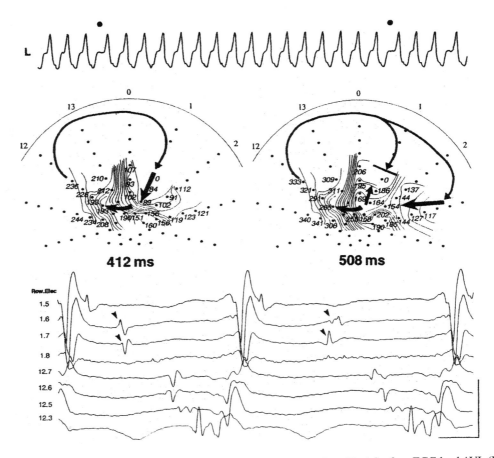

Figure 17.12   Alternate entry tracts into a return path complex. (Top) Surface ECG lead AVL (L) of a monoform ventricular tachycardia with intermittent pauses. The black dots indicate abrupt prolongation of the tachycardia cycle length by almost 100 ms. Although only two examples are shown, many occurred every few seconds, all without change in configuration of the tachycardia. The middle two panels show the diastolic activation maps. (Middle, left) Activation during the more common short cycles. (Middle, right) Diastolic activation during the rare prolonged cycles. (Bottom) Selected local bipolar electrograms from rows 2 and 12 on the return tract. In both cases, the onset of diastole was designated coincident with a second smaller component of the electrogram of electrode 1.5. The more common shorter cycles were sustained by activation from row 1 reaching electrode 13.8 at 190 ms. The longer cycles were due to block of this input forcing a more circuitous input from row three with a consequent delay of 65 ms in reaching electrode 13.8 at 255 ms. Onset of the next systolic activation of electrode 12.6 was further delayed starting at 290 ms instead of 193 ms. (Bottom) Diastolic potentials on electrodes 1.6 and 1.7 are reversed in timing and polarity with long cycles. (Reproduced with permission from Ref. 25.)

order is reversed, with electrode 1.7 leading. Not only is the order reversed but so is the polarity of both potentials. This is consistent with block into the faster tract being associated with retrograde activation of that tract during the long cycles. The diastolic maps show this reversal in activation along row 1.

Although the above case demonstrates the effects of intermittent block in one of two

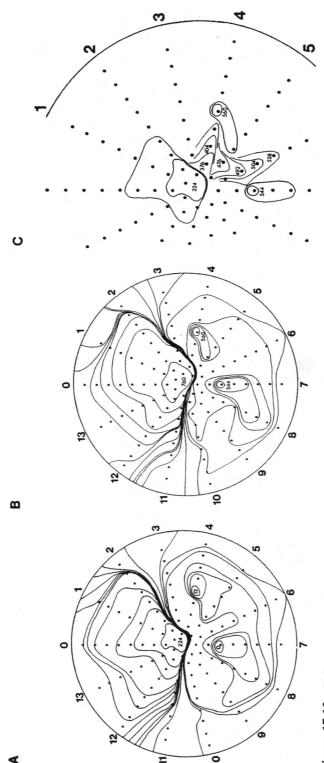

Figure 17.13 Left ventricular endocardial activation of a tachycardia with two separate endocardial origins. (A) and (B) Activation of two consecutive beats. (C) Details of diastolic activation linking the two beats. See text for details. (Reproduced with permission from Ref. 24.)

entry paths, other examples were detailed where the short or long entry path, once engaged, sustained the tachycardia permanently. In such cases, it was possible to initiate two ventricular tachycardias with the same site of origin and the same surface ECG morphology but differing in that one was malignant with a fast rate while the second was more benign, occurring at a slow rate.

Figure 17.13 shows an example of alternative exit paths from a common return complex. It shows the endocardial activation of two beats of a ventricular tachycardia, each beat of which originated at separate sites. Panel (A) shows that the site of earliest activation occurred at the sixth electrode of row 7. Twelve milliseconds later, a second site of activation appeared on the fifth electrode of row 4, some 4 cm distant from the first. At 48 ms, the two activation fronts had merged but were subsequently forced to proceed around an arc of block across the apex. The latest activation occurred at the apex on the distal side of the block at 224 ms. The next beat of the tachycardia (B) started at the same two sites as before at 544 and 560 ms. Both sites were about 3 cm distant from the region of latest activation. Close examination of the local electrograms revealed that this gap was bridged by a meandering reactivation front shown in (C). In a case such as this, if intermittent block were to occur in one of these two exit paths, the result would be a dramatic change in morphology of the ventricular tachycardia on the surface ECG, without termination of the tachycardia. Graded responses in one or both exits could provide an interplay of relative contributions to global cardiac excitation that could present a surface ECG pattern of polymorphic ventricular tachycardia.

Spontaneous block in the return pathway, without an active alternate path, will result in spontaneous termination of the tachycardia. Block occurred at the entrance to or exit from a return tract (Fig. 17.14) and the block was of either a sudden Mobitz type II or a Wenckebach pattern.

Clinical observations based on intraoperative mapping studies such as the above indicate that the substrate for reentrant ventricular tachycardia in a healed infarct is extremely varied. At one extreme of the spectrum is a simple, surviving muscle bundle forming a single return pathway, while at the other extreme is an extensive subendocardial sheetlike structure linked to the surrounding myocardium by multiple connecting bundles. The latter substrate can behave like a switching matrix allowing "multiuse" reentry (Fig. 17.15). The functional significance of such a structure is that:

1. The periodicity of a tachycardia (either sustained or intermittent) will be determined by which input paths are engaged. Provided the exit path is unaltered, the "site of origin" and the surface ECG morphology will remain unchanged.
2. A choice of exit paths will define different "sites of origin" and different ECG morphologies but these tachycardias will all share the same initial portion of a

---

Figure 17.14   Spontaneous block in return tracts with termination of ventricular tachycardia. (A) Spontaneous block at the entrance of the return tract of a nonsustained ventricular tachycardia. The top trace shows surface lead 2 (II) of initiation and spontaneous termination of a monoform tachycardia. (Middle, left) Systolic activation starting at the asterisk and ending at 209 ms on apical electrode rows 0 to 2. (Middle, right) Subsequent diastolic activation with the start of the following beat (asterisk) at 270 ms. (Bottom) bipolar electrograms from the return tract. Note the sudden block between the two components of the electrograms from electrodes 2.8, 3.8, and 4.8 resulting in termination of tachycardia. (B) Spontaneous block near the exit of the return tract with termination

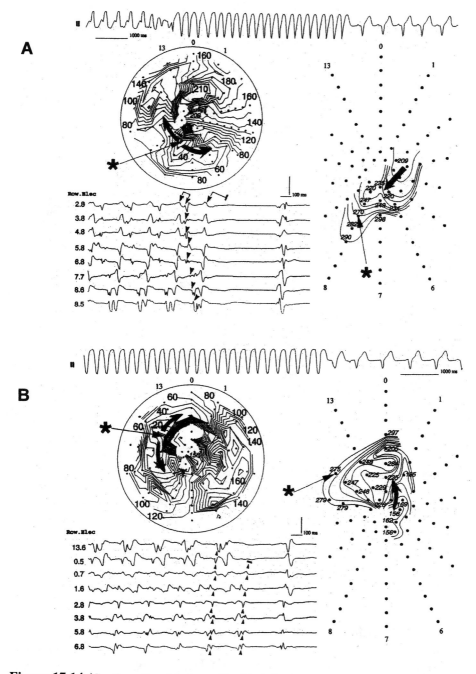

Figure 17.14 (Continued)   of tachycardia. (Top) Surface ECG lead 2 (II) of self-terminating tachycardia. (Middle, left) Systolic activation sequence starting at the asterisk and ending at 156 ms at the apex. (Middle, right) subsequent diastolic activity in the return tract moving back to the site of origin (asterisk) of the next beat at 275 ms. (Bottom) Bipolar electrograms from the return tract. Arrowheads show local components linking one beat of the tachycardia to the next. Note the sudden failure of conduction from electrode 0.7 to electrode 0.5 with abrupt termination of the tachycardia. (Reproduced with permission from Ref. 25.)

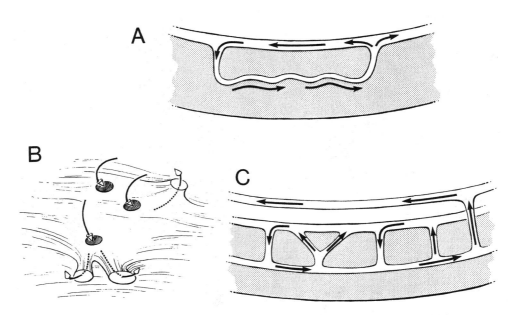

Figure 17.15 Functional schematic showing the range of the return path complexes in infarct-related ventricular tachycardia. Instead of envisioning a single bundle forming a return path (A), an extensive sheet is proposed, linked with the subendocardium through multiple connections (B). Some act as sink holes (shaded ovals), whereas others provide exit points (white ovals) to the subendocardium. The connections may involve portions of the left bundle branch system (C). See text for discussion. (Reproduced with permission from Ref. 24.)

return complex. By corollary, tachycardias of different morphologies (or even polymorphic morphology) are not necessarily derived from different substrates—they may all provide a single successful target for ablation.

3. Functional conduction block can occur in any of the muscle tracts forming the input and exit paths to and from the return complex to produce the above effects. Such block can be spontaneous or pacing-induced (or even tachycardia-induced based on concealed conduction).

4. Spontaneous block with tachycardia termination can occur at the entrance to as well as the exit from the return complex. Traditionally, ablation procedures have targeted the preexit site. However, the best target sites may be those that exhibit spontaneous block. Successful ablation of a tachycardia can be achieved at entrance sites as well (personal observation).

5. Functional tests for identification of a return pathway [such as concealed entrainment (26) and postpacing intervals] may not be sufficiently discriminatory to allow successful disruption of reentry if the return complex is a broad sheet-like structure.

New criteria are needed to identify not only the return path of a tachycardia but also its most vulnerable components. Such criteria may include morphology of local electrograms and the identification of sites of block in response to pacing-induced termination.

Unlike many supraventricular tachycardias, infarct-related ventricular tachycardia

stimuli ↓

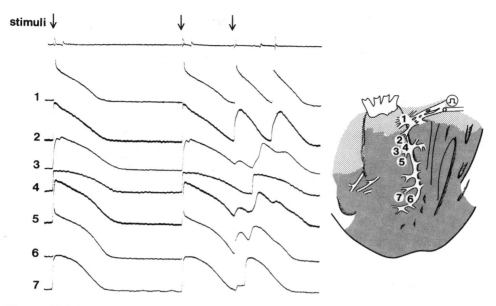

Figure 17.16 Transmembrane potentials from Purkinje fibers (1,2,6) and myocardial fibers (3,4,5,7) on the endocardial surface of a 72-h-old infarct in a canine heart (dark area in diagram is the infarct). Stimuli are delivered to a free-running strand of Purkinje fibers (square wave). The first two arrows are the basic drive. A premature impulse (third arrow) is blocked at the Purkinje–muscle junction at sites 2, 3, and 4, but successful excitation of the surviving muscle layers occurs at sites 6 and 7. From there, the muscle layer is retrogradely activated at site 2. (Janse MJ, Downar E, unpublished results, 1976).

remains a major challenge for clinical interventional electrophysiology. New technical advances are required field to provide high-resolution mapping with minimal morbidity or mortality. New percutanous electrode arrays, such as the basket catheter (27), offer some promise in this regard. New signal-processing techniques may help identify electrograms that are currently undetectable to enhance mapping of diastolic activation of the surviving muscle bundles that contribute to the return complexes.

Figure 17.16 shows a last example of anatomical reentry in an experimental infarction. The inset depicts part of the endocardial surface, including a papillary muscle, a free running strand of Purkinje fibers, and Purkinje fibers on the surface of the papillary muscle of a canine heart with a 72-h-old infarct (the darker area is the infarct). Stimuli were delivered to the free-running Purkinje strand and microelectrode recordings were made at the sites indicated. Recordings at sites 1, 2, and 6 were from Purkinje fibers overlying the infarct, recordings at sites 3, 4, 5, and 7 were from surviving muscle fibers on the endocardial surface of the infarct that were kept alive because oxygen and substrate from the cavitary blood reached the myocytes by diffusion. A premature impulse induced in the free-running strand (second arrow) is followed by a nonstimulated impulse. The recordings at sites 2, 3, and 4 show that at this Purkinje–muscle junction the premature impulse is blocked and does not reach the muscle. However, successful excitation of the muscle layer occurs at the junction between sites 6 and 7. Activity spreads slowly through the surviving muscle layer in a retrograde direction to reexcite the Purkinje layer at site 2.

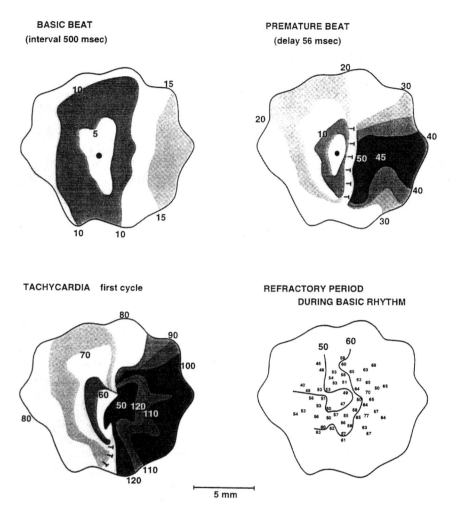

Figure 17.17 Initiation of functional reentry by premature stimulation in an isolated preparation of rabbit atrial muscle (A) Isochronal activation map of basic beat (interval 500 ms). The dot indicates the site of stimulation. Activation times (ms) are given relative to the stimulus onset. (B) Map of a premature beat (interval 56 ms). Double bars indicate conduction block. (C) First cycle of tachycardia. (D) Refractory periods measured during basic rhythm (ms). (Reproduced with permission from Ref. 29.)

VI. FUNCTIONAL REENTRY

In functional reentry, the reentrant impulse propagates around an area with a functional (i.e., unassociated with an anatomical barrier) conduction block. Allessie and coworkers (28–30) induced a rapid tachycardia in isolated preparations of rabbit atrial tissue by application of a critically timed premature stimulus. Figure 17.17 shows an activation map during regular pacing (basic beat), during the premature beat that initiated the tachycardia, and during the first cycle of tachycardia. Also shown is a distribution of refractory periods measured during regular basic rhythm. No anatomical obstacle was detected on the map of the basic beat (i.e., the excitation propagated in all directions from the central stimulating electrode). The premature wave propagated into the areas with shorter refractory periods and was blocked in the direction where refractory periods were longer. The line of conduction block extended across the center of the preparation along a distance of approximately 5 mm. Excitation propagated in two directions around the line of block and the two wavefronts merged behind the line of block. At this time, the tissue proximal to the site of block had recovered from the premature excitation and could be reexcited by the merged wavefront. The original area of conduction block broke up into two new areas and two wavefronts propagated around them in opposite directions: one in clockwise, the other in counterclockwise direction. Subsequently, one wave became extinct at the border of the preparation leaving only a single reentrant circuit. Arrhythmias induced in such a way were often short-lived, terminating spontaneously after one or several beats, or more stable, lasting for many seconds. Figure 17.18 shows an activation map during stable tachycardia, together with recordings of action potentials from the center of the reentrant circuit. Intracellular recordings were made from seven fibers located on a straight line through the zone of functional conduction block. Recordings obtained from one side of the central area (traces A, 1, 2, and 3) demonstrate a gradual decrease of the amplitude, rate of rise, and the duration of the action potentials. The recording from the fiber 3 demonstrates double potentials where the larger voltage deflection is caused by the wavefront propagating from left to right and the smaller voltage deflection is caused by the electrotonic influence of the same wavefront propagating half a cycle length later from left to right. The same sequence

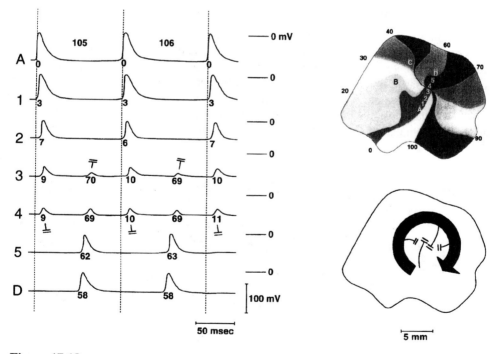

Figure 17.18 Functional reentry and tachycardia. (Right) Activation map and (left) action potential recordings obtained during steady-state tachycardia. Cells in the central area of the reentrant circuit show double potentials of low amplitude (traces 3 and 4). (Lower right) Schematic representation of the activation pattern. Double bars indicate conduction block. (Reproduced with permission from Ref. 30.)

of events takes place on the opposite side of the circuit (traces D, 5, and 4). The stable tachycardias could be reset or terminated by a properly timed stimulus delivered from an electrode located close to the central reentrant circuit that indicated the presence of a partially excitable gap.

After the pioneering experiments of Allessie and coworkers in atrial muscle, functional reentrant circuits with activation patterns of varying complexity were observed in both atrial and ventricular muscle. Because the functional reentrant circuits are not tied to anatomical structures, they can change in location and size. The activation can become even more complex when several rotating wavefronts are present in cardiac muscle (called "random" reentry) (31). Thus, multiple reentrant circuits were observed during stable atrial fibrillation (32) and during ventricular fibrillation in ischemic hearts (33). An example of the latter is shown in Figure 17.19. The activation pattern of part of the left ventricular surface, covered by a multiterminal electrode, is shown during spontaneously occurring ventricular fibrillation, 5 min after complete occlusion of the left anterior descending coronary artery of a porcine heart. Three consecutive time windows are shown. Multiple, apparently independent activation fronts propagate around multiple islets of conduction block (indicated by the shaded areas). Only seldom is a reentrant circuit completed (see middle panel) and usually an area is reexcited by another wavefront rather than by the one which activated it before. (For a more detailed description of atrial and ventricular fibrillation, see Chaps. 18 and 19.)

VII.   THE LEADING CIRCLE

To explain the properties of a single functional reentrant circuit, Allessie et al. formulated the concept of leading circle reentry (30). It was postulated that, during wave rotation in a tissue without unexcitable obstacles, the wavefront impinges on its refractory tail and travels through partially refractory tissue. The interaction between the wavefront and the refractory tail determines the properties of functional reentry. The leading circle was defined as the smallest possible pathway in which the impulse can continue to circulate, and in which the stimulating efficacy of the wavefront is just enough to excite the tissue ahead,

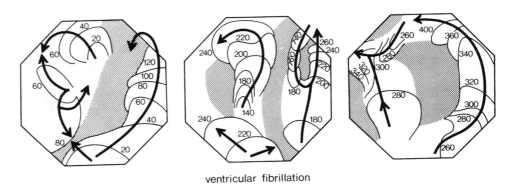

ventricular fibrillation

Figure 17.19   Activation patterns of three successive time windows during ischemia-induced ventricular fibrillation in an isolated, Langendorff-perfused pig heart. Numbers are in milliseconds. Shaded areas are zones of conduction block. Note the presence of multiple independent wavefronts. Both collision and fusion of wavefronts occur, and occasionally a complete reentrant circuit is seen. (Reproduced with permission from Ref. 33.)

which is still in its relative refractory phase. Because the wavefront propagates through partially refractory tissue, the conduction velocity is reduced. The velocity value as well as the length of the circuit depends on the excitability of the partially refractory tissue and on the stimulating efficacy of the wavefront, which is determined by the amplitude and the upstroke velocity of the action potential. The revolution time is confined to the relative refractory period and no fully excitable gap exists according to this mechanism.

VIII. ANISOTROPIC REENTRY

It has long been known that conduction velocity in atrial and ventricular tissue depends on the relationship between the direction of the propagating wave and the alignment of the myocardial cells. Because of the difference in spacing of cell borders and gap junctions in the longitudinal versus the transverse direction, the intracellular transverse resistivity is about nine times higher than the longitudinal resistivity. This accounts for the fact that longitudinal conduction velocity is three times higher than transverse conduction velocity (34).

Spach and colleagues (35) showed that the anisotropic properties of cardiac muscle may provide the spatial nonuniformity required to produce reentry. Premature impulses generated by point stimulation in the crista terminalis of the atrium were blocked in the longitudinal direction, but conducted in the transverse direction. Eventually, a more distal site in the crista terminalis was excited, and activity propagated retrogradely in the longitudinal direction to induce a reentrant premature impulse. Anisotropic reentry has also been observed in surviving epicardial muscle overlying a healed infarct (36). Anisotropic reentry is functional in the sense that no gross anatomical obstacle is present. However, unlike the leading circle model, anisotropic reentry is characterized by the presence of a distinct excitable gap (see Fig. 17.20).

Tachycardias usually are initiated by premature stimulation. The unidirectional block of the premature impulse that initiates the tachycardia is usually more prone to occur in the longitudinal direction, but may occur also in the transverse direction. During the sustained phase of the tachycardia, however, the line of block is always oriented parallel to the long fiber axis. A figure-8 type of reentry also can be observed, where activity propagates in two semicircular wavefronts, one clockwise and the other counterclockwise, around two lines of block to join in a common pathway. The lines of block and the common pathway are oriented also parallel to the fiber direction (Fig. 17.20). Across the line of block, there is electrotonic interaction between the two longitudinal limbs of the circuit. Action potentials from cells within the line of block consist of distinct electrotonic humps caused by activity of the two longitudinal limbs (Fig. 17.20, lower panel). Conduction along the longitudinal limbs parallel to the line of block is about three times as fast as in the two transverse limbs. At the pivoting points, where the impulse changes direction from transverse to longitudinal, the action potentials are preceded by a steplike depolarization, resulting in a local conduction delay in the order of 30 ms. This prolongs the action potential to such a degree that at the pivoting points no diastolic interval is present between successive action potentials. In contrast, in the longitudinal limbs, action potentials are separated by a distinct diastolic interval. Therefore, the excitable gap, so characteristic for anisotropic reentry, appears to be caused by a lengthening of the action potential at the pivoting points of the circuit, related to local delay of the impulse caused by the sudden increase of

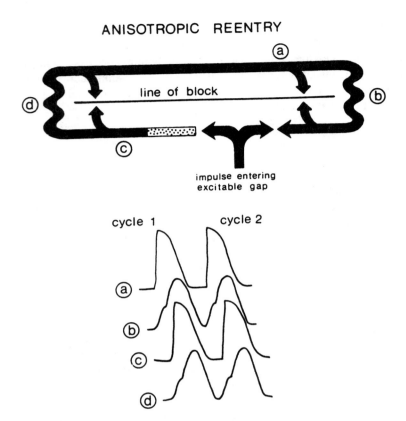

Figure 17.20 Schematic representation of an anisotropic reentrant circuit. The line of block is parallel to fiber orientation. The part of the circuit that is absolutely refractory is black, the part that is relatively refractory is stippled, and the part that is fully excitable is white. Conduction parallel to the long fiber axis (longitudinal conduction) is fast, and conduction perpendicular to it (transverse conduction) is slow. Only the two longitudinal limbs of the circuit have an excitable gap, as shown by the schematic drawings of the action potentials in both longitudinal limbs (a and c) and at the two pivoting points of transverse conduction (b and d). In b and d, there is no diastolic interval separating the repolarization phase and the upstroke of the next action potential, whereas this is present in a and c. Therefore, an impulse originating outside the circuit can penetrate the excitable gap only in the longitudinal limbs. It will conduct retrogradely and collide with the oncoming reentrant wavefront. Depending on the state of recovery of the tissue in the antegrade direction, the penetrating impulse may continue in the circuit (resetting or entraining) or block. (Reproduced with permission from: Task Force of the Working Group on Arrhythmias of the European Society of Cardiology. The Sicilian Gambit. A new approach to the classification of antiarrhythmic drugs based on their actions on arrhythmogenic mechanisms. Circulation 1991; 84:1831–1851 and Eur Heart J 1991; 12:1112–1131.)

electrotonic load as the impulse changes direction from longitudinal to transverse (37, 38), as schematically illustrated in Figure 17.20.

IX. SPIRAL WAVE REENTRY

Rotating waves of excitation have been described in a variety of excitable biological, physical, and/or chemical systems. The best-known example is the Belousov–Zhabotinsky re-

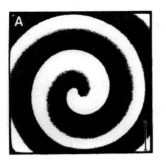 

Figure 17.21 Spiral waves. Spiral waves in (A) Belousov-Zhabotinsky reaction and (B) isolated preparation of canine epicardial muscle. (Reproduced with permission from Refs. 39 and 41, respectively.)

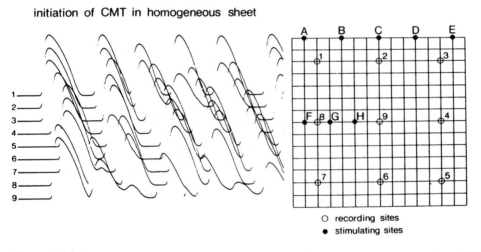

initiation of CMT in homogeneous sheet

O recording sites
● stimulating sites

Figure 17.22 Initiation of circus-movement tachycardia (CMT) in a homogeneous sheet. Filled circles are stimulus sites; unfilled circles correspond to tracings. A stimulus is applied within a short time interval to the five stimulation sites (A–E), located at the upper edge of the sheet. Stimulation of the other three stimulus sites (F, G, H) at the appropriate time results in the onset of circus-movement tachycardia. (Reproduced with permission from Ref. 40.)

action, where malonic acid is reversibly oxidized by bromate in the presence of ferroin. During this process, ferroin changes in color from red to blue and then back to red, allowing visualization of the reaction (39). As shown in Figure 17.21(A), in the center of the rotating wave (core) the tip of the wave moves along the surrounding medium effectively creating a spiral. In the heart, such spiral waves have been implicated in the generation of arrhythmias [Fig. 17.21(B)] (39). Figures 17.22 and 17.23 show how a spiral wave can be initiated in a computer model consisting of 650 identical excitable elements coupled to each other (40). Initially, a plane wavefront was set up by simultaneous stimulation of electrodes A to E in Figure 17.22. The resultant wavefronts merged into a single plane wave that traversed the sheet. Thereafter, a premature stimulus was delivered simultaneously to electrodes F, G, and H at a critical coupling interval. The resulting activation was

Figure 17.23 Isochronal lines corresponding to the initiation of the tachycardia shown in Figure 17.22. The distance between the lines corresponds to 20 time steps. (Left, upper panel) Plane wave caused by the conditioning stimulus. (Right, upper panel) Premature beat, which is blocked in the antegrade direction, starts a clockwise circus movement. (Left, lower panel). First isochronal line of this panel corresponds to the last one of the right, upper panel. (Right, lower panel) Continuation of the left lower panel. (Reproduced with permission from Ref. 40.)

blocked antegradely (see local response in element 7 of Fig. 17.22), but propagated retrogradely, invading the upper part of the sheet (element 1) and turning right and downward at the opposite side of the sheet (elements 2, 3, 4, etc.). The holes in the isochronal maps of Figure 17.23 represent regions where no action potentials of sufficient magnitude were present. For example, the first action potential of element 9 in Figure 17.22 corresponds to the planar wave of the left upper panel of Figure 17.23; the second, depressed, action potential to the premature activation whose path is shown in the right upper panel of Figure 17.23; and the local response corresponds to the "hole" in the left lower panel of Figure 17.23. The vortex of the spiral wave was not stationary but could move through the sheet. Similar spiral waves have been observed in cardiac muscle (41,42) [see also Fig. 17.21(B)].

Whereas in the leading circle model of functional reentry excitability is the crucial factor that determines the reentrant circuit, where the core is kept permanently refractory by centripetal wavelets, it has become apparent that the curvature of the circulating wavefronts is another important factor in maintaining functional reentry (43). A curving wavefront may cease to propagate altogether when a critical curvature is reached, despite the presence of excitable tissue. The difference between leading circle reentry and spiral wave reentry is that in the former the core of the circuit is kept permanently refractory, while in the latter the core is excitable but not excited. Recent studies, employing isolated perfused canine atria, clearly demonstrated spiral wave reentry, where cells in the core sometimes were quiescent at almost normal levels of membrane potential. The excitable gap was larger near the core than in the periphery of the reentrant circuit, which is incompatible with the leading circle concept (44).

In conclusion, spiral wave reentry may be the most important mechanism of functional reentry; it may be modified by structural inhomogeneities, such as tissue anisotropy, blood vessels, or fibrosis, so that obvious spiral patterns of activation need not be present. When looking at reentry on a microscopic level, the difference between anatomical and functional reentry tends to disappear because, obviously, cardiac tissue is not homogeneous.

REFERENCES

1. Mines GR. On dynamic equilibrium in the heart. J Physiol (Lond) 1913; 46:349–382.
2. Mines GR. On circulating excitations in heart muscles and their possible relation to tachycardia and fibrillation. Trans R Soc Can 1914; Sect IV:43–52.
3. Garrey WE. The nature of fibrillary contraction of the heart. Its relation to tissue mass and form. Am J Physiol 1914; 33:397–414.
4. Janse MJ, van Capelle FJL, Freud GE, Durrer D. Circus movement within the AV node as a basis for supraventricular tachycardia as shown by multiple microelectrode recording in the isolated rabbit heart. Circ Res 1971; 28:403–414.
5. MacLean WAH, Plumb VJ, Waldo AL. Transient entrainment and interruption of ventricular tachycardia. PACE 1981; 4:358–365.
6. Kent AFS. Observations on the auriculo-ventricular junction of the mammalian heart. Q J Exp Physiol 1913; 7:193–195.
7. Wolff L, Parkinson J, White PD. Bundle-branch block with short P-R interval in healthy young people prone to paroxysmal tachycardia. Am Heart J 1930; 5:685–704.
8. Holzmann M, Scherf D. Ueber Elektrokardiogramme mit verkürzter Vorhof-Kammer Distanz und positiven P-Zacken. Z Klin Med 1932; 21:404–423.
9. Wolferth CC, Wood FC. The mechanism of production of short PR intervals and prolonged QRS complexes in patients with presumably undamaged hearts. Hypothesis of an accessory pathway of auriculoventricular conduction (bundle of Kent). Am Heart J 1933; 8:297–308.
10. Durrer D, Roos JR. Epicardial excitation of the ventricles in a patient with a Wolff-Parkinson-White syndrome (type B). Circulation 1967; 35:15–21.
11. Burchell HB, Frye RB, Anderson MW, McGoon DC. Atrioventricular and ventriculoatrial excitation in Wolff-Parkinson-White syndrome (type B). Circulation 1967;36:663–672.
12. Durrer D, Schoo L, Schuilenburg RM, Wellens HJJ. The role of premature beats in the initiation and termination of supraventricular tachycardia in the Wolff-Parkinson-White syndrome. Circulation 1967;36:644–662.
13. Gallagher JJ, Selby WC, Kasell J, Wallace AG. Multiple accessory pathways in patients with the pre-excitation syndrome. Circulation 1976;54:571–591.
14. Chung KY, Walsh TI, Messic E. Wolff-Parkinson-White syndrome. Am Heart J 1965;69:116.
15. Coumel P, Attuel P. Reciprocating tachycardia in overt and latent pre-excitation. Influence of functional bundle branch block on the rate of the tachycardia. Eur J Cardiol 1974;1/4:423–436.
16. Janse MJ. The effects of changes in heart rate on the refractory period of the heart. Ph.D. Thesis, University of Amsterdam; Mondeel Offsetdrukkerij, 1971.
17. Akhtar M, Damato AN, Batsford WP, Ruskin JN, Ogunkelu JB, Vargas G. Demonstration of reentry within the His-Purkinje system in man. Circulation 1974;50:1150–1162.
18. Lyons CJ, Burgess MJ. Demonstration of reentry within the canine specialized conduction system. Am Heart J 1979;98:595–603.

19. Caceres J, Jazayeri M, McKinnie J, Avitall B, Denkeer ST, Tchou P, Akhtar M. Sustained bundle branch reentry as a mechanism of clinical tachycardia. Circulation 1989;79:256–270.

20. Guérot C, Valère PE, Castillo-Fenoy A, Tricot R. Tachycardie par ré-entree de branche à branche. Arch Mal Coeur 1974;67:1–11.

21. Tchou P, Jazayeri M, Denker S, Dongas J, Caceres J, Akhtar M. Trans catheter electrical ablation of the right bundle branch. A method of treating macroreentrant ventricular tachycardia attributed to bundle branch reentry. Circulation 1988;78:246–257.

22. Blanck Z, Dhala A, Deshpande S, Sra J, Jazayeri M, Akhtar M. Bundle Branch Reentrant Ventricular Tachycardia: Cumulative Experience in 48 Patients. J Candiouase Electrophysiol 1993;4:253–262.

23. De Bakker JMT, Van Capelle FJL, Janse MJ, Wilde AAM, Coronel R, Becker AE, Dingemans KP, Van Hemel NM, Hauer RNW. Reentry as a cause of ventricular tachycardia in patients with chronic ischemic heart disease: electrophysiologic and anatomic correlation. Circulation 1988;77:589–606.

24. Downar E, Kimber S, Harris L, Mickleborough L, Sevaptsidis E, Masse S, Chen TK, Genga A. Endocardial mapping of ventricular tachycardia in the intact human heart. II. evidence for multiuse reentry in a functional sheet of surviving myocardium. J Am Coll Cardiol 1992;20:869–78.

25. Downar E, Saito J, Doig CJ, Chen TK, Sevaptsidis E, Masse S, Kimber S, Mickleborough L, Harris L. Endocardial mapping of ventricular tachycardia in the intact human ventricle. III. Evidence of multiuse reentry with spontaneous and induced block in portions of reentry path complex. J Am Coll Cardiol 1995;25:1591–600.

26. Stevenson WG, Khan H, Sager P, Saxon LA, Middlekauff HR, Natterson, Wiever I. Identification of reentry circuit sites during catheter mapping and radiofrequency ablation of ventricular tachycardia late after myocardial infarction. Circulation 1993;88:1647–70.

27. Eldar M, Fitzpatrick AP, Ohab D, Smith MF, Hsu S, Whayne JG, Vered Z, Rotstein Z, Kordis T, Swenson DK, Chin M, Scheinman MM, Lesh MD, Greenspan AJ. Percutaneous multielectrode endocardial mapping during ventricular tachycardia in the swine model. Circulation 1996;94:1125–1130.

28. Allessie MA, Bonke FIM, Schopman FJG. Circus movement in rabbit atrial muscle as a mechanism of tachycardia. Circ Res 1973;33:54–62.

29. Allessie MA, Bonke FIM, Schopman FJG. Circus movement in rabbit atrial muscle as a mechanism of tachycardia. II. The role of non-uniform recovery of excitability in the occurrence of unidirectional block as studied with multiple microelectrodes. Circ Res 1976;39:168–177.

30. Allessie MA, Bonke FIM, Schopman FJG. Circus movement in rabbit atrial muscle as a mechanism of tachycardia. III. The "leading circle" concept: A new model of circus movement in cardiac tissue without the involvement of an anatomical obstacle. Circ Res 1977;41:9–18.

31. Hoffman BF, Rosen MR. Cellular mechanisms for cardiac arrhythmias. Circ Res 1981;49:1–15.

32. Allessie MA, Lammers WJEP, Bonke FIM, Hollen J. Experimental evaluation of Moe's multiple wavelet hypothesis of atrial fibrillation. In: Zipes DP, Jalife J, eds. Cardiac Arrhythmias. New York: Grune & Stratton, 1985:265–276.

33. Janse MJ, van Capelle FJL, Morsink H, Kléber AG, Wilms-Schopman FJL, Cardinal R, Naumann d'Alnoncourt C, Durrer D. Flow of "injury" current and patterns of excitation during early ventricular arrhythmias in acute regional ischemia in isolated porcine and canine hearts. Evidence for two different arrhythmogenic mechanisms. Circ Res 1980;47:151–165.

34. Clerc L. Directional differences of impulse spread in trabecular msucle from mammalian heart. J Physiol (Lond) 1976;255:335–346.

35. Spach MS, Dolber PC, Heidlage JF. Influence of the passive anisotropic properties on directional differences in propagation following modification of the sodium conductance in human atrial muscle. A model of reentry based on anisotropic discontinuous propagation. Circ Res 1988;62:811–832.

36. Wit AL. Rentrant excitation in the ventricles. In: Rosen MR, Janse MJ, AL Wit, eds. Cardiac Electrophysiology: A Textbook. Mount Kisco, NY: Futura Publishing Company, 1990:603–622.

37. Van Capelle FJL, Allessie MA. Computer simulation of anisotropic impulse propagation: characteristics of action potentials during re-entrant arrhythmias. In: Goldbeter A, Ed. Cell to Cell Signalling: From Experiments to Theoretical Models. London: Academic Press, 1989:577–588.

38. Allessie MA, Schalij MJ, Kirchhof CJHJ, Boersma L, Huybers M, Hollen J. The role of anisotropic impulse propagation in ventricular tachycardia. In: Goldbeter A, ed. Cell to Cell Signalling: From Experiments to Theoretical Models. London: Academic Press, 1989:565–576.

39. Winfree AT. When time breaks down. Princeton, NJ: Princeton University Press, 1987.

40. Van Capelle FJL, Durrer D. Computer simulation of arrhythmias in a network of coupled excitable elements. Circ Res 1980;47:454–466.

41. Davidenko JM, Pertsov AV, Salomonsz R, Baxter W, Jalife J. Stationary and drifting spiral waves of excitation in isolated cardiac muscle. Nature 1992;355:349–351.

42. Davidenko JM. Spiral wave activity: A possible common mechanism for polymorphic and monomorphic ventricular tachycardia. J Cardiovasc Electrophysiol 1993;4:730–746.

43. Fast VG, Kléber AG. Role of wavefront curvature in propagation of cardiac impulse. Cardiovasc Res 1997;33:258–271.

44. Athill CA, Ikeda T, Kim Y-H, Wu T-J, Fishbein MC, Karagueuzian HS, Chen P-S. Transmembrane potential properties at the core of functional reentrant wavefronts in isolated canine right atria. Circulation 1998;98:1556–1567.

18

Atrial Fibrillation

ALBERT L. WALDO

Case Western Reserve University/University Hospitals of Cleveland, Cleveland, Ohio

DAVID R. VAN WAGONER

Cleveland Clinic Foundation, Cleveland, and Ohio State University, Columbus, Ohio

I. OVERVIEW

Atrial fibrillation (AF) is the most prevalent arrhythmia in the western world, with an incidence that increases significantly with age. It affects about 6% of individuals over age 65, and 10% of those over age 80 (1). AF is characterized by disorganized, high-rate (300–500/min) atrial electrical activity (2). Until recently, AF has largely had an electrocardiographic (ECG) diagnosis, characterized by an irregular ventricular rate in which the baseline between the QRS complexes is characterized by the presence of either fine or coarse fibrillatory atrial complexes (i.e., normal P waves are absent and there is no isoelectric interval between QRS complexes) (Fig. 18.1).

The loss of synchrony in electrical activation leads to severely impaired atrial contractility. As a result, AF increases the risk of stroke four- to five-fold (3), putatively due to thrombus formation in the relatively stagnant blood that pools in the fibrillating atrial appendages. In the absence of depressed atrioventricular nodal conduction (either intrinsic or drug related), the very rapid and irregular atrial activation rate during AF can result in an irregular and increased ventricular rate. Thus, in addition to increasing the risk of stroke, AF can impair diastolic ventricular filling and can have deleterious consequences for patients with poor ventricular function. Further, the rapid ventricular activation rate can lead to degenerative changes in ventricular function. Overall, recent results from the Framingham Heart Study demonstrate an important negative impact of AF on long-term survival (4). In recognition of its clinical significance, as well as the apparent potential for better treatment strategies, AF has received great attention from both clinicians and basic scientists in the past decade.

479

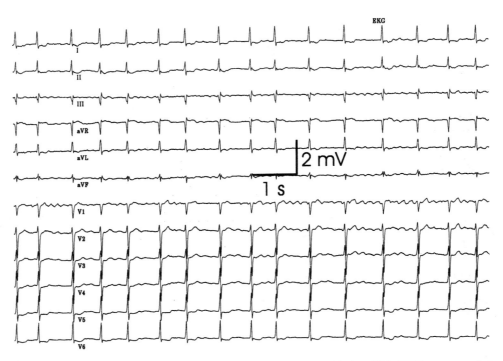

Figure 18.1 Typical 12-lead ECG manifestation of atrial fibrillation. This ECG was recorded from a 59-year-old male with a history of hypertension, suffering from recent-onset atrial fibrillation. The patient became dizzy upon initiation of the arrhythmia and suffered a stroke.

AF is often a progressive disease. Initial episodes are frequently *paroxysmal*, self-terminating within a few minutes to a few hours. Commonly, the duration of the paroxysms lengthens to the point where the arrhythmia becomes *persistent*, and the arrhythmia is continuous, unless the arrhythmia is either pharmacologically or electrically converted back to normal sinus rhythm. After variable periods of time, a combination of significant molecular and structural changes occur in the atria which make it significantly more difficult to achieve and maintain normal sinus rhythm. In some patients, the AF becomes *permanent* (i.e., it is either not possible to restore sinus rhythm or sinus rhythm can be achieved but atrial fibrillation recurs promptly, within a few minutes). When this happens, it is believed that the combination of anatomical and electrophysiological changes caused by the atrial fibrillation has become irreversible.

The major risk factors for the development of AF are advanced age (>65 years), male gender, hypertension, and ischemic heart disease. Other independent risk factors include diabetes mellitus, heart failure, increased vagal tone, congenital heart disease, valvular heart disease, enlarged left or right atrium, pulmonary disease, and hyperthyroidism (5). When AF occurs in younger (<65 years old) patients with none of the above risk factors and no underlying structural heart disease, it is termed *lone* AF.

Although the diagnosis of AF is rarely a clinical problem, with several exceptions, the diagnosis of the *mechanism* of AF is not yet at hand and treatment is still fraught with problems. The principal treatment remains antiarrhythmic drug therapy, but it suffers from unpredictable and, ultimately, relatively poor efficacy. AF recurs in about 50% of patients despite therapy, regardless of the drug used. In addition, important adverse effects may occur, including serious and even lethal proarrhythmia. Other forms of therapy to suppress or cure AF include surgical approaches (the Maze procedure), atrial pacing, and radiofrequency catheter ablation. At present, with the exception of vagally mediated AF, the treatments for AF are largely empiric in their approach.

For a long time, there was much speculation but little detailed understanding of the mechanisms of AF. With improvements in technology, particularly in the last two decades, and the development of several animal models, we now have come to appreciate that there are probably several mechanism of AF. All of these may lead to electrophysiological and structural remodeling of the atria, thereby providing a final common pathway. The structural, cellular, and molecular changes that accompany the different stages of AF are now a subject of intense study and are discussed in detail below.

II. MECHANISMS OF ATRIAL FIBRILLATION

A. Reentry as a Potential Mechanism

AF was first conceptualized as an arrhythmia based on the presence of reentrant circuits of electrical activation (see Fig. 18.2). In papers published in 1906 (6) and 1908 (7), Mayer was the first to elucidate the fundamental principles of reentry. Mayer studied the properties of contractions initiated in rings of tissue cut from the bell of jellyfish (Scyphomedusae). He showed that he could stimulate the tissue in such a way that the contractile wave would flow in only one direction. Once this wave of contraction was started, it could be maintained for hours or days. He realized that for reentry to occur, there a tissue was needed in which conduction was unidirectional, and in which conduction time exceeded the refractory period. In 1913, Mines extended this work, studying the atria of turtle, frog, and electric ray hearts. He showed that refractory period was inversely related to stimulation rate, and characterized the impact of wavelength (conduction velocity × refractory period) on fibrillation. And, just as in the Scyphomedusae of Mayer, he showed that in these atria a "circus movement" of contractile waves could be induced (8).

Initially, a major challenge to the study of AF was the difficulty in creating suitable animal models. As recognized by Lewis et al. (9), there was an inability to produce sustained arrhythmias in the normal canine atria, and a technical inability to map AF even if it lasted for any period. As we now know, to make sustained AF in the canine heart and other mammalian hearts of moderate size (e.g., goats), abnormal (pathological) conditions resulting from interventions such as application of substances, vagal stimulation, inflammation, prolonged rapid pacing, or heart failure usually must first be imposed.

Despite this, early studies produced seminal work. In a paper published in 1914 (10), Garrey established the fundamental concept that a critical mass of tissue was necessary to sustain fibrillation of any sort (atrial or ventricular). Particularly remarkable was the fact that he reached this conclusion based on observations of dying hearts without any electrophysiological or even mechanical recordings. Nevertheless, his conclusions have stood the test of time. He induced AF by introducing Faradic stimulation at the tip of one of the atrial appendages. When he separated the tip of the appendage from the fibrillating atria, he found that "as a result of this procedure, the appendage came to rest, but the auricles invariably continued their delirium unaltered." From such observations, Garrey concluded that "any small auricular piece will cease fibrillating even though the excised pieces retained their normal properties." Also, based on his studies, Garrey proposed (11) that fibrillation was due to ". . . a series of ring-like circuits of shifting location and multiple com-

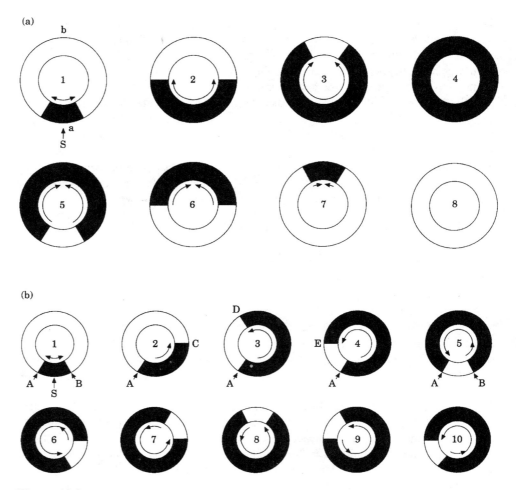

Figure 18.2 Conceptual diagrams depicting the mechanisms of re-entry, and the requirement for an excitable gap. (a) Illustrates the progress of a single reentrant wave passing through a ring of muscle, following stimulation at site *a*. The black portion of the ring corresponds to the tissue in the refractory state, and the figure shows the progress of the wave through the tissue. (b) Illustrates the establishment of circus movement in a ring of muscle tissue. The ring is stimulated in its lower quadrant, and the wave spreads in both directions, to A and B. At A it is blocked, but in B it continues to propagate around the ring. When it arrives at E (4), the refractory state is passing, so the wave continues to travel around the circle (5 to 10). (Reproduced from Ref. 14.)

plexity." Interestingly enough, this is now demonstrated as one of the mechanisms of AF (12). Actually, Lewis (13) proposed a similar mechanism, namely, that

> In fibrillation . . . a single circus movement does exist, but the path changes more grossly; but in general the same broad path is used over and over again. A *priori* it is possible to conceive of circus movements of many types. We might even assume several circuits, completely or transiently independent of each other, and each controlling for a time material sections of the muscle . . .

However, Lewis felt "that [in AF] the most mass of muscle is animated by a single circus movement . . . varying with limits . . ." As we shall see, such a mechanism of AF has now been demonstrated (12).

In the mid 1940s, Wiener and Rosenbleuth (15) calculated on the basis of estimates of the velocity of potential circulating reentrant wavefronts of excitation that some anatomical orifices (entry and exit points of the great vessels and pulmonary veins) were too small to permit sustained reentrant excitation. However, they suggested that the orifice of the inferior vena cava might be large enough to sustain reentrant atrial flutter, and they inferred that the smaller orifice of the superior vena cava would serve for AF. They also mentioned "the possibility that the pulmonary veins, singly or jointly, may provide effective obstacles for flutter or fibrillation." Two important principles were implicit in this work. First, similar to the hypotheses of Garrey (11) and Lewis (16), AF could be due to a single reentrant circuit generating a rhythm of such short cycle length that the remainder of the atria cannot follow 1:1. Second, also implicit in the early studies, including those of Lewis (9), was the apparent assumption that conduction velocity in the reentrant circuit was constant. We now know the former is likely and the latter is most unusual. Concerning the latter, functional or anatomical areas of slow conduction in reentrant circuits probably are the rule.

B. Focal, High-Rate Activity as a Mechanism of Atrial Fibrillation

The concept that AF may result from a single focus firing rapidly initially comes from the work of Scherf and colleagues (17–19), and later repeated by others (20,21). They placed aconitine on the heart and demonstrated that both the organized rhythm of atrial flutter and the disorganized rhythm of AF could be generated from a single focus firing rapidly. When the site of aconitine application was excluded, the tachycardia terminated. Additionally, it was suggested that it was the degree of rapidity of firing at the aconitine site that determined whether the rhythm generated was atrial flutter or AF. This concept is quite important, and is supported by subsequent studies on rabbit atria by Goto et al. (22) and Azuma et al. (23), who found that aconitine causes a very rapid rate, apparently due to abnormal automaticity. These findings demonstrate that a single focus firing rapidly (whatever the cause) is capable of producing AF. It is assumed that the impulses generated from the aconitine site occur so rapidly that the atria cannot follow in a 1:1 fashion. The result is AF. As we begin to understand some of the more contemporary models of AF, this old concept is again quite relevant. Furthermore, the idea that a single focus firing rapidly regardless of the cause (e.g., reentry, automaticity, or rapid pacing) is capable of generating AF will be seen in subsequent animal studies, and has now been demonstrated in patients (24–27).

C. Leading Circle Reentry

Implicit in most of the early studies was an assumption that there was only a single reentrant circuit responsible for either atrial flutter or fibrillation. It was thought that this circuit was "anchored" to an anatomical obstacle, such as a vessel orifice. The presence of an anatomical obstacle as central to reentrant excitation was usually implicit until the 1970s, when Allessie and colleagues (28–30) demonstrated in the isolated left atrium of rabbit hearts that reentrant excitation could occur in which the center of the reentrant circuit was functionally determined. This form of reentry was called "leading circle reentry" (30) and was an important advance in our understanding of reentry and our subsequent appreciation

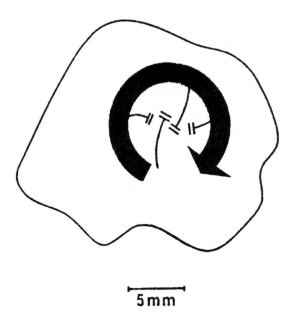

5 mm

Figure 18.3 Leading circle reentry. Diagram shows a circuit consisting of a reentrant wavefront (black arrow) circulating around a functionally refractory center produced by converging wavelets that block in the center. Block is indicated by the double bars. (Reproduced from Ref. 30.)

that functionally determined reentrant circuits of very short cycle length might generate AF. This concept is illustrated in Figure 18.3.

D. The Multiple Wavelet Hypothesis

Since the studies of Hoff (31), it has been known that vagal stimulation can produce AF in the canine heart, either alone or in association with atrial pacing. In 1959, largely based on studies of a vagally mediated model of AF and the aconitine-induced model of AF in the canine heart, Moe and Abildskov proposed the multiple wavelet hypothesis, in which random reentry was the cause of AF (21,23). AF was postulated to consist of multiple, dynamically distributed reentrant wavelets. The pathways of these wavelets were not anatomically determined, but rather were determined by the local atrial refractoriness and excitability. Because of this, the wavelets could collide and annihilate, divide, or fluctuate in size and velocity. They considered the multiple wavelet hypothesis to be one of several competing mechanisms that could explain the observed properties of AF. Alternative mechanisms which they suggested could underlie high-rate atrial activation included: (1) a single ectopic focus firing rapidly; (2) multiple ectopic foci firing rapidly; and (3) a single reentrant impulse around a fixed circuit. Moe favored the multiple wavelet hypothesis because it could best explain the stability of episodes of AF, which he recognized could last for years in some individuals (32). In one of the earliest computer simulations, Moe and colleagues were able to demonstrate that the multiple wavelet model could reproduce many of the features of AF in animals or humans (33). However, his model predicted that a large number (>30) of circulating wavelets were necessary to sustain AF. The experimental work of Allessie et al. (34) (Fig. 18.4) has shown that, in the canine atria, far fewer wavelets (4 to 6) were required to sustain AF.

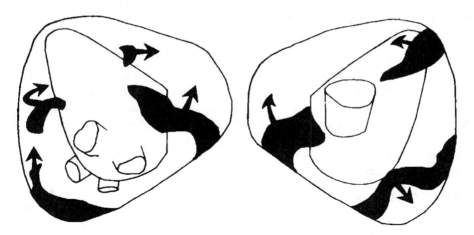

Figure 18.4   The first direct demonstration of the presence of multiple reentrant wavelets during atrial fibrillation. Endocavitary electrodes recorded simultaneous electrical activity from 192 sites during atrial fibrillation initiated during acetylcholine infusion in the canine heart. Activation maps were constructed these recordings. The figure illustrates the simultaneous presence of four waves of reentrant activity in the left atria, and three in the right atria. This was a typical observation during episodes of sustained atrial fibrillation. (Reproduced from Ref. 34).

III.   MULTISITE MAPPING AND EXPERIMENTAL MODELS OF SUSTAINED ATRIAL FIBRILLATION

A.   Multisite Mapping/In Vitro Acetylcholine-Induced Canine Atrial Fibrillation

The next advances in our understanding of AF resulted from the combination of the use of simultaneous multisite mapping techniques to analyze activation of the atria and the development of models in which AF was sustained. Allessie and colleagues (34,35) developed a Langendorff-perfused canine atrial model of AF. In this model, sustained AF was produced by rapid atrial pacing during infusion of acetylcholine. When the pacing was stopped, AF persisted as long as the acetylcholine was infused. They could record simultaneously from 192 of the 960 electrodes present in specially designed electrode arrays inserted through the tricuspid and mitral valve orifices. Thereby, atrial activation was recorded from selected endocardial portions of either or both atria during AF. These studies clearly demonstrated the presence of multiple, simultaneously circulating reentrant wavelets of the random reentry type, although they also occasionally described reentrant circuits with head–tail interaction (i.e., circus movement) [Fig. 18.2(b)]. These studies provided the proof that the multiple wavelet hypothesis proposed by Moe and Abildskov could be an operative mechanism (21).

More recently, several new models of AF have been described. An in vitro canine right atrial model described by Schuessler et al. (36) was shown to have a functionally determined figure-8 reentrant circuit of short cycle length induced by rapid pacing during acetylcholine infusion. The very short cycle length of the reentrant circuit (45 ms) generated an AF rhythm in the remainder of the preparation. Thus, once again, the concept of a single focus producing a rapid rhythm that the atria cannot follow 1:1 (fibrillatory conduction) is operative.

B. In Vivo Canine Mitral Regurgitation Model

A third model of AF was developed by Cox and colleagues (37) by creating mitral regurgitation in the canine heart. This model is difficult to produce and is associated with a high mortality rate. Simultaneous multisite mapping studies ". . . exhibited a spectrum of abnormal patterns ranging from the simplest pattern, in which a single reentrant circuit was present that activated the remainder of the atria, to the most complex cases, in which no consistent pattern of activation could be identified." Although septal activation maps were not obtained, and mapping for longer periods during AF are needed to characterize this model further, the presence of unstable reentrant circuits of short cycle length indicate, once again, that AF can be produced by this mechanism.

C. Canine Sterile Pericarditis Model

A fourth model of AF is the canine sterile pericarditis model (12), suggested by observation of patients with AF in the immediate postoperative period following open-heart surgery (38). AF is induced by rapid atrial pacing or programmed atrial pacing 1 to 2 days after surgically creating the pericarditis. On postoperative days 3 and 4, the inducibility of AF decreases, principally because atrial flutter is induced (12). Simultaneous multisite mapping studies have shown that, in this model, AF is produced by either of two mechanisms. One is due to multiple unstable reentrant circuits of very short cycle length which drive the atria at rates that cannot be followed in a 1:1 fashion. These reentrant circuits are short-lived (mean three to four rotations), but subsequently are reformed so that one to four (mean 1.3 per 100-ms window) unstable reentrant circuits are always present (12) (Fig. 18.5). The other mechanism is a single, stable reentrant circuit of very short cycle length, generally traveling around one or more pulmonary veins, which generates fibrillatory conduction to the remainder of the atria (39).

D. Canine Continuous Vagal Stimulation

Another canine heart model of AF is the continuous vagal stimulation model used by Moe and Abildskov (21), in which AF is initially induced by a burst of rapid atrial pacing (40–42). Although not fully characterized, the mechanism of maintenance of AF in this model has been suggested to be due to unstable reentrant circuits.

E. Chronic Atrial Pacing Models

In a patient with sustained atrial tachycardia and resulting class IV heart failure due to tachycardia-mediated cardiomyopathy, Moreira et al. (43) were the first to show that continuous, long-duration rapid atrial pacing could produce persistent atrial fibrillation. Experimental models of AF were subsequently produced by sustained or intermittent rapid atrial pacing in the canine or goat heart, respectively (44,45). While this approach is still being characterized, demonstration of consistent pathophysiological changes (shortening of the atrial effective refractory period, and histological changes consistent with hibernation) resulting from the persistent rapid atrial rate has already led to the realization that "atrial fibrillation begets atrial fibrillation." This is an important concept with enormous implications for understanding the progressive nature of AF (45). Furthermore, the fact that these changes seem reversible, at least after 2 weeks of AF, also has important implications.

In sheep, AF has been produced simply by pacing the atria rapidly for a brief period (46–48). This results in AF that persists for relatively long periods. This model remains to

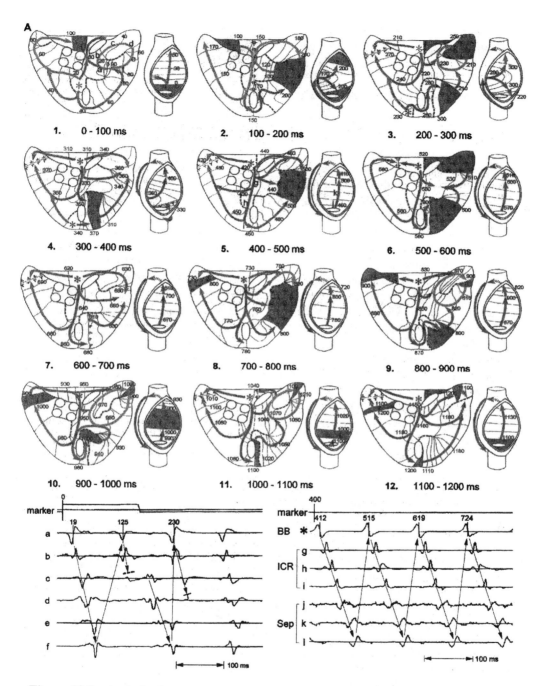

Figure 18.5 Analysis of consecutive atrial activation patterns during an episode of sustained AF in a dog. During this 1.2-s recording period, 16 reentrant circuits in total (mean 1.3 per 100-ms window) were observed. Seven of the circuits involved the septum and atrial epicardium; six circulated around the pulmonary veins, and three were observed in the right atrial free wall. The circuits lasted from 1.5 to 4.5 (mean 2.7) consecutive rotations, with a mean cycle length of 118 ms. (Reproduced from Ref. 12.)

be characterized more fully, but already has been shown to produce either a single rotor (reentrant circuit) in the left atrium, which causes fibrillatory conduction in the right atrium (47), or spiral wave reentry in the right atrium.

F. Heart Failure–Associated Atrial Fibrillation

AF is a frequent complication of heart failure. Several models of chronic AF have been developed following the onset of experimentally induced heart failure. Following high-rate (190/min) continuous pacing of the ventricles of sheep, a pacing-induced cardiomyopathy diminishes ventricular performance (49). In this condition, there is a significant increase in the susceptibility of the atria to the induction of AF. Two other models of AF in conjunction with heart failure have recently been described. One is a canine model in which the ventricles are paced rapidly for several weeks, after which atrial pacing can induce an atrial tachycardia due to a single focus firing rapidly. This focus is usually in the region of the pulmonary veins, but can also be found in the right atrium, and seems to be due to delayed after-depolarizations (50). When the atria can no longer follow this high-rate stimulus, or if an extrastimulus is given, atrial fibrillation is initiated. Another model is one created by infusing microspheres into the coronary arteries (51). This causes global ventricular dysfunction and stable AF, once it has been initiated either spontaneously or with pacing. The mechanism of the latter is not yet characterized.

The atrial electrophysiological changes associated with *ventricular* pacing and heart failure in the canine model may be distinct from those caused by AF induced with rapid *atrial* pacing. In further characterization of this model, it has been reported that the atrial effective refractory period (ERP) is actually increased (as opposed to a reduction in response to rapid atrial pacing) prior to the initiation of AF (52). In addition, the extent of atrial fibrosis was also dramatically increased (0.3% in control, up to ~15% in the animals in congestive heart failure), presumably leading to conduction defects. There was no change in the heterogeneity of refractoriness or conduction velocity at a cycle length of 360 ms (52).

G. Human Atrial Fibrillation Studies

There are several studies in patients which, though limited, have been informative. Cox et al. (37) induced AF in patients who had undergone surgical ablation of an accessory AV connection (Wolff–Parkinson–White syndrome) and showed in limited mapping studies that a single unstable reentrant circuit was present in some instances, and an uncertain mechanism, possibly multiple-reentrant wavelets, was present in other instances. Allessie's group (53) mapped part of the right atrial free wall during AF, also induced in patients after surgical ablation of an accessory AV connection. In the latter studies, three patterns of activation in the right atrial free wall were seen (Fig. 18.6). One included an unstable reentrant circuit. The other two did not really permit a mechanism to be discerned, although one was consistent with multiple-reentrant circuits.

Finally, there is the mechanism of tachycardia-induced tachycardia. Haisseguerre and colleagues have shown in a small cohort of patients with paroxysmal AF but without structural heart disease that AF was often generated by a single focus, principally in one of the pulmonary veins, which fired rapidly and generated AF (27). Ablation of the provoking focus resulted in disappearance of the AF. Similarly, not only has it long been recognized that both atrioventricular reentrant tachycardias (AVRT) and atrioventricular nodal reentrant tachycardias (AVNRT) may initiate AF, but also that suppression or cure of the AVRT or AVNRT is associated with the disappearance of AF.

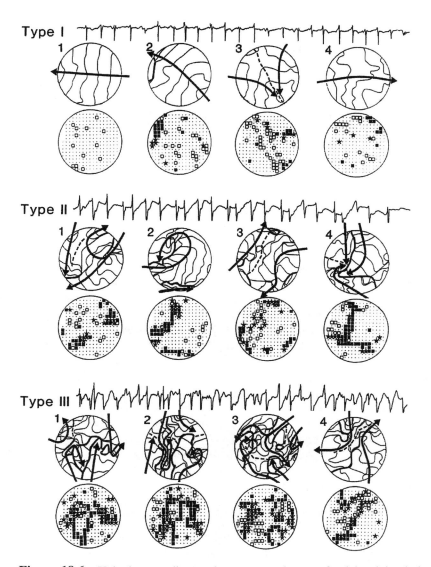

Figure 18.6 Unipolar recordings and reconstructed maps of atrial activity during human atrial fibrillation. Examples show the spatial distribution of unipolar electrogram morphologies during types I, II, and III fibrillation. Each panel documents 4 s of AF. For each type of fibrillation, a right atrial unipolar electrogram is displayed, together with four activation maps made at 1-s intervals. The diagram below each map gives the spatial distribution of the observed electrogram configurations (◆, single potentials; □, short-double potentials; ■, long-double potentials; and *, fragmented electrograms). During type I AF (top), broad activation waves propagated rapidly in different directions across the free wall of the right atrium and most of the electrograms showed single potentials. Fragmented electrograms were extremely rare. During type II AF (middle) a larger number of long-double and fragmented potentials were recorded. However, no preferential areas for these multicomponent electrograms were seen, and the beat-to-beat changing location of intra-atrial conduction block and slow conduction was distributed randomly. During type III AF (bottom), about one-third of the electrograms had multiple components. This was due to the high degree of fragmentation of the fibrillation waves and the associated high incidence of slow conduction, conduction block, and pivoting of wavefronts. Also, during type III AF, no preferential sites for intra-atrial conduction block or slow conduction were found in these patients. (Reproduced from Ref. 53.)

H.   Summary

All of these models, both animal and patient, provide new and important opportunities to understand better the nature of AF and its treatment. Animal models have already provided significant insights into the nature and mechanism of AF, and there is every expectation that they will continue to do so. Further studies and new models will serve to provide valuable new insights that should be expected to have important implications for patient care. Of course, the best animal model to study is the human. Such studies have already begun during open heart surgery (37). With techniques of simultaneous multisite mapping and other new technologies, such studies should continue to provide the requisite insights and understanding of AF to advance patient care.

IV.   ATRIAL FIBRILLATION–INDUCED ATRIAL REMODELING

A.   Overview

The process whereby paroxysmal atrial fibrillation becomes persistent and then permanent involves both structural changes in the atria (with respect to the degree of dilatation, trabeculation, fibrosis, fatty infiltration, and the gap junctional connections between myocytes), as well as biochemical changes in the individual atrial myocytes (e.g., hypertrophy, changes in ion channel density or distribution, etc.). This pathophysiological adaptation of the atria to the fibrillatory rhythm has been termed *remodeling*. More specifically, the changes in chamber size, collagen deposition, and gross structure have been termed *structural remodeling*, and those changes primarily affecting the excitability and electrical activity of the myocytes have been termed *electrophysiological remodeling*. The concept of electrophysiological remodeling was first recognized by Wijffels and colleagues, who noted that there were rapid, yet reversible, changes in atrial effective refractory period in their burst-pacing goat model of AF (45). Studies characterizing AF-induced atrial electrophysiological and structural remodeling are discussed in more detail below.

B.   Electrophysiological Remodeling

1.   Observations from Clinical Studies

For approximately 25 years, it has been known that significant electrophysiological changes occur in the atria of patients with diseased atria. In 1976, Hordof and colleagues were among the first to study the correlation between disease status and the cellular electrophysiological properties of human atrial tissue. They demonstrated that AF was most commonly observed in dilated atria, where the resting potential was frequently depolarized (54). They further noted that verapamil modulated the plateau of action potentials recorded from healthy atria, but not from patients prone to AF. Further, they showed that verapamil could prevent slow automaticity in the diseased atria. These studies suggested that calcium cycling may be significantly altered.

In a clinical electrophysiological study, Attuel and colleagues showed that there was a decrease in the effective refractory period and a loss in the adaptation of the atria to changes in rate in the atria of patients vulnerable to the induction of AF with pacing (55). In 1986, this observation was extended with a microelectrode study performed on tissue removed from a similar group of patients with chronic AF (56). It was shown that, relative to the normal patients, the action potentials recorded from atrial tissue of patients with chronic AF were briefer and more triangular. In addition, some myocytes were relatively depolarized. Similar to the report of Attuel et al., an abbreviation of APD_{90} (action potential duration measured at 90% repolarization) and the effective refractory period (ERP) were reported in the tissue of chronic AF patients. Finally, both of these parameters were shown to have a diminished accommodation to changes in rate. Together, these three studies (54–56) were among the first to demonstrate that AF was associated with significant long-term electrophysiological changes. The reduction in ERP could contribute to a decreased wavelength, assuming that conduction velocity did not change. Thus, these changes reflected an adaptation (electrophysiological remodeling) that facilitated the maintenance of AF once it became established.

2. Studies in the Burst-Pacing Goat Model: Time Course

Results from the burst-pacing goat model of AF (45) brought new attention and focus to studies of the mechanisms underlying the adaptation of the atria to the presence of AF. In this model, sustained AF was created in goats with use of a fibrillation pacemaker, i.e., a pacemaker that automatically distinguishes between sinus rhythm and AF. Whenever sinus rhythm was detected, the pacemaker introduced a 1-second burst of rapid atrial pacing (20 msec cycle length) which then again precipitated AF (45). During control periods, episodes of induced AF lasted a mean of 6 ± 3 s. However, fibrillation pacemaker-induced AF resulted in a progressive prolongation in the duration of AF, until it became sustained after 7.1 ± 4.8 days in 10 of 11 goats. During the first 24 h of AF, the median interval between atrial electrograms (F-F interval) shortened from a mean of 145 ± 18 to a mean of 108 ± 8 ms. This was a reflection of shortening of the atrial ERP. In addition, the inducibility of AF by a single premature atrial beat increased from 24 to 76%. The atrial ERP, as determined by programmed stimulation at a pacing cycle length of 400 ms, decreased 35% from a mean of 146 ± 19 to a mean of 98 ± 20 ms. At higher pacing rates (i.e., pacing at shorter drive cycle lengths), the decrease in atrial ERP was less (−12%), demonstrating a reversal of the normal adaptation of the atrial ERP to heart rate (Fig. 18.7). In five goats, after 2 to 4 weeks of AF, sinus rhythm was restored and all electrophysiological changes reversed in 1 week.

Thus, in this model, sustained AF led to a rapid, marked shortening of the atrial ERP. As the refractory period shortened, the episodes of induced AF became longer. Following approximately 5 days of paced AF, most of the goats remained in persistent AF with no need for additional pacing. From this observation, the phrase "atrial fibrillation begets atrial fibrillation" was coined. If the AF was cardioverted to sinus rhythm following relatively brief periods of AF (days to weeks), and the goats were then kept in normal sinus rhythm, the atrial ERP returned to baseline values within a week. The time course of the changes in atrial ERP was complex.

3. Factors Responsible for Atrial Electrophysiological Remodeling

Several elegant follow-up studies have been performed using the same goat model. It was demonstrated that it is most likely that the fibrillation-induced high-rate electrical activity of the atria, rather than other factors such as ischemia, stretch, or neurohumoral changes that accompany AF were responsible for the electrophysiological remodeling (57). Treatment of the goats with verapamil significantly reduced the extent of the changes in the atrial ERP, while having little effect on the inducibility of the AF (58). This suggests that the electrophysiological remodeling process is a response to cellular calcium overload. Similar results have now also been obtained from patient studies, where the changes in atrial ERP after brief periods (~15 min) of rapid pacing-induced AF were prevented by pretreat-

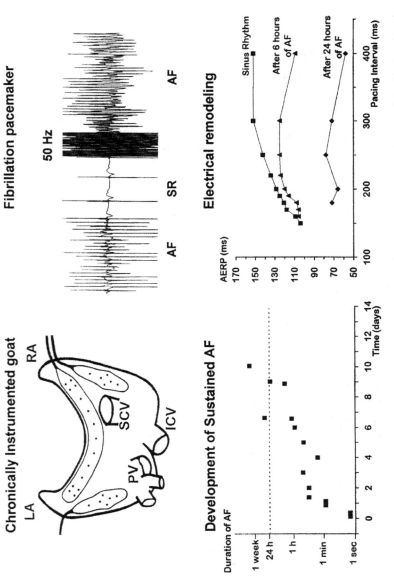

Figure 18.7 The goat model of sustained atrial fibrillation. (Upper left) Schematic diagram of the implanted epicardial unipolar recording and stimulation electrodes. Abbreviations: LA, left atria; RA, right atria; PV, pulmonary veins; SCV, superior caval vein; ICV, inferior caval vein. (Upper right) Three to four weeks after implant, the goats were connected to an external fibrillation pacemaker. The pacemaker detected conversion from AF to normal sinus rhythm, and delivered a 1-s burst of 50-Hz stimulation ($4 \times$ diastolic threshold) upon sensing the conversion. (Lower left) Representative example of the time course of development of sustained AF. Initial episodes are brief after pacemaker is switched on, but become progressively longer with continued maintenance in AF. (Lower right) An example of the electrophysiological remodeling that accompanies the development of sustained AF. The atrial ERP is plotted as a function of pacing interval. Note that significant shortening of the ERP is detectable within 6 h of AF, and that the normal accommodation to changes in rate is reversed by 24 h of AF. (Reproduced from Ref. 57).

ment with verapamil, but not procainamide (59). These short episodes of AF are also characterized by a postepisode contractile dysfunction (60). The contractile dysfunction recovers over ~10 min. Verapamil pretreatment prevented the short-term changes in contractile function as well. More detailed studies characterizing the fundamental mechanism(s) whereby calcium overload results in changes in the atrial ERP and mechanical function are still ongoing.

Whereas the goat studies suggest that ischemia is not a primary factor in the remodeling process, ischemia is likely to have a role in the inducibility and initiation of the arrhythmia, perhaps due to its effects on the heterogeneity of repolarization (see below). A significant reduction in atrial blood flow has been demonstrated in dogs paced at 600 bpm for 6 weeks (61). Note that episodes of AF frequently accompany acute myocardial infarction. The impact of coronary artery disease as a major risk factor for AF strongly suggests that ischemia has some role in the development of AF. An interesting case report notes the resolution of episodes of paroxysmal atrial fibrillation following angioplasty (62). Thus, while ischemia may not directly influence the electrophysiological remodeling process, it is very likely a significant factor in the genesis of AF.

4. The Role of Heterogeneity

The atria are not homogeneous structures, but are composed of areas of complex trabeculae and areas of muscular sheet. Because of the number of blood vessels entering the tissue, there are numerous orifices and areas of transition between atrial myocytes and vascular tissue. There are specialized nodal regions (sinoatrial node and atrioventricular node), and transitional fibers linking the atria to these regions. In view of its innate complexity, it is not surprising that the electrophysiological properties of the tissue vary significantly, depending on the specific location (63). The electrophysiological tasks of the atria include: (1) generation of pacemaker activity, with a stimulus to be spread throughout the atria, and following a variable delay generated at the atrioventricular node, conducted to the ventricles; (2) electrical activation of synchronized contractile activity in the atria. Regional differences in ion channel distribution lead to normal, regional variations in the atrial ERP. In normal sinus rhythm, the regional variations in ERP help to direct the path of atrial excitation in order to coordinate the muscular activity of the atria.

Neural fibers (sympathetic and parasympathetic) are distributed in a nonuniform manner throughout the atria. Neural activity modulates specific ion currents (e.g., I_{KACh}, I_{Ca}), and can alter the normal regional distribution of atrial ERP, thus having a major impact on both the inducibility of AF and the ability of the atria to sustain it (64). In the presence of AF, sympathetic tone is generally increased. In some cases, vagal tone is also increased. All of these factors tend to accentuate regional differences in ERP and wavelength. The increase in heterogeneity leads to the presence of more areas of slow conduction (due to incomplete recovery from inactivation), and tends to promote the maintenance of atrial fibrillation.

5. Review of Atrial Cellular Electrophysiology

Atrial myocytes are morphologically, functionally, biochemically, and electrophysiologically similar to, yet distinct from, their ventricular counterparts. Figure 18.8 illustrates a typical human atrial action potential, along with the major voltage-dependent currents that are believed to contribute to the shape of the human atrial action potential. Activation of any of the upper three currents would depolarize the myocytes, while the remaining currents are responsible for repolarization. The relative contribution and direction of the cur-

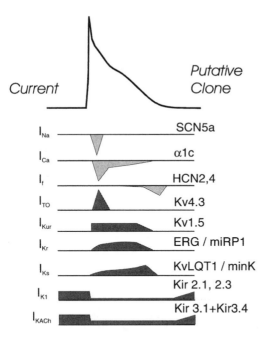

Figure 18.8   Ion currents and ion channels contributing to the human atrial action potential. A representative human atrial action potential is shown in the top of the figure. The measurable currents are listed on the left, with the gene(s) encoding the channel protein(s) shown on the right. The time course and direction of the currents are shown schematically. The top three currents (light shading) depolarize the atrial myocyte, while the lower currents (dark shading) repolarize the myocytes.

rent during the time course of the action potential are illustrated schematically. In the column on the right, the putative clones responsible for each current in human atrial myocytes are listed.

In general, there is a significant overlap in the distribution of ion channels between atrial and ventricular myocytes. In both, the resting potential is maintained by inward rectifier K currents (I_{K1}). In both, the major contributor to the upstroke of the action potential (phase 0) is the voltage-dependent Na current (I_{Na}). To a variable extent, the transient outward K current (I_{TO}) causes an early rapid repolarization of the action potential. The plateau phase of the action potential is a balance between inward L-type Ca current, perhaps a small inward Na current, and the repolarizing delayed rectifier K currents. Atrial myocytes have two additional repolarizing currents that are not present in ventricular myocytes: the ultrarapid delayed rectifier K current, I_{Kur}, and the muscarinic K current, I_{KACh}. While both of these currents are present at high density in healthy myocytes, I_{KACh} is only activated by either vagal activity (which releases acetylcholine) or by circulating adenosine. Thus, I_{KACh} confers significant neurohumoral control over atrial electrical activity. It is the combined presence of I_{Kur}, together with a larger I_{TO} current, which are primarily responsible for the characteristically more triangular shape of atrial action potentials, in comparison to the relatively rectangular mammalian ventricular action potentials.

Whereas outward K currents are greater in atrial than in ventricular myocytes, the current density of I_{K1} is much lower (<20%) than in ventricular myocytes, and there is more pronounced inward rectification. Thus, whereas I_{K1} generates a distinct outward current at potentials of –40 to –20 mV in ventricular myocytes, the current–voltage relationship for I_{K1} is relatively flat in atrial myocytes over this potential range. The net result is that the resting potential in atrial myocytes is typically less hyperpolarized, less stable, and more sensitive to neurohormonal modulation than in ventricular myocytes.

Not shown in Figure 18.8, but present in atrial myocytes from some species (especially canine), is a T-type Ca current. The kinetics and voltage dependence of the T-type current are usually similar to that of the Na current. There is thus far no published evidence for this current in human atrial myocytes.

An additional significant difference between atrial and ventricular myocytes is structural. Ventricular myocytes have a fully developed t-tubular system, consisting of periodic invaginations in the surface membrane into the core of the myocyte, arranged at the Z-line of the myofilaments. The sarcoplasmic reticulum (SR) in ventricular myocytes is in contact with the t-tubule forming the junctional triad structure. A primary mechanism of excitation–contraction coupling depends upon Ca influx from t-tubular Ca channels, which then triggers release of Ca from sarcoplasmic stores in the junctional SR by activating the ryanodine receptor Ca release channels (RyR). Atrial myocytes typically have a much lower density of t-tubules than are present in ventricular myocytes. Atrial myocytes have an additional class of sarcoplasmic reticulum (corbular SR), which is closely apposed to the surface membrane, but has no contact with a t-tubular membrane (65). In view of this, it is quite likely there are fundamental differences between atrial and ventricular myocytes in the process of excitation–contraction coupling (65). Efforts to exploit the structural and electrophysiological differences between atrial and ventricular tissues aimed at identifying specific atrial targets for pharmacological intervention may lead to safer drugs—drugs devoid of either ventricular proarrhythmia or negative inotropic complications.

6.   Cellular Studies of Electrophysiological Remodeling in Animal Models of AF

The most comprehensive series of studies on the functional and biochemical changes accompanying persistent high-rate atrial pacing have been performed in the canine model (44,66–68). The major finding here is that there are specific ion channels whose expression and functional density is reduced following atrial pacing, whereas there are other channels that remain unaffected. Thus, high-rate atrial pacing is associated with a significant reduction in the cellular current density of I_{Ca}, I_{Na}, and I_{TO}, whereas the current density of I_{K1}, $I_{Ca(T)}$, and $I_{Cl(Ca)}$ do not change. In the cases where a change in current density was detected, there was no detectable change in any of the other biophysical characteristics of those currents (with respect to intrinsic voltage dependence or to the kinetics of current activation/inactivation). In addition, the levels of steady-state mRNA expression for Kv4.3 (I_{TO}), the α_{1c}-subunit of the L-type Ca channel, the α-subunit of the cardiac Na channel, the Na/Ca exchanger protein, and the Kir2.3 inward rectifier K channel were quantified. A very close correlation was reported between the changes in mRNA expression and the changes in functional currents (68).

7.   Cellular Studies of Electrophysiological Remodeling in Chronic Human AF

Table 18.1 lists the predominant ion channels that are known to be functional in *human* atria, along with data on the impact of chronic AF on those currents.

Table 18.1 emphasizes that there are still many gaps in our knowledge of the functional distribution of ion currents, as well as in our understanding of the biochemical impact of AF on the distribution of channel subunits in AF. As shown in Figure 18.9, in patients with chronic AF there was consistently a significant reduction in the density of both outward

Table 18.1 Changes in the Density of Ionic Currents and/or Distribution of Ion Channel Subunits Associated with Chronic AF

Current/original characterization papers	Current change in chronic AF	Clone	Change in chronic AF
I_{Ca} (69),(70) (only L-type current demonstrated thus far)	Decreased 60–70% (71)	α_{1c} α_{1g}? (T-type) α_{1h}? (T-type)	α_{1c} mRNA decreased 60% (72)
I_{Na} (73)	Unknown (decreased V_{max}) (54)	hH1 (74)	Unknown
I_{K1} (75)	No change (76)	HIRK HIR	Unknown
I_{to} (77)	Decreased 60% (76)	Kv4.3 (78)	Unchanged (protein) (79)
I_{Kur} (80,81)	Decreased ~50% (76)	Kv1.5 (80)	Decreased ~50% (protein) (76)
I_{Kr} (82)	Unknown	HERG	No change (protein) (76)
I_{Ks} (82)	Unknown	min-K/KvLQT1	Unknown
I_{Kach} (83)	Unknown (decreased in heart failure (83))	GIRK1+CIR (84)	Unknown
I_f (85), (86)	Unknown		
I_{NS} (87)	Unknown	Unknown	
$I_{Na/Ca}$ (88,89)	Unknown	NCX (90)	

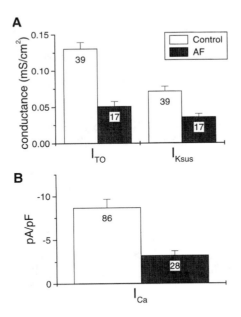

Figure 18.9 The current densities of ITO, IKsus, and ICa are reduced in atrial myocytes isolated from patients with chronic AF. (A) Mean (±SEM) conductance values for I_{TO} and I_{Ksus} (measured from +10 to +70 mV) are shown for myocytes isolated from patients in normal sinus rhythm and from patients in chronic AF. (B) Mean (±SEM) peak calcium current densities are plotted for myocytes isolated from 42 patients in normal sinus rhythm and from 11 patients in chronic AF. For each group, the number of myocytes studied are indicated within the bar. (Summary data are adapted from studies in Refs. 76 and 71.)

K currents (76) and inward L-type calcium currents (71). Studies using the rapidly paced canine atria have documented similar long-term (up to 42 days) functional (66) and biochemical changes (68). With respect to understanding the factors responsible for the changes in excitability and the atrial effective refractory period detected in the fibrillating atria (56), the reduction in Ca current density (by 63%) is likely to be the most significant functional change (71). It is therefore important to note that the changes observed following the onset of AF are similar to those caused by atrial dilation alone in the absence of AF (91).

Several recent studies have analyzed the mRNA levels for the L-type Ca channel, the RyR, and other Ca handling proteins in tissue from patients with AF (72,92). These studies also demonstrated a reduction in the expression of mRNA for the L-type Ca channel, as well as a reduction in the expression of the SR Ca-ATPase (SERCA). No change was detected in the expressed mRNA for the RyR, calsequestrin, phospholamban, or the Na/Ca exchanger (72,92). A reduction in the expression of SERCA would suggest difficulty in re-sequestering Ca into the sarcoplasmic reticulum, and may aggravate the Ca overload experienced by fibrillating atria.

C. Structural Remodeling of the Heart During Atrial Fibrillation

Reentrant activity is enhanced by the presence of either functional or anatomical heterogeneities, which increase the chances of unidirectional block occurring, disrupting the normal pathway of atrial excitation and probably facilitating multiple wavelet formation. Anatomical–pathological studies of atria from patients with mitral valve disease reveal (1) cellular myolysis and hypertrophy; (2) increased interstitial fibrosis; and (3) diminished cell-to-cell contact (93). The extent of these pathological changes is increased in those patients who were in persistent AF. The degree of pathological change is generally correlated with the electrophysiological status of the diseased atria, with the more diseased atria being relatively more depolarized (94). There is no clear evidence of changes in conduction velocity in the atria of patients with persistent AF.

Animal studies of goats (95) and dogs (96) in either persistent AF or rapidly paced reveal subtle and different changes in conduction velocity and in the distribution of gap junctions (connexins) electrically connecting the atrial myocytes. No changes in conduction velocity were detected in the fibrillating goat model. In this model, the overall expression of the important atrial connexins (Cx40, Cx43) was unchanged. However, a redistribution of Cx40 was detected, such that there were small areas of low-density expression surrounded by larger areas of normal expression. This expression profile could conceivably form the anatomical basis for increased micro-reentrant circuit formation. In contrast, studies on the rapidly paced canine atria have documented reductions in both Na channel density (97) and conduction velocity (98). In addition, expression of Cx43 was relatively *enhanced* in the rapidly paced canine atria (96). Clear changes in either conduction velocity or in the distribution of connexins in the fibrillating human atria have not yet been well characterized. Further studies on the quantitative and regional distribution of connexins and fibrosis in fibrillating human atria are needed to determine the role of altered pathways of conduction in human AF.

D. The Role of Remodeling in the Propagation of AF: Conceptual Implications for Therapy

Both the electrophysiological and structural remodeling processes are likely to have a major impact on the propagation of the fibrillatory rhythm, although the time course of these events may be different. Electrophysiological remodeling occurs on both a rapid (seconds to minutes) and slower (days to weeks) time scale. The very rapid changes probably involve modulation of specific ion currents (particularly I_{Ca} and I_{TO}) by increased intracellular Ca, altered pH, phosphorylation state, oxidative state, metabolic demand, or other post-translational protein modifications. The slower changes in atrial electrophysiology (days to weeks) are probably due to changes in the rate of translation, synthesis, or degradation of specific ion channels in the myocyte membrane.

The most rapid change associated with the onset of AF is a decrease in the ERP. This change can be blocked by prior administration of verapamil, suggesting that calcium overload (Ca inactivation of the Ca channel?) is a primary cause of this rapid change in ERP (59). In addition, there are slower and longer lasting reductions in the ERP-associated longer episodes of AF. These, too, can be significantly blunted by preadministration of a Ca channel blocker (58). Thus, with respect to the chronology of the major electrophysiological changes, calcium overload is likely the primary factor mediating the long-term electrophysiological remodeling associated with AF. Therefore, to prevent the early electrophysiological remodeling, minimizing calcium overload should be a major goal. Three distinct approaches to this goal seem apparent. First, direct suppression of Ca channel activity with Ca channel blockers could be utilized. Second, a reduction in the autonomic increase in I_{Ca} using β-adrenergic receptor antagonists, or drugs with mixed K-channel blocking and β-adrenergic receptor antagonist activity (such as sotalol, amiodarone, etc.) are available. Finally, suppression of calcium influx mediated by the sodium–calcium exchanger (in response to sodium overload) is a possibility. From a prevention perspective, the latter two groups seem to have the least risk of proarrhythmia or negative inotropy.

Structural changes occur at several levels, including myocyte hypertrophy, ultrastructural changes (loss of fibrillar structure, myolysis, mitochondrial enlargement, etc.), changes in the number and order of connexin junctions, and in the degree of interstitial fibrosis and fatty infiltration. It is likely that the onset of AF-induced structural remodeling (fibrosis, dilatation, etc.) occurs at a significantly slower rate than the electrophysiological remodeling, and it is not known whether these changes are reversible. Clearly, the progressive nature of AF involves a broad continuum of electrophysiological and structural changes.

V. INDICATIONS AND LIMITATIONS OF CURRENT THERAPIES FOR AF

A. Overview

Although important gains have been made in understanding the treatment of atrial fibrillation, much is still unanswered. Basically, there are two established treatment strategies. One is to make every effort to restore and maintain sinus rhythm. The other is to allow the atrial fibrillation to persist, but control the ventricular response rate. Both strategies rely heavily on antiarrhythmic drugs and/or AV nodal blocking drugs, and also include anticoagulation with warfarin. Both strategies may also involve nonpharmacological therapies. This choice of strategies, rhythm control, or rate control applies only to patients with either paroxysmal or persistent atrial fibrillation. For patients with permanent atrial fibrillation, control of the ventricular response rate is the only therapeutic choice.

What are the potential advantages of suppressing atrial fibrillation and of maintaining sinus rhythm? With this approach, the patient may benefit from more physiological rate control, maintenance of the atrial contribution to cardiac output, better exercise tolerance, reduced thromboembolic risk (logical but unproven), and prevention of rate-related cardiac chamber dilatation. What are the potential advantages of simply treating atrial fibrillation with optimal ventricular rate control and warfarin therapy? With this approach, the patient may benefit from simple control of ventricular rate, avoidance of potentially proarrhythmic effects of antiarrhythmic drug therapy, avoidance of other potentially adverse effects of antiarrhythmic drugs, avoidance of frequent recurrence of atrial fibrillation, fewer compliance problems, and perhaps lower cost.

Resolution of the relative advantages of these approaches is currently being sought in large clinical trials directly comparing the two strategies. For example, the National Heart, Lung, and Blood Institute of the National Institutes of Health has sponsored the Atrial Fibrillation Follow up Investigation of Rhythm Management (AFFIRM) trial (99) that will be completed in 2001. This study and several smaller studies comparing rhythm control versus rate control strategies (PIAF, RACE, SAFE-T) are likely to be quite helpful clinically, but none are likely to provide new insights into the prevention or cure of AF.

B. The Dilemma of Antiarrhythmic Drug Therapy

At the time of this writing, there are eight antiarrhythmic drugs available for use in the United States that have demonstrated efficacy in the suppression of atrial fibrillation, and two more that should be available soon. Using the Vaughn–Williams antiarrhythmic drug classification, the eight "old" drugs include the class IA agents—quinidine, procainamide, and disopyramide; the class IC drugs—flecainide and propafenone; a IC-like drug—moricizine; and the class III drugs—sotalol and amiodarone. The two new drugs, dofetilide and azimilide, are also class III. In addition, some other antiarrhythmic drugs are available for use in other countries (e.g., cibenzoline in Japan and France).

In general, no antiarrhythmic drug has been shown to be clearly better than any other for the suppression of atrial fibrillation. This conclusion is based on analysis of the many, primarily small, studies that have examined antiarrhythmic drug efficacy and have demonstrated that atrial fibrillation recurs in at least one-half of the patients being treated, regardless of the drug selected (see Table 18.2) (101,102). In short, recurrence of AF while on drug therapy should be expected. Fortunately, if the patient is properly anticoagulated, and if the ventricular response rate is clinically acceptable during the recurrence, emergency treatment is rarely required. In fact, from this perspective, recurrence most often

Table 18.2 Efficacy of Antiarrhythmic Drugs for Chronic AF at Late Follow-Up

Drug	Sinus rhythm retention rate (%) mean (range)	Number of studies
No drug	31 (15–56)	10
Quinidine	41 (11–54)	11
Disopyramide	49 (44–54)	3
Propafenone	39 (30–46)	3
Flecainide	62 (49–81)	3
Sotalol	42 (37–49)	3
Amiodarone	53 (36–83)	4

Source: Ref. 100.

would just be a clinically important nuisance. It is from this perspective that the current philosophy of antiarrhythmic drug treatment arises, namely, that recurrence of AF on antiarrhythmic drug therapy should not be considered drug failure per se. The measure of efficacy should be the frequency and severity of recurrence (101,103). Occasional recurrence of AF may be clinically acceptable, and preferable to persistent AF or frequent episodes of paroxysmal AF. This philosophy directs the present approach to therapy of AF, but it is a compromise in that one must accept the inevitability of AF recurrence.

It logically follows that when AF recurs and persists, outpatient cardioversion of the AF to restore sinus rhythm is generally recommended. This prevents needless abandonment of otherwise acceptable antiarrhythmic drug therapy, and avoids hospitalizations for the initiation of new drug therapy, particularly in patients with underlying structural heart disease. In short, acute cardioversion from time to time to restore sinus rhythm is presently considered a part of the expected long-term therapy of AF. Finally, because antiarrhythmic drug selection is largely empiric, and because apparent efficacy is similar for most drugs, the drug selected preferably should be one that is well tolerated, easy to take, and with the fewest side effects, particularly serious adverse effects such as ventricular proarrhythmia. The ideal drug would have no organ toxicity or proarrhythmia, and would be highly effective and given at most only once per day. Unfortunately, at present, there is no such drug.

1. Antiarrhythmic Drug Therapy for Patients with AF in the Absence of Structural Heart Disease (Lone AF)

For these patients, available data demonstrate that if antiarrhythmic drugs are given properly, the risk of serious or life-threatening adverse effects, particularly ventricular proarrhythmia, is quite limited, and therefore, clinically acceptable. Since there is not much difference in efficacy in suppressing AF amongst the several available antiarrhythmic drugs, selection of antiarrhythmic drug therapy currently is guided by the above-mentioned criteria of tolerability and minimization of side effects. It is usually important to use an AV nodal blocking drug as concomitant therapy to control ventricular rate should AF recur. When using class I antiarrhythmic drugs or amiodarone, arrhythmia recurrence may be in the form of atrial flutter with a relatively slow atrial rate (180–220 bpm), making it all the more important to use an AV nodal blocking drug or drugs when one of these agents is used.

2. Antiarrhythmic Drug Therapy in Patients with Underlying Structural Heart Disease

Unfortunately, there are few studies of both efficacy and safety of antiarrhythmic drug therapy to maintain sinus rhythm in these patients. This is especially important because although these patients may be substantially helped by maintaining sinus rhythm and the associated atrial contribution to cardiac output, these same patients, particularly those with congestive heart failure, have the greatest potential for ventricular proarrhythmia (104), and most of the antiarrhythmic drugs are negative inotropic agents. Therefore, selection of antiarrhythmic drug therapy to suppress AF in patients with structural heart disease has been largely influenced by studies in patients with ventricular dysfunction treated for ventricular arrhythmias. Recently, however, there have been studies that demonstrated both the safety and efficacy of dofetilide, azimilide, and sotalol for treatment of atrial fibrillation in patients with structural heart disease (105–109).

3. Safety of Antiarrhythmic Drug Therapy

Background

All antiarrhythmic drugs have some potential to provoke proarrhythmia, and the latter may even be lethal. Since clinically appropriate anticoagulation and clinically appropriate ventricular rate control may reduce AF simply to a clinically important nuisance, the potential risks of antiarrhythmic drug therapy must be weighed relative to the potential benefits. Unfortunately, there are few prospectively obtained data regarding the relative risk of available antiarrhythmic drugs in the primary treatment of AF. For a long time, studies primarily evaluated the efficacy but not the safety of antiarrhythmic drug therapy of AF. Thus, most of the data come from meta-analyses of old prospectively done studies, retrospective analysis of other data, or extrapolation from studies of antiarrhythmic drug therapy in the primary prevention of sudden cardiac death in patients with underlying structural heart disease. The latter usually consisted of ventricular dysfunction, either in the presence or absence of ischemic heart disease (110).

Safety Studies of Antiarrhythmic Drug Therapy for AF

A retrospective meta-analysis study of quinidine therapy (111) demonstrated a 2.9% mortality rate for those assigned to quinidine, but only a 0.8% mortality rate for those assigned to placebo ($p < 0.05$). While many questions have been raised about the six studies comprising the meta-analysis and the meta-analysis itself, the results of this study were most disconcerting. They clearly demonstrated the need to consider safety as well as efficacy of antiarrhythmic drug therapy of AF.

Another retrospective analysis of survival data from the Stroke Prevention in Atrial Fibrillation (SPAF I) trial (104) demonstrated that patients without a history of congestive heart failure had no increased risk of cardiac mortality associated with antiarrhythmic drug therapy with a 2-year follow-up. However, for patients with a history of congestive heart failure, 2-year survival was only 76%, with a relative risk of cardiac death of 4.7 when compared with similar patients not receiving antiarrhythmic medication ($p < 0.001$). These data indicate that antiarrhythmic drug therapy can be given rather safely to patients with no history of congestive heart failure. However, in those patients with a history of congestive heart failure, treatment with antiarrhythmic drug therapy adds significantly to the risk.

Subsequently, several prospective, randomized, placebo-controlled studies of antiarrhythmic drugs such as flecainide and propafenone (112–114) led to their approval by the United States Food and Drug Administration (FDA) for the treatment of paroxysmal AF in patients without structural heart disease. Most recently, prospective, randomized, placebo-controlled studies of patients with AF treated with dofetilide, sotalol, and azimilide have been performed (105–109) in a wide spectrum of patients with AF. Because these studies have demonstrated both safety and efficacy, the FDA recommended a broad indication approval for the use of dofetilide and sotalol in the treatment of patients with AF. Azimilide will also soon seek a broad indication FDA approval for use by patients with AF.

Other Safety Studies

It is now recommended that class I antiarrhythmic drugs not be given to patients with structural heart disease. This recommendation is based on the results of the Cardiac Arrhythmia Suppression Trial (CAST) (115), and the results of a subsequent very large meta-analysis of antiarrhythmic agents used to treat patients with structural heart disease in an effort to prevent sudden cardiac death (116). In particular, caution should be used when treating patients with coronary artery disease, recent myocardial infarction, ventricular dysfunction, and frequent and complex ventricular ectopy. However, not all of the news is adverse when extrapolating from studies of antiarrhythmic drugs in the primary prevention of sudden cardiac death to the use of antiarrhythmic drugs to treat AF. Several clinical trials have had long-term neutral effects, suggesting that the drugs studied can be used safely in patients with underlying structural heart disease. These trials include the sotalol post-myocardial infarction study (117), the DIAMOND studies of dofetilide (118,119), the CHF–STAT, EMIAT, and CAMIAT studies of amiodarone (110), and the CAST II study of moricizine (120). The ALIVE study of azimilide is ongoing (121). A few studies, in patients primarily with nonischemic heart disease (GESICA, EPAMSA, and a CHF–STAT subgroup), suggest that amiodarone may even have a favorable effect on outcome (110). In patients with AF in the presence of substantial underlying congestive heart failure (e.g., New York Heart Association Class IV heart failure or a left ventricular ejection fraction ≤25%), amiodarone may be the only antiarrhythmic drug that can be given safely at the present time.

C. Long-Term Therapy for Ventricular Rate Control

Ventricular rate control during episodes of recurrent AF (paroxysmal or persistent) or during permanent AF is an important, and usually necessary, part of the treatment of AF. Clinically adequate ventricular rate control is generally considered to be a resting ventricular rate ≤80 beats per minute (bpm), with an increase in ventricular rate to no more than 110 bpm with routine activities of daily living, such as walking. Patients with adequate ventricular rate control during AF in the absence of drug therapy to increase conduction block at the AV node, of course, do not require further therapy. Such patients usually have an intrinsic AV conduction abnormality, often in association with sinus node dysfunction. Ultimately, the adequacy of ventricular rate control during AF is best assessed with some form of ECG monitoring [e.g., ambulatory 24-h ECG (Holter) monitor] during AF.

Either a Ca channel blocker that reduces AV nodal conduction or a beta-blocker is usually the treatment of choice for long-term therapy to control ventricular rate response in those patients who do not have congestive heart failure. Digoxin, on the other hand, should be considered for patients who have congestive heart failure secondary to impaired systolic ventricular function. It is not uncommon for patients to be well served by or even to require combination treatment with two or three of these agents to achieve ventricular rate control.

When clinically acceptable ventricular rate control is not obtained for any number of reasons, two therapeutic options are available. These are either AV conduction modification by radiofrequency ablation of the slow pathway input to the AV node (122), or deliberate ablation of the AV node or His bundle to create complete heart block (123), followed by implant of an appropriate pacemaker system.

D. Nonpharmacological Therapy of Atrial Fibrillation

1. Pacemaker Therapy

One important role for pacing (either atrial alone, or both the atria and ventricles) is to permit the administration of drug therapy to patients with either paroxysmal or persistent AF and underlying sinus node dysfunction. Most, if not all, antiarrhythmic drugs (class I, II,

III, and IV) used to treat AF affect sinus node function adversely to varying degrees. Thus, if sinus bradycardia is initially present or develops after initiation of drug therapy, concomitant use of an atrial pacing system that includes mode switching (one that stops sensing or pacing the atria should AF recur) is generally desirable. A potential role for single-site or even dual-site atrial pacing (high and low right atrium, or right atrial and left atrial pacing) at a rate of about 80 to 90 bpm in combination with antiarrhythmic drug therapy is currently being assessed. Preliminary data (124,125) suggest that this combination of pacemaker and drug therapy in patients who were otherwise "resistant" to drug therapy alone can be remarkably effective in suppressing recurrent AF.

A new direction in pacing treatment, still quite early in its investigation, is rapid atrial pacing to terminate AF. To date, it is not clear why this should work. However, if AF is being driven by a rapidly firing focus whose mechanism is amenable to interruption by rapid pacing (e.g., reentry or delayed after-depolarizations), then a reason for the efficacy of rapid atrial pacing would be understandable based on our current understanding of how pacing works to terminate tachycardias (126).

2. The Implantable Low-Energy Atrial Defibrillator

A series of systematic studies of the safety and efficacy of low-energy atrial defibrillation in animal models and patients has led to the development of devices that will detect AF and deliver a low-energy shock (6–12 J) to restore sinus rhythm. If the shock is synchronized to ventricular activation so that it is not delivered during the T wave (the time of ventricular repolarization) of the preceding QRS complex (the time of ventricular depolarization), something the device has been reliably programmed to do, the shock can be delivered safely.

Early experience in patients with the low-energy atrial defibrillation indicates it is safe and effective (127). This suggests it may have a useful role to play in selected patients with AF. It is probably reasonable to anticipate that this form of therapy will usually be combined with either or both antiarrhythmic drug therapy and atrial pacing. Its limitations include the discomfort of the shock; an important incidence of very early recurrence of AF (<1 min); its inapplicability for patients with a history of frequent episodes of AF; and the fact that it does not prevent AF. An additional benefit may be the logical expectation that by converting AF back to sinus rhythm promptly, within minutes to perhaps a few hours, the adverse effects of atrial remodeling are prevented or so quickly reversed that the pathophysiological changes in the atria that would otherwise occur and promote AF (45) are aborted.

3. Surgery for Treatment of AF

Until recently, the only surgical treatment of AF consisted of the Maze operation (128). The primary goal of the surgery was to create lines of conduction block with surgical incisions, thus preventing the propagation of reentrant wavelets. In its recent incarnations, this procedure has been quite effective in achieving a cure. In a long-term follow-up study, 70 to 80% of atrial mechanical function was restored (129). This therapy is something of a "one size fits all" approach, because it assumes that AF is due to multiple reentrant wavelets. Due to its invasive nature, the number of patients who have been treated with the Maze procedure (or variants thereof) is relatively small. Thus, while effective, it is unlikely that surgical interventions will have a major role in the long-term treatment of the majority of patients with AF.

4.   Radiofrequency Ablation to Treat AF

There have been many efforts to reproduce the Maze operation or apply its concept using endocardial catheter radiofrequency ablation techniques (130–132). The latter, very much investigational, have had some success, but have been associated with important morbidity which includes stroke, pulmonary hypertension, and atrial flutter, as well as AF recurrence (130–133). When the several mechanisms of AF are better understood, one or more of them may identify an accessible target for effective ablation, as has happened with the recently described rapidly firing ectopic focus found primarily in one of the pulmonary veins (27).

There is yet another promising, in fact, unique circumstance for the use of radiofrequency ablation in the treatment of AF (134). As discussed earlier, AF may recur as classic atrial flutter in the presence of some antiarrhythmic drugs, particularly those in class IC. This atrial flutter may be successfully ablated using standard catheter electrode techniques. After ablation, the atrial flutter will not recur, and by continuing the antiarrhythmic drug therapy, the AF will not recur either in about 70% of patients, according to early studies (134).

E.   Summary of Current Therapies

Antiarrhythmic drug therapy remains the primary treatment to suppress AF and maintain sinus rhythm. The efficacy of the several drugs now available appears about the same. Because of this, the adverse effect profile as well as the potential for proarrhythmia should guide drug selection. Furthermore, the measure of efficacy of antiarrhythmic drug therapy in most patients should be the frequency and severity of recurrence, not just simple recurrence. Thus, when AF recurs, DC cardioversion to restore sinus rhythm generally should be considered, rather than changing the antiarrhythmic drug therapy because of its "failure." Any of the available antiarrhythmic drugs may be used to treat patients with lone AF. Available antiarrhythmic drug therapy to treat patients with AF in the presence of structural heart disease is more limited because of serious safety concerns. Warfarin remains a critical component of therapy for patients at risk for systemic embolism or stroke. Concomitant device therapy, particularly an atrial pacing system with mode switching, may be helpful or even necessary to provide effective drug therapy. Single- or dual-site atrial pacing and the implantable atrial defibrillator may evolve to have an increasing role in the treatment of AF, most likely in combination with antiarrhythmic drug therapy. Ablative techniques continue to hold hope for cure of AF, and have already accomplished that for patients with lone AF caused by a single-focus firing rapidly.

VI.   DIRECTIONS FOR FUTURE RESEARCH

Significant progress has been made in the past decade to better understand the mechanisms responsible for the initiation and maintenance of AF. Several areas of future basic science and clinical investigation seem particularly promising. At the fundamental level, additional studies are needed to better understand signal transduction pathways that are activated by high-rate atrial activity that results in immediate and delayed changes in refractory period. A better understanding of this process may help identify specific, novel targets for preventive pharmacological therapies. Efforts that simultaneously seek to identify atrial specific targets (with the goal of avoiding ventricular proarrhythmia) are also likely to be fruitful.

One major unresolved clinical question has to do with timing. How long should one wait before attempting to terminate initial episodes of paroxysmal AF? Is it possible to identify more specifically the etiology of the arrhythmia on a patient-by-patient basis to tailor therapies to the specific cause of the arrhythmia? Following either pharmacological or electrical cardioversion, early recurrence of AF postcardioversion is a significant problem. Are there new ways to deal with this problem in a manner that is less painful and more effective in maintaining normal sinus rhythm?

The most vexing and challenging question is to identify the factor(s) that cause the incidence of AF to increase so dramatically as a function of age, with a nearly exponential increase in incidence after the age of 65. If one can clearly identify the crucial age-related changes, it becomes possible to envision new therapies aimed at preventing them.

ACKNOWLEDGMENTS

Supported in part by grants RO1-HL38408 (A.L.W.) and RO1-HL57262 (D.V.W.) from the National Institutes of Health, National Heart, Lung, and Blood Institute, Bethesda, Maryland.

REFERENCES

1. Feinberg WM, Blackshear JL, Laupacis A, Kronmal R, Hart RG. Prevalence, age distribution, and gender of patients with atrial fibrillation. Analysis and implications. Arch Intern Med 1995; 155:469–473.
2. Wells JL, Jr., Karp RB, Kouchoukos NT, MacLean WA, James TN, Waldo AL. Characterization of atrial fibrillation in man: studies following open heart surgery. Pacing Clin Electrophysiol 1978; 1:426–438.
3. Laupacis A, Cuddy TE. Prognosis of individuals with atrial fibrillation. Can J Cardiol 1996; 12 (suppl A):14A–16A.
4. Benjamin EJ, Wolf PA, D'Agostino RB, Silbershatz H, Kannel WB, Levy D. Impact of atrial fibrillation on the risk of death: The Framingham Heart Study. Circulation 1998; 98:946–952.
5. Kannel WB, Wolf PA, Benjamin EJ, Levy D. Prevalence, incidence, prognosis, and predisposing conditions for atrial fibrillation: population-based estimates. Am J Cardiol 1998; 82:2N–9N.
6. Mayer AG. Rhythmical pulsation in scyphomedusae. 47, 1–62. 1906. Washington, D.C., Carnegie Institution of Washington.
7. Mayer AG. Rhythmical pulsation in scyphomedusae. II. Papers from the Tortugas Laboratory of the Carnegie Institution of Washington 1908; I:113–131.
8. Mines GR. On dynamic equilibrium in the heart. J Physiol (Lond) 1913; 46:349–383.
9. Lewis T, Feil HS, Stroud WD. Observations upon flutter and fibrillation. II: The nature of auricular flutter. Heart 1920; 7:191–245.
10. Garrey W. The nature of fibrillatory contraction of the heart: its relation to tissue mass and form. Am J Physiol 1914; 33:397–414.
11. Garrey WE. Auricular fibrillation. Physiol Rev 1924; 4:215–250.
12. Kumagai K, Khrestian C, Waldo AL. Simultaneous multisite mapping studies during induced atrial fibrillation in the sterile pericarditis model. Insights into the mechanism of its maintenance. Circulation 1997; 95:511–521.
13. Lewis T. The Mechanism and Graphical Registration of the Heart Beat, 3rd ed. London: Shaw and Sons, Ltd., 1925:340–342.
14. Lewis T. Observations upon flutter and fibrillation. IV: Impure flutter; theory of circus movement. Heart 1920; 7:293–345.

15. Wiener N, Rosenblueth A. The mathematical formulation of the problem of conduction of impulses in a network of connected excitable elements, specifically in cardiac muscle. Arch Inst Cardiol Mex 1946; 16:205–265.

16. Lewis T. The Mechanism and Graphical Registration of the Heart Beat, 3rd ed. London: Shaw and Sons, Ltd., 1925.

17. Scherf D. Studies on auricular tachycardia caused by aconitine administration. Proc Exp Biol Med 1947; 64:233–239.

18. Scherf D, Romano FJ, Terranova R. Experimental studies on auricular flutter and auricular fibrillation. Am Heart J 1958; 36:241–251.

19. Scherf D, Terranova R. Mechanism of auricular flutter and fibrillation. Am J Physiol 1949; 159:137–142.

20. Kimura E, Kato K, Murao S, Ajisaka H, Kayama S, Omiya Z. Experimental studies on the mechanism of auricular flutter. Tohoku J Exp Med 1954; 60:197–207.

21. Moe GK, Abildskov JA. Atrial fibrillation as a self-sustained arrhythmia independent of focal discharge. Am Heart J 1959; 58:59–70.

22. Goto M, Sakamoto Y, Imanaga I. Aconitine-induced fibrillation of the different muscle tissues of the heart and the action of acetylcholine. In Sano T, Matsuda K, Mizuhira B, eds. Electrophysiology and Ultrastructure of the Heart. New York: Grune & Stratton, 1967:199–209.

23. Azuma K, Iwane H, Ibukiyama C, Watabe Y, Shin-Mura H, Iwaoka M, Wakatsuki T, Saito K, Shimizu K, Takada S, Yasui N. Experimental studies on aconitine-induced atrial fibrillation with microelectrodes. Isr J Med Sci 1969; 5:470–474.

24. Waldo AL, MacLean WA, Karp RB, Kouchoukos NT, James TN. Continuous rapid atrial pacing to control recurrent or sustained supraventricular tachycardias following open heart surgery. Circulation 1976; 54:245–250.

25. Moreira DA, Shepard RB, Waldo AL. Chronic rapid atrial pacing to maintain atrial fibrillation: use to permit control of ventricular rate in order to treat tachycardia induced cardiomyopathy. Pacing Clin Electrophysiol 1989; 12:761–775.

26. Waldo AI, Cooper TB. Spontaneous onset of type I atrial flutter in patients. J Am Coll Cardiol 1996; 28:707–712.

27. Haissaguerre M, Jais P, Shah DC, Takahashi A, Hocini M, Quiniou G, Garrigue S, Mouroux A, Metayer P, Clementy J. Spontaneous initiation of atrial fibrillation by ectopic beats originating in the pulmonary veins. N Engl J Med 1998; 339:659–666.

28. Allessie MA, Bonke FI, Schopman FJ. Circus movement in rabbit atrial muscle as a mechanism of trachycardia. Circ Res 1973; 33:54–62.

29. Allessie MA, Bonke FI, Schopman FJ. Circus movement in rabbit atrial muscle as a mechanism of tachycardia. II. The role of nonuniform recovery of excitability in the occurrence of unidirectional block, as studied with multiple microelectrodes. Circ Res 1976; 39:168–177.

30. Allessie MA, Bonke FI, Schopman FJ. Circus movement in rabbit atrial muscle as a mechanism of tachycardia. III. The "leading circle" concept: a new model of circus movement in cardiac tissue without the involvement of an anatomical obstacle. Circ Res 1977; 41:9–18.

31. Hoff HE, Geddes LA. Cholinergic factor in auricular fibrillation. J Appl Physiol 1955; 8:177–192.

32. Moe GK. A conceptual model of atrial fibrillation. J Electrocardiol 1968; 1:145–146.

33. Moe GK, Rheinbolt WC, Abildskov JA. A computer model of atrial fibrillation. Am Heart J 1964; 67:200–220.

34. Allessie MA, Lammers WJEP, Bonke FIM, Hollen J. Experimental evaluation of Moe's multiple wavelet hypothesis of atrial fibrillation. In Zipes DP, Jalife J, eds. Cardiac Electrophysiology and Arrhythmias. Orlando, FL: Grune & Stratton, 1985:265–276.

35. Allessie MA, Lammers W, Smeets J, Bonke F, Hollen J. Total mapping of atrial excitation during acetylcholine-induced atrial flutter and fibrillation in the isolated canine heart. In Kulbertus HE, Olsson SB, Schlepper M, eds. Atrial Fibrillation. Molndal, Sweden: A.B. Hassle, 1982:44–59.

36. Schuessler RB, Grayson TM, Bromberg BI, Cox JL, Boineau JP. Cholinergically mediated tachyarrhythmias induced by a single extrastimulus in the isolated canine right atrium. Circ Res 1992; 71:1254–1267.

37. Cox JL, Canavan TE, Schuessler RB, Cain ME, Lindsay BD, Stone C, Smith PK, Corr PB, Boineau JP. The surgical treatment of atrial fibrillation. II. Intraoperative electrophysiologic mapping and description of the electrophysiologic basis of atrial flutter and atrial fibrillation. J Thorac Cardiovasc Surg 1991; 101:406–426.

38. Page PL, Plumb VJ, Okumura K, Waldo AL. A new animal model of atrial flutter. J Am Coll Cardiol 1986; 8:872–879.

39. Matsuo K, Tomita Y, Uno K, Khrestian C, Waldo AL. A new mechanism of sustained atrial fibrillation: studies in the canine sterile pericarditis model. Circulation 1998; 98:I-209.

40. Wang Z, Page P, Nattel S. Mechanism of flecainide's antiarrhythmic action in experimental atrial fibrillation. Circ Res 1992; 71:271–287.

41. Wang J, Bourne GW, Wang Z, Villemaire C, Talajic M, Nattel S. Comparative mechanisms of antiarrhythmic drug action in experimental atrial fibrillation. Importance of use-dependent effects on refractoriness. Circulation 1993; 88:1030–1044.

42. Wang J, Liu L, Feng J, Nattel S. Regional and functional factors determining induction and maintenance of atrial fibrillation in dogs. Am J Physiol 1996; 271:H148–H158.

43. Moreira DA, Shepard RB, Waldo AL. Chronic rapid atrial pacing to maintain atrial fibrillation: use to permit control of ventricular rate in order to treat tachycardia induced cardiomyopathy. Pacing Clin Electrophysiol 1989; 12:761–775.

44. Morillo CA, Klein GJ, Jones DL, Guiraudon CM. Chronic rapid atrial pacing. Structural, functional, and electrophysiological characteristics of a new model of sustained atrial fibrillation. Circulation 1995; 91:1588–1595.

45. Wijffels MC, Kirchhof CJ, Dorland R, Allessie MA. Atrial fibrillation begets atrial fibrillation. A study in awake chronically instrumented goats. Circulation 1995; 92:1954–1968.

46. Powell AC, Garan H, McGovern BA, Fallon JT, Krishnan SC, Ruskin JN. Low energy conversion of atrial fibrillation in the sheep. J Am Coll Cardiol 1992; 20:707–711.

47. Skanes AC, Mandapati R, Berenfeld O, Davidenko JM, Jalife J. Spatiotemporal periodicity during atrial fibrillation in the isolated sheep heart. Circulation 1998; 98:1236–1248.

48. Ayers GM, Alferness CA, Ilina M, Wagner DO, Sirokman WA, Adams JM, Griffin JC. Ventricular proarrhythmic effects of ventricular cycle length and shock strength in a sheep model of transvenous atrial defibrillation. Circulation 1994; 89:413–422.

49. Power JM, Beacom GA, Alferness CA, Raman J, Wijffels M, Farish SJ, Burrell LM, Tonkin AM. Susceptibility to atrial fibrillation: a study in an ovine model of pacing-induced early heart failure. J Cardiovasc Electrophysiol 1998; 9:423–435.

50. Stambler BS, Shepard RK, Turner DA, Fenelon G. Evidence of triggered activity as the mechanism of atrial tachycardia in dogs with pacing-induced heart failure. J Am Coll Cardiol 1997; 29:254A(abstr).

51. Nabih MA, Prcevski P, Fromm BS, Lavine SJ, Elnabtity M, Munir A, Steinman RT, Meissner MD, Lehmann MH. Effect of ibutilide, a new class III agent, on sustained atrial fibrillation in a canine model of acute ischemia and myocardial dysfunction induced by microembolization. Pacing Clin Electrophysiol 1993; 16:1975–1983.

52. Li D, Fareh S, Leung TK, Nattel S. Promotion of atrial fibrillation by heart failure in dogs: atrial remodeling of a different sort. Circulation 1999; 100:87–95.

53. Konings KT, Smeets JL, Penn OC, Wellens HJ, Allessie MA. Configuration of unipolar atrial electrograms during electrically induced atrial fibrillation in humans. Circulation 1997; 95:1231–1241.

54. Hordof AJ, Edie R, Malm JR, Hoffman BF, Rosen MR. Electrophysiologic properties and response to pharmacologic agents of fibers from diseased human atria. Circulation 1976; 54: 774–779.

55. Attuel P, Childers R, Cauchemez B, Poveda J, Mugica J, Coumel P. Failure in the rate adapta-

tion of the atrial refractory period: its relationship to vulnerability. Int J Cardiol 1982; 2:179–197.

56. Boutjdir M, Le Heuzey JY, Lavergne T, Chauvaud S, Guize L, Carpentier A, Peronneau P. Inhomogeneity of cellular refractoriness in human atrium: factor of arrhythmia? Pace 1986; 9: 1095–1100.

57. Wijffels MC, Kirchhof CJ, Dorland R, Power J, Allessie MA. Electrical remodeling due to atrial fibrillation in chronically instrumented conscious goats: roles of neurohumoral changes, ischemia, atrial stretch, and high rate of electrical activation. Circulation 1997; 96: 3710–3720.

58. Tieleman RG, De Langen C, Van Gelder IC, de Kam PJ, Grandjean J, Bel KJ, Wijffels MC, Allessie MA, Crijns HJ. Verapamil reduces tachycardia-induced electrical remodeling of the atria. Circulation 1997; 95:1945–1953.

59. Daoud EG, Knight BP, Weiss R, Bahu M, Paladino W, Goyal R, Man KC, Strickberger SA, Morady F. Effect of verapamil and procainamide on atrial fibrillation-induced electrical remodeling in humans. Circulation 1997; 96:1542–1550.

60. Daoud EG, Marcovitz P, Knight BP, Goyal R, Man KC, Strickberger SA, Armstrong WF, Morady F. Short-term effect of atrial fibrillation on atrial contractile function in humans. Circulation 1999; 99:3024–3027.

61. Jayachandran V, Winkle W, Sih HJ, Zipes DP, Hutchins GD, Olgin JE. Chronic atrial fibrillation from rapid atria pacing is associated with reduced atrial blood flow: a positron emission tomography study. Circulation 1998; 98:I-209(abstr).

62. Schoonderwoerd BA, Van Gelder IC, Crijns HJ. Left ventricular ischemia due to coronary stenosis as an unexpected treatable cause of paroxysmal atrial fibrillation. J Cardiovasc Electrophysiol 1999; 10:224–228.

63. Feng J, Yue L, Wang Z, Nattel S. Ionic mechanisms of regional action potential heterogeneity in the canine right atrium. Circ Res 1998; 83:541–551.

64. Liu L, Nattel S. Differing sympathetic and vagal effects on atrial fibrillation in dogs: role of refractoriness heterogeneity. Am J Physiol 1997; 273:H805–H816.

65. Hatem SN, Benardeau A, Rucker-Martin C, Marty I, de CP, Villaz M, Mercadier JJ. Different compartments of sarcoplasmic reticulum participate in the excitation-contraction coupling process in human atrial myocytes. Circ Res 1997; 80:345–353.

66. Yue L, Feng J, Gaspo R, Li G-R, Wang Z, Nattel S. Ionic remodeling underlying action potential changes in a canine model of atrial fibrillation. Circ Res 1997; 81:512–525.

67. Sun H, Gaspo R, Leblanc N, Nattel S. Cellular mechanisms of atrial contractile dysfunction caused by sustained atrial tachycardia. Circulation 1998; 98:719–727.

68. Yue L, Melnyk P, Gaspo R, Wang Z, Nattel S. Molecular mechanisms underlying ionic remodeling in a dog model of atrial fibrillation. Circ Res 1999; 84:776–784.

69. Escande D, Coulombe A, Faivre JF, Coraboeuf E. Characteristics of the time-independent slow inward current in adult human atrial single myocytes. J Mol Cell Cardiol 1986; 18:547–551.

70. Mewes T, Ravens U. L-type calcium currents of human myocytes from ventricle of non-failing and failing hearts and from atrium. J Molec Cell Cardiol 1994; 26:1307–1320.

71. Van Wagoner DR, Pond AL, Lamorgese M, Rossie SS, McCarthy PM, Nerbonne JM. L-type Ca^{2+} currents and human atrial fibrillation. Circ Res 1999; 85:428–436.

72. Lai LP, Su MJ, Lin JL, Lin FY, Tsai CH, Chen YS, Huang SKS, Tseng YZ, Lien WP. Downregulation of L-type calcium channel and sarcoplasmic reticulum Ca^{2+}-ATPase mRNA in human atrial fibrillation without significant change in the mRNA of ryanodine receptor, calsequestrin and phospholamban: an insight into the mechanism of atrial electrical remodeling. J Am Coll Cardiol 1999; 33:1231–1237.

73. Sakakibara Y, Wasserstrom JA, Furukawa T, Jia H, Arentzen CE, Hartz RS, Singer DH. Characterization of the sodium current in single human atrial myocytes. Circ Res 1992; 71:535–546.

74. Gellens ME, George ALJ, Chen LQ, Chahine M, Horn R, Barchi RL, Kallen RG: Primary structure and functional expression of the human cardiac tetrodotoxin-insensitive voltage-dependent sodium channel. Proc Natl Acad Sci U.S.A. 1992; 89:554–558.

75. Koumi S, Backer CL, Arentzen CE. Characterization of inwardly rectifying K^+ channel in human cardiac myocytes. Alterations in channel behavior in myocytes isolated from patients with idiopathic dilated cardiomyopathy. Circulation 1995; 92:164–174.

76. Van Wagoner DR, Pond AL, McCarthy PM, Trimmer JS, Nerbonne JM: Outward K^+ current densities and Kv1.5 expression are reduced in chronic human atrial fibrillation. Circ Res 1997; 80:772–781.

77. Fermini B, Wang Z, Duan D, Nattel S. Differences in rate dependence of transient outward current in rabbit and human atrium. Am J Physiol 1992; 263:H1747–H1754.

78. Wang Z, Feng J, Shi H, Pond A, Nerbonne JM, Nattel S. Potential molecular basis of different physiological properties of the transient outward K^+ current in rabbit and human atrial myocytes. Circ Res 1999; 84:551–561.

79. Pond AL, Nerbonne JM, Rossie S, Van Wagoner DR: Impact of chronic atrial fibrillation (AF) on functional K^+ and Ca^{2+} currents and channel subunit expression. Circulation 1998; 98(17):I-819(abstr).

80. Fedida D, Wible B, Wang Z, Fermini B, Faust F, Nattel S, Brown AM. Identity of a novel delayed rectifier current from human heart with a cloned K^+ channel current. Circ Res 1993; 73:210–216.

81. Wang Z, Fermini B, Nattel S. Sustained depolarization-induced outward current in human atrial myocytes. Evidence for a novel delayed rectifier K^+ current similar to Kv1.5 cloned channel currents. Circ Res 1993; 73:1061–1076.

82. Wang Z, Fermini B, Nattel S. Rapid and slow components of delayed rectifier current in human atrial myocytes. Cardiovasc Res 1995; 28:1540–1546.

83. Koumi S, Arentzen CE, Backer CL, Wasserstrom JA. Alterations in muscarinic K^+ channel response to acetylcholine and to G protein-mediated activation in atrial myocytes isolated from failing human hearts. Circulation 1994; 90:2213–2224.

84. Krapivinsky G, Gordon EA, Wickman K, Velimirovic B, Krapivinsky L, Clapham DE. The G-protein-gated atrial K^+ channel IKACh is a heteromultimer of two inwardly rectifying K^+-channel proteins. Nature 1995; 374:135–141.

85. Porciatti F, Pelzmann B, Cerbai E, Schaffer P, Pino R, Bernhart E, Koidl B, Mugelli A. The pacemaker current I(f) in single human atrial myocytes and the effect of beta-adrenoceptor and A1-adenosine receptor stimulation. Br J Pharmacol 1997; 122:963–969.

86. Hoppe UC, Beuckelmann DJ. Characterization of the hyperpolarization-activated inward current in isolated human atrial myocytes. Cardiovasc Res 1998; 38:788–801.

87. Crumb WJ, Pigott JD, Clarkson CW. Description of a non-selective cation current in human atrium. Circ Res 1995; 77:950–956.

88. Benardeau A, Hatem SN, Rucker-Martin C, Le Grand B, Mace L, Dervanian P, Mercadier JJ, Coraboeuf E. Contribution of Na^+/Ca^{2+} exchange to action potential of human atrial myocytes. Am J Physiol 1996; 271:H1151–H1161.

89. Li GR, Nattel S. Demonstration of an inward Na^+-Ca^{2+} exchange current in adult human atrial myocytes. Ann N Y Acad Sci 1996; 779:525–8:525–528.

90. Wang J, Schwinger RH, Frank K, Muller-Ehmsen J, Martin-Vasallo P, Pressley TA, Xiang A, Erdmann E, McDonough AA. Regional expression of sodium pump subunits isoforms and Na^+-Ca^{++} exchanger in the human heart. J Clin Invest 1996; 98:1650–1658.

91. Le Grand B, Hatem S, Deroubaix E, Couétil J-P, Coraboeuf E. Depressed transient outward and calcium currents in dilated human atria. Cardiovasc Res 1994; 28:548–556.

92. Brundel BJJM, Van Gelder IC, Henning RH, Tuinenburg AE, Deelman LE, Tieleman RG, Grandjean JG, Van Gilst WH, Crijns HJGM. Gene expression of proteins influencing the calcium homeostasis in patients with persistent and paroxysmal atrial fibrillation. Cardiovasc Res 1999; 42:443–454.

93. Thiedemann KU, Ferrans VJ. Ultrastructure of sarcoplasmic reticulum in atrial myocardium of patients with mitral valvular disease. Am J Pathol 1976; 83:1–38.

94. Mary-Rabine L, Albert A, Pham TD, Hordof A, Fenoglio JJ, Jr., Malm JR, Rosen MR. The relationship of human atrial cellular electrophysiology to clinical function and ultrastructure. Circ Res 1983; 52:188–199.

95. van der Velden HMW, Kempen MJ, Wijffels MC, Zijverden M, Groenewegen WA, Allessie MA, Jongsma HJ. Altered pattern of connexin 40 distribution in persistent atrial fibrillation in the goat. J Cardiovasc Electrophysiol 1998; 9:596–607.

96. Elvan A, Huang XD, Pressler ML, Zipes DP. Radiofrequency catheter ablation of the atria eliminates pacing-induced sustained atrial fibrillation and reduces connexin 43 in dogs. Circulation 1997; 96:1675–1685.

97. Gaspo R, Bosch RF, Bou-Abboud E, Nattel S. Tachycardia-induced changes in Na^+ current in a chronic dog model of atrial fibrillation. Circ Res 1997; 81:1045–1052.

98. Gaspo R, Bosch RF, Talajic M, Nattel S: Functional mechanisms underlying tachycardia-induced sustained atrial fibrillation in a chronic dog model. Circulation 1997; 96:4027–4035.

99. Atrial fibrillation follow-up investigation of rhythm management—the AFFIRM study design. The Planning and Steering Committees of the AFFIRM study for the NHLBI AFFIRM investigators. Am J Cardiol 1997; 79:1198–1202.

100. Crijns HJGM, Gosselink AT. Prophylactic drug therapy after cardioversion of atrial fibrillation or flutter. Cardio, July 1994; 31–35.

101. Atrial fibrillation: current understandings and research imperatives. The National Heart, Lung, and Blood Institute Working Group on Atrial Fibrillation. J Am Coll Cardiol 1993; 22:1830–1834.

102. Crijns HJGM, Gosselink ATM, Van Gelder IC. Drugs after cardioversion to prevent relapses of chronic atrial fibrillation or flutter. In: Kingma JH, van Hemel NM, Lie KI, eds. Atrial Fibrillation, A Treatable Disease? Boston: Kluwer Academic Publishers, 1992:105–148.

103. Waldo AL. Measures of drug efficacy in treating atrial fibrillation. Am J Cardiol 1998; 82:7I–8I.

104. Flaker GC, Blackshear JL, McBride R, Kronmal RA, Halperin JL, Hart RG. Antiarrhythmic drug therapy and cardiac mortality in atrial fibrillation. The Stroke Prevention in Atrial Fibrillation Investigators. J Am Coll Cardiol 1992; 20:527–532.

105. Pedersen OD, and The DIAMOND Study Group. Dofetilide in the treatment of atrial fibrillation in patients with impaired left ventricular function: Atrial fibrillation in the DIAMOND study. Circulation 1998; 98:I-632(abstr).

106. Greenbaum RA, Campbell TJ, Channer KS, Dalrymple HW, Kingma JHSM, Theisen K, Toivonnen LK. Conversion of atrial fibrillation and maintenance of sinus rhythm by dofetilide. The EMERALD (European and Australian Multicenter Evaluative Research on Atrial Fibrillation Dofetilide) Study. Circulation 1998; 98:I-633(abstr).

107. Singh SN, Berk MR, Yellen LG, Zoble RG, Abrahamson D, Satler CA, Friedrich T. Efficacy and safety of oral dofetilide in maintaining normal sinus rhythm in patients with atrial fibrillation: A multicenter study. Circulation 1997; 96:I-383(abstr).

108. Pritchett E, Page R, Connolly S, Marcello S, Schnell D, Wilkinson W. Azimilide treatment of atrial fibrillation. Circulation 1998; 98:I-633(abstr).

109. The dl-sotalol AFIB/AFL Multicenter Study Group. Efficacy, safety and dose response of dl-sotalol (Betapace®) for prevention of recurrence of atrial fibrillation/atrial flutter. Pace 1998; 21:8A(abstr).

110. Pratt CM, Waldo AL, Camm AJ. Can antiarrhythmic drugs survive survival trials? Am J Cardiol 1998; 81:24D–34D.

111. Coplen SE, Antman EM, Berlin JA, Hewitt P, Chalmers TC. Efficacy and safety of quinidine therapy for maintenance of sinus rhythm after cardioversion. A meta-analysis of randomized control trials Circulation 1990; 82:1106–1116.

112. Anderson JL, Gilbert EM, Alpert BL, Henthorn RW, Waldo AL, Bhandari AK, Hawkinson

RW, Pritchett EL. Prevention of symptomatic recurrences of paroxysmal atrial fibrillation in patients initially tolerating antiarrhythmic therapy. A multicenter, double-blind, crossover study of flecainide and placebo with transtelephonic monitoring. Flecainide Supraventricular Tachycardia Study Group. Circulation 1989; 80:1557–1570.

113. Henthorn RW, Waldo AL, Anderson JL, Gilbert EM, Alpert BL, Bhandari AK, Hawkinson RW, Pritchett EL. Flecainide acetate prevents recurrence of symptomatic paroxysmal supraventricular tachycardia. The Flecainide Supraventricular Tachycardia Study Group. Circulation 1991; 83:119–125.

114. A randomized, placebo-controlled trial of propafenone in the prophylaxis of paroxysmal supraventricular tachycardia and paroxysmal atrial fibrillation. UK Propafenone PSVT Study Group. Circulation 1995; 92:2550–2557.

115. The Cardiac Arrhythmias Suppression Trial (CAST) Investigators: Preliminary Report. Effect of encainide and flecainide on mortality in a randomized trial of arrhythmia suppression after myocardial infarction. N Engl J Med 1989; 321:406–410.

116. Teo KK, Yusuf S, Furberg CD. Effects of prophylactic antiarrhythmic drug therapy in acute myocardial infarction. An overview of results from randomized controlled trials. JAMA 1993; 270:1589–1595.

117. Julian DG, Prescott RJ, Jackson FS, Szekely P. Controlled trial of sotalol for one year after myocardial infarction. Lancet 1982; 1:1142–1147.

118. Dofetilide in patients with left ventricular dysfunction and either heart failure or acute myocardial infarction: rationale, design, and patient characteristics of the DIAMOND studies. Danish Investigations of Arrhythmia and Mortality ON Dofetilide. Clin Cardiol 1997; 20:704–710.

119. Kober L. A clinical trial of dofetilide in patients with acute myocardial infarction and left ventricular dysfunction: the DIAMOND-MI study. Eur Heart J 1998; 19:90(abstr).

120. Effect of the antiarrhythmic agent moricizine on survival after myocardial infarction. The Cardiac Arrhythmia Suppression Trial II Investigators. N Engl J Med 1992; 327:227–233.

121. Page RL, Connolly SJ, Marcello SR, Schnell DJ, Wilkinson WE, Pritchett ELC: Dose response for azimilide treatment of paroxysmal supraventricular tachycardia. Pacing Clin.Electrophysiol. 1999; 22:770(abstr).

122. Williamson BD, Man KC, Daoud E, Niebauer M, Strickberger SA, Morady F. Radiofrequency catheter modification of atrioventricular conduction to control the ventricular rate during atrial fibrillation. N Engl J Med 1994; 331:910–917.

123. Rosenquvist M, Lee MA, Moulinier L, Springer MJ, Abbott JA, Wu J, Langberg JJ, Griffin JC, Scheinman MM. Long-term follow-up of patients after transcatheter direct current ablation of the atrioventricular junction. J Am Coll Cardiol 1990; 16:1467–1474.

124. Delfaut P, Saksena S, Prakash A, Krol RB. Long-term outcome of patients with drug-refractory atrial flutter and fibrillation after single- and dual-site right atrial pacing for arrhythmia prevention. J Am Coll Cardiol 1998; 32:1900–1908.

125. Saksena S, Prakash A, Hill M, Krol RB, Munsif AN, Mathew PP, Mehra R. Prevention of recurrent atrial fibrillation with chronic dual-site right atrial pacing. J Am Coll Cardiol 1996; 28:687–694.

126. Waldo AL, Wit AL. Mechanism of cardiac arrhythmias and conduction disturbances. In: Schlant RC, Alexander RW, eds: Hurst's the Heart, 9th ed. New York: McGraw-Hill, 1997:825–872.

127. Wellens HJ, Lau CP, Luderitz B, Akhtar M, Waldo AL, Camm AJ, Timmermans C, Tse HF, Jung W, Jordaens L, Ayers G. Atrioverter: an implantable device for the treatment of atrial fibrillation. Circulation 1998; 98:1651–1656.

128. Cox JL, Boineau JP, Schuessler RB, Jaquiss RD, Lappas DG. Modification of the maze procedure for atrial flutter and atrial fibrillation. I. Rationale and surgical results. J Thorac Cardiovasc Surg 1995; 110:473–484.

129. Albirini A, Scalia GM, Murray RD, Chung MK, McCarthy PM, Griffin BP, Arheart KL, Klein

AL. Left and right atrial transport function after the Maze procedure for atrial fibrillation: an echocardiographic Doppler follow-up study. J Am Soc Echocardiogr 1997; 10:937–945.

130. Swartz JF, Pellersels G, Silvers J, Patten L, Cervantez D. A catheter-based curative approach to atrial fibrillation in humans. Circulation 1994; 90:I-335(abstr).

131. Robbins IM, Colvin EV, Doyle TP, Kemp WE, Loyd JE, McMahon WS, Kay GN. Pulmonary vein stenosis after catheter ablation of atrial fibrillation. Circulation 1998; 98:1769–1775.

132. Haissaguerre M, Jais P, Shah DC, Gencel L, Pradeau V, Garrigues S, Chouairi S, Hocini M, Le Metayer P, Roudaut R, Clementy J. Right and left atrial radiofrequency catheter therapy of paroxysmal atrial fibrillation. J Cardiovasc Electrophysiol 1996; 7:1132–1144.

133. Andersen HR, Nielsen JC, Thomsen PE, Thuesen L, Mortensen PT, Vesterlund T, Pedersen AK. Long-term follow-up of patients from a randomised trial of atrial versus ventricular pacing for sick-sinus syndrome. Lancet 1997; 350:1210–1216.

134. Huang DT, Monahan KM, Zimetbaum P, Papageorgiou P, Epstein LM, Josephson ME. Hybrid pharmacologic and ablative therapy: a novel and effective approach for the management of atrial fibrillation. J Cardiovasc Electrophysiol 1998; 9:462–469.

19

Ventricular Fibrillation and Defibrillation

STEPHEN M. DILLON

University of Pennsylvania and Presbyterian Medical Center, Philadelphia, Pennsylvania

RICHARD A. GRAY

University of Alabama at Birmingham, Birmingham, Alabama

I. FIBRILLATION

Ventricular fibrillation (VF) is the leading cause of death in the industrialized world (1) and atrial fibrillation (AF) often leads to stroke (2). Fibrillation has been studied for more than a century, beginning with the finding that spontaneous termination of VF occurred frequently in small hearts but infrequently in large hearts (3,4). In 1914, Garrey demonstrated that the maintenance of fibrillation was independent of the initiating event (5). By cutting fibrillating hearts into various size pieces, Garrey showed that a "critical mass" of tissue was required to sustain fibrillation (~25% of the canine ventricle). Furthermore, Garrey demonstrated that the shape of heart tissue was related to the ability to sustain fibrillation and showed that impulses could circulate around rings of tissue. In 1914, Mines (6) also showed that self-sustaining circulating impulses could be supported in rings of cardiac tissue similar to those recorded earlier in the jellyfish (7). So-called "circus movement" reentry is usually described as an electrical impulse continuously propagating over a unidimensional circuitous pathway (i.e., a ring), as shown in Figure 19.1. The impulse can only be maintained if the circulation time around the circuit is long enough to permit expiration of the refractory period of the previous excitation. The rotation period is determined by the ratio of the length of the circuit (i.e., perimeter) and the conduction velocity along the pathway. In this essentially one-dimensional (1-D) model of reentry, the portion of the circuit that is not refractory can be excited and is called the spatial "excitable gap." Both Garrey and Mines summarized their findings by suggesting that fibrillation resulted from spatially disordered impulses with ringlike circuits.

Although the present reentrant view of VF is similar to that adopted in 1914, techno-

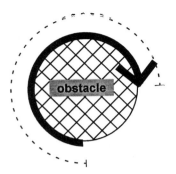

Figure 19.1 Anatomical reentry. Anatomical or circus movement reentry results from an excitation wave (black) continuously propagating around an inexcitable obstacle (hatched circle). The one-dimensional circuit is the perimeter of the obstacle. The wavelength (dashed line) is the extent of the depolarized region. The portion of the circuit that is not depolarized is excitable and this region is called the excitable gap. (Reproduced with permission from Gray RA, Cardiovasc Rev Rep 1999; 20(4):206–215.)

logical advances have allowed us to refine our understanding of the detailed events occurring during VF. The first recordings of the spatiotemporal dynamics of fibrillation were by Wiggers in 1930 (8), who filmed the rapid events that led to his characterization of four "stages" of fibrillation. It would be many years before it was possible to record the electrical activity at many sites simultaneously (i.e., electrical mapping) during fibrillation to determine the events responsible for these complex contractions. Nevertheless, seminal work by Moe involving numerical simulations of electrical activity in a two-dimensional grid led to the idea that a spatial dispersion of refractoriness was required for the initiation and maintenance of fibrillation (9–11). Moe described ". . . numerous vortices, shifting in position and direction like eddies in a turbulent pool, accounted for the sustained activity . . ." (11). This became known as the "multiple/wandering/meandering wavelet" hypothesis of fibrillation.

Before it became possible to make many simultaneous recordings, in 1973 Allessie cleverly used a roving electrode array to record from multiple sites sequentially to "map" stable reentry in the rabbit heart (12). This led to the formulation of the "leading circle" model. According to the leading circle concept (Fig. 19.2), a reentrant wave circulates around a functionally determined area of inexcitable tissue at the highest possible speed (13). The leading circle is the smallest possible pathway in which a rotating impulse can continue to circulate in the absence of an anatomically defined obstacle, and where the stimulating efficacy of the head of the impulse is just sufficient to excite the relatively refractory tissue that lies ahead in the pathway. Consequently, in the leading circle model there is no excitable gap. The well-known strength–interval curve of cardiac threshold can explain the minimum size of the functional reentrant circuit.

With the advent of multiple channel recording arrays, Janse (14), using 60 electrodes, found activation patterns resembling meandering wavelets in ischemic hearts in the early stages of fibrillation. A year later, in 1981, Ideker (15), using 27 electrodes, found that there was a period of organized epicardial activation during the transition to ventricular fibrillation induced by reperfusion following acute ischemia. Soon afterward, a number of investigators, using up to 196 electrodes, found rotating waves during ventricular ar-

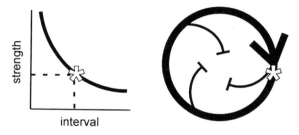

interval

Figure 19.2 Leading circle reentry. The leading circle concept postulates that the propagating depolarization wavefront is just sufficient to excite tissue direct ahead. The size of the reentrant circuit is determined by the strength–interval curve (left). The strength–interval curve provides the relationship between the strength needed to activate cardiac tissue and the duration since the last activation. The cycle length of reentry is the interval from the strength–interval curve where the strength is equal to the ability of the propagating wave front to depolarize downstream regions. The central regions are activated by centripetal wavelets. (Reproduced with permission from Gray RA, Cardiovasc Rev Rep 1999; 20(4):206–215.)

rhythmias (16–20). In the early 1980s, Allessie and coworkers mapped acetylcholine-induced atrial fibrillation using 480 electrodes (21). These experiments were considered confirmation of Moe's multiple wavelet hypothesis of atrial fibrillation. Fully developed VF has been mapped with hundreds of electrodes (22–24) and at over 10,000 sites simultaneously during VF using optical methods (25–28). However, traditional analysis yields only minor improvements to the qualitative description of fibrillation as "meandering wavelets."

In most fields there tends to be an emphasis on improving recording methods with the idea that obtaining more detailed information will lead to increased understanding of the underlying events. While this is undoubtedly true, examining data with nontraditional methods sometimes leads to great advances in our understanding of certain phenomena. The analysis of reentry and fibrillation in the heart has experienced a revolution of sorts due to two fields of research: (1) chaos theory and (2) nonlinear waves in excitable systems. Of course, both fields involve mathematical approaches well beyond the scope of this chapter; nevertheless, we hope to present a brief overview of their importance to the study of fibrillation.

Chaos theory is particularly attractive to those studying fibrillation because of the way it describes complex phenomena. Traditionally, the amount of complexity observed in a system was directly attributed to the degree of randomness in the system. Chaos theory states that very simple deterministic systems can give rise to tremendously complex behavior. The frequency spectra of VF ECGs indicated certain periodic components demonstrating that VF was not a completely random phenomenon (29). Nevertheless, early studies of VF ECGs failed to find evidence of simple chaotic systems (22,30). However, by incorporating spatial information, Damle (31) was able to find evidence of organized electrical activation during VF. Witkowski (32) has also found evidence of determinism during VF. Garfinkel (33) found that the transition to VF occurred via a specific process characteristic of the onset of chaos. Overall, these results demonstrate that VF is not entirely random.

In 1946, pioneering work involving the collaboration of a mathematician (Norbert Wiener) and a cardiologist (Arturo Rosenblueth) provided a quantitative description of reentry in heart tissue that is still much cited today (34). In 1948, Selfridge extended this analysis and proved that reentry could exist in uniform tissue in the absence of a nonconducting obstacle (35). Since the mid-1960s, there has been a vast amount of research (mostly in Russia) involving the mathematical formulation of wave propagation in the heart (see Ref. 36 and 37 for excellent reviews). This work is based on theoretical considerations rooted in the idea that the heart acts as a nonlinear excitable system (36,38–43). Cardiac tissue satisfies the three conditions of excitable systems. First, cardiac cells are "excitable," because, upon receiving an appropriate electrical stimulus that brings their cell membrane from a resting potential to a threshold voltage, they become fully depolarized. Second, cardiac cells have a refractory period. Finally, the depolarization process propagates as a nonlinear wave. In other words, waves propagate in the heart with constant shape because they propagate at the expense of energy taken from heart cells, and the collision of two waves results in mutual annihilation. In contrast, linear waves propagate with decreasing amplitude and "pass through" each other resulting in interference patterns. Nonlinear waves propagate in a variety of excitable media, including chemical, physical, mechanical, and biological systems in a similar manner and are well characterized (36–38). All of these systems exhibit self-organized activity in the form of rotating spiral waves, an apparently ubiquitous feature of excitable systems. In fact, there is mounting evidence supporting the notion that spiral waves underlie many cardiac arrhythmias (40–43).

For nonlinear waves in excitable systems, the conduction velocity of the wavefront is linearly related to its curvature (44). Indeed, there is a critical curvature above which the wavefront does not propagate. During reentry, this critical curvature determines the size of the "core," which is the region around which the tip of the spiral wave rotates. This form of reentry is typically called spiral wave reentry because the reentrant wavefront forms a spiral shape exhibiting decreasing curvature away from the center of rotation (Fig. 19.3). The conduction velocity is slowest at the spiral wave "tip" that rotates around the core; no propagation occurs within the core, only electrotonic diffusion. Together with the refractory period and the excitability of the medium, the critical curvature determines the rotation period of the spiral wave. In the limiting case, when the core size is zero, the refractory period alone determines the rotation period (i.e., leading circle reentry).

Recent experiments with isolated thin pieces of the sheep heart reveal a linear relationship between conduction velocity and wavefront curvature (45) that is in agreement with nonlinear wave theory (44,46). By studying planar waves propagating through various size isthmuses these experiments also indirectly determined the critical curvature for propagation to be between 0.7 and 1.5 mm^{-1} at the spiral wave period (45). These data suggest that if spiral waves in the heart are determined by wavefront curvature alone, the diameter of their core should be approximately 1 to 3 mm. Clever experimental approaches to measure spiral wave parameters indirectly were carried out by Pertsov and Medvinskii in 1986 (47) and later by Ikeda (48). Both studies estimated the diameter of the spiral wave core to be ~7 mm by analyzing the rotation period of waves circulating around holes cut in thin pieces of heart tissue as the size of the hole was varied. When the hole size was larger than the size of the core, the period of rotation was equal to the ratio of the perimeter of the hole and the conduction velocity along the pathway (i.e., anatomical reentry). When the hole was smaller than the core size, the wave rotated at the spiral wave rotation period defined by the heart tissue (i.e., functional reentry). The first direct measurement of spiral wave parameters in heart tissue resulted from high-resolution optical mapping by Davidenko and Pertsov (49,50). In these experiments, spiral wave activity was consistently initiated in thin slices of canine and sheep ventricular muscle. As a result of the anisotropic

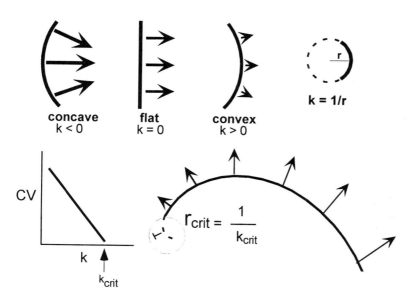

Figure 19.3 Spiral wave reentry. For many nonlinear waves the conduction velocity (CV) is linearly related to the curvature of the wavefront (bottom left). Concave wavefronts propagate faster than planar waves, which propagate faster than convex waves (top). Curvature (k) is defined for circular waves in isotropic tissue as $1/r$, where r is the radius of the circular wavefront (top right). Conduction fails (i.e., CV = 0) at some critical value of curvature, k_{crit} (bottom left). For rotating nonlinear waves (i.e., spiral waves), conduction is slower near the center of rotation compared to the periphery because the curvature of the wavefront decreases radially from the center (bottom right). The wavefront does not propagate near the center because of its high curvature. Therefore, the spiral wave tip circumvents a region called the core shown as gray circle (bottom right). (Reproduced with permission from Gray RA, Cardiovasc Rev Rep 1999; 20(4):206–215.)

properties of heart muscle, propagation along fibers was about four times faster than across fibers, which resulted in elliptical spiral waves. The average rotation period was 180 ms and a large excitable gap was present (~30% of the circuit). The average size of the elongated spiral wave cores was 17 mm². A more recent study of reentry in the atria revealed similar results (albeit with a larger excitable gap): a cycle length of 162 ms, an excitable gap of 67%, and a core area of 12 mm² (51).

Three-dimensional spiral waves, so-called scroll waves, can be conceptualized by stacking two-dimensional spiral waves. The cylinder connecting the cores is called a filament and these scroll-wave filaments come in many shapes (52): they can be linear (straight), L-shaped, U-shaped, or even ring-shaped (the filament forms a closed loop in this case). Scroll-wave filaments can also become twisted, in which case exceedingly complex dynamics may occur (53). These different filament shapes are important because the manifestation of the activity on the surface will depend on the dynamics of the scroll-wave filament (54). These 3-D scroll waves are not just conceptual abstractions, they have been extensively studied numerically as well as in chemical reactions (55–60).

The earliest evidence that 3-D spiral waves occur in the whole heart comes from studies of healthy canine hearts. First, Chen (20) showed that a single premature stimulus delivered within the recovering tail of a propagating wave resulted in a "figure-8" pattern of reentry. The premature stimulus caused nonuniform excitation in a region exhibiting a

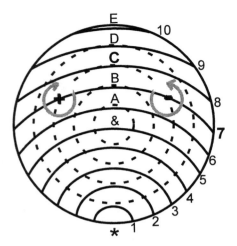

Figure 19.4 Pinwheel stimulation. A wave propagates from the pacing site (*) at the bottom to top revealing a gradient in recovery. A stimulus is applied at a point (&) located within the tail of the propagating wave. The stimulus creates nonuniform excitation of the tissue with the strongest response at the stimulus site (&). Contours of excitation strength are shown as dashed lines (decreasing levels indicated by A–D). A critical combination of stimulus strength and recovery phase (in this case, stimulus strength C and recovery phase 7) are required to induce reentry. Therefore, this point stimulation protocol induces a pair of counter-rotating spiral waves if the critical combination of stimulus strength and recovery phase is achieved. (See Ref. 40 for further details.)

spatial gradient of refractoriness that led to the formation of a pair of counter-rotating spiral waves as predicted by Winfree (40) (see Fig. 19.4). Second, Frazier (61) and Shibata (62) demonstrated that it was possible to induce spiral waves whose chirality and location were consistent with those predicted from theoretical considerations (40,63). In some of these experiments, plunge electrodes attached to needles enabled the recording of electrical activity across the ventricular wall, revealing that the 3-D scroll waves were oriented such that their filaments were perpendicular to the heart surface. These studies were accomplished by analyzing data from extracellular potentials from up to 120 sites. With the introduction of video-imaging technology, it is now possible to record transmembrane signals on the surface of the whole heart from tens of thousands of sites (64). Using this technology, Gray observed patterns on the surface of the rabbit heart consistent with 3-D spiral waves. The movement of the scroll waves correlated with the irregularity of the ECGs, with the episodes displaying the largest movement resulting in the largest changes in the ECG (64). These results confirmed the predictions of Davidenko (42) by showing:(1) stationary scroll waves resulted in ECGs characteristic of monomorphic tachycardia; (2) drifting scroll waves gave rise to ECG patterns typical of polymorphic tachycardia; and (3) a scroll wave moving throughout a large portion of the heart in a cycloid pattern gave rise to undulating ECGs characteristic of the arrhythmia known as "torsade de pointes" [also predicted by Winfree (63)].

It has generally been thought that VF results from a large number of reentrant waves. Recent experimental evidence suggests that possibly only a few reentrant waves are present on the heart at one time during fibrillation (64–66). It has been shown that a single scroll wave can give rise to extremely irregular ECGs if it moves rapidly through the heart

(64). The authors provided a mathematical equation that relates the irregularity of the ECG to the movement of the core of reentrant waves in the heart by invoking the Doppler effect. A drifting rotating wave behaves similar to other moving sources emitting waves and, as such, results in the Doppler effect, which causes an abbreviation or extension of the cycle length ahead or behind the moving organizing center (49,64,67). Gray (64) estimated that a single reentrant wave can give rise to patterns resembling fibrillation if it moves at 30 to 40% of the wave speed. These results suggest that the velocity of the organizing centers of reentrant waves, not their number, is responsible for the irregular ECGs recorded during fibrillation. Although fibrillation can result from a single reentrant wave in the rabbit heart, most often VF results from multiple unstable reentrant circuits. The complex spatiotemporal patterns recorded from the heart during fibrillation are difficult to interpret in regard to the leading circle and spiral wave models of reentry, especially because reentrant waves on the surface are observed only infrequently (23,24,68). Nevertheless, spiral wave parameters such as cycle length, core size, and excitable gap have been calculated. The cycle length during VF is ~100 to 120 ms in dogs, sheep, pigs, and rabbits. Rough measures of core size (25–30 mm^2) (23) and excitable gap (~50%) (23,69) during VF are slightly larger than observed during stable reentry. Finally, Gray (70), using a new signal-processing technique, was able to track the dynamics of rotating waves with a temporal resolution much less than the rotation period. Using this technique coupled with high-resolution video imaging, this group was able to elucidate the mechanisms of rotor formation and termination during sustained VF in rabbit and sheep hearts, which were consistent with the theoretical predictions of Winfree (40).

If VF is indeed maintained via reentry, the irregularity of the ECG reflects the instability of these reentrant waves (i.e., the processes of wave creation and destruction). The study of how reentrant waves are formed and annihilated, as well as the factors responsible for these events, is an active field of research. Here we present an overview of the factors thought to play a role in the stability of reentry that are actively being studied experimentally.

A. Mechanisms of Reentry Formation

The mechanisms listed below are not mutually exclusive, some factors are common to several, and all of them have been proposed to explain the transition from monomorphic VT (MVT) to VF and the maintenance of VF.

1. Inherent Gradients in Refractoriness

Spatial gradients in electrophysiological parameters (e.g., refractory period) were traditionally thought essential for reentry formation and fibrillation. Spatial gradients in the refractory period can cause a wavefront to "break," (i.e., block at one part but conduct at an adjacent segment), thus allowing the wavefront to rotate and curl around the free end(s) formed at the break(s)(38). Moe's (11) cellular automata model showed that "turbulent" unstable reentrant patterns resulted if the tissue had a random distribution of refractory periods, while homogeneous refractory periods resulted in stable reentry. However, now it is known that heterogeneity of electrophysiological parameters is not required to generate unstable reentry and the turbulent patterns (71–73). Using the same model as Moe (11), Winfree (74) showed that turbulent activity resulted in uniform tissue if the restitution function was altered. In principle, therefore, fibrillation does not require electrophysiological inhomogeneities.

2. Doppler-Shift-Induced Wavebreaks

If the organizing center of a reentrant wave moves rapidly enough, it can break into two counter-rotating reentrant waves. This phenomenon may repeat itself, eventually resulting in turbulence, characteristic of fully developed fibrillation. In a numerical model that exhibited rapid movement of the spiral wave tip, Bär and Eiswirth (72) found that sites ahead of the moving spiral wave tip displayed an instantaneous period that was shorter than the true rotation period, due to the Doppler effect. They demonstrated that when these abbreviated periods were less than the minimum period required for wave propagation, conduction block along the direction of motion occurred, leading to the formation of a new pair of spiral waves. This mechanism provides a robust explanation of the transition from MVT to VF and has been observed in animal models of fibrillation (75).

In the whole heart, reentrant waves tend to move (75), although we do not know why. Both stationary and nonstationary spiral waves have been observed in a variety of excitable systems (even homogeneous ones) (76–83). The movement of rotating waves probably depends upon a number of factors. Many numerical and analytical studies have been accomplished for two-dimensional spiral waves (36,84–86). Two parameters that influence spiral wave tip trajectories are the threshold for excitation and the ratio of characteristic rates of recovery and excitation (87–89). Zykov (87) and Winfree (88) varied these two parameters systematically and showed that a variety of spiral wave tip trajectories were formed ranging from small circular trajectories to large complex cycloid trajectories. The motion of spiral wave tips in inhomogeneous excitable media has also been studied by creating gradients in parameters (e.g., refractory period, excitability, etc.). Spiral waves tend to drift in the presence of an inhomogeneity gradient (90) or changes in fiber curvature (91). It has also been shown that spiral waves drift along impermeable boundaries (i.e., no ions flowing across border) with a simultaneous decrease in their rotation period (92). In addition, organizing centers may move due to features that are uniquely three-dimensional. For example, linear filaments tend to move toward areas of minimal thickness (93) and the shape of the scroll-wave filament itself can result in drift (94). Furthermore, spiral waves have been shown to be stationary on a spherical surface, but move if the gradient of the surface curvature is not constant (56,95,96). Typically, in these numerical models, the movement of the spiral wave tip is much smaller than the wave speed and, although the cycle length at each site is not periodic, single spiral waves do not "break up" into multiple spiral waves. However, it is not known how ion currents, gradients, etc., in the heart influence the speed of the organizing center of reentrant waves. Simulations using ion models have resulted in unstable reentry (71); however, it has recently been shown that by changing parameter values, stable spiral waves can also be obtained (88,97–101).

3. Action Potential and Conduction Velocity Instabilities

The action potential duration (APD) is known to be related to the preceding diastolic interval (DI) (102). This relationship is known as the restitution curve and is influenced by the basic cycle length (CL). The dynamics of APD in a single cell can be graphically analyzed using the fact that the APD equals CL-DI for constant cycle lengths (103). This analysis predicts that if the slope of the restitution curve is greater than 1 at a particular CL, then APD will vary when pacing at that CL. The particular dynamics (e.g., alternans, Wenckebach, chaotic, etc.) depend on the shape of the restitution curve. This idea has led to the supposition that tissue exhibiting a restitution curve with a slope greater than 1 will not support stable reentry (i.e., the restitution hypothesis). Accordingly, the stability of propagation around a 1-D ring has been studied theoretically (73,104). These results demonstrate that when the perimeter of the ring is shorter than a critical length, oscillations in APD and conduction velocity occur. These results reproduce similar oscillations observed experimentally (105), and are thought to be caused by alterations in the recovery time resulting from the kinetics of ion currents. Karma (73) was the first to show that the pulse instabilities found in a 1-D ring can cause the destabilization of reentrant waves. The cellular properties of restitution have been implicated as an important factor for the transition to fibrillation (33,106). However, as yet, the details of the complex spatial aspects of the transition to fibrillation have not been incorporated into the restitution hypothesis.

4.   Small Obstacles and Lateral Instabilities

The heart is a complex structure with features of various sizes ranging from angstroms to centimeters. It is not clear what role these features play in the propagation of electrical waves throughout the heart. Although microscopic heterogeneities (less than approximately 1 mm) cause discontinuous conduction at a cellular level (107,108), it has recently been suggested that they play only a minor role in regard to macroscopic wave propagation because they are smaller than the size of the wavefront (109). Conduction abnormalities involving structures of the heart greater than the width of the wavefront can lead to the formation of reentrant waves (110). Structures of scale greater than 1 mm may lead to propagation failure during events that decrease the width of the wavefront (e.g., reduced excitablity, perhaps caused by ischemia and/or rapid heart rate). Propagation failure leads to the formation of wave breaks. The evolution of these wave breaks depends on the properties of the tissue in the nearby region and have been studied in detail theoretically (111,112). In "normal" cardiac tissue, these wave breaks form "pivot points" around which wavefronts rotate. New reentrant waves will be formed if conduction around these pivot points is slow enough to allow the tissue between wave breaks to recover by the time the wave rotates 180 degrees (70). If the tissue near the wave breaks is "depressed" (e.g., reduced dV/dt), the critical curvature for propagation is decreased and wavefronts do not turn sharply around their ends ("lateral instability"). The resulting larger radius turns may allow the tissue in the region of block to recover, allowing reentry to occur. Cabo showed that obstacles and lateral instabilities result in reentry formation in cardiac tissue at high rates, or when sodium conductance had been decreased with tetrodotoxin (110). These results confirmed the idea that obstacle size and lateral instabilities may play a role in the formation of reentrant waves in the heart.

5.   Filament Instabilities

These instabilities are unique to three-dimensional systems. Filaments can expand based solely on their shape (94). A parameter called filament tension, which depends on the dynamics of cellular kinetics, determines how local filament curvature evolves (113). If filament tension is positive, local curvature tends to decrease spontaneously. This means that curved filaments tend to straighten, and that scroll rings will shrink and collapse. If filament tension is negative, however, local curvature tends to increase spontaneously. This means that curved filaments tend to become increasingly curved and, therefore, they lengthen; similarly, scroll rings will expand. Filament expansion leads to the formation of new reentrant waves because the filament becomes increasingly convoluted and eventually hits a boundary that leads to the formation of new rotors (75,113). This mechanism provides a robust explanation, uniquely 3-D, which is consistent with how some investigators view VF (74). However, the few known examples of models that exhibit negative filament tension are characterized by low excitability, and the heart is known to exhibit a high degree of excitability. Another uniquely 3-D instability results from twisting of the filament. Cellular properties vary across the ventricular wall and may be important factors in establishing the stability of reentry in the heart. Gradients across the wall may lead to inherently different rotation periods on the epi- and endocardial surfaces. This difference in rotation rates can cause the filament to twist, evolving into a helical shape (114). If the twist exceeds a critical value, new reentrant waves can form, resulting in turbulent activity (57,59). In addition, parameter gradients can cause scroll rings to drift and expand (60), which may lead to the formation of new reentrant waves either via collision of the filament with a boundary, or the Doppler shift mechanism described above. The rotation of fiber direction across the ventricular wall has also been implicated as a destabilizing factor for reentry (115,116). This fiber rotation causes linear filaments to twist, which can result in propagated local disturbances that lengthen the filament giving rise to the formation of new reentrant waves when the filament collides with a boundary (116).

B. Mechanisms of Spontaneous Reentry Termination

Typically, termination of a reentrant wave occurs when it moves within a critical distance of an impermeable boundary or when a pair of counter-rotating reentrant waves moves too close together. Here, the latter is discussed for simplicity. During a portion of the reentrant cycle, a single wavefront is propagating through the central common pathway between the two wave breaks. Propagation success or failure through this isthmus is determined by the ratio of current available to excite downstream tissue and current required to depolarize the membrane (i.e., safety factor) (117,118). When the safety factor for propagation through this isthmus is less than unity, propagation fails, thus resulting in mutual annihilation of the counter-rotating pair of waves. The safety factor can become less than 1 for a variety of reasons (e.g., the wavebreaks become too close together, critically narrowing the isthmus, or the isthmus is refractory because of cycle length abbreviaton).

II. DEFIBRILLATION

This section focuses on defibrillation including: (1) coupling between the flow of shock current and induced myocardial membrane potential changes; (2) cellular electrophysiological responses; and (3) resulting effect of these responses on the maintenance of fibrillation. The last aspect has been a long-standing topic of interest and the discussion here will focus on the key question, "What causes defibrillation to fail?"

Several assumptions are made here that are unproved (Fig. 19.5). The first is that ventricular fibrillation (VF) is a reentrant arrhythmia and so depends on the presence of multiple propagating wavefronts. The second assumption is that defibrillation shocks ultimately elicit regenerative, depolarizing electrophysiological responses from the myocardium (see later in chapter for opposing view). The last assumption is that the effective strength of a shock will depend upon its local electrical field intensity (i.e., the voltage gradient) (119). The shock voltage gradient is measured in volts per centimeter and denotes how rapidly shock voltage varies in space. How the shock voltage gradient couples to the myocardium to give rise to membrane potential changes is still under investigation. Possibilities include the creation of "secondary sources" due to gap junctions (120) or microscopic discontinuities (121,122), "sawtooth potentials" (123), or heterogeneities in the

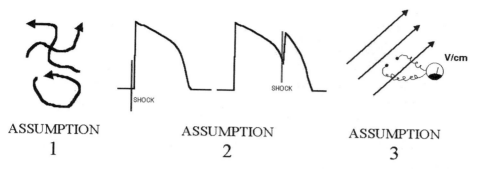

ASSUMPTION
1

ASSUMPTION
2

ASSUMPTION
3

Figure 19.5 Assumptions implicit in the descriptions of defibrillation made in this chapter. Assumption 1 is that VF is maintained by multiple wavefront reentry (arrows represent erratic, multiple wavefront propagation). Assumption 2 is that shocks evoke regenerative membrane depolarizations. (That is, they stimulate schematic transmembrane voltage recordings of shock responses. This assumption is not part of the Efimov hypothesis.) Assumption 3 is that the local effective strength of a shock is a linear function of the shock voltage gradient (arrows represent shock current flow).

proportions of intracellular and extracellular volumes (124,125). This simplification ignores the influence of the orientation of the myocardial "fibers" (126). Theoretical developments also indicate that macroscopic changes in membrane potential may arise as a result of the interaction between the shock electric field and the variations in myocardial fiber orientation and curvature intrinsic to the ventricle (127). A generalized "activating function" concept incorporates all of the recognized means by which a shock can affect the membrane potential (128). It considers both the spatial rate of change in the voltage gradient itself and the spatial rate of change of intracellular conductivity. Despite this variety, in vivo experimental evidence indicates that in the intact, in situ ventricle, the voltage gradient is still a reasonably accurate means of gauging the local effective strength of a shock (126,129–131).

More than 6 decades ago Wiggers attributed failed defibrillation to the failure of a shock to stop all fibrillatory activity (132). Many years later, two groups of investigators showed that the failure to immediately eliminate all fibrillatory activity does not guarantee unsuccessful defibrillation because, in some circumstances, the continuing electrical activity apparently terminates spontaneously (133,134). The critical mass hypothesis arising from this observation stipulated that a shock can defibrillate even if fibrillatory activity is present after the shock, as long as the activity is confined to less than a critical mass of the myocardium needed to sustain fibrillation (134). The fraction of the ventricle needed to sustain VF in the dog was estimated by quenching VF in varying volumes of tissue either through different routes of intracoronary potassium injection or different defibrillation shock current paths through the ventricle (134). On this basis, it was proposed that a shock can successfully defibrillate if it arrests fibrillatory activity in at least 75% of the dog's ventricle. Later studies by Witkowski (130) using epicardial mapping raised this proportion to nearly 100%, but this may be an overestimate owing to the possibility of intramural and endocardial activity not detected by the recording system. The following describes two contemporary schools of thought: the first is not consistent with this picture while the second is. The chapter then concludes with a brief review of other proposed defibrillation mechanisms.

A. Failed Defibrillation Due to Refibrillation Following Total Wavefront Termination

The development of shock-proof large-scale multielectrode recording systems in the years following the publication of the critical mass hypothesis enabled Chen and Ideker to measure more precisely the minimum fraction of the dog's ventricle in which the shock needs to terminate fibrillatory activity in order to still successfully defibrillate (135). The electrical recordings from a canine heart could resume within 20 ms or so after the shock (136). It was assumed that, as the shock energy was increased, fewer and fewer electrodes would show continuing fibrillatory activity after the shock, up until a point was reached when no electrode would register electrical activity following a sufficiently strong shock. The results were strikingly contrary to this expectation. All shocks, from energies that never defibrillated up to those that always did, were followed by a period of electrical silence on all recording channels. In those shocks that failed to defibrillate, the electrical silence was broken by the reappearance of electrical activity that eventually brought the ventricle back into fibrillation (Fig. 19.6). This period of electrical silence was called the isoelectric interval and for unsuccessful defibrillation shocks it averaged 64 ± 22 ms. This value stood in contrast to the fact that during VF the average time between activations across all channels was 1.7 ± 3.2 ms. It was established that even low-energy shocks, as little as 1 J in canines, were able to stop fibrillatory activity on the epicardial surface. It was concluded that even in failed defibrillation the shock terminates all fibrillatory activity in the ventricle. All subsequent electrical mapping studies by this laboratory (129,137–141)—except for one (142)—have also found an isoelectric interval following failed defibrillation shocks.

To account for the ultimate failure to defibrillate, it was proposed that the shock engendered the delayed appearance of new wavefronts that reinduced fibrillation (135). A successful defibrillation shock must therefore be one that does not subsequently reinduce

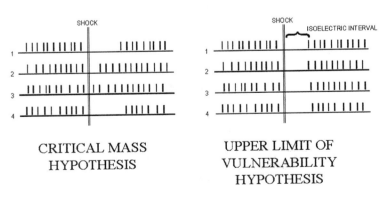

CRITICAL MASS HYPOTHESIS   UPPER LIMIT OF VULNERABILITY HYPOTHESIS

Figure 19.6 Schematic of electrical recordings made during failed defibrillation according to two different hypotheses. These schematized tracings show a tick whenever a recording channel registers electrical activation from the ventricle. Preceding the shock, all channels show the erratic activation thought to typify VF. The recordings shown on the left side portray the shock outcome according to the Critical Mass hypothesis. At least one channel shows continued activity following the shock, while the others undergo varying periods of temporary inactivity. The recordings on the right side portray the shock outcome according to the Upper Limit of Vulnerability hypothesis. All channels show a period of temporary inactivity, the isoelectric interval, followed by the resumption of VF. (The duration of the isoelectric interval is exaggerated for clarity in this figure.)

VF. This unique viewpoint sees failed defibrillation as a consequence of VF reinitiation as opposed to VF continuation (135). This concept was embodied in a new defibrillation hypothesis called the Upper Limit of Vulnerability (143). This hypothesis unified two apparently disparate phenomena: the ability of T-wave shocks (i.e., shocks applied during the vulnerable period) to induce VF and the ability of shocks to terminate fibrillation. It is well known that a shock applied during the vulnerable period can induce VF (144) and that there is an upper limit on the strength of the shock that is able to do so. Too strong a shock will not induce VF, at least until the point where it damages the heart (143,145,146). It was shown in the dog's ventricle that there is a numerical correlation between the minimum shock energy needed to guarantee defibrillation and the maximum shock energy capable of initiating VF (Fig. 19.7) (143). To explain this numerical correlation, the upper limit of vulnerability hypothesis proposed that the mechanism responsible for the initiation of VF by a shock is the same as that which reinitiates VF following an unsuccessful defibrillation shock (Fig. 19.8).

Guided by this hypothesis, electrical mapping studies of VF initiation by shocks were used to discover the causes of failed defibrillation. The first of these studies (62) was conducted in canines and it showed that, as in failed defibrillation, there was an isoelectric interval following shocks that initiated VF. Interestingly, the earliest sites of postshock activation showed a similar dependence on the energy of shocks that failed to defibrillate the same animals (137). The significance of this dependence on shock energy was provided by

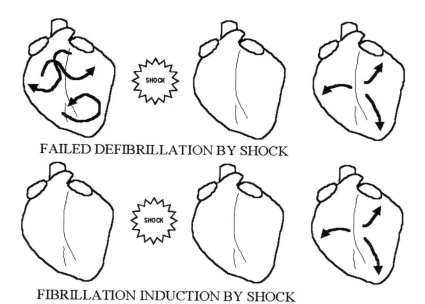

FAILED DEFIBRILLATION BY SHOCK

FIBRILLATION INDUCTION BY SHOCK

Figure 19.7 Failed defibrillation and VF induction by shock according to the Upper Limit of Vulnerability hypothesis. Going from left to right: in failed defibrillation, a shock eliminates all VF wavefronts and renders the ventricle electrically quiescent before, after tens of milliseconds, wavefronts arise spontaneously to induce VF. In VF induction, a shock is applied while the ventricle is repolarizing following activation by a paced or supraventricular wavefront. For tens of milliseconds after the shock, the ventricle is electrically quiescent until wavefronts arise spontaneously to induce VF. The same mechanisms are thought to be responsible for generating the wavefronts that induce VF in either failed defibrillation or VF induction.

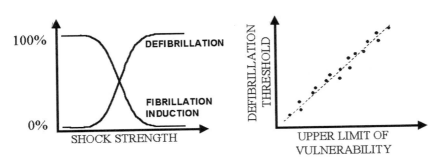

Figure 19.8 Numerical relationships between shock outcome and shock strength. The plot on the left schematically shows the sigmoid-shaped relationships found for the probability of defibrillation or VF induction versus the shock strength. (VF induction is accomplished by a shock in excess of the VF threshold that is applied during the vulnerable period.) In any heart there is a threshold shock strength that can be associated with a given probability (e.g., 50%) of defibrillating the ventricle or inducing VF. The plot on the right side schematically shows the relationships between the threshold value for defibrillation versus that for inducing VF for a number of hearts. The existence of a numerical correlation between the two values is taken as evidence of a mechanistic linkage between the two processes.

a study that mapped the spatial distribution of the shock voltage gradient created by shocks. This study noted that the shock voltage gradient was strongest near the shock electrodes and that it decreased with distance from it. More importantly, these data suggested that the numerical correlation between the upper limit of vulnerability and the defibrillation threshold could be explained if the impulses that fibrillated the ventricle were suppressed by exposure to shock voltage gradient in excess of some critical level. Therefore, as defibrillation or fibrillation induction is attempted with ever-increasing shock energies, successively larger volumes of the ventricle should become unable to generate VF-(re)initiating wavefronts. Thus, to successfully defibrillate or to prevent VF induction, all of the ventricle must be exposed to a minimum, critical shock voltage gradient. This prediction was met by an electrical mapping study of defibrillation that placed this critical value at about 5 V/cm for a 10-ms monophasic shock waveform (139). Another electrical mapping study strongly suggested that the critical shock voltage gradient for preventing VF induction is ~5 V/cm (61). In further validation of the critical shock voltage gradient concept it was found that the critical shock voltage gradient was lower for a biphasic shock waveform than for a monophasic waveform (2.7 vs. 5.4 V/cm) in accordance with the higher defibrillating efficacy of the former (139).

A natural issue arising from the results of these electrical mapping studies was establishing how shocks actually refibrillated ventricles during failed defibrillation. Electrical activation mapping was once again used to address this question, but this time employing a larger number of recording sites spread across a small area of myocardium. In one study, the 3-D propagation pattern was mapped by using plunge electrodes that sampled the endocardium, midwall, and epicardium of the thin-walled right ventricular outflow tract (138). Early postshock activation patterns in failed defibrillation took on one of two forms: unidirectional or focal conduction. In any given episode, the same pattern was seen in all three tissue layers. In some cases, it was possible to relate the unidirectional propagation of a postshock wavefront to the pattern of wavefront propagation present just before the shock. It was noticed that, in some cases, the earliest activation seen after the shock was a wavefront that propagated away from the area that would have been activated by the preshock wavefront, had the shock not been applied. Knowing that shocks are able to depolarize myocardium, it was inferred that the area lying between where the preshock wavefront was last observed, and where the postshock wavefront originated, was stimulated by the shock (i.e., the shock directly excited the myocardium about to be depolarized by the preshock wavefront). The postshock wavefront was therefore presumed to originate at the border of the directly excited area propagated in a unidirectional fashion. This unidirectional propagation pattern, about half of the episodes, contrasts with the focal pattern in which the wavefront originated from a circumscribed area and propagated in all directions. This type of pattern was not easily explained. It was thought possible that focal patterns of postshock activation were due to triggered activity or a microreentrant circuit too small to be resolved by the mapping electrode array. This behavior was not dependent on shock waveform since unidirectional and focal activation patterns were seen in another electrical mapping study in which monophasic and biphasic defibrillation shocks were used (139).

To explain the unidirectional conduction patterns, the shock was proposed to elicit three possible responses from myocardium, depending on both its refractory state and the shock voltage gradient (62,138,147). A shock is able to directly excite fully recovered myocardium as long as the shock voltage gradient exceeds a low threshold value which, in a nonfibrillating canine heart, was estimated to be 0.64 to 1.84 V/cm for a 3-ms rectangular waveform (126). As shown by Kao and Hoffman, a graded response can be elicited from relatively refractory myocardium by a sufficiently strong stimulus (148). In contrast to the all-or-none response, the amplitude and duration of the graded response are an increasing function of stimulus intensity. As envisioned for defibrillation, not only does the graded response not propagate but, by delaying repolarization, it temporarily blocks the propagation of any subsequent depolarizing activity. The last possible outcome of a shock is to have no effect because of an insufficient shock intensity relative to the degree of myocardial refractoriness. This will also occur if the shock finds the myocardium in its absolute refractory phase. Electrical mapping has established that in both atrial (149) and ventricular fibrillation (69,150), there is region of excitable myocardium ahead of the advancing wavefront. Therefore, a weak shock will stimulate an all-or-none action potential in the most excitable myocardium immediately ahead of the wavefront, but will not affect the less recovered myocardium a distance further away. After the shock, the directly excited tissue will border on myocardium that was not affected by the shock. This unaffected myocardium will therefore still be undergoing repolarization. The difference in membrane potential between these zones sets up a flow of depolarizing current such that, once the unaffected myocardium recovers its excitability, it becomes activated and gives rise to a propagated wavefront. The long delay between the delivery of the shock and the later appearance of the unidirectionally conducted wavefront has been attributed to the latent period that can occur when stimulating partially recovered myocardium (138,147). This latent period is imagined to be due to latency in the development of the membrane response and an initially slowed propagation. A stronger shock will prevent postshock propagation because in addition to directly exciting the myocardium immediately ahead of the preshock wavefront, the shock also evokes a graded response in the less well-recovered myocardium bordering the zone of direct excitation. The graded response suppresses the delayed propagation in two ways: (1) by minimizing the membrane potential difference between the zones of direct excitation and graded response and (2) by prolonging the refractoriness of the myocardium bordering the directly excited region. Thus the myocardium directly excited by the shock is unable to evoke a propagating impulse. Supporting evidence for the first proposed process came from a microelectrode study of field-stimulated rabbit papillary muscle that showed that there is little variation in membrane potential present after a sufficiently strong shock because of the production of a graded response (151). The second piece of evidence came from numerical simulations that show that a shock delivered in the repolarizing wake of an impulse will give rise to a propagating impulse if the shock is weak because of a strong difference in membrane potential between the adjacent zones of excited and unexcited myocardium (152). Stronger shocks eliminate the large difference in membrane potential and with it the flow of depolarizing current and the ability to launch an impulse.

According to the Upper Limit of Vulnerability hypothesis, the means by which fibrillation is precipitated in defibrillation or VF induction should be closely related, if not identical. This question had already been addressed by the theoretical work of Winfree, who showed that it is possible for a critically timed stimulus to induce reentrant excitation in myocardium (40). An electrical mapping study of VF induction not only demonstrated shock-induced reentrant circuits, but also the expected dependence on shock timing (62). A unique electrical mapping study pulled together the conceptual threads of this hypothesis in a quantitative manner embodied in the critical point mechanism for VF initiation (Fig. 19.9) (61). The critical point mechanism is an elaboration on the rotor hypothesis of Winfree described above. The critical point itself describes where a reentrant circuit arises in response to a shock. It is a consequence of the fact that neither the recovery of excitability nor the effective shock strength is uniformly distributed throughout the myocardium but rather exhibits smooth gradients. A spatial gradient in refractoriness is created after the passage of a wavefront and it approximately mirrors the differences in activation time. A spatial gradient in shock field strength (i.e., a gradient in the voltage gradient) occurs because shock electrodes do not create a uniform flow of current through the ventricle. The gradients in refractoriness and shock strength are independent of each other and, for a given electrode configuration, will depend upon the timing and strength of the shock. A critical point is formed where: (1) these gradients intersect in a nonparallel fashion; and (2) some point on this intersection has a critical degree of refractoriness and a critical shock strength. Recall that the electrophysiological response of the myocardium to a shock depends on the local effective strength of the shock and the degree of refractoriness. Either direct excitation, a nonpropagating graded response, or no effect will occur. If the gradients in refractoriness and shock strength intersect as described, the shock will create a stereotypical arrangement of these electrophysiological responses in the myocardium arranged around the critical point. The shock will directly excite the most recovered myocardium and will have no effect where the myocardium is absolutely refractory. These two zones will be adjacent to each other. After the shock, a propagating impulse will form at the border between the directly excited myocardium and the zone of no effect. In a thin zone of relative refractoriness and high shock strength, the shock elicits a nonpropagating graded response. The critical point is where the graded response zone meets the border between the zones of direct excitation and no effect as one moves in the direction of decreasing shock strength. It forms the pivot point for the wavefront arising from the zone of direct excitation. By the time this wavefront pivots around the critical point and enters the graded response zone, the myocardium in this area will have recovered excitability and so support passage of the impulse. In this manner, the critical point becomes the center of the shock-induced reentrant circuit. Since such a reentrant circuit requires a gradient in shock strength that spans the critical shock strength, wavefront propagation will not occur if all points on the ventricle are exposed to a shock voltage gradient equal to or exceeding the

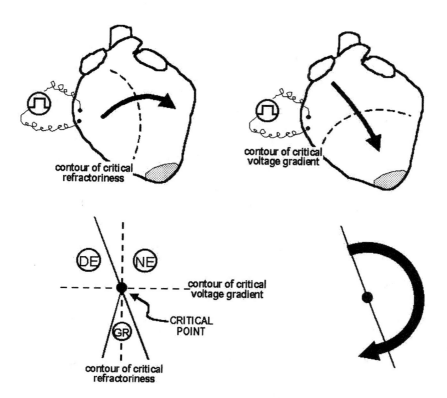

Figure 19.9 Diagram of the critical point mechanism for VF induction by a shock. The top drawings use arrows to depict the direction of the spatial gradients in myocardial refractoriness created by the passage of a wavefront elicited by a pacing stimulus (left) and the drop-off in shock field strength with distance from a shock electrode (right). The shaded area of apex represents the shock electrode. The dashed lines indicate where the myocardium experiences the critical degree of refractoriness and voltage gradient. The bottom drawings indicate the electrophysiological responses arising in the area around the critical point in response to a shock (left) and the reentrant pattern of excitation which follows (right). See text for explanation (DE: direct excitation; GR: graded response; NE: no effect).

critical shock voltage gradient. This critical shock voltage gradient is essentially the threshold local field shock field strength first recognized in the electrical mapping studies of defibrillation. As long as a defibrillation shock is strong enough to expose all of the ventricle to local shock strength of at least the critical voltage gradient, it will not be possible to form critical point reentrant circuits. Thus refibrillation cannot occur and defibrillation will succeed. Further, the critical point concept can explain the probabilistic nature of defibrillation if one considers that the frequency of the required arrangements and orientation of the gradients in myocardial refractoriness will be a function of the essentially random processes governing VF.

B. Failed Defibrillation Due to Insufficient Wavefront Termination

Electrical mapping provided new information in unprecedented quantity and detail about the defibrillation process. However, electrical recording is unavoidably interrupted during and shortly following the application of a shock due to electrical overload. Also, electrogram recordings, particularly bipolar recordings, only register activity when wavefront propagation occurs and so cannot sense stationary electrical activation. Last, slowly propagating wavefronts or wavefronts mediated by small membrane depolarizations are harder for electrical mapping systems to detect than otherwise, especially if the ability to sense low-frequency signals has been reduced in order to dampen postshock electrode polarization. For these reasons, Dillon and Kwaku sought to apply the technique of optical activation mapping to the study of how electrical shocks affect VF wavefront propagation (153). Dillon had previously applied optical recording to a study of the electrophysiological effects of shocks because this technique is inherently immune to overload by high-voltage shocks (154,155). Optical recording uses a voltage-sensitive dye (156) to monitor the changes in membrane potential in myocardial tissue (157,158). Because of this ability, optical recording cannot only be used to map the spread of the cardiac impulse, but also to track the phases of action potential and the membrane potential changes caused by the shocks and their subsequent electrophysiological responses. Although recently developed (159–161), optical mapping is not easily adaptable to in vivo use and so defibrillation was almost wholly studied in isolated rabbit hearts whose mechanical activity was pharmacologically arrested. However, to date all of these data have paralleled those taken from whole animals, and there are no indications as yet of a model-dependent discrepancy in results.

An optical mapping study of defibrillation in isolated rabbit hearts revealed a significantly different picture from that envisioned by the Upper Limit of Vulnerability hypothesis (153). Optical recordings were made at 100 epicardial sites on the epicardium during electrically induced, nonischemic VF. The shock was asynchronous with respect to the wavefront dynamics underway within the field of view of the recording system. On occasion, a VF wavefront was in view at the time of the shock. In these cases, the shock promptly depolarized (i.e., stimulated) the myocardium ahead of the advancing wavefront. Thereafter, there was only one of two possible outcomes: a wavefront either seamlessly propagated away from the zone of shock-depolarized myocardium or no propagation occurred. The difference with respect to the Upper Limit of Vulnerability hypothesis was that delayed wavefront evolution following wavefront termination was never seen. In cases where propagated activity followed the shock, there was no spatial or temporal discontinuity between the shock-induced depolarization and the subsequent regenerative, membrane-based depolarization processes. Whenever a wavefront failed to propagate away from the shock-depolarized area, there was a brief period of electrical inactivity during which the myocardium repolarized. The next membrane depolarizations were almost always due to wavefronts that propagated inward from the perimeter. These first sites of postshock activation could be remote from the area depolarized by the shock. The remaining minority of cases showed the first postshock activation arising wholly within the recording area. There were two possible sources for these types of activation patterns. The first is that wavefronts originated at deeper layers, (e.g., the endocardium) and propagated toward the epicardium. The second is that these wavefronts originated from epicardial sources of focal activation. An early after-depolarization scenario was not possible because the postshock wavefronts did not arise from depolarized levels. Since no diastolic depolarization was evident, an automatic focus or late after-depolarization source was also rejected. The simplest conclusion is that all postshock wavefronts arose as a result of propagation away from areas that were depolarized by the shock. When epicardial activation did not occur immediately after the shock, subsequent activity, if present, could be attributed to a wavefront that was launched by a shock in an area of myocardium not under view by the mapping system. These results were wholly consistent with the traditional view of defibrillation: shocks either stopped VF wavefronts or they did not. For the rabbit heart at least, the Upper Limit of Vulnerability hypothesis did not accurately describe defibrillation.

Except for the crucial discrepancy described above, the optical mapping study yielded a picture that was largely in keeping with the Upper Limit of Vulnerability hypothesis. For instance, except for a shock too weak to have any electrophysiological effects, postshock wavefronts were not continuations of the preshock wavefronts but were rather due to the shock itself. Therefore, VF wavefront termination amounts to being able to prevent the shock from creating new wavefronts. The optical mapping data showed that shocks were able to depolarize (i.e., stimulate or excite) myocardium and that this depolarization either did or did not propagate. If the shock-depolarized region bordered myocardium that was not effectively refractory, then depolarization may propagate away. However, if the shock-depolarized area bordered effectively refractory myocardium, then, by definition, no postshock propagation is possible. The shock-depolarized myocardium would be surrounded by tissue incapable to sustaining regenerative activation on its own. This idea was tested by using the membrane potential as a proxy for myocardial refractoriness. Since optical recording cannot yield actual millivolt values, the optical signal was scaled so that resting potential was represented by 0% depolarization and the maximum amplitude of a paced action potential as 100% depolarization. Thus, higher levels of membrane potential were correlated with higher degrees of refractoriness. The border between the myocardium that was directly excited by the shock and the myocardium into which shock-evoked depolarization could propagate was determined. If excessive refractoriness prevented wavefront propagation away from the border of the shock-depolarized area, then the likelihood of postshock propagation should be correlated with the degree of refractoriness at this border. This was shown by plotting the percentage of times that a wavefront propagated away from a site on the border versus the membrane potential at that site at the time of the shock (Fig. 19.10). This plot showed a decreasing sigmoidal shaped relationship between propagation likelihood and optical membrane potential. For low levels of optical membrane potential, and hence the least degree of refractoriness, the likelihood of postshock propagation was constant and at its highest level. The likelihood of postshock propagation then smoothly decreased as the optical membrane potential at the border sites increased. Eventually an optical membrane potential was reached in which propagation fell to zero. Data points corresponding to progressively higher optical membrane potentials also showed a 0% propagation incidence. This limiting potential could be explained by the onset of effective refractoriness in the myocardium bordering the shock-depolarized area.

Does it make sense to speak of depolarizing or stimulating myocardium that is effectively refractory? Nothing rules out the possibility of a shock producing a stationary depolarization of the membrane. The key difference between these two types of depolarization is that while myocardium can become refractory to cellular sources of depolarization, it is never refractory to a sufficiently strong electrical shock. It had been dogma that, following excitation, myocardium is refractory to further stimulation for a finite period because of membrane-voltage-dependent inactivation of the sodium channel. However, Hoffman indicated long ago that myocardium is never absolutely refractory to the stimulating effects of a sufficiently strong and/or prolonged anodal stimulus (162). This type of stimulation is not typically encountered in clinical or experimental practice and there are no naturally occurring sources of anodal stimulation in the heart. However, it was shown that shocks strong enough to defibrillate the rabbit ventricle were also capable of eliciting a second phase of depolarization out of already depolarized myocardium (154,155). It was subsequently shown that this ability of a shock to stimulate refractory myocardium was de-

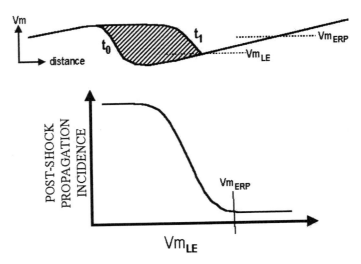

Figure 19.10 Dependence of postshock propagation incidence on membrane potential. The top drawing shows a hypothetical membrane potential profile along the path of VF wavefront immediately before (t_0) and after (t_1) a shock. Application of a shock depolarizes some of the excitable myocardium (shaded) ahead of the wavefront up to level Vm_{LE} along the trailing edge of the preceding wavefront. It is hypothetically possible for propagation to continue away from the shock-depolarized area as long as the leading edge membrane voltage (Vm_{LE}) is less than that associated with effective refractoriness (Vm_{ERP}). The bottom plot schematically reproduces the experimental relationship found between postshock propagation incidence versus the leading edge membrane voltage. It shows propagation falling to 0% when the shock depolarizes the myocardium up to a threshold potential which, by definition, demarcates the onset of effective refractoriness.

pendent on the strength of the shock and the degree of refractoriness (131). If the myocardium had just recently been excited, then a strong shock was needed to elicit a second phase of regenerative depolarization. As the myocardium was allowed to repolarize, it was found that progressively weaker shocks were able to depolarize it. Fully repolarized myocardium could be stimulated by very weak shocks and in this limit the shock actions were indistinguishable from those due to ordinary pacing. Of course, such weak shocks would have little, if any, effect on relatively repolarized myocardium and no effect on effectively refractory myocardium (Fig. 19.11, bottom row). Therefore, stimulation by a shock is not a qualitatively different process than ordinary pacing, but the unfamiliar end of the spectrum of the general process of electrical stimulation (163). Only with the advent of optical recording did it become possible to recognize that myocardium is never absolutely refractory to a sufficiently strong stimulus. At the time it was hypothesized that the shock circumvented the voltage-dependent inactivation of the sodium channel through its ability to cause a bipolar change in membrane potential on a microscopic scale. A discussion of the details of this process is beyond the scope of this chapter. Whatever its mechanism, the ability of a shock to stimulate myocardium any point in the cardiac cycle in a strength- and refractoriness-dependent manner can be used to predict how to prevent a shock from launching a propagating wavefront (131,151).

Another consequence of a shock's ability to regeneratively depolarize myocardium is to delay the ultimate repolarization of the stimulated myocardium if the shock is applied during an action potential already in progress. Dillon (154,155) and others (151,164–166) showed that defibrillation shocks are able to delay repolarization in a shock strength- and

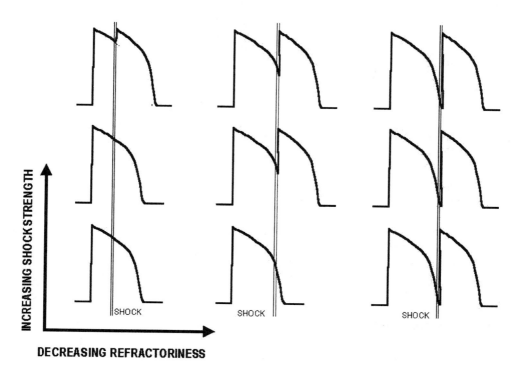

Figure 19.11 Electrophysiological effects of shocks as portrayed by optical recordings. The drawing schematically shows the a range of possible membrane voltage responses that can be evoked by shocks depending on the shock strength and the degree of membrane refractoriness. It has been experimentally demonstrated that the myocardium is not absolutely refractory to excitation of a depolarizing response by a sufficiently strong shock (upper left drawing).

timing-dependent manner (Fig. 19.11). Shocks applied late during the action potential (e.g., during terminal repolarization) were shown to produce a range of shock strength-dependent responses ranging from no effect, to a "bump," to an action potential whose amplitude and duration progressively increased to maximal values. An upper limit on the amount of additional depolarization time produced by a shock is set by the duration of the full amplitude action potential evoked by a sufficiently strong shock. Thus a shock of sufficient amplitude can markedly prolong the time that the myocardium remains depolarized. By prolonging refractoriness in this way, the shock my abet defibrillation by suppressing propagated activity in the ventricle. Another potentially more powerful antiarrhythmic effect is the shock's ability to homogenize repolarization times. Optical recordings made from a fixed site during consecutive defibrillation episodes using a fixed shock strength have experimentally verified this expectation (167). That study, performed in the isolated rabbit heart, showed in fact that there was a constant repolarization time following the shock regardless of the membrane potential at the time of the shock (Fig. 19.12, left). Later, in vivo optical mapping of canine defibrillation showed that defibrillation shocks were able to simultaneously induce action potentials that caused near simultaneous repolarization within an extended area of the epicardium (161). In such an instance, it could be imagined that a postshock wavefront would either be uniformly conducted or uniformly blocked in the myocardium that formerly supported the hypothetical reentrant circuit. Even shocks whose field strength falls short of that required to evoke a full action potential

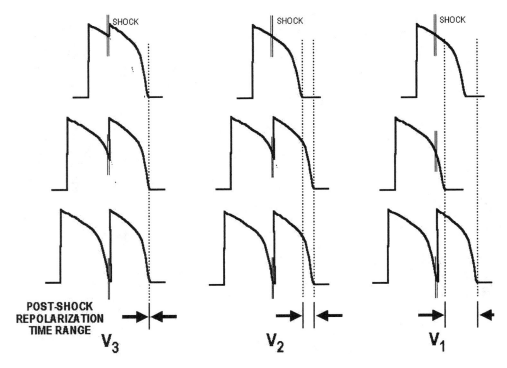

Figure 19.12   Progressive homogenization of postshock repolarization times by progressively stronger shocks. The hypothetical membrane voltage responses to shocks shown in Figure 19.11 are rearranged and time aligned with respect to the shock to illustrate disparities in postshock repolarization times. Three adjacent areas of myocardium are exposed to the same shock field strength but are considered to reside at three levels of refractoriness. The dotted vertical lines indicate the earliest and latest post-shock repolarization times in these three regions. The strongest shock V_3 is followed by near-simultaneous postshock repolarization as indicated by the dotted line. The weakest shock V_1 shows the widest disparity in postshock repolarization times.

from all points during the VF action potential are able to mitigate the spatially inhomogeneous patterns of refractoriness created by VF (Fig. 19.12, middle). This strength-dependent homogenization of postshock repolarization time was demonstrated by Knisley (151) by electrically recording membrane responses elicited in paced frog myocardium by shocks having various strengths and timings. Similarly, optical mapping studies of the responses of fibrillating rabbit hearts to various shock strengths has also shown a progressive homogenization of repolarization times in response to progressively stronger shocks (168).

Taken together, the optical recording data suggest that a single process is involved in defibrillation, namely, the ability of shock to depolarize (i.e., to stimulate) the ventricle. If a shock is strong enough to depolarize all of the excitable myocardium in the ventricle, then it will then fail to launch propagating wavefronts. This should be sufficient to guarantee defibrillation since all propagating activity will come to a halt. Weaker shocks that do not depolarize all of the excitable myocardium may or may not defibrillate the heart, depending upon whether these shocks elicit propagating wavefronts and whether these wavefronts have the capacity to continue reentrant propagation. It is known that the probability of successful defibrillation versus shock strength is a sigmoid-shaped curve that has a smoothly rising segment between the shock strengths associated with 0% and 100% defibrillation likelihood. This regime of shock outcome could be explained by the electrophysiological consequences of the shock being an unpredictable function of the effectively random disposition of the ventricle at the time of the shock. Whereas for a given shock strength the distribution of shock field strength is fixed, the distribution of refractory states in the ventricle, a factor that also strongly influences the ability of the shock to depolarize the myocardium, will vary in an unpredictable way from shock episode to shock episode. Depending upon chance, a shock may generate greater or fewer numbers of propagating wavefronts. The shock may prolong and homogenize postshock refractoriness to differing degrees because of the unpredictability of refractory state in the ventricle at the time of the shock. Therefore, if the ultimate ability of the ventricle to continue VF depends on the prevalence of wavefronts and their likelihood to continue arrhythmic conduction, then the outcome of any given shock will be unpredictable. On average, however, as the shock strength is increased, the likelihood of continued fibrillation will decrease because on average fewer wavefronts will be launched by the shock and these wavefronts will on average find an increasingly larger fraction of the ventricle undergoing increasingly prolonged and increasingly more homogeneous refractoriness (Fig. 19.13). This explanation of defibrillation as well as the related process of VF induction is given in the Progressive Depolarization hypothesis (169).

C. Defibrillation Mechanisms Based on Shock-Induced Prolongation of Refractoriness

Contemporaneous with the optical studies, there were a series of other studies that focused on the role of shock-induced prolongation of refractoriness. In this work, Sweeney (170,171) used standard electrophysiological methods to deduce that shocks could delay the recovery of excitability in the in vivo canine heart. It was found that shocks applied

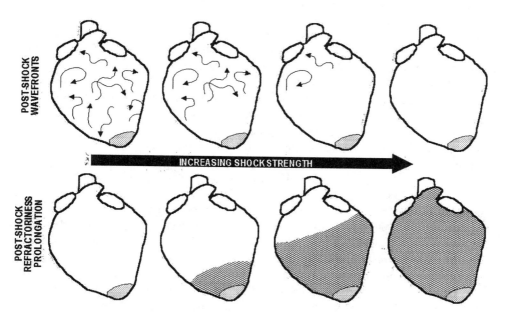

Figure 19.13 Postshock wavefront termination and refractoriness prolongation in a progressively larger fraction of the ventricle following shocks of progressively increasing strength. (See text for explanation.)

when the heart had just been depolarized by stimulation did not affect the effective refractory period (170). The ventricle remained insensitive to shocks delivered later in time up until about 70 ms before the lapse of the effective refractory period. Shocks applied with this timing or later caused a progressive lengthening of the effective refractory period. It was hypothesized that the shock effectively acted as a strong pacing stimulus for the entire ventricle. It was thought that shocks were able to evoke graded responses that prolonged the time that the myocardium remained at least partially depolarized, which in turn delayed the ultimate return of excitability. Subsequent studies using these same methods established that shocks could prolong refractoriness under conditions more closely resembling those of VF (171,172). These basic findings were elaborated into the Refractory Period Extension hypothesis (170,171,173), which assumes that a fibrillating ventricle will have three possible responses to a shock: (1) no effect in recently depolarized myocardium; (2) refractoriness extension in effectively refractory myocardium; and (3) direct depolarization of well-recovered myocardium. This process is largely the same as described previously in the Upper Limit of Vulnerability. The consequences of this mechanism are that both successful and unsuccessful shocks terminate VF wavefronts and that unsuccessful defibrillation is due to the delayed generation of new VF wavefronts owing to the action of the shock. Another aspect of this hypothesis concerns the ability of myocardium that was unaffected by the shock to "remember" the preshock impulse pathway by continuing to undergo spatially nonuniform recovery of excitability. This means that a postshock impulse could be shepherded into an erratic conduction path. Strong shocks will diminish this possibility by extending the refractoriness of effectively refractory myocardium and directly activating the excitable segments. Since the most refractory myocardium would continue to repolarize as before, a strong shock will therefore cause all segments to repolarize together after about 1 VF cycle length. Data consistent with this hypothesis were found in a study that used a high-current stimulus applied after a defibrillation shock to see if the outcome was altered (174). Defibrillation occurs when a shock is able to homogenize and delay repolarization times by extending the refractoriness of effectively refractory myocardium and by directly activating excitable myocardium. More importantly, however, is that such shocks eliminate the spatial gradients in membrane potential that are otherwise created by a shock that is too weak to prolong refractoriness (174). These gradients in membrane potential were postulated to give rise to delayed impulse formation due to the passive flow of depolarizing current from activated regions into adjacent myocardium that has repolarized sufficiently to become excitable once again.

The Extension of Refractoriness hypothesis, another related viewpoint, also envisions a primary role for the shock-induced prolongation of refractoriness in causing defibrillation. It emphasizes the need to homogeneously prolong the extra degree of refractoriness induced by the shock. Swartz and Jones (166) had earlier shown that simulations of defibrillation shocks by means of current injection into chick myocyte aggregates caused timing- and strength-dependent delays in repolarization. Later, Tovar and Jones (175) showed that similar results could be obtained using external stimulation of myocyte aggregates. They knew that defibrillation failed because postshock wavefronts arose in areas of the ventricle exposed to low-shock field strengths. Therefore, they hypothesized that shocks must prolong refractoriness in the low field strength regions of the ventricle in order to block the wavefronts that were not immediately terminated by the shock. To test this, they recorded action potentials from an isolated, perfused rabbit heart in an area that received the lowest shock field strength (176). An overall correlation between the extra refractoriness induced by a shock applied during VF and the likelihood of that shock defibrillating was found. These data were taken to support the idea that successful defibrillation is caused by prolonging refractoriness in the low field strength region of the ventricle. The Extension of Refractoriness hypothesis also emphasized the importance of a uniform prolongation of refractoriness with regard to low-versus high-shock field strength regions of the ventricle. This concept was held to explain the superiority of biphasic shock waveforms over monophasic waveforms that prolonged refractoriness to a greater, but not necessarily more uniform, extent (175). Like the Critical Mass hypothesis, the Extension of Refractoriness hypothesis considered failed defibrillation to be due to a failure to terminate a sufficient number of wavefronts.

D. Failed Defibrillation Due to Accelerated Repolarization

This last section describes a markedly different concept in which a shock can terminate or shorten an action potential. Rather than acting as a stimulus, the shock either raises or lowers the cardiac membrane through the direct action of passive transmembrane current flow, as Efimov demonstrated by optically mapping a paced rabbit heart (177). Shocks applied during the action potential plateau evoked a complex, but consistent, pattern of strong depolarizing and hyperpolarizing membrane potential changes in adjacent regions of myocardium over a length scale of several millimeters. By shortening the action potential through hyperpolarizing effects, a shock could act against defibrillation by restoring the excitability of the myocardium more quickly than otherwise and so condition the ventricle to support postshock wavefronts. Efimov, however, described a more direct means by which the hyperpolarizing effects of the shock led to the formation of reentrant circuits (178). Because of the stereotyped patterns of hyperpolarization and depolarization created around the endocardial electrode, a wavefront has the ability to propagate in a reentrant pathway back to its point of origin and complete one or more circuits. Thus, in a manner significantly different from the critical point mechanism described by the Upper Limit of Vulnerability hypothesis, a shock can immediately induce a closed-loop reentrant circuit. Therefore, a shock could thwart defibrillation by creating a reentrant source of arrhythmic impulse formation called a Virtual Electrode Induced Phase Singularity.

REFERENCES

1. Myerburg RJ, Kessler KM, Kimura S, Bassett AL, Cox MM, Castellanos A. Life-threatening ventricular arrhythmias: the link between epidemiology and pathophysiology. In: Zipes DP, J Jalife, eds. Philadelphia: W.B. Saunders, 1995:723–731.
2. Wolf PA, Dawber TR, Thomas EJ, Kannel WB. Epidemiologic assement of chronic atrial fibrillation and risk of stroke: the Framingham study. Neurology 1978; 28:973–977.
3. McWilliam JA. Fibrillar contraction of the heart. J Physiol 1887; 8:296–310.
4. Porter WT. The recovery of the heart from fibrillatory contractions. Am J Physiol 1898; 1:71–82.
5. Garrey WE. The nature of fibrillary contraction of the heart. Its relation to tissue mass and form. Am J Physiol 1914; 33:397–414.
6. Mines GR. On circulating excitations in heart muscles and their possible relations to tachycardia and fibrillation. Trans R Soc Canada (Biol) 1914; 8:43–53.
7. Mayer AG. Rhythmical pulsation in scyphomedusae. Carnegie Inst Wash Publ 1906; 47.
8. Wiggers CJ. Stusies of ventricular fibrillation caused by eletric shock: Cinematographic and electrocardiographic observations of the natural process in the dog's heart: Its inhibition by potassium and the revival of coordinated beats by calcium. Am Heart J 1930; 5:351–365.
9. Moe GK, Abildskov JA. Experimental and laboratory reports. Am Heart J 1959; 58:59–70.
10. Moe GK. On the multiple wavelet hypothesis of atrial fibrillation. Arch Int Pharmacodyn 1962; 140:183–188.

11. Moe GK, Rheinboldt WC, Abildskov JA. A computer model of atrial fibrillation. Am J Physiol 1964; 67:200–220.

12. Allessie MA, Bonke FIM, Schopman FJG. Circus movement in rabbit atrial muscle as a mechanism of tachycardia. Circ Res 1973; 333:54–62.

13. Allessie MA, Bonke FIM, Schopman FJG. Circus movement in rabbit atrial muscle as a mechanism of tachycardia. III. The "leading circle" concept: a new model of circus movement in cardiac tissue without the involvement of an anatomical obstacle. Circ Res 1977; 41:9–18.

14. Janse MJ, van Cappelle FJL, Morsink H, Kleber AG, Wilms-Schopman F, Cardinal R, d'Alnoncourt CN, Durrer D. Flow of "injury" current and patterns of excitation during early ventricular arrhythmias in acute regional myocardial ischemia in isolated porcine and canine hearts. Circ Res 1980; 47:151–165.

15. Ideker RE, Klein GJ, Harrison L, Smith WM, Kasell J, Reimer KA, Wallace AG, Gallagher JJ. The transition to ventricular fibrillation induced by reperfusion after acute ischemia in the dog: a period of organized epicardial activation. Circ 1981; 63(6):1371–1379.

16. El-Sherif N, Smith RA, Evans K. Canine ventricular arrhythmias in the late myocardial infarction period. 8. Epicardial mapping of reentrant circuits. Circ Res 1981; 49:255–265.

17. Wit AL, Allessie MA, Bonke FM, Lammers W, Smeets J, Fenoglio JJ, Jr. Electrophysiological mapping to determine the mechanisms of experimental ventricular tachycardia initiated by premature impulses Experimental approach and initial results demonstrating reentrant excitation. Am J Cardiol 1982; 49:166–185.

18. El-Sherif N, Mehra R, Gough WB, Zeiler RH. Ventricular activation patterns of spontaneous and induced ventricular rhythms in canine one-day-old myocardial infarction. Evidence for focal and reentrant mechanisms. Circ Res 1982; 51:152–166.

19. Downar E, Parson ID, Mickleborough LL, Camerson DA, Yao LC, Waxman MB. On-line epicardial mapping of intraoperative ventricular arrhythmias: Initial clinical experience. J Am Coll Cardiol 1984; 4:703–714.

20. Chen PS, Wolf PD, Dixon EG, Danieley ND, Frazier DW, Smith WM, Ideker RE. Mechanism of ventricular vulnerability to single premature stimuli in open-chest dogs. Circ Res 1988; 62:1191–1209.

21. Allessie MA, Lammers WJEP, Bonke FIM, Hollen J. Experimental evaluation of Moe's multiple wavelet hypothesis of atrial fibrillation. In: Zipes DP, Jalife J, eds. Cardiac Electrophysiology and Arrhythmias. Orlando: Grune & Stratton., 1985:265–275.

22. Witkowski FX, Penkoske PA. Activation patterns during ventricular fibrillation. Ann NY Acad Sci 1990; 591:219–231.

23. Lee JJ, Kamjoo K, Hough D, Hwang C, Fan W, Fishbein MC, Bonometti C, Ikeda T, Karagueuzian HS, Chen PS. Reentrant wave fronts in Wiggers' stage II ventricular fibrillation. Characteristics and mechanisms of termination and spontaneous regeneration. Circ Res 1996; 78(4):660–675.

24. Rogers JM, Huang J, Smith WM, Ideker RE. Incidence, evolution and spatial distribution of functional reentry during ventricular fibrillation in pigs. Circ Res 1999; 84:945–954.

25. Gray RA, Jalife J, Panfilov AV, Baxter WT, Cabo C, Davidenko JM, Pertsov AM. Mechanisms of cardiac fibrillation: Drifting rotors as a mechanism of cardiac fibrillation. Science 1995; 270:1222–1223.

26. Gray RA, Pertsov AM, Jalife J. Incomplete reentry and epicardial breakthrough patterns during atrial fibrillation in the sheep heart. Circulation 1996; 94(10):2649–2661.

27. Bove RT, Dillon SM. Optically imaging cardiac activation with a laser system. IEEE EMB 1998; 17:84–94.

28. Witkowski FX, Leon LJ, Penkoske PA, Clark RB, Spano ML, Ditto W, Giles WR. A method for visualization of ventricular fibrillation. Chaos 1998; 8:942–102.

29. Goldberger AL, Bhargava V. Some observations on the question: Is ventricular fibrillation "chaos"? Phys D 1986; 19:282–289.

30. Kaplan DT, Cohen RJ. Is fibrillation chaos? Circ Res 1990; 67:862–886.

31. Damle RS, Kanaan NM, Robinson NS, Ge Y, Goldberger JJ, Kadish AH. Spatial and temporal linking of epicardial activation directions during ventricular fibrillation in dogs: evidence for underlying organization. Circulation 1992; 86:1547–1558.

32. Witkowski FX, Kavanagh KM, Penkoske PA, Plonsey R, Spano ML, Ditto WL, Kaplan DT. Evidence for determinism in ventricular fibrillation. Phys Rev Lett 1995; 75:1230–1233.

33. Garfinkel A, Chen PS, Walter DO, Karagueuzian HS, Kogan B, Evans SJ, Karpoukhin M, Hwang C, Uchida T, Gotoh M, Nwasokwa O, Sager P, Weiss JN. Quasiperiodicity and chaos in cardiac fibrillation [see comments]. J Clin Invest 1997; 99(2):305–314.

34. Wiener N, Rosenblueth A. The mathematical formulation of the problem of conduction of impusles in a network of connected excitable elements, specifically in cardiac muscle. Arch Inst Cardiol Mex 1946; 16:1–13.

35. Selfridge O. Studies on flutter and fibrillation: some notes on the theory of flutter. Arch Inst Cardiol Mexico 1948; 18:177–187.

36. Zykov VS. Simulation of Wave Processes in Excitable Media. Manchester: Manchester University Press, 1987.

37. Tyson JJ, Keener JP. Singular perturbation theory of traveling waves in excitable media. Phys Rev D 1988; 32:327–361.

38. Krinsky VI. Mathematical models of cardiac arrhythmias (spiral waves). Pharm Ther B 1978; 3:539–551.

39. Pertsov AM, Ermakova EA, Panfilov AV. Rotating spiral waves in a modified Fitz-Hugh-Nagumo model. Phys Rev D 1984; 14:117–124.

40. Winfree AT. When Time Breaks Down. The Three-Dimensional Dynamics of Electrochemical Waves and Cardiac Arrhythmias. Princeton: Princeton University Press, 1987.

41. Jalife J, Davidenko JM, Michaels DC. A new perspective on the mechanisms of arrhythmias and sudden cardiac death: spiral waves of excitation in heart muscle. J Cardiovasc Electrophysiol 1991; 2 (suppl 3):S133–S152.

42. Davidenko JM. Spiral wave activity: A possible common mechanism for polymorphic and monomorphic ventricular tachycardias. J Cardiovasc Electrophysiol 1993; 4:730–746.

43. Gray RA, Jalife J. Spiral waves and the heart. Int J Bifurc Chaos 1996; 6:415–435.

44. Zykov VS. Analytical evaluation of the dependence of the speed of an excitation wave in a teo-dimensional excitable medium on the curvature of its wavefront. Biophysics 1980; 25:906–911.

45. Cabo C, Pertsov AM, Baxter WT, Davidenko JM, Gray RA, Jalife J. Wave-front curvature as a cause of slow conduction and block in isolated cardiac muscle. Circ Res 1994; 75(6):1014–1028.

46. Fast VG, Kleber AG. Role of wavefront curvature in propagation of cardiac impulse. Cardiovasc Res 1997; 33:258–271.

47. Pertsov AM, Medvinskii AB. Experimental evaluation of the critical size of a reverberator in the myocardium. Biophysics 1986; 31:752–756.

48. Ikeda T, Yashima M, Uchida T, Hough D, Fishbein MC, Mandel WJ, Chen PS, Karaguezian HS. Attachment of meandering reentrant wave fronts to anatomic obstacles in the atrium. Role of the obstacle size. Circ Res 1997; 81:753–764.

49. Davidenko JM, Pertsov AM, Salomonz R, Baxter W, Jalife J. Stationary and drifting spiral waves of excitation in isolated cardiac muscle. Nature 1992; 355:349–351.

50. Pertsov AM, Davidenko JM, Salomonsz R, Baxter WT, Jalife J. Spiral waves of excitation underlie reentrant activity in isolated cardiac muscle. Circ Res 1993; 72:631–650.

51. Ikeda T, Uchida T, Hough D, Lee JJ, Fishbein MC, Mandel WJ, Chen P, Karagueuzian HS. Mechanism of spontaneous termination of functional reentry in isolated canine right atrium. Circulation 1996; 94:1962–1973.

52. Pertsov AM, Jalife J. Three-dimensional vortex-like reentry. In: Zipes DP, Jalife J, eds. Cardiac Electrophysiology: From Cell to Bedside. Philadelphia: W.B. Saunders, 1995:403–410.

53. Keener JP, Tyson JJ. The dynamics of scroll waves in excitable media. SIAM Rev 1992; 34:1–39.

54. Winfree AT. Stable particle-like solutions to the nonlinear wave equations of three-dimensional excitable media. SIAM Rev 1990; 32:1–53.

55. Winfree AT. Scroll-shaped waves of chemical activity in three dimensions. Science 1973; 181:937–939.

56. Gomatam J, Hodson DA. The eikonal equation: stability of reaction-diffusion waves on a sphere. Phys Rev D 1991; 49:82–89.

57. Pertsov AM, Aliev RR, Krinsky VI. Three-dimensional twisted vortices in an excitable media. Nature 1990; 345:419–421.

58. Pertsov AM, Vinson M, Muller SC. Three-dimensional reconstruction of organizing centers in excitable chemical media. Phys Rev D 1993; 63:233–240.

59. Mironov S, Vinson M, Mulvey S, Pertsov A. Destabilization of three-dimensional rotating chemical waves in an inhomogeneous BZ reaction. J Phys Chem 1996; 100:1975–1983.

60. Vinson M. Interactions of spiral waves in inhomogeneous excitable media. Phys Rev D 1997; 116:313–324.

61. Frazier DW, Wolf PD, Wharton JM, Tang AS, Smith WM, Ideker RE. Stimulus-induced critical point. Mechanism for electrical initiation of reentry in normal canine myocardium. J Clin Invest 1989; 83:1039–1052.

62. Shibata N, Chen PS, Dixon EG, Wolf PD, Danieley ND, Smith WM, Ideker RE. Influence of shock strength and timing on induction of ventricular arrhythmias in dogs. Am J Physiol 1988; 255:H891–H901.

63. Winfree AT. Electrical instability in cardiac muscle: Phase singularities and ritors. J Theoret Biol 1989; 138:353–371.

64. Gray RA, Jalife J, Panfilov A, Baxter WT, Cabo C, Davidenko JM, Pertsov AM. Nonstationary vortexlike reentrant activity as a mechanism of polymorphic ventricular tachycardia in the isolated rabbit heart. Circulation 1995; 91:2454–2469.

65. Hillsley RE, Bollacker KD, Simpson EV, Rollins DL, Yarger MD, Wolf PD, Smith WM, Ideker RE. Alteration of ventricular fibrillation by propranolol and isoproterenol detected by epicardial mapping with 506 electrodes. J Cardiovasc Electrophysiol 1995; 6(6):471–485.

66. Janse MJ, Wilms-Schopman FJ, Coronel R. Ventricular fibrillation is not always due to multiple wavelet reentry. J Cardiovasc Electrophysiol 1995; 6:512–521.

67. Fast VG, Pertsov AM. Drift of a vortex in the myocardium. Biophysics 1990; 35:489–494.

68. Kim DT, Kwan Y, Lee JJ, Ikeda T, Uchida T, Kamjoo K, Kim K, Ong JC, Athill CA, Wu TJ, Czer L, Karaguezian, HS, Chen PS. Patterns of spiral tip motion in cardiac tissues. Chaos 1998; 8:137–148.

69. Cha YM, Birgersdotter-Green U, Wolf PL, Peters BB, Chen PS. The mechanism of termination of reentrant activity in ventricular fibrillation. Circ Res 1994; 74:495–506.

70. Gray RA, Pertsov AM, Jalife J. Spatial and temporal organization during cardiac fibrillation. Nature 1998; 392:675–678.

71. Panfilov AV, Holden AV. Spatiotemporal irregularity in a two-dimensional model of cardiac tissue. Int J Bifurc Chaos 1991; 1:219–225.

72. Bar M, Eiswirth M. Turbulence due to spiral breakup in a continuous excitable medium. Physiol Rev 1993; 48:1635–1637.

73. Karma A. Spiral breakup in model equations of action potential propagation in cardiac tissue. Phys Rev Lett 1993; 71:1103–1106.

74. Winfree AT. Electrical turbulence in three-dimensional heart muscle. Science 1994; 266:1003–1006.

75. Gray RA, Jalife J. Ventricular fibrillation and atrial fibrillation are two different beasts. Chaos 1998; 8:65–78.

76. Winfree AT. The Geometry of Biological Time. New York: Springer-Verlag, 1980.

77. Muller SC, Plesser T, Hess B. The structure of the core of the spiral wave in the Belousoc-Zhabotinsky reagent. Science 1985; 230:661–663.

78. Suzuki R. Electrochemical neural model. Adv Biophys 1976; 9:115–156.
79. Coullet P, Frisch T, Gilli JM, Rica S. Excitability in inquid crystal. Chaos 1994; 4:485–489.
80. Gerisch G. Standienpezifische aggregationsmuster bei distyostelium discoideum. Wilhelm Roux Archiv Entwick Org 1965; 156:127–132.
81. Goldbeter A. Mechanism for oscillatory synthesis nof cAMP in Dictyostelium discoideum. Nature 1975; 253:540–542.
82. Gorelova NA, Bures J. Spiral waves of spreading depression in the isolated chicken retina. Neurology 1983; 14:353–360.
83. Lechleiter J, Girard S, Peralta E. Spiral calcium wave propagation and annihilation in Xenopus laevis oocytes. Science 1991; 252:123–125.
84. Mikhailov AS, Zykov VS. Kinematical theory of spiral waves in excitable media: Comparison with numerical simulations. Phys Rev D 1991; 52:379–397.
85. Meron E. The role of curvature and wavefront interactions in spiral wave dynamics. Phys Rev D 1991; 49:98–106.
86. Barkley D. Euclidean symmetry and the dynamics of rotating spiral waves. Phys Rev Lett 1994; 72:164–167.
87. Zykov VS. Cycloid circulation of spiral waves in an excitable medium. Biophysics 1986; 31:940–944.
88. Courtemanche M, Winfree AT. Re-entrant rotating waves in a beeler-reuter based model of two-dimensional cardiac electrical activity. Int J Biol Chem 1991; 1:431–444.
89. Krinsky VI, Efimov IR, Jalife J. Vortices with linear cores in excitable media. Proc R Soc Lond A 1992; 437:645–655.
90. Pertsov AM, Yermakova YA. Mechanism of drift of a helical wave in an inhomogeneous medium. Biophysics 1988; 33:364–369.
91. Rogers JM, McCulloch AD. Nonuniform muscle fiber orientation causes spiral wave drift in a finite element model of cardiac action potential propagation. J Cardiovasc Electrophysiol 1994; 5:496–509.
92. Yermakova YA, Pertsov AM. Interaction of rotating spiral waves with boundary. Biophysics 1986; 31:932–940.
93. Panfilov AV, Vasiev RR, Mushinsky AV. An intergral invariant for scroll rings in a reaction-diffusion system. Phys Rev D 1987; 36:181–188.
94. Panfilov AV, Pertsov AM. Vortex rings in a three-dimensional medium described by reaction-diffusion equations. Doklady Biophys 1984; 274:58–60.
95. Davydov VA, Zykov VS. Kinematics of spiral waves on nonuniformly curved surfaces. Physic Rev D 1991; 49:71–74.
96. Rogers JM, McCulloch AD. The effect of nonuniform anisotropy on wavefront stability in a finite element model of cardiac action potential propagation. Proceedings of the 14th annual international conference of the IEEE engineering in medicine and biology society, 1992.
97. Fishler MG, Thakor NV. A massively parallel computer model of propagation through a two-dimensional cardiac syncytium. PACE 1991; 14:1694–1699.
98. Starmer CF, Biktashev VN, Romashko DN, Stepanov MR, Makarova ON, Krinsky VI. Vulnerability in an excitable medium: analytical and numerical studies of initiating unidirectional propagation. Biophys J 1993; 65:1775–1787.
99. Courtemanche MA. Complex spiral wave dynamics in a spatially distributed ionic model of cardiac activity. Chaos 1996; 6:579–600.
100. Beaumont J, Davidenko N, Davidenko JM, Jalife J. Model study of vortex-like activity in cardiac muscle. PACE 1995; 18:934.
101. Leon LJ, Roberge FA. A model study of reentrant activity in 2D cardiac tissue. PACE 1995; 18:935.
102. Elharrar V, Surawicz B. Cycle length effect on restitution of action potential duration in dog cardiac fibers. Am J Physiol 1983; 244:H782–H792.
103. Guevara MR, Glass L. Phase locking, period-doubling bifurcations and irregular dynamics in periodically stimulated cardiac cells. Science 1981; 214:1350–1353.

104. Courtemanche MA, Glass L, Keener JP. Instabilities of a propagating pulse in a ring of excitable media. Phys Rev Lett 1993; 70:2182–2185.

105. Frame LH, Simson MB. Oscillations of conduction, action potential duration, and refractoriness: A mechanism for spontaneous termination of reentrant tachycardias. Circulation 1988; 78:1277–1287.

106. Koller ML, Riccio ML, Gilmour RF. Dynamic restitution of action potential auration during electrical alternans and ventricular fibrillation. Am J Physiol 1998; 275:H1635–H1642.

107. Spach MS, Miller WT, Geselowitz DB, Barr RC, Kootsey JM, Johnson EA. The discontinuous nature of propagation in normal canine cardiac muscle. Evidence for recurrent discontinuous of intracellular resistance that affect the membrane currents. Circ Res 1981; 48:39–54.

108. Spach MS, Heidlage JF. The stochastic nature of cardiac propagation at a microscopic level. Electrical description of myocardial architecture and its application to conduction. Circ Res 1995; 76(3):366–380.

109. Pertsov A. Scale of geometric structures responsible for discontinuous propagation in myocardial tissue. In: Spooner PM, Joyner RW, Jalife J, eds. Discontinuous conduction in the heart. Armonk, NY: Futura, 1997:273–293.

110. Cabo C, Pertsov AM, Davidenko JM, Baxter WT, Gray RA, Jalife J. Vortex shedding as a precursor of turbulent electrical activity in cardiac muscle. Biophys J 1996; 70(3):1105–1111.

111. Panfilov AV, Hogeweg P. Spiral wave breakup in a modified Fitzhugh-Nagumo model. Phys Lett A 1993; 176:295–299.

112. Starobin JM, Zilberter YI, Rusnak EM, Starmer CF. Wavelet formation in excitable cardiac tissue: The role of wavefront-obstacle interactions in initiating high-frequency fibrillatory-like arrhythmias. Biophys J 1996; 70:581–594.

113. Biktashev V, Holden A, Zhang H. Tension of organizing filaments of scroll waves. Phil Trans R Soc Lond A 1994; 347:611–630.

114. Henze C, Lugosi E, Winfree AT. Helical organizing centers in excitable media. Can J Phys 1990; 68:683–708.

115. Panfilov AV, Keener JP. Reentry in an anatomical model of the heart. Chaos, Solitons, Fractals 1995; 5:681–689.

116. Fenton F, Karma A. Vortex dynamics in three-dimensional continuous myocardium with fiber orientation: Filament instability and fibrillation. Chaos 1998; 8:20–47.

117. Leon LJ, Roberge FA. Directional characteristics of action potential propagation in cardiac muscle: A model study. Circ Res 1991; 69:378–395.

118. Shaw RM, Rudy Y. Electrophysiologic effects of acute myocardial ischemia. Circ Res 1997; 80:124–138.

119. Crampton R. Accepted, controversial, and speculative aspects of ventricular defibrillation. Prog Cardiovasc Dis 1980; 23:167–186.

120. Plonsey R, Barr RC. Effect of microscopic and macroscopic discontinuities on the response of cardiac tissue to defibrillating (stimulating) currents. Med Biol Eng Comput 1986; 24:130–136.

121. Gillis AM, Fast VG, Rohr S, Kleber AG. Spatial changes in transmembrane potential during extracellular electrical shocks in cultured monolayers of neonatal rat ventricular myocytes. Circ Res 1996; 79:676–690.

122. Fast VG, Rohr S, Gillis AM, Kleber AG. Activation of cardiac tissue by extracellular electrical shocks. Formation of "secondary sources" at intercellular clefts in monolayers of cultured myocytes. Circ Res 1998; 82:375–385.

123. Krassowska W, Pilkington TC, Ideker RE. Periodic conductivity as a mechanism for cardiac stimulation and defibrillation. IEEE Trans Biomed Eng 1987; 34:555–560.

124. Fishler MG. Syncytial heterogeneity as a mechanism underlying cardiac far-field stimulation during defibrillation-level shocks. J Cardiovasc Electrophysiol 1998; 9:384–394.

125. Fishler MG, Vepa K. Spatiotemporal effects of syncytial heterogeneities on cardiac far-field

excitations during monophasic and biphasic shocks. J Cardiovasc Electrophysiol 1998; 9:1310–1324.

126. Frazier DW, Krassowska W, Chen PS, Wolf PD, Dixon EG, Smith WM, Ideker RE. Extracellular field required for excitation in three-dimensional anisotropic canine myocardium. Circ Res 1988; 63:147–164.

127. Trayanova N, Skouibine K, Aguel F. The role of cardiac tissue structure in defibrillation. Chaos 1998; 8:221–233.

128. Sobie EA, Susil RC, Tung L. A generalized activating function for predicting virtual electrodes in cardiac tissue. Biophys J 1997; 73:1410–1423.

129. Chen PS, Wolf PD, Claydon FJ, Dixon EG, Vidaillet HJ Jr, Danieley ND, Pilkington TC, Ideker RE. The potential gradient field created by epicardial defibrillation electrodes in dogs. Circulation 1986; 74:626–636.

130. Witkowski FX, Penkoske PA, Plonsey R. Mechanism of cardiac defibrillation in open-chest dogs with unipolar DC-coupled simultaneous activation and shock potential recordings. Circulation 1990; 82:244–260.

131. Dillon SM, Mehra R. Prolongation of ventricular refractoriness by defibrillation shocks may be due additional depolarization of the action potential. J Cardiovasc Electrophysiol 1992; 3:442–456.

132. Wiggers CJ. The physiologic basis for cardiac resuscitation from ventricular fibrillation— Method for serial defibrillation. Am Heart J 1940; 20:413–422.

133. Mower MM, Mirowski M, Spear JF, Moore EN. Patterns of ventricular activity during catheter defibrillation. Circulation 1974; 69:858–861.

134. Zipes DP, Fischer J, King RM, Nicoll AB, Jolly WW. Termination of ventricular fibrillation in dogs by depolarizing a critical amount of myocardium. Am J Cardiol 1975; 36:37–44.

135. Chen PS, Shibata N, Dixon EG, Wolf PD, Danieley ND, Sweeney MB, Smith WM, Ideker RE. Activation during ventricular defibrillation in open-chest dogs. Evidence of complete cessation and regeneration of ventricular fibrillation after unsuccessful shocks. J Clin Invest 1986; 77:810–823.

136. Colavita PG, Wolf P, Smith WM, Bartram FR, Hardage M, Ideker RE. Determination of effects of internal countershock by direct cardiac recordings during normal rhythm. Am J Physiol 1986; 250:H736–H740.

137. Shibata N, Chen PS, Dixon EG, Wolf PD, Danieley ND, Smith WM, Ideker RE. Epicardial activation after unsuccessful defibrillation shocks in dogs. Am J Physiol 1988; 255:H902–H909.

138. Chen PS, Wolf PD, Melnick SD, Danieley ND, Smith WM, Ideker RE. Comparison of activation during ventricular fibrillation and following unsuccessful defibrillation shocks in open-chest dogs. Circ Res 1990; 66:1544–1560.

139. Zhou X, Daubert JP, Wolf PD, Smith WM, Ideker RE. Epicardial mapping of ventricular defibrillation with monophasic and biphasic shocks in dogs. Circ Res 1993; 72:145–160.

140. Usui M, Callihan RL, Walker RG, Walcott GP, Rollins DL, Wolf PD, Smith WM, Ideker RE. Epicardial sock mapping following monophasic and biphasic shocks of equal voltage with an endocardial lead system. J Cardiovasc Electrophysiol 1996; 7:322–334.

141. Cha YM, Peters BB, Chen PS. The effects of lidocaine on the vulnerable period during ventricular fibrillation. J Cardiovasc Electrophysiol 1994; 5(7):571–580.

142. Chattipakorn N, Kenknight BH, Rogers JM, Walker RG, Walcott GP, Rollins DL, Smith WM, Ideker RE. Locally propagated activation immediately after internal defibrillation. Circulation 1998; 97:1401–1410.

143. Chen PS, Shibata N, Dixon EG, Martin RO, Ideker RE. Comparison of the defibrillation threshold and the upper limit of ventricular vulnerability. Circulation 1986; 73:1022–1028.

144. Wiggers CJ, Wegria R. Ventricular fibrillation due to single, localized induction and condenser shocks applied during the vulnerable phase of ventricular systole. Am J Physiol 1940; 128:500–505.

145. Ferris LP, King BG, Spence PW, Williams HB. Effect of electric shock on the heart. AIEE Trans (Elect Eng) 1936; 55:498–515.

146. Fabiato A, Coumel P, Gourgon R, Saumont R. Le seuil de reponse synchrone des fibres myocardiques. Application a la comparison experimentale de l'efficacite des differentes formes de chocs electriques de defibrillation. Arch Mal d Coeur 1967; 60:527–544.

147. Chen PS, Wolf PD, Ideker RE. Mechanism of cardiac defibrillation. A different point of view. Circulation 1991; 84:913–919.

148. Kao CY, Hoffman BF. Graded and decremental response in heart muscle fibers. Am J Physiol 1958; 194:187–196.

149. Kirchhof C, Chorro F, Scheffer GJ, Brugada J, Konings K, Zetelaki Z, Allessie MA. Regional entrainment of atrial fibrillation studied by high-resolution mapping in open-chest dogs. Circulation 1993; 88:736–749.

150. Kenknight BH, Bayly PV, Gerstle RJ, Rollins DL, Wolf PD, Smith WM, Ideker RE. Regional capture of fibrillating ventricular myocardium. Evidence of an excitable gap. Circ Res 1995; 77:849–855.

151. Knisley SB, Smith WM, Ideker RE. Effect of field stimulation on cellular repolarization in rabbit myocardium. Implications for reentry induction. Circ Res 1992; 70:707–715.

152. Fishler MG, Sobie EA, Tung L, Thakor NV. Modeling the interaction between propagating cardiac waves and monophasic and biphasic field stimuli: The importance of the induced spatial excitatory response. J Cardiovasc Electrophysiol 1996; 7:1183–1196.

153. Kwaku KF, Dillon SM. Shock-induced depolarization of refractory myocardium prevents wave-front propagation in defibrillation. Circ Res 1996; 79:957–973.

154. Dillon SM, Wit AL. Use of voltage sensitive dyes to investigate electrical defibrillation. Proc IEEE Eng Med Biol 1988; 10 (Pt 1):215–216.

155. Dillon SM. Optical recordings in the rabbit heart show that defibrillation strength shocks prolong the duration of depolarization and the refractory period. Circ Res 1991; 69:842–856.

156. Davila HV, Cohen LB, Salzberg BM, Waggoner AS. A large change in axon fluorescence that provides a promising method for measuring membrane potential. Nature N Biol 1973; 241:159.

157. Salama G, Morad M. Merocyanine 540 as an optical probe of transmembrane electrical activity in the heart. Science 1976; 191(4226):485–487.

158. Girouard SD, Laurita KR, Rosenbaum DS. Unique properties of cardiac action potentials recorded with voltage-sensitive dyes. J Cardiovasc Electrophysiol 1996; 7:1024–1038.

159. Dillon SM, Wit AL. Optical action potentials recorded in-situ from the normal and epicardial borderzones of healing canine infarcted hearts show comparable upstroke durations. Circulation 1994; 90(4 Pt 2):I-466.

160. Menz V, Li KS, Schwartzman DJ, Dillon SM. In situ optical action potential recordings of defibrillation in swine. PACE 1995; 18(Pt II):877.

161. Dillon SM, Kerner TE, Hoffman J, Menz V, Li KS, Michele JJ. A system for in-vivo cardiac optical mapping. IEEE EMB 1998; 17:95–108.

162. Hoffman BF, Cranefield PF. Electrophysiology of the heart. New York: McGraw Hill, 1960.

163. Dillon SM. The electrophysiology of ventricular defibrillation: Today and Yesterday. In: Alessie MA, Fromer M, eds. Atrial and Ventricular Fibrillation Mechanisms and Device Therapy. Armonk, NY: Futura Publishing, 1997:113–144.

164. Zhou XH, Knisley SB, Wolf PD, Rollins DL, Smith WM, Ideker RE. Prolongation of repolarization time by electric field stimulation with monophasic and biphasic shocks in open-chest dogs. Circ Res 1991; 68:1761–1767.

165. Zhou X, Wolf PD, Rollins DL, Afework Y, Smith WM, Ideker RE. Effects of monophasic and biphasic shocks on action potentials during ventricular fibrillation in dogs. Circ Res 1993; 73:325–334.

166. Swartz JF, Jones JL, Jones RE, Fletcher R. Conditioning prepulse of biphasic defibrilla-

tor waveforms enhances refractoriness to fibrillation wavefronts. Circ Res 1991; 68:438–449.

167. Dillon SM. Synchronized repolarization after defibrillation shocks. A possible component of the defibrillation process demonstrated by optical recordings in rabbit heart. Circulation 1992; 85:1865–1878.

168. Wang T, Kwaku KF, Dillon SM. Repolarization resynchronization may underlie the efficacy of biphasic shocks. Circulation 1994; 90(4 Pt 2):I-446.

169. Dillon SM, Kwaku KF. Progressive depolarization: A unified hypothesis for defibrillation and fibrillation induction by shocks. J Cardiovasc Electrophysiol 1998; 9:529–552.

170. Sweeney RJ, Gill RM, Steinberg MI, Reid PR. Ventricular refractory period extension caused by defibrillation shocks. Circulation 1990; 82:965–972.

171. Sweeney RJ, Gill RM, Reid PR. Characterization of refractory period extension by transcardiac shock. Circulation 1991; 83:2057–2066.

172. Gill RM, Sweeney RJ, Reid PR. Refractory period extension during ventricular pacing at fibrillatory pacing rates. PACE 1997; 20:647–653.

173. Sweeney RJ, Gill RM, Steinberg MI, Reid PR. Effects of flecainide, encainide, and clofilium on ventricular refractory period extension by transcardiac shocks. PACE 1996; 19:50–60.

174. Sweeney RJ, Gill RM, Reid PR. Refractory interval after transcardiac shocks during ventricular fibrillation. Circulation 1996; 94:2947–2952.

175. Tovar OH, Jones JL. Biphasic defibrillation waveforms reduce shock-induced response duration dispersion between low and high shock intensities. Circ Res 1995; 77:430–438.

176. Tovar OH, Jones JL. Relationship between "extension of refractoriness" and probability of successful defibrillation. Am J Physiol 1997; 272:H1011–H1019.

177. Efimov IR, Cheng YN, Biermann M, Van Wagoner DR, Mazgalev TN, Tchou PJ. Transmembrane voltage changes produced by real and virtual electrode during monophasic defibrillation shock delivered by an implantable electrode. J Cardiovasc Electrophysiol 1997; 8:1031–1045.

178. Efimov IR, Cheung Y, Van Wagoner DR, Mazgalev TN, Tchou PJ. Virtual electrode-induced phase singularity. A basic mechanism for defibrillation failure. Circ Res 1998; 82:918–925.

20

Origins, Classification, and Significance of Ventricular Arrhythmias

ROBERT J. MYERBURG and AGUSTIN CASTELLANOS

University of Miami School of Medicine, Miami, Florida

HEIKKI V. HUIKURI

Oulu University Hospital, Oulu, Finland

Ventricular arrhythmias are defined as single or repetitive impulses that originate distal to the bifurcation of the bundle of His and do not require atrial or proximal atrioventricular (AV) junctional tissue for their expression or maintenance. While they commonly occur in the presence of structural heart diseases, and usually identify a risk for potentially fatal arrhythmias in those clinical settings, some forms of ventricular arrhythmias occur in the absence of structural heart disease and may have little or no influence on the risk of fatal events. The interpretation of clinical significance of ventricular arrhythmias is based upon both the form of the arrhythmia and the clinical settings in which they occur.

I. DEFINITIONS OF VENTRICULAR ARRHYTHMIAS

Ventricular arrhythmias may be categorized broadly as premature ventricular contractions (PVCs) and ventricular tachyarrhythmias, the latter including ventricular tachycardia (VT) and ventricular fibrillation (VF). According to strict criteria, premature ventricular contractions are single or paired extrasystoles that originate in the ventricles, and the ventricular tachyarrhythmias are repetitive forms of three or more consecutive ventricular ectopic beats (1). In fact, this definition is somewhat arbitrary because, in clinical circumstances, there is a transition of forms between repetitive extrasystoles and nonsustained or sustained ventricular tachycardias. Figure 20.1 demonstrates a method of classification of ventricular arrhythmias based upon the frequency of ectopic beats and their electrocardiographic forms or patterns (2).

FREQUENCY	FORMS
GRADE 0 - NIL (-)	GRADE 0 - NIL
GRADE 1 - RARE (< 1 PVC / hr)	GRADE A - SINGLE, UNI
GRADE 2 - INFREQUENT (1 - 9 / hr)	GRADE B - SINGLE, MULTI
GRADE 3 - INTERMEDIATE ($\geq$ 10, < 30 / hr)	GRADE C - COUPL, SALVO (2 - 5 PVCs)
GRADE 4 - FREQUENT ($\geq$ 30, < 60 / hr)	GRADE D - NON-SUST VT ($\geq$ 6 PVCs, < 30 sec)
GRADE 5 - DENSE ($\geq$ 60 / hr)	GRADE E - SUSTAINED VT ($\geq$ 30 sec)

Figure 20.1   Classification of ventricular arrhythmias. Ventricular arrhythmias may be classified according to frequency (number of ectopic beats per unit of time) and the forms of the arrhythmias—single, repetitive, ventricular tachycardia. In the presence of heart disease, particularly recent myocardial infarction, frequency of PVCs as a marker of risk begins to be expressed at the level of Grade 2, and seems to plateau at Grade 3 and beyond. In persons with normal hearts, even very frequent single ectopic beats may be benign. Forms of ventricular arrhythmias may be single ectopic beats, either of a single or uniform (UNI) morphology, or of multiple morphologies (MULTI). The latter is commonly referred to as multifocal. Repetitive forms may occur as couplets (COUPL–2 consecutive beats), salvos (3–5 consecutive beats), and nonsustained and sustained ventricular tachycardia (VT). Sustained VT is generally defined as 30 s or more of continuous activity of ventricular origin. (Modified from Ref. 2, with permission.)

Frequency of premature ventricular contractions has received considerable attention because of clinical and epidemiological observations that, in stable settings of clinical heart disease (i.e., unrelated to acute ischemic events or acute heart failure), 10 or more ventricular ectopic impulses per hour during continuous monitoring identifies an increased risk of sudden death and total mortality (3–5). Moreover, there are data that suggest that frequencies as low as three or four ectopic impulses per hour identify the beginning of an ascending slope on the risk curve, which plateaus at 10 or more beats per hour (5). It is not clear that numbers higher than 10 per hour confer any larger risk as an independent determinant.

Interest in forms of electrocardiographic patterns derives from observations that repetitive responses, in association with a low ejection fraction, indicate an increased mortality risk (5,6). There are conflicting data regarding the question as to whether repetitive forms in the range of 3 to 15 consecutive impulses carries any more power than frequent single premature beats. Some observers believe that repetitive forms confer independent increased risk and other data sets challenge that concept.

Sustained ventricular tachycardia (repetitive sequences lasting 30 s or longer) is generally viewed as a marker of increased risk in the presence of heart disease, particularly in its later stages. Such tachyarrhythmias in the absence of structural heart disease may be clinically benign for risk prediction, although they may be troublesome to the patient in

DURATION

- SALVOS
  (3 - 5 IMPULSES)

- NON-SUSTAINED
  (≥6 IMPULSES,
     <30 SECONDS)

- SUSTAINED VT
  (≥30 SECONDS)

MORPHOLOGY

- MONOMORPHIC

- POLYMORPHIC

- Torsade de pointes

- RVOT PATTERN

- BIDIRECTIONAL

Figure 20.2   Categories of ventricular tachycardia. Ventricular tachycardia is classified by duration and by electrocardiographic morphology. An arbitrary distinction, based on limited and not universally accepted stratification of risks, is salvos of 3–5 consecutive impulses, with nonsustained ventricular tachycardia defined as 6 or more impulses up to a maximum of less than 30 s and sustained ventricular tachycardia lasting 30 s or more. Morphologies may be monomorphic or polymorphic (see text), a specific pattern of polymorphic referred to as torsade de pointes, a left bundle branch pattern that is relatively specific for tachycardias originating in the right ventricular outflow tract (RVOT), and bidirectional ventricular tachycardia, in which there is alternation of the vectors of successive beats. (Modified from Ref. 1, with permission.)

terms of symptoms. The categories of ventricular tachyarrhythmias are listed in Figure 20.2. Sustained ventricular tachycardias may be classified as *monomorphic*, in which each consecutive impulse is similar to the others and the rate is relatively constant, and *polymorphic*, in which there is beat-to-beat variation as well as variation in the cycle lengths between beats. When polymorphic ventricular tachycardias occur as spontaneous clinical events, they are generally considered to reflect a higher risk. However, during electrophysiological studies with programmed electrical stimulation, the induction of polymorphic ventricular tachycardia by aggressive pacing protocols paradoxically may be a low-risk, nonspecific phenomenon.

Another characteristic of tachyarrhythmias that influences risk is their hemodynamic stability. Even for monomorphic forms, hemodynamically unstable tachycardias are considered to reflect higher risk than stable monomorphic tachycardias. This concept refers to hemodynamic status after the onset of tachycardia (Fig. 20.3). In contrast, patients who are hemodynamically unstable before the onset of tachycardia—that is, the arrhythmia is secondary to hemodynamic deterioration—have an even worse prognosis. Accordingly, VT of this type has a very high fatality rate (7,8).

The ventricular arrhythmias of greatest concern are ventricular flutter, a monomorphic tachycardia fast enough to generate a waveform that is similar to a sine wave (usually at cycle lengths ≤ 300 ms, rates ≥ 200 beats per min) and ventricular fibrillation, an irregular and uncoordinated form of ventricular activation is incapable of maintaining mechanical function because of its chaotic pattern. Ventricular fibrillation is actually defined by its mechanical features. Certain repetitive electrocardiographic patterns within the spectrum of ventricular fibrillation that generate a pulse (however weak) are defined as polymorphic, ventricular tachycardia. Ventricular fibrillation is uniformly fatal if not treated immediately.

● ELECTROPHYSIOLOGICAL STABILITY

<u>MORPHOLOGY</u>   <u>CYCLE LENGTH</u>

Monomorphic VT   > 400-250 ms [Stable]

Polymorphic VT   < 400-200 ms [Variable]

Ventricular Flutter   240- < 200 ms [Stable >
                                                       Unstable]

Ventricular Fibrillation   ? - << 200 ms
                                  [Mechanical Definition]

● HEMODYNAMIC STABILITY

BP, Cardiac Output   Level of Consciousness

Figure 20.3   Electrical and hemodynamic stability of ventricular tachycardia. There are general associations among the electrocardiographic patterns of ventricular tachyarrhythmias, their rate or cycle lengths, and their electrical and hemodynamic stability. In general, monomorphic ventricular tachycardias tend to be slower and electrically and hemodynamically more stable, whereas polymorphic tachycardias are faster and have a higher likelihood of being hemodynamically unstable. Ventricular flutter may be electrically stable at least for a period of time, but is usually hemodynamically unstable, and ventricular fibrillation is, by definition, unable to support hemodynamic stability.

II.   STRUCTURE, FUNCTION, AND PATHOPHYSIOLOGY

For many years, clinical and epidemiological associations between premature ventricular contractions and ventricular tachycardia or fibrillation have been recognized. This association resulted in an electrogenic pathophysiological concept, in which ambient PVCs were considered to be causally related to ventricular tachycardia or fibrillation. While such an initiating mechanism may exist, to assume a simple cause–effect relationship is to oversimplify pathophysiological mechanisms (9).

PVCs in persons without structural heart disease, which are morphologically indistinguishable from PVCs in the presence of structural heart disease, do not confer risk of initiation of fatal arrhythmias. Thus, the association between PVCs and ventricular tachyarrhythmias requires additional influences that we have referred to as conditioning and triggering or initiating factors (10). Conditioning risk factors refer to those structural or functional abnormalities or variations in the heart which, under appropriate circumstances, will establish a mechanism by which a rhythm disturbance can be sustained. Once the conditioning risk factor is in place, the PVC can be viewed as having the potential to initiate a sustained arrhythmia. However, even in patients having the structural substrate for a sustained arrhythmia, PVCs can be very frequent over long periods of time without initiating arrhythmias, indicating that another factor is required to allow the PVC/VT relationship to become manifest. This is referred to as transient or initiating risk. It can be viewed as a short-term change in the physiology of the electrophysiological systems in the heart that will allow an abnormal, but electrophysiologically inactive, conditioning influence to become unstable for a period of time, resulting in the onset of a tachyarrhythmia (Fig. 20.4). Thus the pathophysiological cascade includes an electrogenic relationship (PVC → VT)

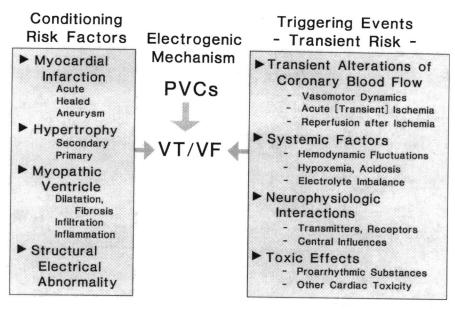

Figure 20.4   Ventricular tachyarrhythmias: The association between structure, function, and on-set. The concept that PVCs induce ventricular tachycardia or fibrillation (VT/VF) is an electrogenic concept that cannot be isolated from the underlying structural heart disease (conditioning risk factors) and triggering events (transient risk factors). PVCs in a normal heart rarely, if ever, induce life-threatening VT/VF; but the presence of various structural abnormalities create a conditioned risk that becomes manifest at the time of transient changes referred to as triggering events. (From Ref. 9, with permission.)

that is usually not manifest unless a defined substrate is present over time, with the nature of that substrate changing at the time when acute transient factors become manifest (Fig. 20.5).

The relationship between conditioning substrates and time requires special considerations. In some circumstances, such as a healed myocardial infarction with a chronically scarred ventricle, the time domain is very long, measured in terms of months or years. In contrast, during an acute myocardial infarction, in which risk of ventricular arrhythmias may occur over a relatively short time period (minutes, hours, or days), the time domain has to be viewed differently. In that circumstance, it is short and dynamic. Nonetheless, a conditioning substrate is present for a defined period of time, which allows initiating influences to express themselves if they are present during the same time period. In the active circumstance of myocardial infarction, in which there is a transition from ischemia to infarction, to healing, and finally to recurrent ischemia in the presence of a scar, the model of arrhythmia risk is very dynamic. Figure 20.6 demonstrates five time periods in the evolution from ischemia to infarction to recurrent ischemia. In each of these time periods, the nature of the anatomical substrate for ventricular arrhythmias is different, and the time frame of their expression (especially for ischemia, acute infarction, and healing infarction) is time-limited. Any approach to the pathophysiology of ventricular arrhythmias, based on the anatomical features of ischemic heart disease, has to take into consideration these variations.

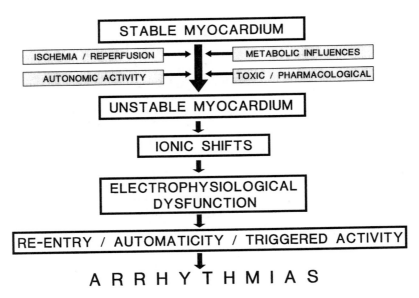

Figure 20.5   Cascade of the evolution from stability to instability in the onset of ventricular tachyarrhythmias. The functional modulations responsible for destabilizing a stable, but structurally abnormal heart include alterations in myocardial perfusion, hemodynamic or metabolic states, autonomic status, and pharmacological effects. With the transition from stable to unstable myocardium, acute changes in membrane properties create a transient substrate for reentry, or of enhanced or abnormal forms of automaticity, which lead to arrhythmias. (From Ref. 25, with permission.)

From an epidemiological point of view, the structural abnormality of greatest importance for ventricular arrhythmias is coronary artery disease and its myocardial consequences. This single entity accounts for approximately 80% of fatal ventricular arrhythmias based upon the prevalence of that category of disease among the Western population (Fig. 20.7). However, other structural abnormalities of the heart also define a risk of ventricular arrhythmogenesis or contribute to that risk in the setting of coronary artery disease. An example of the latter is left ventricular hypertrophy in patients with coronary artery disease. It is well established that global left ventricular hypertrophy secondary to hypertension adds risk of arrhythmias and sudden death to the patient with coronary artery disease (11,12). In addition, it is well established that part of the remodeling process after myocardial infarction involves the development of regional hypertrophy surrounding the area of an infarct scar (13,14), and that the pathophysiology of hypertrophied myocytes is altered, likely favoring the genesis of arrhythmias during subsequent ischemic events (15). In fact, the remodeling phenomenon can be viewed at three different levels in cardiac tissue—the whole heart, cellular interactions, and single cell phenomena (see Fig. 20.8). In a general sense, remodeling can also be viewed as a pathophysiological concept at the same three levels in nonischemic cardiomyopathies and hypertrophic cardiomyopathies as well. At each level, including the molecular level, alterations as a consequence of disease can create the substrates that allow the initiation and maintenance of arrhythmias under appropriate circumstances.

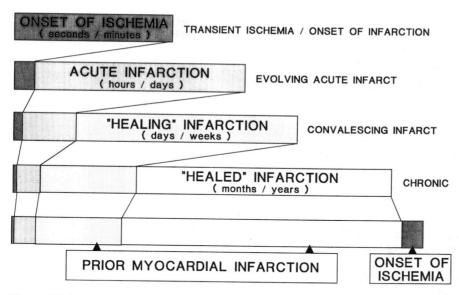

Figure 20.6   Models for study of arrhythmias associated with ischemia and infarction. The complexity of ischemic heart disease and its risk of arrhythmias is a result of the heterogeneity of the disease processes, both structurally and over time. Mechanisms responsible for arrhythmias during the onset of ischemia, which is a period lasting only seconds to minutes, are distinct from those during the acute phase of myocardial infarction or during healing. The healed myocardial infarction is generally stable until a recurrent episode of ischemia creates destabilization in a substrate different than that of the normal heart. These complexities must be considered in the design and evaluation of data from models intended to study mechanisms of arrhythmias in ischemic heart disease. (From Ref. 10, with permission.)

▷ CORONARY HEART DISEASE
- Acute ischemic events
- Chronic ischemic heart disease
▷ CARDIOMYOPATHIES
- Dilated cardiomyopathies
- Hypertrophic cardiomyopathies
▷ INFLAMMATION / INFILTRATION
▷ SUBTLE, POORLY-DEFINED LESIONS
▷ LESIONS OF MOLECULAR STRUCTURE
▷ FUNCTIONAL ABNORMALITIES
▷ "NORMAL" HEARTS - IDIOPATHIC VF

Figure 20.7   Structural or etiological basis of ventricular arrhythmias. Coronary heart disease accounts for approximately 80% of ventricular arrhythmias and sudden cardiac death risk in the western hemisphere. The cardiomyopathies collectively (dilated and hypertrophic) account for an additional 10 to 15%. All other diseases or functional conditions, including hearts thought to be structurally normal, account for only a small percentage of ventricular arrhythmias collectively.

	ISCHEMIC HEART DISEASE	NON-ISCHEMIC CARDIOMYOPATHY	HYPERTROPHIC CARDIOMYOPATHY
MACRO-REMODELING	Scar -path length; Regional hypertrophy; Non-uniform conduction and recovery of excitability	Slowed conduction Non-uniform conduction Stretch-alteration	Global hypertrophy [regional variation?] Path length
MICRO-REMODELING	Altered gap junctions [regional anisotrophy] Fragmented conduction	Interstitial fibrosis Fragmented conduction Non-uniform conduction	Interstitial fibrosis Non-uniformity Vascular mismatch
MOLECULAR REMODELING	Regional depolarization and repolarization	Global or regional repolarization	Global or regional repolarization

Figure 20.8 Mechanisms of remodeling, disease states, and arrhythmia risk. The effect of disease states upon the electrophysiological substrate of the heart as viewed at 3 levels. (1) Macro-, remodeling, or gross effects refer to structural changes at a macroscopic level, including regional diseases and pathways. (2) Micro remodeling involving small region down to microscopic levels. (3) "Molecular remodeling" refers to the molecular structures, channels, and other membrane functions that are altered. The expression of these three forms of remodeling differ in ischemic heart disease, nonischemic cardiomyopathy, and hypertrophic cardiomyopathies. Note that this figure provides examples and is not intended to be all-inclusive.

III. FORMS OF VENTRICULAR ARRHYTHMIAS IN RELATION TO CARDIAC ABNORMALITIES

Ventricular tachyarrhythmias that occur in hearts with defined structural abnormalities are generally conditioned by an anatomical substrate (see Fig. 20.7). The substrate for ventricular tachyarrhythmias may be anatomical and/or physiological. In either case, slow conduction and unidirectional conduction disturbances are required, as are triggering events. In the presence of structural heart disease, anatomical pathways are commonly present. An example is the heart with a healed myocardial infarction and a defined scar, through which (or across which) abnormal pathways of conduction persist. Such an anatomical pattern establishes the properties required for sustaining an intraventricular reentrant loop when an appropriate triggering event initiates that pattern. It is important to recognize that reentry may also occur because of functional physiological changes, but in these cases the pattern is often unstable and less predictable.

Anatomical reentrant pathways can often be demonstrated by programmed electrical stimulation (i.e., electrophysiological testing designed to reproduce an individual patient's arrhythmia under controlled circumstances). Successful induction is particularly predictable for the defined pathways established in healed myocardial infarctions. It is much less predictable for the anatomical substrate caused by nonischemic cardiomyopathies (Fig. 20.8). Moreover, programmed electrical stimulation is usually unsuccessful in functionally based substrates, unless provocations are invoked that will recreate the physiological disturbance responsible for the tachycardia at the time of the programmed electrical stimulation study.

IV. TRIGGERING OF VENTRICULAR ARRHYTHMIAS BY TRANSIENT ALTERATIONS IN PHYSIOLOGY

Transient functional alterations may alter structural conditions in the abnormal heart that are pathophysiologically stable under baseline conditions. It is likely that such functional modulation exerts its adverse influence by creating nonuniform responses in structurally heterogeneous abnormal hearts. The four major categories of functional heterogeneity include: (1) transient ischemia and reperfusion; (2) systemic metabolic and hemodynamic alterations; (3) neurochemical and neurophysiological factors; and (4) exogenous toxic or pharmacological effects (Fig. 20.4).

A. Ventricular Arrhythmias Initiated by Ischemia and Reperfusion

Transient reduction in blood flow causes alterations in the electrophysiology of ventricular myocardium and specialized conducting tissue. The nature of these electrophysiological changes is a function of the duration and severity of ischemia. In addition, the reestablishment of blood flow after transient ischemia may initiate yet other mechanisms for altered ventricular muscle electrophysiology and arrhythmias.

The onset of ischemia is accompanied by altered conduction and changes in the time course of repolarization. Ischemia is responsible for decreasing the resting membrane potential of myocytes from –90 mV toward 0. Given the association between the membrane potential at the time of activation and the conduction velocity of an initiated impulse (16), regional ischemia will cause regional changes in velocity of propagation of impulses. Such slow conduction is one of the requisites for reentrant activation. Ischemia alters the time course of repolarization in a more complex fashion. Whereas alterations in resting membrane potential and conduction velocity are often uniform across the ventricular wall, ion channels may manifest dramatic transmural changes in formation during ischemia. For example, the ATP-activated potassium channel ($I_{K\text{-}ATP}$) is much more densely distributed on the epicardium than on the endocardium (17). This channel is activated in the absence of ATP and, therefore, the depletion of ATP levels, and the increase in ADP associated with ischemia, result in an enhanced outward potassium current in the epicardial tissue generated by this mechanism. This shortens the refractory period of the epicardial muscle markedly compared to endocardium, setting the stage for reentrant activation.

Reperfusion after transient ischemia also appears to play an important role in initiation of ventricular arrhythmias, and does so by mechanisms that appear to be different than those responsible for ischemic arrhythmias. Moreover, specific patterns of reperfusion arrhythmias, and the mechanisms responsible for them, appear to be dependent upon the duration of ischemia (18) (Fig. 20.9). A number of studies have demonstrated that for the most dangerous reperfusion arrhythmias, polymorphic ventricular tachycardia or ventricular fibrillation, a duration of ischemia more than 5 and less than 20 min, is required. Shorter durations of ischemia may generate ischemic-mediated ventricular tachycardia or fibrillation. Arrhythmias are generally not triggered during reperfusion after periods of ischemia lasting less than 5 min. However, with the longer periods of ischemia, fatal arrhythmias may occur during reperfusion. For durations of ischemia greater than 20–30

CLINICAL STATE	ISCHEMIC TIME	ARRHYTHMIA
CORONARY SPASM	SECONDS / MINUTES	VF, VT
CARDIOPULMONARY BYPASS; ISCHEMIC CARDIAC ARREST	MINUTES / HOURS	PVCs, VF
THROMBOLYSIS, ACUTE ANGIOPLASTY	HOURS	AVR, PVCs, VT
MYOCARDIAL REVASCULARIZATION	DAYS / WEEKS	- ? -

Figure 20.9   Duration of ischemia and the characteristics of reperfusion arrhythmias. The duration of ischemia appears to determine both the probability of occurrence and the form of reperfusion arrhythmias. The most predictably life-threatening forms of reperfusion arrhythmias occur after ischemia that has lasted for at least 5 min and no longer than 20 to 30 min. Experimentally this appears to be associated with the triggering of rapid polymorphic tachycardias (AVR = accelerated ventricular rhythm). (Modified from Refs. 18 and 25, with permission.)

min, risk of such fatal arrhythmias decreases, and the more common patterns that emerge are accelerated automatic ventricular rhythms and PVCs. As shown in Figure 8, the major clinical associations with the various forms of reperfusion arrhythmias are coronary vasospasm or other causes of transient occlusion (life-threatening forms of arrhythmias), and thrombolysis and angioplasty in the setting of acute myocardial infarction (the more benign accelerated ventricular rhythms).

Reperfusion related to polymorphic ventricular tachycardia or ventricular fibrillation may occur by mechanisms of triggered activity (15,19) or by focal reentry in the reperfused region (20,21). Some data suggest that hypertrophied myocytes are more vulnerable to the triggered activity mechanism (15,22,23) and, therefore, hearts with healed myocardial infarction and regional hypertrophy may be particularly vulnerable.

B.   Ventricular Arrhythmias Associated with Transient Systemic Factors

A diverse variety of systemic factors may participate in initiation of ventricular arrhythmias in the conditioned heart. These include metabolic factors, such as hypoxemia and acidosis as well as electrolyte disturbances (24–26). However, the factor that is probably most important in general clinical circumstances is transient changes in the hemodynamic status. It is becoming increasingly evident that alterations in myocardial stretch, ventricular filling pressures, and chamber dimensions and functions can be associated with the initiation of transient ventricular arrhythmias (27–29). Both ambient PVCs and sustained ventricular arrhythmias appear to be more common during overt heart failure, and treatment of certain of these arrhythmias is often successful by targeting on the hemodynamic abnormalities rather than the use of antiarrhythmic drugs. Neither experimental nor clinical data on the role of hemodynamic manipulation for the initiation or termination of arrhythmias has been developed to a very great extent, despite the obvious clinical associations. This is a field of opportunity for future research that warrants increasing attention.

C. Autonomic Factors in Ventricular Arrhythmias

Systemic and regional neurophysiological factors are receiving a great deal of attention as markers for identifying risk of ventricular arrhythmias and for elucidating mechanisms of potentially fatal arrhythmias. A number of studies have demonstrated alterations in adrenergic receptors, coupling proteins and adenylyl cyclase activity in hearts with healed myocardial infarction (30,31). When these changes are regional, they are associated with alterations in the regional properties of refractoriness (32,33), a factor that may be arrhythmogenic. Clinical studies of the role of adrenergic stimulation for the initiation of arrhythmias during programmed electrical stimulation also support the concept of a neurogenic role in arrhythmogenesis. At a systemic level, additional data also support this concept (34–38). Observations of measures such as heart rate variability and baroreceptor sensitivity in patients with prior myocardial infarctions are instructive. They indicate that alterations in global systemic autonomic function, as measured by these techniques, are associated with risks of potentially fatal arrhythmias and, at the very least, may have a role for risk stratification. At best, they may even lead to new therapeutic approaches.

D. Effects of Toxic and Pharmacological Substances on the Heart

The classic proarrhythmic response to antiarrhythmic drugs, torsade de pointes, is a particularly useful model for exogenous arrhythmogenic influences (39,40). Through effects on potassium channels, and perhaps other channels and receptors, the Class IA and Class III antiarrhythmic agents (such as quinidine and sotalol) are known to establish a small but finite risk of *torsade de pointes* in susceptible individuals. This risk is higher in the presence of structural heart disease than in individuals with structurally normal hearts. In recent years, there has been increasing attention to this phenomenon, now extending beyond antiarrhythmic drugs to other categories of drugs that may influence channels responsible for the repolarizing properties of the heart and, therefore, risk of torsade de pointes. The most evident among these is the story of the antihistamine, terfenidine, which blocks the delayed rectifier current I_{KR}. Its interaction with the antifungal, ketoconazole, which results in altered metabolism by hepatic P-450 enzymes, has defined a new example of proarrhythmic interactions (41). Ketoconazole blocks the metabolism of terfenadine to a benign metabolite that retains antihistaminic, but not I_{KR}-blocking properties, thereby potentiating the occurrence of proarrhythmia. Similar phenomena occur with a number of other drugs and the list is rapidly expanding to include many others such as macrolide antibiotics and antipsychotics. Clearly this proarrhythmic phenomenon is more general than was previously appreciated.

V. THE EPIDEMIOLOGY OF VENTRICULAR ARRHYTHMIAS

The epidemiology of ventricular arrhythmias parallels the epidemiology of sudden cardiac death. In recent years, it has become increasingly apparent that the epidemiology is dynamic and not restricted to behavior of defined population pools observed over time. Morever, many of the clinical characteristics of patients at risk should be viewed in an epidemiological context. The factors that must be considered are the risks within population

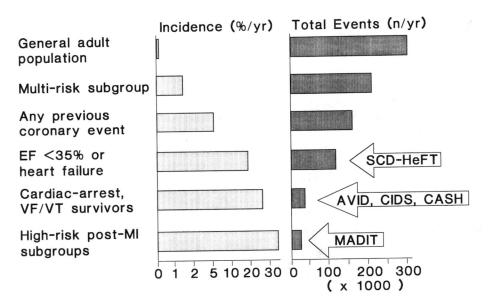

Figure 20.10   The relationship between incidence of life-threatening ventricular arrhythmias in specific population subgroups and the number of events accounted for in those groups. Among the 300,000 sudden cardiac deaths that occur in the general adult population of the United States annually (0.1–0.2% per year), only a small fraction are contained within those subgroups at substantially increased risk (heart failure, cardiac arrest survivors, high-risk post-MI patients). Most of the deaths occur in lower risk subgroups with much larger denominators. This impacts upon the population effect of clinical trials designed to study efficacy in high-risk populations. MADIT (67), AVID (68), and related defibrillator trials (SIDS and CASH) cumulatively demonstrate benefit to a small group of very high-risk patients. Another trial in progress, SCD-HEFT (sudden cardiac death–heart failure trial) addresses another subgroup of somewhat larger population size. (Modified from Ref. 69, with permission.)

subgroups, time-dependent risks, the conditioning and triggering factors (see Sec. IV), and the dynamics of interventions.

A.   Risks Within Population Subgroups

As discussed above, patients with advanced heart disease, defined as ejection fractions of less than 35%, and a history of manifest ventricular arrhythmias, are at high risk for potentially fatal ventricular arrhythmias. However, the populations defined by such criteria are small in the context of total population risk of fatal cardiac arrest. Among the 300,000 sudden cardiac deaths that occur in the United States each year, only a small fraction are associated with prior life-threatening events or very high-risk markers of cardiac arrest (Fig. 20.10) (10). Most of the fatal arrhythmic events occur among patient subgroups with less advanced heart disease, or no previously recognized disease at all. While these patient subgroups represent individual risks that are quite a bit lower, thereby reducing the efficiency of any intervention (42), they paradoxically account for the majority of fatal ventricular arrhythmias in the community. This paradox is a consequence of the denominators of the subgroups. Identification of individuals at risk within the populations with lower case fatality rates, but high total number of events, remains problematic (see Chap. 28). The challenge for the future is to develop approaches or techniques that will allow screening of specific risk for fatal ventricular arrhythmias as the expression of a first or recurrent event in patient populations that cumulatively have a low risk, but generate large numbers of victims (43).

B. Time Dependence of Risk

The risk of fatal ventricular arrhythmias after a conditioning cardiovascular event is not linear over time (10). Data from various sources, representing different kinds of cardiovascular events, demonstrate that conditions such as recent myocardial infarction, recent onset of heart failure, unstable angina pectoris, and survival from out-of-hospital cardiac arrest are all accompanied by a higher risk of recurrent events during the first 6 to 18 months after the index event than is present later. The time dependence of risk within high-risk subgroups puts a limit on the opportunity for intervention strategies beyond the early period after the index event. Thus, for both purposes of clinical trials and clinical practice, it is important to recognize that there is a need to invoke preventive therapeutic strategies early after index events in order to achieve maximum benefits.

C. Identification of Triggers of Life-Threatening Ventricular Arrhythmias

Identification of the trigger mechanisms of life-threatening ventricular arrhythmias represents one of the most challenging issues for understanding the mechanisms onset of arrhythmias in various clinical settings and provides a logical basis for improved therapeutic strategies in the prevention of sudden arrhythmic death (44,45).

Analysis of standard and ambulatory ECGs has a potential to give insight into the factors and mechanisms responsible for the onset and perpetuation of ventricular tachyarrhythmias. Several epidemiological studies have shown that certain risk markers, such as heart rate variability measured over a 24-h period from ambulatory ECG, QT interval dispersion measured from standard 12-lead ECG, and late potentials measured from signal-averaged ECG can predict the occurrence of subsequent arrhythmic events (37,46–49). The positive predictive accuracy of these measures is relatively low for prediction of future arrhythmic death, however, perhaps partly because the risk profile and arrhythmia vulnerability may change significantly during the time course after the identification of the risk. The difficulty of actually documenting the arrhythmic death is also a problem in epidemiological studies. Sudden death commonly has been attributed to arrhythmic death, but data based on stored ECGs of patients with implantable cardioverter defibrillators have documented that a significant proportion of sudden deaths are not due to ventricular tachyarrhythmia (50,51). Therefore, the epidemiological follow-up studies may not be able to give reliable information on the risk markers of ventricular tachyarrhythmias. In case control studies, the arrhythmia mechanism can be documented more reliably, and these studies have provided more accurate information on the differences in risk profile between the patients with and without vulnerability to ventricular tachyarrhythmias. Heart rate variability is reduced and dispersion of QT interval is increased in survivors of ventricular tachyarrhythmic events compared to matched controls (52,53). Tachyarrhythmia patients also show more abundant ventricular ectopic activity and episodes of nonsustained ventricular tachycardia than matched nonarrhythmia patients. There is significant overlap in the individual measures between the arrhythmia and nonarrhythmia patients, however, suggesting that traditional analysis of Holter recordings or standard ECG may not give specific information on the arrhythmia risk for therapeutic decisions.

An important approach in understanding the association between ECG abnormalities as predictors of ventricular tachyarrhythmias is provided by analyzing the ECGs recorded prior to the onset of arrhythmias in patients experiencing a tachyarrhythmic event. Short–long–short sequences of R-R intervals caused by ectopic beats precedes the onset of ventricular tachyarrhythmias in one-third of the cases (54). Abnormalities both in long-term and beat-to-beat R-R interval dynamics are also evident preceding the arrhythmia onset (55). Spectral analysis of heart rate variability has documented that altered heart rate variability frequently precedes the onset of sustained ventricular tachycardia (56,57). Ventricular fibrillation is not commonly preceded by altered heart rate variability measured by traditional nonspectral methods (58), but recent data suggest that analysis of heart rate behavior by nonlinear methods can detect abnormal heart rate dynamics also before the spontaneous onset of ventricular fibrillation (59). Overall, there is increasing evidence that abnormal heart rate behavior precedes the onset of ventricular tachyarrhythmias, but the current knowledge is based on relatively small and selected patient groups. Stored ECGs of patients with implanted cardioverter defibrillators offer the potential to study various features in R-R interval dynamics before the arrhythmia onset in larger patient samples with various underlying substrates for ventricular arrhythmias. More studies in this area would be useful in providing definitive information on the value of continuous ECG monitoring for identifying the onset of life-threatening arrhythmias.

Dynamic analysis of cardiac electrical signals occurring in relation to ventricular tachyarrhythmias has focused mainly on R-R interval variability. Because life-threatening arrhythmias, particularly those initiated or perpetuated by reentrant excitation, usually originate from myocardial tissue, assessment of dynamic changes in ventricular repolarization could provide important new information on the propensity to ventricular arrhythmias. Abnormalities in QT interval adaptation to an increase in heart rate have been observed in patients with the congenital long-QT Syndrome (60). Occurrence of episodes of T-wave alternans in ECG recordings has also been observed in patients with the long-QT syndrome, a phenomenon that often precedes the onset of torsade de pointes. The presence of T-wave alternans at the microvolt level many also predict vulnerability to ventricular tachyarrhythmias in patients with structural heart disease (61), and loss of heart rate dependency of QT interval behavior seems to precede the onset of ventricular fibrillation in patients with a prior myocardial infarction (62). These preliminary observations suggest that, in addition to analysis of heart rate behavior, dynamic analysis of ventricular repolarization may give even more important new information on vulnerability to ventricular tachyarrhythmias (63). Epidemiological data and autopsy findings from victims of sudden death provide evidence that fatal ventricular arrhythmias precipitated by myocardial ischemia, in the setting of subtotal or total coronary occlusion, are a major cause of sudden cardiac death. Acute coronary occlusion does not invariably lead to fatal arrhythmias, but may cause acute coronary events (e.g., unstable angina pectoris or acute myocardial infarction) without occurrence of ventricular arrhythmias.

A key issue in clinical research is to identify the patients with vulnerability to ischemia-induced ventricular tachyarrhythmias. Experimental data show that autonomic mechanisms modify the clinical manifestations of acute coronary occlusion (64). Augmented parasympathetic activity seems to have antifibrillatory effects during myocardial ischemia, while sympathetic stimulation increases the electrical instability of the ischemic myocardium. One clinical tool to study the effects of acute myocardial ischemia on arrhythmic vulnerability is to assess the individual responses to coronary occlusion during percutaneous coronary angioplasty. Abrupt coronary occlusion causes a wide range of autonomic reactions as evidenced by changes in heart rate, blood pressure, and heart rate variability (65). Recent data show that a coronary occlusion-induced increase in heart rate variability protects against occurrence of complex ventricular arrhythmias during the early phase of abrupt coronary occlusion (66), providing evidence that vagal activation has a role in the expression of ischemia-induced clinical arrhythmias. Together, experimental and clinical data support the view that responses of the autonomic nervous system to acute hemodynamic changes or ischemia are important in the genesis of life-threatening arrhythmias. Identification of the factors responsible for wide intersubject differences in autonomic responses, either genetic or acquired, will be of considerable importance in future research.

Mechanistic understanding of the factors predisposing to the onset and perpetuation of ventricular arrhythmias may assist in improving future therapy. Implantable devices having sensors to monitor electrical and hemodynamic parameters, such as R-R intervals, QT intervals, pH, temperature, and myocardial contractility, have been developed. Studies using advanced new techniques, both invasive and noninvasive, and new analytical methods, based on both traditional and nonlinear analytical techniques, aimed at identifying the imminent life-threatening arrhythmic event will be a fruitful area for future research in understanding the pathophysiology of ventricular arrhythmias. If the triggering factors can be identified accurately, it may be possible to develop individualized new therapeutic options to prevent sudden arrhythmic events. The key, of course, is to identify which potential measures of risk have practical, predictive value in specific patient categories.

D. Population Impact of Interventions

Aggressive interventions for the prevention of fatal arrhythmic events have been studied largely in patient populations at very high risk for fatal arrhythmias. The outcomes of adequately powered trials have demonstrated a hierarchy of benefit from various interventions, with the greatest benefit appearing to derive from implantable defibrillators (67,68), and with some potential survival benefit from amiodarone, at least in comparison to other antiarrhythmic drugs. However, the application of these observations is limited by the populations studied and by study design (Fig. 20.10). Specifically, the studies demonstrating benefit are limited to certain very high-risk subgroups (69), and the lower risk subgroups with larger numbers of cumulative events have not yet been adequately evaluated. Furthermore, the efficiency of such strategies for more general populations is open to serious question. Accordingly, higher degrees of risk resolution are needed in order to establish a rationale for aggressive therapy in other populations. Clinical markers, as developed to date, appear not to have the power to discriminate within intermediate-risk groups, and other strategies are needed. Among the latter is identification of genetically based risk factors. A recent epidemiological survey suggests that there is a specific familial risk for sudden cardiac death as a manifestation of acute myocardial infarction within a population of families with a strong history of acute coronary events (70). This implies that there may be genetic factors conditioning the response to ischemia that increase the susceptibility of a patient with a specific genetic pattern to an arrhythmic response to ischemia. Such observations are in their infancy, but offer the opportunity to identify high-risk subjects within the general population and apply preventive strategies to specific high-risk subgroups far in advance of their development of conditioning or triggering substrates.

VI. TREATMENT STRATEGIES FOR VENTRICULAR ARRHYTHMIAS IN THE CONTEXT OF UNDERLYING PATHOLOGY

Sustained ventricular tachyarrhythmias as defined earlier (i.e., VT, VF) are the most common mechanisms for sudden cardiac death, regardless of the specific structural etiology. However, the approaches to therapy may differ according the pathological basis of the arrhythmia (Table 20.1). In coronary heart disease, four general strategies are available (Fig. 20.11). Historically, antiarrhythmic drug therapy, initially prescribed empirically and subsequently guided by the results of electrophysiological studies performed with catheter techniques, was the only approach available. Over time, it was learned that stable and sustained monomorphic ventricular tachycardia was the most likely to respond to membrane-active drugs; but when multiple monomorphic tachycardias were observed in the same patient, success measured by loss of inducibility in the laboratory was less likely.

For the unstable monomorphic tachycardias, membrane-active drugs are generally less effective, with the possible exception of amiodarone given empirically and beta-blockers in some patients. Again, multiple unstable tachycardias were more difficult to treat. For the polymorphic ventricular tachycardias, beta-blockers, and to a lesser extent amiodarone, appear to have some benefit. For certain of the polymorphic tachycardias, especially in the absence of advanced structural heart disease, beta-blockers are still considered useful as primary therapy. For patients who survive ventricular fibrillation not associated with transient causes or an acute ischemic event, the trend recently has been away from using drug therapy as primary treatment in preference to implantable defibrillators. However, it is not yet clear that defibrillators offer an advantage over drug therapy in survivors of cardiac arrest who have ejection fractions greater than 40%.

Drug therapy has always been recognized to be limited by the fact that even a successful outcome in the laboratory left a residual treated risk that was still unacceptably high in patients with low ejection fractions and advanced disease. Surgical approaches emerged in the late 1970s and were popular through the early 1990s. Revascularization surgery is still an important component of therapy, particularly for the various unstable patterns (unstable monomorphic sustained VT, polymorphic VT, and ventricular fibrillation) when it can be reasonably demonstrated that ischemia is the mechanism for the ar-

Table 20.1 Therapeutic Strategy for Prevention of Ventricular Arrhythmias Based on Pathophysiologic Mechanisms of Remodeling.[a]

Therapeutic strategies	Techniques
Block or uncouple abnormal channels	Drugs
Balance regional channel dispersion	Drugs, Revascularization
Modify acquired pathways	Drugs, Surgery, Ablation
Alter loading conditions	Drugs, Surgery (?)
Alter stress/strain relationships	Drugs, Surgery
Control heart rate	Drugs, Ablation

[a]Strategies can be designed to address channel alterations, dispersion of channel properties, anatomical pathways, loading conditions, and heart rate. The techniques available include drugs, various forms of surgery and catheter ablation techniques.

CLINICAL RHYTHM	INDUCED RHYTHM	ANTI-ARR DRUGS	SURGERY or PTCA	CATHETER ABLATION	ICD
SUSTAINED VT: Monomorphic, Stable	Single VT	Membrane active	Anti-arr surg +/- anti-isch	Feasible	Tiered Rx
	Multiple VTs	Secondary	Anti-ischemia, if appropriate	Very low success rate	Tiered Rx
SUSTAINED VT: Monomorphic, Unstable	Single VT	Membrane active ?; β-blockers	Anti-ischemia, if appropriate	No	Defib mode; Tiered Rx ?
	Multiple VTs	Secondary			
POLYMORPHIC VENTRICULAR TACHYCARDIA	Sustained or non-sustained	β-blockers?, amiodarone?	Anti-ischemia, if appropriate	No	Defib mode, Tiered Rx?
VENTRICULAR FIBRILLATION	Sus Mono VT [Stable]	Secondary	Anti-ischemia, if substrate identified	Low success; efficacy?	Tiered Rx
	Polymorphic or unstable VT; Non-Sus VT	β-blockers?		No	Defib mode
	V Fibrillation	Secondary			

Figure 20.11 Therapeutic strategies based on patterns of ventricular tachyarrhythmias in ischemic heart disease. The applicability and success of various strategies are based upon characteristics of spontaneous (clinical) and laboratory-induced arrhythmias. Strategies including antiarrhythmic drugs, surgery, or angioplasty (PTCA), catheter ablation, and implantable defibrillators (ICD). The application of these strategies depends upon the clinical arrhythmia. The options are not mutually exclusive, since there is the potential for both morbidity and mortality benefits by multiple strategies. (Modified from Ref. 1, with permission.)

rhythmia. As sole therapy, it is less useful for stable monomorphic ventricular tachycardias because they are not usually ischemically mediated, even though they may be ischemically triggered. For the latter category, there is still a role for surgical mapping and cryoablation along with aneursymectomy. In the patient with stable monomorphic tachycardia, when the heart can be mapped in the operating room, and revascularized at the same time, very good success rates can be achieved, both acutely and long-term. However, this is a small group of patients and the numerical importance of this procedure has faded from its former preeminence.

For certain forms of stable tachycardias that can be mapped in the electrophysiology laboratory, catheter ablation techniques are feasible and useful. Unfortunately, for most of the categories shown in Figure 20.11, this is not the case; but it is potentially useful for stable monomorphic ventricular tachycardias. As technology and mapping techniques improve, perhaps a larger subset of these patients will be considered candidates. Finally, the use of implantable defibrillators has emerged into a predominant role in the high-risk forms of ventricular arrhythmias. Tiered therapy, in which the defibrillator is programmed to pace-terminate sustained, hemodynamically stable rhythms before delivering a defibrillating shock, is commonly used for the monomorphic tachycardias. For patients with clearly documented, hemodynamically unstable tachycardias, in whom pacing is not expected

CLINICAL RHYTHM	INDUCED RHYTHM	ANTI-ARR DRUGS	CATHETER ABLATION	ICD
SUSTAINED VT: **Monomorphic,** **Stable**	Bundle-branch reentry	**Membrane active**	Yes	**Tiered Rx**
	Other or multiple reentrant VTs	Membrane active [low success rate]	Very low success rate	**Tiered Rx**
	Automatic VTs	Membrane active?; β-blockers ?	Unlikely	No
SUSTAINED VT: **Monomorphic,** **Unstable**	Single VT	Membrane active?; β-blockers ?	No	**Defib mode;** Tiered Rx?
	Multiple VTs			
POLYMORPHIC **VENTRICULAR** **TACHYCARDIA**	Sustained or non-sustained	β-blockers, Amiodarone?	No	**Defib mode**
VENTRICULAR **FIBRILLATION**	Sus Mono VT [Stable]	Membrane active	Low success rate	**Tiered Rx**
	Polymorphic or unstable VT; Non-Sus VT	β-blockers ? Amiodarone ? [Selected cases]	No	**Defib mode**
	V Fibrillation	Amiodarone ?		

Figure 20.12   Therapeutic strategies based on patterns of ventricular tachyarrhythmias in non-ischemic cardiomyopathies. Arrhythmia inducibility and options for therapy differ in nonischemic arrhythmias compared to arrhythmias in ischemic heart disease. In general, arrhythmias are less easy to induce in the laboratory setting, and the data are of uncertain value. Revascularization surgery has no role, but remodeling surgery might prove to be of some value in the future. For now, the tendency is to use reactive therapy (ICDs) rather than preventive therapy (drugs and ablation) in this category of patients. (Modified Ref. 1, with permission.)

to achieve conversion, programming of the devices directly to the defibrillation mode is generally used.

The various approaches outlined in Figure 20.11 are not mutually exclusive. Antiarrhythmic drugs are commonly used in conjunction with defibrillators to reduce the number of events to which the defibrillator responds, while depending on the defibrillator for reverting life-threatening events. Revascularization is not always sufficient for controlling the risk of arrhythmias and some patients who undergo revascularization will also receive either drugs or defibrillators. In addition, catheter ablation, even when not reliable for a primary therapy, is often used to reduce the number of events in patients with defibrillators by interrupting at least some of the circuits responsible for tachycardias.

The approach to the nonischemic cardiomyopathies is somewhat different (Fig. 20.12). For these patients, revascularization and antiarrhythmic surgery have no roles, and the strategies are limited to antiarrhythmic drugs, catheter ablation, and implantable defibrillators. One of the major problems with the nonischemic cardiomyopathies is that the results observed in the electrophysiology laboratory commonly do not correlate well with clinical arrhythmias; and, therefore, electrophysiological testing is less useful. There are some exceptions, however. A specific form of monomorphic ventricular tachycardia, which is usually hemodynamically stable, is bundle branch reentry. In this arrhythmia, an impulse propagating from the atrium conducts down a bundle branch (usually the right), crosses the septum, and reenters the left bundle branch retrogradely, causing a stable reentrant tachycardia. This particular form, when recognized in the electrophysiology laboratory, is very amenable to catheter ablation as the primary therapy. Failing that, either drug therapy or tiered defibrillator therapy are useful. However, when the monomorphic tachycardias are multiple, catheter ablation alone will not be successful. For the remainder of the techniques and specific arrhythmias shown in Figure 20.12, implantable defibrillators have emerged as the preferred therapy. Antiarrhythmic drugs are unpredictable in the setting and not easily guided by electrophysiological studies, and only bundle branch reentry is predictably responsive to catheter ablation techniques. However, as in the case of ventricular arrhythmias in coronary heart disease, antiarrhythmic drugs and, to a lesser extent, catheter ablation, maintain a secondary role for patients who have implantable devices.

Beyond these two categories of diseases, other entities might require still different general approaches. For example, sustained ventricular arrhythmias caused by acute viral myocarditis is generally a transient risk, and although it may be life threatening at the time of the active inflammatory process, it generally does not indicate long-term risk. Accordingly, transient antiarrhythmic therapy may be appropriate, avoiding a commitment to long-term device therapy. In contrast, in the absence of structural disease, a benign form of monomorphic left bundle branch block tachycardia pattern may originate from a focus in the right ventricular outflow tract or left side of the septum in young individuals. These focal tachycardias are generally amenable to catheter ablation therapy; and even if they have a potential to respond to antiarrhythmic or beta-blocker therapy, many consider catheter ablation as the treatment of choice for these unusual arrhythmias.

For the patient at extraordinarily high risk of cardiac arrest due to ventricular tachyarrhythmias, such as survivors of out-of-hospital cardiac arrest and certain other categories of patients with sustained or nonsustained tachyarrhythmias and low ejection fractions, drug therapy has generally been disappointing. As a result of several clinical trials that have been reported recently, studying patients with ejection fractions less than 35% and other markers of risk for life-threatening arrhythmias, the implantable defibrillator has emerged as the preferred long-term therapy. Although the relative risk reductions using such therapy in reported trials have been impressive, the clinical impact of this approach has been limited because the population of individuals at risk for sudden death who fit into the categories described by these trials is quite small.

ACKNOWLEDGMENTS

Dr. Myerburg is supported in part by the American Heart Association Chair in Cardiovascular Research at the University of Miami and a grant from the National Institutes of Health, NHLBI–HL21735. Dr. Huikuri is supported in part by a grant from the Finnish Academy of Science, Helsinki, Finland.

REFERENCES

1. Myerburg RJ, Kessler KM, Castellanos A. Recognition, clinical assessment and management of arrhythmias and conduction disturbances. In: Alexander RW, Schlart RC, Fuster V, eds. Hurst's "The Heart." New York: McGraw Hill, 1998; 873–941.
2. Myerburg RJ, Kessler KM, Luceri RM, Zaman L, Trohman RG, Estes D, Castellanos A. Classification of ventricular arrhythmias based on parallel hierarchies of frequency and form. Am J Cardiol 1984; 54:1355–1358.

3. Chiang B, Perlman HV, Fulton M, Ostrander ID, Epstein RH. Predisposing factor in sudden cardiac death in Tecumseh, Michigan; A prospective study. Circulation 1970; 41:31–37.

4. Ruberman W, Weinblatt M, Goldberg JD, Frank CW, Chaudhary BS, Shapiro S. Ventricular premature complexes and sudden death after myocardial infarction. Circulation 1981; 64:297–305.

5. Bigger JT, Fleiss JL, Kleiger R, Miller JP, Rolnitzky LM, and the Multicenter Post-Infarction Research Group. The relationships among ventricular arrhythmias, left ventricular dysfunction, and mortality in the 2 years after myocardial infarction. Circulation 1984; 69:250–258.

6. Vismara LA, Amsterdam BA, Mason DT. Relation of ventricular arrhythmias in the late-hospital phase of acute myocardial infarction to sudden death after hospital discharge. Am J Med 1975; 5:6–12.

7. Robinson JS, Sloman G, Matthew TH, Goble, AJ. Survival after resuscitation from cardiac arrest in acute myocardial infarction. Am Heart J 1965; 69:740–747.

8. Hinkle LE, Jr, Thaler HT. Clinical classification of cardiac deaths. Circulation 1982; 65:457–464.

9. Myerburg RJ, Kessler KM, Bassett AL, Castellanos A. A biological approach to sudden cardiac death: Structure, function and cause. Am J Cardiol 1989; 63:1512–1516.

10. Myerburg RJ, Kessler KM, Castellanos A. Sudden cardiac death: Structure, function, and time-dependence of risk. Circulation (suppl I) 1992; 85:1–21–10.

11. Cooper RS, Simmons BE, Castaner A, Santhanam V, Ghali J, Mar M. Left ventricular hypertrophy is associated with worse survival independent of ventricular function and number of coronary arteries severely narrowed. Am J Cardiol 1990; 65:441–445.

12. Cupples LA, Gagnon DR, Kannel WB. Long- and short-term risk of sudden coronary death. Circulation 1992; 85(suppl I):I-II–I-18.

13. Ginzton LE, Conant R, Rodrigues DM, Laks MM. Functional significance of hypertrophy of the non-infarcted myocardium after myocardial infarction in humans. Circulation 1989; 80:816–822.

14. Cox MM, Berman I, Myerburg RJ, Smets MJD, Koslovskis PL. Morphometric mapping of regional myocyte diameters after healing of myocardial infarction in cats. J Mol Cell Cardiol 1991; 23:127–135.

15. Furukawa T, Bassett AL, Kimura S, Furukawa N, Myerburg RJ. The ionic mechanism of reperfusion-induced early after depolarizations in feline left ventricular hypertrophy. J Clin Invest 1993; 91:1521–1531.

16. Fozzard HA. The roles of membrane potential and inward Na^+ and Ca^{2+} currents in determining conduction. In: Rosen MR, Janse MJ, Wit AL, eds. Cardiac Electrophysiology: A Textbook. Mt. Kisco, NY: Futura Publishing, 1990:415–425.

17. Furukawa T, Kimura S, Furukawa N, Bassett AL, Myerburg RJ. Role of cardiac ATP-regulated potassium channels in differential responses of endocardial and epicardial cells to ischemia. Circ Res 1991; 68:1693–1702.

18. Manning AS, Hearse DJ. Reperfusion-induced arrhythmias: mechanisms and prevention. J Mol Cell Cardiol 1984; 16:497–518.

19. Priori SG, Mantica M, Napolitano C, Schwartz PJ. Early after depolarization induced in vivo by reperfusion of ischemic myocardium. Circulation 1990; 81:1911–1920.

20. Ideker RE, Klein GJ, Harrison L, Smith WM, Kasell J, Reimer KA, Wallace AG, Gallagher JJ. The transition to ventricular fibrillation induced by reperfusion after ischemia in the dog: A period of organized epicardial activation. Circulation 1981; 63:1371–1379.

21. Coronel R, Wilms-Schopman FJ, Opthof T, Cinca J, Fiolet JW, Janse MJ. Reperfusion arrhythmias in isolated perfused pig hearts: Inhomogeneities in extracellular potassium, ST and TQ potentials, and transmembrane action potentials. Circ Res 1992; 71:1131–1142.

22. Furukawa T, Myerburg RJ, Furukawa N, Kimura S, Bassett AL. Metabolic inhibition of $I_{Ca,L}$ and I_K differs in feline left ventricular hypertrophy. Am J Physiol 1994; 266 (Heart Circ Physiol 35):H1121–H1131.

23. Koyha T, Kimura S, Myerburg RJ, Bassett AL. Susceptibility of hypertrophied rat hearts to ventricular fibrillation during acute ischemia. J Mol Cell Cardiol 1988; 20:159–168.

24. Multiple Risk Factor Intervention Trial Research Group. Multiple-risk factor intervention trial: Risk factor changes in mortality results. J Am Med Assoc 1982; 248:1465–1477.

25. Myerburg RJ, Kessler KM, Castellanos A. Pathophysiology of sudden cardiac death. PACE 1991; 14(Part II):935–943.

26. Gettes LS. Electrolyte abnormalities underlying lethal ventricular arrhythmias. Circulation 1992; 85 (suppl I):1-70–I-76.

27. Lab MJ. Contraction-excitation feedback in myocardium: Physiologic basis and clinical relevance. Circ Res 1982; 50:757–766.

28. Calkins H, Maughan WL, Weissman HF, Surgiura S, Sagawa K, Levine JH. Effect of acute volume load on refractoriness and arrhythmia development in isolated chronically infarcted canine hearts. Circulation 1989; 79:687–697.

29. Franz MR, Burkhoff D, Yue DT, Sagawa K. Mechanically induced action potential changes and arrhythmia in isolated and in situ canine hearts. Cardiovasc Res 1989; 23:213–223.

30. Kammerling JJ, Green FJ, Watanabe AM, Inoue H, Barber MJ, Henry DP, Zipes DP. Denervation supersensitivity of refractoriness in non-infarcted areas apical to transmural myocardial infarction. Circulation 1987; 76:383–393.

31. Kozlovskis PL, Smets MJD, Duncan RC, Bailey BK, Bassett AL, Myerburg RJ. Regional beta-adrenergic receptors and adenylate cyclase activity after healing of myocardial infarction in cats. J Mol Cell Cardiol 1990; 22:311–322.

32. Barber MJ, Mueller TM, Henry DF, Felton SJ, Zipes DP. Transmural myocardial infarction in the dog produces sympathectomy in non-infarcted myocardium. Circulation 1982; 67:787–796.

33. Gaide MS, Myerburg RJ, Kozlovskis PL, Bassett AL. Elevated sympathetic response of epicardium proximal to healed myocardial infarction. Am J Physiol 1983; 14:646–652.

34. Schwartz PJ, Vanoli E, Stramba-Badiale M, De Ferrari GM, Billman GE, Foreman RD. Autonomic mechanisms and sudden death: New insights from analysis of baroreceptor reflexes in conscious dogs with and without a myocardial infarction. Circulation 1988; 78:969–979.

35. Le Rovere MT, Specchia G, Mortara A, Schwartz PJ. Baroreflex sensitivity, clinical correlates and cardiovascular mortality among patients with first myocardial infarction: A prospective study. Circulation 1988; 78:816–824.

36. Huikuri HV, Linnaluoto MK, Seppanen T, Airaksinen KEJ, Kessler KM, Takkunen JT, Myerburg RJ. Heart rate variability and its circadian rhythm in survivors of cardiac arrest. Am J Cardiol 1992; 70:610–615.

37. La Rovere MT, Bigger Jr. JT, Marcus FI, Mortara A, Schwartz PJ, for the ATRAMI Investigators. Baroreflex sensitivity and heart-rate variability in prediction of total cardiac mortality after myocardial infarction. Lancet 1998; 351:478–484.

38. Bigger JT, Fleiss JL, Steinman RC, Rolnitzky LM, Kleiger RE, Rottman JN. Frequency domain measures of heart period variability and mortality after myocardial infarction. Circulation 1992; 85:164–171.

39. Selzer A, Wray HW. Quinidine syncope: Paroxysmal ventricular fibrillation occurring during treatment of chronic atrial arrhythmias. Circulation 1964; 30:17.

40. Roden DM, Bennett PB, Snyders DJ, Balser JR, Hondeghem LM. Quinidine delays I_K activation in guinea pig ventricular myocytes. Circ Res 1988; 62:1055–1058.

41. Woosley RL, Chen Y, Freiman JP, Gillis RA. Mechanism of cardiotoxic actions of terfenadine. J Am Med Assoc 1993:269:1533–1536.

42. Myerburg RJ, Kessler KM, Castellanos A. Sudden cardiac death: Epidemiology, transient risk, and intervention assessment. Ann Intern Med 1993; 119:1187–1197.

43. Myerburg RJ, Kessler KM, Kimura S, Castellanos A. Sudden cardiac death: Future approaches based on identification and control of transient risk factors. J Cardiovasc Electrophysiol 1992; 3:626–640.

44. Muller JE, Tofler GH, Stone PH. Circadian variation and triggers of onset of acute cardiovascular disease. Circulation 1989; 79:733–743.

45. MacLure M. The case-crossover design: A method for studying transient effects on the risk of acute events. Am J Epidemiol 1991; 133:144–153.

46. Kleiger RE, Miller JP, Bigger Jr JT, Moss AJ, and the Multicenter Post-Infarction Research Group. Decrease heart rate variability and it association with increased mortality after acute myocardial infarction. Am J Cardiol 1987; 59:256–262.

47. Hartikainen JEK, Malik M, Staunton A, Poloniecki J, Camm AJ. Distinction between arrhythmic and nonarrhythmic deaths after acute myocardial infarction based on heart rate variability, signal-averaged electrocardiogram, ventricular arrhythmias and left ventricular ejection fraction. J Am Coll Cardiol 1996; 28:296–304.

48. Huikuri HV. Heart rate variability in coronary artery disease. J Intern Med 1995;237:349–357.

49. Zabel M, Klingenheben T, Franz MR, Hohnloser SH. Assessment of QT dispersion for prediction of mortality or arrhythmic events after myocardial infarction. Results of a prospective, long-term follow-up study. Circulation 1998; 97:2543–2550.

50. Epstein AE, Carlson MD, Fogoros RN, Higgins SL, Venditti Jr. FJ. Classification of death in antiarrhythmia trials. J Am Coll Cardiol 1996; 27:433–442.

51. Pratt CM, Greenway PS, Schoenfeld MH, Hibbem ML, Reiffel JA. Exploration of the precision of classifying sudden cardiac death. Implications for the interpretation of clinical trials. Circulation 1996; 93:519–524.

52. Perkiomaki JS, Huikuri HV, Koistinen JM, Mäkikallio T, Castelanos A, Myerburg RJ. Heart rate variability and dispersion of QT interval in patients with vulnerability to ventricular tachycardia and ventricular fibrillation after previous myocardial infarction. J Am Coll Cardiol 1997; 30:1331–8.

53. Valkama JO, Huikuri HV, Koistinen MJ, Yli-Mayry S, Airaksinen KEJ, Myerburg RJ. Relation between heart rate variability and spontaneous and induced ventricular arrhythmias in patients with coronary artery disease. J Am Coll Cardiol 1995; 25:437–443.

54. Schaumann A, Neufert C, Fabian O, Schuenemann S, Herse B. Short-long short sequences are in 33% the trigger mechanism of 1004 analyzed spontaneous ventricular arrhythmias. Circulation 1998; 98(suppl):I164.

55. Huikuri HV, Seppanen T, Koistinen MJ, Airaksinen KEJ, Ikaheimo MJ, Castellanos A, Myerburg RJ. Abnormalities in beat-to-beat dynamics of heart rate before the spontaneous onset of life-threatening ventricular tachyarrhythmias in patients with prior myocardial infarction. Circulation 1996; 93:1836–1844.

56. Huikuri HV, Valkama JO, Airaksinen KEJ, Seppanen T, Castellanos A, Myerburg RJ. Frequency domain measures of heart rate variability before the onset of nonsustained and sustained ventricular tachycardia in patients with coronary artery disease. Circulation 1993;87:1220–1228.

57. Shusterman V, Aysin B, Gottipaty V, Weiss R, Brode S, Schwartzman D, Anderson KP, for the ESVEM Investigators. Autonomic nervous system and spontaneous initiation of ventricular tachycardia. J Am Coll Cardiol 1998; 32:1891–9.

58. Vybirial T, Glaeser DH, Goldberger AL, Rigney DR, Hess KR, Mietus J, Skinner JE, Francis M, Pratt CM. Conventional heart rate variability analysis of ambulatory electrocardiographic recordings fails to predict imminent ventricular fibrillation. J Am Coll Cardiol 1993; 22:557–565.

59. Mäkikallio TH, Koistinen MJ, Tulppo MP, Jordaens L, Wood N, Peng CK, Goldberger AL, Huikuri HV. RR interval dynamics before the spontaneous onset of ventricular fibrillation. J Am Coll Cardiol 1998; 202C(suppl):26–30.

60. Schwartz PJ, Priori SG, Locati EH. Long QT syndrome patients with mutations of the SCN5A and HERG genes have differential responses to Na+ channel blockade and to increase in heart rate. Circulation 1995; 92:3381–3386.

61. Rosenbaum DS, Jackson LE, Smith JM, Garan H, Ruskin JN, Cohen RJ. Electrical alternans and vulnerability to ventricular arrhythmias. N Eng J Med 1994; 330:235–241.

62. Huikuri HV, Saarnivirta A, Tikkanen P, Airaksinen KEJ. QT interval dynamics before the spontaneous onset of ventricular fibrillation. Circulation 1998; 98(suppl):I–26.

63. Huikuri HV. Abnormal dynamics of ventricular repolarization—A new insight into mechanisms of ventricular tachyarrhythmias. Eur Heart J 1997; 18:893–895.

64. Schwartz PJ, Billman GE, Stone HL. Autonomic mechanisms in ventricular fibrillation induced by myocardial ischemia during exercise in dogs with a healed myocardial infarction. An experimental preparation for sudden cardiac death. Circulation 1984; 69:780–790.

65. Airaksinen KEJ, Ikaheimo MJ, Linnaluoto MK, Thavanainen KUO, Huikuri HV. Gender difference in autonomic and hemodynamic reactions to abrupt coronary occlusion J Am Coll Cardiol 1998; 31:301–306.

66. Airaksinen KEJ, Ylitalo A, Niemela M, Tahvanainen KUO, Huikuri HV. Heart rate variability and occurrence of ventricular arrhythmias during abrupt coronary occlusion. Am J Cardiol, in press.

67. Moss AJ, Hall WJ, Cannon DS, Daubert JP, Higgens SL, Klein H, Levine JH, Saksena S, Waldo AL, Wilber D, Brown MW, Heo M, for the Multicenter Automatic Defibrillator Implantation Trial Investigators. Improved survival with an implanted defibrillator in patients with coronary disease at high risk for ventricular arrhythmia. N Engl J Med 1996; 335:1933–1940.

68. The Antiarrhythmics Versus Implantable Defibrillators (AVID) Investigators. A comparison of antiarrhythmic-drug therapy with implantable defibrillators in patients resuscitated from near-fatal ventricular arrhythmias. N Engl J Med 1997; 337:1576–1583.

69. Myerburg RJ, Mitrani R, Interian A, Jr, Castellanos A. Interpretation of outcomes of antiarrhythmic clinical trials: Design features and population impact. Circulation 1998; 97:1514–1521.

70. Friedlander Y, Siscovick DS, Weinmann S, Austin MA, Psaty BM, Lemaitre RN, Arbogast P, Raghunathan TE, Cobb LA. Family history as a risk factor for primary cardiac arrest. Circulation 1998; 97:155–160.

21

Cardiac Pacemakers

ADAM ZIVIN and GUST H. BARDY

University of Washington, Seattle, Washington

RAHUL MEHRA

Medtronic, Inc., Minneapolis, Minnesota

I. PACEMAKERS

A. Technology

Attempts at electrical stimulation of cardiac muscle began early in the nineteenth century, but it was Marmorstein, in 1927, who first demonstrated clearly that the heart rate of a dog could be increased by pacing with an external pulse generator (1). In the early 1930s, Lidwill and Hyman were then first to successfully pace the human ventricle with an external artificial pacemaker (2,3). They used partially insulated needle electrodes that were simply plunged into the ventricle. Hyman argued that since the unit was to be used for emergency applications, it should not require an external source of electrical power, and decided against the use of batteries because the life period of available commercial batteries at that time was only about 6 months. His pacemaker delivered current from a small hand-cranked spring wound motor that could provide 6 min of pacing. Early pacemakers were capable of saving lives, but aroused little interest in the medical community. Jeffrey has argued that early pacemakers were "premature technologies" that did not address a significant medical need, as the issues relating to cardiac arrest were not appreciated at that time (4).

In the subsequent two decades, awareness of the problem of bradycardia-related cardiac arrest intensified because of the improved understanding of cardiac arrhythmias and of Stokes-Adams attacks (syncope occurring as a result of acquired third-degree heart block). Interest in cardiac pacing was reignited in 1952 when Zoll announced that he had revived a patient from cardiac standstill by means of external pacing (4–6). Zoll utilized two electrodes strapped across the chest. Although this approach was noninvasive, it was very painful and resulted in skin burns. Zoll's external power-driven pacemaker maintained the patient's heartbeat for more than 50 h until the patient recovered sufficiently to go home. By 1954, a commercial version of Zoll's externally driven, nonimplantable pacemaker had begun to be used in academic centers around the country. A power outage in 1957 motivated Lillehei at the University of Minnesota to ask a local electrical engineer, Earl Bakken, to develop a compact, battery-operated pacemaker (7). This device was soon in widespread use for postoperative heart block following cardiac surgery.

The advent of the implantable pacemaker began in 1958 when a Swedish team led by Sennings (a physician) and Elmqvist (an engineer) implanted the first internal pacemaker (8,9). A thoracotomy was required and pacing was done through electrodes sutured on the epicardium. In these early systems, significant problems with changes in pacing threshold, lead infection, and lead breakage were common occurrences. Furman subsequently argued that transvenous lead implantation would resolve many of these issues and the advent of transvenous pacing was heralded in 1958 when he successfully paced an elderly patient with a catheter electrode inserted transvenously (10). Other investigators, including Greatbach and Bakken continued to refine the implantable pacemaker and took on the challenge of solving various technical problems, notably device miniaturization, longer life batteries, and stable, reliable lead material (11,12).

The need for implantable pacemakers grew as the indications for implantation expanded from atrioventricular (AV) conduction disturbances to management of sinus node dysfunction. Technology evolved rapidly with the development of lithium-iodide batteries that had greater longevity (12). Early pacemakers were encased in silicone rubber, but metallic encapsulation was found to be biocompatible and facilitated device extraction. The need to noninvasively change pacemaker rate and other stimulation and sensing parameters led to the development of pacemakers whose functions could be altered with an external programmer using radio-frequency energy to communicate with the pacemaker, a technology used to this day. Electronic advances then led to major miniaturization using integrated circuits (ICs) as opposed to discrete components—a real breakthrough. Continued improvement of pacing leads has also been a major technical challenge, as these need to endure literally millions of flexures during their lifetime. Lead materials used today rely on silicone and/or polyurethane, which are more biocompatible and reliable than earlier materials. With all these technical refinements, present-day pacemakers are relatively small (10–15 cc) and can pace reliably for up to 8–10 years before generator replacement is needed. Figure 21.1 shows in graphic terms the marked reduction in device size that has occurred over the past 30 years.

The primary functional challenge for contemporary pacemakers is to maintain heart rate based on circulatory needs, pacing in a manner that mimics the natural physiology of excitation and conduction. In a healthy heart, the sinus node is modulated by the autonomic nervous system and its rate is determined by a multiplicity of factors, such as physical activity, emotion, blood pressure, etc. Not only the rate, but also the activation sequence and AV conduction time vary with demand and these requirements must also be considered. Rate is controlled by pacemaker discharge rate and the excitation and conduction sequence are dependent on the placement of the pacing electrodes.

1. NBG Code

To simplify and standardize the functional descriptions of all pacemakers, a five-letter code was developed by the North American Society of Pacing and Electrophysiology in conjunction with the British Pacing and Electrophysiology Group (NBG Code). This code

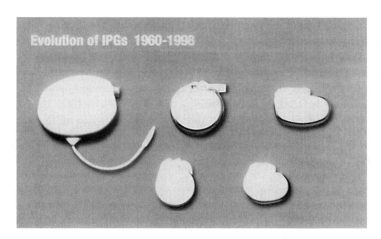

Figure 21.1 Evolution of implantable pacemaker generators. Early pacemaker devices were large and have reduced markedly in size over recent years. This reduction in size occurred as a result of improved packing density and use of integrated circuits and batteries with higher energy density such as the lithium-iodine battery. Shown (upper left to right) are representative models from 1960, 1978, and (bottom row) 1985, 1988, and 1999 models. Around 1980, microprocessors began to be incorporated into pacemakers; 1985 saw the first rate-responsive pacemaker widely used in patients. It used a piezoelectric crystal to modulate the rate of the pacemaker output. Size has continued to decrease and contemporary models occupy volumes of about 10–11 ccs.

Table 21.1 NBG Pacemaker Code

1st position Chamber(s) paced	2nd position Chamber(s) sensed	3rd position Pacing response to sensed activity	4th position Rate modulation
V = **V**entricle	**V** = **V**entricle	**O** = No sensing (asynchronous)	**O** or none = not rate modulated
A = **A**trium	**A** = **A**trium	**I** = Output **I**nhibited	**R** = Activity **R**esponsive
D = **D**ual (A and V)	**D** = **D**ual	**T** = Output **T**riggered	
O = Neither	**O** = Neither	**D** = **D**ual (inhibition and triggering[a])	

[a]In dual-chamber "D" mode, "A" sense triggers "V" pace. See text for discussion.

is used to indicate device pacing and sensing functions, as well as rate response and programming capabilities (Table 21.1).

Nearly all current applications of pacemakers place only one pacing/sensing lead in a given cardiac chamber; consequently, pacemakers are usually termed "single-chamber" or "dual-chamber," depending on the number of leads they can accept. The simplest type of pacemaker, therefore, is a single-lead, asynchronous, "VOO" type, which has the ability to pace in the ventricle only ("V" in 1st position), and is unable to sense in either chamber ("O" in 2nd position) or, by extension, respond to sensed activity ("O" in 3rd position). The next level of complexity is a single-lead system capable of sensing in the ventricle, with the ability to inhibit pacing output in response to intrinsic activity. This type is coded "VVI." The lead of a single-chamber pacemaker could also be placed in the atrium for isolated sinus node dysfunction with "AOO" and "AAI" being the appropriate designations. Dual-chamber pacemakers allow pacing and sensing in both the atrium and ventricle and are the most common type implanted in the United States. A common mode notation for such a dual-chamber pacemaker is DDD, indicating pacing in both chambers (D in the 1st position), sensing in both chambers (D in the 2nd position), and the ability to inhibit and trigger output (atrial sense/ventricular pace; D in the 3rd position).

2. Single-Chamber Pacemakers

The most basic pacemaker design is a single-chamber "asynchronous" pacemaker that paces the atrium or the ventricle at a fixed rate. The term asynchronous means this pacemaker has no capacity to sense intrinsic atrial or ventricular activity when there is intermittent spontaneous activity, the pacing stimuli from these pacemakers compete with the intrinsic activity and such stimuli could fall in the relative refractory period of the myocardium, potentially initiating atrial or ventricular tachyarrhythmias. Such a pacemaker would also have compromised longevity by pacing more frequently than needed.

To prevent such problems, the "demand" pacemaker was developed. An atrial demand pacemaker (AAI), for example, can sense intrinsic atrial excitation and alter the timing of each pacing stimulus accordingly. Pacing and sensing occur through the same lead electrode. The response of an atrial demand pacemaker and the shape of a typical sensed signal in the atrium is shown in Figure 21.2. The goal of such a pacemaker is to pace the heart at a minimum rate and reset the timing of the next scheduled pacing stimulus if an atrial beat is sensed. Although this concept is simple, technical limitations of the device make its operation complex. In all pacemakers, after a sensed atrial signal (the atrial electrogram) or delivery of a pacing stimulus, a "refractory period" is initiated. The initial segment of the refractory period is the "blanking" period. The difference between the "blanking" and "refractory" period is that during the blanking period, the pacemaker sensing function is completely shut off, whereas during the refractory period, sensing can occur, but it is not used for timing of the next pacing stimulus. Detection of atrial events during

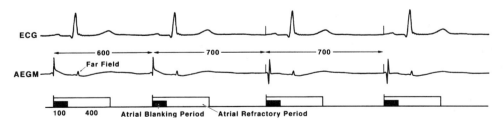

Figure 21.2 Electrical function of an AAI pacemaker programmed to pace at an interval of 700 ms. An electrode in the atrium senses sinus beats at 600 ms, which inhibits the pacing output. This is followed by two atrial-paced beats at 700 ms since no additional spontaneous atrial activity is detected. The AEGM (atrial electrogram) channel illustrates the signal recorded by the atrial lead. It is composed of a signal from the atrium (large sharp deflection) followed by a smaller "far-field" signal that represents ventricular depolarization. The bottom trace shows that the atrial blanking and the atrial refractory periods begin at the onset of detection of the atrial beat or the pacing stimulus, and last for 100 and 400 ms, respectively.

the refractory period is used for detection of atrial arrhythmias and for "mode-switching," a feature discussed below.

There are several reasons for incorporating a blanking period. Following application of a pacing stimulus, an electrical charge develops at the electrode–tissue interface due to polarization of the electrode, regardless of whether the pacing stimulus successfully stimulates the tissue. This polarization voltage can be relatively large, but decays with time. To avoid detection of this voltage, which is not a representation of true intrinsic atrial activity, a blanking period of about 100 ms is used. The blanking period following a sensed atrial electrogram prevents multiple sensing of the same beat when the atrial signal is fragmented. In some pacemakers, the duration of the atrial blanking period following a paced beat is longer than that following a sensed beat, to accommodate for polarization.

A refractory period is incorporated into the timing cycle because during this period a sensed signal is most likely "noise." The most common source of noise is ventricular depolarization detected by the atrial lead as a "far-field" signal (Fig. 21.2). Other sources are externally generated electromagnetic interference and myopotentials. The key disadvantage of the refractory period mode of operation is that an atrial premature beat occurring during this period is ignored. Following the refractory period, the pacemaker is capable of sensing intrinsic electrical activity and responding accordingly. A premature beat that is sensed after the refractory period, but prior to the anticipated delivery of the next pacing stimulus, resets the timing cycle.

A single-chamber ventricular "demand" pacemaker has the same features as the single-chamber, atrial demand pacemaker, except that the pacing electrode is in the ventricle (Fig. 21.3). Detection of far-field signals (i.e., signals arising from the unpaced chamber) does not occur due to the inability of atrial mass to generate enough voltage to be sensed. Therefore, the duration of the refractory period can be shorter for a ventricular pacemaker than for an atrial pacemaker, but the blanking period is similar because of the polarization factor discussed above. This refractory period helps avoid oversensing of the T wave in the ventricle.

3. Dual-Chamber Pacemakers

If a patient has an intact AV conduction pathway, an atrial demand pacemaker can simulate normal excitation and conduction of the heart by taking advantage of conduction through

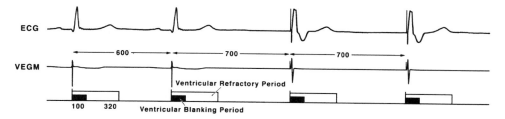

Figure 21.3 Electrical function of a VVI pacemaker set to pace at an interval of 700 ms. The electrode in the ventricle senses ventricular activations at 600 ms and suppresses pacemaker output. This is followed by two ventricular paced beats at 700 ms as no additional spontaneous activity is detected until that interval. The VEGM (ventricular electrogram) trace shows the signal recorded by the lead and consists primarily of ventricular depolarizations. The lower channel trace shows that the ventricular blanking and ventricular refractory periods begin at the onset of detection of ventricular depolarizations or the pacing stimulus and last 100 and 320 ms, respectively.

the His-Purkinje system. Such a pacemaker, however, lacks the ability to change its rate based on metabolic demand, and in case of AV conduction abnormalities, cannot pace the ventricles. Primary developments that help to address these two issues have been the incorporation of protocols to incorporate rate responsiveness and to perform dual-chamber pacing.

Dual-chamber pacemakers are significantly more complex than single-chamber units. The simplest operation of a dual-chamber DDD pacemaker is described in Figure 21.4. After pacing in the atrium, if a ventricular beat is not sensed within a designated AV interval, the ventricle is paced to maintain AV synchrony. If no further intrinsic activity is sensed, the next atrial pacing stimulus occurs to maintain a preprogrammed minimum heart rate. Additional technical challenges arise with these designs. First, a pacing stimulus creates such a large electrical field that it could be sensed by the electrode in the other chamber as cardiac activity were it not for a "cross-chamber blanking period" (Fig. 21.4). The sense amplifier for the nonpaced chamber is shut off (blanked) for 20 to 30 ms to prevent sensing of this pacing stimulus. Although the cross-chamber blanking period solves a technical problem, any true cardiac signal that occurs during this period would be ignored. Typically, this is not an issue as true cardiac signals last longer than 20 to 30 ms and extend beyond the blanking period.

An additional problem can arise if conduction does not occur through the AV node following atrial pacing. The paced ventricular beat could potentially conduct retrograde

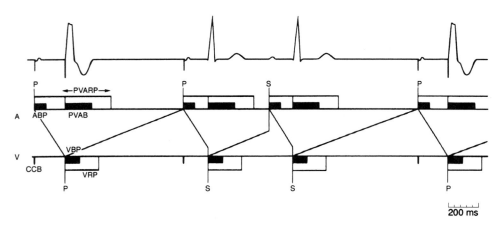

200 ms

Figure 21.4 Electrical function of DDD pacemaker function. The top channel trace simulates the surface ECG, "A" the atrial channel, and "V" the ventricular channel. The first beat is an atrial and ventricular paced beat, the second an atrial paced beat that occurs at the lower pacing rate with normal AV conduction. The third is a premature atrial beat that also conducts to the ventricle and inhibits atrial and ventricular pacing. The atrial pacing stimulus "P" initiates an atrial blanking period (ABP) in the atrium and a cross chamber blanking (CCB) period in the ventricle. After an AV delay, since no spontaneous ventricular activity is sensed, the ventricular pacing stimulus is delivered. This initiates the ventricular blanking period (VBP) and ventricular refractory period (VRP). In the atrial channel, the ventricular stimulus also initiates a post ventricular blanking period (PVAB) and post ventricular atrial refractory period (PVARP). The functions of each of these periods are discussed in the text. After the ventricular pacing stimulus, a VA delay is initiated, resulting in atrial pacing for the second beat. This atrial beat conducts to the ventricle and is sensed "S" by the ventricular lead prior to the expiration of the pacemaker's AV delay. Hence, no ventricular stimulus is delivered for this beat.

through the AV node and reactivate the atria soon after the previous atrial activation. If this retrograde atrial activation is outside the atrial refractory period, it would be sensed by the atrial sense amplifier and the next ventricular pacing stimulus would be scheduled after the programmed AV interval. With relatively short VA and AV intervals, this can result in inappropriate high-rate pacing in the ventricle, and is referred to as pacemaker-mediated tachycardia (PMT). This does not occur normally because after AV node–His–Purkinje conduction, ventricular activation cannot immediately reactivate the atria because the AV node is refractory. To solve this problem, a post-ventricular atrial refractory period (PVARP) is programmed into dual-chamber pacemakers. A retrograde atrial beat would fall within this period and be ignored. This refractory period in the atrial channel begins at the onset of the ventricular pacing stimulus. The initial part of the PVARP is called the post-ventricular atrial blanking period (PVAB), during which time no sensing occurs in the atrial channel. The PVAB is typically long enough to avoid both sensing of the ventricular pacing stimulus and the far-field QRS signal. Note that although the function of these refractory and blanking periods is to solve problems, they create unique issues of their own, as true cardiac signals occurring during these periods are ignored. The ideal scenario would be for the pacemaker to have enough intelligence to eliminate the need for all refractory and blanking periods.

With this elementary background it is possible to predict the behavior of dual-chamber pacemakers during premature atrial and ventricular beats, as well as during supraventricular tachyarrhythmias. In the DDD mode, no detection of atrial premature beats can occur during the atrial blanking and the PVAB periods. If the atrial premature beat falls between PVAB and end of PVARP, its detection is not used for timing of ventricular events, but for determining if the pacemaker should switch its mode of operation, a feature used to prevent high-rate tracking of atrial arrhythmias. Only atrial beats that fall outside the PVARP will result in a ventricular pacing stimulus after an AV delay. The fastest rate at which the ventricle can follow the atrium corresponds to an interval that is the sum of the AV interval and the PVARP and is typically less than 150 beats per minute. If an atrial tachyarrhythmia such as atrial fibrillation occurs, this is the maximum rate at which the ventricle will be paced. For patients with supraventricular tachyarrhythmias, the DDD pacing mode is compromised by this potential for rapid ventricular pacing. To circumvent this problem, algorithms are incorporated into pacemakers that change the pacemaker's response to sensed atrial events from a "tracking" (DDD) to a "nontracking mode" (DDI), if the atrial rate exceeds a programmable detection rate. DDI mode avoids rapid tracking of atrial tachyarrhythmias and behaves like VVI during atrial fibrillation. During DDI function, AV synchrony is maintained during atrial pacing, but not following native P waves.

4. Rate-Responsive Pacing

To mimic physiological changes in heart rate during exercise, emotional swings, and other demands of normal daily life, it is desirable to be able to change the pacing rate fairly rapidly. Under conditions of increased metabolic demand, the total oxygen uptake (cardiac output × arteriovenous oxygen difference) may increase from 0.3 to 3 L/per min. Although the arteriovenous oxygen difference increases somewhat to meet this demand, increases in cardiac output are the primary contributor and can increase from 5 to 20 L/min, largely as a result of increases in heart rate, and secondarily by an increase in stroke volume. The relationship between cardiac output and heart rate has been well investigated and typically consists of three phases (13). In the first, the increase in cardiac output is proportional to heart rate. In the second, cardiac output reaches a plateau with respect to heart rate due to inadequate ventricular filling time. In the third, cardiac output decreases with increasing heart rate due to markedly reduced time for diastolic filling. The heart rates for the three phases can vary and also depends on the extent of autonomic tone, myocardial disease, posture, etc.

Atrioventricular synchrony contributes to the hemodynamic performance of the heart. It has been shown that atrial systole can increase cardiac output in some cases by 20 to 30% (14). The extent of improvement is influenced by both pathological and normal physiological parameters. In the normal heart, this contribution decreases at higher heart rates, which explains the small difference found in the hemodynamic performance of patients with normal hearts with high-rate ventricular (VVI) pacing, compared with DDD pacing. Cardiac pumping efficiency is higher with AAI pacing than with VVI pacing, with DDD pacing falling in between, even when the AV interval is optimized (15–18). Preserving the normal activation sequence through the AV node and His-Purkinje system improves atrial contractility and reduces preload dependence, in comparison with dual-chamber pacing (16). It has been proposed that in a patient with a DDD pacemaker and intact AV conduction, long AV intervals are required to promote normal AV conduction and avoid ventricular stimulation. However, due to the PVARP interval, this could restrict the maximum atrial rate a pacemaker could track. Some of the more recent generations of pacemakers attempt to minimize ventricular pacing by automatically prolonging the AV interval, if intrinsic AV conduction is observed. Another way of avoiding stimulation in the ventricular apex would be to change the site of ventricular pacing. Despite promising early experiments, recent results with pacing the proximal septum or right ventricular outflow tract have not been encouraging (19,20). Chronic His bundle pacing has also been shown to be possible in a limited series of patients, but it is technically demanding (19–21).

There are several different types of artificial sensors available to adjust heart rate, with the ideal sensor being one that is sensitive to physiological demands, such as exercise and emotion, by sensing circulating catecholamines, neural activity, or some other activity-related parameter. To date, it has not been possible technically to develop such sensors and most present-day sensors detect consequences of basic physiological variables with high specificity (i.e., unresponsive to environmental noise), and an appropriate speed of response (22), but not in any sense with the same degree of sensitivity as the intact, sinus node.

The most commonly used artificial sensor is an activity sensor that detects body movement and increases heart rate accordingly. This sensor responds to movement and its key disadvantage is that it has an inadequate response to activities with little movement, such as isometric exercise and emotional changes. Another commonly used sensor measures respiratory minute ventilation, as determined by cyclical changes in impedance between the pacemaker housing and the pacing lead. Other sensors detect central venous temperature, pH, oxygen saturation, QT interval, right ventricular pressure gradient, pre-ejection interval, and endocardial-evoked responses. Most of these are incorporated within special leads and each has been evaluated with varying degrees of rigor in clinical studies (23). Recently, there has been a trend toward using more than one sensor type within a single pacemaker to maximize the advantages of each.

5. Electrical Stimulation

To excite a single cardiac cell, the transmembrane potential must depolarize to a critical threshold value. The normal resting potential of a cardiac cell varies, depending on the tissue studied, from –50 to –60 mV in the SA and AV nodes, to –80 mV in working myocar-

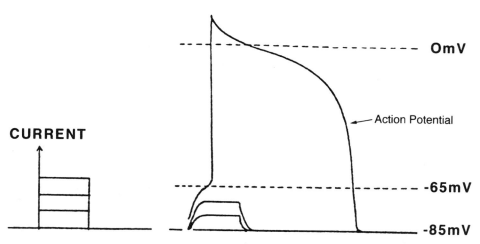

Figure 21.5   The transmembrane potential of a cell is shown at increasing levels of depolarizing current. At the lowest stimulus amplitude, the cell depolarizes, but does not reach threshold (–65 mV) for activation. Although the stimulus current changes instantaneously, the transmembrane potential of the cell changes slowly due to membrane capacitance. The stimulus of the next magnitude also fails to excite the cell. However, when the stimulus amplitude is increased further, the transmembrane potential reaches threshold and an action potential is elicited.

dium, and –90 mV in the Purkinje system. Also, enough transmembrane current must be delivered by a pacemaker for each cell to reach its threshold potential, which also varies, being more negative in Purkinje fibers and less negative in the SA node. Figure 21.5 shows the effect of varying depolarizing current in a single cell. When a rectangular stimulus is turned on, the transmembrane potential does not reach a stable value instantaneously. Rather, because the cell membrane can store charge (due to its intrinsic capacitance), the potential change occurs more slowly than the instantaneous change shown in the stimulus pulse. Similarly, when the stimulus is turned off, the transmembrane potential decays more slowly. As the stimulus amplitude is increased, threshold transmembrane potential is reached and the ion mechanisms responsible for the action potential are initiated. Due to membrane capacitance, if the stimulus duration were too short, the threshold potential would not be reached before the stimulus was turned off, or an extremely high stimulus amplitude would be required to elicit the action potential. If the stimulus duration was very long and the transmembrane potential reached threshold before completion of the stimulus, the electrical energy in the remainder of the electrical stimulus would be wasted. Due to inherent cell capacitance, there is a defined relationship between stimulus duration and threshold for excitation. This relationship is the "strength–duration" curve of excitation (Fig. 21.6), which defines the effect of stimulus or pulse duration on excitation threshold (24–28). Within this relationship there is an optimal pulse duration that can be used to minimize the energy required for excitation. The energy dissipated per pulse is the product of $V*I*d$ ($V*V*d/R$ or $I*I*R*d$), where V is the voltage threshold, I is the current threshold, R is the resistance of the pacing circuit, and d is the pulse duration. Figure 21.6 illustrates how, by assuming a certain resistance, the voltage threshold can be converted to an energy threshold, resulting in an energy strength–duration curve. The pulse duration at which the energy threshold reaches a minimum defines the optimal pulse duration.

CARDIAC PACING

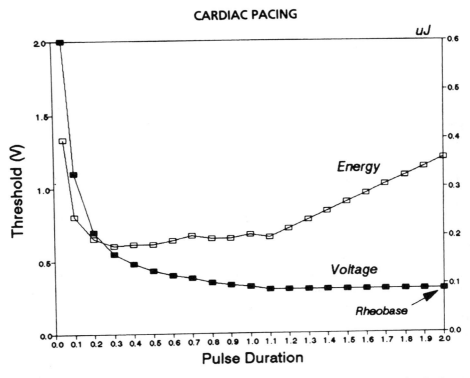

Figure 21.6 Typical voltage and energy thresholds for electrode cardiac pacing is shown as a function of pulse duration, expressed in milliseconds. Voltage threshold is expressed in volts and energy threshold in microjoules. With increasing duration, the voltage threshold decreases and reaches a plateau called the Rheobase. On the other hand, energy threshold has a minimum value between 0.3 and 0.5 ms.

Not only should the pulse duration be optimized, but also pacemaker electrode designs should be designed to provide the lowest energy threshold. It is postulated that to elicit myocardial excitation, a certain minimum number of cells need to be excited. Because the heart is a functional syncytium (i.e., an aggregate of electrically interconnected cells), an impulse will propagate once this critical volume of cells is excited. Therefore, for pacing, a minimum amount of current is needed per unit area of tissue (current density) to excite this aggregate of cardiac cells (29,30). The typical current density required for excitation is about 2.5 mA/cm^2 (31). If one assumes a simple spherical electrode geometry, the current threshold would be proportional to the product of current density threshold times the square of the electrode radius (since current density threshold = current threshold/electrode surface area). This implies that if the electrode surface area is small, a smaller amount of current is required to reach this threshold current density and excite the tissue. The resistance of a spherical electrode is inversely proportional to its radius because, as discussed above, the energy used by the pacing pulse is $I*I*R*d$ and therefore it proportional to $(r)^3$—the cube of the electrode radius. Hence, the energy threshold can be minimized by reducing the radius of the electrode: the smaller the electrode area, the lower the energy requirement. For this reason, decreasing the surface areas of pacing electrodes is a continuously improving goal within the pacing industry.

Unfortunately, the pacing threshold represents a variety of dynamic physiological factors and changes with daily activity, metabolic activity, pH, lead maturity, etc. (32). Therefore, if one were to set the stimulus magnitude of a pacing device at just this threshold, consistent pacing capture might not be achieved. There are two methods for resolving this issue: one is to program the pacemaker output at a multiple of the pacing threshold called a "safety factor," and the second is to develop systems that measure the threshold and adjust the output of the pacemaker accordingly. The value of the "safety factor" is based on the type of lead and the expected changes in the threshold. The larger the safety factor, the higher the pacing output and the greater the battery drain. Systems that automatically adjust output, based on frequent measurement of the capture threshold, must assess capture from the pacing electrode itself. This usually requires special low-polarization leads, special detection algorithms, or unique pacing waveforms; it has been shown to be technically effective in the ventricle, but continues to be a challenge in the atrium (33).

It is important to note that pacing threshold is also affected by other variables, such as the polarity of the stimulus. For current to flow, one needs a negative and positive electrode. If the pacing electrode at the tip of the catheter is negative with respect to the housing of the pacemaker electrode (the "can"), it is termed "cathodal." If the electrode is positive, pacing is "anodal." Cathodal pacing thresholds are lower than anodal pacing thresholds and therefore all unipolar pacing is done in the cathodal mode (34). During bipolar pacing, the lead tip electrode is made negative with respect to the proximal ring electrode to achieve a lower threshold. Unfortunately, it has been found that after leads have been in place for a period of time, a fibrotic capsule frequently forms around the electrode and separates it from excitable tissue (35,36). For the same current stimulus, the current density in the excitable tissue decreases and therefore the stimulus current must be increased to reach the threshold current density. In general, this fibrotic tissue is a good conductor of electricity and does not lead to a dramatic increase in pacing resistance. "Steroid-eluting" electrodes are used routinely to minimize the size of this fibrotic capsule and minimize increases in pacing threshold over time (37).

6. Pacemaker Sensing

Sensing intrinsic cardiac electrical activity is required in pacing devices to determine the timing of the pacing stimulus and help decide which chamber to pace in a multielectrode system. The first generation of pacemakers lacked sensing capability and paced the heart at a fixed rate. This mode of pacing had two problems. First, if the patient's intrinsic rhythm increases, the device has no way to suppress output to prevent unnecessary battery drain. Second, if the stimulus were to fall during relative refractoriness in the vulnerable period, there is a theoretical possibility one could initiate fibrillation. In practice, this is extremely rare because the pacing stimulus strength is usually inadequate to capture the tissue, except perhaps during periods of acute ischemia.

When a wave of cardiac depolarization travels around an electrode, it gives rise to an electrical signal with three components: an initial R wave, the intrinsic deflection (ID), and the S wave. The R wave reflects the signal recorded as the electrical wavefront approaches; the ID occurs as the wavefront passes; and the S wave is recorded as the wavefront recedes from the electrode (Fig. 21.7). Pacemaker sensing circuits detect signals that have an adequate amplitude and slope, or "slew rate." Slew rate is usually defined as the rate of change of the signal amplitude expressed in V/s. The intrinsic deflection typically has the largest slew rate. It is advantageous to detect only signals with high slew rates for two reasons. First, this minimizes detection by atrial electrodes of far-field ventricular signals that

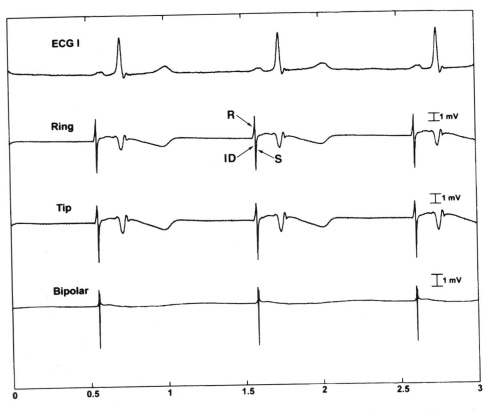

Figure 21.7   Signals recorded from a bipolar atrial electrode located in the right atrial appendage during sinus rhythm. The unipolar signals recorded from the ring and tip electrodes show an atrial activation signal followed by a smaller far field signal due to ventricular activation. The "R", ID (intrinsic deflection), and "S" waves represent the wavefront approaching, passing underneath, and receding from the electrode, respectively. The bipolar signal, which is a subtraction of the two unipolar signals, shows distinct atrial activation with no far-field "noise" signal.

may be of high amplitude, but typically a low slew rate (Fig. 21.2). Second, for ventricular electrodes, this prevents overdetection of ventricular T waves that can also have high amplitudes but low slew rates.

Another method of minimizing far-field signal detection for atrial electrodes is to sense with a "bipolar" electrode (between lead tip and ring) rather than with a "unipolar" electrode (between tip and housing). Bipolar electrodes record smaller far-field signals, but even here it is desirable to reduce the spacing between the tip and the ring electrodes and to place the atrial lead as far away from the ventricle as possible (38). The interelectrode spacing should not be too small because reduction in size of, and therefore ability to detect, the intrinsic deflection may occur. Another disadvantage of unipolar sensing is that because of the widely spaced sensing "antenna," myopotential signals can be detected during maneuvers that exercise the pectoral muscles. This problem is eliminated with bipolar sensing as the pacemaker electrode housing is not used for sensing. There are many other variables in addition to electrode spacing that can affect the size of the detected signal: the size of the electrode (smaller electrodes typically give rise to larger signals), the distance between the electrode and viable tissue once a fibrotic capsule has formed (signal size decreases with increasing distance from viable tissue), and respiration (39).

7. Batteries

The longevity and, hence, value of a pacemaker is significantly influenced by the energy drained by the pacing stimulus. However, this is only one of many functions that deplete battery energy. Both the analog sense amplifiers and the digital circuits that control device timing require power. These semiconductor circuits are typically of CMOS (complementary metallic oxide semiconductor) design, which facilities miniaturization and very low power consumption. Batteries that power these devices have undergone significant evolution since the introduction of early pacemakers, which used a rechargeable nickel–cadmium formulation (40). These pacemakers were inductively recharged by an external transmitter and an implanted receiver. This technology soon fell into disfavor, however, because physicians were uncomfortable placing the responsibility of charging pacemakers in the hands of patients, and the technology was not particularly efficient. Subsequently, mercury–zinc batteries were used that were cast in epoxy to permit dissipation of the evolved hydrogen. Unfortunately, battery longevity was still limited to about 2 years. Although nuclear batteries (plutonium-238) were used for a brief period, the major advance came with the introduction of lithium batteries which, since 1972, have become the primary supply for most pacemaker systems (11).

A critical safety issue in pacemaker therapy is the determination of pacemaker batteries' "end-of-life" date, which augurs complete system failure. In most pacemakers, the rate of pacing is used as an indicator of battery voltage and, as the battery reaches its end of life (almost 6–10 years), the pacing rate decreases slightly, although usually not enough to be clinically significant. This slight change in pacing rate can be monitored by a transtelephonic transmission allowing early detection and remote safety evaluations.

8. Leads and Electrodes

An ideal pacemaker electrode lead would be one that lasts the life of the patient without mechanical or electrical degradation. Electrical performance is primarily influenced by the electrode design and mechanical performance by the conducting and insulation materials. As discussed above, electrodes are usually designed to have the lowest energy thresholds and a minimal increase in threshold over time. Presently, small surface area electrodes are the favored approach to this issue. It is also important that there be no energy loss in polarization at the electrode–tissue interface, as this tends to dissipate and waste energy. New materials, such as porous and platinized electrodes, have been developed to reduce polarization. One of the major advances in lead technology has been the development of electrodes that elute small quantities of the corticosteroid dexamethasone sodium phosphate (37,41). Such steroid-eluting leads demonstrate minimal change in threshold over several years. The peak in stimulation threshold that occurs 4 to 8 weeks postimplantation because of inflammation and edema adjacent to the pacing electrode is also reduced dramatically. Although it has been hypothesized that the steroid reduces the size of the fibrotic layer around the electrode, the mechanisms by which steroid leads improve pacing efficiency are not well understood.

One of the continuing challenges with lead design has been ensuring stable positioning. There are two mechanisms for lead fixation: active and passive (Fig. 21.8). Passive fixation refers to the use of "tines" that are extensions of the silicone or polyurethane insulation and lodge within the trabeculae. This technology limits the placement of such leads

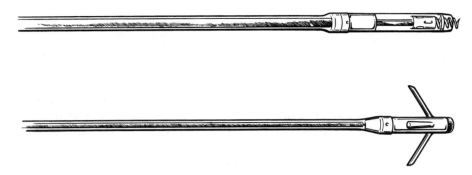

Figure 21.8   Illustrations are of active fixation (top) and passive fixation (bottom) pacing leads. The active fixation lead has a "screw-in" electrode to anchor the lead to the tissue and the passive fixation lead has tines that become lodged in the trabeculae of the endocardium to hold the lead in

to the trabeculated regions of the right atrial appendage or the apex of the ventricle. With the advent of active fixation leads that have a helical screw that is secured in the endocardium, it has been possible to place electrodes virtually anywhere in the right atrium or ventricle. This technology has facilitated evaluation of other pacing sites to optimize performance. With this new approach ventricular pacing is now being attempted using septal sites as well as the His bundle to optimize ventricular function, and in the atrium at Bachmann's bundle or proximal coronary sinus in an attempt to prevent atrial fibrillation.

Pacing lead insulation is typically made of silicone or polyurethane (36). Silicone leads have performed well historically, but have a high coefficient of friction and can be difficult to manipulate. New surface treatments of silicone rubber have helped address this problem. Polyurethane was introduced as an insulating material because of its superior strength and lower coefficient of friction (42). This property has allowed polyurethane leads to have smaller diameters than silicone leads and newer formulations have proven to be as durable as silicone.

B.   Clinical Applications

1.   Current Indications for Pacemaker Implantation

Approximately 120,000 pacemakers are implanted each year in the United States. This number has increased only moderately since 1982, following concerns about overuse. The first ACC/AHA guidelines were released in 1981 and have since been updated in 1998 (43). While the overall prevalence of pacemaker use is 2.6 per 1,000 in the U. S, the majority are implemented in persons over the age of 65. Use reaches 26 per 1,000 over age 75 (44,45). With advances in generator and lead technology, the proportion of dual-chamber devices has increased in the last 10 years and now represents about two-thirds to three-quarters of all new implants. For new implementations, the indication for implantation is sick-sinus syndrome in 40 to 55% of cases, AV block in 30 to 35%, and the remainder for miscellaneous conditions, including carotid sinus hypersensitivity, malignant vasodepressor syndrome, hypertrophic cardiomyopathy, etc. (46,47).

2.   Sinus Bradycardia

Sinus bradycardia, defined as a sinus rate of less than 50 bpm, has been estimated to occur in 6 to 25% of all individuals, but true incidence is difficult to ascertain as most people with "electrocardiographic" sinus bradycardia are probably asymptomatic. Resting heart rates of 40 to 50 bpm are relatively common particularly in young, healthy individuals (48), and lower rates can be seen in conditioned athletes and during sleep. In subjects who are not young or conditioned with a high resting vagal tone, profound daytime or nocturnal sinus bradycardia should obviously prompt investigation of comorbid possibilities, such as coronary disease, obstructive sleep apnea, etc., before contemplating pacemaker therapy (49). In persons with underlying structural heart disease, the presence of asymptomatic sinus bradycardia, without other conduction abnormalities, does not have independent clinical implications and does not warrant pacemaker implantation, as prognosis remains dependent on the underlying form of heart disease. In patients with such comorbidities, however, sinus node dysfunction that would be asymptomatic in healthy individuals may be clinically relevant and subtly manifest as worsening of congestive heart failure, fatigue, irritability, etc. The burden of proof nonetheless rests on the physician to establish that sinus bradycardia is, in fact, of sufficient clinical importance to consider pacemaker therapy (50). Moreover, when an acute reversible or treatable trigger can be identified (e.g., vagotonic maneuvers or bradycardiac drugs such as digitalis, Ca^{2+} channel blockers, and beta-blockers), even those with symptomatic sinus bradycardia may not require a permanent pacemaker. Pacemaker implantation for drug-induced symptomatic bradycardia is usually reserved for patients who require treatment with the offending agent for another condition. One important use for pacemaker treatment of asymptomatic bradycardia is in patients with documented pause-dependent torsade de pointes not due to a reversible cause, such as concurrent treatment with a QT-prolonging drug.

3. Heart Block

In general, the incidence of bundle branch block increases with patient age, reaching 11% after age 60 (51). Isolated bundle branch block [either right (RBBB) or left (LBBB)], without any history of symptoms suggesting intermittent complete heart block, is not an indication for permanent pacemaker implantation. Longitudinal studies on subjects with LBBB have demonstrated that cardiac mortality is related to the presence of other underlying heart disease, not the progression to third-degree AV block (52–57). RBBB as an isolated finding is common in patients with pulmonary disease and has a benign prognosis. Similarly, first-degree AV block is rarely symptomatic and as such is not an indication for permanent pacemaker implantation (58).

Mobitz I (Wenckebach) AV block is a form of second-degree heart block that most often reflects decremental conduction within the AV node. It may be either a physiological response to high parasympathetic tone, or a "toxic" effect of drugs such as Ca^{2+} channel or β-adrenoreceptor blocking agents, or exogenously administered adenosine (59). Mobitz I AV block is usually transient or physiological, is rarely symptomatic, and, as such, is not an indication for permanent pacing. Mobitz I block may, however, be clinically significant under certain circumstances when the trigger is recurrent and unavoidable (e.g., in the ventilator-dependent patient or when it is a consequence of medication or trauma to the compact AV node).

In contrast, Mobitz II heart block reflects infranodal conduction pathology and is more likely to be symptomatic because the degree of block will often worsen rather than improve. As with sinus bradycardia and Mobitz I AV block (AVB), permanent pacing for Mobitz II AVB has been used for alleviation of documented associated symptoms, including dizziness or exercise intolerance, or worsening of congestive heart failure (60). Indications for permanent pacing for asymptomatic Mobitz II AVB include unwitnessed syncope of no known cause in a patient with second-degree AVB, and second-degree AVB in the presence of LBBB, bifascicular block or so-called "trifascicular block" (61–63).

Third-degree (complete) AV block may be congenital or acquired. Inherited forms are often asymptomatic because of an autonomically sensitive junctional escape rhythm. In this case, prophylactic pacemaker implantation is usually not necessary early in life. However, patients with complete congenital AVB often become symptomatic as they age and chronotropic response becomes impaired (64). Causes of acquired complete AV block are manifold and include idiopathic degenerative conduction system disease, myocardial infarction, postcardiac surgery (particularly following aortic or mitral valve surgery), cardiomyopathy (ischemic or nonischemic), bacterial endocarditis, drugs, etc. Patients with sleep apnea may manifest high-degree heart block, but this appears to be vagally mediated and has been reported to respond well to positive airway pressure (65). The incidence of conduction defects leading to pacemaker implantation after cardiac surgery ranges from 0.8 to 4.0%, but long-term follow-up indicates that 40% of these patients will eventually become pacemaker-independent (66). With rare exception, acquired complete heart block, unless due to a reversible or transient cause, is an accepted indication for permanent pacing.

4. Hypertrophic Cardiomyopathy

Experimental and clinical trials have been conducted in patients with hypertrophic cardiomyopathy to determine the value of AV pacing with a short AV delay in this population. The principle is that RV apical pacing will preexcite the interventricular septum, resulting in its paradoxical motion away from the LV outflow tract during systole. Acute hemodynamic studies have shown that significant reduction of the outflow gradient can be achieved with this technique, although careful optimization of intervals is required. More interesting is the observation that, after chronic pacing, transient electrical remodeling of the septum may occur with persistent reduction of the gradient after pacing is discontinued (67,68). Some studies have suggested that this type of pacing may also provide a mortality benefit, but patient selection bias has made this difficult to interpret. Recent randomized studies have come to diverging conclusions about the true benefit of AV sequential pacing, even when maximal preexcitation of the interventricular septum can be achieved (69,70). Placebo effect also appears to play a significant role in the symptomatic benefits achieved in many patients with pacemakers (71).

5. Neurocardiogenic Syncope

Neurocardiogenic syncope, sometimes referred to as vasodepressor syncope, is characterized by paradoxical peripheral vasodilatation and bradycardia in the face of hypotension. It appears, in part, to be a parasympathetic overreaction to sympathetic surge. The spectrum of hemodynamic profiles ranges from primarily "vasodilatory" to primarily "cardioinhibitory," depending on which component dominates. In patients with a profound cardioinhibitory component, both sinus arrest and complete heart block may be observed. Recently completed trials indicate that in selected patients with severe symptoms, implantation of a pacemaker with a "rate-drop" feature reduces the number of episodes (72). Rate drop refers to a pacemaker algorithm that detects a rapid fall in sensed heart rate, and responds by AV pacing at a relatively high (e.g., 120 bpm) rate to correct the bradycardia and attempt to compensate for peripheral vasodilatation (73). Pacing for vasodepressor syncope is usually reserved for patients who are symptomatic despite maximal medical therapy.

6. Dilated Cardiomyopathy

Initial investigations focused on potential uses of AV synchronous pacing in patients with dilated cardiomyopathy in the hopes of optimizing the ventricular filling period. In patients with first-degree AVB, left ventricular diastolic filling becomes compromised because of loss of a so-called synchronized "atrial kick" and increased diastolic mitral regurgitation. Although initial studies were promising, long-term follow-up failed to confirm these results (74,75). The most common conduction defect in patients with dilated cardiomyopathy is LBBB, and RV apical pacing even with optimal AV delay does nothing to help late activation of the left ventricle. Pacing from the RV outflow tract, or basal septum, has not proven better than apical pacing (76). In patients with a primary indication for pacing (sinus bradycardia, AV block, etc.) careful tailoring of programmed AV intervals may be needed to avoid worsening of hemodynamics. AV delays of 120 to 140 ms have been optimal in most acute hemodynamic studies (77). At this point, in the absence of a primary indication for permanent pacing, dual-chamber pacing for dilated cardiomyopathy is not a therapy of choice.

C. Ongoing Clinical Investigations

1. Biventricular Pacing for Congestive Heart Failure

Because of failure of RV pacing alone to provide hemodynamic benefit in dilated cardiomyopathy, attention has recently shifted to using biventricular or left ventricular pacing for patients in this category (78–83). In particular, for patients with intraventricular conduction delays and first-degree AV block, resynchronization of atrial and ventricular contraction could be expected to improve ventricular mechanical performance and experimental and early clinical studies have been promising. LV pacing has been accomplished both with epicardial leads and with transvenous leads placed into the middle cardiac vein via the coronary sinus. Whether these techniques will result in long-term benefit, however, awaits further study.

2. Pacing for Atrial Tachyarrhythmias

Recent observations into the electrophysiology and epidemiology of atrial fibrillation (AF) suggest that this disorder is much more heterogeneous than previously appreciated. The spectrum of atrial fibrillation patients ranges from those with significant structural heart disease and atrial myopathy (ischemic, valvular, hypertensive) to "idiopathic" AF in normal hearts. A significant number of these patients also have symptomatic bradycardia as a comorbidity. Because cardiac pacing is routinely conducted for management of bradycardia, several studies have attempted to compare the effects of atrial versus ventricular pacing on development of atrial fibrillation. A prospective randomized trial in patients with sick sinus syndrome followed for 8 years showed that patients with atrial-based pacing had a lower incidence of atrial fibrillation, improved survival, and fewer thromboembolic complications (84). It was difficult to assess from this work whether atrial pacing was antiarrhythmic or ventricular pacing proarrhythmic. Although it has been suggested that these beneficial effects of atrial pacing may carry over to dual-chamber pacing, in a small prospective study in tachy–brady-syndrome patients (Pac-A-Tach), DDDR pacing did not improve atrial tachyarrhythmia incidence, as compared to ventricular pacing, but showed a significant improvement in survival at a 2-year follow-up (85,86). Another study in patients undergoing AV nodal ablation also concluded that atrial pacing does not result in a lower recurrence of AF than dual-chamber pacing over a short-term follow-up (87). Additional large, mortality-based trials, such as the Canadian Physiological Pacing study, comparing DDD and ventricular pacing are ongoing and will likely shed new light on the subject.

The ability of atrial pacing to prevent AF in patients without structural heart disease, but with vagally mediated paroxysmal atrial fibrillation, has been suggested by early results in a small trial (88). In these patients, the arrhythmia typically occurred at night and was preceded by significant bradycardia. It is important to recognize, though, that in patients without clinical bradycardia, the onset of AF is preceded by bradycardia in only 8 to 10% of such episodes, and the bradycardia may not always be causative (89). "Short–long sequences" more frequently precede the onset of atrial fibrillation and unique pacing algorithms that prevent these sequences have the potential of preventing AF episodes, although this idea remains untested clinically.

Initial attempts at conventional atrial pacing for atrial fibrillation were based on the concept that pacing may prevent the premature beats that initiate AF by suppressing them with overdrive pacing, or by reducing the probability of reentrant excitation in the presence of premature beats by preexciting the abnormal substrate and increasing the relative homogeneity of repolarization (90,91). Dual-site atrial pacing, or atrial septal pacing, attempts to accomplish the latter and the results from initial nonrandomized studies have been promising (92–94). Prospective randomized studies are presently underway to determine the efficacy of these novel approaches.

Termination of atrial fibrillation has been even a greater challenge than its prevention. Experimentally, Allessie and coworkers developed some rapid pacing algorithms to terminate atrial fibrillation in goats (95). Although they were able to capture and control a small region around the pacing electrode, they were not able to terminate atrial fibrillation. This finding has since been confirmed in humans (96) and the long-term value of this approach remains questionable.

Atrial antitachycardia pacing has been much more successful for termination of atrial flutter. Atrial pacing is conducted at a rate slightly greater than the flutter rate, and following the end of the pacing sequence, termination has been observed in more than 60% of cases, including both typical and atypical flutter rhythms (97). One thought has been that flutter consists of a single reentrant rhythm and a paced activation wavefront might block this form of reentry. Flutter episodes that are not amenable to termination by overdrive pacing may be terminated by pacing at an even higher rate (e.g., high-frequency pacing at 3000 stimuli/min). This technique has been successfully used for termination of rapid atypical flutter which frequently was seen to degenerate into an unstable atrial tachyarrhythmia that spontaneously converts to sinus rhythm (98).

D. Pacemakers

1. Device Selection

Figure 21.9 depicts some general principles regarding pacemaker selection. For patients with an intact sinus mechanism (whether chronotropically competent or not), inclusion of an atrial lead is appropriate. In isolated sinus node dysfunction with intact AV conduction, single-chamber atrial pacing (AAIR) can be used, and the rate of subsequent development of AV block in patients without structural heart disease or following cardiac transplantation appears to be low (99–101). On the other hand, for patients with heart disease and sinus node dysfunction, the development of AV conduction disorders is higher and dual-

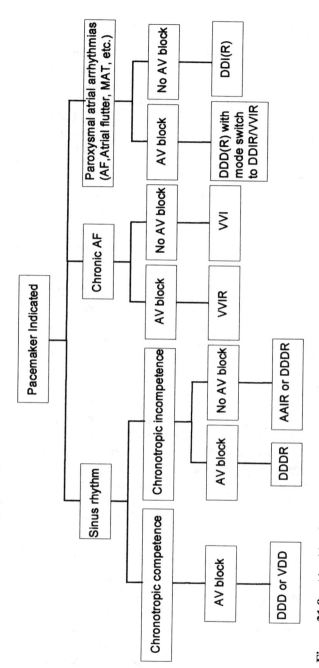

Figure 21.9 Algorithm for pacemaker selection and programming mode are shown based on the underlying rhythm and presenting arrhythmia.

chamber pacemaker implantation is often prudent. Dual-chamber pacing is being used for patients with intact atrial function for several reasons: for those with an intact AV conduction system, asynchronous ventriculoatrial conduction from ventricular pacing (VVI) increases the likelihood of development of atrial fibrillation (84). Ventricular pacing in AV block may result in "pacemaker syndrome," due to asynchronous contraction of the atrium against a closed tricuspid valve during pacemaker-initiated ventricular systole. Retrospective studies comparing mortality in elderly recipients with dual-chamber vs. single-chamber ventricular pacing have been inconclusive because of selection bias, with frailer, sicker patients receiving single-chamber pacemakers more often for the same indications (84,102–104). A recent study comparing symptomatic benefit of dual-chamber vs. single-chamber pacing in patients with intact atrial function, while showing no benefit of dual-chamber pacing, was compromised by a 25% crossover rate to dual-chamber pacing because of intolerance (105). Nonetheless, ventricular pacing alone may be appropriate in specific circumstances (e.g., because of patient frailty or difficulties with venous access).

In patients with permanent AF or with persistent, refractory, atrial arrhythmias of other types, an atrial lead is not of value, and single-chamber ventricular pacing is appropriate. For patients with pure AV block, but intact sinus node function, development of sinus node dysfunction is rare and single-lead "dual-chamber" VDD pacing systems are available that allow only sensing in the atrium, and both pacing and sensing in the ventricle (106). This permits ventricular tracking of the intrinsic sinus rate with a single lead. Single-pass leads that allow atrial pacing as well as sensing are under development, but have been plagued by problems with diaphragmatic capture (107).

2.   Programming and Sensors

As with the selection of pacemaker type, programming should generally favor use of whatever conduction properties remain intact in an individual patient. Single-chamber atrial pacemakers are usually programmed AAIR because the primary indication is symptomatic sinus node dysfunction. In patients in chronic atrial fibrillation, the decision to program VVI vs. VVIR depends on the appropriateness of the patient's intrinsic ventricular response rate. Those patients who only require backup for periods of extreme bradycardia may do well programmed to VVI mode, to minimize competition with their intrinsic ventricular response rate. On the other hand, if AV conduction is impaired or absent because of underlying disease, drugs, or AV node ablation, VVIR will result in better exercise tolerance after optimizing rate response parameters. Single-chamber ventricular pacemakers are generally contraindicated in patients with the capability for sinus rhythm, but if used, rate response is usually activated only in the setting of complete heart block or sinus node dysfunction to avoid competition with intrinsic sinus rate.

With dual-chamber pacing for isolated AV block, either chronic or intermittent DDD or VDD mode is frequently chosen, and the programmed AV delay is extended to allow intrinsic conduction to help maximize pacemaker longevity and more natural pacing patterns (108). With the exception of hypertrophic cardiomyopathy (discussed above), there is no evidence that artificially shortening the AV delay at the expense of normal ventricular activation via the His-Purkinje system provides any hemodynamic or symptomatic benefit (74,77). Rate-responsive modes are usually programmed only in patients with sinus node dysfunction, with the upper sensor-driven rate based on patient age and underlying cardiac disease.

No particular activity sensor technology has demonstrated superiority for all patient categories. Accelerometers are the most common type, and are generally reliable, but can be fooled by motion unrelated to activity (e.g., riding in a vehicle). Piezoelectric crystal detection of muscle activity may be more appropriate for patients who do a lot of upper body exertion, but these sensors may oversense muscle activity, especially if placed subpectorally. Minute ventilation monitors detect respiratory rate by measurement of undulations in transthoracic impedance. This is probably the most physiological method available, but is usually avoided in patients who hyperventilate (e.g., COPD or anxiety disorders). QT interval measurements are based on autonomically mediated alterations in repolarization, and can be used if the patient does not need rate response during ventricular pacing. Pacemakers with dual sensors (e.g., both accelerometer and minute ventilation detection) are available and may be helpful particularly in active patients.

Paroxysmal supraventricular arrhythmias including atrial fibrillation and atrial flutter present a problem for dual-chamber pacemakers. Device selection and programming need to be tailored to avoid rapid ventricular tracking during atrial arrhythmias. If AV block is present, this can be done by restricting the pacemaker's "upper tracking rate," but may compromise exercise tolerance in sinus rhythm or lead to pacemaker syndrome due to pacemaker-mediated Wenckebach block ("upper rate behavior") at physiological sinus rates. Implantation of a device with the capability to "mode switch" to VVIR or DDIR is more appropriate and will allow atrial tracking to higher sinus rates, but limit the ventricular rate to the upper sensor rate during atrial arrhythmias (109). In patients with intact AV conduction and paroxysmal atrial arrhythmias, DDI(R) mode is usually ideal. Because this is a nontracking mode, atrial arrhythmias will not cause rapid ventricular rates, but atrial pacing in sinus rhythm will still be available without the need for the pacemaker to mode switch. DDI(R) is not used in the presence of persistent AV block because it does not allow programming of the AV delay after sensed atrial events (the "sensed AV delay").

3. Electromagnetic Interference and Other Complications

Interference from electrocautery equipment is a well-recognized cause of pacemaker dysfunction, but is usually limited to the duration of the electrocautery pulse. There have also been case reports of permanent, or at least persistent, pacemaker malfunction after exposure to electrocautery, but these remain rare, and the etiologies were probably multifactorial (110). The high-strength radio-frequency output from this type of equipment (typically 500–750 kHz) apparently can generate electrical noise that can be interpreted by the pacemaker as intrinsic myocardial activity. With unipolar electrocautery, the most common interactions are ventricular output inhibition due to sensed ventricular noise, or ventricular tracking of atrial noise. High-level noise can also cause a pacemaker to change to a noise "reversion mode," which is usually asynchronous pacing. One way to prevent either noise reversion mode—which will require reprogramming after surgery—or inappropriate pacing or inhibition is to change the pacemaker to a nonsensing (VOO or DOO) mode using telemetry to reprogram the device prior to the use of this type of equipment.

Ultrasonic dental equipment has also been reported to cause pacemaker malfunctions, and their use is generally avoided in pacemaker-dependent patients (111,112). Interesting interactions have been reported between cardiac monitoring equipment and pacemakers that use minute ventilation sensing to adjust rate response. The signal used in the measurement of transthoracic impedance and respiratory rate may result in inappropriate behavior of the rate-response circuitry, usually manifesting as upper rate pacing. This resolves after the monitoring equipment is disconnected (113). Likewise, use of external radiotherapy with pacemaker patients also needs to be closely monitored. Ionizing radiation applied to the chest (e.g., for lung or breast cancer) can cause permanent damage to the pacemaker circuitry, and the current recommendation is that the dose received by the pace-maker should not exceed 2 Gy, although malfunction is rarely reported at less than 10 Gy (114).

Magnetic Resonance Imaging (MRI) is generally regarded as contraindicated in pa-tients with either pacemakers or defibrillators, and pacemaker manufacturers uniformly advise against it. The first problem is the large static magnetic field, which will close the pacemaker reed switch and put the device in magnet mode. The oscillating component of the magnetic field can induce measurable voltages in the lead, which can be interpreted by the pacemaker as intrinsic activity. The radio-frequency pulses themselves may similarly lead to inappropriate sensing. Permanent reprogramming or malfunction has not been ob-served. Modern pacemakers are largely nonferrous, and pacemaker movement generally does not occur during imaging pulses. If MRI must be performed on a patient with a pace-maker, asynchronous pacing can usually be anticipated and appropriate cardiac monitoring is usually indicated (115).

Wireless phones have come under much scrutiny in the past few years regarding pos-sible interactions with pacemakers and implantable defibrillators. In vitro experiments have documented transient pacemaker malfunctions triggered by digital phones in ex-tremely close proximity. Complications have included inappropriate and erratic pacing and, occasionally, inhibition of pacemaker output. Interactions were seen only with digital wireless phones. Because of the higher sensitivity of atrial sensing circuitry, dual-chamber pacemakers are more likely to be affected. Unlike other sources of electromagnetic inter-ference, the high-frequency output of wireless phones is detected at the pacemaker header, rather than in the leads, and no difference in susceptibility has been noted between unipo-lar and bipolar systems. All pacing alterations seen to date appear limited to the duration of exposure to the phone and only at distances of less than 6 in. (116). A recent multicenter study investigating the clinical significance of these interactions found that "clinically sig-nificant" interactions occurred only rarely (1.7%).

Theft-prevention systems or electronic article surveillance systems installed in many stores are of three types: magnetic audiofrequency, swept radio frequency, and acoustom-agnetic. Initial trials indicate that only the acoustomagnetic system commonly causes pacemaker malfunction, usually manifesting as inhibition of output and limited to the du-ration of exposure. As patients usually will not know what type of theft gate they are walk-ing through, it is usually prudent to limit such exposures (117). There are also many other sources of electromagnetic interference and local magnetic fields are ubiquitous; therefore, patients are usually cautioned to limit exposure to any source of potentially strong inter-ference.

II. SUMMARY

The initial, and still primary, purpose of implantable pacemakers is for treatment of symp-tomatic bradycardia. With advances in technology, pacemakers have progressed from large fixed-rate single-chamber devices to multiprogrammable, multichamber devices with the ability to respond to changing hemodynamic demands. Large amounts of diagnostic data including patient heart rate and activity levels, Holter and "event" monitoring capabilities, and generator battery and lead status can be stored in many current generation pacemakers. Even with these additional features, device size continues to shrink and longevity and reli-ability of both generators and leads continues to improve. One of the challenges faced by manufacturers is to ensure that these features can be used to enhance patient care without unnecessarily complicating device implantation, programming, or follow-up. In spite of increasing levels of automation, an understanding of the indications and limitations of pacemaker therapy remains essential; pacemakers are not the solution for every patient with a slow heart rate or syncope.

With the extraordinary developments that have occurred in pacemaker therapy for the traditional indication, bradycardia, new uses are now beginning to be explored. Under investigation are pacing for treatment of vasodepressor syndrome, atrial pacing for prevention or treatment of atrial arrhythmias, and pacing for various forms of cardiomyopathy, including biventricular pacing in the treatment of congestive heart failure. As the technology improves, other possible uses are also likely to be conceived. Which of these will withstand rigorous scientific and clinical validation remains the challenge for the future.

REFERENCES

1. Marmorstein M. Contribution a l'etude des excitations electriques localisees sur le coeur en rapport avec al topographie de l'innervation du coeur chez le chien. J Physiol Pathol 1927; 25:617–625.
2. Lidwill M. Cardiac disease in relation to anaesthesia. Transactions of the third session, australasian medical congress (British Medical Association). Sydney, Australia, 1929.
3. Hyman A. Resuscitation of the stopped heart by intracardial therapy. II. Experimental use of an artificial pacemaker. Arch Intern Med 1932; 50:283.
4. Jeffrey K. The invention and reinvention of cardiac pacing. Cardiol Clin 1992; 10:561–571.
5. Zoll P, Linenthal A, Normal L, et al. Treatment of Stokes-Adams disease by external electric stimulation of the heart. Circulation 1954; 9:482.
6. Zoll P. Development of electric control of cardiac rhythm. JAMA 1973; 226:881.
7. Nelson G. A brief history of cardiac pacing. Texas Heart Inst J 1993; 20:12–18.
8. Elmqvist R, Landegren J, Pettersson S, Senning A, William-Olson G. Artificial pacemaker for treatment of Adams-Stokes syndrome and slow heart rate. Am Heart J 1963; 65:731–748.
9. Senning A. Discussion of a paper by Stephenson SS, Edwards WH, Jolly PC, Scott AW. Physiologic P-wave cardiac stimulator. J Thorac Surg 1959; 38:604–609. J Thorac Surg 1959; 38:639.
10. Furman S, Robinson G. The use of an intracardiac pacemaker in the correction of total heart block. Surg Forum 1958; 9:245.
11. Greatbach W, Lee J, Mathias W, Eldridge M, Moser J, Schneider A. The solid state lithium battery: a new improved chemical power source for implantable cardiac pacemakers. IEEE Trans Biomed Eng 1971; 18:317.
12. Greatbach W, Holmes C. History of implantable devices: entrepreneurs, bioengineers and the medical profession, in a unique collaboration, build an important new industry. IEEE Eng Med Biol 1991; 10:38–49.
13. Geddes L, Fearnot N, Smith H. The exercise-responsive cardiac pacemakers. Biomed Eng 1984; 31:763–770.
14. Martin R, Cobb L. Observations on the effect of atrial systole in man. J Lab Clin Med 1966; 68:224.
15. Bedotto J, Grayburn P, Black W, et al. Alteration in left ventricular relaxation during atrioventricular pacing in humans. J Am Coll Cardiol 1990; 15:658–664.
16. Rosenqvist M, Isaaz K, Botvinick E, et al. Relative importance of activation sequence compared to atrioventricular synchrony in left ventricular function. Am J Cardiol 1991; 67:148–156.
17. Wirtzfeld A, Himmler F, Klein G, et al. Atrial pacing in patients with sick sinus syndrome: acute and long-term hemodynamic effects. In: Feruglio G, ed. Cardiac Pacing: Electrophysiology and Pacemaker Technology. Padova, Italy: Piccin Medical Books, 1982:651–654.

18. El Gamal M, van Gelder L. Preliminary experience with the Helifix electrode for transvenous atrial implantion. Pacing Clin Electrophysiol 1981; 4:100–105.

19. Barin E, Jones S, Ward D, Camm A, Nathan A. The right ventricular outflow tract as an alternative permanent pacing site: long-term follow-up. Pacing Clin Electrophysiol 1991; 14:3–5.

20. Karpawich P, Justice C, Chang C, Gause C, Kuhns L. Septal ventricular pacing in the immature canine heart: a new perspective. Am Heart J 1991; 121:827–833.

21. Deshmukh P, Anderson K. Direct His bundle pacing: novel approach to permanent pacing in patients with severe left ventricular dysfunction and atrial fibrillation (abstr). Pacing Clin Electrophysiol 1996; 19:700.

22. Chirife R. Physiological principles of a new method for rate responsive pacing using the pre-ejection interval. Pacing Clin Electrophysiol 1988; 11:1545–1554.

23. Lau C. Sensors and algorithms for rate adaptive pacing. Rate Adaptive Cardiac pacing: Single and Dual Chamber. Mount Kisco, NY: Futura Publishing Company, 1993: 63–228.

24. Lapicque L. Definition experimentale de l'excitabilite. Soc Biol 1909; 77:280–283.

25. Hoorweg L. Uber die elektrische nerve merregung. Pflugers Arch Physiol 1892; 52:87–99.

26. Weiss G. Sur la possibilite de rendre comparable entre eux les apparels servant a l'excitation electrique. Arch It Biol 1901; 35:413–446.

27. Blair H. On the intensity-time relations for stimulation by electric currents. J Gen Physiol 1932; 15:709–729.

28. Furman S, Hurzeler P, Mehra R. Cardiac pacing and pacemakers IV. Threshold of cardiac stimulation. Am Heart J 1977; 94:115–124.

29. Furman S, Parker B, Escher D, Solomon N. Endocardial threshold of cardiac response as a function of electrode surface area. J Surg Res 1968; 8:161–166.

30. Furman S, Parker B, Escher D. Decreasing electrode size and increasing efficiency on cardiac stimulation. J Surg Res 1971; 11:105–110.

31. Irnich W. Engineering concepts of pacemaker electrodes. In: Schaldach M, Furman S, eds. Engineering in Medicine I. Advances in Pacemaker Technology. Berlin: Springer-Verlag, 1975:241–272.

32. Doenecke P, Flothner R, Rettig G, Bette L. Studies of short and long term threshold changes. In: Schaldach M, Furman S, eds. Engineering in Medicine I. Advances in Pacemaker Technology. Berlin: Springer-Verlag, 1975:283–296.

33. Vonk B, Oort G. New method of atrial and ventricular capture detection. Pacing Clin Electrophysiol 1998; 21:217–222.

34. Mehra R, Furman S. A comparison of cathodal, anodal and bipolar strength interval curves with temporary and permanent electrodes. Br Heart J 1979; 41:468.

35. Beyersdorf F, Schneider M, Kreuzer J, Falk S, Zegelman M, Satter P. Studies of the tissue reaction induced by transvenous pacemaker electrodes. I. Microscopic examination fo the extent of connective tissue around the electrode tip in the human right ventricle. Pacing Clin Electrophysiol 1988; 11:1753–1759.

36. Kay G. Basic aspects of cardiac pacing. In: Ellenbogen K, ed. Cardiac Pacing. Boston, MA: Blackwell Scientific Publications, 1992:32–119.

37. Stokes K, Graf J, Wiebusch W. Drug-eluting electrodes improved pacemaker performance. Fourth Annual Conference IEEE Engineering in Medicine and Biology Society. New York, 1982:499.

38. Adler S, Brown M, Nelson L, Mehra R. Interelectrode spacing and right atrial location both influence farfield ventricular electrogram amplitude. Pacing Clin Electrophysiol 1999; 22:877.

39. DeCaprio V. Endocardial electrograms with transvenous pacemaker electrodes. 1977.

40. Fischell R, Schulman J. A rechargeable power system for cardiac pacemakers. Proc 11th Intersoc Energy Conv Eng Conf 1976; 1:163.

41. Stokes K, Bornzin G. The electrode-biointerface: Stimulation. In: Barold S, ed. Modern Cardiac Pacing. Mount Kisco, NY: Futura Publishing Company, 1985:37–77.

42. Mond H, Stokes K, Helland J, et.al. The porous titanium steroid eluting electrode: A double blind study assessing the stimulation threshold effects of steroid. Pacing Clin Electrophysiol 1988; 11:214–219.

43. Gregoratos G, Cheitlin M, Conill A, et al. ACC/AHA guidelines for implantation of cardiac pacemakers and antiarrhythmia devices: a report of the American College of Cardiology/ American Heart Association Task Force on Practice Guidelines (Committee on Pacemaker Implantation). J Am Coll Cardiol 1998; 31:1175–1209.

44. Silverman BG, Gross TP, Kaczmarek RG, Hamilton P, Hamburger S. The epidemiology of pacemaker implantation in the United States. Pub Health Rep 1995; 110:42–46.

45. Bernstein AD, Parsonnet V. Survey of cardiac pacing and defibrillation in the United States in 1993. Am J Cardiol 1996; 78:187–196.

46. Daley WR, Kaczmarek RG. The epidemiology of cardiac pacemakers in the older US population. J Am Geriatr Soc 1998; 46:1016–1019.

47. Daley WR. Factors associated with implantation of single- versus dual-chamber pacemakers in 1992. Am J Cardiol 1998; 82:392–395.

48. Bjornstad H, Storstein L, Meen HD, Hals O. Ambulatory electrocardiographic findings in top athletes, athletic students and control subjects. Cardiology 1994; 84:42–50.

49. Stegman SS, Burroughs JM, Henthorn RW. Asymptomatic bradyarrhythmias as a marker for sleep apnea: appropriate recognition and treatment may reduce the need for pacemaker therapy. Pacing Clin Electrophysiol 1996; 19:899–904.

50. Taylor IC, Stout RW. The significance of cardiac arrhythmias in the aged. Age Ageing 1983; 12:21–28.

51. Kreger BE, Anderson KM, Kannel WB. Prevalence of intraventricular block in the general population: the Framingham Study. Am Heart J 1989; 117:903–10.

52. Gil VM, Almeida M, Ventosa A, et al. Prognosis in patients with left bundle branch block and normal dipyridamole thallium-201 scintigraphy. J Nucl Cardiol 1998; 5:414–417.

53. Heinsimer JA, Irwin JM, Basnight LL. Influence of underlying coronary artery disease on the natural history and prognosis of exercise-induced left bundle branch block. Am J Cardiol 1987; 60:1065–1067.

54. Schneider JF, Thomas HE, Jr., McNamara PM, Kannel WB. Clinical-electrocardiographic correlates of newly acquired left bundle branch block: the Framingham Study. Am J Cardiol 1985; 55:1332–1338.

55. McAnulty JH, Kauffman S, Murphy E, Kassebaum DG, Rahimtoola SH. Survival in patients with intraventricular conduction defects. Arch Intern Med 1978; 138:30–35.

56. McAnulty JH, Rahimtoola SH. Chronic bundle-branch block. Clinical significance and management. JAMA 1981; 246:2202–2204.

57. McAnulty JH, Rahimtoola SH, Murphy E, et al. Natural history of "high-risk" bundle-branch block: final report of a prospective study. N Engl J Med 1982; 307:137–143.

58. Mymin D, Mathewson F, Tate R, Manfreda J. The natural history of primary first-degree atrioventricular heart block. N Engl J Med 1986; 315:1183–1187.

59. Bergfeldt L, Rosenqvist M, Vallin H, Edhag O. Disopyramide induced second and third degree atrioventricular block in patients with bifascicular block. An acute stress test to predict atrioventricular block progression. Br Heart J 1985; 53:328–334.

60. Strasberg B, Amat-y-Leon F, Dhingra R, et.al. Natural history of chronic second-degree atrioventricular nodal block. Circulation 1981; 63:1043–1049.

61. Morady F, Higgins J, Peters R, et al. Electrophysiologic testing in bundle branch block and unexplained syncope. Am J Cardiol 1984; 54:587–591.

62. Dhingra R, Denes P, Wu D, Chuquimia R, Rosen K. The significance of second degree atrioventricular block and bundle branch block: observations regarding site and type of block. Circulation 1974; 49:638–646.

63. Ezri M, Lerman BB, Marchlinski FE, Buxton AE, Josephson ME. Electrophysiologic evaluation of syncope in patients with bifascicular block. Am Heart J 1983; 106:693–697.

64. Michaelsson M, Jonzon A, Riesenfeld T. Isolated congenital complete heart block in adult life: a prospective study. Circulation 1995; 92:442–449.

65. Koehler U, Fus E, Grimm W, et al. Heart block in patients with obstructive sleep apnoea: pathogenetic factors and effects of treatment. Eur Respir J 1998; 11:434–439.

66. Glikson M, Dearani JA, Hyberger LK, Schaff HV, Hammill SC, Hayes DL. Indications, effectiveness, and long-term dependency in permanent pacing after cardiac surgery. Am J Cardiol 1997; 80:1309–1313.

67. Fananapazir L, Epstein N, Curiel R, et al. Long term results of dual chamber (DDD) pacing in obstructive hypertrophic cardiomyopathy. Circulation 1994; 90:2731–2742.

68. McAreavey D, Fananapazir L. Altered cardiac hemodynamic and electrical state in normal sinus rhythm after chronic dual-chamber pacing for relief of left ventricular outflow obstruction in hypertrophic cardiomyopathy. Am J Cardiol 1992; 70:651–656.

69. Ommen SR, Nishimura RA, Squires RW, Schaff HV, Danielson GK, Tajik AJ. Comparison of dual-chamber pacing versus septal myectomy for the treatment of patients with hypertrophic obstructive cardiomyopathy: a comparison of objective hemodynamic and exercise end points. J Am Coll Cardiol 1999; 34:191–196.

70. Maron BJ, Nishimura RA, McKenna WJ, Rakowski H, Josephson ME, Kieval RS. Assessment of permanent dual-chamber pacing as a treatment for drug-refractory symptomatic patients with obstructive hypertrophic cardiomyopathy. A randomized, double-blind, crossover study (M-PATHY). Circulation 1999; 99:2927–33.

71. Nishimura R, Trusty J, Hayes D, et al. Dual-chamber pacing for hypertrophic cardiomyopathy: A randomized, double-blind, crossover trial. J Am Coll Cardiol 1997; 29:435–441.

72. Connolly SJ, Sheldon R, Roberts RS, Gent M. The North American Vasovagal Pacemaker Study (VPS). A randomized trial of permanent cardiac pacing for the prevention of vasovagal syncope. J Am Coll Cardiol 1999; 33:16–20.

73. Benditt DG, Sutton R, Gammage M, et al. "Rate-drop response" cardiac pacing for vasovagal syncope. Rate-Drop Response Investigators Group. J Interv Card Electrophysiol 1999; 3:27–33.

74. Shinbane JS, Chu E, DeMarco T, et al. Evaluation of acute dual-chamber pacing with a range of atrioventricular delays on cardiac performance in refractory heart failure. J Am Coll Cardiol 1997; 30:1295–300.

75. Auricchio A, Salo RW. Acute hemodynamic improvement by pacing in patients with severe congestive heart failure. Pacing Clin Electrophysiol 1997; 20:313–24.

76. Victor F, Leclercq C, Mabo P, et al. Optimal right ventricular pacing site in chronically implanted patients. A prospective randomized crossover comparison of apical and outflow tract pacing. J Am Coll Cardiol 1999; 33:311–316.

77. Gold M, Feliciano Z, Gottlieb S, Fisher M. Dual-chamber pacing with a short atrioventricular delay in congestive heart failure: A randomized study. J Am Coll Cardiol 1995; 26:967–973.

78. Cazeau S, Ritter P, Lazarus A, et al. Multisite pacing for end-stage heart failure: early experience. Pacing Clin Electrophysiol 1996; 19:1748–1757.

79. Daubert JC, Ritter P, Le Breton H, et al. Permanent left ventricular pacing with transvenous leads inserted into the coronary veins. Pacing Clin Electrophysiol 1998; 21:239–245.

80. Gras D, Mabo P, Tang T, et al. Multisite pacing as a supplemental treatment of congestive heart failure: preliminary results of the Medtronic Inc. InSync Study. Pacing Clin Electrophysiol 1998; 21:2249–2255.

81. Leclercq C, Cazeau S, Le Breton H, et al. Acute hemodynamic effects of biventricular DDD pacing in patients with end-stage heart failure. J Am Coll Cardiol 1998; 32:1825–1831.

82. Saxon L, Boehmer J, Hummel J, et al. Biventricular pacing in patients with congestive heart failure: Two prospective randomized trials. Am J Cardiol 1999; 83:120D–123D.

83. Auricchio A, Stellbrink C, Sack S, et al. The pacing therapies for congestive heart failure (PATH-CHF) study: Rationale, design, and endpoints of a prospective randomized multicenter study. Am J Cardiol 1999; 83:130D–135D.

84. Andersen HR, Nielsen JC, Thomsen PE, et al. Long-term follow-up of patients from a randomized trial of atrial versus ventricular pacing for sick-sinus syndrome. Lancet 1997; 350:1210–1216.

85. Sgarbossa E, Pinski S, Maloney J, et al. Chronic atrial fibrillation and stroke in patients with Sick Sinus Syndrome. Relevance of clinical characteristics and pacing modalities. Circulation 1993; 88:1045–1053.

86. Wharton J, Sorrentino R, Campbell P, for the P-A-T Investigators. Effect of pacing modality on atrial tachyarrhythmia recurrences in the Tachycardia-Bradycardia Syndrome: Preliminary results of the Pacemaker Atrial Tachycardia trial. Circulation 1998; 98:I-494.

87. Gillis A, Wyse D, Connolly S, et al. Atrial pacing periablation for prevention of paroxysmal atrial fibrillation. Circulation 1999; 99:2553–2558.

88. Coumel P, Friocourt P, Mugica J, Attuel P, LeClerq J. Long term prevention of vagal atrial arrhythmias by atrial pacing at 90/minute: Experience with cases. Pacing Clin Electrophysiol 1983; 6:552–560.

89. Hill M, Hammill S, Mehra R. Onset and termination of spontaneous episodes of atrial fibrillation. Pacing Clin Electrophysiol 1995; 18:89.

90. Mehra R, Hill M. Prevention of atrial fibrillation/flutter by pacing techniques. In: Saksena S, Luderitz B, eds. Interventional Electrophysiology: A Textbook, 2nd ed. Armonk, NY: Futura Publishing Company, 1996:521–540.

91. Mehra R. How might pacing prevent atrial fibrillation? In: Murgatroyd F, Camm A, eds. Non-pharmacologic Management of Atrial Fibrillation. Armonk, NY: Futura Publishing Company, 1998.

92. Saksena S, Prakash A, Hill M, et. al. Prevention of recurrent atrial fibrillation with chronic dual site right atrial pacing. J Am Coll Cardiol 1996; 28:687–694.

93. Daubert C, Mabo P, Berder V, Gras D, LeClerq C. Atrial tachyarrhythmias associated with high degree interatrial conduction block; prevention by permanent atrial resynchronization. Eur J Card Pacing Electrophysiol 1994; 4:35–44.

94. Padaletti L, Porciani M, Michelucci A, et al. Interatrial septum pacing: a new approach to prevent recurrent atrial fibrillation. J Interv Card Electrophysiol 1999; 3:35–43.

95. Kirchoff C, Chorro F, Scheffer G, et al. Regional entrainment of atrial fibrillation studied by high-resolution mapping in open-chest dogs. Circulation 1993; 88:736–749.

96. Paladino W, Bahu M, Knight BP, et al. Failure of single- and multisite high-frequency atrial pacing to terminate atrial fibrillation. Am J Cardiol 1997; 80:226–227.

97. Wharton M, Santini M. Treatment of spontaneous atrial tachyarrhythmias with the Medtronic 7250 Jewel AF: Worldwide clinical experience. Circulation 1998; 98:I-190.

98. Giorgberidze I, Saksena S, Mongeon L, et al. Effects of high-frequency atrial pacing in atypical atrial flutter and atrial fibrillation. 1997.

99. Andersen HR, Nielsen JC, Thomsen PE, et al. Atrioventricular conduction during long-term follow-up of patients with sick sinus syndrome. Circulation 1998; 98:1315–1321.

100. Woodard DA, Conti JB, Mills RM, Jr., Williams RA, Curtis AB. Permanent atrial pacing in cardiac transplant patients. Pacing Clin Electrophysiol 1997; 20:2398–2404.

101. Heinz G, Kratochwill C, Hirschl M, et al. Normal AV node function in patients with sinus node dysfunction after cardiac transplantation. J Card Surg 1993; 8:417–424.

102. Brady PA, Shen WK, Neubauer SA, Hammill SC, Hodge DO, Hayes DL. Pacing mode and long-term survival in elderly patients with congestive heart failure: 1980–1985. J Interv Card Electrophysiol 1997; 1:193–201.

103. Andersen HR, Thuesen L, Bagger JP, Vesterlund T, Thomsen PE. Prospective randomized trial of atrial versus ventricular pacing in sick-sinus syndrome. Lancet 1994; 344:1523–1528.

104. Jahangir A, Shen WK, Neubauer SA, et al. Relation between mode of pacing and long-term survival in the very elderly. J Am Coll Cardiol 1999; 33:1208–1216.

105. Lamas G, Orav E, Stambler B, et al. Quality of life and clinical outcomes in elderly patients

treated with ventricular pacing as compared with dual-chamber pacing. N Engl J Med 1998; 338:1097–1104.

106. Morsi A, Lau C, Nishimura S, Goldman BS. The development of sinoatrial dysfunction in pacemaker patients with isolated atrioventricular block. Pacing Clin Electrophysiol 1998; 21:1430–1434.

107. Tse H, Lau C, Leung S, Leung Z, Mehta N. Single lead DDD system: a comparative evaluation of unipolar, bipolar, and overlapping biphasic stimulation and the effects of right atrial floating electrode location on atrial pacing and sensing thresholds. Pacing Clin Electrophysiol 1996; 19:1758–1763.

108. Harper G, Pina I, Kutalek S. Intrinsic conduction maximizes cardiopulmonary performance in patients with dual chamber pacemakers. Pacing Clin Electrophysiol 1991; 14:1787–1791.

109. Provenier F, Boudrez H, Deharo JC, Djiane P, Jordaens L. Quality of life in patients with complete heart block and paroxysmal atrial tachyarrhythmias: a comparison of permanent DDIR versus DDDR pacing with mode switch to DDIR. Pacing Clin Electrophysiol 1999; 22:462–468.

110. Peters B, Gold M. Reversible prolonged pacemaker failure due to electrocautery. J Interv Card Electrophysiol 1998; 2:343–344.

111. Vlay S. Electromagnetic interference and ICD discharge related to chiropractic treatment. Pacing Clin Electrophysiol 1998; 21:2009.

112. Miller CS, Leonelli FM, Latham E. Selective interference with pacemaker activity by electrical dental devices. Oral Surg Oral Med Oral Pathol Oral Radiol Endod 1998; 85:33–36.

113. Chew EW, Troughear RH, Kuchar DL, Thorburn CW. Inappropriate rate change in minute ventilation rate responsive pacemakers due to interference by cardiac monitors. Pacing Clin Electrophysiol 1997; 20:276–282.

114. Last A. Radiotherapy in patients with cardiac pacemakers. Br J Radiol 1998; 71:4–10.

115. Lauck G, von Smekal A, Wolke S, et al. Effects of nuclear magnetic resonance imaging on cardiac pacemakers. Pacing Clin Electrophysiol 1995; 18:1549–1555.

116. Schlegel R, Grant F, Shivakumar R, Reynolds D. Electromagnetic compatibility study of the in-vitro interaction of wireless phones with cardiac pacemakers. Biomed Inst Tech 1998; 32:645–655.

117. McIvor ME, Reddinger J, Floden E, Sheppard RC. Study of Pacemaker and Implantable Cardioverter Defibrillator Triggering by Electronic Article Surveillance Devices (SPICED TEAS). Pacing Clin Electrophysiol 1998; 21:1847–1861.

22

Implantable Cardioverter Defibrillators

ADAM ZIVIN and GUST H. BARDY

University of Washington, Seattle, Washington

RAHUL MEHRA

Medtronic, Inc., Minneapolis, Minnesota

The concept of an automatic internal defibrillator (AID) or implantable cardioverter defibrillator (ICD) arose in the 1960s as external cardiac defibrillation was increasingly being used in coronary care units for treatment of ventricular fibrillation, and sudden cardiac death (SCD) was gaining attention as a major public health problem. Although the idea of automatic external defibrillation had been discussed initially by Zucoto, Mirowski was the first to champion and begin practical development of an automatic internal device (1,2). Mirowski had realized that bouts of recurrent ventricular tachyarrhythmia were life threatening, but was frustrated by an inability to deliver therapeutic electrical interventions on a continuously monitored basis. In 1969, he and Mower developed a prototype of today's automatic internal defibrillator (3). It performed reliably in animal experiments and human testing of internal defibrillation was successfully performed during a bypass operation using a catheter electrode in the right ventricle and a second indifferent electrode (4). To bring the concept to widespread public availability, they collaborated with various commercial organizations interested in its development and in 1980 the first human implant of the defibrillator was performed at the Johns Hopkins University Hospital largely with private support.

I. DEFIBRILLATION USING ICDs

The primary goal of all defibrillators is to terminate ventricular tachyarrhythmias by delivering high-voltage shocks to the ventricle. As with implantable pacemakers, implantable defibrillating devices need to be small, reliable, and to have adequate longevity. Since their introduction, implantable ICDs have evolved to not only perform this function, but to take on additional tasks such as antitachycardia pacing of the ventricle, dual-chamber (i.e., atrial and ventricular) pacing, and even termination of atrial tachyarrhythmias. Many of the dual-chamber pacing features, including rate-responsive pacing, are very similar to those in pacemakers and the reader is referred to the previous chapter for additional information on these issues. Due to the requirement of ICDs to deliver high-voltage shocks of 700 to 800 V following detection of ventricular tachyarrhythmias, defibrillators need an electrical capacitor to store energy and appropriate circuitry to charge the capacitors to this high voltage using a low-voltage battery. The capacitors required, along with their electronic circuitry, result in an increase in the size of these devices in comparison to pacemakers, yet their development and increasing miniaturization has, as with pacemakers, been remarkable.

A key difference between pacing and defibrillation of the heart is that for pacing, only a very small mass of the myocardium needs to be stimulated, while for defibrillation most, if not all, of the myocardium must be stimulated. As the myocardium is easily excitable throughout diastole, a small single wave of depolarization during pacing can readily propagate throughout the whole heart. In contrast, during ventricular fibrillation, there are usually multiple reentrant wavefronts that are continuously changing in location and size that must be quelled. To defibrillate successfully, most of these wavefronts have to be interrupted simultaneously. To achieve this, one needs to capture most of the tissue that is in a state of relative refractoriness. Pacing during diastole can be achieved with a current density of 2.5 mA/cm^2, but to excite tissue in its relative refractory period, a stimulus at least five times larger is needed (5). Pacing at a single site with such a magnitude could excite the tissue directly underneath the electrode, but the wavefront would not propagate far due to the refractoriness of the adjacent tissue. Theoretically, if electrodes could be placed in every region of the heart, defibrillation might be possible with low-energy stimuli. Clearly, this is impractical. In clinical practice, two to three high-voltage electrodes are used, with one of them located in the right ventricle, in addition to normal sensing electrodes which provide input into detection circuits of the device. A block, electrical schematic that represents in simplified form most contemporary devices is shown in Figure 22.1. Clearly, a major aspect in device design is the high-voltage circuit that can provide shocks at multiple sites within the myocardium. The magnitude of the electrical stimulus needed to achieve defibrillation must be high, because these electrodes must stimulate tissue at some distance from the site of electrode placement. In comparison to ventricular pacing where about 1 mA of current is frequently sufficient, electrical defibrillation in patients can frequently require up to 15 A (i.e., more than 1000 times higher for a finite period of time). Although electrical current is the fundamental property that stimulates tissue, defibrillation strength is usually expressed in units of energy (Joules), which reflect the product of the voltage and current delivered, as well as the duration of the shock.

One unique property of defibrillation success is that it is probabilistic (6,7). The same energy that can defibrillate the heart on one occasion, may be unsuccessful at another time. One explanation for this is that during fibrillation multiple reentrant wavefronts move across the heart continuously and their proximity to the defibrillating electrodes as well as their state of refractoriness at the time of the shock may vary. As the energy of the shock increases, the probability that it will successfully defibrillate also increases. This characteristic of defibrillation is illustrated in Figure 22.2. The term "defibrillation threshold" is used to describe a shock strength that is successful in defibrillating in a certain percentage of attempts (e.g., 50%).

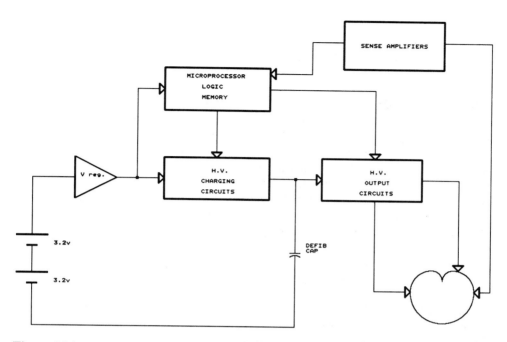

Figure 22.1   ICD system block diagram. An ICD typically has six basic components: a battery as the power source; a sense amplifier that senses the electrical activity of the heart; control circuits composed of the microprocessor, logic, and memory; a high-voltage charging circuit that converts the low voltage of the battery to high voltage and charges the capacitor; a capacitor that stores the defibrillation energy and a switching circuit that creates the appropriate waveform for defibrillating the heart. The voltage regulator (V reg) supplies a stable voltage source to control circuits that compensates for fluctuations in the battery voltage. The battery, storage capacitor, and the electronic circuitry each consume about one-third of the volume of the device.

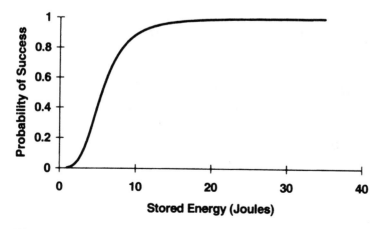

Figure 22.2   An example of a defibrillation success curve showing that the probability of successful defibrillation increases as the stored energy of the shock increases. The probability of successful defibrillation is about 50% at 6 J, 90% at 10 J, and reaches a plateau thereafter.

Generally, it is desirable to work with low defibrillation thresholds, so that the size of the electrical capacitors within the device can be small, minimizing device size and maximizing battery longevity. Apart from basic electrophysiological properties of the myocardium being defibrillated, variables that affect defibrillation threshold include the size, shape, and location of the electrodes and the shock waveform itself. In early implantable defibrillators, shocking electrodes were large patches that were sutured onto the epicardial surface of the heart during thoracotomy. Early defibrillator generators were also so large that subcutaneous or subrectus implantation in the abdomen was required. Defibrillation threshold was reduced by increasing the size of the patches, but there was high morbidity and mortality associated with this approach (5). This led to the development of transvenous defibrillation electrodes that consist of 4- to 5-cm-long coils wound around the lead body. These do not need to be in direct contact with the myocardium to defibrillate and are remarkably effective. Defibrillation thresholds with electrodes placed in the right ventricle (RV), superior vena cava (SVC), and the coronary sinus were initially evaluated along with subcutaneous electrodes and various combinations thereof. The leads were frequently tunneled under the skin down to the abdominally placed generator. With advancements in technology, it has been possible to reduce the size of the ICD itself so that they could be implanted in the pectoral region.

This led to a critical advance in defibrillation in 1993, in which the metal defibrillator body itself was first used as one of the electrodes, in conjunction with an electrode placed in the right ventricle (8). This advance resulted in a system with a low defibrillation threshold that requires only one lead. The lowest defibrillation thresholds have been reported when the body or "can" of such a device, which serves as one electrode, and a coil in the superior vena cava are electrically tied together with the current being delivered to the myocardium between this pair (9) (Fig. 22.3). This configuration, and defibrillation using the right ventricle and can alone, are the systems of choice today. Pacing and detection

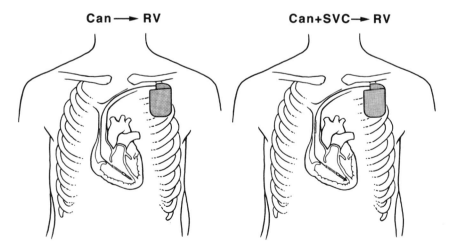

Figure 22.3   Two commonly used configurations for ventricular defibrillation are shown. During can to RV defibrillation, the shock is delivered across a coil in the right ventricle and the defibrillator can, located in the left pectoral region. During can+SVC to RV defibrillation, the coil electrode in the superior vena cava and the can electrode are at the same voltage and a shock is delivered between them and the coil electrode in the right ventricle.

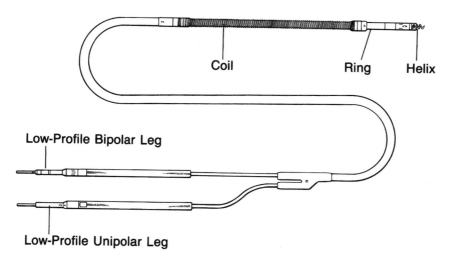

Figure 22.4 A typical ventricular fibrillation lead, which consists of a metallic helix that is screwed into the myocardium for fixation. The helix and the ring are used for pacing and sensing of ventricular activity. The long coil behind the ring is 4 to 5 cm in length and is used for fibrillation. One method of defibrillating the heart is to apply the high voltage across this coil in the right ventricle and the ICD can (active can). The lead terminates in two legs, the bipolar leg that has the electrical connection to the helix tip and the ring and the unipolar leg that is connected to the coil. These two legs are inserted into the header block of the ICD.

of ventricular tachyarrhythmias when incorporated in these designs are conducted using more complex electrodes, such as that shown in Figure 22.4.

The shock waveform used with the first generation of implantable defibrillators was a monophasic "capacitive" discharge waveform" (Fig. 22.5, upper panel). This exponentially decaying waveform is created as the capacitor is allowed to discharge. Use of this waveform was based on the technical ease of waveform generation. Since that time, however, animal and clinical experimentation has indicated that a waveform in which the polarity of the terminal portion of the waveform was reversed (i.e., a "biphasic" waveform) (Fig. 22.5, lower panel) resulted in significantly lower thresholds (10–13). Biphasic waveforms have now become the standard for all defibrillators today and can be generated relatively efficiently. The duration of the waveform pulse is typically 5 to 10 ms, 10 to 20 times that used for cardiac pacing.

II. ANTITACHYCARDIA PACING

The original concept addressed by ICD development was that all ventricular tachyarrhythmias would require defibrillation (i.e., application of a high-energy shock). However, it has been observed that in the case of monomorphic ventricular tachycardias many episodes can also be terminated by rapid ventricular pacing [antitachycardia pacing (ATP)] instead of a high-voltage shock (8,14). The advantages of terminating ventricular tachyarrhythmias by pacing rather than defibrillation are elimination of shock-associated pain and reduced battery drain. ATP is generally performed by pacing the ventricle at a rate slightly faster than the ventricular tachycardia rate and then abruptly terminating the pacing sequence (Fig. 22.6) (15). This pacing sequence entrains and interrupts the reentrant tachycar-

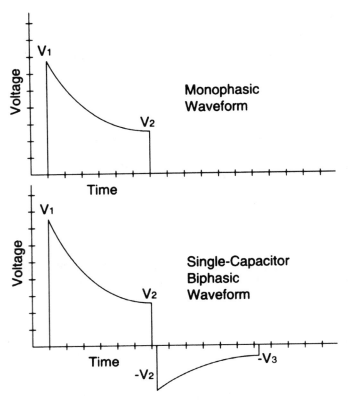

Figure 22.5 Monophasic and biphasic waveforms used for defibrillation are shown. The monophasic waveform is generated by discharging a charged capacitor and truncating the pulse when it reaches a certain voltage, V2. The single-capacitor biphasic waveform is generated by discharging the capacitor and, when the voltage reaches V2, reversing the direction of current flow. The pulse is eventually truncated at voltage V3.

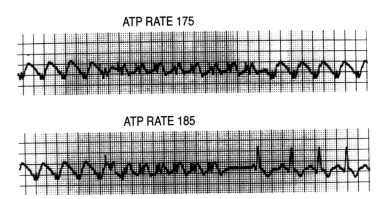

Figure 22.6 An ECG of a monomorphic ventricular tachycardia at 150 beats per minute is shown being terminated by antitachycardia pacing delivered through an electrode in the right ventricle. Pacing at a rate of 175 beats/min is unsuccessful, but increasing the rate to 185 beats/min terminates the tachycardia and results in sinus rhythm.

dia circuit. If the VT cycle length is very short, it is often difficult to terminate VT by ATP and high-energy shocks may be required. The precise pacing rate for termination of VT is difficult to predict and therefore a sequence of rates and ATP modalities are typically used.

III. ARRHYTHMIA DETECTION

Prior to delivering electrical therapy to terminate VF or VT, tachyarrhythmias must first be detected. Single-chamber defibrillators have a single lead in the ventricle that must first be capable of detecting VF and, ideally, differentiating ventricular from supraventricular tachyarrhythmias. It is critical that the sensitivity for detection of VT/VF be very high to ensure that all potentially lethal episodes can be treated. To achieve these very high sensitivities, the specificity of detection is frequently compromised technically, resulting in occasional episodes of inappropriate therapy during sinus tachycardia, or rapid conduction during atrial fibrillation (AF) or flutter. If detection is based solely on ventricular rate, it is difficult to distinguish VT from sinus tachycardia or other supraventricular tachyarrhythmias (e.g., atrial fibrillation). In the detection of true arrhythmias, various algorithms have been used to help discriminate VT from SVT; that is, to increase specificity without compromising sensitivity (16). Most are based on two concepts: stability of ventricular intervals (stability criterion), or the rate of onset of the tachycardia (onset criterion). Rapid conduction, for example, during atrial fibrillation, results in irregular ventricular intervals and the presence of variable intervals in adjacent beats can be used to help differentiate rapidly conducting AF from VT. This discrimination technique is only used for ventricular tachycardias, because during VF the ventricular intervals are also irregular, but of much shorter duration. The onset criterion is used to help discriminate sinus tachycardia from ventricular tachycardia, based on the assumption that VT initiates abruptly, whereas sinus tachycardia has a gradual warm-up period. The primary problem with this algorithm has been that sinus tachycardia can transition to VT and an algorithm based on detection of the onset of the arrhythmia can compromise sensitivity of VT detection. Other techniques, including measurement of electrogram width and morphological comparison between sinus rhythm and VT electrograms, have been used to improve the specificity of single-lead systems (17–19). Often, these detection features are activated only if there is clinical evidence that the incidence of inappropriate electrical therapy is abnormally high.

With the advent of dual-chamber ICDs, the addition of an atrial lead presents the potential to significantly improve the specificity of VT detection. Various dual-chamber algorithms have been developed with most focused on analyzing the relationship between atrial and ventricular activity to discriminate VT from SVT (20). Under highly controlled circumstances, these algorithms have been reported to reduce the incidence of inappropriate ventricular electrical therapy to less than 10%, without loss of sensitivity for VT or VF detection, whereas in regular practice false positive rates of up to one-third are commonplace. Discriminating sinus tachycardia from 1:1 retrograde conduction during VT remains a key challenge and, in the future, algorithms that also incorporate analysis of the morphology of the ventricular complex may help resolve this more difficult issue.

IV. TECHNOLOGICAL DEVELOPMENTS

Developments in ICD technology, as well as the results of various large, multicenter trials demonstrating their clinical value in specific populations have been directly responsible for the recent dramatic growth in use of ventricular ICDs over the last decade (see Chap.

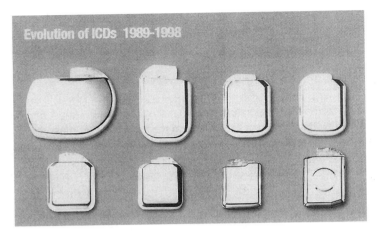

Figure 22.7 Evolution of implantable defibrillators. As with the development of cardiac pacemakers, miniaturization of these devices has progressed markedly over the past 30 years as their capabilities have expanded. Each of the devices shown here is capable of providing output shocks between 30 to 34 J. Shown (upper row, left to right), are representative models from 1989, 1990, 1994, 1995, and (bottom row, left to right), 1996, 1997, 1998, and 1999. Beginning in 1995, these devices have employed the metallic housing of the defibrillator itself as one of the defibrillator electrodes to help reduce defibrillation threshold. Beginning in 1998, the models shown incorporated rate-responsive pacing and sufficient memory to store more than 20 min of electrogram information. Typical longevity of recent models, depending on extent of pacing and shocks required, runs between 6 and 11 years. The latest 1999 model provides dual-chamber pacing in both the atria and ventricles.

29). The size of the devices has been reduced from 150 cc in the early 1980s to less than 40 ccs at present, and typical device lifetimes are between 4 to 8 years (Fig. 22.7) (21). Improvements in capacitor and battery technology, which together occupy about half the ICD volume (Fig. 22.8), have been primarily responsible for this reduction in size. The weight of ICDs has also been dramatically reduced from almost 300 to less than 75 g. Present-day ICDs are also capable of storing vast amounts of information useful in patient management, as well as monitoring device efficacy, safety, and optimizing programming. In the early 1980s, these devices typically stored 256 bytes of information, whereas most of to-

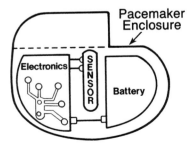

Figure 22.8 Location of the battery, sensor, and electronics within a pacemaker. In this example, the sensor is a piezoelectric crystal that is welded to the pacemaker enclosure and is used for rate-responsive pacing. The battery and the electronics occupy most of the space within the device. The unit above the dotted line is the "header block" into which the pacing lead is inserted.

day's devices can store at least 256 Kbytes, a thousand-fold increase. One of the key challenges in the future is to determine more optimal methods to display these data in ways that are both easy to use and clinically more valuable. With the considerable experience achieved so far, implantation procedures have developed to a state where they are straightforward and rarely provide a challenge except in patients with structural disease. Figure 22.9 shows the placement of one contemporary "active can" device, including placement of atrial and ventricular leads within the heart and chest cavity.

V. POTENTIAL NEW APPLICATIONS

Although the original impetus for ICD development was the management of ventricular tachyarrhythmias, interest in managing atrial tachyarrhythmias has also grown significantly in recent years. It is now recognized that between 20 to 30% of patients with ventricular tachyarrhythmias also have atrial tachyarrhythmias (21–24). Such atrially initiated tachyarrhythmias can worsen patient symptoms, result in inappropriate ventricular shocks, and may be responsible for initiating ventricular tachyarrhythmias that can exacerbate other pathologies, such as heart failure. Therefore, it makes great sense to incorporate new strategies for treatment and prevention of atrial tachyarrhythmias into devices that are capable of defibrillation and antitachycardia pacing in the atrium as well as the ventricle, in addition to combined dual-chamber pacing (25). The clinical benefits of managing atrial

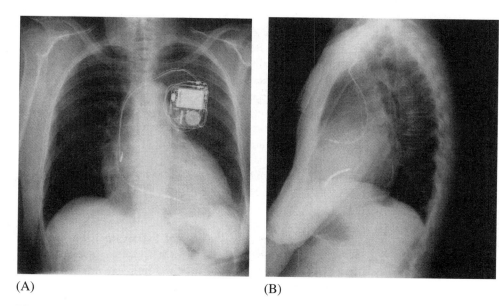

(A)   (B)

Figure 22.9   (A) Front and (B) lateral chest radiographs showing placement of a "dual-chamber ICD." In addition to providing dual-chamber pacing, the ability of these devices to sense atrial activity can help improve the specificity of VT and VF detection and minimize inappropriate shocks for supraventricular arrhythmias or sinus tachycardia. This patient underwent ICD implantation for recurrent VF and high-degree AV block in the setting of nonischemic cardiomyopathy. The atrial lead is a standard pacemaker lead placed in the right atrial appendage. The ventricular lead is an active fixation single-coil defibrillator lead positioned in the right ventricular apex. Care needs to be taken during implantation of dual-chamber devices to avoid crosstalk, which could affect not only the pacing functions but also ventricular arrhythmia discrimination.

tachyarrhythmias in patients with VT/VF is presently being investigated and we should have better indications for this use in the near future. In principle, these devices should be able to provide significant benefit in managing patients with highly symptomatic atrial tachyarrhythmias, without the comorbidity due to VT/VF. Early results indicate that about 50% of the atrial tachyarrhythmias in this patient population can be terminated by pacing. The key issue with this device is the level of discomfort associated with an internal atrial defibrillation shock for termination of AF. Patient-initiated, rather than automatic activation of the atrial defibrillation shock, and shock during sleep, is preferred by some patients, and both modes of therapy are being evaluated. An especially attractive aspect is that prompt treatment of AF may retard the progression and electrical remodeling that accompanies chronic AF, prevent the onset of heart failure, and improve the quality of life of these patients. Last, but not least, it should be mentioned that a very important potential application of dual-chamber ICDs is in the management of patients with congestive heart failure. New technologies allow both right and left ventricular pacing in an attempt to enhance contractile efficiency. Long-term benefits of this technology to improve quality of life, as well as reduce mortality, are under investigation (26).

VI. CLINICAL APPLICATIONS

The purpose of the implantable ICD is to prevent death due to hemodynamically unstable ventricular tachyarrhythmias. While advances in technology have made these devices much more flexible in terms of arrhythmia detection and electrical therapy options, their main purpose is to reduce sudden cardiac death, a threat that claims approximately 300,000 lives in the United States annually. Recent (1998–1999), rates of ICD placement in the United States are about 30,000 annually, corresponding to a frequency of 120 per million population (27). The majority of these are implanted in patients with documented sustained ventricular arrhythmias, or aborted episodes of sudden death. Results from on-going primary prevention trials of ICDs that are now underway (e.g., in heart failure), however, could cause a tremendous escalation in these numbers based on new indications for their use. The most recent ACC/AHA guidelines outline accepted indications for ICD implantation as well as for pacemakers and the reader is referred to that valuable guide and its expected updates in the future for a review of current recommendations (28).

Secondary prevention of sudden cardiac death in patients who have survived cardiac arrest remains the best-studied indication for ICD use in the United States. In such patients, and especially in those of this group for whom no reversible or curable cause can be found, implantation of an ICD has been repeatedly documented to provide a major mortality benefit. Three large-scale trials have compared the value of ICDs with conventional antiarrhythmic drug therapies in the treatment of ischemic and nonischemic cardiomyopathies, with documented ventricular arrhythmias. CIDS (the Canadian Implantable Defibrillator Study) enrolled patients with an ejection fraction (EF) $\leq$ 35%, presenting with VF or hemodynamically unstable VT (29,30). Patients were randomized to undergo implantation of an ICD or empirical treatment with amiodarone. After 3-year follow-up, there was a nonsignificant 20% reduction in all-cause mortality in the ICD group. AVID (Antiarrhythmics Versus Implantable Defibrillator trial) randomized patients with an EF $\leq$ 40% and VF or sustained VT to ICD vs. empirical use of a class III antiarrhythmic drug (primarily amiodarone) (31). After 3-year follow-up, there was a statistically significant 31% reduction in total deaths in the group who received an ICD. CASH (Cardiac Arrest Study Hamburg) randomized SCD survivors to ICD versus drug therapy with propafenone, amiodarone, or beta-blockade (32). The propafenone arm of the trial was terminated prematurely because of excess mortality. CASH did not restrict enrollment to patients with impaired LV function, but again demonstrated 37% lower mortality in the ICD group, with no difference observed between the beta-blocker metoprolol and amiodarone. Despite various flaws and concerns in both design and analysis, these studies remain the best available data on secondary prevention of sudden death. As of this writing the consensus opinion of these studies is that they strongly support the use of ICDs in high-risk patients who have survived sudden death or have hemodynamically destabilizing VT when no treatable trigger can be identified.

Another high-risk group in whom use of ICDs is currently being explored is patients with various forms of the congenital long-QT (LQT) syndrome. LQT patients who have already survived an episode of cardiac arrest, or documented polymorphic VT, especially if on pharmacological therapy at the time, are increasingly being evaluated as ICD candidates. Brugada syndrome is a related disease, also due to abnormalities in ion channel function, with a constellation of findings that includes an anatomically structurally normal heart, an ECG showing right bundle branch block with characteristic ST elevation in the right precordium, and predisposition to sudden death (33). Factors associated with an increased risk of sudden death events include a family history of sudden death events and history of syncope. Additional methods of risk stratification for this group, including sensitivity to pharmacological challenge with sodium channel blockers, like mexiletine, and EP testing are also under investigation. At this point, use of ICDs is indicated for those patients who also have had actual episodes of syncope or a strong family history of sudden death (34). Also, young patients with familial forms of hypertrophic cardiomyopathy who have survived an episode of VF or who are judged to be high risk or proven not responsive to standard medical therapies may be additional candidates.

Generally accepted indications for ICD implantation in patients with hypertrophic cardiomyopathies (HCM) and without a history of sudden death include sustained ventricular arrhythmias, nonexertional syncope, and strong family history of sudden death with early age at presentation (35,36). Patients, especially younger ones, with primary VF without underlying structural heart disease and without a definable trigger for arrhythmias (e.g., those with drug intoxication, metabolic abnormalities, or clear evidence of acute ischemia, etc.) may also benefit (37–39).

The use of ICDs in patients without a documented history of sustained ventricular arrhythmias or aborted sudden death is controversial. The area of greatest focus has been patients with ischemic cardiomyopathies. Here, the risk of sudden death is known to be higher in patients with demonstrated coronary disease who have impaired LV function. Several trials have focused on identifying patients in this category who might benefit from prophylactic ICD implantation. Most of these have tried to use some risk-stratification technique to identify higher risk populations (40–44) in addition to a reduced LVEF. An abnormal signal-averaged ECG (SAECG), inducibility of sustained arrhythmias during programmed electrical stimulation (PES), runs of nonsustained ventricular tachycardias (NSVT), have all been employed with varying degrees of success in different groups. MADIT (the Multicenter Automatic Defibrillator Implantation Trial) studied patients with coronary artery disease (CAD) and an EF $\leq 40\%$, who were inducible and nonsuppressible on PES (45). Patients were randomized to receive "conventional" therapy with antiarrhythmic drugs (most often amiodarone), or an ICD. This trial was terminated when a predetermined statistical endpoint was reached and showed a relative mortality hazard ratio of 0.46 in the ICD-treated group. The CABG-Patch Trial selected patients with a positive

SAECG who were scheduled to undergo clinically indicated coronary bypass surgery (46). Half of these patients underwent simultaneous epicardial ICD implantation at the time of surgery. In contrast to MADIT, no survival advantage was seen in the ICD group in this study although this could have been related to high rates of nonarrhythmic mortality. The recently completed MUSTT study (Multicenter UnSustained Tachycardia Trial) randomized patients with coronary artery disease (CAD), an EF ≤ 40%, and NSVT who were inducible at PES to no antiarrhythmic therapy versus PES-guided antiarrhythmic treatment (including ICD implantation as last resort) (47). This study was primarily designed to test the value of PES testing in guiding therapy, but also has had interesting implications for ICD use. A 27% reduction in the primary endpoint of arrhythmic death or cardiac arrest in the PES-guided group was entirely due to the benefit of ICD implantation. Nonetheless, reduction in total mortality did not reach statistical significance after a mean follow-up period of 39 months. Although the results of MADIT were sufficient to obtain FDA approval for prophylactic ICD implantation in patients fitting the enrollment criteria used in that trial, the results of this study have been controversial in some quarters and it should be noted none of these studies has yet conclusively supported prophylactic defibrillator implantation in ischemic cardiomyopathy, regardless of the technique used for risk stratification. Whether this conclusion will be addressed by future trials and which approaches will need to be employed in particular populations remains a clinical challenge of first order. Woosley and Singh deal extensively with the strengths and weaknesses of many antiarrhythmia trials in great detail (48).

Last, we should note that another area where primary prevention with ICD therapy is being evaluated is in patients awaiting cardiac transplant, who have a high risk of sudden death. Studies have generally been positive with regard to survival to transplant in the ICD-treated group (49). However, patients included in these types of studies have been heterogeneous with respect to functional class and substrate, and because of the retrospective, nonrandomized nature of the studies, it is difficult to determine the validity of this application. Data from primary prevention trials will also hopefully clarify management for this subgroup in the not too distant future.

VII. CONTRAINDICATIONS TO ICD IMPLANTATION

As the purpose of an implantable defibrillator is to facilitate outpatient management of malignant ventricular arrhythmias, implantation of an ICD is not appropriate for a hospitalized patient who is not otherwise ready for discharge. Firm contraindications include hemodynamic or pulmonary instability, evidence of ongoing infection or ischemia, incessant ventricular arrhythmias not controllable with oral antiarrhythmic agents, and concomitant illnesses from which the patient is not expected to recover. Relative contraindications might include planned thoracic surgery, contraindications to the use of fluoroscopy (e.g., pregnancy), or unsuitability for VF induction for intraoperative testing of the device (e.g., severe aortic stenosis or inoperable left-main or left-main equivalent coronary artery disease). Although implantation of an ICD without intraoperative VF testing is technically feasible, it can be argued that a patient who would not tolerate VF induction and defibrillation under controlled circumstances, would be unlikely to survive an out-of-hospital episode, even if successfully defibrillated (50). Exceptions to this position exist, and individual cases have to be considered following a full review of risk/benefit ratios for various conditions.

VIII. ONGOING INVESTIGATIONS

Two major clinical ICD trials are presently underway which may further clarify the value of primary prevention in patients at high risk for arrhythmic death. The NHLBI supported, SCD-HeFT (the Sudden Cardiac Death Heart Failure Trial) is a placebo-controlled study of amiodarone versus ICD in patients with NYHA class II-III heart failure who have an EF $\leq 35\%$ (ischemic and nonischemic), and no history of sustained ventricular arrhythmias (51). The primary endpoint is total mortality, with specific guidelines for maximizing other therapies known to improve survival (e.g., β-blockers, ACE inhibitors, and cholesterol-reducing statin drugs). The MADIT II trial is an ongoing attempt to extend the results of MADIT I by removing the criteria of inducibility at PES. Its goal is to randomize 1200 post-MI patients with an EF $\leq 30\%$ to ICD versus "conventional" therapy (52).

IX. CURRENT CLINICAL STATUS

Single-chamber ventricular defibrillators represent the majority of ICD devices in clinical use today. Their primary purpose is the termination of malignant ventricular arrhythmias and they are the preferred mode of therapy for most clinical indications (see Fig. 22.10). Such devices also usually have the capability of back-up pacing for bradycardia that may occur after a shock, or because of the underlying cardiac pathology. Empirical programming of postshock bradycardia pacing is appropriate for most patients.

For patients who require ICD implantation because of hemodynamically unstable fast VT or VF only, or for prophylactic indications, a single-lead device programmed with a cutoff rate of 180–200 bpm (320–300 ms) is frequently optimal. The risk of inappropriate shocks for supraventricular arrhythmias at this detection rate is low unless the patient is known to have PSVT or AF with rapid ventricular response. Particularly for prophylactic indications, minimizing both system complexity and the chance of malfunction, which creates the potential for inappropriate shocks due to low-programmed cutoff rate, is of utmost importance for patient acceptance. Patients who have slower ventricular arrhythmias, which may overlap in rate with sinus rhythm during exercise, or with PSVTs, such as atrial flutter conducting 2:1, present more of a device-selection problem, but single-chamber ICDs are still frequently preferable because of their comparative simplicity. Often detection enhancements, including onset (to distinguish sinus tachycardia) or stability criteria (to distinguish atrial fibrillation) are generally adequate to reduce inappropriate shocks, especially in older patients. While specific programming strategies are beyond the scope of this chapter, it should be evident that any of these enhancements could potentially result in underdetection of VT. For this reason, enhanced detection criteria such as R-R stability onset, EGM width, or template matching should be programmed only for ventricular rates known to be hemodynamically stable in a particular patient. Familiarity with a patient's history, including history of atrial fibrillation, rate of clinically documented tachycardias, and sinus rate during activity are very important considerations in tailoring programming.

The term "dual-chamber" defibrillators is usually used to refer to ventricular defibrillators that have an additional pace/sense lead in the atrium (Fig. 22.9). The ability to sense in the atrium, while complicating hardware and software design considerably, has the potential to reduce the likelihood of inappropriate shocks for supraventricular arrhythmias. The criteria may be simple, such as recognition of ventriculoatrial dissociation (ventricular rate greater than atrial rate), measurement of atrial electrogram rate to detect atrial

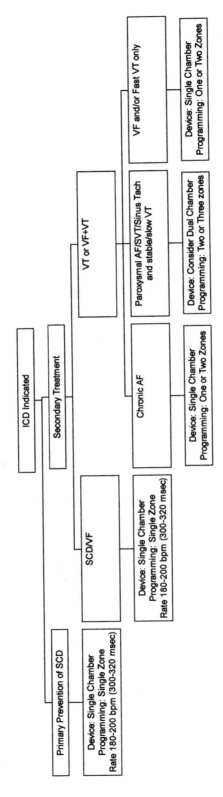

Figure 22.10 Indications for ICD selection and typical programming are shown based on the presenting clinical condition and presenting arrhythmia. Algorithm for ICD selection and programming based on primary or secondary prevention and underlying arrhythmia.

fibrillation, or more complex examinations of relative atrial and ventricular electrogram timing (20,53).

The other value of a dedicated atrial lead is the capability for dual chamber pacing. This eliminates the potential for device–device interactions between a separate pacemaker and ICD in the same patient. Despite their intrinsic appeal, dual-chamber ICDs are not appropriate for all patients needing a defibrillator. The addition of a second lead greatly increases system complexity; cost and size must be weighed against the value of dual-chamber pacing and sensing. Dual-chamber devices are generally reserved for patients in need of a defibrillator, who have a coexisting indication requiring a permanent pacemaker. In patients with documented slower ventricular tachycardias that overlap in rate with expected sinus rhythm or atrial arrhythmias, the additional specificity of dual-chamber sensing has also been reported to be of benefit. Nonetheless, the potential for underdetection of VT remains, and use of diagnostic enhancements is usually limited to ventricular rates that are hemodynamically stable, regardless of the arrhythmia. In the younger patient with slower monomorphic VT, the additional detection specificity provided by the dual-chamber system must be weighed against the increased likelihood of lead failure, or other complications, over the course of the patient's lifetime. Ablation of slower, stable tachycardias may be a viable alternative strategy to minimize overlap between sinus tachycardia and ventricular tachycardia (54).

Lead selection is generally based on inherent properties of those that may be available at any particular time and physician preference. For biphasic waveforms, slightly lower defibrillation thresholds have been demonstrated for dual-coil (SVC+RV) versus single-coil leads, but this may not be clinically significant (55). The routine addition of a separate SVC coil is not necessary (9). With the development of nonthoracotomy lead systems, implantation of ICDs can usually be safely performed in the EP laboratory under conscious conditions with local anesthesia or following deep sedation. With attention paid to closely monitoring vital signs and oxygenation, even patients with severely compromised LV function usually undergo implantation without the attendant risks of endotracheal intubation and ventilation (56,57). Current generations of ICDs are small enough to permit pectoral implantation in the vast majority of patients and implantation of an ICD differs little from pacemaker implantation with regards to surgical techniques, venous access, and fluoroscopic visualization. Left pre-pectoral implantation is generally preferred with unipolar ICDs, using the defibrillator case as one electrode ("active can" or "hot can" systems). The shock vector between RV coil and can is better with left pectoral implantation, and while right-sided implantation is usually successful, the defibrillation threshold may be higher (58). This may be clinically important in patients with high defibrillation threshold as is sometimes the case with patients with dilated cardiomyopathy.

Unless contraindicated for reasons of patient safety, testing is generally performed at the time of implantation. Induction of VF can be done with synthesized AC current delivered through the lead or with T-wave synchronized low-energy shock. The purpose is twofold: to verify device sensing of ventricular fibrillation and to determine an adequate safety margin between the patient's defibrillation energy requirement and the maximum output of the device. Although full step-down defibrillation threshold testing to failure is not standard, it may permit programming of a lower first shock energy, minimizing not only the time to shock (quicker capacitor charge time), but also limiting the hemodynamic compromise associated with defibrillator discharge (59,60). Stepdown threshold testing will usually identify a shock strength that will be successful 70% of the time. By programming the first delivered therapy to the defibrillation threshold value plus 10 J, a 98% first shock success rate is expected (61,62). In patients with known monomorphic VT, anti-tachycardia pacing (ATP) is often programmed in a VT zone as the first therapy, followed by lower and high energy shocks. In patients with known VT, ATP can generally be safely programmed empirically, that is without specific VT induction and testing. In patients who already have a permanent pacemaker at the time of defibrillator implantation, device–device interactions need to be tested. Complications are often due to oversensing of pace-maker spikes or "double counting" of spikes and myocardial potentials. These can mani-fest in two ways: (1) double counting of pacemaker spikes and myocardial potentials leading to inappropriate shocks; or (2) underdetection of VF due to oversensing of pace-maker spikes. In order to challenge the detection capabilities of the defibrillator in a "worst-case" scenario, the pacemaker can be programmed to an asynchronous mode (VOO or DOO) (see Chap. 21 for NBG code abbreviations) during threshold testing.

Routine follow-up on patient-device status after implantation is generally performed every 3 to 4 months, depending on age of the device and expected battery status. Routine device interrogation usually includes read-outs on battery capacity, delivered therapies, and verification of pacing threshold and sensing functions. A yearly chest x-ray has been advocated by some to look for unsuspected lead fractures, but this may become unneces-sary with the advent of automatic shocking-coil impedance measurements in recent de-vices. The issue of routine defibrillation threshold testing has been debated, and while some studies have documented occasional instances of a rise in threshold, potentially clin-ically important changes have been rare (63–65). Unless a patient with a marginal defibril-lation safety margin is placed on an antiarrhythmic known to increase defibrillation thresh-old (e.g., amiodarone), the risk of routine postimplant VF testing is generally difficult to justify.

For a patient who previously had been "quiet," the occurrence of frequent shocks can cause considerable emotional distress and potential hemodynamic decompensation. It is therefore important to investigate whether previous shocks have been "appropriate" (i.e., for true VT or VF), or "inappropriate." Inappropriate device therapies have been re-ported in up to 20 to 30% of implanted patients and review of recorded electrograms or R-R interval plots, along with clinical history, are often adequate to make this diagnosis. In the patient who presents with syncope or near-syncope accompanying device therapy, in-appropriate shocks are less likely. An investigation of the trigger (ischemia, electrolyte im-balance, etc.) and use of other antiarrhythmic therapies may be helpful in making this de-cision. Routine initiation of antiarrhythmic drug therapy is usually not advocated in the patient with relatively few discharges per year. Many antiarrhythmic drugs may also ele-vate the defibrillation threshold above the output capability of the ICD generator. Patients who experience five or six shocks in succession, without warning and without loss of con-sciousness, may be receiving "inappropriate" therapies for supraventricular arrhythmias or because of a "make/break" artifact occurring as a result of a fractured lead.

X. ELECTROMAGNETIC INTERFERENCE AND ICDs

Artifacts leading to inappropriate device therapy include extraneous electromagnetic inter-ference from radio-controlled toys, theft-deterrent systems, high-voltage generators, etc., or myopotential oversensing that results from automatic sense-amplifier adjustment during periods of bradycardia (i.e., "bradycardia-dependent oversensing"). Precautions to prevent electromagnetic interference and strong magnetic fields that may interfere with ICD func-tions are the same as for cardiac pacemakers, with differences related to the severity of the underlying condition or, in some cases, peculiarities in defibrillator response circuitry. As the magnet mode for ICDs is inhibition of detection and therapy for tachyarrhythmias, prolonged exposure to magnetic fields can be potentially catastrophic. In the case of electromagnetic "noise," the response of the ICD will be detection and therapy (ATP and/or shock) for a nonexistent arrhythmia. These may be repeated until the patient moves away from the source of interference. Inappropriate defibrillator therapy has been reported from sources as varied as slot machines, radio-controlled models, TENS units, arc welders, and electrocautery machines (66–70). With wireless phones, sensing of the high-frequency transmission occurs in the defibrillator header, so interference, as with pacemakers, can be avoided by maintaining a distance of at least 8 cm between phone units and the defibrillator generator.

XI. CONCLUSIONS

The advent of transvenous defibrillation systems has reduced the morbidity and mortality of ICD implantation devices significantly compared with early procedures that required a thoracotomy for placement of epicardial defibrillation patches. Current generation unipolar ICDs do not differ substantially from pacemakers in terms of implantation technique. Nonetheless, management and follow-up of a patient with an ICD is significantly more involved because of device complexity and the greater seriousness of the potential for malignant ventricular arrhythmias. An ICD that is not appropriately programmed for the specific patient can lead to significant morbidity and potentially mortality as a result of either inadequate or unnecessary device therapy. Some patients need only a "shock box" for treatment of VF, whereas others require the full complement of options including dual-chamber pacing, management of atrial arrhythmias, antitachycardia pacing, and VF therapy.

From a technological standpoint, ICD size and reliability continue to improve. Generator size is now approximately 35 cc, limited in large part by battery, capacitor, and connector size. Despite defibrillation thresholds that are commonly only 5 J, the probabilistic nature of defibrillation and the requirement for a safety margin may impose a lower limit to ICD size, barring a dramatic change in defibrillation technique. Enormous development costs, short product life cycles, and patient and physician demand for "only the latest" will challenge significant cost reductions even as the ICD population expands. As with pacemakers, ICD automation and data-gathering abilities have become simplified and improved clinical management procedures have begun to be clear. Even more sophisticated automation and data-gathering capabilities, such as remote device follow-up and reprogramming, continuous hemodynamic and metabolic monitoring, etc., are under development and should continue to expand device utility in the future.

The relative ease of implantation and longevity of current generation defibrillators has also made ICDs a valuable new tool in primary prevention. No longer must patients survive a cardiac arrest to justify the risk of ICD implantation. In the case of SCD survivors or patients with hemodynamically destabilizing ventricular arrhythmias, the benefit from defibrillator treatment is relatively clear-cut, and in most cases has become the new standard of care in the United States. From a public health standpoint, preventing sudden death before the first event—"primary prevention"—is a major area of emphasis and ongoing investigation. This is potentially an enormous patient population. Risk-stratification techniques are the weak link, and none have been proven particularly sensitive and specific above and beyond relatively broad clinical characteristics, such as ejection fraction or

New York Heart Association functional class for heart failure candidates. With other major public health issues competing for available dollars in a health-care industry with skyrocketing costs, large-scale trials become even more critical in helping to clarify the most cost-efficient uses for these devices. The potential for misuse or overuse will continue to be a serious concern—one that can only be answered with future clinical evaluation.

REFERENCES

1. Mirowski M, Mower M, Staewen W, et al. Standby automatic defibrillator: An approach to prevention of sudden coronary death. Arch Intern Med 1970; 126:158–161.
2. Mirowski M, Mower M, Staewen W, et al. The development of the transvenous automatic defibrillator. Arch Intern Med 1972; 129:773–779.
3. Mower M. In the beginning: From dogs to humans. Pacing Clin Electrophysiol 1995; 18:506–511.
4. Mirowski M, Mower M, Staewen W, et al. Ventricular defibrillation through a single intravascular catheter system. Clin Res 1971; 19:328.
5. Mehra R, Cybulski Z. Tachyarrhythmia termination: Lead systems and hardware design. In: Singer I, ed. Implantable Cardioverter-Defibrillator. Armonk, NY: Futura Publishing Co, 1994:109–133.
6. Davy J, Fain E, Dorian P, et al. The relationship between successful defibrillation and delivered energy in open-chest dogs: Reappraisal of the "defibrillation threshold" concept. Am Heart J 1987; 113:77.
7. McDaniel W, Schuder J. The cardiac ventricular defibrillation threshold-inherent limitations in its application and interpretation. Med Instrum 1987; 21:170.
8. Bardy G, Johnson G, Poole J, et al. A simplified, single lead unipolar transvenous cardioverter-defibrillator. Circulation 1993; 88:543–547.
9. Bardy G, Dolack G, Kudenchuk P, Poole J, Mehra R, Johnson G. Prospective, randomized comparison in humans of a unipolar defibrillation system with that using an additional superior vena cava electrode. Circulation 1994; 89:1090–1093.
10. Gurvich N, Markarychev V. Defibrillation of the heart with biphasic electrical impulses. Kardiologiia 1967; 7:109–112.
11. Schuder J, McDaniel W, Stoeckle H. Defibrillation fo 100-kg calves with asymmetrical, bidirectional, rectangular pulses. Cardiovasc Res 1984; 18:419–426.
12. Bardy G, Ivey T, Allen M, Johnson G, Mehra R, Greene L. A prospective randomized evaluation of biphasic versus monophasic waveform pulses on defibrillation efficacy in humans. J Am Coll Cardiol 1989; 14:728–733.
13. Blanchard S, Knisley S, Walcott G, Ideker R. Defibrillation waveforms. In: Singer I, ed. Implantable Cardioverter-Defibrillator. Armonk, NY: Futura Publishing Co, 1994:153–178.
14. Fromer M, Brachmann J, Block M, et al. Efficacy of automatic multimodal device therapy for ventricular tachyarrhythmias as delivered by a new implantable pacing cardioverter-defibrillator: results of a European multicenter study of 102 implants. Circulation 1992; 86:363.
15. Fischer J, Mehra R, Furman S. Termination of ventricular tachycardia with bursts of rapid ventricular pacing. Am J Cardiol 1978; 21:94.
16. Olson W. Tachyarrhythmia sensing and detection. In: Singer I, ed. Implantable Cardioverter-Defibrillator. Armonk, NY: Futura Publishing Co, 1994:71–107.
17. Klingenheben T, Sticherling C, Skupin M, Hohnloser S. Intracardiac QRS Electrogram Width-an Arrhythmia detection feature for implantable cardioverter defibrillators: Exercise induced variation as a base for device programming. PACE 1998; 21:1609–1617.
18. Brugada J, Mont L, Figueiredo M, Valentino M, Matas M, Navarro-Lopez F. Enhanced detection criteria in implantable defibrillators. J Cardiovasc Electrophysiol 1998; 9:261–268.
19. Gold M, Hsu W, Marcovecchio A, Olsovsky M, Lang D, Shorofsky S. A new defibrillator dis-

crimination algorithm utilizing electrogram morphology analysis. Pacing Clin Electrophysiol 1999; 22:179–182.

20. Korte T, Jung W, Wolpert C, et al. A new classification algorithm for discrimination of ventricular from supraventricular tachycardia in a dual chamber implantable cardioverter defibrillator. J Cardiovasc Electrophysiol 1998; 9:70–73.

21. Center C-SM. Comprehensive Evaluation of Defibrillators and Resuscitative Shocks (CEDARS) Study: Does atrial fibrillation increase the incidence of inappropriate shock by implanted defibrillators? J Am Coll Cardiol 1993; 21:201A.

22. Higgins S, Williams S, Pak J, Meyer D. Indication for a dual chamber pacemaker combined with an implantable cardioverter-defibrillator. Am J Cardiol 1998; 81:1360–1362.

23. Schmitt C, Montero M, Melichercik J. Significance of supraventricular tachyarrhythmias in patients with implanted pacing cardioverter defibrillators. Pacing Clin Electrophysiol 1998; 17:295–302.

24. Neuzner J, Petschner H. Managing the problem of atrial tachyarrhythmias in patients with ICDs. Practice and Progress in Cardiac Pacing and Electrophysiology: Dordrecht, The Netherlands, Kluwer Academic Publishers, 1996:353–360.

25. Wharton M, Santini M. Treatment of spontaneous atrial tachyarrhythmias with the Medtronic 7250 Jewel AF: Worldwide clinical experience. Circulation 1998; 98:I-190.

26. Saxon L, Boehmer J, Hummel J, et al. Biventricular pacing in patients with congestive heart failure: Two prospective randomized trials. Am J Cardiol 1999; 83:120D–123D.

27. Bernstein AD, Parsonnet V. Survey of cardiac pacing and defibrillation in the United States in 1993. Am J Cardiol 1996; 78:187–96.

28. Gregoratos G, Cheitlin M, Conill A, et al. ACC/AHA guidelines for implantation of cardiac pacemakers and antiarrhythmia devices: a report of the American College of Cardiology/American Heart Association Task Force on Practice Guidelines (Committee on Pacemaker Implantation). J Am Coll Cardiol 1998; 31:1175–1209.

29. Connolly SJ, Gent M, Roberts RS, et al. Canadian Implantable Defibrillator Study (CIDS): study design and organization. CIDS Co-Investigators. Am J Cardiol 1993; 72:103F–108F.

30. Connolly S. Results from the Canadian Implantable Defibrillator Study (CIDS). Late Breaking Clinical Trials. ACC 47th Annual Scientific Sessions, 1998.

31. AVID Investigators. A comparison of antiarrhythmic-drug therapy with implantable defibrillators in patients resuscitated from near-fatal ventricular arrhythmias. The Antiarrhythmics Versus Implantable Defibrillators (AVID) Investigators. N Engl J Med 1997; 337:1576–83.

32. Kuck K. Cardiac Arrest Study Hamburg. Late Breaking Clinical Trials I. ACC 47th Annual Scientific Sessions, 1998.

33. Brugada J, Brugada R, Brugada P. Right bundle-branch block and ST-segment elevation in leads V1 through V3: A marker for sudden death in patients without demonstrable structural heart disease. Circulation 1998; 97:457–460.

34. Groh W, Silka M, Oliver R, Halperin B, McAnulty J, Kron J. Use of implantable cardioverter-defibrillators in the congenital long QT syndrome. Am J Cardiol 1996; 78:703–706.

35. Primo J, Geelen P, Brugada J, et al. Hypertrophic cardiomyopathy: role of the implantable cardioverter- defibrillator. J Am Coll Cardiol 1998; 31:1081–1085.

36. Zhu DW, Sun H, Hill R, Roberts R. The value of electrophysiology study and prophylactic implantation of cardioverter defibrillator in patients with hypertrophic cardiomyopathy. Pacing Clin Electrophysiol 1998; 21:299–302.

37. Viskin S, Belhassen B. Idiopathic ventricular fibrillation. Am Heart J 1990; 120:661–671.

38. Wellens H, Lemery R, Smeets J, et. al. Sudden arrhythmic death without overt heart disease. Circulation 1992; 85(suppl I):I92–I97.

39. Yetman AT, Hamilton RM, Benson LN, McCrindle BW. Long-term outcome and prognostic determinants in children with hypertrophic cardiomyopathy. J Am Coll Cardiol 1998; 32:1943–1950.

40. Armoundas AA, Rosenbaum DS, Ruskin JN, Garan H, Cohen RJ. Prognostic significance of

electrical alternans versus signal averaged electrocardiography in predicting the outcome of electrophysiological testing and arrhythmia-free survival. Heart 1998; 80:251–256.

41. Armoundas AA, Osaka M, Mela T, et al. T-wave alternans and dispersion of the QT interval as risk stratification markers in patients susceptible to sustained ventricular arrhythmias. Am J Cardiol 1998; 82:1127–1129.

42. Das SK, Morady F, DiCarlo L, Jr., et al. Prognostic usefulness of programmed ventricular stimulation in idiopathic dilated cardiomyopathy without symptomatic ventricular arrhythmias. Am J Cardiol 1986; 58:998–1000.

43. Farrell T, Bashir Y, Cripps T, etal. Risk stratification for arrhythmic events in postinfarction patients based on heart rate variability, ambulatory electrocardiographic variables and the signal-averaged electrocardiogram. J Am Coll Cardiol 1991; 18:687–697.

44. Grimm W, Glaveris C, Hoffmann J, et al. Noninvasive arrhythmia risk stratification in idiopathic dilated cardiomyopathy: design and first results of the Marburg Cardiomyopathy Study. Pacing Clin Electrophysiol 1998; 21:2551–2556.

45. Moss A, Hall W, Cannom D, et al. Improved survival with an implanted defibrillator in patients with coronary disease at high risk for ventricular arrhythmia. N Engl J Med 1996; 335:1933–1940.

46. Bigger J, for the Coronary Artery Bypass Graft (CABG) Patch Trial Investigators. Prophylactic use of implanted cardiac defibrillators in patients at high risk for ventricular arrhythmias after coronary artery bypass graft surgery. N Engl J Med 1997; 337:1569–1575.

47. Buxton AE, Leek L, Fisher JD, Josephson ME, Prystowsky EN, Hafley G. Randomized study of the prevention of sudden death in patients with coronary artery disease. N Engl J Med 1999; 841:1882–1890.

48. Woosley RL, Singh SN, eds. Arrhythmia Treatment and Therapy: Evaluation of Clinical Trial Evidence. New York: Marcel Dekker, 2000.

49. Schmidinger H. The implantable cardioverter defibrillator as a "bridge to transplant": a viable clinical strategy? Am J Cardiol 1999; 83:151D–157D.

50. Weiss D, Zilo P, Luceri R, Platt S, Rosenbaum M. Predischarge arrhythmia induction testing of implantable defibrillators may be unnecessary in selected cases. Am J Cardiol 1997; 80:1562–1565.

51. Bardy G, Lee K, Mark D, et al. SCD-HeFT: Prevention of Sudden Cardiac Death in Patients with Congestive Heart Failure Trial. In: Woosley R, Singh S, eds. Arrhythmia Treatment and Therapy: Evaluation of Clinical Trial Evidence. New York: Marcel Dekker, 2000.

52. Moss A, Cannom D, Daubert J, et al. Multicenter Automatic Defibrillator Implantation Trial II (MADIT II): design and clinical protocol. Ann Noninvas Electrocardiol 1999; 4:83–91.

53. Nair M, Saoudi N, Kroiss D, Letac B. Automatic arrhythmia identification using analysis of the atrioventricular association. Application to a new generation of implantable defibrillators. Participating centers of the automatic recognition of arrhythmia study group. Circulation 1997; 95:967–973.

54. Strickberger SA, Man KC, Daoud EG, et al. A prospective evaluation of catheter ablation of ventricular tachycardia as adjuvant therapy in patients with coronary artery disease and an implantable cardioverter-defibrillator. Circulation 1997; 96:1525–31.

55. Gold M, Olsovsky M, Pelini M, Peters R, Shorofsky S. Comparison of single- and dual-coil pectoral defibrillation lead systems. J Am Coll Cardiol 1998; 31:1391–1394.

56. Craney JM, Gorman LN. Conscious sedation and implantable devices. Safe and effective sedation during pacemaker and implantable cardioverter defibrillator placement. Crit Care Nurs Clin North Am 1997; 9:325–334.

57. Jones G, Bardy G. Implantation of ICDs in the electrophysiology laboratory. In: Singer I, ed. Interventional Electrophysiology, 1st ed. Baltimore, MD: Williams and Wilkins, 1997:725–740.

58. Flaker G, Tummala T, Wilson J, Investigators ftWWJ. Comparison of right- and left-sided pectoral implantation parameters with the Jewel active can cardiodefibrillator. Pacing Clin Electrophysiol 1998; 21:447–451.

59. Zivin A, Souza J, Pelosi F, et al. Relationship between shock energy and postdefibrillation ventricular arrhythmias in patients with implantable defibrillators. J Cardiovasc Electrophysiol 1999; 10:370–377.

60. Tokano T, Bach D, Chang J, et al. Effect of ventricular shock strength on cardiac hemodynamics. J Cardiovasc Electrophysiol 1998; 9:791–797.

61. Strickberger SA, Man KC, Souza J, et al. A prospective evaluation of two defibrillation safety margin techniques in patients with low defibrillation energy requirements. J Cardiovasc Electrophysiol 1998; 9:41–46.

62. Strickberger SA, Daoud EG, Davidson T, et al. Probability of successful defibrillation at multiples of the defibrillation energy requirement in patients with an implantable defibrillator. Circulation 1997; 96:1217–1223.

63. Daoud EG, Man KC, Morady F, Strickberger SA. Rise in chronic defibrillation energy requirements necessitating implantable defibrillator lead system revision. Pacing Clin Electrophysiol 1997; 20:714–719.

64. Daoud EG, Man KC, Horwood L, Morady F, Strickberger SA. Relation between amiodarone and desethylamiodarone plasma concentrations and ventricular defibrillation energy requirements. Am J Cardiol 1997; 79:97–100.

65. Goyal R, Harvey M, Horwood L, et al. Incidence of lead system malfunction detected during implantable defibrillator generator replacement. Pacing Clin Electrophysiol 1996; 19:1143–1146.

66. Vlay S. Electromagnetic interference and ICD discharge related to chiropractic treatment. Pacing Clin Electrophysiol 1998; 21:2009.

67. Glotzer TV, Gordon M, Sparta M, Radoslovich G, Zimmerman J. Electromagnetic interference from a muscle stimulation device causing discharge of an implantable cardioverter defibrillator: epicardial bipolar and endocardial bipolar sensing circuits are compared. Pacing Clin Electrophysiol 1998; 21:1996–8.

68. Madrid A, Sanchez A, Bosch E, Fernandez E, Moro Serrano C. Dysfunction of implantable defibrillators caused by slot machines. Pacing Clin Electrophysiol 1997; 20:212–214.

69. Santucci PA, Haw J, Trohman RG, Pinski SL. Interference with an implantable defibrillator by an electronic antitheft-surveillance device. N Engl J Med 1998; 339:1371–1374.

70. Man K, Davidson T, Langberg J, Morady F, Kalbfleisch S. Interference from a hand held radiofrequency remote control causing discharge of an implantable defibrillator. Pacing Clin Electrophysiol 1993; 16:1756–1758.

23

Remodeling
Structural and Electrical Considerations

GORDON F. TOMASELLI

The Johns Hopkins University, Baltimore, Maryland

KARL T. WEBER

University of Tennessee Health Science Center, Memphis, Tennessee

I. INTRODUCTION

Despite remarkable improvements in medical therapy, the prognosis of patients with myocardial failure remains poor, with over 15% of patients dying within 1 year of initial diagnosis and up to an 80% 6-year mortality (1, 2). Of the deaths in patients with heart failure, 50% are sudden and unexpected.

The failing heart undergoes a complex series of changes in both myocyte and non-myocyte elements. In an attempt to compensate for the reduction in cardiac function, the sympathetic nervous system (SNS) and the renin-angiotensin-aldosterone system (RAAS), as well as other neurohumoral mechanisms, are activated but ultimately prove to be maladaptive. Neurohumoral activation predisposes to myocyte loss and ventricular remodeling, including interstitial hyperplasia, which results in a progressive reduction in force development and impairment of ventricular relaxation.

Intrinsic cardiac and neurohumoral responses to myocardial failure adversely alter the electrophysiology of the heart, prediposing patients with heart failure to an increase in arrhythmic death. With progression of heart failure, there is an increase in the frequency and complexity of ventricular ectopy (3, 4). Total mortality in heart failure patients correlates with left ventricular function and the presence of complex ventricular ectopy (5–7). However, there is no clear correlation between sudden cardiac death (SCD) and ventricular function or ventricular ectopy, and death can be disproportionately sudden in patients with more modest myocardial dysfunction (8). A major caveat is that the mechanism of sudden death is highly heterogeneous and includes ventricular tachycardia, ventricular fibrillation, profound bradycardia, asystole, and pulseless electrical activity.

This chapter considers the changes, both structural and electrical, that have been observed in myocardial hypertrophy and heart failure that predispose to cardiac arrhythmias.

II.   CARDIAC STRUCTURE AND REMODELING IN HEART FAILURE

A.   Structural Remodeling

The myocardium is composed of cardiac myocytes, which are highly differentiated and specialized cells responsible for the heart's contractile behavior, and nonmyocytes. Myocytes occupy two-thirds of the myocardium's structural space; however, they represent only one-third of all cells. Nonmyocyte cardiac cells include fibroblasts responsible for turnover of an extracellular matrix that consists predominantly of fibrillar type I and III collagens. The fibrillar collagen scaffolding provides for cardiac myocyte alignment and coordinated transmission of contractile force to the ventricular chamber. Other nonmyocyte cells include endothelial and smooth muscle cells of the intramural vasculature.

Hypertrophy per se does not predispose to arrhythmias. Hypertrophy is adaptive when myocyte and matrix compartments undergo balanced growth that preserves their respective proportionalities. Such remodeling appears with athletic training. Increments in myocardial mass that appear in trained athletes are comparable to levels achieved in patients with cardiovascular disease. It is the patient, however, in whom left ventricular hypertrophy is associated with enhanced cardiovascular risk, including sudden cardiac death (9). Thus, it is not the quantity but rather the quality of cardiac tissue that determines risk. Sudden cardiac death can appear in the athlete when hypertrophy is pathological, as represented by a disproportionate growth of myocyte and nonmyocyte compartments. This can occur, for example, when exercise training is combined with anabolic steroids or some agent injurious to cardiac myocytes or when systemic hypertension is accompanied by a chronic activation of the RAAS (10).

The failing human heart undergoes considerable structural remodeling (11). Such remodeling provides a structural basis for ventricular diastolic and systolic dysfunction and arrhythmogenesis. In ischemic cardiomyopathy following a previous myocardial infarction(s), a segmental loss of cardiac myocytes is replaced by macroscopic scar tissue. This infarct scar, which consists of type I and III collagens, is anchored into viable myocardium at the border zone of the infarction by collagen fibers that encircle myocytes. Throughout noninfarcted myocardium of infarcted and noninfarcted ventricles there are microscopic scars that have replaced lesser numbers of lost myocytes. Such replacement (or reparative) fibrosis accompanies myocyte necrosis of diverse causality. Remaining viable myocytes may hypertrophy; those encircled by collagen often atrophy. In contrast, an inflammatory cell response and subsequent fibrous tissue formation do not follow myocyte apoptosis.

Another prominent feature of the structural remodeling of myocardium in ischemic cardiomyopathy is an adverse accumulation of fibrillar collagen that is not related to myocyte necrosis. Such a reactive interstitial fibrosis is found remote to the infarct site and involves noninfarcted myocardium of both right and left ventricles, including the interventricular septum (12–15). Not unlike hepatic cirrhosis or renal tubulointerstitial sclerosis, interstitial fibrosis in the myocardium compromises the function of its parenchyma (or cardiac myocytes). It augments tissue stiffness and electrical dispersion. Interstitial fibrosis predisposes to reentrant arrhythmias (9, 16–18).

B. Determinants of Remodeling

The hypertrophy of cardiac myocytes residing in a given ventricle or atrium is determined by their mechanical loading conditions. Myocyte loading is a function of the prevailing pressure and volume events and the size of the corresponding ventricular or atrial chamber. Certain hormonal stimuli, such as growth and thyroid hormones, can also promote myocyte hypertrophy.

Fibroblast-like cells regulate matrix remodeling. Collagen turnover at sites of tissue repair in the infarcted heart is governed by myofibroblasts, a phenotypically transformed fibroblast that expresses alpha smooth muscle actin. Actin microfilaments confer myofibroblasts with contractile behavior that contributes to scar tissue remodeling (e.g., thinning) and whose tonic contraction could alter myocardial stiffness. Myofibroblasts are persistent and active at an infarct site for years after myocardial infarction (19). They are nourished by a neovasculature. Thus, as opposed to considerable earlier thinking, the infarct scar is anything but inert tissue.

Myofibroblast collagen turnover is not regulated by hemodynamic factors, but instead by signals found in the biochemical milieu of the interstitial space. Transforming growth factor-β_1 (TGF-β_1) is a cytokine released into this tissue fluid where it serves to regulate myofibroblast collagen synthesis in a paracrine and autocrine manner. TGF-β_1 is initially derived from inflammatory cells that have invaded the infarct site; it subsequently is derived from myofibroblasts themselves. This fibrogenic peptide contributes importantly to fibrous tissue formation by promoting myofibroblast transcription of type I and III collagen genes and their expression of tissue inhibitors of matrix metalloproteinases. Myofibroblast TGF-β_1 expression is regulated in an autocrine manner by angiotensin II generated de novo at sites of repair (11). Whether locally generated angiotensin II contributes to the arrhythmogenic potential of the failing heart is unknown, but its receptors are found on cardiac myocytes and myofibroblasts.

Chronic inappropriate (relative to dietary sodium and intravascular volume) elevations in angiotensin II are associated with myocardial fibrosis. The same is true for chronic elevations in aldosterone, the other effector hormone of the circulating RAAS. Each of these circulating hormones contributes to cardiac myocyte necrosis and therefore is associated with microscopic scarring. Likewise, all are associated with a reactive perivascular/interstitial fibrosis that appears in right and left ventricles, right and left atria, and systemic organs (including kidneys) (20, 21). Pathophysiological mechanisms involving angiotensin II and aldosterone in the appearance of cardiac fibrosis have been reviewed elsewhere (10). A chronic activation of the RAAS in heart failure is pathological and contributes to the progressive structural remodeling of cardiac tissue and the progressive nature of heart failure.

III. CELLULAR ELECTROPHYSIOLOGY IN HYPERTROPHY AND HEART FAILURE

A. Changes in the Action Potential Profile and Duration in Hypertrophy and Heart Failure

An elementary and distinctive signature of any excitable tissue is its action potential profile. Cardiac myocytes possess a characteristically long action potential (Fig. 23.1): after an initial rapid upstroke, there is a plateau of maintained depolarization before repolarization. The duration of the action potential is primarily responsible for the time course of re-

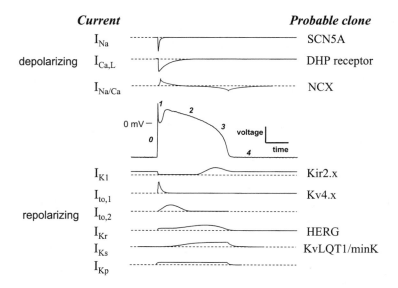

Figure 23.1   Schematic of the depolarizing and repolarizing currents that underlie the action potential in the mammalian ventricle. A schematic of the time course of each of the currents is shown as well as the gene product that underlies the current. The phases of the action potential are labeled. (Modified from Ref. 198.)

polarization of the heart; prolongation of the action potential produces delays in cardiac repolarization.

Changes in action potential duration and profile result from alterations in the functional expression of depolarizing and repolarizing currents. Prolongation of the action potential is characteristic of cells and tissues isolated from ventricles of animals with heart failure independent of the mechanism (Fig. 23.2), including pressure and/or volume over-

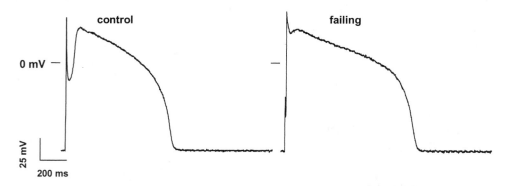

Figure 23.2   Action potentials from ventricular myocytes isolated from normal and failing canine hearts. The action potential on the left was recorded from a ventricular myocyte isolated from a control heart using the whole-cell configuration of the patch clamp at 37°C, stimulation frequency of 0.5 Hz, and minimal intracellular Ca^{2+} buffering. On the right is a similarly recorded action potential from a ventricular myocyte from a terminally failing heart.

load (22–36), genetic (37–39), metabolic (40), ischemia/infarction (41–44), and chronic pacing tachycardia models (45–47). Similarly, tissues (48–50) and cells (51, 52) from failing human ventricles exhibit action potential prolongation.

The pathophysiological significance of action potential prolongation in myocytes isolated from hypertrophied and failing hearts has been questioned on several grounds. First, most action potential recordings from ventricular myocytes are made at unphysiologically slow rates and indeed the difference in duration between cells from failing and control ventricles converges at high stimulation frequency (47, 50). However, slow heart rates and pauses after premature contractions are common in heart failure, and the postpause prolongation of the action potential duration may be highly significant. Second, isolated myocytes are no longer electrically coupled to other cells in the cardiac syncytium; however, intact muscle preparations from failing hearts (e.g., 49) and monophasic action potential recordings in whole hearts (53) also exhibit action potential prolongation. Finally, the duration of the action potential is quite sensitive to mechanical load and increasing the load tends to shorten action potential duration and refractoriness more in failing than in normal hearts (54).

An important and understudied question is the effect of hypertrophy and failure on regional differences in action potential duration. Action potential durations vary across the ventricular wall (55–58) and in different regions (59) of the mammalian heart. Data from experimental animal models of hypertrophy suggest regional inhomogeneity in action potential prolongation (25,28). The finding of enhanced spatial and temporal dispersion of monophasic action potential duration, refractoriness, and electrocardiographic QT intervals in humans (60,61) and animals with heart failure (53) is consistent with an exaggerated dispersion of action potential duration that may predispose to ventricular arrhythmias.

B. Downregulation of Potassium Currents

The duration and shape of the action potential is the result of a delicate balance between the depolarizing and repolarizing currents that are active during the plateau phase (Fig. 23.1). Repolarization in the mammalian heart is achieved primarily by the activity of potassium-selective ion currents, although the exact molecular composition of the channels carrying these currents varies from species to species. Reduced density (functional downregulation) of K currents is a recurring theme in hypertrophied and failing ventricular myocardium. Ventricular myocytes contain several distinct classes of voltage-dependent K channels. The inward rectifier K^+ current, I_{K1}, sets the resting membrane potential and contributes to the terminal phase of repolarization. Another important K current is the calcium-independent transient outward current, I_{to}. Unlike the inward rectifier, I_{to} is expressed in heart cells in a species- and cell-type-specific fashion. This current plays a crucial role in the early phase of repolarization. The delayed rectifier K current (I_K), composed of molecularly distinct rapid (I_{Kr}) and slow (I_{Ks}) components, is important in phase 3 repolarization. I_K density varies regionally in the hearts of some species (62,63) and I_{Kr} is the target of several antiarrhythmic drugs with Vaughan–Williams class III action.

A reduction in the current density of I_{to} is arguably the most consistent ion current change in cardiac hypertrophy and failure (Table 23.1). Several notable exceptions are studies of compensated pressure overload hypertrophy that were associated with either no change (24) or an increase in I_{to} density (30,64). Downregulation of I_{to}, without a significant change in the voltage dependence or kinetics of the current has also been observed in cells isolated from terminally failing human hearts (52,65,66).

Table 23.1   K Currents in Hypertrophy and Heart Failure

Ref.	Model	APD	I_{to}	I_{K1}	I_K	Comment
Pressure/volume overload						
29	cat/RV			↑	↓	I_K slowed activation, faster deact
23	rat/Ao		↓			I_{to} density LVFW > apex > septum
24	SHR	↑	↔	↓	↔	
81	cat/Ao	↑			↓	Slowed act and faster deact of I_K
33	gp/Ao	↑		↔	↔	
26	SHR	↑	↓	↔		
27	DOCA/salt	↑	↓			Small negative shift gating
30	RVH rat	↑	↑			Slowed I_{to} decay
34	rat/Ao	↑	↓			No change in Ito gating/kinetics
35	ferret/RV		↓			Slowed TTP, decay and recovery
36	RVH rat					↓ Kv4.2/4.3 mRNA
						No change Kv1.2, 1.4, 1.5, 2.1,LQT1
Pacing-tachycardia						
45	dog	↑	↓		↓	No change in I_{to} gating/kinetics
46	rabbit	↑	↓		↔	No change in I_{to} recovery
Ischemia/infarction						
41	dog	↓	↓			Infarct vs. noninfarct zone
42	rat	↑	↓			No change: RMP, Ito kinetics/gating
44	rat					↓ Kv1.4, 2.1, 4.2 mRNA
						↓ Kv2.1, 4.2 protein
Genetic/misc.						
38	hamster	↑	↓			Slowed Ito recovery
40	rat/GH	↑	↓			No change Ito gating/kinetics
Human						
52		↑	↓	↓	↔	
65			↓ subendo			
66			↓ subepi			Slow recovery of I_{to} in subendo
77						↓ Kv4.3, ↔ Kv1.4, Kvβ1, Kir2.1, herg

RV, right ventricle; Ao, aorta; LVFW, left ventricular free wall; SHR, spontaneously hypertensive rat; DOCA, deoxycorticosterone acetate; RVH, right ventricular hypertrophy; TTP, time to peak; RMP, resting membrane potential; GH, growth hormone.

The density of I_{to} varies regionally and transmurally in the heart, and there is some evidence that the density of I_{to} may be reduced differentially in heart failure (65,66). The mechanisms underlying regional and transmural differences in I_{to} current density in the heart are not clear. Some data suggest that there are differences in the level of expression of the same K channel gene; alternatively, distinct gene products may underlie I_{to} in different regions of the heart and at various stages of development (67,68). In humans, it has been hypothesized that the K channel Kv1.4 is the predominant gene that encodes endocardial I_{to}, while Kv4.3 underlies midmyocardial and epicardial I_{to} (69). Interestingly, these two K channels (Kv1.4 and Kv4.3 or Kv4.2) exhibit distinct kinetic behavior when heterologously expressed, with Kv1.4 having much slower inactivation recovery kinetics than Kv4.x (70–73). Preferential expression of Kv1.4 in the endocardium may underlie the different electrophysiological behavior of I_{to} in human myocytes isolated from the subendocardium and subepicardium (65,66).

The molecular mechanisms of I_{to} downregulation in heart failure are likely to be multifactorial. I_{to} is regulated by neurohumoral mechanisms; specifically, α_1-adrenergic stimulation reduces the current size (74–76). In animal models (44) and human heart failure (77), a reduction in the steady-state level of Kv4 mRNA has been associated with downregulation of I_{to}. In the rat, reduction in the steady-state level of mRNA is associated with a commensurate decrease in the level of immunoreactive Kv4 protein (44). The reduction in mRNA level results from a change in the balance between transcription and mRNA degradation, the precise molecular mechanism of which is unknown. It is interesting to note that regulated expression of I_{to} and Kv4 mRNA and protein occurs during development (67) and exposure to thyroid hormone (68,78).

Because I_{to} is brief, its role in setting the action potential duration in larger animals and humans remains controversial. I_{to} does profoundly influence rapid early repolarization (phase 1) and the level of the plateau (Fig. 23.2), thereby affecting all of the currents that are active later in the action potential. However, most of the studies examining I_{to} in heart failure have used Ca^{2+}-buffered internal solutions, thus eliminating any possible role of calcium-dependent processes. Under such conditions, there are several lines of evidence (45,46,52) that I_{to} can influence overall action potential duration. Nevertheless, it is not clear whether this conclusion would also apply under more physiological conditions. Winslow et al. have simulated the role of I_{to} in setting the action potential duration in a novel canine ventricular action potential model (79). They find that I_{to} has a sizable effect when intracellular Ca^{2+} is buffered, but not when Ca^{2+} cycling occurs unimpeded. More experiments will be required to resolve the physiological importance of I_{to} downregulation in heart failure.

Changes in other K currents in hypertrophy and heart failure have been reported, but not with the consistency of downregulation of I_{to} (Table 23.1). The inward rectifier K current (Kir2 family of genes) maintains the resting membrane potential and contributes to the terminal phase of repolarization in ventricular myocytes. The important component of I_{K1} for action potential repolarization is the outward current at voltages positive to the equilibrium potential for K^+ (E_K). In mild ventricular hypertrophy, increased (29), decreased (but no change at voltages positive to E_K) (24), and unchanged (26,33) I_{K1} density has been reported. Similar inconsistencies have been observed in pacing tachycardia models: reduced I_{K1} density has been demonstrated in the dog (45) and both decreased (J. Rose et al., unpublished) and unchanged current density have been found in the rabbit (46). In human heart failure, ventricular myocytes exhibit significantly reduced current density at negative voltages. The underlying basis of the downregulation of I_{K1} in human heart failure is uncertain, but one report indicates no change in the steady-state level of Kir2.1 mRNA in failing compared to control hearts (77). In studies of human ventricular I_{K1}, a differential reduction in the current was noted between cells isolated from failing hearts with dilated versus ischemic cardiomyopathy. The whole-cell slope conductance at the Nernst potential for K^+ was significantly smaller in cells from hearts with dilated cardiomyopathy; these cells also had longer action potential durations with slower terminal (phase 3) repolarization (80). Ventricular myocytes isolated from controls and hearts with ischemic cardiomyopathy exhibited a change in the likelihood of single-channel opening of I_{K1} at different voltages, a response that was absent in cells isolated from hearts of patients with dilated cardiomyopathy (80).

Studies of the delayed rectifier K current in hypertrophic and failing hearts are sparse. Myocytes isolated from hypertrophied right (29) and left ventricles (81,82) of the cat have reduced I_K current density with slowed activation and faster deactivation. The reduction in the outward current over the plateau voltage range in cells from the hypertrophied feline left ventricle exhibit a greater predisposition to developing potentially arrhythmogenic early after-depolarizations (EADs) (81). In contrast, studies of cells isolated from pressure-overload guinea pig (33) and spontaneously hypertensive rat (24) ventricles demonstrate no change in I_K. There are no studies comparing I_K in control and failing human hearts. The rapid component of the delayed rectifier current is encoded by HERG (*human ether a go-go related gene*); we found no change in the steady-state level of HERG mRNA (77) but others have reported a decrease in failing compared to control hearts (83).

The ATP-gated potassium channel ($I_{K\text{-ATP}}$) is the principal mediator of action potential shortening in response to ischemia in the heart. $I_{K\text{-ATP}}$ is activated under conditions of intracellular ATP depletion or ADP accumulation. Differences in the activity of $I_{K\text{-ATP}}$ in hypertrophied or failing hearts may have profound implications for susceptibility to arrhythmias induced by myocardial ischemia. Human ventricular $I_{K\text{-ATP}}$ in cells isolated from failing ventricles is fundamentally similar to that observed in myocytes from control ventricles but less sensitive to ATP inhibition (84). Action potential shortening that occurs in response to ischemia or metabolic inhibition is exaggerated in cells from hypertrophied compared to normal ventricles (82). The differential sensitivity of the action potential duration to ischemic stress may be a result of altered $I_{K\text{-ATP}}$ sensitivity to intracellular [ATP]; however, the L-type Ca current in myocytes from hypertrophied hearts is also more profoundly suppressed by metabolic inhibition than the current in cells from control hearts (82).

C. Alterations in Ca²⁺ Homeostasis

Heart failure is characterized by depression of developed contractile force, prolongation of relaxation, and blunting of the frequency-dependent facilitation of contraction. The fundamental changes in Ca^{2+} handling that attend ventricular failure are thought to account for the abnormalities in excitation–contraction coupling; however, the cellular and molecular basis of the Ca^{2+} handling deficits in ventricular hypertrophy and failure remain controversial.

The voltage-dependent L-type Ca channel is a multisubunit protein that is ubiquitous in the heart and is often referred to as the dihydropyridine (DHP) receptor for the clinically important class of calcium channel blocking drugs that inhibit this channel. The L-type Ca current is the primary source of Ca^{2+} entry, triggering release of additional Ca^{2+} from the sarcoplasmic reticulum, thereby initiating actin-myosin cross-bridge cycling. The density of Ca current has been studied in a number of animal models of ventricular hypertrophy and failure (85). The severity of hypertrophy or failure appears to influence the density of the L-type current (24,26,33,42,45,82,86–99) or the number of DHP binding sites (93,100–107). In general, when a difference in L-type Ca current density has been detected, the current is increased in mild–moderate hypertrophy and decreased in more severe hypertrophy and heart failure (Table 23.2). Studies of L-type Ca current in cells isolated from failing human hearts parallel the findings in animal models with severe hypertrophy or failure; human cells exhibit either no change (51,108–110), or a decrease in current density (111) or DHP binding sites (112,113) (Table 23.2). Ventricular myocytes isolated from failing hearts exhibit attenuated augmentation of the L-type Ca current by β-adrenergic stimulation (109,111) and depression of rate-dependent potentiation (114) compared to cells isolated from control hearts.

The basic electrophysiological features of the L-type current are altered in some studies of hypertrophy and failure. The most common change is a slowing of the decay of the whole-cell current (e.g., 24,33,87,92), a change that could alter excitation–contraction coupling and would tend to prolong action potential duration. Prolongation of the decay of the Ca current is curious, particularly in view of the common association of elevation of intracellular [Ca^{2+}] in cardiac hypertrophy and failure, a change that should promote calcium-induced inactivation of the current (115,116). However, the slowed decay of the L-type current may reflect deficiencies in Ca^{2+} handling as exemplified by the reduction in the peak of the Ca^{2+} transient causing less Ca^{2+}-induced inactivation of the Ca current. The underlying mechanism of the prolonged whole-cell current decay is unknown; however, a recent single-channel comparison of Ca current in human ventricular myocytes has identified an increase in open probability of channels consistent with a dephosphorylation defect in myocytes isolated from failing hearts (117).

The molecular basis of changes in the density of the L-type Ca current is unknown. In failing human hearts, the steady-state level of α_{1C} mRNA (the gene that encodes the major subunit of the L-type current) has been reported to decrease by Northern blot (112,117) but was unchanged by ribonuclease protection assay (77). It is not known whether there is a change in the level of immunoreactive protein, although a reduction in the number of DHP binding sites has been reported in various studies (Table 23.2). Hypertrophy after myocardial infarction in the rat is associated with reemergence of expression of the fetal isoform of the α_{1C} gene (98). Northern blots of samples from the left ventricle of terminally failing hearts revealed no change in β-subunit (or α_{1C}) mRNA (117). In contrast, samples from right ventricular endomyocardial biopsies revealed an inverse relationship between β-subunit mRNA levels measured by competitive PCR and LV end-diastolic pressure in transplanted hearts (118).

Increased density (upregulation) or reexpression of the T-type Ca current is a prominent feature of some animal models of ventricular hypertrophy (97). The T-type current activates at hyperpolarized voltages and may participate in automaticity in some cells and tissues in the heart. The distribution of the T-type current is more restricted in the heart (119) than the L-type current, particularly in the adult ventricle. Normal maturation of cardiomyocytes is associated with loss of the T-type current, but myocytes grown in primary culture (120), exposed to insulin-like growth factor (IGF-1) in short-term culture (121), or isolated from the atria of rats with growth-hormone-secreting tumors (122) reexpress this current. The T-type current has not been detected in cells isolated from either normal or failing human ventricles (51,109,123); therefore, a role for this current in progression of human heart failure or associated arrhythmogenesis is unlikely.

The amplitude of the intracellular Ca^{2+} transient and its rate of decay are reduced in intact muscles (49) and cells (47,51,109,124) isolated from failing ventricles compared with normal controls (Fig. 23.3). The changes in the Ca^{2+} transient are the result of defective function of the sarcoplasmic reticulum, but the precise molecular mechanism(s) of this defect are controversial. The sarcoplasmic reticulum (SR) Ca^{2+}-ATPase (SERCA2a) and the Na^+-Ca^{2+} exchanger (NCX) are primary mediators of Ca^{2+} removal from the cytoplasm. SERCA2a is inhibited by unphosphorylated phospholamban (PLB) by a direct protein–protein interaction (125), when PLB is phosphorylated, SERCA2a inhibition is re-

Table 23.2 Calcium Current Changes in Hypertrophy and Failure

Ref.	Model	Current density/binding sites	Comment
Mild-moderate hypertrophy			
100	PO/rat	↑DHP-binding sites	
87	RVH rat	↑I_{Ca-L}	Slowed decay
88	PO/rat	↔ I_{Ca-L}	↓β-adrenergic responsiveness
42	post-MI/rat	↔ I_{Ca-L}	No change in kinetics
89	post-MI/rat	↓ I_{Ca-L}	No significant change in kinetics
90	SHR	↔ I_{Ca-L}	No change in kinetics
91	SHR	↑ I_{Ca-L} (10 weeks)	
92	PA banding/cat	↑ I_{Ca-L}	No change in kinetics
101	Syrian hamster	↑ DHP binding sites (early)	
102	Syrian hamster	↑ DHP binding sites (early)	
93	chick PDA	↑ I_{Ca-L} no change DHP binding sites	
33	PO/gp	↑ I_{Ca-L}	Slowed decay, depolarizing shift of SS inactivation
94	pacing pig (1 week)	↓ I_{Ca-L}	No change in kinetics, ↓ β-adrenergic responsiveness
Severe hypertrophy and failure			
95	DOCA-salt/rat	↔ I_{Ca-L}	No change in kinetics
90	SHR	↔ I_{Ca-L}	No change in kinetics
24	SHR	↑ I_{Ca-L}	Slowed decay
26	SHR	↔ I_{Ca-L}	
104	post MI/rat	↓ DHP binding sites	
103	post-MI/rat	↓ PN200-110 binding sites	
82	post-MI/cat	↔ I_{Ca-L}	Slowed decay
101	Syrian hamster	↓ DHP binding sites (late)	
102	Syrian hamster	↓ DHP binding sites (late)	
96	PA banding/ferret	↓ I_{Ca-L}	

Animal Models

Ref.	Model	Current density/binding sites	Comment
97	PA banding/cat	↓I_{Ca-L}	Reexpression of I_{Ca-T}
98	post-MI/gp	↓I_{Ca-L}	Increase in the fetal splice variant of α1C
99	RV HTN/gp	↓I_{Ca-L}	No change in kinetics, ↓β-adrenergic responsiveness
105	pacing rabbit	↓DHP binding sites	
106	pacing dog	↓DHP binding sites	
45	pacing dog	↔I_{Ca-L}	No change in kinetics
107	MI dog	↓DHP binding sites	
86	MI dog	↓I_{Ca-L}	No change in kinetics, ↓β-AR sensitivity
94	pacing pig (3 week)	↓I_{Ca-L}, ↓DHP binding sites	No change in kinetics, ↓β-adrenergic responsiveness

Human studies

Ref.	Current density/binding sites	Comment
109	↔I_{Ca-L}	No change in kinetics, ↓β-AR sensitivity
110	↔I_{Ca-L}	No change in kinetics or voltage dependence
111	↓I_{Ca-L}	↓β-AR sensitivity
114		Blunted upregulation of I_{Ca-L} by rapid stimulation in hearts with EF < 40%
108	↔DHP binding sites	
112	↓DHP binding sites, α1C mRNA	
113	↔DHP binding sites	Normal vs. DCM LV
	↓DHP binding sites	Normal vs. ICM LV
117	↑ensemble average I_{Ca-L}	↑P_{open}, single-channel availability
	↔α_{1C}, β mRNA	

PO, pressure overload; RVH, right ventricular hypertrophy; MI, myocardial infarction; SHR, spontaneously hypertensive rat; PA, pulmonary artery; PDA, patent ductus arteriosus; gp, guinea pig; DOCA, deoxycorticosterone acetate; DHP, dihydropyridine; DCM, dilated cardiomyopathy; ICM, ischemic cardiomyopathy; LV, left ventricle; P_{open}, single-channel open probability.

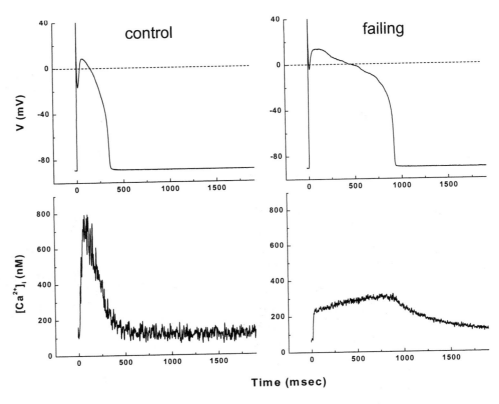

Figure 23.3 Action potentials (upper) and Ca^{2+} transients (lower) recorded in canine ventricular myocytes isolated from a normal heart (left) and failing heart (right). The action potentials are recorded at 37°C with indo-1 in the pipet for determination of the intracellular Ca^{2+} concentration. The transient of the failing cell is smaller, with a slowly rising phase during the action potential plateau and delayed decay following repolarization.

lieved. Ca^{2+} entry into the cell through the L-type Ca channel stimulates release of Ca^{2+} from the SR by the ryanodine receptor (RyR) in a process known as Ca^{2+}-induced Ca^{2+} release. The level of ventricular RyR mRNA decreases in some studies of terminal human heart failure (126,127) but no change in RyR protein level has been demonstrated (128).

Many studies have demonstrated a reduction in SERCA2a mRNA (112,129–137), but fewer studies have shown a reduction in immunoreactive protein (47,137–139). Despite unanimity of opinion that Ca^{2+} sequestration by the SR is defective in failing myocardium, there is controversy about whether there is a change in the pump function of SERCA (for discussion, see Refs. 140 and 141). SERCA2a function may also be altered in hypertrophy and heart failure by changes in the relative expression or function of PLB. PLB mRNA is consistently reduced in failing human hearts (129,132–134,142), but this has not translated into a decrease in PLB protein (Table 23.3) (132–134,143).

Intracellular Ca^{2+} concentration ($[Ca^{2+}]_i$) is an important modulator of the cellular electrophysiology of the heart, affecting the function of a number of ion channels and transporters and increasing the resistance between cells by reducing gap junctional conductance (144). The NCX importantly contributes to control of $[Ca^{2+}]_i$, extruding cytoplasmic Ca^{2+} by electrogenically exchanging it for extracellular Na^+. Most studies from hypertrophied and failing hearts have demonstrated an increase in both NCX mRNA and protein (47,131,145,146) (Table 23.3), suggesting that enhanced NCX function compensates for defective SR removal of Ca^{2+} from the cytoplasm in the failing heart. Others (47) have reported functional evidence for a prominent role of NCX in myocytes from failing canine hearts, which was interpreted as being compensatory for a decrease in Ca^{2+} reuptake by the SR. However, direct studies of NCX function in failing hearts are limited; Na^+-dependent $^{45}Ca^{2+}$ flux into sarcolemmal vesicles has been shown to be increased in a human sarcolemmal preparation (146). In contrast, no change in the Ni^{2+}-sensitive exchanger current was observed in cells isolated from failing hearts compared with normal controls in the rabbit pacing tachycardia heart failure model (147). This, however, does not imply that the current carried by the NCX is the same in cells from failing hearts compared with controls. In the context of a prolonged Ca^{2+} transient, the NCX is likely to play a significant role in reshaping the action potential profile. Forward-mode exchanger function (Na^+ in and Ca^{2+} out) compensates for defective SR Ca^{2+} removal at the expense of depletion of the releasable pool of Ca^{2+} with repetitive stimulation (flat or negative force-frequency relation) (148–150), increasing depolarizing current. Reverse-mode exchange (Na^+ out and Ca^{2+} in) has been suggested to provide inotropic support to the failing heart (145,151). Computer simulations based on the canine pacing tachycardia model suggest that augmentation of reverse-mode exchanger function during the early plateau will tend to shorten the action potential duration. However, with exaggerated forward-

Table 23.3 Alterations in Ca^{2+} Regulatory Proteins in Human Heart Failure

Ref.	SERCA mRNA	SERCA protein	NCX mRNA	NCX protein	PLB mRNA	PLB protein	RYR mRNA	RYR protein
130	↓							
142					↑			
112	↓							
126							↓ ICM ↔ DCM	
129	↓				↓		↓	
131	↓ DCM, ICM		↑ DCM, ICM	↑ DCM, ICM				
143		↔ function				↔		
138		↓ function						
127							↓ DCM, ICM	
128		↔				↔		↔
145			↑ DCM, ICM	↑ DCM, ICM				
133	↓	↔			↓	↔		
146			↑ trend	↑ trend				
132	↓	↔			↓	↔		
134	↓	↔			↓	↔		
149		↓				↑		

SERCA, sarcoplasmic reticulum Ca^{2+}; NCX, Na^+-Ca^{2+} exchanger; PLB, phospholamban; RYR, ryanodine receptor ATPase; DCM, dilated cardiomyopathy; ICM, ischemic cardiomyopathy.

mode function and changes in the decay rate of the L-type Ca current the net effect is prolongation of the action potential (79).

Calsequestrin is an abundant low-affinity, high-capacity Ca^{2+} binding protein in the SR (152). Calsequestrin colocalizes in the junctional SR with the ryanodine receptor. Alterations in the Ca^{2+} binding capacity of the SR could significantly alter Ca^{2+} homeostasis; however, the level of calsequestrin appears to be unchanged in human heart failure (112,128,129,143,153).

The function of the SR is disabled in cardiac hypertrophy and failure. The peak of the Ca^{2+} transient is reduced out of proportion to the change in the Ca current, and the removal of Ca^{2+} from the cytoplasm is retarded. In addition to the elevated $[Ca^{2+}]_i$, other changes in the intracellular milieu (e.g., increased $[Na^+]_i$) may profoundly influence the function of Ca^{2+} homeostatic proteins. Because of interactions between the different components of the system, alteration in SR function in heart failure is complex and not simply explained by changes in the level of proteins that are involved in Ca^{2+} homeostasis.

D. Other Currents and Transporters

The hyperpolarization-activated "funny" or pacemaker current (I_f) in the heart is a nonselective cation current that was originally described in tissues that spontaneously depolarize such as the sinoatrial node (154–156). More recently, I_f has been demonstrated in ventricular cells from animal (157,158) and human hearts (159,160), activating at very negative voltages outside the physiological range. The channel gene underlying this current has recently been cloned (161). I_f generates an inward current that drives the membrane voltage toward threshold, thus significantly contributing to diastolic depolarization in automatic cells. In rat ventricular myocytes, I_f density increases with the severity of cardiac hypertrophy (162). In contrast, although I_f is found in higher density in ventricular myocytes from failing human hearts, the difference from controls did not reach statistical significance. Furthermore, no differences in the voltage dependence, kinetics, or isoproterenol-induced gating shift were noted in cells from failing compared to control hearts (160). Nonetheless, the trend toward an increase in I_f in the setting of reduced I_{K1} current density could predispose ventricular myocytes isolated from failing hearts to enhanced automaticity.

E. Na⁺-K⁺ ATPase

The Na^+-K^+ ATPase (Na, K pump) transports K^+ into the cell and Na^+ out with a stoichiometry of 2:3 generating an outward repolarizing current. The Na^+-K^+ ATPase is dimeric consisting of α- and β-subunits, each of which has three isoforms. The α-subunit determines the glycoside sensitivity of the pump. The majority of experimental data suggest that the expression and function of the Na^+-K^+ ATPase are reduced in failing compared with control hearts (163–167). Decreased Na^+, K^+ pump function in heart failure has several consequences that might be relevant to production of arrhythmias. First is the reduction in the outward repolarizing current that would tend to prolong action potential duration. Second, all else being equal, reduced pump function would lead to an increase in intracellular $[Na^+]$ and enhanced reverse-mode Na^+-Ca^{2+} exchange, increasing depolarizing current. Finally, cells with less Na^+-K^+ ATPase activity have greater difficulty handling changes in extracellular $[K^+]$. Low $[K^+]_0$ itself tends to inhibit the ATPase, while increases in $[K^+]_0$ would tend to be cleared less rapidly in the setting of relative pump inhibition.

F. Modulation of Channel and Transporter Function

In the face of impaired left ventricular pump function with impaired systemic perfusion, the body attempts to maintain circulatory homeostasis through a complex series of neurohumoral changes. Prominently, the SNS and RAAS systems are activated. Activation of the SNS increases heart rate and contractility and redistributes blood flow centrally by vasoconstriction. The RAAS similarly causes vasoconstriction and increases circulatory volume. These neurohumoral changes contribute to progression of the heart failure phenotype. Chronic elaboration of catecholamines can be directly cardiotoxic and result in a series of changes in adrenergic receptor densities that are maladaptive. The volume overload and vasoconstriction produced by chronic activation of both the SNS and RAAS increases myocardial wall stress resulting in increased oxygen demand and the possibility of progressive myocyte damage. The combination of neurohumoral activation and mechanical stress activates signal transduction cascades that produce myocyte hypertrophy and result in the elaboration of trophic factors that increase the interstitial content of collagen; both effects combine to impair systolic and diastolic function of the heart. The changes in neurohumoral signaling have prominent effects on the electrophysiology of the failing heart.

Since the early 1980s (168), adrenergic signaling in human heart failure has been the subject of extensive study (for reviews see Refs. 169 and 170). β_1-, β_2-, and α_1-adrenergic receptors (AR) mediate the effects of increased catecholamines (both circulating epinephrine and norepinephrine released from cardiac nerve terminals) in the heart. These receptor subtypes are coupled to different signaling systems. The β_1- and β_2-receptors are coupled by stimulatory G-proteins to adenylyl cyclase; activation results in increased cellular levels of cAMP, which may be quite local in the case of β_2-receptors (171). The α_1-receptor is coupled by a G-protein to phospholipase C (PLC), which hydrolyzes inositol phospholipids increasing cellular inositol 1,4,5-trisphosphate (IP3) and diacylglycerol (DAG). Angiotensin II (AT1) receptors are similarly coupled to PLC. Activation of the AT1 receptor or the α-adrenergic pathway initiates a kinase cascade triggering cell growth and altering the level of intracellular Ca^{2+}. Indeed, a byproduct of local catecholamine excess in the heart is an increase in cellular Ca^{2+} load. The possible adverse consequences of increased Ca^{2+} load is activation of phospholipases, proteases, and endonucleases culminating in cell necrosis or apoptosis and progression of the failing phenotype.

β- and α-AR signaling pathways significantly affect the function of a number of ion channels and transporters. The net effect of β-AR stimulation is to shorten the ventricular action potential duration due to an increase in the outward current density and a hyperpolarizing shift of the activation of I_K (172), despite β_1-receptor stimulation of depolarizing current through the L-type Ca channel. α_1-AR stimulation inhibits several K currents in the mammalian heart, including I_{to}, I_{K1}, and I_K in rat ventricle with the effect of prolonging action potential duration (173).

Chronic catecholamine stimulation of the failing heart decreases responsiveness of the β-adrenergic signaling system in a manner resembling agonist-induced receptor desensitization (174). A consistent finding is a decrease in the number of β_1-receptors (169, 175) accompanied by a reduction in β_1-receptor mRNA (176, 177), suggesting a decline in the de novo synthesis of receptors in the failing heart. There is controversy as to whether the number of β_2-ARs is decreased in failing human hearts (175); however, there is no change in the steady-state β_2-AR mRNA levels (176, 177). It is generally agreed that any change in β_2-receptor number is less than the change in β_1-receptor number—thus there is a change in the ratio of β-receptor subtypes in the failing heart compared with normal controls. The ratio of β1:β2 AR in the normal heart is approximately 80%:20%, in the failing heart the ratio is closer to 60%:40% (169). β1-AR receptor down regulation in heart failure involves phosphorylation by β-AR kinase and enhanced binding by the inhibitory protein β-arrestin to the receptor. This phosphorylation pathway is likely to be operative in post-translational inhibition of β_2-AR signaling in heart failure. There is also evidence for uncoupling of the β_2-receptor from its intracellular signaling pathway (175).

The desensitization of the β-adrenergic signaling system observed in heart failure is not confined to a reduction in receptor number or uncoupling the receptors from stimulatory G-proteins (170). Bypassing the receptor and G-proteins, for example, by treatment with phosphodiesterase isoenzyme III inhibitors such as milrinone, does not produce equivalent inotropic effects in normal and failing myocardium, suggesting an additional defect in adenylyl cyclase signaling in the failing heart (178). Adenylyl cyclase is under complex regulatory control by G-protein α- and βγ-subunits in the heart. Membrane preparations from human hearts exhibit reduced basal and guanine nucleotide-stimulated adenylyl cyclase activity (179,180). Furthermore, activation of adenylyl cyclase through a G-protein-mediated pathway (e.g., GTP, forskolin, or NaF) is depressed in failing hearts but direct activation of the catalytic subunit by Mn^{2+} is no different in control and failing hearts (181). These data are consistent with a defect in guanine nucleotide binding protein regulation of adenylyl cyclase. The content of the stimulatory G-protein, Gsα, is unchanged in failing hearts (179,180,182,183). The family of inhibitory G-proteins, Giα ($Gi\alpha_2$ is the predominant form in the human heart) are involved in inhibitory regulation of adenylyl cyclase. The level of Giα is increased from 40 to 90% in failing compared with control human hearts (179,180). Pretreatment of failing human myocytes with pertussis toxin to inactivate inhibitory G-proteins restores the positive inotropic effect of isoproterenol (184). These data suggest that in addition to a reduction in β_1-adrenergic receptor number, an increase in Giα is responsible for desensitization of the β-adrenergic receptor signaling in the failing heart.

Mechanical load is another important modulator of excitability in the heart. The effect of altered hemodynamic load may be exaggerated in the failing compared with the normal ventricle. In doxorubicin-induced heart failure in the rabbit, increased load produced exaggerated shortening of the action potential duration and enhanced arrhythmia susceptibility in failing compared to control hearts (54). The effect of load is not likely to be distributed uniformly across the ventricular wall or throughout the myocardium, and thus has the potential to increase dispersion in action potential duration with arrhythmogenic consequences.

G. Electrical Remodeling and Arrhythmia Mechanisms in Cardiac Hypertrophy and Failure

Sudden death due to ventricular arrhythmias associated with cardiac hypertrophy and failure is heterogeneous and it is probable that multiple mechanisms are involved. The variability in the reported electrophysiological changes are certainly in part methodological, but also reflect a high degree of heterogeneity in the pathobiology of hypertrophy and heart failure in animal models and human disease. The stage of disease is crucial in determining the degree and character of electrical remodeling and arrhythmic risk. Data from human heart failure trials support this concept; in Veterans Administration heart failure therapy trial (VHeFT) the risk of sudden and presumed arrhythmic death was proportionately greater in patients with less severe heart failure (8). Changes in the risk of sudden death with progression of heart disease are likely to be a reflection of changes in electrophysiological substrate. The great challenge remains to use the understanding provided by in vitro studies to more completely understand arrhythmic mechanisms in the intact heart and to prevent sudden death in patients.

Toward this end, it is useful to consider the possible mechanisms of ventricular arrhythmias in terms of cellular electrophysiological parameters and the molecular changes in hypertrophy and heart failure that modulate these parameters (Table 23.4) (185). Abnormal automaticity may arise in hypertrophied and failing hearts in the setting of a reduction in resting membrane potential or acceleration of phase 4 diastolic depolarization, such that the threshold for activation of the Na current is reached rapidly. Reexpression of I_{Ca-T} in ventricular cells, changes in the voltage dependence, β-adrenergic sensitivity, or an increase in the density of I_f, and reduced I_{K1} density could conspire to enhance automaticity in ventricular myocytes in failing hearts.

Triggered activity arising from after-depolarizations could be enhanced by several electrophysiological changes described in the failing and hypertrophied heart. Cells isolated from failing animal and human hearts consistently reveal a significant prolongation of action potentials compared to those in normal hearts, independent of the mechanism of heart failure (Table 23.1). The plateau phase of the action potential is known to be quite labile: this is a time of high membrane resistance, during which small changes in current can easily tip the balance either toward repolarization or maintained depolarization. As a rule, the longer the action potential, the more labile is the repolarization process (186). Action potential lability may be manifest as variability in duration and/or secondary depolarizations, such as EADs that interrupt repolarization, which then initiate triggered arrhythmias including torsade de pointes ventricular tachycardia. Indeed, enhanced susceptibility to af-

Table 23.4 Arrhythmia Mechanisms in Cardiac Hypertrophy and Heart Failure

	Arrhythmogenic mechanism	Molecular changes in hypertrophy/HF
abnormal automaticity		
↓ RMP-V$_{threshold}$	Phase 4 diastolic depolarization (enhanced) Maximum diastolic potential (reduced)	↑I_{Ca-T}, ↓I_{K1}, ↑I_f
triggered automaticity		
EAD-mediated	AP duration (↑ AP duration & altered profile)	↓ K currents, ↑ NCX, Altered I_{Ca-L} density and kinetics,
late EAD or DAD-mediated	[Ca^{2+}]$_i$ (increased)	Slowed Ca^{2+} transient, ↑ NCX
reentry		
reactivation (short excitable gap)	APD (prolonged)	
↓ conduction and block (long excitable gap)	Anisotropic conduction (altered)	Microfibrosis in the interstitium

RMP, resting membrane potential; AP, action potential; APD, action potential duration; NCX, Na$^+$-Ca^{2+} exchanger; EAD, early after-depolarization; DAD, delayed afterdepolarization.

ter-depolarization-mediated ventricular arrhythmias has been demonstrated experimentally. Prolongation of repolarization (187–189), enhanced dispersion of repolarization and susceptibility to cesium-induced action potential prolongation have been demonstrated in the canine pacing-tachycardia heart failure model, a preparation with a high incidence of sudden death (53). Ventricular myocytes isolated from the failing canine heart exhibit more spontaneous EADs than cells from control hearts and have an exaggerated response (more frequent and complex EADs) to reduction of $[K^+]_0$ and the addition of the nonspecific K-channel blocker, cesium chloride (190). Complex after-depolarizations and triggered arrhythmias are more common in hypertrophied rat ventricular myocardium exposed to K-channel blockers (191) and dogs with left ventricular hypertrophy (LVH) exposed to the Ca channel agonist BayK 8644 (192). Alterations in Ca current density or kinetics can predispose to EAD- or DAD-mediated arrhythmias (186). Changes in the cellular environment such as hypokalemia, hypomagnesemia, and elevated levels of catecholamines may further increase the susceptibility to after-depolarization-mediated triggered arrhythmias (50).

Changes in Ca^{2+} handling in the hypertrophic and failing heart may also contribute to electrical instability. The characteristic slow decay of the Ca^{2+} transient and increased diastolic $[Ca^{2+}]_i$ can predispose to oscillatory release of Ca^{2+} from the SR and DAD-mediated triggered arrhythmias. The slow decay of the Ca^{2+} transient will influence ion flux through the NCX and may also predispose to late phase 3 EAD-mediated triggered arrhythmias.

The most common mechanism of ventricular arrhythmias is reentry due to abnormal impulse conduction. There are a number of changes characteristic of failing myocardium, in both the myocyte and interstitial compartments that increase the liklihood of reentry. In hearts that are failing as the result of myocardial infarction, a macroreentry circuit may exist in the border zone of the infarction. Normally there is a dispersion of action potential duration in the ventricle of both humans (58) and animals (193); it is possible that changes in the expression of K currents (and other currents) could enhance this dispersion of action potential duration. There is clinical evidence that spatial (60) and temporal dispersion of repolarization (61) is enhanced in the failing human heart. Such dispersion of repolarization may predispose to nonexcitable gap reentry, such as that proposed to underlie polymorphic ventricular tachycardia and ventricular fibrillation.

Alterations in anisotropic conduction may also contribute to the production of arrhythmias in hypertrophic and failing hearts. Alterations in intracellular $[Ca^{2+}]$ (144,194) and redistribution of gap junctions (195,196) will affect intercellular conduction, and microfibrosis will alter anisotropic conduction (197) leading to spatial nonuniformities of electrical loading resulting in conduction block and reentry.

IV. CONCLUSIONS

The increased risk of sudden cardiac death in patients with myocardial hypertrophy and heart failure is the result of remodeling processes that occur in both the myocyte and interstitial compartments of the heart. The key components of ventricular myocyte remodeling are the functional expression of a number of ion channels, transporters, and receptors that result in action potential prolongation, abnormal Ca^{2+} handling, and aberrant adrenergic signaling. The remodeling process creates a substrate that is highly sensitive to triggers for potentially lethal ventricular arrhythmias.

ACKNOWLEDGMENTS

This work was supported by a National Heart, Lung and Blood Institute, Specialized Center of Research in Sudden Cardiac Death Award (NIH P50 H252307). We thank Brad Nuss for Figure 23.1 and Brian O'Rourke for Figure 23.3. We gratefully acknowledge Eduardo Marbán, Brian O'Rourke, Brad Nuss, Rai Winslow, Dirk Beuckelmann, Stefan Kääb, Jochen Rose, and Michael Näbauer for helpful discussions.

REFERENCES

1. Changes in mortality from heart failure—United States, 1980–1995. MMWR 1998; 47:633–637.
2. Konstam MA, Remme WJ. Treatment guidelines in heart failure. Prog Cardiovasc Dis 1998; 48:65–72.
3. Kjekshus J. Arrhythmias and mortality in congestive heart failure. Am J Cardiol 1990; 65:42I–48I.
4. Chakko CS, Gheorghiade M. Ventricular arrhythmias in severe heart failure: incidence, significance, and effectiveness of antiarrhythmic therapy. Am Heart J 1985; 109:497–504.
5. Wilson JR, Schwartz JS, Sutton MS, Ferraro N, Horowitz LN, Reichek N, Josephson ME. Prognosis in severe heart failure: relation to hemodynamic measurements and ventricular ectopic activity. J Am Coll Cardiol 1983; 2:403–410.
6. von Olshausen K, Schafer A, Mehmel HC, Schwarz F, Senges J, Kubler W. Ventricular arrhythmias in idiopathic dilated cardiomyopathy. Br Heart J 1984; 51:195–201.
7. Califf RM, McKinnis RA, Burks J, Lee KL, Harrell FE, Jr., Behar VS, Pryor DB, Wagner GS, Rosati RA. Prognostic implications of ventricular arrhythmias during 24 hour ambulatory monitoring in patients undergoing cardiac catheterization for coronary artery disease. Am J Cardiol 1982; 50:23–31.
8. Cohn JN, Archibald DG, Ziesche S, Franciosa JA, Harston WE, Tristani FE, Dunkman WB, Jacobs W, Francis GS, Flohr KH, et al. Effect of vasodilator therapy on mortality in chronic congestive heart failure. Results of a Veterans Administration Cooperative Study. N Engl J Med 1986; 314:1547–1552.
9. Weber KT. Enhanced cardiovascular risk in hypertensive heart disease. Cardiovasc Risk Factors 1995; 5:87–92.
10. Weber KT, Brilla CG. Pathological hypertrophy and cardiac interstitium. Fibrosis and renin-angiotensin-aldosterone system. Circulation 1991; 83:1849–1865.
11. Weber KT. Extracellular matrix remodeling in heart failure: a role for de novo angiotensin II generation. Circulation 1997; 96:4065–4082.
12. van Krimpen C, Schoemaker RG, Cleutjens JP, Smits JF, Struyker-Boudier HA, Bosman FT, Daemen MJ. Angiotensin I converting enzyme inhibitors and cardiac remodeling. Basic Res Cardiol 1991; 86:149–155.
13. Smits JF, van Krimpen C, Schoemaker RG, Cleutjens JP, Daemen MJ. Angiotensin II receptor blockade after myocardial infarction in rats: effects on hemodynamics, myocardial DNA synthesis, and interstitial collagen content. J Cardiovasc Pharmacol 1992; 20:772–778.
14. Volders PG, Willems IE, Cleutjens JP, Arends JW, Havenith MG, Daemen MJ. Interstitial collagen is increased in the non-infarcted human myocardium after myocardial infarction. J Mol Cell Cardiol 1993; 25:1317–1323.
15. Beltrami CA, Finato N, Rocco M, Feruglio GA, Puricelli C, Cigola E, Quaini F, Sonnenblick EH, Olivetti G, Anversa P. Structural basis of end-stage failure in ischemic cardiomyopathy in humans. Circulation 1994; 89:151–163.
16. Assayag P, Carre F, Chevalier B, Delcayre C, Mansier P, Swynghedauw B. Compensated car-

diac hypertrophy: arrhythmogenicity and the new myocardial phenotype. I. Fibrosis. Cardiovasc Res 1997; 34:439–444.

17. Swynghedauw B, Chevalier B, Charlemagne D, Mansier P, Carre F. Cardiac hypertrophy, arrhythmogenicity and the new myocardial phenotype. II. The cellular adaptational process. Cardiovasc Res 1997; 35:6–12.

18. Swynghedauw B. Molecular mechanisms of myocardial remodelling. Physiol Rev, 1999.

19. Willems IE, Havenith MG, De Mey JG, Daemen MJ. The alpha-smooth muscle actin-positive cells in healing human myocardial scars. Am J Pathol 1994; 145:868–875.

20. Hall CE, Hall O. Hypertension and hypersalimentation. II. Deoxycorticosterone Hypertens Lab Invest 1965; 14:1727–1735.

21. Sun Y, Ramires FJ, Weber KT. Fibrosis of atria and great vessels in response to angiotensin II or aldosterone infusion. Cardiovasc Res 1997; 35:138–147.

22. Bassett AL, Gelband H. Chronic partial occlusion of the pulmonary artery in cats. Change in ventricular action potential configuration during early hypertrophy. Circ Res 1973; 32:15–26.

23. Benitah JP, Gomez AM, Bailly P, Da Ponte JP, Berson G, Delgado C, Lorente P. Heterogeneity of the early outward current in ventricular cells isolated from normal and hypertrophied rat hearts. J Physiol (Lond) 1993; 469:111–138.

24. Brooksby P, Levi AJ, Jones JV. The electrophysiological characteristics of hypertrophied ventricular myocytes from the spontaneously hypertensive rat. J Hypertens 1993; 11:611–622.

25. Bryant SM, Shipsey SJ, Hart G. Regional differences in electrical and mechanical properties of myocytes from guinea-pig hearts with mild left ventricular hypertrophy. Cardiovasc Res 1997; 35:315–323.

26. Cerbai E, Barbieri M, Li Q, Mugelli A. Ionic basis of action potential prolongation of hypertrophied cardiac myocytes isolated from hypertensive rats of different ages. Cardiovasc Res 1994; 28:1180–1187.

27. Coulombe A, Momtaz A, Richer P, Swynghedauw B, Coraboeuf E. Reduction of calcium-independent transient outward potassium current density in DOCA salt hypertrophied rat ventricular myocytes. Pfluegers Arch 1994; 427:47–55.

28. Keung EC, Aronson RS. Non-uniform electrophysiological properties and electrotonic interaction in hypertrophied rat myocardium. Circ Res 1981; 49:150–158.

29. Kleiman RB, Houser SR. Outward currents in normal and hypertrophied feline ventricular myocytes. Am J Physiol 1989; 256:H1450–H1461.

30. Li Q, Keung EC. Effects of myocardial hypertrophy on transient outward current. Am J Physiol 1994; 266:H1738–H1745.

31. Gulch RW. Alterations in excitation of mammalian myocardium as a function of chronic loading and their implications in the mechanical events. Basic Res Cardiol 1980; 75:73–80.

32. Nordin C, Siri F, Aronson RS. Electrophysiologic characteristics of single myocytes isolated from hypertrophied guinea-pig hearts. J Mol Cell Cardiol 1989; 21:729–739.

33. Ryder KO, Bryant SM, Hart G. Membrane current changes in left ventricular myocytes isolated from guinea pigs after abdominal aortic coarctation. Cardiovasc Res 1993; 27:1278–1287.

34. Tomita F, Bassett AL, Myerburg RJ, Kimura S. Diminished transient outward currents in rat hypertrophied ventricular myocytes. Circ Res 1994; 75:296–303.

35. Potreau D, Gomez JP, Fares N. Depressed transient outward current in single hypertrophied cardiomyocytes isolated from the right ventricle of ferret heart. Cardiovasc Res 1995; 30:440–448.

36. Takimoto K, Li D, Hershman KM, Li P, Jackson EK, Levitan ES. Decreased expression of Kv4.2 and novel Kv4.3 K$^+$ channel subunit mRNAs in ventricles of renovascular hypertensive rats. Circ Res 1997; 81:533–539.

37. Li GR, Ferrier GR, Howlett SE. Calcium currents in ventricular myocytes of prehypertrophic cardiomyopathic hamsters. Am J Physiol 1995; 268:H999–H1005.

38. Thuringer D, Coulombe A, Deroubaix E, Coraboeuf E, Mercadier JJ. Depressed transient out-

ward current density in ventricular myocytes from cardiomyopathic Syrian hamsters of different ages. J Mol Cell Cardiol 1996; 28:387–401.

39. Thuringer D, Deroubaix E, Coulombe A, Coraboeuf E, Mercadier JJ. Ionic basis of the action potential prolongation in ventricular myocytes from Syrian hamsters with dilated cardiomyopathy. Cardiovasc Res 1996; 31:747–757.

40. Xu XP, Best PM. Decreased transient outward K^+ current in ventricular myocytes from acromegalic rats. Am J Physiol 1991; 260:H935–942.

41. Lue WM, Boyden PA. Abnormal electrical properties of myocytes from chronically infarcted canine heart. Alterations in Vmax and the transient outward current. Circulation 1992; 85:1175–1188.

42. Qin D, Zhang ZH, Caref EB, Boutjdir M, Jain P, el-Sherif N. Cellular and ionic basis of arrhythmias in postinfarction remodeled ventricular myocardium. Circ Res 1996; 79:461–473.

43. Bril A, Forest MC, Gout B. Ischemia and reperfusion-induced arrhythmias in rabbits with chronic heart failure. Am J Physiol 1991; 261:H301–307.

44. Gidh-Jain M, Huang B, Jain P, el-Sherif N. Differential expression of voltage-gated K^+ channel genes in left ventricular remodeled myocardium after experimental myocardial infarction. Circ Res 1996; 79:669–675.

45. Kääb S, Nuss HB, Chiamvimonvat N, O'Rourke B, Pak PH, Kass DA, Marban E, Tomaselli GF. Ionic mechanism of action potential prolongation in ventricular myocytes from dogs with pacing-induced heart failure. Circ Res 1996; 78:262–273.

46. Rozanski GJ, Xu Z, Whitney RT, Murakami H, Zucker IH. Electrophysiology of rabbit ventricular myocytes following sustained rapid ventricular pacing. J Mol Cell Cardiol 1997; 29:721–732.

47. O'Rourke B, Kass DA, Tomaselli GF, Kääb S, Tunin R, Marban E. Mechanisms of altered excitation-contraction coupling in canine tachycardia-induced heart failure: 1 Experimental studies. Circ Res, 1999.

48. Coltart DJ, Meldrum SJ. Intracellular action potential in hypertrophic obstructive cardiomyopathy. Br Heart J 1972; 34:71127497.

49. Gwathmey JK, Copelas L, MacKinnon R, Schoen FJ, Feldman MD, Grossman W, Morgan JP. Abnormal intracellular calcium handling in myocardium from patients with end-stage heart failure. Circ Res 1987; 61:70–76.

50. Vermeulen JT, McGuire MA, Opthof T, Coronel R, de Bakker JM, Klopping C, Janse MJ. Triggered activity and automaticity in ventricular trabeculae of failing human and rabbit hearts. Cardiovasc Res 1994; 28:1547–1554.

51. Beuckelmann DJ, Näbauer M, Erdmann E. Intracellular calcium handling in isolated ventricular myocytes from patients with terminal heart failure. Circulation 1992; 85:1046–1055.

52. Beuckelmann DJ, Näbauer M, Erdmann E. Alterations of K^+ currents in isolated human ventricular myocytes from patients with terminal heart failure. Circ Res 1993; 73:379–385.

53. Pak PH, Nuss HB, Tunin RS, Kääb S, Tomaselli GF, Marban E, Kass DA. Repolarization abnormalities, arrhythmia and sudden death in canine tachycardia-induced cardiomyopathy. J Am Coll Cardiol 1997; 30:576–584.

54. Pye MP, Cobbe SM. Arrhythmogenesis in experimental models of heart failure: the role of increased load. Cardiovasc Res 1996; 32:248–257.

55. Litovsky SH, Antzelevitch C. Rate dependence of action potential duration and refractoriness in canine ventricular endocardium differs from that of epicardium: role of the transient outward current. J Am Coll Cardiol 1989; 14:1053–1066.

56. Fedida D, Giles WR. Regional variations in action potentials and transient outward current in myocytes isolated from rabbit left ventricle. J Physiol (Lond) 1991; 442:191–209.

57. Lukas A, Antzelevitch C. Differences in the electrophysiological response of canine ventricular epicardium and endocardium to ischemia. Role of the transient outward current. Circulation 1993; 88:2903–2915.

58. Drouin E, Charpentier F, Gauthier C, Laurent K, Le Marec H. Electrophysiologic characteris-

tics of cells spanning the left ventricular wall of human heart: evidence for presence of M cells. J Am Coll Cardiol 1995; 26:185–192.

59. Di Diego JM, Sun ZQ, Antzelevitch C. I(to) and action potential notch are smaller in left vs. right canine ventricular epicardium. Am J Physiol 1996; 271:H548–H561.

60. Barr CS, Naas A, Freeman M, Lang CC, Struthers AD. QT dispersion and sudden unexpected death in chronic heart failure. Lancet 1994; 343:327–329.

61. Berger RD, Kasper EK, Baughman KL, Marban E, Calkins H, Tomaselli GF. Beat-to-beat QT interval variability: novel evidence for repolarization lability in ischemic and nonischemic dilated cardiomyopathy. Circulation 1997; 96:1557–1565.

62. Furukawa T, Kimura S, Furukawa N, Bassett AL, Myerburg RJ. Potassium rectifier currents differ in myocytes of endocardial and epicardial origin. Circ Res 1992; 70:91–103.

63. Liu DW, Antzelevitch C. Characteristics of the delayed rectifier current (IKr and IKs) in canine ventricular epicardial, midmyocardial, and endocardial myocytes. A weaker IKs contributes to the longer action potential of the M cell. Circ Res 1995; 76:351–365.

64. Ten Eick RE, Zhang K, Harvey RD, Bassett AL. Enhanced functional expression of transient outward current in hypertrophied feline myocytes. Cardiovasc Drugs Ther 1993; 3:611–619.

65. Wettwer E, Amos GJ, Posival H, Ravens U. Transient outward current in human ventricular myocytes of subepicardial and subendocardial origin. Circ Res 1994; 75:473–482.

66. Näbauer M, Beuckelmann DJ, Uberfuhr P, Steinbeck G. Regional differences in current density and rate-dependent properties of the transient outward current in subepicardial and subendocardial myocytes of human left ventricle. Circulation 1996; 93:168–177.

67. Xu H, Dixon JE, Barry DM, Trimmer JS, Merlie JP, McKinnon D, Nerbonne JM. Developmental analysis reveals mismatches in the expression of K$^+$ channel alpha subunits and voltage-gated K$^+$ channel currents in rat ventricular myocytes. J Gen Physiol 1996; 108:405–419.

68. Wickenden AD, Kaprielian R, Parker TG, Jones OT, Backx PH. Effects of development and thyroid hormone on K$^+$ currents and K$^+$ channel gene expression in rat ventricle. J Physiol (Lond) 1997; 504:271–286.

69. Näbauer M, Barth A, Kääb S. A second calcium-independent transient outward current present in human left ventricular myocardium. Circulation 1998; 98:I–231.

70. Blair TA, Roberds SL, Tamkun MM, Hartshorne RP. Functional characterization of RK5, a voltage-gated K$^+$ channel cloned from the rat cardiovascular system. FEBS Lett 1991; 295:211–213.

71. Po S, Snyders DJ, Baker R, Tamkun MM, Bennett PB. Functional expression of an inactivating potassium channel cloned from human heart. Circ Res 1992; 71:732–736.

72. Dixon JE, Shi W, Wang HS, McDonald C, Yu H, Wymore RS, Cohen IS, McKinnon D. Role of the Kv4.3 K$^+$ channel in ventricular muscle. A molecular correlate for the transient outward current. Circ Res 1996; 79:659–668.

73. Kong W, Po S, Yamagishi T, Ashen MD, Stetten G, Tomaselli GF. Isolation and characterization of the human gene encoding the transient outward potassium current: further diversity by alternative mRNA splicing. Am J Physiol 1998; 275:H1963–H1970.

74. Apkon M, Nerbonne JM. Alpha 1-adrenergic agonists selectively suppress voltage-dependent K$^+$ current in rat ventricular myocytes. Proc Natl Acad Sci USA 1988; 85:8756–8760.

75. Fedida D, Shimoni Y, Giles WR. A novel effect of norepinephrine on cardiac cells is mediated by alpha 1-adrenoceptors. Am J Physiol 1989; 256:H1500–H1504.

76. Braun AP, Fedida D, Clark RB, Giles WR. Intracellular mechanisms for alpha 1-adrenergic regulation of the transient outward current in rabbit atrial myocytes. J Physiol (Lond) 1990; 431:689–712.

77. Kääb S, Dixon J, Duc J, Ashen MD, Näbauer M, Beuckelmann DJ, Steinbeck G, Tomaselli GF. Molecular basis of transient outward potassium current downregulation in human heart failure: A decrease in Kv4.3 mRNA correlates with a reduction in current density. Circulation 1998; 98:1383–1393.

78. Shimoni Y, Fiset C, Clark RB, Dixon JE, McKinnon D, Giles WR. Thyroid hormone regulates

postnatal expression of transient K[+] channel isoforms in rat ventricle. J Physiol (Lond) 1997; 500:65–73.

79. Winslow R, Rice J, Jafri S, Marban E, O'Rourke B. Mechanisms of altered excitation-contraction coupling in canine tachycardia-induced heart failure. II. Model studies. Circ Res, 1999.

80. Koumi S, Backer CL, Arentzen CE. Characterization of inwardly rectifying K[+] channel in human cardiac myocytes. Alterations in channel behavior in myocytes isolated from patients with idiopathic dilated cardiomyopathy. Circulation 1995; 92:164–174.

81. Furukawa T, Bassett AL, Furukawa N, Kimura S, Myerburg RJ. The ionic mechanism of reperfusion-induced early afterdepolarizations in feline left ventricular hypertrophy. J Clin Invest 1993; 91:1521–1531.

82. Furukawa T, Myerburg RJ, Furukawa N, Kimura S, Bassett AL. Metabolic inhibition of ICa,L and IK differs in feline left ventricular hypertrophy. Am J Physiol 1994; 266:H1121–H1131.

83. Choy A-M, Kuperschmidt S, Lang CC, Pierson RN, Roden DM. Regional expression of HERG and KvLQT1 in heart failure. Circulation 1996; 94:164.

84. Koumi SI, Martin RL, Sato R. Alterations in ATP-sensitive potassium channel sensitivity to ATP in failing human hearts. Am J Physiol 1997; 272:H1656–H1665.

85. Hart G. Cellular electrophysiology in cardiac hypertrophy and failure. Cardiovasc Res 1994; 28:933–946.

86. Aggarwal R, Boyden PA. Altered pharmacologic responsiveness of reduced L-type calcium currents in myocytes surviving in the infarcted heart. J Cardiovasc Electrophysiol 1996; 7:20–35.

87. Keung EC. Calcium current is increased in isolated adult myocytes from hypertrophied rat myocardium. Circ Res 1989; 64:753–763.

88. Scamps F, Mayoux E, Charlemagne D, Vassort G. Calcium current in single cells isolated from normal and hypertrophied rat heart. Effects of beta-adrenergic stimulation. Circ Res 1990; 67:199–208.

89. Santos PE, Barcellos LC, Mill JG, Masuda MO. Ventricular action potential and L-type calcium channel in infarct-induced hypertrophy in rats. J Cardiovasc Electrophysiol 1995; 6:1004–1014.

90. Gomez AM, Benitah JP, Henzel D, Vinet A, Lorente P, Delgado C. Modulation of electrical heterogeneity by compensated hypertrophy in rat left ventricle. Am J Physiol 1997; 272.

91. Xiao YF, McArdle JJ. Elevated density and altered pharmacologic properties of myocardial calcium current of the spontaneously hypertensive rat. J Hypertens 1994; 12:783–790.

92. Kleiman RB, Houser SR. Calcium currents in normal and hypertrophied isolated feline ventricular myocytes. Am J Physiol 1988; 255:H1434–H1442.

93. Creazzo TL. Reduced L-type calcium current in the embryonic chick heart with persistent truncus arteriosus. Circ Res 1990; 66:1491–1498.

94. Mukherjee R, Hewett KW, Walker JD, Basler CG, Spinale FG. Changes in L-type calcium channel abundance and function during the transition to pacing-induced congestive heart failure. Cardiovasc Res 1998; 37:432–444.

95. Momtaz A, Coulombe A, Richer P, Mercadier JJ, Coraboeuf E. Action potential and plateau ionic currents in moderately and severely DOCA-salt hypertrophied rat hearts. J Mol Cell Cardiol 1996; 28:2511–2522.

96. Bouron A, Potreau D, Raymond G. The L type calcium current in single hypertrophied cardiomyocytes isolated from the right ventricle of ferret heart. Cardiovasc Res 1992; 26:662–670.

97. Nuss HB, Houser SR. T-type Ca[2+] current is expressed in hypertrophied adult feline left ventricular myocytes. Circ Res 1993; 73:777–782.

98. Gidh-Jain M, Huang B, Jain P, Battula V, el-Sherif N. Reemergence of the fetal pattern of L-type calcium channel gene expression in non infarcted myocardium during left ventricular remodeling. Biochem Biophys Res Comm 1995; 216:892–897.

99. Ming Z, Nordin C, Siri F, Aronson RS. Reduced calcium current density in single myocytes isolated from hypertrophied failing guinea pig hearts. J Mol Cell Cardiol 1994; 26:1133–1143.

100. Mayoux E, Callens F, Swynghedauw B, Charlemagne D. Adaptational process of the cardiac Ca2+ channels to pressure overload: biochemical and physiological properties of the dihydropyridine receptors in normal and hypertrophied rat hearts. J Cardiovasc Pharmacol 1988; 12:390–396.

101. Finkel MS, Marks ES, Patterson RE, Speir EH, Steadman KA, Keiser HR. Correlation of changes in cardiac calcium channels with hemodynamics in Syrian hamster cardiomyopathy and heart failure. Life Sci 1987; 41:153–159.

102. Wagner JA, Weisman HF, Snowman AM, Reynolds IJ, Weisfeldt ML, Snyder SH. Alterations in calcium antagonist receptors and sodium-calcium exchange in cardiomyopathic hamster tissues. Circ Res 1989; 65:205–214.

103. Gopalakrishnan M, Triggle DJ, Rutledge A, Kwon YW, Bauer JA, Fung HL. Regulation of K^+ and Ca^{2+} channels in experimental cardiac failure. Am J Physiol 1991; 261:H1979–H1987.

104. Dixon IM, Lee SL, Dhalla NS. Nitrendipine binding in congestive heart failure due to myocardial infarction. Circ Res 1990; 66:782–788.

105. Colston JT, Kumar P, Chambers JP, Freeman GL. Altered sarcolemmal calcium channel density and Ca(2+)-pump ATPase activity in tachycardia heart failure. Cell Calcium 1994; 16:349–356.

106. Vatner DE, Sato N, Kiuchi K, Shannon RP, Vatner SF. Decrease in myocardial ryanodine receptors and altered excitation-contraction coupling early in the development of heart failure. Circulation 1994; 90:1423–1430.

107. Gengo PJ, Sabbah HN, Steffen RP, Sharpe JK, Kono T, Stein PD, Goldstein S. Myocardial beta adrenoceptor and voltage sensitive calcium channel changes in a canine model of chronic heart failure. J Mol Cell Cardiol 1992; 24:1361–1369.

108. Rasmussen RP, Minobe W, Bristow MR. Calcium antagonist binding sites in failing and nonfailing human ventricular myocardium. Biochem Pharmacol 1990; 39:691–696.

109. Beuckelmann DJ, Erdmann E. Ca(2+)-currents and intracellular $[Ca^{2+}]_i$-transients in single ventricular myocytes isolated from terminally failing human myocardium. Basic Res Cardiol 1992; 87:235–243.

110. Mewes T, Ravens U. L-type calcium currents of human myocytes from ventricle of non-failing and failing hearts and from atrium. J Mol Cell Cardiol 1994; 26:1307–1320.

111. Ouadid H, Albat B, Nargeot J. Calcium currents in diseased human cardiac cells. J Cardiovasc Pharmacol 1995; 25:282–291.

112. Takahashi T, Allen PD, Lacro RV, Marks AR, Dennis AR, Schoen FJ, Grossman W, Marsh JD, Izumo S. Expression of dihydropyridine receptor (Ca^{2+} channel) and calsequestrin genes in the myocardium of patients with end-stage heart failure. J Clin Invest 1992; 90:927–935.

113. Gruver EJ, Morgan JP, Stambler BS, Gwathmey JK. Uniformity of calcium channel number and isometric contraction in human right and left ventricular myocardium. Bas Res Cardiol 1994; 89:139–148.

114. Piot C, Lemaire S, Albat B, Seguin J, Nargeot J, Richard S. High frequency-induced upregulation of human cardiac calcium currents. Circulation 1996; 93:120–128.

115. Cavalie A, Pelzer D, Trautwein W. Fast and slow gating behaviour of single calcium channels in cardiac cells. Relation to activation and inactivation of calcium-channel current. Pflugers Arch 1986; 406:241–258.

116. Yue DT, Backx PH, Imredy JP. Calcium-sensitive inactivation in the gating of single calcium channels. Science 1990; 250:1735–1738.

117. Schroeder F, Handrock R, Beuckelmann DJ, Hirt S, Hullin R, Priebe L, Schwinger RHG, Weil J, Herzig S. Increased availability and open probability of single L-type calcium channels from failing compared with nonfailing human ventricle. Circulation 1998; 98:969–976.

118. Hullin RA, Asmus F, Berger HJ, Boekstegers P. Differential expression of the subunits of the

cardiac L-type calcium channel in diastolic failure of the transplanted heart. Circulation 1997; 96:I–55.

119. Vassort G, Alvarez J. Cardiac T-type calcium current: pharmacology and roles in cardiac tissues. J Cardiovasc Electrophysiol 1994; 5:376–393.

120. Fares N, Gomez JP, Potreau D. T-type calcium current is expressed in dedifferentiated adult rat ventricular cells in primary culture. C R Acad Sci III 1996; 319:569–576.

121. Chen CC, Best PM. Effects of IGF-1 on T-type calcium currents in cultured atrial myocytes. FASEB J 1996; 10:A310.

122. Xu XP, Best PM. Increase in T-type calcium current in atrial myocytes from adult rats with growth hormone-secreting tumors. Proc Natl Acad Sci USA 1990; 87:4655–4659.

123. Beuckelmann DJ, Näbauer M, Erdmann E. Characteristics of calcium-current in isolated human ventricular myocytes from patients with terminal heart failure. J Mol Cell Cardiol 1991; 23:929–937.

124. Beuckelmann DJ, Näbauer M, Kruger C, Erdmann E. Altered diastolic $[Ca^{2+}]i$ handling in human ventricular myocytes from patients with terminal heart failure. Am Heart J 1995; 129:684–689.

125. James P, Inui M, Tada M, Chiesi M, Carafoli E. Nature and site of phospholamban regulation of the Ca^{2+} pump of sarcoplasmic reticulum. Nature 1989; 342:90–92.

126. Brillantes AM, Allen P, Takahashi T, Izumo S, Marks AR. Differences in cardiac calcium release channel (ryanodine receptor) expression in myocardium from patients with end-stage heart failure caused by ischemic versus dilated cardiomyopathy. Circ Res 1992; 71:18–26.

127. Go L, Moschella MC, Watras J, Handa KK, Fyfe BS, Marks AR. Differential regulation of two types of intracellular calcium release channels during end-stage heart failure. J Clin Invest 1995; 95:888–894.

128. Meyer M, Schillinger W, Pieske B, Holubarsch C, Heilmann C, Posival H, Kuwajima G, Mikoshiba K, Just H, Hasenfuss G. Alterations of sarcoplasmic reticulum proteins in failing human dilated cardiomyopathy. Circulation 1995; 92:778–784.

129. Arai M, Alpert NR, MacLennan DH, Barton P, Periasamy M. Alterations in sarcoplasmic reticulum gene expression in human heart failure. A possible mechanism for alterations in systolic and diastolic properties of the failing myocardium. Circ Res 1993; 72:463–469.

130. Mercadier JJ, Lompre AM, Duc P, Boheler KR, Fraysse JB, Wisnewsky C, Allen PD, Komajda M, Schwartz K. Altered sarcoplasmic reticulum Ca2(+)-ATPase gene expression in the human ventricle during end-stage heart failure. J Clin Invest 1990; 85:305–309.

131. Studer R, Reinecke H, Bilger J, Eschenhagen T, Bohm M, Hasenfuss G, Just H, Holtz J, Drexler H. Gene expression of the cardiac Na^+-Ca^{2+} exchanger in end-stage human heart failure. Circ Res 1994; 75:443–453.

132. Schwinger RH, Bohm M, Schmidt U, Karczewski P, Bavendiek U, Flesch M, Krause EG, Erdmann E. Unchanged protein levels of SERCA II and phospholamban but reduced Ca^{2+} uptake and Ca(2+)-ATPase activity of cardiac sarcoplasmic reticulum from dilated cardiomyopathy patients compared with patients with nonfailing hearts. Circulation 1995; 92:3220–3228.

133. Flesch M, Schwinger RH, Schnabel P, Schiffer F, van Gelder I, Bavendiek U, Sudkamp M, Kuhn-Regnier F, Bohm M. Sarcoplasmic reticulum Ca^{2+} ATPase and phospholamban mRNA and protein levels in end-stage heart failure due to ischemic or dilated cardiomyopathy. J Mol Med 1996; 74:321–332.

134. Linck B, Boknik P, Eschenhagen T, Muller FU, Neumann J, Nose M, Jones LR, Schmitz W, Scholz H. Messenger RNA expression and immunological quantification of phospholamban and SR-Ca(2+)-ATPase in failing and nonfailing human hearts. Cardiovasc Res 1996; 31:625–632.

135. Kuo TH, Tsang W, Wang KK, Carlock L. Simultaneous reduction of the sarcolemmal and SR calcium ATPase activities and gene expression in cardiomyopathic hamster. Biochim Biophys Acta 1992; 1138:343–349.

136. Feldman AM, Weinberg EO, Ray PE, Lorell BH. Selective changes in cardiac gene expression during compensated hypertrophy and the transition to cardiac decompensation in rats with chronic aortic banding. Circ Res 1993; 73:184–192.

137. Zarain-Herzberg A, Afzal N, Elimban V, Dhalla NS. Decreased expression of cardiac sarcoplasmic reticulum Ca(2+)-pump ATPase in congestive heart failure due to myocardial infarction. Mol Cell Biochem 1996; 164:285–290.

138. Hasenfuss G, Reinecke H, Studer R, Meyer M, Pieske B, Holtz J, Holubarsch C, Posival H, Just H, Drexler H. Relation between myocardial function and expression of sarcoplasmic reticulum Ca(2+)-ATPase in failing and nonfailing human myocardium. Circ Res 1994; 75:434–442.

139. Kiss E, Ball NA, Kranias EG, Walsh RA. Differential changes in cardiac phospholamban and sarcoplasmic reticular Ca(2+)-ATPase protein levels. Effects on Ca2+ transport and mechanics in compensated pressure-overload hypertrophy and congestive heart failure. Circ Res 1995; 77:759–764.

140. Hasenfuss G. Alterations of calcium-regulatory proteins in heart failure. Cardiovasc Res 1998; 37:279–289.

141. Movsesian MA, Schwinger RH. Calcium sequestration by the sarcoplasmic reticulum in heart failure. Cardiovasc Res 1998; 37:352–359.

142. Feldman AM, Ray PE, Silan CM, Mercer JA, Minobe W, Bristow MR. Selective gene expression in failing human heart. Quantification of steady-state levels of messenger RNA in endomyocardial biopsies using the polymerase chain reaction. Circulation 1991; 83:1866–1872.

143. Movsesian MA, Karimi M, Green K, Jones LR. Ca(2+)-transporting ATPase, phospholamban, and calsequestrin levels in nonfailing and failing human myocardium. Circulation 1994; 90:653–657.

144. Noma A, Tsuboi N. Dependence of junctional conductance on proton, calcium and magnesium ions in cardiac paired cells of guinea-pig. J Physiol (Lond) 1987; 382:193–211.

145. Flesch M, Schwinger RH, Schiffer F, Frank K, Sudkamp M, Kuhn-Regnier F, Arnold G, Bohm M. Evidence for functional relevance of an enhanced expression of the Na(+)-Ca2+ exchanger in failing human myocardium. Circulation 1996; 94:992–1002.

146. Reinecke H, Studer R, Vetter R, Holtz J, Drexler H. Cardiac Na$^+$/Ca^{2+} exchange activity in patients with end-stage heart failure. Cardiovasc Res 1996; 31:48–54.

147. Rose J, O'Rourke B, Kass DA, Tomaselli GF. Na$^+$-Ca^{2+} exchange (NCX) current density is unchanged in heart failure despite an increase in NCX protein. Circulation 1998; 98:I–679.

148. Gwathmey JK, Slawsky MT, Hajjar RJ, Briggs GM, Morgan JP. Role of intracellular calcium handling in force-interval relationships of human ventricular myocardium. J Clin Invest 1990; 85:1599–1613.

149. Hasenfuss G, Reinecke H, Studer R, Pieske B, Meyer M, Drexler H, Just H. Calcium cycling proteins and force-frequency relationship in heart failure. Basic Res Cardiol 1996; 91:17–22.

150. Pieske B, Sutterlin M, Schmidt-Schweda S, Minami K, Meyer M, Olschewski M, Holubarsch C, Just H, Hasenfuss G. Diminished post-rest potentiation of contractile force in human dilated cardiomyopathy. Functional evidence for alterations in intracellular Ca^{2+} handling. J Clin Invest 1996; 98:764–776.

151. Mattiello JA, Margulies KB, Jeevanandam V, Houser SR. Contribution of reverse-mode sodium-calcium exchange to contractions in failing human left ventricular myocytes. Cardiovasc Res 1998; 37:424–431.

152. Yano K, Zarain-Herzberg A. Sarcoplasmic reticulum calsequestrins: structural and functional properties. Mol Cell Biochem 1994; 135:61–70.

153. Schillinger W, Meyer M, Kuwajima G, Mikoshiba K, Just H, Hasenfuss G. Unaltered ryanodine receptor protein levels in ischemic cardiomyopathy. Mol Cell Biochem 1996; 160–161:297–302.

154. Yanagihara K, Irisawa H. Inward current activated during hyperpolarization in the rabbit sinoatrial node cell. Pflugers Arch 1980; 385:11–19.

155. Brown HF, DiFrancesco D, Noble SJ. How does adrenaline accelerate the heart? Nature 1979; 280:235–236.

156. Brown H, Difrancesco D. Voltage-clamp investigations of membrane currents underlying pace-maker activity in rabbit sino-atrial node. J Physiol (Lond) 1980; 308:331–351.

157. Yu H, Chang F, Cohen IS. Pacemaker current exists in ventricular myocytes. Circ Res 1993; 72:232–236.

158. Ranjan R, Chiamvimonvat N, Thakor NV, Tomaselli GF, Marban E. Mechanism of anode break stimulation in the heart. Biophys J 1998; 74:1850–1863.

159. Cerbai E, Pino R, Porciatti F, Sani G, Toscano M, Maccherini M, Giunti G, Mugelli A. Characterization of the hyperpolarization-activated current, I(f), in ventricular myocytes from human failing heart. Circulation 1997; 95:568–571.

160. Hoppe UC, Jansen E, Sudkamp M, Beuckelmann DJ. Hyperpolarization-activated inward current in ventricular myocytes from normal and failing human hearts. Circulation 1998; 97:55–65.

161. Ludwig A, Zong X, Jeglitsch M, Hofmann F, Biel M. A family of hyperpolarization-activated mammalian cation channels. Nature 1998; 393:587–591.

162. Cerbai E, Barbieri M, Mugelli A. Occurrence and properties of the hyperpolarization-activated current If in ventricular myocytes from normotensive and hypertensive rats during aging. Circulation 1996; 94:1674–1681.

163. Dhalla NS, Dixon IM, Rupp H, Barwinsky J. Experimental congestive heart failure due to myocardial infarction: sarcolemmal receptors and cation transporters. Basic Res Cardiol 1991; 86:13–23.

164. Houser SR, Freeman AR, Jaeger JM, Breisch EA, Coulson RL, Carey R, Spann JF. Resting potential changes associated with Na-K pump in failing heart muscle. Am J Physiol 1981; 240:H168–H176.

165. Kjeldsen K, Bjerregaard P, Richter EA, Thomsen PE, Norgaard A. Na$^+$,K$^+$-ATPase concentration in rodent and human heart and skeletal muscle: apparent relation to muscle performance. Cardiovasc Res 1988; 22:95–100.

166. Zahler R, Gilmore-Hebert M, Sun W, Benz EJ. Na, K-ATPase isoform gene expression in normal and hypertrophied dog heart. Basic Res Cardiol 1996; 91:256–266.

167. Spinale FG, Clayton C, Tanaka R, Fulbright BM, Mukherjee R, Schulte BA, Crawford FA, Zile MR. Myocardial Na$^+$,K$^+$-ATPase in tachycardia induced cardiomyopathy. J Mol Cell Cardiol 1992; 24:277–294.

168. Bristow MR, Ginsburg R, Minobe W, Cubicciotti RS, Sageman WS, Lurie K, Billingham ME, Harrison DC, Stinson EB. Decreased catecholamine sensitivity and beta-adrenergic-receptor density in failing human hearts. N Engl J Med 1982; 307:205–211.

169. Bristow MR. Changes in myocardial and vascular receptors in heart failure. J Am Coll Cardiol 1993; 22:61A–71A.

170. Bohm M, Flesch M, Schnabel P. Beta-adrenergic signal transduction in the failing and hypertrophied myocardium. J Mol Med 1997; 75:842–848.

171. Zhou YY, Cheng H, Bogdanov KY, Hohl C, Altschuld R, Lakatta EG, Xiao RP. Localized cAMP-dependent signaling mediates beta 2-adrenergic modulation of cardiac excitation-contraction coupling. Am J Physiol 1997; 273:H1611–1618.

172. Hartzell HC, Duchatelle-Gourdon I. Regulation of the cardiac delayed rectifier K current by neurotransmitters and magnesium. Cardiovasc Drugs Ther 1993; 7(suppl 3):547–554.

173. Fedida D, Braun AP, Giles WR. Alpha 1-adrenoceptors in myocardium: functional aspects and transmembrane signaling mechanisms. Physiol Rev 1993; 73:469–487.

174. Lohse MJ, Engelhardt S, Danner S, Bohm M. Mechanisms of beta-adrenergic receptor desensitization: from molecular biology to heart failure. Basic Res Cardiol 1996; 91:29–34.

175. Brodde OE. Beta-adrenergic receptors in failing human myocardium. Basic Res Cardiol 1996; 91:35–40.

176. Bristow MR, Minobe WA, Raynolds MV, Port JD, Rasmussen R, Ray PE, Feldman AM. Re-

duced beta-1 receptor messenger RNA abundance in the failing human heart. J Clin Invest 1993; 92:2737–2745.

177. Ungerer M, Bohm M, Elce JS, Erdmann E, Lohse MJ. Altered expression of beta-adrenergic receptor kinase and beta1-adrenergic receptors in the failing human heart. Circulation 1993; 87:454–463.

178. Feldman MD, Copelas L, Gwathmey JK, Phillips P, Warren SE, Schoen FJ, Grossman W, Morgan JP. Deficient production of cyclic AMP: pharmacologic evidence of an important cause of contractile dysfunction in patients with end-stage heart failure. Circulation 1987; 75:331–339.

179. Feldman AM, Cates AE, Veazey WB, Hershberger RE, Bristow MR, Baughman KL, Baumgartner WA, Van Dop C. Increase of the 40,000-mol wt pertussis toxin substrate (G protein) in the failing human heart. J Clin Invest 1988; 82:189–197.

180. Bohm M. Alterations of beta-adrenoceptor-G-protein-regulated adenylyl cyclase in heart failure. Mol Cell Biochem 1995; 147:147–160.

181. Bristow MR, Anderson FL, Port JD, Skerl L, Hershberger RE, Larrabee P, O'Connell JB, Renlund DG, Volkman K, Murray J, et al. Differences in beta-adrenergic neuroeffector mechanisms in ischemic versus idiopathic dilated cardiomyopathy. Circulation 1991; 84:1024–1039.

182. Schnabel P, Bohm M, Gierschik P, Jakobs KH, Erdmann E. Improvement of cholera toxin-catalyzed ADP-ribosylation by endogenous ADP-ribosylation factor from bovine brain provides evidence for an unchanged amount of Gsalpha in failing human myocardium. J Mol Cell Cardiol 1990; 22:73–82.

183. Eschenhagen T, Mende U, Nose M, Schmitz W, Haverich A, Hirt S, Doring V, Kalmar P, Hoppner W, Seitz H-J. Increased messenger RNA level of the inhibitory G-protein alpha subunit Gialpha-2 in human end-stage heart failure. Circ Res 1992; 70:688–696.

184. Brown LA, Harding SE. The effect of pertussis toxin on beta-adrenoceptor responses in isolated cardiac myocytes from noradrenaline-treated guinea-pigs and patients with cardiac failure. Br J Pharmacol 1992; 106:115–122.

185. Marban E. Molecular approaches to arrhythmogenesis. In: Chien K, ed. Molecular Basis of Heart Disease. New York: W.B. Saunders, 1998.

186. Aronson RS, Ming Z. Cellular mechanisms of arrhythmias in hypertrophied and failing myocardium. Circulation 1993; 87:76–83.

187. Li HG, Jones DL, Yee R, Klein GJ. Arrhythmogenic effects of catecholamines are decreased in heart failure induced by rapid pacing in dogs. Am J Physiol 1993; 265:H1654–1662.

188. Li HG, Jones DL, Yee R, Klein GJ. Electrophysiologic substrate associated with pacing-induced heart failure in dogs: potential value of programmed stimulation in predicting sudden death. J Am Coll Cardiol 1992; 19:444–449.

189. Wang Z, Taylor LK, Denney WD, Hansen DE. Initiation of ventricular extrasystoles by myocardial stretch in chronically dilated and failing canine left ventricle. Circulation 1994; 90:2022–2031.

190. Nuss HB, Kääb S, Kass DA, Tomaselli GF, Marban E. Increased susceptibility to arrhythmogenic early after depolarization and oscillatory prepotentials in failing canine ventricular myocytes. Circulation 1995; 92:434.

191. Aronson RS. Afterpotentials and triggered activity in hypertrophied myocardium from rats with renal hypertension. Circ Res 1981; 48:720–727.

192. Ben-David J, Zipes DP, Ayers GM, Pride HP. Canine left ventricular hypertrophy predisposes to ventricular tachycardia induction by phase 2 early afterdepolarizations after administration of BAY K 8644. J Am Coll Cardiol 1992; 20:1576–1584.

193. Sicouri S, Antzelevitch C. A subpopulation of cells with unique electrophysiological properties in the deep subepicardium of the canine ventricle. The M cell. Circ Res 1991; 68:1729–1741.

194. Maurer P, Weingart R. Cell pairs isolated from adult guinea pig and rat hearts: effects of $[Ca^{2+}]_i$ on nexal membrane resistance. Pflugers Arch 1987; 409:394–402.

195. Peters NS. New insights into myocardial arrhythmogenesis: distribution of gap-junctional coupling in normal, ischaemic and hypertrophied human hearts. Clin Sci (Colch) 1996; 90:447–452.

196. Severs NJ. Gap junction alterations in the failing heart. Eur Heart J 1994; 15(suppl D):53–57.

197. Spach MS, Boineau JP. Microfibrosis produces electrical load variations due to loss of side-to-side cell connections: a major mechanism of structural heart disease arrhythmias. Pacing Clin Electrophysiol 1997; 20:397–413.

198. The Task Force of the Working Group on Arrhythmias of the European Society of Cardiology. The Sicilian Gambit. A new approach to the classification of antiarrhythmic drugs based on their actions on arrhythmogenic mechanisms. Circulation 1991; 84:1831–1851.

24

Transgenic Electrophysiology and Small Animal Models

AMIT RAKHIT and CHARLES I. BERUL

Children's Hospital and Harvard Medical School, Boston, Massachusetts

I. INTRODUCTION

Based on recent advances in genetics and molecular biology, genetically manipulated animals have become increasingly relevant for understanding human cardiovascular disease (1–3), offering the potential to supplement and extend the information obtained over the years using traditional animal models. In addition to standard transgenic models that either include or exclude specific genes of interest, these advances now allow the transfer of whole-cell nuclei into unfertilized cells without a nucleus to produce live clonal offspring (4). Such techniques ultimately may be extended to develop large animal models with important cardiovascular and electrophysiological applications.

For technical and economic reasons, the mouse is currently the model of choice for many genetic manipulations. The development of reproducible and stable genetic mutations with identifiable phenotypes allows researchers to study phenotypic changes whose primary genetic cause is known. The limitations of murine cardiovascular physiology must be recognized, however. The length of the cardiac cycle is 1/10 that of the human, with heart rates of 500 to 700 beats per minute in the conscious, freely moving mouse (3). Whereas some murine models may display phenotypes similar to those of human disease states, they do not always manifest the full spectrum of any one disease found in humans (3,5). For example, differences in contractile proteins exist between human and mouse cardiac tissues which probably play an important role in the differing responses of the two species to similar pathological conditions.

In this chapter, we place particular emphasis on new small animal models, novel techniques, and the major advances being offered by utilizing direct transgenic and genetically manipulated models. We also review some of the approaches and techniques involved in studying specific arrhythmias with these small animal models.

II. SMALL ANIMAL PREPARATIONS AND ELECTROBIOLOGY

Molecular biological advances have provided exciting new tools and approaches in identifying the genetic basis of electrophysiological disorders. The mouse has become the principal mammalian species for transgenic studies and several mouse models for cardiac electrophysiological diseases have now been developed. Murine models are available as models for familial hypertrophic cardiomyopathy (6), long-QT syndrome (7–9), and myotonic dystrophy (10). Other pertinent models are presently being engineered as more experience with the techniques is obtained. These include mice with defects in genes regulating cardiac connexin proteins, ion channels, G-proteins, and other proteins critical in intra- and intercellular signaling processes. These murine models display particular electrophysiological abnormalities that can be characterized using both ex vivo and in vivo techniques.

The Langendorff perfused whole-heart technique was first described in 1895 as a means to study ex vivo cardiovascular hemodynamics. Modern electrophysiological techniques using this method involve removing the heart rapidly and connecting it to an apparatus that maintains perfusion via the coronary arteries. A physiological salt solution saturated with oxygen and minimal carbon dioxide is perfused with constant perfusion pressure through a cannula in the aortic root and thus to the coronary arteries. A stable temperature and coronary artery flow are maintained. The heart can then be immersed in the salt solution and electrodes can be positioned to monitor a summation single-lead ECG. Endocardial and epicardial electrodes can also be placed in multiple sites to monitor action potentials and activation mapping through vector analysis (11). Such isolated heart techniques can also be used for detailed optical mapping of action potentials using voltage-sensitive fluorescent dyes (12,13). Light from a high-intensity light source, such as tungsten-halogen, is passed through heat and interference filters and used to illuminate the epicardial surface. Epicardial and endocardial mapping can then be performed. A lens is used to collect emitted light and then transferred to an emission filter, coupled to a video camera that records the images. Computers process the high-density optical images to digital video data. Background fluorescence can be subtracted from each frame and the true signal images intensified. By this means, beat-to-beat changes in wave propagation patterns as well as patterns of excitation can be studied.

Using such preparations, information on a variety of experimental influences can be obtained under well-controlled conditions. For example, recent studies evaluated the effects of calcium and potassium ion channel blockade on atrial conduction (14). Isolated rabbit hearts were perfused using the Langendorff technique and recordings obtained from 12 epicardial atrial electrodes. These data were used to determine activation time, effective refractory period (ERP), action potential duration, and conduction time both before and after 2 h of rapid atrial pacing. Action potential duration was found to be shortened after rapid atrial pacing. The study demonstrated that even short periods of rapid atrial rhythms can alter atrial conduction properties. This "remodeling" of the atrial conduction system appeared related to calcium in that verapamil, a calcium channel blocker, prevented the pacing-induced decrease in action potential duration.

High-resolution optical mapping techniques have also been used to map repolarization and depolarization patterns in the heart (13). In one study, the effects of atrial defibrillation shocks were studied and found to demonstrate excitation of the ventricles through local activation near the base. Appropriate timing of the shock in the vulnerable period of relative refractoriness could also induce ventricular fibrillation. Although extraneous variables, such as autonomic influences and circulating regulators, are mostly eliminated, in this preparation one limitation of ex vivo studies is that extended exposure to artificial environments does not replicate the complex hemodynamic properties, tissue electrolyte balance, metabolic variables, and cardiac electrophysiological properties seen in vivo.

Before detailing some of the most interesting studies being done on small mammals, it is worth noting here another recent development not usually considered in the context of cardiac electrophysiology. That is the zebra fish as a model for studying the embryonic development of the vertebrate heart and its electrophysiology (15). Recent studies have shown that the gene *Nkx2.5*, along with other transcription factor–related gene products, is important in regulating the development of normal cardiac progenitors including myocardium and conduction tissues (16). Nevertheless, no genetically manipulated cardiovascular models of electrophysiological disease are known in fish. However, it seems clear this preparation will be quite valuable in the study of various genetic elements in development and may provide important clues to understanding congenital structural and conduction abnormalities that do contribute substantially to pediatric arrhythmias.

Another preparation that has been very useful in these efforts has involved study of cardiac conduction development in avian embryos. For example, studies have shown a close link between pacemaker development and polarity and the anteroposterior axis of the heart (17). It is also known from studies in other organisms that an essential element in the development of the conduction system is the cardiac endothelium, a cellular layer unique to vertebrate hearts. Conduction tissue is thought to develop from myocardial cells that lie adjacent to endothelium (18), although there has been some controversy over its precise origin. Other studies have shown that Purkinje cells in embryonic chicks develop from specialized myocytes that lie close to developing coronary arteries (19). Gap junction proteins also appear to be crucial to conduction system development, and this question can also be studied in these models. Purkinje fibers of the atrioventricular bundle and the bundle branches of bovine embryonic hearts express abundant connexin43, especially on the plasma membrane facing other Purkinje cells, but not that facing surrounding connective tissue (20). Future studies may benefit from the use of these models to study development of conduction system abnormalities, and may have important bearing in tissue-engineering efforts directed at cardiac pacemaking or conductive cells. Many aspects of cardiac development from regulatory proteins, myosin, actin, and cellular structures to conduction properties and characteristics of the neuroelectrophysiological system can be explained using these approaches.

A.   Normal Mouse Electrophysiology

In vivo mouse cardiac electrophysiology studies now permit measurements of ECG and electrophysiological data similar to those performed on humans (21). In the first reported in vivo experiments, mice were anesthetized, intubated, and mechanically ventilated. Epicardial electrodes were placed on the surfaces of the right ventricle, left ventricle, and right atrium [Fig. 24.1(top)] and a 12-lead electrocardiogram and electrophysiological data recorded. Programmed stimulation can also be performed to assess conduction characteristics and arrhythmia inducibility in the in vivo mouse model. The few difficulties with this model are those associated with using an open-chest procedure and the alterations in hemodynamics and conduction associated with this intervention.

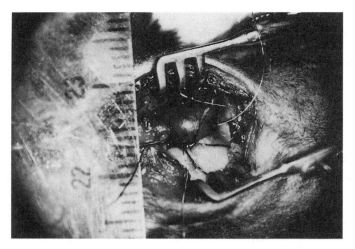

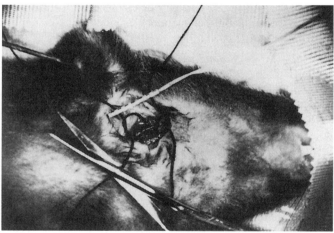

Figure 24.1 Depiction of mouse electrophysiology (EP) study illustrating both epicardial and endocardial methods. (Top) A photograph taken through an operating microscope of an epicardial approach. A midline stereotomy is performed on an anesthetized, sedated mouse. The pericardium is incised and insulated pacing wires are sutured on to the left ventricular, right ventricular, and atrial epicardial surfaces in a bipolar fashion. The wires are connected to an amplifier for stimulation and electrogram recording. (Bottom) An octapolar mouse EP catheter (NuMed, Inc.) is placed into the right heart via a right jugular vein cutdown approach. The right jugular vein is exposed between the two suture ties and is cannulated with the endocardial electrophysiology catheter. The catheter is positioned with the tip at the right ventricular apex so that a pair of electrodes can pace the RV, a second pair can record the RV intracardiac electrogram, a third pair can pace the right atrium, and the proximal pair can record the RA electrogram. A central lumen is utilized for administration of medications during the EP procedure. (Adapted from Refs. 21 and 29.)

Recent advances have further allowed for an endocardial approach to pacing, pharmacological intervention, blood sampling, and pressure transduction measurements. Venous access is obtained under microscopic guidance via a right jugular vein cutdown approach [Fig. 24.1(bottom)]. An octapolar electrode catheter specifically designed for simultaneous atrial and ventricular endocardial recording and pacing is used (3). By this approach, measurement of surface ECG and simultaneous electrophysiological parameters can be obtained by pacing and recording directly from the endocardium of an intact closed-chest mouse.

Methods for studying conscious, freely moving mice with implanted telemetry devices have also been developed that can accurately measure normal electrocardiographic intervals (3,22). Using sterile techniques, mice are anesthetized and an incision made on the back along the spine. A subcutaneous pocket is formed and a 3.5-g radiofrequency transmitter is implanted under the muscle. The cathodal lead is looped forward to the scapula and anchored in place. The anodal lead is then brought near the heart apex. Electrocardiographic information from a single lead is then monitored by placing the animals near a receiver that records transmitted signals. Using this technology, research on conscious untethered mice can include many ECG parameters, such as heart rate and QT-interval variability and other pertinent electrocardiographic data.

B. Genetic Approaches

An exciting approach to the study of human cardiovascular diseases today is the development of transgenic and other genetically manipulated animal models. An animal that gains new genetic information from the addition of foreign DNA is said to be transgenic. Any change in a nucleotide or codon sequence has the potential to cause an altered protein product. Such changes can result in loss, gain, or dysfunction of specific proteins and affect the overall working properties of the heart. If such changes are incorporated into the germ line, the mutation will be transmitted indefinitely to future progeny. In contrast, changes in coding sequences that occur solely in somatic cells are typically not inherited.

The physical site of each gene on a particular chromosome is known as its chromosomal locus. Each gene is present in two copies, one on each of the pairs of chromosomes, and are known as alleles. The sites for each allele, located on the two homologous chromosomes, can carry an identical gene from each parent (homozygous) or different information (heterozygous). Patterns of inheritance of specific diseases depend on whether different alleles are dominant or recessive and on whether the mode of transmission to succeeding generations is autosomal, X-linked, or mitochondrial.

Intentional changes in gene sequences form the basis of molecular biological techniques used to create transgenic models of human disease. With the identification of a great many of the genes involved in causing human disease, researchers are presently developing animal models to study genotype–phenotype relationships. Rapid advances in technology have led to refinement of older methods and newer, more effective transgenic solutions. Initial studies involved inserting plasmid vectors carrying genes of interest into the nucleus of an unfertilized oocyte or into the pronucleus of a fertilized cell. The egg would then be planted into a pseudopregnant animal, such as a mouse, and after birth, the recipient could be examined for expression of the newly transmitted DNA (23).

Expression of donor genes can be extremely variable using plasmid vectors because expression depends on several factors: the integration site in the host DNA, the number and site of DNA promoter regions, possible inactivation of tandem sequences, and the influence of multiple transcriptional regulatory proteins. The plasmid vector may cause rearrangement of host-cell DNA because insertion sites into the chromosome are random, potentially leading to a nonfunctional cell and cell death. These challenges using plasmid vectors have led to development of newer techniques. One procedure that has emerged is to use retroviral vectors to introduce DNA into the host cell. Such vectors cause minimal rearrangement of host DNA and allow targeting of specific somatic mutations. However, it is difficult to use this procedure to affect germ line cells, making it impossible to propagate induced phenotypes using this method (23).

One of the newest techniques in the transgenic armamentarium is the use of mouse embryonic stem cells derived from blastocysts shortly after fertilization and before implantation in the uterine wall. DNA of interest is obtained from a donor mouse or from recombinant techniques and mouse embryonic stem cells innoculated with this DNA by microinjection. Resulting hybrid cells are screened for expression of the new DNA using techniques such as polymerase chain reaction (PCR). Identification of a particular marker known to be included on the transfected DNA is done by seeing whether the cells express this new genomic material. Once identified, cells expressing the transfected sequences are injected into a new host blastocyst, which is then implanted into a foster mother. Transfected stem cells then develop normally within the hybrid animal. Each animal is now chimeric, having tissue derived from both the implanted as well as endogenous stem cells. Such models provide stable lines that can then be further genetically selected by crossbreeding (24).

Genes can also be "knocked out" using genetic engineering techniques by means of targeted gene disruption. Thus, a "wild-type" gene can be modified by interrupting a specific gene exon with marker sequences, such as an exogenous gene like "neo," which confers drug resistance to G418, providing a convenient method of selecting affected cells. An exon is that portion of the gene that contains specific DNA base pairs that directly encode a given protein. Other parts of genes, known as introns, comprise the intervening sequences between exons and may act as punctuation in reading DNA sequences and in identifying specific regions to be actively transcribed into mRNA. When foreign DNA is inserted into an embryonic stem cell, it can either undergo homologous or nonhomologous combination with host DNA. Homologous recombination requires that the entire gene of interest be incorporated into the chromosome without intervening flanking regions that may disrupt its expression. Cells can then be screened using PCR to assess whether an intact recombination event has occurred. By using gene targeting via homologous recombination in murine embryonic stem cells, a specific gene can either be ablated or modified (25). If the process results in lack of expression of the modified sequences, the gene is said to be "knocked out" from the genome. Such changes can indicate either lack of transcription or dysfunctional transcription. Once obtained, heterozygous "knock-out" animals can then be bred to obtain those homozygous for the absence of a particular gene. Homozygous recombinants will then display the phenotype associated with complete absence of this protein product. Genes can be removed or modified in a particular domain, or even a single amino acid residue, to produce animals in which the mutation can be propagated indefinitely through the germ line (26).

C. Mouse Models of Electrogenic Diseases

There is an expanding number of mouse models that have been created in attempts to replicate specific forms of human arrhythmic diseases (Table 24.1). For example, genetically manipulated mice having an α-myosin heavy-chain missense mutation (Arg403Gln) display a phenotype strikingly similar to familial hypertrophic cardiomyopathy (FHC) caused by the identical mutations in humans (6). Hypertrophic cardiomyopathy is a heterogeneous group of disorders characterized by ventricular hypertrophy, arrhythmias, and sudden cardiac death. The usual course of the disease is slow progression of arrhythmia symptoms with a significant incidence of sudden death. Most clinical studies have reported arrhythmias as the mode of sudden death in these patients (27). The genetically manipulated mice

Table 24.1 Human Electrophysiological Disorders with Mouse Models

Clinical syndrome	Murine gene product
Familial hypertrophic cardiomyopathy	Myosin heavy-chain, sarcomeric proteins, binding proteins
Familial dilated cardiomyopathy	Tropomodulin, muscle LIM protein
Duchenne's muscular dystrophy	Dystrophin
Myotonic muscular dystrophy	Myotonic dystrophy protein kinase (DMPK)
Congenital long-QT syndrome	KVLQT1, MERG, K^+ channel subtypes
Gap junction connexin mutations	Connexin 40 and 43

into which this Arg403Gln mutation have been engineered have many changes characteristic of human FHC, including alterations in myocyte and ventricular structure and a predisposition to both atrial and ventricular arrhythmias. In vivo mouse electrophysiology studies performed on FHC mice with an α-myosin heavy-chain substitution have revealed cardiac conduction abnormalities and arrhythmia inducibility (28). Gender differences in the effects of this mutation were also seen, with male mice more likely to have abnormalities in sinus node function, right axis deviation, and inducibility of ventricular tachycardia, as shown in Figure 24.2 (29). Heterogeneity of ventricular repolarization was also demonstrated in these α-myosin heavy-chain mutant FHC mice (30), and further studies have evaluated the cardiovascular responses to exercise and beta blockade. Murine models of human FHC also include mutations in other sarcomeric proteins, such as cardiac troponins, tropomodulin, myosin light chain, myosin binding proteins, and tropomyosin. These studies may be relevant in the further understanding of the genetic mechanisms and pathophysiology of hypertrophy and ventricular arrhythmia vulnerability in hypertrophic cardiomyopathy and other inherited cardiac disorders.

Mouse models of familial dilated cardiomyopathy (FDC) have also been developed by creating "knock-out" mice lacking the gene encoding for the muscle LIM protein (MLP). FDC in humans accounts for up to 30% of all cases of idiopathic dilated cardiomyopathy (31). The majority of cases are transmitted in an autosomal dominant fashion; however, multiple modes of transmission have been described including an X-linked recessive pattern, X-linked recessive with myocyte mitochondria aberration, and various linkage sites on somatic chromosomes, specifically chromosomes 3, 9, and 1 (3). These MLP-deficient mice develop a phenotype similar to FDC with disrupted cytoskeletal architecture and neonatal onset of dilated cardiomyopathy. Transgenic tropomodulin overexpression mutations also have been utilized in murine models of dilated cardiomyopathy. Disease expression has been shown to involve calcineurin signaling pathways, and dilated cardiomyopathy appears to be prevented experimentally with calcineurin inhibitors, such as cyclosporin (32).

Congenital long-QT syndrome (LQTS) is another example of an inherited human arrhythmic syndrome studied using transgenic animal counterparts. LQTS is a disorder characterized in humans by a prolonged QT interval on the electrocardiogram, syncope, seizures, and sudden death from ventricular arrhythmias, especially torsades de pointes (33,34). There is significant genetic heterogeneity, and patients having the same mutation may have varying degrees of disease severity secondary to variable penetrance and expressivity (35). With new approaches to genetic analysis and molecular techniques, more than 150 mutations for LQTS have been described in five different sites (described in Chap. 25).

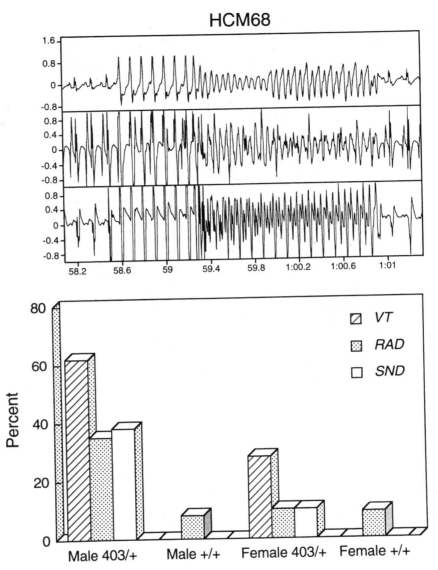

Figure 24.2 Illustration of ventricular tachycardia and other electrophysiological abnormalities induced in familial hypertrophic cardiomyopathy (FHC) mice during in vivo electrophysiology studies. The top panel is an example of inducible ventricular tachycardia (VT) with programmed ventricular stimulation in a heterozygous mutant FHC mouse. The three channel recordings are from surface ECG, right atrial intracardiac electrogram, and right ventricular intracardiac electrogram (data displayed in volts). The electrogram begins in normal sinus rhythm, followed by ventricular programmed electrical stimulation that induces VT. The bottom panel is a summary of the electrophysiological findings in male and female mutant FHC mice, illustrating the percentage of mice with VT, right axis deviation (RAD) on surface ECG, and sinus node dysfunction (SND) during atrial pacing studies. Male mice had a higher frequency of each of these abnormalities, implying gender differences in electrophysiological phenotype severity. (Adapted from Ref. 29.)

Mutations in the LQT1 gene, which encodes a potassium ion channel subunit important in myocyte repolarization, cause the chromosome 11 variant of LQTS. Alterations in the LQT2 gene encoding the HERG ion channel subunit cause the chromosome 7 form of the disease. Mutations in either LQT1 or LQT2 lead to prolongation of repolarization time. Interestingly, drug-induced long-QT syndrome caused by agents such as terfenadine, sotalol, erythromycin, cisapride, and quinidine do so by blocking the same potassium channels that are mutated in congenital LQTS (36–39). The LQT3 gene encodes a myocyte sodium channel involved in the influx of sodium ions during phase 0 depolarization. Mutations in this gene lead to another form of LQTS as well as a different arrhythmic condition known as Brugada syndrome (24,33), both due to similar sodium channel defects.

Genetic animal models of LQTS have been developed which have defects in those ion channel proteins important in repolarization. For example, mutations in the HERG gene have been developed using the closely related bovine ether-a-go-go gene (BEAG). Several chimeric and mutant channels were created in Xenopus oocytes using complementary RNA/DNA constructs (40). Patch-clamp electrophysiological studies were then performed to assess susceptibility of the mutant channels to class III antiarrhythmic drugs. The different mutants were studied to determine which specific amino acid mutations led to changes in drug binding and activity.

1. Mouse LQTS Models with Potassium Channel Defects

Transgenic mice with mutations in KvLQT1, HERG, and minK have been developed that cause prolonged repolarization times and consequently prolonged QT intervals on the surface ECG. Using transgene constructs with viral promoter and expression sequences, recombinant plasmid DNA was injected into fertilized mouse oocytes (41). Heterozygous and littermate control mice were studied using surface electrocardiograms to assess QT interval. Murine cardiac myocytes were then assessed for whole-cell action potential and potassium currents using patch-clamp techniques. Findings included prolongation of the QT interval as well as ion channel abnormalities that altered cardiac myocyte repolarization. Mice that overexpressed the N-terminus of a voltage-gated potassium channel (Kv1.1N206Tag) have QT prolongation and inducible ventricular tachycardia (41). Mouse lines have been developed which lack the potassium channel gene Kv1.4, which was demonstrated not to be primarily responsible for the transient outward current and do not have an apparent electrocardiographic phenotype (42). Conversely, mice expressing a dominant negative form of Kv4, leading to functional inactivation of the transient outward current, were developed and shown to be the molecular species coding for the transient outward current (43). Mice with this mutation have QT prolongation on the surface ECG, and increased action potential duration in ventricular myocytes. Whole animal electrophysiology studies have not yet been performed in this mouse model.

Mutations in the HERG gene in humans are responsible for LQT2, and MERG is the mouse homologue of this channel gene. Several different mutations have been designed in transgenic mice, including overexpression of a human HERG mutation (44) and a different isoform of MERG that is expressed selectively in the murine heart (45). Interestingly, both of these mutations lead to alterations in potassium current regulation, with inhibition of the rapidly activating component of the delayed rectifier potassium channel (I_{Kr}) at the cellular level, but without QT prolongation in whole animal ECG recordings. A murine model of the Jervell and Lange-Nielsen autosomal recessive variant of LQTS was engineered by targeted disruption of the slowly activating component of the delayed rectifier current (I_{Ks}), with a phenotype of deafness and QT prolongation (46).

2.   Mouse LQTS Models with Sodium Channel or Sodium–Potassium
     Pump Defects

Mutations in genes that encode proteins in the voltage-gated sodium channel are responsible for the LQT3 form of LQTS (33,47). Mouse lines with sodium channel mutations are currently being developed as models for LQT3 and a related condition, the Brugada syndrome, which in humans is also caused by mutations in the cardiac sodium channel. A murine model of LQTS has been described which has a mutation in the sodium–potassium ATPase pump (48). These transgenic mice express a human isoform of the α3-subunit, and have a long-QT interval on ECG, electrical T-wave alternans, and inducible ventricular arrhythmias during programmed stimulation studies (48).

Further studies including single-cell analyses, tissue conduction studies, and whole animal in vivo EP studies may prove useful in elucidating the mechanisms of these disease processes and the relationship of long-QT syndrome to sudden death and ventricular tachyarrhythmias.

D.   Mouse Models of Conduction Disturbances

No naturally occurring small-animal models of inherited conduction defects have been described, but transgenic animals with specific electrophysiological phenotypes have been engineered and can provide a very useful tool in studying the cardiac conduction system.

Myotonic muscular dystrophy is the most common adult-onset type of muscular dystrophy. It affects all muscle types and causes varying degrees of debilitation. Cardiac manifestations of myotonic dystrophy include AV block and bradycardia secondary to involvement of the conduction system. The disease is inherited in an autosomal dominant fashion that progressively worsens in subsequent generations as a result of the repetitive expansion of a trinucleotide repeat sequence mutation in a protein kinase gene (DMPK) on chromosome 19 (49). A murine model of homozygous DMPK deficiency has been developed, which appears quite relevant in understanding the human disease (10). Extensive in vivo electrophysiology studies of these animals have been performed and distinct abnormalities in atrioventricular (AV) conduction properties were discerned (50). Both heterozygote and homozygote mice had prolongation of the PR interval on surface ECG, and longer AV conduction times. The homozygous mutant mice had more severe AV conduction defects, including second- and third-degree AV block (50). Interestingly, the AV conduction abnormalities appeared to be developmentally progressive, with PR prolongation and AV block only evident in adult mice, similar to the human phenotype. The heterozygote mice displayed a milder phenotype, but clearly different from wild-type littermates, linking DMPK haploinsufficiency with cardiac conduction pathophysiology.

Transgenic mice with connexin protein defects constitute another group of mice with cardiac conduction disorders. Connexins are proteins that form the gap junctions between myocytes, providing a pathway for transcellular electrical communication (see Chap. 8). Cardiac myocytes contain different types of connexin isoforms, and these are selectively expressed in a species-specific manner in different parts of the myocardium. Disorders of gap junction proteins have not yet been confirmed as responsible for human disease, although they may be implicated in some congenitally inherited structural heart abnormalities and conduction disorders (51). Human ventricular myocytes express primarily connexin 43 and 45, whereas atrial myocytes express primarily connexin 43, 45, and 40. Targeted mutagenesis in different connexin proteins has been performed and electrophysiology studies have been reported for connexin 43 mutated mice (52,53). Homozygous connexin 43–deficient mice did not live, but heterozygotes survived, albeit with distinct ventricular conduction abnormalities. These studies revealed prolonged intraventricular conduction delays in single myocytes, epicardial whole heart preparations, and whole-animal in vivo electrophysiology studies. The QRS duration on ECG was significantly longer in connexin 43–deficient mice, without atrial electrophysiological abnormalities or inducible arrhythmias (53). Structurally, major abnormalities in the normal looping of the heart during embryonic development occurred in connexin 43–deficient mice, leading to congenital heart abnormalities (54).

Connexin 40–deficient mice have also been recently engineered and bred to homozygosity. These mice have a severe lack of connexin 40, without compensation from other connexin isoforms in the heart (55). Connexin 40–deficient mice have been shown to have distinct atrioventricular and distal conduction system disturbances using surface ECG and intracardiac electrophysiological studies (55,56). Site-specific AV conduction delay was demonstrated both during sinus rhythm and with atrial pacing, as was increased atrioventricular refractoriness (56). Alterations in cardiac gap junction isoform-specific density, and distribution, may also play an important role in arrhythmogenesis. Mouse models of connexin subtype deficiencies and studies of the effects of chronic arrhythmias on connexin isoform distribution may provide further information linking gap junction proteins with human clinical arrhythmias (57).

III. CONCLUSIONS

With the identification of hundreds of new genes potentially modulating electrophysiological characteristics, specific genetic etiologies of cardiovascular electrical disease can now be identified. Collectively, the development of animal models to delineate the genetic underpinnings represents a major progression in our understanding of human arrhythmias and sudden cardiac death. Such advances permit individuals that have specific genetic defects to be identified and risk factors related to inheritable electrophysiological consequences determined. As more is learned about the biology of genetic defects involved in isolated "pure" forms of disease and genetic patterns of susceptibility to acquired arrhythmia conditions, emphasis will shift from diagnosis to treatment using pharmacogenetic and gene therapy strategies. Gene replacement therapy is just beginning to be utilized in animal models of cardiovascular disease (58), and potential for curative therapy in human patients is high, but unlikely to become commonplace for arrhythmic diseases in the near future. To date, gene transfer in the cardiovascular field has been useful in creating animal models of disease in which the in vivo functions of gene products (or lack of products) can be studied (59). As technology improves, gene-to-gene transfer and gene replacement modalities will become increasingly important areas of study and may provide rational and exciting new therapeutic approaches to human cardiovascular pathophysiology.

REFERENCES

1. Schwartz P. Do animals have clinical value? Am J Cardiol 1998; 81(6A)14D–20D.
2. Paigen K. A miracle enough: the power of mice. Nat Med 1995; 1:215–220.
3. Berul CI, Mendelsohn ME. Molecular biology and genetics of cardiac disease associated with sudden death: electrophysiologic studies in mouse models of inherited diseases. In: Estes NAM, Salem DN, Wang PJ, eds. Sudden Cardiac Death in the Athlete. Armonk, NY: Futura Publishing Co, 1998:465–481.

4. Wilmut I, Schnieke AE, McWhir J, Kind AJ, Campbell KH. Viable offspring derived from fetal and adult mammalian cells. Nature 1997; 385:810–813.

5. Mukerjee M. Trends in animal research. Sci Am 1997; Feb:86–93.

6. Geisterfer-Lowrance AA, Christe M, Conner DA, Ingwall JS, Schoen FJ, Seidman CE, Seidman JG. A mouse model of familial hypertrophic cardiomyopathy. Science 1996; 272:731–734.

7. Wang Q, Shen J, Splawski I, Atkinson D, Li Z, Robinson JL, Moss AJ, Towbin JA, Keating MT. SCN5A mutation associated with an inherited cardiac arrhythmia-long QT syndrome. Cell 1995; 80:805–811.

8. Curran ME, Splawski I, Timothy KW, Vincent GM, Green ED, Keating MT. A molecular basis of cardiac arrhythmias: HERG mutations cause long QT syndrome. Cell 1995; 80:795–803.

9. Grace AA, Chien KR. Congenital long QT syndromes: Toward molecular dissection of arrhythmia substrates. Circulation 1995; 92:2786–2789.

10. Reddy S, Smith DB, Rich MM, Leferovich JM, Reilly P, Davis BM, Tran K, Rayburn H, Bronson R, Cros D, Balice-Gordon RJ, Housman D. Mice lacking the myotonic dystrophy protein kinase develop a late-onset progressive myopathy. Nat Genet 1996; 13:325–335.

11. Gottwald M, Gottwald E, Dhein S. Age-related electrophysiological and histological changes in rabbit hearts: age-related changes in electrophysiology. Int J Cardiol 1997; 62:97–106.

12. Asano Y, Davidenko JM, Baxter WT, Gray RA, Jalife J. Optical mapping of drug-induced polymorphic arrhythmias and torsades de pointes in the isolated rabbit heart. J Am Coll Cardiol 1997; 29:831–842.

13. Gray RA, Jalife J. Effects of atrial defibrillation shocks on the ventricles in isolated sheep hearts. Circulation 1998; 97:1613–1622.

14. Wood MA, Caponi D, Sykes AM, Wenger EJ. Atrial electrical remodeling by rapid pacing in the isolated rabbit heart: effects of Ca++ and K+ channel blockade. J Intervent Cardiac Electrophysiol 1998; 2:15–23.

15. Fishman MC, Olson EN. Parsing the heart: genetic modules for organ assembly. Cell 1997; 91:153–156.

16. Serbedzija GN, Chen J, Fishman MC. Regulation of the heart field in zebrafish. Development 1998; 125:1095–1101.

17. Moorman AFM, de Jong F, Denyn MMFJ, Lamers WH. Development of the cardiac conduction system. Circ Res 1998; 82:629–644.

18. Gourdie RG, Mima T, Thompson RP, Mikawa T. Terminal diversification of the myocyte lineage generates Purkinje fibers of the cardiac conduction system. Development 1995; 121(5):1423–31.

19. Gourdie RG, Wei Y, Kim D, Klatt SC, Mikawa T. Endothelin-induced conversion of embryonic heart muscle cells into impulse-conducting Purkinje fibers. Proc Natl Acad Sci USA 1998; 95(12):6815–8.

20. Oosthoek PW, Viragh S, Lamers WH, Moorman AF. Immunohistochemical delineation of the conduction system. II: The atrioventricular node and Purkinje fibers. Circ Res 1993; 73(3):482–91.

21. Berul CI, Aronovitz MJ, Wang PJ, Mendelsohn ME. In vivo cardiac electrophysiology studies in the mouse. Circulation 1996; 94:2641–2648.

22. Mitchell GF, Jeron A, Koren G. Measurement of heart rate and Q-T interval in the conscious mouse. Am J Physiol 1998 (Heart Circ Physiol 43); 274:H747–H751.

23. Lewin B. Genes VI. Oxford: Oxford University Press, 1997:980–987.

24. Brugada R, Roberts R. The molecular genetics of arrhythmias and sudden death. Clin Cardiol 1998; 21:553–560.

25. Melton DW. Gene targeting in the mouse. Bioessays 1994; 16:633–638.

26. James JF, Hewett TE, Robbins J. Cardiac physiology in transgenic mice. Circ Res 1998; 82:407–415.

27. Colan SD, Spevak PJ, Parness IA, Nadas AS. Cardiomyopathies. In: Fyler DC, ed. Nadas' Pediatric Cardiology. Philadelphia: Hanley & Belfus, Inc., 1992:338–344.

28. Berul CI, Christe M, Aronovitz MA, Seidman CE, Seidman JG, Mendelsohn ME. Electrophysiological abnormalities and arrhythmias in MHC mutant familial hypertrophic cardiomyopathy mice. J Clin Invest 1997; 99:570–576.

29. Berul CI, Christe ME, Aronovitz MJ, Maguire CT, Seidman CE, Seidman JG, Mendelsohn ME. Familial hypertrophic cardiomyopathy mice display gender differences in electrophysiological abnormalities. J Intervent Cardiac Electrophysiol 1998; 2:7–14.

30. Bevilacqua LM, Maguire CT, Seidman CE, Seidman JG, Berul CI. QT dispersion in MHC familial hypertrophic cardiomyopathy mice. Pediatr Res 1999; 45:190–194.

31. Michels VV, Moll PP, Miller FA, Tajik AJ, Chu JS, Driscoll DJ, Burnett JC, Rodeheffer RJ, Chesebro JH, Tazelaar HD. The frequency of familial dilated cardiomyopathy in a series of patients with idiopathic dilated cardiomyopathy. N Engl J Med 1992; 326:77–82.

32. Sussman MA, Lim HW, Gude N, taigen T, Olson EN, Robbins J, Colbert MC, Gualberto A, Wieczorek DF, Molkentin JD. Prevention of cardiac hypertrophy in mice by calcineurin inhibition. Science 1998; 281:1690–1693.

33. Towbin JA. New revelations about the long QT-syndrome. N Engl J Med 1995; 333:384–385.

34. Berul CI, Sweeten TL, Hill SL, Vetter VL. Provocative testing in children with suspect congenital long QT syndrome. Ann Noninvas Electrocardiol 1998; 3:3–11.

35. Vincent GM. The molecular genetics of the long QT syndrome: genes causing fainting and sudden death. Annu Rev Med 1998; 49:263–274.

36. Hill SL, Evangelista JK, Pizzi AM, Mobasseleh M, Fulton DR, Berul CI. Potential proarrhythmia associated with cisapride in children. Pediatrics 1998; 101:1053–1056.

37. Honig PK, Wortham DC, Zamani K, Conner DP, Mullin JC, Cantilena LR. Terfenadine-ketoconazole interaction. J Am Med Assoc 1993; 269:1513–1518.

38. Nattel S, Ranger S, Talajic M, Lemery R, Roy D. Erythromycin-induced long QT syndrome: Concordance with quinidine and underlying cellular electrophysiologic mechanism. Am J Med 1990; 89:235–238.

39. Berul CI, Morad M. Regulation of potassium channels by nonsedating antihistamines. Circulation 1995; 91:2220–2225.

40. Ficker E, Jarolimek W, Kiehn J, Baumann A, Brown AM. Molecular determinants of dofetilide block of HERG K+ channels. Circ Res 1998; 82:386–395.

41. London B, Jeron A, Zhou J, Buckett P, Han X, Mitchell GF, Koren G. Long QT and ventricular arrhythmias in transgenic mice expressing the N terminus and first transmembrane segment of a voltage-gated potassium channel. Proc Natl Acad Sci USA 1998; 95:2926–2931.

42. London B, Wang DW, Hill JA, Bennett PB. The transient outward current in mice lacking the potassium channel gene Kv1.4. J Physiol 1998; 509:171–182.

43. Barry DM, Xu H, Schuessler RB, Nerbonne JM. Functional knockout of the transient outward current, long QT syndrome, and cardiac remodeling in mice expressing a dominant-negative Kv4 alpha subunit. Circ Res 1998; 83:560–567.

44. Babij P, Askew GR, Nieuwenhuijsen B, Su CM, Bridal TR, Jow B, Argentieri TM, Kulik J, DeGennaro LJ, Spinelli W, Colatsky TJ. Inhibition of cardiac delayed rectifier K+ current by overexpression of the long QT syndrome HERG G628S mutation in transgenic mice. Circ Res 1998; 83:668–678.

45. Lees-Miller JP, Kondo C, Wang L, Duff HJ. Electrophysiological characterization of an alternatively processed ERG K+ channel in mouse and human hearts. Circ Res 1997; 81:719–726.

46. Drici MD, Arrighi I, Chouabe C, Mann JR, Lazdunski M, Romey G, Barhanin J. Involvement of IsK-associated K+ channel in heart rate control of repolarization in a murine engineered model of Jervell and Lange-Nielsen syndrome. Circ Res 1998; 83:95–102.

47. An RH, Wang XL, Kerem B, Benhorin J, Medina A, Goldmit M, Kass RS. Novel LQT-3 muta-

tion affects Na+ channel activity through interactions between alpha- and beta 1-subunits. Circ Res 1998; 83:141–146.

48. O'Brien SE, Apkon MA, Berul CI, Patel HT, Zahler R. Electrocardiographic phenotype of long QT syndrome in transgenic mouse model expressing human 3 Na,K-ATPase isoform in heart. Circulation 1997; 96:I–422.

49. Housman DE, Shaw DJ. Expansion of an unstable DNA region and phenotypic variation in myotonic dystrophy. Nature 1992; 355:545–546.

50. Berul CI, Maguire CT, Aronovitz MJ, Greenwood J, Miller C, Gehrmann J, Housman DE, Mendelsohn ME, Reddy S. DMPK dosage alterations result in atrioventricular conduction abnormalities in a mouse myotonic dystrophy model. J Clin Invest 1999; 103:R1–R7.

51. Davis LM, Kanter HL, Beyer EC, Saffitz JE. Distinct gap junction protein phenotypes in cardiac tissues with disparate conduction properties. J Am Coll Cardiol 1994; 24:1124–1132.

52. Guerrero PA, Schuessler RB, Davis LM, Beyer EC, Johnson CM, Yamada KA, Saffitz JE. Slow ventricular conduction in mice heterozygous for a connexin43 null mutation. J Clin Invest 1997; 99:1991–1998.

53. Thomas SA, Schuessler RB, Berul CI, Beardslee MA, Beyer EC, Mendelsohn ME, Saffitz JE. Disparate effects of deficient expression of connexin43 on atrial and ventricular conduction: evidence for chamber-specific molecular determinants of conduction. Circulation 1998; 97:686–691.

54. Ya J, Erdstieck-Ernste EBHW, de Boer PAJ, van Kempen MJA, Jongsma H, Gros D, Moorman AFM, Lamers WH. Heart defects in connexin 43 deficient mice. Circ Res 1998; 82:360–366.

55. Simon AM, Goodenough DA, Paul DL. Mice lacking connexin40 have cardiac conduction abnormalities characteristic of atrioventricular block and bundle branch block. Curr Biol 1998; 8:295–298.

56. Bevilacqua LM, Simon AM, Maguire CT, Gehrmann J, Rakhit A, Paul DL, Berul CI. A targeted disruption of connexin40 causes distinct atrioventricular conduction abnormalities. J Am Coll Cardiol 1999; 33:1A (abstr).

57. Thomas SA, Schuessler RB, Saffitz JE. Connexins, conduction, and atrial fibrillation. J Cardiovasc Electrophysiol 1998; 9:608–611.

58. Akhter SA, Skaer CA, Kypson AP, McDonald PH, Peppel KC, Glower DD, Lefkowitz RJ, Koch WJ. Restoration of β-adrenergic signaling in failing cardiac ventricular myocytes via adenoviral-mediated gene transfer. Proc Natl Acad Sci USA 1997; 94:12100–12105.

59. Nabel EG. Gene therapy for cardiovascular disease. Circulation 1995; 91:541–548.

25

Genetic Approaches and Familial Arrhythmias

JEFFREY A. TOWBIN

Baylor College of Medicine, Houston, Texas

KETTY SCHWARTZ

Institut de Myologie, Paris, France

I. INTRODUCTION

In this chapter, we describe our current understanding of the genetic basis of familial arrhythmias and the approaches used to clarify these genetic diseases. The molecular bases of primary ventricular arrhythmias (long-QT syndrome, Brugada syndrome), primary atrial arrhythmias (atrial fibrillation), and secondary ventricular arrhythmias (hypertrophic cardiomyopathy) are discussed.

Sudden cardiac death (SCD) is a significant problem in the United States, with an incidence reported to be greater than 300,000 persons per year (1,2). Interest in identifying the underlying cause of death has been focused on cases of unexpected arrhythmogenic death, estimated to represent 5% of all sudden deaths. In cases in which no structural heart disease can be identified, the long-QT syndrome and ventricular preexcitation are most commonly considered as likely causes. Recently, Brugada syndrome [also known by some investigators as idiopathic ventricular fibrillation (IVF)], a disease associated with an electrocardiographic abnormality of right bundle branch block with ST-elevation in the right precordial leads (V1–V3), has been added to the list of possible causes of sudden death in otherwise healthy, young individuals (3). In addition, hypertrophic cardiomyopathy has been considered to be the most common cause of SCD in the United States in young adults and athletes, with ventricular arrhythmias playing a central role. This chapter describes the current understanding of the molecular genetics of inherited diseases in which arrhythmias are prominent features.

II. THE LONG-QT SYNDROMES

A.   Disease Classification

The long-QT syndromes (LQTS) are disorders of repolarization identified by the electro-cardiographic (ECG) abnormalities of prolongation of the QT interval corrected for heart rate (QTc), relative bradycardia, T-wave abnormalities, and episodic ventricular tachyarrhythmias (2), particularly *torsade de pointes*. The diagnosis usually relies on a QTc measurement of >460–480 ms using Bazzet's formula (QTc = QT/$\sqrt{\text{R-R}}$) associated with the T-wave abnormalities. A diagnostic point system has been developed which takes these features into account (4) (Table 25.1). LQTS occurs either as an inherited disorder, a sporadic disorder, or it may be acquired (usually after use of various medications or electrolyte abnormalities, such as hypokalemia) (Tables 25.2 and 25.3). The clinical presentation is similar in most forms of LQTS, but two distinct forms with differing patterns of transmission have been described: the Romano–Ward syndrome (5,6) and the Jervell and Lange-Nielsen syndrome (7).

Romano–Ward syndrome is the most common of the inherited forms of LQTS and

Table 25.1   Diagnostic Criteria in LQTS

Clinical finding	Points
Electrocardiographic findings[a]	
$\quad$ QT$_c$[b]	
$\qquad$ >480 ms$^{1/2}$	3
$\qquad$ 460–470 ms$^{1/2}$	2
$\qquad$ 450 (male) ms$^{1/2}$	1
$\quad$ Torsade de pointes[c]	2
$\quad$ T-wave alterans	1
$\quad$ Notched T wave in three leads	1
$\quad$ Low heart rate for age[d]	0.5
Clinical history	
$\quad$ Syncope[e]	
$\qquad$ With stress	2
$\qquad$ Without stress	1
$\quad$ Congenital deafness	0.5
Family history[e]	
$\quad$ Family members with definite LQTS[f]	1
$\quad$ Unexplained sudden cardiac death below age	
$\qquad$ 30 among immediate family members	0.5

Scoring: <1 point = low probability of LQTS; 2 to 3 points = intermediate probability of LQTS; >4 points = high probability of LQTS.
[a]In the absence of medications or disorders known to affect these ECG features.
[b]QTc calculated by Bazett's formula, where QTc = QT/$\sqrt{\text{RR}}$.
[c]Mutually exclusive.
[d]Resting heart rate below the second percentile for age.
[e]The same family member cannot be counted twice.
[f]Definite LQTS is defined by an LQTS score >4.
Source: Ref. 4.

Table 25.2   Causes of Acquired LQTS

Drug name	Chemical name
Anesthetics/asthma	
Adrenaline	Epinephrine
Antihistamines	
Seldane	Terfenedine
Hismanol	Astemizole
Benadryl	Diphenhydramine
Antibiotics	
E-Mycin, EES, EryPeds, PCE, etc.	Erythromycin
Bactrim, Septra	Trimethoprim and Sulfamethoxazole
Pentam intravenous	Pentamidine
Heart medications	
Heart rhythm drugs	
Quinidine, Quinidex, Duraquin, Quiniglute, etc.	Quinidine
Pronestyl	Procainamide
Norpace	Disopyramide
Betapace	Sotalol
Lipid-lowering drugs	
Lorelco	Probucol
Antianginal drugs	
Vascor	Bepridil
Gastrointestinal	
Propulsid	Cisapride
Antifungal drugs	
Nizoral	Ketoconazole
Diflucan	Fluconazole
Sporanox	Itraconazole
Psychotropic drugs	
Elavil, Norpramine, Viractil Compazine, Stelazine, Thorazine, Mellaril, Etrafon, Trilafon, others	Amitriptyline (Tricyclics)
Haldol	Haloperidol
Risperdal	Risperidone
Orap	Pimozide
Diuretics	
Lozol	Indapamide
Potassium loss	

appears to be transmitted as an autosomal dominant trait (1,2,5,6). In this disorder, gene carriers are expected to be clinically affected and have a 50% likelihood of transmitting the disease-causing gene to their offspring. However, low penetrance has been described and therefore gene carriers may, in fact, have no clinical features of disease (8). When clinically affected, individuals with Romano–Ward syndrome have the pure syndrome of a prolonged QT interval on ECG, with the associated symptom complex of syncope, sudden death, and, in some patients, seizures (9,10). Occasionally, other noncardiac abnormalities such as diabetes mellitus (11,12), asthma (13), or syndactyly (14) may also be associated with QT prolongation. LQTS may also be involved in some cases of sudden infant death syndrome (SIDS) (15–17).

Table 25.3   Drugs Causing Torsade de Pointes

I.   Antiarrhythmic agents
     Quinidine
     Disopyramide
     Procainamide (*N*-Acetyl-procainamide)
     Sotalol
     Ibutilide
     Amiodarone
II. Calcium channel blocking agents
     Bepridil
     Lidoflazine
III. Central nervous system active agents
     Thioridazine
     Tricyclic antidepressants
     Pimozide
IV. Antibiotics
     Macrolides (e.g., Erythromycin)
     Pentamidine
     Trimethoprim-sulfa
V.   Antihistamines
     Terfenadine
     Astemizole
VI. Miscellaneous
     Terodiline
     Liquid protein diets
     Organophosphorous insecticide poisoning
     Ketanserin
     Cisapride
     Probucol

The Jervell and Lange-Nielsen syndrome (JLNS) is a relatively uncommon inherited form of LQTS. Classically, this disease has been described as having apparent autosomal recessive transmission (7). Patients have an identical clinical presentation as those with Romano–Ward syndrome, but also have associated sensorineural deafness (7,18,19). Individuals with JLNS usually have longer QT intervals as compared to individuals with Romano–Ward syndrome as well as a more malignant course. Recently this distinction has been blurred, as autosomal recessive cases of Romano–Ward syndrome have been described (20).

B. Mapping of LQTS Genes in Romano–Ward Syndrome

The first gene for autosomal dominant LQTS was mapped by Keating et al. (21) to chromosome 11p15.5 *(LQT1)* (Fig. 25.1) using genome-wide linkage analysis in a large Utah family. Soon afterward, Keating et al. (22) reported consistent linkage of several other LQTS families to chromosome 11p15.5. Linkage analysis is essentially a statistical method of localizing (i.e., mapping) an affected locus or gene to a particular region of a chromosome using a set of genetic markers at known positions within the genome. LQTS locus heterogeneity was subsequently reported by Towbin and colleagues and others

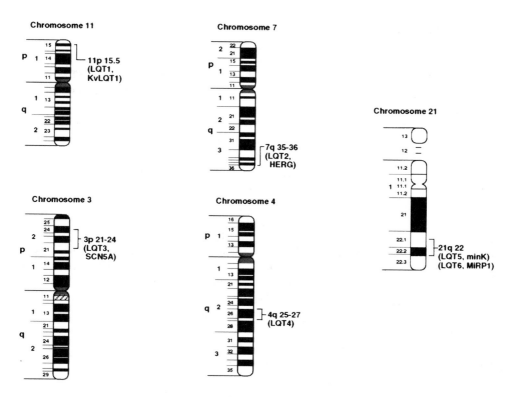

Figure 25.1 Ideograms of chromosome 11, 7, 3, 4, and 21 showing approximate locations of *LQT1 (KVLQT1)*, *LQT2 (HERG)*, *LQT3 (SCN5A)*, *LQT4*, *LQT5 (minK)*, and *LQT6 (MiRP1)*, respectively.

(23–26) and confirmed by the mapping of the second LQTS locus to chromosome 7q35-36 *(LQT2)*, and the third LQTS locus to chromosome 3p21-24 *(LQT3)* (27). Schott et al. (28) mapped the fourth LQTS locus to chromosome 4q25-57 *(LQT4)*, while a fifth gene *(minK)* located on chromosome 21q22 (29,30) was shown to be *LQT5*. More recently, a sixth gene, *minK*-related peptide 1 *(MiRP1)*, localized to 21q22 as well, was identified (31) (Fig. 25.1). Several other families with autosomal dominant LQTS are not linked to any known LQTS loci, indicating the existence of additional LQTS-causing genes.

C. Gene Identification in Romano–Ward Syndrome

1. Cardiac Potassium Channel Gene *KVLQT1* and *LQT1*

The *LQT1* gene required 5 years from the time that mapping was first reported to gene cloning. The positional cloning method was used to identify this gene, *KVLQT1*, a novel potassium channel with six membrane-spanning segments (32). *KVLQT1* (also called *KCNQ1*) consists of 16 exons, spans approximately 400kb, and is widely expressed in multiple human tissues including heart, kidney, lung, placenta, and pancreas, but not skeletal muscle, liver, or brain. In the original report, 11 different types of *KVLQT1* mutations (deletion and missense mutations) were identified in 16 LQTS families, establishing *KVLQT1* as *LQT1*. More than 100 families with *KVLQT1* mutations have since been described. There is at least one frequently mutated region (called a "hot spot") of *KVLQT1*

(32,33). This gene is believed to be the most commonly mutated gene causing LQTS (33,34).

Analysis of the predicted amino acid sequence of *KVLQT1* suggests that it encodes a potassium channel α-subunit with a conserved potassium-selective signature pore sequence, flanked by six membrane-spanning segments (29,30,32) (Fig. 25.2). The putative voltage sensor domain is found in the fourth membrane-spanning region (S4) and the selective pore loop is located between the fifth and sixth membrane-spanning regions (S5,S6). Electrophysiological characterization of the *KVLQT1* protein in various in vitro heterologous expression systems confirmed that *KVLQT1* is a voltage-gated potassium channel protein subunit that requires an accessory β-subunit to function properly (29,30). This β-subunit, which coassembles with *KVLQT1*, is called *minK* (Fig. 25.2); *minK* (IsK), which is coded by the *KCNE1* gene, is a short protein, with only 130 amino acids and only one transmembrane-spanning segment (35). At the time of its initial identification, *minK* did not have any sequence or structural homology to any of the other cloned channels, but it is now known to be part of a family of similar proteins (31). When *minK* and *KVLQT1* are coexpressed in either mammalian cell lines, or *Xenopus* oocytes, a potassium current that is similar to the slowly activating delayed rectifier potassium current [I_{Ks}] in cardiac myocytes appears (29,30). Immunoprecipitation experiments have confirmed the physical interaction between *KVLQT1* and *minK* (29), and the formation of the cardiac slowly activating delayed rectifier I_{Ks} current. Combination of normal and mutant *KVLQT1* subunits results in abnormal I_{Ks} channels and, hence, LQTS-associated mutations of *KVLQT1* are believed to act through a "dominant-negative" mechanism (a mutant subunit of *KVLQT1* interferes with the function of paired normal wild-type subunits resulting in dysfunction of the intact tetramer complex). This "poison pill"-type mechanism results in what has been termed a loss-of-function mutation since the entire complex becomes inoperative (36).

Since mutations in *KVLQT1* were shown to cause chromosome 11–linked LQTS

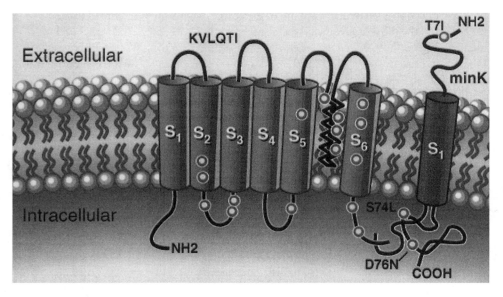

Figure 25.2   The molecular architecture of the cardiac I_{Ks} potassium channel encoded by *KVLQT1* and *minK*. Mutations are noted.

(LQT1), mutations in *minK* were also expected to cause LQTS, and this idea was subsequently demonstrated to be correct (see Sec. II.C.4) (37).

2. Cardiac Potassium Channel Gene *HERG* and *LQT2*

Both *LQT2* and *LQT3* (described below) were identified by the candidate gene approach. This relies on a defined mechanistic hypothesis based on knowledge of the physiology of the disease of interest. Since LQTS is considered to be largely a disorder of abnormal repolarization, genes encoding ion channels or proteins modulating channel function were considered the most likely candidates. After initial localization of *LQT2* to chromosome 7q35-36, candidate genes in this region were analyzed. *HERG* (*h*uman *e*ther-a-go-go-*re*lated *g*ene), a cardiac potassium channel gene with six transmembrane segments, which was originally cloned from a brain cDNA library (38) and found to be expressed in neural-crest-derived neurons (39), microglia (40), a wide variety of tumor cell lines (41), and the heart (42) was evaluated. Curran et al. (42) demonstrated linkage of *HERG* to the *LQT2* locus on chromosome 7q35-36 and six LQTS-associated mutations were identified in *HERG*, including missense mutations, intragenic deletions, and a splicing mutation. Later, Schulze-Bahr et al. (43) reported a single base pair deletion and a stop codon mutation in *HERG*, conforming this gene to be a common cause of LQTS when mutated. Currently, this gene is thought to be the second most common gene mutated in LQTS (second to *KVLQT1*), and mutations scattered throughout its entire sequence without preferential clustering have been seen (Table 25.4).

HERG consists of 16 exons and spans 55 kb of genomic sequence (42). The predicted topology of *HERG* is shown in Figure 25.3, and is similar to *KVLQT1*. Unlike *KVLQT1*, *HERG* has extensive intracellular amino and carboxyl termini, with a region in the carboxyl terminal domain having sequence similarity to nucleotide binding domains.

Electrophysiological and biophysical characterization of expressed *HERG* in *Xenopus* oocytes established that *HERG* encodes the rapidly activating, delayed rectifier potassium current I_{Kr} (44,45). Electrophysiological studies of LQTS-associated mutations showed that they act through either "loss-of-function" or a "dominant-negative" mechanism (46). In addition, protein trafficking abnormalities have been shown to occur (47). This channel has been shown to coassemble with β-subunits for normal function, similar to that seen in I_{Ks}. McDonald et al. (48) initially suggested that the complexing of *HERG* with *minK* is needed to regulate the I_{Kr} potassium current. More recently, Abbott et al. (31) more definitively identified a *M*inK related protein *(MiRP1)* as an activity-modifying β-subunit that coassembles with *HERG* (see below).

3. Cardiac Sodium Channel Gene *SCN5A* and *LQT3*

The positional candidate gene approach was also used to establish that the gene responsible for chromosome 3–linked LQTS *(LQT3)* is the cardiac sodium channel gene *SCN5A* (49,50). *SCN5A* is highly expressed in human myocardium, but not in skeletal muscle, liver, or uterus (51,52). It consists of 28 exons that span 80 kb and encodes a protein of 2016 amino acids with a putative structure that consists of four homologous domains (I–IV), each of which contains six membrane-spanning segments (S1–S6), each similar to the structure of a potassium channel α-subunit (32,42) (Fig. 25.4). Linkage studies with *LQT3* families and *SCN5A* initially demonstrated linkage to the *LQT3* locus on chromosome 3p21-24 (51,52) and three types of mutations, one 9-bp intragenic deletion ($\Delta DK_{1505}P_{1506}Q_{1507}$) and two missense mutations ($R_{1644}H$ and $N_{1325}S$), were also identified in six LQTS families (51,52) (Table 25.4). All three mutations were expressed in *Xenopus*

Table 25.4   Gene Mutations Causing LQTS

Gene	Function	Type of mutation	Mutation	Result[a]
KVLQT1 (KCNQ1)	I_{Ks}	Intragenic deletion (3bp)	F38W/G39Δ	Tryptophan for conserved phenylalanine at position 38, and deletion of glycine at position 39 in S2
		Missense mutation	G39R	Arginine for conserved glycine, S2
		Missense mutation	R45C	Cysteine for conserved arginine, S2-S3 intercellular loop
		Missense mutation	A49P	Proline for conserved alanine, S2-S3 cytoplasm loop
		Missense mutation	A49T	Threonine for conserved alanine, S2-S3 cytoplasm loop
		Intragenic deletion (5bp)	Δ58	Disrupts, S2-S3
		Missense mutation	G60R	Arginine for conserved glycine, S2-S3
		Insertion (1bp)	$G_{60}R_{61}L_{62}G_{63}$ to $G_{60}A_{61}A_{62}A_{63}$	Frameshift, S2-S3
		Missense mutation	R61Q	Glutamine for conserved arginine, S2-S3
		Missense mutation	V125M	Methionine for conserved valine, S4-S5
		Missense mutation	E132L	Leucine for conserved glutamic acid, S4-S5
		Missense mutation	V135M	Methionine for valine, S4-S5
		Missense mutation	G140D	Aspartic acid for conserved glycine, S5
		Missense mutation	L144F	Phenylalanine for conserved leucine, S5
		Missense mutation	W176S	Serine for conserved tryptophan, S5-pore extracelluar loop
		Missense mutation	G177R	Arginine for conserved glycine, pore
		Intragenic deletion	SP/V178 (ΔGGT)	Disrupts pore
		Missense mutation	T180R	Arginine for conserved threonine, pore
		Missense mutation	T183I	Isoleucine for conserved threonine, pore
		Missense mutation	I184M	Methionine for conserved isoleucine, pore
		Missense mutation	G185S	Serine for conserved glycine, pore
		Missense mutation	Y186S	Serine for conserved tyrosine, pore
		Missense mutation	G186C	Cysteine for conserved glycine, pore
		Missense mutation	K189N	Lysine for conserved asparagine, pore
		Missense mutation	P191A	Alanine for conserved proline, pore
		Missense mutation	W193C	Cysteine for conserved tryptophan, pore-S6
		Missense mutation	G196S	Serine for conserved glycine, S6

Mutation type	Mutation	Description
Missense mutation	G196R	Arginine for conserved glycine, S6
Missense mutation	F211R	Arginine for conserved phenylalanine, S6
Missense mutation	A212E	Glutamate for conserved alanine, S6
Missense mutation	A212V	Valine for conserved alanine, S6
Missense mutation	L213F	Glutamate acid for conserved phenylalanine, S6
Missense mutation	A215V	Valine for conserved alanine, S6
Missense mutation	G216E	Glutamate acid for conserved glycine, S6
Missense mutation	I217T	Threonine for conserved isoleucine, S6
Splicing mutation	SP/A215 (G-A)	Disrupts S6
Missense mutation	G219S	Serine for conserved glycine, S6
Missense mutation	D222N	Asparagine for conserved aspartic acid, S6
Missense mutation	L224P	Proline for conserved leucine, S6
Missense mutation	R237W	Trytophan for conserved arginine, S6-C-terminus
Missense mutation	R237P	Proline for conserved arginine, S6-C-terminus
Missense mutation	A242T	Threonine for conserved alanine, S6-C-terminus
Deletion mutation	ΔC318	Frameshift mutation, C-terminus
Missense mutation	R366W	Tryptophan for conserved arginine (C-terminus, S6) leading to frameshift
Missense mutation	R410W	Tryptophan for conserved arginine, S6-C-terminus
Deletion + Insertion	$Q_{415}Y_{416}S_{417}Q_{418}G_{419}$ to $V_{415}E_{416}I_{417}A_{419}X_{522}$	Disrupts S6-C-terminus
Missense mutation	R426C	Cysteine for conserved arginine, C-terminus
Missense mutation	T458M	Methionine for conserved threonine, C-terminus
Missense mutation	G460D	Aspartic acid for conserved glycine, cytoplasmic tail
Missense mutation	R462H	Histidine substitution for arginine, C-terminus
Insertion mutation	InsC502	Frameshift mutation, C-terminus
Deletion mutation	ΔC502	Frameshift mutation, C-terminus

Table 25.4 Continued

Gene	Function	Type of mutation	Mutation	Result[a]
HERG	I_{Kr}	Intragenic deletion	Δ1261	Frameshift in S1, sequences → premature stop codon
		Intragenic deletion	Δ1500-F508	Disrupts S3
		Missense mutation	N470D	Aspartic acid for conserved asparagine, S2-intracellular loop
		Missense mutation	T474I	Isoleucine for conserved threonine, S2-S3
		Nonessens mutation	Y493ter	Stop codon for conserved tyrosine, S2-S3
		Missense mutation	R534C	Cysteine for conserved arginine, S4
		Intragenic deletion	ΔT557	Disrupts S3
		Missense mutation	A561V	Valine for conserved alanine, S5
		Missense mutation	G572C	Cysyeine for conserved glycine, S5
		Missense mutation	N588D	Aspartic acid for conserved asparagine, S5-pore
		Missense mutation	A561T	Threonine for conserved alanine, S5
		Missense mutation	I593R	Arginine for conserved isoleucine, pore extracellular loop
		Missense mutation	G601S	Serine for conserved glycine, S5-pore extra cellular loop
		Missense mutation	Y611E	Glutamic acid for conserved tyrosine, S5-pore
		Missense mutation	Y611H	Histidine for conserved tyrosine, S5-pore
		Missense mutation	A614V	Valine for conserved alanine, pore
		Missense mutation	G628S	Serine for highly conserved glycine, pore
		Missense mutation	V630L	Leucine for conserved valine, pore
		Missense mutation	V630A	Alanine for conserved valine, pore
		Intragenic deletion (2 bp)	Δ711	Frameshift, S6
		Nonessens mutation	Q725ter	Stop codon for glutamine, S6
		Missense mutation	V822M	Methionine for conserved valine, C-terminus
		Duplication (31 bp)	Dup845-856	Frameshift, affects nucleotide-binding domain (NBD)
		Missense mutation	I_{III}	Disrupts splice-donor sequence of intron III, affects NBD

Gene	Current	Mutation	Type	Description
SCN5A	I_{Na}	N1325S	Missense mutation	Serine for conserved asparagine, S4-S5 of DIII
		R1623Q	Missense mutation	Glutamine for conserved arginine, S4 DIV
		E1784K	Missense mutation	Lysine for glutamic acid in acidic domain of C-terminus (S6DIV)
		R1644H	Missense mutation	Histidine for highly conserved arginine, S4 DIV
		D1790G	Missense mutation	Glycine for conserved aspartic acid, disrupts interaction of α + β subunits
		Δ1505–1507	Intragenic deletion (9bp)	In-frame deletion of 3 conserved amino acids (KPQ) in the cytoplasmic linker between DIII and DIV
MinK	I_{Ks}	T71	Missense mutation	Isoleucine for conserved threonine
		V47F	Missense mutation	Phenylalanine for conserved valine
		L51H	Missense mutation	Histidine for conserved leucine
		T59P, L60P	Substitution ACCCTG → CCCCG	Proline at codons 59 + 60 substitution for threonine at codon 59 and leucine at codon 60
		S74L	Missense mutation	Serine for leucine substitution at codon 74
		D76N	Missense mutation	Aspartic acid substitution for asparagine at codon 76
MiRP1	I_{Kr}	Q9E	Missense mutation	Glutamic acid for glutamine in extracellular domain
		M54T	Missense Mutation	Threonine substitution for methionine
		I57T	Missense Mutation	Threonine substitution for isoleucine

The mutation notation for KVLQT1 is based on the original sequence of KVLQT1 (Ref. 26). A new notation scheme based on the newly described complete sequence (Ref. 52) adds 129 additional amino acid sequences (386 bp) to the 5′ end of the original sequence (i.e., R45C = R174C, etc.). [a]Structural features that are mutated are indicated by letters (S2, S3, etc.) referring to specific ion channel protein transmembrane segments illustrated in Figures 9–11. The sodium channel (SCN5A) protein domains are indicated by DI-DIV (Domain I-Domain IV).

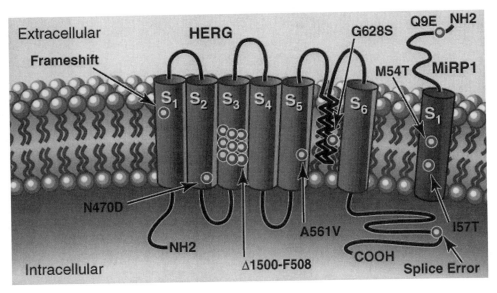

Figure 25.3   The molecular architecture of the cardiac I_{Kr} potassium channel encoded by *HERG* and *MiRP1*. Mutations are noted.

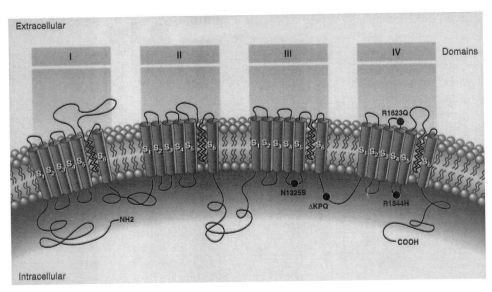

Figure 25.4   The molecular architecture of the cardiac I_{Na} sodium channel encoded by *SCN5A*. Mutations are noted.

oocytes and it was found that all mutations generated a late phase of inactivation-resistant, mexiletine-, and tetrodotoxin-sensitive, whole-cell current through multiple mechanisms (53,54). Two of the three mutations showed dispersed channel reopenings after the initial transient, but the other mutation showed both dispersed reopening and long-lasting bursts (54). These results suggested that *SCN5A* mutations act through a "gain-of-function" mechanism (the mutant channel functions normally, but because of delayed inactivation there is an additional influx of sodium current). The mechanism of chromosome 3–linked LQTS thus appears to be persistent noninactivated sodium current occurring in the plateau phase of the action potential. Later An et al. (55) showed that not all mutations in *SCN5A* are associated with persistent current. Further, mutations in *SCN5A* were identified in patients with Brugada syndrome and idiopathic VF (56); these mutations result in more rapid recovery from inactivation of the mutant channels or loss of function, causing the Brugada-syndrome-type phenotype.

4.   Cardiac Potassium Channel Gene *minK* and *LQT5*

The *minK* gene (*IsK*, or *KCNE1*), was initially localized to chromosome 21 (21q22.1) and found to consist of three exons that span approximately 40 kb. It encodes a short protein consisting of 130 amino acids and has only one transmembrane-spanning segment with small extracellular and intercellular regions (Fig. 25.2) (35–37) When expressed in *Xenopus* oocytes, it produces a potassium current that closely resembles the slowly activating delayed-rectifier potassium current, I_{Ks}, in cardiac cells (35,37). The fact that the *minK* clone was only expressed in *Xenopus* oocytes, and not in mammalian cell lines, raised the question whether *minK* is in fact a human channel protein. With the cloning of *KVLQT1* and coexpression of *KVLQT1* and *minK* in both mammalian cell lines and *Xenopus* oocytes, the molecular mystery was revealed; *KVLQT1* interacts with *minK* to form the cardiac slowly activating delayed-rectifier I_{Ks} current (29,30). *MinK* alone cannot form a functional channel but induces the I_{Ks} current by interacting with endogenous *KVLQT1* protein in *Xenopus* oocytes and mammalian cells. Immunoprecipitation experiments also demonstrated the physical interaction between *KVLQT1* and *minK* (29). Since mutations in *KLVQT1* cause chromosome 11–linked LQTS and *HERG* mutations cause *LQT2*, mutations in *minK* alone were expected to cause LQTS. This was confirmed when Splawski *et al.* (37) identified mutations in two families with LQTS. In both cases, missense mutations (S74L, D76N) were identified (Table 25.4), which reduced I_{Ks} by shifting the voltage dependence of activation and accelerating channel deactivation. The functional consequences of these mutations included delayed cardiac repolarization and, hence, an increased risk of arrhythmias.

5.   Cardiac Potassium Channel *MiRP1* and *LQT6*

MiRP1, the *KCNE2* gene product, is a novel potassium channel cloned and characterized by Abbott and colleagues (31). This small integral membrane subunit protein assembles with *HERG (LQT2)* to alter its function, enabling full development of the I_{Kr} current. *MiRP1* is a 123-amino-acid channel protein with a single predicted transmembrane segment similar to that described for *minK* (31,35). Chromosomal localization studies mapped this *KCNE2* gene to chromosome 21q22.1, within 79 kb of *KCNE1 (minK)* and arrayed in opposite orientation (31). The open reading frames of these two genes share 34% identity and both are contained in a single exon, suggesting that they are related through gene duplication and divergent evolution.

Three missense mutations associated with LQTS and ventricular fibrillation were identified in *KCNE2* by Abbott et al. (31) and biophysical analysis demonstrated that these mutants form channels that open slowly and close rapidly, thus diminishing potassium currents. In one case, the missense mutation, a C to G transversion at nucleotide 25, which produced a glutamine (Q) to glutamic acid (E) substitution at codon 9 (Q9E) in the putative extracellular domain of *MiRP1* (Table 25.4), led to the development of *torsade de pointes* and ventricular fibrillation after intravenous clarithromycin infusion (i.e., a drug-induced arrhythmia).

Therefore, like *minK*, this channel protein acts as a β-subunit but, by itself, leads to increased ventricular arrhythmia risk when mutated. These similar channel proteins (i.e., *minK* and *MiRP1*) suggest that a family of channels exist which regulate ion channel α-subunits. The specific roles of these proteins remain unclear and are currently hotly debated.

D.   Genetics and Physiology of Autosomal Recessive LQTS (Jervell and Lange-Nielsen Syndrome)

Neyroud et al. (58) discovered the first molecular abnormality in patients with Jervell and Lange-Nielsen syndrome when they studied two families with three affected children with a novel homozygous deletion–insertion mutation in *KVLQT1*. A deletion of 7 bp and an insertion of 8 bp at the same location led to premature termination at the C-terminal end of the *KVLQT1* channel. In addition, Splawski et al. (59) also identified a homozygous insertion of a single nucleotide that caused a frameshift in the coding sequence after the second putative transmembrane domain of *KVLQT1*. Together, these data strongly suggested that at least one form of JLNS is caused by homozygous mutations in *KVLQT1*. This has since been confirmed by others (36,60,61).

It is interesting that, in general, heterozygous mutations in *KVLQT1* cause Romano–Ward syndrome (LQTS only), while homozygous mutations in *KVLQT1* cause JLNS (LQTS and deafness). The likely explanation is as follows: although heterozygous *KVLQT1* mutations act by a dominant-negative mechanism, some functional *KVLQT1* potassium channels still exist in the stria vascularis of the inner ear. Therefore, congenital deafness is averted in patients with heterozygous *KLVQT1* mutations. For patients with homozygous *KVLQT1* mutations, the idea is that no functional *KVLQT1* potassium channels can be formed. It has been showed by in situ hybridization that *KVLQT1* is expressed in the inner ear (58), suggesting that homozygous *KVLQT1* mutations can cause the dysfunction of potassium secretion in the inner ear and lead to deafness. However, because of incomplete penetrance, not all heterozygous or homozygous mutations follow this rule (8,20).

Schulze-Bahr et al. (62) showed that mutations in *minK* result in JLNS syndrome as well. Hence, abnormal I_{Ks} current, whether it occurs due to homozygous mutations in *KVLQT1* or the *minK* subunit, results in LQTS and deafness.

E.   Genotype–Phenotype Correlations

Zareba et al. (63) have recently shown that the particular gene that is mutated results in characteristic phenotypes whose identity can help predict outcome. For instance, these authors suggested that mutations in the *LQT1* and *LQT2* genes result in early symptoms (i.e., syncope) but the risk of sudden death seems relatively low. In contrast, mutations in *LQT3* result in a few symptoms, but when symptoms occur they are associated with a high likelihood of sudden death. In addition, mutations in *LQT1* and *LQT2* appear associated with stressed-induced symptoms, including response to auditory triggers. Events in *LQT3* patients, on the other hand, appear associated with sleep-associated symptoms. Coupled with the findings by Moss et al. (64) that differences in ECG patterns could be identified based on the gene mutated, it could be suggested that understanding the specific mutated gene that causes LQTS in any individual could be used to improve diagnosis and survival.

F. Genetic Testing

Currently six LQTS-causing genes have been identified and more than 100 mutations described. This genetic heterogeneity makes genetic testing more difficult than if a single gene defect were responsible for the disease. However, under certain conditions genetic testing can be performed. In large families in which linkage analysis can be performed, identification of the gene of interest (if the linkage is to one of the known genes) can be discerned relatively rapidly and screening of mutations undertaken. Once a mutation is identified in one affected family member (usually the initially presenting proband patient), remaining family members can be screened for this mutation quickly. In small families or sporadic cases, mutation screening for all known genes is usually required. *KVLQT1 (LQT1)* mutations are usually screened first since this appears to be the most common disease-causing gene (65). If no mutation is uncovered in *KVLQT1*, *HERG* (the next most common gene mutated) is screened before *SCN5A*, *KCNE1*, and *KCNE2* (31,34,61,66). If no mutation is found in any of these genes, however, one cannot conclude that the individual does not have LQTS, since other disease-causing genes remain to be discovered. In the future, technological advances are likely to enable more rapid, automated screening to proceed. These advances, such as "chip" technology, could take genetic testing from the research laboratory to the diagnostic laboratory ultimately.

G. Therapy

Currently, the standard therapeutic approach with suspected or confirmed LQTS patients is the initiation of β-blockers at the time of diagnosis (1,2). In cases in which β-blockers cannot be used, such as in patients with asthma, other drugs such as mexiletine, have been tried. When medical therapy fails, left sympathectomy or implantation of an automatic cardioverter defibrillator (ICD) have been utilized.

Recently, genetic-based therapy has been attempted. One group (67) showed that sodium channel blocking agents (i.e., mexiletine) shorten the QTc in patients with *LQT3*, while others (68,69) have demonstrated that exogenous potassium supplementation or potassium-channel openers, respectively, may be useful in patients with potassium channel defects. However, no definitive evidence that these approaches improve survival has yet been published.

H. LQTS Summary

The long-QT syndromes are genetically and clinically heterogeneous. The affected gene in any patient can lead to a wide spectrum of clinical outcomes depending on its specific mutation. These mutations, however, remain difficult to identify, but once known can be useful in therapy. Gene-specific therapy may be an option in the future.

III. BRUGADA SYNDROME

A. Clinical Aspects of Brugada Syndrome

Multiple other causes of ventricular arrhythmias and sudden cardiac death exist (70). Although in many instances the underlying disorder remains unknown, new information concerning some of the potential etiologies is emerging. One of these disorders is idiopathic ventricular fibrillation (IVF) or Brugada syndrome. The first identification of ECG pattern of right bundle branch block (RBBB) with ST elevation in leads V1–V3 was reported in three apparently healthy males (71). Shortly thereafter, persistent ST elevation without RBB in another 10 asymptomatic males and ST elevation in the right chest leads and conduction block in the right ventricle in patients with severe hyperkalemia were reported. Although multiple other reports of patients with variations of this ECG pattern exist (74–78), the association of the ECG abnormality with sudden death was largely ignored until Martini et al. (79) and Aihara et al. (80) focused attention on the possible link; it was confirmed in 1991 by Pedro and Josep Brugada (81) who described four patients with sudden and aborted sudden death with ECGs demonstrating RBBB and persistent ST elevation in leads V1–V3 (Fig. 25.5). In 1992, these authors characterized what they believed to be a distinct clinical and electrocardiographic syndrome (3).

The finding of ST elevation in the right chest leads has been observed in a variety of clinical and experimental settings and is not unique or diagnostic of Brugada syndrome by itself. Situations in which these ECG findings occur include electrolyte or metabolic disor-

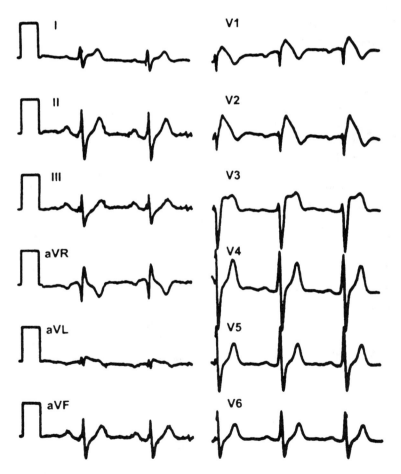

Figure 25.5   ECG manifestations of Brugada syndrome on the 12-lead ECG. Note the ST-segment elevation in leads V1–V3. Right bundle branch block can also occur. The symptoms in Brugada syndrome occur due to ventricular fibrillation.

ders, pulmonary or inflammatory diseases, abnormalities of the central or peripheral nervous system. In the absence of these abnormalities, the term "idiopathic ST elevation" is often used and may identify Brugada syndrome patients. The prevalence of idiopathic ST elevation varies from 2.1 to 2.65% (82,83), with elevation of the ST segment limited to the right precordial leads occurring in less than 1% of all cases of ST elevation.

ECGs with these characteristics associated with unexpected sudden deaths have been reported commonly in Japan and Southeast Asia, frequently affecting men during sleep (83). This disorder, known as sudden and unexpected death syndrome (SUDS) or sudden unexpected nocturnal death syndrome (SUNDS), has many names in Southeast Asia, including "bangungut" (to rise and moan in sleep) in the Philippines; "non-laitai" (sleep-death) in Laos; "lai-tai" (died during sleep) in Thailand; and "pokkuri" (sudden and unexpectedly ceased phenomena) in Japan. General characteristics of SUDS patients include young, healthy males in whom death occurs suddenly with a groan, usually during sleep late at night. No precipitating factors are identified and autopsy findings are generally negative (84). Life-threatening ventricular tachyarrhythmias as a primary cause of SUDS has been demonstrated, with VF occurring in many cases (85).

B. Brugada Syndrome and Arrhythmogenic Right Ventricular Dysplasia

Controversy exists concerning the possible association of Brugada syndrome and arrhythmogenic right ventricular dysplasia (ARVD), with some investigators arguing that these are the same disorder or that one is a forme-fruste of the other (86–91). However, the classic echocardiographic, angiographic, and magnetic resonance imaging findings of ARVD are not seen in Brugada syndrome patients. In addition, Brugada syndrome patients typically are without the histopathological findings of ARVD. Further, the morphology of the VT/VF pattern appears to differ (92).

C. Clinical Genetics of Brugada Syndrome

Like ARVD, most of the families thus far identified with Brugada syndrome appear to have an autosomal dominant pattern of inheritance (88,92–94). In these families, approximately 50% of offspring of affected patients develop the disease. Although the number of families reported has been small, it is likely that this is due to under-recognition, as well as premature and unexpected death (94,95).

D. Molecular Genetics of Brugada Syndrome

To identify the gene(s) responsible for Brugada syndrome, moderate-size families with excellent clinical surveillance are required for gene-mapping studies that can then guide subsequent positional cloning or positional candidate gene cloning. In Brugada syndrome, several good candidate genes exist. In animal studies, blockade of the calcium-independent 4-aminopyridine-sensitive, transient outward potassium current (I_{to}) results in surface ECG findings of elevated, downsloping ST segments (95,96). This appears to be due to greater prolongation in the epicardial action potential compared to the endocardium, which lacks a plateau phase (97). Loss of the action potential plateau (or dome) in the epicardium, but not endocardium, would be expected to cause ST-segment elevation because loss of the dome is caused by an outward shift in the balance of currents active at the end of phase 1 of the action potential (principally I_{to} and I_{Ca}). Autonomic neurotransmitters, like acetylcholine, can facilitate loss of the action potential dome by suppressing calcium current and augmenting potassium current, whereas β-adrenergic agonists (i.e., isoproterenol, dobutamine) restore the dome by augmenting I_{Ca} (98,99). Sodium channel blockers also facilitate loss of the canine right ventricular action potential dome as a result of a negative shift in the voltage at which phase 1 begins (100,101). Hence, genes underlying I_{to}, I_{Ca}, and I_{Na} are excellent candidates to study. Since defects in I_{Na} (*SCN5A*) have previously been shown to cause VT/VF in humans with the long-QT syndrome (51,52), this gene became the subject of much active investigation.

The laboratory of one of the authors (JAT) reported findings on six families and several sporadic cases of Brugada syndrome (56). The families were initially studied by linkage analysis using markers to known ARVD loci on chromosomes 1 (102) and 14 (103,104), and linkage to these sites was excluded. More recently, seven other families were also studied who did not show any linkage to known ARVD loci. This suggests that many of the families with Brugada syndrome, may indeed be affected by an entity distinct from ARVD. Candidate gene screening using the mutation analysis approach of single-strand conformation polymorphism (SSCP) analysis and DNA sequencing was performed. In three families, mutations in *SCN5A* were identified (56), including: (1) a missense mutation (C → T base substitution) causing a substitution of a highly conserved threonine by methionine at codon 1620 (T1620M) in the extracellular loop between transmembrane segments S3 and S4 of domain IV (see Fig. 25.4, IVS3–IVS4), an area important for coupling of channel activation to fast inactivation; (2) a two nucleotide insertion (AA) that disrupts the splice-donor sequence of intron 7; and (3) a single nucleotide deletion (A) at position 1397, which results in an in-frame stop codon that eliminates IIIS6, IVS1–IVS6, and the carboxy-terminus of *SCN5A* (Fig. 25.6).

Electrophysiological analysis of these mutant proteins in *Xenopus* oocytes demonstrated a reduction in the number of functional sodium channels for both the splicing mutation and one-nucleotide deletion mutation, which could promote development of reentrant arrhythmias. Examination of the missense mutation showed sodium channels that recover from inactivation more rapidly than normal. In this case, the presence of both normal and mutant channels in the same tissue would increase heterogeneity of the refractory

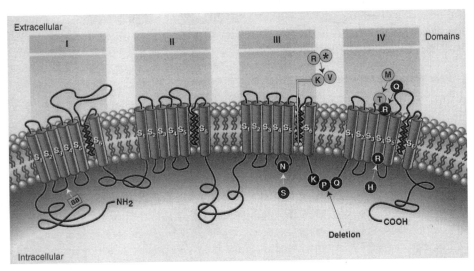

Figure 25.6   The molecular architecture of the cardiac I_{Na} sodium channel encoded by *SCN5A* and the mutations causing *LQT3* (black) and Brugada syndrome (gray).

period, a well-established mechanism of arrhythmogenesis. Inhibition of sodium channel current causes heterogeneous loss of the action potential dome in the right ventricular epicardium, leading to a marked dispersion of depolarization and refractoriness, an ideal substrate for development of reentrant arrhythmias. Phase 2 reentry produced by the same substrate is believed to provide the premature beat necessary for initiation of the VT and VF responsible for symptoms in these patients.

E. Brugada Syndrome Summary

It is important to note that both Brugada syndrome patients and LQT3 patients studied to date have abnormalities in the sodium ion channel gene *SCN5A*. In Brugada syndrome, mutations in this gene result in a loss in function of the channels or rapid recovery from inactivation. Unlike Brugada syndrome, *LQT3* occurs due to a gain of function in *SCN5A*, where persistence of inactivation is seen. The differences in the clinical findings between *LQT3* and Brugada syndrome are thus the phenotypically different biophysical result of a difference in the position of the mutations within this gene. Important similarities should be noted, however. In particular, both disorders cause life-threatening ventricular tachyarrhythmias. The molecular similarities involved should be helpful to identify any remaining genes responsible for both disorders.

IV. ATRIAL FIBRILLATION

Atrial fibrillation (AF) is the most common sustained cardiac arrhythmia, affecting more than 2 million persons in the United States alone (105). The overall prevalence of AF is 0.89% but rises with age from 0.05% among people aged 25 to 35 years to more than 5% among those over age 65 years (106,107). In addition, AF is more common in men than women. The racial differences are unknown.

A. Clinical Aspects of Atrial Fibrillation

AF usually presents with palpitations, exercise intolerance, and heart failure symptoms. In one study of hospitalized patients, 52% had dyspnea, 34% presented with chest pain, 26% had palpitations, and 19% had dizziness or syncope. When associated with preexisting cardiac disease, serious clinical deterioration may occur due to AF. In certain patients, such as those with Wolff–Parkinson–White syndrome, the accessory pathway may enable to high rate of atrial activation during AF to be directly transmitted to the ventricles and lead to ventricular fibrillation (109). In patients with hypertrophic cardiomyopathy, AF occurs in 15% and results in significant clinical worsening due to associated cardiac failure and thromboembolic risk. Similarly, AF can cause clinical deterioration in patients with restrictive cardiomyopathy or following the Fontan operation.

B. Disease Classification

The nomenclature used to describe the clinical syndromes of AF is confusing. AF can be broadly classified as "acute" when it exists for a few days. "Paroxysmal" AF is characterized by recurrent episodes that resolve spontaneously or require intervention; this form represents nearly one-fourth of nonacute cases and progresses to "chronic" AF within 2 years in over 10% of patients (110). Chronic AF is of long duration. "Lone" AF includes a subset of patients lacking clinical, ECG, or echocardiographic evidence of cardiovascular risk factors (e.g., rheumatic valve disease, structural heart disease, heart failure, hypertensive heart disease, pericarditis, etc.) and occurs in up to 17% of all cases of AF. Another group of patients have an inherited form of AF; up until recently, this group has been considered quite rare. The inheritance of this disorder is autosomal dominant (111).

C. Molecular Aspects of Familial Atrial Fibrillation

Brugada et al. (111) studied three families with familial AF (FAF) and using linkage analysis, were able to localize a gene for FAF on a region of the long arm of chromosome 10 (10q22–q24). Interestingly, this region corresponds to a genetic locus for autosomal dominant dilated cardiomyopathy described by Bowles et al. (112). However, the gene for both disorders remains elusive. Using positional cloning and candidate gene positional cloning approaches, the region of interest for AF has been narrowed significantly; candidate genes under study include those encoding channel proteins and those encoding molecular components of the sympathetic or parasympathetic nervous system that influence cardiac conduction or impulse initiation. Since the burdens imposed by long-term medical therapy and associated complications associated with AF, such as strokes, are immense, with a total annual cost upward of $9 billion, identification of the gene(s) responsible for this disorder can be expected to have a major impact on health care.

IV. HYPERTROPHIC CARDIOMYOPATHY

Hypertrophic cardiomyopathy is a complex cardiac disease with unique pathophysiological characteristics and a great diversity of morphological, functional, and clinical features (113,114). Although hypertrophic cardiomyopathy has been regarded largely as a relatively uncommon cardiac disease, the prevalence of echocardiographically defined hypertrophic cardiomyopathy in a large cohort of apparently healthy young adults selected from a community-based general population was reported recently to be as high as 0.2% (115). Familial disease with autosomal dominant inheritance predominates.

A. Clinical Aspects of Familial Hypertrophic Cardiomyopathy

Observations of myocardial diseases that can reasonably be interpreted as hypertrophic cardiomyopathy were made in the middle of the last century at the Hospital La Salpêtrière in Paris by Vulpian, who called what he saw at the macroscopic level a "rétrécissement de l'orifice ventriculo aortique" (subaortic stricture) (116). It was, however, only in the late 1950s that the unique clinical features of hypertrophic cardiomyopathy were systematically described. It is characterized by left and/or right ventricular hypertrophy, which is usually asymmetric and which can affect different regions of the ventricle. The interventricular septum is most commonly affected, with or without involvement of either the anterior wall or the posterior wall in continuity. A particular form of regional involvement affects the apex, but spares the upper portion of the septum (apical hypertrophy) (113). Typically, left ventricular volume is normal or reduced. Systolic gradients are common. Typical morphological changes include myocyte hypertrophy and disarray surrounding the areas of increased loose connective tissue. Patients with HCM frequently report a reduced exercise capacity and functional limitations. Though the pathophysiological features of the disease that contribute to this limitation are complex and not fully understood, left ventricular outflow tract obstruction, if present, is believed to contribute to increased filling pressures and a failure to augment cardiac output during exercise, leading to exertional symptoms. Arrhythmias and premature sudden deaths are common (114,115).

B. Mapping of Familial Hypertrophic Cardiomyopathy Genes

The first gene for familial hypertrophic cardiomyopathy (FHC) was mapped to chromosome 14q11.2–q12 using genome-wide linkage analysis in a large Canadian family (117). Soon afterward, FHC locus heterogeneity was reported (118,119) and subsequently confirmed by the mapping of the second FHC locus to chromosome 1q3 and of the third locus to chromosome 15q2 (120,121). Carrier et al. (122) mapped the fourth FHC locus to chromosome 11p11.2. Four other loci were subsequently reported, located on chromosomes 7q3 (123), 3p21.2–3p21.3 (124), 12q23–q24.3 (125), and 15q14 (126). Several other families are not linked to any known FHC loci, indicating the existence of additional FHC-causing genes.

C. Gene Identification in Familial Hypertrophic Cardiomyopathy

All known FHC disease genes encode proteins that are part of the sarcomere, the elemental unit of contraction and a highly complex structure. The sarcomere is composed of multiple proteins organized in an exact stoichiometry as shown in Table 25.5 (see review, Ref. 127). These sites affected in FHC consist of three myofilament proteins, the β-myosin heavy chain (β-MyHC), the ventricular myosin essential light chain 1 (MLC-1s/v), and the ventricular myosin regulatory light chain 2 (MLC-2s/v); four thin filament proteins—cardiac actin, cardiac troponin T (cTnT), cardiac trooponin 1 (Ctnl), and α-tropomyosin (α-TM); and, finally, the cardiac myosin binding protein C (cMyBP-C). Each of these proteins is encoded by multigene families that exhibit tissue-specific, developmental, and physiologically regulated patterns of expression.

1. Thick Filament Proteins

Myosin Subunits

Myosin is the molecular motor that transduces energy from the hydrolysis of ATP into directed muscle fiber movement and, in so doing, drives sarcomere shortening and muscle contraction. Cardiac myosin consists of two heavy chains (MyHC) and two pairs of light chains (MLC), referred to as essential (or alkali) light chains (MLC-1) and regulatory (or phosphorylatable) light chains (MLC-2), respectively (see review, Ref. 128). The myosin molecule is highly asymmetrical, consisting of two globular heads joined to a long rodlike

Table 25.5   FHC Loci and Disease Genes

Locus	Gene	Protein	Expression in adult striated muscles
14q11-12	*Myh7*	β-myosin heavy chain	Cardiac and slow-twitch
1q3	*TNNT2*	Cardiac troponin T	Cardiac and slow-twitch
15q2	*TPM1*	α-tropomyosin	Cardiac and fast-twitch
11p11.2	*MYBPC3*	Cardiac myosin binding protein C	Cardiac
12q23q24.3	*MYL2*	Cardiac myosin regulatory light chain	Cardiac and slow-twitch
3p	*MYL3*	Cardiac myosin essential light chain	Cardiac and slow-twitch
19p13.2-q13.2	*TNNC1*	Cardiac troponin I	Cardiac
15q14	ACTC	Cardiac actin	Cardiac
7q3	?	?	?

tail. The light chains are arranged in tandem in the head–tail junction. Their function is not fully understood. Neither myosin light chain type is required for the adenosine triphosphatase (ATPase) activity of the myosin head, but they probably modulate it in the presence of actin and contribute to the rigidity of the neck, which is hypothesized to function as a lever arm for generating an effective power stroke. Mutations have been found in the heavy chains and in the two types of ventricular light chains (127).

Concerning the heavy chains, the β-isoform (β-MyHC) is the major isoform of the human ventricle and slow-twitch skeletal fibers. It is encoded by *MYH7*. At least 50 mutations have been found in unrelated families with FHC (Fig. 25.7), and three "hot spots" or areas with a high frequency of mutation were identified in codons 403, 719, and 741. Most of the mutations are missense mutations located either in the head or in the head–rod junction of the molecule. Three exceptions to this are two 3-bp deletions that do not disrupt the reading frame, one of codon 10 and the other of codon 930, and a 2.4-kb deletion in the 3′ region. In kindred with the latter mutation, only the proband had developed clinically diagnosed hypertrophic cardiomyopathy at a very late age of onset (59 years).

As for the light chains, the isoforms expressed in the ventricular myocardium and in the slow-twitch muscles are the so-called ventricular myosin regulatory light chains (MLC-2s/v) encoded by *MYL2*, and the ventricular myosin essential light chain (MLC-1s/v) encoded by *MYL3*. They both belong to the superfamily of EF-hand proteins. Two missense mutations have been reported in *MYL3*, and five in *MYL2* (124) (Fig. 25.8).

Myosin Binding Protein C (MyBP-C)

MyBP-C is part of the thick filaments of the sarcomere, being located at the level of the transverse stripes, 43 nm apart, seen by electron microscopy in the sarcomere A band. Its function is uncertain, but, for a decade, evidence has existed to indicate potential structural and regulatory roles. Partial extraction of cMyBP-C from rat skinned cardiac myocytes and rabbit skeletal muscle fibers alters Ca^{2+}-sensitive tension (129), and it was shown that phosphorylation of cMyBP-C alters myosin cross-bridges in native thick filaments, suggesting that cMyBP-C can modify force production in activated cardiac muscles.

The cardiac isoform of this protein is encoded by the *MYBPC3* gene that was analyzed extensively in one of the author's (KS) laboratories recently determined its. Subsequently, Gautel et al. (131) showed that three distinct regions are specific to the cardiac isoform: the NH_2-terminal domain C0 Ig-I containing 101 residues, the MyBP-C motif (a 105-residue stretch linking the C1 and C2 Ig-I domains), and a 28-residue loop inserted in the C5 Ig-I domain (131). It was also shown that cMyBP-C is specifically expressed in the heart during human and murine development (132,133).

At least 27 *MYBPC3* mutations have been identified in unrelated families with FHC (Fig. 25.9). Seventeen of these result in aberrant transcripts that are predicted to encode COOH-terminal truncated cardiac MyBP-C polypeptides lacking at least the myosin-binding domain. Seven others result in mutated or deleted proteins without disruption of the reading frame: five are missense mutations in exons 6, 17, 21, and 23, one is a splice donor site mutation in intron 27, and one is an 18-residue duplication in exon 33. Finally, three mutations are predicted to produce either a mutated protein or a truncated one: two are missense mutations in exons 15 and 17 and one is a branch point mutation in intron 23.

2. Thin-Filament Proteins

The thin filament contains actin, the troponin (Tn) complex, and tropomyosin (TM). The troponin complex and tropomyosin constitute the Ca^{2+}-sensitive switch that regulates the

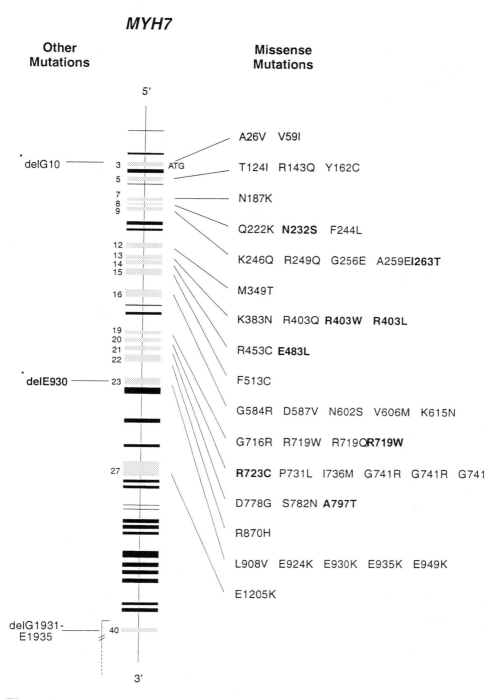

Figure 25.7 Genomic structure of *MYH7*. Mutations are noted. Bold characters indicate mutations found within the French INSERM collaborative network.

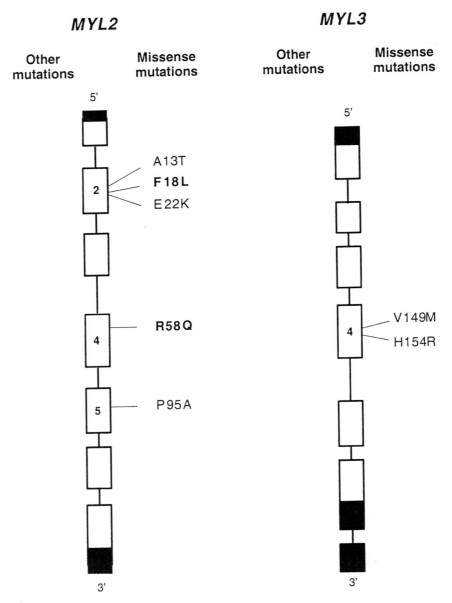

Figure 25.8 Genomic structure of *MYL2* and *MYL3*. Mutations are noted. Bold characters indicate mutations found within the French INSERM collaborative network.

contraction of cardiac muscle fibers. Mutations have been found in α-TM and in two of the subunits of the cardiac troponin complex: cTnl, the inhibitory subunit, and cTnT, the tropomyosin-binding subunit. Recently, actin mutations have also been reported (126).

α-TM is encoded by the *TPM1* gene. The cardiac isoform is expressed both in the ventricular myocardium and in fast-twitch skeletal muscles (134). It shares the overall structure of other tropomyosins that are rodlike proteins that possess a simple dimeric α-coiled coil structure in parallel orientation along their entire length (134). Four missense

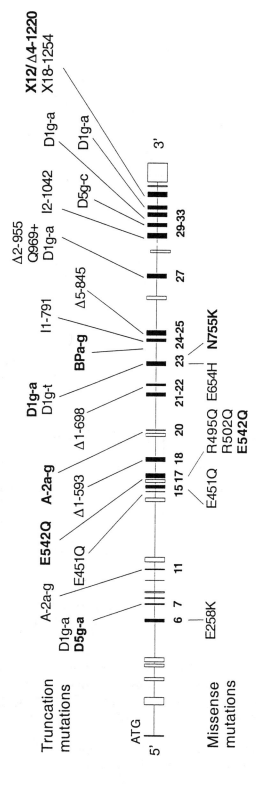

Figure 25.9 Genomic structure of *MYBPC3*. Mutations are noted. Bold characters indicate mutations found within the French INSERM collaborative network.

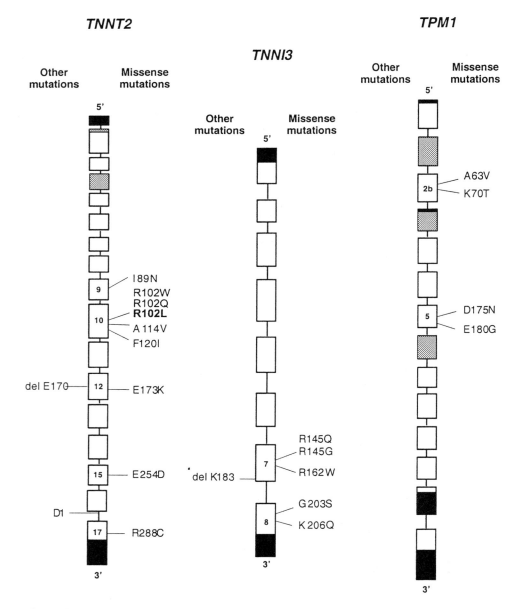

Figure 25.10   Genomic structure of *TNNT₂*, *TNNI₃*, and *TPM1*. Mutations are noted. Bold characters indicate mutations found within the French INSERM collaborative network.

mutations were found in unrelated FHC families (Fig. 25.10). Two of them, A63V and K70T, are located in exon 2b within the consensus pattern of sequence repeats of α-TM and could alter tropomyosin binding to actin. Mutations D175N and E180G are both located within constitutive exon 5, in a region near the C190 and near the calcium-dependent TnT binding domain.

cTnT is encoded by the *TNNT2* gene. In human cardiac muscle, multiple isoforms of cTnT have been described which are expressed in the fetal, adult, and diseased heart, and which result from alternative splicing of the single gene *TNNT2* (135,136). The precise physiological relevance of these isoforms is currently poorly understood, but the organization of the human gene has been partially established, thus allowing identification of the position of those mutations within exons, including those alternatively spliced during development; it also enables an amino acid numbering that reflects the full coding potential of human *TNNT2* (137,138). Eleven mutations have been found in unrelated FHC families, three of which are located in a hot spot (codon 102) (Fig. 25.10). Ten mutations are missense types located between exons 9 and 17, one mutation is a 3-bp deletion located in exon 12 that does not disrupt the coding frame, and the last is located in the intron 16 splice donor site and is predicted to produce a truncated protein in which the C-terminal binding sites are disrupted.

cTnI is encoded by the *TNNI3* gene. The cTnI isoform is expressed only in cardiac muscle (139). Cooperative binding of cTnI-actin-tropomyosin is a unique property of the cardiac variant (see review, Ref. 140). Six mutations were recently identified (Fig. 25.10). Five are missense mutations located in exons 7 and 8, and one is a K183D mutation that does not disrupt the coding frame.

Alpha-cardiac actin (ACTC) was recently identified as another cause of FHC. Mogensen et al. (126) studied a family with heterogeneous phenotypes, ranging from asymptomatic with mild hypertrophy, to pronounced septal hypertrophy and left ventricular outflow tract obstruction. Using linkage analysis and mutation screening, the gene was mapped to chromosome 15q14, identified as a missense mutation (G → T in position 253, exon 5), and resulted in an Ala295Ser amino acid substitution. The mutation is localized at the surface of actin in proximity to a putative myosin-binding site. This mutation, which causes FHC, differs from the two mutations reported to cause dilated cardiomyopathy (DCM) (141), which were localized in the immobilized end of actin that cross-binds to the anchor polypeptides in the Z bands.

D. Genotype–Phenotype Relations in Familial Hypertrophic Cardiomyopathy

The pattern and extent of left ventricular hypertrophy in patients with HCM vary greatly even in first-degree relatives with similar mutations, and a high incidence of sudden deaths has been reported in many selected families. An important issue, therefore, is to determine whether the genotype heterogeneity observed in FHC accounts for the phenotypic diversity of the disease. However, present conclusions must be viewed as preliminary because the available data relate to only a few hundred individuals, and it is obvious that although a given phenotype may be apparent in a small family, examining large or multiple families with the same mutation is required before drawing unambiguous conclusions. Nevertheless, several concepts have begun to emerge, at least for mutations in the *MYH7*, *TNNT2*, and *MYBPC3* genes. For *MYH7*, it is clear that prognosis for patients with different mutations varies considerably (reviewed in Ref. 142). For example, the R403Q mutation appears to be associated with markedly reduced survival (143), whereas some others, such as V606M, appear more benign (144). The disease caused by *TNNT2* mutations is usually associated with a 20% incidence of nonpentrance, a relatively mild and sometimes subclinical hypertrophy, but a high incidence of sudden death that can occur even in the absence of significant clinical left ventricular hypertrophy (120,145,146). In one family with a *TNNT2* mutation with complete penetrance, echocardiographic data showed a wide range of hypertrophy, yet there was no sudden cardiac death (138). Mutations in *MYBPC3* seem to be characterized by specific clinical features with a mild phenotype in young subjects, a delayed age at the onset of symptoms, and a favorable prognosis before the age of 40 years (147–150).

Genetic studies have also revealed the presence of apparently clinically healthy individuals who carry a mutant allele associated in first-degree relatives with a typical deleterious phenotype. A number of mechanisms have been discussed that could account for the large variability in phenotypic expression of these mutations: environmental differences, for example, or acquired traits (e.g., differences in lifestyle, risk factors, and exercise), or the existence of modifier genes or polymorphisms in other proteins that could modulate expression of the disease. The only significant results obtained so far concern the influence of an angiotensin-I converting enzyme insertion/deletion (ACE I/D) polymorphism. Association studies show that, compared to a control population, the D allele is more common in patients with hypertrophic cardiomyopathy and in patients with a high incidence of sudden cardiac death (151,152). An association between the D allele and hypertrophy has been observed in the case of *MYH7* R403 codon mutations, but not with *MYBPC3* mutation carriers (153), raising the concept of multiple genetic modifiers in FHC.

V. CONCLUSIONS

Genetic studies of inherited diseases in which arrhythmias are prominent features have deepened our understanding of these disorders by many orders of magnitude and have provided completely new insights into their pathogenesis and the heterogeneity of their clinical features. In all these diseases, the clinical phenotype, including outcome, appears to differ based on the gene mutated and the specific mutation. Similar findings are emerging in studies on familial dilated cardiomyopathy (FDCM) as well (154). From all these, it appears that affecting a particular protein at any point within the molecular pathways required for a specific cardiac function (i.e., contractile apparatus resulting in mechanical function; ion channels resulting in cardiac rhythm; cytoskeletal proteins resulting in cardiac structural organization) results in a spectrum of similar diseases. Thus, contractile apparatus mutations cause FHC; ion channel mutations result in LQTS and Brugada syndrome; mutations in cytoskeletal protein genes result in FDCM. This "final common pathway" hypothesis should be valuable in helping to identify the remaining genes contributing to these and other cardiac disorders. In addition, understanding the molecular bases of these conditions has raised new and compelling questions about the optimal management of genotypically affected individuals who are considered to be at particularly high risk for life-threatening events. Future therapies based on these discoveries offer great promise for the prevention of sudden death over years to come.

REFERENCES

1. Priori SG, Barhanin J, Hauer RNW, et al. Genetic and molecular basis of cardiac arrhythmias: impact on clinical management (Parts I and II). Circulation 1999; 99:518–528.
2. Schwartz PJ, Locati EH, Napolitano C, Priori SG. The long QT syndrome. In: Zipes DP, Jalife J, eds. Cardiac Electrophysiology: From Cell to Bedside. Philadelphia: WB Saunders Co., 1996; 72:788–811.
3. Brugada P, Brugada J. Right bundle-branch block, persistent ST segment elevation and sudden cardiac death: a distinct clinical and electrocardiographic syndrome. A multicenter report. J Am Coll Cardiol 1992; 20:1391–1396.
4. Schwartz PJ, Moss AJ, Vincent GM, Crampton RS. Diagnostic criteria for the long QT syndrome: an update. Circulation 1993; 88:782–784.

5.   Romano C, Gemme G, Pongiglione R. Antmie cardiache rare in eta pediatrica. Clin Pediatr 1963; 45:656–683.

6.   Ward OC. A new familial cardiac syndrome in children. J Ir Med Assoc 1964; 54:103–106.

7.   Jervell A, Lange-Nielsen F. Congenital deaf-mutism, function heart disease with prolongation of the Q-T interval and sudden death. Am Heart J 1957; 54:59–68.

8.   Priori SG, Napolitano C, Schwartz PJ. Low penetrance in the long-QT syndrome. Clinical impact. Circulation 1999; 99:529–533.

9.   Ratshin RA, Hunt D, Russell RO Jr., Rackley CE. QT-interval prolongation, paroxysmal ventricular arrhythmias, and convulsive syncope. Ann Intern Med 1971; 75:19–24.

10. Singer PA, Crampton RS, Bass NH. Familial Q-T prolongation syndrome: Convulsive seizures and paroxysmal ventricular fibrillation. Arch Neurol 1974; 31:64–66.

11. Bellavere F, Ferri M, Fuarini L, Bax F, Piccoli A, Fedele D. Prolonged QT period in diabetic autonomic neuropathy: A possible role in sudden cardiac death. Br Heart J 1988; 59:379–383.

12. Ewing DJ, Boland O, Neilson JMM, Cho CG, Clarke BF. Autonomic neuropathy, QT interval lengthening, and unexpected deaths in male diabetic patients. Diabetologia 1991; 34:182–185.

13. Weintraub G, Gow RM, Wilkinson JL. The congenital long QT syndromes in children. J Am Coll Cardiol 1990; 16:674–680.

14. Marks ML, Trippel DL, Keating MT. Long QT syndrome associated with syndactyly identified in females. Am J Cardiol 1995; 10:744–745.

15. Schwartz PJ, Segantini A. Cardiac innervation, neonatal electrocardiography and SIDS. A key for a novel preventive strategy? Ann NY Acad Sci 1988; 338:1709–1714.

16. Schwartz PJ, Stramba-Badiale M, Segantini A, et al. Prolongation of the QT interval and the Sudden Infant Death Syndrome. N Engl J Med 1998; 338:1709–1714.

17. Towbin JA, Friedman RA. Prolongation of the Long QT Syndrome and Sudden Infant Death Syndrome. N Engl J Med 1998; 338:1760–1761.

18. Jervell A. Surdocardiac and related syndromes in children. Adv Intern Med 1971; 17:425–438.

19. James TN. Congenital deafness and cardiac arrhythmias. Am J Cardiol 1967; 19:627–643.

20. Priori SG, Schwartz PJ, Napolitano C, et al. A recessive variant of the Romano-Ward long-QT syndrome. Circulation 1998; 97:2420–2425.

21. Keating MT, Atkinson D, Dunn C, Timothy K, Vincent GM, Leppert M. Linkage of a cardiac arrhythmia, the long QT syndrome, and the Harvey ras-1 gene. Science 1991; 252:704–706.

22. Keating MT, Atkinson D, Dunn C, Timothy K, Vincent GM, Leppert M. Consistent linkage of the long QT syndrome to the Harvey ras-1 locus on chromosome 11. Am J Hum Genet 1991; 49:1335–1339.

23. Towbin JA, Pagotto L, Siu B, et al. Romano-Ward long QT syndrome (RWLQTS): evidence of genetic heterogeneity. Pediatr Res 1992; 31:23A.

24. Benhorin J, Kalman YM, Madina A, et al. Evidence of genetic heterogeneity in the long QT syndrome. Science 1993; 260:1960–1962.

25. Curran ME, Atkinson D, Timothy K, et al. Locus heterogeneity of autosomal dominant long QT syndrome. J Clin Invest 1993; 92:799–803.

26. Towbin JA, Li H, Taggart T, et al. Evidence of genetic heterogeneity in Romano-Ward long QT syndrome: analysis of 23 families. Circulation 1994; 90:2635–2644.

27. Jiang C, Atkinson D, Towbin JA, et al. Two long QT syndrome loci map to chromosome 3 and 7 with evidence for further heterogeneity. Nature Genet 1994; 8:141–147.

28. Schott J, Charpentier F, Peltier S, et al. Mapping of a gene for long QT syndrome to chromosome 4q25–27. Am J Hum Genet 1995; 57:1114–1122.

29. Barhanin J, Lesage F, Guillemare E, Finc M, Lazunski M, Romey G. KVLQT1 and IsK (minK) proteins associate to form the I_{Ks} cardiac potassium current. Nature 1996; 384:78–80.

30. Sanguinetti MC, Curran ME, Zou A, et al. Coassembly of KVLQT1 and minK (IsK) proteins to form cardiac I_{Ks} potassium channel. Nature 1996; 384:809–813.

31. Abbott GW, Sesti F, Splawski I, et al. MiRP1 forms I_{Kr} potassium channels with HERG and is associated with cardiac arrhythmia. Cell 1999; 97:175–187.

32. Wang Q, Curran ME, Splawski I, et al. Positional cloning of a novel potassium channel gene: KVLQT1 mutations cause cardiac arrhythmias. Nature Genet 1996; 12:17–23.

33. Li H, Chen Q, Moss AJ, et al. New mutations in the KVLQT1 potassium channel that cause long QT syndrome. Circulation 1998; 97:1264–1269.

34. Choube C, Neyroud N, Guicheney P, Lazdunski M, Romey G, Barhanin J. Properties of KVLQT1 K^+ channel mutations in Romano-Ward and Jervell and Lange-Nielsen inherited cardiac arrhythmias. EMBO J 1997; 16:5472–5479.

35. Honore E, Attali B, Heurteaux C, et al. Cloning, expression, pharmacology and regulation of a delayed rectifier K^+ channel in mouse heart. EMBO J 1991; 10:2805–2811.

36. Wollnick B, Schreeder BC, Kubish C, Esperer HD, Wieacker P, Jensch TJ. Pathophysiological mechanisms of dominant and recessive KVLQT1 K^+ channel mutations found in inherited cardiac arrhythmias. Hum Molec Genet 1997; 6:1943–1949.

37. Splawski I, Tristani-Firouzi M, Lehmann MH, Sanguinetti MC, Keating MT. Mutations in the minK gene cause long QT syndrome and suppress I_{Ks} function. Nature Genet 1997; 17:338–340.

38. Warmke JE, Ganetzky B. A family of potassium channel genes related to eag in Drosophila and mammals. Proc Natl Acad Sci USA 1994; 91:3438–3442.

39. Arcangeli A, Rosati B, Cherubini A, et al. HERG- and IRK-like inward rectifier currents are sequentially expressed during neuronal crest cells and their derivatives. Eur J Neurosci 1997; 9:2596–2604.

40. Pennefather PS, Zhou W, DeCoursey TE. Idiosyncratic gating of HERG-like K^+ channels in microglia. J Gen Physiol 1998; 111:795–805.

41. Bianchi L, Wible B, Arcangeli A, et al. HERG encodes a K^+ current highly conserved in tumors of different histogenesis: a selective advantage for cancer cells? Cancer Res 1998; 58:815–822.

42. Curran ME, Splawski I, Timothy KW, Vincent GM, Green ED, Keating MT. A molecular basis for cardiac arrhythmia: HERG mutations cause long QT syndrome. Cell 1995; 80:795–803.

43. Schulze-Bahr E, Haverkamp W, Funke H. The long-QT syndrome. N Engl J Med 1995; 333:1783–1784.

44. Sanguinetti MC, Jiang C, Curran ME, Keating MT. A mechanistic link between an inherited and an acquired cardiac arrhythmia: HERG encodes the I_{Kr} potassium channel. Cell 1995; 81:299–307.

45. Trudeau MC, Warmke J, Ganetzky B, Robertson G. HERG, a human inward rectifier in the voltage-gated potassium channel family. Science 1995; 269:92–95.

46. Sanguinetti MC, Curran ME, Spector PS, Keating MT. Spectrum of HERG K^+-channel dysfunction in an inherited cardiac arrhythmia. Proc Natl Acad Sci USA 1996; 93:2208–2212.

47. Furutani M, Trudeau MC, Hagiwara N, et al. Novel mechanism associated with an inherited cardiac arrhythmia. Defective protein trafficking by the mutant HERG (G601S) potassium channel. Circulation 1999; 99:2290–2294.

48. McDonald TV, Yu Z, Ming Z, et al. A MinK-HERG Complex regulates the cardiac potassium current I_{Kr}. Nature 1997; 388:289–292.

49. Gellens M, George AL, Chen L, et al. Primary structure and functional expression of the human cardiac tetrodotoxin-insensitive voltage-dependent sodium channel. Proc Natl Acad Sci USA 1992; 89: 554–558.

50. George AL, Varkony TA, Drakin HA, et al. Assignment of the human heart tetrodotoxin-resistant voltage-gated Na channel α-subunit gene (SCN5A) to band 3p21. Cytogenet Cell Genet 1995; 68:67–70.

51. Wang Q, Shen J, Splawski I, et al. SCN5A mutations associated with an inherited cardiac arrhythmia, long QT syndrome. Cell 1995; 80:805–811.

52. Wang Q, Shen J, Li Z, et al. Cardiac sodium channel mutations in patients with long QT syndrome: an inherited cardiac arrhythmia. Hum Mol Genet 1995; 4:1603–1607.

53. Bennett PB, Yazawa K, Makita N, George AL Jr. Molecular mechanism for an inherited cardiac arrhythmia. Nature 1995; 376:683–685.

54. Bennett PB, Yazawa K, Makita N, George AL Jr. Molecular mechanisms of sodium channel-linked long QT syndrome. Circ Res 1996; 78:916–924.

55. An RH, Wang XL, Kerem B, Benhorin J, Medina A, Goldsmit M, Kass RS. Novel LQT3 mutation affects Na$^+$ channel activity through interactions between alpha- and beta-1-subunits. Circ Res 1998; 83:141–146.

56. Chen Q, Kirsch GE, Zhang D, et al. Genetic basis and molecular mechanism for idiopathic ventricular fibrillation. Nature 1998; 392:293–296.

57. Arena JP, Kass RS. Block of heart potassium channels by clofilium and its tertiary analogs: relationship between drug structure and type of channel blocked. Mol Pharmacol 1988; 34:60–66.

58. Neyroud N, Tesson F, Denjoy I, et al. A novel mutation on the potassium channel gene KVLQT1 causes the Jervell and Lange-Nielsen cardioauditory syndrome. Nature Genet 1997; 15:186–189.

59. Splawski I, Timothy KW. Vincent GM, Atkinson DL, Keating MT. Brief report: molecular basis of the long-QT syndrome associated with deafness. N Engl J Med 1997; 336:1562–1567.

60. Chen Q, Zhang D, Gingell RL, et al. Homozygous deletion in KLVQT1 associated with Jervell and Lange-Nielsen syndrome. Circulation 1999; 99:1344–1347.

61. Tyson J, Tranebjaerg L, Bellman S, et al. IsK and KVLQT1: mutation in either of the two subunits of the slow component of the delayed rectifier potassium channel can cause Jervell and Lange-Nielsen syndrome. Hum Molec Genet 1997; 12:2179–2185.

62. Schulze-Bahr E, Wang Q, Wedekind H, et al. KCNE1 mutations cause Jervell and Lange-Nielsen syndrome. Nat Genet 1997; 17:267–268.

63. Zareba W, Moss AJ, Schwartz PJ, et al. Influence of the genotype on the clinical course of the long-QT syndrome. N Engl J Med 1998; 339:960–965.

64. Moss AJ, Zareba W, Benjorin J, et al. ECG T-wave patterns in genetically distinct forms of the hereditary long QT syndrome. Circulation 1995; 92:2929–2934.

65. Li H, Schwartz P, Locati E, Moss A, Robinson J, Towbin JA. Chromosome 11-linked long QT syndrome LQT1 is the most common from of long QT syndrome. Pediatrics 1996; 98:534.

66. Splawski I, Shen J, Timothy KW, Vincent GM, Lehmann MH, Keating MT. Genomic structure of three long QT syndrome genes: KVLQT1, HERG, and KCNE1. Genomics 1998; 51:86–97.

67. Schwartz PJ, Priori SG, Locati EH, et al. Long-QT syndrome patients with mutations of the SCN5A and HERG genes have differential responses to Na$^+$ channel blockade and to increases in heart rate: Implications for gene-specific therapy. Circulation 1995; 92:3381–3386.

68. Compton SJ, Lux RL, Ramsey MR, et al. Genetically defined therapy of inherited long-QT syndrome: Correction of abnormal repolarization by potassium. Circulation 1996; 94:1018–1022.

69. Shimizu W, Kurita T, Matsuo K, et al. Improvement of repolarization abnormalities by a K$^+$ channel opener in the LQTQ1 form of congenital long-QT syndrome. Circulation 1999; 97:1581–1588.

70. Myerburg RJ. Sudden cardiac death in persons with normal (or near normal) heart. Am J Cardiol 1997; 79(6A):3–9.

71. Osher HL, Wolff L. Electrocardiographic pattern simulating acute myocardial injury. Am J Med Sci 1953; 226:541–545.

72. Edeiken J. Elevation of RS-T segment, apparent or real in right precordial leads as probable normal variant. Am Heart J 1954; 48:331–339.

73. Levine HD, Wanzer SH, Merrill JP. Dialyzable currents of injury in potassium intoxication resembling acute myocardial infarction or pericarditis. Circulation 1956; 13:29–36.

74. Roesler H. An electrocardiographic study of high take-off of the R (R)-T segment in right pre-cordial leads. Altered repolarization. Am J Cardiol 1960; 6:920–928.

75. Calo AA. The triad secondary R wave, RS-T segment elevation and T waves inversion in right precordial leads: a normal electrocardiographic variant. G Ital Cardiol 1975; 5:955–960.

76. Parisi AF, Beckmann CH, Lancaster MC. The spectrum of ST elevation in the electrocardio-grams of healthy adult men. J Electrocardiol 1971; 4:137–144.

77. Wasserburger RH, Alt WJ, Lloyd CJ. The normal RS-T segment elevation variant. Am J Car-diol 1971; 8:184–192.

78. Goldman MJ. RS-T segment elevation in mid- and left precordial leads as a normal variant. Am Heart J 1953; 46:817–820.

79. Martini B, Nava A, Thiene G, et al. Ventricular fibrillation without apparent heart disease. De-scription of six cases. Am Heart J 1989; 118:1203–1209.

80. Aihara N, Ohe T, Kamakura S, et al. Clinical and electrophysiologic characteristics of idio-pathic ventricular fibrillation. Shinzo 1990; 22(suppl 2):80–86.

81. Brugada P, Brugada J. A distinct clinical and electrocardiographic syndrome: right bundle-branch block, persistent ST segment elevation with normal QT interval and sudden cardiac death (Amstar). PACE 1991; 14:746.

82. Sumita S, Yoshida K, Ishikawa T, et al. ST level in healthy subjects with right bundle branch block in relation to Brugada syndrome (abstr). Eur J Card Pacing Electrophysiol 1996; 6(1):270.

83. Nademanee K, Veerakul G, Nimmannit S, et al. Arrhythmogenic marker for the sudden unex-plained death syndrome in Thai men. Circulation 1997; 96:2595–2600.

84. Gotoh K. A histopathological study on the conduction system of the so-called Pokkuri dis-ease (sudden unexpected cardiac death of unknown origin in Japan). Jpn Circ J 1976; 40:753–768.

85. Hayashi M, Murate A, Satoh M, Aizawa Y, Oda E, Oda Y, Watanabe T, Shibata A. Sudden nocturnal death in young males from ventricular flutter. Jpn Heart J 1985; 26:585–591.

86. Naccarella F. Malignant ventricular arrhythmias in patients with a right bundle-branch block and persistent ST segment elevation in V1–V3; a probable arrhythmogenic cardiomyopathy of the right ventricle (Editorial comment.). G Ital Cardiol 1993; 23:1219–1222.

87. Fontaine G. Familial cardiomyopathy associated with right bundle branch block, ST segment elevation and sudden death (Letter). J Am Coll Cardiol 1996; 28:540.

88. Corrado D, Nava A, Buja G, et al. Familial cardiomyopathy underlies syndrome of right bun-dle branch block, ST segment elevation and sudden death. J Am Coll Cardiol 1996; 27:443–448.

89. Scheinman MM. Is Brugada syndrome a distinct clinical entity? J Cardiovasc Electrophysiol 1997; 8:332–336.

90. Ohe T. Idiopathic ventricular fibrillation of the Brugada type—an atypical form of arrhythmo-genic right ventricular cardiomyopathy (Editorial). Intern Med 1996; 35:595.

91. Fontaine G, Piot O, Sohal P, et al. Right precordial leads and sudden death. Relation with ar-rhythmogenic right ventricular dysplasia. Arch Mal Coeur Vaiss 1996; 89:1323–1329.

92. Brugada J, Brugada P. Further characterization of the syndrome of right bundle branch block, ST segment elevation, and sudden death. J Cardiovasc Electrophysiol 1997; 8:325–331.

93. Kobayashi T, Shintani U, Tamamoto T, et al. Familial occurrence of electrocardiographic ab-normalities of the Brugada-type. Intern Med 1996; 35:637–640.

94. Gussak I, Antzelevitch C, Bjerregaard P, Towbin JA, Chaitman BR. The Brugada syndrome; clinical, electrophysiological, and genetic considerations. J Am Coll Cardiol 1999; 33:5–15.

95. Antzelevitch C. The Brugada Syndrome. J Cardiovasc Electrophys 1998; 9:513–516.

96. Suzuki J, Tsubone H, Sugano S. Characteristics of ventricular activation and recovery patterns in the rat. J Vet Med Sci 1992; 54:711–716.

97. Lukas A, Antzelevitch C. Differences in the electrophysiological response of canine ventricu-lar epicardium and endocardium to ischemia: Role of the transient outward current. Circula-tion 1993; 88:2903–2915.

98. Antzelevitch C, Sicouri S, Lukas A, et al. Clinical implications of electrical heterogeneity in the heart: The electrophysiology and pharmacology of epicardial, M and endocardial cells. In: Podrid PJ, Kowey PR, eds. Cardiac Arrhythmia: Mechanism, Diagnosis and Management. Baltimore, MD: William & Wilkins, 1995:88–107.

99. Litovsky SH, Antzelevitch C. Differences in the electrophysiological response of canine ventricular subendocardium and subepicardium to acetylcholine and isoproterenol. A direct effect of acetylcholine in the ventricular myocardium. Circ Res 1990; 67:615–627.

100. Krishnan SC, Antzelevitch C. Flecainide-induced arrhythmia in canine ventricular epicardium: Phase 2 Reentry? Circulation 1990; 67:615–627.

101. Krishnan SC, Antzelevitch C. Sodium channel blockade produces electrophysiologic effects in canine ventricular epicardium and endocardium. Circ Res 1991; 69:277–291.

102. Rampazzo A, Nava A, Erne P, Eberhand M, Vian E, Slomp P, Tiso N, Thiene G, Daniele GA. A new locus for arrhythmogenic right ventricular cardiomyopathy (ARVD 2) maps to chromosome 1q42-q43. Hum Molec Genet 1995; 4:2151–2154.

103. Rampazzo A, Nava A, Danieli GA, Buja G, Daliento L, Fasoli G, Scognamiglio R, Corrado D, Thiene G. The gene for arrhythmogenic right ventricular cardiomyopathy maps to chromosome 14q23-q24. Hum Molec Genet 1994; 3:959–962.

104. Severini GM, Krajinovic M, Pinamonti B, Sinagra G, Fioretti P, Brunazzi MC, Flashchi A, Camerini F, Giacca M, Mestroni L. A new locus for arrhythmogenic right ventricular dysplasia on the long arm of chromosome 14. Genomics 1996; 31:193–200.

105. The National Heart, Lung, and Blood Institute Working Group on Atrial Fibrillation. Atrial fibrillation: current understanding and research imperatives. J Am Coll Cardiol 1993; 22:1830–1834.

106. Feinberg WM, Blackshear JL, Laupacis A, Kronmal R, Hart RG. Prevalence, age distribution, and gender of patients with atrial fibrillation: analysis and implications. Arch Intern Med 1995; 155:469–473.

107. Kannel WB, Abbott RD, Savage DD, McNamara PM. Epidemiological features of chronic atrial fibrillation. N Engl J Med 1982; 306:1018–1022.

108. Lip GYH, Tean KN, Dunn FG. Treatment of atrial fibrillation in a district general hospital Br Heart J 1994; 71:92–95.

109. Narayan SM, Cain ME, Smith JM. Atrial fibrillation. Lancet 1997; 350:943–950.

110. Flaker GC, Fletcher KA, Rothbart RM, Halperin JL, Hart RG. Clinical and echocardiographic features of intermittent atrial fibrillation that predict recurrent atrial fibrillation. Am J Cardiol 1995; 76:355–358.

111. Brugada R, Tapscott T, Czernuszewicz GZ, et al. Identification of a genetic locus for familial atrial fibrillation. N Engl J Med 1997; 336:905–911.

112. Bowles KR, Gajarski R, Porter P, et al. Gene mapping of familial autosomal dominant dilated cardiomyopathy to chromosome 10q21-23. J Clin Invest 1996; 98:1355–1360.

113. Wigle ED, Sasson Z, Henderson MA, et al. Hypertrophic cardiomyopathy. The importance of the site and the extent of hypertrophy. A review. Prog Cardiovasc Dis 1985; 28:1–83.

114. Maron BJ, Bonow RO, Cannon RO, Leon MB, Epstein SE. Hypertrophic cardiomyopathy: interrelations of clinical manifestations, pathophysiology, and therapy. N Engl J Med 1987; 316:780–789, 844–852.

115. Maron BJ, Gardin JM, Flack JM, Gidding SS, Kurosaki TT, Bild DE. Prevalence of hypertrophic cardiomyopathy in a general population of young adults: echocardiographic analysis of 4111 subjects in the CARDIA study. Circulation 1995; 92:785–789.

116. Vulpian A. Contribution à l'étude des rétrécissements de l'orifice ventriculo-aortique. Arch Physiol 1868; 3:220–222.

117. Jarcho JA, McKenna W, Pare JAP, et al. Mapping a gene for familial hypertrophic cardiomyopathy to chromosome 14q1. N Engl J Med 1989; 321:1372–1378.

118. Solomon SD, Jarcho JA, McKenna WJ, et al. Familial hypertrophic cardiomyopathy is a genetically heterogeneous disease. J Clin Invest 1990; 86:993–999.

119. Schwartz K, Dufour C, Fougerousse F, et al. Exclusion of myosin heavy chain and cardiac

actin gene involvement in hypertrophic cardiomyopathies of several French families. Circ Res 1992; 71:3–8.

120. Watkins H, MacRae C, Thierfelder L, et al. A disease locus for familial hypertrophic cardiomyopathy maps to chromosome 1q3. Nature Genet 1993; 3:333–337.

121. Thierfelder L, MacRae C, Watkins H, et al. A familial hypertrophic cardiomyopathy locus maps to chromosome 15q2. Proc Natl Acad Sci USA 1993; 90:6270–6274.

122. Carrier L, Hengstenberg C, Beckmann JS, et al. Mapping of a novel gene for familial hypertrophic cardiomyopathy to chromosome 11. Nature Genet 1993; 4:311–313.

123. MacRae CA, Ghaisas N, Kass S, et al. Familial hypertrophic cardiomyopathy with Wolff-Parkinson-White Syndrome maps to a locus on chromosome 7q3. J Clin Invest 1995; 96:1216–1220.

124. Poetter K, Jiang H, Hassanzadeh S, et al. Mutation in either the essential regulatory light chains of myosin are associated with a rare myopathy in human heart and skeletal muscle. Nature Genet 1996; 13:63–69.

125. Kimura A, Harada H, Park JE, et al. Mutations in the cardiac troponin I gene associated with hypertrophic cardiomyopathy. Nature Genet 1997; 16:379–382.

126. Mogensen J, Klausen IC, Pederson AK, et al. α-Cardiac actin is a novel disease gene in familial hypertrophic cardiomyopathy. J Clin Invest 1999; 103:T39–R43.

127. Bonne G, Carrier L, Richard P, Hainque B, Schwartz K. Familial hypertrophic cardiomyopathy from mutations to functional defects. Circ Res 1998; 83:380–593.

128. Schiaffino S, Reggiani C. Molecular diversity of myofibrillar proteins: gene regulation and functional significance. Physiol Rev 1996; 76:371–423.

129. Hofmann PA, Hartzell HC, Moss RL. Alterations in Ca^{2+} sensitive tension due to partial extraction of C-protein from rat skinned cardiac myocytes and rabbit skeletal muscle fibers. J Gen Physiol 1991; 97:1141–1163.

130. Carrier L, Bonne G, Bährend E, et al. Organization and sequence of human cardiac myosin binding protein C gene (MYBPC3) and identification of mutations predicted to produce truncated proteins in familial hypertrophic cardiomyopathy. Circ Res 1997; 80:427–434.

131. Gautel M, Zuffardi O, Freiburg A, Labeit S. Phosphorylation switches specific for the cardiac isoform of myosin binding protein C: a modulator of cardiac contraction? EMBO J 1995; 14:1952–1960.

132. Fougerousse F, Delezoide AL, Fiszman MY, Schwartz K, Beckman JS, Carrier L. Cardiac myosin binding protein C gene is specifically expressed in heart during murine and human development. Circ Res 1998; 82:130–133.

133. Gautel M, Fürst DO, Cocco A, Schiaffino S. Isoform transitions of the myosin-binding protein C family in developing human and mouse muscles. Lack of isoform transcomplementation in cardiac muscle. Circ Res 1998; 82:124–129.

134. Lees-Miller JP, Helfman DM. The molecular basis for tropomyosin isoform diversity. Bioessays 1991; 13:429–437.

135. Mesnard L, Logeart D, Taviaux S, Diriong S, Mercadier JJ, Samson F. Human cardiac troponin T: cloning and expression of new isoforms in the normal and failing heart. Circ Res 1995; 76:687–692.

136. Townsend P, Barton P, Yacoub M, Farza H. Molecular cloning of human cardiac troponin T isoforms: expression in developing and failing heart. J Mol Cell Cardiol 1995; 27:2223–2236.

137. Forissier JF, Carrier L, Farza H, et al. Codon 102 of the cardiac troponin T gene is a putative hot sot for mutations in familial hypertrophic cardiomyopathy. Circulation 1996; 94:3069–3073.

138. Farza H, Townsend PJ, Carrier L, et al. Genomic organisation, alternative splicing and polymorphisms of the human cardiac troponin T gene. J Mol Cell Cardiol 1998; 3:1247–1253.

139. Hunkeler NM, Kullman J, Murphy AM. Troponin I isoform expression in human heart. Circ Res 1991; 69:1409–1414.

140. Solaro RJ, Van Eyk J. Altered interactions among thin filament proteins modulate cardiac function. J Mol Cell Cardiol 1996; 28:217–230.

141. Olson T, Michels VV, Thibodeau SN, Tai YS, Keating MT. Actin mutations in dilated cardiomyopathy, a heritable form of heart failure. Science 1998; 280:750–752.

142. Spirito P, Seidman CE, McKenna WJ, Maron BJ. The management of hypertrophic cardiomyopathy. N Engl J Med 1997; 336:775–785.

143. Watkins H, Rosenzweig T, Hwang DS, et al. Characteristic and prognostic implications of myosin missense mutations in familial hypertrophic cardiomyopathy. N Engl J Med 1992; 326:1106–1114.

144. Fananapazir L, Epstein ND. Genotype-phenotype correlations in cardiomyopathy: Insights provided by comparisons of kindreds with distinct and identical β-myosin heavy chain mutations. Circulation 1994; 89:22–32.

145. Moolman JC, Corfield VA, Posen B, et al. Sudden death due to troponin T mutations. J Am Coll Cardiol 1997; 29:549–555.

146. Nakajima-Taniguchi C, Matsui H, Fujio Y, Nagata S, Kishimoto T, Yamauchi-Takihara K. Novel missense mutation in cardiac troponin T gene found in Japanese patient with hypertrophic cardiomyopathy. J Mol Cell Cardiol 1997; 29:839–843.

147. Bonne G, Carrier L, Bercovici J, et al. Cardiac myosin binding protein-C gene splice acceptor site mutation is associated with familial hypertrophic cardiomyopathy. Nature Genet 1995; 11:438–440.

148. Watkins H, Conner D, Thierfelder L, et al. Mutations in the cardiac myosin binding protein-C gene on chromosome 11 cause familial hypertrophic cardiomyopathy. Nature Genet 1995; 11:434–437.

149. Nimura H, Bachinski LL, Sangwatanaroh S, et al. Mutations in the gene for cardiac myosin-binding protein C and late-onset familial hypertrophic cardiomyopathy. N Engl J Med 1998; 33:1248–1257.

150. Charron P, Dubourg O, Desnos M, et al. Clinical features and prognostic implications of familial hypertrophic cardiomyopathy related to cardiac myosin binding protein C gene. Circulation 1998; 97:2230–2236.

151. Marian AJ, Yu Q-T, Workman R, Greve G, Roberts R. Angiotensin-converting enzyme polymorphism in hypertrophic cardiomyopathy and sudden cardiac death. Lancet 1993; 342:1085–1086.

152. Yonega K, Okamoto H, Machida M, et al. Angiotensin-converting enzyme gene polymorphism in Japanese patients with hypertrophic cardiomyopathy. Am Heart J 1995; 130:1089–1093.

153. Tesson F, Dufour C, Moolman JC, et al. The influence of the angiotensin I converting enzyme genotype in familial hypertrophic cardiomyopathy varies with the disease gene mutation. J Mol Cell Cardiol 1997; 29:831–838.

154. Towbin JA. The role of cytoskeletal proteins in cardiomyopathies. Curr Opin Cell Biol 1998; 10:131–139.

26

New Directions for Antiarrhythmic Drug Development Based on Molecular Approaches to Arrhythmogenesis

EDUARDO MARBÁN

The Johns Hopkins University, Baltimore, Maryland

MICHAEL SANGUINETTI

University of Utah, Salt Lake City, Utah

I. INTRODUCTION

The last two decades of the twentieth century have witnessed a revolution in our understanding of the biology of excitable membranes. Currents flowing through individual channel molecules were resolved for the first time, using a technique known as "patch clamp," which earned its creators, Erwin Neher and Bert Sakmann, the Nobel Prize in 1991. Such single-channel recordings enabled physiologists to fingerprint each of the many classes of ion channels in a given cell (1). These investigators also introduced the technique of tight-seal, whole-cell recording that allowed the study of currents in isolated cardiac myocytes with a precision not previously possible with the commonly used two-microelectrode or sucrose gap voltage clamp of multicellular cardiac preparations. Before the development of whole-cell and single-channel recording, quantitative biophysical studies of ion currents were only possible in certain neuronal preparations like the giant squid axon. In addition, a variety of new and powerful approaches made it possible to measure ion concentrations within living cells with excellent spatiotemporal resolution, expanding our understanding of the interactions between the surface membrane and the cell interior. Fortuitously, advances in the molecular genetics of channels and transporters kept full pace with those in cellular physiology. The laboratory of Shosaku Numa was the first to clone and to deduce the general structure of a voltage-dependent ion channel—the sodium channel of the electric eel (2). Cloning of mammalian homologs followed quickly. The parallel explosion of potassium channel biology was sparked by positional cloning of the first such channel by Jan and coworkers (3). Important electrogenic transporters such as the Na–Ca exchanger were also isolated and cloned (4). Expression in heterologous systems coupled with application of modern physiological approaches enabled remarkable insights into the structure-function relationships of the newly cloned ion transport proteins. All these methods were brought to bear upon cells and gene products of cardiovascular significance.

These astounding advances have opened up novel opportunities to address the persistent problem of cardiac arrhythmias. In a few instances, such opportunities have already begun to be realized. Linkage analysis and positional cloning by Keating and associates has pinpointed several genetic mutations responsible for heritable long-QT syndrome (see Chap. 25). The identification of the culprit genes and their functional alterations has enabled the elaboration of rational, gene-specific therapeutic strategies for this fascinating, but uncommon, disorder. Unfortunately, such progress has not yet been achieved for the common, often lethal, arrhythmias that complicate ischemic heart disease and heart failure. The major advances of the last two decades in this area have been limited to mechanical therapeutic strategies such as automatic defibrillators and catheter ablation. Many paroxysmal supraventricular tachycardias have become increasingly rare due to these potentially curative modalities. In contrast, the pharmacological treatment of ventricular arrhythmias is a problematic situation that was highlighted by the discouraging outcome of the Cardiac Arrhythmia Suppression Trial (CAST). In CAST, class I antiarrhythmic drugs that effectively suppressed ectopic beats unexpectedly increased mortality in survivors of myocardial infarction, particularly in patients with poor ventricular function (5). Until then, conventional wisdom dictated that ambient ectopy predisposed to lethal arrhythmias; by challenging this truism, the CAST results exposed the extent of our ignorance regarding the mechanisms of sudden cardiac death. More than 400,000 Americans still die annually of arrhythmias. The most important lesson to be learned from the failure of many antiarrhythmic drugs in large-scale clinical trials is not the hopelessness of the pharmacological approach, but rather that no single drug can be expected to treat a disorder with such diverse etiology.

There is no small irony in the fact that we know much more about the workings of the individual molecules that underlie excitability than we do about common disorders of excitability, notably ventricular arrhythmias. The problem represents not so much an unbridgeable chasm between basic biology and clinical practice as the fact that much remains to be learned at the fundamental level. A fresh look at mechanisms of arrhythmia is necessary to transcend the classical concepts of fixed wiring abnormalities. Perhaps the right questions have yet to be asked. For example, the spectrum of arrhythmogenic mechanisms is poorly defined, and it is likely that arrhythmias that appear similar based on ECG recordings have very different molecular and cellular mechanisms. Moreover, we know precious little about the relationship between the genome and the rich tapestry of gene expression in the heart in general, and even less so about the genes that shape excitability. Do changes in channel or transporter gene expression occur in arrhythmogenic disorders (e.g., heart failure)? Are such changes adaptive or maladaptive? Can gene transfer methods be used in selected patients to treat or to prevent arrhythmias?

It is our central thesis that the next phase of antiarrhythmic drug development must emerge from modern concepts of excitation and how it is altered in disease states. If this knowledge can be paired with mechanism-based clinical diagnosis, then it should be possible to effectively treat the wide spectrum of life-threatening arrhythmias. The molecular and cellular basis of arrhythmia is complex. Although classification of arrhythmia types has some clinical utility, it is important to dismiss the possibility of discovering only a few "magic bullet" drugs for treatment of these disorders. The present chapter will review existing concepts of arrhythmogenesis, discuss emerging new insights, and highlight the emerging opportunities for drug development.

II. PATHOPHYSIOLOGICAL COMPONENTS OF CARDIAC ARRHYTHMIAS

The heart is a network of myocardial cells surrounded by a complex matrix of extracellular fluid, fibrous tissue, and nonmyocardial cells. Myocardial cells, which are interconnected by gap junctions (6,7), are the only ones that contribute directly to electrogenesis and conduction. The other constituents of the heart play important modulatory functions but do not actually generate or transmit the cardiac impulse. Thus, cardiac excitation can be logically divided into two components: the processes that shape excitability in individual cardiac myocytes, and those that govern the coupling among myocytes. Both of these components are subject to modulation by intracellular and extracellular factors.

To discuss the mechanisms of arrhythmia, we must first review the basic properties of several transport pathways not covered elsewhere in this volume. Arrhythmogenesis involves not only classic voltage-dependent ion channels (Chap. 3) but also mechanosensitive channels, electrogenic transporters, and molecules that connect cells to each other electrically.

A. Integrative Considerations

Figure 26.1 depicts the various contributors to arrhythmias and the interactions among them. Proteins contributing to excitability reside within the surface membrane of the cell (Fig. 26.1, left). The human genetics of the congenital long-QT syndrome illustrate that primary abnormalities of channel proteins suffice to produce arrhythmias. For example, faulty inactivation of sodium channels due to a nine-base-pair deletion within the SCN5A cardiac sodium channel gene, or mutation of HERG, KVLQT1, or hminK potassium channel subunits retard repolarization and predispose to torsade de pointes (Chap. 25). Such rare disorders serve as valuable prototypes for understanding more common arrhythmias. They emphasize that apparently complex rhythms can arise from discrete lesions in individual genes that encode signaling molecules. However, it is also clear that mutation of an ion channel gene is not sufficient to provoke arrhythmia. Many known gene carriers have few or no documented incidences or arrhythmia, clearly indicating that other factors are important to trigger the arrhythmias associated with this inherited disorder. Primary genetic abnormalities in cell–cell coupling (Fig. 26.1, right) may well turn out to be capable of producing arrhythmias, but this has not yet been established. Recent studies of connexin proteins have emphasized the very dynamic nature of the manner in which ion channel subunits are modulated at levels of transcription, translation, and protein trafficking and turnover (8). Linkage analysis has pinpointed the chromosomal location of the lesion in various kindreds with hereditary conduction disorders such as familial Wolff–Parkinson–White syndrome (Chap. 25). It will be interesting to determine whether any such genetic disorders involve lesions of connexin genes or of other genes intimately involved in forming cell–cell junctions.

More commonly, rhythm disturbances arise as a consequence of modulatory factors acting either on the surface membrane or on cell–cell coupling (or both). Figure 26.1 shows these modulatory factors in the center. The effects can be acute, producing changes

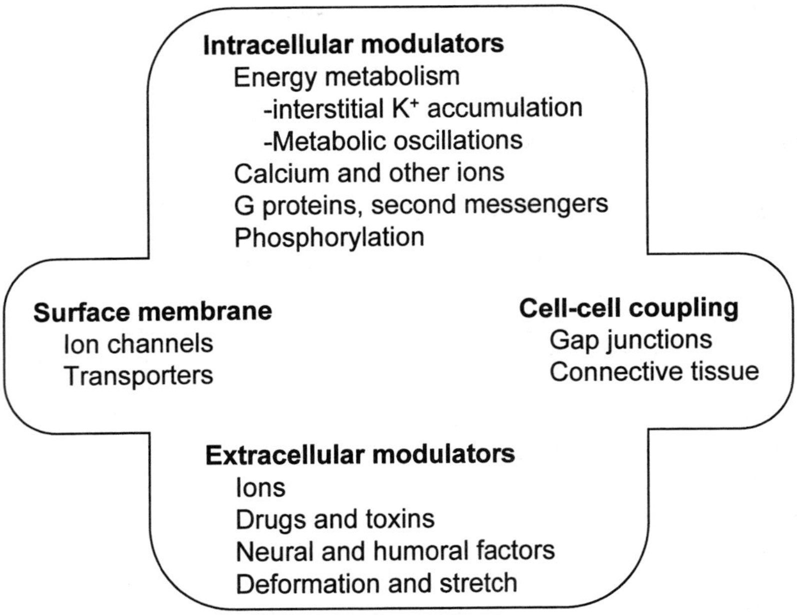

Figure 26.1 Pathophysiological components of arrhythmias. Components within the surface membrane (left) interact with those involved in cell–cell coupling (right). Both sets of components are subject to modulation by various intra- and extracellular factors (center). (Adapted from Ref. 137, by permission.)

in excitation on the time scale of seconds to minutes. Such acute changes generally involve post-translational modifications of excitability proteins or direct biophysical consequences of the concentrations of permeant ions. Changes in gene expression presumably underlie the longer term effects, although the actual mediators may be identical. Beta-adrenergic stimulation exemplifies such a dual effect. Acutely, the increase in intracellular cyclic AMP concentration will activate protein kinase A and stimulate the phosphorylation of calcium channels, a post-translational modification (9). Sustained elevation of cyclic AMP acts at the transcriptional level by activating CREB and increasing the expression of its target genes (10). Little work has been reported on the transcriptional regulation of ion channels, but at least some K-channel genes are believed to be regulated by cyclic AMP-dependent mechanisms (11). Likewise, the expression of cardiac connexins increases dramatically when ventricular myocytes are cultured in the presence of dibutyryl cyclic AMP (12).

III. INTRACELLULAR FACTORS THAT INFLUENCE EXCITABILITY

A. Energy Metabolism

Intracellular ATP concentrations normally approximate 5 mM, far in excess of the levels required to support hydrolysis and other ATP-dependent processes (13). During ischemia, ATP is rapidly consumed, reaching submillimolar levels within several minutes (14); lethal ventricular arrhythmias occur most frequently during the early minutes of acute myocardial infarction (15). The decrease in ATP is accompanied by accumulation of its degradation products (ADP, AMP, adenosine, and inorganic phosphate). Cessation of perfusion under anaerobic conditions also leads to retention of lactate and thus to acidosis (16).

These conditions are ideal for the activation of ATP-dependent potassium channels. These channels have a high conductance and are expressed so densely that even the activation of a small fraction suffices to render heart cells inexcitable (17). The opening of such channels also favors interstitial K^+ accumulation, although other mechanisms also play an important role (18). Thus, the opening of ATP-dependent K channels will decrease excitability both by direct effects on membrane conductance and indirectly due to the rise of interstitial $[K^+]$. The decrease in cellular excitability associated with activation of ATP-dependent K channels has been shown to be antiarrhythmic in some animal models (19). Drugs that activate these K channels can be cardioprotective in models of ischemia. The doses required for protection do not alter electrical properties (no change in action potential duration) (20), and may be mediated by modulation of ATP-dependent K channels in the mitochondria.

Activation of ATP-dependent K channels may also be arrhythmogenic under conditions that are much less severe than those associated with total global ischemia (21,22). We now appreciate that energy metabolism oscillates when cells are stressed (23). Such primary metabolic oscillations drive oscillations of ATP-dependent channel activity that suffice to render cells cyclically inexcitable (24). This behavior, which has only recently been recognized, forms the basis for a novel arrhythmogenic principle. More investigation will be required to place this mechanism into its proper context.

Energy metabolism influences a variety of other ion transport pathways within the cell, including L-type calcium channels, calcium release channels in the sarcoplasmic reticulum, and Na–Ca exchange (25). Gap junction channels are also sensitive to the intracellular concentrations of adenine nucleotides. In general, the activity of these transport molecules decreases as the cellular concentration of MgATP falls. Interestingly, such modulatory effects do not necessarily reflect a simple shortage of ATP as a fuel: nonhydrolyzable ATP analogs can often substitute for ATP in maintaining ion transport (25).

B. Ions

Changes in various physiologically relevant inorganic cations importantly influence cellular excitability and cell–cell communication. Calcium is the best-recognized example. The diastolic intracellular free Ca^{2+} concentration $([Ca^{2+}]_i)$ is tightly regulated at approximately 100 nM under physiological conditions (26,27). Excessive accumulation of cellular calcium can lead to "calcium overload," a condition characterized by abnormal diastolic bursts of calcium release from the sarcoplasmic reticulum (28,29). Calcium overload can occur as a consequence of drug exposure (e.g., as a toxic effect of digitalis or of phosphodiesterase inhibitors), rapid pacing, or ischemia (27,30). The tight regulation of $[Ca^{2+}]_i$ is lost; myocardial cells then undergo cyclical elevations of $[Ca^{2+}]_i$ during diastole (31). These $[Ca^{2+}]_i$ oscillations activate an inward current that leads to delayed after-depolarizations (DADs), displacements of the diastolic membrane potential that can reach threshold and initiate premature beats (32). Figure 26.2 shows action potentials recorded from a strip of ventricular muscle. Panel (A) shows a normal action potential. Repolarization is monotonic, and there is no secondary electrical activity after repolarization has been completed. Panel (B) is a record of transmembrane potential from a preparation that had been exposed to toxic concentrations of digitalis. Here, the stimulated action potential is followed by a series of damped oscillations during diastole: these are DADs. Calcium oscillations and DADs are widely believed to underlie the ventricular arrhythmias associated with digitalis

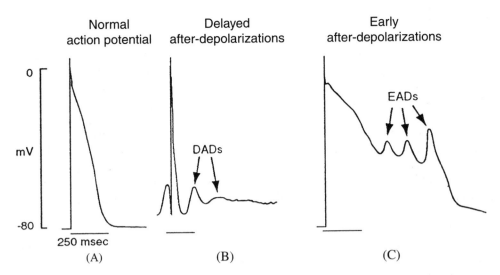

Figure 26.2 Action potentials illustrating delayed and early afterdepolarizations. (A) Normal action potential recorded from ferret ventricular muscle. (B) Action potential recorded from digitalis-intoxicated ferret ventricular muscle. Note the damped oscillations of diastolic membrane potential that follow the action potential; these are delayed after-depolarizations (DADs). (C) Action potential recorded from ferret ventricular muscle exposed to cesium, a potassium channel blocker. Note the marked prolongation of the action potential, which leads to an oscillations of the membrane potential during the late plateau phase. These are early afterdepolarizations (EADs). (Adapted from Ref. 32.)

toxicity (33); they have also been proposed to mediate multifocal atrial tachycardia (34). The identity of the inward current activated by diastolic $[Ca^{2+}]_i$ oscillations remains controversial, with three candidates vying for prominence: a nonselective Ca^{2+}-activated cation channel (35); Na–Ca exchange current triggered by the rise in $[Ca^{2+}]_i$ (36,37); and a Ca^{2+}-activated chloride current (38). The lack of selective pharmacological inhibitors has hampered efforts to distinguish among these possibilities. DADs, by definition, are diastolic events that occur after complete repolarization of the preceding action potential (39). In this sense, they are readily distinguishable from early after-depolarizations (EADs) [Fig. 26.2(C)], which have a fundamentally different mechanism (see below). A drug that could prevent intracellular Ca overload would prevent DADs and presumably arrhythmias associated with this trigger mechanism.

The major arrhythmogenic effects of intracellular $[Na^+]$ are attributable to its direct relationship to $[Ca^{2+}]_i$, mediated by the Na–Ca exchanger (40,41). Any maneuver that increases $[Na^+]_i$ (e.g., rapid pacing, digitalis-mediated or hypokalemic inhibition of the Na-K ATPase, or Na channel toxins that increase Na^+ influx) will blunt the transmembrane Na^+ gradient and decrease the driving force for Ca^{2+} extrusion via the exchanger. Cellular Ca loading will therefore increase whenever $[Na^+]_i$ rises. Although Na^+-activated K^+ channels have been described (42), their sensitivity to $[Na^+]_i$ is so low that it is unclear whether they come into play under physiological (or even under common pathophysiological) conditions.

Elevations of proton concentration have several consequences that may contribute to arrhythmias (43). Heart cells contain a Na–H exchanger that reacts to acidosis by extruding H^+ from the cell in exchange for Na^+ (44); the enhanced Na^+ influx will increase calcium loading via the Na–Ca exchanger and thus promote calcium overload. Protons also directly inhibit a variety of membrane transport proteins including sodium and calcium channels (43).

Ca^{2+} and H^+ act not only on surface membrane pathways but also on cell–cell coupling. Elevations of either ion have been shown to decrease the open probability of gap junction channels (7,45). By increasing the resistance to intercellular current flow, such changes have been proposed to underlie the decrease in conduction velocity that characterizes ischemic ventricular muscle. Although the relative importance of Ca^{2+} and H^+ is still debated (45), calcium accumulation during ischemia correlates temporally much better with uncoupling than does acidosis: calcium accumulation and uncoupling begin only after 10 to 20 min of severe ischemia, whereas acidosis is virtually immediate (46). Phospholipid breakdown products such as lysophosphatidylcholine have also been proposed to decrease cell–cell coupling during ischemia (47).

Cells contain appreciable concentrations of various organic cations. Among these, polyamines, such as spermidine, have assumed special importance since the recognition that they block outward current through inwardly rectifying K channels (48). Simulations reveal that significant depletion of intracellular polyamines would be capable of altering repolarization of the cardiac action potential, but it is still unclear whether such depletion ever occurs in vivo.

This section has focused exclusively on cations, which historically have been overwhelmingly acknowledged more than anions as signaling molecules. Nevertheless, it is important to recognize the emerging understanding of the roles of anions. Several types of chloride channels have been identified in heart cells, including one that is activated by cyclic AMP and another that is calcium-dependent (49). The expression pattern of such channels varies widely among species. Organic anions such as taurine figure prominently as intracellular osmolytes, and a putative pathway for their transport has been identified (50). Anions and their transporters have yet to be implicated in arrhythmogenesis (with the exception of their possible involvement in DADs, as cited above), but a contribution to some disorders of excitation is certainly plausible.

C. G-Proteins, Second Messengers, and Phosphorylation

Most of the proteins involved in excitability are subject to direct modulation by G-proteins or second messengers, and/or to post-translational modification by kinases and/or phosphatases. Only two examples of special relevance to arrhythmias will be presented here.

β-Adrenergic receptor stimulation leads to activation of G_s, elaboration of cyclic AMP, and stimulation of protein kinase A. These effects can alter excitability by a variety of mechanisms, including upregulation of L-type calcium channels (9). Arrhythmias initiated by recruitment of this pathway would logically be expected to be precipitated by catecholamine infusion or by exercise, and inhibited by vagotonic maneuvers, adenosine, or β-blockers. Several forms of atrial tachycardia conform to these predictions (51), as do a subset of ventricular tachycardias (52). Some patients without evidence of anatomical heart disease are susceptible to monomorphic ventricular tachycardia that originates in the right ventricular outflow tract. Such tachycardias are often precipitated by exercise and inhibited by adenosine and β-blockers (53). The basis for these arrhythmias is presently unknown; as one possibility, it is intriguing to wonder whether such patients have a relatively benign genetic lesion (e.g., a polymorphism) in one or more components of the β-adrenergic signaling cascade that increases cardiac excitability.

IV. EXTRACELLULAR MODULATORS

A. Ions

Cells expend energy to maintain ion gradients across the surface membrane. Those ion gradients are then used by channels and exchangers for signaling. The major physiological cations that can influence excitation are Na^+, K^+, Ca^{2+}, Mg^{2+}, and polyamines. Among these, K^+ and Mg^{2+} are the ones most commonly implicated in clinical arrhythmias.

Extracellular $[K^+]$ is normally tightly regulated within the range of 3.5 to 5.0 mM by renal mechanisms. As $[K^+]$ drops below 3.5 mM, action potentials become progressively longer; eventually, repolarization becomes unstable and ventricular tachycardia results. This effect of lowering $[K^+]$ is paradoxical, because consideration of the driving force alone would predict the opposite result (54). It turns out that the effects of $[K^+]_0$ reflect a specialized property of several potassium-selective channels in which the probability that the channels are open is regulated by $[K^+]_0$: as external potassium concentrations rise, the open probability increases. At least two cardiac K currents (I_{K1} and I_{Kr}) are subject to this type of regulation (54–56), which may arise from K binding to a specific regulatory site on the outer face of the channel (57). An increase in extracellular K shifts the voltage dependence of I_{Kr} activation to more positive potentials and thereby increases the probability of channel opening at any given membrane potential (58). Changes in $[K^+]_0$ also affect drug binding to a variety of channels and transporters. For example, blockade of the delayed rectifier I_{Kr} by dofetilide is exquisitely sensitive to the extracellular K^+ concentration (59). Thus, the anti- (and pro-) arrhythmic effects of this drug would vary with changes in potassium homeostasis.

Changes in free magnesium ion concentration ($[Mg^{2+}]$), either intra- or extracellular, influence excitability indirectly by screening the negative charges that stud the phospholipid bilayer (60). Hypomagnesemia lowers the threshold for excitation because fewer of these surface charges are screened; as a result, ion channels sense that transmembrane voltage is more depolarized than predicted by bulk electrochemical equilibrium alone. Conversely, an increase of $[Mg^{2+}]_0$ raises the threshold for channel activation and thus exerts a generally depressant effect on excitability. The antiarrhythmic effects of magnesium are probably not due to block of Ca current (61), as has often been hypothesized.

Magnesium has recently attracted considerable attention as adjunctive treatment during acute myocardial infarction (62). Some (but by no means all) clinical trials have shown a dramatic benefit of intravenous magnesium supplementation on outcomes as diverse as pump failure and arrhythmias (63). One study notes that the magnesium can be either antiarrhythmic or proarrhythmic depending on dosage (64). Catecholamines induce acute cellular magnesium loss by as-yet unclear mechanisms (65,66). This effect may help to explain the apparent benefit of magnesium in acute MI, a setting in which sympathetic stimulation is markedly accentuated. Likewise, there is good reason to expect benefit in those patients on chronic diuretic therapy, given that such therapy commonly depletes magnesium.

B. Drugs

A broad variety of drugs influence cardiac excitability directly by altering the activity of channels, or indirectly by acting on modulatory pathways. Let us briefly consider those drugs which act directly on channels to influence cardiac excitation. Drugs such as lidocaine that block Na channels (also known as "class I" agents) produce local anesthesia when infiltrated at high concentrations around a nerve. When given systemically, these drugs act on the heart at much lower concentrations. Lidocaine blocks cardiac Na channels more potently than those of skeletal muscle and nerve due to intrinsic differences in the structure of the pore-forming α-subunits (67). Drug-induced block of cardiac Na channels slows conduction throughout the atria and ventricles. This effect can be either antiarrhythmic or proarrhythmic, depending on the mechanism of the rhythm disturbance. Local anesthetics are often effective in suppressing the ventricular arrhythmias encountered during ischemia, but they do not decrease mortality in high-risk survivors of acute myocardial infarction (5). This example emphasizes the need to match the use of a specific drug to a particular arrhythmia. A formidable challenge for the future will be to distinguish the many types of cardiac electrical dysfunctions from one another and then to tailor the discovery and development of specific therapies for each type.

Calcium channel blockers ("class IV" agents) are specific antagonists of ion flux through L-type calcium channels. Arguably the most useful channel-specific drugs, the major indications for which they are prescribed (hypertension and coronary artery disease) target the L-type channels in smooth muscle to produce vasodilatation. Functional differences in the α_1-subunits render the L-type channels of smooth muscle more sensitive to dihydropyridine blockers (e.g., nifedipine) relative to the closely related cardiac channels (68). Other families of calcium channel blockers (epitomized by diltiazem and verapamil) are more cardioselective and thus more useful for treating rhythm disturbances. Such compounds have emerged as the drugs of choice for rate control in atrial fibrillation (51). This suppressive effect on atrioventricular conduction is entirely consistent with the fact that AV nodal conduction is dependent upon calcium channels. Conversely, calcium channel blockers are only rarely effective in ventricular arrhythmias (53).

A number of long-available drugs, including quinidine and amiodarone, block voltage-dependent K channels and thereby prolong cardiac refractoriness (69). Such drugs block several other classes of ion channels as well, making it difficult to assign their effects solely to their action on K channels. So-called "class III" drugs that block specific K channels are now available, and new agents are being aggressively developed. The most popular target has been the delayed rectifier I_{Kr}, which is blocked by sotalol and by dofetilide (59). While such drugs are often effective in terminating atrial fibrillation, their track record in ventricular arrhythmias is spotty. K-channel blockade has been particularly problematic given its potential to delay repolarization excessively and thus to produce iatrogenic long-QT syndrome (69). The block of K channels that are specifically expressed in the atria (e.g., Kv1.5 channels) (70,71) may provide a novel mechanism for the treatment of atrial fibrillation that is not associated with proarrhythmic activity in the ventricle. However, this approach is complicated by the recent observation that Kv1.5 as well as transient outward K-channel expression are downregulated during atrial fibrillation (72). It is unclear if block of a repolarizing current that is already depressed in the disease state would have the desired effect of reducing atrial refractory period.

Mutagenesis and expression studies have begun to define the molecular sites of interaction between drugs and ion channels. A number of common features are emerging, as shown in Figure 26.3. Panel (A) shows the linear structure of a single α-subunit of a voltage-dependent K channel. The molecule begins with a cytoplasmic N terminus, then crosses the membrane six times (S1–S6). X-ray analysis of the atomic structure of a two-transmembrane domain K channel of *Streptomyces lividans* has confirmed and remarkably refined the structural features of the S5–S6 regions of voltage-gated K channels deduced from earlier biophysical and pharmacological studies, and identified the likely molecular

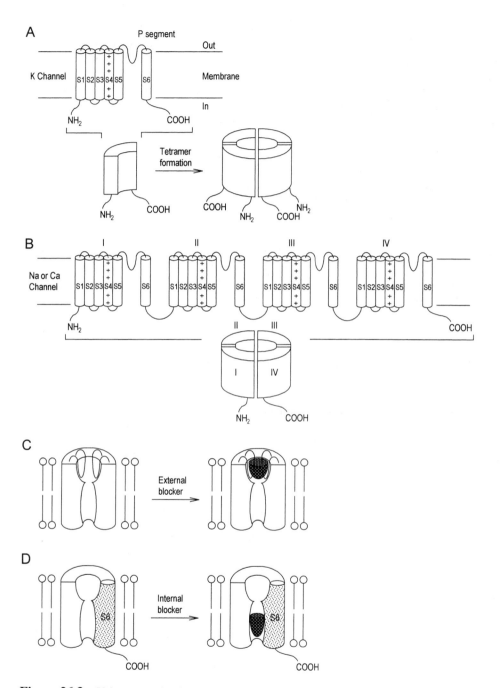

Figure 26.3 Voltage-gated cation channels and their drug-binding sites. (A) The top panel depicts the linear structure of a single α-subunit from a K channel, with its presumed transmembrane folding pattern. Each α-subunit consists of six transmembrane segments, depicted as cylinders. The fourth such segment (S4) is studded with positively charged arginine and lysine residues that form the voltage sensor for activation. A loop of protein between S5 and S6 (the P segment) dips back into the transmembrane region to form the lining of the pore. Four such α-subunits unite to form a functional channel. (B) Presumed folding pattern of Na- and Ca-channel α-subunits. Each domain (denoted

basis for many of the unique properties of K channels (73). The linker between S5 and S6, known as the P segment, dips back into the membrane to form the external vestibule of the pore and the selectivity filter. The carbonyl oxygen atoms from three amino acids (GYG) line the selectivity filter. The filter is stabilized in such a way as to specifically coordinate K^+, but not other ions such as Na^+. Four α-subunits assemble to create a functional channel, with one P segment contributed by each (74). Part of S6 lines the internal vestibule. Figure 26.3(B) shows that the α-subunits of Na and Ca channels consist of four internally homologous domains. Each domain topologically resembles a single K-channel α-subunit. The P segment in each domain is distinctive, but otherwise Na and Ca channels resemble a fused K-channel tetramer.

Drugs that inhibit current flow through ion channels generally do so by occluding the pore. Two major classes of pore-blocking sites have been identified. The first involves P-segment residues; compounds that bind here occlude ion flux by plugging the external vestibule [Fig. 26.3(C)]. This mechanism is favored by naturally occurring toxins; the pufferfish-derived Na-channel blocker tetrodotoxin (75,76) and the scorpion venom–derived (60) K-channel blocker charybdotoxin represent well-characterized examples. In contrast to toxins, clinically useful molecules bind preferentially to internal sites, particularly the S6 segment [Fig. 26.3(D)]. This general feature is curiously conserved among channels of very different pharmacology (77). Several residues in S6 of domain IV figure critically in Na channel block by local anesthetics (78); residues at comparable positions in Ca channels mediate block by dihydropyridines (79), and in K channels mediate block by quinidine (80). What is so special about S6? The answer is not yet clear. A clue may be found in the fact that all of the clinically useful pore-blocking drugs exhibit a strong dependence on the particular gating state of their target channel. For example, local anesthetics "bind" much more potently to inactivated Na channels than to those which are either resting or open. It may be that S6 forms a part of the ion flux pathway that is particularly sensitive to the gating conformation of the channel, as has been suggested for the Shaker K channel (81,82). If so, small perturbations of S6 structure could translate into large changes in the efficacy of channel block, even by compounds that are only weakly attracted to the internal vestibule (83). More experiments will be required to evaluate this idea and other detailed mechanisms of drug action.

Most drugs that decrease channel current do so by blocking the pore of the channel. One problem with this mechanism is that high doses of the drug can fully block channel conductance, rarely a desirable effect. Another mechanism is modulation of channel gating. For example, spider toxins such as hannatoxins (84,85) and heteropodatoxins (86) shift the voltage dependence of activation for specific K channels to more positive potentials, thereby reducing current magnitude. The potential advantage of such a mechanism is that the effect is self-limiting because shifts in gating are saturable. Even at high toxin concentrations, channels can open if the membrane is sufficiently depolarized. A small mole-

---

by a roman numeral from I to IV) resembles a K-channel α-subunit topologically, although there is extensive divergence among the domains (particularly in the P segments). Each α-subunit folds into a functional channel. (C) Cross-sectional schematic of an ion channel. The P segments (highlighted) form the outer pore lining. External blockers bind to the P segments and occlude the pore. (D) Cross-sectional schematic of an ion channel highlighting one of the S6 segments. Internal blockers interact with the surface of S6 accessible from the inner pore lumen and thereby block the channel. (Adapted from Ref. 137, by permission.)

cule that affected gating of a target channel without causing overt block might be a useful antiarrhythmic agent.

C. Humoral Factors

A number of neurohumoral factors affect ion transport pathways and cellular ion homeostasis. Among these, catecholamines and angiotensin II merit special mention since their circulating levels are markedly increased in heart failure and in other cardiovascular disorders (87). It is important to note that β-adrenergic receptor blockers are the only drug therapy that has been shown to reduce the risk of sudden cardiac death. Various cytokines, growth factors, and immune factors have cellular effects that may render them arrhythmogenic (88). Modulation of cytokines has received little or no attention as a mechanism to reduce the risk of arrhythmia. Regional inhomogeneities of cardiac autonomic innervation have also been implicated in the pathogenesis of arrhythmias. The reader is referred elsewhere for discussion of these emerging areas (89,90).

D. Deformation and Stretch

Mechanical events can clearly influence cardiac excitation. Sudden distension of the ventricles induces extrasystoles in a process known as mechanoelectrical feedback (91–96). While the possible correlates of this effect in patients remain unclear, diseases complicated by ventricular dilatation (e.g., dilated cardiomyopathy) exhibit a high incidence of sudden cardiac death (97). Hypertrophic hearts are also prone to arrhythmias (98). One study has noted that a swelling-activated cation current was persistently activated in cells isolated from hearts with pacing-induced heart failure (99). Myocytes in the midventricular wall undergo extensive deformation during systole; such deformation would logically be accentuated in cardiac hypertrophy. Arrhythmias related to stretch and to hypertrophy undoubtedly reflect long-term adaptive changes in various ion transport pathways, which have been well characterized in various animal models (100–102). It seems likely that mechanosensitive ion channels will turn out to play an equally important role on a beat-to-beat basis (103). Modeling studies predict that stretch-induced increase in cellular excitability may lead to an increased dispersion of refractoriness throughout the heart and therefore to an increased risk of arrhythmia (104,105). There are likely to be many types of channels that are activated by membrane stretch of cardiac myocytes. Nonselective, K^+-selective, and Cl-selective channels have been described from cardiac and other cell types (106–109). Selective inhibitors of such channels will help to elucidate their contribution to arrhythmogenesis, and may represent a novel antiarrhythmic mechanism.

V. MECHANISMS OF ARRHYTHMIAS

A strictly molecular approach provides very limited insight into the mechanisms of arrhythmias. Cellular and network properties must be considered in order to understand the initiation and propagation of rhythm disturbances. Table 26.1 represents an attempt to classify fundamental arrhythmic mechanisms in a biologically intuitive manner. The first column depicts the basic processes of excitability at various levels of integration. The second column lists the known or presumed molecular components at each level of integration. The last two columns list the corresponding arrhythmogenic mechanisms and relevant examples of clinical arrhythmias (the latter listing is not meant to be all-inclusive). Two levels of integration must be considered: processes that are intrinsic to individual cells (the

Table 26.1 Mechanisms of Arrhythmias

Level of integration	Key molecular components	Arrhythmogenic mechanism	Prototypical arrhythmias
Myocyte level			
Impulse initiation	Pacemaker current (I_f or I_{st}) T-type calcium channels	Suppression or acceleration of the physiological pacemaker	Sinus tachycardia or sinus bradycardia
	Unclear Na–Ca exchange Ca-activated chloride channel Ca-activated nonspecific cation channel	Abnormal automaticity Triggered diastolic activity (DADs)	Ectopic atrial tachycardia Ventricular tachycardia of digitalis toxicity; possibly some reperfusion arrhythmias
Excitation	Na channels ATP-sensitive K channels	Conduction slowing or block in atria or ventricles	Ischemic arrhythmias with slow conduction resulting from interstitial K accumulation
	L-type Ca channels	Conduction slowing or block in AV node	Iatrogenic AV block caused by calcium channel blocker
Repolarization	Voltage-dependent K channels Sodium channels L-type calcium channels Na-Ca exchange	Action potential prolongation (EADs)	Torsade de pointes Polymorphic VT
Multicellular level			
Cell-cell coupling	Connexins	Conduction delay or block caused by cellular uncoupling	Acute ischemic arrhythmias
Network properties	Collagen and other extracellular matrix proteins	Reentry Impedance mismatch Discontinuous conduction	Inherited: Wolff–Parkinson–White syndrome Acquired: monomorphic VT around an infarct

Source: Adapted from Ref. 137, by permission.

"myocyte level") and processes that by their very nature can occur only in networks of coupled cells (the "multicellular level").

A. Myocyte Processes and Related Arrhythmic Mechanisms

At the myocyte level, the electrical impulse can logically be divided into three components: first, impulse initiation, which encompasses normal pacemaker mechanisms; second, excitation, in which high-density inward currents depolarize the cell; and, finally, repolarization.

Each of these physiological processes (Table 26.1, left) can contribute to rhythm disturbances (Table 26.1, right). Let us first consider impulse initiation. Alterations of the rate of the physiological sinus pacemaker lead to sinus arrhythmias (51). Other rhythm disturbances can be initiated by nonphysiological pacemakers in a mechanism known as abnormal automaticity. These can either be subsidiary pacemakers (in the AV node or Purkinje fibers) or pathological excitatory mechanisms that produce automatic activity in normally quiescent tissues such as atrial muscle. The cellular mechanisms for abnormal automaticity are generally obscure (110). The final manifestation of abnormal impulse initiation is in diastolic triggered activity due to calcium overload; the cellular events underlying this form of activity are delayed afterdepolarizations (111) (DADs), whose basis has been discussed above.

The second set of myocyte processes are those that underlie excitation. In atrial and ventricular muscle, excitation occurs with the activation of sodium channels; in contrast, nodal tissues depend upon L-type calcium channels for excitation (112). When excitation is depressed, the sinus node can fail to depolarize, leading to asystole, or to failure of conduction within the AV node ("AV block"). The latter occurs commonly as a consequence of pharmacological calcium channel blockade (51). Within the ventricles, ischemia is the most common cause of lethal arrhythmias (113). Slow conduction of the electrical impulse, or complete block, can occur due to depression of sodium channels (60). These channels tend to inactivate with modest depolarization of the resting membrane potential; ischemia produces such depolarization by enabling the accumulation of potassium in the interstitial spaces surrounding the ischemic myocytes (114). Such depression can occur within minutes of the onset of ischemia, well before the onset of necrosis (114). Thus, cells that may still be viable can be electrically silenced by a failure of excitation due to sodium channel inactivation.

The final process intrinsic to each myocyte is that of repolarization. The plateau of the action potential reflects a delicate balance between inward and outward currents so that relatively little current is required to tip the balance either to repolarization or to maintained depolarization. Prolongation of the action potential, as occurs in heart failure, makes repolarization even more unstable and predisposes toward secondary depolarizations (early afterdepolarizations [EAD]), (32,33). These are the cellular events that initiate long-QT-related arrhythmias (115–117). Figure 26.3(C) shows an example of EADs precipitated by cesium-induced blockade of potassium currents in a strip of ventricular muscle. A broad variety of interventions (including hypokalemia, hypomagnesemia, and various drugs) favor the development of EADs; no single ionic mechanism can be implicated (115). The common link is perturbation of the balance between inward and outward currents during the plateau of the action potential. An increase in inward current (e.g., by incomplete inactivation of sodium channels) would have the same consequences as a net reduction of outward currents. Although EADs have been closely associated with conditions that favor the polymorphic ventricular tachycardias of the long-QT syndrome, the precise mechanism of the arrhythmia remains unclear. Spatiotemporal inhomogeneity of repolarization must play an important role (116): if EADs were to occur homogeneously throughout the myocardium, the worst that might happen would be a particularly prolonged QT interval. On the other hand, if one region of the heart were to experience EADs while others repolarize fully, the repolarized areas could recover their excitability and be reexcited by the depolarized EAD-containing region. This could lead to the opportunistic spread of depolarization wavefronts throughout the heart; without a fixed wiring abnormality to favor a stereotyped pattern of repetitive activation, the surface electrogram would register the characteristic polymorphic appearance (118). This general idea and others are being tested both experimentally and numerically.

Heart failure, whatever the initiating pathology, ends up being a highly lethal syndrome in which fully half of the deaths are due to arrhythmias (119). The mechanism of sudden death in heart failure is only now beginning to be clarified. In human and animal

(120) models of heart failure, action potentials are markedly prolonged. The primary culprit appears to be disease-related attenuation of the voltage-dependent transient outward K current (120). This observation and other lines of evidence support the hypothesis that heart failure is a common, acquired form of the long-QT syndrome (119). Thus, abnormalities of repolarization are probably of much greater clinical importance than previously realized in the pathogenesis of common lethal ventricular arrhythmias.

B. Multicellular Processes and Related Arrhythmic Mechanisms

Arrhythmias are of little consequence at the unicellular level; only when the gross pattern of cardiac excitation is altered is the function of the heart compromised. To understand how the pattern of excitation can be disturbed, it is necessary to consider how groupings of cells contribute to arrhythmias.

Two categories of factors, cell–cell coupling and network properties, must be considered. The first simply acknowledges that cardiac cells are coupled to each other. The electrical behavior of one cell will influence, and be influenced by, that of its neighbors (45). The extent of coupling determines the ease with which the impulse can travel from cell to cell. Coupling is inherently asymmetrical: heart cells are long structures that are predominantly coupled end to end rather than side to side (7). Thus, current spreads more easily along the long axis of cells than in the transverse direction, a feature known as anisotropy (121). In quantitative models of cardiac conduction, spatial irregularities make it easier to generate arrhythmias (122). For example, current can follow circuitous paths so as to reexcite tissue that is no longer refractory, producing so-called reentrant arrhythmias (39–112). Nevertheless, spatial inhomogeneity is not required for the induction or maintenance of arrhythmias. Perfectly homogeneous excitable media can exhibit complex patterns of periodic and chaotic excitation (118,123). Thus, a sufficiently large sheet of cardiac cells, coupled to each other with perfect symmetry, would still be capable of supporting arrhythmias. This important insight has arisen from the application of nonlinear dynamics to the study of cardiac excitation.

While recognizing that homogenous media can support arrhythmias, the additional complexity of spatial inhomogeneity makes arrhythmias even easier to generate (121). The factors that contribute to this spatial inhomogeneity are called network properties (Table 26.1); these include not only asymmetries of cell–cell coupling but also barriers to conduction. Such barriers can arise from naturally occurring structures such as valves and multicellular bundles (122), or from pathological remodeling of the myocardium as in infarcted tissue (113). Fibrosis, denervation, and atrial dilatation are important factors besides "electrophysiological remodeling" in the occurrence and maintenance of atrial fibrillation (124). Network properties have been the major focus of traditional work on arrhythmias; this work is extensively reviewed elsewhere (113,121). The remainder of this chapter will focus on the emerging areas in which molecular approaches offer immediate promise.

VI. OPPORTUNITIES FOR DRUG DEVELOPMENT

The most dazzling contributions of molecular genetics to the field of cardiac excitability have been in ion channel structure–function analysis. Most of the genes that encode specific transport pathways have been identified (notable exceptions include the calcium-activated transient outward current and mechanosensitive channels). A remarkable amount is known about which parts of a given protein confer specific functional properties: we can now inspect the sequence of a novel voltage-dependent channel and confidently predict which residues line the pore and which mediate activation and inactivation. Nevertheless, we are only beginning to scratch the surface of ion channel molecular biology and to put the principles learned to practical use. The following areas represent just a sample of those that are ripe for exploitation.

Rational drug design. The cloning of ion channel genes and the functional characterization of their products has identified several novel therapeutic targets. Such genes can be exploited to create empirical drug screening assays against existing libraries of compounds. A more intellectually pleasing alternative would be to use rational drug design (i.e., to progress from the known structure of a binding site to create a drug that will bind selectively to that site). Such a site might be the pore of a targeted channel. This will be impossible until we move from the cartoon phase of channel structure to that of atomic resolution. We need to become more successful at applying standard structural approaches, such as x-ray crystallography, to ion channel proteins. There is no inherent reason why this should be impossible; recently, the structures of a number of complex membrane proteins, including a potassium channel (73) and cytochrome c oxidase (125) have been resolved. More importantly, we need to broaden our horizons when searching for potential therapeutic targets. Most antiarrhythmic drugs act by blocking sodium, calcium, or potassium channels. The rationale behind this approach is obvious, but has severe limitations. Future efforts should be directed toward defining new targets that include modulation of intracellular signaling pathways.

Somatic gene transfer. We already know enough about ion channel genetics that we can design strategies to tinker with excitability by overexpressing functional ion channels. These can be tailor-made to alter specific features of excitability; for example, HERG might be overexpressed in an effort to increase refractoriness without excessively abbreviating the action potential (126). Conversely, antisense or dominant negative strategies could be applied to suppress selected gene products and thereby to probe their roles in the process of excitability. Connexin genes would also be obvious targets for overexpression or knockout in a bid to manipulate cell–cell coupling. Such approaches have already proven to be realizable for probing model systems, such as cultured cells (127,128). The realistic application of such approaches in vivo will require improved methods for somatic gene transfer.

Transcriptional regulation of genes encoding excitability proteins. Much less is known about the regulation of ion channel gene expression than is known about the structure–function relationships of the encoded proteins. Nevertheless, the expression of channels changes dramatically with development, and is quite specific in different regions of the heart. How is this temporal and spatial specificity achieved? How does expression change in arrhythmia-prone disease states such as heart failure? What nontranscriptional processes control the levels of functional excitability proteins? These problems are amenable to existing, albeit laborious, methods of genetic analysis.

Coupling of energy metabolism to excitability. Heart cells have evolved a variety of pathways whereby changes in energy metabolism feed back upon and regulate excitability and cell-cell communication. It is increasingly apparent that energy metabolism itself is far from static, and that primary oscillations in energetics may constitute a novel arrhythmogenic principle (23,24). An improved understanding of these coupling processes at the molecular and cellular levels will go a long way toward elucidating the pathogenesis of ischemic arrhythmias, which remain the biggest killers in Western society.

Improved animal models. Existing animal models of arrhythmias are far from perfect. In the case of ventricular tachyarrhythmias, individual laboratories have developed approaches with singular advantages and limitations, but no single model is widely accepted. It is understandably difficult to motivate extensive molecular characterization of any model whose relevance is questionable. Almost all efforts have been directed at large animal models with some element of superimposed ischemia. Conventional wisdom holds that large animal models are necessary to maintain arrhythmias, particularly those in which abnormal conduction patterns are a prominent feature. The physical dimensions of an electrical circuit must be large compared to the characteristic space constant of the tissue; otherwise, activation will be too nearly synchronous for the heart to be able to support arrhythmias (112). Given such considerations, transgenic and knockout approaches to arrhythmias may be more fruitfully directed at larger animals despite the obvious technical advantages of working with mice. Alternatively, somatic gene transfer methods may offer a means to alter the electrophysiological substrate in large animals without manipulation of the germ line.

Interactions with quantitative models of cardiac excitability. Readers of this chapter will readily concede that arrhythmias present a difficult challenge. While it is certain that we will continue to learn much about individual pieces of the puzzle, putting it all together will require more than simple intuition. Quantitative approaches offer unique promise in this regard. Models of excitation and contraction in individual cardiac cells are already fairly sophisticated (129,130); perturbations of individual ion conductances reproduce and predict experimental behavior reasonably well. Given the remarkable advances in numerical methods and in computing power in our time, virtually infinite refinements of any given cellular model are possible. More importantly, individual cells can be tied together into large networks using supercomputers (131–133). Such networks can be made geometrically realistic guided by anatomical and histological data. To the extent that fruitful interactions can be developed between groups of investigators from quite different cultures (computer scientists and experimental biologists), molecular advances can be used to constrain and to improve quantitative models (133). The development of reliable quantitative models of cardiac electrical activity (104, 134–136) will also facilitate the critical evaluation of molecular interventions. We will be able to begin to address a number of questions more intelligently. What will be the consequences of knocking out a particular ion channel? What is the physiological role of the transmural gradients of transient outward potassium current? In any given arrhythmia, what are the relative contributions of the various factors listed in Table 1? Can the arrhythmia be terminated or prevented by alterations of one particular ion transport pathway?

VI. CONCLUDING REMARKS

It must be emphasized that molecular approaches alone will never solve the problem of arrhythmogenesis. We need a concerted effort at various levels of integration. In order for this goal to be realized, molecular and cellular scientists need to understand the complexity of arrhythmias, to look past the daunting jargon of arrhythmias, and then to break down the problem of arrhythmogenesis into simple elements. Even if the big picture appears elusive, individual pieces are certainly amenable to reductionist approaches. Conversely, electrophysiologists need to gain an appreciation for the vast potential, as well as the limitations, of molecular genetic approaches to arrhythmogenesis. As our understanding of the molecular and cellular basis of arrhythmias advances, novel drug targets will be defined and it will likely be necessary to devise new paradigms for discovery of drugs to treat this multifactorial disorder.

ACKNOWLEDGMENTS

We thank the National Heart, Lung and Blood Institute for supporting our research on cardiac excitability and contractility. This chapter was loosely adapted from E. Marbán, Molecular approaches to arrhythmogenesis. In: K. Chien et al., eds. *Molecular Basis of Heart Disease*. Philadelphia: W.B. Saunders, 1998.

REFERENCES

1. Sakmann B, Neher E, eds. Single-Channel Recording. New York: Plenum Press, 1995.
2. Noda M, Shimizu S, Tanabe T, Takai T, Kayano T, Ikeda T, Takahashi H, Nakayama H, Kanaoka Y, Minamino N, Kangawa K, Matsuo H, Raftery M, Hirose T, Inayama S, Hayashida H, Miyata T, Numa S. Primary structure of electrophorus electricus sodium channel deduced from cDNA sequence. Nature 1984; 312:121–127.
3. Papazian DM, Schwarz TL, Tempel BL, Jan YN, Jan LY. Cloning of genomic and complementary DNA from Shaker, a putative potassium channel gene from Drosophila. Science 1987; 237:749–753.
4. Nicoll DA, Longoni S, Philipson KD. Molecular cloning and functional expression of the cardiac sarcolemmal Na^+-Ca^{2+} exchanger. Science 1990; 250:562–565.
5. Echt DS, Liebson PR, Mitchell LB. Mortality and morbidity in patients receiving encainide, flecainide, or placebo: the Cardiac Arrhythmia Suppression Trial. N Engl J Med 1991; 324:781–788.
6. Severs NJ. Pathophysiology of gap junctions in heart disease. J Cardiovasc Electrophysiol 1994; 5:462–475.
7. Saffitz JE, Davis LM, Darrow BJ. The molecular basis of anisotropy: role of gap junctions. J Cardiovasc Electrophysiol 1995; 6:498–510.
8. Beardslee MA, Laing JG, Beyer EC, Saffitz JE. Rapid turnover of connexin43 in the adult rat heart. Circ Res 1998; 83:629–635.
9. Tsien RW, Bean BP, Hess P, Lansman JB, Nilius B, Nowycky MC. Mechanisms of calcium channel modulation by beta-adrenergic agents and dihydropyridine calcium agonists. J Mol Cell Cardiol 1986; 18:691–710.
10. Leonard J, Montminy MR. Control of gene expression by transmitters and peptide hormones. Curr Opin Neurobiol 1991; 1:74–78.
11. Mori Y, Matsubara H, Folco E, Siegel A, Koren G. The transcription of a mammalian voltage-gated potassium channel is regulated by cAMP in a cell-specific manner. J Biol Chem 1993; 268:26482–26493.
12. Darrow BJ, Fast VG, Kleber AG, Beyer EC, Saffitz JE. Functional and structural assessment of intercellular communication: increased conduction velocity and enhanced connexin expression in dibutyryl cAMP-treated cultured cardiac myocytes. Circ Res 1996; 79:174–183.
13. Allen DG, Orchard CH. Myocardial contractile function during ischemia and hypoxia. Circ Res 1987; 60:153–168.
14. Kusuoka H, Porterfield JK, Weisman HF, Weisfeldt ML, Marban E. Depressed Ca^{2+}-activa-

tion as a consequence of reperfusion-induced cellular calcium overload in ferret hearts. J Clin Invest 1987; 79:950–961.

15. Myerburg RJ, Castellanos A. Cardiac Arrest and Sudden Cardiac Death. In: Braunwald E, ed. Heart Disease: A Textbook fof Cardiovascular Medicine. Philadelphia: W.B. Saunders Co., 1997:742–779.

16. Jennings RB. Symposium on the pre-hospital phase of acute myocardial infarction. Part II. Early phase of myocardial ischemic injury and infarction. Am J Cardiol 1969; 24:753–765.

17. Nichols CG, Ripol C, Lederer WJ. ATP-sensitive potassium channel modulation of the guinea pig ventricular action potential and contraction. Circ Res 1991; 68:280–287.

18. Weiss JN, Shieh RC. Potassium loss during myocardial ischaemia and hypoxia: does lactate efflux play a role? Cardiovasc Res 1994; 28:1125–1132.

19. Carlsson L, Abrahamsson C, Drews L, Duker G. Antiarrhythmic effects of potassium channel openers in rhythm abnormalities related to delayed repolarization. Circulation 1992; 85:1491–1500.

20. Grover GJ, D'Alonzo AJ, Parham CS, Darbenzio RB. Cardioprotection with the KATP opener cromakalim is not correlated with ischemic myocardial action potential duration. J Cardiovasc Pharmacol 1995; 26:145–152.

21. Wilde AA. Role of ATP-sensitive K+ channel current in ischemic arrhythmias. Cardiovasc Drugs Ther 1993; 7(suppl 3):521–526.

22. Bellemin-Baurreau J, Poizot A, Hicks PE, Rochette L, Armstrong JM. Effects of ATP-dependent K+ channel modulators on an ischemia-reperfusion rabbit isolated heart model with programmed electrical stimulation. Eur J Pharmacol 1994; 256:115–124.

23. Marban E, Ramza BM, Romashko DN, O'Rourke B. Nonjunctional channels of cardiac cells and metabolic oscillation as contributors to discontinuous conduction. In: Spooner PS, ed. Discontinuous Cardiac Conduction. Mt. Kisco, NY: Futura Press, 1996.

24. O'Rourke B, Ramza BM, Marban E. Oscillations of membrane current and excitability driven by metabolic oscillations in heart cells. Science 1994; 265:962–966.

25. O'Rourke B. Ion channels as sensors of cellular energy. Mechanisms for modulation by magnesium and nucleotides. Biochem Pharmacol 1993; 46:1103–1112.

26. Endoh M. Physiological and pathophysiological modulation of calcium signaling in myocardial cells. Jpn Circ J 1991; 55:1108–1117.

27. Barry WH, Bridge JH. Intracellular calcium homeostasis in cardiac myocytes. Circulation 1993; 87:1806–1815.

28. Kort AA, Lakatta EG, Marban E, Stern MD, Wier WG. Fluctuations in intracellular $[Ca^{2+}]$ and their effect on tonic tension in canine cardiac Purkinje fibres. J Physiol 1985; 367:291–308.

29. Wier WG, Kort AA, Stern M, Lakatta E, Marban E. Cellular calcium fluctuations in mammalian heart: direct evidence from noise analysis of aequorin signals in Purkinje fibers. Proc Natl Acad Sci 1983; 80:7367–7371.

30. Marban E, Koretsune Y, Kusuoka H. Disruption of intracellular Ca^{2+} homeostatsis in hearts reperfused after prolonged episodes of ischemia. In: Das D, ed. Cellular, Biochemical and Molecular Aspects of Reperfusion Injury. New York: Academy of Sciences, 1994:38–50.

31. Wier WG, Hess P. Excitation-contraction coupling in cardiac Purkinje fibers. Effects of cardiotonic steroids on the intracellular $[Ca^{2+}]$ transient, membrane potential, and contraction. J Gen Physiol 1984; 83:395–415.

32. Marban E, Robinson SW, Wier WG. Mechanisms of arrhythmogenic delayed and early afterdepolarizations in ferret ventricular muscle. J Clin Invest 1986; 78:1185–1192.

33. Rosen MR. Cellular electrophysiology of digitalis toxicity. J Am Coll Cardiol 1985; 5:22A–34A.

34. Levine JH, Michael JR, Guarnieri T. Treatment of multifocal atrial tachycardia with verapamil. N Engl J Med 1985; 312:21–25.

35. Kass RS, Tsien RW, Weingart R. Ionic basis of transient inward current induced by strophanthidin in cardiac Purkinje fibres. J Physiol 1978; 281:209–226.

36. Mechmann S, Pott L. Identification of Na-Ca exchange current in single cardiac myocytes. Nature 1986; 319:597–599.

37. Han X, Ferrier GR. Contribution of Na^+-Ca^{2+} exchange to stimulation of transient inward current by isoproterenol in rabbit cardiac Purkinje fibers. Circ Res 1995; 76:664–674.

38. Zygmunt AC. Intracellular calcium activates a chloride current in canine ventricular myocytes. Am J Physiol 1994; 267:H1984–H1995.

39. Hoffman BF, Rosen MR. Cellular mechanisms for cardiac arrhythmias. Circ Res 1981; 49:1–15.

40. Eisner DA. Intracellular sodium in cardiac muscle: effects on contraction. Exp Physiol 1990; 75:437–457.

41. Eisner DA, Lederer WJ. Na-Ca exchange: stoichiometry and electrogenicity. Am J Physiol 1985; 248:C189–C202.

42. Wang Z, Kimitsuki T, Noma A. Conductance properties of the Na(+)-activated K^+ channel in guinea-pig ventricular cells. J Physiol 1991; 433:241–257.

43. Orchard CH, Cingolani HE. Acidosis and arrhythmias in cardiac muscle. Cardiovasc Res 1994; 28:1312–1319.

44. Cingolani HE, Koretsune Y, Marban E. Recovery of contractility and intracellular pH during respiratory acidosis in ferret hearts: role of Na^+/H^+ exchange. Am J Physiol 1990; 259:H843–H848.

45. DeMello WC. Gap junctional communication in excitable tissues; the heart as a paradigma. Progr Biophys Mol Biol 1994; 61:1–35.

46. Marban E, Koretsune Y, Kusuoka H. Disruption of intracellular Ca2+ homeostasis in hearts reperfused after prolonged episodes of ischemia. In: Das D, ed. Cellular, Biochemical and Molecular Aspects of Reperfusion Injury. New York: New York Academy of Sciences, 1994:38–50.

47. Yamada KA, McHowat J, Yan GX, Donahue K, Peirick J, Kleber AG, Corr PB. Cellular uncoupling induced by accumulation of long-chain acylcarnitine during ischemia. Circ Res 1994; 74:83–95.

48. Nichols CG, Makhina EM, Pearson WL, Sha Q, Lopatin AN. Inward rectification and implications for cardiac excitability. Circ Res 1996; 78:1–7.

49. Hume JR, Horowitz B. A plethora of cardiac chloride conductances: molecular diversity or a related gene family. J Cardiovasc Electrophysiol 1995; 6:325–331.

50. Moorman JR, Ackerman SJ, Kowdley GC, Griffin MP, Mounsey JP, Chen Z, Cala SE, O'Brian JJ, Szabo G, Jones LR. Unitary anion currents through phospholemman channel molecules. Nature 1995; 377:737–740.

51. Ganz LI, Friedman PL. Supraventricular tachycardia. N Engl J Med 1995; 332:162–173.

52. Lerman BB, Stein K, Engelstein ED, Battleman DS, Lippman N, Bei D, Catanzaro D. Mechanism of repetitive monomorphic ventricular tachycardia. Circulation 1995; 92:421–429.

53. Lerman BB. Response of nonreentrant catecholamine-mediated ventricular tachycardia to endogenous adenosine and acetylcholine. Evidence for myocardial receptor-mediated effects. Circulation 1993; 87:382–390.

54. Noble D. The initiation of the heartbeat. Oxford, UK: Clarendon Press, 1975.

55. Sanguinetti MC, Jiang C, Curran ME, Keating MT. A mechanistic link between an inherited and an acquired cardiac arrhythmia: HERG encodes the I_{Kr} potassium channel. Cell 1995; 81:299–307.

56. Trudeau MC, Warmke JW, Ganetzky B, Robertson GA. HERG, a human inward rectifier in the voltage-gated potassium channel. Science 1995; 269:92–95.

57. Pardo LA, Heinemann SH, Terlau H, Ludewig U, Lorra C, Pongs O, Stuhmer W. Extracellular K^+ specifically modulates a rat brain K^+ channel. Proc Natl Acad Sci 1992; 89:2466–2470.

58. Zou A, Xu Q, Sanguinetti MC. A mutation in the pore region of HERG K^+ channels reduces rectification by shifting the voltage dependence of inactivation. J Physiol 1998; 509:129–138.

59. Yang T, Roden DM. Extracellular potassium modulation of drug block of I_{Kr}. Implications for torsade de pointes and reverse use-dependence. Circulation 1996; 93:407–411.

60. Hille B, ed. Ionic Channels of Excitable Membranes. Sunderland, MA: Sinauer Associates, 1992.

61. Song Y, Liu QY, Vassalle M. On the antiarrhythmic actions of magnesium in single guinea-pig ventricular myocytes. Clin Exp Pharmacol Physiol 1996; 23:830–838.

62. Orlov MV, Brodsky MA, Douban S. A review of magnesium, acute myocardial infarction and arrhythmia. J Am Coll Nutr 1994; 13:127–132.

63. Arsenian MA. Magnesium and cardiovascular disease. Progr Cardiovasc Dis 1993; 35:271–310.

64. Aupetit JF, Freysz M, Faucon G, Loufoua-Moundanga J, Coquelin H, Timour Q. Magnesium—a profibrillatory or antifibrillatory drug depending on plasma concentration, heart rate and myocardial perfusion. Acta Anaesthesiol Scand 1997; 41:516–523.

65. Romani A, Marfella C, Scarpa A. Regulation of magnesium uptake and release in the heart and in isolated ventricular myocytes. Circ Res 1993; 72:1139–1148.

66. Altschuld RA, Jung DW, Phillips RM, Narayan P, Castillo LC, Whitaker TE, Hensley J, Hohl CM, Brierley GP. Evidence against norepinephrine-stimulated efflux of mitochondrial Mg^{2+} from intact cardiac myocytes. Am J Physiol 1994; 266:H1103–H1111.

67. Nuss HB, Tomaselli GF, Marban E. Cardiac sodium channels (hH1) are intrinsically more sensitive to block by lidocaine than are skeletal muscle (μl) channels. J Gen Physiol 1996; 106:1193–1209.

68. Welling A, Kwan YW, Bosse E, Flockerzi V, Hofmann F, Kass RS. Subunit-dependent modulation of recombinant L-type calcium channels. Molecular basis for dihydropyridine tissue selectivity. Circ Res 1993; 73:974–980.

69. Roden DM. Risks and benefits of antiarrhythmic therapy. N Engl J Med 1994; 331:785–791.

70. Li GR, Feng J, Wang Z, Fermini B, Nattel S. Adrenergic modulation of ultrarapid delayed rectifier K+ current in human atrial myocytes. Circ Res 1996; 78:903–915.

71. Feng J, Wible B, Li GR, Wang Z, Nattel S. Antisense oligodeoxynucleotides directed against Kv1.5 mRNA specifically inhibit ultrarapid delayed rectifier K^+ current in cultured adult human atrial myocytes. Circ Res 1997; 80:572–579.

72. Van Wagoner DR, Pond AL, McCarthy PM, Trimmer JS, Nerbonne JM. Outward K^+ current densities and Kv1.5 expression are reduced in chronic human atrial fibrillation. Circ Res 1997; 80:772–781.

73. Doyle DA, Cabral JM, Pfuetzner RA, Kuo A, Gulbis JM, Cohen SL, Chait BT, MacKinnon R. The structure of the potassium channel: molecular basis of K^+ conduction and selectivity. Science 1998; 280:69–77.

74. Tomaselli GF, Backx P, Marban E. Molecular basis of permeation in voltage-gated ion channels. Circ Res 1993; 72:491–496.

75. Backx P, Yue DT, Lawrence JH, Marban E, Tomaselli GF. Molecular localization of an ion-binding site within the pore of mammalian sodium channels. Science 1992; 257:248–251.

76. Lipkind GM, Fozzard HA. A structural model of the tetrodotoxin and saxitoxin binding site of the Na^+ channel. Biophys J 1994; 66:1–13.

77. Colatsky TJ. Antiarrhythmic drug binding sites in cardiac K^+ channels. Circ Res 1996; 78:1115–1116.

78. Ragsdale DS, McPhee JC, Scheuer T, Catterall WA. Molecular determinants of state-dependent block of Na^+ channels by local anesthetics. Science 1994; 265:1724–1728.

79. Hockermann GH, Johnson BD, Scheuer T, Catterall WA. Molecular determinants of high affinity phenylalkylamine block of L-type calcium channels. J Biol Chem 1995; 270:22119–22122.

80. Yeola SW, Rich TC, Uebele VN, Tamkun MM, Snyders DJ. Molecular analysis of a binding

site for quinidine in a human cardiac delayed rectifier K^+ channel: role of S6 in antiarrhythmic drug binding. Circ Res 1996; 78:1105–1114.

81. Liu Y, Holmgren M, Jurman ME, Yellen G. Gated access to the pore of a voltage-dependent K^+ channel. Neuron 1997; 19:175–184.

82. Holmgren M, Shin KS, Yellen G. The activation gate of a voltage-gated K^+ channel can be trapped in the open state by an intersubunit metal bridge. Neuron 1998; 21:617–621.

83. Balser JR, Nuss HB, Orias DW, Johns DC, Marban E, Tomaselli GF, Lawrence JH. Local anesthetics as effectors of allosteric gating: lidocaine effects on inactivation-deficient rat skeletal muscle Na channels. J Clin Invest 1996; 98:2874–2886.

84. Swartz KJ, MacKinnon R. Hanatoxin modifies the gating of a voltage-dependent K+ channel through multiple binding sites. Neuron 1997; 18:665–673.

85. Swartz KJ, MacKinnon R. Mapping the receptor site for hanatoxin, a gating modifier of voltage-dependent K^+ channels. Neuron 1997; 18:675–682.

86. Sanguinetti MC, Johnson JH, Hammerland LG, Kelbaugh PR, Volkmann RA, Saccomano NA, Mueller AL. Heteropodatoxins: peptides isolated from spider venom that block Kv4.2 potassium channels. Mol Pharmacol 1997; 51:491–498.

87. Packer M. The neurohormonal hypothesis: a theory to explain the mechanism of disease progression in heart failure (editorial). J Am Coll Cardiol 1992; 20:248–254.

88. Schneider MD, McLellan WR, Black FM, Parker TG: Growth factors, growth factor response elements, and the cardiac phenotype. Basic Res Cardiol 1992; 87:33–48.

89. Zipes DP. Sudden cardiac death. Future approaches. Circulation 1992; 85:I160–I166.

90. Priori SG, Napolitano C, Schwartz PJ. Cardiac receptor activation and arrhythmogenesis. Eur Heart J 1993; 14:20–26.

91. Lab MJ. Contraction-excitation feedback in myocardium. Physiological basis and clinical relevance. Circ Res 1982; 50:757–766.

92. Franz MR, Burkhoff D, Yue DT, Sagawa K. Mechanically induced action potential changes and arrhythmia in isolated and in situ canine hearts. Cardiovasc Res 1989; 23:213–223.

93. Hansen DE, Craig CS, Hondeghem LM. Stretch-induced arrhythmias in the isolated canine ventricle. Evidence for the importance of mechanoelectrical feedback. Circulation 1990; 81:1094–1105.

94. Franz MR, Cima R, Wang D, Profitt D, Kurz R. Electrophysiological effects of myocardial stretch and mechanical determinants of stretch-activated arrhythmias. Circulation 1992; 86:968–978.

95. Dick DJ, Lab MJ. Mechanical modulation of stretch-induced premature ventricular beats: induction of mechanoelectric adaptation period. Cardiovasc Res 1998; 38:181–191.

96. Sarubbi B, Ducceschi V, Santangelo L, Iacono A. Arrhythmias in patients with mechanical ventricular dysfunction and myocardial stretch: role of mechano-electric feedback. Can J Cardiol 1998; 14:245–252.

97. Dean JW, Lab MJ. Arrhythmia in heart failure: role of mechanically induced changes in electrophysiology. Lancet 1989; 1:1309–1312.

98. Myerburg RJ, Kessler KM, Castellanos A. Sudden cardiac death. Structure, function, and time-dependence of risk. Circulation 1992; 85:I2–I10.

99. Clemo HF, Stambler BS, Baumgarten CM. Persistent activation of a swelling-activated cation current in ventricular myocytes from dogs with tachycardia-induced congestive heart failure. Circ Res 1998; 83:147–157.

100. Cameron JS, Bassett AL, Gaide MS, Lodge NJ, Wong SS, Kozlovskis PL, Myerburg RJ. Cellular electrophysiology of coronary artery ligation in chronic pressure overload. Int J Cardiol 1987; 14:155–168.

101. Nuss HB, Houser SR. Effect of duration of depolarisation on contraction of normal and hypertrophied feline ventricular myocytes. Cardiovasc Res 1994; 28:1482–1489.

102. Siri FM, Nordin C, Factor SM, Sonnenblick E, Aronson R. Compensatory hypertrophy and

failure in gradual pressure-overloaded guinea pig heart. Am J Physiol 1989; 257:H1016–H1024.

103. Taggart P, Sutton P, Lab M. Interaction between ventricular loading and repolarisation: relevance to arrhythmogenesis. Br Heart J 1992; 67:213–215.

104. Rice JJ, Winslow RL, Dekanski J, McVeigh E. Model studies of the role of mechano-sensitive currents in the generation of cardiac arrhythmias. J Theor Biol 1998; 190:295–312.

105. Riemer TL, Sobie EA, Tung L. Stretch-induced changes in arrhythmogenesis and excitability in experimentally based heart cell models. Am J Physiol 1998; 275:H431–442.

106. Bustamante JO, Ruknudin A, Sachs F. Stretch-activated channels in heart cells: relevance to cardiac hypertrophy. J Cardiovasc Pharmacol 1991; 17(suppl 2):S110–113.

107. Filipovic D, Sackin H. A calcium-permeable stretch-activated cation channel in renal proximal tubule. Am J Physiol 1991; 260:F119–129.

108. Hu H, Sachs F. Mechanically activated currents in chick heart cells. J Membr Biol 1996; 154:205–216.

109. Hu H, Sachs F. Stretch-activated ion channels in the heart. J Mol Cell Cardiol 1997; 29:1511–1523.

110. Binah O, Rosen MR. Mechanisms of ventricular arrhythmias. Circulation 1992; 85:I25–I31.

111. Ferrier GR, Saunders JH, Mendez C. A cellular mechanism for the generation of ventricular arrhythmias by acetylstrophanthidin. Circ Res 1973; 32:600–609.

112. Cranefield PF, ed. The Conduction of the Cardiac Impulse. The Slow Response and Cardiac Arrhythmias. Mount Kisco, NY: Futura Publishing Company, 1975.

113. Wit AL, Janse MJ, eds. The Ventricular Arrhythmias of Ischemia and Infarction. Electrophysiological Mechanisms. Mount Kisco, NY: Futura Publishing Company, 1993.

114. Cascio WE, Johnson TA, Gettes LS. Electrophysiologic changes in ischemic ventricular myocardium: I. Influence of ionic, metabolic and energetic changes. J Cardiovasc Electrophysiol 1995; 6:1039–1062.

115. January CT, Chau V, Makielski JC. Triggered activity in the heart: cellular mechanisms of early after-depolarizations. Eur Heart J 1991; 12:4–9.

116. Leichter D, Danilo PJ, Boyden P, Rosen TS, Rosen MR. A canine model of torsades de pointes. Pacing Clin Electrophysiol 1988; 11:2235–2245.

117. Levine JH, Spear JF, Guarnieri T, Weisfeldt ML, deLangen CD, Becker LC, Moore EN. Cesium chloride-induced long QT syndrome: demonstration of afterdepolarizations and triggered activity in vivo. Circulation 1985; 72:1092–1103.

118. Gray RA, Jalife J, Panfilov A, Baxter WT, Cabo C, Davidenko JM, Pertsov AM. Nonstationary vortexlike reentrant activity as a mechanism of polymorphic ventricular tachycardia in the isolated rabbit heart. Circulation 1995; 91:2454–2469.

119. Tomaselli GF, Beuckelmann DJ, Calkins HG, Berger RD, Kessler PD, Lawrence JH, Kass D, Feldman AM, Marban E. Sudden cardiac death in heart failure: the role of abnormal repolarization. Circulation 1994; 90:2534–2539.

120. Beuckelmann D, Nabauer E, Erdmann E. Alterations of K^+ currents in isolated human ventricular myocytes from patients with terminal heart failure. Circ Res 1993; 73:379–385.

121. Kaab S, Nuss HB, Chiamvimonvat N, O'Rourke B, Kass DA, Marban E, Tomaselli GF. Ionic mechanism of action potential prolongation in ventricular myocytes from dogs with pacing-induced heart failure. Circ Res 1996; 78:262–273.

122. Spach MS. Alignment of myocardial cells and its role in the genesis of cardiac arrhythmias. Pacing Clin Electrophysiol 1990; 13:1535–1540.

123. Spach MS, Heidlage JF. The stochastic nature of cardiac propagation at a microscopic level. Electrical description of myocardial architecture and its application to conduction. Circ Res 1995; 76:366–380.

124. Janse MJ. Why does atrial fibrillation occur? Eur Heart J 1997; 18(suppl C):C12–18.

125. Iwata S, Ostermeier C, Ludwig B, Michel H. Structure at 2.8A resolution of cytochrome c oxidase from Paracoccus denitrificans. Nature 1995; 376:660–669.

126. Miller C. The inconstancy of the human heart. Nature 1996; 379:767–768.

127. Johns DC, Nuss HB, Chiamvimonvat N, Ramza BM, Marban E, Lawrence JH. Adenovirus-mediated expression of a voltage-gated potassium channel *in vitro* (rat cardiac myocytes and *in vivo* (rat liver): a novel strategy for modifying excitability. J Clin Invest 1995; 96:1152–1158.

128. Nuss HB, Johns DC, Kaab S, Tomaselli GF, Kass DA, Lawrence JH, Marban E. Reversal of potassium channel deficiency in cells from failing hearts by adenoviral gene transfer: a prototype for gene therapy for disorders of cardiac excitability and contractility. Gene Therapy 1993; 3:900–912.

129. DiFrancesco D, Noble D. A model of cardiac electrical activity incorporating ionic pumps and concentration changes. Phil Trans R Soc Lond Ser B: Biol Sci 1985; 307:353–398.

130. Luo CH, Rudy Y. A dynamic model of the cardiac ventricular action potential. II. Afterdepolarization, triffered activity and potentiation. Circ Res 1994; 74:1097–1113.

131. Winslow RL, Kimball A, Varghese A, Noble D. Simulating cardiac sinus and atrial network dynamics on the Connection Machine. Phys D: Nonlin Phenom 1993; 64:281–298.

132. Winslow RL, Noble D, Varghese T, Adlakha C, Hothyta A. Generation and propagation of ectopic beats induced by Na-K pump inhibition in atrial network models. Proc R Soc London B 1993; 254:55–61.

133. Fast VG, Kleber AG. Cardiac tissue geometry as a determinant of unidirectional conduction block: assessment of microscopic excitation spread by optical mapping in patterned cell cultures and in a computer model. Cardiovasc Res 1995; 29:697–707.

134. Gibb WJ, Wagner MB, Lesh MD. Modeling triggered cardiac activity: an analysis of the interactions between potassium blockade, rhythm pauses, and cellular coupling. Math Biosci 1996; 137:101–133.

135. Keener JP, Panfilov AV. A biophysical model for defibrillation of cardiac tissue. Biophys J 1996; 71:1335–1345.

136. Bub G, Glass L, Publicover NG, Shrier A. Bursting calcium rotors in cultured cardiac myocyte monolayers. Proc Natl Acad Sci USA 1998; 95:10283–10287.

137. Marbán E. Molecular approaches to arrhythmogenesis. In: Chien K et al., eds. Molecular Basis of Heart Disease. Philadelphia: W.B. Saunders Co., 1998:313–328.

27

Ablation Therapy of Cardiac Arrhythmias

DOUGLAS P. ZIPES

Indiana University School of Medicine, Indianapolis, Indiana

I. INTRODUCTION

The goal of catheter ablation is to destroy myocardial tissue responsible for a tachyarrhythmia by delivering energy through a catheter that is usually placed at an endocardial position in the heart. Multiple electrodes are inserted into the leg, arm or neck veins, and the femoral artery (when left ventricular approach is required) to map (locate) the arrhythmia and ablate it. Usually the diagnostic portion (i.e., determining the mechanism and site of the tachycardia) and the therapeutic portion (i.e., the actual ablation) of the study are done during the same procedure. The ablation catheter has electrodes on its tip that are manipulated by electroanatomical mapping techniques to a position adjacent to cardiac cells determined by the map to be integrally related to the onset and/or maintenance of the arrhythmia. Most commonly, radiofrequency (RF) energy (Fig. 27.1) is delivered from an external generator through the catheter to the tip electrode. The RF energy irreversibly destroys tissue in contact with the electrodes by controlled heat production, somewhat similar to electrocautery, when the tissue temperature reaches about 50° C (Fig. 27.2). Lasers, cryothermy, focused ultrasound, and microwave energy sources have been used, but much less often than RF energy because of greater technical complexities and inability to regulate the amount of tissue destruction as finely as with RF ablation. Because of the precise delivery and minimal extent of tissue damage, RF ablation is very "forgiving." The safety, efficacy, and cost effectiveness of RF ablation, as well as an end result that actually cures the patient of the arrhythmia as noted above, make it an attractive therapeutic choice for treating many arrhythmias. Presently, the cure rates and low incidence of complications have made RF ablation the initial treatment of choice for patients with virtually all symp-

725

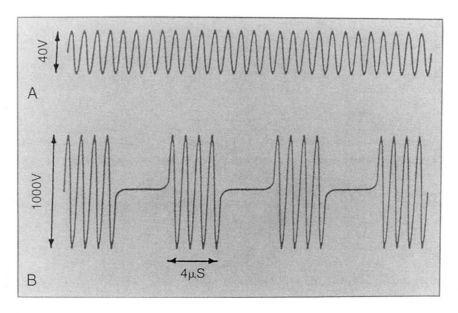

Figure 27.1 Comparison of output waveforms used for radiofrequency (A) catheter ablation and (B) electrosurgical cutting. Resistive heating during ablation is produced by a relatively low voltage (40 to 70 V) delivered in a continuous unmodulated fashion. The brief, high-voltage pulses used during electrosurgery promote arcing and coagulum formation. (From Kalbfleisch SJ, Langberg JJ. Catheter ablation with radiofrequency energy: biophysical aspects and clinical applications. J Cardiovasc Electrophysiol 1992; 3:173.)

tomatic supraventricular tachycardia, including atrial flutter, atrial tachycardia, atrioventricular (AV) nodal reentry, and AV reentry. Atrial fibrillation is also being treated with RF ablation, with varying success, depending on the type of atrial fibrillation. Many ventricular tachycardias, especially those occurring in patients with structurally normal hearts, can be ablated as well.

Application of RF ablation for a number of specific arrhythmias is discussed briefly in the sections that follow.

II. RADIOFREQUENCY MODIFICATION OF THE AV NODE TO TREAT AV NODAL REENTRANT TACHYCARDIA

AV node reentrant tachycardia (AVNRT) accounts for almost two-thirds of the patients presenting with a paroxysmal supraventricular tachycardia (PSVT). It is due to a circulating wave of excitation (reentry) over two (or more) AV nodal pathways. The two pathways that comprise the usual tachycardia circuit are called the fast and slow pathways, so named because of fast and slow conduction capabilities. These pathways are the anterosuperior and inferoposterior atrial approaches to the AV node (Fig. 27.3). Since the dual AV nodal pathways are the basis for this tachycardia, interruption of conduction in either pathway can eliminate the reentry responsible for AVNRT (1–3).

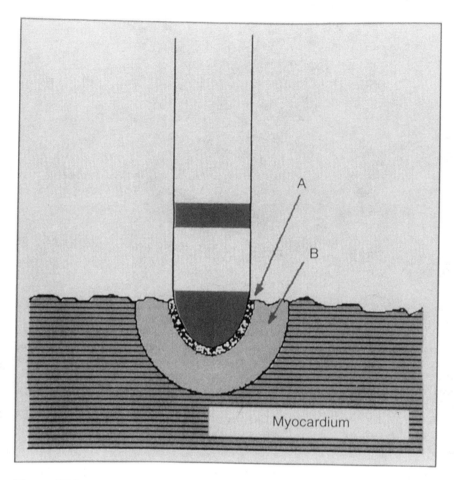

Figure 27.2 Mechanism of heating during radiofrequency catheter ablation. Because current density drops off rapidly as a function of distance from the electrode surface, only a small shell of myocardium adjacent to the distal electrode (A) is heated directly. The major portion of the lesion (B) is produced by conduction of heat away from the electrode–tissue interface into surrounding tissue. (From Langberg JJ, Leon A. Energy sources for catheter ablation. In: Zipes DP, Jalife J, eds. Cardiac Electrophysiology: From Cell to Bedside, 2nd ed. Philadelphia: W.B. Saunders Company, 1994:1434–1441.)

A. Fast Pathway Ablation

Ablation of the slow pathway is preferred because the complication of heart block is minimized and patients with an unusual form of AVNRT that involves two or more slow pathways, rather than the usual type of reentry that uses the slow pathway to conduct to the ventricles and the fast pathway back to the atria, can be treated effectively. On occasion, fast pathway ablation can be attempted if slow pathway ablation (see below) has been unsuccessful. For fast pathway ablation, the electrode tip is positioned along the AV node–His bundle axis in the anterosuperior portion of the tricuspid annulus. During energy delivery, the ECG is monitored for excessive PR prolongation and/or the occurrence of AV

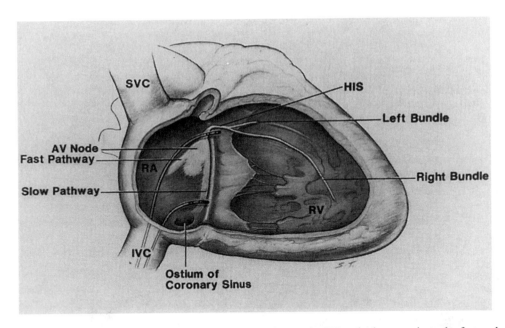

Figure 27.3 Schematic diagram of atrial approaches to the AV node that constitute the fast and slow pathways. AV node and pathways greatly enlarged for graphic purposes. Lower catheter lies posteroinferiorly over the slow pathway, while the upper catheter lies anterosuperiorly over the His bundle. SVC, superior vena cava; RA, right atrium; IVC, inferior vena cava; and RV, right ventricle. (Reproduced with permission from Zipes DP. Management of cardiac arrhythmias: Pharmacological, electrical and surgical techniques. In: Braunwald E, ed. Heart Disease. A textbook of Cardiovascular Medicine, 5th ed. Philadelphia: W.B. Saunders Company, 1997:593–639.)

block (i.e., the development of unwanted AV conduction disturbances). The initial RF pulse is delivered at 15 to 20 W for 10 to 15 s and the duration is gradually increased. Endpoints are PR prolongation, elimination of retrograde fast pathway conduction, and noninducibility of AVNRT. RF current should be discontinued if the PR interval prolongs by more than 50% or if AV block results.

The major electrophysiological effects of fast pathway ablation are elimination or marked attenuation of ventriculoatrial conduction, an increase in the A-H interval, which is a measure of AV nodal conduction time, and elimination of dual AV nodal physiology (Fig. 27.4). Starting with low energies and gradually increasing the output may reduce the risk of complete AV block, which is the most important complication associated with ablation of the fast pathway. High-degree AV block, requiring pacemaker implantation, occurs in up to 10% of patients after the ablation. This complication generally occurs during the ablation procedure, but some episodes have presented 24 h or more later, possibly as a result of the extension of the RF lesion over time. Successful elimination of AVNRT by fast pathway ablation occurs in about 80 to 95%, with a recurrent tachycardia rate of about 5 to 15%.

B. Slow Pathway Ablation

The slow pathway can be located by mapping along the posteromedial tricuspid annulus close to the coronary sinus os. Using an anatomical approach, target sites can be selected

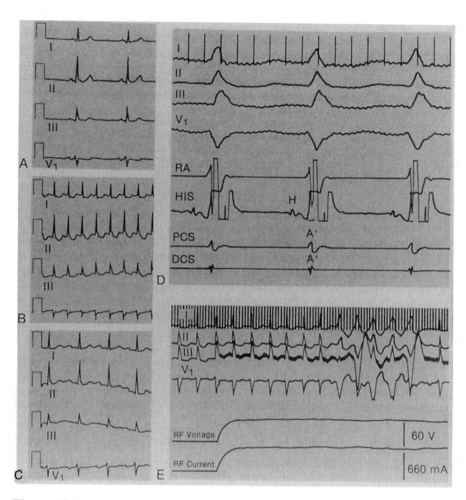

Figure 27.4 Radiofrequency (RF) fast pathway AV nodal modification for AV nodal reentrant tachycardia. (A) Normal sinus rhythm. (B) AV nodal reentrant tachycardia. (C) Normal sinus rhythm following AV nodal ablation. Note prolonged PR interval. (D) AV nodal reentrant tachycardia with intracavitary recordings. Note virtually simultaneous activation of atria and ventricles, consistent with AV nodal reentrant tachycardia. (E) Radiofrequency ablation with catheter placed in the anterior region of the AV node producing selective ablation of the anterogradely conducting fast pathway. Leads I, II, III, and V₁, scalar recordings. RA, right atrial electrogram; His, His bundle electrogram; PCS, electrogram recorded from the proximal electrodes of the coronary sinus catheter; DCS, electrogram recorded from the distal electrode of the coronary sinus catheter. Large time lines 50 ms; small time lines 10 ms. Vertical bars, calibration for RF voltage and current. Square wave for ECG = 1 mV, 200 ms. (Reproduced with permission from Zipes DP. Management of cardiac arrhythmias: Pharmacological, electrical and surgical techniques. In: Braunwald E, ed. Heart Disease. A Textbook of Cardiovascular Medicine, 5th ed. Philadelphia: W.B. Saunders Company, 1997:593–639.)

fluoroscopically by dividing the level of the coronary sinus os and the His bundle electrogram recording position into six anatomical regions. Serial RF lesions are created in each region, starting at the most posterior site and progressing to the more anterior locus. The success rate with the anatomical or electrogram mapping approach is equivalent, and most often, combinations of both are used, yielding successful elimination of AVNRT in almost 100% of patients, with less than a 1% chance of complete heart block, and a recurrence rate of 0 to 2%. Significant complications occur in less than 1 to 2%, and death from the procedure is quite rare (2).

Slow pathway ablation results in an increase in the AV nodal effective refractory period without a change in the anterograde or retrograde conduction time through the AV node. Approximately 40% of patients may have evidence of residual slow pathway function after successful elimination of sustained AVNRT, but no tachycardia. The endpoint for slow pathway ablation is the elimination of sustained AVNRT both with and without an infusion of isoproterenol and that can occur despite the presence of continued slow pathway function. At present, the fast pathway ablation approach is appropriate when the slow pathway approach has been found unsuccessful and perhaps for some patients in whom the induction of AVNRT is not reproducible because fast-pathway ablation provides a reliable endpoint of PR prolongation. In slow pathway ablation, the only reliable endpoint is elimination of tachycardia.

C. Indications

Radiofrequency catheter ablation for AV nodal reentrant tachycardia can be considered in patients with symptomatic sustained AVNRT that is drug resistant or when the patient is drug intolerant or does not desire long-term drug treatment. Since AVNRT occurs commonly in young females, the issue of subsequent pregnancy and chronic drug treatment supports the elimination of the tachycardia by RF ablation. The procedure also can be considered in patients with sustained AVNRT identified during electrophysiological study or catheter ablation of another arrhythmia.

III. RADIOFREQUENCY CATHETER ABLATION OF ACCESSORY PATHWAYS

Almost one-third of patients presenting with PSVT have an atrioventricular reentrant tachycardia (AVRT). This tachycardia is characterized in its usual (orthodromic) form by anterograde conduction to the ventricle over the normal AV conduction system, and retrograde conduction to the atrium over the accessory pathway. The latter is a muscular bridge connecting atrium with ventricle, thus bypassing the normal AV node. When the accessory pathway conducts in both anterograde and retrograde directions (it can conduct only retrogradely and still cause AVRT, so-called concealed accessory pathway), it can produce a typical Wolff–Parkinson–White (WPW) electrocardiographic abnormality during sinus rhythm. This is characterized by a short PR interval and slurring (delta wave) of the upstroke of the QRS. Patients can also have atrial tachycardia, atrial flutter, and atrial fibrillation with conduction to the ventricles over the normal or accessory pathway. These arrhythmias can result in very rapid ventricular rates due to the fast conduction over the accessory pathway. Other tachycardia variants and types of pathways exist (3).

The safety, efficacy, and cost effectiveness of RF catheter ablation of accessory atrioventricular pathways have made RF ablation the initial treatment of choice in most adult

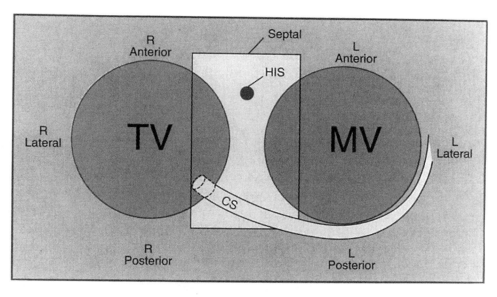

Figure 27.5 Schematic of free wall accessory pathway locations around the mitral and tricuspid annuli as visualized in the left anterior oblique projection. TV, tricuspid valve; MV, mitral valve; CS, coronary sinus; and His, His bundle. (From Miles WM, Zipes DP, Klein LS. Ablation of free wall accessory pathways. In: Zipes DP, ed. Catheter Ablation of Arrhythmias. Armonk, NY: Futura Publishing Company, 1994:211–230.)

and many pediatric patients who have AVRT or atrial tachyarrhythmias associated with a rapid ventricular response due to conduction over the accessory pathway. An electrophysiological study is performed initially to determine that the accessory pathway is an integral part of the tachycardia circuit and to locate the optimal site for ablation. Pathways can be located in the right or left free wall or septum of the heart (Fig. 27.5). Septal accessory pathways are classified as anteroseptal, midseptal, and posteroseptal (4,5). Pathways at all locations and in all age groups can be ablated successfully. Multiple pathways are present in about 5% of patients. Some pathways with epicardial locations may be more easily approached from within the coronary sinus. Direct recordings of the accessory pathway potential (Fig. 27.6) can be used to establish the optimal ablation site. In other instances, the site of the shortest conduction time between atrial and ventricular activation can be used to locate the site of insertion of the accessory pathway. Reproducible mechanical inhibition of accessory pathway conduction and subthreshold stimulation also have been used to determine the optimal site. Accidental catheter trauma should be avoided, however. Intracardiac echocardiography can be helpful at times. Accessory pathways often cross the left atrioventricular groove obliquely and therefore may not be directly across the AV groove from each other. Ablation of the accessory pathway can be accomplished from the atrial aspect of the mitral annulus or from the ventricular side.

Successful ablation sites should exhibit stable fluoroscopic and electrical characteristics. When thermistor-tipped ablation catheters are used, a stable rise in catheter tip temperature is a helpful adjunct to insure catheter stability and adequate catheter–tissue contact. In such an instance, the peak temperature generally exceeds 59°C. The retrograde transaortic and transseptal approaches have been used with equal success to ablate acces-

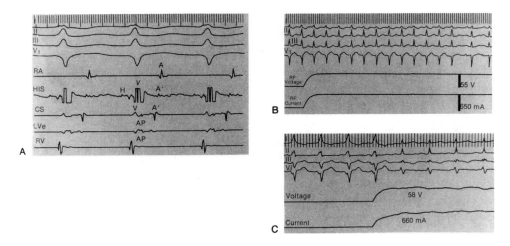

Figure 27.6   (A) and (B) Radiofrequency (RF) ablation of a left free-wall accessory pathway. (A) Depicts atrioventricular reentrant tachycardia with anterograde conduction over the normal pathway and retrograde conduction over the left free-wall accessory pathway. The electrodes in the coronary sinus (CS) record activation over the accessory pathway (AP), which is apposed by the catheter positioned in the left ventricular endocardium (LV$_e$). (B) RF energy is delivered during the tachycardia and produces termination after 3.8 ss. The delta wave has disappeared and tachycardia can no longer be initiated (not shown). (C) Radiofrequency catheter ablation of a right free-wall accessory pathway. Elimination of accessory pathway conduction almost immediately after deliver of radiofrequency energy indicates that the catheter is positioned virtually on the accessory pathway and best insures a successful ablation. Leads I, II, III, V scalar recordings. (From Zipes DP, et al. Nonpharmacologic therapy: Can it replace antiarrhythmic drug therapy? J Cardiovasc Electrophysiol 1991; 2:S255.)

sory pathways located on the left side of the heart, and the choice should reflect the experience of the electrophysiologist. Routine electrophysiological study performed weeks after the ablation procedure is generally not indicated unless the patient has a recurrent delta wave on the ECG or symptoms of tachycardia.

Patients can have atriofascicular accessory pathways that connect the right atrium with the right ventricular endocardium or right bundle branch (Fig. 27.7). These connections consist of a proximal nodelike portion responsible for conduction delay and decremental conduction properties and a long distal segment located along the endocardial surface of the right ventricular free wall that has electrophysiological properties similar to the right bundle branch. The distal end of the right atriofascicular accessory pathway can insert into the apical region of the right ventricular free wall close to the distal right bundle branch or can actually fuse with the latter. Right atriofascicular accessory pathways actually may represent a duplication of the AV conduction system and can be localized for ablation by recording potentials from the rapidly conducting distal component extending from the tricuspid annulus to the apical region of the right ventricular free wall. Ablation attempts should be performed more proximally to avoid inadvertently ablating the distal right bundle branch, which can create incessant tachycardia by lengthening the reentrant circuit.

The results of one survey by a large group of heart rhythm experts (6), showed that successful ablation of left free wall accessory pathways was obtained in 2312 of 2527

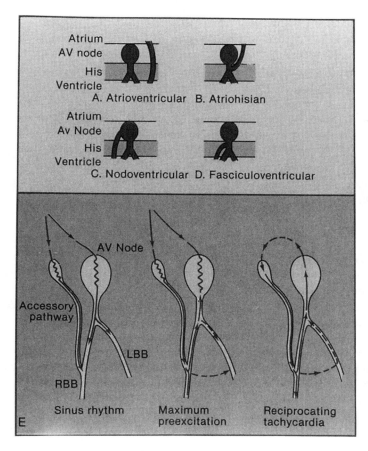

Figure 27.7 Schematic representation of accessory pathways. (A) Demonstrates the "usual" atrioventricular accessory pathway giving rise to most clinical presentations of tachycardia associated with Wolff–Parkinson–White syndrome. (B) Illustrates the very uncommon atriohisian accessory pathway. If the Lown Ganong Levine syndrome exists, it would have this type of anatomy, which has been demonstrated on occasion histopathologically. (C) Nodoventricular pathways, original concept, in which anterograde conduction travels down the accessory pathway with retrograde conduction in the bundle branch–His bundle–AV node (see below). (D) Demonstrates the fasciculoventricular connections, not thought to play an important role in the genesis of tachycardias. (E) Illustrates the current concept of nodoventricular accessory pathway in which the accessory pathway is an atrioventricular communication with AV nodal-like properties. Sinus rhythm results in a fusion QRS complex, as in the usual form of WPW shown in panel (A). Maximum preexcitation results in ventricular activation over the accessory pathway and the His bundle is activated retrogradely. During reciprocating tachycardia, anterograde conduction occurs over the accessory pathway with retrograde conduction over the normal pathway. [Panels (A)–(D) reproduced with permission from Zipes DP. Management of cardiac arrhythmias: Pharmacological, electrical and surgical techniques. In: Braunwald E, ed. Heart Disease. A Textbook of Cardiovascular Medicine, 5th ed. Philadelphia: W.B. Saunders Company, 1997:593–639. Panel (E) reproduced with permission from Benditt DG, Milstein S. Nodoventricular accessory connection: A misnomer or a structural/functional spectrum. J Cardiovasc Electrophysiol 1990; 1:231.]

(91%) patients; for septal accessory pathways, 1115 of 1279 (87%); and for right free wall accessory pathways, 585 or 715 (82%). As with ablation for AVNRT noted above, results today can be expected to be better, but the differential success rates, depending on the location of the accessory pathway, still exist. Today, competent electrophysiologists can achieve long-term elimination of tachycardia in 90 to 100% of patients, with a recurrence rate of less than 5%. Significant complications were reported in 94 of 4521 patients (2.1%) and there were 13 procedure-related deaths in 4521 patients studied (0.2%). In a European survey, the complication rate was 4.4%, with three deaths in 2222 patients (7). Complication rates not exceeding 1 to 2% are to be expected in most high-volume electrophysiology labs today. Most common complications include heart block during ablation for septal pathways and nonfatal cardiac tamponade. A death rate of 0.1% or less is in the range of the 0.05 to 0.5% annual risk of sudden death in WPW patients (2).

A. Indications

Ablation of accessory pathways is indicated in patients with symptomatic AVRT that is drug resistant or when the patient is drug intolerant or does not desire long-term drug therapy. It is also indicated in patients with atrial fibrillation (or other atrial tachyarrhythmias) and a rapid ventricular response via the accessory pathway when the tachycardia is drug resistant or when the patient is drug intolerant or does not desire long-term drug therapy. Other candidates might include patients with AVRT or atrial fibrillation, both with and without rapid ventricular preexcitation rates, whose livelihood, profession, important activities, insurability, and mental well-being, or the public safety, would be affected by spontaneous tachyarrhythmias or by the presence of the electrocardiographic abnormality, patients with atrial fibrillation and a controlled ventricular response via the accessory pathway, and patients with a family history of sudden cardiac death (8).

IV. RADIOFREQUENCY CATHETER ABLATION OF ATRIAL TACHYCARDIA, SINUS NODAL REENTRY/INAPPROPRIATE SINUS TACHYCARDIA, AND JUNCTIONAL TACHYCARDIA

Atrial arrhythmias amenable to catheter ablation include atrial tachycardias that are caused by virtually any electrophysiological mechanism including automatic, triggered, or reentrant mechanisms. In addition, sinus node reentry, inappropriate sinus tachycardia, and junctional tachycardias can be ablated. These arrhythmias account for only 5 to 10% of patients presenting with PSVTs (2).

Activation mapping is used to determine the site of the atrial tachycardia by recording the earliest onset of local atrial activation. Ten to 15% of patients may have multiple atrial foci. Sites tend to cluster near the pulmonary veins in the left atrium and the mouths of the atrial appendages and along the crista terminalis on the right. Reentrant atrial tachycardia appears to occur more commonly in the setting of structural heart disease, specifically following prior atrial surgery. The region of slow conduction varies from patient to patient depending on the atrial incisions and operation performed, such as a Fontan or Mustard procedure. Therefore, careful review of operative reports and electrophysiological mapping is essential because the atriotomy scar often plays an important role in the genesis of the tachycardia. When the sinus node area is to be ablated, it can be identified anatomically such as with intracavitary echocardiography, as well as electrophysiologically, and ablation lesions are usually placed initially between the superior vena cava and crista terminalis. Lesions can be extended lower on the crista, depending on the sinus rate response (9,10).

Results from an earlier survey showed that of 371 patients who underwent ablation for tachycardia and atrial flutter, there was a 75% success rate, with three significant complications (0.8%) and no deaths (6). The complication rate was 5% in a European survey, and there were no deaths in 141 patients (7). Today, success rates range from 80 to 90%, with recurrence rates of 5 to 10%, depending on the arrhythmia, complications are about 2% (2).

A. Indications

Candidates for RF catheter ablation include patients with atrial tachyarrhythmias that are drug resistant, patients who are drug intolerant, or those who do not desire long-term drug therapy (8).

V. RADIOFREQUENCY CATHETER ABLATION OF ATRIAL FLUTTER

Understanding the reentrant pathway for typical atrial flutter (negative sawtooth waves in leads II, III, and aV_f at a rate of about 300/min), has been essential in developing an anatomically directed ablation approach (3). Reentry in the right atrium, with the left atrium passively activated, constitutes the mechanism of typical atrial flutter with a caudocranial activation along the right atrial septum and a craniocaudal activation of the right atrial free wall, as the tachycardia circulates around the tricuspid annulus [Fig. 27.8(A)]. A zone of relatively slow conduction in the low right atrium, typically bounded by the tricuspid annulus, the inferior vena cava, and the coronary sinus, exists in the region of the slow pathway (11). Placing an ablation lesion across this zone (isthmus of tissue between the inferior vena cava orifice and tricuspid annulus) and ensuring the presence of bidirectional block at that site abolishes the typical atrial flutter (Fig. 27.9). This can be accomplished near the entrance of the slow zone in the low inferolateral right atrium, at the midpoint of the slow zone in the inferior right atrium, or near the exit at the inferomedial right atrium. Lesions can be guided anatomically or electrophysiologically. Successful elimination of typical atrial flutter today can be achieved in 90% or more patients, with a recurrence rate of 10% or less, by creating bidirectional block at the isthmus. Atypical or unusual atrial flutter can be ablated as well and has a clockwise rotation, cephalad up the right atrial free wall and caudad down the septum, with upright flutter waves in the inferior leads [Fig. 27.8(B)]. In addition, some atrial flutters can be due to surgical scars, as described for atrial tachycardias, and each location has to be determined for the individual patient (12,13). Indications are as for atrial tachycardia (8).

VI. RADIOFREQUENCY CATHETER ABLATION OF ATRIAL FIBRILLATION

Several surgical procedures involving incision and isolation of atrial myocardium have been devised to eliminate atrial fibrillation, and their feasibility has been demonstrated. Based on the surgical successes, catheter techniques for eliminating atrial fibrillation have also been developed and include AV nodal modification/ablation for rate control, ablation of focal atrial fibrillation, and ablation using linear atrial lesions (14).

To achieve RF catheter ablation of AV conduction, and rate control in patients with atrial fibrillation, a catheter is placed across the tricuspid valve and positioned to record the largest His bundle electrogram associated with the largest atrial electrogram. RF energy is

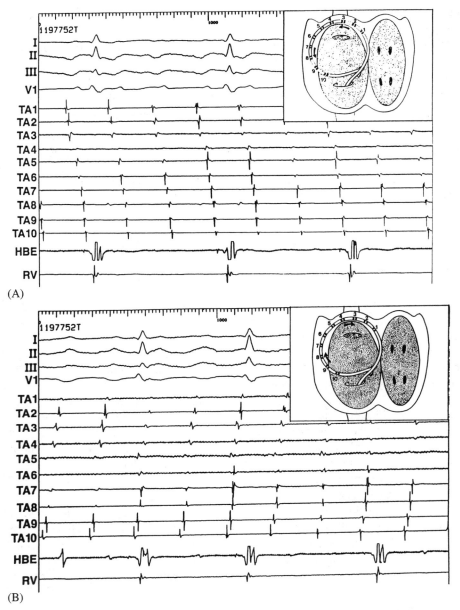

Figure 27.8 Atrial flutter. (A) records typical atrial flutter, with negative flutter waves in leads II and III. The insert (upper right) is a schematic of the right and left atria. A catheter has been inserted through the inferior vena cava and loops around the right atrium. Electrodes 1 to 10 are marked and correspond to electrogram recordings TA1 to TA10. Note that the atrial activation sequence proceeds in a counterclockwise direction from TA1 to TA10, cephaled up the septum and caudally down the right atrial free wall. HBE, His bundle electrogram; RV, right ventricular electrogram; TA, tricuspid anulus I, II, III; and V$_1$, scalar recordings. (B) Atypical atrial flutter in the same patient, with flutter waves positive in leads II and III. Recordings as in panel (A). Note that the activation sequence travels in a clockwise direction in this example from the same patient. (Reproduced with permission from Zipes DP. Management of cardiac arrhythmias: Pharmacological, electrical and surgical techniques. In: Braunwald E, ed. Heart Disease. A Textbook of Cardiovascular Medicine, 5th ed. Philadelphia: W.B. Saunders Company, 1997:593–639.)

During Isthmus Ablation

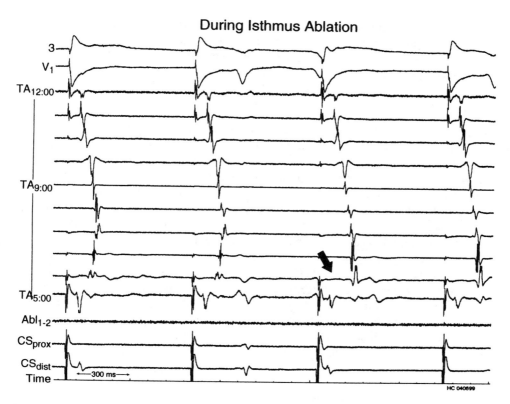

Figure 27.9   Ablation of isthmus conduction for atrial flutter. A halo catheter with multiple electrodes is positioned around the tricuspid annulus (TA) from a position registered at 12:00 through 9:00 and then 5:00. During coronary sinus (CS_{prox} and CS_{dist}) pacing conduction proceeds in the first two complexes around the tricuspid annulus in both directions, giving a "bow-shaped" distribution of the electrograms. However, at the arrow, conduction block is achieved across the isthmus by RF ablation (ABL_{1-2}), so that conduction proceeds around the TA only in one direction, from 12:00 to 9:00 to 5:00. (Tracing courtesy of John M. Miller, M.D.)

applied until complete AV block is achieved and is continued for an additional 30 s. If no change in AV conduction is observed after 15 to 20 of RF ablation, the catheter is repositioned and the attempt is repeated. Patients who fail conventional RF ablation attempts from the right ventricle can undergo an attempt from the left ventricle with a catheter positioned along the posterior interventricular septum to record a large sharp His bundle electrogram. Energy is applied between the catheter electrode and the skin patch or between catheters in the left and right ventricles. Success rates approach 100% in most studies today, with recurrence of AV conduction in less than 5%. Improved left ventricular function can result following control of the ventricular rate.

The AV junction can also be modified to slow the ventricular rate without producing complete AV block by ablating in the region of the slow pathway, as described under AV nodal modification for AV nodal reentry. Long-term rate control without need for pacing can be achieved in about 75% of patients, with the remaining requiring a pacemaker for AV block (2). This procedure can be tried prior to producing complete AV block (15).

Results from a large United States survey (6) indicated that AV junctional ablation was successful in producing complete AV block in 95% of 1600 patients, with significant complications occurring in 21 (1.3%) and two procedure-related deaths (0.1%) (6). In Europe, the complication rate was 3.2%, and there was one death in 900 patients (7). Today, successful interruption of AV conduction can be achieved in almost 100% of patients with minimal complications (2).

A. Indications

Ablation and modification of AV conduction can be considered in patients with symptomatic atrial fibrillation or other atrial tachyarrhythmias who have inadequately controlled ventricular rates, despite optimal drug therapy, unless primary ablation of the atrial tachyarrhythmia is possible, or when drugs are not tolerated or the patient does not wish to take them, even though the ventricular rate can be controlled. Other patients include those with symptomatic nonparoxysmal junctional tachycardia that is drug resistant or when drugs are not tolerated or are not desired, patients resuscitated from sudden cardiac death due to atrial flutter or atrial fibrillation with a rapid ventricular response in the absence of an accessory pathway, and patients with a dual-chamber pacemaker and a pacemaker-mediated tachycardia that cannot be treated effectively by drugs or by reprogramming the pacemaker. Modification, when successful, eliminates the need for a permanent pacemaker, but has less reliable rate control than does producing complete AV block and may predispose some patients to bradycardia-related complications, especially during the first few weeks after the procedure. In both approaches, bradycardia-dependent sudden death due to development of a long-QT interval and polymorphic ventricular tachyarrhythmias must be prevented (8).

B. Ablation of Atrial Fibrillation

Linear lesions placed in the right and left atrial endocardium and epicardium by RF catheter ablation have been used in an attempt to replicate the success of the surgical Maze procedure. The underlying hypothesis is based on the concept that reentrant wavelets maintain atrial fibrillation, and reduction of the size of the anatomical substrate by compartmentalization of the atria would eliminate the atrial fibrillation. Preliminary experience indicates that this approach is successful in abolishing atrial fibrillation in about 50% of patients, with a higher success when antiarrhythmic drugs are administered. Success rates are also higher when left atrial ablation lesions are administered than when only the right atrium is ablated. Residual atrial flutter, perhaps due to incomplete ablation lesions, is a problem, and this procedure must still be regarded as experimental (14).

An unknown number of patients have a focal source of atrial fibrillation (14, 16–18). Patients with this type of atrial fibrillation are frequently young, without other evidence of heart disease, and have salvos of atrial premature complexes, some of which can be seen to induce atrial fibrillation. One or more rapidly discharging foci, often originating in one of the pulmonary veins, can conduct to the atrium and cause disorganized atrial discharge consistent with atrial fibrillation (Fig. 27.10). Of 97 such patients, Haissaguerre et al. (14) found one focus in 50 patients, two foci in 22, three in 19, and four in six patients, totaling 175 foci, of which 164 (95%) originated from the pulmonary veins. Ablation of the initiating trigger eliminated the atrial fibrillation and depended on the number of foci, with 90% success in patients with one focus, 67% with two, and 25% when more are involved. Ablation procedures can be thwarted by lack of the presence of atrial premature complexes at the time of study.

The implications of these observations are rather important from an electrophysio-

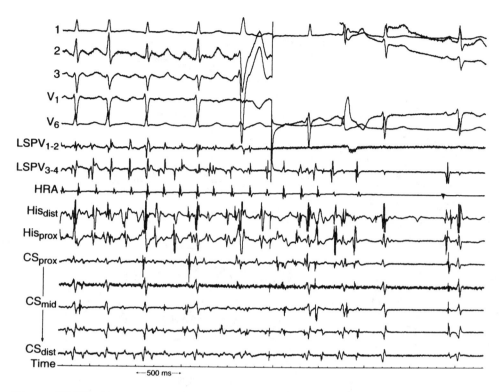

Figure 27.10 Focal atrial fibrillation from left superior pulmonary vein (LSPV). Frequent episodes of paroxysmal atrial fibrillation were all precipitated by premature atrial complexes of the same morphology originating in the LSPV in this 18-year-old female. Ablation at that site eliminated all episodes of atrial fibrillation. Conventions as in previous figure. (Tracing courtesy of John M. Miller, M.D.)

logical as well as a clinical standpoint. Electrophysiologically, whether a rapidly discharging focus can produce atrial fibrillation, long thought to be due to multiple wavelet reentry, has been the subject of much debate over many years. The question has been answered convincingly by these recent observations (14). Clinically, it is now clear that a relatively simple (compared with creating linear ablation lines) ablation approach can effectively eliminate atrial fibrillation in appropriate patients. How often focal discharge causes "routine" atrial fibrillation has yet to be established. A potential complication of ablation in or around the pulmonary veins is the subsequent development of pulmonary venous stenosis and pulmonary hypertension. Thus, the procedure has to be performed carefully and by skilled electrophysiologists. New approaches to isolating the pulmonary veins without producing pulmonary venous stenosis are being investigated.

VII. RADIOFREQUENCY CATHETER ABLATION OF VENTRICULAR TACHYCARDIA

In general, the success rate for ablation of ventricular tachycardias is less than for AV nodal or AV reentry, in part due to very difficult mapping and ablation requirements in the thick-walled ventricles (Fig. 27.11). Further, the ventricular tachycardia must be reproducibly

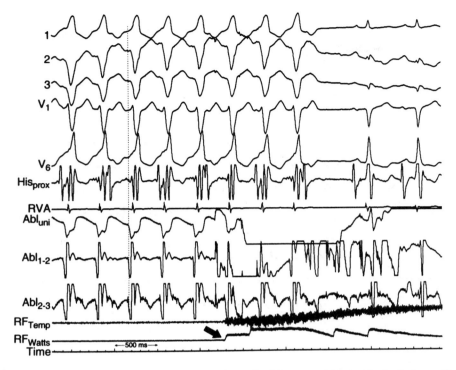

Figure 27.11   Ablation of ventricular tachycardia. Monomorphic ventricular tachycardia following an inferior wall myocardial infarction in a 57-year-old man is ablated shortly after the onset of
delivery of radiofrequency energy (arrow). Leads I, II, III, V1, and V6 of the scalar ECG are displayed. His$_{prox}$, proximal His electrogram; RVA, right ventricular apical electrogram; ABL$_{uni}$, unipolar recording from the ablation catheter; ABL$_{1-2}$ and ABL$_{2-3}$ bipolar recordings from the ablation
catheter; RF$_{temp}$ and RF$_{watts}$, temperature and energy recordings during ablation. Dashed line indicates onset of QRS complex. (Tracing courtesy of John M. Miller, M.D.)

inducible, monomorphic, sustained, and hemodynamically stable so that the patient can
tolerate the ventricular tachycardia during the procedure. Also, the origin of the ventricular
tachycardia must be fairly circumscribed and, preferably, endocardially situated, although
epicardial origins have been ablated. Very rapid ventricular tachycardias, polymorphic
ventricular tachycardias, and infrequent nonsustained episodes are not amenable to this
form of therapy at this time. Newer mapping technologies presently being investigated,
such as noncontact balloon electrodes that are used to plot immediate activation sequences
using mathematical techniques, may change those restrictions in the near future.

Radiofrequency catheter ablation of ventricular tachycardia is divided into three major categories: (1) idiopathic ventricular tachycardias that occur in patients with essentially normal hearts; (2) ventricular tachycardias that present in a variety of disease settings
but without coronary artery disease; and (3) ventricular tachycardia in patients with coronary artery disease (19). In the first group, the ventricular tachycardias originate most commonly in the right ventricular outflow tract and less often in the inflow tract or in the left
ventricular outflow tract (20). Left ventricular tachycardias are characteristically septal in
origin (21). Ventricular tachycardias in abnormal hearts without coronary artery disease

can be due to bundle branch reentry, usually a characteristic ventricular tachycardia of dilated cardiomyopathies. In these patients, tachycardia is maintained by excitation circulating over the bundle branches and ablation of the right bundle branch, to produce right bundle branch block, eliminates the tachycardia. Ventricular tachycardia can occur in patients with right ventricular dysplasia, hypertrophic cardiomyopathy, and a host of other noncoronary disease problems. Frequent ventricular complexes can be ablated (22).

Activation mapping to plot the actual spread of excitation, and pace mapping, are effective in patients with idiopathic ventricular tachycardias to locate the site of origin of the ventricular tachycardia. Pace mapping involves stimulation of various ventricular sites to initiate a QRS contour that duplicates the QRS contour of the spontaneous ventricular tachycardia, thus establishing the apparent site of origin of the arrhythmia. This technique is limited by several methodological problems, but may be useful when the tachycardia cannot be initiated and when a 12-lead ECG has been obtained during the spontaneous ventricular tachycardia. Localization of the site of origin of ventricular tachycardia in patients with coronary artery disease is more difficult than in patients with structurally normal hearts because of the altered anatomy and electrophysiology. Pace mapping is not as helpful as it is for idiopathic ventricular tachycardia. Further, reentry circuits can sometimes be large and resistant to the relatively small lesions produced by RF catheter ablation. Finding the area of slow conduction used as part of the reentrant circuit is helpful, since ablation at this site has a good chance of eliminating the tachycardia. Purkinje potentials can be recorded in some patients with left ventricular tachycardias (21). New, non-contact-mapping approaches, as mentioned earlier, appear useful (23).

In patients without structural heart disease, only a single ventricular tachycardia is usually present, and catheter ablation of that ventricular tachycardia is curative. In patients with extensive structural heart disease, especially those with prior myocardial infarction, multiple ventricular tachycardia pathway are often present. Catheter ablation of a single ventricular tachycardia in such patients may only be palliative and may not eliminate the need for further antiarrhythmic therapy, such as an implanted cardioverter defibrillator.

In one U.S. survey (6), 429 patients underwent ablation with an overall success rate of 71%. In 224 patients with structurally normal hearts, the success rate was 85%. The success rate was 54% in 115 patients with ventricular tachycardia due to ischemic heart disease and 61% in 90 patients with idiopathic cardiomyopathy. There were 13 significant complications (3.0%) and, interestingly, considering the nature of the disease, no procedure-related deaths reported. The complication rate was 7.5% in a European survey, and there was one death in 320 patients (7). Today, successful elimination of ventricular tachycardias ranges from 67% to 96%, with minimal complications (2).

A. Indications

Patients considered for RF catheter ablation of ventricular tachycardia are those with symptomatic sustained monomorphic ventricular tachycardia when the tachycardia is drug resistant, when the patient is drug intolerant, or when the patient does not desire long-term drug therapy, patients with bundle branch reentrant ventricular tachycardia, and patients with sustained monomorphic ventricular tachycardia and an implantable cardioverter-defibrillator (ICD) who are receiving multiple shocks not manageable by reprogramming or concomitant drug therapy. In some patients, the ventricular tachycardia may be too slow to be recognized by the ICD and still cause symptoms. Occasionally, nonsustained ventricular tachycardia or even severely symptomatic premature ventricular complexes can be eliminated by RF catheter ablation.

VIII. CONCLUSIONS

RF catheter ablation has revolutionized the treatment of cardiac arrhythmias, offering cures to many thousands of patients. The virtual immediate restoration to full health by such a procedure in many patients, without a need for continuing medical and drug therapy is unprecedented in the history of cardiology. Drawbacks include the costs and the risk of complications (24). However, given the totality of the issues involved, for most patients with symptomatic tachyarrhythmias that are amenable to elimination with RF ablation, the choice for such therapy should be considered early in the treatment course.

ACKNOWLEDGMENT

Supported in part by the Herman C. Krannert Fund, by grant HL 52323 from the National Heart, Lung and Blood Institute of the National Institutes of Health.

REFERENCES

1. Langberg JJ, Leon A, Borganelli M, et al. A randomized, prospective comparison of anterior and posterior approaches to radiofrequency catheter ablation of atrioventricular nodal reentry tachycardia. Circulation 1993; 87:1551.
2. Morady F. Radio-frequency ablation as treatment for cardiac arrhythmias. N Engl J Med 1999; 340:534–544.
3. Zipes DP. Specific arrhythmias: diagnosis and treatment. In: Braunwald E, ed. Heart Disease. A Textbook of Cardiovascular Medicine, 5th ed. Philadelphia: W.B. Saunders, 1997; 640–704.
4. Scheinman MM, Wang YS, Van Hare GF, et al. Electrocardiographic and electrophysiologic characteristics of anterior, midseptal and right anterior free wall accessory pathways. J Am Coll Cardiol 1992; 20:1220.
5. Haissaguerre M, Marcus F, Poquet F, et al. Electrocardiographic characteristics and catheter ablation of parahissian accessory pathways. Circulation 1994; 90:1124.
6. Scheinman MM. Patterns of catheter ablation practice in the United States: Results of the 1992 NASPE survey. PACE 1994; 17:873.
7. Hindricks G, on behalf of the Multicentre European Radiofrequency Survey (MERFS) Investigators of the Working Group on Arrhythmias of the European Society of Cardiology: The Multicentre European Radiofrequency Survey. Complications of radiofrequency catheter ablation of arrhythmias. Eur Heart J 1993; 14:1644.
8. Zipes DP, DiMarco JP, Gillette PC, et al. ACC/AHA guidelines for clinical intracardiac electrophysiologic procedures. Circulation 1995; 92:673. J Am Coll Cardiol 1995; 26:555. J Cardiovasc Electrophysiol 1995; 6:652.
9. Chen SA, Tai CT, Chiang CE, Ding YA, Chang MS. Focal atrial tachycardia: reanalysis of the clinical and electrophysiologic characteristics and prediction of successful radiofrequency ablation. J Cardiovasc Electrophysiol 1998; 9:355–365.
10. Markowitz SM, Stein KM, Mittal S, Slotwiner DJ, Lerman BB. Differential effects of adenosine on focal and macroreentrant atrial tachycardia. J Cardiovasc Electrophysiol 1999; 10:489–502.
11. Shah DC, Takahashi A, Jais P, Hocini M, Clementy J, Haissaguerre M. Local electrogram-based criteria of cavotricuspid isthmus block. J Cardiovasc Electrophysiol 1999; 10:662–669.
12. Gandhi SK, Bromberg BI, Schuessler RB, Turken BJ, Boineau JP, Cox JL, Huddleston CB.

Characterization and surgical ablation of atrial flutter after the classic Fontan repair. Ann Thorac Surg 1996; 61:1666–1678.

13. Gatzoulis MA, Freeman MA, Siu, SC, Webb, GD, Harris L. Atrial arrhythmia after surgical closure of atrial septal defects in adults. N Engl J Med 1999; 340:839–846.

14. Haissaguerre M, Shah DC, Jais P. Atrial fibrillation: Mapping insights and curative approaches. In: Califf RM, Isner J, Prystowsky EN, Serruys P, Swain J, Thomas J, Thompson P, Young JH, eds. Textbook of Cardiovascular Medicine, Updates. Cedar Knolls, NJ: Lippincott-Raven, 1999:1–11.

15. Narasimhan C, Blanck Z, Akhtar M. Atrioventricular nodal modification and atrioventricular junctional ablation for control of ventricular rate in atrial fibrillation. J Cardiovasc Electrophysiol 1998; 9:S146–150.

16. Jais P, Haissaguerre M, Shah DC, Chouairi S, Gencel L, Hocini M, Clementy J. A focal source of atrial fibrillation treated by discrete radiofrequency ablation. Circulation 1997; 95:572–576.

17. Chen SA, Tai CT, Yu WC, Chen YJ, Tsai CF, Hsieh MH, Chen CC, Prakash VS, Ding YA, Chang MS. Right atrial focal atrial fibrillation: electrophysiologic characteristics and radiofrequency catheter ablation. J Cardiovasc Electrophysiol 1999; 10:328–335.

18. Hwang C, Karagueuzian HS, Chen PS. Idiopathic paroxysmal atrial fibrillation induced by a focal discharge mechanisms in the left superior pulmonary vein: possible roles of the ligament of Marshall. J Cardiovasc Electrophysiol 1999; 10:636–648.

19. El-Shalakany A, Hadjis T, Papageorgiou P, Monahan K, Epstein L, Josephson ME. Entrainment/mapping criteria for the prediction of termination of ventricular tachycardia by single radiofrequency decision in patients with coronary artery disease. Circulation 1999; 99:2283–2289.

20. Shimoike E, Ohnishi Y, Ueda N, Maruyama T, Kaji Y. Radiofrequency catheter ablation of left ventricular outflow tract tachycardia from the coronary cusp: a new approach to the tachycardia focus. J Cardiovasc Electrophysiol 1999; 10:1005–1009.

21. Tsuchiya T, Okumura K, Honda T, Honda T, Iwasa A, Yasue H, Tabuchi T. Significance of late diastolic potential preceding Purkinje potential in verapamil-sensitive idiopathic left ventricular tachycardia. Circulation 1999; 99:2408–2413.

22. Seidl K, Schumacher B, Hauer B, Jung W, Drogemuller A, Senges J, Luderitz B. Radiofrequency catheter ablation of frequent monomorphic ventricular ectopic activity. J Cardiovasc Electrophysiol 1999; 10:924–934.

23. Schilling RJ, Peters NS, Davies DW. Feasibility of a noncontact catheter for endocardial mapping of human ventricular tachycardia. Circulation 1999; 99:2543–2552.

24. Wellens, HJJ. Catheter ablation of cardiac arrhythmias. Usually cure, but complications can occur. Circulation 1999; 99:195–197.

28

Risk Identification by Noninvasive Markers of Cardiac Vulnerability

RICHARD L. VERRIER

Harvard Medical School and Beth Israel Deaconess Medical Center, Boston, Massachusetts

RICHARD J. COHEN

Harvard–MIT Division of Health Sciences and Technology, Massachusetts Institute of Technology, Cambridge, Massachusetts

I. INTRODUCTION

Sudden cardiac death is a preeminent public health problem (1). Effective therapy to prevent this event today primarily involves placement of an implantable cardioverter defibrillator (ICD) in survivors of cardiac arrest (2) and postmyocardial infarction patients identified to be at high risk (3), which has resulted in highly significant reductions in total mortality. However, the advent of the ICD and other therapies has not significantly reduced the number of sudden cardiac deaths, because only a small fraction of the large group of patients who will ultimately die of sudden cardiac death are prospectively identified and treated. The reason is that most of the sudden cardiac deaths occur in large patient populations that have a relatively low sudden death rate (Fig. 28.1) (4). For example, the sudden death rate of survivors of cardiac arrest is approximately 25% per year, but accounts for only about 50,000 sudden deaths per year in the United States. In contrast, for patients with any prior coronary event, the sudden death rate is approximately 5% per year, accounting for about 150,000 sudden deaths. The disproportion in the total number of sudden cardiac deaths in these groups occurs because the group of patients with prior coronary events is so much larger than the cardiac arrest survivor group. Furthermore, approximately 30% of sudden cardiac deaths occur in patients without a prior history of heart disease.

Prevalence and success of prophylactic treatment by ICDs are limited in low-risk populations primarily due to the cost of ICD therapy. The cost-effectiveness ratio increas-

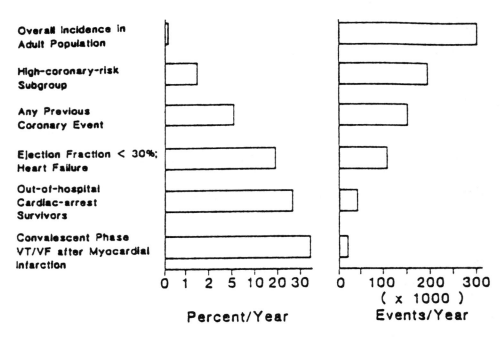

Figure 28.1 Sudden cardiac deaths among population subgroups. Estimates of incidence (percent per year) and total number of sudden cardiac deaths per year are shown for the overall adult population in the United States and for higher-risk subgroups. The overall estimated incidence is 0.1% or 0.2% per year, totaling more than 300,000 deaths per year. Within subgroups identified by increasingly powerful risk factors, the increasing incidence is accompanied by progressively decreasing total numbers. Practical interventions for the larger subgroups will require identification of higher-risk clusters within the groups. VT/VF = ventricular tachycardia/ventricular fibrillation. The horizontal axis for the incidence figures is nonlinear. (From Ref. 4.)

es if therapy is applied to progressively lower risk groups, because more devices are implanted in more patients who will not need them. For example, a medical care system might support ICD implantation in patients who have a 2-year probability of sudden death of 25%; in this case, four ICDs could result in up to two patient-years of life saved. However, if the 2-year sudden death rate is 2.5%, it would require 40 ICDs to achieve at most two patient-years of life saved. The cost of prophylactic treatment of this latter group is not likely to be acceptable, and ICD therapy here might not be justified in view of the limited benefit.

The challenge is how to apply effective treatments in the low-risk patient groups in which the large majority of sudden deaths occur. The ideal solution to this problem lies in developing effective noninvasive systems of risk stratification. With accurate screening of a moderate-risk group, such as patients with heart failure, medically and economically feasible treatment strategies may become possible. A complementary approach would be to develop lower cost and lower morbidity therapies that can be applied in lower risk groups. Here, effective risk stratification could also play a role by guiding physicians in choosing appropriate therapy. For example, recent studies (5,6) suggest the drug amiodarone may have a beneficial effect in reducing arrhythmic mortality. However, the benefit in moderate-risk populations appears limited; if risk stratification could be used to assess effectiveness of treatment in different subgroups, then pharmacological therapy could be applied only to those patients most likely to benefit.

Recent clinical studies provide some important implications for the development of effective stratification for risk. The relevant studies can be divided into two categories—natural history and therapeutic trials. The natural history studies involve evaluation of various diagnostic tests as predictors of outcome events. These might be documented arrhythmias or sudden death, but also may be the result of a reference diagnostic test, such as programmed electrophysiological study. One special category of natural history investigation is the case control study. Here, tests are performed on matched patients and the endpoint, a prior event, selects the patients for the study. The results of this type of study are far less reliable than prospective studies not only because of the possibility of selection bias, but also because of pathophysiological considerations. Specifically, the index event may alter the underlying physiological substrate, potentially introducing bias into later evaluations. In addition, the occurrence of an event in the past may not be a good predictor of the occurrence of a future event. For example, patients who have an episode of sustained ventricular tachycardia or fibrillation due to an unstable electrical substrate are probably at high risk of a future episode. However, patients who have an episode of sustained ventricular tachycardia or fibrillation due to a myocardial infarction might have a stable electrical substrate and, short of a recurrent infarction, might be at lesser risk. In addition, even though a natural history study may indicate that a given patient group is at high risk for sudden cardiac death, this result does not necessarily mean that a given treatment will eliminate that risk. For example, with patients with extremely poor ventricular function at increased risk for sudden cardiac death, an ICD might have only limited effectiveness if such patients develop ventricular tachycardia or fibrillation that is refractory to countershock. Or, even if such patients are successfully cardioverted by the ICD, their lives might only be extended for a short period as they may die of heart failure due to reduced ventricular function. Early data from the CASH study (7) revealed, for example, that three of eight patients with appropriate ICD discharges died within 2 months after this event, whereas another study found that the death rate among ICD recipients with New York Heart Association class III heart failure was 13.8%, 28.6%, and 44.1% at 1, 3 and 5 years (8).

Another important issue in the evaluation of risk stratifiers is that because classification of deaths as arrhythmic is difficult, the tendency has been to use total mortality as the endpoint of the study. In many patient populations, approximately half of the deaths are due to tachyarrhythmia and, of those, some events may be due to processes, such coronary artery plaque disruption, which would not be predicted by most arrhythmic risk stratifiers. Thus, arrhythmic risk stratifiers would be expected to predict only a fraction of the deaths that could be prevented by antiarrhythmic therapy.

The second type of study, a therapeutic study, directly tests the hypothesis that a given treatment benefits a selected group of patients. For purposes of evaluating risk, the most direct approach is one in which patients in a specific category are randomized for treatment. This category may be identified in part in terms of the outcome of a risk stratification test. Then, event rates are determined among patients in the category with and without prophylactic treatment. Alternatively, in a substudy design, all patients who meet the entry criteria, not including the risk stratification test, are randomized to treatment and also undergo the risk stratification test. Then, event rates are determined among four groups of patients (i.e., those with a positive and a negative test result, with or without prophylactic treatment). This design requires a substantially larger number of randomized patients, whereas the former approach requires a larger number of patients to be screened for entry into the study. Analysis of endpoint data thus enables determination of the utility of the risk stratifier in identifying a patient population that benefits from the specific prophylactic treatment.

Natural history studies are primarily useful in identifying risk stratification tests for evaluation in prospective therapeutic trials. Only therapeutic trials can determine whether a therapy directed by a risk stratification test results in an improved outcome. Therapeutic trials tend to be much more difficult and expensive to conduct than natural history trials, but in their absence, natural history studies may be useful if they identify populations at extraordinarily high risk that warrants medical intervention, such as placement of an ICD, even prior to a trial.

Therapeutic trials of arrhythmia therapies have been fewer in number than natural history studies relevant to the individual risk stratification. Recent ones have focused on evaluating ICD treatment. The MADIT trial, for example (3), followed 196 postmyocardial infarction patients with New York Heart Association functional class I, II, or III disease, with a left ventricular ejection fraction ≤0.30 and the presence of nonsustained ventricular tachycardia on Holter monitoring. In order to be eligible for randomization to ICD therapy, patients also must have had inducible ventricular tachyarrhythmia at programmed electrophysiological stimulation testing that was not suppressible with administration of an antiarrhythmic agent. ICD therapy resulted in a significant improvement in survival (approximately 84% versus 70%) at 2 years. In contrast, the CABG Patch trial (9) involved patients with a left ventricular ejection fraction ≤0.40 and a positive signal-averaged ECG who were scheduled for coronary artery bypass graft surgery. Nine hundred patients were randomized at the time of the surgery. Over the course of a 5-year follow-up, no significant difference in survival between ICD and non-ICD treatment was detected (10).

Clinical characteristics of CABG Patch trial patients and MADIT patients were quite similar in terms of the presence of coronary artery disease and low left ventricular ejection fraction, and yet the outcomes of the studies were extremely different (11). One basis for such differences in the effectiveness of ICD therapy as assessed by the two trials is that patients were assigned to ICD therapy using quite differing risk criteria but with equivalent effective risk stratification. An alternative explanation is that bypass graft surgery was effective in preventing arrhythmias.

Many drug therapy trials for the prevention of arrhythmic events have also been conducted in postmyocardial infarction patients and heart failure patients with mixed results (12–18).

Substantial benefit for prophylactic therapies directed specifically at preventing sudden cardiac death, even in fairly high-risk populations, appears to have been demonstrated only in patients treated according to risk stratification based either on a prior cardiac arrest (2) or on invasive electrophysiological study (3). Large studies of prophylactic ICD and amiodarone therapy (SCD-Heft and MADIT II) are now underway in heart failure patients with no additional risk stratification but results will not be available for some time and it will be interesting to see if improvements in survival are demonstrated without effective stratification approaches.

II. CRITERIA FOR EVALUATION OF AN EXISTING RISK STRATIFIER

Risk stratification tests must pass certain benchmarks to have practical utility in clinical decision making. Below we have incorporated and adapted some of the criteria proposed by Surawicz (19).

1. The method of measurement of the risk stratifier must be sufficiently standardized so that the measured values can be reproduced at different treatment sites. Generally, at the inception of a new method, multiple variations in the method of measurement are developed by different investigators. For tests involving signal processing, the development of commercial equipment may provide a means for implementing standardized measurements.

2. "Normal" results of the standardized test must be established for various patient populations. Normal values obtained from healthy 25-year-old individuals or older patients with different pathology may not be relevant for risk stratification of heart failure patients.

3. The measured parameter must predict endpoint events in different patient populations. These data come from natural history studies. Common statistical indices of performance include sensitivity, specificity, positive predictive value, negative predictive value, and relative risk. The values of each of these must be specified with respect to the time elapsed since the initial test. This information generally evolves with the duration of follow-up, as additional endpoint events occur. Statistical indices can then be computed with confidence intervals using actuarial Kaplan-Meier curves (e.g., Figs. 3, 5, and 6). *Sensitivity* is the fraction of patients with events who were identified by the test. *Specificity* is the fraction of nonevent patients who were correctly identified by the test. *Positive predictive value (PPV)* is the fraction of patients with a positive test who had an endpoint event. *Negative predictive value (NPV)* is the fraction of patients with a negative test who did not have an endpoint event. *Relative risk (RR)* is the ratio of the probability of the occurrence of an endpoint event for a patient who tests positive, compared to the probability of that event for a patient who tests negative (Table 28.1). Sensitivity and specificity in principle do not depend on the fraction of patients in the population who will have events, whereas the positive and negative predictive value and relative risk are highly dependent on this fraction. In practice, all of these statistical indices are population-dependent; thus it is important that values of these indices be established in each patient population studied.

The indices that are most useful clinically are: sensitivity, positive predictive value, and relative risk. While extremely high sensitivity is not required for a test to be clinically useful, a very low sensitivity may render administration of the test costly or not worthwhile. It may also be desirable to adjust the cut-point of a parameter (the point above which the test is either positive or negative) to increase sensitivity at the expense of specificity in cases of serious disease, especially if treatment is available (20). If the treatment is either expensive or has significant morbidity, a high positive predictive value is required to justify treatment. A test with a low positive predictive value may have limited application if the treatment cannot be justified. However, an inexpensive test with high sensitivity and moderate positive predictive value could be used for screening and referral for invasive or

Table 28.1 Definition of Statistical Measures

	Event	No Event
Test Positive	True Positive (TP)	False Positive (FP)
Test Negative	False Negative (FN)	True Negative (TN)

Sensitivity = TP/(TP + FN); specificity = TN/(TN + FP); positive predictive value = TP/(TP + FP); negative predictive value = TN/(FN + TN); relative risk = positive predictive value/(1 − negative predictive value).

more expensive testing. Relative risk indicates whether or not the test provides information beyond the natural survival statistics of the population in question.

In most natural history studies, the various survival groups are usually compared and statistical significance established by computing "p value." A significant p value indicates simply that the sets of survival curves are statistically different, not that the difference is necessarily large enough to be meaningful on a clinical level. If there is any difference at all between survival rates for test-positive and test-negative subjects, a significant p value can usually be achieved if the number of subjects in the study is sufficiently large. By way of illustration, consider Figure 28.2 with a hypothetical distribution of values of a test among subjects with and without an endpoint event after 2 years of follow-up. Although there is a slight difference in the means of the two distributions, which would be significant if the number of subjects in the study were sufficiently large, the test is likely completely useless from a clinical perspective as the relative risk is nearly equal to 1.

4.   The most conclusive demonstration of the utility of a stratification test is proof that therapy guided by the test improves survival. As discussed above, there are a multiplicity of reasons that a test which is predictive in a natural history trial may not lead to a therapeutic benefit when coupled with treatment. Prophylactic treatment trials generally occur quite late in the development of a new risk stratifier, after it is strongly believed on

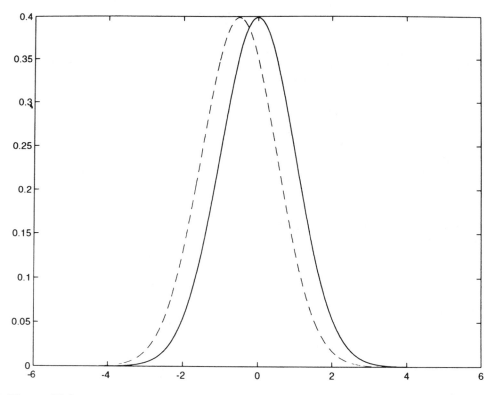

Figure 28.2 Two probability distributions that overlap considerably but from which one could draw samples which, if sufficiently large, would result in a statistically significant difference in the means of the samples.

clinical grounds that a test is useful or even after the test is incorporated into clinical practice, especially if it has a high positive predictive value.

III. REVIEW OF INDIVIDUAL RISK STRATIFIERS

Because of the absence of prodromes and the lack of screening tools, the occurrence of arrhythmic sudden cardiac death events is frequently the initial manifestation of cardiac disease in up to 30% of patients (21). Risk stratification for the general cardiac patient population is limited because ECG markers at best detect susceptibility for primary arrhythmic events (Table 28.2), which account for less than 70% of sudden cardiac deaths. Approximately 30% of sudden deaths are attributed to plaque disruption (22), and currently there are no reliable methods to sense a plaque that is vulnerable to disruption. Some small fraction of sudden deaths may be due to bradyarrhythmias rather than tachyarrhythmias and overall unrecognized prior myocardial infarction in 75% of sudden death victims (21). This has stimulated an intensive search for predictors of sudden death in survivors of myocardial infarction. In patients with heart failure and cardiomyopathies and those evaluated in the electrophysiology laboratory, it is likely that the vast majority of sudden deaths are due to an arrhythmogenic substrate (23,24).

The following is a brief summary of various risk stratifiers that have been used for predicting sudden cardiac death in such populations.

A. Stratifiers Reflective of Impaired Mechanical Function

Because of experimental and clinical evidence that the size and extent of myocardial injury following an infarction are related to propensity for ventricular tachyarrhythmias, early attention was focused on markers of impaired mechanical function as potential risk stratifiers (25). The most commonly studied parameter has been left ventricular ejection fraction, defined as the fraction of blood ejected from the left ventricle with each heart beat. This parameter can readily be measured by echocardiography and other noninvasive as well as invasive tests (26). The basic rationale is that depressed ejection fraction reflects either weakened heart muscle or extensive area of injury. The precise reason that depressed ejection fraction is associated with enhanced susceptibility to life-threatening arrhythmias is unknown. However, it is likely that several factors are involved, including a substrate for reentry, excessive stretch of Purkinje fibers, enhanced background level of cate-

Table 28.2 Pathophysiological Factors and Corresponding Markers for Prediction of Cardiac Events

Event	Marker
Plaque disruption	None suitable[a]
Autonomic nervous system activity	HRV, BRS
Active myocardial ischemia	ST-segment deviation, especially with T-wave alternans
Primary arrhythmia resulting from altered electrical substrate	Programmed electrical stimulation, T-wave alternans, SAECG, nonsustained ventricular tachycardia, QT dispersion

[a]This is an area of intense research with significant potential for breakthrough. In any case, the markers listed above could help to determine the prognosis and risk of sudden cardiac death secondary to plaque disruption.

cholamines, and a sizable heterogeneous matrix in terms of impulse formation and wave-front fractionation. Several studies have shown that a marked depression in left ventricular ejection fraction (LVEF) is a powerful predictor of sudden cardiac death in individuals with chronic heart disease (26). The incidence of nonarrhythmic cardiac death also appears to increase with decreasing ejection fraction although with low specificity. The utility of LVEF testing is underscored by the fact that in many investigations, LVEF is an important frame of reference and is often used as a trial entry or rejection criterion.

B. Risk Stratification Based on Spontaneous Ventricular Arrhythmia

There has been longstanding interest in the use of ventricular ectopic activity also referred to as premature ventricular contractions (PVCs), ventricular premature beats (VPBs), and ventricular premature discharges (VPDs) as a marker of vulnerability to life-threatening ventricular arrhythmias (27). In the absence of heart disease, VPDs are usually prognostically benign. After individuals reach the age of 30 years, frequent VPDs appear to select a subgroup with a higher incidence of coronary artery disease and sudden cardiac death. VPDs in asymptomatic patients are a well-established independent risk factor for sudden death after myocardial infarction and other cardiac causes (25,28), particularly if the VPD pattern is frequent, multiform, or repetitive. Most investigations employ a frequency cutoff of 10 VPDs/h as a threshold level for increased risk (29). High risk is thought to be associated with multiform VPDs, bigeminy, short-coupling intervals (the R-on-T phenomenon), and salvos of three or more ectopic beats (nonsustained ventricular tachycardia). Lown and Wolf (27) proposed a grading system based on these features and Ruberman and coworkers found that complex forms of VPDs indicated increased risk following myocardial infarction (30) while nonsustained ventricular tachycardia alone has been used as an indicator of arrhythmic risk (31).

Approximately a decade ago, the Cardiac Arrhythmia Suppression Trial (CAST) (15,32) tested the hypothesis that VPD suppression by antiarrhythmic agents decreases risk for SCD following myocardial infarction (15,16). The results were both surprising and disappointing, primarily because the rate of cardiac events and death among patients in the group treated with antiarrhythmic drugs (encainide, flecainide, or moricizine) exceeded placebo by more than threefold. CAST determined that effective arrhythmia suppression with these agents increased incidence of death from both arrhythmic and nonarrhythmic causes. These results have been interpreted as confirmation that VPD suppression, at least by these agents, does not confer protection against cardiac events including SCD. Whether this was a drug-specific effect or a demonstration that the premature beat hypothesis is not valid remains to be determined. When VPDs are evaluated in conjunction with derangements in left ventricular dysfunction, the predictive ability appears to be more robust. There is a particularly high risk of sudden cardiac death following myocardial infarction in the presence of high-grade VPDs when LVEF is significantly reduced (Fig. 28.3). Finally, the risk associated with complex VPDs in postinfarction patients is higher in individuals with non-Q-wave infarction than in those with transmural infarctions (29).

C. Stratifiers Based on Autonomic Function

Over the past two decades, a sizable body of evidence has been obtained that associates altered autonomic nervous system activity with the development of life-threatening arrhythmias (33–35). Excessive sympathetic nerve activity appears to be deleterious, while enhanced vagal activity appears to oppose the proarrhythmic influences of adrenergic

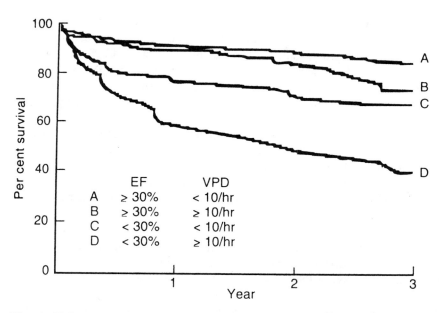

Figure 28.3 Survival during 3 years of follow-up after acute myocardial infarction as a function of left ventricular dysfunction (ejection fraction, EF) and ventricular arrhythmias (VPDs/h as measured by Holter monitoring). The survival curves were calculated as Kaplan-Meier estimates. With higher VPD frequencies and lower ejection fractions, the mortality rates increase. The number of patients in groups A, B, C, and D were 536, 136, 80, and 37, respectively. (From Ref. 29.)

stimuli. The two autonomic markers that have been most explored in terms of their potential clinical utility are heart rate variability (HRV), a marker of autonomic tone, and baroreflex sensitivity (BRS), an indicator that reflects the capability to activate potent cardioprotective reflexes (34,36). Recently, both HRV and BRS were compared in a sizable prospective clinical study of postinfarction patients. It concluded that not only were both markers useful indicators of risk for cardiac events, but also that they appeared to provide complementary information (37–39).

1. Heart Rate Variability (HRV)

HRV can be defined as "the interbeat oscillations that reflect the influence of tonic autonomic activity on the sinoatrial node" (40,41). This parameter is based on the concept that subtle fluctuations in interbeat dynamics provide insight into neurohumoral control of the heart. It should not be considered alone, however, because control of heart rate is also influenced by other factors, notably respiration, hemodynamic factors, and the renin-angiotensin system. Extensive experimental studies have shown that the time constants for cardiac control differ considerably between sympathetic and parasympathetic nerve activity (40,42–47). Adrenergic factors generally act with a relatively slower time course, whereas vagal influences are considerably more rapid.

Two basic approaches have been employed clinically to analyze HRV from surface ECG recordings (40,48,49). The most straightforward, time-domain analysis, involves determining the overall variance of the interbeat interval signal in the 24-h ECG record. Among parameters frequently measured are the standard deviation of all successive R–R intervals (SDNN), or portions thereof, and standard deviation of the averages of R–R in-

tervals in 5-min segments (SDANN). The other frequently employed time-domain techniques that provide information primarily about parasympathetic neural influences include
the "root-mean-square standard deviation" of successive intervals (rMSSD) and the proportion of successive interbeat intervals that diverge more than 50 ms (pNN50). The advantages of time-domain techniques are simplicity, reproducibility, and demonstrated
prognostic utility. Their shortcomings include the requirements for adequate sampling frequency and a stationary ambulatory signal (40).

Frequency domain analysis provides information concerning the amount of overall
variance in heart rate due to oscillations of heart rate at various frequencies. The power
spectrum is usually divided into four bands. High-frequency (HF) power is generally defined as within the 0.15–0.4 Hz band and appears to reflect parasympathetically mediated
respiratory variation. Low-frequency (LF) power, in the 0.04- to 0.15-Hz band, reflects
sympathetically and parasympathetically mediated heart-rate responses to oscillations in
arterial pressure (Fig. 28.4) (50,51). Together, the HF and LF bands account for approximately 6% of the total power in the 24-h HRV spectrum. Many investigators obtain a ratio
of the low:high frequency variability, which is thought to yield information concerning the
relative sympathetic–parasympathetic balance (40,52). A higher LF:HF ratio indicates
sympathetic preponderance, and a lower LF:HF ratio is reflective of parasympathetic dominance (40). The remaining two bands have been designated "very low frequency" (VLF,
0.0033- to 0.04-Hz band) and "ultralow frequency" (ULF, 0.0000115- to 0.0033-Hz band).
Although together these low-frequency spectra account for more than 90% of the total
power in the 24-h HRV spectrum and are independent predictors of mortality, the physiological mechanisms responsible for these two components are not well understood. Some
evidence suggests that they reflect the influence of thermoregulatory (45), peripheral vaso-

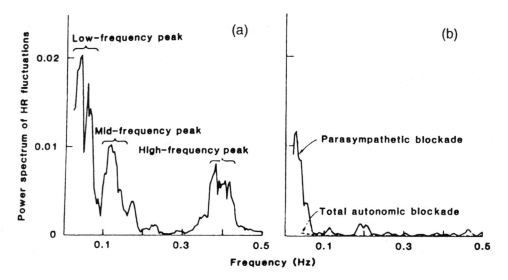

Figure 28.4 (a) Power spectrum of heart rate in an adult conscious dog, demonstrating three discrete spectral peaks. The low- and mid-frequency peaks are often not distinct. The high-frequency
peak is associated with respiration. (b) Power spectrum of heart rate fluctuations under parasympathetic nerve blockade and combined parasympathetic and sympathetic (beta-adrenergic) nerve
blockade. (From Ref. 50.)

motor (54), or renin-angiotensin systems (50). Bigger and coworkers (53) have shown that the ULF and VLF bands are excellent univariate predictors of mortality, independent of the five main clinical post-MI risk factors: age, New York Heart Association functional class, rales in the coronary care unit, left ventricular ejection fraction, and ventricular arrhythmias detected in a 24-h ambulatory ECG recording.

The clinical utility of HRV in risk stratification for sudden cardiac death, cardiac death, and all-cause mortality appears promising for patients with myocardial infarction (54), heart failure, diabetic neuropathies, and following cardiac transplantation (40).

Post-MI risk stratification has been perhaps the most extensively explored application, starting with the pioneering study of Kleiger and coworkers (54), who demonstrated a significant predictive power of HRV in survivors enrolled in the Multicenter Post-Infarction Research Group. This study entailed retrospective analysis of 24-h ECG recordings in a series of 808 individuals in whom a cycle-length variability (SDNN) of <50 ms showed a relative mortality risk 5.3 times higher than those with a SDNN >100 ms (Fig. 28.5). Numerous groups have confirmed this basic finding but, while HRV appears to be as reliable a predictor of all-cause mortality as ejection fraction, it is not a specific marker of arrhythmic susceptibility. For sudden cardiac death, several studies have provided evidence of an association between depressed HRV and propensity to life-threatening ventricular arrhythmias (40). However, ability of HRV to predict onset of sustained ventricular tachycardia or fibrillation has met with limited success (49), as no significant alterations in HRV have been observed immediately prior to the initiation of sustained VT. Patients with sustained VT have a lower HRV than those with repetitive premature beats, but HRV does not differ between patients with and without nonsustained VT. In general, it appears that HRV

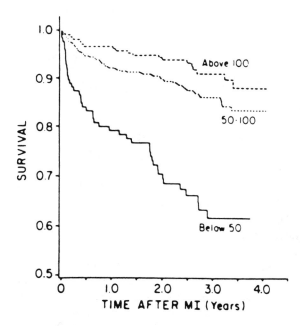

Figure 28.5 Survival after myocardial infarction for patients grouped according to the standard deviation of all normal R-R intervals (SDNN) computed over a 24-h period. The data demonstrate a strong association between this parameter and all-cause mortality after myocardial infarction. (From Ref. 54.)

does not have sufficient predictive accuracy to be employed as an independent screening tool for SCD.

In interpreting HRV data, it is essential to keep in mind that none of the various parameters used, either in isolation or collectively, constitutes a direct measure of sympathetic or parasympathetic nerve activity. At best, they provide inferences regarding the state of autonomic balance as it impacts on the sinoatrial node. Because of the relatively low predictive value of HRV measures, they are considered less useful for screening unselected populations and may be best suited as an adjunctive measure, useful in conjunction with others, such as LV ejection fraction, arrhythmias, or markers of cardiac electrical instability.

2.   Baroreceptor Sensitivity (BRS)

BRS can be defined as the gain of the baroreflex response to a pressor stimulus (36–38,55). Baroreceptors in the aortic arch and carotid arteries respond to changes in arterial blood pressure and can elicit an appropriate reciprocal change in parasympathetic and sympathetic nerve activity to counter this change. This in turn results in an appropriate change in heart rate, cardiac output, and peripheral vascular resistance to restore arterial blood pressure to normal. The extent to which heart rate is reduced in response to an increase in blood pressure indicates the briskness of autonomic activation and constitutes a measure of BRS. Baroreflex mechanisms are thus central to cardiovascular homeostasis, as they exert potent influences on vagal and sympathetic neural outflow to the heart and peripheral vessels (56,57). BRS is most commonly measured by the ability of a small intravenous injection of a pressor drug, typically phenylephrine, to increase arterial blood pressure. For BRS determinations, arterial blood pressure and heart rate are recorded on a beat-to-beat basis and plotted against each other. The slope of the resulting plot, expressed as milliseconds of increase in R–R interval per mmHg rise in systolic blood pressure, measures the sensitivity of arterial baroreflex control. The steeper the slope, the more potent is the individual's baroreceptor system in activating the vagus nerve to modulate sympathetic nerve activity. In normal individuals, a representative mean value is approximately 13 ms/mmHg, whereas following myocardial infarction, individuals at high risk for sudden cardiac death may exhibit values in the range of 3 ms/mmHg. Until recently, a major limitation of the BRS assessment technique has been the need for continuous invasive arterial blood pressure measurement. However, finger plethysmography allows arterial blood pressure measurement through a small finger cuff equipped with an infrared digital plethysmograph and permits adequate tracking of blood pressure during baroreceptor testing (58).

In the 1980s an extensive series of experiments in dogs undergoing coronary artery occlusion determined that myocardial infarction could impair vagal reflexes and that attendant changes in BRS predicted likelihood of ventricular fibrillation during an exercise stress test when acute ischemia was superimposed (59). These studies also proved that exercise conditioning increases BRS gain and vagus nerve response to an ischemic stimulus. Such alterations in autonomic tone and reflexes have been implicated as important mechanisms by which physical fitness can decrease arrhythmic events following myocardial infarction (37,38). LaRovere et al. (37,38) examined whether these experimental results applied in human subjects. They prospectively evaluated BRS in 78 postinfarction patients who were followed for 2 years during which seven cardiovascular deaths occurred, including four that were sudden. BRS was significantly lower in the seven deceased patients compared to the survivors. There was no apparent relationship between reduced BRS and low left ventricular ejection fraction (LVEF). Later, Farrell and coworkers (60) found that

BRS was depressed in 10 of 122 infarct patients who experienced arrhythmic events. The most extensive study of BRS as a risk stratifier occurred in the ATRAMI study, a multi-center prospective investigation in 1284 patients with a recent (<28 day) myocardial infarction (38). During the 21-month follow-up period, 44 cardiac deaths and 5 nonfatal cardiac arrests occurred. Low BRS values (<3.0 ms/mmHg) were found to confer significant multivariate risk of cardiac mortality, with a relative risk of 3.2 (Fig. 28.6). These results provide good evidence that following a myocardial infarction, depression of vagal reflexes, as determined by BRS, discloses prognostic information independent of LVEF and of the presence of ventricular arrhythmias. The study also suggests that BRS adds to the prognostic utility of HRV and that the two reflect different facets of autonomic activity, BRS being an indicator of reflexes and HRV an indicator of autonomic tone.

The main limitations in the use of BRS testing have centered around the need to administer a pressor drug. However, newer methods have recently been developed to measure baroreceptor function noninvasively (61). While BRS is a measure of autonomic reflexes, an established factor in arrhythmogenesis, its most important strength may be as a complementary endpoint to other indices (60). Depressed BRS has been associated with increased total mortality but not necessarily arrhythmic events. Importantly, it has been determined that neither HRV nor BRS, nor, for that matter, any noninvasive risk stratifier other than T-wave alternans and LVEF, can predict ventricular tachycardia or fibrillation in ICD recipients with coronary artery disease or dilated cardiomyopathy (24).

D. Stratifiers Reflective of Cardiac Electrophysiological Status

Several parameters have been explored as potential markers of cardiac electrical instability. These include assessment of late potentials by signal-averaged ECG technology, measurement of high-grade, nonsustained ventricular tachycardia on ambulatory ECG, ar-

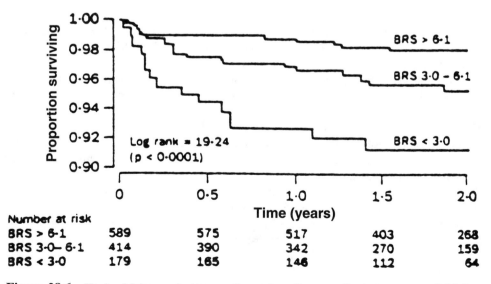

Figure 28.6 Kaplan-Meier survival curves for total cardiac mortality in postmyocardial infarction patients according to baroreceptor sensitivity (BRS) in the ATRAMI study. P value refers to differences in event rates among subgroups. (From Ref. 38.)

rhythmia inducibility by programmed electrical stimulation, and measurement of repolarization status by QT-interval dispersion and, most recently, T-wave alternans.

1. Signal-Averaged ECG (SAECG)

Experimental studies in animals have revealed that the presence of microvolt levels of "late potentials" that occur at the end of the QRS complex are indicative of areas of slowed conduction in the ventricular myocardium, where reentrant ventricular arrhythmias are likely to arise (62). In the early 1980s, the appearance of late potentials in the ECG of patients, determined by signal-averaging spectral analytical and time-domain methods, demonstrated a correlation with the occurrence of spontaneous, sustained ventricular tachycardia events. SAECG testing has since been employed in diverse groups of patients, including those with syncope, sustained VT, myocardial infarction, and cardiomyopathy (63). Notwithstanding a considerable effort, definitive evidence that this endpoint is useful to guide therapy is lacking. Thus, its major limitation appears to be a low positive predictive value, at ~16%, despite sensitivity and specificity in the range of 60% and 80%, respectively. Therefore, current interest in the method has been primarily as an adjunctive tool to other techniques (63–65).

2. Programmed Electrical Stimulation (PES)

This parameter represents a more aggressive approach to risk prediction than simply recording rhythm abnormalities. It entails invasive, provocative testing using electrical stimuli delivered to the heart via transvenous catheters to disclose intrinsic electrical instabilities (66,67). Extensive information accrued over 20 years suggests that PES can readily disclose latent propensities for arrhythmias (66,68,69). The established endpoint for PES testing is considered to be the induction of a sustained ventricular tachycardia following no more than three extrastimuli. The method is based on studies in animals in which various protocols were employed. In particular, it was shown that the electrical threshold for inducing fibrillation is generally predictive of intrinsic risk under diverse physiological conditions including ischemia, reperfusion, and changes in autonomic activity, and following administration of some antiarrhythmic agents (70–72). In animals recovering from myocardial infarction, PES can be useful in exposing residual electrical instability, which manifest as ventricular tachyarrhythmias (73).

PES has become the clinically accepted standard for arrhythmic risk classification in patients with coronary artery disease. Numerous studies have determined it to be useful in identifying individuals susceptible to spontaneous ventricular tachycardia and sudden cardiac death. Richards and coworkers (68) examined 361 patients within 1 to 2 weeks after MI and continued follow-up for at least 1 year. There were 34 eye-witnessed, instantaneous deaths, and 26 other deaths, with 9 patients surviving one or more episodes of VF or VT. Patients with PES-inducible VT were 15.2 times more likely to experience serious electrical events than patients without inducible VT. Sensitivity for electrical events was 58% within 1 year, specificity was 95%, positive predictive value 30%, and negative predictive value 98%. More recent data (74) from the MUSTT trial suggest that in patients with prior MI, a LVEF of ≤0.40, and nonsustained ventricular tachycardia, the relative risk of arrhythmic events associated with a positive versus negative PES is moderate (1.5 at two years and 1.33 at five years). This relative risk is much less than the relative risk of 15.2 reported from data collected a decade earlier by Richards et al. (68). One possible reason for this discrepancy is that PES may be a better predictor of arrhythmic risk in postmyocardial infarction patients in the prethrombolytic era, when the occurrence of spontaneous postinfarction monomorphic ventricular tachycardia was much more common than at present. For other patients with conditions such as nonischemic dilated cardiomyopathy, the utility of PES has not been established. In the MADIT (3) and MUSTT (8a,74) trials, which studied postmyocardial infarction patients with reduced LVEF and nonsustained ventricular tachycardia, the criteria of PES-inducible VT successfully identified a patient group who benefited from ICDs.

The major limitations of PES are that it is highly invasive, costly, and certainly not risk-free. Thus, it is not readily adaptable for widespread screening and is generally reserved for in-hospital evaluation of patients thought to be at high risk.

3. QT Dispersion

The arrhythmogenic potential of heterogeneity of repolarization is one of the most fundamental and enduring concepts in modern cardiac electrophysiology, and has frequently been invoked to explain the onset of life-threatening ventricular arrhythmias. The basic principle is that when adjoining myocardial cells repolarize at differing rates, the stage is set for electrical gradients that lead to an irregularly excitable matrix, disruption of the wavefront, and increased potential for arrhythmias. Elegant mapping techniques are now available that provide repolarization maps that readily detect repolarization inhomogeneities (73). In 1990, Day and colleagues (75) introduced the technique of measuring interlead differences in recorded QT intervals in the 12-lead ECG to obtain another measure of heterogeneity of repolarization. This they termed "QT dispersion." Basically, this index is calculated as the maximum minus the minimum QT interval of a single beat recorded in the 12-lead ECG.

QT dispersion has gained popularity as a result of its simplicity and putative electrophysiological underpinnings. Nevertheless, its utility in clinical decision making remains uncertain, as was pointed out in a recent review by Surawicz (19). He compared QT-dispersion levels from normal subjects with those of patients with cardiac disease (Fig. 28.7). Dispersion levels were greatest in the long-QT syndrome, where they can be clearly dis-

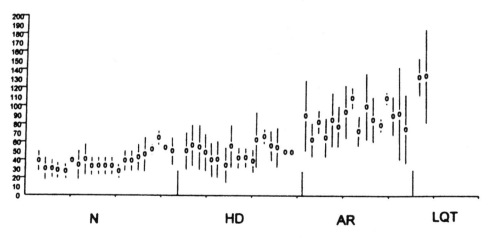

Figure 28.7   Average values (0) and standard deviations (vertical bars) of QT dispersion in milliseconds (ordinate) in 21 studies of normal control subjects (N); 12 studies (17 patient groups) of patients who had either heart disease (HD) or diabetes mellitus but no serious symptomatic ventricular tachyarrhythmias, or ventricular tachyarrhythmias in the presence of structurally normal heart; 13 studies (16 patient groups) of patients at risk of serious ventricular tachyarrhythmias (AR) or sudden cardiac death; and 2 groups of patients with congenital long-QT syndrome (LQT). (From Ref. 19.)

tinguished from normal values. Unfortunately, there was considerable overlap among the average values and a wide standard deviation in normal, heart disease, and arrhythmia-prone individuals, which weakens predictive power. One key to unlocking the potential of QT dispersion may reside in the recent finding by Sporton and coworkers (76), who determined that, at rest, QT dispersion is identical in normal subjects and patients with significant, angiographically demonstrated coronary disease. However, when heart rate is elevated above 100 beats/min with right atrial pacing, the degree of QT dispersion became significantly different (Fig. 28.8).

Methodological difficulties in accurately measuring QT dispersion are manifold. These involve establishing an exact point for the termination of the T wave and, when present, the confounding influences of a U wave. The need for rate correction introduces further ambiguity. Perhaps the most fundamental issue, however, is whether surface QT-interval differences actually provide an adequate picture of temporospatial dispersion of repolarization as it occurs within the heart (19,77).

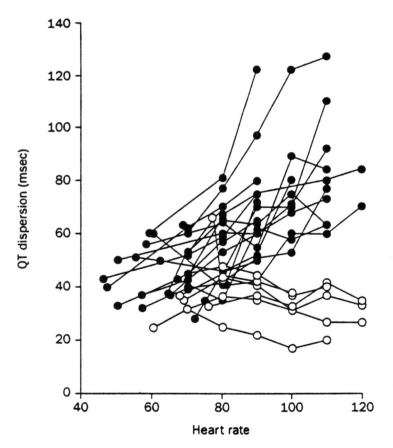

Figure 28.8 Effect of myocardial ischemia induced by incremental atrial pacing on QT dispersion in 18 patients with coronary artery disease (●) and 6 normal controls (○). Symbols connected by lines indicate the same patient, illustrating an increased predictive power over resting QT dispersion. (From Ref. 76.)

4. T-Wave Alternans

Electrical alternans refers to an alternating pattern in either the shape or presence of component forms within the recorded electrocardiographic complexes, often occurring with an ABABAB type variability (Figs. 28.9–11). T-wave alternans specifically refers to alternation in the morphology of the T wave, although the term alternans is often applied to alternation in the ST segment as well. Thus, T-wave alternans and repolarization alternans are synonymous terms. Electrical alternans in the electrocardiogram was described as early as 1909 (78,79). One setting in which alternans is common is that of pericardial effusion, a condition in which the heart may rotate in the pericardial sac on an every-other-beat basis, leading to apparent alternation in the morphology of all electrocardiographic complexes resulting from rotation of the electrical axis. This is not true electrical alternans as it is not associated with alternation of electrical processes within the heart. We will restrict the discussion here to true electrical alternans.

Historically, T-wave alternans has been anecdotally associated with a variety of pathophysiological conditions such as electrolyte abnormalities (80–82), Prinzmetal's angina (83–85), ischemia (86–94), and the inherited or acquired long-QT syndrome (95,96). Interestingly, all of these conditions are associated with an increased incidence of ventricular arrhythmias. The inherited long-QT syndrome, in particular, is associated with a high incidence of sudden cardiac death, primarily in young patients (Fig. 28.10) (95). Onset of myocardial infarction is also characterized by episodes of T-wave alternans (Fig.

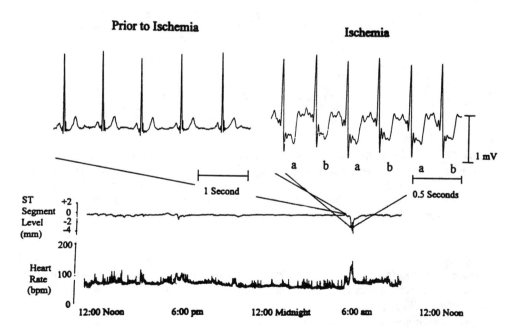

Figure 28.9   Representative ambulatory ECG recordings of ST-segment level, heart rate, and ECG in a patient from the Angina and Silent Ischemia Study (ASIS) using an ACS ambulatory recorder. In the absence of significant ST-segment depression, there was no visible T-wave alternans. During a bout of ischemia, as indicated by a drop in ST-segment level, there is marked T-wave alternans in the absence of R-wave alternans or other notable changes in the activation waveform. (From Ref. 93.)

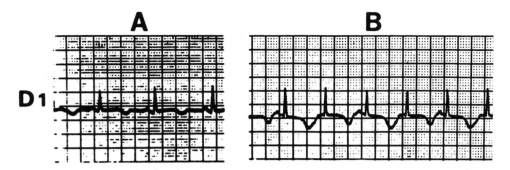

Figure 28.10   Electrocardiogram of 9-year-old patient affected by the long-QT syndrome. Left panel: at rest; right panel: T-wave alternans appeared during unintentionally induced fear. (From Ref. 34.)

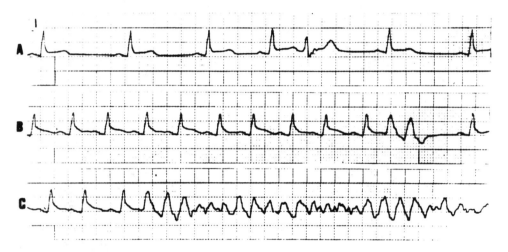

Figure 28.11   Sinus tachycardia after atropine leading to ventricular fibrillation. Patient with anterior infarction, lead I. Upper panel: Sinus bradycardia (55 beats/min) with ventricular ectopics. Middle panel: Record 2 min after atropine (0.6 mg, i.v.) Sinus tachycardia (110 beats/min) with consecutive ventricular ectopics. Lower panel: Record 3 min after atropine showing development of T-wave alternans and ventricular fibrillation. (From Ref. 97.)

28.11) (97). In 1948, Kalter and Schwartz (98) reviewed all published cases of electrical alternans and reported finding a total of 46. They calculated an incidence of visible alternans of 0.08%, with a mortality rate in these cases of 61%. Despite this remarkable finding, electrical alternans was widely considered, until recently, largely an electrocardiographic curiosity. With improvements in methods to measure T-wave alternans (88,91,99), and in particular of alternans that occurs at microvolt levels which cannot be detected by visual inspection of the electrocardiogram because its amplitude is too low, much renewed interest in this indicator has developed. Microvolt level T-wave alternans has been demonstrated to be a sensitive and specific marker of susceptibility for ventricular tachyarrhythmias and sudden cardiac death in patients referred for electrophysiological testing (23) and those implanted with ICDs (24) (Table 28.3). In the experimental laboratory, visible

Table 28.3 Clinical Studies on Predictive Power of T-Wave Alternans

Patient Group	Number of patients	Endpoint	Findings	Investigators (Ref.)
High-risk patients				
EP patients (64% CAD, 75% of whom had MI; 8% dilated cardiomyopathy; 24% no organic heart disease)	83	Inducible during PES	sensitivity = 81%, specificity = 84%, PPV = 76%, NPV = 88%, RR = 5.2	23
EP patients	45	Inducible during PES	sensitivity = 85%, specificity = 68%	127
EP patients (64% CAD, 75% of whom had MI; 8% dilated cardiomyopathy; 24% no organic heart disease)	66	Ventricular tachyarrhythmic events	sensitivity = 89%, specificity = 89%, PPV = 80%, NPV = 94% RR = 13.3	23
EP patients (48% CAD)	201	Ventricular tachyarrhythmic events	RR = 11.0	120
Consecutive ICD patients with ventricular arrhythmia (75% CAD, 17% dilated cardiomyopathy)	95	Ventricular tachyarrhythmic events	sensitivity = 78%, specificity = 61%, PPV = 67%, NPV = 73%, RR = 2.5	24
Consecutive ICD patients with ventricular arrhythmia and CAD	71	Ventricular tachyarrhythmic events	sensitivity = 85%, specificity = 68%, PPV = 73%, NPV = 81%, RR = 3.9	24
Moderate-risk patients				
Congestive heart failure patients without MI or VT/VF (63% CAD)	107	Ventricular tachyarrhythmic events	All events during followup occurred in patients with TWA	121
Post-MI patients	102	Ventricular tachyarrhythmic events	sensitivity = 93%, specificity = 59%, PPV = 28%, NPV = 98%, RR = 16.8	122

EP patients = patients undergoing electrophysiological testing; ICD = implantable cardioverter/defibrillator; CAD = coronary artery disease; MI = myocardial infarction; VT/VF = ventricular tachycardia/fibrillation; SCD = sudden cardiac death; PPV = positive predictive value; NPV = negative predictive value; RR = relative risk; TWA = T-wave alternans.

T-wave alternans in the millivolt range has been demonstrated consistently during episodes of myocardial ischemia and reperfusion, and, in these states, its magnitude is strongly correlated with onset of ventricular tachyarrhythmias (88,91,92).

There are two hypotheses linking T-wave alternans to enhanced susceptibility to ventricular tachyarrhythmias. Both involve a coupling between alternans and dispersion of recovery. The first hypothesis is that increased dispersion of refractoriness leads to a sub-population of cells that have a refractory period that exceeds the interbeat interval. These cells can at most depolarize normally only on an every-other-beat basis. This absent, or abnormal, depolarization on an every-other-beat basis leads to alternans on the surface ECG. In turn, the dispersion in recovery promotes wavefront fractionation and reentry leading to the development of reentrant ventricular arrhythmias. One weakness of this hypothesis is that it would be expected to lead equally to depolarization (QRS) alternans, as well as to repolarization alternans. In fact, QRS alternans tends to be far less prominent than T-wave alternans and is more weakly correlated with susceptibility to ventricular arrhythmias (88).

The second hypothesis (100,101) is that action potential alternans, in particular alternation in the duration of the action potential, is the primary event. Action potential alternans may result in a wide variety of conditions as a result of alterations in the shape of the restitution curve (i.e., in the relationship between action potential duration and the preceding diastolic interval). If the heart rate is fixed, a longer action potential duration results in a shorter subsequent diastolic interval, which in turn leads to a shorter subsequent action potential duration. A sufficiently steep restitution curve results in stable action potential alternans (100,102) and causes a dynamic spatial dispersion of recovery. Discordant action potential alternans results when the phase of action potential alternans varies from one region of the myocardium to another. That is, in one region the action potential may alternate on a long-short-long-short basis, whereas in an adjacent region the alternation is short-long-short-long. The spatial dispersion of recovery in turn increases susceptibility to ventricular arrhythmias. The alternation in action potential, as a result of alternation in the duration of repolarization in individual cells, is directly reflected in the surface ECG as T-wave alternans. This second hypothesis indicates that repolarization alternans may not just be a "fellow traveler" of a substrate more susceptible to ventricular tachyarrhythmias, but that it is mechanistically involved in the arrhythmogenic process. Direct experimental evidence that action potential alternans leading to dispersion of recovery and to the development of reentrant arrhythmias has recently been provided (100). The weight of the evidence thus supports this second hypothesis.

Alterations in transmembrane or intracellular calcium movements appear to play a major, but not exclusive, role in the initiation of action potential alternans, and likely T-wave alternans, during acute myocardial ischemia (88,99,103–105). This conclusion is based on investigations involving calcium channel blockers, such as verapamil, diltiazem, and nexopamil, which have been found to suppress T-wave alternans and prevent ventricular tachyarrhythmias during ischemia (106–108). In patients with Prinzmetal's variant angina, Salerno and colleagues (84) reported that pretreatment with diltiazem abolished both ischemia-induced T-wave alternans and ventricular arrhythmias. Whereas these agents affect myocardial perfusion as well as cardiac excitable properties, TQ-segment data suggest that the capability of these agents to suppress T-wave alternans is independent of the influence of the agent on the extent of the ischemic insult. Lee and colleagues (109) demonstrated that alternans in an ischemic area is associated with spatial heterogeneity of calcium transients. In addition, ryanodine, an agent that blocks calcium release from the sarcoplasmic reticulum, was noted to suppress ischemia-induced alternans and accompanying transients in calcium ion flux. The precise mechanisms whereby calcium induces alternans, the specific intracellular compartments involved, and the possible role of the sodium-calcium exchanger in this process require further exploration.

Potassium may also play an important role in ischemia-induced T-wave alternans. Extracellular K^+ accumulation is thought to be responsible for postrepolarization refractoriness and is likely to contribute to the well-established depression of restitution of action potential duration. Lukas and Antzelevitch (110) suggested that the transient outward current, I_{to}, is prominent mainly in epicardial and not endocardial tissue and suggested that this divergence may contribute to differences in action potential morphology. Electrical alternans of the epicardial action potential with a 2:1 or 3:1 pattern was evident during ischemia. This finding further emphasizes that the pattern of T-wave alternans may result from various changes in ion channel kinetics (111).

Although computer simulations suggest that electrical alternans should be a reliable marker of increased susceptibility to ventricular arrhythmias (112), clinical reports generally indicated that visible T-wave alternans is a relatively rare phenomenon (98). This disparity led to the development of the "microvolt-level T-wave alternans hypothesis," which states that T-wave alternans is commonly associated with a substrate more susceptible to ventricular tachyarrhythmias, but that its amplitude is so small it is generally not seen on the surface ECG. This idea led to the development of signal-processing methods to detect levels of T-wave alternans in the microvolt range, which is below the noise level of the ECG.

Microvolt-level T-wave alternans is generally measured by a spectral method (88). The T wave is first sampled at corresponding points in sequential beats (Fig. 28.12). The amplitudes of these points in successive beats constitute a beat series. This beat series is subjected to Fourier analysis to create a power spectrum. The power spectrum indicates the energy of the beat-to-beat fluctuations in the amplitudes, as a function of frequency, measured in units of cycles per beat. The power spectra corresponding to different offset points spanning the ST segment and the T wave are averaged, resulting in a composite power spectrum. The presence of alternans is indicated by a peak at the last point in the power spectrum corresponding to a frequency of one cycle every two beats, the Nyquist frequency. The alternans voltage V_{alt} represents the square root of the amplitude of this peak above background noise, and the alternans ratio k indicates the number of standard deviations by which this peak exceeds background noise. Thus, the alternans voltage reflects the amplitude of the peak and the alternans ratio represents its statistical significance. The ability to detect microvolt-level T-wave alternans may be further enhanced by the use of multicontact electrodes that combine multiple signals from one anatomical location to reduce noise (99).

One characteristic of microvolt-level T-wave alternans is that it appears to be highly rate-dependent, evident only above a threshold heart rate. Figure 28.13 illustrates the effect of increasing heart rate during an exercise stress test. T-wave alternans appears abruptly when the heart rate exceeds a patient-specific threshold and is then sustained until the rate drops below that value. Thus, heart rate threshold seems to be a critically important variable and tends to be low in patients at risk for ventricular tachyarrhythmias or sudden cardiac death (24,113). In contrast, normal individuals generally develop sustained T-wave alternans only at heart rates near their maximum predicted heart rate if at all (92,113). In ischemic patients, extent of ischemia rather than heart rate is probably the overriding factor in the development of T-wave alternans, as it can occur during the recovery phase of an exercise stress test, when heart rate is decreasing (87,114). In such patients, a general cor-

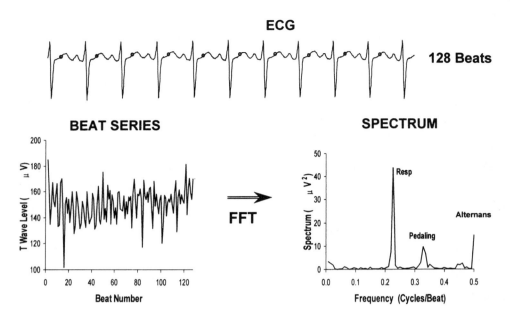

Figure 28.12   Illustration of spectral method of measuring T-wave alternans. Amplitude of corresponding points on successive T-waves is measured to construct a beat series. Fourier analysis of the beat series is used to construct a power spectrum that indicates the energy of the fluctuations in amplitude as a function of frequency measured in cycles per beat. Power spectra corresponding to different points in the T wave are then averaged. The alternans voltage V_{alt} is the square root of the amplitude of the spectral peak at 0.5 cycles/beat above the mean level in a predefined noise band located near 0.5 cycles per beat. The statistical significance of the measured alternans is determined by computing the ratio of the amplitude of the alternans peak above the background noise level to the standard deviation of the noise in the predefined noise band. Despite the absence of any visible T-wave alternans on the ECG, a clearly definable peak is present at the alternans frequency.

respondence between heart rate and T-wave alternans magnitude may result from the fact that elevated heart rates exacerbate ischemia by compromising diastolic coronary perfusion time and by increasing the degree of action potential alternans leading to enhanced heterogeneity of repolarization (105,115). Patients with Prinzmetal's angina who exhibit visible ECG T-wave alternans during ischemia were also found to be at greater risk for ventricular tachyarrhythmias (84,85). Visible and quantifiable T-wave alternans averaging >55 µV was identified during spontaneous ischemic events or exercise in stable coronary artery disease patients (93). Thus, in patients with ischemic disease, T-wave alternans magnitude appears readily detectable with ambulatory precordial lead ECG monitoring, providing appropriate attention is given to methodology considerations (115). Because the phenomenon is regionally specific, it may not be evident in limb leads even when it is sizable in the precordial leads overlying the ischemic zone (90,92). Further investigation is required to establish whether T-wave alternans can be detected using conventional ambulatory ECG technology in other patient populations and whether it can be successfully used to predict life-threatening events in prospective trials.

Association of microvolt-level T-wave alternans with increased susceptibility to ventricular arrhythmias was demonstrated in patients referred for electrophysiological testing because of high risk for ventricular arrhythmias (23) (Table 28.3). Microvolt T-

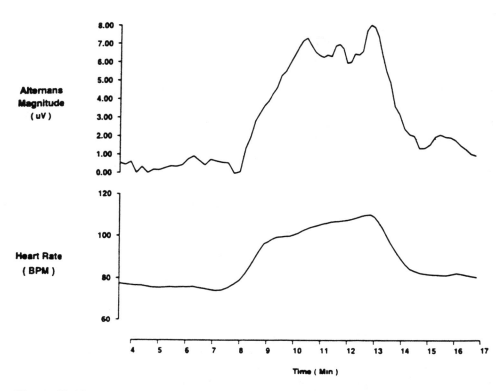

Figure 28.13 Plot of alternans magnitude and heart rate during the course of bicycle exercise test. (From Ref. 99.)

wave alternans measured during right atrial pacing at 100 bpm predicted the outcome of electrophysiological testing (relative risk 5.4; $p < 0.001$) and also predicted arrhythmia-free survival ($p < 0.001$) as well as PES testing. Patients without significant levels of T-wave alternans had a 96% rate of arrhythmia-free survival after 20 months of follow-up, whereas patients with significant levels of alternans had only a 19% rate of arrhythmia-free survival at 20 months (Fig. 28.14). Subsequent analyses confirmed that neither QT dispersion nor SAECG was predictive in patients at high risk for ventricular arrhythmias referred for electrophysiological testing (116,117).

Important progress in the use of microvolt-level T-wave alternans involved the development of a completely noninvasive test that eliminates the need for right atrial pacing. Exercise stress testing, a standard test for measuring risk, was explored as a means for inducing microvolt T-wave alternans (112,118,119). To use the technique during exercise required major advances in signal processing and development of multicontact electrodes to reduce noise (99). Hohnloser and colleagues (113) observed an 84% concurrence rate of T-wave alternans induced by exercise as compared to right atrial pacing in patients without evidence of active ischemia referred for PES. When the two techniques were compared, it was found that onset heart rates were indistinguishable whether heart rate was elevated by exercise or pacing. Estes and coworkers (118) then confirmed that exercise-induced microvolt T-wave alternans predicted the outcome of electrophysiological testing. More recently, Coch and colleagues (119) demonstrated that microvolt levels of T-wave alternans can be measured equally well during pharmacological elevation of heart rate with atropine.

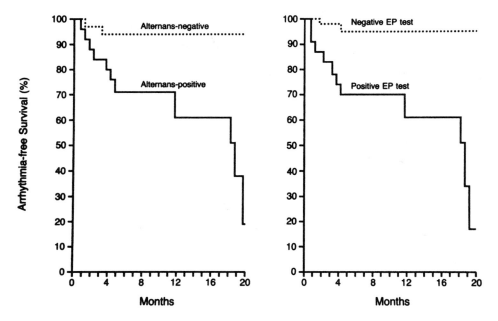

Figure 28.14 T-wave alternans and results of electrophysiological testing in relation to arrhythmia-free survival among 66 patients. In the left panel, arrhythmia-free survival according to Kaplan-Meier life-table analysis is compared in patients with T-wave alternans (alternans ratio >3.0) and without it (ratio ≤3.0). Note that the presence of T-wave alternans is a strong predictor of reduced arrhythmia-free survival. In the right panel, arrhythmia-free survival among patients with positive EP tests is compared with that among patients in whom ventricular arrhythmias were not induced on EP testing (negative EP test). The predictive accuracy of EP testing and T-wave alternans in these patients is essentially the same. (From Ref. 23.)

Hohnloser and coworkers (24) conducted the first prospective study in which microvolt-level T-wave alternans was compared with other risk stratifiers including PES, LVEF, BRS, HRV, SAECG, QT dispersion, and the presence of nonsustained ventricular tachycardia on Holter monitoring (Table 28.3). Seventy-five percent of the patients receiving an ICD because of high risk for arrhythmia had coronary disease and 17% had dilated cardiomyopathy; those with evidence of active ischemia were excluded. The measured endpoint was the first appropriate discharge of the ICD. Of all of the indices monitored, only T-wave alternans ($p < 0.006$) and LVEF ($p < 0.04$) predicted event-free survival over an 18-month follow-up period. Both were also the only predictors of event-free survival in a coronary artery disease subgroup. On multivariate analysis, T-wave alternans was the only significant independent predictor of event-free survival (Fig. 28.15). In one preliminary report on 201 patients referred for PES (120), it was reported that exercise-induced T-wave alternans predicted 1-year event-free survival with a relative risk of 11.0 ($p < 0.001$) compared to a relative risk of 3.1 ($p < 0.01$) for PES and 1.8 (not significant) for SAECG. In a separate study of 107 patients with congestive heart failure and no prior history of ventricular arrhythmias (121), all arrhythmic events during followup occurred in patients with T-wave alternans (Table 28.3). Recently, T-wave alternans was demonstrated to be a highly effective predictor of arrhythmic events in patients after myocardial infarction (122). In 102 patients, T-wave alternans predicted arrhythmic events with a relative risk of 16.8 ($p <$

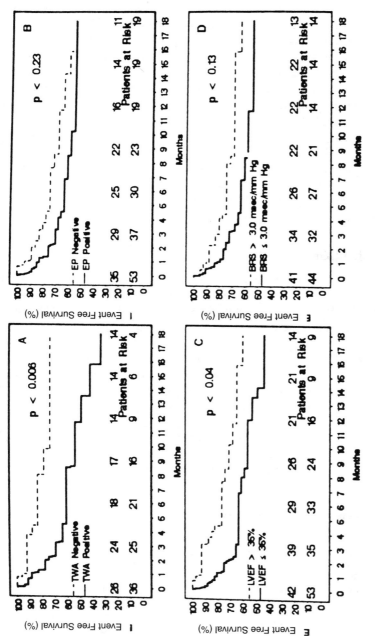

Figure 28.15 Comparison of the predictive value of T-wave alternans, electrophysiological testing, left ventricular ejection fraction, and baroreflex sensitivity in predicting appropriate ICD discharge in patients with ICDs with coronary artery disease or dilated cardiomyopathy (Table 28.3). (A) Survival in 62 patients with determinate T-wave alternans results. (B) Survival in 88 patients according to results of electrophysiological testing. (C) Survival in 95 patients according to left ventricular ejection fraction. (D) Survival in 85 patients according to determination of baroreflex sensitivity. (From Ref. 24.)

0.006) (Table 28.3). Finally, T-wave alternans has been reported to be predictive of ventricular arrhythmias in patients with nonischemic dilated cardiomyopathy (123,124), a result believed not to be predictable with PES. Thus, present indications are that microvolt T-wave alternans may be superior to other noninvasive arrhythmic risk stratifiers and at least equivalent to invasive PES. Nevertheless, there is a need for additional demonstration of the utility of T-wave alternans in stratification of lower risk patient populations. Among the most sizable and clinically significant populations are those with a prior myocardial infarction and those with heart failure. Because of the size of these populations and their vulnerability to life-threatening arrhythmias, it is quite important to establish the utility of measuring this electrophysiological phenomenon in those populations. Also, because the ultimate measure of a noninvasive risk stratifier is suitability to direct therapy, prospective randomized trials will be required to demonstrate that the risk of sustained ventricular arrhythmias and sudden death in patients identified by T-wave alternans testing can be reduced by ICDs or pharmacological therapy.

IV. SUMMARY AND CONCLUSIONS

PES is the currently accepted clinical standard for arrhythmic risk stratification for patients with coronary artery disease. However, because PES is invasive, costly, and not risk-free, it is not suitable for widespread screening of patients potentially at risk for sudden cardiac death. As illustrated in Table 28.4, conventional noninvasive arrhythmic risk stratifiers tend to be characterized by moderate sensitivity, but low positive predictive value in postmyocardial infarction patients. This low positive predictive value limits their use in guiding treatments that can be quite costly or associated with patient risk. Accordingly, natural history studies may be very useful in identifying effective noninvasive risk stratifiers, but only therapeutic trials can be relied on to test these concepts with respect to treatment outcome.

T-wave alternans is a promising noninvasive risk stratifier at this time and in high-risk patients it appears to identify individuals at risk for arrhythmic events and cardiac death as well as PES (Table 28.3). Its suitability and reliability when used with exercise

Table 28.4 Predictors of Arrhythmic Events in Postmyocardial Infarction Patients

Risk factor	Ref.	Follow-up period	Sensitivity	PPV
LVEF < 0.4	(125)	15 mos	79%	26%
LVEF ≤ 0.4	(68)	12 mos	71%	11%
LVEF ≤ 0.4	(64)	17 mos	46%	10%
SAECG	(125)	15 mos	79%	17%
SAECG	(64)	17 mos	63%	17%
HRV	(125)	15 mos	89%	15%
HRV	(64)	17 mos	92%	17%
NSVT	(125)	15 mos	42%	25%
NSVT	(64)	17 mos	54%	16%
PES	(68)	12 mos	58%	30%

LVEF = left ventricular ejection fraction; SAECG = signal-averaged electrocardiogram; HRV = heart rate variability; NSVT = nonsustained ventricular tachycardia; PES = programmed electrical stimulation; PPV = positive predictive value.

stress testing appear to constitute a distinct advantage (125) over other alternatives. The use of exercise testing in conjunction with other risk stratifiers has the potential to improve their performance as well, but may be hampered by significant technical obstacles like interlead morphological changes, data nonstationarity, muscle noise and artifacts, such as those associated with motion or repositioning during testing. To date, few of the other markers (e.g., QT dispersion, HRV, or SAECG) have been explored with provocative exercise. It will therefore be critical to determine the success of exercise-T-wave alternans in direct clinical trial evaluations now underway in several centers.

Because of the multifactorial nature of sudden cardiac death (Table 28.2), it is likely that no single parameter will adequately reflect the complex factors that lead to lethal arrhythmias in different categories of patients. One optimum strategy for the future, therefore, would seem to be to examine whether combinations of several risk stratification parameters may be more effective than any individual one. Farrell and associates (64) and Pedretti and coworkers (126) demonstrated that various combinations of markers may be more effective than any single index. In the recent ATRAMI study, the combination of HRV and BRS together appeared to increase predictive accuracy (38). Based on current understanding of factors that alter and modulate the myocardial substrate and triggers of life-threatening arrhythmias, a rational approach would involve combining a measure of electrical stability, such as T-wave alternans, QT dispersion, SAECG, or PES (24) with a measure of mechanical function (LVEF) and a measure of autonomic nervous system activity (HRV or BRS). A remaining challenge is thus to evaluate which particular combination of predictors will be most successful in large-scale controlled evaluations of specific patient categories.

ACKNOWLEDGMENTS

Supported by Grants P01 ES08129 and P01 ES09825 from the National Institutes of Health, NAG5-4989 from the U.S. National Aeronautics and Space Administration, and a grant from the National Space Biomedical Research Institute. The authors thank Murray A. Mittleman, M.D., Dr.P.H., for an excellent discussion and review of the manuscript. Serene Yeam and Sandra S. Verrier provided valuable editorial assistance.

REFERENCES

1. Heart and Stroke Statistical Update. Dallas: American Heart Association, 1999.
2. Antiarrhythmics Versus Implantable Defibrillators (AVID) Investigators. A comparison of antiarrhythmic-drug therapy with implantable defibrillators in patients resuscitated from near-fatal ventricular arrhythmias. N Engl J Med 1997; 337:1576–1583.
3. Moss AJ, Hall J, Cannom DS, Daubert JP, Higgins SL, Klein H, Levine JH, Saksena S, Waldo AL, Wilber D, Brown MW, Heo M. Improved survival with an implanted defibrillator in patients with coronary disease at high risk for ventricular arrhythmia. N Engl J Med 1996; 335:1933–1940.
4. Myerburg RJ, Kessler KM, Castellanos A. Sudden cardiac death. Structure, function, and time-dependence of risk. Circulation 1992; 85 (suppl I):2–10.
5. Cairns JA, Connolly SJ, Roberts R, Gent M. Randomised trial of outcome after myocardial infarction in patients with frequent or repetitive ventricular premature depolarisations: CAMIAT. Lancet 1997; 349:675–682.
6. Julian DG, Camm AJ, Frangin G, Janse MJ, Munoz A, Schwartz PJ, Simon P. Randomised tri-

al of effect of amiodarone on mortality in patients with left ventricular dysfunction after recent myocardial infarction: EMIAT. Lancet 1997; 349:667–674.

7. Siebels J, Cappato R, Rupel R, Schneider MAE, Kuck KH. Preliminary results of the cardiac arrest study Hamburg (CASH). Am J Cardiol 1993; 72:109F–113F.

8. Bocker D, Bansch D, Heinecke A, Weber M, Brunn J, Hammel D, Borggrefe M, Breithardt G, Block M. Potential benefit from implantable cardioverter-defibrillator therapy in patients with and without heart failure. Circulation 1998; 98:1636–1643.

9. Bigger JT Jr. Prophylactic use of implanted cardiac defibrillators in patients at high risk for ventricular arrhythmias after coronary-artery bypass graft surgery. Coronary Artery Bypass Graft (CABG) Patch Trial Investigators. N Engl J Med 1997; 337:1569–1575.

10. Bigger JT, Whang W, Rottman JN, Kleiger RE, Gottlieb CD, Namerow PB, Steinman RC, Estes III NAM. Mechanisms of death in the CABG Patch Trial. Circulation 1999; 99:1416–1421.

11. Friedman PL, Stevenson WG. Unsustained ventricular tachycardia—to treat or not to treat? N Engl J Med 1996; 335:1984–1985.

12. Pfeffer MA, Braunwald E, Moye LA, Basta L, Brown EJ Jr, Cuddy TE, Davis BR, Geltman EM, Goldman S, Flaker GC. Effect of captopril on mortality and morbidity in patients with left ventricular dysfunction after myocardial infarction. Results of the survival and ventricular enlargement trial. The SAVE Investigators. N Engl J Med 1992; 327:669–677.

13. Gottlieb SS, McCarter RJ, Vogel RA. Effect of beta-blockade on mortality among high-risk and low-risk patients after myocardial infarction. N Engl J Med 1998; 339:489–497.

14. Pitt B, Zannad F, Remme WJ, Cody R, Castaigne A, Perez A, Palensky J, Wittes J, for the Randomized Aldactone Evaluation Study Investigators. Effect of spironolactone on morbidity and mortality in patients with severe heart failure. N Engl J Med 1999; 341:709–717.

15. Cardiac Arrhythmia Suppression Trial (CAST) Investigators. Preliminary report: effect of encainide and flecainide on mortality in a randomized trial of arrhythmia suppression after myocardial infarction. N Engl J Med 1989; 321:406–412.

16. Ruskin JN. The cardiac arrhythmia suppression trial (CAST) [editorial]. N Engl J Med 1989; 321:386–388.

17. Echt DS, Liebson PR, Mitchell LB, Peters RW, Obias-Mannon D, Barker AH, Arensberg D, Baker A, Friedman L, Greene L, Huther ML, Richardson DW. Mortality and morbidity in patients receiving encainide, flecainide, or placebo. N Engl J Med 1991; 324:781–788.

18. Singh SN, Fletcher RD, Gisher SG, Singh BN, Lewis HD, Deedwania PC, Massie BM, Colling C, Lazzeri D. Amiodarone in patients with congestive heart failure and asymptomatic ventricular arrhythmia. N Engl J Med 1995; 333:77–82.

19. Surawicz B. Will QT dispersion play a role in clinical decision-making? J Cardiovasc Electrophysiol 1996; 7:777–784.

20. Hennekens CH, Buring JE. Epidemiology in Medicine. Boston: Little, Brown, 1987.

21. Myerburg RJ, Castellanos A. Cardiac arrest and sudden cardiac death. In: Braunwald E, ed. Heart Disease: A Textbook of Cardiovascular Medicine. Philadelphia: WB Saunders, 1997:742–779.

22. Thomas AC, Davies MJ, Kulbertus HE, Wellens HJJ. A pathologist's view of sudden cardiac death. In: Kulbertus HE, Wellens HJJ, eds. Sudden Death. The Hague: Martinus Nijhoff, 1980: 34–48.

23. Rosenbaum DS, Jackson LE, Smith JM, Garan H, Ruskin JN, Cohen RJ. Electrical alternans and arrhythmia vulnerability in man. N Engl J Med 1994; 330:235–241.

24. Hohnloser SH, Klingenheben T, Li Y-G, Zabel M, Peetermans J, Cohen RJ. T-wave alternans as a predictor of recurrent ventricular tachyarrhythmias in ICD recipients: prospective comparison with conventional risk markers. J Cardiovasc Electrophysiol 1998; 9:1258–1268.

25. Bigger JT Jr, Fleiss JL, Kleiger R, Miller JP, Rolnitzky MN, Multicenter Post-Infarction Research Group. The relationship among ventricular arrhythmias, left ventricular dysfunction and mortality in the 2 years after myocardial infarction. Circulation 1984; 69:250–258.

26. Bigger JT Jr. Role of left ventricular ejection fraction. In: Akhtar M, Myerburg RJ, Ruskin JN, eds. Sudden Cardiac Death: Prevalence, Mechanisms, and Approaches to Diagnosis and Management. Wayne, PA: Andover Medical Publishers, 1994:190–201.

27. Lown B, Wolf M. Approaches to sudden death from coronary heart disease. Circulation 1971; 44:130–142.

28. Mukharji J, Rude RE, Poole KE, Gustafson N, Thomas LJ Jr, Strauss HW, Jaffe AS, Muller JE, Roberts R, Raabe DS Jr, and the MILIS Study Group. Risk factors for sudden death after acute myocardial infarction: Two-year follow-up. Am J Cardiol 1984; 54:31–36.

29. Bigger JT. Relation between left ventricular dysfunction and ventricular arrhythmias after myocardial infarction. Am J Cardiol 1986; 57:8B–14B.

30. Ruberman W, Weinblatt E, Frank CW, Goldberg JD, Shapiro S. Repeated 1-hour electrocardiographic monitoring of survivors of myocardial infarction at 6-month intervals: Arrhythmia detection and relation to prognosis. Am J Cardiol 1981; 47:1197–1204.

31. Hohnloser SH, Klingenheben T, Zabel M, Schopperl M, Mauss O. Prevalence, characteristics and prognostic value during long-term follow-up of nonsustained ventricular tachycardia after myocardial infarction in the thrombolytic era. J Am Coll Cardiol 1999; 33:1895–1902.

32. Greene HL, Roden DM, Katz RJ, Woosley RL, Salerno DM, Henthorn RW. The Cardiac Arrhythmia Suppression Trial: first CAST ... then CAST-II. J Am Coll Cardiol 1992; 19:894–898.

33. Lown B, Verrier RL. Neural activity and ventricular fibrillation. N Engl J Med 1976; 294:1165–1170.

34. Schwartz PJ, Zaza A, Locati E, Moss AJ. Stress and sudden death. The case of the long QT syndrome. Circulation 1991; 83(suppl 4):1171–1190.

35. Zipes DP. Autonomic modulation of cardiac arrhythmias. In: DP Zipes, J Jalife, eds. Cardiac Electrophysiology: From Cell to Bedside. Philadelphia: WB Saunders, 1995:441–453.

36. Schwartz PJ, La Rovere MT, Vanoli E. Autonomic nervous system and sudden cardiac death. Circulation 1992; 85(suppl I):77–91.

37. La Rovere MT, Specchia G, Mortara A, Schwartz PJ. Baroreflex sensitivity, clinical correlates, and cardiovascular mortality among patients with a first myocardial infarction. Circulation 1988; 78:816–824.

38. La Rovere MT, Bigger JT Jr, Marcus FI, Mortara A, Schwartz PJ. Baroreflex sensitivity and heart-rate variability in prediction of total cardiac mortality after myocardial infarction. ATRAMI (Autonomic Tone and Reflexes After Myocardial Infarction) Investigators. Lancet 1998; 351:478–484.

39. Barron HV, Viskin S. Autonomic markers and prediction of cardiac death after myocardial infarction [comment]. Lancet 1998; 351:461–462.

40. Task Force of the European Society of Cardiology and the North American Society of Pacing and Electrophysiology. Heart rate variability: standards of measurement, physiological interpretation and clinical use. Circulation 1996; 93:1043–1065.

41. Berntson GG, Bigger JT, Eckberg DL, Grossman P, Kaufmann PG, Malik M, Nagaraja HN, Porges SW, Saul JP, Stone PH, van der Molen MW. Heart rate variability: origins, methods, and interpretive caveats. Psychophysiology 1997; 34:623–648.

42. Jalife J, Slenter VAJ, Salata JJ. Dynamic vagal control of pacemaker activity in the mammalian sinoatrial node. Circ Res 1983; 52:642–656.

43. Baselli G, Cerutti S, Civardi S, Lombardi F, Malliani A, Merri M, Pagani M, Rizzo G. Heart rate variability signal processing: a quantitative approach as an aid to diagnosis in cardiovascular pathologies. Int J BioMed Comput 1987; 20:51–70.

44. Lombardi F, Sandrone G, Pernpruner S, Sala R, Garimoldi M, Cerutti S, Baselli G, Pagani M, Malliani A. Heart rate variability as an index of sympathetic interaction after acute myocardial infarction. Am J Cardiol 1987; 60:1239–1245.

45. Fallen EL, Kamath MV, Ghistra DN. Power spectrum of heart rate variability: a non-invasive test of integrated neurocardiac function. Clin Invest Med 1988; 11:331–340.

46. Appel ML, Berger RD, Saul JP, Smith JM, Cohen RJ. Beat-to-beat variability in cardiovascular parameters: noise or music? J Am Coll Cardiol 1989; 14:1139–1148.

47. Malliani A, Pagani M, Lombardi F, Cerutti S. Cardiovascular neural regulation explored in the frequency domain. Circulation 1991; 84:482–492.

48. Bigger JT, Schwartz PJ. Markers of vagal activity and the prediction of cardiac death after myocardial infarction. In: Levy MN, Schwartz PJ. Vagal Control of the Heart. Mt. Kisco, NY: Futura, 1994:481–508.

49. Fallen EL. Clinical utility of heart rate variability. Cardiac Electrophysiol Rev 1997; 3:347–351.

50. Akselrod S, Gordon D, Ubel FA, Shannon DC, Barger AC, Cohen RJ. Power spectrum analysis of heart rate fluctuation: a quantitative probe of beat-to-beat cardiovascular control. Science 1981; 213:220–222.

51. Saul JP, Rea RF, Eckberg DL, Berger RD, Cohen RJ. Heart rate and muscle sympathetic nerve variability during reflex changes in autonomic activity. Am J Physiol 1990; 258:H713–H721.

52. Pagani M, Lombardi F, Guzzetti S, Rimoldi O, Furlan R, Pizzinelli P, Sandrone G, Malfatto G, Dell'Orto S, Piccaluga E, Turiel M, Baselli G, Cerutti S, Malliani A. Power spectral analysis of heart rate and arterial pressure variabilities as a marker of sympathovagal interaction in man and conscious dog. Circ Res 1986; 59:178–193.

53. Bigger JT, Fleiss JL, Steinman RC, Rolnitzky LM, Kleiger RE, Rottman JN. Frequency domain measures of heart period variability and mortality after myocardial infarction. Circulation 1992; 85:164–171.

54. Kleiger RE, Miller JP, Bigger JT Jr, Moss AJ. Decreased heart rate variability and its association with increased mortality after acute myocardial infarction. Am J Cardiol 1987; 59:256–262.

55. Hohnloser SH, Klingheben T. Clinical utility of baroreflex sensitivity measurements. Cardiac Electrophysiol Rev 1997; 3:354–356.

56. Thames MD, Dibner-Dunlap ME, Smith ML. Mechanisms of arterial baroreflex control. In: Levy MN, Schwartz PJ, eds. Vagal Control of the Heart. Mt. Kisco, NY: Futura, 1994.

57. Somers VK, Abboud FM. Baroreflexes in health and disease. In: Levy MN, Schwartz PJ, eds. Vagal Control of the Heart. Mt. Kisco, NY: Futura, 1994.

58. Silke B, McAuley D. Accuracy and precision of blood pressure determination with the Finapres: an overview using re-sampling statistics. J Hum Hypertens 1998; 12:403–409.

59. Schwartz PJ, Vanoli E, Stramba-Badiale M, DeFerrari GM, Billman GE, Foreman RD. Autonomic mechanisms and sudden death. New insights from analysis of baroreceptor reflexes in conscious dogs with and without a myocardial infarction. Circulation 1988; 78:969–979.

60. Farrell TG, Odemuyiwa O, Bashir Y, Cripps TR, Malik M, Ward DE, Camm AJ. Prognostic value of baroreflex sensitivity testing after acute myocardial infarction. Br Heart J 1992; 67:129–137.

61. Mullen TJ, Appel ML, Mukkamala R, Mathias JM, Cohen RJ. System identification of closed-loop cardiovascular control: effects of posture and autonomic blockade. Am J Physiol 1997; 272:H448–H461.

62. Breithardt G, Cain ME, El-Sherif N, Flowers NC, Hombach V, Janse M, Sinson MB, Steinbeck G. Standards for analysis of ventricular late potentials using high-resolution or signal-averaged electrocardiography: a statement by a Task Force Committee of the European Society of Cardiology, the American Heart Association, and the American College of Cardiology. J Am Coll Cardiol 1991; 17:999–1006.

63. Kulakowski P. Clinical utility of signal-averaged electrocardiography. Cardiac Electrophysiol Rev 1997; 3:321–324.

64. Farrell TG, Bashir Y, Cripps T, Malik M, Poloniecki J, Bennett ED, Ward DE, Camm AJ. Risk stratification of arrhythmic events in postinfarction patients based on heart rate variability, ambulatory electrocardiographic variables, and the signal-averaged electrocardiogram. J Am Coll Cardiol 1991; 18:687–697.

65. Cain ME, Anderson JL, Arnsdorf MF, Mason JW, Scheinman MM, Waldo AL. Signal averaged electrocardiography. J Am Coll Cardiol 1996; 27:238–249.
66. Josephson MA. Clinical Cardiac Electrophysiology. Philadelphia: Williams & Wilkins, 1993.
67. Wellens HJJ, Schuilenburg RM, Durrer D. Electrical stimulation of the heart in patients with ventricular tachycardia. Circulation 1972; 46:216–226.
68. Richards DAB, Byth K, Ross DL, Uther JB. What is the best predictor of spontaneous ventricular tachycardia and sudden death after myocardial infarction? Circulation 1991; 83:756–763.
69. Zipes DP, Wellens HJJ. Sudden cardiac death. Circulation 1998; 98:2334–2351.
70. Moore EN, Spear JF. Ventricular fibrillation threshold. Its physiological and pharmacological importance. Arch Intern Med 1975; 135:446–453.
71. Verrier RL, Brooks WW, Lown B. Protective zone and the determination of vulnerability to ventricular fibrillation. Am J Physiol 1978; 234:H592–H596.
72. Verrier RL. Autonomic modulation of arrhythmias in animal models. In: Rosen MR, Wit AL, Janse MJ, eds. Cardiac Electrophysiology: A Textbook in Honor of Brian Hoffman. Mt. Kisco, NY: Futura Press, 1990:933–949.
73. Janse MJ, Wit AL. Electrophysiological mechanism of ventricular arrhythmias resulting from myocardial ischemia and infarction. Physiol Rev 1989; 69:1049–1169.
74. Buxton AE, Lee KL, DiCarlo L, Gold MR, Greer GS, Prystowsky EN, O'Toole MF, Tang A, Fisher JD, Coromilas J, Talajic M, Hafley G. Electrophysiologic testing to identify patients with coronary artery disease who are at risk for sudden death. Multicenter UnSustained Tachycardia Trial Investigators. N Engl J Med 2000; 342:1937–1945.
75. Day CP, McComb JM, Campbell RWF. QT dispersion: an indication of arrhythmia risk in patients with long QT intervals. Br Heart J 1990; 63:342–344.
76. Sporton SC, Taggart P, Sutton PM, Walker JM, Hardman SM. Acute ischaemia: a dynamic influence on QT dispersion. Lancet 1997; 349:306–309.
77. Elming H, Jun L, Torp-Pedersen C, Kober L, Malik M. Measurement of QT interval dispersion. Cardiac Electrophysiol Rev 1997; 3:372–376.
78. Hering HE. Experimentelle Studien an Saeugethieren ueber das Elektrocardiogramm. Z Exper Pathol Therap 1909; 7:363–378.
79. Lewis T. Notes upon alternation of the heart. Q J Med 1910; 4:141–144.
80. Ricketts HH, Denosin EK, Haywood IJ. Unusual T-wave abnormality. Repolarization alternans associated with hypomagnesemia, acute alcoholism, and cardiomyopathy. JAMA 1969; 207:365–366.
81. Reddy CVR, Kiok JP, Khan RG, El-Sherif N. Repolarization alternans associated with alcoholism and hypomagnesemia. Am J Cardiol 1984; 53:390–391.
82. Shimoni Z, Flatau E, Schiller D, Barzilay E, Kohn D. Electrical alternans of giant U waves with multiple electrolyte deficits. Am J Cardiol 1984; 54:920–921.
83. Kleinfeld MJ, Rozanski JJ. Alternans of the ST segment in Prinzmetal's angina. Circulation 1977; 55:574–577.
84. Salerno JA, Previtali M, Panciroli C, Klersy C, Chimienti M, Regazzi Bonora M, Maragoni E, Falcone C, Guasti L, Campana C, Rondanelli R. Ventricular arrhythmias during acute myocardial ischaemia in man. The role and significance of R-ST-T alternans and the prevention of ischaemic sudden death by medical treatment. Eur Heart J 1986; (7 Suppl A):63–75.
85. Turitto G, El-Sherif N. Alternans of the ST segment in variant angina. Incidence, time course and relation to ventricular arrhythmias during electrocardiographic recording. Chest 1988; 93:587–591.
86. Joyal M, Feldman RL, Pepine CJ. ST-segment alternans during percutaneous transluminal coronary angioplasty. Am J Cardiol 1984; 54:915–916.
87. Ring ME, Fenster PE. Exercise-induced ST segment alternans. Am Heart J 1986; 111:1009–1011.
88. Smith JM, Clancy EA, Valeri CR, Ruskin JN, Cohen RJ. Electrical alternans and cardiac electrical instability. Circulation 1988; 77:110–121.

89. Gilchrist IC. Prevalence and significance of ST-segment alternans during coronary angioplasty. Am J Cardiol 1991; 68:1534–1535.

90. Sutton PMI, Taggart P, Lab M, Runnalls ME, O'Brien W, Treasure T. Alternans of epicardial repolarization as a localized phenomenon in man. Eur Heart J 1991; 12:70–78.

91. Nearing BD, Huang AH, Verrier RL. Dynamic tracking of cardiac vulnerability by complex demodulation of the T wave. Science 1991; 252:437–440.

92. Nearing BD, Oesterle SN, Verrier RL. Quantification of ischaemia-induced vulnerability by precordial T-wave alternans analysis in dog and human. Cardiovasc Res 1994; 28:1440–1449.

93. Verrier RL, Nearing BD, MacCallum G, Stone PH. T-wave alternans during ambulatory ischemia in patients with stable coronary disease. Ann Noninvas Electrocardiol 1996; 1:113–120.

94. Verrier RL, Zareba W, Nearing BD. T-wave alternans monitoring to assess risk for ventricular tachycardia and fibrillation. In: Moss AJ, Stern S, eds. Noninvasive Electrocardiology: Clinical Aspects of Holter Monitoring. London: WB Saunders, 1996: 445–464.

95. Schwartz PJ, Malliani A. Electrical alteration of the T-wave: Clinical and experimental evidence of its relationship with the sympathetic nervous system and with long QT syndrome. Am Heart J 1975; 89:45–50.

96. Platt SB, Vijgen JM, Albrecht P, Hare GFV, Carlson MD, Rosenbaum DS. Occult T-wave alternans in long QT syndrome. J Cardiovasc Electrophysiol 1996; 7:144–148.

97. Pantridge JF. Autonomic disturbance at the onset of acute myocardial infarction. In: Schwartz PJ, Brown AM, Malliani A, Zanchetti A, eds. Neural Mechanisms in Cardiac Arrhythmias. New York: Raven Press, 1978:7.

98. Kalter HH, Schwartz ML. Electrical alternans. NY State J Med 1948; 1:1164–1166.

99. Rosenbaum DS, Albrecht P, Cohen RJ. Predicting sudden cardiac death from T-wave alternans of the surface electrocardiogram: promise and pitfalls. J Cardiovasc Electrophysiol 1996; 7:1095–1111.

100. Pastore JM, Girourard SD, Laurita KR, Akar FG, Rosenbaum DS. Mechanism linking T-wave alternans to the genesis of cardiac fibrillation. Circulation 1999; 99:1385–1394.

101. Chinushi M, Restivo M, Caref EB, El-Sherif N. Electrophysiological basis of arrhythmogenicity of QT/T alternans in the long-QT syndrome. Tridimensional analysis of the kinetics of cardiac repolarization. Circ Res 1998; 83:614–628.

102. Konta T, Ikeda K, Yamaki M, Nakamura K, Honma K, Kubota I, Yasui S. Significance of discordant ST alternans in ventricular fibrillation. Circulation 82:2185–2189, 1990.

103. Surawicz B, Fisch C. Cardiac alternans: diverse mechanisms and clinical manifestations. J Am Coll Cardiol 1992; 20:483–499.

104. Verrier RL, Nearing BD. Electrophysiologic basis for T-wave alternans as an index of vulnerability to ventricular fibrillation. J Cardiovasc Electrophysiol 1994; 5:445–461.

105. Shimizu W, Antzelevitch C. Cellular and ionic basis for T-wave alternans under long QT conditions. Circulation 1999; 99:1499–1507.

106. Hashimoto H, Suzuki K, Miyake S, Nakashima M. Effects of calcium antagonists on the electrical alternans of the ST segment and on associated mechanical alternans during acute coronary occlusion in dogs. Circulation 1983; 68:667–672.

107. Nearing BD, Hutter JJ, Verrier RL. Potent antifibrillatory effect of combined blockade of calcium channels and 5-HT$_2$ receptors with nexopamil during myocardial ischemia and reperfusion in canines: comparison to diltiazem. J Cardiovasc Pharmacol 1996; 27:777–787.

108. Wu Y, Clusin WT. Calcium transient alternans in blood-perfused ischemic hearts: observations with fluorescent indicator fura red. Am J Physiol 1997; 273:H2161–H2169.

109. Lee H-C, Mohabir R, Smith N, Franz MR, Clusin WT. Effect of ischemia on calcium-dependent fluorescence transients in rabbit hearts containing indo 1. Correlation with monophasic action potentials and concentration. Circulation 1988; 78:1047–1059.

110. Lukas A, Antzelevitch C. Differences in the electrophysiologic response of canine ventricular epicardium and endocardium to ischemia: Role of the transient outward current. Circulation 1993; 88:2903–2915.

111. Coronel R, Wilms-Schopman FJ, Opthof T, Cinca J, Fiolet JW, Janse MJ. Reperfusion arrhythmias in isolated perfused pig hearts. Inhomogeneities in extracellular potassium, ST and TQ potentials, and transmembrane action potentials. Circ Res 1992; 71:1131–1142.

112. Smith JM, Cohen RJ. Simple finite-element model accounts for wide range of cardiac dysrhythmias. Proc Natl Assoc Sci 1984; 81:233–237.

113. Hohnloser SH, Klingenheben T, Zabel M, Li Y-G, Albrecht P, Cohen RJ. T-wave alternans during exercise and atrial pacing in humans. J Cardiovasc Electrophysiol 1997; 8:987–993.

114. Belic N, Gardin JM. ECG manifestations of myocardial ischemia. Arch Intern Med 1980; 140:1162–1165.

115. Verrier RL, Nearing BD. T-wave alternans as a harbinger of ischemia-induced sudden cardiac death. In: Zipes DP, Jalife J, eds. Cardiac Electrophysiology: From Cell to Bedside. Philadelphia: WB Saunders, 1995:467–477.

116. Armoundas AA, Osaka M, Mela T, Rosenbaum DS, Ruskin JN, Garan H, Cohen RJ. T-wave alternans and dispersion of the QT interval as risk stratification markers in patients susceptible to sustained ventricular arrhythmias. Am J Cardiol 1998; 82:1127–1129.

117. Armoundas AA, Rosenbaum DS, Ruskin JN, Garan H, Cohen RJ. Prognostic significance of electrical alternans versus signal averaged electrocardiography in predicting the outcome of electrophysiological testing and arrhythmia-free survival. Heart 1998; 80:251–256.

118. Estes NAM, Michaud G, Zipes DP, El-Sherif N, Venditti FJ, Rosenbaum DS, Albrecht P, Wang PJ, Cohen RJ. Electrical alternans during rest and exercise as predictors of vulnerability to ventricular arrhythmias. Am J Cardiol 1997; 80:1314–1318.

119. Coch M, Weber S, Buck L, Waldecker B. Assessment of T-wave alternation using atropine [abstract]. Circulation 1998; 98:I–442.

120. Gold MR, Bloomfield DM, Anderson KP, Wilber DJ, El-Sherif N, Estes NAM II, Groh WJ, Kaufman ES, Greenberg ML, Cohen RJ. A comparison of T-wave alternans, signal averaged electrocardiography, and electrophysiology study to predict arrhythmia vulnerability [abstract]. J Am Coll Cardiol 1999; 33(suppl A): 145A.

121. Klingenheben T, Zabel M, D'Agostino RB, Cohen RJ, Hohnloser SH. Predictive value of T-wave alternans for arrhythmic events in patients with congestive heart failure. Lancet, in press.

122. Ikeda T, Sakata T, Takami M, Kondo N, Tezuka N, Nakae T, Noro M, Enjoji Y, Abe R, Sugi K, Yamaguchi T. Combined assessment of T-wave alternans and late potentials used to predict arrhythmic events after myocardial infarction. J Am Coll Cardiol 2000; 35:722–730.

123. Adachi K, Ohnishsi Y, Shima T, Yamashiro K, Takei A, Tamura N, Yokoyama M. Determinant of microvolt-level T-wave alternans in patients with dilated cardiomyopathy. J Am Coll Cardiol 1999; 34:374–80.

124. Klingenheben T, Credner SC, Bender B, Cohen RJ, Hohnloser SH. Exercise-induced microvolt T-wave alternans identifies patients with non-ischemic dilated cardiomyopathy at high risk of ventricular tachyarrhythmic events [abstract]. PACE 1999; 22:860.

125. Verrier RL, Stone PH. Exercise stress testing for T-wave alternans to expose latent electrical instability. J Cardiovasc Electrophysiol 1997; 8:994–997.

126. Pedretti R, Etro MD, Laporta A, Braga SS, Caru B. Prediction of late arrhythmic events after acute myocardial infarction from combined use of noninvasive prognostic variables and inducibility of sustained monomorphic ventricular tachycardia Am J Cardiol 1993; 71:1131–1141.

127. Kavesh NG, Shorofsky SR, Sarang SE, Gold MR. The effect of heart rate on T-wave alternans. J Cardiovasc Electrophysiol 1998; 9:703–708.

29

Sudden Cardiac Death and Its Prevention

CHRISTINE M. ALBERT and JEREMY N. RUSKIN

Massachusetts General Hospital, Boston, Massachusetts

LEONARD A. COBB

University of Washington, Seattle, Washington

Sudden cardiac death (SCD) is a clinical entity characterized by abrupt and unexpected cardiac arrest in the absence of a noncardiac precipitating event such as trauma, exsanguination, electrocution, poisoning, etc. SCD typically occurs outside the hospital. A cardiac arrhythmia is generally considered to be the initiating mechanism and, although most victims have underlying heart disease, few have a documented history of prior arrhythmias. Earlier definitions of SCD have often invoked a consideration of associated cardiac symptoms—i.e., symptoms from 1 to 24 h duration. In view of the imprecision of such estimates, it is not surprising that opinions regarding the classification of mode of death may differ, particularly when the collapse was not witnessed.

I. PREDISPOSING CARDIAC DISORDERS

In contemporary Western societies, the incidence of SCD is largely set by the prevalence of coronary heart disease (CHD). Although SCD may be associated with virtually any cardiac disorder, the high prevalence of CHD typically overshadows the importance of other conditions. Since SCD events are rarely monitored by ECG or directly observed by medical personnel, the true incidence of arrhythmic SCD has not been established. Prospective and longitudinal observations from the National Heart, Lung, and Blood Institute's Framingham study indicate that more than half of all deaths from CHD can be considered to be SCD [1]. A typical victim is male (75% of cases), in his sixties, has recognized heart disease or hypertension, and smokes cigarettes. It is estimated that there are at least 250,000 episodes of SCD each year in the U.S. (2). Not unexpectedly, there is beginning to be evidence that the incidence of SCD has recently declined, as the age-specific mortality attributed to CHD has fallen in many countries (2–5).

The linkage between SCD and CHD may be considered within three general settings: patients with acute myocardial infarction, those with ischemia without infarction, and the setting of structural alterations secondary to infarction or chronic ischemia (e.g., mechanical dysfunction due to scar formation or ventricular dilatation). Although most episodes of SCD are unrelated to acute infarction (5,6), SCD is often described in the lay press as a "massive heart attack." Similarly, the oft-repeated, but incorrect, statement that most patients with acute infarction die before reaching the hospital is based on the faulty premise that sudden death in the setting of CHD is due to acute infarction.

Although CHD is by far the predominant cause of SCD, virtually all structural cardiac disorders may lead to SCD, particularly cardiomyopathies and hemodynamically significant aortic valve abnormalities. Other structural conditions that can serve as a basis for SCD include arrhythmogenic right ventricular dysplasia, atrial septal defect, nonatherosclerotic coronary artery disease, and tetralogy of Fallot (7,8). Important as they may be for the afflicted patient, disorders other than CHD contribute little to the total SCD burden in developed, Western countries.

II. IDIOPATHIC SCD

Whereas underlying cardiac disorders can usually be identified in SCD victims, there are puzzling situations in which sudden cardiac arrest, apparently arrhythmic in nature, occurs in the absence of any significant cardiac abnormality (9,10). In a report of survivors who had experienced out-of-hospital episodes of ventricular fibrillation, 8% of patients were found to have no evident structural cardiac abnormality, or only what was considered insignificant coronary atherosclerosis (10).

Another example of idiopathic SCD occurs among young Southeast Asian men who die unexpectedly while asleep. This syndrome has been given several different names, depending on the region where it is found (e.g., it is termed "nonlaitai" in Laos) (11,12). An outbreak of such cases appeared in the United States during the 1970s and 1980s subsequent to an influx of refugees following the Vietnam War. When direct observation of these cases was possible, for example, during attempted resuscitation, VF was frequently discovered to be the first recorded rhythm (11). Interestingly, examples of this syndrome have become increasingly rare in the United States, most likely reflecting a smaller number of susceptible immigrants. Resuscitated victims of this Southeast Asian "sudden unexplained nocturnal sudden death syndrome" as well as other forms of idiopathic VF have been shown to experience a high rate of recurrent cardiac arrest, documented to be VF in those few cases where direct observations have been made (10,11). Some of these cases of idiopathic VF are quite possibly indicative of a genetically determined susceptibility, and as indicated in Chapter 25, familial aggregation of cases of this syndrome have been described in several Southeast Asian families (13,14) presumably due to cardiac Na channel defects.

Another well-recognized form of SCD that does not involve structural abnormalities is that which occurs with higher incidence in children and young adults, in association with prolongation of the QT interval, at times accompanied by congenital deafness. The identification of the molecular bases for this spectrum of long-QT syndrome (LQTS) disorders represents a fundamental advancement in the understanding of at least some facets of arrhythmogenesis (15). Additionally, several other primarily electrophysiological abnormalities are occasionally responsible for SCD (16). Many of these disorders are discussed in Chapter 28.

III. ARRHYTHMIAS AND SCD

A. Ventricular Fibrillation (VF)

Although SCD is generally considered to be the consequence of VF, that belief is backed only by fragmentary clinical observations. The initiation of episodes has rarely been documented by ECG monitoring, and those that have been monitored are almost always highly selected cases in which histories were suggestive of cardiac arrhythmia. About half of apparent SCD victims demonstrate asystole or pulseless electrical activity (PEA) when first examined by rescue teams; however, this observation is of limited value because of the progressive degradation of VF to asystole over the course of several minutes (17). In cases where there has been a relatively short delay between collapse and the initial determination of rhythm (i.e., witnessed cases in a public location), the proportion of cases with VF increases to over 80% (Fig. 29.1). Thus, VF appears to be the underlying mechanism in the majority of events in which instantaneous collapse has occurred. However, it is also likely that asystole or PEA underlies a proportion of episodes classified as SCD (18).

While VF may be initiated with a few organized cardiogram complexes that resemble ventricular tachycardia (VT), the clinical syndrome of recurrent VT does not appear to contribute appreciably to the huge numbers of SCD victims. Additionally, VT is uncommonly identified in cardiac arrest victims examined a few minutes after collapse (5). Al-

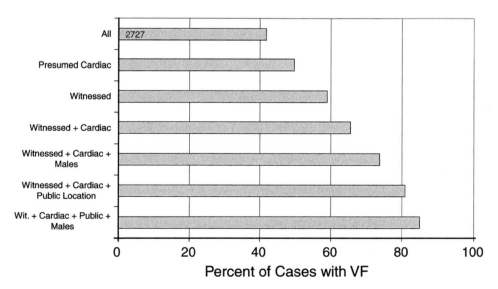

Figure 29.1   The bars indicated the initially recorded heart rhythms (shown as percent of subgroups with ventricular fibrillation), in 2727 consecutive cases of out-of-hospital cardiac arrest treated by Seattle paramedics between 1992 and 1996. Conditions associated with a high prevalence of VF are surrogates of shortened time from collapse to documentation of rhythm (witnessed collapse, public locations) and male gender. (LA Cobb, personal observations, 1999.)

though there is no evidence to support the use of electronic cardiac pacing for the treatment of cardiac arrest (19), the probability remains that AV block or primary disorders of impulse formation may be implicated in SCD, at least in a small proportion of cases (18,20,21).

IV. INFLUENCE OF THE AUTONOMIC NERVOUS SYSTEM

There is substantial experimental and clinical evidence supporting involvement of the autonomic nervous system (ANS) in the genesis of life-threatening ventricular tachyarrhythmias (22,23). These interactions are discussed in detail in Chapters 20 and 28. In general, environmental or physiological perturbations leading to sympathetic stimulation facilitate development of VT or VF; conditions impeding sympathetic neural traffic tend to have the opposite effect. Examples of the former include the arrhythmogenic properties of epinephrine and other catecholamines, the stimulus of an alarm in the LQT syndrome (24) and vigorous physical exertion (25–27). Furthermore, beta-sympathetic antagonists have been shown to reduce the incidence of SCD in post-MI patients (28). The circadian variation in the occurrence of out-of-hospital cardiac arrest is likely due to activation of the ANS after arising from sleep (29,30), as illustrated in Figure 29.2, and may be modified by β-adrenergic blockade (31).

Risk stratifiers indicative of enhanced sympathetic over vagal imbalance have been suggested as indicators of persons at risk for SCD. These include heart rate variability (32,33) and baroreceptor sensitivity to acute changes in blood pressure in post-MI patients at relatively high risk for death, including SCD (34).

V. ROLE OF CORONARY ARTERY THROMBOSIS

As detailed above, the importance of CAD as a causal factor for SCD is indisputable. However, the relationship of coronary artery thrombosis to sudden death is less clear. In a group of patients who could not be resuscitated from out-of-hospital VF, Reichenbach et

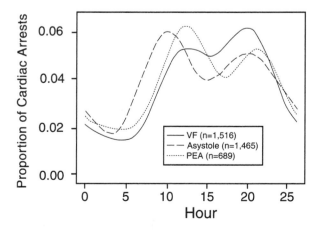

Figure 29.2   The relationship between time of day and frequency of out-of-hospital cardiac arrest in patients whose first recorded rhythm was VF or asystole; 3670 consecutive cases of treated cardiac arrest in Seattle. (Reproduced with permission from Ref. 30.)

al. (6) described acute thrombotic occlusion (assessed by post-mortem angiography) in 10% of 87 patients, 55% of whom had healed myocardial infarction. On the other hand, Davies and Thomas (35) reported intraluminal thrombosis in 74 of 100 cases of SCD; 18% of cases had thrombotic occlusions of at least 75% of the vessel lumen. They also found intraintimal thrombosis in an additional 21 patients and concluded that virtually all patients who die suddenly of CHD have actively progressive arterial lesions that are predominantly thrombotic. Others have described acute arterial thrombi in 4% to 59% of autopsied SCD victims (6). A confounding variable in virtually all of the autopsy studies is the matter of case identification and selection, the exact quantitative significance of nonocclusive arterial thrombosis, and the potential contribution of downstream embolization awaits further clarification.

VI. RACIAL AND SOCIOECONOMIC ASPECTS

Racial and socioeconomic factors have long been associated with health status and known to impact mortality rates (36). Most often, blacks fare less well than whites; such appears to be the case also for SCD. Blacks in two U.S. studies experienced out-of-hospital cardiac arrest at an average age several years younger than whites, and the age-specific incidence of cardiac arrest among blacks was twice that of whites (37,38). In one study in Chicago, the survival rate after an out-of-hospital cardiac arrest among blacks was only 31% of that among whites. Blacks were significantly less likely to have a witnessed cardiac arrest, bystander-initiated cardiopulmonary resuscitation, a favorable initial rhythm, or to be admitted to the hospital. When they were admitted, blacks were half as likely to survive. However, the association between race and survival persisted even when these and other recognized risk factors were taken into account. It is unclear whether such racial differences reflect genetic predisposition or a disadvantage related to access and adherence to good health care, or perhaps both influences. It is relevant to note that Hallstrom et al. (39) have described a direct correlation between apparent economic status (assessed value of housing) and survival from sudden cardiac arrest.

VII. AGE AND GENDER

A major public health reason for the concern about SCD is the sometimes unstated issue of "premature" death (i.e., a fatal event for an individual who otherwise would be expected to live for several years). Whereas half of SCDs occur in persons under approximately age 65, the age-specific annual incidence increases markedly as age rises (i.e., about 100 per 100,000 population for men aged 50, compared to 800/100,000 aged 75) (Fig. 29.3) (40).

SCD is approximately three times more common in men as compared to women. Even after adjustment for known cardiac risk factors, women have only 32% the rate of SCD as men (41) and the rate is lower at all ages (1). Consistent with this, only a minority (26%) of cardiac arrest victims are women. This suggests that women have a lower incidence of arrhythmic death (42). One possible explanation for this female advantage may be the lower prevalence of CHD in women, although it is unlikely to be the entire explanation. Presence of CHD does increase the risk of SCD in women; however, it does not completely eliminate the female advantage over men in susceptibility to SCD (43). Women with symptomatic myocardial infarction still have half the incidence of SCD (44). It appears then there is something about being female that is protective against SCD, even

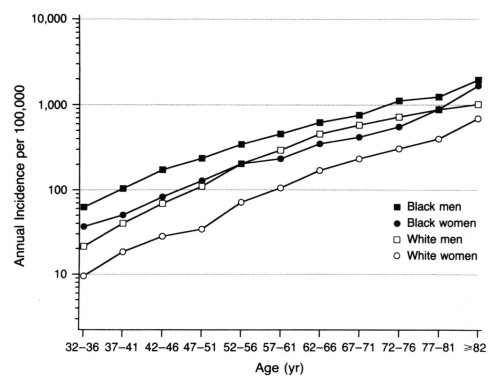

Figure 29.3   The incidence of out-of-hospital cardiac arrest according to age, sex, and race in 6451 patients with presumed cardiac basis for arrest in Chicago. (Reproduced with permission from Ref. 38.)

when overt CHD is present and, since the incidence of SCD becomes a more prominent feature of CHD mortality after menopause, (45) sex hormones may be partly responsible.

VIII.   PREDICTORS OF SCD

As discussed above and later in this chapter, there are a number of clinical correlates and predictors of sudden cardiac death. These population-based characteristics, which include CHD risk factors, dietary habits, genetic influences, and acute triggering events, serve to emphasize the multifactorial nature of SCD. While these predictors all share a statistical correlation with the subsequent development of SCD, they have inadequate sensitivity and specificity for predicting SCD within a reasonable period of time. Presently, the presence and severity of underlying heart disease is the condition most predictive of SCD.

IX.   PREVENTION IN THE GENERAL POPULATION

Nevertheless, the majority of SCDs occur in the general population and not in recognized high-risk subsets, such as patients with a history of MI, severe LV dysfunction, or a prior cardiac arrest. Fifty-five percent of male and 64% of female SCD victims have no previous history of heart disease, and therefore SCD is frequently the initial manifestation of heart disease (1). Therefore, SCD in this segment of the population must be addressed to impact significantly the overall incidence of SCD, even though the overall incidence is only 0.1% per year (46). This raises the obvious problem with respect to primary prevention; any broad-based intervention would have to be applied to 1000 people to prevent one sudden death. Several potential strategies might be employed to deal with this dilemma. First, since the majority of SCDs occur in the setting of CHD, focusing preventive efforts on the modification of known CHD risk factors seems a productive approach. Second, searching for more specific markers of risk in the general population to identify subgroups at high enough risk to justify an invasive and expensive intervention, such as a prophylactic ICD, could be quite effective. Third, developing harmless, low-cost interventions that could be applied to the population at large would be most useful. The latter would require identification of dietary or lifestyle risk factors specific to SCD, not just to CHD. Even if identification of those at risk remains poor, preventive strategies based on the modification of these risk factors could be usefully applied to the entire population at low cost and little risk.

A. CHD Risk Factors

To date, most efforts have focused on the first approach to primary prevention, modification of traditional CHD risk factors. Modifiable CHD risk factors that have been demonstrated to predict SCD in diverse cohorts include hypertension, hypercholesterolemia, diabetes, obesity, and smoking (1,41,43,45,47,48). Although initial data from the Framingham study suggested that these CHD risk factors might not be as predictive in women, recent data with longer follow-up suggests that they do predict SCD in both sexes (49). Smoking, in particular, appears to be a very important risk factor for SCD and predisposes men to acute coronary thrombosis (50) and women to plaque erosion with resultant acute thrombus (51). Elevated total cholesterol (TC), and in particular the TC/HDL (high-density lipoprotein) ratio, predisposes to acute plaque rupture in men and in postmenopausal women (50,51).

Population reductions in SCD incidence and in overall CHD mortality occurring since the mid-1960s provides indirect evidence of the success of CHD risk factor modification efforts by public health officials. Over the past 30 years, total mortality from CHD has declined more than 50% (52). Temporal declines in the incidence rates of all manifestations of CHD, including SCD, acute myocardial infarction, and angina pectoris, would suggest that the mortality decline is due, at least in part, to measures of primary prevention and changes in the prevalence and levels of major CHD risk factors. More recently, data from the NHLBI's ARIC study (53) suggest that the incidence of myocardial infarction did not decrease from 1987 to 1994, and that recent declines in CHD death rates may be due, at least in part, to improvements in secondary prevention.

Despite the decline in CHD mortality over the last decade, little proportionate change has been seen in the nature of these deaths, suggesting that modification of known risk factors impacts CHD in general and not SCD specifically. The majority of these events are still sudden, occurring out-of-hospital and in the emergency room (2,3). Thus, SCD is still the most common cause of death in adults less than 65 years old (2,3) and a major public health problem in the United States and all other industrialized countries, despite declining prevalence of CHD risk factors.

Although CHD risk factors are predictive of SCD, no single risk factor predicts with any accuracy who will die suddenly, and therefore the ability to identify specific individuals who are at high risk of SCD is limited. In an attempt to identify high-risk individuals, multivariate risk indexes have been created based on coronary risk factor status and electrocardiographic abnormalities. Those in the upper quintile of multivariate risk have a tenfold increased incidence of SCD compared to the lower quintile (1). Although it seems possible to identify 42% and 39% of SCDs in the upper decile of multivariate risk for men and women, respectively (1), only 4.5% of the men classified as high risk actually died suddenly over 15 years (54), and therefore the positive predictive value for any specific individual still remains poor. Clearly, continued research is needed to delineate both the factors that specifically predispose to SCD and the pathogenic mechanisms that precipitate it in both the general population and in persons with clinically manifest coronary artery disease.

B. Triggers

The risk of sudden cardiac death is not only a function of the underlying cardiac substrate and its vulnerability to arrhythmias, but also the frequency of exposure to acute precipitants or "triggers." These triggers may increase sympathetic activity, which in turn may precipitate coronary vasospasm, arrhythmias, and SCD. Two of the most prevalent triggers of SCD are vigorous exercise and psychosocial stress.

1. Physical Activity

The impact of physical activity on SCD seems paradoxical. On one hand, case reports suggest that cardiac arrhythmias and SCD occur with unusually high frequency during or directly after vigorous exercise (55). Exertion-related cardiac arrests have been reported in 10 to 20% of SCDs. Case-control and case-crossover studies have demonstrated that vigorous exertion can trigger myocardial infarction (27,56), cardiac arrest (25), and SCD (57). One autopsy study found that plaque rupture was much more common in men with severe CHD who died suddenly during exertion than in those who died at rest, suggesting that this may be the mechanism underlying the increase in risk (58). Despite the increase in risk associated with an episode of vigorous exertion, SCD during exercise is a rare event (16) and several epidemiological studies have suggested that habitual vigorous exercise is associated with an overall reduction in the risk of SCD (59,60) and total mortality (61,62). It appears that habitual vigorous exertion lowers the risk of cardiac arrest or myocardial infarction associated with an acute episode of exertion, and the risk associated with physical activity was highest in those who were usually sedentary (25,56,57). Experimentally, regular exercise has been demonstrated to prevent ischemia-induced ventricular fibrillation and death in dogs by increasing vagal activity (63). Thus, it appears regular conditioning exercise may decrease cardiovascular morbidity and mortality, whereas acute vigorous periods of exercise particularly in untrained individuals may have an adverse effect. More research on factors that might modify the excess risk of SCD associated with exertion is needed so that the benefits of regular conditioning exercise can be obtained without significant accompanying risk.

2. Psychosocial Stress

In one case-controlled study involving 64 SCDs, psychiatric illness, educational incongruity with spouse, and having no children were independently associated with increased risk of SCD in women (64). High levels of stress and social isolation are associated with an increase in both SCD and total mortality after myocardial infarction (65). Socially isolated men had an increased risk of cardiovascular mortality, but not of myocardial infarction in the Health Professionals Follow-up Study, suggesting that social support is protective against death in those with established disease (66). Social supports may alleviate anxiety, which has also been linked to fatal CHD, but not to nonfatal events (67). Individuals with high levels of anxiety have reduced heart rate variability compared to normal subjects (68), which has been shown to independently predict mortality after myocardial infarction (69) and perhaps SCD in the general population (70). It also appears that acute mental stress can trigger SCD, although determination of the role of mental stress is much more difficult than determination of the role of physical stress because of the difficulties inherent in assessing mental state. However, a convincing example of mental stress triggering SCD can be found in the "natural experiment" of the 1996 Northridge earthquake in Southern California, one of the strongest earthquakes ever recorded in a major city in North America (71). On the day of the earthquake, there was a sharp increase in the number of SCDs from cardiac causes that were related to CHD and not accounted for by physical exertion. These findings, along with the unusually low incidence of such deaths in the week after the earthquake, suggest that emotional stress may precipitate cardiac events in people who have underlying increased susceptibility.

3. Dietary Factors

In addition to the traditional CHD risk factors, nutritional factors also appear associated with CHD mortality to a greater extent than nonfatal CHD, and some have even been associated with SCD specifically. These risk factors may be specific for SCD as opposed to other manifestations of CHD, and therefore may have a selective effect on susceptibility to arrhythmias. Such factors may modify the electrophysiological properties of the myocardium, rendering it less vulnerable to arrhythmogenic triggers, thereby decreasing the risk of sustained ventricular arrhythmias and SCD. Experimental data in dogs (72) and primates (73) suggest that the n-3 polyunsaturated fatty acids (PUFA) found in cold water fish have antiarrhythmic properties. Animals fed a diet enriched with or given intravenous infusions of fish oil also appear to have a reduced risk of ventricular fibrillation during coronary artery ligation and reperfusion and in response to programmed ventricular stimulation (72–76). Plausible mechanisms for these antiarrhythmic effects include modulation of Na, K^+, and Ca^{2+} channels and there is in vitro evidence for such effects on channel activity in cardiac myocytes (76). Last, it is worth noting that consumption of one fish meal per week as compared to no fish was associated with an increase in heart rate variability (77) in patients, a predictor of a lower risk for arrhythmic death after myocardial infarction (78).

There are also data to support an antiarrhythmic property of n-3 fatty acids in normal subjects. Small amounts of fish intake, such as one meal per week (79), have been inversely associated with cardiac mortality in several prospective cohorts (79–82). Based on these promising observational data, a randomized trial was conducted. The Diet and Reinfarction Trial (DART) randomized 2033 men after myocardial infarction to receive or not receive advice to eat at least two portions of fatty fish per week. The intervention group had a significant 29% reduction in total mortality (primarily composed of ischemic heart disease deaths) at 2 years and no difference in the incidence of nonfatal myocardial infarctions (83). From these results, the authors hypothesized that fish consumption may reduce the risk of fatal arrhythmias and therefore mortality after myocardial infarction without affecting the incidence of repeat myocardial infarction.

Two epidemiological studies and one randomized trial have examined the relationship between fish consumption and SCD specifically. In a retrospective, population-based case-control study, both dietary intake of n-3 fatty acids and red-blood-cell n-3 fatty acid composition were associated with reduced risks of primary cardiac arrest. A 50% reduction in risk was associated with n-3 fatty acid levels equivalent to one fish meal per week (84). Similar findings were reported in the prospective Physicians Health Study (85). Among 20,551 U.S. male physicians, dietary fish intake was associated with a reduced risk of subsequent SCD, with an apparent threshold at a consumption level of one fish meal per week. For men who consumed fish at least once per week, the multivariate relative risk of sudden death was a significant value of 0.48 compared to those who consumed fish less than monthly. These findings are consistent with those reported from the GISSI-Prevenzione Trial, a randomized trial of n-3 polyunsaturated fatty acid supplements (1 g daily) in patients surviving recent myocardial infarction (86). Patients who received treatment with n-3 fatty acids had a significantly reduced risk of the primary endpoint (death, nonfatal MI, and nonfatal stroke) primarily due to a statistically significant 45% reduction in the risk of sudden death resulting in a significant, 17% reduction in risk of death. As in prior studies, there was no benefit on nonfatal events.

In addition to eicosapentanoic acid (EPA) and docosahexaenoic acid (DHA) found in fish, there are other dietary sources of unsaturated n-3 fatty acids. Alpha-linolenic acid is a long-chain n-3 fatty acid found in tofu, soybean and canola oils, nuts, and some other foods of plant origin that humans can metabolize to EPA and DHA, the n-3 fatty acids found in fish (76). Therefore, α-linolenic acid found in these important dietary oils may also have antiarrhythmic properties. In populations where cardiovascular mortality is very low, the traditional diet frequently has a high content of α-linolenic acid (87). A randomized secondary prevention trial (88) that evaluated the effect of a Mediterranean α-linolenic acid-rich diet found a significant 76% reduction in cardiac death in the group assigned to the Mediterranean diet despite no changes in serum lipids. There were no sudden deaths in the intervention group compared to eight in the control group (88).

In addition to n-3 fatty acids, alcohol is another dietary factor that may have a selective effect on risk of SCD. It has long been known that heavy alcohol consumption (>5 drinks/day) is associated with an increased risk of SCD (89,90); however, the results of studies addressing light-to-moderate alcohol consumption and SCD have not been consistent. Prospective studies have found either no (91,92) or a positive (93) association, while case-control studies have reported inverse linear associations (94). In the prospective Physicians Health Study (95), a significant U-shaped association between moderate alcohol consumption and subsequent SCD was found. After controlling for multiple confounding variables, men who consumed 2 to 6 drinks per week at baseline had significant 60 to 79% reductions in risk of SCD compared to those who rarely or never consumed alcohol, and compared to those who consumed two or more drinks per day. In contrast, the relationship of alcohol intake and non-sudden-CHD death was L-shaped or linear. Presumably, the differential effect on SCD is due to the superimposed effect of alcohol on arrhythmogenesis on a background effect on atherogenesis and thrombosis.

4.   Genetic Factors

Two recent studies have suggested that there may be a very significant genetic component to risk of SCD. In a recently reported retrospective case-control study of primary cardiac arrest victims (i.e., cardiac arrest not secondary to noncardiac causes), a family history of MI or primary cardiac arrest in a first-degree relative was associated with a 57% increase in the risk of primary cardiac arrest independent of other risk factors (96). This study suggests that in addition to known CHD risk factors, there are genetic or unknown environmental factors that are partially responsible for the familial aggregation of primary cardiac arrest. These findings were confirmed and expanded in a prospective study. In the Paris Prospective Study I, 118 SCDs were documented in 7746 men employed by the Paris Civil Service over an average of 23 years of follow up (47). Parental history of SCD was a risk factor for occurrence of SCD (relative risk, 1.95), but was not associated with the occurrence of fatal MI. Conversely, parental history of myocardial infarction was a risk factor for the occurrence of fatal MI, but was not associated with risk of SCD. The effect was cumulative, with a relative risk for those with one parental history of SCD equal to 1.89, and the RR of having two parents with a history of SCD being 9.44. In addition, there was a significant direct correlation between the age of the parent and age of the progeny at the time of death.

The selectivity of the effect of family history of SCD on risk for SCD suggests that genetic or environmental factors responsible for the familial aggregation may predispose to fatal arrhythmia rather than to CHD in general. Currently, the factors involved are unknown, but perhaps recent advances in our understanding of the genetics of primary arrhythmic disorders such as LQTS and idiopathic ventricular fibrillation (97) may shed light on the genetics of the more common syndrome of SCD. At the present time, it is unclear whether these advances will translate into improved detection and treatment of the broader population of patients at risk for ventricular arrhythmia and SCD and more research in this area is a high priority.

X.   PREVENTION IN PATIENTS WITH KNOWN CARDIAC DISEASE

A.   Postmyocardial Infarction Patients

As discussed above, the incidence of SCD is substantially higher in both men and women who have overt CHD and increases significantly after any CHD event and specifically after myocardial infarction (98,99). SCD accounts for 30 to 60% of the mortality in the first year after infarction (99,100). However, the absolute risk in unselected patients is still low (1–3% per year). Multiple techniques have been devised to stratify patients into those who are at increased risk of SCD. Those explored extensively in previous studies include assessment of left ventricular (LV) function, ambulatory electrocardiographic recordings for the detection of ventricular arrhythmias, measurement of Q-T dispersion, T-wave alternans, signal-averaged electrocardiograms, measurement of heart rate variability or baroreflex sensitivity, and invasive electrophysiology studies. The aim is to identify a high-risk subset of patients who might benefit from specific aggressive prevention strategies such as an implantable cardioverter defibrillator (ICD). Each of these tests has been demonstrated to predict increased risk of SCD in one or another subgroup of patients; however, when viewed in isolation, each has relatively low positive predictive value and, therefore, many patients with positive tests will not suffer SCD. This has led to the strategy of combining different tests to increase predictive value.

B.   Risk Stratification and Implantable Defibrillators

This was the strategy that was successfully employed in the Multicenter Automatic Defibrillator Implantation Trial (MADIT I) (101) and Multicenter Unsustained Tachycardia Trial (MUSTT) Trial (102). Both trials combined information on LV function, ventricular ectopic activity, and the results of electrophysiological testing to identify a subgroup of patients with a high risk of arrhythmic death. MADIT I randomized 196 patients with a history of prior MI and LV dysfunction [LV ejection fraction (LVEF) $\leq$ .35], a documented episode of asymptomatic, unsustained ventricular tachycardia, and inducible, nonsuppressible ventricular tachyarrhythmia or fibrillation on electrophysiological study to receive either an ICD or conventional drug therapy. The conventional therapy group primarily received the antiarrhythmic drug amiodarone (75%) and only 10% of patients were treated with Na channel blocking antiarrhythmic agents. MADIT was terminated when a highly significant 54% reduction in the risk of death emerged in the group treated with a defibrillator, resulting primarily from a decrease in arrhythmic deaths. On analysis, however, there was also a lower incidence of nonarrhythmic deaths in the ICD treatment group, suggesting either some misclassification in the cause of death or the possibility that antiarrhythmic drugs in the conventional therapy treatment group resulted in an increase in nonarrhythmic cardiac mortality. Also, β-receptor blocking agents were used significantly more frequently in the ICD arm, although the main result still persisted after statistical controls were used to adjust for their use. Based on the results of this trial, the use of the ICD in patients with nonsustained VT, prior MI, LV dysfunction, and inducible VF or sustained VT at electrophysiological study not suppressed by procainamide has been accepted as a Class I indication for ICD use by the American College of Cardiology/American Hearth Association Task Force on Practice Guidelines.

MUSTT randomized a similar group of patients (CAD patients, LVEF <0.40, unsustained VT, and inducible ventricular arrhythmia at electrophysiological study) either to treatment with conservative drugs (ACE inhibitors and/or beta-blockers) or to "electrophysiologically guided" therapy, followed by an ICD if antiarrhythmic drugs did not suppress the arrhythmia (102). The strengths of this trial include its large size (704 patients) and the use of a control group that did not receive antiarrhythmic drugs. The results of MUSTT confirm there is a significant mortality benefit of the ICD, compared to both conventional therapy and EP-guided antiarrhythmic drug therapy in this subgroup of patients. The MUSTT results lend further support to the recommendation made by the Taskforce on Practice Guidelines and may serve to broaden this recommendation to include patients with ventricular tachycardia induced at EP study, regardless of whether the arrhythmia is suppressible.

Contrary to the previous two examples of successful risk stratification, results from the Coronary Artery Bypass Graft (CABG) Patch trial demonstrate how risk stratification sometimes fails. This trial randomized 900 patients who were undergoing coronary artery bypass grafting with a LVEF of less than 0.36 and abnormal signal-averaged electrocardiograms to therapy with an ICD or to a control group (103). After an average of 32 months of follow-up, there was no difference in mortality between the two treatment arms. A subsequent analysis (104) demonstrated that ICD use did reduce arrhythmic death. However, only 29% of the total deaths were arrhythmic (compared to 50% in the control group of MADIT) and, therefore, total mortality was not significantly affected. The population selected was apparently not at high enough risk for arrhythmic death to derive an overall mortality benefit from the ICD. A number of possible explanations exist to explain the low arrhythmic mortality in this trial. First, the signal-averaged electrocardiogram may have a lower positive predictive value than the combination of unsustained ventricular tachycardia and inducible ventricular arrhythmias at EP study in patients with CAD and LV dysfunction. Second, revascularization may markedly decrease the risk of arrhythmic death (105) and, perhaps, this population regardless of the risk stratification approach employed, would not benefit from ICD implantation.

C. Risk Stratification and Antiarrhythmic Drugs

The results of MADIT and the MUSTT trials discussed above argue against the strategy of using prophylactic antiarrhythmic drug therapy in high-risk patients with LV dysfunction. Several other trials have also examined the role of prophylactic antiarrhythmic drug therapy in patients with prior MI; all have been disappointing. The hypothesis that long-term antiarrhythmic drug suppression of asymptomatic ventricular premature beats (PVCs) after myocardial infarction would decrease mortality was tested in the NHLBI's Cardiac Arrhythmia Suppression Trial (CAST) (106). Patients in whom ventricular ectopy could be suppressed with Na channel blockers (encainide, flecainide or moricizine) were assigned to active drug or placebo. After a mean follow-up time of 10 months, the Data and Safety Monitoring Board recommended that the encainide and flecainide arms be discontinued because of an excess mortality in patients receiving these drugs. The relative risk of death or cardiac arrest due to all causes was 2.38 in the encainide and flecainide arms compared to placebo (106). A possible, but nonsignificant trend toward lower mortality, was observed with moricizine treatment, and this arm was continued as the so-called CAST II study, but this, too, was terminated early due to a higher mortality in the moricizine arm during the initial 14-day run-in period (107). A subsequent analysis of the CAST data suggested that the higher mortality rate in the flecainide–encainide arm might have been due to an interaction between ischemia and proarrhythmia, with resultant conversion of nonfatal ischemic events to fatal events (108). In addition, a meta-analysis of 18 long-term trials of Na channel blockers post-MI, involving more than 6300 patients, showed a significant excess of mortality associated with active treatment (relative risk, 1.21) (109). As a result of these studies, it seems likely that those antiarrhythmia agents which block Na channels (e.g., encainide, flecainide, etc.) are associated with increased risks of SCD in patients with ischemic heart disease.

Agents that prolong repolarization, largely due to block of potassium channels, have also been evaluated in postmyocardial infarction patients, but with mixed results. D-sotalol, a blocker of the delayed rectifier K current, was tested in infarction patients with depressed left ventricular function in the Survival With ORal D-sotalol (SWORD) trial. This study of 3119 patients was terminated early because of a statistically significant excess mortality in the D-sotalol treatment group (4.6 vs. 2.7%) (110). Unlike the mixture of d- and 1-sotalol isomers, the d-sotalol isomer has minimal β-adrenergic blocking properties and its predominant activity is as a K channel blocker; thus, it is inappropriate to extrapolate the SWORD data to the d,l-sotalol mixture. This supposition was supported when the d,l-isomer was tested in a large trial of patients in the United Kingdom and did not significantly increase or decrease risk of mortality or sudden death (111). In addition to sotalol, prophylactic trials in post-MI patients with ejection fraction <35% have been completed with the I_{Kr} blocker dofetilide (112). A total of 8272 patients were screened and 1510 (63% of those eligible based on left ventricular function) participated in the study. No difference in overall survival was seen with this drug (230 vs. 243 deaths, relative risk = 0.94), nor was there a statistical difference in cardiac or arrhythmic mortality rate.

Several other trials have evaluated the efficacy of using the drug amiodarone to prevent mortality after acute myocardial infarction. The two largest, the European Myocardial Infarct Amiodarone Trial (EMIAT) of 1486 patients (113) and the Canadian Amiodarone Myocardial Infarction (CAMIAT) of 1202 patients (114) both found reductions in arrhythmic death, but no effect on total mortality. In an attempt to increase the risk for arrhythmic death, EMIAT enrolled only patients with a LVEF <40% and CAMIAT enrolled patients with frequent or repetitive premature depolarizations (VPDs). Risk reductions of approximately 35% in arrhythmic death were seen in both trials, but this did not translate into an overall mortality benefit. However, these trials were not designed or powered to detect small-to-moderate benefits (<35%) on mortality. A subsequent meta-analysis involving 6553 post-MI or heart failure patients reported a trend toward a decrease in mortality with amiodarone (odds ratio = 0.85; p = 0.081); however, when the subgroup of post-MI trials was examined alone the difference was not significant (odds ratio, 0.92) (115).

In summary, no antiarrhythmic drug has yet demonstrated conclusive mortality benefits when given prophylactically to patients who have suffered prior infarction. Amiodarone, d,l-sotalol, and perhaps dofetilide appear not to be associated with an increase in total mortality when used in post-MI patients. This neutral effect on mortality makes these agents preferable to sodium channel blocking agents for the treatment of other symptomatic arrhythmias (e.g., atrial fibrillation), in patients with coronary artery disease.

1. Conventional Therapy Applied to All Patients

Despite results using channel-directed drugs, there are pharmacological therapies that have proven effects in helping to reduce the risk of SCD in post-MI survivors. Beta-adrenergic antagonist therapy initiated 5 to 28 days after myocardial infarction significantly reduces mortality, primarily due to a decrease in SCD (28,116). The benefits appear to be consistent across various subgroups and serious side effects are uncommon in patients able to tolerate beta blockade. In a meta-analysis of 26 trials, β-adrenergic antagonist therapy after myocardial infarction reduced mortality by 23% (109). Despite the clear efficacy of this therapy, only 36 to 42% of the 240,989 patients enrolled in the National Registry of Myocardial Infarction from 1990 to 1993 actually received oral β-adrenergic antagonist therapy (117) and the continued underutilization of these highly effective agents, especially in the elderly (118), is a subject of national concern.

Angiotensin-converting-enzyme inhibitors initiated within the first few days to weeks after myocardial infarction have also shown benefits in patients with left ventricular dysfunction. The TRACE trial (119) demonstrated that trandolapril produced a significant reduction of 24% in SCD in patients with echocardiographic evidence of LV dysfunction (LVEF ≤35%), 3 to 7 days after myocardial infarction. More importantly, this trial, the Survival and Ventricular Enlargement (SAVE) Trial (120), and the Acute Infarction Ramipril Efficacy (AIRE) study (121) demonstrated significant mortality benefits of ACE-inhibitor therapy in post-MI patients with LV dysfunction. Given these findings, this therapy should be considered standard of care in patients where ACE inhibitors are tolerated, though again there are data suggesting these agents are underutilized (122).

Another important strategy in reducing the risk for SCD among patients with prior MI or angina and elevated LDL cholesterol levels is aggressive cholesterol reduction with lipid-lowering agents. Although none of the lipid-lowering trials has examined incidence of SCD, there is some evidence that part of the benefit on cardiovascular mortality may be through a reduction in the risk of SCD. In the Scandinavian Simvastatin Survival Study, or so-called (4S study), the number of deaths occurring within 1 h of symptoms was lower among study patients assigned to the lipid-lowering agent Simvastatin as compared to placebo (37 vs. 63) (123). As with ACE inhibitors, lipid-lowering agents have been proven to reduce mortality in patients with elevated cholesterol levels and documented CHD or CHD risk factors (123,124).

2. Congestive Heart Failure

Congestive heart failure (CHF) markedly increases the risk for SCD. In the Framingham Study, symptomatic CHF was associated with a two- to fivefold increase in the risk of SCD (41). The absolute risk of SCD increases with severity of CHF; however, the impact of SCD is heightened in patients with symptomatically mild CHF where the majority of deaths are sudden and not due to progressive heart failure (125). Results applying to post-MI patients with CHF have been described above. However, it is worth noting that the risk of SCD is increased in CHF regardless of its etiology, and therefore another major strategy being employed to prevent SCD is to target all patients with CHF for intervention. Again, the major modalities being tested include antiarrhythmic agents (primarily amiodarone), conventional agents (β-adrenergic antagonists), and ICD therapy. The data regarding use of amiodarone in patients with CHF are limited, and currently the only positive results are from a single trial, the Grupo de Estudio de la Sobrevida en la Insuficiencia Cardiaca en Argentina (GESICA) trial. Of 516 patients with LVEF <35% and symptomatic CHF enrolled in this trial, those assigned to amiodarone had a significant, 28% reduction in mortality (126). Both SCD and death due to progressive CHF were found to be reduced. Contrary to these results, in the Veteran's Administration cooperative study, or Survival Trial of Antiarrhythmic Therapy in Congestive Heart Failure trial (STAT-CHF) of 674 patients with symptomatic CHF and LVEF <0.40, there was no beneficial effect of amiodarone on either total mortality or SCD (127). Reasons for these disparate results are unclear, but may be due to differences in patient populations or in compliance. In addition, the total mortality in the placebo group was much higher in the GESICA Trial than STAT-CHF (55% vs. 29%) and, therefore, presumably the severity of CHF differed. Also the proportion of patients with nonischemic cardiomyopathy differed significantly in the two studies, 61% in GESICA compared with only 30% in STAT-CHF. Subgroup analyses of these results have generated the hypothesis that amiodarone may reduce mortality in nonischemic cardiomyopathy; however, this hypothesis has not been tested in rigorous prospective randomized trials.

Strong data supporting use of β-adrenergic receptor antagonists in patients with heart failure has been accumulated in three large-scale trials showing highly significant reductions in total mortality and SCD. The Carvedilol Heart Failure Study Group reported a significant 65% reduction in death and a 45% reduction in SCD among 1094 patients with mild-to-moderate CHF treated with carvedilol (128). Since carvedilol is not a pure β-adrenergic antagonist and also blocks α_1-receptors and exerts antioxidant effects, it was unclear whether beneficial effects would be seen with other β-adrenergic antagonists. Two randomized trials of selective antagonists of β_1-receptors, the Cardiac Insufficiency Bisoprolol Study (CIBIS-II) (129) and the Metoprolol CR/XL Randomized Intervention Trial in Heart Failure (MERIT-HF) (130) have resolved this uncertainty. Both trials enrolled patients with depressed systolic function and NYHA Class III or IV heart failure. Both trials were stopped early due to highly significant 34% reductions in mortality among patients assigned to β_1-receptor antagonists, largely due to significant reductions in SCD. Although newer generation β-adrenergic antagonists such as carvedilol and bucindolol have theoretical advantages for the treatment of heart failure, CIBIS-II and MERIT-HF demonstrate that older generation β_1-selective agents without vasodilatory properties also produce substantial survival benefits in patients with CHF. Trials comparing these newer agents to metoprolol should provide important information regarding relative efficacy and cost-effectiveness.

Given the results of the MADIT and MUSTT studies, there is also much current enthusiasm that prophylactic ICD therapy will prevent SCD among patients with heart failure. As discussed above, these two prophylaxis trials effectively utilized the presence of nonsustained ventricular tachycardia and inducible, sustained ventricular tachycardias at electrophysiological study to identify a subgroup of patients with CAD and CHF at high risk of SCD. However, the arrhythmic event rate and overall mortality among patients without inducible VT at electrophysiological study who were therefore excluded from these trials is disturbingly high. A 24% arrhythmic event rate among the patients without inducible VT was reported in the MUSTT trial. This raises the possibility that at least some of these patients could benefit from ICD therapy. Several ongoing trials are examining the effectiveness of ICD therapy in unselected patients with LV dysfunction secondary to both ischemic and nonischemic cardiomyopathy (131). The largest one evaluating defibrillator use as primary prevention of SCD in CHF is the Sudden Cardiac Death in Heart Failure Trial (SCD-HeFT). This trial will randomize 2500 patients with an LVEF <0.36 and NYHA Class II–III CHF to three treatment groups: ICD therapy, amiodarone therapy, and "optimal" medical therapy alone. When completed, results of this trial will not only supply important information regarding the efficacy of ICD therapy in the prevention of SCD in patients with CHF, but crucial data regarding cost-effectiveness of widespread use of this therapy in patients with CHF. Such cost-effectiveness data are important in making policy decisions regarding expenditure of health resources for this very expensive, but potentially equally valuable, therapy.

XI. COMMUNITY RESUSCITATION PROGRAMS

Nearly 40 years ago, following the development and introduction of closed chest compression, mouth-to-mouth ventilation, and transthoracic defibrillation, the ability to resuscitate victims of cardiac arrest was finally realized (132). In the mid-1960s the first organized effort to provide critical cardiac care outside the hospital was undertaken in Northern Ireland (133). Subsequently, the concept of rapidly responding prehospital emergency care by paramedical personnel was formulated and introduced in the United States (132). With that development, there also arose an opportunity for community-based interventions to facilitate the resuscitation of victims of out-of-hospital cardiac arrest. By the mid-1980s, most medium-to-large U.S. cities had developed rapid response paramedic programs. Recognized determinants of survival in patients treated in these types of programs are summarized in Table 29.1.

A. CPR Training Programs

More than 50 million Americans are estimated to have received instruction in providing CPR. Both personal training classes and programmed instruction approaches using, for example, television (134) have been used. Most important, there is convincing evidence that the survival rate for out-of-hospital cardiac arrest can be improved when bystanders initiate CPR prior to the arrival of professional medical personnel (135,136). Timing in use of CPR therapy is absolutely critical and the window where it is effective is relatively short. Initiation within the first 4 to 6 min after collapse and continuation until the arrival of

Table 29.1 Determinants of Survival for Out-of-Hospital Cardiac Arrest

Variable	Factors favoring survival
Initial cardiac rhythm	Ventricular fibrillation (149) (few survivors with other rhythms)
Rapidity of response	Quick response is mandatory (survival declines by 3 to 10% for each minute delay in providing treatment for VF)(150)
Comorbidity	Absence of major comorbidity (151)
Race	Survival greater in whites than blacks (37,38)
Socioeconomic status	Greater financial resources (39)
Age	Slightly improved survival if <70 Yrs (152)

skilled rescuers (136) is necessary. Nevertheless, CPR as a stand-alone procedure is rarely a definitive, life-saving procedure. Rather, for victims of sudden cardiac arrest, CPR represents a temporary life-saving measure, providing a few critical minutes of reduced circulation to vital structures until advanced life support, particularly defibrillation, is made available (137). In view of the demonstrated effectiveness of bystander-initiated CPR, the community investment in CPR training is a sound (137) strategy indeed, although it should be recognized that due to the frequency of these events, many trained providers are not likely to need these skills.

B. Public Access Defibrillation

An impressive example of recent technological advancement in this area is the development of automated external defibrillators (AEDs) capable of recognizing VF in patients and delivering a remedial precordial shock. AEDs are now extensively and safely used by first-response emergency medical technicians. With the declining cost of these devices, consideration is being directed toward widespread deployment of AEDs and to the training in their use by the general public [i.e., "public access defibrillation" (PAD)] (138).

To a limited extent, PAD is already in place. Some airlines, airports, sports arenas, and a few other public sites have installed automated defibrillators for use, principally, by trained employees (139). It is a virtual certainty that lives can be saved by the proper use of these devices. When defibrillation is provided within the first minute or two following cardiac arrest, the initial rhythm is most often VF, and the survival rate is excellent—100%— in one series of patients who developed cardiac arrest during the course of a cardiac rehabilitation program (140). Over a 17-year experience in a Seattle professional sports facility where paramedics were available in about 2 mins, 90% of 29 cardiac arrests had VF when first monitored, and had a 62% survival rate (141).

Enthusiasm for PAD is, however, somewhat dampened by the fact that most cardiac arrests occur at home rather than in public places (142) and the likelihood of utilizing defibrillators in any given public place is quite small. Based on a retrospective analysis of cardiac arrest locations within Seattle and surrounding King County, Washington, investigators concluded that the most efficient placement of devices for public access would have been accomplished by the strategic deployment of 276 AEDs to cover a total of 134 cardiac arrests during the 5-year study period (40). Of this population, only 60% actually had VF when rescuers arrived on the scene. Although it is not known how much improvement in survival could be achieved over that provided by the existing emergency medical services, it is reasonable to consider that PAD programs will have their greatest impact in communities with relatively low survival rates.

C. Secondary Prevention

Despite improved cardiopulmonary resuscitation and the beginnings of PAD, currently only 6 to 32% of cardiac arrest victims who reach the hospital survive to hospital discharge (143). If a cardiac arrest victim is fortunate enough to survive, the risk for recurrence of a life-threatening ventricular arrhythmia is quite high (20–30% at 2 years) (144). Understandably, this group of patients is the first to which ICD therapy was directed. However, until recently, no randomized trial data were available and, although observational studies suggested benefits (145), there were concerns regarding patient selection and use of historical controls.

Three randomized trials of ICD efficacy have recently reached completion. The Antiarrhythmics Versus Implantable Defibrillator (AVID) trial randomized 1016 patients resuscitated from near-fatal ventricular fibrillation, or with symptomatic, sustained ventricular tachycardia and hemodynamic compromise, to treatment with an ICD as first-line therapy versus antiarrhythmic drug therapy, primarily amiodarone (146). If patients did not have syncope and had only symptoms suggestive of hemodynamic compromise (presyncope, angina, or CHF), or if patients had undergone revascularization, a LVEF <40% was required for enrollment. Patients assigned to ICD therapy had significant 39%, 27%, and 31% reductions in all-cause mortality at 1, 2, and 3 years, respectively, compared with antiarrhythmic drug therapy. Patient groups in this study were well balanced, except for treatment with β-adrenergic antagonists, which was significantly more frequent in the ICD group (42.3% vs. 16.5%); however, the beneficial effect of the defibrillator persisted after statistically adjusting for use of β-adrenergic antagonists using a multivariate Cox analysis model.

The Cardiac Arrest Hamburg Study (CASH) was another clinical trial directed at this question and, as in the AVID study, ICD therapy was associated with significant reductions in patient mortality (147). This trial randomized 288 cardiac arrest survivors to either ICD therapy, amiodarone, propafenone, or metoprolol alone. Propafenone treatment was stopped early due to an excess mortality rate (29% at 2 years). The power to detect significant differences between each group was limited by small sample size; therefore, the amiodarone and metoprolol groups were combined and their combined outcome compared to results obtained with the ICD. Two-year mortality was 12.1% in the ICD arm and 19.6% in the combined amiodarone and metoprolol arm. In this analysis, ICD use was associated with a significant 37% reduction in total mortality at 2 years, similar to the mortality reduction seen in AVID.

Yet a third secondary prevention study, the Canadian Implantable Defibrillator Study (CIDS), also assessed the efficacy of the ICD versus amiodarone among 659 patients with a history of cardiac arrest or hemodynamically unstable ventricular tachycardia (148). This trial showed a less impressive reduction in mortality (19.7%), which was not significant. The preliminary results of a meta-analysis combining the three secondary prevention trials discussed here suggests there is a significant mortality benefit of the ICD compared to antiarrhythmic drug therapy in patients who have survived life-threatening ventricular arrhythmias. As a result of these trials, the ICD has now become the standard of care in this group of patients.

XII. SUMMARY

On reflection of the findings discussed in this chapter, it is evident that the most clinically relevant aspect of studies on SCD lies in its prevention. To this end, substantial progress has been made—mostly through indirect efforts directed at the prevention of coronary heart disease and its complications. Reduction of risk factors and employment of procedures aimed at myocardial revascularization have both had a significant impact. Additionally, we have seen real success with the long-term administration of β-adrenergic antagonists, angiotensin-converting-enzyme inhibitors, and cholesterol-lowering agents following myocardial infarction. Further, the development of implantable cardioverter defibrillators has provided an effective means to abort SCD in patients at very high risk. Notably absent from a listing of effective preventive measures are the classic antiarrhythmic drugs. The potential value of community resuscitation programs is evident, but, to date, the implementation of emergency medical services is remarkable for its lack of consistent benefits in different communities.

An encouraging development is the recognition of dietary measures that appear to have antifibrillatory properties. The mechanism whereby omega-3 and related polyunsaturated fatty acids may provide protection from SCD represents an important opportunity for further investigation where more research is clearly needed. The influence of this and other dietary modifications or supplementation should begin to be examined in appropriately designed, rigorous trials.

Last, it should be noted that our abilities to identify patients at high risk for SCD remain unsatisfactory, aside from recognizing those with manifest CAD and poor ventricular function. In the future, it is possible that efforts to identify genetic factors that increase vulnerability to VF may lead to novel methods of risk identification and hence beneficial intervention. Meanwhile, it is important to recognize that we now have effective, albeit indirect, tools that help to reduce the burden imposed by SCD and the challenge here is to ensure that these are applied to today's patients.

REFERENCES

1. Kannel WB, Schatzkin A. Sudden death: Lessons from subsets in population studies. J Am Coll Cardiol 1985; 5:141B–149B.
2. Gillum RF. Sudden death in the United States: 1980–1985. Circulation 1989; 79:756–765.
3. Traven ND, Kuller LH, Ives DG, Rutan GH, Perper JA. Coronary heart disease mortality and sudden death: Trends and patterns in 35- to 44-year-old white males, 1970–1990. Am J Epidemiol 1995; 142:45–52.
4. Goldberg RJ. Declining out-of-hospital sudden coronary death rates. Additional pieces of the epidemiologic puzzle. Circulation 1989; 79:1369–1372.
5. Cobb LA, Weaver WD, Fahrenbruch CE, Hallstrom AP, Copass MK. Community-based interventions for sudden cardiac death: Impact, limitations, and changes. Circulation 1992; 85 [suppl I]:I98–I-102.
6. Reichenbach DD, Moss NS, Meyer E. Pathology of the heart in sudden cardiac death. Am J Cardiol 1977; 39:865–872.
7. Tabib A, Miras A, Taniere P, Loire R. Undetected cardiac lesions cause unexpected sudden cardiac death during occasional sport activity. A report of 80 cases. Eur Heart J 1999; 20(12):900–903.
8. Silka MJ, Hardy BG, Menashe VD, Morris CD. A population-based prospective evaluation of

risk of sudden cardiac death after operation for common congenital heart defects. J Am Coll Cardiol 1998; 32:245–251.

9. Marcus FI. Idiopathic ventricular fibrillation. J Cardiovasc Electrophysiol 1997; 8:1075–1083.

10. Kudenchuk PJ, Cobb LA, Greene L, Fahrenbruch CE, Sheehan FH. Late outcome of survivors of out-of-hospital cardiac arrest with left ventricular ejection fractions <50% and without significant coronary arterial narrowing. Am J Cardiol 1991; 67:704–708.

11. Otto CM, Tauxe RV, Cobb LA, Greene HL, Gross BW, Werner JA, Burroughs RW, Samson WE, Weaver WD, Trobaugh GB: Ventricular fibrillation causes sudden death in southeast Asian immigrants. Ann Intern Med 1984; 101:45–47.

12. Baron RC, Thacker SB, Gorelkin L, Vernon AA, Taylor WR, Choi K. Sudden death among Southeast Asian refugees: an unexplained nocturnal phenomenon. JAMA 1983; 250:2947–2951.

13. Otto CM, Tauxe RV, Cobb LA, Greene HL, Gross BW, Werner JA, Burroughs RW, Samson WE, Weaver WD, Trobaugh GB: Ventricular fibrillation causes sudden death in southeast Asian immigrants. Ann Intern Med 1984; 101:45–47.

14. Goh KT, Chao TC, Chew CH. Sudden nocturnal deaths among Thai construction workers in Singapore. Lancet 1990; 335:1154.

15. Zareba W, Moss AJ, Schwartz PJ, Vincent GM, Robinson JL, Priori SG, Benhorin J, Locati EH, Towbin JA, Keating MT, Lehmann MH, Hall WJ. Influence of genotype on the clinical course of the long-QT syndrome. International Long-QT Syndrome Registry Research Group. N Engl J Med 1998; 339:960–965.

16. Zipes DP, Wellens HJJ. Sudden cardiac death. Circulation 1998; 98:2334–2351.

17. Weaver WD, Cobb LA, Dennis D, Ray R, Hallstrom AP, Copass MK. Amplitude of ventricular waveform and outcome after cardiac arrest. Ann Intern Med 1985; 102:53–55.

18. Roelandt J, Klootwijk P, Lubsen J, Janse MJ. Sudden death during long-term ambulatory monitoring. Eur Heart J 1984; 5:7–20.

19. Cummins RO, Graves JR, Larsen MP, Hallstrom AP, Hearne TR, Ciliberti J, Nicola RM, Horan S. Out-of-hospital transcutaneous pacing by emergency medical technicians in patients with asystolic cardiac arrest. N Engl J Med 1993; 328:1377–1382.

20. James TN, St Martin E, Willis PW 3rd, Lohr TO. Apoptosis as a possible cause of gradual development of complete heart block and fatal arrhythmias associated with absence of the AV node, sinus node, and internodal pathways. Circulation 1996; 93:1424–1438.

21. Michaelsson M, Riesenfeld T, Jonzon A. Natural history of congenital complete atrioventricular block. Pacing Clin Electrophysiol 1997; 20:2098–2101.

22. Schwartz PJ. The autonomic nervous system and sudden death. Eur Heart J 1998; 19:F72–80.

23. Barron HV, Lesh MD. Autonomic nervous system and sudden cardiac death. J Am Coll Cardiol 1996; 27:1053–1060.

24. Wellens HJJ, Vermeulen A, Durrer D. Ventricular fibrillation occurring on arousal from sleep by auditory stimuli. Circulation 1972; 46:661–665.

25. Siscovick DS, Weiss NS, Fletcher RH, Lasky T. The incidence of primary cardiac arrest during vigorous exercise. N Engl J Med 1984; 311:874–877.

26. Cobb LA, Weaver WD. Exercise: A risk for sudden death in patients with coronary heart disease. J Am Coll Cardiol 1986; 7:215–219.

27. Willich SN, Lewis M, Lowel H, et al. Physical exertion as a trigger of acute myocardial infarction. N Engl J Med 1993; 329:1684–1690.

28. Beta Blocker Heart Attack Trial Research Group. A randomized trial of propranolol in patients with acute myocardial infarction. JAMA 1992; 247:1707–1714.

29. Muller JE, Ludmer PL, Willich SN, Tofler GH, Aylmer G, Klangos I, Stone PH. Circadian variation in the frequency of sudden cardiac death. Circulation 1987; 75:131–138.

30. Peckova M, Fahrenbruch CE, Cobb LA, Hallstrom AP. Circadian variations in the occurrence of cardiac arrest. Initial and repeat episodes. Circulation 1998; 98:31–39.

31. Peters RW. Propranolol and the morning increase in sudden cardiac death: (The beta-blocker heart attack trial experience). Am J Cardiol 1990; 66:57G–59G.

32. Kleiger RE, Miller P, Bigger JT, Moss AJ. Multi-center post-infarction research group. Decreased heart rate variability and its association with increased mortality after acute myocardial infarction. Am J Cardiol 1987; 59:256–262.

33. Schwartz PJ, Rovere MT, Vanoli E. Autonomic nervous system and sudden cardiac death. Experimental basis and clinical observations for post-myocardial infarction risk stratification Circulation 1992; 85(suppl 1):177–191.

34. La Rovere MT, Specchia G, Mortara A, Schwartz PJ. Baroreflex sensitivity, clinical correlates, and cardiovascular mortality among patients with a first myocardial infarction. A prospective study. Circulation 1988; 78:816–824.

35. Davies MJ, Thomas A. Thrombosis and acute coronary artery lesions in sudden cardiac ischemic death. N Engl J Med 1984; 310:1138–1140.

36. Pappas G, Queen S, Hadden W, Fisher G. The increasing disparity in mortality between socioeconomic groups in the United States, 1960 and 1986. N Engl J Med 1993; 329:103–109.

37. Cowie M, Fahrenbruch C, Cobb L, Hallstrom A. Out-of-hospital cardiac arrest: racial differences in outcome in Seattle. Am J Publ Health 1993; 83:955–959.

38. Becker LB, Han BH, Meyer PM, Wright FA, Rhodes KV, Smith DW, Barrett J, The CPR Chicago Project. Racial differences in the incidence of cardiac arrest and subsequent survival. N Engl J Med 1993; 329:600–606.

39. Hallstrom A, Boutin P, Cobb L, Johnson E. Socioeconomic status and prediction of ventricular fibrillation survival. Am J Publ Health 1993; 83:245–248.

40. Becker L, Eisenberg M, Fahrenbruch C, Cobb L. Public locations of cardiac arrest, implications for public access defibrillation. Circulation 1998; 97:2106–2109.

41. Cupples LA, Gagnon DR, Kannel WB. Long- and short-term risk of sudden coronary death. Circulation 1992; 85 (suppl I):11–50.

42. Cobb LA, Weaver WD, Fahrenbruch CE, Hallstrom AP, Copass MK. Community-based interventions for sudden cardiac death. Circulation 1992; 82 (suppl I):I:98–I:102.

43. Kannel WB, Cupples LA, D'Agostino RB. Sudden death risk in overt coronary heart disease: The Framingham Study. Am Heart J 1987; 113:799–804.

44. Kannel WB, Abbott RD. Incidence and prognosis of myocardial infarction in women: The Framingham Study. In: Eaker ED, Packard B, Wenger NK, Carlson TB, Tyroler HA, eds. Coronary Heart Disease in Women. Bethesda: National Heart, Lung and Blood Institute, National Institutes of Health, 1987:208–214.

45. Kannel WB, Thomas HE Jr. Sudden coronary death: the Framingham Study. Ann NY Acad Sci 1982; 382:3–20.

46. Myerburg RJ, Kessler KM, Castellanos A. Sudden cardiac death: epidemiology, transient risk, and intervention assessment. Ann Intern Med 1993; 119:1187–1197.

47. Jouven X, Desnos M, Guerot C, Ducimetiere P. Predicting sudden death in the population: the Paris Prospective Study I. Circulation 1999; 99(15):1978–1983.

48. Wannamethee G, Shaper AG, Macfarlane PW, Walker M. Risk factors for sudden cardiac death in middle-aged British men. Circulation 1995; 91:1749–1756.

49. Kannel WB, Wilson PWF, D'Agostino RB, Cobb J. Sudden Coronary death in women. Am Heart J 1998; 136:205–212.

50. Burke AP, Farb A, Malcom GT, Llang Y-H, Smialek J, Virmani R. Coronary risk factors and plaque morphology in men with coronary disease who died suddenly. N Engl J Med 1997; 336:1276–1282.

51. Burke AP, Farb A, Malcom GT, Llang Y-H, Smialek J, Virmani R. Effect of risk factors on the mechanism of acute thrombosis and sudden coronary death in women. Circulation 1998; 97:2110–2116.

52. National Center for Health Statistics, Division of Vital Statistics. Public use data tapes for

U.S. mortality, 1970 to 1995 and provisional tabulations for 1996. Hyattsville, MD.: National Center for Health Statistics, 1997.

53. Rosamond WD, Chambless LE, Folsom AR, Cooper LS, Conwill DE, Clegg L, Wang C-H, Heiss G. Trends in the incidence of myocardial infarction and in mortality due to coronary heart disease, 1987 to 1994. N Engl J Med 1988; 339:861–867.

54. Kreger BE, Cupples A, Kannel WB. The electrocardiogram in prediction of sudden death: The Framingham Study experience. Am Heart J 1987; 2:377–382.

55. Thompson PD, Stern WP, Williams P, et al. Death during jogging or running: a study of 18 cases. JAMA 1979; 242:2535–2538.

56. Mittleman MA, Maclure M, Tofler GH, et al. Triggering of acute myocardial infarction by heavy physical exertion: protection against triggering by regular exertion. N Engl J Med 1993; 329:1677–1683.

57. Albert CM, Mittleman MA, Chae CU, Lee I-M, Manson JE, Hennekens CH. Habitual vigorous exercise diminishes the excess risk of sudden cardiac death during vigorous exertion [Abstract]. Circulation 1997; 96(suppl):I-1688.

58. Burke AP, Farb A, Malcom GT, Llang Y-H, Smialek J, Virmani R. Plaque rupture and sudden death related to exertion in men with coronary artery disease. JAMA 1999; 281:921–926.

59. Morris JN, Everitt MG, Pollard R, Chave SPW. Vigorous exercise in leisure-time: protection against coronary heart disease. Lancet 1980; 2:1207–1210.

60. Paffenberger RS Jr, Hale WE. Work activity and coronary heart mortality. N Engl J Med 1975; 292:545–550.

61. Sandovik L, Erikssen J, Thaulow E, et al. Physical fitness as a predictor of mortality among healthy middle-aged norwegian men. N Engl J Med 1993; 328:533–537.

62. Paffenberger RS Jr, Hyde RT, Wing AL, Hsieh C. Physical activity, all cause mortality, and longevity of college alumni. N Engl J Med 1986; 314:605–613.

63. Hull SS, Vanoli E, Adamson PB, Verrier RL, Foreman RD, Schwartz PJ. Exercise training confers anticipatory protection fron sudden death during acute myocardial ischemia. Circulation 1994; 89:548–552.

64. Talbott E, Kuller LH, Detre K, Perper J. Biologic and psychosocial risk factors of sudden death from coronary disease in white women. Am J Cardiol 1977; 39:858–864.

65. Ruberman W, Weinblatt E, Goldberg J, Chaudhary BS. Psychosocial influences on mortality after myocardial infarction. N Engl J Med 1984; 311:552–559.

66. Kawachi I, Colditz GA, Ascherio A, et al. Prospective study of phobic anxiety and risk of coronary heart disease in men. Circulation 1994; 89:1992–1997.

67. Haines AP, Imeson JD, Meade TW. Phobic anxiety and ischemic heart disease. Br Med J 1987; 295:297–299.

68. Offerhaus RE. Heart rate variability in psychiatry. In: Kitney RJ, Rompelman O, eds. The Study of Heart Rate Variability. Oxford: Oxford University Press, 1980:225–238.

69. Bigger JT Jr, Fleiss JL, Steinman RC, et al. Frequency domain measures of heart rate period variability and mortality after myocardial infarction. Circulation 1992; 85:164–171.

70. Fei L, Anderson MH, Katrisis D, et al. Decreased heart rate variability in survivors of sudden cardiac death not associated with coronary artery disease. Br Med J 1994; 71:16–21.

71. Leor J, Poole WK, Kloner RA. Sudden cardiac death triggered by an earthquake. N Engl J Med 1996; 334:413–419.

72. Billman GE, Hallaq H, Leaf A. Prevention of ischemia-induced ventricular fibrillation by ω3 fatty acids. Proc Natl Acad Sci 1994; 91:4427–4430.

73. McLennan PL, Bride TM, Abeywardena MY, Charnock JS. Dietary lipid modulation of ventricular fibrillation threshold in the marmoset monkey. Am Heart J 1992; 123:1555–1561.

74. McLennan PL, Abeywardena MY, Charnock JS. Dietary fish oil prevents ventricular fibrillation following coronary artery occlusion and reperfusion. Am Heart J 1988; 116:709–717.

75. McLennan PL, Bridle TM, Abeywardena MY, Charnock JS. Comparative efficacy of n-3 and n-6 polyunsaturated fatty acids in modulating ventricular fibrillation threshold in marmoset monkeys. Am J Clin Nutr 1993; 58:666–669.

76. Kang JX, Leaf A. Antiarrhythmic effects of polyunsaturated fatty acids: recent studies. Circulation 1996; 94:1774–1780.

77. Christensen JH, Korup E, Aaroe J, Toft E, Moller J, Rasmussen K, Dyerberg J, Schmidt EB. Fish consumption, n-3 fatty acids in cell membranes, and heart rate variability in survivors of myocardial infarction with left ventricular dysfunction. Am J Cardiol 1997; 79:1670–1673.

78. Hartikainen JK, Malik M, Staunton A, Poloniecki J, Camm J. Distinction between arrhythmic and nonarrhythmic death after acute myocardial infarction based on heart rate variability, signal-average electrocardiogram, ventricular arrhythmias and left ventricular ejection fraction. J Am Coll Cardiol 1996; 28:296–304.

79. Kromhout D, Bosschieter EB, de Lezenne Coulander C. The inverse relation between fish consumption and 20-year mortality from coronary heart disease. N Engl J Med 1985; 312:1205–1209.

80. Norell SE, Ahlbom A, Feychting M, Pederson NL. Fish consumption and mortality from coronary heart disease. Br Med J 1986; 293:426.

81. Shekelle RB, Missell LV, Paul O, Shyrock AM, Stamler J. Fish consumption and mortality from coronary heart disease. N Engl J Med 1985; 313:820.

82. Dolecek TA, Epidemiological evidence of relationships between dietary polyunsaturated fatty acids and mortality in the Multiple Risk Factor Intervention Trial. Proc Soc Exp Biol Med 1992; 200:177–182.

83. Burr ML, Fehily AM, Gilbert JF, et al. Effects of changes in fat, fish, and fibre intakes on death and myocardial reinfarction: Diet and Reinfarction Trial (DART). Lancet 1989; 2:757–761.

84. Siscovick DS, Raghunathan TE, King I, et al. Dietary intake and cell membrane levels of long-chain n-3 polyunsaturated fatty acids and the risk of primary cardiac arrest. JAMA 1995; 274:1363–1367.

85. Albert CM, Hennekens CH, O'Donnell CJ, Ajani UA, Carey VC, Willett WC, Ruskin JN, Manson JE. Fish consumption and decreased risk of sudden cardiac death. JAMA 1998; 279 (1):23–28.

86. GISSI-Prevenzione Investigators. Dietary supplementation with n-3 polyunsaturated fatty acids and vitamin E after myocardial infarction: results of the GISSI-Prevenzione trial. Lancet 1999; 354:447–455.

87. Kardinal AFM, Kok FJ, Ringstad J, et al. Antioxidants in adipose tissue and risk of myocardial infarction: the Euramic study. Lancet 1993; 342:1379–1384.

88. de Logreril M, Renaud S, Mamelle N, Salen P, Martin J-L, Monjaud I, Guidollet J, Touboul P, Delaye J. Mediterranean alpha-linolenic acid-rich diet in secondary prevention of coronary heart disease. Lancet 1989; 334:757–761.

89. Wannamethee G, Shaper AG. Alcohol and sudden cardiac death. Br Heart J 1992; 68:443–448.

90. Dyer AR, Stamler J, Paul O, Berkson DM, Lepper MH, McKean H, Shekelle RB, Lindberg HA, Garside D. Alcohol consumption, cardiovascular risk factors, and mortality in two Chicago epidemiologic studies. Circulation 1977; 56:1067–1074.

91. Kozarevic D, Demirovic J, Gordon T, Kaelber CT, McGee D, Zukel WJ. Drinking habits and coronary heart disease: the Yugoslavia Cardiovascular Disease Study. Am J Epidemiol 1982; 116:748–758.

92. Gordon T, Kannel WB. Drinking habits and cardiovascular disease: the Framingham Study. Am Heart J 1983; 105:667–673.

93. Suhonen O, Aromaa A, Reunanen A, Knekt P. Alcohol consumption and sudden coronary death in middle-aged Finnish men. Acta Med Scand 1987; 221:335.

94. Siscovick DS, Weiss NS, Fox N. Moderate alcohol consumption and primary cardiac arrest. Am J Epidemiol 1986; 123:499–503.

95. Albert CM, Manson JE, Cook NR, Ajani UA, Gaziano JM, Hennekens CH. Moderate Alcohol Consumption and the Risk of Sudden Cardiac Death among U.S. Male Physicians. Circulation, 1999; 100:944–950.

96. Friedlander Y, Siscovick DS, Weinmann S, Austin MA, Psaty BM, Lemaitre RN, Arbogast P,

Raghunathan TE, Cobb LA. Family history as a risk factor for primary cardiac arrest. Circulation 1998; 97:155–160.

97. Priori SG, Barhanin J, Hauer RNW, Haverkamp W, Jongsma HJ, Kleber AG, McKenna WJ, Roden DM, Rudy Y, Schwartz K, Schwartz PJ, Towbin JA, Wilde AM. Genetic and molecular basis of cardiac arrhythmias: impact on clinical management, Parts I and II. Circulation 1998; 99:518–528.

98. Moss AJ, Davis HT, DeCamilla J, et al. Ventricular ectopic beats and their relation to sudden and non-sudden cardiac death after myocardial infarction. Circulation 1979; 60:998–1003.

99. Daly LE, Hickey N, Graham IM, Mulcahy R. Predictors of sudden death up to 18 years after a first attack of unstable angina or myocardial infarction. Br Heart J 1987; 58:567–571.

100. The Multicenter Postinfarction Research Group. Risk stratification and survival after myocardial infarction. N Engl J Med 1983; 309:331–336.

101. Moss AJ, Hall WJ, Cannom DS, Daubert JP, Higgins SL, Levine JH, Saksena S, Waldo AL, Wilber D, Brown MW, Heo M. Improved survival with an implanted defibrillator in patients with coronary disease at high risk for ventricular arrhythmia. N Engl J Med 1996; 335:1933–1940.

102. Buxton BE, Lee KL, Fisher JD, Josephson ME, Prystowsky EN, Hafley G, for the Multicenter Unsustained Tachycardia Trial Investigators. A Randomised Study of the Prevention of Sudden Death in Patients with Coronary Artery Disease. N Engl J Med 1999; 341:1882–1890.

103. Bigger JT Jr, for the CABG Patch Trial Investigators. Prophylactic use of implanted cardiac defibrillators in patients at high risk for ventricular arrhythmias after coronary artery bypass graft surgery. N Engl J Med 1997; 337:1569–1575.

104. Bigger JT Jr, Whang W, Rottman JN, Kleiger RE, Gottlieb CD, Namerow PB, Steinman RC, Estes NA, 3rd. Mechanisms of death in the CABG Patch trial: a randomized trial of implantable cardiac defibrillator prophylaxis in patients at high risk of death after coronary artery bypass graft surgery. Circulation 1999; 99:1416–1421.

105. Kelly P, Ruskin JN, Vlahakes GJ, Buckley MJ Jr, Freeman CS, Garan H. Surgical coronary revascularization in survivors of prehospital cardiac arrest: its effect on inducible ventricular arrhythmias and long-term survival. J Am Coll Cardiol 1990; 15:267–273.

106. The Cardiac Arrhythmia Suppression Trial (CAST) Investigators. Preliminary report: effect of encainide and flecainide on mortality in a randomized trial of arrhythmia suppression after myocardial infarction. N Engl J Med 1989; 321:406–412.

107. The Cardiac Arrhythmia Suppression Trial II investigators. Effect of the antiarrhythmic agent moricizine on survival after myocardial infarction. N Engl J Med 1992; 327:227–233.

108. Greenberg HM, Dwyer EM, Hochman JS, Steinberg JS, Echt DS, Peters RW. Interaction of ischaemia and encainide/flecainide treatment: A proposed mechanism for the increased mortality in CAST1. Br Heart J 1995; 74:631–635.

109. Teo KK, Yusuf S, Furberg CD. Effects of prophylactic antiarrhythmic drug therapy in acute myocardial infarction. An overview of results from randomized control trials. JAMA 1993; 270:1589–1595.

110. Waldo AL, Camm AJ, deRuyter H, Friedman PL, MacNeil DJ, Pauls JF, Pitt B, Pratt CM, Schwartz PJ, Veltri EP. Effect of d-sotalol on mortality in patients with left ventricular dysfunction after recent and remote myocardial infarction. The SWORD Investigators. Survival With Oral d-Sotalol. Lancet 1996; 348(9019):7–12.

111. Julian DG, Jackson FL, Prescott RJ, Szekely P. Controlled trial of sotalol for one year after myocardial infarction. Lancet 1982; i:1142–1150.

112. Lars Kober, The Diamond Study Group. A clinical trial of dofetilide in patients with acute myocardial infarction and left ventricular dysfunction (abstr). Circulation 1998; 98(suppl):I-471.

113. Julian DG, Camm AJ, Frangin G, et al. Randomised trial of effect of amiodarone on mortality in patients with left ventricular dysfunction after recent myocardial infarction: EMIAT. Lancet 1997; 349:667–674.

114. Cairns JA, Connolly SJ, Roberts R, Gent M. Randomised trial of outcome after myocardial infarction in patients with frequent or reptitive ventricular premature depolarisations: CAMIAT. Lancet 1997; 349:675–682.

115. Amiodarone Trials Meta-Analysis Investigators. Effect of prophylactic amiodarone on mortality after acute myocardial infarction and in congestive heart failure: meta-analysis of individual data from 6500 patients in randomised trials. Lancet 1997; 350:1417–1424.

116. The Norwegian Multicenter Study Group. Timolol-induced reduction in mortality and reinfarction in patients surviving acute myocardial infarction. N Engl J Med 1981; 304:801–807.

117. Rogers WJ, Bowlby LJ, Chandra NC et al. for the Participants in the National Registry of Myocardial Infarction. Treatment of myocardial infarction in the United States (1990–1993). Observations from the National Registry of Myocardial Infarction. Circulation 1994; 90:2103–2114.

118. Krumholz HM, Radford MJ, Wang Y, Chen J, Heiat A, Marciniak TA. National use and effectiveness of beta-blockers for the treatment of elderly patients after acute myocardial infarction: National Cooperative Cardiovascular Project [published erratum appears in JAMA 1999; 281(1):37]. JAMA 1998; 280:623–629.

119. Kober L, Torp-Pedersen C, Carlsen J, et al. A clinical trial of the angiotension-converting-enzyme inhibitor trandolapril in patients with left ventricular dysfunction after myocardial infarction. N Engl J Med 1995; 333:1670–1676.

120. Pfeffer MA, Braunwald E, Moyé LA, et al. on behalf of the SAVE investigators. Effect of Captopril on mortality and morbidity in patients with left ventricular dysfunction after myocardial infarction. N Engl J Med 1992; 327:669–677.

121. The Acute Infarction Ramipril Efficacy (AIRE) Study Investigators. Effect of ramipril on mortality and morbidity of survivors of acute myocardial infarction with clinical evidence of heart failure. Lancet 1993; 342:821–828.

122. Barron HV, Michaels AD, Maynard C, Every NR. Use of angiotensin-converting enzyme inhibitors at discharge in patients with acute myocardial infarction in the United States: data from the National Registry of Myocardial Infarction 2. J Am Coll Cardiol 1998; 32:360–367.

123. Scandinavian Simvastatin Survival Study Group. Randomised trial of cholesterol lowering in 4444 patients with coronary heart disease: the Scandinavian Simvastatin Survival Study (4S). Lancet 1994; 344:1383–1389.

124. Byington RP, Jukema JW, Salonen JT, Pitt B, Bruschke AV, Hoen H, Furberg CD, Mancini GB. Reduction in cardiovascular events during pravastatin therapy: pooled analysis of clinical events of the Pravastatin Atherosclerosis Intervention Program. Circulation 1995; 92:2419–2425.

125. Goldman S, Johnson G, Cohn JN, Cintron G, Smith R, Francis G. Mechanism of death in heart failure. The Vasodilator-Heart Failure Trials. The V-HeFT VA Cooperative Studies Group. Circulation 1993; 87:VI24–31.

126. Doval H, Nul D, Grancelli H, Perrone S, Bortman G, Curiel R. Randomised trial of low-dose amiodarone in severe congestive heart failure. Lancet 1994; 1:493–498.

127. Singh SN, Fletcher RD, Fisher SG, et al. Amiodarone in patients with congestive heart failure and asymptomatic ventricular arrhythmia. N Engl J Med 1995; 333:77–82.

128. Packer M, Bristow MR, Cohn JN, Colucci WS, Fowler MB, Gilbert EM, Shusterman NH. The effect of carvedilol on morbidity and mortality in patients with chronic heart failure. N Engl J Med 1996; 334(21):1349–1355.

129. CIBIS-II Investigators and Committees. The Cardiac Insufficiency Bisoprolol Study II (CIBIS II). Lancet 1999; 353:9–13.

130. MERIT-HF Study Group. Effect of metoprolol CR/XL in chronic heart failure: Metoprolol CR/XL Randomised Intervention Trial in Congestive Heart Failure (MERIT-HF). Lancet 1999; 353:2001–2007.

131. Klein H, Auricchio A, Reek S, Geller C. New primary prevention trials of sudden cardiac

death in patients with left ventricular dysfunction: SCD-HEFT and MADIT-II. Am J Cardiol 1999; 83:91D–97D.

132. Eisenberg MS, Pantridge F, Cobb LA, Geddes JS. The revolution and evolution of prehospital cardiac care. Arch Intern Med 1996; 156:1611–1620.

133. Pantridge JF, Geddes JS. A mobile intensive-care unit in the management of myocardial infarction. Lancet 1967; 2:271–273.

134. Becker L, Vath J, Eisenberg M, Meischke H. The impact of television public service announcements on the rate of bystander CPR. Prehosp Emerg Care 1999; 3:353–356.

135. Thompson RG, Hallstrom AP, Cobb LA. Bystander-initiated cardiopulmonary resuscitation in the management of ventricular fibrillation. Ann Intern Med 1979; 90:737–740.

136. Cummins RO, Eisenberg MS. Prehospital cardio-pulmonary resuscitation: is it effective? JAMA 1985; 253:2408–2412.

137. Mandel LP, Cobb LA. Reinforcing CPR skills without mannequin practice. Ann Emerg Med 1987; 16:1117–1120.

138. Nichol G, Hallstrom AP, Ornato JP, Riegel B, Stiell IG, Valenzuela T, Wells GA, White RD, Weisfeldt ML. Potential cost-effectiveness of public access defibrillation in the United States. Circulation 1998; 7:1315–1320.

139. O'Rourke MF, Donaldson E, Geddes JS. An airline cardiac arrest program. Circulation 1997; 4:2849–2853.

140. Hossack KF, Hartwig R. Cardiac arrest associated with supervised cardiac rehabilitation. J Cardiac Rehab 1982; 2:402–408.

141. Cobb LA, Personal communication, 1999.

142. Litwin PE, Eisenberg MS, Hallstrom AP, Cummins RO. The location of collapse and its effect on survival from cardiac arrest. Ann Emerg Med 1987; 16:787–791.

143. Valenzuela TD, Spaite DW, Meislin HW, Clark LL, Wright AL, Ewy GA. Case and survival definitions in out-of-hospital cardiac arrest: effect on survival rate calculation. JAMA 1992; 267:272–274.

144. Schaffer WA, Cobb LA. Recurrent ventricular fibrillation and modes of death in survivors of out-of-hospital ventricular fibrillation. N Engl J Med 1975; 293:259–262.

145. Powell AC, Fuchs T, Finkelstein DM, Garan H, Cannom DS, McGovern BA, Kelly E, Vlahakes GJ, Torchiana DF, Ruskin JN. Influence of implantable cardioverter-defibrillators on the long-term prognosis of survivors of out-of-hospital cardiac arrest. Circulation 1993; 88:1083–1092.

146. The Antiarrhythmics Versus Implantable Defibrillators (AVID) Investigators. A comparison of antiarrhythmic-drug therapy with implantable defibrillators in patients resuscitated from near-fatal ventricular arrhythmias. N Engl J Med 1997; 337:1576–1583.

147. Cappato R. Secondary prevention of sudden death: the Dutch Study, the Antiarrhythmics Versus Implantable Defibrillator Trial, the Cardiac Arrest Study Hamburg, and the Canadian Implantable Defibrillator Study. Am J Cardiol 1999; 83:68D–73D.

148. Connolly SJ, Gent M, Roberts RS, Dorian P, Sheldon RS, Mitchell B, Green MS, Klein GJ, O'Brien B. Canadian Implantable Defibrillator Study (CIDS): A randomized trial of the implantable cardioverter defibrillator against amiodarone. Circulation 2000; 101:1297–1302.

149. Weaver WD, Cobb LA, Hallstrom AP, Copass MK, Ray R, Emery M, Fahrenbruch C. Considerations for improving survival from out-of-hospital cardiac arrest. Ann Emerg Med 1986; 15:1181–1186.

150. Weaver WD, Cobb LA, Hallstrom AP, Fahrenbruch C, Copass MK, Ray R. Factors influencing survival following out-of-hospital cardiac arrest. J Am Coll Cardiol 1986; 7:752–757.

151. Eisenberg MS, Horwood BT, Cummins RO, Reynolds-Haertle R, Hearne TR. Cardiac arrest and resuscitation: A tale of 29 cities. Ann Emerg Med 1990; 19:179–186.

152. Longstreth WT, Cobb LA, Fahrenbruch CE, Copass MK. Does age affect outcomes of out-of-hospital cardiopulmonary resuscitation? JAMA 1990; 264:2109–2110.

Index

About the Editors

PETER M. SPOONER is Director of the Arrhythmias, Ischemia and Sudden Cardiac Death Research Program; Division of Heart and Vascular Disease; National Heart, Lung, and Blood Institute, Bethesda, Maryland, where he has held positions in program development in support of fundamental research in the arrhythmia sciences for more than a dozen years. The author of journal articles and earlier volumes on arrhythmias and their causes, he has been involved with these issues as well with the American Heart Association Councils on Clinical Cardiology and on Basic Cardiovascular Sciences. He has held memberships in the American Society of Biochemistry and Molecular Biology, the Endocrine Society, and the American Physiological Society, and has worked as a consultant to the European Society of Cardiology and other European health care organizations. He has held appointments at the Uniformed Services University of Health Sciences and at the ICRF in London. Dr. Spooner received the B.S. degree (1964) from Bates College, Lewiston, Maine, and the M.S. (1966) and Ph.D. (1969) degrees from the University of Illinois. He completed postdoctoral work at the Cardiovascular Research Institute, University of California San Francisco Medical Center, before joining the NIH in 1974.

MICHAEL R. ROSEN is Gustavus A. Pfeiffer Professor of Pharmacology, Professor of Pediatrics, and Director of the Center for Molecular Therapeutics at Columbia University College of Physicians and Surgeons, New York, New York; Adjunct Professor of Physiology and Biophysics at the State University of New York, Stonybrook School of Medicine; and Attending Physician at the New York Presbyterian Hospital. The author, coauthor, editor, or coeditor of over 500 journal publications, book chapters and books, he is a Diplomate of the American Board of Internal Medicine, a Fellow of the American College of Physicians, the American College of Clinical Pharmacology, the American College of Cardiology, and the European Society of Cardiology, and a member of numerous societies and associations. He has received a number of awards, including the Honorary Regent for Life of the American College of Clinical Pharmacology and the Award of Merit of the American Heart Association. He is consulting editor for *Circulation Research and Cardiovascular Research* and co-Editor-in-Chief of the *Journal of Cardiovascular Pharmacology.* Dr. Rosen received the B.A. degree (1960) from Wesleyan University, Middletown, Connecticut, and the M.D. degree (1964) from the State University of New York, Downstate Medical Center, New York City.